# Table of Atomic Masses

Atomic masses are taken from Phys. Rev. D *50* 1243 (1994).
A bracketed value denotes the mass of the longest-lived isotope.

| | Symbol | Z | Atomic Mass | | Symbol | Z | Atomic Mass |
|---|---|---|---|---|---|---|---|
| Actinium | Ac | 89 | [227.0278] | Mercury | Hg | 80 | 200.59 |
| Aluminum | Al | 13 | 26.981539 | Molybdenum | Mo | 42 | 95.94 |
| Americium | Am | 95 | [243.0614] | Neodymium | Nd | 60 | 144.24 |
| Antimony | Sb | 51 | 121.757 | Neon | Ne | 10 | 20.1797 |
| Argon | Ar | 18 | 39.948 | Neptunium | Np | 93 | [237.0482] |
| Arsenic | As | 33 | 74.92159 | Nickel | Ni | 28 | 58.6934 |
| Astatine | At | 85 | [209.9871] | Nielsbohrium | Ns | 107 | [262.1231] |
| Barium | Ba | 56 | 137.327 | Niobium | Nb | 41 | 92.90638 |
| Berkelium | Bk | 97 | [247.0703] | Nitrogen | N | 7 | 14.00674 |
| Beryllium | Be | 4 | 9.012182 | Nobelium | No | 102 | [259.1011] |
| Bismuth | Bi | 83 | 208.98037 | Osmium | Os | 76 | 190.23 |
| Boron | B | 5 | 10.811 | Oxygen | O | 8 | 15.9994 |
| Bromine | Br | 35 | 79.904 | Palladium | Pd | 46 | 106.42 |
| Cadmium | Cd | 48 | 112.411 | Phosphorus | P | 15 | 30.973762 |
| Calcium | Ca | 20 | 40.078 | Platinum | Pt | 78 | 195.08 |
| Californium | Cf | 98 | [251.0796] | Plutonium | Pu | 94 | [244.0642] |
| Carbon | C | 6 | 12.011 | Polonium | Po | 84 | [208.9824] |
| Cerium | Ce | 58 | 140.115 | Potassium | K | 19 | 39.0983 |
| Cesium | Cs | 55 | 132.90543 | Praseodymium | Pr | 59 | 140.90765 |
| Chlorine | Cl | 17 | 35.4527 | Promethium | Pm | 61 | [144.9127] |
| Chromium | Cr | 24 | 51.9961 | Protactinium | Pa | 91 | 231.03588 |
| Cobalt | Co | 27 | 58.93320 | Radium | Ra | 88 | [226.0254] |
| Copper | Cu | 29 | 63.546 | Radon | Rn | 86 | [222.0176] |
| Curium | Cm | 96 | [247.0703] | Rhenium | Re | 75 | 186.207 |
| Dysprosium | Dy | 66 | 162.50 | Rhodium | Rh | 45 | 102.90550 |
| Einsteinium | Es | 99 | [252.0830] | Rubidium | Rb | 37 | 85.4678 |
| Erbium | Er | 68 | 167.26 | Ruthenium | Ru | 44 | 101.07 |
| Europium | Eu | 63 | 151.965 | Rutherfordium | Rf | 104 | [261.1089] |
| Fermium | Fm | 100 | [257.0951] | Samarium | Sm | 62 | 150.36 |
| Fluorine | F | 9 | 18.9984032 | Scandium | Sc | 21 | 44.955910 |
| Francium | Fr | 87 | [223.0197] | Seaborgium | Sg | 106 | [263.1186] |
| Gadolinium | Gd | 64 | 157.25 | Selenium | Se | 34 | 78.96 |
| Gallium | Ga | 31 | 69.723 | Silicon | Si | 14 | 28.0855 |
| Germanium | Ge | 32 | 72.61 | Silver | Ag | 47 | 107.8682 |
| Gold | Au | 79 | 196.96654 | Sodium | Na | 11 | 22.989768 |
| Hafnium | Hf | 72 | 178.49 | Strontium | Sr | 38 | 87.62 |
| Hahnium | Ha | 105 | [262.1144] | Sulfur | S | 16 | 32.066 |
| Hassium | Hs | 108 | [265.1306] | Tantalum | Ta | 73 | 180.9479 |
| Helium | He | 2 | 4.002602 | Technetium | Tc | 43 | [97.9072] |
| Holmium | Ho | 67 | 164.93032 | Tellurium | Te | 52 | 127.60 |
| Hydrogen | H | 1 | 1.00794 | Terbium | Tb | 65 | 158.92534 |
| Indium | In | 49 | 114.818 | Thallium | Tl | 81 | 204.3833 |
| Iodine | I | 53 | 126.90447 | Thorium | Th | 90 | 232.0381 |
| Iridium | Ir | 77 | 192.22 | Thulium | Tm | 69 | 168.93421 |
| Iron | Fe | 26 | 55.847 | Tin | Sn | 50 | 118.710 |
| Krypton | Kr | 36 | 83.80 | Titanium | Ti | 22 | 47.88 |
| Lanthanum | La | 57 | 138.9055 | Tungsten | W | 74 | 183.84 |
| Lawrencium | Lr | 103 | [262.1098] | Uranium | U | 92 | 238.0289 |
| Lead | Pb | 82 | 207.2 | Vanadium | V | 23 | 50.9415 |
| Lithium | Li | 3 | 6.941 | Xenon | Xe | 54 | 131.29 |
| Lutetium | Lu | 71 | 174.967 | Ytterbium | Yb | 70 | 173.04 |
| Magnesium | Mg | 12 | 24.3050 | Yttrium | Y | 39 | 88.90585 |
| Manganese | Mn | 25 | 54.93805 | Zinc | Zn | 30 | 65.39 |
| Meitnerium | Mt | 109 | [266.1378] | Zirconium | Zr | 40 | 91.224 |
| Mendelevium | Md | 101 | [258.0984] | | | | |

# CHEMISTRY
## Principles & Reactions
*Third Edition*

# CHEMISTRY
## Principles & Reactions

*Third Edition*           *A Core Text*

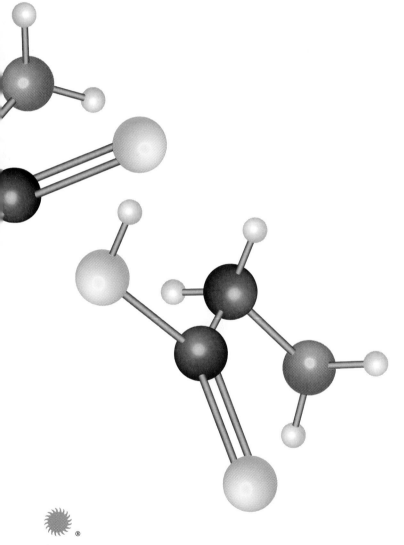

**William L. Masterton**
University of Connecticut

**Cecile N. Hurley**
University of Connecticut

**Saunders Golden Sunburst Series**
Saunders College Publishing
Harcourt Brace College Publishers

FORT WORTH   PHILADELPHIA   SAN DIEGO   NEW YORK   ORLANDO   AUSTIN

SAN ANTONIO   TORONTO   MONTREAL   LONDON   SYDNEY   TOKYO

Requests for permission to make copies of any part of the work should be mailed to: Permissions Department, Harcourt Brace & Company, 6277 Sea Harbor Drive, Orlando, Florida 32887-6777.

Text Typeface: Stone Serif
Compositor: York Graphic Services
Vice President/Publisher: John Vondeling
Senior Developmental Editor: Beth C. Rosato
Managing Editor: Carol Field
Project Editor: Laura Shur
Copy Editor: Janis Moore
Manager of Art & Design: Carol Bleistine
Senior Art Director: Joan Wendt
Cover Designer: Ruth A. Hoover
Text Artwork: Rolin Graphics, Inc.
Manager of Production: Joanne Cassetti
Senior Production Manager: Charlene Catlett Squibb
Manager, Photo Research & Permissions: Dena Digilio-Betz
Product Manager: Angus McDonald

Cover Credit: Charles D. Winters
Cover Image: Burning in methanol. Sodium, Strontium, Boric Acid.

Printed in the United States of America

Chemistry: Principles & Reactions, Third Edition

ISBN: 0-03-005889-9

Library of Congress Catalog Card Number: 95-072523

789  048  9876543

**William L. Masterton** received his Ph.D. in physical chemistry from the University of Illinois in 1953. Two years later, he arrived at the University of Connecticut, where he taught general chemistry and a graduate course in chemical thermodynamics. He received numerous teaching awards; the one of which he is most proud came from the Student Senate at UConn. Bill wrote, with co-author Emil Slowinski, the all-time best selling general chemistry textbook, CHEMICAL PRIN-CIPLES, which sold well over one and a half million copies. Between editions, Bill continues to work on the definitive account of the most celebrated crime of the 19th century, the Borden case. Bill's field of research, solution thermodynamics, prepared him well for making maple syrup each March at the family farmhouse in New Hampshire.

**Cecile Nespral Hurley** received her M.S. at the University of California, Los Angeles. Since 1979, she has served as Lecturer and Coordinator of Freshman Chemistry at the University of Connecticut, where she has directed a National Science Foundation supported project on study groups in general chemistry. She also coordinates the University's High School Cooperative Program in Chemistry, through which superior Connecticut high school students take the University's general chemistry course at their high schools. In her spare time, Cecile roots for the UConn Husky Women's Basketball Team and roots out weeds from her country garden that she likes to imagine rivals Monet's garden at Giverny.

*To Loris and Jim*
*For energy, enthalpy, and equilibrium*

# Contents Overview

For many years our colleagues told us that general chemistry texts were too long, too repetitive, and too heavy. We listened, we agreed, and we did something about it. Four years ago, our second edition weighed in at 606 pages. It was well received; reviewers and, more important, instructors were enthusiastic about it. Moreover, the incidence of hernias among college freshmen taking general chemistry dropped by 6.4%*.

This edition, at 640 pages, is only slightly longer than its predecessor. Our goal throughout has been to write a book that can be covered in two semesters without deleting any fundamental topics. Beyond that we were determined to include enough explanatory material and applications to make chemistry intelligible and interesting to students. Perhaps the best way to describe how we did this is to try to answer some questions that instructors have asked about the book.

## What Did We Leave Out?

We have attempted to eliminate repetition and duplication wherever possible. For example, this book contains

— one and only one method of balancing redox equations, the half-equation approach introduced in Chapter 4.
— one and only one way of working gas law problems, using the ideal gas law in all cases (Chapter 5).
— one and only one way of calculating $\Delta H$ (Chapter 8), using enthalpies of formation.
— one and only one equilibrium constant for gas phase reactions (Chapter 12), the thermodynamic constant $K$, commonly referred to as $K_p$. This simplifies not only the treatment of gaseous equilibrium, but also the discussion of reaction spontaneity (Chapter 17) and electrochemistry (Chapter 18).

Nearly all of the topics covered in the typical general chemistry text are found in this book, with a few exceptions, including

— *biochemistry,* traditionally covered in the last chapter of every general chemistry text. Fascinating as this material is, it requires an understanding of organic chemistry that few students have. Our last chapter (Chapter 22) is devoted to organic chemistry, with emphasis on practical applications such as synthetic polymers.
— *molecular orbitals.* Many people have suggested deleting this material from general chemistry texts; we relegated it to an appendix. To be honest, we're not sure whether molecular orbitals *should* be covered. Our experience suggests that MO theory doesn't go over very well with beginning students. On the other hand, this is the approach to bonding that chemists use today.

Several topics discussed at considerable length in many general chemistry texts are condensed in this one. For example,

*Not really; you can't believe everything you read in prefaces.

— *Lewis acids* and bases are discussed quite briefly when complex ion formation is introduced in Chapter 15. Two acid-base models (Arrhenius in Chapter 4, Brønsted-Lowry in Chapter 13) and probably enough for the general chemistry student.
— *qualitative analysis* is summarized in a few pages in section 16.3, which is flexible enough to be assigned when students start cation analysis in the laboratory.
— *descriptive inorganic chemistry* is covered formally in two chapters (Chapter 20 on metals, Chapter 21 on nonmetals). This text is short enough so that both of these chapters can be covered. To help you do that, we've reorganized the descriptive material to make it easier to teach, more interesting, and less encyclopedic.

## How Is This Book Organized?

As you can see from the Table of Contents (p. xix), the organization is more or less conventional. A few features are worthy of comment:

— Chapter 4 (Reactions in Aqueous Solution) shows students how to write, balance, and apply net ionic equations for precipitation, acid-base, and redox reactions. It seems reasonable to introduce this fundamental material early; among other things it establishes a background for a meaningful first-semester laboratory.
— the two factors that determine reaction feasibility, rate and equilibrium, are presented back-to-back in Chapters 11 and 12. This seems a logical starting point for the second semester.
— coordination chemistry (Chapter 15) is presented somewhat earlier than is customary. This is our subtle way of encouraging you to spend time on an area of inorganic chemistry that students find fascinating albeit challenging.

## What Are the Changes in This Edition?

When we lopped off nearly 400 pages from our text four years ago, we sometimes overdid it. Students told us that in a few areas the treatment was too terse; we agree. Accordingly we have expanded several discussions in the interest of clarity and/or added more in-text examples. The modest increase in the number of pages in this edition is due almost entirely to this factor.

Beyond that we have

— rewritten about 75% of the end-of-chapter problems. We added conceptual problems, many of them in the "unclassified" category (see, for example, #66 in Chapter 3). As before, the "classified" problems are paired; the second member of each pair is answered in Appendix 6.
— split acid-base and solubility equilibria into two chapters. Our students convinced us that the topic density here was too high in the Second Edition. Interspersing coordination chemistry between

Chapters 14 and 16 makes it possible to discuss in a rational way the reactions used to bring insoluble solids into solution.

## What Learning Aids Are Available to Students?

Each chapter

— starts with a brief summary of topics to be covered. This way the student (and the instructor) knows what's coming.
— contains in-text examples featuring a "Strategy" section which outlines the reasoning behind the solution. This helps distinguish the "why" from the "how" of problem solving.
— ends with a "Highlights" section which includes key concepts (indexed to corresponding examples and problems), key equations, and key terms (defined in a glossary/index). The section ends with a "Summary Problem" which ties together all of the ideas introduced in the chapter.

## How Is Student Interest Maintained?

It took us a long time to realize that the function of a chemistry teacher and hence of a textbook goes well beyond explaining chemical principles. Students must be convinced that chemistry is so important and so exciting that they should devote the time and effort required to master these principles. To help here we have pointed out, wherever appropriate, the practical applications of chemical principles and the way they relate to current research. This is done throughout the text but especially in two feature sections:

— *"Chemistry Beyond the Classroom"* occurs near the end of each chapter, where chemical principles covered in the chapter are applied to the world around us. New to this edition are discussions of forensic chemistry (Chapters 1 and 4), fullerenes (Chapter 2), weather phenomena (Chapter 5), the aurora (Chapter 6), the chemistry of selenium (Chapter 21), and the structure of proteins (Chapter 22).
— *"Chemistry: The Human Side"* provides biographical sketches of famous chemists. Here we have tried to illuminate the personality as well as the accomplishments of the individual. Dorothy Hodgkin, Henry Eyring and Alfred Werner have been added to this edition.

## What Ancillaries Are Available?

The following materials were developed specifically for use with the text. Other ancillaries of a more general type are listed after the preface.

- **Study Guide/Workbook** by Cecile N. Hurley. Worked examples and problem-solving techniques help the student understand the principles of general chemistry. Each chapter is outlined for the student with fill-in-the-blanks, and exercises and self-tests allow the students to gauge their mastery of the chapter.

- **Instructor's Manual** by William L. Masterton. Included are lecture outlines and lists of demonstrations for each chapter. Worked-out solutions are provided for all of the end-of-chapter problems that do not have answers in the appendix.
- **Student Solutions Manual** by Cassandra T. Eagle and David G. Farrar (both of Appalachian State University). Complete solutions to all the problems answered in the text, including the Challenge Problems. References to the main text's sections and tables are provided as a guide for problem-solving techniques employed by the authors.
- **Lecture Outline** by Ronald O. Ragsdale (University of Utah). Organized to follow class lectures to free students from extensive note-taking during lectures.
- **Cooperative Learning Worksheets** by Cecile N. Hurley. A collection of worksheets (about three per chapter) that students work on in groups. Questions are designed to stimulate group activity and discussion. Intended for group work, the questions provided on each worksheet are equally conceptually-oriented and quantitatively-oriented. The booklet includes instructions for use, how to guide student discussion, and supporting data on the success of cooperative learning at University of Connecticut.
- **Printed Test Bank** by Ronald O. Ragsdale (University of Utah). Over 1000 class-tested multiple-choice, five-part questions.
- **ExaMaster™ Computerized Test Bank** is the software version of the printed Test Bank. Instructors can create thousands of questions in a multiple-choice format. A command reformats the multiple-choice question into a short-answer question. Adding or modifying existing problems, as well as incorporating graphics, can be done. ExaMaster has gradebook capabilities for recording and graphing students' grades.
- **Chemical Principles in the Laboratory,** Sixth Edition, by Emil Slowinski and Wayne Wolsey (both of Macalester College) and William Masterton (University of Connecticut) provides detailed directions and advance study assignments. This manual contains 42 experiments that have been selected with regard to cost and safety. All the experiments have been thoroughly class-tested. Alternatively, the *Chemical Principles in the Laboratory with Qualitative Analysis,* Sixth Edition, Alternate Version is available, offering 30 of the 42 experiments in the former book with eight new experiments covering qualitative analysis.
- An **Instructor's Manual** is available for each version of *Chemical Principles in the Laboratory,* and each IM provides lists of equipment and chemicals needed for each experiment.
- **Overhead Transparencies** set includes 120 full-color acetates with labels enlarged for easy viewing.

Saunders College Publishing may provide complimentary instructional aids and supplements or supplemental packages to those adopters qualified under our adoption policy. Please contact your sales representative for more information. If, as an adopter or potential user, you receive supplements you do not need, please return them to your sales representative or send them to:

Attn: Returns Department
  Troy Warehouse
  465 South Lincoln Drive
  Troy, MO 63379

Through the services of the **Harcourt Brace Custom Publishing Group,** portions of *Chemistry: Principles & Reactions,* Third Edition, can be packaged according to individual needs. Instructors who wish to augment *Chemistry: Principles & Reactions,* Third Edition with their own material, make selected chapters available in courses with a different focus than that of the textbook as a whole, or package *Chemistry: Principles & Reactions,* Third Edition with select chapters from other Saunders College Publishing textbooks should contact their local Saunders College Publishing sales representative.

## Who Helped Produce This Text?

The short answer is "a great many people." Many of our revisions reflect the comments of students and colleagues. Beyond that we are indebted to the many reviewers who devoted so much time to the project; many of them earned somewhat less than the minimum wage. Reviewers of the Third Edition include:

Linda Atwood, *California Polytechnic State University, San Luis Obispo*
Janice Bradley, *Lake City Community College*
James Carroll, *University of Nebraska, Omaha*
Cassandra Eagle, *Appalachian State University*
William B. Euler, *University of Rhode Island*
Sammye Sue Harrill, *Mount St. Mary Academy*
Daniel Haworth, *Marquette University*
Barbara Hopkins, *Xavier University*
Donald Kleinfelter, *The University of Tennessee*
Carol Martinez, *Albuquerque Technical, Vocational Institute*
Gregory Neyhart, *North Carolina State University*
George Patterson, *Suffolk University*
Lawrence Potts, *Gustavus Adolphus College*
Ronald Ragsdale, *University of Utah*
Al Shina, *San Diego City College*
Steven Socol, *Southern Utah University*
Joseph Stenson, *Delaware Valley College*
Richard Treptow, *Chicago State University*
Robert A. Wilkins, *Andrews University*
Stanley Williamson, *University of California, Santa Cruz*

We are particularly pleased to acknowledge the contributions of Donald Kleinfelter, Professor of Chemistry at The University of Tennessee. In particular, we adopted his method of balancing redox half-equations (Chapter 4); we're convinced it is an improvement on the traditional method.

Reviewers of the Second Edition include:

Peter Baine, *California State University, Long Beach*
John E. Bauman, *University of Missouri, Columbia*
Jesse Binford, *University of South Florida*
James D. Cherry, *Enrico Fermi High School*
Coran Cluff, *Brigham Young University*
Robert Conley, *New Jersey Institute of Technology*
M. Davis, *University of Texas, El Paso*
Elizabeth S. Friedman, *Los Angeles Valley College*
Steven D. Gammon, *University of Idaho*
Frederick A. Grimm, *The University of Tennessee, Knoxville*
Wyman K. Grindstaff, *Southwest Missouri State University*
Daniel T. Haworth, *Marquette University*
Douglas W. Hensley, *Louisiana Tech University*
David Hilderbrand, *South Dakota State University*
Grant N. Holder, *Appalachian State University*
Pushkar Kaul, *Boston College*
Paul B. Kelter, *The University of Wisconsin, Oshkosh*
Michael Kenney, *Marquette University*
Leslie Kinsland, *University of Southwestern Louisiana*
Deborah M. Nycz, *Broward Community College*
William E. Ohnesorge, *Lehigh University*
Virgil L. Payne, *Florida Atlantic University*
Paul S. Poskozim, *Northeastern Illinois University*
Ronald Ragsdale, *University of Utah*
T. W. Sottery, *University of Southern Maine*
Richard L. Snow, *Brigham Young University*
Donald Titus, *Temple University*
William Van Doorne, *Calvin College*
Charles A. Wilkie, *Marquette University*
Stanley M. Williamson, *University of California*
Sidney H. Young, *University of South Alabama*

Many people at Saunders made major contributions to this book. Three individuals were particularly helpful:

**John Vondeling**  After 30 years, it's hard to find anything new to say about John. His title keeps changing, but fortunately his competence and his personality remain unchanged.

**Beth Rosato,**  our developmental editor. Again and again, she went well beyond her job description to help us. She is truly a joy to work with.

**Laura Shur,**  our talented, friendly and overworked project editor.

> William L. Masterton
> Cecile N. Hurley
> University of Connecticut
> Storrs
> July, 1996

## Multimedia Materials

Available for Use with *Chemistry: Principles & Reactions,* Third Edition

- **Saunders Interactive General Chemistry CD-ROM** is a revolutionary interactive tool. This multimedia presentation serves as a useful companion to the text. Divided into chapters, the CD-ROM presents ideas and concepts with which the user can interact in several different ways, for example, by watching a reaction in progress, changing a variable in an experiment and observing the results, and listening to tips and suggestions for understanding concepts or solving problems. Students navigate through the CD-ROM using original animation and graphics, interactive tools, pop-up definitions, over 100 video clips of chemical experiments, which are enhanced by sound effects and narration, and over 100 molecular models and animations. A chapter-by-chapter guide/workbook for the students accompanies the CD-ROM. Important technological information, as well as suggestions for accessing some of the CD-ROM's unique teaching features is available for professors.
- **Cambridge Scientific ChemDraw and Chem3D** are available shrink-wrapped with the text for a nominal fee. These software packages enable students to draw molecular structures using ChemDraw. Users draw with ChemDraw and then can transfer their structures into Chem3D, which allows them to create, manipulate, and view three-dimensional color models for a clearer image of a molecule's shape and reaction sites. Cambridge Scientific provides an accompanying User's Guide and Quick Reference Card, written exclusively for Saunders College Publishing.
- **CalTech Chemistry Animation Project (CAP)** is a set of five video units of unmatched quality and clarity that cover the chemical topics of Atomic Orbitals, Valence Shell Electron Pair Repulsion Theory, Crystals, Molecular Orbitals, and Periodic Trends.
- **Chemistry 1997 MediaActive™** CD-ROM provides still imagery from Masterton and other quality Saunders textbooks. Available as a presentation tool, this CD-ROM can be used in conjunction with commercial presentation packages, such as Powerpoint™, Persuasion™, and Podium™, as well as the Saunders LectureActive™ presentation software. Available in both Macintosh and Windows platforms.
- Lecture outlines for the professor are also available created with **Powerpoint™**. Relating to the text on a per chapter basis, Powerpoint™ users can edit the content with their own material or import material from our Chemistry 1997 MediaActive™.
- The **Chemistry in Perspective Videodisc** contains over 110 minutes of motion footage, including molecular model animations, chemical reaction videos, animated principles of chemistry, and videos demonstrating chemical principles at work in everyday life, as well as 2000 still images from several of Saunders College Publishing's chemistry titles. Excerpts from *World of Chemistry*

video series, published by Annenberg/CPB Project and corporate sponsors, also appear on this videodisc.

- **LectureActive™ Software for Macintosh and IBM formats** accompanies both the Chemistry 1997 MediaActive™ CD-ROM and Chemistry in Perspective videodisc and contains references to all video clips and still images. Instructors can create custom lectures quickly and easily. Lectures can be read directly from the computer screen or printed with barcodes that contain videodisc instructions.

- **Barcode Manuals** for the MediaActive 1997 CD-ROM and Chemistry in Perspective videodisc contain complete descriptions, barcode labels, and reference numbers for every still image or video clip. The manual also provides practical advice about MediaActive 1997 CD-ROM and Chemistry in Perspective videodisc set-up instructions for first-time users.

- **KC?Discoverer Software (JCE:Software),** developed by a team of chemists, is an extensive database allowing users to explore 48 different properties of the chemical elements, such as atomic radii, density, ionization energy, color, reactivity with air, water, and acids and bases. The HELP menu provides a reference for the source of data for each of the properties and the database. KC?Discoverer Software has the capability to correlate with the Periodic Table Videodisc. This program is available to qualified adopters or it may be purchased directly from JCE: Software Department of Chemistry, University of Madison, Wisconsin, 1101 University Avenue, Madison, WI 53706.

- **Periodic Table Videodisc: Reactions of the Elements (JCE: Software),** by Alton Banks, North Carolina State University, features still and live footage of the elements, their uses, and their reactions with air, water, acids, and bases. Users operate the videodisc from a videodisc player with a hand-controlled keypad, a barcode reader, or an interface to a computer running KC?Discoverer Software. It is particularly useful as a way to demonstrate chemical reactions in a large lecture room. Available to qualified adopters.

- **Shakhashiri Demonstration Videotapes** feature well-known instructor Bassam Shakhashiri, University of Wisconsin, performing 50 three- to five-minute chemical demonstrations. An accompanying Instructor's Manual describes each demonstration and includes discussion questions.

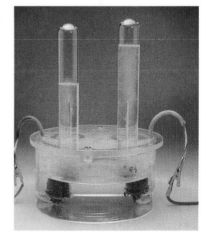

# Contents

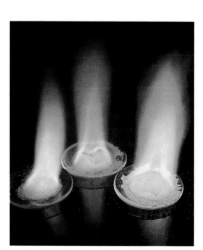

# 7  Covalent Bonding ................................... 173

# 8  Thermochemistry ................................... 207

# 9  Liquids and Solids ................................... 238

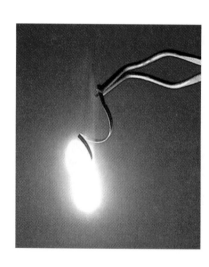

An attractive ore (orpiment, $As_2S_3$) of a poisonous element (see p. 20) (Paul Silverman/ Fundamental Photos, NYC)

# Matter and Measurement     1

**I**nchworm, inchworm, measuring
the marigolds,
You and your arithmetic will
probably go far.
Inchworm, inchworm, measuring
the marigolds,
Seems to me you'd stop and see
how beautiful they are.

—NURSERY RHYME

Almost certainly, this is your first college course in chemistry; perhaps it is your first exposure to chemistry at any level. Unless you are a chemistry major, you may wonder why you are taking this course and what you can expect to gain from it. To address that question, it will be helpful to look at some of the ways in which chemistry contributes to other disciplines.

If you're planning to be an engineer, you can be sure that many of the materials you will work with have been synthesized by chemists. Some of these materials are organic (carbon-containing). They could be familiar plastics like polyethylene (Chapter 22) or the more esoteric plastics used in unbreakable windows and nonflammable clothing. Other materials, including metal alloys (Chapter 20) and semiconductors (Chapter 9), are inorganic in nature.

Perhaps you are a health science major, looking forward to a career in medicine or pharmacy. If so, you will want to become familiar with the properties of aqueous solutions (Chapters 4, 10, 14, and 16), which include blood and other body fluids. Chemists today are involved in the synthesis of a variety of life-saving products. These range from drugs used in chemotherapy (Chapter 15) to new antibiotics used against resistant microorganisms.

Beyond career preparation, an objective of a college education is to stimulate your curiosity about the world around you or, to abuse a cliché, to make you a "better informed citizen." In this text, we will look at some of the chemistry-related issues facing society today, including

— forensic science in the courtroom (Chapters 1, 4)
— depletion of the ozone layer (Chapter 11)
— the phenomenon of acid rain (Chapter 13)
— fluoridation of water supplies (Chapter 16)
— the greenhouse effect (Chapter 17)

These topics are covered in end-of-chapter essays entitled, "Chemistry Beyond the Classroom."

We hope that when you complete this course you too will be convinced of the importance of chemistry in today's world. We should, however, caution you upon one point. Although we will talk about many of the applications of chemistry, *our main concern will be with the principles that govern chemical reactions*. Only by mastering these principles will you understand the basis of the applications referred to in the preceding paragraphs.

Taxol, a complex organic molecule used in chemotherapy, was synthesized by two groups of chemists in 1994. (Charles D. Winters)

This chapter begins the study of chemical principles by

— considering the different types of matter: pure substances versus mixtures, elements versus compounds (Section 1.1)
— looking at the nature of measurements (Section 1.2), the uncertainties associated with them (Section 1.3), and the method used to convert measured quantities from one set of units to another (Section 1.4)
— focusing on certain physical properties, including density and water solubility, used to identify substances (Section 1.5)

## 1.1  Types of Matter

Matter is anything that has mass and occupies space. It exists in three phases: solid, liquid, and gas. A solid has a rigid shape and a fixed volume. A liquid has a fixed volume but is not rigid in shape; it takes on the shape of the container. A gas has neither a fixed volume nor a rigid shape; it takes on both the volume and the shape of the container.

Matter can be classified into two categories:

— pure substances, each of which has a fixed composition and a unique set of properties.
— mixtures, composed of two or more substances.

Pure substances are either elements or compounds (Fig. 1.1), whereas mixtures can be either homogeneous or heterogeneous.

*Most materials you encounter are mixtures*

### Elements

An **element** is a type of matter which cannot be broken down into two or more pure substances. There are 112 known elements, of which 91 occur naturally.

Many elements are familiar to all of us. The charcoal used in outdoor grills is nearly pure carbon. Electrical wiring, jewelry, and water pipes are often made from copper, a metallic element. Another such element, aluminum, is used in many household utensils. The shiny liquid in the thermometers you use is still another metallic element, mercury.

Some elements come in and out of fashion, so to speak. Fifty years ago, elemental silicon was a chemical curiosity. Today, ultrapure silicon has become

**Figure 1.1**
Classification of matter.

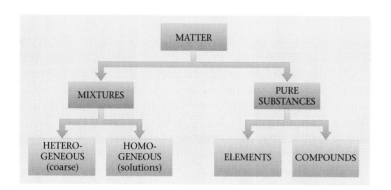

Elemental silicon rests atop silicon wafer used in the integrated circuits of semiconductors. (Charles D. Winters)

| TABLE 1.1    **Names and Symbols of Some of the More Familiar Elements** | | | | | | | |
|---|---|---|---|---|---|---|---|
| Aluminum | Al | Chlorine | Cl | Lithium | Li | Rubidium | Rb |
| Antimony | Sb | Chromium | Cr | Magnesium | Mg | Selenium | Se |
| Argon | Ar | Cobalt | Co | Manganese | Mn | Silicon | Si |
| Barium | Ba | Copper | Cu | Mercury | Hg | Silver | Ag |
| Beryllium | Be | Fluorine | F | Neon | Ne | Sodium | Na |
| Bismuth | Bi | Gold | Au | Nickel | Ni | Strontium | Sr |
| Boron | B | Helium | He | Nitrogen | N | Sulfur | S |
| Bromine | Br | Hydrogen | H | Oxygen | O | Tin | Sn |
| Cadmium | Cd | Iodine | I | Phosphorus | P | Uranium | U |
| Calcium | Ca | Iron | Fe | Platinum | Pt | Xenon | Xe |
| Carbon | C | Krypton | Kr | Plutonium | Pu | Zinc | Zn |
| Cesium | Cs | Lead | Pb | Potassium | K | | |

the basis for the multibillion-dollar semiconductor industry. Lead, on the other hand, is an element moving in the other direction. A generation ago it was widely used to make paint pigments, plumbing connections, and gasoline additives. Today, because of the toxicity of lead compounds, all of these applications have been banned in the United States.

In chemistry, an element is identified by its symbol. This consists of one or two letters, usually derived from the name of the element. Thus the symbol for carbon is C; that for aluminum is Al. Sometimes the symbol comes from the Latin name of the element or one of its compounds. The two elements copper and mercury, which were known in ancient times, have the symbols Cu *(cuprum)* and Hg *(hydrargyrum)*. Table 1.1 lists the names and symbols of the elements that we will be most concerned with in this text.

## Compounds

A **compound** is a pure substance that contains more than one element. Water is a compound of hydrogen and oxygen. The compounds methane, acetylene, and naphthalene all contain the elements carbon and hydrogen, in different proportions.

Compounds have fixed compositions. That is, a given compound always contains the same elements in the same percentages by mass. A sample of pure water contains precisely 11.19% hydrogen and 88.81% oxygen. In contrast, mixtures can vary in composition. For example, a mixture of hydrogen and oxygen might contain 5, 10, 25, or 60% hydrogen, along with 95, 90, 75, or 40% oxygen.

The properties of compounds are very different from those of the elements they contain. Ordinary table salt, sodium chloride, is a white, unreactive solid. As you can guess from its name, it contains the two elements sodium and chlorine. Sodium (Na) is a shiny, extremely reactive metal. Chlorine (Cl) is a poisonous, greenish-yellow gas. Clearly, when these two elements combine to form sodium chloride, a profound change takes place (Fig. 1.2).

Many different methods can be used to resolve compounds into their elements. Sometimes, but not often, heat alone is sufficient. Mercury(II) oxide, a

**Figure 1.2**
Sodium is a shiny, highly reactive metal, ordinarily stored under toluene (bottle) to prevent reaction with air and water. Chlorine is a greenish-yellow gas, shown here in a high-pressure cylinder. Sodium chloride is a white, nontoxic solid (table salt). (Charles D. Winters)

compound of mercury and oxygen, decomposes to its elements when heated to 600°C. Joseph Priestley, an English chemist, discovered oxygen 200 years ago when he carried out this reaction by exposing a sample of mercury(II) oxide to an intense beam of sunlight focused through a powerful lens. Another method of resolving compounds into elements is called electrolysis. This involves passing an electric current through a compound, usually in the liquid state. Through the process of electrolysis, it is possible to separate water into the two elements hydrogen and oxygen.

**Figure 1.3**
Mixtures can be *homogeneous,* as with brass, which is a solid solution of copper and zinc. Alternatively, they can be *heterogeneous,* as with granite, which contains discrete regions of different minerals (feldspar, mica, and quartz). (Charles D. Winters)

## Mixtures

A **mixture** contains two or more pure substances combined in such a way that each substance retains its chemical identity. When you shake iron filings with powdered sulfur, a mixture is formed; the two elements are chemically unchanged. In contrast, when sodium is exposed to chlorine gas, a compound, sodium chloride, is formed; the two elements lose their chemical identity.

There are two types of mixtures.

**1. Homogeneous** or uniform mixtures, in which the composition of the mixture is the same throughout. Another name for a homogeneous mixture is a **solution.** A solution is made up of a solvent, the substance present in largest amount, and one or more solutes. Most commonly, the solvent is a liquid, while solutes may be solids, liquids, or gases. Soda water is a solution of carbon dioxide (solute) in water (solvent). Seawater is a more complex solution in which there are several solid solutes, including sodium chloride; the solvent is water. It is also possible to have solutions in the solid state. Brass (Fig. 1.3) is a solid solution containing the two metals copper (67%–90%) and zinc (10%–33%).

**2. Heterogeneous** or nonuniform mixtures are those in which the composition varies throughout. Most rocks fall into this category. In a piece of granite (Fig. 1.3), several components can be distinguished, differing from one another in color.

Many different methods can be used to separate the components of a mixture from one another. A couple of methods that you may have carried out in the laboratory are

— *filtration,* used to separate a heterogeneous solid–liquid mixture. The mixture is passed through a barrier with fine pores such as filter paper.
— *distillation* (Fig. 1.4, p. 6), used to resolve a homogeneous solid–liquid mixture. The liquid vaporizes, leaving a residue of the solid in the distilling flask. Pure liquid is obtained by condensing the vapor.

A more complex but more versatile separation method is **chromatography,** a technique widely used in teaching, research, and industrial laboratories to separate all kinds of mixtures. This method takes advantage of differences in solubility and/or extent of adsorption on a solid surface. In *gas-liquid chromatography,* a mixture of volatile liquids and gases is introduced into one end of a heated glass tube. As little as one microliter ($10^{-6}$ L) of sample may be used. The tube is packed with an inert solid whose surface is coated with a viscous liquid. An unreactive "carrier gas," often helium, is passed through the tube.

All gaseous mixtures, including air, are solutions

A heterogeneous mixture of copper sulfate crystals (blue) and sand. (Charles D. Winters)

**Figure 1.4**

The two components of a water solution of potassium chromate can be separated from each other by distillation. Water is vaporized by heating; cooling the vapor causes it to condense to a pure liquid in the flask at the lower right. Because solid potassium chromate is not volatile, it remains in the distillation flask as a yellow residue.

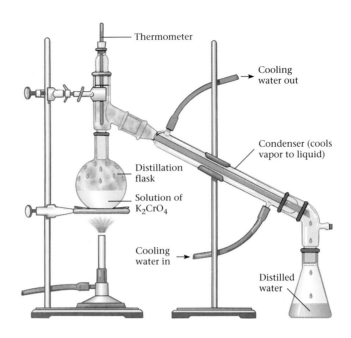

Thermometer

Cooling water out

Condenser (cools vapor to liquid)

Distillation flask

Solution of $K_2CrO_4$

Cooling water in

Distilled water

The components of the sample gradually separate as they vaporize into the helium or condense into the viscous liquid. Usually the more volatile fractions move faster and emerge first; successive fractions activate a detector and recorder. The end result is a plot such as that shown in Figure 1.5.

Gas-liquid chromatography (GLC) finds many applications outside the chemistry laboratory. If you've ever had an emissions test on the exhaust sys-

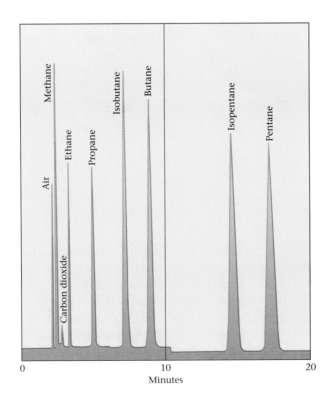

Air

Methane

Carbon dioxide

Ethane

Propane

Isobutane

Butane

Isopentane

Pentane

0          10          20

Minutes

**Figure 1.5**

The components of natural gas (mostly methane) can be separated by gas-liquid chromatography. With some volatile mixtures, the sample can be as small as $10^{-6}$ L.

tem of your car, GLC was almost certainly the analytical method used. Pollutants such as carbon monoxide and unburned hydrocarbons appear as peaks on a graph such as that shown in Figure 1.5. A computer sums the areas under these peaks, which are proportional to the concentrations of pollutants, and prints out a series of numbers that tells you whether your car passed or failed the test. Many of the techniques used to test people for drugs (marijuana, cocaine, and others) or alcohol also make use of gas-liquid chromatography.

## 1.2 Measurements: Quantities and Units

When you measure a quantity such as length or mass you express your result as a number attached to a unit, e.g., "6.02 *inches*," "148 *pounds*." Scientists ordinarily use a set of units, developed in France in the eighteenth century, known as the **metric system.** This, as you probably know, is a decimal-based system in which all the units of a particular quantity are related to each other by powers of ten.

Surprising as it may seem, all the quantities that scientists measure can be expressed in terms of seven base quantities (Table 1.2). Putting it another way, if base units can be established for each of these quantities (length, mass, and so forth), then units for all other quantities (e.g., volume, force, pressure, energy) can be related to them. Recognizing this fact, the General Conference of Weights and Measures in 1960 recommended an International System of Units **(SI)** constructed upon the seven base units listed in Table 1.2. Decimal multiples of these units are shown using the prefixes listed at the right of the table.

To see how this all works out, let us examine three base quantities commonly measured in chemistry: length, mass, and temperature. We will consider how these quantities are measured as well as the units in which they are expressed.

### Length

The meter stick found in every general chemistry laboratory reproduces the SI base unit of length, the meter. The meter was originally intended to be 1/40,000,000 of the Earth's meridian that passes through Paris. It is now defined in terms of the speed of light, which travels a distance of one meter in 1/299,792,458 second.

#### TABLE 1.2    SI Base Units and Selected Prefixes

| Base Quantity | Base Unit | Factor* | Prefix |
|---|---|---|---|
| Length | meter (m) | $10^6$ | mega (M) |
| Mass | kilogram (kg) | $10^3$ | kilo (k) |
| Temperature | kelvin (K) | $10^{-1}$ | deci (d) |
| Time | second (s) | $10^{-2}$ | centi (c) |
| Amount of substance | mole (mol) | $10^{-3}$ | milli (m) |
| Electric current | ampere (A) | $10^{-6}$ | micro ($\mu$) |
| Luminous intensity | candela (cd) | $10^{-9}$ | nano (n) |
| | | $10^{-12}$ | pico (p) |

*Exponential notation is discussed in Appendix 3.

Picometers are also used; 1 nm = $10^3$ pm

**Figure 1.6**
Single-pan analytical balance with digital readout; the solid sample and container weigh 46.289 g. (Marna G. Clarke)

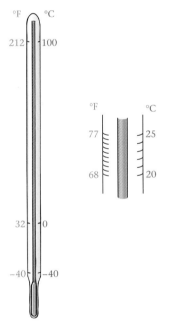

**Figure 1.7**
On the Celsius scale, the distance between the freezing and boiling points of water is 100°; on the Fahrenheit scale it is 180°. This means that the Celsius degree is $\frac{9}{5}$ as large as the Fahrenheit degree, as is evident from the magnified section of the thermometer at the right.

Other units of length are expressed in terms of the meter, using the prefixes listed in Table 1.2. You are familiar with the centimeter, the millimeter, and the kilometer:

$$1 \text{ cm} = 10^{-2} \text{ m} \qquad 1 \text{ mm} = 10^{-3} \text{ m} \qquad 1 \text{ km} = 10^3 \text{ m}$$

The dimensions of very tiny particles are often expressed in nanometers:

$$1 \text{ nm} = 10^{-9} \text{ m}$$

## Mass

In the metric system, mass may be expressed in grams, kilograms, or milligrams; the base SI unit is the kilogram.

$$1 \text{ g} = 10^{-3} \text{ kg} \qquad 1 \text{ mg} = 10^{-6} \text{ kg}$$

The megagram, more frequently called the **metric ton,** is

$$1 \text{ Mg} = 10^6 \text{ g} = 10^3 \text{ kg}$$

Properly speaking, there is a distinction between mass and weight. **Mass** is a measure of the amount of matter in an object; *weight* is a measure of the gravitational force acting on the object. Chemists often use these terms interchangeably; we determine the mass of an object by "weighing" it on a balance (Fig. 1.6).

## Temperature

Temperature is the factor that determines the direction of heat flow. When two objects at different temperatures are placed in contact with one another, heat flows from the one at the higher temperature to the one at the lower temperature.

Temperature is measured indirectly, by observing its effect upon the properties of a substance. A mercury-in-glass thermometer takes advantage of the fact that mercury, like other substances, expands as temperature increases. When the temperature rises, the mercury in the thermometer expands up a narrow tube. The total volume of the tube is only about 2% of that of the bulb at the base. In this way, a rather small change in volume is made readily visible.

Thermometers used in chemistry are marked in degrees *Celsius* (referred to as degrees centigrade until 1948). On this scale, named after the Swedish astronomer Anders Celsius (1701–1744), the freezing point of water is taken to be 0°C. The normal boiling point of water is 100°C. Household thermometers in the United States are commonly marked in *Fahrenheit* degrees. Daniel Fahrenheit (1686–1736) was a German instrument maker who was the first to use the mercury-in-glass thermometer. On this scale, the normal freezing and boiling points of water are taken to be 32° and 212°, respectively (Fig. 1.7). It follows that (212°F − 32°F) = 180°F covers the same temperature interval as (100°C − 0°C) = 100°C. This leads to the general relation between the two scales:

$$t_{°F} = 1.8 \, t_{°C} + 32°$$

Notice (Fig. 1.7) that the two scales coincide at −40°, as can readily be seen from the equation just written:

$$\text{At } -40°C: \, t_{°F} = 1.8(-40°) + 32° = -72° + 32° = -40°$$

The SI base unit for temperature is the **kelvin (K);** note the absence of the degree sign. The kelvin is defined to be 1/273.16 of the difference between the lowest attainable temperature (0 K) and the triple point of water* (0.01°C). The relationship between temperature in K and in °C is

$$T_K = t_{°C} + 273.15$$

This scale is named after Lord Kelvin, a British scientist who showed in 1848, at the age of 24, that it is impossible to reach a temperature lower than 0 K.

---

**Example 1.1**   Express normal body temperature, 98.60°F, in °C and K.

*Strategy*   Use the relations $t_{°F} = 1.8t_{°C} + 32°$ and $T_K = t_{°C} + 273.15$. Solve algebraically for the desired quantity, $t_{°C}$ in the first case, $T_K$ in the second.

*Solution*

To find $t_{°C}$:      $98.60° = 1.8t_{°C} + 32°$

Solving:   $t_{°C} = \dfrac{98.60° - 32°}{1.8} =$   37.00°C

To find $T_K$:   $T_K = 37.00 + 273.15 =$   310.15 K

---

## Derived Quantities and Units

As pointed out earlier, units for any measured quantity can be derived from the base units listed in Table 1.2. We will consider the units used to express four derived quantities: volume, force, pressure, and energy.

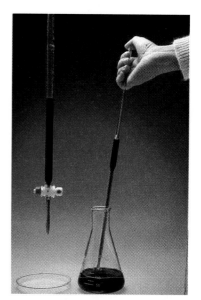

**Volume**   The SI unit for volume is the *cubic meter,* $m^3$. This is a very large unit; a cubic meter of water weighs about a ton. We will more often express volumes in

— cubic centimeters    $1\ cm^3 = (10^{-2}\ m)^3 = 10^{-6}\ m^3$
— liters (L)         $1\ L = 10^{-3}\ m^3 = 10^3\ cm^3$
— milliliters (mL)     $1\ mL = 10^{-3}\ L = 10^{-6}\ m^3$

Notice that a milliliter is equal to one cubic centimeter: $1\ mL = 1\ cm^3$.

The device most commonly used to measure volumes in general chemistry is the graduated cylinder. A pipet or buret (Fig. 1.8) is used when greater accuracy is required. A pipet is calibrated to deliver a fixed volume of liquid, e.g., 25.00 mL, when filled to the mark and allowed to drain. Variable volumes can be delivered accurately by a buret, perhaps to ± 0.01 mL.

**Force**   The SI unit of force is the *newton* (N); it is the force required to give a mass of one kilogram an acceleration of one meter per second squared. That is,

$$1\ N = 1\ kg \cdot m/s^2$$

The newton, by itself, is not widely used in general chemistry. It is, however, involved in the definitions of the SI units for pressure and energy.

**Figure 1.8**
A buret (left) is used to deliver a variable volume of liquid, accurately measured. A pipet (right) is used to deliver a fixed volume (e.g., 25.00 mL) of liquid. (Charles D. Winters)

---

*The triple point of water (Chapter 9) is the one, unique temperature at which ice, liquid water, and water vapor can coexist in contact with one another.

**Pressure**   The SI unit for pressure is the **pascal (Pa),** which is the pressure exerted by a force of one newton operating on an area of one square meter.

$$1 \text{ Pa} = 1 \text{ N/m}^2 = 1 \text{ kg/m} \cdot \text{s}^2$$

A pascal is a very small unit; a film of water 0.1 mm thick exerts a pressure of about 1 Pa on a surface with which it is in contact. The *kilopascal* is used more frequently; atmospheric pressure is of the order of 100 kPa (or $10^5$ Pa).

Chemists use a variety of units to express pressure, as you will see in Chapter 5. Perhaps the most common is the **standard atmosphere (atm),** defined by the relation

$$1 \text{ atm} = 1.01325 \times 10^5 \text{ Pa}$$

**Energy**   The SI unit of energy is the **joule (J),** which is the work done when a force of one newton acts through a distance of one meter.

$$1 \text{ J} = 1 \text{ N} \cdot \text{m} = 1 \text{ kg} \cdot \text{m}^2/\text{s}^2$$

In the past, chemists used the calorie,* which is the amount of heat required to raise the temperature of one gram of water one degree Celsius. The calorie is now defined by the relation

$$1 \text{ cal} = 4.184 \text{ J}$$

A joule is a rather small energy unit. One joule of electrical energy would keep a 10-watt light bulb burning for only a tenth of a second. For that reason, we will often express energy changes in kilojoules.

$$1 \text{ kJ} = 10^3 \text{ J}$$

> Another pressure unit is the millimeter of mercury (1 atm = 760 mm Hg)

# Chemistry: *The Human Side*

The discussion in Section 1.2 emphasizes the importance of numerical measurements, involving such quantities as mass, temperature, and volume. When a chemist carries out an experiment in the laboratory, he or she almost always makes quantitative measurements. Consider, for example, the following directions for the preparation of aspirin, abstracted from a laboratory manual in organic chemistry.

> Add *2.0 g* of salicylic acid, *5.0 mL* of acetic anhydride, and *5 drops* of 85% $H_3PO_4$ to a 50-mL Erlenmeyer flask. Heat in a water bath at *75°C* for *15 minutes*. Add cautiously *20 mL* of water and transfer to an ice bath at *0°C*. Scratch the inside of the flask with a stirring rod to initiate crystallization. Separate aspirin from the solid-liquid mixture by filtering through a Buchner funnel *10 cm* in diameter.

Chemistry was not always so quantitative. The following recipe for finding the philosopher's stone was recorded about 300 years ago.

> Take all the mineral salts there are, also all salts of animal and vegetable origin. Add all the metals and minerals, omitting none. Take two parts of the salts and grate in one part of the metals and minerals. Melt this in a crucible, forming a mass that reflects the essence of the world in all its colors. Pulverize this and pour vinegar over it. Pour off the red liquid into English wine bottles, filling them half-full. Seal them with the bladder of an ox (*not* that of a pig). Punch a hole in the top with a coarse needle. Put the bottles in hot sand for three months. Vapor will escape through the hole in the top, leaving a red powder. . . .

**Antoine Lavoisier**

(1743–1794)

(Northwind Picture Archive)

---

*The "calorie" referred to by nutritionists is actually a kilocalorie (1 kcal = $10^3$ cal). On a "2000-calorie" per day diet, you eat food capable of producing 2000 kcal = $2 \times 10^3$ kcal = $2 \times 10^6$ cal of energy.

One man more than any other transformed chemistry from an art to a science. Antoine Lavoisier was born in Paris; he died on the guillotine during the French Revolution. Above all else, Lavoisier understood the importance of carefully controlled, quantitative experiments. These were described in his book *Elements of Chemistry.* Published in 1789, it is illustrated with diagrams by his wife.

The results of one of Lavoisier's quantitative experiments are shown in Table 1.A; the data are taken directly from Lavoisier. If you add up the masses of reactants and products (expressed in arbitrary units), you find them to be the same, 510. As Lavoisier put it, "In all of the operations of men and nature, nothing is created. An equal quantity of matter exists before and after the experiment."

Lavoisier was executed because he was a tax collector; chemistry had nothing to do with it

**TABLE 1.A** **Quantitative Experiment on the Fermentation of Wine (Lavoisier)**

| Reactants | Mass (Relative) | Products | Mass (Relative) |
|---|---|---|---|
| Water | 400 | carbon dioxide | 35 |
| Sugar | 100 | alcohol | 58 |
| Yeast | 10 | acetic acid | 3 |
| | | water | 409 |
| | | sugar (unreacted) | 4 |
| | | yeast (unreacted) | 1 |

This was the first clear statement of the law of conservation of mass (Chapter 2), which was the cornerstone for the growth of chemistry in the nineteenth century. Again, to quote Lavoisier, "It is on this principle that the whole art of making experiments is founded."

# 1.3 Uncertainties in Measurements; Significant Figures

Every measurement carries with it a degree of uncertainty whose magnitude depends upon the nature of the measuring device and the skill with which it is used. Suppose, for example, you measure out 8 mL of liquid using the 100-mL graduated cylinder shown in Figure 1.9. Here the volume is uncertain to perhaps ±1 mL. With such a crude measuring device, you would be lucky to obtain a volume between 7 and 9 mL. To obtain greater precision, you could use a narrow 10-mL cylinder, which has divisions in small increments. You might now measure a volume within 0.1 mL of the desired value, in the range of 7.9 to 8.1 mL. Using a buret, the uncertainty could be reduced to ±0.01 mL.

Anyone making a measurement has a responsibility to indicate the uncertainty associated with it. Such information is vital to anyone who wants to repeat the experiment or judge its precision. The three volume measurements referred to above could be reported as

| | |
|---|---|
| 8 ± 1 mL | (large graduated cylinder) |
| 8.0 ± 0.1 mL | (small graduated cylinder) |
| 8.00 ± 0.01 mL | (buret) |

**Figure 1.9**
The uncertainty associated with a measurement depends upon the nature of the measuring device. To measure out small volumes of liquid precisely, a 10-mL graduated cylinder is much more effective than one with a volume of 100 mL. (Marna G. Clarke)

In this text, we will drop the ± notation and simply write

8 mL     8.0 mL     8.00 mL

When we do this, it is understood that there is an ***uncertainty of at least one unit in the last digit,*** that is, 1 mL, 0.1 mL, 0.01 mL, respectively. This method of citing the degree of confidence in a measurement is often described in terms of **significant figures,** the meaningful digits obtained in a measurement. In 8.00 mL there are three significant figures; each of the three digits has experimental meaning. Similarly, there are two significant figures in 8.0 mL and one significant figure in 8 mL.

*There's a big difference between 8 mL and 8.00 mL, perhaps as much as half a milliliter*

## Counting Significant Figures

Frequently, you need to know the number of significant figures in a measurement that someone else has reported. To do this, you apply the following common-sense rules.

1. *All nonzero digits are significant.* There are three significant figures in 5.37 cm and four significant figures in 4.293 cm.
2. *Zeros between nonzero digits are significant.* There are three significant figures in 106 g or in 1.02 g.
3. *Zeros beyond the decimal point at the end of a number are significant.* When the volume of a liquid is reported as 8.00 mL, this implies that the two zeros are experimentally meaningful. The quantity 8.00 mL carries the same degree of precision as 8.13 mL or 7.98 g; all of these quantities have three significant figures.
4. *Zeros preceding the first nonzero digit in a number are not significant.* In a mass measurement of 0.002 g, there is only one significant figure—the "2" at the end. The zeros serve only to fix the position of the decimal point. This becomes obvious if the mass is expressed in **exponential** (scientific) **notation** (Appendix 3). In that case, 0.002 g is written as

$$2 \times 10^{-3} \text{ g}$$

Now, clearly, there is only one significant figure. The uncertainty is $\pm 1 \times 10^{-3}$ g.

Sometimes the number of significant figures in a reported measurement is ambiguous. Suppose that a piece of metal is reported to weigh "500 g." You cannot be sure how many of these digits are meaningful. Perhaps the metal was weighed to the nearest gram ($500 \pm 1$ g). If so, the "5" and the two zeros are significant; there are three significant figures. Then again, the metal might have been weighed only to the nearest 10 g ($500 \pm 10$ g). In this case, only the "5" and one zero are known accurately; there are two significant figures. About all you can do in such cases is to wish the person who carried out the weighing had used exponential notation. The mass should have been reported as

*The number of significant figures is the number of digits shown when a quantity is expressed in exponential notation*

$5.00 \times 10^2$ g     (3 significant figures)

or                $5.0 \times 10^2$ g     (2 significant figures)

or                $5 \times 10^2$ g     (1 significant figure)

In general, *any ambiguity concerning the number of significant figures in a measurement can be resolved by using exponential notation.*

---

**Example 1.2**    Three different students weigh the same object, using different balances. They report the following masses:

(a) 15.02 g     (b) 15.0 g     (c) 0.01502 kg

How many significant figures are there in each value?

**Strategy**    Follow the rules for counting significant figures. If you have trouble deciding whether a zero is significant, put the number in exponential notation.

**Solution**

(a)   4.

(b)   3.   The zero after the decimal point is significant. It indicates that the object was weighed to the nearest 0.1 g.

(c)   4.   The zeros at the left are not significant. They are there only because the mass was expressed in kilograms rather than grams. Note that "15.02 g" and "0.01502 kg" represent the same mass.

---

## Significant Figures in Multiplication and Division

Most measured quantities are not end results in themselves. Instead, they are used to calculate other quantities, often by multiplication or division. The precision of any such derived result is limited by that of the measurements on which it is based. *When measured quantities are multiplied or divided, the number of significant figures in the result is the same as that in the quantity with the smallest number of significant figures.*

---

**Example 1.3**    An overseas flight leaves New York in the late afternoon and arrives in London 8.50 hours later. The airline distance from New York to London is about $5.6 \times 10^3$ km, depending to some extent upon the flight path followed. What is the average speed of the plane, in kilometers per hour?

**Strategy**    Calculate the average speed by taking the quotient

$$\text{speed} = \frac{\text{distance traveled}}{\text{time elapsed}}$$

Count the number of significant figures in the numerator and in the denominator; the smaller of these two numbers is the number of significant figures in the quotient.

**Solution**    The average speed will appear on your calculator as

$$\frac{5.6 \times 10^3 \text{ km}}{8.50 \text{ h}} = 658.8235294 \text{ km/h}$$

It makes no sense to report all these numbers, so don't do it

There are three significant figures in the denominator and two in the numerator. The answer should have two significant figures; round off the average speed to

$6.6 \times 10^2$ km/h.

---

Expanding upon the calculation in Example 1.3, we can see the reason behind the rule for the number of significant figures retained in multiplication or division. When we say that the distance is "$5.6 \times 10^3$ km," we mean that it lies between $5.5 \times 10^3$ and $5.7 \times 10^3$ km. Similarly, if the time is quoted to $\pm 0.01$ h, its true value should lie between 8.49 and 8.51 h. We see then that the average speed might be as large as

$$\frac{5.7 \times 10^3 \text{ km}}{8.49 \text{ h}} = 6.7 \times 10^2 \text{ km/h}$$

On the other hand, the average speed could be as little as

$$\frac{5.5 \times 10^3 \text{ km}}{8.51 \text{ h}} = 6.5 \times 10^2 \text{ km/h}$$

Looking at the results of these calculations, we see that it is entirely reasonable to report the average speed to be $6.6 \times 10^2$ km/h, implying an uncertainty of $\pm 0.1 \times 10^2$ km/h.

## Uncertainties in Addition and Subtraction

When measured quantities are added or subtracted, the uncertainty in the result is found in a quite different way than in multiplication and division. It is determined by counting the number of "decimal places," i.e., the number of digits to the right of the decimal point for each measured quantity. *When measured quantities are added or subtracted, the number of decimal places in the result is the same as that in the quantity with the smallest number of decimal places.*

To illustrate this rule, suppose you want to find the total mass of a solution made up of 10.21 g of instant coffee, 0.2 g of sugar, and 256 g of water.

|  | Mass | Uncertainty |  |
| --- | --- | --- | --- |
| Instant coffee | 10.21 g | $\pm$ 0.01 g | 2 decimal places |
| Sugar | 0.2  g | $\pm$ 0.1 g | 1 decimal place |
| Water | 256  g | $\pm$ 1 g | 0 decimal places |
| Total mass | 266 g |  |  |

Since there are no digits after the decimal point in the mass of water, there are none in the total mass. Looking at it another way, we can say that the total mass, 266 g, has an uncertainty of $\pm 1$ g, as does the mass of water, the quantity with the greatest uncertainty.

---

**Example 1.4**  A beaker containing lead pellets has a mass of 185.36 g. The empty beaker has a mass of 75.681 g. What is the mass of the lead pellets?

*Strategy*  Count the number of decimal places in both measurements. Your answer should have the same number of decimal places as the measurement with the smallest number of decimal places.

*In adding and subtracting, we count decimal places, not significant figures*

*Solution*  The mass of the lead pellets is

$$\begin{array}{ll} 185.36 \text{ g} & \text{2 decimal places} \\ \underline{75.681 \text{ g}} & \text{3 decimal places} \\ 109.679 \text{ g} & \end{array}$$

The answer should have two decimal places and is   109.68 g.

---

## Exact Numbers

In applying the principles just described, keep in mind one important point. Some numbers involved in calculations are exact rather than approximate because they are defined rather than measured quantities. Exact numbers never limit the precision of any calculated result. To illustrate the situation, consider the equation relating Fahrenheit and Celsius temperatures:

$$t_{°F} = 1.8t_{°C} + 32°$$

The numbers 1.8 and 32 are exact. Hence, they do not limit the number of significant figures in a temperature conversion; that limit is determined only by the precision of the thermometer used to measure temperature.

A different type of exact number arises in certain calculations. Suppose you are asked to determine the amount of heat evolved when *one kilogram* of coal burns. The implication is that since "one" is spelled out, *exactly* one kilogram of coal burns. The uncertainty in the answer should be independent of the amount of coal.

A number which is spelled out (one, two, --) does not affect the number of significant figures

## 1.4 Conversion of Units

It is often necessary to convert measurements expressed in one unit (e.g., grams) to another unit (milligrams or kilograms). To do this, we follow what is known as a **conversion factor** approach. For example, to convert a volume of 536 cm$^3$ to liters, the relation

$$1 \text{ L} = 1000 \text{ cm}^3$$

is used. Dividing both sides of this equation by 1000 cm$^3$ gives a quotient equal to one:

$$\frac{1 \text{ L}}{1000 \text{ cm}^3} = \frac{1000 \text{ cm}^3}{1000 \text{ cm}^3} = 1$$

The quotient 1 L/1000 cm$^3$, which is called a *conversion factor,* is multiplied by 536 cm$^3$. Since the conversion factor equals one, this does not change the actual volume. However, it does accomplish the desired conversion of units. The cm$^3$ in the numerator and denominator cancel to give the desired unit—liters.

$$536 \text{ cm}^3 \times \frac{1 \text{ L}}{1000 \text{ cm}^3} = 0.536 \text{ L}$$

The relation 1 L = 1000 cm$^3$ can be used equally well to convert a volume in liters, say, 1.28 L, to cubic centimeters. In this case, the necessary conversion factor is obtained by dividing both sides of the equation by 1 L:

$$\frac{1000 \text{ cm}^3}{1 \text{ L}} = \frac{1 \text{ L}}{1 \text{ L}} = 1$$

Multiplying 1.28 L by the quotient 1000 cm$^3$/1 L converts the volume from liters to cubic centimeters:

$$1.28 \text{ L} \times \frac{1000 \text{ cm}^3}{1 \text{ L}} = 1280 \text{ cm}^3 = 1.28 \times 10^3 \text{ cm}^3$$

Notice that a single relation (1 L = 1000 cm$^3$) gives two conversion factors:

$$\frac{1 \text{ L}}{1000 \text{ cm}^3} \quad \text{and} \quad \frac{1000 \text{ cm}^3}{1 \text{ L}}$$

## TABLE 1.3 Relations Between Length, Volume, and Mass Units

| Metric | | English | | Metric-English | |
|---|---|---|---|---|---|
| **Length** | | | | | |
| 1 km | $= 10^3$ m | 1 ft | $= 12$ in | 1 in | $= 2.54$ cm* |
| 1 cm | $= 10^{-2}$ m | 1 yd | $= 3$ ft | 1 m | $= 39.37$ in |
| 1 mm | $= 10^{-3}$ m | 1 mile | $= 5280$ ft | 1 mile | $= 1.609$ km |
| 1 nm | $= 10^{-9}$ m $= 10$Å | | | | |
| **Volume** | | | | | |
| 1 m$^3$ | $= 10^6$ cm$^3 = 10^3$ L | 1 gallon | $= 4$ qt $= 8$ pt | 1 ft$^3$ | $= 28.32$ L |
| 1 cm$^3$ | $= 1$ mL $= 10^{-3}$ L | 1 qt (U.S. liq.) | $= 57.75$ in$^3$ | 1 L | $= 1.057$ qt (U.S. liq.) |
| **Mass** | | | | | |
| 1 kg | $= 10^3$ g | 1 lb | $= 16$ oz | 1 lb | $= 453.6$ g |
| 1 mg | $= 10^{-3}$ g | 1 short ton | $= 2000$ lb | 1 g | $= 0.03527$ oz |
| 1 metric ton | $= 10^3$ kg | | | 1 metric ton | $= 1.102$ short ton |

*This conversion factor is exact; the inch is defined to be exactly 2.54 cm. The other factors listed in this column are approximate, quoted to four significant figures. Additional digits are available if needed for very accurate calculations. For example, the pound is defined to be 453.59237 g.

To go from cubic centimeters to liters, use the ratio 1 L/1000 cm$^3$; to go from liters to cubic centimeters, use the ratio 1000 cm$^3$/1 L. In general, when you make a conversion, *choose the factor that cancels out the initial unit.*

$$\text{initial quantity} \times \text{conversion factor(s)} = \text{desired quantity}$$

Conversions between English and metric units can be made using Table 1.3.

A highway sign in Missouri.
(Photograph by Beverly March)

**Example 1.5** According to a highway sign, the distance from St. Louis to Chicago is 295 miles. Express this distance in kilometers.

***Strategy*** Use Table 1.3 to find a relation between miles and kilometers. Write the conversion factor in such a way that miles cancel out and are replaced by kilometers.

***Solution*** From Table 1.3, the required relation is

$$1 \text{ mile} = 1.609 \text{ km}$$

Since the initial unit, miles, is in the numerator, the conversion factor must have miles in the denominator, i.e., 1.609 km/1 mile:

$$295 \text{ miles} \times \frac{1.609 \text{ km}}{1 \text{ mile}} = \boxed{475 \text{ km}}$$

*Later (Ch. 3, 4) we'll use conversion factors with chemical units*

Table 1.3 can be used to find certain conversion factors that are not directly listed. Suppose, for example, that you need a relation between cubic meters (m$^3$) and cubic inches (in$^3$). The table gives a relation between meters (m) and inches (in):

$$1 \text{ m} = 39.37 \text{ in}$$

Cubing both sides of this equation:

$$1 \text{ m}^3 = (39.37 \text{ in})^3 = 6.102 \times 10^4 \text{ m}^3$$

This relation can be used to convert a volume of $0.100 \, m^3$ to cubic inches:

$$0.100 \, \cancel{m^3} \times \frac{6.102 \times 10^4 \, in^3}{1 \, \cancel{m^3}} = 6.10 \times 10^3 \, in^3$$

Frequently, it is necessary to carry out more than one conversion to work a problem. This can be done by setting up successive conversion factors (Example 1.6).

---

**Example 1.6**   A certain U.S. car has a fuel efficiency rating of 36.2 miles per gallon. Convert this to kilometers per liter.

*Strategy*   Use Table 1.3 to find a relation between kilometers and miles and between gallons and liters. Sometimes there is no direct relation shown in the table; in that case, use two relations to accomplish the desired conversion. Set up the arithmetic in a single expression, writing conversion factors in such a way that, after cancellation, only the desired unit remains.

*Solution*   The relations to be used are

$$1 \text{ mile} = 1.609 \text{ km}$$

$$1 \text{ gallon} = 4 \text{ quarts} \qquad \text{(This is an exact relation.)}$$

$$1 \text{ L} = 1.057 \text{ quart}$$

Using these relations as conversion factors gives

$$36.2 \frac{\cancel{miles}}{\cancel{gallon}} \times \frac{1.609 \text{ km}}{1 \, \cancel{mile}} \times \frac{1 \, \cancel{gallon}}{4 \, \cancel{qt}} \times \frac{1.057 \, \cancel{qt}}{1 \text{ L}} = \boxed{15.4 \text{ km/L}}$$

Notice that three conversion factors are required. First miles are converted to kilometers to obtain the fuel efficiency in kilometers per gallon. Then gallons are converted to quarts, and, finally, quarts to liters.

---

The conversion factor approach shown in Examples 1.5 and 1.6 will be used throughout this text. If this is your first contact with it, it may seem awkward or artificial. You will find, however, that it is the best way to solve a wide variety of problems in chemistry. It is particularly useful when multiple conversions are required (Example 1.6).

## 1.5   Properties of Substances

Every pure substance has its own unique set of properties that serve to distinguish it from all other substances. A chemist most often identifies an unknown substance by measuring its properties and comparing them to the properties recorded in the chemical literature for known substances.

The properties used to identify a substance must be **intensive;** that is, they must be independent of amount. The fact that a sample weighs 4.02 g or has a volume of 229 mL tells us nothing about its identity; mass and volume are **extensive** properties; that is, they depend on amount. Beyond that, substances may be identified on the basis of their

— **chemical properties,** observed when the substance takes part in a **chemical reaction,** a change that converts it to a new substance. For example,

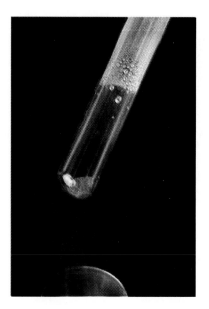

A red oxide of mercury, HgO, can be decomposed to the elements by heating to about 500°C.

(Leon Lewandowski)

the fact that mercury(II) oxide decomposes to mercury and oxygen upon heating to 600°C can be used to identify it.

— **physical properties,**  observed without changing the chemical identity of a substance. One such property is color; the fact that potassium chromate is yellow serves to distinguish it from a great many other substances.

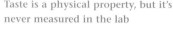
Taste is a physical property, but it's never measured in the lab

In this section, we consider four physical properties that you may well have occasion to measure in the general chemistry laboratory.

## Density

The density of a substance is the ratio of mass to volume:

$$\text{density} = \frac{\text{mass}}{\text{volume}} \qquad d = \frac{m}{V}$$

Note that even though mass and volume are extensive properties, the ratio of mass to volume is intensive. Samples of copper weighing 1.00 g, 10.5 g, 264 g, . . . all have the same density, 8.94 g/mL at 25°C.

For liquids or gases, density can be found in a straightforward way by measuring, independently, the mass and volume of a sample (Example 1.7). For liquids, density is most often expressed in g/mL; for gases, g/L is more common.

---

**Example 1.7**    To determine the density of ethyl alcohol, a student pipets a 5.00-mL sample into an empty flask weighing 15.246 g. He finds that the mass of the flask + ethyl alcohol = 19.171 g. Calculate the density of ethyl alcohol.

***Strategy***    Determine the mass of the alcohol by subtracting the mass of the empty flask from the mass of the flask and the alcohol. The volume is given. Take the quotient of mass/volume as the density.

*Solution*

mass of ethyl alcohol = 19.171 g − 15.246 g = 3.925 g

volume of ethyl alcohol = 5.00 mL

density = 3.925 g/5.00 mL =    0.785 g/mL

Note that the density is expressed to three significant figures since the volume (5.00 mL) contains only three significant figures.

---

The three layers are made up of three liquids with different densities. Gasoline is the top layer, water is the middle layer, and mercury, the densest of the three, is at the bottom. Cork floats on gasoline, oak wood sinks in gasoline but floats on water, while brass sinks in gasoline and water but floats on mercury. (Charles Steele)

For solids, density is a bit more difficult to determine. A common approach (for insoluble solids) is shown in Figure 1.10. The mass of the solid sample is found in the usual way; its volume is found indirectly.

## Melting Point and Boiling Point

The **melting point** is the temperature at which a substance changes from the solid to the liquid state. If the substance is pure, the temperature stays constant during melting. This means that, for a pure substance, the melting point is identical with the freezing point. Ice melts at 0°C; pure water freezes at that same temperature.

**Figure 1.10**
To determine the density of a solid, you first find its mass, *m*. To find its volume, add the solid to a flask of known volume, *V*, and determine the volume of water, $V_m$, required to fill the flask. The density of the solid is $m/(V - V_m)$. (Marna G. Clarke)

The **boiling point** of a liquid is the temperature at which bubbles filled with vapor form within the liquid. For reasons to be discussed in Chapter 9, boiling point depends upon the pressure above the liquid. The normal boiling point is the temperature at which a liquid boils when the pressure above it is one atmosphere. For a pure liquid, temperature remains constant during the boiling process.

If you find that a colorless liquid freezes at 0°C and boils at 100°C (at 1 atm pressure), chances are the liquid is water. To be sure, you might check the density, which should be 0.997 g/mL at 25°C. Melting point and boiling point, like density, are intensive properties. The melting point of ice is 0°C whether you're dealing with a single ice cube or a skating rink.

## Solubility

The extent to which a substance dissolves in a particular solvent can be expressed in various ways. A common method is to state the number of grams of the substance that dissolves in 100 g of solvent at a given temperature. At 20°C, about 32 g of potassium nitrate dissolves in 100 g of water. At 100°C, the solubility of this solid is considerably greater, about 246 g/100 g of water.

---

**Example 1.8**  Taking the solubility of potassium nitrate, $KNO_3$, to be 246 g/100 g of water at 100°C and 32 g/100 g of water at 20°C, calculate

(a) the mass of water required to dissolve one hundred grams of $KNO_3$ at 100°C.
(b) The amount of $KNO_3$ that remains in solution when the mixture in (a) is cooled to 20°C.

***Strategy*** The solubility at a particular temperature gives you a relationship between grams of solute ($KNO_3$) and grams of solvent (water). This in turn leads to the conversion factor required to calculate the mass of water in (a) or that of $KNO_3$ in (b). Note that the temperature is 100°C in (a), 20°C in (b).

*Solution*

(a) Mass of water required = $100 \text{ g } KNO_3 \times \dfrac{100 \text{ g water}}{246 \text{ g } KNO_3} =$ | 40.7 g water |

(b) Since the solution contains 40.7 g of water,

mass of $KNO_3$ in solution = $40.7 \text{ water} \times \dfrac{32 \text{ g } KNO_3}{100 \text{ g water}} =$ | 13 g $KNO_3$ |

In effect, we use solubility as a conversion factor

The remaining 87 g of potassium nitrate crystallizes out of solution when the temperature drops from 100°C to 20°C.

---

# CHEMISTRY

## *Beyond the Classroom*

**Figure 1.A**
Arsenic is found in nature as a yellow sulfide orpiment ($As_2S_3$). (Charles D. Winters)

# Arsenic

An element everyone has heard about but almost no one has ever seen is arsenic, symbol As. It is a gray solid with some metallic properties, melts at 816°C, and has a density of 5.78 g/mL. Among the elements, arsenic ranks fifty-first in abundance; it is about as common as tin or beryllium. Two brightly colored sulfides of arsenic, realgar and orpiment (Figure 1.A), were known to the ancients. The element is believed to have been isolated for the first time by Albertus Magnus in the thirteenth century; he heated orpiment with soap.

The principal use of elemental arsenic is in its alloys with lead. The "lead" storage battery contains a trace of arsenic along with 3% antimony. Lead shot, which are formed by allowing drops of molten metal to fall through air, contains from 0.5 to 2.0% arsenic. The presence of arsenic raises the surface tension of the liquid and hence makes the shot more nearly spherical.

In the early years of this century, several thousand organic compounds of arsenic were synthesized and tested for medicinal use, mainly in the treatment of syphilis. Inorganic compounds, including lead arsenate and arsenious acid, were widely used as insecticides. Nowadays, the use of arsenic compounds in medicine has virtually disappeared, and their use in agriculture has declined sharply. More effective and/or environmentally friendly products have been developed for these purposes. Today, by far the most important compound of arsenic is gallium arsenide, which is widely used in semiconductor devices. Although more expensive than silicon, GaAs is more efficient for certain applications, including solar batteries.

The "arsenic poison" referred to in true crime dramas is actually the oxide of arsenic, $As_2O_3$, rather than the element itself. Less than 0.1 g of this white, slightly soluble powder can be fatal. The classic symptoms of acute arsenic poisoning involve various unpleasant gastrointestinal disturbances, severe abdominal pain, and burning of the mouth and throat.

In the modern forensic chemistry laboratory (Figure 1.B), arsenic is detected by analysis of hair samples, where the element tends to concentrate in chronic arsenic poisoning. A single strand of hair is sufficient to establish the presence or absence of the element. The technique most commonly used is neutron activation analysis, described in Chapter 19. If the concentration found is greater than about 0.0003%, poisoning is indicated; normal arsenic levels are much lower than this.

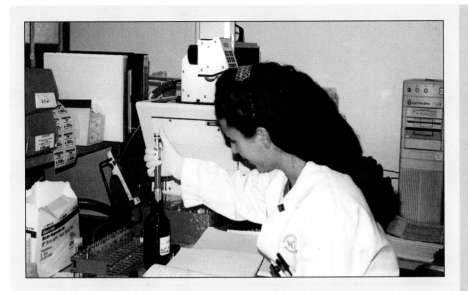

**Figure 1.B**
A chemistry graduate student at the University of Connecticut, Gina Cassella, works on a research project at the forensic toxicology laboratory of Hartford Hospital.
(Photo courtesy of Professor S. Ruven Smith)

This technique was applied in the early 1960s to a lock of hair taken from Napoleon Bonaparte (1769–1821) on St. Helena. Arsenic levels of up to 50 times normal suggested he may have been a victim of poisoning, perhaps on orders from the French royal family. More recently (1991), U.S. President Zachary Taylor (1785–1850) was exhumed on the unlikely hypothesis that he had been poisoned by Southern sympathizers concerned about his opposition to the extension of slavery. The results indicated normal arsenic levels. Apparently "Old Rough and Ready" died of cholera morbus, brought on by overindulgence in overripe fruit.

# CHAPTER HIGHLIGHTS

## *Key Concepts*

**1.** Convert between °F, °C, and K
  (Example 1.1; Problems 9–12, 57)
**2.** Determine the number of significant figures in a measured quantity
  (Example 1.2; Problems 17, 18)
**3.** Determine the number of significant figures in a calculated quantity
  (Examples 1.3, 1.4; Problems 19–24)
**4.** Use conversion factors to change the units of a measured quantity
  (Examples 1.5, 1.6, 1.8; Problems 7, 8, 15, 16, 25–36, 52, 53, 58)
**5.** Relate density to mass and volume
  (Example 1.7; Problems 37–44, 56)

## *Key Equations*

$$t_{°F} = 1.8t_{°C} + 32° \qquad T_K = t_{°C} + 273.15$$

## Key Terms

boiling point
centi-
compound
conversion factor
density
element
kelvin (K)

kilo-
melting point
milli-
mixture
—homogeneous
—heterogeneous
nano-

property
—chemical
—extensive
—intensive
—physical
significant figure
solution

## Summary Problem

Potassium permanganate is a purple compound containing three elements: potassium, manganese, and oxygen. It has a density of 2.703 $g/cm^3$; its melting point is $2.40 \times 10^2$ °C. At 20°C its solubility is 6.38 g/100 g of water; at 60°C the solubility is 25 g/100 g of water.

(a) What are the symbols of the three elements in potassium permanganate?
(b) List the physical properties of potassium permanganate given above.
(c) What is the mass of a sample of potassium permanganate with a volume of 48.7 $cm^3$?
(d) Express the density in pounds per cubic foot.
(e) Express the melting point of potassium permanganate in °F and K.
(f) How much potassium permanganate can be dissolved in 38.5 g of water at 20°C?
(g) How much potassium permanganate can be dissolved in 38.5 g of water at 60°C?
(h) Can one dissolve 10.0 g of potassium permanganate in 55.0 g of water at 60°C? If so, how much potassium permanganate remains dissolved when the solution (10.0 g of potassium permanganate/55.0 g of water) is cooled from 60°C to 20°C? How much, if any, crystallizes out?

Potassium permanganate crystals.
(Charles D. Winters)

Express all your answers to the correct number of significant figures; use the conversion factor approach throughout.

### Answers

(a) K, Mn, O    (b) density, melting point, solubility    (c) 132 g
(d) 168.7 $lb/ft^3$    (e) 464°F; 513 K    (f) 2.46 g    (g) 9.6 g
(h) yes; 3.51 g will dissolve; 6.5 g will crystallize out

## Questions & Problems

The questions and problems listed here are typical of those at the end of each chapter. Some are conceptual (these are marked by an asterisk). Most require calculations, writing equations, or other quantitative work. The topic emphasized in each question or problem is indicated in the heading, such as "Symbols and Formulas" or "Significant Figures." Those in the "Unclassified" category may involve more than one concept, perhaps including topics from a preceding chapter.

"Challenge Problems," listed at the end of the set, require extra skill and/or effort.

The "classified" questions and problems (Problems 1 to 46 in this set) are arranged in matched pairs, one below the other, and illustrate the same concept. For example, Questions 1 and 2 are nearly identical in nature; the same is true of Questions 3 and 4, and so on. Problems numbered in color are answered in Appendix 6.

## Types of Matter

**\*1.** Classify each of the following as element, compound, or mixture.

(a) air ∾

(b) table salt c

(c) vinegar ∿

(d) mercury e

**\*2.** Classify each of the following as element, compound, or mixture.

(a) gold c

(b) milk ∿

(c) sugar c

(d) mayonnaise ∿

**\*3.** Classify the following as solution or heterogeneous mixture.

(a) wine ∩

(b) vinaigrette dressing ∿

(c) batter for chocolate chip cookies ∿

**\*4.** Classify the following as solution or heterogeneous mixture.

(a) iron ore ∿

(b) seawater S

(c) gasoline S

## Measurements

**\*5.** What quantity and unit are given for

(a) an aspirin tablet mg

(b) a bottle of fruit juice oz

(c) the air in a bicycle tire

**\*6.** What quantity and unit are given for

(a) a bag of fertilizer

(b) a sphygmomanometer (reads blood pressure)

(c) the output of a water heater

**7.** Write the appropriate symbol in the blank ( $>$ , $<$ , or $=$ )

(a) 303 m _____ $303 \times 10^3$ km

(b) 500 g _____ 0.500 kg

(c) $1.50 \ cm^3$ _____ $1.50 \times 10^3 \ nm^3$

**8.** Write the appropriate symbol in the blank ( $>$ , $<$ , or $=$ )

(a) 37.12 g _____ 0.3712 kg

(b) $28 \ m^3$ _____ $28 \times 10^2 \ cm^3$

(c) 525 mm _____ $525 \times 10^6$ nm

**9.** Comfortable room temperature is 68°F. Express this in

(a) °C    (b) K

**10.** Lukewarm water has to be used when dissolving yeast. This translates to a temperature of 52°C. What is the temperature of lukewarm water in degrees Fahrenheit?

**11.** Computers are not supposed to be in very warm rooms. The highest temperature tolerated for maximum performance is 308 K. What is this temperature in °C? in °F?

**12.** "Dry Ice" is prepared by freezing carbon dioxide at −56.5°C and 5 atm. What is the freezing point of carbon dioxide in Kelvin?

**13.** Express the following derived units in terms of the base SI units listed in Table 1.2 (e.g., $1 \ cm^3 = 10^{-6} \ m^3$).

(a) liter    (b) joule    (c) kilopascal

**14.** Follow the directions in Question 13 for the following derived units.

(a) newton    (b) atmosphere    (c) $cm^3$

**15.** Convert

(a) $3.65 \times 10^5$ kPa to atmospheres

(b) 375 J to kcal

**16.** Convert

(a) 12.6 atm to Pa

(b) 15 nutritional calories to kJ

## Significant Figures

**\*17.** How many significant figures are there in each of the following?

(a) 0.285 km    (b) 0.003010 g    (c) $3.600 \times 10^{12}$ nm

(d) 4102 mL    (e) $5 \times 10^{-6}$ tons

**\*18.** How many significant figures are there in each of the following?

(a) 136.509 g    (b) 100.20 m    (c) 0.07302 atm

(d) $6.000 \times 10^4$ J    (e) 1500 min

**19.** A graduated cylinder has a circular cross section with a radius of 2.500 cm. What is the volume of water in the graduated cylinder with a measured height of 1.20 cm? (The volume of a cylinder is $\pi r^2 \ell$, where $r$ is the radius and $\ell$ is the length.)

**20.** The volume of a sphere is $\frac{4}{3}\pi r^3$, where $r$ is the radius. One student measured the radius to be 4.30 cm. Another measured the radius to be 4.33 cm. What is the difference in volume between the two measurements?

**21.** Calculate the following to the correct number of significant figures.

(a) $x = \dfrac{2.63 \ g}{4.982 \ cm^3}$    (b) $x = \dfrac{13.54 \ miles}{5.00 \ hours}$

(c) $x = 13.2 \ g + 1468 \ g + 0.04 \ g$

(d) $x = \dfrac{2 \ g + 0.127 \ g + 459 \ g}{6.2 \ cm^3 - 0.567 \ cm^3}$

**22.** How many significant figures are there in the values of $x$ obtained from

(a) $x = \dfrac{34.0300 \ g}{12.09 \ cm^3}$

(b) $x = (0.00630 \ cm)(2.003 \ cm)(200.0 \ cm)$

(c) $x = 32.647 \ in - 32.327 \ in$

(d) $x = \dfrac{236.45 \ g - 1.3 \ g}{(3.4561 \ cm)(32.567 \ cm^2)}$

**\*23.** Round off the following quantities to the indicated number of significant figures.

(a) 7.4855 g (3 significant figures)

(b) 298.693 cm (5 significant figures)

(c) 13.452 lb (2 significant figures)

(d) 346 oz (2 significant figures)

**\*24.** Round off the following quantities to the indicated number of significant figures.

(a) 132.505 g (4 significant figures)

(b) 17.2509 cm (3 significant figures)

(c) 168.51 lb (3 significant figures)

(d) 4389 oz (2 significant figures)

## Conversion Factors

**25.** Convert 22.3 mL to
   **(a)** liters   **(b)** in$^3$   **(c)** quarts

**26.** Convert 6743 nm to
   **(a)** Å   **(b)** inches   **(c)** miles

**27.** During earlier times in England, land was measured in units such as fardells, nookes, yards, and kides.

$$2 \text{ fardells} = 1 \text{ nooke} \qquad 4 \text{ nookes} = 1 \text{ yard}$$

$$4 \text{ yards} = 1 \text{ kide}$$

   **(a)** How many kides are there in 27 fardells?
   **(b)** What is the area (in square yards) of land that measures 15 fardells by 12 kides?

**28.** In 1618, the Pharmacopoeia of London defined the following mass units to be used in the preparation of drugs.

$$20 \text{ grains} = 1 \text{ scruple} \qquad 3 \text{ scruples} = 1 \text{ drachm}$$
$$8 \text{ drachms} = 1 \text{ ounce} \qquad 12 \text{ ounces} = 1 \text{ pound}$$

   **(a)** How many grains were present in 16 ounces?
   **(b)** How many pounds were there in 78.6 drachms?

**29.** The unit of land measure in the metric system is the hectare. In the English system, it is the acre. A hectare is equal to exactly 100 meters on a side. If one hectare is equal to 2.47 acres, how many square feet are there in one acre?

**30.** A light year is defined to be the distance traveled by light in one year. If the speed of light is $3.0 \times 10^{10}$ cm/s, how many miles are equal to one light year?

**31.** An average adult has 6.0 L of blood. The Red Cross usually takes one pint of blood from its donors at each donation. What percent (by volume) of her blood does a blood donor give in one donation?

**32.** Cholesterol in blood is measured in mg of cholesterol/dL of blood. If the unit of measurement was changed to g of cholesterol/mL of blood, what would a cholesterol reading of 185 translate to?

**33.** In Germany, nutritional information is given in kilojoules instead of in nutritional calories (1 nutritional calorie = 1 kcal). A packet of Rindfleisch-Suppe (meat soup) has the following information:

$$250 \text{ mL of prepared soup} = 235 \text{ kJ}$$

A packet of the same soup sold in the United States would have the same nutritional information in nutritional calories/cup. If there are two cups to a pint, what nutritional information would appear on the packet?

**34.** The legal limit for alcohol sobriety is 0.10% alcohol by volume in blood plasma. How many mL of alcohol in 3.0 qt of blood plasma does this represent?

**35.** Silver dollars must contain 90.0% silver. A silver dollar has a mass of 27.0 g. In September 1995 silver sold for $5.18 an ounce. In September 1995 did the silver dollar have more value as currency or as a source for silver?

**36.** When the *Exxon Valdez* ran aground off the coast of Alaska, $2.5 \times 10^5$ barrels of crude oil were spilled. There are exactly 42 gallons to a barrel. If the oil was allowed to flow and fill a two-car garage with dimensions $8 \times 25 \times 25$ feet, how many of these garages would be filled by the oil spilled from the tanker?

## Physical and Chemical Properties

**37.** The "cup" is a measure of volume widely used in cookbooks. One cup is equivalent to 225 mL. What is the density of clover honey (g/mL) if one-third cup has a mass of 112 grams?

**38.** The volume occupied by egg whites from four "large" eggs is 112 mL. The mass of the egg whites from these four eggs is $1.20 \times 10^2$ g. What is the average density of the egg white from one "large" egg?

**39.** A solid with an irregular shape and a mass of 12.65 g is added to a graduated cylinder filled with water (d = 1.00 g/mL) to the 29.7-mL mark. After the solid sinks to the bottom, the water level is read to be at the 36.6-mL mark. What is the density of the solid?

**40.** A metal slug weighing 16.32 g is added to a flask with a volume of 52.6 mL. It is found that 40.2 g of methanol (d = 0.791 g/mL) must be added to the metal to fill the flask. What is the density of the metal?

**41.** A sheet of aluminum foil that is 11 inches wide and 12 inches long weighs 8.9 g. If aluminum has a density of 2.70 g/cm$^3$, what is the thickness, in mm, of the foil? (Assume that the foil is pure aluminum.)

**42.** A cube of ice (d = 0.917 g/mL) is 2.00 inches on a side. How many mL of water (d = 1.00 g/mL) are obtained when the ice melts?

**43.** Vinegar contains 5.00% acetic acid by mass and has a density of 1.01 g/mL. What mass (in grams) of acetic acid is present in 10.0 L of vinegar?

**44.** Air is 21% oxygen by volume. Oxygen has a density of 1.31 g/L. What is the volume, in liters, of a room that holds enough air to contain 75 kg of oxygen?

**45.** The solubility of potassium chloride is 37.0 g/100 g of water at 30°C. Its solubility at 70°C is 48.3 g/100 g of water.
   **(a)** Calculate the mass of potassium chloride that dissolves in 46.5 g of water at 30°C.
   **(b)** Calculate the mass of water required to dissolve 27.9 g of potassium chloride at 70°C.
   **(c)** If 30.0 g of KCl were added to 75.0 g of water at 30°C, would it all dissolve? If the temperature were increased to 70°C, would it all dissolve?

**46.** The solubility of ammonium bromide in water at 20°C is 75.5 g/100 g of water. Its solubility at 50°C is 99.2 g/100 g of water. Calculate
   **(a)** the mass of ammonium bromide that dissolves in 65.0 g of water at 20°C.
   **(b)** the mass of water required to dissolve 50.0 g of ammonium bromide at 50°C.
   **(c)** the mass of ammonium bromide that would not remain in solution if a solution made up of 29.0 g of ammonium bromide in 35.0 g of water at 50°C is cooled to 20°C.

## Unclassified

**\*47.** The following data refer to the element phosphorus. Classify each as a physical or a chemical property.

(a) It exists in several forms, e.g., white, black, and red phosphorus.

(b) It is a solid at 25°C and 1 atm.

(c) It is insoluble in water.

(d) It burns in chlorine to form phosphorus trichloride.

**\*48.** The following data refer to the compound water. Classify each as a chemical or a physical property.

(a) It is a colorless liquid at 25°C and 1 atm.

(b) It reacts with sodium to form hydrogen gas as one of the products.

(c) Its melting point is 0°C.

(d) It is insoluble in carbon tetrachloride.

**\*49.** How do you distinguish

(a) chemical properties from physical properties?

(b) distillation from filtration?

(c) a solute from a solution?

**\*50.** How do you distinguish

(a) density from solubility?

(b) an element from a compound?

(c) a solution from a heterogeneous mixture?

**51.** Lead has a density of 11.34 g/cm$^3$ and oxygen has a density of $1.81 \times 10^{-3}$ g/cm$^3$ at room temperature. How many cm$^3$ are occupied by one gram of lead? By one gram of oxygen? Comment on the difference in volume for the two elements.

**52.** A cup of brewed coffee is made with about 9.0 grams of ground coffee beans. If a student drinks four cups of gourmet coffee a day, how much does the student spend on a year's supply of gourmet coffee that sells for $8.99 a pound?

**53.** The Kohinoor Diamond (d = 3.51 g/cm$^3$) is 108 carats. If one carat has a mass of $2.00 \times 10^2$ mg, what is the mass of the Kohinoor Diamond in pounds? What is the volume of the diamond in cubic inches?

**54.** A roll of aluminum foil sold in the supermarket is $66\frac{2}{3}$ yards by 12 inches. It has a mass of 0.83 kg. If the density of aluminum foil is 2.70 g/cm$^3$, what is the thickness of the foil in inches?

**55.** Titanium is used in airplane bodies because it is strong and light. It has a density of 4.55 g/cm$^3$. If a cylinder of titanium is 6.18 cm long and has a mass of 147.3 g, calculate the diameter of the cylinder. ($V = \pi r^2 \ell$, where $V$ is the volume of the cylinder, $r$ is its radius, and $\ell$ is the length.)

**56.** A pycnometer is a device used to measure density. It weighs 20.455 g empty and 31.486 g when filled with water (d = 1.00 g/cm$^3$). Pieces of an alloy are put into the empty, dry pycnometer. The mass of the alloy and pycnometer is 28.695 g. Water is added to the alloy to exactly fill the pycnometer. The mass of the pycnometer, water, and alloy is 38.689 g. What is the density of the alloy?

## Challenge Problems

**57.** At what point is the temperature in °F exactly twice that in °C?

**58.** Oil spreads on water to form a film about 100 nm thick (two significant figures). How many square kilometers of ocean will be covered by the slick formed when one barrel of oil is spilled (1 barrel = 31.5 U.S. gallons)?

**59.** A laboratory experiment requires ten grams of aluminum wire (d = 2.70 g/cm$^3$). The diameter of the wire is 0.200 in. Determine the length of the wire, in centimeters, to be used for this experiment. The volume of a cylinder is $\pi r^2 \ell$, where $r$ = radius and $\ell$ = length.

**60.** An average human male breathes about $8.50 \times 10^3$ L of air per day. The concentration of lead in highly polluted urban air is $7.0 \times 10^{-6}$ grams of lead per m$^3$ of air. Assume that 75% of the lead is present as particles less than $1.0 \times 10^{-6}$ m in diameter and that 50% of the particles smaller than that are retained in the lungs. Calculate the mass of lead absorbed in this manner in one year by an average male living in this environment.

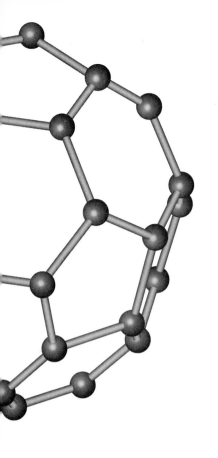

What does a geodesic dome (Epcot Center) have to do with carbon chemistry? (See p. 46) (Joach Messerschmidt/Tony Stone Images)

# 2  Atoms, Molecules, and Ions

**A**tom from atom yawns as far
As moon from earth, or star
from star.

—RALPH WALDO EMERSON

*Atoms*

**CHAPTER OUTLINE**

To learn chemistry, you must become familiar with the "building blocks" that chemists use to describe the structure of matter. These include

— *atoms* (Section 2.1), which in turn are composed of *electrons* and *nuclei* (Section 2.2). An atomic nucleus contains *neutrons* and *protons*. The stability of a nucleus depends upon the neutron-to-proton ratio; unstable nuclei decompose by a process called *radioactivity* (Section 2.4).

— *molecules,* the building blocks of several elements and a great many compounds. Molecular substances can be identified by their formulas (Section 2.5) or their names (Section 2.7).

— *ions.* An ionic compound is made up of two different kinds of ions of opposite charge. Using relatively simple principles, it is possible to derive the formulas (Section 2.6) and names (Section 2.7) of ionic compounds.

Early in this chapter (Section 2.3) we will introduce a classification system for elements known as the *periodic table*. It will prove useful in this chapter and throughout the remainder of this text.

## 2.1 Atoms and the Atomic Theory

In 1808, an English scientist and schoolteacher, John Dalton, developed the atomic model of matter that underlies modern chemistry. Three of the main postulates of modern atomic theory, all of which Dalton suggested, are stated below and illustrated in Figure 2.1.

1. *An element is composed of tiny particles called atoms.* All atoms of a given element show the same chemical properties. Atoms of different elements show different properties.
2. *In an ordinary chemical reaction, no atom of any element disappears* or is changed into an atom of another element.
3. *Compounds are formed when atoms of two or more elements combine.* In a given compound, the relative numbers of atoms of each kind are definite and constant. In general, these relative numbers can be expressed as integers or simple fractions.

On the basis of Dalton's theory, the atom can be defined as the smallest particle of an element that can enter into a chemical reaction.

Atoms of element 1

Atoms of element 2

—Compound 1

—Compound 2

Different combinations produce different compounds

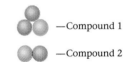
Atoms of different elements have different masses

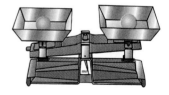
No atom disappears or is changed in a chemical reaction

**Figure 2.1**
Some features of Dalton's theory.

# Chemistry: *The Human Side*

**John Dalton**
(1766–1844)

(The E.F. Smith Memorial Collection in the History of Chemistry, Department of Special Collections, Van Pel-Dietrich Library, University of Pennsylvania)

**Figure 2.A**
Chromium forms two different compounds with oxygen, as shown by their different colors. In the green compound on the left, there are two chromium atoms for every three oxygen atoms (2 Cr : 3 O) and 2.167 g of chromium per gram of oxygen. In the red compound on the right, there is one chromium atom for every three oxygen atoms (1 Cr : 3 O) and 1.083 g of chromium per gram of oxygen. The ratio of the chromium masses, 2.167 : 1.083, is that of two small whole numbers, 2.167 : 1.083 = 2 : 1, an illustration of the law of multiple proportions. (Marna G. Clarke)

Dalton was a quiet, unassuming man and a devout Quaker. When presented to King William IV of England, Dalton refused to wear the colorful court robes because of his religion. His friends persuaded him to wear the scarlet robes of Oxford University, from which he had a doctor's degree. Dalton was color blind, so he saw himself clothed in gray.

Dalton was a prolific scientist who made contributions to biology and physics as well as chemistry. At a college in Manchester, England, he did research and spent as many as 20 hours a week lecturing in mathematics and the physical sciences. Dalton never married; he said once, "my head is too full of triangles, chemical properties, and electrical experiments to think much of marriage."

Dalton's atomic theory explained three of the basic laws of chemistry:

The **law of conservation of mass:** This states that *there is no detectable change in mass in an ordinary chemical reaction.* If atoms are "conserved" in a reaction (Postulate 2 of the atomic theory), mass will also be conserved.

The **law of constant composition:** This tells us that *a compound always contains the same elements in the same proportions by mass.* If the atom ratio of the elements in a compound is fixed (Postulate 3), their proportions by mass must also be fixed.

The **law of multiple proportions:** This law, formulated by Dalton himself, was crucial to the establishment of atomic theory. It applies to situations in which two elements form more than one compound. The law states that in these compounds, *the masses of one element that combine with a fixed mass of the second element are in a ratio of small whole numbers.*

The validity of this law depends upon the fact that atoms combine in simple, whole-number ratios (Postulate 3). Its relation to atomic theory is further illustrated in Figure 2.A.

## 2.2 Components of the Atom

Like any useful scientific theory, the atomic theory raised more questions than it answered. Scientists wondered whether atoms, tiny as they are, could be broken down into still smaller particles. Nearly 100 years passed before the existence of subatomic particles was confirmed by experiment. Two future Nobel laureates did pioneer work in this area. J.J. Thomson was an English physicist working at the Cavendish Laboratory at Cambridge. Ernest Rutherford, at one time a student of Thomson's (Fig. 2.2), was a native of New Zealand. Rutherford carried out his research at McGill University in Montreal and at Manchester and Cambridge in England. He was clearly the greatest experimental physicist of his time, one of the greatest of all time.

## Electrons

The first evidence for the existence of subatomic particles came from studies of the conduction of electricity through gases at low pressures. When the glass tube shown in Figure 2.3 is partially evacuated and connected to a spark coil, an electric current flows through it. Associated with this flow are colored rays of light called *cathode rays,* which are bent by both electric and magnetic fields. From a careful study of this deflection, J.J. Thomson showed in 1897 that the rays consist of a stream of negatively charged particles, which he called **electrons.** Electrons are common to all atoms, carry a unit negative charge ($-1$), and have a very small mass, roughly 1/2000 of that of the lightest atom.

Every atom contains a definite number of electrons. This number, which runs from 1 to more than 100, is characteristic of a neutral atom of a particular element. All atoms of hydrogen contain one electron; all atoms of the element uranium contain 92 electrons. We will have more to say in Chapter 6 about how these electrons are arranged relative to one another. Right now, you need only know that they are found in the outer regions of the atom, where they form what amounts to a cloud of negative charge.

## Protons and Neutrons; the Atomic Nucleus

A series of experiments carried out under the direction of Ernest Rutherford in 1911 shaped our ideas about the nature of the atom. He and his students bombarded a piece of thin gold foil (Fig. 2.4) with $\alpha$-particles (helium atoms minus their electrons). With a fluorescent screen, they observed the extent to which the $\alpha$-particles were scattered. Most of the particles went through the foil unchanged in direction; a few, however, were reflected back at acute angles. This was a totally unexpected result, inconsistent with the model of the atom in vogue at that time. In Rutherford's words, "It was as though you had fired a 15-inch shell at a piece of tissue paper and it had bounced back and hit you." By a mathematical analysis of the forces involved, Rutherford showed that the scattering was caused by a small, positively charged **nucleus** at the center of the gold atom.

**Figure 2.2**
J.J. Thomson and Ernest Rutherford (right) talking, perhaps, about nuclear physics, but more likely about yesterday's cricket match. (AIP Niels Bohr Library, Bainbridge Collection)

Before this experiment, it was believed that + and − particles were more or less uniformly distributed through the atom

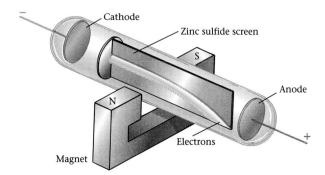

**Figure 2.3**
Cathode ray tube. The ray, shown here as a yellow beam, is made up of fast-moving electrons. In an electric or magnetic field, the beam is deflected in such a way as to indicate that it carries a negative charge.

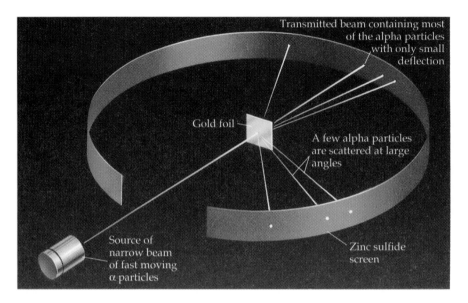

**Figure 2.4**

Rutherford's scattering experiment. Most of the α-particles are essentially undeflected, but a few are scattered at large angles. In order to cause the large deflections, atoms must contain heavy, positively charged nuclei.

A cathode ray tube used in the laboratory. Note the beam deflected on the right tube, which is subjected to a magnetic field. (Richard Megna/Fundamental Photographs, NY)

Since the time of Rutherford, scientists have learned a great deal about the properties of atomic nuclei. For our purposes in chemistry, the nucleus of an atom can be considered to consist of two different types of particles (Table 2.1).

1. The **proton,** which has a mass nearly equal to that of an ordinary hydrogen atom. The proton carries a unit positive charge ($+1$), equal in magnitude to that of the electron ($-1$).
2. The **neutron,** an uncharged particle with a mass slightly greater than that of a proton.

## Atomic Number

All the atoms of a particular element have the same number of protons in the nucleus. This number is a basic property of an element, called its **atomic number** and given the symbol $Z$:

$$Z = \text{number of protons}$$

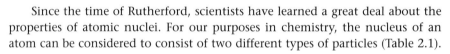

| TABLE 2.1 | **Properties of Subatomic Particles** | | |
|-----------|----------|-----------------|----------------|
| Particle  | Location | Relative Charge | Relative Mass* |
| proton    | nucleus  | $+1$            | 1.00728        |
| neutron   | nucleus  | 0               | 1.00867        |
| electron  | outside nucleus | $-1$     | 0.00055        |

*These are expressed in atomic mass units (Chap. 3).

In a neutral atom, the number of protons in the nucleus is exactly equal to the number of electrons outside the nucleus. Consider, for example, the elements hydrogen ($Z = 1$) and uranium ($Z = 92$). All hydrogen atoms have one proton in the nucleus; all uranium atoms have 92. In a neutral hydrogen atom there is one electron outside the nucleus; in a uranium atom there are 92.

| H atom: | 1 proton, 1 electron | $Z = 1$ |
| U atom: | 92 protons, 92 electrons | $Z = 92$ |

## Mass Numbers; Isotopes

The **mass number** of an atom, given the symbol $A$, is found by adding up the number of protons and neutrons in the nucleus:

$$A = \text{number of protons} + \text{number of neutrons}$$

No. of neutrons = A − Z

All atoms of a given element have the same number of protons, hence the same atomic number. They may, however, differ from one another in mass and hence in mass number. This can happen because, although the number of protons in an atom of an element is fixed, the number of neutrons is not. It may vary and often does. Consider the element hydrogen ($Z = 1$). There are three different kinds of hydrogen atoms. They all have one proton in the nucleus. A "light" hydrogen atom (the most common type) has no neutrons in the nucleus ($A = 1$). Another type of hydrogen atom (deuterium) has one neutron ($A = 2$). Still a third type (tritium) has two neutrons ($A = 3$).

Atoms that contain the same number of protons but a different number of neutrons are called **isotopes.** The three kinds of hydrogen atoms just described are isotopes of that element. They have masses that are very nearly in the ratio 1:2:3. Among the isotopes of the element uranium are the following:

| Isotope | $Z$ | $A$ | Number of Protons | Number of Neutrons |
|---------|-----|-----|-------------------|--------------------|
| uranium-235 | 92 | 235 | 92 | 143 |
| uranium-238 | 92 | 238 | 92 | 146 |

The composition of a nucleus is shown by its **nuclear symbol.** Here, the atomic number appears as a subscript at the lower left of the symbol of the element. The mass number is written as a superscript at the upper left.

$$\text{mass number} \longrightarrow A$$
$$X \longleftarrow \text{element symbol}$$
$$\text{atomic number} \longrightarrow Z$$

The nuclear symbols for the isotopes of hydrogen and uranium referred to above are

$$_1^1H, \ _1^2H, \ _1^3H \qquad _{92}^{235}U, \ _{92}^{238}U$$

Quite often, isotopes of an element are distinguished from one another by writing the mass number after the symbol of the element. The isotopes of uranium are often referred to as U-235 and U-238.

"Ordinary" ice (right) floats because the solid phase of water, $_1^1H_2O$, is less dense than the liquid. In contrast, "heavy" ice sinks because the solid form of deuterium oxide, $_1^2H_2O$ is more dense than the liquid. (Charles D. Winters)

---

### Example 2.1

(a) An isotope of cobalt (Co) is used in radiation therapy for certain types of cancer. Write nuclear symbols for three isotopes of cobalt ($Z = 27$) in which there are 29, 31, and 33 neutrons, respectively.

(b) One of the most harmful components of nuclear waste is a radioactive isotope of strontium, $^{90}_{38}Sr$; it can be deposited in your bones, where it replaces calcium. How many protons are there in the nucleus of Sr-90? How many neutrons?

**Strategy**   Remember the definitions of atomic number and mass number and where they appear in the nuclear symbol.

**Solution**

(a) The mass numbers are

$$27 + 29 = 56 \qquad 27 + 31 = 58 \qquad 27 + 33 = 60$$

Thus the nuclear symbols are   $^{56}_{27}Co, \; ^{58}_{27}Co, \; ^{60}_{27}Co.$

(b) The number of protons is given by the atomic number (left subscript) and is 38. The mass number (left superscript) is 90. The number of neutrons is $90 - 38 =$ 52

---

## 2.3   Introduction to the Periodic Table

From a structural point of view, an element is a substance all of whose atoms have the same number of protons, i.e., the same atomic number. The chemical properties of elements depend upon their atomic numbers, which can be read from the **periodic table** (inside front cover of this text). The atomic number is given directly above the symbol of the element. For example, sulfur (S) has an atomic number of 16; lead (Pb) has an atomic number of 82.

The periodic table is useful for a great many purposes. Here, we will take a brief look at some of the characteristics of the table. Later, in Chapter 6, the periodic table will be examined in greater detail.

### Periods and Groups

The horizontal rows in the table are referred to as **periods.** The first period consists of the two elements hydrogen (H) and helium (He). The second period starts with lithium (Li) and ends with neon (Ne).

The vertical columns are known as **groups** or **families.** Historically, many different systems have been used to designate the different groups. Both Arabic and Roman numerals have been used in combination with the letters A and B. The system used in this text is the one recommended by the International Union of Pure and Applied Chemistry (IUPAC) in 1985. The groups are numbered from 1 to **18**, starting at the left.

Elements falling in Groups 1, 2, **13**, **14**, **15**, **16**, **17** and **18**\* are referred to as **main-group elements.** The ten elements in the center of periods 4 through

*Other periodic tables label the groups differently, but the elements have the same positions*

---

\*Prior to 1985, Groups **13–18** were commonly numbered 3–8 or 3A–8A in the United States.

6 are called **transition elements;** they fall in Groups 3 through 12. The first transition series (period 4) starts with Sc (Group 3) and ends with Zn (Group 12).

Certain main groups are given special names. The elements in Group 1, at the far left of the periodic table, are called *alkali metals;* those in Group 2 are referred to as *alkaline earth metals*. Moving to the right, the elements in Group 17 are called *halogens;* at the far right, the *noble* (unreactive) *gases* constitute Group 18.

Elements in the same main group show very similar chemical properties. For example,

— lithium (Li), sodium (Na), and potassium (K) in Group 1 all react vigorously with water to produce hydrogen gas.

— helium (He), neon (Ne), and argon (Ar) in Group 18 do not react with any other substances.

On the basis of observations such as these, we can say that *the periodic table is an arrangement of elements, in order of increasing atomic number, in horizontal rows of such a length that elements with similar chemical properties fall directly beneath one another in vertical groups.*

## Metals and Nonmetals

The diagonal line or stairway that starts to the left of boron in the periodic table separates metals from nonmetals. The more than 80 elements to the left and

The reaction of the alkali metals with water (potassium is shown here) generates enough heat to ignite the hydrogen formed.
(Charles D. Winters)

below that line, shown in blue in the table, have the properties of **metals;** in particular they have high electrical conductivities. Elements above and to the right of the stairway are **nonmetals** (yellow); about 20 elements fit in that category.

Along the diagonal line in the periodic table are several elements that are difficult to classify exclusively as metals or nonmetals. They have properties in between those of elements in the two classes. In particular, their electrical conductivities are intermediate between those of metals and nonmetals. The six elements

These elements, particularly Si, are used in semiconductors

| B | Si | Ge | As | Sb | Te |
|---|----|----|----|----|----|
| boron | silicon | germanium | arsenic | antimony | tellurium |

are often called **metalloids.**

In this chapter, we will use the periodic table to

— relate the symbol of an element to its atomic number. This is particularly useful in balancing nuclear equations (Section 2.4).
— deduce the charge of a monatomic ion formed by an element. From this information, it is possible to obtain the formula of an ionic compound (Section 2.6).

## 2.4  Nuclear Stability; Radioactivity

There are eight known isotopes of carbon:

$$^{9}_{6}C \quad ^{10}_{6}C \quad ^{11}_{6}C \quad ^{12}_{6}C \quad ^{13}_{6}C \quad ^{14}_{6}C \quad ^{15}_{6}C \quad ^{16}_{6}C$$

Of these, two ($^{12}_{6}C$, $^{13}_{6}C$) are stable in the sense that they do not decompose over time. In contrast, the three lighter nuclei ($^{9}_{6}C$, $^{10}_{6}C$, $^{11}_{6}C$) and the three heavier nuclei ($^{14}_{6}C$, $^{15}_{6}C$, $^{16}_{6}C$) are unstable; as time passes they decompose to other nuclei. These isotopes are said to be *radioactive;* the process by which they decompose (*decay*) is referred to as **radioactivity.**

There is no simple way to predict whether a given isotope will be radioactive. About all we can say is that if an isotope lies within the "belt of stability" shown in blue in Figure 2.5, it is likely to be stable; $^{12}_{6}C$ and $^{13}_{6}C$ fall in that category. Isotopes falling outside that belt because they have too few neutrons ($^{9}_{6}C$, $^{10}_{6}C$, $^{11}_{6}C$) or too many neutrons ($^{14}_{6}C$, $^{15}_{6}C$, $^{16}_{6}C$) are radioactive.

As you can see from Figure 2.5, the neutron-to-proton ratio required for stability varies with atomic number. For light elements ($Z < 20$), this ratio is close to one. For example, the isotopes $^{12}_{6}C$, $^{14}_{7}N$, and $^{16}_{8}O$ are stable. As atomic number increases, the ratio increases; the "belt of stability" shifts to higher numbers of neutrons. With very heavy isotopes such as $^{206}_{82}Pb$, the stable neutron-to-proton ratio is about 1.5:

$$(206 - 82)/82 = 124/82 = 1.51$$

### Radioactive Decay Processes; Nuclear Equations

The process of radioactivity was discovered, almost accidentally, by Henri Becquerel at the Sorbonne in Paris in 1896. Early investigators in the field, who included Marie Curie in France and Ernest Rutherford in England, detected three different types of radiation produced when unstable nuclei decomposed (Figure 2.6).

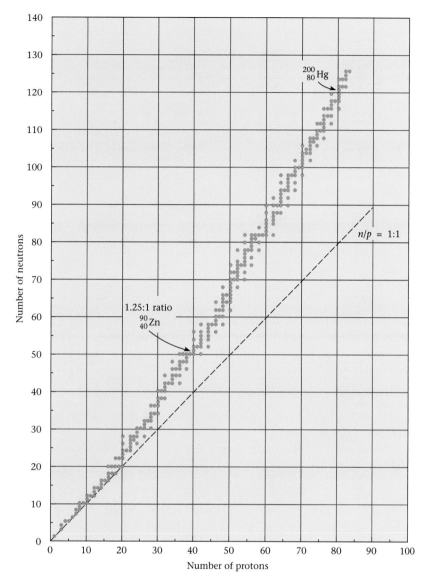

**Figure 2.5**
Stable isotopes (blue dots) have neutron-to-proton ratios that fall within a narrow range, referred to as a "belt of stability." For light isotopes of small atomic number, the stable ratio is 1.0; with heavier isotopes it increases to about 1.5. There are no stable isotopes for elements of atomic number greater than 83 (Bi).

**1. Alpha radiation** is made up of a stream of positively charged particles (alpha particles). These are $^4_2\text{He}$ nuclei, which carry a +2 charge because the two extranuclear electrons of the helium atom are missing. When an alpha particle is emitted from an unstable nucleus, another nucleus is formed. That nucleus has a mass number four less and an atomic number two less than the original nucleus. For example, when a $^{238}_{92}\text{U}$ nucleus decomposes, emitting a $^4_2\text{He}$ particle, an isotope of thorium, $^{234}_{90}\text{Th}$, is formed. This process can be represented by a **nuclear equation:**

$$^{238}_{92}\text{U} \longrightarrow {}^4_2\text{He} + {}^{234}_{90}\text{Th}$$

The *reactant* is the U-238 nucleus; the *products* are the alpha particle and the Th-234 nucleus. The arrow separates reactants on the left from products on the right. Here, as in all nuclear equations, there is a balance of both atomic number (90 + 2 = 92) and mass number (4 + 234 = 238) on the two sides.

**2. Beta radiation** is made up of a stream of negatively charged particles (beta particles) identical in their properties to electrons. In a nuclear

**Figure 2.6**
The direction in which beta particles are deflected shows that they are negatively charged. Alpha particles move so as to indicate that they carry a positive charge. Gamma rays are undeflected and so must be uncharged.

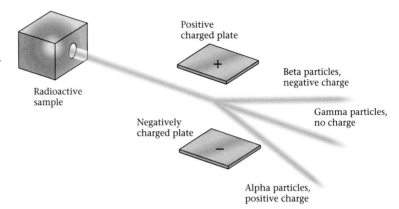

equation, a beta particle is shown as $_{-1}^{0}e$. The ejection of a beta particle (mass ≈ 0, charge = −1) leaves the mass number unchanged but increases the atomic number by one unit. An example of beta emission is the radioactive decay of thorium-234:

$$_{90}^{234}\text{Th} \longrightarrow _{-1}^{0}e + _{91}^{234}\text{Pa}$$

Again, there is a balance of mass number (234) and atomic number (90) on both sides of the equation.

Beta emission comes about when a neutron ($_{0}^{1}n$) in the nucleus is converted to a proton ($_{1}^{1}\text{H}$):

$$_{0}^{1}n \longrightarrow _{1}^{1}\text{H} + _{-1}^{0}e$$

*Effectively, an electron has an atomic number of −1*

This type of reaction occurs with unstable nuclei containing "too many" neutrons, that is, nuclei lying above the belt of stability shown in Figure 2.5.

**3. Gamma radiation** consists of high-energy photons. The emission of gamma radiation accompanies most nuclear reactions. Since gamma radiation changes neither the atomic number nor the mass number, it is ordinarily omitted in writing nuclear equations.

---

**Example 2.2**    Thorium-232 undergoes radioactive decay in a three-step process:

(a) $_{90}^{232}\text{Th} \longrightarrow Q + _{2}^{4}\text{He}$    (b) $Q \longrightarrow R + _{-1}^{0}e$    (c) $R \longrightarrow T + _{90}^{228}\text{Th}$

Write a balanced nuclear equation for each step, identifying Q, R, and T by their nuclear symbols.

**Strategy**    The basic principle is that mass number and atomic number must be "conserved," i.e., they must have the same sum on both sides of the equation. This enables you to determine the mass numbers and atomic numbers of Q, R, and T. You can use the periodic table to find the symbols of the elements Q, and R.

**Solution**

(a) Alpha emission by $_{90}^{232}\text{Th}$ decreases the mass number by four and the atomic number by two. The mass number of Q is 232 − 4 = 228; its atomic number is 90 − 2 = 88. Referring to the periodic table, we see that element 88 is radium, symbol Ra.

$$_{90}^{232}\text{Th} \longrightarrow _{2}^{4}\text{He} + _{88}^{228}\text{Ra}$$

(b) When $^{228}_{88}Ra$ emits an electron, the mass number is unchanged, but the atomic number increases by 1 to 89 (element 89 = actinium, Ac).

$$^{228}_{88}Ra \longrightarrow {}^{0}_{-1}e + {}^{228}_{89}Ac$$

(c) The unbalanced equation in this case is

$$^{228}_{89}Ac \longrightarrow T + {}^{228}_{90}Th$$

For the equation to balance, T must have a mass number of 0 and an atomic number of $-1$; T must be an electron. The balanced equation is

$$^{228}_{89}Ac \longrightarrow {}^{0}_{-1}e + {}^{228}_{90}Th$$

Nuclear reactions are "extraordinary" in that they usually involve the transmutation of elements

## 2.5 Molecules and Ions

Isolated atoms rarely occur in nature; only the noble gases (He, Ne, Ar, . . .) consist of individual, nonreactive atoms. Atoms tend to combine with one another in various ways to form more complex structural units. Two such units, which serve as building blocks for a great many elements and compounds, are molecules and ions.

### Molecules

Two or more atoms may combine with one another to form an uncharged **molecule.** The atoms involved are usually those of nonmetallic elements. Within the molecule, atoms are held to one another by strong forces called *covalent bonds,* which consist of shared pairs of electrons (Chapter 7). Forces between neighboring molecules, in contrast, are quite weak.

The structures of molecules are sometimes represented by **structural formulas,** which show the bonding pattern within the molecule. The structural formulas of hydrogen chloride, water, ammonia, and methane are

"Superglue" is weak compared to covalent bonds

$$H—Cl \qquad H—O—H \qquad H—\overset{\displaystyle |}{\underset{\displaystyle |}{N}}—H \qquad H—\overset{\displaystyle H}{\underset{\displaystyle H}{\overset{|}{\underset{|}{C}}}}—H$$

The dashes represent covalent bonds. The three-dimensional geometries of these molecules are shown in Figure 2.7.

Most commonly, molecular substances are represented by **molecular formulas,** in which the number of atoms of each element is indicated by a subscript written after the symbol of the element. The molecular formulas of the substances just described are

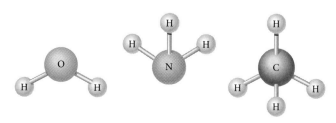

**Figure 2.7**
Ball-and-stick models of $H_2O$, $NH_3$, and $CH_4$. The "sticks" represent covalent bonds between H atoms and O, N, or C atoms.

hydrogen chloride:  HCl   (1 H atom, 1 Cl atom per molecule)
water:  $H_2O$   (2 hydrogen atoms, 1 oxygen atom per molecule)
ammonia:  $NH_3$   (3 hydrogen atoms, 1 nitrogen atom per molecule)
methane:  $CH_4$   (4 hydrogen atoms, 1 carbon atom per molecule)

Sometimes we represent a molecular substance with a formula intermediate between a structural formula and a molecular formula. A **condensed structural formula** suggests the bonding pattern in the molecule and highlights the presence of a reactive group of atoms within the molecule. Consider, for example, the organic compounds commonly known as methyl alcohol and methylamine. Their structural formulas are

*Structural formulas help us predict chemical properties*

$$
\begin{array}{ccc}
& H & \\
& | & \\
H-C-O-H & & \\
& | & \\
& H &
\end{array}
\qquad
\begin{array}{ccc}
& H & \\
& | & \\
H-C-N-H & \\
& | \; | & \\
& H \; H &
\end{array}
$$

methyl alcohol                methylamine

The condensed structural formulas of these compounds are written as

$$CH_3OH \qquad CH_3NH_2$$

These formulas take up considerably less space and emphasize the presence in the molecule of

— the OH group found in all *alcohols,* including ethyl alcohol, $C_2H_5OH$.
— the NH₂ group found in one class of *amines.* Ethylamine has the condensed structural formula $C_2H_5NH_2$.

---

**Example 2.3**    Give the molecular formulas of ethyl alcohol and ethylamine.

*Strategy*    To find the molecular formulas, simply add up the atoms of each type and use the sums as subscripts in the formulas.

*Solution*

(a)   $C_2H_6O$    (b)   $C_2H_7N$

Note that although these formulas give the composition of the molecule, they reveal nothing about the way the atoms fit together.

---

Elements as well as compounds can exist as discrete molecules. In hydrogen gas, the basic building block is a molecule consisting of two hydrogen atoms joined by a covalent bond:

$$H-H$$

Other molecular elements are shown in Figure 2.8.

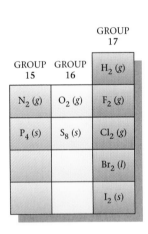

**Figure 2.8**
Molecular elements and their physical states in the periodic table.

| GROUP 15 | GROUP 16 | GROUP 17 |
|---|---|---|
| | | $H_2$ (g) |
| $N_2$ (g) | $O_2$ (g) | $F_2$ (g) |
| $P_4$ (s) | $S_8$ (s) | $Cl_2$ (g) |
| | | $Br_2$ (l) |
| | | $I_2$ (s) |

## Ions

When an atom loses or gains electrons, charged particles called **ions** are formed. Typically, metal atoms tend to lose electrons to form positively charged ions called **cations** (pronounced CAT-i-ons). Examples include the $Na^+$ and $Ca^{2+}$ ions, formed from atoms of the metals sodium and calcium:

*Ions are formed when metals react with nonmetals: Na + Cl → Na⁺ + Cl⁻*

$$
\begin{array}{ccc}
\text{Na atom} & \longrightarrow & Na^+ \text{ ion} \quad + e^- \\
(11p^+,\ 11e^-) & & (11p^+,\ 10e^-)
\end{array}
$$

$$\begin{array}{ccc} \text{Ca atom} & \longrightarrow & \text{Ca}^{2+} \text{ ion} + 2\ e^- \\ (20p^+,\ 20e^-) & & (20p^+,\ 18e^-) \end{array}$$

(The arrows separate *reactants,* Na and Ca atoms, from *products,* cations and electrons.)

Nonmetal atoms form negative ions (**anions**—pronounced AN-i-ons) by gaining electrons. Consider, for example, what happens when atoms of the nonmetals chlorine and oxygen acquire electrons:

$$\begin{array}{ccc} \text{Cl atom} + e^- & \longrightarrow & \text{Cl}^- \text{ ion} \\ (17p^+,\ 17e^-) & & (17p^+,\ 18e^-) \end{array}$$

$$\begin{array}{ccc} \text{O atom} + 2e^- & \longrightarrow & \text{O}^{2-} \text{ ion} \\ (8p^+,\ 8e^-) & & (8p^+,\ 10e^-) \end{array}$$

*Notice that when an ion is formed, the number of protons in the nucleus is unchanged.* It is the number of electrons that increases or decreases. In other words, these are not nuclear reactions. They involve extranuclear electrons, as do all "ordinary" chemical reactions, discussed in Chapter 3 and subsequent chapters.

---

**Example 2.4**　　Give the number of protons and electrons in $Al^{3+}$, a cation suspected of playing a role in Alzheimer's disease.

*Strategy*　　The atomic number (from the periodic table) gives the number of protons and electrons in the neutral atom. The positive charge tells you how many electrons have been lost.

*Solution*　　The atomic number of aluminum is 13; the charge of the cation is +3. Hence:

$$\text{no. protons} = \boxed{13} \quad \text{no. electrons} = 13 - 3 = \boxed{10}$$

---

The ions dealt with to this point (e.g., $Na^+$, $Cl^-$) are **monatomic;** that is, they are derived from a single atom by the loss or gain of electrons. Many of the most important ions in chemistry are **polyatomic,** containing more than one atom. Examples include the hydroxide ion ($OH^-$) and the ammonium ion ($NH_4^+$). In these and other polyatomic ions, the atoms are held together by covalent bonds, e.g.,

$$(\text{O—H})^- \qquad \left(\begin{matrix} & \text{H} & \\ & | & \\ \text{H—N—H} \\ & | & \\ & \text{H} & \end{matrix}\right)^+$$

In a very real sense, you can think of a polyatomic ion as a "charged molecule."

Since a bulk sample of matter is electrically neutral, ionic compounds always contain both cations (positively charged particles) and anions (negatively charged particles). Ordinary table salt, sodium chloride, is made up of an equal number of $Na^+$ and $Cl^-$ ions. The structure of sodium chloride is shown in Figure 2.9, p. 40. Notice that

You can't buy a bottle of $Na^+$ ions

— there are two kinds of structural units in NaCl, the $Na^+$ and $Cl^-$ ions.
— there are no discrete molecules; $Na^+$ and $Cl^-$ ions are bonded together in a continuous network.

**Figure 2.9**
Two different ways of showing the structure of NaCl. The small spheres represent $Na^+$ ions, the large spheres $Cl^-$ ions. Notice that there are equal numbers of $Na^+$ and $Cl^-$ ions, but no NaCl molecules.

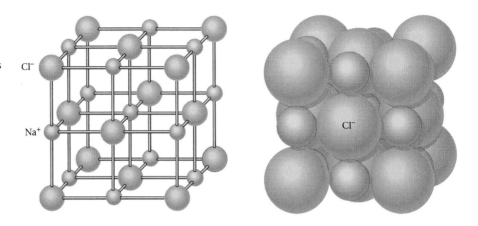

Ionic compounds are held together by strong electrical forces between oppositely charged ions (e.g., $Na^+$, $Cl^-$). These forces are referred to as **ionic bonds.** Typically, ionic compounds are solids at room temperature and have relatively high melting points (mp NaCl = 801°C, $CaCl_2$ = 772°C). To melt an ionic compound requires that oppositely charged ions be separated from one another, thereby breaking ionic bonds.

## 2.6 Formulas of Ionic Compounds

The formula of an ionic compound (e.g., NaCl, $CaCl_2$) shows the simplest ratio between cations and anions (1 $Na^+$ ion for 1 $Cl^-$ ion; 1 $Ca^{2+}$ ion for 2 $Cl^-$ ions). In that sense, the formulas of ionic compounds are *simplest formulas*.

To determine the formula of an ionic compound, we apply the **principle of electrical neutrality,** which requires that the total positive charge of the cations equal the total negative charge of the anions. Consider, for example, the ionic compound calcium chloride. The ions present are $Ca^{2+}$ and $Cl^-$. Clearly, for the compound to be electrically neutral, there must be two $Cl^-$ ions for every $Ca^{2+}$ ion. The formula of calcium chloride must be $CaCl_2$, indicating that the simplest ratio of $Cl^-$ to $Ca^{2+}$ ions is 2:1.

To predict the formulas of ionic compounds, you must know the charges of the ions involved. The remainder of this section is devoted to that topic.

### Monatomic Ions

Figure 2.10 shows the charges of a large number of monatomic ions superimposed upon the periodic table. The charges of ions formed by main-group atoms, shown in blue in the figure, can be predicted by applying a simple principle:

*Atoms that are close to a noble gas* (Group 18) *in the periodic table tend to form ions that contain the same number of electrons as the neighboring noble gas atom.*

This is reasonable; noble gas atoms must have an extremely stable electronic structure, since they are so unreactive. Other atoms might be expected to acquire noble gas electronic structures by losing or gaining electrons.

Applying this principle, you can deduce the charges of ions formed by main-group atoms:

**Figure 2.10**

| 1 | 2 | 3 | 4 | 5 | 6 | 7 | 8 | 9 | 10 | 11 | 12 | 13 | 14 | 15 | 16 | 17 | 18 |
|---|---|---|---|---|---|---|---|---|---|---|---|---|---|---|---|---|---|
| | | | | | | | | | | | | | | | | $H^-$ | He |
| $Li^+$ | | | | | | | | | | | | | | $N^{3-}$ | $O^{2-}$ | $F^-$ | Ne |
| $Na^+$ | $Mg^{2+}$ | | | | | | | | | | | $Al^{3+}$ | | | $S^{2-}$ | $Cl^-$ | Ar |
| $K^+$ | $Ca^{2+}$ | | | | $Cr^{3+}$ | $Mn^{2+}$ | $Fe^{2+}$ $Fe^{3+}$ | $Co^{2+}$ | $Ni^{2+}$ | $Cu^+$ $Cu^{2+}$ | $Zn^{2+}$ | | | | $Se^{2-}$ | $Br^-$ | Kr |
| $Rb^+$ | $Sr^{2+}$ | | | | | | | | | $Ag^+$ | $Cd^{2+}$ | | $Sn^{2+}$ | | $Te^{2-}$ | $I^-$ | Xe |
| $Cs^+$ | $Ba^{2+}$ | | | | | | | | | | | | $Pb^{2+}$ | $Bi^{3+}$ | | | Rn |

Charges of ions found in solid ionic compounds. The step-like diagonal line separates metals from nonmetals and cations from anions. Ions shown in blue have the same number of electrons as the neighboring noble gas atom. For example, $S^{2-}$, $Cl^-$, $K^+$, and $Ca^{2+}$ all have 18 electrons, as does the Ar atom. The cations shown in red do not have a noble gas structure.

| Group | No. of Electrons in Atom | Charge of Ion Formed |
|---|---|---|
| 1 | 1 more than noble gas atom | +1 |
| 2 | 2 more than noble gas atom | +2 |
| 16 | 2 less than noble gas atom | −2 |
| 17 | 1 less than noble gas atom | −1 |

Several metals that are farther removed from the noble gases in the periodic table form positive ions. The structures of these ions are not related in any direct way to those of noble gas atoms. Indeed, there is no simple way to predict their charges. Some of the more important of these ions are listed in Figure 2.10. They are derived from the **transition metals** (those in the groups near the center of the periodic table) and the *post-transition* metals in Groups **14** and **15**. The most common charge among these ions, you will note, is +2. Charges of +1 ($Ag^+$) and +3($Cr^{3+}$, $Bi^{3+}$) are less common. Several of these metals form more than one cation:

$$Fe^{2+} \text{ and } Fe^{3+} \qquad Cu^+ \text{ and } Cu^{2+}$$

Using Figure 2.10 and the principle of electrical neutrality, it is possible to predict the formulas of a large number of ionic compounds. Note that, *in writing the formula of an ionic compound, the positive ion is always placed first.*

Metals form cations (+1, +2,+3); nonmetals in Groups **16** and **17** form anions (−1, −2)

---

**Example 2.5**   Predict the formulas of the ionic compounds formed by

(a) magnesium and sulfur    (b) cobalt and chlorine
(c) aluminum and oxygen    (d) bismuth and fluorine

***Strategy***   First, identify the charges of the cation and anion, using Figure 2.10. Then balance positive with negative charges to arrive at the formula.

Copper (II) Oxide (Black) and Copper (I) Oxide (red). (Charles D. Winters)

*Solution*

(a)    MgS;   one $Mg^{2+}$ ion requires one $S^{2-}$ ion.

(b)    $CoCl_2$;   one $Co^{2+}$ ion requires two $Cl^-$ ions.

(c)    $Al_2O_3$;   two $Al^{3+}$ ions (total charge = +6) require three $O^{2-}$ ions (total charge = −6).

(d)    $BiF_3$;   one $Bi^{3+}$ ion requires three $F^-$ ions.

## Polyatomic Ions

Table 2.2 lists some of the polyatomic ions that you will need to know, along with their names and charges. Notice that

Nearly all cations are monatomic; the majority of anions are polyatomic

— there is only one common polyatomic cation, $NH_4^+$. ***All other cations considered in this text are derived from metal atoms*** (e.g., $Na^+$ from Na, $Ca^{2+}$ from Ca, . . .).
— most of the polyatomic anions contain one or more oxygen atoms; collectively these species are called **oxoanions.**

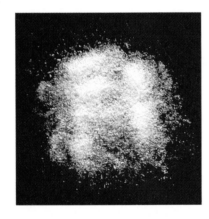

Sodium carbonate crystals. (Charles D. Winters)

**Example 2.6**   Using Figure 2.10 and Table 2.2, predict the formulas of (a) strontium hydroxide (b) sodium carbonate (c) ammonium phosphate.

***Strategy***   The reasoning here is entirely similar to that in Example 2.5. The only difference is that you must know the formulas and charges of the polyatomic ions in Table 2.2.

*Solution*

(a) One $Sr^{2+}$ ion requires two $OH^-$ ions. The formula is   $Sr(OH)_2$.   Parentheses are used to indicate that there are two polyatomic $OH^-$ ions for every $Sr^{2+}$.

(b) Two $Na^+$ ions require one $CO_3^{2-}$ ion. The formula is   $Na_2CO_3$.

(c) Three $NH_4^+$ ions are required for one $PO_4^{3-}$ ion. The formula is   $(NH_4)_3PO_4$.

TABLE 2.2   **Some Common Polyatomic Ions**

| +1 | −1 | −2 | −3 |
|---|---|---|---|
| $NH_4^+$ (ammonium) | $OH^-$ (hydroxide) | $CO_3^{2-}$ (carbonate) | $PO_4^{3-}$ (phosphate) |
| | $NO_3^-$ (nitrate) | $SO_4^{2-}$ (sulfate) | |
| | $ClO_3^-$ (chlorate) | $CrO_4^{2-}$ (chromate) | |
| | $ClO_4^-$ (perchlorate) | $Cr_2O_7^{2-}$ (dichromate) | |
| | $CN^-$ (cyanide) | $HPO_4^{2-}$ (hydrogen phosphate) | |
| | $C_2H_3O_2^-$ (acetate) | | |
| | $MnO_4^-$ (permanganate) | | |
| | $HCO_3^-$ (hydrogen carbonate) | | |
| | $H_2PO_4^-$ (dihydrogen phosphate) | | |

# 2.7 Names of Compounds

A compound can be identified either by its formula (e.g., NaCl) or its name (sodium chloride). In this section, you will learn the rules used to name ionic and simple molecular compounds. To start with, it will be helpful to show how individual ions within ionic compounds are named.

## Ions

Monatomic cations take the name of the metal from which they are derived. Examples include

$$Na^+ \text{ sodium} \qquad K^+ \text{ potassium}$$

There is one complication: Certain metals, notably those in the transition series, form more than one type of cation. An example is iron, which forms both $Fe^{2+}$ and $Fe^{3+}$. To distinguish between these cations, the charge must be indicated in the name. This is done by putting the charge as a Roman numeral in parentheses after the name of the metal:

We don't use Roman numerals with cations of Groups 1 and 2 metals; they always have charges of +1 and +2 respectively

$$Fe^{2+} \text{ iron(II)} \qquad Fe^{3+} \text{ iron(III)}$$

(An older system used the suffixes *-ic* for the ion of higher charge and *-ous* for the ion of lower charge. These were added to the stem of the Latin name of the metal, so that the $Fe^{3+}$ ion was referred to as ferric and the $Fe^{2+}$ ion as ferrous.)

Monatomic anions are named by adding the suffix *-ide* to the stem of the name of the nonmetal from which they are derived.

|  |  |  |  |  |
|---|---|---|---|---|
|  |  |  | $H^-$ | hydride |
| $N^{3-}$ nitride | $O^{2-}$ | oxide | $F^-$ | fluoride |
|  | $S^{2-}$ | sulfide | $Cl^-$ | chloride |
|  | $Se^{2-}$ | selenide | $Br^-$ | bromide |
|  | $Te^{2-}$ | telluride | $I^-$ | iodide |

Polyatomic ions, as you have seen (Table 2.2), are given special names. Certain nonmetals in Groups **15–17** of the periodic table form more than one polyatomic ion containing oxygen (oxoanions). The names of several such **oxoanions** are shown in Table 2.3. From the entries in the table, you should be able to deduce the following rules.

1. When a nonmetal forms two oxoanions, the suffix *-ate* is used for the anion with the larger number of oxygen atoms. The suffix *-ite* is used for the anion containing fewer oxygen atoms.
2. When a nonmetal forms more than two oxoanions, the prefixes *per-* (largest number of oxygen atoms) and *hypo-* (fewest oxygen atoms) are used as well.

Cl, Br and I form more than two oxyanions

| TABLE 2.3 | **Oxoanions of Nitrogen, Sulfur, and Chlorine** | |
|---|---|---|
| **Nitrogen** | **Sulfur** | **Chlorine** |
|  |  | $ClO_4^-$ *perchlorate* |
|  |  | $ClO_3^-$ *chlorate* |
| $NO_3^-$ nit*rate* | $SO_4^{2-}$ sulf*ate* | $ClO_2^-$ *chlorite* |
| $NO_2^-$ nit*rite* | $SO_3^{2-}$ sulf*ite* | $ClO^-$ *hypochlorite* |

Potassium dichromate, $K_2Cr_2O_7$, is an ionic compound. (Marna G. Clarke)

## Ionic Compounds

The name of an ionic compound consists of two words. The first word names the cation and the second names the anion. This is, of course, the same order in which the ions appear in the formula.

---

**Example 2.7**   Name the following ionic compounds:

(a) CaS     (b) $Al(NO_3)_3$     (c) $FeCl_2$

***Strategy***   To name an ionic compound, you must know the rules for naming individual ions, as discussed above. Where transition metals are involved, it is customary to show the charge of the cation; main-group metals typically form only one cation.

***Solution***

(a)   calcium sulfide     (b)   aluminum nitrate     (c)   iron(II) chloride

---

## Binary Molecular Compounds

When a metal combines with a nonmetal, the product is ordinarily an ionic compound. As you have just seen, the formulas and names of these compounds can be deduced in a straightforward way. When two nonmetals combine with each other, the product is most often a binary molecular compound. There is no simple way to deduce the formulas of such compounds. There is, however, a systematic way of naming molecular compounds that differs considerably from that used with ionic compounds.

The systematic name of a binary molecular compound, which contains two different nonmetals, consists of two words.

**1.** The first word gives the name of the element that appears first in the formula; a Greek prefix (Table 2.4) is used to show the number of atoms of that element in the formula.

**2.** The second word consists of

— the appropriate Greek prefix designating the number of atoms of the second element
— the stem of the name of the second element
— the suffix *-ide*

To illustrate these rules, consider the names of the several oxides of nitrogen:

| | | | |
|---|---|---|---|
| $N_2O_5$ | dinitrogen *pentaoxide* | $N_2O_3$ | dinitrogen *trioxide* |
| $N_2O_4$ | dinitrogen *tetraoxide* | NO | nitrogen *oxide* |
| $NO_2$ | nitrogen *dioxide* | $N_2O$ | dinitrogen *oxide* |

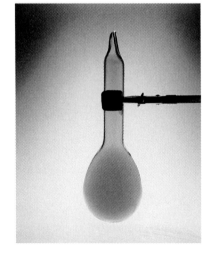

Nitrogen dioxide, $NO_2$, is a reddish-brown gas at 25°C and 1 atm.
(Charles D. Winters)

TABLE 2.4   **Greek Prefixes Used in Nomenclature**

| Number* | Prefix | Number | Prefix | Number | Prefix |
|---|---|---|---|---|---|
| 2 | di | 5 | penta | 8 | octo |
| 3 | tri | 6 | hexa | 9 | nona |
| 4 | tetra | 7 | hepta | 10 | deca |

*The prefix mono (1) is seldom used

**Example 2.8**   Give the names of

(a) $SO_2$   (b) $SO_3$   (c) $PCl_3$   (d) $Cl_2O_7$

**Strategy**   Start with the prefix denoting the number of atoms of the first element, followed by the name of that element. Repeat for the second element, ending with the suffix *-ide*.

*Solution*

(a)   sulfur dioxide

(b)   sulfur trioxide

(c)   phosphorus trichloride

(d)   dichlorine heptaoxide

---

Many of the best-known binary compounds of the nonmetals have acquired common names. These are widely and, in some cases, exclusively used. Examples include

| | | | |
|---|---|---|---|
| $H_2O$ | water | $PH_3$ | phosphine |
| $H_2O_2$ | hydrogen peroxide | $AsH_3$ | arsine |
| $NH_3$ | ammonia | $NO$ | nitric oxide |
| $N_2H_4$ | hydrazine | $N_2O$ | nitrous oxide |
| $C_2H_2$ | acetylene | $CH_4$ | methane |

## Acids

A few binary molecular compounds containing H atoms ionize in water to form $H^+$ ions. These are called **acids.** One such compound is hydrogen chloride, HCl; in water solution it exists as aqueous $H^+$ and $Cl^-$ ions. The water solution of hydrogen chloride is given a special name; it is referred to as hydrochloric acid. A similar situation applies with HBr and HI:

Water, a molecular compound, is never called "dihydrogen oxide." (Charles D. Winters)

| **Pure Substance** | | **Water Solution** | |
|---|---|---|---|
| $HCl(g)$ | hydrogen chloride | $H^+(aq)$, $Cl^-(aq)$ | hydrochloric acid |
| $HBr(g)$ | hydrogen bromide | $H^+(aq)$, $Br^-(aq)$ | hydrobromic acid |
| $HI(g)$ | hydrogen iodide | $H^+(aq)$, $I^-(aq)$ | hydriodic acid |

Most acids contain oxygen in addition to hydrogen atoms. Such species are referred to as **oxoacids.** Two oxoacids that you are likely to encounter in the general chemistry laboratory are

$$HNO_3 \quad \text{nitric acid} \qquad H_2SO_4 \quad \text{sulfuric acid}$$

The names of oxoacids are simply related to those of the corresponding oxoanions. The *-ate* suffix of the anion is replaced by *-ic* in the acid. Similarly, the suffix *-ite* is replaced by the suffix *-ous.* The prefixes *per-* and *hypo-* found in the name of the anion are retained in the name of the acid.

| | | | |
|---|---|---|---|
| $ClO_4^-$ | *perchlorate* ion | $HClO_4$ | *perchloric* acid |
| $ClO_3^-$ | chlor*ate* ion | $HClO_3$ | chlor*ic* acid |
| $ClO_2^-$ | chlor*ite* ion | $HClO_2$ | chlor*ous* acid |
| $ClO^-$ | *hypo*chlor*ite* ion | $HClO$ | *hypo*chlor*ous* acid |

**Example 2.9**    Give the names of

(a) $HNO_2$      (b) $H_2SO_3$      (c) HIO

***Strategy***    In (a) and (b), refer back to Table 2.3 for the name of the oxoanion. In (c), reason by analogy with chlorine oxoacids.

***Solution***

(a)  nitrous acid      (b)  sulfurous acid      (c)  hypoiodous acid

---

## CHEMISTRY

### *Beyond the Classroom*

Richard E. Smalley

**Figure 2.B**
The Structure of $C_{60}$, buckminsterfullerene, (left) is identical to that of a soccer ball (right) with a carbon atom at each vertex. It resembles that of a geodesic dome (chapter opener) of the type originally designed by R. Buckminster Fuller. (Photo by George Semple)

## Fullerenes

Until recently, it was generally supposed that the elements forming discrete polyatomic molecules were limited to those in Groups **15** to **17** of the Periodic Table (recall Fig. 2.8, p. 38). Then, in 1985, a molecular form of carbon with the formula $C_{60}$ was discovered by a group of chemists led by Richard E. Smalley at Rice University and Harry W. Kroto at the University of Sussex, U. K. (For a fascinating account of the history of the $C_{60}$ molecule, see the article "Great Balls of Carbon" by Smalley, published in the March/April 1991 issue of *The Sciences*.)

The structure of this molecule is shown in Figure 2.B. As you can see, it forms a nearly spherical cage. More exactly, the geometry of $C_{60}$ is that of a polygon with 32 faces. Of these, 12 are pentagons. These are separated from each other by hexagons, 20 in all. There is a carbon atom at the corner of each pentagon ($5 \times 12 = 60$). From a slightly different point of view, each carbon atom is located at the intersection of a pentagon with two hexagons. In this sense, all of the carbon atoms in the molecule are geometrically equivalent, which explains in part the unusual stability of this large molecule.

If you're a sports fan, you've almost certainly seen this structure before; it is that of a soccer ball. Taking this into account, Smalley and his co-workers considered names such as "carbosoccer" or "soccerene" for this molecular form of carbon. In the end, though, they called it "buckminsterfullerene" in honor of the controversial American

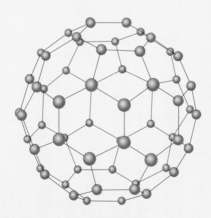

architect, engineer, and philosopher R. Buckminster Fuller. In the 1950s, Fuller started building geodesic domes (p. 26) with structures resembling that of the $C_{60}$ molecule. Less formally, $C_{60}$ is referred to as "buckyball."

Macroscopic quantities of $C_{60}$ did not become available until 1990, when a group of physicists worked out a process for its synthesis. They vaporized graphite electrodes in a helium atmosphere. When an arc is struck between the electrodes, a black soot containing up to 20% $C_{60}$ is formed. The process is slow and relatively expensive. Buckminsterfullerene sells today for $100 to $200 a gram, roughly ten times the price of gold.

Ever since $C_{60}$ was discovered, scientists have speculated on its possible existence in nature. Recently (August 1994) Luann Becker, a geochemistry graduate student, found small amounts of buckminsterfullerene in a meteor crater formed in Sudbury, Ontario, nearly two billion years ago. As Smalley put it, "buckyballs have been with us even before there was an us."

Buckminsterfullerene is not a unique molecular form of carbon. Molecules ranging in size from $C_{32}$ to $C_{300}$ have been identified. Collectively, these species are called "fullerenes." One of them, $C_{70}$, is found in relatively good yield in the soot referred to above, accounting for about 5% of the total mass. It has a structure similar to that of the $C_{60}$ molecule, except that there are 25 hexagons rather than 20. The extra hexagons are arranged in a band around the equator of the molecule, which looks somewhat like a tiny rugby ball.

Over the past few years, one of the hottest areas of chemical research has been the synthesis of fullerene derivatives. The simplest of these are "metal fullerenes," in which one or more metal atoms are trapped within a $C_{60}$ or $C_{70}$ cage. For some curious reason, all the metals undergoing this reaction are from the left side of the periodic table (e.g., K, Cs, La, U).

Certain more complex fullerene derivatives show biological activity. Specifically, a water-soluble compound formed by $C_{60}$ has been found to inhibit HIV-1 and HIV-2 the *h*uman *i*mmunodeficiency *v*iruses that cause AIDS. It appears that the $C_{60}$ molecule has the correct size and geometry to block the active sites of key enzymes associated with these viruses.

# CHAPTER HIGHLIGHTS

## *Key Concepts*

1. Relate a nuclear symbol to the number of protons and neutrons in the nucleus
   (Examples 2.1, 2.4; Problems 13–24).
2. Write balanced nuclear equations
   (Example 2.2; Problems 31–38).
3. Relate structural, condensed structural, and molecular formulas
   (Example 2.3; Problems 43, 44).
4. Predict formulas of ionic compounds
   (Examples 2.5, 2.6; Problems 49–52).
5. Name
   — ionic compounds
   (Example 2.7; Problems 53, 54).
   — binary molecular compounds
   (Example 2.8; Problems 45–48).
   — oxoacids
   (Example 2.9; Problems 55, 56).

# Key Terms

| | | |
|---|---|---|
| atom | —anion | nonmetal |
| atomic number | —cation | nuclear equation |
| electron | isotope | nuclear symbol |
| formula | mass number | periodic table |
| —condensed structural | metal | —group |
| —molecular | metalloid | —period |
| —structural | molecule | proton |
| ion | neutron | radioactivity |

# Summary Problem

Iodine crystals (Charles Steele)

Iodine is a dark purple solid. It was once popularly used as an antiseptic.

(a) What is its molecular formula?

(b) How many protons and electrons are there in a molecule of iodine? In an iodide ion?

(c) What is the atomic number for iodine?

(d) Write the nuclear symbol for the iodine atom with 53 protons and 78 neutrons.

(e) In what group and period does iodine belong in the periodic table? Is it a metal, non-metal, or metalloid?

(f) I-131 is a radioactive isotope used to destroy thyroid tumors. It decays by emitting a beta particle. Write an equation to represent the decay.

(g) I-123 is a radioactive isotope used as a diagnostic imaging tool. How many neutrons are there in I-123?

(h) A compound is made up of one atom of iodine and three atoms of chlorine. What is the molecular formula of this compound? What is the name of this compound?

(i) When iodine and barium combine, an ionic compound is formed. Give its name and formula.

(j) Iodine and oxygen combine to form oxoanions entirely analogous to the oxoanions of chlorine. Give the name and the formula of the oxoanion of iodine made up of one atom of oxygen and one atom of iodine. Give the name and formula of the corresponding oxoacid.

## Answers

(a) $I_2$

(b) 106 protons, 106 electrons; 53 protons, 54 electrons

(c) 53

(d) $^{131}_{53}I$

(e) period 5, group **17**, nonmetal

(f) $^{131}_{53}I \longrightarrow {}^{0}_{-1}e + {}^{131}_{54}Xe$

(g) 70

(h) $ICl_3$; iodine trichloride

(i) $BaI_2$; barium iodide

(j) $IO^-$, hypoiodite; HIO, hypoiodous acid

# Questions & Problems

## Atomic Theory and Laws

**\*1.** State in your own words the law of conservation of mass. State the law in its modern form.

**\*2.** State in your own words the law of constant composition.

**\*3.** Which of the three laws (if any) listed on p. 28 is illustrated by the following statements?

**(a)** Hydrogen has three isotopes. One has a mass number $A$ equal to its atomic number $(Z)$; in another $A = 2Z$, and in a third, $A = 3Z$.

**(b)** When copper(II) oxide decomposes, the total mass of the copper and oxygen formed equals the mass of copper oxide decomposed.

**(c)** Analysis of water in the rain collected in the Amazon rain forest and of that formed in a test tube by combining hydrogen and oxygen gives the same value for the percent oxygen.

**(d)** The atom ratio of hydrogen to carbon is twice as large in one compound as it is in another compound of the two elements.

*4. Which of the three laws (if any) listed on p. 28 is illustrated by the following statement?

**(a)** A sealed bag of popcorn has the same mass before and after it is put in a microwave oven. (Assume no breaks in the bag after it has been put into the microwave oven.)

**(b)** A teaching assistant writes "highly improbable" in a student's report that states that his unknown is $S_{1.2}O_{2.7}$.

**(c)** The mass of phosphorus, P, combined with one gram of hydrogen, H, in the highly toxic gas phosphine, $PH_3$, is a little more than twice the mass of nitrogen, N, combined with one gram of hydrogen in ammonia gas, $NH_3$.

*5. Using the laws of constant composition and the conservation of mass, complete the molecular picture of hydrogen molecules ( $\bigcirc\!-\!\bigcirc$ ) reacting with chlorine molecules ( $\square\!-\!\square$ ) to give hydrogen chloride ( $\square\!-\!\bigcirc$ ) molecules.

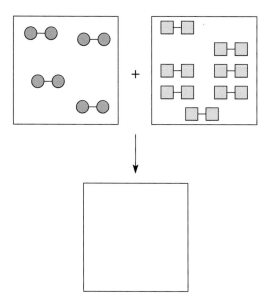

**6.** Using the law of conservation of mass, which numbered box(es) represent the product mixture after the substances in the box at the upper right, undergo a reaction?

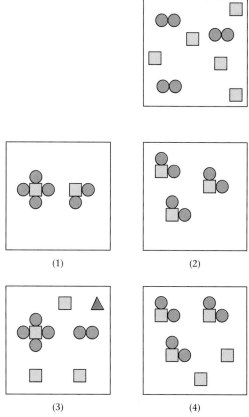

(1)          (2)

(3)          (4)

**7.** Mercury(II) oxide, a red powder, can be decomposed by heating to produce liquid mercury and oxygen gas. When a sample of this compound is decomposed, 3.87 g of oxygen and 48.43 g of mercury are produced. In a second experiment, 15.68 g of mercury is allowed to react with an excess of oxygen; 16.93 g of red mercury(II) oxide is produced. Show that these results are consistent with the law of constant composition.

**8.** When magnesium ribbon, Mg(*s*), is heated in oxygen gas, magnesium oxide, a white powder, is produced. In one experiment, 3.56 g of magnesium ribbon is completely consumed in reacting with 7.00 g of oxygen to produce 5.93 g of magnesium oxide; some oxygen remains unreacted. In a second experiment, 2.50 g of magnesium ribbon reacts with 1.10 g of oxygen gas. This time, all the oxygen is consumed; some unreacted magnesium remains, and 2.75 g of magnesium oxide is produced. Show that these results are consistent with the law of constant composition.

**9.** A series of compounds containing only nitrogen and oxygen was prepared and analyzed. Given the following results

| Compound | Mass of Nitrogen (g) | Mass of Oxygen (g) |
|----------|----------------------|--------------------|
| A | 12.6 | 14.4 |
| B | 23.5 | 53.7 |
| C | 18.2 | 10.4 |

**(a)** calculate the mass of nitrogen per gram of oxygen in each compound.

**(b)** show that the results in part (a) follow the law of multiple proportions.

10. Three compounds containing only carbon and hydrogen are analyzed. The results for the analysis of the first two compounds are given below:

| Compound | Mass of Carbon (g) | Mass of Hydrogen (g) |
|----------|--------------------|-----------------------|
| X | 28.5 | 2.39 |
| Y | 34.7 | 11.6 |
| Z | 16.2 | |

Which, if any, of the following results for the mass of hydrogen in compound Z follows the law of multiple proportions?

**(a)** 5.84 g    **(b)** 3.47 g    **(c)** 2.72 g

## Nuclear Symbols and Isotopes

**\*11.** Who discovered the nucleus? Describe the experiment that led to the discovery.

**\*12.** Who discovered the electron? Describe the experiment that led to the deduction that electrons are negatively charged particles.

**\*13.** Selenium is widely sold as a dietary supplement. It is advertised to "protect" women from breast cancer. Write the nuclear symbol for naturally occurring selenium. It has 34 protons and 46 neutrons.

**\*14.** Radon is a radioactive gas that can cause lung cancer. How many protons are there in a Rn-222 atom? How many neutrons?

**\*15.** Answer the following questions about the bromine atom.

**(a)** Do the symbols $_{35}^{79}Br$ and Br-79 convey the same information?

**(b)** Do the symbols $_{35}^{79}Br$ and $_{35}Br$ convey the same information?

**\*16.** How do the two isotopes N-14 and N-15 differ from each other? Write nuclear symbols for both.

**\*17.** Nuclei with the same mass number but different atomic numbers are called isobars. Consider Ca-40, Ca-41, K-41, and Ar-41.

**(a)** Which of these are isobars? Which are isotopes?

**(b)** What do Ca-40 and Ca-41 have in common?

**(c)** Correct the statement (if it is incorrect): Ca-41, K-41, and Ar-41 have the same number of neutrons.

**\*18.** Consider Na-21 and review the definition of isobars given in Question 17. Assuming that they exist, write the nuclear symbol for

**(a)** an isotope of Na-21 with one more neutron than Na-21 has.

**(b)** an isobar of Na-21 with atomic number 10.

**(c)** a nucleus with 11 protons and 12 neutrons. Is this nucleus an isotope or an isobar of Na-21?

**\*19.** A large amount of the element iron is present in steel. Fe-56 is one of its isotopes. How many

**(a)** protons are in its nucleus?

**(b)** neutrons are in its nucleus?

**(c)** electrons are in an iron atom?

**(d)** neutrons, protons, and electrons are in the $Fe^{3+}$ ion formed from this isotope?

**\*20.** Lithium is an element that is used by physicians in the treatment of some mental disorders. Lithium-7 is one of its isotopes. How many

**(a)** protons are in its nucleus?

**(b)** neutrons are in its nucleus?

**(c)** electrons are in a lithium atom?

**(d)** neutrons, protons, and electrons are in the $Li^+$ ion formed from this isotope?

**\*21.** Complete the table below using the periodic table if necessary.

| Nuclear Symbol | Charge | Number of Protons | Number of Neutrons | Number of Electrons |
|----------------|--------|-------------------|--------------------|--------------------| 
| $_{7}^{14}N$ | ____ | ____ | ____ | ____ |
| ____ | ____ | 17 | 20 | 18 |
| ____ | +2 | 27 | 29 | ____ |
| $_{34}^{82}Se^{2-}$ | ____ | ____ | ____ | ____ |

**\*22.** Complete the table below using the periodic table if necessary.

| Nuclear Symbol | Charge | Number of Protons | Number of Neutrons | Number of Electrons |
|----------------|--------|-------------------|--------------------|--------------------| 
| $_{37}^{85}Rb$ | ____ | ____ | ____ | ____ |
| ____ | +3 | 13 | 14 | ____ |
| ____ | −1 | 9 | 10 | ____ |

**\*23.** Give the number of protons and electrons in

**(a)** a $C_{60}$ molecule.

**(b)** a $C^{4-}$ ion.

**(c)** a $CH_4$ molecule.

**(d)** an $H^+$ ion.

**\*24.** Give the number of protons and electrons in

**(a)** an $N_2$ molecule.

**(b)** an $N^{3-}$ ion.

**(c)** an $N_2O_4$ molecule.

**(d)** an $H^-$ ion.

## Elements and the Periodic Table

**\*25.** Give the symbols for

**(a)** cesium    **(b)** tungsten    **(c)** antimony

**(d)** phosphorus    **(e)** potassium

**\*26.** Name the elements whose symbols are

**(a)** C    **(b)** Co    **(c)** Cd    **(d)** Cl    **(e)** Cu

**\*27.** Classify the elements in Question 25 as metals, non-metals, or metalloids.

**\*28.** Classify the elements in Question 26 as metals, non-metals, or metalloids.

**\*29.** How many elements are there in the following groups?
 **(a)** Group 11 **(b)** Group 2
 **(c)** Group 17 **(d)** Group 5

**\*30.** How many elements are in the following periods?
 **(a)** period 1 **(b)** period 2 **(c)** period 3
 **(d)** period 4 **(e)** period 5

## Nuclear Equations

**\*31.** Lead-210 is used to prepare eyes for corneal transplants. Its decay product is bismuth-210. Identify the emission from lead-210.

**\*32.** Smoke detectors contain small amounts of americium-241. It decays by emitting an alpha particle. Write the equation for the nuclear decay.

**\*33.** Write balanced nuclear equations for
 **(a)** the loss of an alpha particle by Th-230.
 **(b)** the loss of an electron by lead-210.
 **(c)** the formation of osmium-187 by beta decay.
 **(d)** the formation of californium-243 by alpha decay.

**\*34.** Write balanced nuclear equations for
 **(a)** the alpha emission resulting in the formation of Pa-233.
 **(b)** the beta emission resulting in the formation of yttrium-90.
 **(c)** the beta decay of Ar-37.
 **(d)** the alpha decay of astatine-218.

**\*35.** Thorium-231 is the product of alpha emission and is radioactive, emitting beta radiation. Determine
 **(a)** the parent nucleus of Th-231.
 **(b)** the product of Th-231 decay.

**\*36.** Radon-206 is the product of alpha emission and is radioactive, emitting beta radiation. Determine
 **(a)** the parent nucleus of Rn-206.
 **(b)** the product of Rn-206 decay.

**\*37.** Radon-222 decays by a five-step series of three alpha emissions followed by two beta emissions. Identify each intermediate and the final product.

**\*38.** Americium-241 has a sequential decay, part of which is alpha, alpha, beta, alpha. Identify each intermediate.

## Nuclear Stability

**\*39.** Which isotope in each of the following pairs should be more stable?
 **(a)** $^{12}_{6}C$ or $^{13}_{6}C$ **(b)** $^{6}_{3}Li$ or $^{8}_{3}Li$ **(c)** $^{14}_{7}N$ or $^{16}_{7}N$

**\*40.** Which isotope in each of the following pairs should be more stable?
 **(a)** $^{28}_{14}Si$ or $^{29}_{14}Si$ **(b)** $^{19}_{9}F$ or $^{20}_{9}F$ **(c)** $^{20}_{11}Na$ or $^{23}_{11}Na$

**\*41.** Indicate whether each of the following nuclei lies within the belt of stability in Figure 2.5.
 **(a)** Na-24 **(b)** Cf-251 **(c)** Ne-20 **(d)** Th-225

**\*42.** Indicate whether each of the following nuclei lies within the belt of stability in Figure 2.5.
 **(a)** Rn-228 **(b)** P-34 **(c)** He-6 **(d)** Po-209

## Names and Formulas of Ionic and Molecular Compounds

**\*43.** Write the condensed structural formulas and molecular formulas for the following molecules. The reactive groups are shown in red.

 **(a)** $H-\underset{\underset{\displaystyle H}{|}}{\overset{\overset{\displaystyle H}{|}}{C}}-\underset{\underset{\displaystyle O-H}{|}}{C}=O$ (acetic acid) **(b)** $H-\underset{\underset{\displaystyle H}{|}}{\overset{\overset{\displaystyle H}{|}}{C}}-\underset{\underset{\displaystyle H}{|}}{\overset{\overset{\displaystyle H}{|}}{C}}-Cl$ (ethyl chloride)

**\*44.** Given the following condensed formulas, write the molecular formulas for the molecules.
 **(a)** dimethylamine $(CH_3)_2NH$
 **(b)** *n*-propyl alcohol $CH_3(CH_2)_2OH$

**\*45.** Write the formulas for the following molecules.
 **(a)** ammonia **(b)** carbon dioxide **(c)** acetylene
 **(d)** xenon tetrafluoride
 **(e)** germanium tetrachloride

**\*46.** Write the formulas for the following molecules.
 **(a)** methane **(b)** hydrogen peroxide
 **(c)** dinitrogen pentaoxide **(d)** boron trifluoride
 **(e)** diselenium dichloride

**\*47.** Write the names of the following molecules.
 **(a)** $XeO_3$ **(b)** $N_2H_4$ **(c)** $S_4N_4$ **(d)** $CCl_4$
 **(e)** $BrF_5$

**\*48.** Write the names of the following molecules.
 **(a)** $NF_3$ **(b)** $PCl_3$ **(c)** $PH_3$ **(d)** $IF_7$
 **(e)** $SiC$

**\*49.** Give the formulas of all the compounds containing no ions other than $Na^+$, $Ba^{2+}$, $I^-$, and $O^{2-}$.

**\*50.** Give the formulas of compounds in which
 **(a)** the cation is $Mg^{2+}$ and the anion is $Br^-$ or $N^{3-}$.
 **(b)** the anion is $S^{2-}$ and the cation is $Co^{2+}$ or $Co^{3+}$.

**\*51.** Write formulas for the following ionic compounds.
 **(a)** potassium dichromate **(b)** sodium nitrate
 **(c)** magnesium phosphate **(d)** chromium(II) oxide
 **(e)** nickel(III) chloride

**\*52.** Write the formulas of the following ionic compounds.
 **(a)** aluminum sulfate **(b)** sodium acetate
 **(c)** rubidium bromide **(d)** sodium nitride
 **(e)** ammonium chloride

**\*53.** Write the names of the following ionic compounds.
 **(a)** $Fe(C_2H_3O_2)_3$ **(b)** $NaClO_3$ **(c)** $KHCO_3$
 **(d)** $CdS$ **(e)** $K_2O$

**\*54.** Write the names of the following ionic compounds.
 **(a)** $Fe_2(CO_3)_3$ **(b)** $CaSO_4$ **(c)** $Cu_2S$
 **(d)** $PbO_2$ **(e)** $H_2SO_4(aq)$

**\*55.** Complete the following table.

| Name | Formula |
|---|---|
| bromine trichloride | _____ |
| _____ | $HClO_4(aq)$ |
| _____ | $NF_3$ |
| potassium permanganate | _____ |
| _____ | $Ni_2(HPO_4)_3$ |
| sulfur dioxide | _____ |

**\*56.** Complete the following table.

| Name | Formula |
|---|---|
| hypochlorous acid | _____ |
| _____ | $Ba(NO_2)_2$ |
| _____ | $Au_2S_3$ |
| nitric oxide | _____ |
| nickel(II) periodate | _____ |
| _____ | $S_2Cl_2$ |

## Unclassified

**\*57.** If squares represent Cl atoms and spheres represent K atoms, make a representation of a KCl crystal.

**\*58.** If squares represent carbon and spheres represent chlorine, make a representation of liquid $CCl_4$.

**\*59.** Which of the following statements are always true? Never true? Usually true?

(a) Compounds containing carbon atoms are molecular.

(b) A molecule is made up of nonmetal atoms.

(c) An ionic compound has at least one metal atom.

**\*60.** Criticize each of the following statements.

(a) In an ionic compound, the number of cations is always the same as the number of anions.

(b) The molecular formula for strontium bromide is $SrBr_2$.

(c) The mass number is always equal to the atomic number.

(d) For any ion, the number of electrons is always more than the number of protons.

**\*61.** Write the formulas and names of the following.

(a) A molecule made up of one atom of an element with 14 protons and four atoms of a halogen in period 3

(b) The major inorganic component of bones, made up of calcium and an anion containing four oxygen atoms and one phosphorus atom

(c) The aluminum salt of the anion derived from the sulfate ion by the loss of one oxygen atom

**\*62.** Identify each of the following elements.

(a) A halogen with 35 protons in its nucleus

(b) A member of the same period as germanium (Ge) whose anion has a $-2$ charge and 36 electrons

(c) A member of Group 10 with 28 protons in the nucleus

(d) A metalloid in Group 13

## Challenge Problems

**63.** Ethane and ethene are two gases containing only hydrogen and carbon atoms. In a certain sample of ethane, 4.53 g of hydrogen is combined with 18.0 g of carbon. In a sample of ethene, 7.25 g of hydrogen is combined with 43.20 g of carbon.

(a) Show how these data illustrate the law of multiple proportions.

(b) Suggest reasonable formulas for the two compounds.

**64.** Calculate the average density of a single Al-27 atom by assuming that it is a sphere with a radius of 0.143 nm. The masses of a proton, electron, and neutron are $1.6727 \times 10^{-24}$ g, $9.1095 \times 10^{-28}$ g, and $1.6750 \times 10^{-24}$ g, respectively. The volume of a sphere is $4\pi r^3/3$, where $r$ is its radius. Express the answer in grams per cubic centimeter. The density of aluminum is experimentally found to be 2.70 g/cm³. What does that suggest about the packing of aluminum atoms in the metal?

**65.** The mass of a beryllium atom is $1.4965 \times 10^{-23}$ g. Using that fact and other information in this chapter, find the mass of a $Be^{2+}$ ion.

**66.** Each time you inhale, you take in about 500 mL (two significant figures) of air; each mL contains $2.5 \times 10^{19}$ molecules. It has been estimated that Abraham Lincoln, in delivering the Gettysburg Address, inhaled about 200 times.

(a) How many molecules did Lincoln take in?

(b) In the entire atmosphere, there are about $1.1 \times 10^{44}$ molecules. What fraction of the molecules in the Earth's atmosphere was inhaled by Lincoln at Gettysburg?

(c) In the next breath that you take, estimate the number of molecules that were inhaled by Lincoln at Gettysburg.

Gypsum crystals making up ordinary plaster (p. 72) look like this under magnification. (Martin Land/Science Photo Library/Photo Researchers)

# Mass Relations in Chemistry; Stoichiometry

3

To this point, our study of chemistry has been largely qualitative, involving very few calculations. However, chemistry is a quantitative science. Atoms of elements differ from one another not only in composition (number of protons, electrons, neutrons), but also in mass. Chemical formulas of compounds tell us not only the atom ratios in which elements are present, but also the mass ratios.

The general topic of this chapter is stoichiometry (stoy-key-OM-e-tree), the study of mass relations in chemistry. Whether dealing with atoms and molecules (Section 3.1), molar masses (Section 3.2), chemical formulas (Section 3.3), or chemical reactions (Section 3.4), you will be answering some very practical questions that ask "how much—" or "how many—," e.g.,

— How many molecules are there in a glass of water? (Section 3.1)
— How much iron can be obtained from a ton of iron ore? (Section 3.3)
— How much nitrogen gas is required to form a kilogram of ammonia? (Section 3.4)

## 3.1  Atomic and Formula Masses

Individual atoms are too small to be seen, let alone weighed. However, as you will soon see, it is possible to determine quite accurately the relative masses of different atoms and molecules. Indeed, it is possible to go a step further and calculate the actual masses of these tiny building blocks of matter.

### Atomic Masses; the Carbon-12 Scale

Relative masses of atoms of different elements are expressed in terms of their **atomic masses** (often referred to as atomic weights). The atomic mass of an element indicates how heavy, on the average, one atom of that element is compared with an atom of another element.

In order to set up a scale of atomic masses, it is necessary to establish a standard value for one particular species. For many years, two different standards that differed slightly from one another were in common use. Chemists took the atomic mass of the element oxygen to be exactly 16, whereas physicists assigned that value to the most common isotope of oxygen, $^{16}_{8}O$. In 1961, this confusing situation was resolved by adopting a single scale based on the most common isotope of carbon, $^{12}_{6}C$. This isotope is assigned a mass of exactly 12 *atomic mass units (amu)*.

1 amu = 1/12 mass C atom ≈ mass H atom

$$\text{mass of C-12 atom} = 12 \text{ amu (exactly)}$$

It follows that an atom half as heavy as a C-12 atom would weigh 6 amu, an atom twice as heavy as C-12 would have a mass of 24 amu, and so on.

Atomic masses are readily obtained from the periodic table, where they are listed directly below the symbol of the element. In the table on the inside front cover of this text, atomic masses are listed to four significant figures although more precise values are available (e.g., atomic mass of F = 18.998403 amu). Notice that hydrogen has an atomic mass of 1.008 amu; helium has an atomic mass of 4.003 amu. This means that, on the average, a helium atom has a mass that is about one third that of a C-12 atom:

$$\frac{4.003 \text{ amu}}{12.00 \text{ amu}} = 0.3336$$

or about four times that of a hydrogen atom:

$$\frac{4.003 \text{ amu}}{1.008 \text{ amu}} = 3.971$$

In general, for two elements X and Y:

$$\frac{\text{at. mass X}}{\text{at. mass Y}} = \frac{\text{mass of atom of X}}{\text{mass of atom of Y}}$$

*Atomic masses give relative masses of different atoms*

## Atomic Masses and Isotopic Abundances

Relative masses of individual atoms can be determined using a mass spectrometer (Fig. 3.1). Here, gaseous atoms or molecules at very low pressures are ionized by removing one or more electrons. The cations formed are accelerated by a potential of 500 to 2000 V toward a magnetic field, which deflects the ions from their straight-line path. The extent of deflection is inversely related to the mass of the ion. By measuring the voltages required to bring two ions of different mass to the same point on the collector, it is possible to determine their relative masses. For example, using a mass spectrometer, it is found that a $^{19}_{9}\text{F}$ atom is 1.583 times as heavy as a $^{12}_{6}\text{C}$ atom and so has a mass of

$$1.583 \times 12.00 \text{ amu} = 19.00 \text{ amu}$$

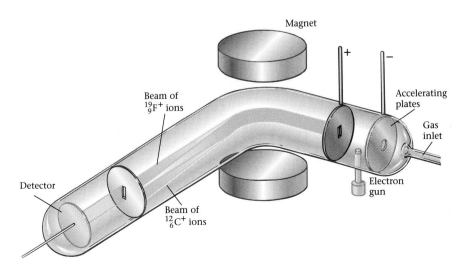

Magnet

Beam of $^{19}_{9}\text{F}^+$ ions

Accelerating plates

Gas inlet

Detector

Electron gun

Beam of $^{12}_{6}\text{C}^+$ ions

**Figure 3.1**
The mass spectrometer. A beam of gaseous ions is deflected in the magnetic field toward the collector plate. Light ions are deflected more than heavy ones. By comparing the accelerating voltages required to bring the two ions to the same point, it is possible to determine the relative masses of the ions.

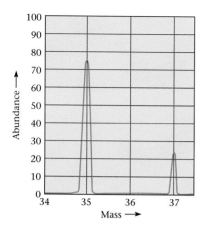

**Figure 3.2**
Mass spectrum of chlorine, which contains isotopes with atomic masses of 34.97 amu (75.53%) and 36.97 amu (24.47%).

As it happens, naturally occurring fluorine consists of a single isotope, $^{19}_{9}F$. It follows that the atomic mass of the element fluorine must be the same as that of F-19, 19.00 amu. The situation with most elements is more complex, since they occur in nature as a mixture of two or more isotopes. In order to determine the atomic mass of such an element, it is necessary to know not only the masses of the individual isotopes but also their percentages *(abundances)* in nature.

Fortunately, isotopic abundances as well as isotopic masses can be determined by mass spectroscopy. The situation with chlorine, which has two stable isotopes, Cl-35 and Cl-37, is shown in Figure 3.2. The atomic masses of the two isotopes are determined in the usual way. The relative abundances of these isotopes are proportional to the heights of the recorder peaks or, more accurately, to the areas under these peaks. For chlorine, the data obtained from the mass spectrometer are

| | Atomic Mass | Abundance |
|---|---|---|
| Cl-35 | 34.97 amu | 75.53% |
| Cl-37 | 36.97 amu | 24.47% |

Using these data and the general equation

$$\text{atomic mass } Y = (\text{atomic mass } Y_1) \times \frac{\%Y_1}{100\%} + (\text{atomic mass } Y_2) \times \frac{\%Y_2}{100\%} + \cdots$$

where Y is the element in question and $Y_1$, $Y_2$, ... are its stable isotopes, the atomic mass of the element is readily calculated.

---

**Example 3.1**   Using the data cited for chlorine, calculate its atomic mass.

*Strategy*   Substitute the data into the general relation given above and solve for atomic mass.

*Solution*

You can also calculate abundances, knowing atomic and isotopic masses (cf. Problem 8)

$$\text{atomic mass Cl} = 34.97 \text{ amu} \times \frac{75.53\%}{100\%} + 36.97 \text{ amu} \times \frac{24.47\%}{100\%} = \boxed{35.46} \text{ amu}$$

The atomic mass of chlorine is a weighted average of those of the two isotopes; it is closer to that of the more abundant isotope, Cl-35.

---

Calculations of the type shown in Example 3.1, using data obtained with a mass spectrometer, can give answers precise to seven or eight significant figures. The accuracy of tabulated atomic masses is limited mostly by variations in natural abundances. Sulfur is an interesting case in point. It consists largely of two isotopes, $^{32}_{16}S$ and $^{34}_{16}S$. The abundance of sulfur-34 varies from about 4.18% in sulfur deposits in Texas and Louisiana to 4.34% in volcanic sulfur from Italy. This leads to an uncertainty of 0.006 amu in the atomic mass of sulfur.

Sulfur deposit on a Hawaiian island volcano. (D. Cavagnaro)

## Masses of Individual Atoms; Avogadro's Number

For most purposes in chemistry, it is sufficient to know the relative masses of different atoms. Sometimes, however, it is necessary to go one step further and calculate the mass in grams of individual atoms. Let us consider how this can be done.

To start with, consider the elements helium and hydrogen. A helium atom is about four times as heavy as a hydrogen atom (He = 4.003 amu, H = 1.008 amu). It follows that a sample containing 100 helium atoms weighs four times as much as a sample containing 100 hydrogen atoms. Again, comparing samples of the two elements containing a million atoms each, the masses will be in a 4 (helium) to 1 (hydrogen) ratio. Turning this argument around, it follows that a sample of helium weighing four grams must contain the same number of atoms as a sample of hydrogen weighing one gram. More exactly:

> no. of He atoms in 4.003 g helium = no. of H atoms in 1.008 g hydrogen

This reasoning is readily extended to other elements. A sample of an element with a mass in grams equal to its atomic mass contains a certain definite number of atoms, $N_A$, regardless of the identity of the element.

The question now arises as to the numerical value of $N_A$; that is, how many atoms are there in 4.003 g of helium, 1.008 g of hydrogen, 32.07 g of sulfur, and so on? As it happens, this problem is one that has been studied for at least a century. Several ingenious experiments have been designed to determine this number, known as **Avogadro's number** and given the symbol $N_A$ (see Problem 82, end of chapter). As you can imagine, it is huge. (Remember that atoms are tiny. There must be a lot of them in 4.003 g of He, 1.008 g of H, and so on.) To four significant figures,

$$N_A = 6.022 \times 10^{23}$$

To get some idea of how large this number is, suppose the entire population of the world were assigned to counting the atoms in 4.003 g of helium. If each person counted one atom per second and worked a 48-hour week, the task would take more than ten million years.

The importance of Avogadro's number in chemistry should be clear. *It represents the number of atoms of an element in a sample whose mass in grams in numerically equal to the atomic mass of the element.* Thus, there are

| | |
|---|---|
| $6.022 \times 10^{23}$ H atoms in 1.008 g H | atomic mass H = 1.008 amu |
| $6.022 \times 10^{23}$ He atoms in 4.003 g He | atomic mass He = 4.003 amu |
| $6.022 \times 10^{23}$ S atoms in 32.07 g S | atomic mass S = 32.07 amu |

Knowing Avogadro's number and the atomic mass of an element, it is possible to calculate the mass of an individual atom (Example 3.2a). You can also determine the number of atoms in a weighed sample of any element (Example 3.2b).

---

**Example 3.2**  When selenium (Se) is added to glass, it gives the glass a brilliant red color. Taking Avogadro's number to be $6.022 \times 10^{23}$, calculate

(a) the mass of a selenium atom.
(b) the number of selenium atoms in a 1.000-g sample of the element.

*Strategy*  The atomic mass of Se, from the periodic table, is 78.96 amu. It follows that

$$6.022 \times 10^{23} \text{ Se atoms} = 78.96 \text{ g Se}$$

*If a nickel weighs twice as much as a dime, there are equal numbers of coins in 1000 g of nickels and 500 g of dimes*

*Most people have better things to do*

Glass is made red by adding a compound of selenium to sand. (Charles D. Winters)

This relation yields the required conversion factors.

*Solution*

(a) mass of Se atom = 1 Se atom $\times \dfrac{78.96 \text{ g Se}}{6.022 \times 10^{23} \text{ Se atoms}} = \boxed{1.311 \times 10^{-22} \text{ g}}$

(b) no. of Se atoms = 1.000 g $\times \dfrac{6.022 \times 10^{23} \text{ Se atoms}}{78.96 \text{ g}}$

$= \boxed{7.627 \times 10^{21} \text{ Se atoms}}$

## Formula Masses

The development we have gone through to this point has been restricted to atoms and, at least by implication, to elements made up of individual atoms. To extend these ideas to other types of particles and to all types of substances, it is helpful to define a quantity called **formula mass.** Quite simply, the formula mass is the sum of the atomic masses in the formula of a substance. Thus,

| Formula | Formula Mass |
|---------|--------------|
| O | 16.00 amu |
| $O_2$ | 2(16.00 amu) = 32.00 amu |
| $H_2O$ | 2(1.008 amu) + 16.00 amu = 18.02 amu |
| NaCl | 22.99 amu + 35.45 amu = 58.44 amu |

The formula mass represents the mass, on the carbon-12 scale, of the unit represented by the formula. The "formula unit" may be an atom (O), a molecule ($O_2$ or $H_2O$), or a set of ions (1 $Na^+$, 1 $Cl^-$ ion). When the formula unit is a molecule, the formula mass is often referred to as the *molecular mass*. From the data above, you can conclude that an $H_2O$ molecule is about 18/16 as heavy as an O atom; an $O_2$ molecule is exactly twice as heavy as an O atom.

The statement made earlier relating Avogadro's number to atomic masses can now be generalized:

*A sample of a substance that has a mass in grams numerically equal to its formula mass contains Avogadro's number of formula units.*

The formula unit of $CaCl_2$ consists of one $Ca^{2+}$ ion and two $Cl^-$ ions

$6.022 \times 10^{23}$ O atoms weigh 16.00 g

$6.022 \times 10^{23}$ $O_2$ molecules weigh 32.00 g

$6.022 \times 10^{23}$ $H_2O$ molecules weigh 18.02 g

$6.022 \times 10^{23}$ ($Na^+$ ions + $Cl^-$ ions) weigh 58.44 g

This relationship is particularly useful for elements and molecular compounds (Fig. 3.3).

---

**Example 3.3**   How many molecules are there in a drop of water weighing 0.050 g?

*Strategy*   Use the relation $6.022 \times 10^{23}$ $H_2O$ molecules = 18.02 g $H_2O$ to find the appropriate conversion factor.

*Solution*

$$0.050 \text{ g H}_2\text{O} \times \frac{6.022 \times 10^{23} \text{ H}_2\text{O molecules}}{18.02 \text{ g H}_2\text{O}} = \boxed{1.7 \times 10^{21} \text{ H}_2\text{O molecules}}$$

Since molecules are so small, it takes a lot of them to make up a sample large enough to be seen and weighed.

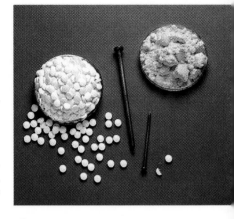

**Figure 3.3**
One mole of iron, sulfur, and aspirin. The iron nails weigh 55.85 g and contain $6.022 \times 10^{23}$ Fe atoms. The pile of yellow sulfur weighs 32.07 g and contains $6.022 \times 10^{23}$ S atoms. The aspirin tablets weigh 180.15 g and contain $6.022 \times 10^{23}$ $C_9H_8O_4$ molecules.  (Marna G. Clarke)

## 3.2   The Mole

The quantity represented by Avogadro's number is so important that it is given a special name, the **mole.** A mole represents $6.022 \times 10^{23}$ items, whatever they may be.

$$1 \text{ mol H atoms} = 6.022 \times 10^{23} \text{ H atoms}$$

$$1 \text{ mol O atoms} = 6.022 \times 10^{23} \text{ O atoms}$$

$$1 \text{ mol H}_2 \text{ molecules} = 6.022 \times 10^{23} \text{ H}_2 \text{ molecules}$$

$$1 \text{ mol H}_2\text{O molecules} = 6.022 \times 10^{23} \text{ H}_2\text{O molecules}$$

$$1 \text{ mol electrons} = 6.022 \times 10^{23} \text{ electrons}$$

$$1 \text{ mol pennies} = 6.022 \times 10^{23} \text{ pennies}$$

(One mole of pennies is a lot of money. It's enough to pay all the expenses of the United States for the next billion years or so.)

A mole represents not only a specific number of particles but also a definite mass of a substance. In general, ***the molar mass, $\mathcal{M}$, in grams per mole, is numerically equal to the formula mass.*** Thus:

The size of the mole was chosen to make this relation true

| Formula | Formula Mass | Molar Mass, $\mathcal{M}$ |
|---------|--------------|---------------------------|
| O       | 16.00 amu    | 16.00 g/mol               |
| $O_2$   | 32.00 amu    | 32.00 g/mol               |
| $H_2O$  | 18.02 amu    | 18.02 g/mol               |
| NaCl    | 58.44 amu    | 58.44 g/mol               |

Notice that the formula of a substance must be known to find its molar mass. It would be ambiguous, to say the least, to refer to the "molar mass of hydrogen." One mole of hydrogen atoms, represented by the symbol H, weighs 1.008 g; the molar mass of H is 1.008 g/mol. One mole of hydrogen molecules, represented by the formula $H_2$, weighs 2.016 g; the molar mass of $H_2$ is 2.016 g/mol.

### Mole-Gram Conversions

As you will see later in this chapter, it is often necessary to convert from moles of a substance to mass in grams or vice versa. Such conversions are readily made by using the general relation

$$m = \mathcal{M} \times n$$

where $m$ is the mass in grams, $\mathcal{M}$ is the molar mass (g/mol), and $n$ is the amount in moles.

There are two crystalline forms of $CaCO_3$, calcite (right) and aragonite (left). (Charles D. Winters)

**Example 3.4** Calcium carbonate is the principal ingredient of the chalk used in most classrooms. Determine the number of moles of calcium carbonate in a stick of chalk containing 14.8 g of calcium carbonate.

***Strategy*** Find the molar mass of calcium carbonate and use it to convert 14.8 g to moles. Before calculating the molar mass, you must come up with the formula of calcium carbonate (Chapter 2).

***Solution*** The formula is $CaCO_3$, so the molar mass is

$$\mathcal{M} = [40.08 + 12.01 + 3(16.00)] \text{ g/mol} = 100.09 \text{ g/mol}$$

$$n = 14.8 \text{ g CaCO}_3 \times \frac{1 \text{ mol CaCO}_3}{100.09 \text{ g CaCO}_3} = \boxed{0.148 \text{ mol CaCO}_3}$$

**Example 3.5** Acetylsalicylic acid, $C_9H_8O_4$, is the principal ingredient of aspirin. What is the mass in grams of 0.287 mol of acetylsalicylic acid?

***Strategy*** Find the molar mass of $C_9H_8O_4$ and use it to obtain the conversion factor required to convert 0.287 mol to mass in grams.

***Solution*** The molar mass of $C_9H_8O_4$ is

$$\mathcal{M} = [9(12.01) + 8(1.008) + 4(16.00)] \text{ g/mol} = 180.15 \text{ g/mol}$$

Hence,

$$\text{mass } C_9H_8O_4 = 0.287 \text{ mol } C_9H_8O_4 \times \frac{180.15 \text{ g } C_9H_8O_4}{1 \text{ mol } C_9H_8O_4} = \boxed{51.7 \text{ g } C_9H_8O_4}$$

Conversions of the type we have just carried out come up over and over again in chemistry. They will be required in nearly every chapter of this text. Clearly, you must know what is meant by a mole. Remember, a mole always represents a certain number of items, $6.022 \times 10^{23}$. Its mass, however, differs with the substance involved: A mole of $H_2O$, 18.02 g, weighs considerably more than a mole of $H_2$, 2.016 g, even though they both contain the same number of molecules. In the same way, a dozen bowling balls weigh a lot more than a dozen eggs, even though each involves the same number of items.

## 3.3 Mass Relations in Chemical Formulas

As you will see shortly, the formula of a compound can be used to determine the mass percents of the elements present. Conversely, if the percentages of the elements are known, the simplest formula can be determined. Knowing the molar mass of the compound, it is possible to go one step further and find the molecular formula. In this section we will consider how these three types of calculations are carried out.

### Percent Composition from Formula

The "percent composition" of a compound is specified by citing the mass percents of the elements present. For example, in a 100-gram sample of water there are 11.19 g of hydrogen and 88.81 g of oxygen. Hence, the percentages of the two elements are

$$\frac{11.19 \text{ g H}}{100.00 \text{ g}} \times 100\% = 11.19\% \text{ H} \qquad \frac{88.81 \text{ g O}}{100.00 \text{ g}} \times 100\% = 88.81\% \text{ O}$$

We would say that the percent composition of water is 11.19% H, 88.81% O.

Knowing the formula of a compound, you can readily calculate the mass percents of its constituent elements. It is convenient to start with one mole of the compound (Example 3.6).

Sodium hydrogen carbonate, $NaHCO_3$, is sometimes called "baking soda." (Charles D. Winters)

---

**Example 3.6** Sodium hydrogen carbonate, commonly called "bicarbonate of soda," is used in many commercial products to relieve an upset stomach. It has the formula $NaHCO_3$. What are the mass percents of Na, H, C, and O in sodium hydrogen carbonate?

**Strategy** Find the mass in grams of each element in one mole of $NaHCO_3$. Then find

$$\% \text{ element} = \frac{\text{mass element}}{\text{total mass compound}} \times 100\%$$

**Solution** It is convenient to set up a table to determine the mass of each element in one mole of $NaHCO_3$.

|     | $n$ | $\times$ | $\mathcal{M}$ | $=$ | $m$ |
|-----|------|-----|-------------|-----|------|
| Na | 1 mol | $\times$ | 22.99 g/mol | $=$ | 22.99 g |
| H | 1 mol | $\times$ | 1.008 g/mol | $=$ | 1.008 g |
| C | 1 mol | $\times$ | 12.01 g/mol | $=$ | 12.01 g |
| O | 3 mol | $\times$ | 16.00 g/mol | $=$ | 48.00 g |
|    |       |          |             |     | 84.01 g $NaHCO_3$ |

Since 84.01 g of $NaHCO_3$ contains 22.99 g of Na, 1.008 g of H, 12.01 g of C, and 48.00 g of O,

$$\text{mass \% Na} = \frac{22.99 \text{ g}}{84.01 \text{ g}} \times 100\% = \boxed{27.36\%} \qquad \text{mass \% H} = \frac{1.008 \text{ g}}{84.01 \text{ g}} \times 100\%$$
$$= \boxed{1.200\%}$$

$$\text{mass \% C} = \frac{12.01 \text{ g}}{84.01 \text{ g}} \times 100\% = \boxed{14.30\%} \qquad \text{mass\% O} = \frac{48.00 \text{ g}}{84.01 \text{ g}} \times 100\%$$
$$= \boxed{57.14\%}$$

The percentages add up to 100, as they should:

$$27.36\% + 1.200\% + 14.30\% + 57.14\% = 100.00\%$$

---

The calculations in Example 3.6 illustrate an important characteristic of formulas. In one mole of $NaHCO_3$, there is 1 mol Na (22.99 g), 1 mol H (1.008 g), 1 mol C (12.01 g), and 3 mol O (48.00 g). In other words, the mole ratio is 1 mol Na:1 mol H:1 mol C:3 mol O. This is the same as the atom ratio in $NaHCO_3$, 1 atom Na:1 atom H:1 atom C:3 atoms O. In general, ***the subscripts in a formula represent not only the atom ratio in which the different elements are combined, but also the mole ratio.*** For example,

Can you explain why the mole ratio must equal the atom ratio?

| Formula | Atom Ratio | Mole Ratio |
|---|---|---|
| $H_2O$ | 2 atoms H:1 atom O | 2 mol H:1 mol O |
| $KNO_3$ | 1 atom K:1 atom N:3 atoms O | 1 mol K:1 mol N:3 mol O |
| $C_{12}H_{22}O_{11}$ | 12 atoms C:22 atoms  H:11 atoms O | 12 mol C:22 mol  H:11 mol O |

The formula of a compound can also be used in a straightforward way to find the mass of an element in a weighed sample of the compound (Example 3.7).

---

**Example 3.7**   An iron-containing mineral responsible for the red color of soils in many parts of the country is limonite, which has the formula $Fe_2O_3 \cdot \frac{3}{2}H_2O$. What mass of iron in grams can be obtained from a metric ton ($10^3$ kg = $10^6$ g) of limonite?

***Strategy***   First find the mass percent of iron. Then, using the mass percent, determine how much iron there is in one metric ton of the ore.

*Solution*   In one mole of limonite, there is

$$2 \text{ mol Fe} \times \frac{55.85 \text{ g Fe}}{1 \text{ mol Fe}} = 111.7 \text{ g Fe}$$

The molar mass of limonite is

$$(111.7 + 3(16.00) + \frac{3}{2}(18.02)) \text{ g/mol} = 186.7 \text{ g/mol}$$

Thus,

$$\% \text{ Fe} = \frac{111.7 \text{ g}}{186.7 \text{ g}} \times 100\% = 59.83\%$$

In one metric ton of limonite

$$\text{mass Fe} = 1.000 \times 10^6 \text{ g limonite} \times \frac{59.83 \text{ g Fe}}{100.0 \text{ g limonite}} = \boxed{5.983 \times 10^5 \text{ g Fe}}$$

---

Limonite ore. (Paul Silverman/Fundamental Photographs, New York)

## Simplest Formula from Chemical Analysis

A major task of chemical analysis is to determine the formulas of compounds. The formula found by the approach described here is the simplest formula, which gives the simplest whole-number ratio of the atoms present. For an ionic compound, the simplest formula is ordinarily the only one that can be written (e.g., $CaCl_2$, $CrBr_3$). For a molecular compound, the molecular formula is a whole-number multiple of the simplest formula, where that number may be 1, 2, . . . .

| Compound | Simplest Formula | Molecular Formula | Multiple |
|---|---|---|---|
| Water | $H_2O$ | $H_2O$ | 1 |
| Hydrogen peroxide | HO | $H_2O_2$ | 2 |
| Propylene | $CH_2$ | $C_3H_6$ | 3 |

The analytical data leading to the simplest formula may be expressed in various ways. You may know

— the masses of the elements in a weighed sample of the compound
— the mass percents of the elements in the compound
— the masses of products obtained by reaction of a weighed sample of the compound

The strategy used to calculate the simplest formula depends to some extent upon which of these types of information is given. The basic objective in each case is to find the number of moles of each element, then the simplest mole ratio, and finally the simplest formula.

---

**Example 3.8**   A 25.0-g sample of an orange compound contains 6.64 g of potassium, 8.84 g of chromium, and 9.52 g of oxygen. Find the simplest formula.

**Strategy**   First (1), convert the masses of the three elements to moles. Knowing the number of moles (*n*) of K, Cr, and O, you can then (2) calculate the mole ratios. Finally (3), equate the mole ratio to the atom ratio, which gives you the simplest formula.

**Solution**

(1) $n_K = 6.64 \text{ g K} \times \dfrac{1 \text{ mol K}}{39.10 \text{ g K}} = 0.170 \text{ mol K}$

$n_{Cr} = 8.84 \text{ g Cr} \times \dfrac{1 \text{ mol Cr}}{52.00 \text{ g Cr}} = 0.170 \text{ mol Cr}$

$n_O = 9.52 \text{ g O} \times \dfrac{1 \text{ mol O}}{16.00 \text{ g O}} = 0.595 \text{ mol O}$

(2) To find the mole ratios, divide by the smallest number, 0.170 mol K:

$$\dfrac{0.170 \text{ mol Cr}}{0.170 \text{ mol K}} = \dfrac{1.00 \text{ mol Cr}}{1 \text{ mol K}} \qquad \dfrac{0.595 \text{ mol O}}{0.170 \text{ mol K}} = \dfrac{3.50 \text{ mol O}}{1 \text{ mol K}}$$

The mole ratio is 1 mol of K : 1 mol of Cr : 3.50 mol of O.
(3) As pointed out earlier, the mole ratio is the same as the atom ratio. To find the simplest whole-number atom ratio, multiply throughout by 2:

$$2 \text{ K} : 2 \text{ Cr} : 7 \text{ O}$$

The simplest formula of the orange compound is   $K_2Cr_2O_7$.

---

Sometimes you will be given the mass percents of the elements in a compound. If that is the case, one extra step is involved. ***Assume a 100-g sample and calculate the mass of each element in that sample.***

Suppose, for example, you are told that the percentages of K, Cr, and O in a compound are 26.6%, 35.4%, and 38.0%, respectively. It follows that in a 100.0-g sample there are

$$26.6 \text{ g K}, \; 35.4 \text{ g Cr}, \; 38.0 \text{ g O}$$

Working with these masses, you can go through the same procedure followed in Example 3.8 to arrive at the same answer (try it!).

The most complex problem of this type requires you to determine the simplest formula of a compound, given only the raw data obtained from its analysis. Here, an additional step is involved; you have to determine the masses of the elements present in a fixed mass of the compound (Example 3.9).

Mass data ⟶ mole ratio = atom ratio ⟶ simplest formula

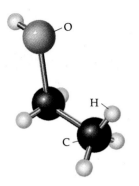

The structural formula of ethyl alcohol, shown above, is often abbreviated as $CH_3CH_2OH$.

**Example 3.9** The compound that gives fermented grape juice, malt liquor, and vodka their intoxicating properties is ethyl alcohol, which contains the elements carbon, hydrogen, and oxygen. When a sample of ethyl alcohol is burned in air, it is found that

$$\text{5.00 g ethyl alcohol} \longrightarrow \text{9.55 g } CO_2 + \text{5.87 g } H_2O$$

What is the simplest formula of ethyl alcohol?

**Strategy** The first step here is to calculate the masses of C, H, and O in the 5.00-g sample. To do this, note that all of the carbon has been converted to carbon dioxide and that there is 12.01 g (1 mol) of C in 44.01 g (1 mol) of $CO_2$. Hence, to find the mass of carbon in 9.55 g of $CO_2$, you use the conversion factor

$$\text{12.01 g C/44.01 g } CO_2$$

By the same token, all the hydrogen ends up as water. There is 2.016 g (2 mol) of H in 18.02 g (1 mol) of $H_2O$. The conversion factor to go from mass of water to mass of hydrogen is

$$\text{2.016 g H/18.02 g } H_2O$$

The mass of oxygen is found by difference:

$$\text{mass O} = \text{mass of sample} - (\text{mass of C} + \text{mass of H})$$

Once the masses of the elements are obtained, it's all downhill; follow the same path as in Example 3.8.

**Solution**

(1) Find the mass of each element in the sample.

$$\text{mass C} = \text{9.55 g } CO_2 \times \frac{\text{12.01 g C}}{\text{44.01 g } CO_2} = \text{2.61 g C}$$

$$\text{mass H} = \text{5.87 g } H_2O \times \frac{\text{2.016 g H}}{\text{18.02 g } H_2O} = \text{0.657 g H}$$

$$\text{mass O} = \text{5.00 g sample} - (\text{2.61 g C} + \text{0.657 g H}) = \text{1.73 g O}$$

(2) Find the number of moles of each element.

$$n_C = \text{2.61 g C} \times \frac{\text{1 mol C}}{\text{12.01 g C}} = \text{0.217 mol C}$$

$$n_H = \text{0.657 g H} \times \frac{\text{1 mol H}}{\text{1.008 g H}} = \text{0.652 mol H}$$

$$n_O = \text{1.73 g O} \times \frac{\text{1 mol O}}{\text{16.00 g O}} = \text{0.108 mol O}$$

(3) Find the mole ratios and then the simplest formula.

$$\frac{\text{0.217 mol C}}{\text{0.108 mol O}} = \frac{\text{2.01 mol C}}{\text{1 mol O}} \qquad \frac{\text{0.652 mol H}}{\text{0.108 mol O}} = \frac{\text{6.04 mol H}}{\text{1 mol O}}$$

Sometimes it's not so easy to convert the atom ratio to simplest formula (cf. Problem 40)

Rounding off to whole numbers, the mole ratios are

$$\text{2 mol C} : \text{6 mol H} : \text{1 mol O}$$

The simplest formula of ethyl alcohol is $C_2H_6O$.

## Molecular Formula from Simplest Formula

Chemical analysis always leads to the simplest formula of a compound because it gives only the simplest atom ratio of the elements. As pointed out earlier, the molecular formula is a whole-number multiple of the simplest formula. That multiple may be 1 as in $H_2O$, 2 as in $H_2O_2$, 3 as in $C_3H_6$, or some other integer. To find the multiple, one more piece of data is needed: the molar mass.

---

**Example 3.10**    Vitamin C may (or may not) help prevent the common cold. Its simplest formula is found by analysis to be $C_3H_4O_3$. From another experiment, the molar mass is found to be about 180 g/mol. What is the molecular formula of vitamin C?

*Strategy*    Calculate the molar mass corresponding to the simplest formula, i.e., $\mathcal{M}_{C_3H_4O_3}$. Then find the multiple by dividing the observed molar mass, 180 g/mol, by $\mathcal{M}_{C_3H_4O_3}$.

*Solution*

$$\mathcal{M}_{C_3H_4O_3} = 3(12.01 \text{ g/mol}) + 4(1.008 \text{ g/mol}) + 3(16.00 \text{ g/mol}) = 88.06 \text{ g/mol}$$

The ratio of the observed molar mass to that of $C_3H_4O_3$ is

$$\frac{180 \text{ g/mol}}{88.06 \text{ g/mol}} = 2.04$$

The multiple is 2; the molecular formula is   $C_6H_8O_6$.

---

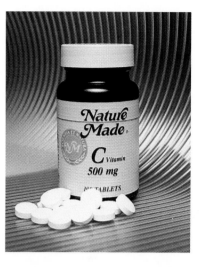

Vitamin C tablets. (Charles D. Winters)

## 3.4  Mass Relations in Reactions

A chemist who carries out a reaction in the laboratory needs to know how much **product** can be obtained from a given amount of starting materials (**reactants**). To do this, he or she starts by writing a balanced chemical equation.

## Writing and Balancing Equations

Chemical reactions are represented by chemical equations, which identify reactants and products. Formulas of reactants appear on the left side of the equation; those of products are written on the right. In a balanced chemical equation, there are the same number of atoms of a given element on both sides. The same situation holds for a chemical reaction that you carry out in the laboratory; atoms are conserved. For that reason, *any calculation involving a reaction must be based upon the balanced equation for that reaction.*

Beginning students are sometimes led to believe that writing a chemical equation is a simple, mechanical process. Nothing could be further from the truth. One point that seems obvious is often overlooked. *You cannot write an equation unless you know what happens in the reaction that it represents.* All the reactants and all the products must be identified. Moreover, you must know their formulas and physical states.

To illustrate how a relatively simple equation can be written and balanced, consider a reaction used in the Titan rocket motor to launch the Gemini spacecraft (Fig. 3.4). The reactants were two liquids, hydrazine and dinitrogen tetraoxide, whose molecular formulas are $N_2H_4$ and $N_2O_4$, respectively. The products of the reaction are gaseous nitrogen, $N_2$, and water vapor. To write a balanced equation for this reaction, proceed as follows:

**1.** *Write a "skeleton" equation in which the formulas of the reactants appear on the left and those of the products on the right.* In this case,

$$N_2H_4 + N_2O_4 \longrightarrow N_2 + H_2O$$

**2.** *Indicate the physical state of each reactant and product, after the formula, by writing*

(*g*)   for a gaseous substance
(*l*)   for a pure liquid
(*s*)   for a solid
(*aq*)  for an ion or molecule in water (aqueous) solution

In this case

$$N_2H_4(l) + N_2O_4(l) \longrightarrow N_2(g) + H_2O(g)$$

**3.** *Balance the equation.* To accomplish this, start by writing a coefficient of 4 for $H_2O$, thus obtaining 4 oxygen atoms on both sides:

$$N_2H_4(l) + N_2O_4(l) \longrightarrow N_2(g) + 4H_2O(g)$$

Now consider the hydrogen atoms. There are $4 \times 2 = 8$ H atoms on the right. To obtain 8 H atoms on the left, write a coefficient of 2 for $N_2H_4$:

$$2N_2H_4(l) + N_2O_4(l) \longrightarrow N_2(g) + 4H_2O(g)$$

Finally, consider nitrogen. There are a total of $(2 \times 2) + 2 = 6$ nitrogen atoms on the left. To balance nitrogen, write a coefficient of 3 for $N_2$:

$$2N_2H_4(l) + N_2O_4(l) \longrightarrow 3N_2(g) + 4H_2O(g)$$

This is the final balanced equation for the reaction of hydrazine with dinitrogen tetraoxide.

Three points concerning the balancing process are worth noting.

**1.** Equations are balanced by adjusting coefficients in front of formulas, never by changing subscripts within formulas. On paper, the equation discussed above could have been balanced by writing $N_6$ on the right, but that would have been absurd. Elemental nitrogen exists as diatomic molecules, $N_2$; there is no such thing as an $N_6$ molecule.
**2.** In balancing an equation, it is best to start with an element that appears in only one species on each side of the equation. In this case, either oxygen or hydrogen are good starting points. Nitrogen would have been a poor choice, however, since there are nitrogen atoms in both reactant molecules, $N_2H_4$ and $N_2O_4$.
**3.** In principle, there are an infinite number of balanced equations that can be written for any reaction. The equations

$$4N_2H_4(l) + 2N_2O_4(l) \longrightarrow 6N_2(g) + 8H_2O(g)$$

$$N_2H_4(l) + \tfrac{1}{2}N_2O_4(l) \longrightarrow \tfrac{3}{2}N_2(g) + 2H_2O(g)$$

**Figure 3.4**
Titan rocket motor. (NASA)

Using the letters (*s*), (*l*), (*g*), (*aq*) lends a sense of physical reality to the equation

are balanced in that there are the same number of atoms of each element on both sides. Ordinarily, the equation with the simplest whole-number co-efficients

$$2N_2H_4(l) + N_2O_4(l) \longrightarrow 3N_2(g) + 4H_2O(g)$$

is preferred.

Frequently, when you are asked to balance an equation, the formulas of products and reactants are given. Sometimes, though, you will have to derive the formulas, given only the names (Example 3.11).

---

**Example 3.11**   Crystals of sodium hydroxide (lye) react with carbon dioxide of the air to form a white powder, sodium carbonate, and a colorless liquid, water. Write a balanced equation for this chemical reaction.

*Strategy*   To translate names into formulas, recall the discussion in Section 2.7, Chapter 2. The physical states are given or implied. To balance the equation, you could start with either sodium or hydrogen.

*Solution*   The "skeleton" equation is

$$NaOH(s) + CO_2(g) \longrightarrow Na_2CO_3(s) + H_2O(l)$$

Since there are two Na atoms on the right, a coefficient of 2 is written in front of NaOH:

$$2NaOH(s) + CO_2(g) \longrightarrow Na_2CO_3(s) + H_2O(l)$$

Careful inspection shows that all other atoms are now balanced.

Can you write an equation for the reaction of aluminum with oxygen? If not, recall Fig. 2.10

---

## Mass Relations from Equations

The principal reason for writing balanced equations is to make it possible to re-late the masses of reactants and products. Calculations of this sort are based upon a very important principle:

**The coefficients of a balanced equation represent numbers of moles of reactants and products.**

To show that this statement is valid, recall the equation

$$2N_2H_4(l) + N_2O_4(l) \longrightarrow 3N_2(g) + 4H_2O(g)$$

The coefficients in this equation represent numbers of molecules; that is,

2 molecules $N_2H_4$ + 1 molecule $N_2O_4 \longrightarrow$
3 molecules $N_2$ + 4 molecules $H_2O$

A balanced equation remains valid if each coefficient is multiplied by the same number, including Avogadro's number, $N_A$:

$2N_A$ molecules $N_2H_4$ + $N_A$ molecules $N_2O_4 \longrightarrow$
$3N_A$ molecules $N_2$ + $4N_A$ molecules $H_2O$

Recall from Section 3.2 that a mole represents Avogadro's number of items, $N_A$. Thus the equation above can be written

$$2 \text{ mol } N_2H_4 + 1 \text{ mol } N_2O_4 \longrightarrow 3 \text{ mol } N_2 + 4 \text{ mol } H_2O$$

The quantities 2 mol $N_2H_4$, 1 mol $N_2O_4$, 3 mol $N_2$, and 4 mol $H_2O$ are chemically equivalent to each other in this reaction. Hence they can be used in conversion factors such as

$$\frac{2 \text{ mol } N_2H_4}{3 \text{ mol } N_2}, \ \frac{3 \text{ mol } N_2}{1 \text{ mol } N_2O_4}, \text{ or a variety of other combinations}$$

If you wanted to know how many moles of hydrazine were required to form 1.80 mol of elemental nitrogen, the conversion would be

$$n_{N_2H_4} = 1.80 \text{ mol } N_2 \times \frac{2 \text{ mol } N_2H_4}{3 \text{ mol } N_2} = 1.20 \text{ mol } N_2H_4$$

To find the number of moles of nitrogen produced from 2.60 mol of $N_2O_4$,

$$n_{N_2} = 2.60 \text{ mol } N_2O_4 \times \frac{3 \text{ mol } N_2}{1 \text{ mol } N_2O_4} = 7.80 \text{ mol } N_2$$

Simple mole relationships of the type just discussed are readily extended to relate

— moles of one substance to grams of another (Example 3.12a)
— grams of one substance to grams of another (Examples 3.12b and c)

---

**Example 3.12**   Ammonia used to make fertilizers for lawns and gardens is made by reacting nitrogen of the air with hydrogen. The balanced equation for the reaction is

$$N_2(g) + 3H_2(g) \longrightarrow 2NH_3(g)$$

Determine

(a) the mass in grams of ammonia, $NH_3$, formed when 1.34 mol of $N_2$ reacts.
(b) the mass in grams of $N_2$ required to form 1.00 kg of $NH_3$.
(c) the mass in grams of $H_2$ required to react with 6.00 g of $N_2$.

*Strategy*   In each case, you use the mole ratios given by the coefficients of the balanced equation to relate moles of one substance to moles of another. Beyond that, use molar mass to relate moles of one substance to mass in grams of that substance. Before starting, decide upon the path you will follow to go from the quantity given to that required, i.e.,

To convert between moles of A and grams of A, use molar mass. To convert between moles of A and moles of B, use the balanced equation

(a) $n_{N_2} \longrightarrow n_{NH_3} \longrightarrow$ mass of $NH_3$
(b) mass of $NH_3 \longrightarrow n_{NH_3} \longrightarrow n_{N_2} \longrightarrow$ mass of $N_2$
(c) mass of $N_2 \longrightarrow n_{N_2} \longrightarrow n_{H_2} \longrightarrow$ mass of $H_2$

Conversions indicated by colored arrows involve mole ratios given by the coefficients of the balanced equation. The other conversions require only molar masses and are essentially identical to those carried out in Examples 3.4 and 3.5.

*Solution*

(a) mass of $NH_3 = 1.34 \text{ mol } N_2 \times \frac{2 \text{ mol } NH_3}{1 \text{ mol } N_2} \times \frac{17.03 \text{ g } NH_3}{1 \text{ mol } NH_3} = \boxed{45.6 \text{ g } NH_3}$

(b)   mass of $N_2 = 1000 \text{ g } NH_3 \times \frac{1 \text{ mol } NH_3}{17.03 \text{ g } NH_3} \times \frac{1 \text{ mol } N_2}{2 \text{ mol } NH_3} \times \frac{28.02 \text{ g } N_2}{1 \text{ mol } N_2}$

$= \boxed{823 \text{ g } N_2}$

(c)   mass of $H_2 = 6.00 \text{ g } N_2 \times \frac{1 \text{ mol } N_2}{28.02 \text{ g } N_2} \times \frac{3 \text{ mol } H_2}{1 \text{ mol } N_2} \times \frac{2.016 \text{ g } H_2}{1 \text{ mol } H_2} = \boxed{1.30 \text{ g } H_2}$

---

## Limiting Reactant and Theoretical Yield

When the two elements antimony and iodine are heated in contact with one another (Fig. 3.5), they react to form antimony(III) iodide.

$$2Sb(s) + 3I_2(s) \longrightarrow 2SbI_3(s)$$

The coefficients in this equation show that two moles of Sb (243.6 g) react with exactly three moles of $I_2$ (761.4 g) to form two moles of $SbI_3$ (1005.0 g). Put another way, the maximum quantity of $SbI_3$ that can be obtained under these conditions, assuming the reaction goes to completion and no product is lost, is 1005.0 g. This quantity is referred to as the **theoretical yield** of $SbI_3$.

Ordinarily, in the laboratory, reactants are not mixed in exactly the ratio required for reaction. Instead, an excess of one reactant, usually the cheaper one, is used. For example, 3.00 mol of Sb could be mixed with 3.00 mol of $I_2$. In that case, after the reaction is over, 1.00 mol of Sb remains unreacted.

$$\text{excess Sb} = 3.00 \text{ mol Sb originally} - 2.00 \text{ mol Sb consumed}$$

$$= 1.00 \text{ mol Sb}$$

The 3.00 mol of $I_2$ should be completely consumed in forming the 2.00 mol of $SbI_3$:

$$n_{SbI_3} \text{ formed} = 3.00 \text{ mol } I_2 \times \frac{2 \text{ mol SbI}_3}{3 \text{ mol I}_2} = 2.00 \text{ mol SbI}_3$$

After the reaction is over, the solid obtained would be a mixture of product, 2.00 mol of $SbI_3$, with 1.00 mol of unreacted Sb.

In situations such as this, a distinction is made between the *excess reactant* (Sb) and the **limiting reactant,** $I_2$. The amount of product formed is determined (limited) by the amount of limiting reactant. With 3.00 mol of $I_2$, only 2.00 mol of $SbI_3$ is obtained, regardless of how large an excess of Sb is used.

Under these conditions, the theoretical yield of product is the amount produced if the limiting reactant is completely consumed. In the case just cited, the theoretical yield of $SbI_3$ is 2.00 mol, the amount formed from the limiting reactant, $I_2$.

Often you will be given the amounts of two different reactants and asked to determine which is the limiting reactant and to calculate the theoretical yield of product. To do this, it helps to follow a systematic, three-step procedure:

1. *Calculate the amount of product that would be formed if the first reactant were completely consumed.*
2. *Repeat this calculation for the second reactant; that is, calculate how much product would be formed if all of that reactant were consumed.*
3. *Choose the smaller of the two amounts calculated in (1) and (2).* **This is the theoretical yield of product; the reactant that produces the smaller amount is the limiting reactant.** The other reactant is in excess; only part of it is consumed.

**Figure 3.5**
When antimony (gray powder) comes in contact with iodine (violet vapor), a vigorous reaction takes place (middle picture). The product is a red solid, $SbI_3$. (Charles D. Winters)

**Example 3.13** Consider the reaction

$$2Sb(s) + 3I_2(s) \longrightarrow 2SbI_3(s)$$

Determine the limiting reactant and the theoretical yield when

(a) 1.20 mol of Sb and 2.40 mol of $I_2$ are mixed.

(b) 1.20 g of Sb and 2.40 g of $I_2$ are mixed. What mass of excess reactant is left when the reaction is complete?

**Strategy** Follow the three steps outlined above. In steps (1) and (2), follow the strategy described in Example 3.12 to calculate the amount of product formed. In (a), a simple one-step conversion is required; in (b), the path is longer because you have to go from mass of reactant to mass of product. To find the mass of reactant left over in (b), calculate how much is required to give the theoretical yield of product. Subtract that from the starting amount to find the amount left.

**Solution**

(a) (1) $n_{SbI_3}$ from Sb = 1.20 mol Sb $\times \dfrac{2 \text{ mol SbI}_3}{2 \text{ mol Sb}}$ = 1.20 mol SbI$_3$

(2) $n_{SbI_3}$ from $I_2$ = 2.40 mol $I_2 \times \dfrac{2 \text{ mol SbI}_3}{3 \text{ mol I}_2}$ = 1.60 mol SbI$_3$

(3) Since $\boxed{1.20 \text{ mol}}$ is the smaller amount of product, that is the theoretical yield of SbI$_3$. This amount of SbI$_3$ is produced by the antimony, so $\boxed{\text{Sb is the}}$ $\boxed{\text{limiting reactant.}}$

You have 62 buns, 95 hamburgers, and 51 slices of cheese. What is the "limiting reactant" for making double cheeseburgers?

(b) (1) mass of SbI$_3$ from Sb = 1.20 g Sb $\times \dfrac{1 \text{ mol Sb}}{121.8 \text{ g Sb}} \times \dfrac{2 \text{ mol SbI}_3}{2 \text{ mol Sb}} \times \dfrac{502.5 \text{ g SbI}_3}{1 \text{ mol SbI}_3}$

$= 4.95$ g SbI$_3$

(2) mass of SbI$_3$ from $I_2$ = 2.40 g $I_2 \times \dfrac{1 \text{ mol I}_2}{253.8 \text{ g I}_2} \times \dfrac{2 \text{ mol SbI}_3}{3 \text{ mol I}_2} \times \dfrac{502.5 \text{ g SbI}_3}{1 \text{ mol SbI}_3}$

$= 3.17$ g SbI$_3$

(3) The reactant that yields the smaller amount (3.17 g) of SbI$_3$ is $I_2$. Hence $\boxed{I_2 \text{ is the limiting reactant.}}$ The smaller amount, $\boxed{3.17 \text{ g}}$ of SbI$_3$, is the theoretical yield.

(4) mass of Sb required = 3.17 g SbI$_3 \times \dfrac{121.8 \text{ g Sb}}{502.5 \text{ g SbI}_3}$ = 0.768 g Sb

mass of Sb left = 1.20 g − 0.768 g = $\boxed{0.43 \text{ g}}$

Remember that, in deciding upon the theoretical yield of product, you *choose the smaller of the two calculated amounts.* To see why this must be the case, refer back to Example 3.13b. There 1.20 g of Sb was mixed with 2.40 g of $I_2$. Calculations show that the theoretical yield of SbI$_3$ is 3.17 g, and 0.43 g of Sb is left over. Thus

$$1.20 \text{ g Sb} + 2.40 \text{ g I}_2 \longrightarrow 3.17 \text{ g SbI}_3 + 0.43 \text{ g Sb}$$

This makes sense: 3.60 g of reactants yields a total of 3.60 g of "products," including the unreacted antimony. Suppose, however, that 4.95 g of $SbI_3$ was chosen as the theoretical yield. The following nonsensical situation would arise:

$$1.20 \text{ g Sb} + 2.40 \text{ g } I_2 \longrightarrow 4.95 \text{ g } SbI_3$$

This violates the law of conservation of mass; 3.60 g of reactants cannot form 4.95 g of product.

## Experimental Yield; Percent Yield

The theoretical yield is the maximum amount of product that can be obtained. In calculating the theoretical yield, it is assumed that the limiting reactant is 100% converted to product. In the real world, that is unlikely to happen. Some of the limiting reactant may be consumed in competing reactions. Some of the product may be lost in separating it from the reaction mixture. For these and other reasons, the experimental yield is ordinarily less than the theoretical yield. Put another way, the *percent yield* is expected to be less than 100%:

$$\text{percent yield} = \frac{\text{experimental yield}}{\text{theoretical yield}} \times 100\%$$

*You never do quite as well as you hoped, so you never get 100%*

---

**Example 3.14**   When methylamine, $CH_3NH_2$, is treated with acid, the following reaction occurs:

$$CH_3NH_2(aq) + H^+(aq) \longrightarrow CH_3NH_3^+(aq)$$

When 3.00 g of $CH_3NH_2$ reacts with 0.100 mol of $H^+$, 2.60 g of $CH_3NH_3^+$ is formed. What is the theoretical yield of product? The percent yield?

***Strategy***   Start by determining the limiting reactant; the theoretical yield is the amount of product that would be formed if all of the limiting reactant were completely consumed. Then use the definition above to find the percent yield.

*Solution*

mass of $CH_3NH_3^+$ from $CH_3NH_2$:

$$3.00 \text{ g } CH_3NH_2 \times \frac{1 \text{ mol } CH_3NH_2}{31.06 \text{ g } CH_3NH_2} \times \frac{1 \text{ mol } CH_3NH_3^+}{1 \text{ mol } CH_3NH_2} \times \frac{32.07 \text{ g } CH_3NH_3^+}{1 \text{ mol } CH_3NH_3^+}$$

$$= 3.10 \text{ g}$$

mass of $CH_3NH_3^+$ from 0.100 mol $H^+$:

$$0.100 \text{ mol of } H^+ \times \frac{1 \text{ mol } CH_3NH_3^+}{1 \text{ mol } H^+} \times \frac{32.07 \text{ g } CH_3NH_3^+}{1 \text{ mol } CH_3NH_3^+}$$

$$= 3.21 \text{ g}$$

The limiting reactant is $CH_3NH_2$; the $H^+$ ion is (slightly) in excess.   The theoretical yield is 3.10 g.

$$\text{percent yield} = \frac{2.60 \text{ g}}{3.10 \text{ g}} \times 100\% = 83.9\%$$

# CHEMISTRY

## *Beyond the Classroom*

# Hydrates

Ionic compounds often separate from water solution with molecules of water incorporated into the solid. Such compounds are referred to as **hydrates.** An example is hydrated copper sulfate, which contains five moles of $H_2O$ for every mole of $CuSO_4$. Its formula is $CuSO_4 \cdot 5H_2O$; a dot is used to separate the formulas of the two compounds $CuSO_4$ and $H_2O$. A Greek prefix is used to show the number of moles of water; the systematic name of $CuSO_4 \cdot 5H_2O$ is copper(II) sulfate pentahydrate.

As you can perhaps guess from Table 3.A, hydrates have been around for a long time. The compound $Na_2SO_4 \cdot 10H_2O$ was first obtained from Hungarian spring waters by the alchemist Johann Glauber in the seventeenth century. At about the same time, $MgSO_4 \cdot 7H_2O$ was isolated from mineral springs at Epsom in England. Both Glauber's salt and Epsom salts were once used as mild laxatives.

**TABLE 3.A    Some Familiar Hydrates**

| Composition | Systematic Name | Common Name |
|---|---|---|
| $Na_2SO_4 \cdot 10H_2O$ | sodium sulfate decahydrate | Glauber's salt |
| $Na_2CO_3 \cdot 10H_2O$ | sodium carbonate decahydrate | washing soda |
| $CaSO_4 \cdot 2H_2O$ | calcium sulfate dihydrate | gypsum |
| $CaSO_4 \cdot \frac{1}{2}H_2O$ | calcium sulfate hemihydrate | plaster of Paris |
| $MgSO_4 \cdot 7H_2O$ | magnesium sulfate heptahydrate | Epsom salts |
| $FeSO_4 \cdot 6H_2O$ | iron(II) sulfate hexahydrate | green vitriol |
| $CuSO_4 \cdot 5H_2O$ | copper(II) sulfate pentahydrate | blue vitriol |
| $CoSO_4 \cdot 6H_2O$ | cobalt(II) sulfate hexahydrate | red vitriol |

Certain hydrates, notably $Na_2CO_3 \cdot 10H_2O$ and $FeSO_4 \cdot 7H_2O$, lose all or part of their water of hydration when exposed to dry air. This process is referred to as *efflorescence;* the glassy (vitreous) hydrate crystals crumble to a powder. Frequently, dehydration is accompanied by a color change (Fig. 3.A). When $CoCl_2 \cdot 6H_2O$ is exposed to dry air or heated, it loses water and changes color from red to purple or blue. Crystals of this compound are used as humidity indicators and as an ingredient of "invisible ink." Writing becomes visible only when the paper is heated, driving off water and leaving a blue residue.

If you're an inexpert skier or horseback rider, you may have come in contact (literally) with the hydrates of calcium sulfate. When plaster of Paris, $CaSO_4 \cdot \frac{1}{2}H_2O$, is ground to a fine powder and mixed with water, it is converted to gypsum, $CaSO_4 \cdot 2H_2O$. This process takes place with an increase in volume, so the solid expands to fill completely any space to which it is confined. This explains its use in forming casts for broken bones.

Molecular as well as ionic substances can form hydrates, but of an entirely different nature. In these crystals, sometimes referred to as *clathrates*, a molecule (such as $CH_4$, $CHCl_3$) is quite literally trapped in an icelike cage of water molecules (Fig. 3.B). Perhaps the best-known molecular hydrate is that of chlorine, which has the approximate composition $Cl_2 \cdot 7.3H_2O$. This compound was discovered by the great English physicist and electrochemist Michael Faraday in 1823. You can make it by bubbling chlorine gas through calcium chloride solution at 0°C; the hydrate comes down as feathery white crystals. In the winter of 1914, the German army used chlorine in chemical warfare on the Russian front against the soldiers of the Tsar. They were puzzled by

**Figure 3.A**
$CuSO_4 \cdot 5H_2O$ (upper left) is blue; the anhydrous solid (lower left) is white. $CoCl_2 \cdot 6H_2O$ (upper right) is pink; lower hydrates such as $CoCl_2 \cdot 4H_2O$ are purple (lower right) or blue. (Marna G. Clarke)

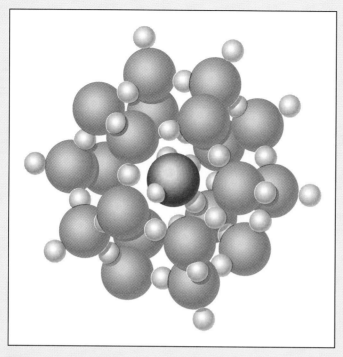

**Figure 3.B**
In the hydrate formed by methane, a $CH_4$ molecule is trapped within a cage of $H_2O$ molecules bonded to one another to give an ice-like structure.

its ineffectiveness; not until spring was deadly chlorine gas liberated from the hydrate, which is stable at cold temperatures.

Recently, geologists have discovered huge deposits of gas hydrates at the bottom of the ocean, where the pressure is high and the temperature slightly above 0°C. Trapped within these solids are methane, $CH_4$, and other combustible hydrocarbons found in natural gas. One hydrate field in a deep region of the Atlantic Ocean is the size of the state of Vermont and is estimated to contain enough methane to produce 350 times as much energy as is annually consumed in the United States. Whether this remarkable discovery will prove practical as a source of energy is another question. Getting the hydrate crystals to the surface and utilizing the methane they contain poses a formidable problem in engineering.

# CHAPTER HIGHLIGHTS

## *Key Concepts*

1. Relate the atomic mass of an element to isotopic masses and abundances
   (Example 3.1; Problems 3–10).
2. Use Avogadro's number to calculate the mass of an atom or molecule
   (Example 3.2, 3.3; Problems 11–18; 77, 78).

**3.** Use molar mass to relate
—moles to mass of a substance
(Examples 3.4, 3.5; Problems 19–26).
—molecular formula to simplest formula
(Example 3.10; Problems 43, 44).
**4.** Use the formula of a compound to find percent composition or mass of an element in a sample
(Examples 3.6, 3.7; Problems 27–34, 80).
**5.** Find the simplest formula of a compound from chemical analysis
(Examples 3.8, 3.9; Problems 35–46, 85).
**6.** Balance chemical equations by inspection
(Example 3.11; Problems 47–54).
**7.** Use a balanced equation to
—relate masses of reactants and products
(Example 3.12; Problems 55–64, 79, 86).
—find the limiting reactant, theoretical yield, and percent yield
(Example 3.13; 3.14; Problems 65–74, 76)

## Key Terms

| | | |
|---|---|---|
| atomic mass unit | molar mass | percent yield |
| Avogadro's number | mole | simplest formula |
| isotopic abundance | molecular formula | theoretical yield |
| limiting reactant | percent composition | |

## Summary Problem

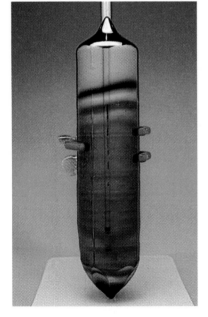

Pure silicon. (Charles D. Winters)

Consider silicon, an element widely used in semi-conductors.

(a) It consists of three isotopes with atomic masses 27.977 amu, 28.977 amu, and 29.974 amu. Their abundances are 92.18%, 4.71%, and 3.12%, respectively. What is the atomic mass of silicon?
(b) What is the mass, in grams, of a silicon atom?
(c) How many moles are there in 57.00 g of silicon?
(d) When silicon reacts with oxygen, silicon dioxide is formed. Write a balanced equation for this reaction.
(e) What is the molar mass of silicon dioxide?
(f) How many grams of silicon will react with 0.287 mol of oxygen?
(g) If 25.00 g of silicon reacts with an excess of oxygen, how many grams of silicon dioxide are formed?
(h) If 25.0 g of silicon react with 25.0 g of oxygen, how many grams of silicon dioxide are formed?
(i) If 42.50 g of silicon dioxide are produced from the reaction in (h), what is the percent yield?
(j) Talc is a common mineral that contains silicon. It is made up of 19.2% Mg, 50.6% O, 0.529% H, and 29.6% Si. What is the simplest formula for talc?

### Answers

| | |
|---|---|
| (a)  28.09 amu | (f)  8.06 |
| (b)  $4.665 \times 10^{-23}$ | (g)  53.48 |
| (c)  2.029 | (h)  46.9 |
| (d)  $Si(s) + O_2(g) \longrightarrow SiO_2(s)$ | (i)  90.6% |
| (e)  60.09 g | (j)  $Mg_3O_{12}H_2Si_4$ |

## *Questions & Problems*

### *Atomic Masses*

**1.** Arrange the following in order of increasing mass.
(a) an oxygen molecule    (b) a sodium atom
(c) a sulfur atom    (d) a sulfide ion

**2.** Calculate the mass ratio of an argon atom to an atom of
(a) helium    (b) arsenic    (c) silver

**3.** Oxygen consists of three isotopes with atomic masses 16.00, 17.00, and 18.00 amu. Their abundances are 99.76%, 0.04%, and 0.20%, respectively. What is the atomic mass of oxygen?

**4.** Strontium consists of four isotopes with masses of 83.9134 amu (0.5%), 85.9094 amu (9.9%), 86.9089 amu (7.0%), and 87.9056 amu (82.6%). Calculate the atomic mass of strontium.

**5.** Elemental bromine has only two naturally occurring isotopes. Br-79 has an atomic mass of 78.9440 amu and an abundance of 50.57%. What is the atomic mass of the second isotope, Br-81?

**6.** Naturally occurring europium (Eu) consists of two isotopes. One of the isotopes has a mass of 152.9212 amu and 51.97% abundance. What is the atomic mass of the other isotope?

**7.** Silicon (atomic mass = 28.0855 amu) consists of three isotopes with masses 27.977 amu, 28.977 amu, and 29.974 amu. The abundance of the last isotope is 2.96%. Estimate the abundance of the other two isotopes.

**8.** Chromium (atomic mass = 51.9961 amu) has four isotopes. Their masses are 49.941 amu, 51.9405 amu, 52.9407 amu, and 53.9389 amu. The first two isotopes have abundances of 4.41% and 83.46%, respectively. Estimate the other two abundances.

**\*9.** Magnesium consists of three isotopes with masses of 23.98 amu, 24.98 amu, and 25.98 amu. Their abundances are 78.6%, 10.1%, and 11.3%, respectively. Sketch the mass spectrum for magnesium.

**\*10.** Chlorine has two isotopes, Cl-35 and Cl-37. Their abundances are 75.53% and 24.47%, respectively. Assume that the only hydrogen isotope present is H-1.
(a) How many different HCl molecules are possible?
(b) What is the sum of the mass numbers of the two atoms in each molecule?
(c) Sketch the mass spectrum for HCl if all the positive ions are obtained by removing a single electron from an HCl molecule.

### *Avogadro's Number and the Mole*

**11.** One chocolate chip used in making chocolate chip cookies has a mass of 0.385 g.
(a) How many chocolate chips are there in one mole of chocolate chips?
(b) If a cookie needs 20 chocolate chips, how many

cookies can one make with a millionth of a mole of chocolate chips? (A millionth of a mole is scientifically known as a micromole.)

**12.** The meat from one hazelnut has a mass of 0.985 g.
(a) What is the mass of a billionth of a mole ($10^{-9}$) of hazelnut meats? (A billionth of a mole is also called a nanomole.)
(b) How many moles are there in a pound of hazelnut meats?

**13.** The atomic mass of silver, Ag, is 107.9 amu. Calculate
(a) the mass, in grams, of one silver atom.
(b) the number of atoms in a picogram of silver (pico = $10^{-12}$).

**14.** Cobalt has an atomic mass of 58.93 amu. Calculate
(a) the mass of a trillion ($10^{12}$) cobalt atoms.
(b) the number of atoms in 0.010 oz of cobalt.

**15.** Determine
(a) the mass of 0.285 mol of lead.
(b) the number of atoms in 0.285 g of lead.
(c) the number of moles of electrons in 0.285 g of lead.

**16.** Consider Ni-60 (59.948 amu). For one gram of Ni-60, calculate
(a) the number of moles.
(b) the number of atoms.
(c) the total number of protons, neutrons, and electrons.

**17.** How many protons are there in
(a) one xenon atom?
(b) one mole of xenon atoms?
(c) $5.0 \times 10^3$ g of xenon?

**18.** How many electrons are there in
(a) ten sulfur atoms?
(b) ten sulfide ions?
(c) one mole of sulfide ions?
(d) one mole of sulfur atoms?

### *Molar Mass and Mole-Gram Conversions*

**19.** Calculate the molar masses (g/mol) of
(a) tungsten, W, a metal used in incandescent light bulbs.
(b) ozone, $O_3$.
(c) vitamin A, $C_{20}H_{30}O$.

**20.** Calculate the molar masses (g/mol) of
(a) naturally occurring sulfur, $S_8$.
(b) plaster of Paris, $CaSO_4 \cdot \frac{1}{2}H_2O$, used in making casts for broken bones.
(c) vitamin C, $C_6H_8O_6$.

**21.** Convert the following to moles.
(a) one hundred milligrams of aspirin ($C_9H_8O_4$).
(b) one milligram of cholesterol, $C_{27}H_{46}O$.
(c) 2.95 grams of $CCl_4$, once used as a dry-cleaning agent.

22. Convert the following to moles.
    **(a)** a metric ton ($10^3$ kg) of hydrazine, $N_2H_4$, a rocket propellant.
    **(b)** one hundred milligrams of caffeine, $C_4H_5N_2O$.
    **(c)** 15.6 g of tin(II) fluoride, the active ingredient in "fluoride" toothpastes.

23. Calculate the mass in grams of 12.00 mol of
    **(a)** phosphorus atoms   **(b)** $P_4(s)$
    **(c)** phosphine gas

24. Calculate the mass in grams of 9.28 mol of
    **(a)** vinyl chloride, $C_2H_3Cl$, the starting material for a plastic.
    **(b)** nitrogen dioxide, the brown gas responsible for the color of smog.
    **(c)** aspartame, $C_{14}H_{18}N_2O_5$, the artificial sweetener.

25. Complete the following table for acetone, $C_3H_6O$, the main component of nail polish remover.

| | Number of Grams | Number of Moles | Number of Molecules | Number of C Atoms |
|---|---|---|---|---|
| **(a)** | 0.3500 | ——— | ——— | ——— |
| **(b)** | ——— | 2.685 | ——— | ——— |
| **(c)** | ——— | ——— | $7.9 \times 10^{12}$ | ——— |
| **(d)** | ——— | ——— | ——— | $14.0 \times 10^{14}$ |

26. Complete the following table for ethylene glycol, $C_2H_6O_2$, an antifreeze used in cars.

| | Number of Grams | Number of Moles | Number of Molecules | Number of O Atoms |
|---|---|---|---|---|
| **(a)** | 13.68 | ——— | ——— | ——— |
| **(b)** | ——— | 0.4952 | ——— | ——— |
| **(c)** | ——— | ——— | $6.0 \times 10^{25}$ | ——— |
| **(d)** | ——— | ——— | ——— | $17 \times 10^{20}$ |

## Percent Composition from Formula

27. Malathion, a commonly used insecticide, has the following chemical formula: $C_{10}H_{19}O_6PS_2$. Calculate the mass percent of each element in malathion.

28. Emerald has the following chemical formula: $Be_3Al_2Si_6O_{18}$. Calculate the mass percent of each element in emerald.

29. Cisplatin, an antitumor ingredient, has the formula $Pt(NH_3)_2Cl_2$. How many milligrams of platinum are present in 250.0 mg of cisplatin?

30. One of the most common rocks on Earth is feldspar. One type of feldspar has the formula $CaAl_2Si_2O_8$. How many grams of silicon can be obtained from one kilogram of feldspar?

31. A tablet of Tylenol has a mass of 0.611 g. It contains 250.0 mg of its "active ingredient," acetominophen, $C_8H_9NO_2$.
    **(a)** What is the mass percent of acetominophen in a tablet of Tylenol?

    **(b)** Assume that all the nitrogen in the tablet is in the acetominophen. How many grams of nitrogen are present in a tablet of Tylenol?

32. The active ingredient in Pepto-Bismol (an over-the-counter remedy for upset stomach) is bismuth subsalicylate, $C_7H_5BiO_4$. Analysis of a 1.50-g sample of Pepto-Bismol yields 346 mg of bismuth. What percent by mass is bismuth subsalicylate? (Assume that there is no other bismuth-containing compound in Pepto-Bismol.)

33. Hexachlorophene, a compound made up of atoms of carbon, hydrogen, chlorine, and oxygen, is an ingredient in germicidal soaps. Combustion of a 1.000-gram sample yields 1.407 g of carbon dioxide, 0.134 g of water, and 0.5228 g of chlorine gas. What are the mass percents of carbon, hydrogen, oxygen, and chlorine in hexachlorophene?

34. Combustion analysis of 1.00 g of the male sex hormone testosterone yields 2.90 g of $CO_2$ and 0.875 g of $H_2O$. What are the mass percents of carbon, hydrogen, and oxygen in testosterone?

## Simplest Formula from Analysis

35. Antimony reacts with oxygen to form an oxide. If 2.643 g of antimony reacts with 0.521 g of oxygen, what is the simplest formula of the oxide?

36. Boron reacts with hydrogen to form a borane. If 4.389 grams of boron reacts with 1.228 g of hydrogen, what is the simplest formula of the compound obtained?

37. Determine the simplest formulas of the following compounds.
    **(a)** aluminum chlorohydrate, the active ingredient of some antiperspirants, which has the composition 30.92% Al, 45.87% O, 2.889% H, and 20.32% Cl.
    **(b)** the food enhancer monosodium glutamate (MSG), which has the composition 35.51% C, 4.77% H, 37.85% O, 8.29% N, and 13.60% Na.
    **(c)** oxalic acid, commonly used to remove rust and blood stains, which has the composition 2.24% H, 26.68% C, and 71.08% O.

38. Determine the simplest formulas of the following compounds.
    **(a)** tetraethyl lead, the gasoline additive, which has the composition 29.71% C, 6.234% H, and 64.07% Pb.
    **(b)** acetylsalicylic acid, which has the composition 60.0% C, 4.48% H, and 35.5% O.
    **(c)** allicin, the compound that gives garlic its characteristic odor, which has the composition 6.21% H, 44.4% C, 9.86% O, and 39.51% S.

39. Methyl salicylate is a common "active ingredient" in liniments such as Ben-Gay. It is also known as oil of wintergreen. It is made up of carbon, hydrogen, and oxygen atoms. When a sample of methyl salicylate weighing 5.287 g is burned in excess oxygen, 12.24 g of carbon dioxide and 2.505 g of water are formed. What is the simplest formula for oil of wintergreen?

40. Ibuprofen, the active ingredient in Advil, is made up of carbon, hydrogen, and oxygen atoms. When a sample of

Ibuprofen weighing 5.000 g burns in oxygen, 13.86 g of $CO_2$ and 3.926 g of water are obtained. What is the simplest formula of ibuprofen?

**41.** Sevin is a common plant insecticide, particularly effective against Japanese beetles. It is made up of carbon, hydrogen, nitrogen, and oxygen atoms. A 10.94-gram sample of Sevin burned in oxygen yields 28.73 g of $CO_2$ and 5.386 g of $H_2O$. Another experiment determines that the insecticide contains 6.963% nitrogen by mass. What is the simplest formula for Sevin?

**42.** Halothane is a commonly used gaseous anesthetic. It is made up of carbon, hydrogen, bromine, fluorine, and chlorine atoms. A ten gram sample of halothane when burned in oxygen gas yields 4.459 g of $CO_2$ and 0.4559 g of $H_2O$. A second ten gram sample is burned in hydrogen and yields 1.847 g of HCl and 4.099 g of HBr. What is the simplest formula for halothane?

**43.** Hexamethylenediamine ($\mathcal{M} = 116.2$ g/mol), a compound made up of carbon, hydrogen, and nitrogen atoms, is used in the production of nylon. When 6.315 g of hexamethylenediamine is burned in oxygen, 14.36 g of carbon dioxide and 7.832 g of water are obtained. What are the simplest and molecular formulas of this compound?

**44.** Dimethylhydrazine, the fuel used in the Apollo lunar descent module, has a molar mass of 60.10 g/mol. It is made up of carbon, hydrogen, and nitrogen atoms. The combustion of 2.859 g of the fuel in excess oxygen yields 4.190 g of carbon dioxide and 3.428 g of water. What are the simplest and molecular formulas for dimethylhydrazine?

**45.** Epsom salts are hydrates of magnesium sulfate. The formula for Epsom salts is $MgSO_4 \cdot 7H_2O$. A 7.834-g sample is heated until a constant mass is obtained, indicating that all the water has been evaporated off. What is the mass of the anhydrous magnesium sulfate? What percentage of the hydrate is water?

**46.** A certain hydrate of potassium aluminum sulfate (alum) has the formula $KAl(SO_4)_2 \cdot xH_2O$. When a hydrate sample weighing 5.459 g is heated to remove all the water, 2.583 g of $KAl(SO_4)_2$ remains. What is the mass percent of water in the hydrate? What is $x$?

## Balancing Equations

**\*47.** The reaction between compounds made up of M (squares) and X (circles) is pictorially shown below. Write a balanced equation to represent the picture shown below using smallest whole-number coefficients.

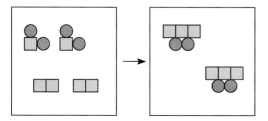

**\*48.** Represent the following equation pictorially (see Question 47 to see a pictorial drawing) using squares to represent A and circles to represent B.

$$A_3 + B_2 \longrightarrow AB_4$$

After you have "drawn" the equation, balance it using your "drawing" as guide.

**\*49.** Balance the following equations:
(a) $Al(s) + Fe_2O_3(s) \longrightarrow Al_2O_3(s) + Fe(s)$
(b) $CO_2(g) + H_2O(l) \longrightarrow C_6H_{12}O_6(s) + O_2(g)$
(c) $C_2H_8N_2(s) + N_2O_4(g) \longrightarrow N_2(g) + CO_2(g) + H_2O(g)$

**\*50.** Balance the following equations:
(a) $Br_2(l) + I_2(s) \longrightarrow IBr_3(g)$
(b) $H_2(g) + N_2(g) \longrightarrow NH_3(g)$
(c) $BF_3(s) + H_2O(g) \longrightarrow B_2O_3(s) + HF(g)$

**\*51.** Write balanced equations for the reaction of barium metal with the following nonmetals:
(a) sulfur     (b) bromine     (c) iodine
(d) nitrogen     (e) oxygen (forming $O^{2-}$ ions)

**\*52.** Write balanced equations for the reaction of oxygen with the following metals to form solids where the anion is $O^{2-}$:
(a) sodium     (b) calcium     (c) aluminum
(d) strontium     (e) cobalt (forming $CO^{3+}$ ions)

**\*53.** Write a balanced equation for
(a) the reaction between fluorine gas and water to give oxygen difluoride and hydrogen fluoride gases.
(b) the reaction between oxygen and ammonia gases to give nitrogen dioxide gas and water.
(c) the decomposition of ammonium nitrate to its elements.
(d) the burning of magnesium in nitrogen to give magnesium nitride.
(e) the burning of gold(III) sulfide in hydrogen to give gold metal and dihydrogen sulfide gas.

**\*54.** Write a balanced equation for
(a) the burning of iron(III) oxide in hydrogen to give iron metal and water.
(b) the reaction between uranium(IV) oxide with liquid hydrogen fluoride to give uranium(IV) fluoride and water.
(c) the combustion of propane ($C_3H_8$) gas to give carbon dioxide and water.
(d) the reaction between xenon tetrafluoride gas and water to give xenon, oxygen, and hydrogen fluoride gases.
(e) the decomposition of potassium chlorate to potassium chloride and oxygen.

## Mole-Mass Relations in Reactions

**55.** Phosphine gas reacts with oxygen according to the following equation:

$$4PH_3(g) + 8O_2(g) \longrightarrow P_4O_{10}(s) + 6H_2O(g)$$

Fill in the blanks below:
(a) 16.3 mol of $PH_3$ reacts with _____ mol of $O_2$.

**(b)** 22.6 mol of $PH_3$ yields_____mol of $H_2O$.
**(c)** 3.48 mol of $P_4O_{10}$ requires_____mol of $PH_3$.
**(d)** 1.76 mol of oxygen produces_____mol of $P_4O_{10}$.

**56.** The combustion of chloroethene, $C_2H_3Cl$, is represented as follows:

$$2C_2H_3Cl(l) + 5O_2(g) \longrightarrow 4CO_2(g) + 2H_2O(g) + 2HCl(g)$$

Fill in the blanks below:
**(a)** 5.893 mol of $C_2H_3Cl$ requires_____mol of $O_2$.
**(b)** 9.683 mol of $C_2H_3Cl$ yields_____mol of $CO_2$.
**(c)** 0.457 mol of $O_2$ reacts with_____mol of $C_2H_3Cl$.
**(d)** 0.734 mol of HCl requires_____mol of $O_2$.

**57.** Using the equation given in Problem 55, calculate
**(a)** the mass of tetraphosphorus decaoxide produced from 7.88 mol of phosphine.
**(b)** the mass of oxygen required to form 7.65 mol of water.
**(c)** the mass of phosphine gas that yields 4.69 g of water.
**(d)** the mass of oxygen required to react with 12.42 g of phosphine.

**58.** Using the equation given in Problem 56, calculate
**(a)** the mass of hydrogen chloride produced from 17.24 mol of chloroethene.
**(b)** the moles of oxygen required to form 28.42 mol of steam.
**(c)** the mass of carbon dioxide formed from 16.32 g of oxygen.
**(d)** the mass of chloroethene required to yield 19.28 g of steam.

**59.** Sand consists mainly of silicon dioxide. When sand is heated with an excess of coke (carbon), pure silicon and carbon monoxide are produced.
**(a)** Write a balanced equation for the reaction.
**(b)** How many moles of silicon dioxide are required to form 25.00 g of silicon?
**(c)** How many grams of carbon monoxide are formed when 32.55 g of silicon are produced?

**60.** The U.S. space shuttle's resuable booster rockets use aluminum and ammonium perchlorate for fuel. The reaction between the two substances yields aluminum oxide, aluminum chloride, nitrogen oxide and steam.
**(a)** Write a balanced equation for the reaction.
**(b)** How many moles of aluminum are required to react with 56.34 g of ammonium perchlorate?
**(c)** If 12.00 g of ammonium perchlorate reacts with an excess of aluminum, how many grams of each product are obtained?

**61.** A wine cooler contains about 4.5% ethyl alcohol, $C_2H_5OH$, by mass. Assume that the alcohol in 7.50 kg of wine cooler is produced by the fermentation of glucose in grapes in the reaction

$$C_6H_{12}O_6(aq) \longrightarrow 2C_2H_5OH(aq) + 2CO_2(g)$$

**(a)** How many grams of glucose are needed to produce the ethyl alcohol in the wine?

**(b)** What volume of carbon dioxide gas ($d = 1.80$ g/L) is produced at the same time?

**62.** When corn is allowed to ferment, ethyl alcohol is produced from the glucose in corn according to the fermentation reaction given in Problem 61.
**(a)** What volume of ethyl alcohol ($d = 0.789$ g/cm$^3$) is produced from ten pounds of glucose?
**(b)** Gasohol is a mixture of 10 cm$^3$ of ethyl alcohol per 90 cm$^3$ of gasoline. How many grams of glucose are required to produce the ethyl alcohol in one liter of gasohol?

**63.** A crude oil burned in electrical generating plants contains about 1.2% sulfur by mass. When the oil burns, the sulfur forms sulfur dioxide gas:

$$S(s) + O_2(g) \longrightarrow SO_2(g)$$

How many liters of $SO_2$ ($d = 2.60$ g/L) are produced when one metric ton ($10^3$ kg) of oil burns?

**64.** Oxygen masks for producing $O_2$ in emergency situations contain potassium superoxide, $KO_2$. It reacts with $CO_2$ and $H_2O$ in exhaled air to produce oxygen:

$$4KO_2(s) + 2H_2O(g) + 4CO_2(g) \longrightarrow 4KHCO_3(s) + 3O_2(g)$$

If a person wearing such a mask exhales 0.702 g $CO_2$/min, how many grams of $KO_2$ are consumed in one hour?

### Limiting Reactant; Theoretical Yield

**\*65.** Nitrogen reacts with hydrogen to form ammonia. Represent each nitrogen atom by a square and each hydrogen atom with a circle. Starting with five molecules of both hydrogen and nitrogen, show pictorially what you have after the reaction is complete.

**\*66.** Consider the following diagram in which atom X is represented by a circle and atom Y is represented by a square.

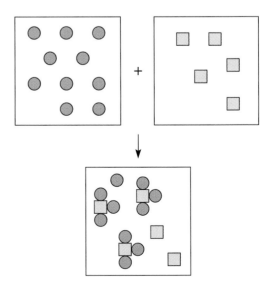

**(a)** Write a balanced equation for the reaction represented by the diagram.
**(b)** If each circle stands for a mole of X and each square a mole of Y, how many moles of X did one start with? How many moles of Y?
**(c)** Using the same representation described in part (b), how many moles of product are formed? How many moles of X and Y are left unreacted?

**67.** Iron reacts with chlorine gas to form iron(III) chloride. If you start with 2.75 mol of iron and 3.50 mol of chlorine,
    **(a)** write a balanced equation for the reaction.
    **(b)** what is the limiting reactant?
    **(c)** what is the theoretical yield of iron(III) chloride in moles?
    **(d)** how many moles of excess reactant remain unreacted?

**68.** A gaseous mixture containing 6.45 mol of hydrogen gas and 8.00 mol of bromine react to form hydrogen bromide gas.
    **(a)** Write a balanced equation for the reaction.
    **(b)** What is the limiting reactant?
    **(c)** What is the theoretical yield of hydrogen bromide in moles?
    **(d)** How many moles of the excess reactant remain unreacted?

**69.** Silicon nitride ($Si_3N_4$) is a ceramic. It is made by reacting silicon and nitrogen at high temperatures.
    **(a)** Write a balanced equation for the reaction.
    **(b)** How much silicon must react with excess nitrogen to prepare 185 g of silicon nitride if the reaction yield is 87%?

**70.** Titanium dioxide ($TiO_2$) is the substance used as the white pigment in paint. It is prepared by the combustion of titanium(IV) chloride. Chlorine gas is also produced.
    **(a)** Write a balanced equation for the reaction.
    **(b)** How much titanium(IV) chloride must react with excess oxygen to prepare 225 g of the pigment? The reaction is found to be 93% efficient.

**71.** Oxacetylene torches used for welding reach temperatures near 2000°C. The reaction involved in the combustion of acetylene is:

$$2C_2H_2(g) + 5O_2(g) \longrightarrow 4CO_2(g) + 2H_2O(g)$$

    **(a)** Starting with 125 g of both acetylene and oxygen, what is the theoretical yield, in grams, of carbon dioxide?
    **(b)** If 60.1 L of carbon dioxide ($d = 1.80$ g/L) are produced, what is the percent yield?
    **(c)** How much of the reactant in excess is unused?

**72.** The space shuttle uses aluminum metal and ammonium perchlorate in its reusable booster rockets. The products of the reaction are aluminum oxide, aluminum chloride, nitrogen oxide gas and steam. The reaction mixture contains 7.00 g of aluminum and 9.32 g of ammonium perchlorate.
    **(a)** Write a balanced equation for the reaction.
    **(b)** What is the theoretical yield of aluminum chloride?

**(c)** If 2.53 g of aluminum chloride are formed, what is the percent yield?
**(d)** How many grams of excess reactant are unused?

**73.** A student prepares phosphorous acid, $H_3PO_3$, by reacting solid phosphorus triiodide with water.

$$PI_3(s) + 3H_2O(l) \longrightarrow H_3PO_3(l) + 3HI(g)$$

The student needs to obtain 0.100 L of $H_3PO_3$ ($d = 1.651$ g/cm$^3$). The procedure calls for a 40.0% excess of water and a yield of 55.0%. How much phosphorus triiodide should she weigh out? What volume of water ($d = 1.00$ g/cm$^3$) should she use?

**74.** Aspirin, $C_9H_8O_4$, is prepared by reacting salicylic acid, $C_7H_6O_3$, with acetic anhydride, $C_4H_6O_3$, in the reaction

$$2C_7H_6O_3(s) + C_4H_6O_3(l) \longrightarrow 2C_9H_8O_4(s) + H_2O$$

A student is told to prepare 25.0 g of aspirin. He is also told that he should use a 50.0% excess of acetic anhydride and expect to get a 65.0% yield in the reaction. How many grams of each reactant should he use?

## Unclassified

***75.** Magnesium ribbon reacts with acid to produce hydrogen gas and magnesium ions. Different masses of magnesium ribbon are added to 10 mL of the acid. The volume of the hydrogen gas obtained is a measure of the number of moles of hydrogen produced by the reaction. Various measurements are given in the following table.

| Expt. | Mass of Mg Ribbon (g) | Volume of Acid Used (mL) | Volume of H$_2$ Gas (mL) |
|---|---|---|---|
| (1) | 0.020 | 10.0 | 21 |
| (2) | 0.040 | 10.0 | 42 |
| (3) | 0.080 | 10.0 | 82 |
| (4) | 0.120 | 10.0 | 122 |
| (5) | 0.160 | 10.0 | 122 |
| (6) | 0.200 | 10.0 | 122 |

    **(a)** Draw a graph of the results plotting the mass of Mg vs. the volume of the hydrogen gas.
    **(b)** What is the limiting reactant in experiment (1)?
    **(c)** What is the limiting reactant in experiment (3)?
    **(d)** What is the limiting reactant in experiment (6)?
    **(e)** Which experiment uses stoichiometric amounts of each reactant?
    **(f)** What volume of gas would be obtained if 0.300 g of Mg ribbon were used? If 0.010 grams were used?

***76.** Iron reacts with oxygen. Different masses of iron are burned in a constant amount of oxygen. The product, an oxide of iron, is weighed. The graph on p. 80 is obtained when the mass of product obtained is plotted against the mass of iron used.
    **(a)** How many grams of product are obtained when 0.50 g of iron are used?

**(b)** What is the limiting reactant when 2.00 g of iron are used?

**(c)** What is the limiting reactant when 5.00 g of iron are used?

**(d)** How many grams of iron react exactly with the amount of oxygen supplied?

**(e)** What is the simplest formula of the product?

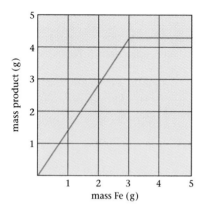

**77.** Some brands of salami contain 0.090% sodium benzoate ($NaC_7H_5O_2$) by mass as a preservative. If you eat 3.00 oz of this salami, how many atoms of sodium will you consume, assuming salami contains no other source of that element?

**78.** Carbon tetrachloride, $CCl_4$, was a popular dry-cleaning agent until it was shown to be carcinogenic. It has a density of 1.594 $g/cm^3$. What volume of carbon tetrachloride will contain a total of $6.00 \times 10^{22}$ molecules of $CCl_4$?

**79.** Most wine is prepared by the fermentation of the glucose in grape juice by yeast:

$$C_6H_{12}O_6(aq) \longrightarrow 2C_2H_5OH(aq) + 2CO_2(g)$$

How many grams of glucose should there be in grape juice to produce 750.0 mL of wine that is 11.0% ethyl alcohol, $C_2H_5OH$ (d = 0.789 $g/cm^3$), by volume?

**80.** All the fertilizers listed below contribute nitrogen to the soil. If all these fertilizers are sold for the same price per gram of nitrogen, which will cost the least per 50-lb bag?

urea, $(NH_2)_2CO$
ammonia, $NH_3$
ammonium nitrate, $NH_4NO_3$
guanidine, $HNC(NH_2)_2$

## Challenge Problems

**81.** Chlorophyll, the substance responsible for the green color of leaves, has one magnesium atom per chlorophyll molecule and contains 2.72% magnesium by mass. What is the molar mass of chlorophyll?

**82.** By x-ray diffraction, it is possible to determine the geometric pattern in which atoms are arranged in a crystal and the distances between atoms. In a crystal of silver, four atoms effectively occupy the volume of a cube 0.409 nm on an edge. Taking the density of silver to be 10.5 $g/cm^3$, calculate the number of atoms in one mole of silver.

**83.** A 5.025-g sample of calcium is burned in air to produce a mixture of two ionic compounds, calcium oxide and calcium nitride. Water is added to this mixture. It reacts with calcium oxide to form 4.832 g of calcium hydroxide. How many grams of calcium oxide are formed? How many grams of calcium nitride?

**84.** A mixture of potassium chloride and potassium bromide weighing 3.595 g is heated with chlorine, which converts the mixture completely to potassium chloride. The total mass of potassium chloride after the reaction is 3.129 g. What percent of the original mixture is potassium bromide?

**85.** A sample of an oxide of vanadium weighing 4.589 g was heated with hydrogen gas to form water and another oxide of vanadium weighing 3.782 g. The second oxide was treated further with hydrogen until only 2.573 g of vanadium metal remained.

**(a)** What are the simplest formulas of the two oxides?

**(b)** What is the total mass of water formed in the successive reactions?

**86.** A sample of cocaine, $C_{17}H_{21}O_4N$, is diluted with sugar, $C_{12}H_{22}O_{11}$. When a 1.00-mg sample of this mixture is burned, 1.00 mL of carbon dioxide (d = 1.80 g/L) is formed. What is the percentage of cocaine in this mixture?

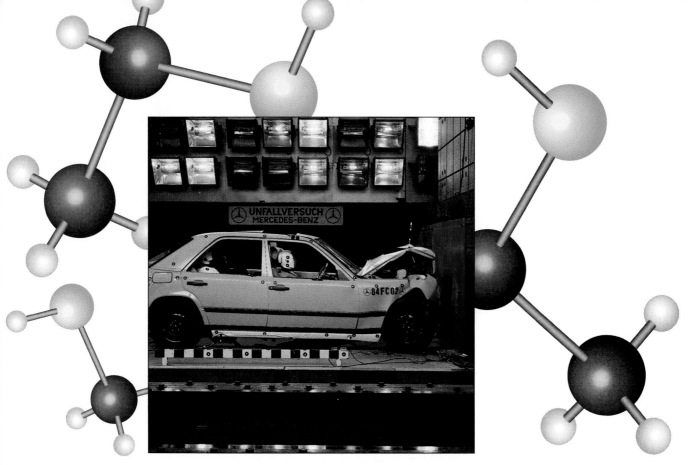

For under $100.00, you can purchase a hand held breathalyzer (p. 102) that will help prevent disasters like this. (Courtesy of Mercedes-Benz)

# Reactions in Aqueous Solution

4

He takes up the waters of the sea in his
hand, leaving the salt;
He disperses it in mist through the skies;
He recollects and sprinkles it like grain in
six-rayed snowy stars over the earth,
There to lie till he dissolves the bonds again.

**—HENRY DAVID THOREAU**

*Journal* **January 5, 1856**

Most of the reactions considered in Chapter 3 involved pure substances reacting with each other. However, most of the reactions you will carry out in the laboratory or hear about in lecture take place in water (aqueous) solution. Beyond that, most of the reactions that occur in the world around you involve ions or molecules dissolved in water. For these reasons, among others, you need to become familiar with some of the more important types of aqueous reactions. These include

— precipitation reactions (Section 4.1)
— acid-base reactions (Section 4.2)
— oxidation-reduction reactions (Section 4.3)

The emphasis in these sections is upon writing and balancing chemical equations for these reactions.

The last half of this chapter is devoted to mass relations in solution reactions. We consider in turn

— the unit of *molarity,* used to express concentrations of solution species (Section 4.4)
— stoichiometric calculations for solution reactions (Section 4.5)
— the use of solution reactions in quantitative analysis (Section 4.6)

## 4.1 Precipitation Reactions

When an ionic solid dissolves in water, the cations and anions separate from one another. As you can see from Figure 4.1, water molecules are intimately involved in this process. However, they are not ordinarily included in the chemical equation written to represent the solution reaction. For the dissolution of sodium chloride, we write

$$NaCl(s) \longrightarrow Na^+(aq) + Cl^-(aq)$$

Similarly, for sodium carbonate and calcium chloride:

$$Na_2CO_3(s) \longrightarrow 2Na^+(aq) + CO_3^{2-}(aq)$$

$$CaCl_2(s) \longrightarrow Ca^{2+}(aq) + 2Cl^-(aq)$$

It is important to realize that compounds such as these are completely ionized in water; there are no "molecules" of NaCl, Na$_2$CO$_3$, or CaCl$_2$. Ionic solids are commonly described as **strong electrolytes.** The ions they produce are good conductors of electricity (Figure 4.2).

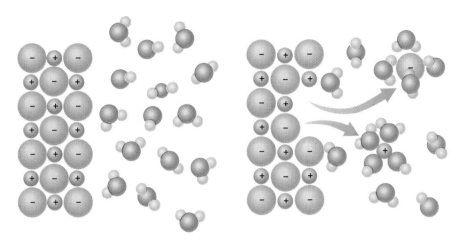

**Figure 4.1**
Dissolving NaCl in water. The attraction between the oxygen atoms in $H_2O$ molecules and $Na^+$ ions in NaCl tends to bring the $Na^+$ ions into solution. At the same time, $Cl^-$ ions, attracted to the hydrogen atoms of $H_2O$ molecules, also enter the solution. The $H_2O$ molecules that surround the ions tend to prevent oppositely charged ions from recombining.

Sometimes when water solutions of two different ionic compounds are mixed, an insoluble solid separates out of solution. The **precipitate** that forms is itself ionic; the cation comes from one solution, the anion from the other. To predict the occurrence of reactions of this type, you must know which ionic substances are insoluble in water.

Ionic solids cover an enormous range of solubilities. At one extreme is lithium chlorate, $LiClO_3$, which dissolves to the extent of 35 mol (3200 g) per liter of water at 25°C. Mercury(II) sulfide, HgS, is at the opposite extreme; the

(a)

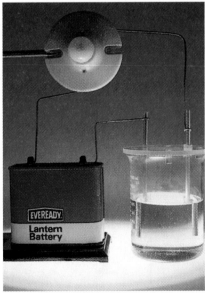
(b)

(c)

**Figure 4.2**
An apparatus for testing electrical conductivity. For an electric current to flow, the solution must contain ions, which carry electric charge. The liquids used are (a) pure water, (b) a solution of pure water and sucrose, and (c) a solution of pure water and NaCl.
(Marna G. Clarke)

You must learn these rules to work with precipitation reactions

(a)

(b)

LiClO₃ (a) is extremely soluble. In contrast, HgS (b) is about as insoluble as it is possible for a compound to be. (Marna G. Clarke)

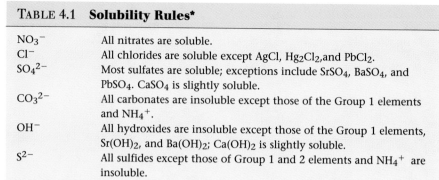

## TABLE 4.1　Solubility Rules*

| | |
|---|---|
| $NO_3^-$ | All nitrates are soluble. |
| $Cl^-$ | All chlorides are soluble except AgCl, $Hg_2Cl_2$, and $PbCl_2$. |
| $SO_4^{2-}$ | Most sulfates are soluble; exceptions include $SrSO_4$, $BaSO_4$, and $PbSO_4$. $CaSO_4$ is slightly soluble. |
| $CO_3^{2-}$ | All carbonates are insoluble except those of the Group 1 elements and $NH_4^+$. |
| $OH^-$ | All hydroxides are insoluble except those of the Group 1 elements, $Sr(OH)_2$, and $Ba(OH)_2$; $Ca(OH)_2$ is slightly soluble. |
| $S^{2-}$ | All sulfides except those of Group 1 and 2 elements and $NH_4^+$ are insoluble. |

*Insoluble compounds are those that precipitate when we mix equal volumes of ~ 0.1 mol/L solutions of the corresponding ions.

calculated solubility at 25°C is $10^{-26}$ mol/L. This means, in principle at least, about 200 L of a solution of HgS would be required to contain a single pair of $Hg^{2+}$ and $S^{2-}$ ions.

## Solubility Rules

Using the solubility rules listed in Table 4.1, it is possible to make qualitative predictions concerning solubilities of literally hundreds of different ionic compounds. These rules are quite simple to interpret. For example, the following facts should be evident from the table:

— $Ni(NO_3)_2$ is soluble. (All nitrates are soluble.)
— $BaCl_2$ is soluble. ($BaCl_2$ is not one of the three insoluble chlorides listed.)
— PbS is insoluble. (Pb is not a Group 1 or Group 2 element.)

Figure 4.3 shows how these rules apply to some common ionic compounds.

## Figure 4.3

Of the compounds of $Pb^{2+}$, $Pb(NO_3)_2$ (tube 2) is soluble; $PbCl_2$ and $PbSO_4$ (tubes 1 and 3) are insoluble white solids. The brownish solids $FeCl_3$ and $Fe(NO_3)_3$ (tubes 4 and 6) are water soluble; $Fe(OH)_3$ (tube 5) is a reddish-brown insoluble solid. Two insoluble compounds of $Ni^{2+}$ are black NiS (tube 7) and green $Ni(OH)_2$ (tube 8); $NiSO_4$ (tube 9) dissolves in water to give a deep green solution. All of these observations are consistent with the solubility rules in Table 4.1 (Charles D. Winters)

Using the solubility rules, you can predict whether a precipitate will form when two solutions of ionic compounds are mixed. For precipitation to occur, *the cation of one solution must combine with the anion of the other solution to form an insoluble solid.*

---

**Example 4.1**   Using the solubility rules in Table 4.1, predict what will happen when the following pairs of aqueous solutions are mixed.

(a) $CuSO_4$ and $NaNO_3$     (b) $Na_2CO_3$ and $CaCl_2$

*Strategy*   First decide what cation and anion are present in each solution. Then write the formulas of the two possible precipitates, combining a cation of one solution with the anion of the other solution. Check the solubility rules to see if one or both of these compounds is insoluble. If so, a precipitation reaction occurs.

*Solution*

(a) Ions present in first solution: $Cu^{2+}$, $SO_4^{2-}$; second solution: $Na^+$, $NO_3^-$
Possible precipitates: $Cu(NO_3)_2$, $Na_2SO_4$
From Table 4.1, both of these compounds must be soluble, so

no precipitate forms.

(b) Ions present: $Na^+$, $CO_3^{2-}$; $Ca^{2+}$, $Cl^-$
Possible precipitates: $NaCl$, $CaCO_3$
Sodium chloride is soluble, but calcium carbonate is not. When these two solutions are mixed,   $CaCO_3$ precipitates   (Fig. 4.4).

---

**Figure 4.4**
When solutions of sodium carbonate, $Na_2CO_3$, and calcium chloride, $CaCl_2$, are mixed, a white precipitate is obtained. (Charles D. Winters)

## Net Ionic Equations

The precipitation reaction that occurs when solutions of $Na_2CO_3$ and $CaCl_2$ are mixed can be represented by a simple equation. The product of the reaction is solid $CaCO_3$, formed by the reaction between $Ca^{2+}$ and $CO_3^{2-}$ ions in aqueous solution. The equation for the reaction is

$$Ca^{2+}(aq) + CO_3^{2-}(aq) \longrightarrow CaCO_3(s)$$

Notice that this equation includes only those ions that participate in the reaction. To be specific, $Na^+$ and $Cl^-$ ions do not appear. They are "spectator ions," which are present in solution before and after the precipitation of calcium carbonate.

Equations such as this, which involve ions and exclude any species that do not take part in the reaction, are referred to as **net ionic equations.** We will use net ionic equations throughout this chapter and indeed the entire text to represent a wide variety of reactions in aqueous solution. Like all equations, net ionic equations must show

— *atom balance.* There must be the same number of atoms of each element on both sides. In the preceding equation, there is one Ca atom, one C atom, and three oxygen atoms on both sides.

— *charge balance.* There must be the same total charge on both sides. In this equation, the total charge is zero on both sides.

$$Ca^{2+}(aq) + CO_3^{2-}(aq) \longrightarrow CaCO_3(s)$$

When solutions of $BaCl_2$ ($Cl^-$ ions) and $Ag_2SO_4$ ($Ag^+$ ions) are mixed, a white precipitate of AgCl forms. (Charles Steele)

To write a net ionic equation, you first have to identify the ions

**Example 4.2** Write a net ionic equation for any precipitation reaction that occurs when solutions of the following ionic compounds are mixed.

(a) NaOH and $Cu(NO_3)_2$    (b) $BaCl_2$ and $Ag_2SO_4$    (c) $(NH_4)_2S$ and $K_2CO_3$

*Strategy*   Follow the procedure of Example 4.1 to decide whether a precipitate will form. If it does, write its formula, followed by (*s*), on the right side of the equation. On the left (reactant) side, write the formulas of the ions (*aq*) required to produce the precipitate. Finally, balance the equation.

*Solution*

(a) Ions present: $Na^+$, $OH^-$; $Cu^{2+}$, $NO_3^-$
Possible precipitates: $NaNO_3$, $Cu(OH)_2$
$NaNO_3$ is soluble, but $Cu(OH)_2$ is not.

Equation:   $Cu^{2+}(aq) + 2OH^-(aq) \longrightarrow Cu(OH)_2(s)$

(b) Ions present: $Ba^{2+}$, $Cl^-$; $Ag^+$, $SO_4^{2-}$
Possible precipitates: $BaSO_4$, AgCl
Both compounds are insoluble, so two reactions occur.

Equations:   $Ba^{2+}(aq) + SO_4^{2-}(aq) \longrightarrow BaSO_4(s)$
$Ag^+(aq) + Cl^-(aq) \longrightarrow AgCl(s)$

(c) Ions present: $NH_4^+$, $S^{2-}$; $K^+$, $CO_3^{2-}$
Possible precipitates: $(NH_4)_2CO_3$, $K_2S$

Both compounds are soluble, so there is   no precipitation reaction and no equation.

---

Although we have introduced net ionic equations to represent precipitation reactions, they have a much wider application. Indeed, we will use them for all kinds of reactions in water solution. In particular, *all of the chemical equations written throughout this chapter are net ionic equations.*

## 4.2   Acid-Base Reactions

You are probably familiar with a variety of aqueous solutions that are either acidic or basic (Fig. 4.5). Acidic solutions have a sour taste and affect the color of certain organic dyes known as acid-base indicators. For example, litmus turns from blue to red in acidic solution. Basic solutions have a slippery feeling and change the colors of indicators (e.g., red to blue for litmus).

The species that give these solutions their characteristic properties are called acids and bases. In this chapter, we use the definitions first proposed by Svante Arrhenius more than a century ago.

*An acid is a species that produces $H^+$ ions in water solution.*

*A base is a species that produces $OH^-$ ions in water solution.*

Later, we will consider more general definitions of acids and bases in Chapter 13 (Brönsted-Lowry) and Chapter 15 (Lewis).

**Figure 4.5**
Many common household items, including vinegar, lemon juice, and cola drinks, are acidic. Others, including ammonia and most detergents and cleaning agents, are basic.
(Marna G. Clarke)

## Strong and Weak Acids and Bases

There are two types of acids, strong and weak, which differ in the extent of their ionization in water. **Strong acids** ionize completely, forming $H^+$ ions and anions. A typical strong acid is HCl. It undergoes the following reaction upon addition to water:

$$HCl(aq) \longrightarrow H^+(aq) + Cl^-(aq)$$

In a solution prepared by adding 0.1 mol of HCl to water, there is 0.1 mol of $H^+$ ions, 0.1 mol of $Cl^-$ ions, and no HCl molecules. There are six common strong acids, whose names and formulas are listed in Table 4.2.

All acids other than those listed in Table 4.2 can be taken to be weak. A **weak acid** is only partially ionized to $H^+$ ions in water. All of the weak acids considered in this chapter are *molecules containing an ionizable hydrogen atom.* Their general formula can be represented as HB; the general ionization reaction in water is

$$HB(aq) \rightleftharpoons H^+(aq) + B^-(aq)$$

The double arrow implies that this reaction does not go to completion. Instead, a mixture is formed containing significant amounts of both products and reactants. With the weak acid hydrogen fluoride

$$HF(aq) \rightleftharpoons H^+(aq) + F^-(aq)$$

Sulfuric acid ionization:
$$H_2SO_4(aq) \rightarrow H^+(aq) + HSO_4^-(aq)$$

TABLE 4.2   **Common Strong Acids and Bases**

| Acid | Name of Acid | Base | Name of Base |
|------|--------------|------|--------------|
| HCl | hydrochloric acid | LiOH | lithium hydroxide |
| HBr | hydrobromic acid | NaOH | sodium hydroxide |
| HI | hydriodic acid | KOH | potassium hydroxide |
| $HNO_3$ | nitric acid | $Ca(OH)_2$ | calcium hydroxide |
| $HClO_4$ | perchloric acid | $Sr(OH)_2$ | strontium hydroxide |
| $H_2SO_4$ | sulfuric acid | $Ba(OH)_2$ | barium hydroxide |

You need to know the strong acids and bases to work with acid-base reactions

a solution prepared by adding 0.1 mol of HF to a liter of water contains about 0.01 mol of $H^+$ ions, 0.01 mol of $F^-$ ions, and 0.09 mol of HF molecules.

Bases, like acids, are classified as strong or weak. A **strong base** in water solution is completely ionized to $OH^-$ ions and cations. As you can see from Table 4.2, the strong bases are the hydroxides of the Group 1 and Group 2 metals. These are typical ionic solids, completely ionized in both the solid state and in water solution. The equations written to represent the processes by which NaOH and $Ca(OH)_2$ dissolve in water are

$$NaOH(s) \longrightarrow Na^+(aq) + OH^-(aq)$$

$$Ca(OH)_2(s) \longrightarrow Ca^{2+}(aq) + 2OH^-(aq)$$

In a solution prepared by adding 0.1 mol of NaOH to water, there is 0.1 mol of $Na^+$ ions, 0.1 mol of $OH^-$ ions, and no NaOH molecules.

**Weak bases** produce $OH^-$ ions in a quite different manner. They react with $H_2O$ molecules, acquiring $H^+$ ions and leaving $OH^-$ ions behind. The reaction of ammonia, $NH_3$, is typical:

$$NH_3(aq) + H_2O \rightleftharpoons NH_4^+(aq) + OH^-(aq)$$

As with all weak bases, this reaction does not go to completion. In a solution prepared by adding 0.1 mol of ammonia to a liter of water, there is about 0.001 mol of $NH_4^+$, 0.001 mol of $OH^-$, and nearly 0.099 mol of $NH_3$.

A common class of weak bases consists of the organic molecules known as *amines*. An amine can be considered to be a derivative of ammonia in which one or more hydrogen atoms have been replaced by hydrocarbon groups (Table 4.3). In the simplest case, methylamine, a hydrogen atom is replaced by a $-CH_3$ group to give the $CH_3NH_2$ molecule, which reacts with water in a manner very similar to $NH_3$:

$$CH_3NH_2(aq) + H_2O \rightleftharpoons CH_3NH_3^+(aq) + OH^-(aq)$$

As we have pointed out, strong acids and bases are completely ionized in water. As a result, compounds such as HCl and NaOH are strong electrolytes like NaCl. In contrast, molecular weak acids and weak bases are poor conductors because their water solutions contain relatively few ions. Hydrofluoric acid and ammonia are commonly described as **weak electrolytes.**

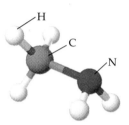

Methylamine, $CH_3NH_2$

When an H atom of $NH_3$ is replaced by a $CH_3$ group, methylamine, $CH_3NH_2$, results.

| TABLE 4.3   **Types of Amines** | | |
| --- | --- | --- |
| **Type** | **General Formula** | **Example** |
| Primary | $RNH_2$ | $CH_3-\overset{\displaystyle \,}{\underset{\displaystyle H}{N}}-H$ |
| Secondary | $R_2NH$ | $CH_3-\overset{\displaystyle \,}{\underset{\displaystyle H}{N}}-CH_3$ |
| Tertiary | $R_3N$ | $CH_3-\overset{\displaystyle \,}{\underset{\displaystyle CH_3}{N}}-CH_3$ |
| where R = $CH_3$, $C_2H_5$, . . . | | |

## Equations for Acid-Base Reactions

When an acidic water solution is mixed with a basic solution, an acid-base reaction takes place. The nature of the reaction and hence the equation written for it depend upon whether the acid and base involved are strong or weak.

**1. Strong acid–strong base.** Consider what happens when a solution of a strong acid such as $HNO_3$ is added to a solution of a strong base such as NaOH. Since $HNO_3$ is a strong acid, it is completely converted to $H^+$ and $NO_3^-$ ions in solution. Similarly, with the strong base NaOH, the solution species are the $Na^+$ and $OH^-$ ions. When the solutions are mixed, the $H^+$ and $OH^-$ ions react with each other to form $H_2O$ molecules. This reaction, referred to as **neutralization,** is represented by the net ionic equation

$$H^+(aq) + OH^-(aq) \longrightarrow H_2O$$

The $Na^+$ and $NO_3^-$ ions take no part in the reaction and so do not appear in the equation.

There is considerable evidence to indicate that the neutralization reaction occurs when any strong base reacts with any strong acid in water solution. It follows that the neutralization equation written above applies to any strong acid–strong base reaction.

**2. Weak acid–strong base.** When a strong base such as NaOH is added to a solution of a weak acid, HB, a two-step reaction occurs. The first step is the ionization of the HB molecule to $H^+$ and $B^-$ ions; the second is the neutralization of the $H^+$ ions produced in the first step by the $OH^-$ ions of the NaOH solution.

$$(1) \ HB(aq) \rightleftharpoons H^+(aq) + B^-(aq)$$

$$(2) \ H^+(aq) + OH^-(aq) \longrightarrow H_2O$$

The equation for the overall reaction is obtained by adding the two equations just written:

$$HB(aq) + OH^-(aq) \longrightarrow B^-(aq) + H_2O$$

where HB stands for any weak acid. For the reaction between solutions of sodium hydroxide and hydrogen fluoride, the equation is

$$HF(aq) + OH^-(aq) \longrightarrow H_2O + F^-(aq)$$

Here, as always, spectator ions such as $Na^+$ are not included in the net ionic equation.

**3. Strong acid–weak base.** As an example of this type of reaction, consider what happens when an aqueous solution of a strong acid like HCl is added to an aqueous solution of ammonia, $NH_3$. Again, we consider the reaction to take place in two steps. The first step is the reaction of $NH_3$ with $H_2O$ to form $NH_4^+$ and $OH^-$ ions. Then, in the second step, the $H^+$ ions of the strong acid neutralize the $OH^-$ ions formed in the first step.

$$(1) \ NH_3(aq) + H_2O \rightleftharpoons NH_4^+(aq) + OH^-(aq)$$

$$(2) \ H^+(aq) + OH^-(aq) \longrightarrow H_2O$$

The overall equation is obtained by summing those for the individual steps. Cancelling species ($OH^-$, $H_2O$) that appear on both sides, we obtain the simple equation

$$H^+(aq) + NH_3(aq) \longrightarrow NH_4^+(aq)$$

When $HClO_4$ reacts with $Ca(OH)_2$, the equation is: $H^+ (aq) + OH^-(aq) \rightarrow H_2O$

When an acid is weak, like HF, its formula appears in the equation

When a base is weak, like $NH_3$, its formula appears in the equation

| TABLE 4.4 | **Types of Acid-Base Reactions** | |
|-----------|-----------------|-----------------|
| **Reactants** | **Reacting Species** | **Net Ionic Equation** |
| Strong acid– | $H^+$ | $H^+(aq) + OH^-(aq) \longrightarrow H_2O$ |
| Strong base | $OH^-$ | |
| Weak acid– | HB | $HB(aq) + OH^-(aq) \longrightarrow H_2O + B^-(aq)$ |
| Strong base | $OH^-$ | |
| Strong acid– | $H^+$ | $H^+(aq) + B(aq) \longrightarrow BH^+(aq)$ |
| Weak base | B | |

In another case, for the reaction of a strong acid such as $HNO_3$ with methyl-amine, $CH_3NH_2$, the equation is

$$H^+(aq) + CH_3NH_2(aq) \longrightarrow CH_3NH_3{}^+(aq)$$

Table 4.4 summarizes the equations written for the three types of acid-base reactions just discussed. You should find it helpful in writing the equations called for in Example 4.3.

---

**Example 4.3**    Write a net ionic equation for each of the following reactions in dilute water solution.

    (a) Nitrous acid ($HNO_2$) with sodium hydroxide (NaOH)
    (b) Ethylamine ($C_2H_5NH_2$) with perchloric acid ($HClO_4$)
    (c) Hydrobromic acid (HBr) with potassium hydroxide (KOH)

*Strategy*    Decide whether the acid and base are strong or weak. Then decide which of the three types of acid-base reactions is involved. Finally, use Table 4.4 to derive the proper equation.

*Solution*

    (a) $HNO_2$ is weak; NaOH is strong. This is a weak acid–strong base reaction.

$$HNO_2(aq) + OH^-(aq) \longrightarrow H_2O + NO_2{}^-(aq)$$

    (b) $HClO_4$ is strong (Table 4.2); $C_2H_5NH_2$ is weak. This is a strong acid–weak base reaction.

$$H^+(aq) + C_2H_5NH_2(aq) \longrightarrow C_2H_5NH_3{}^+(aq)$$

    (c) HBr is a strong acid; KOH is a strong base.

$$H^+(aq) + OH^-(aq) \longrightarrow H_2O$$

---

**Figure 4.6**
When zinc is added to a strong acid, Zn atoms are oxidized to $Zn^{2+}$ ions in solution; $H^+$ ions are reduced to $H_2$ molecules. (Charles D. Winters)

## 4.3    Oxidation-Reduction Reactions

Another common type of reaction in aqueous solution involves an exchange of electrons between two species. Such a reaction is called an **oxidation-reduction** or **redox reaction.** Many familiar reactions fit into this category, including the reaction of metals with acid.

# Chemistry: *The Human Side*

For reasons that are by no means obvious, Sweden produced a disproportionate number of outstanding chemists in the eighteenth and nineteenth centuries. Jons Jakob Berzelius (1779–1848) determined with amazing accuracy the atomic masses of virtually all the elements known in his time. In his spare time, he "invented" such modern laboratory tools as the beaker, the flask, the pipet, and the ringstand.

Svante Arrhenius, like Berzelius, was born in Sweden and spent his entire professional career there. According to Arrhenius, the concept of strong and weak acids and bases came to him on May 13, 1883, when he was 24 years old. He added, "I could not sleep that night until I had worked through the entire problem."

Almost exactly one year later, Arrhenius submitted his Ph.D. thesis at the University of Uppsala. He proposed that salts, strong acids, and strong bases are completely ionized in dilute water solution. Today, it seems quite reasonable that solutions of NaCl, HCl, and NaOH contain, respectively, $Na^+$ and $Cl^-$ ions, $H^+$ and $Cl^-$ ions, and $Na^+$ and $OH^-$ ions. It did not seem nearly so obvious to the chemistry faculty at Uppsala in 1884. Arrhenius's dissertation received the lowest passing grade "approved not without praise."

Arrhenius sent copies of his Ph.D. thesis to several well-known chemists in Europe and America. Most ignored his ideas; a few were openly hostile. A pair of young chemists gave positive responses: Jacobus Van't Hoff (32) at Amsterdam (Holland) and Wilhelm Ostwald (32) at Riga (Latvia). For some years, these three young men were referred to, somewhat disparagingly, as "ionists" or "ionians." As time passed, the situation changed. The first Nobel Prize in chemistry was awarded to Van't Hoff in 1901. Two years later, in 1903, Arrhenius became a Nobel laureate; Ostwald followed in 1909.

Among other contributions of Arrhenius, the most important were probably in chemical kinetics (Chap. 11). In 1889 he derived the relation for the temperature dependence of reaction rate. In quite a different area in 1896 Arrhenius published an article, "On the Influence of Carbon Dioxide in the Air on the Temperature of the Ground." Here he presented the basic idea of the greenhouse effect, discussed in Chapter 17.

In his later years, Arrhenius turned his attention to popularizing chemistry. He wrote several different textbooks that were well received. In 1925, under pressure from his publisher to submit a manuscript, Arrhenius started getting up at 4 AM to write. As might be expected, rising at such an early hour had an adverse effect on his health. Arrhenius suffered a physical breakdown in 1925, from which he never really recovered, dying two years later.

**Svante August Arrhenius**
(1859–1927)

(The E. F. Smith Memorial Collection in the History of Chemistry, Dept. of Special Collections, Van Pelt-Dietrich Library, University of Pennsylvania)

Publishers are like that

In a redox reaction, one species *loses* electrons and is said to be *oxidized.* The other species, which *gains electrons,* is *reduced.* To illustrate this situation, consider the redox reaction that takes place when zinc pellets are added to hydrochloric acid (Fig. 4.6). The net ionic equation for the reaction is

$$Zn(s) + 2H^+(aq) \longrightarrow Zn^{2+}(aq) + H_2(g)$$

Here, zinc atoms are oxidized to $Zn^{2+}$ ions by *losing* electrons; the *half-reaction* is

$$\text{oxidation:} \qquad Zn(s) \longrightarrow Zn^{2+}(aq) + 2e^-$$

while $H^+$ ions are reduced to $H_2$ molecules by *gaining* electrons; the *half-reaction* is

$$\text{reduction:} \qquad 2H^+(aq) + 2e^- \longrightarrow H_2(g)$$

From this example, it should be clear that

— *oxidation and reduction occur together,* in the same reaction; you can't have one without the other.
— *there is no net change in the number of electrons in a redox reaction.* Those given off in the oxidation half-reaction are taken on by another species in the reduction half-reaction.

The oxidizing agent is reduced; the reducing agent is oxidized

The two species that exchange electrons in a redox reaction are given special names. The ion or molecule that accepts electrons is called the **oxidizing agent;** by accepting electrons it brings about the oxidation of another species. Conversely, the species that donates electrons is called the **reducing agent;** when reaction occurs it reduces the other species.

To illustrate these concepts consider the reaction

$$Zn(s) + 2H^+(aq) \longrightarrow Zn^{2+}(aq) + H_2(g)$$

The $H^+$ ion is the oxidizing agent; it brings about the oxidation of zinc. By the same token, zinc acts as a reducing agent; it furnishes the electrons required to reduce $H^+$ ions.

In earlier sections of this chapter, we showed how to write and balance equations for precipitation reactions (Section 4.1) and acid-base reactions (Section 4.2). Here our goal is more modest; we concentrate upon balancing redox equations, given the identity of reactants and products. To do that, it is convenient to introduce a new concept, oxidation number.

## Oxidation Number

The concept of **oxidation number** was introduced to simplify the electron bookkeeping in redox reactions. For a monatomic ion (e.g., $Na^+$, $S^{2-}$), the oxidation number is, quite simply, the charge of the ion ($+1$, $-2$). In a molecule or polyatomic ion, the oxidation number of an element is a "pseudo-charge" obtained in a rather arbitrary way, assigning bonding electrons to the atom with the greater attraction for electrons.

Oxidation numbers are calculated, not determined experimentally

In practice, oxidation numbers in all kinds of species are assigned according to a set of four arbitrary rules.

**1. *The oxidation number of an element in an elementary substance is 0.*** For example, the oxidation number of chlorine in $Cl_2$ or of phosphorus in $P_4$ is 0.

**2. *The oxidation number of an element in a monatomic ion is equal to the charge of that ion.*** In the ionic compound NaCl, sodium has an oxidation number of $+1$, chlorine an oxidation number of $-1$. The oxidation numbers of aluminum and oxygen in $Al_2O_3$ ($Al^{3+}$, $O^{2-}$ ions) are $+3$ and $-2$, respectively.

**3. *Certain elements have the same oxidation number in all or almost all their compounds.*** The Group 1 metals always exist as $+1$ ions in their compounds and hence are assigned an oxidation number of $+1$. By the same token, Group 2 elements always have oxidation numbers of $+2$ in their compounds. Fluorine always has an oxidation number of $-1$.

Oxygen is ordinarily assigned an oxidation number of $-2$ in its compounds. (An exception arises in compounds containing the peroxide ion, $O_2^{2-}$, where the oxidation number of oxygen is $-1$).

In $H_2O_2$, the oxidation numbers of H and O are $+1$ and $-1$ respectively

Hydrogen in its compounds ordinarily has an oxidation number of +1. (The major exception is in metal hydrides such as NaH and $CaH_2$, where hydrogen is present as the $H^-$ ion and hence is assigned an oxidation number of −1.)

**4. *The sum of the oxidation numbers in a neutral species is 0; in a polyatomic ion, it is equal to the charge of that ion.*** The application of this very useful principle is illustrated in Example 4.4.

---

**Example 4.4**   What is the oxidation number of sulfur in $Na_2SO_4$? Of manganese in $MnO_4^-$?

***Strategy***   First look for elements whose oxidation number is always or almost always the same (Rule 3). Then solve for the oxidation number of the remaining element by applying Rule 4.

***Solution***   In $Na_2SO_4$, the oxidation numbers of sodium and oxygen are +1 and −2, respectively. Sodium sulfate, like all compounds, is neutral, so the sum of the oxidation numbers is zero.

$$0 = 2(+1) + \text{oxid. no. S} + 4(-2) \qquad \text{oxid. no. S} = \boxed{+6}$$

In the $MnO_4^-$ ion, oxygen has an oxidation number of −2. The ion has a charge of −1, so the sum of the oxidation numbers must be −1 (Rule 4).

$$-1 = \text{oxid. no. Mn} + 4(-2) \qquad \text{oxid. no. Mn} = \boxed{+7}$$

---

The concept of oxidation number leads directly to a working definition of the terms oxidation and reduction. **Oxidation** is defined as ***an increase in oxidation number and reduction as a decrease in oxidation number.*** Consider once again the reaction of zinc with a strong acid:

$$Zn(s) + 2H^+(aq) \longrightarrow Zn^{2+}(aq) + H_2(g)$$

Zn is oxidized (oxid no.: 0 $\longrightarrow$ +2)

$H^+$ is reduced (oxid no.: +1 $\longrightarrow$ 0)

These definitions are of course compatible with the interpretation of oxidation and reduction in terms of loss and gain of electrons. An element that loses electrons must increase in oxidation number. The gain of electrons always results in a decrease in oxidation number.

## Balancing Half-Equations (Oxidation or Reduction)

Before you can balance an overall redox equation, you have to be able to balance the two half-equations involved. Sometimes that's easy. Given the oxidation half-equation

$$Fe^{2+}(aq) \longrightarrow Fe^{3+}(aq) \qquad \text{(oxid no. Fe: } +2 \longrightarrow +3)$$

it is clear that mass and charge balance can be achieved by adding an electron to the right:

$$Fe^{2+}(aq) \longrightarrow Fe^{3+}(aq) + e^-$$

In another case, this time a reduction half-equation,

$$Cl_2(g) \longrightarrow Cl^-(aq) \qquad \text{(oxid no. Cl: } 0 \longrightarrow -1)$$

mass balance is obtained by writing a coefficient of 2 for $Cl^-$; charge is then balanced by adding two electrons to the left. The balanced half-equation is

$$Cl_2(g) + 2e^- \longrightarrow 2Cl^-(aq)$$

(Throughout this discussion we will show balanced oxidation half-equations in blue, balanced reduction half-equations in red).

Sometimes, though, it is by no means obvious how a given half-equation is to be balanced. This commonly happens when elements other than those being oxidized or reduced take part in the reaction. Most often, these elements are oxygen (oxid no. = −2) and hydrogen (oxid no. = +1). Consider, for example, the half-equation for the reduction of the permanganate ion,

$$MnO_4^-(aq) \longrightarrow Mn^{2+}(aq) \qquad \text{(oxid no. Mn: } +7 \longrightarrow +2)$$

or the oxidation of chromium(III) hydroxide,

$$Cr(OH)_3(s) \longrightarrow CrO_4^{2-}(aq) \qquad \text{(oxid no. Cr: } +3 \longrightarrow +6)$$

To balance half-equations such as these, proceed as follows:

It's important to carry out these steps in the order indicated

(a) **Balance the atoms of the element being oxidized or reduced.**
(b) **Balance oxidation number by adding electrons.** For a reduction half-equation, add the appropriate number of electrons to the left; for an oxidation half-equation, add electrons to the right.
(c) **Balance charge by adding H$^+$ ions in acidic solution, OH$^-$ ions in basic solution.**
(d) **Balance hydrogen by adding H$_2$O molecules.**
(e) Check to make sure that oxygen is balanced. If it is, the half-equation is almost certainly balanced correctly with respect to both mass and charge.

Example 4.5 shows how these rules are applied to balance half-equations.

---

**Example 4.5**    Balance the following half-equations:

1. $MnO_4^-(aq) \longrightarrow Mn^{2+}(aq)$    (acidic solution)
2. $Cr(OH)_3(s) \longrightarrow CrO_4^{2-}(aq)$    (basic solution)

***Strategy***    Follow the rules outlined above, step-by-step in the proper order, and see what happens.

***Solution***

You have to assign oxidation numbers to decide how many electrons to use

**1.** (a) Since there is one atom of Mn on both sides, no adjustment is required here.

$$MnO_4^-(aq) \longrightarrow Mn^{2+}(aq)$$

(b) Since manganese is reduced from an oxidation number of +7 to +2, five electrons must be added to the left.

$$MnO_4^-(aq) + 5e^- \longrightarrow Mn^{2+}(aq)$$

(c) There is a total charge of −6 on the left versus +2 on the right. To balance, add 8H$^+$ to the left to give a charge of +2 on both sides.

$$MnO_4^-(aq) + 8H^+(aq) + 5e^- \longrightarrow Mn^{2+}(aq)$$

(d) To balance the 8H$^+$ ions on the left, add four H$_2$O molecules to the right.

$$MnO_4^-(aq) + 8H^+(aq) + 5e^- \longrightarrow Mn^{2+}(aq) + 4H_2O$$

(e) Note that there are the same number of oxygen atoms, four, on both sides, as there should be. The equation shown in red is the correctly balanced reduction half-equation.

**2.** (a) Again, there is one chromium atom on both sides.

$$Cr(OH)_3(s) \longrightarrow CrO_4{}^{2-}(aq)$$

(b) Since the oxidation number of chromium increases from +3 to +6, add three electrons to the right.

$$Cr(OH)_3(s) \longrightarrow CrO_4{}^{2-}(aq) + 3e^-$$

(c) There is a charge of zero on the left, −5 on the right. To balance charge, add 5 $OH^-$ ions to the left.

$$Cr(OH)_3(s) + 5OH^-(aq) \longrightarrow CrO_4{}^{2-}(aq) + 3e^-$$

(d) There are 8 hydrogens on the left, none on the right. Add 4 $H_2O$ molecules to the right.

$$Cr(OH)_3(s) + 5OH^-(aq) \longrightarrow CrO_4{}^{2-}(aq) + 4H_2O + 3e^-$$

(e) There are 8 oxygen atoms on both sides; the oxidation half-equation is properly balanced.

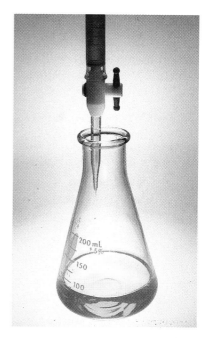

## Balancing Redox Equations

The process used to balance an overall redox equation is relatively straightforward, provided you know how to balance half-equations. Follow a systematic, four-step procedure.

1. **Split the equation into two half-equations,** one for reduction, the other for oxidation.
2. **Balance one of the half-equations** with respect to both atoms and charge as described above (steps a through e).
3. **Balance the other half-equation.**
4. **Combine the two half-equations in such a way as to eliminate electrons.** The final equation should show both atom and charge balance. It should contain no electrons.

**Example 4.6**   Balance the following redox equations.

   (a) $Fe^{2+}(aq) + MnO_4{}^-(aq) \longrightarrow Fe^{3+}(aq) + Mn^{2+}(aq)$   (acidic solution) (Fig. 4.7)
   (b) $Cl_2(g) + Cr(OH)_3(s) \longrightarrow Cl^-(aq) + CrO_4{}^{2-}(aq)$   (basic solution)

***Strategy***   Follow the four-step procedure described above. Actually, if you look carefully, you'll find that all the half-equations have already been balanced! The color-coding should help you find them.

*Solution*

   (a) (1) oxidation: $Fe^{2+}(aq) \longrightarrow Fe^{3+}(aq)$
        reduction: $MnO_4{}^-(aq) \longrightarrow Mn^{2+}(aq)$
     (2), (3) The balanced half-equations, as obtained previously, are

$$Fe^{2+}(aq) \longrightarrow Fe^{3+}(aq) + e^-$$

$$MnO_4{}^-(aq) + 8H^+(aq) + 5e^- \longrightarrow Mn^{2+}(aq) + 4H_2O$$

**Figure 4.7**
When potassium permanganate (buret) is added to an acidic solution containing $Fe^{2+}$ ions (flask), a redox reaction occurs. The equation for the reaction is derived and balanced in the text. As reaction takes place, the purple color characteristic of $MnO_4{}^-$ ion fades. (Charles D. Winters)

(4) To eliminate electrons, multiply the oxidation half-equation by five and add to the reduction half-equation.

$$5[Fe^{2+}(aq) \longrightarrow Fe^{3+}(aq) + e^-]$$
$$\underline{MnO_4^-(aq) + 8H^+(aq) + 5e^- \longrightarrow Mn^{2+}(aq) + 4H_2O}$$
$$5Fe^{2+}(aq) + MnO_4^-(aq) + 8H^+(aq) \longrightarrow 5Fe^{3+}(aq) + Mn^{2+}(aq) + 4H_2O$$

Multiply the half-equations by coefficients (1, 2, —) so that there are equal numbers of electrons on both sides

(b) (1) reduction: $Cl_2(g) \longrightarrow Cl^-(aq)$

oxidation: $Cr(OH)_3(s) \longrightarrow CrO_4^{2-}(aq)$

(2), (3) Earlier, the balanced half-equations were found to be

$$Cl_2(g) + 2e^- \longrightarrow 2Cl^-(aq)$$

$$Cr(OH)_3(s) + 5\ OH^-(aq) \longrightarrow CrO_4^{2-}(aq) + 4H_2O + 3e^-$$

(4) Multiply the reduction half-equation by three, the oxidation half-equation by two, then add. This will produce $6e^-$ on both sides, so they will cancel.

$$3[Cl_2(g) + 2e^- \longrightarrow 2Cl^-(aq)]$$
$$\underline{2[Cr(OH)_3(s) + 5\ OH^-(aq) \longrightarrow CrO_4^{2-}(aq) + 4H_2O + 3e^-]}$$
$$3Cl_2(g) + 2Cr(OH)_3(s) + 10\ OH^-(aq) \longrightarrow 6Cl^-(aq) + 2CrO_4^{2-}(aq) + 8H_2O$$

## 4.4 Solute Concentrations; Molarity

Before discussing solution stoichiometry, it is necessary to consider how to express the **concentrations** of species in solution. That is, we need to specify how much solute is present in a given volume of solution. It would be meaningless to talk about "250 mL of NaOH solution" without knowing how many grams or moles of NaOH there are per liter of solution.

So far as solution stoichiometry is concerned, it is most convenient to express the concentration of solute in terms of **molarity:**

$$\text{molarity } (M) = \frac{\text{moles of solute}}{\text{liters of solution}}$$

[NH₃] means "concentration of ammonia in moles per liter"

The symbol [  ] is commonly used to represent the molarity of a species in solution. For a solution containing 1.20 mol of substance A in 2.50 L of solution,

$$[A] = \frac{1.20 \text{ mol}}{2.50 \text{ L}} = 0.480 \text{ mol/L} = 0.480\ M$$

One liter of such a solution would contain 0.480 mol of A; 100 mL of solution would contain 0.0480 mol of A, and so on.

To prepare a solution to a desired molarity, you first calculate the amount of solute required. This is then dissolved in enough solvent to form the required volume of solution. Suppose, for example, you want to make one liter of 0.100 $M$ $K_2CrO_4$ solution. You first weigh out 19.4 g (0.100 mol) of $K_2CrO_4$ ($M = 194.20$ g/mol). Then stir with enough water (Fig. 4.8) to form one liter (1000 mL) of solution.

The molarity of a solution can be used to calculate

— the number of moles of solute in a given volume of solution
— the volume of solution containing a given number of moles of solute

Here, as in so many other cases, a conversion factor approach is used (Example 4.7).

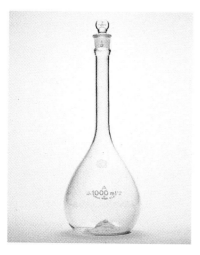

**Figure 4.8**
To prepare one liter of 0.100 *M* $K_2CrO_4$, you would start by weighing out 19.4 g of potassium chromate. The solid is transferred to a 1000-mL volumetric flask. Enough water is added to dissolve, by swirling, all of the potassium chromate. More water is then added to bring the level up to the mark on the neck. Finally, the flask is shaken repeatedly until a homogeneous solution is obtained. (Marna G. Clarke)

---

**Example 4.7** The bottle labeled "concentrated hydrochloric acid" in the lab contains 12.0 mol of HCl per liter of solution. That is, [HCl] = 12.0 *M*.

(a) How many moles of HCl are there in 25.0 mL of this solution?
(b) What volume of concentrated hydrochloric acid must be taken to contain 1.00 mol of HCl?

*Strategy* The required conversion factors are

$$\frac{12.0 \text{ mol HCl}}{1 \text{ L}} \quad \text{or} \quad \frac{1 \text{ L}}{12.0 \text{ mol HCl}}$$

*Solution*

(a) $n_{HCl} = 25.0 \text{ mL} \times \dfrac{1 \text{ L}}{1000 \text{ mL}} \times \dfrac{12.0 \text{ mol HCl}}{1 \text{ L}} = $ 0.300 mol HCl

(b) $V = 1.00 \text{ mol HCl} \times \dfrac{1 \text{ L}}{12.0 \text{ mol HCl}} = $ 0.0833 L (83.3 mL)

---

In all the reactions considered in this chapter, at least one of the reactants is an ion in solution. The concentration of that ion frequently differs from that of the corresponding ionic compound. Consider, for example, what happens when magnesium chloride dissolves in water.

$$MgCl_2(s) \longrightarrow Mg^{2+}(aq) + 2Cl^-(aq)$$

Since 1 mol of $MgCl_2$ yields 1 mol of $Mg^{2+}$ and 2 mol of $Cl^-$, it follows that a 1 *M* solution of $MgCl_2$ is 1 *M* in $Mg^{2+}$ but 2 *M* in $Cl^-$. More generally, in any solution of magnesium chloride, the molarity of $Cl^-$ is twice that of $MgCl_2$.

Dissolving an ionic compound means splitting the compound into its cation and anion.

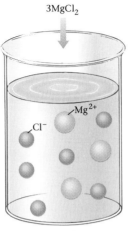
3MgCl$_2$

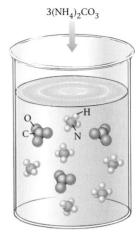
3(NH$_4$)$_2$CO$_3$

(a) $3MgCl_2(s) \rightarrow 3Mg^{2+}(aq) + 6Cl^-(aq)$    (b) $3(NH_4)_2CO_3(s) \rightarrow 6NH_4^+(aq) + 3CO_3^{2-}(aq)$

**Example 4.8**    Give the concentration, in moles per liter, of each ion in

(a) 0.080 *M* K$_2$SO$_4$    (b) 0.40 *M* LaBr$_3$

***Strategy***    To go from concentration of solute to concentration of an individual ion, you must know the conversion factor relating moles of ions to moles of solute. To find this conversion factor, it is helpful to write the equation for the solution process.

***Solution***

(a) $K_2SO_4(s) \longrightarrow 2K^+(aq) + SO_4{}^{2-}(aq)$

The conversion factors are 2 mol K$^+$/1 mol K$_2$SO$_4$ and 1 mol SO$_4{}^{2-}$/1 mol K$_2$SO$_4$.

$$[K^+] = \frac{0.080 \text{ mol K}_2\text{SO}_4}{1 \text{ L}} \times \frac{2 \text{ mol K}^+}{1 \text{ mol K}_2\text{SO}_4} = \boxed{0.16 \ M \ K^+}$$

$$[SO_4{}^{2-}] = \frac{0.080 \text{ mol K}_2\text{SO}_4}{1 \text{ L}} \times \frac{1 \text{ mol SO}_4{}^{2-}}{1 \text{ mol K}_2\text{SO}_4} = \boxed{0.080 \ M \ SO_4{}^{2-}}$$

For aluminum sulfate: $[Al^{3+}] = 2[Al_2(SO_4)_3]$, $[SO_4{}^{2-}] = 3[Al_2(SO_4)_3]$

(b) $LaBr_3(s) \longrightarrow La^{3+}(aq) + 3Br^-(aq)$

$$[La^{3+}] = \frac{0.40 \text{ mol LaBr}_3}{1 \text{ L}} \times \frac{1 \text{ mol La}^{3+}}{1 \text{ mol LaBr}_3} = \boxed{0.40 \ M \ La^{3+}}$$

$$[Br^-] = \frac{0.40 \text{ mol LaBr}_3}{1 \text{ L}} \times \frac{3 \text{ mol Br}^-}{1 \text{ mol LaBr}_3} = \boxed{1.2 \ M \ Br^-}$$

## 4.5   Solution Stoichiometry

Mass relations in solution reactions are very similar to those considered in Chapter 3. The principal difference is that, for species in solution, the number of moles is calculated by multiplying volume by molarity.

**Example 4.9**    When aqueous solutions of sodium hydroxide and iron(III) nitrate are mixed, a red gelatinous precipitate forms. Calculate the mass of precipitate formed when 50.00 mL of 0.200 *M* NaOH and 30.00 mL of 0.125 *M* Fe(NO$_3$)$_3$ are mixed.

***Strategy***   Data are given for two reactants, so this is a limiting-reactant problem, similar to those worked in Chapter 3. There are, however, a couple of preliminary steps. First, decide upon the net ionic equation for the precipitation reaction. Second, calculate the number of moles of each reactant. Then apply the three-step procedure described on p. 69.

## Solution

**1.** Following the procedure outlined in Section 4.1, you should arrive at the net ionic equation

$$Fe^{3+}(aq) + 3OH^-(aq) \longrightarrow Fe(OH)_3(s).$$

**2.** $n_{Fe^{3+}} = 0.03000 \text{ L Fe(NO}_3)_3 \times \dfrac{0.125 \text{ mol Fe(NO}_3)_3}{1 \text{ L Fe(NO}_3)_3} \times \dfrac{1 \text{ mol Fe}^{3+}}{1 \text{ mol Fe(NO}_3)_3}$

$$= 3.75 \times 10^{-3} \text{ mol Fe}^{3+}$$

Note from the formula of iron(III) nitrate that there is one mole of $Fe^{3+}$ per mole of $Fe(NO_3)_3$.

$n_{OH^-} = 0.05000 \text{ L NaOH} \times \dfrac{0.200 \text{ mol NaOH}}{1 \text{ L NaOH}} \times \dfrac{1 \text{ mol OH}^-}{1 \text{ mol NaOH}} = 1.00 \times 10^{-2} \text{ mol OH}^-$

**3.** If $Fe^{3+}$ is limiting,

$$n_{Fe(OH)_3} = 3.75 \times 10^{-3} \text{ mol Fe}^{3+} \times \dfrac{1 \text{ mol Fe(OH)}_3}{1 \text{ mol Fe}^{3+}} = 3.75 \times 10^{-3} \text{ mol Fe(OH)}_3$$

**4.** If $OH^-$ is limiting,

$$n_{Fe(OH)_3} = 1.00 \times 10^{-2} \text{ mol OH}^- \times \dfrac{1 \text{ mol Fe(OH)}_3}{3 \text{ mol OH}^-} = 3.33 \times 10^{-3} \text{ mol Fe(OH)}_3$$

**5.** Since $3.33 \times 10^{-3}$ is less than $3.75 \times 10^{-3}$, $OH^-$ is the limiting reactant. The theoretical yield of $Fe(OH)_3$ is $3.33 \times 10^{-3}$ mol. The molar mass of $Fe(OH)_3$ is 106.87 g/mol, so

$$\text{mass Fe(OH)}_3 = 3.33 \times 10^{-3} \text{ mol Fe(OH)}_3 \times \dfrac{106.87 \text{ g Fe(OH)}_3}{1 \text{ mol Fe(OH)}_3} = \boxed{0.356 \text{ g Fe(OH)}_3}$$

$Fe^{3+}(aq) + 3OH^-(aq) \longrightarrow Fe(OH)_3(s)$
When aqueous solutions of sodium hydroxide and iron (III) nitrate are mixed, a red gelatinous precipitate forms. (Charles D. Winters)

In many solution reactions, you need to know what volume of one solution is required to react with a known amount of the other (Example 4.10).

---

**Example 4.10**   As you found in Example 4.6, the balanced equation for the reaction between $MnO_4^-$ and $Fe^{2+}$ in acidic solution is

$$MnO_4^-(aq) + 8H^+(aq) + 5Fe^{2+}(aq) \longrightarrow 5Fe^{3+}(aq) + Mn^{2+}(aq) + 4H_2O$$

What volume of 0.684 *M* KMnO$_4$ solution is required to react completely with 27.50 mL of 0.250 *M* Fe(NO$_3$)$_2$?

***Strategy***   Start by calculating the number of moles of $Fe^{2+}$. Then use the coefficients of the balanced equation to find the number of moles of $MnO_4^-$. Finally, use molarity as a conversion factor to find the volume of KMnO$_4$ solution.

Calculations such as these require a balanced net ionic equation

## Solution

**1.** $n_{Fe^{2+}} = 0.02750 \text{ L} \times \dfrac{0.250 \text{ mol Fe(NO}_3)_2}{1 \text{ L}} \times \dfrac{1 \text{ mol Fe}^{2+}}{1 \text{ mol Fe(NO}_3)_2}$

$$= 6.88 \times 10^{-3} \text{ mol Fe}^{2+}$$

**2.** $n_{MnO_4^-} = 0.00688 \text{ mol Fe}^{2+} \times \dfrac{1 \text{ mol MnO}_4^-}{5 \text{ mol Fe}^{2+}} = 0.00138 \text{ mol MnO}_4^-$

**3.** $n_{KMnO_4} = n_{MnO_4^-} = 1.38 \times 10^{-3} \text{ mol KMnO}_4$

$$V = 1.38 \times 10^{-3} \text{ mol KMnO}_4 \times \dfrac{1 \text{ L}}{0.684 \text{ mol KMnO}_4} = \boxed{2.02 \times 10^{-3} \text{ L} \quad (2.02 \text{ mL})}$$

## 4.6 Solution Reactions in Quantitative Analysis

Reactions taking place in aqueous solution are commonly used in quantitative analysis to determine the concentration of a dissolved species or its percentage in a solid mixture. To do this, you carry out a titration, measuring the volume of a "standard" solution of known concentration required to react with a measured amount of sample. The standard solution is delivered from a buret (Fig. 4.9). The point in the titration at which the reaction is complete is called the **equivalence point.** This point is commonly detected using an indicator that changes color at the *end point,* telling you it's time to quit. If the indicator has been chosen properly, the equivalence point and the end point coincide.

Calculations for titrations are very similar to those described in Section 4.5. You may be asked to calculate the concentration of a species in solution (Example 4.11) or the percentage of a species in a solid mixture (Example 4.12).

---

**Example 4.11**   One way to determine blood alcohol level is to titrate with a solution containing $Ce^{4+}$, a powerful oxidizing agent. The reaction is

$$C_2H_5OH(aq) + Ce^{4+}(aq) \longrightarrow CO_2(g) + Ce^{3+}(aq) \qquad \text{(acidic solution)}$$

**Figure 4.9**
Titration of a base (flask) with an acid (buret). Originally, the indicator has the color characteristic of basic solution (blue). At the end point, the color changes sharply to green. With excess acid, we get the acid color, yellow. (Charles D. Winters)

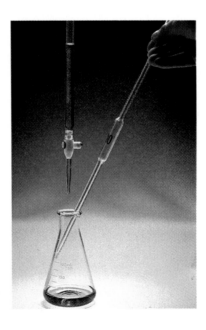
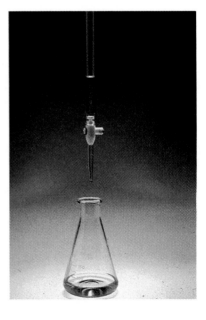

(a) Balance this redox equation.

(b) If 10.0 mL of 0.100 $M$ $Ce^{4+}$ is required to titrate a 5.00-mL sample of blood plasma, what is the molarity of ethyl alcohol, $C_2H_5OH$?

**Strategy**  The equation is balanced in the usual way, following the four-step procedure outlined on p. 95. (To balance the half-equation for the oxidation of $C_2H_5OH$, you will need to use steps a through e, p. 94). Part (b) is very similar to Example 4.10, except that you are asked to calculate concentration rather than volume. Follow the path:

$$nCe^{4+} \longrightarrow nC_2H_5OH \longrightarrow [C_2H_5OH]$$

**Solution**

(a) (1) The half-equations are

> To analyze for alcohol, we take advantage of its ease of oxidation

reduction: $Ce^{4+}(aq) \longrightarrow Ce^{3+}(aq)$     (oxid no. Ce: $+4 \longrightarrow +3$)

oxidation: $C_2H_5OH(aq) \longrightarrow CO_2(g)$     (oxid no. C: $-2 \longrightarrow +4$)

(2) The balanced reduction half-equation is

$$Ce^{4+}(aq) + e^- \longrightarrow Ce^{3+}(aq)$$

(3) For the oxidation half-equation, follow the process described on p. 94.

(a) $C_2H_5OH(aq) \longrightarrow 2CO_2(g)$

(b) $C_2H_5OH(aq) \longrightarrow 2CO_2(g) + 12e^-$

(c) $C_2H_5OH(aq) \longrightarrow 2CO_2(g) + 12e^- + 12H^+(aq)$

(d) $C_2H_5OH(aq) + 3H_2O \longrightarrow 2CO_2(g) + 12e^- + 12H^+(aq)$

(4)     $12[Ce^{4+}(aq) + e^- \longrightarrow Ce^{3+}(aq)]$

$\underline{\quad\quad C_2H_5OH(aq) + 3H_2O \longrightarrow 2CO_2(g) + 12e^- + 12H^+(aq)\quad\quad}$

$12Ce^{4+}(aq) + C_2H_5OH(aq) + 3H_2O \longrightarrow 12Ce^{3+}(aq) + 2CO_2(g) + 12H^+(aq)$

(b) $nCe^{4+} = 0.0100\ L \times \dfrac{0.100\ mol\ Ce^{4+}}{1\ L} = 1.00 \times 10^{-3}\ mol\ Ce^{4+}$

$nC_2H_5OH = 1.00 \times 10^{-3}\ mol\ Ce^{4+} \times \dfrac{1\ mol\ C_2H_5OH}{12\ mol\ Ce^{4+}} = 8.33 \times 10^{-5}\ mol\ C_2H_5OH$

$[C_2H_5OH] = \dfrac{8.33 \times 10^{-5}\ mol}{5.00 \times 10^{-3}\ L} = \boxed{0.0167\ M}$

(A concentration of 0.022 $M$ corresponds to 0.10% alcohol, which is the level judged to be evidence of intoxication in many states.)

---

**Example 4.12**  The principal ingredient of certain commercial antacids is calcium carbonate, $CaCO_3$. A student titrates an antacid tablet weighing 0.542 g with hydrochloric acid; the reaction is

$$CaCO_3(s) + 2H^+(aq) \longrightarrow Ca^{2+}(aq) + CO_2(g) + H_2O$$

If 38.5 mL of 0.200 $M$ HCl is required for complete reaction, what is the percentage of $CaCO_3$ in the antacid tablet?

**Strategy**  From the titration data, the number of moles of $H^+$ is readily calculated. Then follow the path

$$n_{H^+} \longrightarrow n_{CaCO_3} \longrightarrow \text{mass of } CaCO_3 \longrightarrow \%CaCO_3$$

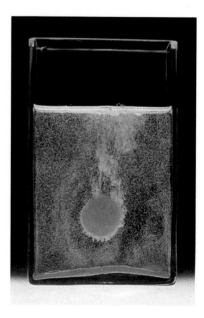

The fizz of the antacid is due to $CO_2$ gas being released.
(Charles D. Winters)

*Solution*

(1) $n_{H^+} = 0.0385 \text{ L} \times \dfrac{0.200 \text{ mol HCl}}{1 \text{ L}} \times \dfrac{1 \text{ mol H}^+}{1 \text{ mol HCl}} = 7.70 \times 10^{-3} \text{ mol H}^+$

(2) $n_{CaCO_3} = 7.70 \times 10^{-3} \text{ mol H}^+ \times \dfrac{1 \text{ mol CaCO}_3}{2 \text{ mol H}^+} = 3.85 \times 10^{-3} \text{ mol CaCO}_3$

(3) mass of $CaCO_3 = 3.85 \times 10^{-3} \text{ mol CaCO}_3 \times \dfrac{100.09 \text{ g CaCO}_3}{1 \text{ mol CaCO}_3} = 0.385 \text{ g CaCO}_3$

(4) $\%CaCO_3 = \dfrac{0.385 \text{ g}}{0.542 \text{ g}} \times 100\% = \boxed{71.0\%}$

A variety of chemical compounds are found in commercial antacids used to relieve "excess stomach acidity." As you can see from Table 4.5, their effectiveness, as measured by the amount of strong acid neutralized per gram of antacid, increases in the order

$$NaHCO_3 < CaCO_3 < MgCO_3 < Mg(OH)_2 < Al(OH)_3$$

TABLE 4.5    **Commercial Antacids**

| Compound | $\mathcal{M}$ (g/mol) | Product* | $\dfrac{n_{H^+}}{n \text{ antacid}}$ | $\dfrac{n_{H^+}}{g \text{ antacid}}$ |
|---|---|---|---|---|
| $NaHCO_3$ | 84.01 | Alka-Seltzer | 1 | 0.0119 |
| $CaCO_3$ | 100.09 | Rolaids, Tums | 2 | 0.0200 |
| $MgCO_3$ | 84.31 | Gaviscon | 2 | 0.0237 |
| $Mg(OH)_2$ | 58.32 | Mylanta, Maalox | 2 | 0.0343 |
| $Al(OH)_3$ | 78.00 | Maalox | 3 | 0.0385 |

*Many commercial products contain a mixture of antacids.

## CHEMISTRY

### *Beyond the Classroom*

## Breath Analysis for Ethyl Alcohol

The first instruments used by police to detect alcohol on the breath of motorists were developed in the 1930s. Until about ten years ago, nearly all of these used a chemical oxidizing agent, most often $K_2Cr_2O_7$, to react with ethyl alcohol:

$$2Cr_2O_7^{2-}(aq) + 16H^+(aq) + 3C_2H_5OH(aq) \longrightarrow 4Cr^{3+}(aq) + 3CH_3COOH(aq) + 11H_2O$$

ethyl alcohol                    acetic acid

Since the dichromate and chromium(III) ions have different colors (Figure 4.A), the quantity of ethyl alcohol present can be determined by measuring the color change. Actually, the dichromate solution is so dilute ($<0.001$ M) that it has a yellow color; the intensity of that color decreases as reaction proceeds.

Knowing the amount of ethyl alcohol present and the volume of the breath sample collected, one can calculate the breath alcohol concentration (BrAC) of the suspect. This in turn can be converted to blood alcohol concentration (BAC) by multiplying by a factor of 2100; one volume of blood contains about 2100 times as much alcohol as the same volume of breath. In practice, all of these calculations are done automatically; an instrument such as the breathalyzer shown in Figure 4.B displays the BAC value directly. If that exceeds the legal limit (0.10%, 0.08%, 0.05% depending on the state), the motorist is considered to be intoxicated (DWI, DUI).

In recent years, many other approaches have been used to determine BrAC values. The newer instruments determine alcohol by gas chromatography (Chap. 1), electrochemical oxidation (Chap. 18), or absorption of infrared radiation. Typically, the portable instruments carried in police cruisers weigh only a few pounds and give quite accurate BrAC values. To be sure, none of these instruments are specific for ethyl alcohol; other organic compounds in the breath can, in principle, interfere. In practice, however, these interferences are too small to be of concern. For example, even though the acetone present in the breath of a diabetic may register as alcohol, it raises the apparent BAC by less than 0.01%.

The greatest uncertainty in the use of instruments like the breathalyzer lies in the conversion from BrAC to BAC values. The factor of 2100 referred to previously can vary considerably depending upon the circumstances under which the sample is taken. Simultaneous measurements of BrAc and BAC values suggest that the factor can be anywhere between 1800 and 2400. This means that a calculated BAC value of 0.100% could be as low as 0.086% or as high as 0.114%.

After drinking an alcoholic beverage, a person's BAC rises to a maximum, typically in 30 to 90 minutes, and then drops steadily at the rate of about 0.02% per hour. The maximum BAC depends upon the amount of alcohol consumed and the person's body weight. The data cited in Table 4.A are for male subjects; females show maximum BACs about 20% higher. Thus, a man weighing 60 kg who takes three drinks may be expected to reach a maximum BAC of 0.09%; a woman of the same weight consuming the same amount of alcohol may show a BAC of

$$0.09\% \times 1.2 = 0.11\%$$

**Figure 4.A**
When alcohol is oxidized by $Cr_2O_7^{2-}$ (right), there is a color change as $Cr^{3+}$ (left) is formed. (Charles D. Winters)

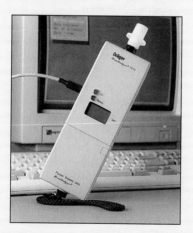

**Figure 4.B**
A late-model breathalyzer using solid-state electronics. (Courtesy of National Draeger, Inc., Breathalyzer Division)

TABLE 4.A **Maximum Blood Alcohol Concentrations (Percents)\***

| Body Weight | No. Drinks (1 drink = 1 oz 100-proof liquor or a 12-oz beer) | | | | | |
|---|---|---|---|---|---|---|
| | 1 | 2 | 3 | 4 | 5 | 6 |
| 50 kg = 110 lb | 0.04 | 0.07 | **0.10** | **0.14** | **0.17** | **0.21** |
| 60 kg = 132 lb | 0.03 | 0.06 | 0.09 | **0.12** | **0.14** | **0.17** |
| 70 kg = 154 lb | 0.02 | 0.05 | 0.07 | **0.10** | **0.12** | **0.15** |
| 80 kg = 176 lb | 0.02 | 0.04 | 0.06 | 0.09 | **0.11** | **0.13** |
| 90 kg = 198 lb | 0.02 | 0.04 | 0.06 | 0.08 | **0.10** | **0.11** |

\*Values are for men; values for women are about 20% higher; BAC values of 0.10% or higher are shown in bold type.

# CHAPTER HIGHLIGHTS

## Key Concepts

1. Apply the solubility rules (Table 4.1) to predict precipitation reactions and write net ionic equations for them
   (Examples 4.1, 4.2; Problems 3–12)
2. Write net ionic equations for acid-base reactions
   (Example 4.3; Problems 21–24)
3. Determine oxidation numbers of elements in compounds
   (Example 4.4; Problems 25–28)
4. Balance redox equations by the half-equation method
   (Examples 4.5, 4.6; Problems 31–44, 76)
5. Relate molarity of a solute to
   — number of moles and volume
   (Example 4.7; Problems 47–50)
   — molarities of ions
   (Example 4.8; Problems, 51, 52)
6. Use balanced net ionic equations to determine
   — the amount of product formed in a solution reaction
   (Example 4.9; Problems 55, 56, 61)
   — the volume of solution required for reaction
   (Example 4.10; Problems 53–55; 58–60, 62, 64, 66, 70)
   — the concentration of a reactant
   (Examples 4.11, 4.12; Problems 61, 67, 68, 71–74, 77, 78)

## Key Terms

| | | |
|---|---|---|
| acid | molarity | precipitate |
| —strong | mole | reducing agent |
| —weak | net ionic equation | reduction |
| base | neutralization | strong electrolyte |
| —strong | oxidation | theoretical yield |
| —weak | oxidation number | titration |
| limiting reactant | oxidizing agent | weak electrolyte |

## Summary Problem

Nitric acid and barium hydroxide are a strong acid and a strong base, respectively. They participate in a variety of acid-base reactions with other species. In addition, nitric acid is often involved in redox reactions because the nitrate ion is readily reduced. Finally, barium hydroxide can form precipitates involving either the barium ion or the hydroxide ion.

(a) Write net ionic equations for the reaction between aqueous solutions of
   (1) barium hydroxide and iron(III) nitrate.
   (2) nitric acid and methylamine ($CH_3NH_2$).
   (3) barium hydroxide and nitric acid.
   (4) barium hydroxide and acetic acid ($HC_2H_3O_2$).
   (5) nitric acid and tin(II) ions forming nitrogen oxide gas and tin(IV) ions.

(b) How would you prepare 500.0 mL of 0.275 $M$ barium hydroxide solution? What is the concentration of each ion in the prepared solution?

(c) How many milliliters of a concentrated (12.0 $M$) nitric acid solution are needed for a reaction that requires 3.84 moles of $H^+$ ions?

(d) When 25.00 mL of 0.1875 $M$ barium hydroxide solution react with 30.00 mL of 0.2228 $M$ nickel(II) sulfate, two precipitates form. Calculate the mass of each precipitate and the concentration of all ions after the reaction, assuming that the final volume is the sum of the initial volumes.

(e) In the analysis of an alloy, copper is oxidized to copper(II) ions while nitrate ions are reduced to nitrogen oxide gas. It is found that 26.06 mL of a solution that is 0.4500 $M$ in nitrate is required to react exactly with a 1.500-g sample of the alloy. Calculate the mass percent of copper in the alloy.

(f) To determine the molarity of a solution of hydrochloric acid, a student titrates it with 0.1750 $M$ barium hydroxide. He finds that 15.75 mL of barium hydroxide is required to react with 20.00 mL of hydrochloric acid. What is the molarity of the hydrochloric acid?

### Answers

(a) (1) $Fe^{3+}(aq) + 3OH^-(aq) \longrightarrow Fe(OH)_3(s)$
    (2) $H^+(aq) + CH_3NH_2(aq) \longrightarrow CH_3NH_3^+(aq)$
    (3) $H^+(aq) + OH^-(aq) \longrightarrow H_2O$
    (4) $HC_2H_3O_2(aq) + OH^-(aq) \longrightarrow C_2H_3O_2^-(aq) + H_2O$
    (5) $8H^+(aq) + 2NO_3^-(aq) + 3Sn^{2+}(aq) \longrightarrow 2NO(g) + 4H_2O + 3Sn^{4+}(aq)$

(b) Dissolve 23.6 g of $Ba(OH)_2$ in enough water to form 500.0 mL of solution; 0.275 $M$ $Ba^{2+}$; 0.550 $M$ $OH^-$.

(c) $3.20 \times 10^2$ mL

(d) 1.094 g $BaSO_4$; 0.4346 g $Ni(OH)_2$; 0.03629 $M$ $Ni^{2+}$; 0.03629 $M$ $SO_4^{2-}$; $[Ba^{2+}] = [OH^-] \approx 0$

(e) 74.53%

(f) 0.2756 $M$

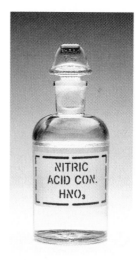

An aqueous solution of nitric acid.
(Marna G. Clarke)

## Questions & Problems

### Precipitation Reactions and Solubility

**\*1.** Write the formulas of the following compounds and decide which are soluble in water.
    **(a)** aluminum chloride
    **(b)** ammonium carbonate
    **(c)** strontium sulfate
    **(d)** chromium(III) hydroxide

**\*2.** Follow the instructions for Question 1 for the following compounds.
    **(a)** mercury(II) chloride    **(b)** cerium(III) nitrate
    **(c)** aluminum sulfate     **(d)** calcium hydroxide

**\*3.** Name the reagent, if any, that you would add to a solution of cobalt(II) nitrate to precipitate
    **(a)** cobalt(II) sulfide    **(b)** cobalt(II) carbonate
    **(c)** cobalt(II) hydroxide

**\*4.** Describe how you would prepare
    **(a)** lead sulfate from a solution of lead nitrate.
    **(b)** aluminum hydroxide from a solution of sodium hydroxide.
    **(c)** barium carbonate from a solution of barium chloride.

**\*5.** Write net ionic equations to explain the formation of
    **(a)** a red precipitate when solutions of iron(III) chloride and sodium hydroxide are mixed.
    **(b)** two different precipitates, one yellow and the other white, when solutions of cadmium(II) sulfate and strontium sulfide are mixed.

**\*6.** Write net ionic equations for the formation of
    **(a)** a green precipitate when solutions of nickel(II) nitrate and sodium hydroxide are mixed.
    **(b)** a white precipitate (that is soluble in acid) when potassium hydroxide and magnesium chloride are mixed.

**\*7.** Decide whether a precipitate will form when the following solutions are mixed. If a precipitate forms, write a net ionic equation for the reaction.
    **(a)** copper(II) sulfate and sodium chloride.
    **(b)** mercury(I) nitrate and hydrochloric acid.
    **(c)** ammonium sulfide and potassium hydroxide.
    **(d)** silver nitrate and lithium carbonate.
    **(e)** aluminum sulfate and strontium hydroxide.

*8. Follow the directions of Question 7 for solutions of
   (a) copper(II) chloride and lead nitrate.
   (b) cobalt(II) sulfate and barium hydroxide.
   (c) potassium nitrate and magnesium sulfate.
   (d) ammonium carbonate and cobalt(III) chloride.
   (e) strontium nitrate and sodium sulfide.

*9. Write a net ionic equation for any precipitation reaction that occurs when 0.1 $M$ solutions of the following are mixed.
   (a) potassium carbonate and strontium chloride.
   (b) zinc sulfate and potassium sulfide.
   (c) ammonium chloride and mercury(I) nitrate.
   (d) barium sulfide and sodium hydroxide.

*10. Follow the directions for Question 9 for the following pairs of solutions.
   (a) scandium(III) nitrate and nickel(II) chloride.
   (b) potassium carbonate and calcium nitrate.
   (c) lead(II) nitrate and sodium sulfide.
   (d) iron(III) nitrate and barium hydroxide.

*11. Using circles to represent cations and squares to represent anions, draw the reactions that occur between
   (a) $Na^+$ and $Cl^-$   (b) $Ag^+$ and $Cl^-$

*12. Assume that the circles represent cations and the squares represent anions, match the incomplete net ionic equations to their pictorial representations below
   (1) $2Na^+ + S^{2-} \longrightarrow$
   (2) $Sr^{2+} + 2OH^- \longrightarrow$
   (3) $Ba^{2+} + CO_3^{2-} \longrightarrow$

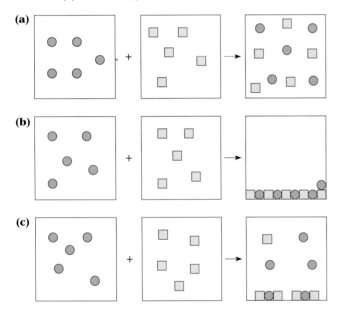

### Acid–Base Reactions

*13. Using squares to represent atoms of one element (or cations) and circles to represent the atoms of the other element (or anions), represent the following solutions pictorially. You may represent the hydroxide anion as a single circle.

   (a) a solution of HCl       (b) a solution of HF
   (c) a solution of KOH       (d) a solution of $NH_3$

*14. The following figures represent species before and after they are dissolved in water. Classify each species as weak electrolyte, strong electrolyte, or nonelectrolyte. You can assume that species that dissociate during solution break up as ions.

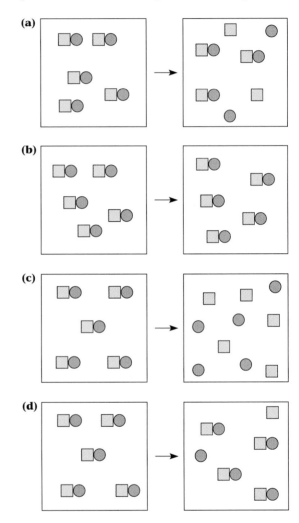

*15. For an acid-base reaction, what is the reacting species (i.e., the ion or molecule that appears in the chemical equation) in the following acids?
   (a) nitrous acid            (b) sulfuric acid
   (c) perchloric acid         (d) chlorous acid
   (e) formic acid ($HCHO_2$)

*16. For an acid-base reaction, what is the reacting species in the following acids?
   (a) nitric acid
   (b) hydrobromic acid
   (c) propionic acid ($HC_3H_5O_2$)
   (d) sulfurous acid
   (e) lactic acid ($HC_3H_5O_3$)

**\*17.** For an acid-base reaction, what is the reacting species in the following bases?

    **(a)** lithium hydroxide

    **(b)** ammonia

    **(c)** trimethylamine, $(CH_3)_3N$

    **(d)** strontium hydroxide

**\*18.** For an acid-base reaction, what is the reacting species in the following bases?

    **(a)** potassium hydroxide    **(b)** aniline, $C_6H_5NH_2$

    **(c)** barium hydroxide    **(d)** pyridine, $C_5H_5N$

**\*19.** Classify the following compounds as acids or bases, weak or strong.

    **(a)** perchloric acid

    **(b)** cesium hydroxide

    **(c)** carbonic acid, $H_2CO_3$

    **(d)** ethylamine, $C_2H_5NH_2$

**\*20.** Follow the directions of Question 19 for

    **(a)** sulfurous acid    **(b)** ammonia

    **(c)** barium hydroxide    **(d)** hydriodic acid

**\*21.** Write a balanced net ionic equation for each of the following acid-base reactions in water.

    **(a)** butyric acid $(HC_4H_7O_2)$ with potassium hydroxide.

    **(b)** methylethylamine $(C_3H_8NH)$ with hydrochloric acid.

    **(c)** aqueous hydrogen cyanide (HCN) with lithium hydroxide.

**\*22.** Write a balanced net ionic equation for each of the following acid-base reactions in water.

    **(a)** nitrous acid and sodium hydroxide.

    **(b)** cesium hydroxide and sulfuric acid.

    **(c)** aniline $(C_6H_5NH_2)$ and nitric acid.

**\*23.** Consider the equation $H^+(aq) + OH^-(aq) \longrightarrow H_2O$. For which of the following pairs would this be the correct equation for the acid-base reaction in solution? If it is not correct, write the proper equation for the acid-base reaction between the pair.

    **(a)** nitric acid and $C_2H_5NH_2$

    **(b)** perchloric acid and cesium hydroxide

    **(c)** $HC_2H_3O_2$ and LiOH

    **(d)** sulfuric acid and calcium hydroxide

    **(e)** barium hydroxide and hydriodic acid

**\*24.** Follow the directions of Question 23 for the following pairs.

    **(a)** hydrochloric acid and calcium hydroxide

    **(b)** HBr and $CH_3NH_2$

    **(c)** nitric acid and ammonia

    **(d)** $H_2SO_4$ and KOH

    **(e)** HF and barium hydroxide

## Redox Reactions

**\*25.** Assign oxidation numbers to each element in

    **(a)** nitrogen dioxide    **(b)** sulfur trioxide

    **(c)** permanganate ion    **(d)** chlorate ion

**\*26.** Assign oxidation numbers to each element in

    **(a)** nitrate ion

    **(b)** sulfate ion

    **(c)** sodium peroxide $(Na_2O_2)$

    **(d)** oxalate ion $(C_2O_4^{2-})$

**\*27.** Assign oxidation numbers to each element in

    **(a)** $SF_6$    **(b)** $Sb_4O_{10}$    **(c)** $N_2H_4$

    **(d)** $S_2O_3^{2-}$    **(e)** $Cr(OH)_4^-$

**\*28.** Assign oxidation numbers to each element in

    **(a)** $HIO_2$    **(b)** $Na_2MoO_4$    **(c)** $Fe_2O_3$

    **(d)** NOF    **(e)** $KO_2$

**\*29.** Classify each of the following half-reactions as oxidation or reduction.

    **(a)** $Ti^{3+}(aq) \longrightarrow TiO_2(s)$

    **(b)** $Sn^{4+}(aq) \longrightarrow Sn^{2+}(aq)$

    **(c)** $H_2O_2(aq) \longrightarrow O_2(g)$

    **(d)** $CH_3OH(aq) \longrightarrow CH_2O(aq)$

**\*30.** Classify each of the following half-reactions as oxidation or reduction.

    **(a)** $O_2(g) \longrightarrow OH^-(aq)$

    **(b)** $MnO_4^-(aq) \longrightarrow MnO_4^{2-}(aq)$

    **(c)** $Cr^{3+}(aq) \longrightarrow Cr_2O_7^{2-}(aq)$

    **(d)** $NH_4^+(aq) \longrightarrow NO_3^-(aq)$

**\*31.** Balance the half-equations in Question 29. Balance (a) and (b) in acidic medium, (c) and (d) in basic medium.

**\*32.** Balance the half-equations in Question 30. Balance (a) and (b) in basic medium, (c) and (d) in acidic medium.

**\*33.** Classify each of the following half-equations as oxidation or reduction and balance.

    **(a)** (acidic)  $Mn^{2+}(aq) \longrightarrow MnO_4^-(aq)$

    **(b)** (basic)  $CrO_4^{2-}(aq) \longrightarrow Cr(OH)_4^-(aq)$

    **(c)** (basic)  $Bi^{3+}(aq) \longrightarrow BiO_3^-(aq)$

    **(d)** (acidic)  $VO^{2+}(aq) \longrightarrow V^{3+}(aq)$

**\*34.** Classify each of the following half-equations as oxidation or reduction and balance.

    **(a)** (basic)  $ClO^-(aq) \longrightarrow Cl^-(aq)$

    **(b)** (acidic)  $NO_3^-(aq) \longrightarrow NO(g)$

    **(c)** (basic)  $Ni(OH)_2(s) \longrightarrow Ni_2O_3(s)$

    **(d)** (acidic)  $Mn^{2+}(aq) \longrightarrow MnO_2(s)$

**\*35.** For each unbalanced equation given below

— write unbalanced half-equations.

— identify the species oxidized and the species reduced.

— identify the oxidizing and reducing agents.

    **(a)** $H_2O_2(aq) + Fe^{2+}(aq) \longrightarrow Fe^{3+}(aq) + H_2O$

    **(b)** $C_2H_4(g) + O_2(g) \longrightarrow CO_2(g) + H_2O$

**\*36.** Follow the directions of Question 35 for the following unbalanced equations.

    **(a)** $Te(s) + NO_3^-(aq) \longrightarrow TeO_2(s) + NO(g)$

    **(b)** $Cr_2O_7^{2-}(aq) + Sn^{2+}(aq) \longrightarrow Cr^{3+}(aq) + Sn^{4+}(aq)$

**\*37.** Balance the equations in Question 35 in acid.

**\*38.** Balance the equations in Question 36 in acid.

**\*39.** Write balanced equations for the following reactions in acid solution.

    **(a)** $Fe^{2+}(aq) + IO_4^-(aq) \longrightarrow Fe^{3+}(aq) + I^-(aq)$

    **(b)** $PbO_2(s) + Br^-(aq) \longrightarrow PbBr_2(s) + O_2(g)$

    **(c)** $Ca(s) + VO_4^{3-}(aq) \longrightarrow Ca^{2+}(aq) + V^{2+}(aq)$

    **(d)** $IO_3^-(aq) + I^-(aq) \longrightarrow I_3^-(aq)$

**\*40.** Write balanced equations for the following reactions in acid solution.

**(a)** $P_4(s) \longrightarrow PH_3(g) + HPO_3{}^{2-}(aq)$

**(b)** $MnO_4{}^-(aq) + C_2H_5OH(aq) \longrightarrow$
$$Mn^{2+}(aq) + C_2H_4O(aq)$$

**(c)** $H_3AsO_3(aq) + BiO_3{}^-(aq) \longrightarrow H_3AsO_4(aq) + Bi(s)$

**(d)** $CrO_4{}^{2-}(aq) + HSO_3{}^-(aq) \longrightarrow$
$$Cr^{3+}(aq) + SO_4{}^{2-}(aq)$$

**\*41.** Write balanced equations for the following reactions in basic solution.

**(a)** $Ni(OH)_2(s) + N_2H_4(aq) \longrightarrow Ni(s) + N_2(g)$

**(b)** $Fe(OH)_3(s) + Cr(OH)_4{}^-(aq) \longrightarrow$
$$Fe(OH)_2(s) + CrO_4{}^{2-}(aq)$$

**(c)** $MnO_4{}^-(aq) + BrO_3{}^-(aq) \longrightarrow$
$$MnO_2(s) + BrO_4{}^-(aq)$$

**(d)** $H_2O_2(aq) + IO_4{}^-(aq) \longrightarrow IO_2{}^-(aq) + O_2(g)$

**\*42.** Write balanced equations for the following reactions in basic solution.

**(a)** $S_2O_3{}^{2-}(aq) + I_2(aq) \longrightarrow S_4O_6{}^{2-}(aq) + I^-(aq)$

**(b)** $Zn(s) + NO_3{}^-(aq) \longrightarrow NH_3(aq) + Zn(OH)_4{}^{2-}(aq)$

**(c)** $ClO^-(aq) + CrO_2{}^-(aq) \longrightarrow Cl^-(aq) + CrO_4{}^{2-}(aq)$

**(d)** $Al(s) + H_2O \longrightarrow Al(OH)_4{}^-(aq) + H_2(g)$

**\*43.** Write balanced net ionic equations for the following reactions in acid solution.

**(a)** Silver is dissolved in nitric acid, forming aqueous silver nitrate and nitrogen dioxide gas.

**(b)** Solid copper(II) sulfide is dissolved in nitric acid, forming copper(II) nitrate (*aq*), sulfur, and nitrogen oxide gas.

**(c)** Tin(II) ion reacts with periodate ion, $IO_4{}^-$, yielding iodide ion and tin(IV) ion.

**\*44.** Write balanced net ionic equations for the following reactions in acid solution.

**(a)** Solid phosphorus ($P_4$) reacts with hypochlorous acid, HClO, to form phosphoric acid, $H_3PO_4$, and chloride ion.

**(b)** Tellurium, Te, is oxidized by nitrate ion to form solid tellurium dioxide and nitrogen oxide gas.

**(c)** An aqueous solution of bromine is reduced to bromide; at the same time iodide ions are oxidized to iodate ions, $IO_3{}^-$.

**\*45.** Use the following reactions to arrange the elements W, X, Y, and Z in order of their increasing ability as reducing agents.

$$W^+ + Z \longrightarrow \text{no reaction}$$
$$X^+ + W \longrightarrow \text{no reaction}$$
$$Y^+ + X \longrightarrow X^+ + Y$$
$$Y^+ + Z \longrightarrow Z^+ + Y$$

**\*46.** Based on the relative strengths of the elements W, X, Y, and Z obtained in Question 45, which of the following reactions would you expect to occur?

**(a)** $Y^+ + W \longrightarrow$

**(b)** $Z^+ + X \longrightarrow$

**(c)** $W^+ + X \longrightarrow$

## Molarity

**47.** How would you prepare 615 mL of 0.483 *M*

**(a)** $KMnO_4$  **(b)** NaCl  **(c)** $C_6H_8O_6$ (vitamin C)

**48.** How would you prepare from the solid and pure water

**(a)** 0.300 L of 6.00 *M* NaOH

**(b)** 2.55 L of 0.750 *M* $NH_4NO_3$

**49.** A reagent bottle is labeled 0.450 *M* $K_2CO_3$.

**(a)** How many moles of $K_2CO_3$ are present in 45.6 mL of this solution?

**(b)** How many mL of this solution are required to furnish 0.800 mol of $K_2CO_3$?

**(c)** Assuming no volume change, how many grams of $K_2CO_3$ do you need to add to 2.00 L of this solution to obtain a 1.000 *M* solution of $K_2CO_3$?

**(d)** If 50.0 mL of this solution is added to enough water to make 125 mL of solution, what is the molarity of the diluted solution?

**50.** You are asked to prepare a 1.21 *M* solution of barium nitrate. You find that you have only 60.00 g of the solid.

**(a)** What is the maximum volume of solution that you can prepare?

**(b)** How many mL of this prepared solution are required to furnish 0.0355 mol of barium nitrate to a reaction?

**(c)** If 1.000 L of the prepared solution is required, how much more barium nitrate would you need?

**(d)** Five hundred mL of a 0.750 *M* solution of barium nitrate is needed. How would you prepare the required solution from the solution prepared in (a)?

**51.** How many moles of ions are present in water solutions prepared by dissolving 0.625 mol of

**(a)** magnesium chloride  **(b)** aluminum sulfate

**(c)** cobalt(III) nitrate  **(d)** iron(II) permanganate

**52.** How many moles of ions are present in water solutions prepared by dissolving 0.0855 mol of

**(a)** strontium hydroxide  **(b)** perchloric acid

**(c)** aluminum chloride  **(d)** calcium sulfide

## Solution Stoichiometry

**53.** What volume of 0.250 *M* lead nitrate is required to react completely with

**(a)** 15.0 mL of 0.0356 *M* cobalt(III) sulfate?

**(b)** 38.2 mL of 0.458 *M* hydrochloric acid?

**(c)** 35.0 mL of 0.297 *M* potassium hydroxide?

**54.** What volume of 0.333 *M* nickel(II) sulfate is required to react completely with

**(a)** 35.6 mL of 0.600 *M* lithium hydroxide?

**(b)** 89.5 mL of 1.35 *M* strontium nitrate?

**(c)** 15.0 mL of 0.896 *M* ammonium sulfide?

**55.** A 50.00-mL sample of 0.0250 *M* mercury(I) nitrate, $Hg_2(NO_3)_2$, is mixed with 0.0500 *M* scandium chloride.

**(a)** What is the minimum volume of scandium chloride required to completely precipitate mercury(I) chloride (also known as calomel)?

**(b)** How many grams of calomel are produced from (a)?

**56.** Aluminum ions react with carbonate ions to form an insoluble compound, aluminum carbonate.

(a) Write the net ionic equation for this reaction.

(b) What is the molarity of a solution of aluminum chloride if 25.0 mL is required to react with 35.5 mL of 0.155 $M$ sodium carbonate?

(c) How many grams of aluminum carbonate are formed?

**57.** What is the molarity of a solution of nitric acid if 0.216 g of barium hydroxide is required to neutralize 20.00 mL of nitric acid?

**58.** What volume of 0.185 $M$ strontium hydroxide is required to neutralize 35.00 mL of 0.175 $M$ hydrogen fluoride?

**59.** What is the volume of 0.550 $M$ hydrochloric acid required to react with

(a) 25.00 mL of 0.418 $M$ lithium hydroxide?

(b) 10.00 g of calcium hydroxide?

(c) 15.0 mL of a solution ($d = 0.928$ g/cm$^3$) containing 10.0% by mass of methylamine ($CH_3NH_2$)?

**60.** What is the volume of 1.050 $M$ potassium hydroxide required to react with

(a) 22.2 mL of 0.755 $M$ nitrous acid?

(b) 3.00 g of oxalic acid $H_2C_2O_4$ (both $H^+$ ions react)?

(c) 20.0 g of concentrated acetic acid ($HC_2H_3O_2$) that is 75% $HC_2H_3O_2$ by mass?

**61.** Potassium permanganate reacts with oxalic acid, $H_2C_2O_4$, to form carbon dioxide and solid manganese(IV) oxide ($MnO_2$).

(a) Write a balanced net ionic equation for the reaction.

(b) If 25.0 mL of 0.500 $M$ potassium permanganate is required to react with 15.0 mL of oxalic acid, what is the molarity of the oxalic acid?

(c) What is the mass of manganese(IV) oxide formed?

**62.** Iodine reacts with thiosulfate ion, $S_2O_3^{2-}$ to give iodide ion and the tetrathionate ion, $S_4O_6^{2-}$.

(a) Write a balanced net ionic equation for the reaction.

(b) If 15.0 g of iodine is dissolved in enough water to make 0.750 L of solution, what volume of 0.187 $M$ sodium thiosulfate will be needed for complete reaction?

**63.** Consider the reaction between silver and nitric acid for which the unbalanced equation is

$$Ag(s) + H^+(aq) + NO_3^-(aq) \longrightarrow Ag^+(aq) + NO_2(g) + H_2O$$

(a) Balance the equation.

(b) If 35.00 mL of 12.0 $M$ nitric acid furnishes enough $H^+$ to react with silver, how many grams of silver react?

**64.** Zinc metal reacts with nitrate ion in basic solution. The unbalanced equation for the reaction is

$$OH^-(aq) + Zn(s) + H_2O + NO_3^-(aq) \longrightarrow$$
$$NH_3(aq) + Zn(OH)_4^{2-}(aq)$$

(a) Balance the equation.

(b) What volume of 0.250 $M$ barium hydroxide is required to react completely with 2.50 g of zinc?

### Chemical Analysis

**65.** Some hydrochloric acid is spilled on the lab floor. Sodium hydrogen carbonate is sprinkled on the spill to neutralize the acid. The balanced equation for the reaction that takes place is

$$NaHCO_3(s) + H^+(aq) \longrightarrow Na^+(aq) + CO_2(g) + H_2O$$

If 100.0 mL of 3.00 $M$ HCl was spilled, what is the minimum amount (in grams) of $NaHCO_3$ that must be sprinkled to neutralize all the acid in the spill?

**66.** Boric acid ($H_3BO_3$) can be used to neutralize bases. The equation for the reaction is

$$H_3BO_3(s) + 3OH^-(aq) \longrightarrow 3H_2O + BO_3^{3-}(aq)$$

What volume of 0.165 $M$ barium hydroxide can be neutralized by 6.69 g of boric acid?

**67.** The percentage of sodium hydrogen carbonate, $NaHCO_3$, in a powder for stomach upsets is found by titrating with 0.215 $M$ hydrochloric acid. If 18.7 mL of hydrochloric acid is required to react with 0.400 g of the sample, what is the percentage of sodium hydrogen carbonate in the sample? The reaction is given in Problem 65.

**68.** A vitamin C capsule is analyzed by titrating it with 0.250 $M$ sodium hydroxide. It is found that 10.3 mL of base is required to react with a capsule weighing 0.518 g. What is the percentage of vitamin C ($C_6H_8O_6$) in the capsule? (One mole of vitamin C reacts with one mole of hydroxide ion.)

**69.** Lactic acid, $C_3H_6O_3$, is the acid present in sour milk. A 0.100-g sample of pure lactic acid requires 12.95 mL of 0.0857 $M$ sodium hydroxide for complete reaction. How many moles of hydroxide ion are required to neutralize one mole of lactic acid?

**70.** An artificial fruit beverage contains 12.0 g of tartaric acid, $H_2C_4H_4O_6$, to provide tartness. It is titrated with a basic solution that has a density of 1.045 g/cm$^3$ and contains 5.00 mass % KOH. What volume of the basic solution is required? (One mole of tartaric acid reacts with two moles of hydroxide ion.)

**71.** Hair bleaching solutions contain hydrogen peroxide, $H_2O_2$. The hydrogen peroxide content can be determined by reacting $H_2O_2$ with a potassium dichromate acidic solution. The unbalanced equation for the reaction is

$$H_2O_2(aq) + Cr_2O_7^{2-}(aq) + H^+(aq) \longrightarrow$$
$$O_2(g) + Cr^{3+}(aq) + H_2O$$

A 15.0-g bleach solution needed 75.8 mL of 0.388 $M$ $K_2Cr_2O_7$ to react completely with the $H_2O_2$ in the hair bleach. What is the mass % of $H_2O_2$ in the bleach?

**72.** A wire weighing 0.250 g and containing 92.50% Fe is dissolved in HCl. The iron is completely oxidized to $Fe^{3+}$ by bromine water. The solution is then treated with tin(II) chloride to bring about the reaction

$$Sn^{2+}(aq) + 2Fe^{3+}(aq) \longrightarrow Sn^{4+}(aq) + 2Fe^{2+}(aq)$$

If 22.0 mL of tin(II) chloride solution is required for complete reaction, what is its molarity?

**73.** Laundry bleach is a solution of hypochlorous acid (HClO). To determine the HClO content of bleach, sulfide ion is added in basic solution. The balanced equation for the reaction is

$$HClO(aq) + S^{2-}(aq) \longrightarrow Cl^{-}(aq) + OH^{-}(aq) + S(s)$$

The chloride ion resulting from the reduction of HClO is precipitated as AgCl. When 25.0 mL of laundry bleach ($d = 1.02$ g/cm$^3$) is treated as described above, 3.49 g of AgCl are obtained. What is the mass % of HClO in the bleach?

**74.** Laws passed in some states define a drunk driver as one who drives with a blood alcohol level of 0.10% by mass or higher. The level of alcohol can be determined by titrating blood plasma with potassium dichromate according to the unbalanced equation

$$H^{+}(aq) + Cr_2O_7{}^{2-}(aq) + C_2H_5OH(aq) \longrightarrow$$
$$Cr^{3+}(aq) + CO_2(g) + H_2O$$

Assuming that the only substance that reacts with dichromate in blood plasma is alcohol, is a person legally drunk if 45.02 mL of 0.05000 $M$ potassium dichromate is required to titrate a 50.00 g sample of blood plasma?

## Unclassified

**\*75.** Classify each of the following as a precipitation, acid-base, or redox reaction.

   **(a)** the reaction between solutions of sulfuric acid and barium nitrate.
   **(b)** the reaction between solutions of sulfuric acid and calcium hydroxide.
   **(c)** the reaction of HCl with Al to evolve H$_2$.
   **(d)** the reaction of a solution of tin(II) chloride with air to form SnO$_2$.

**76.** Gold metal will dissolve only in *aqua regia,* a mixture of concentrated hydrochloric acid and concentrated nitric acid. The products of the reaction between gold and the concentrated acids are AuCl$_4{}^{-}$(aq), NO(g), and H$_2$O.

   **(a)** Write a balanced net ionic equation for the redox reaction, treating HCl and HNO$_3$ as strong acids.
   **(b)** What ratio of hydrochloric acid to nitric acid is the stoichiometric?
   **(c)** What volumes of 12 $M$ HCl and 16 $M$ HNO$_3$ are required to furnish the Cl$^-$ and NO$_3{}^-$ ions to react with 25.0 g of gold?

**77.** The iron content of hemoglobin is determined by destroying the hemoglobin molecule and producing small water-soluble ions and molecules. The iron in the aqueous solution is reduced to iron(II) ion and then titrated against potassium permanganate. In the titration, iron(II) is oxidized to iron(III), and permanganate is reduced to manganese(II) ion. A 5.00-g sample of hemoglobin requires 32.3 mL of a 0.002100 $M$ solution of potassium permanganate. What is the mass percent of iron in hemoglobin?

**78.** A sample of limestone weighing 0.145 g is dissolved in 50.00 mL of 0.100 $M$ hydrochloric acid. The following reaction occurs:

$$CaCO_3(s) + 2H^{+}(aq) \longrightarrow Ca^{2+}(aq) + CO_2(g) + H_2O$$

It is found that 13.05 mL of 0.175 $M$ NaOH is required to titrate the excess HCl left after reaction with the limestone. What is the mass percent of CaCO$_3$ in the limestone?

## Challenge Problems

**79.** Calcium in blood or urine can be determined by precipitation as calcium oxalate, CaC$_2$O$_4$. The precipitate is titrated with potassium permanganate in strong acid. The products of the reaction are carbon dioxide and manganese(II) ion. A 24-hour urine sample is collected from an adult patient, reduced to a small volume, and titrated with 26.2 mL of 0.0946 $M$ KMnO$_4$. How many grams of calcium oxalate are in the sample? Normal range for Ca$^{2+}$ output for an adult is 100 to 300 mg per 24 hours. Is the sample within the normal range?

**80.** Stomach acid is approximately 0.020 $M$ HCl. What volume of this acid is neutralized by an antacid tablet that weighs 330 mg and contains 41.0% Mg(OH)$_2$, 36.2% NaHCO$_3$, and 22.8% NaCl? The reactions involved are

$$Mg(OH)_2(s) + 2H^{+}(aq) \longrightarrow Mg^{2+}(aq) + 2H_2O$$
$$HCO_3{}^-(aq) + H^{+}(aq) \longrightarrow CO_2(g) + H_2O$$

**81.** Copper metal can reduce silver ions to metallic silver. The copper is oxidized to copper ions according to the reaction

$$2Ag^{+}(aq) + Cu(s) \longrightarrow Cu^{2+}(aq) + 2Ag(s)$$

A copper strip with a mass of 2.00 g is dipped into a solution of AgNO$_3$. After some time has elapsed, the copper strip is coated with silver. The strip is removed from the solution, dried, and weighed. The coated strip has a mass of 4.18 g. What are the masses of copper and silver metals in the strip? (Hint: Remember that the copper metal is being used up as silver metal forms.)

**82.** A solution contains both iron(II) and iron(III) ions. A 50.00-mL sample of the solution is titrated with 35.0 mL of 0.0280 $M$ KMnO$_4$, which oxidizes Fe$^{2+}$ to Fe$^{3+}$. The permanganate ion is reduced to manganese(II) ion. Another 50.00-mL sample of the solution is treated with zinc, which reduces all the Fe$^{3+}$ to Fe$^{2+}$. The resulting solution is again titrated with 0.0280 $M$ KMnO$_4$; this time 48.0 mL is required. What are the concentrations of Fe$^{2+}$ and Fe$^{3+}$ in the solution?

Winters are long and icy in New England (see p. 150)
(©1991 Ron Goulet/Dembinsky Photo Assoc.)

# Gases

5

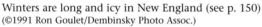

**R**eligious faith is a most filling vapor.

It swirls occluded in us under tight

Compression to uplift us out of weight—

As in those buoyant bird bones thin as paper,

To give them still more buoyancy in flight.

Some gas like helium must be innate.

—ROBERT FROST

*Innate Helium*

## CHAPTER OUTLINE

B y far the most familiar gas to all of us is the air we breathe. The Greeks considered air to be one of the four fundamental elements of nature, along with earth, water, and fire. Late in the eighteenth century, Cavendish, Priestley, and Lavoisier studied the composition of air, which is primarily a mixture of nitrogen and oxygen with smaller amounts of argon, carbon dioxide, and water vapor. Today, it appears that the concentrations of some of the minor components of the atmosphere may be changing, with adverse effects on the environment. The depletion of the ozone layer and increases in the amounts of "greenhouse" gases are topics for the evening news.

*Ozone is a vital but minor component of the upper atmosphere (Ch. 11)*

All gases resemble one another closely in their physical behavior. Their volumes respond in almost exactly the same way to changes in pressure, temperature, or amount of gas. In fact, it is possible to write a simple equation relating these four variables that is valid for all gases. This equation, known as the ideal gas law, is the central theme of this chapter; it is introduced in Section 5.2. The law is applied to

— pure gases in Section 5.3
— gases in chemical reactions in Section 5.4
— gas mixtures in Section 5.5

Section 5.6 considers the kinetic theory of gases, the molecular model upon which the ideal gas law is based. Finally, in Section 5.7, we describe the extent to which real gases deviate from the law.

## 5.1    Measurements on Gases

To completely describe the state of a gaseous substance, its volume, amount, temperature, and pressure are specified. The first three of these quantities were discussed in earlier chapters and will be reviewed briefly here. Pressure, a somewhat more abstract quantity, will be examined in more detail.

### Volume, Amount, and Temperature

A gas expands uniformly to fill any container in which it is placed. This means that the volume of a gas is the volume of its container. Volumes of gases can be expressed in liters, cubic centimeters, or cubic meters:

$$1 \, L = 10^3 \, cm^3 = 10^{-3} \, m^3$$

Most commonly, the amount of matter in a gaseous sample is expressed in terms of the number of moles *(n)*. In some cases, the mass *(m)* is given instead. These two quantities are related through the molar mass, $\mathcal{M}$

$$n = m/\mathcal{M}$$

The temperature of a gas is ordinarily measured using a thermometer marked in degrees Celsius. However, *in any calculation involving the physical behavior of gases, temperatures must be expressed on the Kelvin scale.* To convert between °C and K, use the relation introduced in Chapter 1:

$$T_K = t_{°C} + 273.15$$

Typically, in gas-law calculations, temperatures are expressed only to the nearest degree. In that case, the Kelvin temperature can be found by simply adding 273 to the Celsius temperature.

## Pressure

Pressure is defined as force per unit area. The SI unit of pressure (Chap. 1) is the *pascal* (Pa), the pressure exerted by a force of one newton on an area of one square meter. Atmospheric pressure is about $10^5$ Pa, or 100 *kilopascals* (kPa).

A device commonly used to measure atmospheric pressure is the mercury barometer, first constructed by Evangelista Torricelli in the seventeenth century. This consists of a closed glass tube filled with mercury and inverted over a pool of mercury. The pressure exerted by the mercury column exactly equals that of the atmosphere. Hence the height of the column is a measure of the atmospheric pressure. At or near sea level, it typically varies from 740 to 760 mm, depending upon weather conditions.

Because of the way in which gas pressure is measured, it is often expressed in **millimeters of mercury (mm Hg)**\*. Thus, we might say that the atmospheric pressure on a certain day is 752 mm Hg. This means that the pressure of the air is equal to that exerted by a column of mercury 752 mm high.

Another unit commonly used to express gas pressure is the standard atmosphere, or simply **atmosphere (atm)**. This is the pressure exerted by a column of mercury 760 mm high with the mercury at 0°C. If we say that a gas has a pressure of 0.98 atm, we mean that the pressure is 98% of that exerted by a mercury column 760 mm high.

A representation of Torricelli's barometer. The height of the mercury column gives the atmospheric pressure.

---

**Example 5.1**    A balloon with a volume of $2.36 \times 10^4$ m³ contains $4.68 \times 10^6$ g of helium at 18°C and 120.0 kPa. Express the volume of the balloon in liters, the amount in moles, the temperature in K, and the pressure in both atmospheres and millimeters of mercury.

*Strategy*    Use the following conversion factors:

$$\frac{1\,L}{10^{-3}\,m^3} \qquad \frac{1\,mole\ He}{4.003\,g\ He} \qquad \frac{760\,mm\ Hg}{101.3\,kPa} \qquad \frac{1\,atm}{101.3\,kPa}$$

For the temperature conversion, use the relation: $T_K = t_{°C} + 273$

---

\*The pressure exerted by a column of mercury depends upon its density, which varies slightly with temperature. To get around this ambiguity, the *torr* was defined to be the pressure exerted by 1 mm of mercury at certain specified conditions, notably 0°C. Over time, the unit torr has become a synonym for millimeter of mercury. Throughout this text, we will use millimeter of mercury rather than torr because the former has a clearer physical meaning.

In an aneroid barometer, a change in air pressure moves a diaphragm in or out. This in turn moves a needle across a scale. (Runk/Schoenberger/Grant Heilman, Inc.)

*Solution*

$$V = 2.36 \times 10^4 \, \text{m}^3 \times \frac{1 \, \text{L}}{10^{-3} \, \text{m}^3} = \boxed{2.36 \times 10^7 \, \text{L}}$$

$$n_{\text{He}} = 4.68 \times 10^6 \, \text{g He} \times \frac{1 \, \text{mol He}}{4.003 \, \text{g He}} = \boxed{1.17 \times 10^6 \, \text{mol He}}$$

$$T = 18 + 273 = \boxed{291 \, \text{K}}$$

$$P = 120.0 \, \text{kPa} \times \frac{1 \, \text{atm}}{101.3 \, \text{kPa}} = \boxed{1.185 \, \text{atm}}$$

$$P = 120.0 \, \text{kPa} \times \frac{760 \, \text{mm Hg}}{101.3 \, \text{kPa}} = \boxed{900.3 \, \text{mm Hg}}$$

## 5.2   The Ideal Gas Law

All gases closely resemble each other in the dependence of volume on amount, temperature, and pressure.

**1. Volume is directly proportional to amount.** Figure 5.1a shows a typical plot of volume ($V$) versus number of moles ($n$) for a gas. Notice that the graph is a straight line passing through the origin. The general equation for such a plot is

$$V = k_1 n \quad \text{(constant } T, P\text{)}$$

where $k_1$ is a "constant"; that is, it is independent of individual values of $V$ and $n$ and of the nature of the gas. This is the equation of a direct proportionality.

**2. Volume is directly proportional to absolute temperature.** The dependence of volume ($V$) on the Kelvin temperature ($T$) is shown in Figure 5.1b. Here again, the graph is a straight line through the origin. The equation of the line is

$$V = k_2 T \quad \text{(constant } n, P\text{)}$$

where $k_2$ is a constant. This relationship was first suggested, in a different form, by two French scientists, Jacques Charles (1746–1823) and Joseph Gay-Lussac (1778–1850), both of whom were balloonists.

Gay Lussac isolated boron and was the first to prepare HF

**Figure 5.1**

At constant pressure, the volume of a gas is directly proportional to the number of moles (a) and to the absolute temperature (b).

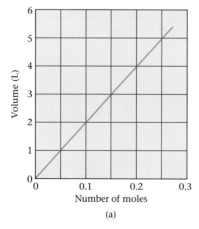

(a)

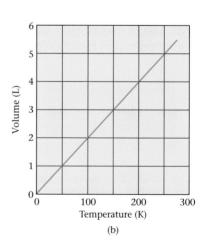

(b)

An illustration of Charles' law. When the air-filled balloons are placed in liquid nitrogen ($T = 77K$), the volume of the air decreases. When the balloons are removed from the liquid nitrogen, they reinflate to their original volume as the air in the balloons returns to room temperature.
(Charles D. Winters)

**3. *Volume is inversely proportional to pressure.*** Figure 5.2 shows a typical plot of volume (V) versus pressure *(P)*. Notice that *V* decreases as *P* increases. The graph is a hyperbola. The general relation between the two variables is

$$V = k_3/P \quad \text{(constant } n, T)$$

The quantity $k_3$, like $k_1$ and $k_2$, is a constant. This is the equation of an inverse proportionality. The fact that volume is inversely proportional to pressure was first established in 1660 by Robert Boyle, an Irish experimental scientist. The equation above is one form of *Boyle's Law*.

The three equations relating the volume, pressure, temperature, and amount of a gas can be combined into a single equation. Since *V* is directly proportional to both *n* and *T*

$$V = k_1 n \qquad V = k_2 T$$

and inversely proportional to *P*

$$V = \frac{k_3}{P}$$

it follows that

$$V = \text{constant} \times \frac{n \times T}{P}$$

Ordinarily, the constant is represented by the symbol *R*. Both sides of the equation are multiplied by *P* to give the **ideal gas law**

$$PV = nRT$$

where *P* is the pressure, *V* the volume, *n* the number of moles, and *T* the Kelvin temperature. The quantity *R* appearing in the ideal gas law is a true constant, independent of *P, V, n, T,* or the identity of the gas. Experimentally, it is found that the ideal gas law predicts remarkably well the experimental behavior of real gases (e.g., $H_2$, $N_2$, $O_2$, . . .) at ordinary temperatures and pressures.

The value of the gas constant *R* can be calculated from experimental values of *P, V, n,* and *T.* Consider, for example, the situation that applies at 0°C and 1 atm. These conditions are often referred to as *standard temperature and pressure* for a gas **(STP).** At STP, one mole of any gas occupies a volume of 22.4 L. Solving the ideal gas law for *R,*

$$R = \frac{PV}{nT}$$

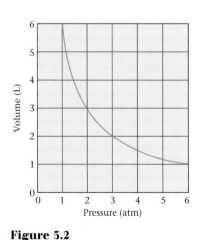

**Figure 5.2**

At constant temperature, the volume of a gas sample is inversely proportional to pressure. In this case, the volume decreases from 6 L to 1 L when the pressure increases from 1 atm to 6 atm.

TABLE 5.1 **Values of $R$ in Different Units**

| Value | Where Used | How Obtained |
|---|---|---|
| $0.0821\dfrac{\text{L} \cdot \text{atm}}{\text{mol} \cdot \text{K}}$ | Gas law problems with $V$ in liters, $P$ in atm | From known values of $P, V, T, n$ |
| $8.31\dfrac{\text{J}}{\text{mol} \cdot \text{K}}$ | Equations involving energy in joules | $1\,\text{L} \cdot \text{atm} = 101.3\,\text{J}$ |
| $8.31 \times 10^3 \dfrac{\text{g} \cdot \text{m}^2}{\text{s}^2 \cdot \text{mol} \cdot \text{K}}$ | Calculation of molecular velocity (p. 125) | $1\,\text{J} = 10^3\,\dfrac{\text{g} \cdot \text{m}^2}{\text{s}^2}$ |

Substituting $P = 1.00$ atm, $V = 22.4$ L, $n = 1.00$ mol, and $T = 0 + 273 = 273$ K,

$$R = \frac{1.00\ \text{atm} \times 22.4\ \text{L}}{1.00\ \text{mol} \times 273\ \text{K}} = 0.0821\ \text{L} \cdot \text{atm}/(\text{mol} \cdot \text{K})$$

The value of $R$ obtained under the most precise conditions at low pressures is $0.082058\ \text{L} \cdot \text{atm}/(\text{mol} \cdot \text{K})$. Note that $R$ has the units of atmospheres, liters, moles, and K. These units must be used for pressure, volume, amount, and temperature in any problem in which this value of $R$ is used.

R is a constant, independent of $P$, $V$, $n$, and $T$, but its numerical value depends upon the units used

Throughout most of this chapter, we will use $0.0821\ \text{L} \cdot \text{atm}/(\text{mol} \cdot \text{K})$ as the value of $R$. For certain purposes, however, $R$ must be expressed in different units (Table 5.1).

## 5.3  Gas Law Calculations

The ideal gas law can be used to solve a variety of problems. We will show how you can use it to find

— the final state of a gas, knowing its initial state and the changes in $P$, $V$, $n$, or $T$ that occur
— one of the four variables, $P$, $V$, $n$, or $T$, given the values of the other three
— the molar mass or density of a gas

### Final and Initial State Problems

Commonly, a gas undergoes a change from an "initial" to a "final" state. Typically, you are asked to determine the effect upon $V$, $P$, $n$, or $T$ of a change in one or more of these variables. For example, starting with a sample of gas at 25°C and 1.00 atm, you might be asked to calculate the pressure developed upon heating to 95°C at constant volume.

The ideal gas law is readily applied to problems of this type. A relationship between the variables involved is derived from this law. In this case, pressure and temperature change, while $n$ and $V$ remain constant.

$$\text{initial state:} \qquad P_1 V = nRT_1$$

$$\text{final state:} \qquad P_2 V = nRT_2$$

To obtain a two-point equation, write the gas law twice and divide to eliminate constants

Dividing the second equation by the first cancels $V$, $n$, and $R$, leaving the relation

$$P_2/P_1 = T_2/T_1 \qquad (\text{constant } n, V)$$

Applying this general relation to the problem described above,

$$P_2 = P_1 \times \frac{T_2}{T_1} = 1.00 \text{ atm} \times \frac{368 \text{ K}}{298 \text{ K}} = 1.23 \text{ atm}$$

Similar "two-point" equations can be derived from the ideal gas law to solve any problem of this type.

---

**Example 5.2**  A 250-mL flask, open to the atmosphere, contains 0.0110 mol of air at 0°C. Upon heating, part of the air escapes; how much remains in the flask at 100°C?

**Strategy**  The key to working this problem is to realize that *pressure and volume remain constant*. The pressure is that of the atmosphere; the volume is that of the flask, 250 mL. Look for a "two-point" relation between $n$ and $T$ at constant $P$ and $V$. Then substitute for $n_1$, $T_1$ and $T_2$; solve for $n_2$.

**Solution**

$$\text{Initial state:} \quad n_1 T_1 = PV/R$$

$$\text{Final state:} \quad n_2 T_2 = PV/R$$

It follows that

$$n_2 T_2 = n_1 T_1; \quad n_2 = n_1 \times \frac{T_1}{T_2}$$

Since $n_1 = 0.0110$ mol; $T_1 = 0 + 273 = 273$ K; and $T_2 = 100 + 273 = 373$ K,

$$n_2 = 0.0110 \text{ mol} \times \frac{273 \text{ K}}{373 \text{ K}} = \boxed{0.00805 \text{ mol}}$$

We conclude that 0.00805/0.0110, or about 73%, of the air remains in the flask at 100°C. About 27% of the air escapes when the temperature is raised.

---

## Calculation of *P, V, n,* or *T*

Frequently, values are known for three of these quantities (perhaps $V$, $n$, and $T$); the other one ($P$) must be calculated. This is readily done by direct substitution into the ideal gas law.

---

**Example 5.3**  The "ozone-friendly" compound now used as a refrigerant in car air conditioners has the molecular formula $C_2F_4H_2$. If 2.50 g of this compound is introduced into an evacuated 500.0-mL container at 10°C, what pressure in atmospheres is developed?

Freon, $CF_2Cl_2$, used to be used

**Strategy**  Substitute directly into the ideal gas law and solve for $P$. Note that $V$, $n$, and $T$ have to be in units consistent with $R = 0.0821$ L · atm/(mol · K).

**Solution**  Converting to the appropriate units,

$$V = 500.0 \text{ mL} \times \frac{1 \text{ L}}{1000 \text{ mL}} = 0.5000 \text{ L}$$

$$T = 10 + 273 = 283 \text{ K}$$

The molar mass of $C_2F_4H_2$ is 102.04 g/mol. Hence

$$n = 2.50 \text{ g } C_2F_4H_2 \times \frac{1 \text{ mol } C_2F_4H_2}{102.04 \text{ g } C_2F_4H_2} = 0.0245 \text{ mol}$$

Substituting into the ideal gas law:

$$P = \frac{nRT}{V} = \frac{0.0245 \text{ mol} \times 0.0821 \text{ L} \cdot \text{atm/(mol} \cdot \text{K)} \times 283 \text{ K}}{0.5000 \text{ L}} = 1.14 \text{ atm}$$

## Molar Mass and Density

The ideal gas law offers a simple approach to the experimental determination of the molar mass of a gas. Indeed, this approach can be applied to volatile liquids like acetone (Example 5.4). All you need to know is the mass of a sample confined to a container of fixed volume at a particular temperature and pressure.

**Example 5.4**   Acetone is widely used as a nail polish remover. A sample of liquid acetone is placed in a 3.00-L flask and vaporized by heating to 95°C at 1.02 atm. The vapor filling the flask at this temperature and pressure weighs 5.87 g. Calculate the molar mass of acetone.

***Strategy***   Perhaps the simplest approach here is to derive from the ideal gas law an expression for the molar mass, $\mathcal{M}$. This is readily done by using the relation $n = m/\mathcal{M}$, where $m$ is the mass in grams.

$$PV = \frac{mRT}{\mathcal{M}}$$

***Solution***   Solving for $\mathcal{M}$,

$$\mathcal{M} = \frac{mRT}{PV}$$

All the quantities required to calculate $\mathcal{M}$ are given in the statement of the problem.

$$\mathcal{M} = \frac{5.87 \text{ g} \times 0.0821 \text{ L} \cdot \text{atm/(mol} \cdot \text{K)} \times 368 \text{ K}}{3.00 \text{ L} \times 1.02 \text{ atm}} = \boxed{58.0 \text{ g/mol}}$$

The density of a gas is dependent upon

— *pressure.* Compressing a gas increases its density by reducing its volume ($d = m/V$).
— *temperature.* Hot air rises because a gas becomes less dense when its temperature is increased.
— *molar mass.* Hydrogen ($\mathcal{M} = 2.016$ g/mol) has the lowest molar mass and the lowest density (at a given $T$ and $P$) of all gases.

The general expression for density is $d = \mathcal{M}P/RT$

The ideal gas law can be used to derive a relation for gas density consistent with these observations (Example 5.5).

**Example 5.5**   Taking the molar mass of dry air to be 29.0 g/mol, calculate the density of air at 27°C and 1 atm.

***Strategy***   Use the ideal gas law to derive a relation between density (*d*), molar mass (*M*), pressure (*P*), and temperature (*T*). It may be simplest to start with the relation derived in Example 5.4:

$$\mathcal{M} = \frac{mRT}{PV}$$

Substituting *d* for *m/V*

$$\mathcal{M} = \frac{dRT}{P}$$

Solving for density

$$d = \frac{\mathcal{M}P}{RT}$$

***Solution***

$$d = \frac{\mathcal{M}P}{RT} = \frac{29.0 \text{ g/mol} \times 1.00 \text{ atm}}{0.0821 \text{ L} \cdot \text{atm/(mol} \cdot \text{K)} \times 300 \text{ K}} = \boxed{1.18 \text{ g/L}}$$

As promised, the density equation shows that gas density increases with increasing molar mass and decreases with increasing temperature. Balloons, which must contain a gas less dense than the surrounding air, take advantage of one or the other of these effects. "Hot air" balloons are filled with air at a temperature higher than that of the atmosphere, making the air inside less dense than that outside. First used in France in the eighteenth century, they are now seen in balloon races and other sporting events. Heat is supplied on demand by a propane burner.

The other type of balloon uses a gas with a molar mass less than that of air. Hydrogen (*M* = 2.016 g/mol) has the greatest lifting power, because it has the lowest density of all gases. However, it has not been used in manned balloons since 1937, when the *Hindenberg,* a hydrogen-filled airship, exploded and burned. Helium (*M* = 4.003 g/mol) is slightly less effective than hydrogen but a lot safer to work with, since it is nonflammable. It is used in a variety of balloons, ranging from the small ones used at parties to meteorological balloons with volumes of a billion liters.

## 5.4  Stoichiometry of Gaseous Reactions

As pointed out in Chapter 3, a balanced equation can be used to relate moles or grams of substances taking part in a reaction. Where gases are involved, these relations can be extended to include volumes. To do this, we use the ideal gas law and the conversion factor approach described in Chapter 3.

*This is the most important application of the ideal gas law to chemistry*

---

**Example 5.6**   Hydrogen peroxide is the active ingredient in commercial preparations for bleaching hair. What mass of hydrogen peroxide must be used to produce 1.00 L of oxygen gas at 25°C and 1.00 atm? The equation for the reaction is

$$2H_2O_2(aq) \longrightarrow O_2(g) + 2H_2O$$

***Strategy***   First convert volume of $O_2(g)$ to moles, using the ideal gas law. Then find the mass of $H_2O_2$ by the conversion-factor approach described in Chapter 3; the path is $n_{O_2} \rightarrow n_{H_2O_2} \rightarrow m_{H_2O_2}$

***Solution***

(1) $n_{O_2} = \dfrac{PV}{RT} = \dfrac{1.00 \text{ atm} \times 1.00 \text{ L}}{0.0821 \text{ L} \cdot \text{atm/(mol} \cdot \text{K)} \times 298 \text{ K}} = 0.0409 \text{ mol } O_2$

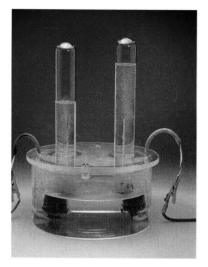

**Figure 5.3**
When water is electrolyzed, the volume of hydrogen gas formed in the tube at the left is twice that of oxygen (right tube), in accordance with the equation $2H_2O(l) \longrightarrow 2H_2(g) + O_2(g)$. (Charles D. Winters)

(2)  mass of $H_2O_2$ = 0.0409 mol $O_2 \times \dfrac{2 \text{ mol } H_2O_2}{1 \text{ mol } O_2} \times \dfrac{34.02 \text{ g } H_2O_2}{1 \text{ mol } H_2O_2}$ = $\boxed{2.78 \text{ g } H_2O_2}$

---

**Example 5.7**    Octane, $C_8H_{18}$, is one of the hydrocarbons in gasoline. Upon combustion (burning in oxygen), octane produces carbon dioxide and water. How many liters of oxygen, measured at 0.974 atm and 24°C, are required to burn 1.00 g of octane?

***Strategy***    First write a balanced equation for the reaction. Then calculate the number of moles of oxygen required to burn 1.00 g of $C_8H_{18}$ ($\mathcal{M}$ = 114.22 g/mol). Finally, use the ideal gas law to find the volume of oxygen.

*Solution*

(1) The balanced equation for the reaction is:

$$2C_8H_{18}(l) + 25\ O_2(g) \longrightarrow 16CO_2(g) + 18H_2O(l)$$

(2)  $n_{O_2}$ = 1.00 g $C_8H_{18} \times \dfrac{1 \text{ mol } C_8H_{18}}{114.22 \text{ g } C_8H_{18}} \times \dfrac{25 \text{ mol } O_2}{2 \text{ mol } C_8H_{18}}$ = 0.109 mol $O_2$

(3)  $V_{O_2} = \dfrac{nRT}{P} = \dfrac{0.109 \text{ mol} \times 0.0821 \text{ L} \cdot \text{atm/(mol} \cdot \text{K)} \times 297 \text{ K}}{0.974 \text{ atm}}$ = $\boxed{2.73 \text{ L}}$

Since air is only 21% $O_2$ by volume, a large volume of air must pass through an automobile engine during combustion.

---

Perhaps the first stoichiometric relationship to be discovered was the **law of combining volumes,** proposed by Gay-Lussac in 1808: ***The volume ratio of any two gases in a reaction at constant temperature and pressure is the same as the reacting mole ratio.***

To illustrate the law, consider the reaction

$$2H_2O(l) \longrightarrow 2H_2(g) + O_2(g)$$

Similarly, 2 L of $H_2$ reacts with 1 L of $O_2$ to form water

As you can see from Figure 5.3, the volume of hydrogen produced is twice that of the other gaseous product, oxygen.

The law of combining volumes, like so many relationships involving gases, is readily explained by the ideal gas law. At constant temperature and pressure, volume is directly proportional to number of moles ($V = k_1n$). It follows that for gaseous species involved in reactions, the volume ratio must be the same as the mole ratio given by the coefficients of the balanced equation.

# Chemistry: *The Human Side*

The equation relating the volume of a gas to the number of moles:

$$V = k_1n$$

can be interpreted to mean that equal volumes of different gases (constant $T$ and $P$) contain equal numbers of moles. This relationship is a modern form of Avogadro's law, proposed in 1811 by an Italian physicist at the University of Turin with the improbable name of Lorenzo Romano Amadeo Carlo Avogadro di Quarequa e di Cerreto. Avogadro (1776–1856) suggested that ***equal volumes of all gases at the same temperature and pressure contain the same number of molecules.***

Avogadro suggested this relationship to explain the law of combining volumes. Today it seems obvious that, since in the reaction

$$2H_2O(l) \longrightarrow 2H_2(g) + O_2(g)$$

the volume of hydrogen, like the number of moles, is twice that of oxygen, equal volumes of these gases must contain the same number of moles or molecules. This was by no means obvious to Avogadro's contemporaries. Berzelius, among others, dismissed Avogadro's ideas because he did not believe diatomic molecules composed of identical atoms ($H_2$, $O_2$) could exist. Dalton went a step further; he refused to accept the law of combining volumes because he thought it implied splitting atoms.

As a result of arguments like these, Avogadro's ideas lay dormant for nearly half a century. They were revived by a fellow countryman, Stanislao Cannizzaro, professor of chemistry at the University of Genoa. At a conference held in Karlsruhe in 1860, he persuaded the chemistry community of the validity of Avogadro's law and showed how it could be used to determine molar and atomic masses.

The quantity now called "Avogadro's number" ($6.02 \times 10^{23}$/mol) was first estimated in 1865, nine years after Avogadro died. Not until well into the twentieth century did it acquire its present name. It seems appropriate to honor Avogadro in this way for the contributions he made to chemical theory.

**Amadeo Avogadro**
(1776–1856)
Courtesy of E. F. Smith Memorial
Collection, Van Pelt Library,
University of Pennsylvania

## 5.5 Gas Mixtures: Partial Pressures and Mole Fractions

The ideal gas law applies not only to pure gases but also to gas mixtures. For a mixture of two gases A and B, the total pressure is given by the expression

$$P_{tot} = n_{tot}\frac{RT}{V} = (n_A + n_B)\frac{RT}{V}$$

Separating the two terms on the right,

$$P_{tot} = n_A\frac{RT}{V} + n_B\frac{RT}{V}$$

The terms $n_ART/V$ and $n_BRT/V$ are, according to the ideal gas law, the pressures that gases A and B would exert if they were alone. These quantities are referred to as **partial pressures,** $P_A$ and $P_B$.

$$P_A = \text{partial pressure A} = n_ART/V$$

$$P_B = \text{partial pressure B} = n_BRT/V$$

Substituting for $n_ART/V$ and $n_BRT/V$ in the equation for $P_{tot}$,

$$P_{tot} = P_A + P_B$$

> Partial pressure is the pressure a gas would exert if it occupied the entire volume by itself

The relation just derived was first proposed by John Dalton in 1801; it is often referred to as **Dalton's law** of partial pressures:

*The total pressure of a gas mixture is the sum of the partial pressures of the components of the mixture.*

To illustrate Dalton's law, consider a gaseous mixture of hydrogen and helium in which

$$P_{H_2} = 2.46 \text{ atm} \qquad P_{He} = 3.69 \text{ atm}$$

It follows from Dalton's law that

$$P_{tot} = 2.46 \text{ atm} + 3.69 \text{ atm} = 6.15 \text{ atm}$$

**Figure 5.4**

When a gas is collected by displacing water, it becomes saturated with water vapor; the partial pressure of $H_2O(g)$ is equal to the vapor pressure of liquid water at the temperature of the system. (Marna G. Clarke)

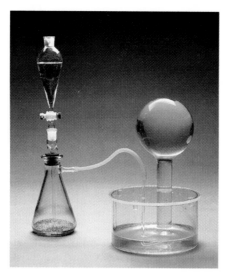 

## Wet Gases; Partial Pressure of Water

When a gas such as hydrogen is collected by bubbling through water (Fig. 5.4), it picks up water vapor; molecules of water escape from the liquid and are carried along with the gas. Dalton's law can be applied to the resulting gas mixture:

$$P_{tot} = P_{H_2O} + P_{H_2}$$

Vapor pressure is further discussed in Chapter 9

In this case, $P_{tot}$ is the measured pressure. The partial pressure of water vapor, $P_{H_2O}$, is equal to the **vapor pressure** of liquid water. It has a fixed value at a given temperature (see Appendix 1). The partial pressure of hydrogen, $P_{H_2}$, can be calculated by subtraction. The number of moles of hydrogen in the wet gas, $n_{H_2}$, can then be determined using the ideal gas law.

---

**Example 5.8**    A student prepares a sample of hydrogen gas by electrolyzing water at 25°C. She collects 152 mL of $H_2$ at a total pressure of 758 mm Hg. Using Appendix 1 to find the vapor pressure of water, calculate

    (a) the partial pressure of hydrogen.
    (b) the number of moles of hydrogen collected.

### Strategy

    (a) Use Dalton's law to find the partial pressure of hydrogen, $P_{H_2}$.
    (b) Use the ideal gas law to calculate $n_{H_2}$, with $P_{H_2}$ as the pressure.

### Solution

    (a) From Appendix 1, $P_{H_2O} = 23.76$ mm Hg at 25°C. The total pressure, $P_{tot}$, is 758 mm Hg.

$$P_{H_2} = P_{tot} - P_{H_2O} = 758 \text{ mm Hg} - 23.76 \text{ mm Hg} = \boxed{734 \text{ mm Hg}}$$

    (b) $n_{H_2} = \dfrac{(P_{H_2})V}{RT} = \dfrac{(734/760 \text{ atm})(0.152 \text{ L})}{(0.0821 \text{ L} \cdot \text{atm/mol} \cdot \text{K})(298 \text{ K})} = \boxed{0.00600 \text{ mol } H_2}$

---

## Partial Pressure and Mole Fraction

As pointed out earlier, the following relationship applies to a mixture containing gas A (and gas B):

$$P_A = \frac{n_A RT}{V} \qquad P_{tot} = \frac{n_{tot} RT}{V}$$

Dividing $P_A$ by $P_{tot}$ gives

$$\frac{P_A}{P_{tot}} = \frac{n_A}{n_{tot}}$$

The fraction $n_A/n_{tot}$ is referred to as the **mole fraction** of A in the mixture. It is the fraction of the total number of moles that is accounted for by gas A. Using $X_A$ to represent the mole fraction of A (i.e., $X_A = n_A/n_{tot}$),

$$P_A = X_A P_{tot}$$

In other words, ***the partial pressure of a gas in a mixture is equal to its mole fraction multiplied by the total pressure.*** This relation is commonly used to calculate partial pressures of gases in a mixture when the total pressure and the composition of the mixture are known (Example 5.9).

If a mixture contains equal numbers of A and B molecules, $X_A = X_B = 0.50$ and $P_A = P_B = \frac{1}{2}P_{tot}$

---

**Example 5.9**   Chemical analysis of dry air shows that the mole fractions of nitrogen, oxygen, and argon are 0.781, 0.210, and 0.009, respectively. Calculate the partial pressure of each of these gases on a day when the barometric pressure is 747 mm Hg.

**Strategy**   The barometric pressure is the total pressure; use the equation $P_A = X_A P_{tot}$ to find the partial pressures.

*Solution*

$$P_{N_2} = 0.781 \times 747 \text{ mm Hg} = \boxed{583 \text{ mm Hg}}$$

$$P_{O_2} = 0.210 \times 747 \text{ mm Hg} = \boxed{157 \text{ mm Hg}}$$

$$P_{Ar} = 0.009 \times 747 \text{ mm Hg} = \boxed{7 \text{ mm Hg}}$$

The partial pressures add to the total pressure, 747 mm Hg, as they should by Dalton's law.

---

## 5.6   Kinetic Theory of Gases

The fact that the ideal gas law applies to all gases indicates that the gaseous state is a relatively simple one from a molecular standpoint. Gases must have certain common properties that cause them to follow the same natural law. Between about 1850 and 1880, James Maxwell, Rudolf Clausius, Ludwig Boltzmann, and others developed the **kinetic theory** of gases. They based it on the idea that all gases behave similarly so far as particle motion is concerned. Since that time, the kinetic theory has had to be modified only slightly. It is one of the most successful scientific theories, ranking in stature with the atomic theory of matter.

## Postulates of the Kinetic Theory

The kinetic theory of gases is based on several assumptions, including the following.

1. Gases consist of atoms or molecules in continuous, random motion. These particles undergo frequent collisions with each other and with the walls of their container. Gas pressure is caused by collisions with the walls.

In air, a molecule undergoes about 10 billion collisions per second

2. Collisions between gas particles are elastic; there is no change in total energy when a collision occurs. No kinetic energy is converted to heat, which explains why the temperature of an insulated gas does not change with time.

3. The volume occupied by gas particles is negligibly small compared with that of their container. This assumption, like (4) below, is valid at ordinary temperatures and pressures, where a gas is mostly "empty space."

4. Attractive forces between particles have a negligible effect on their behavior. The atoms or molecules in a gas can be treated as independent particles.

The two most important postulates for our purposes are:

A baseball in motion has translational energy

5. *The average translational kinetic energy, $E_t$, of a gas particle is directly proportional to the absolute temperature.* That is,

$$E_t = cT$$

where $c$ is a constant.

6. *At a given temperature, all gases have the same average translational kinetic energy.* In other words, the quantity $c$ in the above equation is a universal constant, which has the same value for all gases.

## Average Speeds of Gas Particles

A basic law of physics states that the translational kinetic energy (energy of motion in a straight line) of a particle is one half of the product of the particle's mass times the square of its speed. Applied to gas particles, this means that

$$E_t = mu^2/2$$

where $E_t$ is the average translational energy and $u$ is the corresponding speed, which we will call the average speed.* Combining this relation with Postulate 5 of the kinetic theory, $E_t = cT$, it follows that

$$mu^2/2 = cT$$

Solving this equation for $u$:

$$u = \left(\frac{2cT}{m}\right)^{1/2}$$

where $c$ is a universal constant, with the same value for all gases (Postulate 6).

The constant $c$ in this equation can be evaluated from kinetic theory; it turns out to be

$$c = 3R/2N_A$$

*More rigorously, $u^2$ is the average of the squares of the speeds of all molecules.

where $R$ is the gas constant in the proper units and $N_A$ is Avogadro's number. Substituting for $c$ in the expression for $u$ and realizing that the product $N_A \times m$ is simply the molar mass, $\mathcal{M}$, we obtain:

$$u = \left(\frac{3RT}{\mathcal{M}}\right)^{1/2}$$

$E_t$ depends only upon T; u depends on both T and $\mathcal{M}$

From the equation just written, you can see that the average speed, $u$, is

— *directly proportional to the square root of the absolute temperature.* For a given gas at two different temperatures, $T_2$ and $T_1$, the quantity $\mathcal{M}$ is constant, and we can write

$$\frac{u_2}{u_1} = \left(\frac{T_2}{T_1}\right)^{1/2}$$

— *inversely proportional to the square root of molar mass ($\mathcal{M}$).* For two different gases A and B at the same temperature ($T$ constant):

$$\frac{u_B}{u_A} = \left(\frac{\mathcal{M}_A}{\mathcal{M}_B}\right)^{1/2}$$

---

**Example 5.10**    Calculate the average velocity, $u$, of an $N_2$ molecule at 25°C.

**Strategy**    Use the equation $u = (3RT/\mathcal{M})^{1/2}$; remember to use the proper value of $R = 8.31 \times 10^3$ g · m²/(s² · mol · K). Be careful about units!

**Solution**

$$u = \left(\frac{3RT}{\mathcal{M}}\right)^{1/2} = \left[\frac{3 \times 8.31 \times 10^3 \, \dfrac{\text{g} \cdot \text{m}^2}{\text{s}^2 \cdot \text{mol} \cdot \text{K}} \times 298 \text{ K}}{28.02 \, \dfrac{\text{g}}{\text{mol}}}\right]^{1/2} = 515 \text{ m/s}$$

Notice that

— all of the units within the large brackets cancel except m²/s², leading to the units m/s for $u$.
— the average velocity is very high. In miles per hour, it is

$$515 \, \frac{\text{m}}{\text{s}} \times \frac{1 \text{ mi}}{1.609 \times 10^3 \text{ m}} \times 3.600 \times 10^3 \, \frac{\text{s}}{\text{hr}} = 1.15 \times 10^3 \text{ mi/hr}$$

---

## Effusion of Gases; Graham's Law

One way to check the validity of calculations made from kinetic theory is to study the process of **effusion,** the flow of gas particles through tiny pores or pinholes. The relative rates of effusion of different gases depend upon two factors: the pressures of the gases and the relative speeds of their particles. If two different gases A and B are compared at the same pressure, only their speeds are of concern, and

$$\frac{\text{rate of effusion B}}{\text{rate of effusion A}} = \frac{u_B}{u_A}$$

where $u_A$ and $u_B$ are average speeds. As pointed out earlier, at a given temperature

**Figure 5.5**
When ammonia gas, injected into the left arm of the U-tube, comes in contact with hydrogen chloride (right arm), they react to form a white deposit of ammonium chloride: $NH_3(g) + HCl(g) \longrightarrow NH_4Cl(s)$. Since $NH_3$ ($\mathcal{M} = 17$ g/mol) effuses faster than HCl ($\mathcal{M} = 36.5$ g/mol), the deposit forms closer to the HCl end of the tube. (Marna G. Clarke)

$$\frac{u_B}{u_A} = \left(\frac{\mathcal{M}_A}{\mathcal{M}_B}\right)^{1/2}$$

It follows that, at constant pressure and temperature,

$$\frac{\text{rate of effusion B}}{\text{rate of effusion A}} = \left(\frac{\mathcal{M}_A}{\mathcal{M}_B}\right)^{1/2}$$

This relation in a somewhat different form was discovered experimentally by the Scottish chemist Thomas Graham in 1829. Graham was interested in a wide variety of chemical and physical problems, among them the separation of the components of air. Graham's law can be stated as

***At a given temperature and pressure, the rate of effusion of a gas is inversely proportional to the square root of its molar mass.***

Graham's law gives a way of determining molar masses of gases. All that needs to be done is to compare the rate of effusion of the gas in question with that of another gas of known molar mass. Usually, either the distances moved by the two gases in equal times (Fig. 5.5) or the times required to effuse are measured. Since time is inversely related to rate,

$$\text{rate} = \frac{\text{distance}}{\text{time}}$$

it follows that

If you drive 60 mph instead of 30 mph, you (probably) get there in half the time

$$\frac{\text{time A}}{\text{time B}} = \left(\frac{\mathcal{M}_A}{\mathcal{M}_B}\right)^{1/2}$$

In other words, the time required for effusion increases with molar mass; heavy molecules take longer to effuse.

---

**Example 5.11**    In an effusion experiment, 45 s was required for a certain number of moles of an unknown gas X to pass through a small opening into a vacuum. Under the same conditions, it took 28 s for the same number of moles of Ar to effuse. Find the molar mass of the unknown gas.

*Strategy*    Use the relation between time of effusion and molar mass.

*Solution*

$$\frac{\text{time for Ar}}{\text{time for X}} = \left(\frac{\text{molar mass of Ar}}{\text{molar mass of X}}\right)^{1/2}$$

Substituting numbers into this equation and squaring both sides gives

$$\left(\frac{28 \text{ s}}{45 \text{ s}}\right)^2 = \frac{39.95 \text{ g/mol}}{\text{molar mass of X}}$$

$$\text{molar mass of X} = \frac{39.95 \text{ g/mol}}{(28 \text{ s}/45 \text{ s})^2} = \boxed{1.0 \times 10^2 \text{ g/mol}}$$

A practical application of Graham's law arose during World War II, when scientists were studying the fission of uranium atoms as a source of energy. It became necessary to separate $^{235}_{92}$U, which is fissionable, from the more abundant isotope of uranium, $^{238}_{92}$U, which is not fissionable. Since the two isotopes have almost identical chemical properties, chemical separation was not feasible. Instead, an effusion process was worked out using uranium hexafluoride, $UF_6$. This compound is a gas at room temperature and low pressures. Preliminary experiments indicated that $^{235}_{92}UF_6$ could indeed be separated from $^{238}_{92}UF_6$ by effusion. The separation factor is very small, since the rates of effusion of these two species are nearly equal:

$$\frac{\text{rate of effusion of } ^{235}_{92}UF_6}{\text{rate of effusion of } ^{238}_{92}UF_6} = (352.0/349.0)^{1/2} = 1.004$$

so a great many repetitive separations are necessary. An enormous plant was built for this purpose in Oak Ridge, Tennessee. In this process, $UF_6$ effuses many thousands of times through porous barriers. The lighter fractions move on to the next stage while heavier fractions are recycled through earlier stages. Eventually, a nearly complete separation of the two isotopes is achieved.

## Distribution of Molecular Speeds and Energies

As pointed out earlier, the average speed of an $N_2$ molecule at 25°C is 515 m/s, while that of $H_2$ is even higher, 1920 m/s. However, not all molecules in these gases have these speeds. The motion of particles in a gas is utterly chaotic. In the course of a second, a particle undergoes millions of collisions with other particles. As a result, the speed and direction of motion of a particle are constantly changing. Over a period of time, the speed will vary from almost zero to some very high value, considerably above the average.

In 1860 James Clerk Maxwell, a Scottish physicist and one of the greatest theoreticians the world has ever known, showed that different possible speeds are distributed among particles in a definite way. Indeed, he developed a mathematical expression for this distribution. His results are shown graphically in Figure 5.6 (p. 128) for $O_2$ at 25 and 1000°C. On the graph, the relative number of molecules having a certain speed is plotted against that speed. At 25°C, this number increases rapidly with the speed, up to a maximum of about 400 m/s. This is the most probable speed of an oxygen molecule at 25°C. Above about 400 m/s, the number of molecules moving at any particular speed decreases. For speeds in excess of about 1200 m/s, the fraction of molecules drops off to nearly zero. In general, most molecules have speeds rather close to the average value.

*In a gas sample at any instant, gas molecules are moving at a variety of different speeds*

**Figure 5.6**
The distribution of molecular velocities in oxygen gas at two different temperatures, 25°C and 1000°C. At the higher temperature, the fraction of molecules moving at very high speeds is much greater.

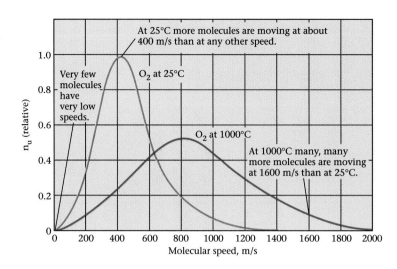

As temperature increases, the speed of the molecules increases. The distribution curve for molecular speeds (Fig. 5.6) shifts to the right and becomes broader. The chance of a molecule having a very high speed is much greater at 1000°C than it is at 25°C. Note, for example, that a large number of molecules have speeds greater than 1200 m/s at 1000°C.

## 5.7 Real Gases

In this chapter, the ideal gas law has been used in all calculations, with the assumption that it applies exactly. Under ordinary conditions, this assumption is a good one; however, all real gases deviate at least slightly from the ideal gas law. Table 5.2 shows the extent to which two gases, $O_2$ and $CO_2$, deviate from ideality at different temperatures and pressures. The data compare the experimentally observed molar volume, $V_m$

The molar volume is V when n = 1

$$\text{molar volume} = V_m = V/n$$

with the molar volume calculated from the ideal gas law $V_m^\circ$:

$$V_m^\circ = RT/P$$

TABLE 5.2  **Real versus Ideal Gases; Percent Deviation\* in Molar Volume**

|  | $O_2$ | | | $CO_2$ | | |
|---|---|---|---|---|---|---|
| P(atm) | 50°C | 0°C | −50°C | 50°C | 0°C | −50°C |
| 1 | −0.0% | −0.1% | − 0.2% | − 0.4% | −0.7% | −1.4% |
| 10 | −0.4% | −1.0% | − 2.1% | − 4.0% | −7.1% | |
| 40 | −1.4% | −3.7% | − 8.5% | −17.9% | | |
| 70 | −2.2% | −6.0% | −14.4% | −34.2% | Condenses to liquid. | |
| 100 | −2.8% | −7.7% | −19.1% | −59.0% | | |

\*% dev. $= \dfrac{(V_m - V_m^\circ)}{V_m^\circ} \times 100\%$

It should be obvious from Table 5.2 that deviations from ideality become larger at *high pressures and low temperatures.* Moreover, the deviations are larger for $CO_2$ than for $O_2$. All of these effects can be correlated in terms of a simple, common-sense observation:

**In general, the closer a gas is to the liquid state, the more it will deviate from the ideal gas law.**

A gas is liquefied by going to low temperatures and/or high pressures. Moreover, as you can see from Table 5.2, carbon dioxide is much easier to liquefy than oxygen.

From a molecular standpoint, deviations from the ideal gas law arise because it neglects two factors (recall Postulates 3 and 4 of the kinetic theory):

**1.** attractive forces between gas particles
**2.** the finite volume of gas particles

We will now consider in turn the effect of these two factors on the molar volumes of real gases.

## Attractive Forces

Notice that in Table 5.2 all the deviations are negative; the observed molar volume is less than that predicted by the ideal gas law. This effect can be attributed to attractive forces between gas particles. These forces tend to pull the particles toward one another, reducing the space between them. As a result, the particles are crowded into a smaller volume, just as if an additional external pressure were applied. The observed molar volume, $V_m$, becomes less than $V_m^\circ$, and the deviation from ideality is *negative:*

*Attractive forces make the molar volume smaller than expected*

$$\frac{V_m - V_m^\circ}{V_m^\circ} < 0$$

The magnitude of this effect depends upon the strength of the attractive forces and hence upon the nature of the gas. Intermolecular attractive forces are stronger in $CO_2$ than they are in $O_2$, which explains why the deviation from ideality of $V_m$ is greater with carbon dioxide and why carbon dioxide is more readily condensed to a liquid than is oxygen.

## Particle Volume

Figure 5.7 shows a plot of $V_m/V_m^\circ$ versus pressure for methane at 25°C. Up to about 150 atm, methane shows a steadily increasing negative deviation from ideality, as might be expected on the basis of attractive forces. At 150 atm, $V_m$ is only about 70% of $V_m^\circ$.

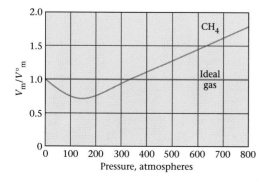

**Figure 5.7**
Below about 350 atm, attractive forces between $CH_4$ molecules cause the observed molar volume of methane gas at 25°C to be less than that calculated from the ideal gas law. At 350 atm, the effect of the attractive forces is just balanced by that of the finite volume of $CH_4$ molecules, and the gas appears to behave ideally. Above 350 atm, the effect of finite molecular volume predominates and $V_m > V_m^\circ$.

At very high pressures, methane behaves quite differently. Above 150 atm, the ratio $V_m/V_m^\circ$ *increases,* becoming 1 at about 350 atm. Above that pressure, methane shows a *positive* deviation from the ideal gas law:

Particle volume makes the molar volume larger than expected

$$\frac{V_m - V_m^\circ}{V_m^\circ} > 0$$

This effect is by no means unique to methane; it is observed with all gases. If the data in Table 5.2 are extended to very high pressures, oxygen and carbon dioxide behave like methane; $V_m$ becomes larger than $V_m^\circ$.

An increase in molar volume above that predicted by the ideal gas law is related to the finite volume of gas particles. These particles contribute to the observed volume, making $V_m$ greater than $V_m^\circ$. Ordinarily, this effect becomes evident only at high pressures, where the particles are quite close to one another.

## CHEMISTRY

### *Beyond the Classroom*

## Water Vapor in the Atmosphere

Of the five major components of the Earth's atmosphere, four (Table 5.A) have essentially constant concentrations. In contrast, the concentration of water vapor varies over a wide range depending, among other things, upon the temperature. The mole fraction of $H_2O$ in the air is typically about 0.05 in the tropics; in polar regions it may be as low as 0.0002.

**TABLE 5.A  Composition of Clean, Dry Air at Sea Level**

| Component | $N_2$ | $O_2$ | Ar | $CO_2$ |
|---|---|---|---|---|
| Mole Fraction | 0.7808 | 0.2095 | 0.00934 | 0.00034 |

To understand why the concentration of water vapor in the atmosphere is temperature dependent, consider Table 5.B. The *equilibrium vapor pressure* listed there is the pressure exerted by water vapor in intimate contact with liquid water (above 0°C) or ice (below 0°C). The value of $P_{H_2O}^\circ$ sets an upper limit to the partial pressure of water vapor in the air. At −40°C air saturated with water vapor contains very little water ($P_{H_2O}^\circ = 0.13$ mm Hg); at 40°C, air can contain a great deal more water ($P_{H_2O}^\circ = 55.3$ mm Hg).

**TABLE 5.B  Equilibrium Vapor Pressure of Water ($P_{H_2O}^\circ$) versus Temperature**

| T(°C) | −40 | −20 | 0 | 10 | 20 | 30 | 40 |
|---|---|---|---|---|---|---|---|
| $P_{H_2O}^\circ$ (mmHg) | 0.13 | 0.90 | 4.6 | 9.2 | 17.5 | 31.8 | 55.3 |

The actual partial pressure of water vapor in the air, $P_{H_2O}$, is ordinarily less than the saturated value, $P_{H_2O}^\circ$. The *relative humidity,* defined as

$$R.\ H. = \frac{P_{H_2O}}{P_{H_2O}^\circ} \times 100\%$$

has a lower limit of 0% (completely dry air) and an upper limit of 100% (air saturated with water vapor). On a day when the temperature is 30°C ($P_{H_2O}^\circ = 31.8$ mm Hg) and the actual partial pressure is 17.5 mm Hg,

$$R.\ H. = \frac{17.5}{31.8} \times 100\% = 55.0\%$$

Our comfort depends upon relative humidity as well as temperature. At relative humidities below about 30%, evaporation of water from body surfaces is extensive and rapid. Mouth and nasal membranes dry out, allowing viruses to enter the lungs more readily. This may be why colds are so common in the winter months, when the relative humidity indoors is often quite low. Most of us are familiar with the uncomfortable effects of high relative humidities, above 80%. Perspiration fails to evaporate, leaving us feeling clammy and hot.

The *dew point* is the temperature to which an air mass must be cooled to raise the relative humidity to 100%. For the air mass referred to above with $P_{H_2O} = 17.5$ mm Hg, the dew point must be 20°C, which is the temperature at which $P^\circ_{H_2O} = 17.5$ mm Hg (Table 5.B). At 20°C,

$$R.\ H. = \frac{P_{H_2O}}{P^\circ_{H_2O}} \times 100\% = \frac{17.5}{17.5} \times 100\% = 100\%$$

Below 20°C, liquid water would condense out of the air, keeping the relative humidity at 100%. When the temperature drops below the dew point on a cool summer night, dew forms on blades of grass and other surfaces. If the "dew point" is below 0°C, the product is solid rather than liquid water (i.e., frost instead of dew).

Clouds form when a moist air mass rises to a height, perhaps 1 to 10 km above the Earth's surface, where atmospheric pressure is much less than it is at sea level. Air, like most "real" gases, cools on expansion. When the temperature of the air mass drops below the dew point, water vapor condenses to form a cloud of tiny water droplets with an average diameter of about 0.01 mm. Clouds formed at very high altitudes where the temperature is below 0°C are made up of tiny ice crystals (Figure 5.A).

The tiny water droplets that make up most clouds are kept aloft by upward air currents. To fall to Earth, they must grow to a certain critical size. This can occur through condensation of water vapor on the droplets or by collision and fusion with other droplets. The resultant "precipitation" is called drizzle if the drop diameter is less than 0.5 mm; the drop size of rain varies from 0.5 to 6 mm. Total rainfall is usually expressed either in inches or centimeters. Annual rainfall in the United States varies from 5 cm in Death Valley, California, to 250 cm (about 100 in) on the coast of the state of Washington.

If you live north of latitude 35° (in the northern hemisphere) you are well aware that water can fall from clouds in solid as well as liquid form. There are several different kinds of frozen precipitation:

— snow is formed when water vapor condenses directly to the solid without passing through the liquid state. Although most snowflakes have a hexagonal pattern, it has been said that no two flakes are exactly alike. The water equivalent of a snowfall can be estimated by dividing by 10; i.e., 10 cm of snow melts down to 1 cm of liquid water.
— sleet is formed when raindrops freeze as they descend through a layer of air at a temperature below 0°C. The resulting ice pellets are less than 0.5 cm in diameter. Larger pellets are referred to as hail. Hailstorms formed in summer thunderstorms can have diameters up to 10 cm, considerably larger than golfballs.

Sometimes, when raindrops fall through a layer of cold air, they "supercool" to a temperature below 0°C rather than turning to ice. When these drops strike the ground, they freeze to form a thin layer of ice that covers the surfaces of trees, power lines, and highways. This phenomenon, called freezing rain or ice rain, is a nightmare for winter drivers in Connecticut (and probably other places as well).

**Figure 5.A**
High cirrus clouds over Yellowstone National Park. (Stan Osolinski/Dembinsky Photo Associates)

**Figure 5.B**
A glacier, in front of which CNH is standing, is a layer of ice, formed from snow compacted under pressure.

# CHAPTER HIGHLIGHTS

## Key Concepts

1. Convert between different units of pressure, volume, temperature, and amount of gas
   (Example 5.1; Problems 1–4)
2. Use the ideal gas law to
   —solve initial and final state problems
   (Example 5.2; Problems 5–16)
   —calculate *P*, *V*, *T*, or *n*
   (Example 5.3; Problems 17–24)
   —calculate density or molar mass
   (Examples 5.4, 5.5; Problems 25–34)
   —relate amounts and volumes of gases in reactions
   (Examples 5.6, 5.7; Problems 35–40, 59, 65, 66, 72)
3. Use Dalton's law to relate partial pressures to total pressures and mole fractions
   (Examples 5.8, 5.9; Problems 41–48)
4. Use the relation $u = (3RT/\mathcal{M})^{1/2}$ to calculate velocities of gas molecules
   (Example 5.10; Problems 53, 54)
5. Use Graham's law to relate rate or time of effusion to molar mass
   (Example 5.11; Problems 49–52, 68, 70)

## Key Equations

| | | |
|---|---|---|
| Ideal gas law | $PV = nRT$ | $n = m/\mathcal{M}$ |
| Dalton's law | $P_{tot} = P_A + P_B + \cdots$ | $P_A = X_A P_{tot}$ |
| Average speed | $u = (3RT/\mathcal{M})^{1/2}$ | |
| Graham's law | $\dfrac{\text{rate}_A}{\text{rate}_B} = \dfrac{\text{time}_B}{\text{time}_A} = \left(\dfrac{\mathcal{M}_B}{\mathcal{M}_A}\right)^{1/2}$ | |

## Key Terms

| | | |
|---|---|---|
| atmosphere | kilopascal | mole fraction |
| density | millimeter of mercury | partial pressure |
| effusion | molar mass | *R* (gas constant) |
| Kelvin scale | mole | STP |

## Summary Problem

When calcium carbonate, the main component of marble and chalk, is added to a solution of hydrochloric acid, the following reaction takes place:

$$CaCO_3(s) + 2H^+(aq) \longrightarrow CO_2(g) + H_2O + Ca^{2+}(aq)$$

(a) If 428 mL of $CO_2$ is produced at 23°C and 755 mm Hg, what is the yield of $CO_2$ in moles? in grams?

(b) The gas collected in (a) is transferred without loss at 23° C to a flask with twice the volume of that in (a). What is the pressure in the flask? At what temperature would the pressure in the larger flask return to the original pressure in (a)?

(c) What is the density of $CO_2$ at 23°C and 755 mm Hg?

(d) Suppose the carbon dioxide is collected at 23°C in a flask that contains nitrogen gas with a pressure of 127 mm Hg. After $CO_2$ is collected, the total pressure is 798 mm Hg. What is the partial pressure of $CO_2$? What mass of $CO_2$ is collected in the 428-mL flask?

(e) What is the mole fraction of nitrogen in the mixture in (d)?

(f) What volume of 0.200 $M$ HCl is required to generate 0.500 L of $CO_2$ at 27°C and 1.20 atm?

(g) What volume of $CO_2$ can be generated at 27°C and 1.20 atm if 2.00 g of $CaCO_3$ is added to 25.00 mL of 1.00 $M$ HCl?

(h) Compare the rate of effusion of $CO_2$ to that of $N_2$ at the same temperature and pressure. Compare the time required for equal numbers of moles of $CO_2$ and $N_2$ to effuse.

(i) The average speed of a carbon dioxide molecule at 25°C is 411 m/s. What is the average speed of a nitrogen molecule at the same temperature?

### Answers

(a) 0.0175 moles; 0.770 g

(b) 378 mm Hg; 319°C

(c) 1.80 g/L

(d) 671 mm Hg; 0.684 g

(e) 0.159

(f) 0.244 L

(g) 257 mL

(h) 0.798; 1.25

(i) 515 m/s

A bed of calcium carbonate (limestone) along the Verde River in Arizona. (Visuals Unlimited/ Ken Wagner)

## Questions & Problems

### Measurements on Gases

**1.** A five-gallon propane tank contains 0.784 mol of propane ($C_3H_8$) at 68°F. Express the volume of the tank in liters, the amount of propane in the tank in grams, and the temperature of the tank in Kelvin.

**2.** An 8.00-ft cylinder has a radius of 32 inches. It contains 235 lb of nitrogen at 27°C. Express the volume of the cylinder ($V = \pi r^2 h$) in liters, the amount of nitrogen in moles, and the temperature in K.

**3.** Complete the following table of pressure conversions.

| mm Hg | Atmospheres | Kilopascals |
|-------|-------------|-------------|
| 913   | ——          | ——          |
| ——    | 0.833       | ——          |
| ——    | ——          | 122         |

**4.** Complete the following table of pressure conversions.

| mm Hg | Atmospheres | Kilopascals |
|-------|-------------|-------------|
| 432   | ——          | ——          |
| ——    | 1.33        | ——          |
| ——    | ——          | 75.2        |

### Initial and Final States

**\*5.** A rigid sealed cylinder has seven molecules of neon (Ne).

(a) Make a sketch of the cylinder with the neon molecules at 25°C. Make a similar sketch of the same seven molecules in the same cylinder at −80°C.

(b) Make the same sketches asked for in part (a), but this time attach a pressure gauge to the cylinder.

**\*6.** Sketch a cylinder with ten molecules of helium (He) gas. The cylinder has a movable piston. Label this sketch BEFORE. Make an AFTER sketch to represent

(a) a decrease in temperature at constant pressure.

(b) a decrease in pressure from 1000 mm Hg to 500 mm Hg at constant temperature.

(c) five molecules of $H_2$ gas added at constant temperature and pressure.

**7.** A five-liter cylinder is filled with butane at 27°C until the pressure in the tank is 1.00 atm. What is the pressure in the tank

(a) on a winter day when the temperature is −10°C?

(b) on a warm day when the temperature is 40°C?

**8.** A 2.00-L balloon is filled with helium gas at 756 mm Hg and 25°C. What is the pressure in another 2.00-L balloon filled with the same amount of helium at 50°C?

**9.** A sample of NO gas at 15°C and 1.00 atm has a volume of 1.50 L. Determine the ratio of the original volume to the final volume when

(a) the pressure and amount of gas remain unchanged and the Celsius temperature is doubled.

(b) the pressure and amount of gas remain unchanged and the Kelvin temperature is doubled.

**10.** A sample of air is originally at 32°C. If $P$ and $n$ are kept constant, to what temperature must the air be cooled to

(a) decrease its volume to a third of its original volume?

(b) decrease its volume by a third?

**11.** A tire is inflated to a gauge pressure of 28.0 psi at 71°F. Gauge pressure is the pressure above atmospheric pressure, which is 14.7 psi. After several hours of driving, the air in the tire has a temperature of 115°F. What is the gauge pressure of the air in the tire? What is the actual pressure of the air in the tire? Assume that the tire volume changes are negligible.

**12.** A basketball is inflated in a garage at 25°C to a gauge pressure of 8.0 psi. Gauge pressure is the pressure above atmospheric pressure, which is 14.7 psi. The ball is used on the driveway at a temperature of −7°C and feels "flat." What is the actual pressure of the air in the ball? What is the gauge pressure?

**13.** On a warm summer day, a person takes in a breath of 500 mL (three significant figures) of air at 751 mm Hg and 39°C. What is the volume of this air in the lungs at 37°C and 1.00 atm pressure?

**14.** A 3.50-cm$^3$ air bubble forms in a deep lake at a depth where the temperature is 6°C at a total pressure of 2.50 atm. The bubble rises to a depth where the temperature and pressure are 13°C and 1.75 atm, respectively. Assuming that the amount of air in the bubble has not changed, calculate its new volume.

**15.** A balloon filled with helium has a volume of $1.28 \times 10^3$ L at sea level, where the pressure is 0.998 atm and the temperature is 31°C. The balloon is taken to the top of a mountain, where the pressure is 0.753 atm and the temperature is −25°C. What is the volume of the balloon at the top of the mountain?

**16.** A 10.00-L flask has 0.475 mol of hydrogen gas at 25°C and a pressure of 1.16 atm. Oxygen gas is added to the flask at the same temperature until the pressure rises to 1.78 atm. How many moles of oxygen gas are added?

### Ideal Gas Law; Calculation of One Variable

**17.** Arterial blood contains about 0.25 mg of oxygen per mL. What is the pressure exerted by the oxygen in one liter of arterial blood at normal body temperature of 37°C?

**18.** A two-liter plastic soft-drink bottle can withstand a pressure of 5 atm. Half a cup (approximately 120 mL) of ethyl alcohol, $C_2H_5OH$ ($d = 0.789$ g/mL), is poured into a soft-drink bottle at room temperature. The bottle is then heated to 100°C, changing the liquid alcohol to a gas. Will the soft-drink bottle withstand the pressure, or will it explode?

**19.** A ball filled with air ($M = 29.0$ g/mol) at a pressure of 2.5 atm at 25°C weighs 522.3 g. All the air in the ball is removed. The "empty" ball weighs 500.1 g. If helium is pumped into the ball to a pressure of 2.5 atm at the same temperature, what is the mass of the helium-filled ball?

**20.** Compressed-air tanks used by scuba divers have a volume of 8.0 L and are filled with air to a pressure of 135 atm at 20°C. How many grams of helium are required to fill a tank under the above conditions?

**21.** Use the ideal gas law to complete the following table for methane gas.

| Pressure | Volume | Temperature | Moles | Grams |
|---|---|---|---|---|
| ——— | 2.50 L | 12°C | 3.70 | ——— |
| 0.769 atm | ——— | 22°C | ——— | 1.000 |
| 639 mm Hg | 5.52 L | ——— | ——— | 7.29 |
| 188 kPa | 977 mL | 45°C | ——— | ——— |

**22.** Complete the following table for carbon monoxide gas.

| Pressure | Volume | Temperature | Moles | Grams |
|---|---|---|---|---|
| 493 mm Hg | 3.75 L | 36°C | ——— | ——— |
| 1.28 atm | 6.39 L | ——— | 0.500 | ——— |
| 125 kPa | ——— | 99°C | ——— | 43.2 |
| ——— | 2.98 L | 125°C | ——— | 0.827 |

**\*23.** Two tanks have the same volume and are kept at the same temperature. Compare the pressure in both tanks if
  **(a)** Tank A has 2.00 moles of carbon dioxide and Tank B has 2.00 moles of helium.
  **(b)** Tank A has 2.00 g of carbon dioxide and Tank B has 2.00 g of helium. (Remember, this should be done without a calculator!)

**\*24.** Tank A contains ammonia at 300 K. Tank B contains nitrogen gas at 150 K. Tanks A and B have the same volume. Compare the pressures in Tanks A and B if
  **(a)** Tank B has twice as many moles of nitrogen as Tank A has of ammonia.
  **(b)** Tank A has the same number of moles of ammonia as Tank B has of nitrogen. (Remember, this should be done without a calculator!)

### Ideal Gas Law; Density and Molar Mass

**25.** Calculate the densities (in g/L) of the following gases at 97°C and 755 mm Hg.
  **(a)** hydrogen chloride
  **(b)** sulfur dioxide
  **(c)** butane ($C_4H_{10}$)

**26.** Calculate the densities (in g/L) of the following gases at 105°F and 112 kPa.
  **(a)** krypton      **(b)** dichlorine oxide      **(c)** ammonia

**27.** Recent measurements show that the atmosphere on Venus consists mostly of carbon dioxide. The temperature at the surface of Venus is 460°C; the pressure is 75 atm. Compare the density of carbon dioxide on Venus's surface with that on the Earth's surface at 25°C and one atmosphere.

**28.** Assuming water vapor behaves ideally at 100°C and 1.00 atm, what is the ratio of the density of $H_2O(l)$ to $H_2O(g)$ under these conditions? The density of liquid water at 100°C is 0.958 g/cm$^3$.

**29.** Freon is a gas made up of carbon, fluorine, and chlorine atoms. It was used as a refrigerant in car air conditioners. It is also one of the culprits in the depletion of the ozone layer. It has a density of 4.65 g/L at 735 mm Hg and 33°C.
  **(a)** What is the molar mass of Freon?
  **(b)** Freon is made up of 9.92% C, 58.6% Cl, and 31.4% F. What is its molecular formula?

**30.** Phosgene is a highly toxic gas made up of carbon, oxygen, and chlorine atoms. Its density at 1.05 atm and 25°C is 4.24 g/L.

(a) What is the molar mass of phosgene?

(b) Phosgene is made up of 12.1% C, 16.2% O, and 71.7% Cl. What is the molecular formula of phosgene?

**31.** Exhaled air contains 74.5% $N_2$, 15.7% $O_2$, 3.6% $CO_2$, and 6.2% $H_2O$ (mole percent).

(a) Calculate the molar mass of exhaled air.

(b) Calculate the density of exhaled air at 37°C and 757 mm Hg and compare the value obtained with that for ordinary air ($\mathcal{M}$ = 29.0 g/mol).

**32.** To prevent a condition called the "bends," deep-sea divers breathe a mixture containing, in mole percent, 10.0% $O_2$, 10.0% $N_2$, and 80.0% He.

(a) Calculate the molar mass of this mixture.

(b) What is the ratio of the density of this gas to that of pure oxygen?

**33.** A 1.58-g sample of $C_2H_3X_3(g)$ has a volume of 297 mL at 769 mm Hg and 35°C. Identify the element X.

**34.** A 2.00-g sample of $SX_6(g)$ has a volume of 329.5 cm³ at 1.00 atm and 20°C. Identify X. Name the compound.

## Gases in Reactions

**35.** Dichlorine oxide is used as a bacteriocide to purify water. It is produced by the reaction

$$SO_2(g) + 2Cl_2(g) \longrightarrow SOCl_2(l) + Cl_2O(g)$$

How many liters of dichlorine oxide can be produced from 6.0 L of chlorine? Assume 100% yield and that all the gases are measured at the same temperature and pressure.

**36.** Hydrogen sulfide gas ($H_2S$) is responsible for the foul odor of rotten eggs. When it reacts with oxygen, sulfur dioxide gas and steam are produced.

(a) Write a balanced equation for the reaction.

(b) How many liters of oxygen would be required to react with 12.0 L of hydrogen sulfide? Assume 100% yield and constant temperature and pressure.

**37.** Hydrogen cyanide, HCN, is a poisonous gas that is used in the gas chamber for the execution of people sentenced to death. It can be formed by the reaction

$$NaCN(s) + H^+(aq) \longrightarrow HCN(g) + Na^+(aq)$$

What volume of 6.00 M HCl is required to react with excess NaCN to produce enough HCN to fill a room with volume 27 m³ at a pressure of 0.987 atm and 23°C?

**38.** When hydrogen peroxide decomposes, oxygen is produced.

$$2H_2O_2(aq) \longrightarrow 2H_2O + O_2(g)$$

What volume of oxygen gas at 25°C and 1.00 atm is produced from the decomposition of 25.00 mL of a solution of hydrogen peroxide ($d$ = 1.05 g/mL) that is 30.0% $H_2O_2$ by mass?

**39.** Ammonium nitrate can be used as an effective explosive because it decomposes into a large number of gaseous products. At a sufficiently high temperature, ammonium nitrate decomposes into nitrogen, oxygen, and steam.

(a) Write a balanced equation for the decomposition of ammonium nitrate.

(b) If 100.0 g of ammonium nitrate is sealed into a 10.0-L steel drum and heated to 315°C, what is the pressure in the drum, assuming 100% decomposition?

**40.** Nitroglycerine is an explosive used by the mining industry. It detonates according to the following equation:

$$4C_3H_5N_3O_9(l) \longrightarrow 12CO_2(g) + 6N_2(g) + 10H_2O(g) + O_2(g)$$

What volume is occupied by the gases produced when 2.00 g of nitroglycerine explodes? The total pressure is 755 mm Hg at 455°C.

## Dalton's Law

**41.** Some chambers used to grow bacteria that thrive on $CO_2$ have a gas mixture made up of 95.0% $CO_2$ and 5.0% $O_2$ (mole percent). What is the partial pressure of each gas if the total pressure is 735 mm Hg?

**42.** A certain laser uses a gas mixture made up of 9.00 g of HCl, 2.00 g of $H_2$, and 165.0 g of Ne. What pressure is exerted by the mixture in a 75.0-L tank at 22°C? Which gas has the smallest partial pressure?

**43.** A sample of gas collected over water at 42°C occupies a volume of one liter. The wet gas has a pressure of 0.986 atm. When dried, the gas occupies 1.04 L with a pressure of 1.00 atm at 90°C. Using this information, calculate the vapor pressure of water at 42°C.

**44.** A sample of oxygen gas is collected over water at 25°C (vp $H_2O$ = 23.8 mm Hg). The wet gas occupies a volume of 15.33 L at a total pressure of 758 mm Hg. If all the water is removed, what volume will the dry oxygen occupy at a pressure of 778 mm Hg and a temperature of 48°C?

**\*45.** The following figure shows three 1.00-L bulbs connected by valves. Each bulb contains neon gas with amounts proportional to the number of atoms pictorially represented in each chamber. All three bulbs are maintained at the same temperature. Unless stated otherwise, assume that the valves connecting the chambers are closed and seal the gases in their respective chambers. Assume also that the volume between chambers is negligible.

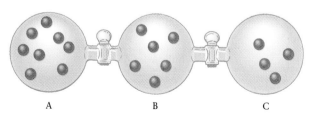

A          B          C

(a) Which bulb has the lowest pressure?

(b) If the pressure in Bulb A is 2.00 atm, what is the pressure in Bulb C?

(c) If the pressure in Bulb A is 2.00 atm, what is the sum of $P_A + P_B + P_C$?

*(continued)*

**(d)** If the pressure in Bulb A is 2.00 atm and the valve between Bulbs A and B is opened, redraw the figure to accurately represent the gas atoms in all the bulbs. What is $P_A + P_B$? What is $P_A + P_B + P_C$? Compare your answer with your answer in part (c).

**(e)** Follow the same instructions as in (d) when all the valves are open.

*46. Consider the sketch below. Each square in Bulb A represents a mole of X atoms. Each circle in Bulb B represents a mole of Y atoms. The bulbs have the same volume, and the temperature is kept constant. When the valve is opened, atoms of X react with atoms of Y according to the following equation:

$$2X(g) + Y(g) \longrightarrow X_2Y(g)$$

The gaseous product is represented as □○□, and each □○□, represents one mole of product.

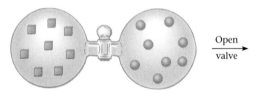

**(a)** If $P_A = 2.00$ atm, what is $P_B$ before the valve is opened and the reaction is allowed to occur? What is $P_A + P_B$?

**(b)** Redraw the sketch to show what happens after the valve is opened and reaction occurs.

**(c)** What is $P_A$? What is $P_B$? What is $P_A + P_B$? Compare your answer with the answer obtained in part (a).

**47.** Consider two bulbs separated by a valve (See Questions 45 and 46). When the valve between the two bulbs is closed, the following data apply:

|        | Bulb A   | Bulb B    |
| ------ | -------- | --------- |
| Gas:   | Ar       | $CO_2$    |
| $V$:   | 3.00 L   | 2.00 L    |
| $P$:   | 2.50 atm | 0.575 atm |

Assuming no temperature change, determine the final pressure inside the system after the valve connecting the two bulbs is opened. Ignore the volume of the tube connecting the two bulbs.

**48.** Follow the instructions for Problem 47 for the following set-up:

|        | Bulb A   | Bulb B   |
| ------ | -------- | -------- |
| Gas:   | $N_2$    | $Cl_2$   |
| $V$:   | 5.00 L   | 1.00 L   |
| $P$:   | 3.50 atm | 1.25 atm |

### Kinetic Theory

**49.** A gas effuses 1.22 times faster than ethane ($C_2H_6$) at the same temperature and pressure.

**\*(a)** Is the gas heavier or lighter than ethane?
**(b)** What is the molar mass of the gas?

**50.** What is the ratio of the rate of effusion of the most abundant gas, nitrogen, to the lightest gas, hydrogen?

**51.** A balloon filled with air ($M = 29$ g/mol) has a small leak. Another balloon filled with helium has an identical leak. How much faster will the helium balloon deflate?

*52. Rank the following gases

$$CO, \quad Ne, \quad CO_2, \quad O_2$$

in order of
**(a)** increasing speed of effusion through a tiny opening.
**(b)** increasing time of effusion.

**53.** What is the average speed of
**(a)** a chlorine molecule at 35°C?
**(b)** a xenon atom at −35°C?

**54.** At what temperature will a molecule of uranium hexafluoride, the densest gas known, have the same average speed as a molecule of the lightest gas, hydrogen, at 25°C?

### Real Gases

**\*55.** A sample of methane gas ($CH_4$) is at 50°C and 20 atm. Would you expect it to behave more or less ideally if
**(a)** the pressure were reduced to 1 atm?
**(b)** the temperature were reduced to −50°C?

*56. The normal boiling points of CO and $SO_2$ are −192°C and −10°C, respectively.
**(a)** At 25°C and 1 atm, which gas would you expect to have a molar volume closest to the ideal value?
**(b)** If you wanted to reduce the deviation from ideal gas behavior, in which direction would you change the temperature? The pressure?

*57. Using Figure 5.7,
**(a)** estimate the density of methane gas at 100 atm.
**(b)** compare the value obtained in (a) with that calculated from the ideal gas law.
**(c)** At what pressure will the calculated density and the actual density of methane be the same?

*58. Using Figure 5.7,
**(a)** estimate the density of methane gas at 200 atm.
**(b)** compare the value obtained in (a) with that calculated from the ideal gas law.

### Unclassified

**59.** Assume that an automobile burns octane, $C_8H_{18}$ ($d = 0.692$ g/cm³).
**(a)** Write the equation for the combustion of octane to carbon dioxide and water.
**(b)** A car has a fuel efficiency of 26 miles/gallon of octane. What volume of carbon dioxide at 25°C and 1.00 atm is generated when that car goes on a fifty-mile trip?

**60.** The pressure exerted by a column of liquid is directly proportional to its density and height. If you wanted to fill a barometer with water ($d = 1.00$ g/cm$^3$), how long a tube should you use? ($d$ of Hg = 13.6 g/cm$^3$)

**\*61.** For an ideal gas, sketch graphs of
   **(a)** $V$ versus $P$ at constant $T$, $n$.
   **(b)** $P$ versus $n$ at constant $T$, $V$.
   **(c)** $u$ versus $T$ at constant $P$, $V$.
   **(d)** $E_{trans}$ versus $P$ at constant $T$, $n$.

**\*62.** For an ideal gas, sketch graphs of
   **(a)** $V$ versus $T$ at constant $P$, $n$.
   **(b)** $P$ versus $T$ at constant $n$, $V$.
   **(c)** $n$ versus $T$ at constant $P$, $V$.
   **(d)** $E_{trans}$ versus $T$ at constant $P$, $n$.

**\*63.** A mixture of 3.5 mol of Ar and 3.9 mol of Ne occupies a 10.00-L container at 300 K. Which gas has the larger
   **(a)** average translational energy?
   **(b)** partial pressure?
   **(c)** mole fraction?
   **(d)** effusion rate?

**64.** Given that 1.00 mol of carbon dioxide and 1.00 mol of sulfur dioxide are in separate containers at the same temperature and pressure, calculate each of the following ratios.
   **(a)** volume of $CO_2$/volume of $SO_2$.
   **(b)** density of $CO_2$/density of $SO_2$.
   **(c)** average translational energy of $CO_2$/average translational energy of $SO_2$.
   **(d)** number of C atoms/number of S atoms.

**65.** A Porsche 928 S4 engine has a cylinder volume of 618 cm$^3$. The cylinder is full of air at 75°C and 1.00 atm.
   **(a)** How many moles of oxygen are in the cylinder? (Mole percent of oxygen in air = 21.0)
   **(b)** Assume that the hydrocarbons in gasoline have an average molar mass of $1.0 \times 10^2$ g/mol and react with oxygen in a 1:12 mole ratio. How many grams of gasoline should be injected into the cylinder to react with the oxygen?

**66.** An intermediate reaction used in the production of nitrogen-containing fertilizers is that between ammonia and oxygen:

$$4NH_3(g) + 5O_2(g) \longrightarrow 4NO(g) + 6H_2O(g)$$

A 150.0-L reaction chamber is charged with reactants to the following partial pressures at 500°C: $P_{NH3} = 1.5$ atm, $P_{O_2} = 0.97$ atm. What is the limiting reactant?

**67.** Glycine is an amino acid made up of carbon, hydrogen, oxygen, and nitrogen atoms. Combustion of a 0.2036-g sample gives 132.9 mL of $CO_2$ at 25°C and 1.00 atm and 0.122 g of water. What are the percentages of carbon and hydrogen in glycine? Another sample of glycine weighing 0.2500 g is treated in such a way that all the nitrogen atoms are converted to $N_2(g)$. This gas has a volume of 40.8 mL at 25°C and 1.00 atm. What is the percentage of nitrogen in glycine? What is the percentage of oxygen? What is the empirical formula of glycine?

**68.** At 25°C and 380 mm Hg, the density of sulfur dioxide is 1.31 g/L. The rate of effusion of sulfur dioxide through an orifice is 4.48 mL/s. What is the density of a sample of gas that effuses through an identical orifice at the rate of 6.78 mL/s under the same conditions? What is the molar mass of the gas?

**69.** A research laboratory has two steel cylinders of equal volume; they are at the same temperature. Cylinder I contains chlorine gas, while cylinder II contains methane gas.
   **(a)** If each cylinder contains one kilogram of gas, which cylinder has the greater pressure?
   **(b)** Suppose cylinder I has a pressure of 7.0 atm at 50°C, while cylinder II has a pressure of 5.5 atm at 10°C. Which cylinder contains the larger number of molecules?

## Challenge Problems

**70.** A tube 5.0 ft long is evacuated. Samples of $NH_3$ and HCl, at the same temperature and pressure, are introduced simultaneously at opposite ends of the tube. When the two gases meet, a white ring of $NH_4Cl(s)$ forms. How far from the end at which ammonia was introduced will the ring form?

**71.** The Rankine temperature scale resembles the Kelvin scale in that 0° is taken to be the lowest attainable temperature (0°R = 0 K). However, the Rankine degree is the same size as the Fahrenheit degree, whereas the Kelvin degree is the same size as the Celsius degree. What is the value of the gas constant in L·atm/(mol·°R)?

**72.** A 0.2500-g sample of an Al-Zn alloy reacts with HCl to form hydrogen gas:

$$Al(s) + 3H^+(aq) \longrightarrow Al^{3+}(aq) + 3/2H_2(g)$$
$$Zn(s) + 2H^+(aq) \longrightarrow Zn^{2+}(aq) + H_2(g)$$

The hydrogen produced has a volume of 0.147 L at 25°C and 755 mm Hg. What is the percentage of zinc in the alloy?

**73.** The buoyant force on a balloon is equal to the mass of air it displaces. The gravitational force on the balloon is equal to the sum of the masses of the balloon, the gas it contains, and the balloonist. If the balloon and the balloonist together weigh 168 kg, what would the diameter of a spherical hydrogen-filled balloon have to be in meters if the rig is to get off the ground at 22°C and 758 mm Hg? (Take $\mathcal{M}_{air} = 29.0$ g/mol.)

**74.** A mixture in which the mole ratio of hydrogen to oxygen is 2:1 is used to prepare water by the reaction

$$2H_2(g) + O_2(g) \longrightarrow 2H_2O(g)$$

The total pressure in the container is 0.950 atm at 25°C before the reaction. What is the final pressure in the container at 125°C after the reaction, assuming an 88.0% yield and no volume change?

**75.** The volume fraction of a gas, A, in a mixture is defined by the equation

$$\text{volume fraction } A = \frac{V_A}{V}$$

where $V$ is the total volume and $V_A$ is the volume that gas A would occupy alone at the same temperature and pressure. Show that, assuming ideal gas behavior, the volume fraction is the same as the mole fraction. Explain why the volume fraction differs from the mass fraction.

Nightlife at Barrow, Alaska (see p. 166) (©1992 Ken Scott/Dembinsky Photo Assoc.)

# 6 Electronic Structure and the Periodic Table

Not chaos-like crush'd and bruis'd,

But, as the world, harmoniously confus'd,

Where order in variety we see,

And where, though all things differ, all agree.

—ALEXANDER POPE

*Windsor Forest*

## CHAPTER OUTLINE

I n Chapter 2 we briefly considered the structure of the atom. You will recall that every atom has a tiny, positively charged nucleus, made up of protons and neutrons. The nucleus is surrounded by negatively charged electrons. The number of protons in the nucleus is characteristic of the atoms of a particular element and is referred to as the atomic number. In a neutral atom, the number of electrons is equal to the number of protons and hence to the atomic number.

In this chapter, we focus upon electron arrangements in atoms, paying particular attention to the relative energies of different electrons (*energy levels*) and their spatial locations (*orbitals*). Specifically, we consider the nature of the energy levels and orbitals available to

Chemical properties of atoms and molecules depend upon their electronic structures

— the single electron in the hydrogen atom (Section 6.2)
— the several electrons in more complex atoms (Section 6.3)

With this background, we show how electron arrangements in multi-electron atoms and the monatomic ions derived from them can be described in terms of

— *electron configurations,* which show the number of electrons in each energy level (Sections 6.4, 6.6)
— *orbital diagrams,* which show the arrangement of electrons within orbitals (Section 6.5, 6.6)

The electron configuration or orbital diagram of an atom of an element can be deduced from its position in the periodic table. Beyond that, position in the table can be used to predict (Section 6.7) the relative sizes of atoms and ions (*atomic radius, ionic radius*) and the relative tendencies of atoms to give up or acquire electrons (*ionization energy, electronegativity*).

Before dealing with electronic structures as such, it will be helpful to examine briefly the experimental evidence upon which such structures are based (Section 6.1). In particular, we need to look at the phenomenon of *atomic spectra*.

## 6.1   Light, Photon Energies, and Atomic Spectra

Fireworks displays are fascinating to watch. Neon lights and sodium vapor lamps can transform the skyline of a city with their brilliant colors. The eerie phenomenon of the aurora borealis is an unforgettable experience when you see it

The colored lights are generated by Sr, Cu, and Na atoms. (Richard Megna/FUNDAMENTAL PHOTOGRAPHS, New York)

for the first time. All of these events relate to the generation of light and its transmission through space.

## The Wave Nature of Light; Wavelength and Frequency

Light travels through space as a wave, which is made up of successive crests, which rise above the midline, and troughs, which sink below it. Waves have three primary characteristics (Fig. 6.1), two of which are of particular interest at this point:

1. **Wavelength** ($\lambda$), the distance between two consecutive crests or troughs, most often measured in meters or nanometers (1 nm = $10^{-9}$ m).
2. **Frequency** ($\nu$), the number of wave cycles (successive crests or troughs) that pass a given point in unit time. If $10^8$ cycles pass a particular point in one second,

$$\nu = 10^8/s = 10^8 \text{ Hz}$$

The frequency unit *hertz* (Hz) represents one cycle per second.

The speed at which a wave moves through space can be found by multiplying the length of a wave cycle ($\lambda$) by the number of cycles passing a point in unit time ($\nu$). For light,

$$\lambda\nu = c$$

where $c$, the speed of light, is a constant, $2.998 \times 10^8$ m/s. To use this equation with this value of $c$

— $\lambda$ should be expressed in meters
— $\nu$ should be expressed in reciprocal seconds (hertz)

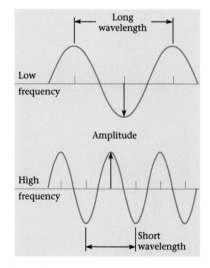

**Figure 6.1**
Three characteristics of a wave are its amplitude, wavelength, and frequency. The *amplitude* ($\psi$) is the height of a crest or the depth of a trough. The *wavelength* ($\lambda$) is the distance between successive crests or troughs. The *frequency* ($\nu$) is the number of wave cycles (successive crests or troughs) that pass a given point in a given time.

Red light (~700 nm) has a longer wavelength than violet (~400 nm)

---

**Example 6.1**   The green light associated with the aurora borealis is emitted by excited (high-energy) oxygen atoms at 557.7 nm. What is the frequency of this light?

**Strategy**   Use the equation: $\lambda\nu = c$, taking $c = 2.998 \times 10^8$ m/s. Note that $\lambda$ must be expressed in meters.

**Solution**

$$\lambda = 557.7 \text{ nm} \times \frac{1 \text{ m}}{10^9 \text{ nm}} = 557.7 \times 10^{-9} \text{ m}$$

$$\nu = \frac{2.998 \times 10^8 \text{ m/s}}{557.7 \times 10^{-9} \text{ m}} = 5.376 \times 10^{14}/s = \boxed{5.376 \times 10^{14} \text{ Hz}}$$

---

Light visible to the eye is a tiny portion of the entire electromagnetic spectrum (Fig. 6.2), covering only the narrow wavelength region 400 to 700 nm. You can see the red glow given off by charcoal on a barbecue grill, but the heat given off is largely in the infrared (IR) region, above 700 nm. Microwave ovens produce radiation at even longer wavelengths. At the other end of the spectrum, below 400 nm, are ultraviolet (UV) radiation, responsible for sunburn, and x-rays, used for diagnostic purposes in medicine.

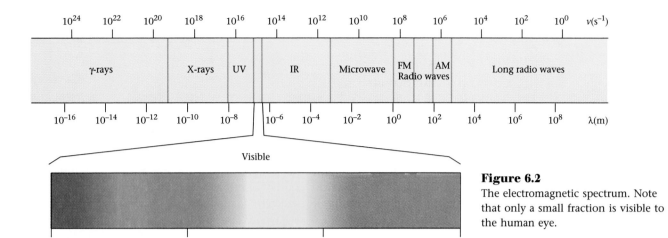

**Figure 6.2**
The electromagnetic spectrum. Note that only a small fraction is visible to the human eye.

## The Particle Nature of Light; Photon Energies

A hundred years ago it was generally supposed that all the properties of light could be explained in terms of its wave nature. A series of investigations carried out between 1900 and 1910 by Max Planck (blackbody radiation) and Albert Einstein (photoelectric effect) discredited that notion. Today we consider light to be generated as a stream of particles called **photons,** whose energy $E$ is given by the Einstein equation

$$E = h\nu = hc/\lambda$$

The quantity $h$ appearing in this equation is referred to as Planck's constant.

$$h = 6.626 \times 10^{-34} \text{ J} \cdot \text{s}$$

Notice from this equation that energy is *inversely* related to wavelength. This explains why you put on a sunscreen to protect yourself from UV solar radiation (<400 nm) and a "lead apron" when dental x-rays are being taken. Conversely, IR (>700 nm) and microwave radiation are of relatively low energy (but don't try walking on hot coals).

---

**Example 6.2**    Referring back to Example 6.1, calculate

   (a) the energy, in joules, of a photon emitted by an excited oxygen atom.
   (b) the energy, in kilojoules, of a mole of such photons.

***Strategy***    Use the equation $E = hc/\lambda$. In part (b), remember that 1 mol = $6.022 \times 10^{23}$ particles.

*Solution*

   (a) $E = \dfrac{hc}{\lambda} = \dfrac{(6.626 \times 10^{-34} \text{ J} \cdot \text{s})(2.998 \times 10^8 \text{ m/s})}{557.7 \times 10^{-9} \text{ m}} = \boxed{3.562 \times 10^{-19} \text{ J}}$

This seems like a tiny amount of energy, but remember that it's coming from a single oxygen atom.

   (b) $E = 3.562 \times 10^{-19} \text{ J} \times \dfrac{1 \text{ kJ}}{10^3 \text{ J}} \times \dfrac{6.022 \times 10^{23}}{1 \text{ mol}} = \boxed{2.145 \times 10^2 \text{ kJ/mol}}$

Lithium flame test (Larry Cameron)

This is roughly comparable to energy effects in chemical reactions; about 240 kJ of heat is evolved when a mole of hydrogen burns.

## Atomic Spectra

Atomic spectroscopy can identify metals at concentrations as low as $10^{-7}$ mol/L

In the seventeenth century, Sir Isaac Newton showed that visible (white) light from the sun can be broken down into its various color components by a prism. The *spectrum* obtained is continuous; it contains essentially all wavelengths between 400 and 700 nm.

The situation with high-energy atoms of gaseous elements is quite different (Fig. 6.3). Here, the spectrum consists of discrete lines given off at specific wavelengths. Each element has a characteristic spectrum that can be used to identify it. In the case of sodium, there are two strong lines in the yellow region at 589.0 nm and 589.6 nm. These lines account for the yellow color of sodium vapor lamps used to illuminate highways.

The fact that the photons making up atomic spectra have only certain discrete wavelengths implies that they can only have certain discrete energies, since

$$E = h\nu = hc/\lambda$$

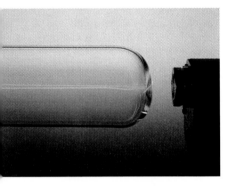

The He-Ne laser gives off orange-red light at 640.2 nm, corresponding to an electronic transition in the neon atom (Marna G. Clarke).

These photons are produced when an electron in an atom moves from one energy level to another. It follows that the electronic energy levels in an atom must be *quantized;* that is, they are limited to particular values. Moreover, it suggests that by measuring the spectrum of an element it should be possible to unravel its electronic energy levels. This is indeed possible, but it isn't easy. Gaseous atoms typically give off hundreds, even thousands, of spectral lines.

One of the simplest of atomic spectra, and the most important from a theoretical standpoint, is that of hydrogen. When energized by a high-voltage discharge, gaseous hydrogen atoms emit radiation at wavelengths that can be grouped into several different series (Table 6.1). The first of these to be discovered, the Balmer series, lies in the visible region. It consists of a strong line at 656.28 nm followed by successively weaker lines, closer and closer together, at lower wavelengths.

**Figure 6.3**

*Emission spectra.* The continuous spectrum at the top contains all visible wavelengths. The atomic spectra of gaseous Na, H, Ca, Hg, and Ne are quite different. They consist of discrete lines at certain definite wavelengths.

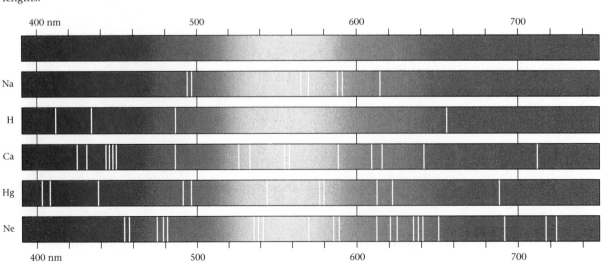

TABLE 6.1   **Wavelengths (nm) of Lines in the Atomic Spectrum of Hydrogen**

| Ultraviolet (Lyman Series) | Visible (Balmer Series) | Infrared (Paschen Series) |
|---|---|---|
| 121.53 | 656.28 | 1875.09 |
| 102.54 | 486.13 | 1281.80 |
| 97.23 | 434.05 | 1093.80 |
| 94.95 | 410.18 | 1004.93 |
| 93.75 | 397.01 | |
| 93.05 | | |

## 6.2   The Hydrogen Atom

The hydrogen atom, containing a single electron, has played a major role in the development of models of electronic structure. In 1913 Niels Bohr, a Danish physicist, offered a theoretical explanation of the atomic spectrum of hydrogen. His model was based largely upon classical mechanics. In 1922 this model won him the Nobel Prize in physics. By that time, Bohr had become director of the Institute of Theoretical Physics at Copenhagen. There he helped develop the new discipline of quantum mechanics, used by other scientists to construct a more sophisticated model for the hydrogen atom.

### Bohr Model

Bohr assumed that a hydrogen atom consists of a central proton about which an electron moves in a circular orbit. He related the electrostatic force of attraction of the proton for the electron to the centrifugal force due to the circular motion of the electron. In this way, Bohr was able to express the energy of the atom in terms of the radius of the electron's orbit. To this point, his analysis was purely classical, based on Coulomb's law of electrostatic attraction and Newton's laws of motion. To progress beyond this point, Bohr boldly and arbitrarily assumed, in effect, that the electron in the hydrogen atom can have only certain definite energies. Using arguments that we will not go into, Bohr obtained the following equation for the energy of the hydrogen electron:

$$E_\mathbf{n} = -R_H/\mathbf{n}^2$$

where $E_\mathbf{n}$ is the energy of the electron, $R_H$ is a quantity called the Rydberg constant (modern value = $2.180 \times 10^{-18}$ J), and $\mathbf{n}$ is an integer called the principal quantum number. Depending upon the state of the electron, $\mathbf{n}$ can have any positive, integral value, i.e.,

$$\mathbf{n} = 1, 2, 3, \ldots$$

Before proceeding with the Bohr model, let us make three points:

**1.** In setting up his model, Bohr designated zero energy as the point at which the proton and electron are completely separated. Energy has to be absorbed to reach that point. This means that the electron, in all its allowed energy states within the atom, must have an energy below zero, i.e., must be negative, hence the minus sign in the equation:

$$E_\mathbf{n} = -R_H/\mathbf{n}^2$$

**2.** Ordinarily the hydrogen electron is in its lowest state, referred to as the **ground state** or ground level, for which $n = 1$. When an electron absorbs enough energy, it moves to a higher, **excited state.** In a hydrogen atom, the first excited state has $n = 2$, the second $n = 3$, and so on.

**3.** When an excited electron gives off energy as a photon of light, it drops back to a lower energy state. The electron can return to the ground state, (from $n = 2$ to $n = 1$, for example) or to a lower excited state (from $n = 3$ to $n = 2$). At any rate, the energy of the photon ($h\nu$) evolved is equal to the difference in energy between the two states:

$$\Delta E = h\nu = E_{hi} - E_{lo}$$

where $E_{hi}$ and $E_{lo}$ are the energies of the higher and lower states, respectively.

Using this expression for $\Delta E$ and the equation $E_n = -R_H/n^2$, it is possible to relate the frequency of the light emitted to the quantum numbers, $n_{hi}$ and $n_{lo}$, of the two states:

$$h\nu = -R_H\left[\frac{1}{(n_{hi})^2} - \frac{1}{(n_{lo})^2}\right]$$

$$\nu = \frac{R_H}{h}\left[\frac{1}{(n_{lo})^2} - \frac{1}{(n_{hi})^2}\right]$$

The last equation written is the one Bohr derived in applying his model to the hydrogen atom. Given

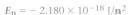

$E_n = -2.180 \times 10^{-18}\ \text{J}/n^2$

$$R_H = 2.180 \times 10^{-18}\ \text{J} \qquad h = 6.626 \times 10^{-34}\ \text{J} \cdot \text{s}$$

you can use the equation to find the frequency or wavelength of any of the lines in the hydrogen spectrum.

---

**Example 6.3**  Calculate the wavelength in nanometers of the line in the Balmer series that results from the transition $n = 3$ to $n = 2$.

***Strategy***  Use the above equation to find the frequency; $n_{lo} = 2$, $n_{hi} = 3$. Then use the equation $\lambda = c/\nu$ to find the wavelength.

*Solution*

$$\nu = \frac{2.180 \times 10^{-18}\ \text{J}}{6.626 \times 10^{-34}\ \text{J} \cdot \text{s}}\left[\frac{1}{4} - \frac{1}{9}\right] = 4.570 \times 10^{14}/\text{s}$$

$$\lambda = \frac{c}{\nu} = \frac{2.998 \times 10^8\ \text{m/s}}{4.570 \times 10^{14}/\text{s}} \times \frac{10^9\ \text{nm}}{1\ \text{m}} = \boxed{656.0\ \text{nm}}$$

Compare this value with that listed in Table 6.1 for the first line in the Balmer series.

---

## Quantum Mechanical Model

Bohr's theory for the structure of the hydrogen atom was highly successful. Scientists of the day must have thought they were on the verge of being able to predict the allowed energy levels of all atoms. However, the extension of Bohr's ideas to atoms with two or more electrons gave, at best, only qualitative agreement with experiment. Consider, for example, what happens when Bohr's theory is applied to the helium atom. Here, the errors in calculated energies and wavelengths are of the order of 5% instead of the 0.1% error with hydrogen.

There appeared to be no way the theory could be modified to make it work well with helium or other atoms. Indeed, it soon became apparent that there was a fundamental problem with the Bohr model. The idea of an electron moving about the nucleus in a well-defined orbit at a fixed distance from the nucleus had to be abandoned.

Scientists in the 1920s, speculating on this problem, became convinced that an entirely new approach was required to treat electrons in atoms, ions, and molecules. A new discipline, called *quantum mechanics,* was developed to describe the motion of small particles confined to tiny regions of space. In the quantum mechanical atom, no attempt is made to specify the position of an electron at a given instant, nor does quantum mechanics concern itself with the path that an electron takes about the nucleus. (After all, if we can't say where the electron is, we certainly don't know how it got there). Instead, quantum mechanics deals only with the *probability* of finding a particle within a given region of space.

In 1926 Erwin Schrödinger, an Austrian physicist working in Zurich, made a major contribution to quantum mechanics. He wrote down a rather complex differential equation to express the wave properties of an electron in an atom. This equation can be solved, at least in principle, to find the amplitude (height) $\psi$ of the electron wave at various points in space. The quantity $\psi$ (psi) is known as the *wave function.* Although we will not use the Schrödinger wave equation in any calculations, you should realize that much of our discussion of electronic structure is based upon solutions to that equation for the electron in the hydrogen atom.

In case you're curious, the equation is:

$$\frac{d^2\psi}{dx^2} + \frac{8\pi^2 m(E - V)}{h^2}\psi = 0$$

For the hydrogen electron, the square of the wave function, $\psi^2$, is directly proportional to the probability of finding the electron at a particular point. If $\psi^2$ at point A is twice as large as at point B, then we are twice as likely to find the electron at A as at B. Putting it another way, over time the electron will turn up at A twice as often as at B.

Figure 6.4 shows, in two different ways, how $\psi^2$ for the hydrogen electron in its ground state (**n** = 1) varies moving out from the nucleus. In Figure 6.4a, an *electron cloud* diagram, the depth of the color is directly proportional to $\psi^2$ and hence to the probability of finding the electron at a point. As you can see,

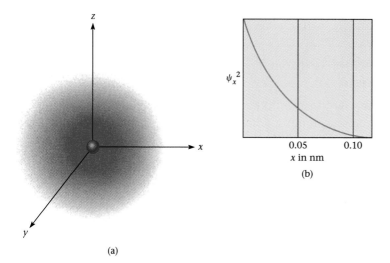

(a)

(b)

**Figure 6.4**
Two different ways of showing the relative probability of finding the hydrogen electron in its ground state at various distances from the nucleus. In (a), the depth of color is proportional to $\psi^2$. In (b), the probability, $\psi_x^2$, of finding the electron at a given distance along the *x*-axis is plotted.

the color fades moving out from the nucleus in any direction; the value of $\psi^2$ drops accordingly. This feature is emphasized in Figure 6.4b, where $\psi_x^2$ is plotted against distance along the *x*-axis. Notice that $\psi_x^2$ has its greatest value at the nucleus ($x = 0$) and decreases smoothly and rapidly as *x* increases.

## 6.3   Quantum Numbers, Energy Levels, and Orbitals

Although the Schrödinger equation is a rather complex one, it can be solved exactly for the hydrogen atom. Indeed, it turns out that there are many solutions for $\psi$, each associated with a set of numbers called **quantum numbers.** There are three such numbers, given the symbols **n**, $\ell$, and $\mathbf{m}_\ell$. A wave function corresponding to a particular set of three quantum numbers (e.g., **n** = 2, $\ell$ = 1, $\mathbf{m}_\ell = 0$) is referred to as an **atomic orbital.** Orbitals differ from one another in their energy and in the shape and spatial orientation of their electron cloud.

For reasons we will discuss later, a fourth quantum number is required to completely describe a specific electron in a multi-electron atom. The fourth quantum number is given the symbol $\mathbf{m}_s$. Each electron in an atom has a set of four quantum numbers, **n**, $\ell$, $\mathbf{m}_\ell$, and $\mathbf{m}_s$. We will now discuss the quantum numbers of electrons as they are used in atoms beyond hydrogen.

### First Quantum Number, n; Principal Energy Levels

The first quantum number, **n**, comes from the Bohr model of the hydrogen atom, where the energy depends only upon **n** ($E_\mathbf{n} = -R_H/\mathbf{n}^2$). In other atoms, the energy of each electron depends mainly, but not completely, upon the value of **n**. As **n** increases, the energy of the electron increases, and, on the average, it is found farther out from the nucleus. The quantum number **n** can take on only integral values, starting with 1:

$$\mathbf{n} = 1, 2, 3, 4, \ldots$$

In an atom, the value of **n** designates what we call a **principal energy level.** Thus, an electron for which **n** = 1 is said to be in the first principal level. If **n** = 2, we are dealing with the second principal level, and so on.

### Second Quantum Number, $\ell$; Sublevels (s, p, d, f)

Each principal level includes one or more **sublevels.** The sublevels are denoted by the second quantum number, $\ell$. As we shall see later, the general shape of the electron cloud associated with an orbital is determined by $\ell$. Larger values of $\ell$ produce more complex shapes.

*This relation between* **n** *and* $\ell$ *comes from the Schrödinger equation*

The quantum numbers **n** and $\ell$ are related; $\ell$ can take on any integral value starting with 0 and going up to a maximum of (**n** − 1). That is,

$$\ell = 0, 1, 2, \ldots, (\mathbf{n} - 1)$$

If **n** = 1, there is only one possible value of $\ell$, namely 0. This means that, in the first principal level, there is only one sublevel, for which $\ell = 0$. If **n** = 2, two

| TABLE 6.2 | Sublevel Designations for the First Four Principal Levels | | | | | | | | |
|---|---|---|---|---|---|---|---|---|---|
| n | 1 | 2 | | 3 | | | 4 | | |
| ℓ | 0 | 0 | 1 | 0 | 1 | 2 | 0 | 1 | 2 | 3 |
| Sublevel | 1s | 2s | 2p | 3s | 3p | 3d | 4s | 4p | 4d | 4f |

values of $\ell$ are possible, 0 and 1. In other words, there are two sublevels ($\ell = 0$ and $\ell = 1$) within the second principal energy level. Similarly,

if **n** = 3:    $\ell$ = 0, 1, or 2    (three sublevels)

if **n** = 4:    $\ell$ = 0, 1, 2, or 3    (four sublevels)

In general, *in the nth principal level, there are* **n** *different sublevels.*

Another method is commonly used to designate sublevels. Instead of giving the quantum number $\ell$, the letters s, p, d, or f* indicate the sublevels $\ell$ = 0, 1, 2, or 3, respectively. That is,

quantum number, $\ell$    0  1  2  3

type of sublevel        s   p   d   f

Usually, in designating a sublevel, a number is included (Table 6.2) to indicate the principal level as well. Thus reference is made to a 1s sublevel (**n** = 1, $\ell$ = 0), a 2s sublevel (**n** = 2, $\ell$ = 0), a 2p sublevel (**n** = 2, $\ell$ = 1), and so on.

For the hydrogen atom, the energy of an electron is independent of the value of $\ell$. (This is implied by the equation $E = -R_H/\mathbf{n}^2$, which involves only one quantum number). For multi-electron atoms, the situation is quite different. Here, the energy is dependent on $\ell$ as well as on **n**. Within a given principal level (same value of **n**), sublevels increase in energy in the order

$$\mathbf{n}s < \mathbf{n}p < \mathbf{n}d < \mathbf{n}f$$

Thus, a 2p sublevel has a slightly higher energy than a 2s sublevel. By the same token, when **n** = 3, the 3s sublevel has the lowest energy, the 3p is intermediate, and the 3d has the highest energy.

Which would have the higher energy, 4p or 3s?

## Third Quantum Number, $m_\ell$; Orbitals

Each sublevel contains one or more **orbitals,** which differ from one another in the value assigned to the third quantum number, $\mathbf{m}_\ell$. This quantum number determines the direction in space of the electron cloud surrounding the nucleus. The value of $\mathbf{m}_\ell$ is related to that of $\ell$. For a given value of $\ell$, $\mathbf{m}_\ell$ can have any integral value, including 0, between $\ell$ and $-\ell$; that is,

$$\mathbf{m}_\ell = \ell, \ldots, +1, 0, -1, \ldots, -\ell$$

To illustrate how this rule works, consider an s sublevel ($\ell = 0$). Here $\mathbf{m}_\ell$ can have only one value, 0. This means that an s sublevel contains only one or-

---

*These letters come from the adjectives used by spectroscopists to describe spectral lines: *sharp, princi*pal, *diffuse, and *fundamental.

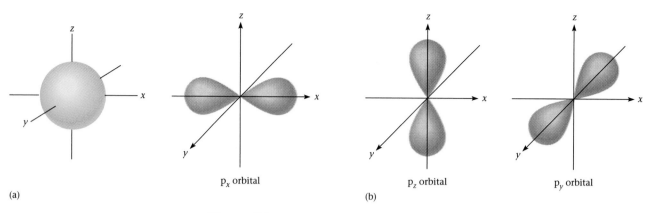

(a)

(b)

$p_x$ orbital

$p_z$ orbital

$p_y$ orbital

**Figure 6.5**
All s orbitals (a) are spherically symmetrical; the density of the electron cloud is independent of direction ($x$, $y$, or $z$). In contrast, the electron density in p orbitals (b) is concentrated along the $x$, $y$, or $z$ axis. The three p orbitals are directed at 90° angles to each other.

bital, referred to as an *s orbital*. The electron cloud associated with an s orbital is spherically symmetrical; the density of the cloud varies with distance from the nucleus but is independent of direction. Most commonly, an s orbital is shown as a simple sphere (Fig. 6.5a). The radius of the sphere indicates the region within which there is a specified probability of finding the electron. If we agree that the sphere should contain 90% of the density of the electron cloud, then a 1s orbital in a hydrogen atom has a radius of 0.14 nm. As the value of **n** increases, the radius of the orbital becomes larger. This means that an electron in a 2s orbital is more likely to be found far out from the nucleus than one in a 1s orbital.

For a p sublevel ($\ell = 1$), $\mathbf{m}_\ell = 1$, 0, or $-1$. Within a given p sublevel, there are three different orbitals described by the quantum numbers $\mathbf{m}_\ell = 1$, 0, and $-1$. Commonly, p orbitals are referred to as $p_x$, $p_y$, and $p_z$ orbitals (Fig. 6.5b).

For the d and f sublevels:

| | | | |
|---|---|---|---|
| d sublevel: | $\ell = 2$ | $\mathbf{m}_\ell = 2, 1, 0, -1, -2$ | 5 orbitals |
| f sublevel: | $\ell = 3$ | $\mathbf{m}_\ell = 3, 2, 1, 0, -1, -2, -3$ | 7 orbitals |

(The shapes and orientations of d orbitals are shown on p. 436).

In general, for a sublevel of quantum number $\ell$, there are a total of $2\ell + 1$ orbitals. All of the orbitals in a given sublevel (e.g., $2p_x$, $2p_y$, $2p_z$) have essentially the same energy.

## Fourth Quantum Number, $m_s$; Electron Spin

The fourth quantum number, $\mathbf{m}_s$, is associated with the spin of the electron. An electron has magnetic properties that correspond to those of a charged particle spinning on its axis. Either of two spins are possible, clockwise or counterclockwise (Figure 6.6).

The quantum number $\mathbf{m}_s$ was introduced to make theory consistent with experiment. In that sense, it differs from the first three quantum numbers, which

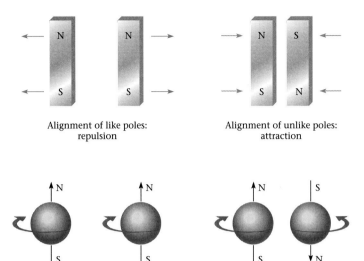

Alignment of like poles:
repulsion

Alignment of unlike poles:
attraction

**Figure 6.6**
In some respects, an electron behaves as if it were a spherical particle spinning about its axis. There is an analogy between the alignment of electron spins and the alignment of bar magnets (top). Within an orbital, the more stable arrangement is the one in which two electrons have opposed spins (lower right).

Repulsion

Some attraction

came from the solution to the Schrödinger wave equation for the hydrogen atom. This quantum number is not related to **n**, **ℓ**, or **m**$_\ell$. It can have either of two possible values:

$$\mathbf{m}_s = +1/2 \quad \text{or} \quad -1/2$$

Electrons that have the same value of **m**$_s$ (i.e., both $+\frac{1}{2}$ or both $-\frac{1}{2}$) are said to have *parallel* spins. Electrons that have different **m**$_s$ values (i.e., one $+\frac{1}{2}$ and the other $-\frac{1}{2}$) are said to have *opposed* spins.

## Pauli Exclusion Principle

The four quantum numbers that characterize an electron in an atom have now been considered. There is an important rule, called the Pauli exclusion principle, that relates to these numbers. It requires that ***no two electrons in an atom can have the same set of four quantum numbers.*** This principle was first stated in 1925 by Wolfgang Pauli, a colleague of Bohr, again to make theory consistent with the properties of atoms.

The Pauli exclusion principle has an implication that is not obvious at first glance. It requires that no more than two electrons can fit into an orbital. Moreover, if two electrons occupy the same orbital they must have opposed spins. To see that this is the case, consider the 2s orbital. Any electron in this orbital must have

$$\mathbf{n} = 2 \quad \mathbf{\ell} = 0 \quad \mathbf{m}_\ell = 0$$

To satisfy the Pauli exclusion principle, the electrons in this orbital must have different **m**$_s$ values. But there are only two possible values of **m**$_s$. Hence, only two electrons can enter the orbital. If the orbital is filled, one electron must have **m**$_s = +\frac{1}{2}$ and the other **m**$_s = -\frac{1}{2}$. In other words, the two electrons must have opposed spins.

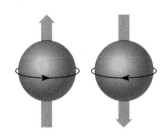

Electrons with opposed spins have different values of $m_s$, $+\frac{1}{2}$ and $-\frac{1}{2}$.

Our model for electronic structure is a pragmatic blend of theory and experiment

## Capacities of Principal Levels, Sublevels, and Orbitals

The rules that we have given for quantum numbers fix the capacities of principal levels, sublevels, and orbitals. In summary,

**1.** Each principal level of quantum number **n** contains a total of **n** sublevels.
**2.** Each sublevel of quantum number $\ell$ contains a total of $2\ell + 1$ orbitals; that is,

an s sublevel ($\ell = 0$) contains 1 orbital

a p sublevel ($\ell = 1$) contains 3 orbitals

a d sublevel ($\ell = 2$) contains 5 orbitals

an f sublevel ($\ell = 3$) contains 7 orbitals

**3.** Each orbital can hold two electrons, which must have opposed spins.

Applying these rules to the first three principal energy levels, we obtain Table 6.3. At the bottom of the table, each arrow indicates an electron. In each orbital, there are two electrons with opposed spins. The number of electrons in a sublevel is found by adding up the electrons in the orbitals within that sublevel. For example, in a p sublevel ($\ell = 1$) there are six electrons, two in each of three orbitals. To find the total number of electrons in a principal level, we add the electrons in the sublevels within that principal level.

---

### Example 6.4

(a) What is the capacity for electrons of the 3d sublevel?
(b) How many electrons can fit into the principal level for which **n** = 4?

***Strategy*** Apply the three rules just cited. Notice particularly that the capacity of a sublevel is found by multiplying the number of orbitals in that sublevel by two.

### *Solution*

(a) Each d sublevel contains five orbitals, so its capacity is

$$5 \times 2e^- = \boxed{10e^-}$$

(b) By rule (1), there must be four sublevels. These are 4s, 4p, 4d, 4f. Remember that an s sublevel has one orbital, a p sublevel three, a d sublevel five, and an f sublevel seven:

$$1(2e^-) + 3(2e^-) + 5(2e^-) + 7(2e^-) = \boxed{32e^-}$$

---

### TABLE 6.3 Allowed Sets of Quantum Numbers for Electrons in Atoms

| Level **n** | 1 | 2 | | | 3 | | | | | | |
|---|---|---|---|---|---|---|---|---|---|---|---|
| Sublevel $\ell$ | 0 | 0 | 1 | | 0 | 1 | | | 2 | | |
| Orbital $m_\ell$ | 0 | 0 | 1 | 0 | -1 | 0 | 1 | 0 | -1 | 2 | 1 | 0 | -1 | -2 |
| Spin $m_s$ $\uparrow = +\frac{1}{2}$ $\downarrow = -\frac{1}{2}$ | ↑↓ | ↑↓ | ↑↓ | ↑↓ | ↑↓ | ↑↓ | ↑↓ | ↑↓ | ↑↓ | ↑↓ | ↑↓ | ↑↓ | ↑↓ | ↑↓ |

| TABLE 6.4   **Capacities of Electronic Levels and Sublevels in Atoms** | | | | | |
|---|---|---|---|---|---|
| | | **Maximum Number of Electrons in Sublevels, $2(2\ell + 1)$** | | | |
| **Level n** | **Total Number of Electrons in Level, $2n^2$** | s | p | d | f |
| 1 | 2 | 2 | — | — | — |
| 2 | 8 | 2 | 6 | — | — |
| 3 | 18 | 2 | 6 | 10 | — |
| 4 | 32 | 2 | 6 | 10 | 14 |

Using the approach in Example 6.4 you can readily obtain the electron capacity of any principal level or sublevel. In Table 6.4, these capacities are listed through **n** = 4. Notice that

— *a sublevel of quantum number $\ell$ has a capacity of* $2(2\ell + 1)$. For example, a 3d sublevel ($\ell = 2$) has a capacity of $2(2 \times 2 + 1) = 10e^-$.
— *a principal level of quantum number n has a capacity of* $2\mathbf{n}^2$. For example, the total capacity of the fourth principal energy level is $2(4)^2 = 32e^-$.

## 6.4   Electron Configurations in Atoms

Given the rules referred to in Section 6.3, it is possible to assign quantum numbers to each electron in an atom. Beyond that, electrons can be assigned to specific principal levels, sublevels, and orbitals. There are several ways to do this. Perhaps the simplest way to describe the arrangement of electrons in an atom is to give its **electron configuration,** which shows the number of electrons, indicated by a superscript, in each sublevel. For example, a species with the electron configuration

$$1s^2 2s^2 2p^5$$

has two electrons in the 1s sublevel, two electrons in the 2s sublevel, and five electrons in the 2p sublevel.

In this section, you learn how to predict the electron configurations of atoms of elements. There are a couple of different ways of doing this, which we consider in turn. It should be emphasized that, throughout this discussion, we refer to *isolated gaseous atoms in the ground state.*

### Electron Configuration from Sublevel Energies

Electron configurations are readily obtained if the order of filling sublevels is known. Electrons enter the available sublevels in order of increasing sublevel energy. Ordinarily, a sublevel is filled to capacity before the next one starts to fill. The relative energies of different sublevels can be obtained from experiment. Figure 6.7 is a plot of these energies for atoms through the **n** = 4 principal level.

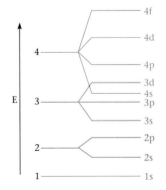

**Figure 6.7**
Arrangement of sublevels in order of increasing energy. This is also the order of filling, starting at the bottom with 1s, as atomic number increases.

From Figure 6.7 it is possible to predict the electron configurations of atoms of elements with atomic numbers 1 through 36. Since an s sublevel can hold only two electrons, the 1s is filled at helium ($1s^2$). With lithium ($Z = 3$), the third electron has to enter a new sublevel: this is the 2s, the lowest sublevel of the second principal energy level. Lithium has one electron in this sublevel ($1s^22s^1$). With beryllium ($Z = 4$), the 2s sublevel is filled ($1s^22s^2$). The next six elements fill the 2p sublevel. Their electron configurations are

| | | | |
|---|---|---|---|
| $_5$B | $1s^22s^22p^1$ | $_8$O | $1s^22s^22p^4$ |
| $_6$C | $1s^22s^22p^2$ | $_9$F | $1s^22s^22p^5$ |
| $_7$N | $1s^22s^22p^3$ | $_{10}$Ne | $1s^22s^22p^6$ |

Beyond neon, electrons enter the third principal level. The 3s sublevel is filled at magnesium:

$$_{12}\text{Mg} \qquad 1s^22s^22p^63s^2$$

Six more electrons are required to fill the 3p sublevel with argon:

$$_{18}\text{Ar} \qquad 1s^22s^22p^63s^23p^6$$

After argon, an "overlap" of principal energy levels occurs. The next electron enters the *lowest* sublevel of the fourth principal level (4s) instead of the *highest* sublevel of the third principal level (3d). Potassium ($Z = 19$) has one electron in the 4s sublevel; calcium ($Z = 20$) fills it with two electrons:

$$_{20}\text{Ca} \qquad 1s^22s^22p^63s^23p^64s^2$$

Now, the 3d sublevel starts to fill with scandium ($Z = 21$). Recall that a d sublevel has a capacity of ten electrons. Hence the 3d sublevel becomes filled at zinc ($Z = 30$):

The order of filling, though Z = 36 is 1s, 2s, 2p, 3s, 3p, 4s, 3d, 4p

$$_{30}\text{Zn} \qquad 1s^22s^22p^63s^23p^64s^23d^{10}$$

The next sublevel, 4p, is filled at krypton ($Z = 36$):

$$_{36}\text{Kr} \qquad 1s^22s^22p^63s^23p^64s^23d^{10}4p^6$$

---

**Example 6.5**   Find the electron configuration of the sulfur atom and the nickel atom.

*Strategy*   Determine the number of electrons in the atom from its atomic number. Fill the appropriate sublevels using the energy diagram in Figure 6.7.

*Solution*   The sulfur atom has atomic number 16, so there are $16e^-$. Two electrons go into the 1s sublevel, two into the 2s sublevel, and six into the 2p sublevel. Two more go into the 3s sublevel, and the remaining four go into the 3p sublevel. The electron configuration of the S atom is

$$1s^22s^22p^63s^23p^4$$

The nickel atom has 28 electrons. Proceeding as above, the electron configuration for nickel is

$$1s^22s^22p^63s^23p^64s^23d^8$$

The 4s sublevel fills before the 3d sublevel.

---

Often, to save space, electron configurations are shortened; the **abbreviated electron configuration** starts with the preceding noble gas. For the elements sulfur and nickel,

| | Electron Configuration | Abbreviated Electron Configuration |
|---|---|---|
| $_{16}$S | $1s^22s^22p^63s^23p^4$ | [Ne] $3s^23p^4$ |
| $_{28}$Ni | $1s^22s^22p^63s^23p^64s^23d^8$ | [Ar] $4s^23d^8$ |

The symbol [Ne] indicates that the first 10 electrons in the sulfur atom have the neon configuration $1s^22s^22p^6$; similarly, [Ar] represents the first 18 electrons in the nickel atom.

## Filling of Sublevels and the Periodic Table

In principle, a diagram such as Figure 6.7 can be extended to include all sublevels occupied by electrons in the 112 known elements. As a matter of fact, that is a relatively simple thing to do; such a figure is in effect incorporated into the periodic table introduced in Chapter 2.

To understand how position in the periodic table relates to the filling of sublevels, consider the metals in the first two groups. Atoms of the Group 1 elements all have one s electron in the outermost principal energy level (Table 6.5. In each Group 2 atom, there are two s electrons in the outermost level. A similar relationship applies to the elements in any group:

*The atoms of elements in a group of the periodic table have the same distribution of electrons in the outermost principal energy level.*

The periodic table "works" because an element's chemical properties depend upon the number of the outer electrons

This means that the order in which electron sublevels are filled is determined by position in the periodic table. Figure 6.8 (p. 155) shows how this works. Notice the following points:

**1. *The elements in Groups 1 and 2 are filling an s sublevel.*** Thus, Li and Be in the second period fill the 2s sublevel. Na and Mg in the third period fill the 3s sublevel, and so on.

**2. *The elements in Groups 13 through 18 (six elements in each period) fill p sublevels,*** which have a capacity of six electrons. In the second period, the 2p sublevel starts to fill with B ($Z = 5$) and is completed with Ne ($Z = 10$). In the third period, the elements Al ($Z = 13$) through Ar ($Z = 18$) fill the 3p sublevel.

**3. *The transition metals, in the center of the periodic table, fill d sublevels.*** Remember that a d sublevel can hold ten electrons. In the fourth period,

---

TABLE 6.5 **Abbreviated Electron Configurations of Group 1 and 2 Elements**

| Group 1 | | Group 2 | |
|---|---|---|---|
| $_3$Li | [He]**2s$^1$** | $_4$Be | [He]**2s$^2$** |
| $_{11}$Na | [Ne]**3s$^1$** | $_{12}$Mg | [Ne]**3s$^2$** |
| $_{19}$K | [Ar]**4s$^1$** | $_{20}$Ca | [Ar]**4s$^2$** |
| $_{37}$Rb | [Kr]**5s$^1$** | $_{38}$Sr | [Kr]**5s$^2$** |
| $_{55}$Cs | [Xe]**6s$^1$** | $_{56}$Ba | [Xe]**6s$^2$** |

the ten elements Sc ($Z = 21$) through Zn ($Z = 30$) fill the 3d sublevel. In the fifth period, the 4d sublevel is filled by the elements Y ($Z = 39$) through Cd ($Z = 48$). The ten transition metals in the sixth period fill the 5d sublevel. Elements 103 to 112 in the seventh, incomplete period are believed to be filling the 6d sublevel.

**4. *The two sets of 14 elements listed separately at the bottom of the table are filling f sublevels*** with a principal quantum number two less than the period number. That is,

— 14 elements in the sixth period ($Z = 57$ to 70) are filling the 4f sublevel. These elements are sometimes called rare earths or, more commonly nowadays, **lanthanides,** after the name of the first element in the series, lanthanum (La). Modern separation techniques, notably chromatography, have greatly increased the availability of compounds of these elements. A brilliant red phosphor used in color TV receivers contains a small amount of europium oxide, $Eu_2O_3$. This is added to a base of yttrium oxide, $Y_2O_3$, or gadolinium oxide, $Gd_2O_3$. Cerium(IV) oxide is used to coat interior surfaces of "self-cleaning" ovens, where it prevents the buildup of tar deposits.

— 14 elements in the seventh period ($Z = 89$ to 102) are filling the 5f sublevel. The first element in this series is actinium (Ac); collectively, these elements are referred to as **actinides.** All these elements are radioactive; only thorium and uranium occur in nature. The other actinides have been synthesized in the laboratory by nuclear reactions. Their stability decreases rapidly with increasing atomic number. The longest-lived isotope of nobelium ($_{102}$No) has a half-life of about 3 min; i.e., in 3 minutes half of the sample decomposes. Nobelium and the preceding element, mendeleevium ($_{101}$Md), were identified in samples containing one to three atoms of No or Md.

## Electron Configuration from the Periodic Table

Figure 6.8 (or any periodic table) can be used to deduce the electron configuration of any element. It is particularly useful for heavier elements such as iodine (Example 6.6).

---

**Example 6.6**   Write the electron configuration of iodine.

***Strategy***   Using Figure 6.8, go across each period in succession, noting the sublevels filled, until you get to iodine.

***Solution***

To obtain electron configurations from the periodic table, consider what sublevels are filled going across each period

Period 1:   the 1s sublevel fills ($1s^2$)

Period 2:   the 2s and 2p sublevels fill ($2s^2 2p^6$)

Period 3:   the 3s and 3p sublevels fill ($3s^2 3p^6$)

Period 4:   the 4s, 3d, and 4p sublevels fill ($4s^2 3d^{10} 4p^6$)

Period 5:   the 5s and 4d sublevels fill; five electrons enter the 5p ($5s^2 4d^{10} 5p^5$)

Putting it all together, the electron configuration of iodine must be

$$1s^2 2s^2 2p^6 3s^2 3p^6 4s^2 3d^{10} 4p^6 5s^2 4d^{10} 5p^5$$

---

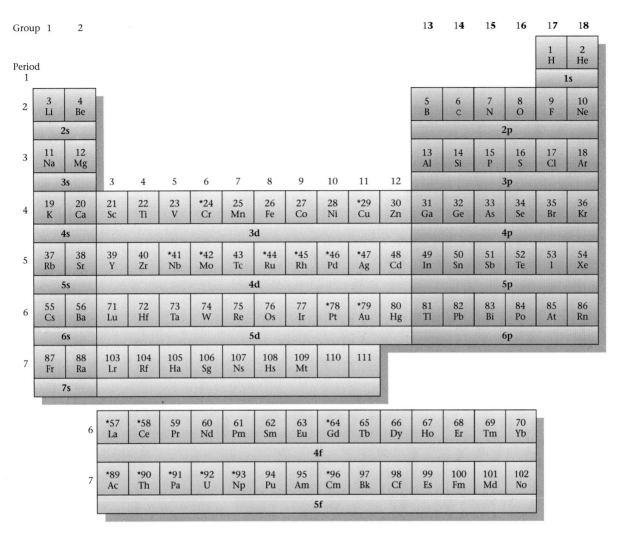

**Figure 6.8**

The periodic table can be used to deduce the electron configurations of atoms. Elements marked with an asterisk have electron configurations slightly different from that predicted by the periodic table.

As you can see from Figure 6.8, the electron configurations of several elements (marked*) differ slightly from those predicted. In every case, the difference involves a shift of one or, at the most, two electrons from one sublevel to another of very similar energy. For example, in the first transition series, two elements, chromium and copper, have an "extra" electron in the 3d as compared with the 4s orbital.

| | Predicted | Observed |
|---|---|---|
| $_{24}$Cr | [Ar] $4s^2 3d^4$ | [Ar] $4s^1 3d^5$ |
| $_{29}$Cu | [Ar] $4s^2 3d^9$ | [Ar] $4s^1 3d^{10}$ |

A similar transposition occurs with six metals (Nb, Mo, Ru, Rh, Pd, Ag) in the second transition series and two more (Pt, Au) in the third.

# Chemistry: *The Human Side*

**Dmitri Mendeleev**
(1836–1907)

Lothar Meyer, a German chemist, also made important contributions to the periodic table

The periodic table, based on chemical similarities among elements, came long before electron configurations. Indeed, the table predated the discovery of the electron itself by 30 years. The person whose name is most closely associated with the periodic table is Dmitri Mendeleev, Professor of Chemistry at the University of St. Petersburg in Tsarist Russia. In writing a textbook of general chemistry, Mendeleev devoted separate chapters to families of elements with similar chemical properties, including the alkali metals (Group 1 in the modern periodic table), the alkaline earth metals (Group 2), and the halogens (Group 17). Reflecting on the behavior of these elements and others, Mendeleev was struck by the fact that their chemical properties vary periodically with atomic mass. Mendeleev went a step further and, in a paper presented to the Russian Chemical Society in March 1869, proposed a primitive version of today's periodic table.

The acclaim that greeted this and successive papers lifted Mendeleev from obscurity to fame. His textbook, *Principles of Chemistry,* first appeared in 1870 and was widely adopted through eight editions.

The genius of Mendeleev was that he shrewdly left empty spaces in his periodic table for new elements yet to be discovered. Indeed, he went further than that, predicting detailed properties for three such elements: "ekaboron" (scandium), "ekaaluminum" (gallium), and "ekasilicon" (germanium). By 1886, all the elements predicted by Mendeleev had been isolated. They were shown to have properties very close to those he predicted (Table 6.A). The spectacular agreement between prediction and experiment removed any doubts about the validity and value of Mendeleev's periodic table.

Mendeleev, like most of us, was fallible. He predicted the existence of two elements with atomic masses smaller than that of hydrogen. Beyond that, convinced that atoms were indivisible, he refused to accept the existence of electrons or ions.

Two remarkable women played a major role in Mendeleev's life. His mother, Maria Korniloff Mendeleev, raised Dmitri and seven siblings on her own; their father became blind shortly after Dmitri was born. In her spare time, she operated a glass factory to support the family. Dmitri's second wife, Anna Popova, was a talented artist who filled his study with sketches of prominent scientists, including Lavoisier, Newton, and Faraday. More important, she aroused in Dmitri an interest in art and literature to complement his devotion to science.

TABLE 6.A **Predicted and Observed Properties of Germanium (Ekasilicon)**

| Property | Predicted by Mendeleev (1871) | Observed (1886) |
|---|---|---|
| Atomic mass | 73 amu | 72.3 amu |
| Density | 5.5 g/cm$^3$ | 5.47 g/cm$^3$ |
| Specific heat | 0.31 J/(g · °C) | 0.32 J/(g · °C) |
| Melting point | very high | 960°C |
| Formula of oxide | RO$_2$ | GeO$_2$ |
| Formula of chloride | RCl$_4$ | GeCl$_4$ |
| Density of oxide | 4.7 g/cm$^3$ | 4.70 g/cm$^3$ |
| Boiling point of chloride | 100°C | 86°C |

# 6.5  Orbital Diagrams of Atoms

For many purposes, electron configurations are sufficient to describe the arrangements of electrons in atoms. Sometimes, however, it is useful to go a step further and show how electrons are distributed among orbitals. In such cases, **orbital diagrams** are used. Each orbital is represented by parentheses ( ), and electrons are shown by arrows written ↑ or ↓, depending upon spin.

To show how orbital diagrams are obtained from electron configurations, consider the boron atom ($Z = 5$). Its electron configuration is $1s^2 2s^2 2p^1$. The pair of electrons in the 1s orbital must have opposed spins ($+\frac{1}{2}, -\frac{1}{2}$ or ↑↓). The same is true of the two electrons in the 2s orbital. There are three orbitals in the 2p sublevel. The single 2p electron in boron could be in any one of these orbitals. Its spin could be either "up" or "down." The orbital diagram is ordinarily written

$$\begin{array}{cccc} & 1s & 2s & 2p \\ \\ _5B & (\uparrow\downarrow) & (\uparrow\downarrow) & (\uparrow)(\ )(\ ) \end{array}$$

with the first electron in an orbital arbitrarily designated by an "up" arrow, ↑.

With the next element, carbon, a complication arises. In which orbital should the sixth electron go? It could go in the same orbital as the other 2p electron, in which case it would have to have the opposite spin, ↓. It could go into one of the other two orbitals, either with a parallel spin, ↑, or an opposed spin, ↓. Experiment shows that there is an energy difference among these arrangements. The most stable is the one in which the two electrons are in different orbitals with parallel spins. The orbital diagram of the carbon atom is

$$\begin{array}{cccc} & 1s & 2s & 2p \\ \\ _6C & (\uparrow\downarrow) & (\uparrow\downarrow) & (\uparrow)(\uparrow)(\ ) \end{array}$$

Similar situations arise frequently. There is a general principle that applies in all such cases; **Hund's rule** (Frederick Hund 1896– ) predicts that

*When several orbitals of equal energy are available, as in a given sublevel, electrons enter singly with parallel spins.*

Only after all the orbitals are half-filled do electrons "pair up" in orbitals.

Following this principle, the orbital diagrams for the elements boron through neon are shown in Figure 6.9 p. 158. Notice that

— in all filled orbitals, the two electrons have opposed spins. Such electrons are often referred to as being *paired*. There are four paired electrons in the B, C, and N atoms, six in the oxygen atom, eight in the fluorine atom, and ten in the neon atom.

— in accordance with Hund's rule, within a given sublevel there are as many half-filled orbitals as possible. Electrons in such orbitals are said to be *unpaired*. There is one unpaired electron in atoms of B and F, two unpaired electrons in C and O atoms, and three unpaired electrons in the N atom. When

Professor Hund celebrated his 100th birthday on February 4, 1996

**Figure 6.9**
Orbital diagrams showing electron arrangements for atoms with five to ten electrons. Electrons entering orbitals of equal energy remain unpaired as long as possible.

| Atom | Orbital diagram | | | Electron configuration |
|------|-----------------|---|---|------------------------|
| B | $(\uparrow\downarrow)$ $(\uparrow\downarrow)$ | $(\uparrow\ )\ (\ \ )\ (\ \ )$ | | $1s^22s^22p^1$ |
| C | $(\uparrow\downarrow)$ $(\uparrow\downarrow)$ | $(\uparrow\ )\ (\uparrow\ )\ (\ \ )$ | | $1s^22s^22p^2$ |
| N | $(\uparrow\downarrow)$ $(\uparrow\downarrow)$ | $(\uparrow\ )\ (\uparrow\ )\ (\uparrow\ )$ | | $1s^22s^22p^3$ |
| O | $(\uparrow\downarrow)$ $(\uparrow\downarrow)$ | $(\uparrow\downarrow)\ (\uparrow\ )\ (\uparrow\ )$ | | $1s^22s^22p^4$ |
| F | $(\uparrow\downarrow)$ $(\uparrow\downarrow)$ | $(\uparrow\downarrow)\ (\uparrow\downarrow)\ (\uparrow\ )$ | | $1s^22s^22p^5$ |
| Ne | $(\uparrow\downarrow)$ $(\uparrow\downarrow)$ | $(\uparrow\downarrow)\ (\uparrow\downarrow)\ (\uparrow\downarrow)$ | | $1s^22s^22p^6$ |
|  | 1s  2s | 2p | | |

there are two or more unpaired electrons, as in C, N, and O, those electrons have parallel spins.

Hund's rule, like the Pauli exclusion principle, is based upon experiment. It is possible to determine the number of unpaired electrons in an atom. With solids, this is done by studying their behavior in a magnetic field. If there are unpaired electrons present, the solid will be attracted into the field. Such a substance is said to be *paramagnetic*. If the atoms in the solid contain only paired electrons, it is slightly repelled by the field. Substances of this type are called *diamagnetic*. With gaseous atoms, the atomic spectrum can also be used to establish the presence and number of unpaired electrons.

---

**Example 6.7**  Construct orbital diagrams for atoms of sulfur and iron.

***Strategy***  Start with the electron configuration, obtained as in Section 6.4. Then write the orbital diagram, recalling the number of orbitals per sublevel, putting two electrons of opposed spin in each orbital within a completed sublevel, and applying Hund's rule where sublevels are partially filled.

***Solution***  Recall from Example 6.5 that the electron configuration of sulfur is $1s^22s^22p^63s^23p^4$. Its orbital diagram is

| | 1s | 2s | 2p | 3s | 3p |
|------|----|----|----|----|----|
| $_{16}$S | $(\uparrow\downarrow)$ | $(\uparrow\downarrow)$ | $(\uparrow\downarrow)(\uparrow\downarrow)(\uparrow\downarrow)$ | $(\uparrow\downarrow)$ | $(\uparrow\downarrow)(\uparrow)(\uparrow)$ |

The atomic number of iron is 26; its electron configuration is $1s^22s^22p^63s^23p^64s^23d^6$. All the orbitals are filled except those in the 3d sublevel, which is populated according to Hund's rule to give four unpaired electrons.

| | 1s | 2s | 2p | 3s | 3p | 4s | 3d |
|------|----|----|----|----|----|----|----|
| $_{26}$Fe | $(\uparrow\downarrow)$ | $(\uparrow\downarrow)$ | $(\uparrow\downarrow)(\uparrow\downarrow)(\uparrow\downarrow)$ | $(\uparrow\downarrow)$ | $(\uparrow\downarrow)(\uparrow\downarrow)(\uparrow\downarrow)$ | $(\uparrow\downarrow)$ | $(\uparrow\downarrow)(\uparrow)(\uparrow)(\uparrow)(\uparrow)$ |

# 6.6  Electron Arrangements in Monatomic Ions

The discussion so far in this chapter has focused upon electron configurations and orbital diagrams of neutral atoms. It is also possible to assign electronic structures to monatomic ions, formed from atoms by gaining or losing electrons. In general, when a monatomic ion is formed from an atom, *electrons are added to or removed from sublevels in the highest principal energy level.*

### Ions with Noble Gas Structures

As pointed out in Chapter 2, elements close to a noble gas in the periodic table form ions that have the same number of electrons as the noble gas atom. This means that these ions have noble gas electron configurations. Thus the three elements preceding Ne (N, O, and F) and the three elements following neon (Na, Mg, and Al) all form ions with the neon configuration, $1s^2 2s^2 2p^6$. The three nonmetal atoms achieve this structure by gaining electrons to form anions:

$$_7N\ (1s^2 2s^2 2p^3) + 3e^- \longrightarrow {}_7N^{3-}\ (1s^2 2s^2 2p^6)$$

$$_8O\ (1s^2 2s^2 2p^4) + 2e^- \longrightarrow {}_8O^{2-}\ (1s^2 2s^2 2p^6)$$

$$_9F\ (1s^2 2s^2 2p^5) + e^- \longrightarrow {}_9F^-\ (1s^2 2s^2 2p^6)$$

The three metal atoms acquire the neon structure by losing electrons to form cations:

$$_{11}Na\ (1s^2 2s^2 2p^6 3s^1) \longrightarrow {}_{11}Na^+\ (1s^2 2s^2 2p^6) + e^-$$

$$_{12}Mg\ (1s^2 2s^2 2p^6 3s^2) \longrightarrow {}_{12}Mg^{2+}\ (1s^2 2s^2 2p^6) + 2e^-$$

$$_{13}Al\ (1s^2 2s^2 2p^6 3s^2 3p^1) \longrightarrow {}_{13}Al^{3+}\ (1s^2 2s^2 2p^6) + 3e^-$$

The species $N^{3-}$, $O^{2-}$, $F^-$, Ne, $Na^+$, $Mg^{2+}$ and $Al^{3+}$ are said to be *isoelectronic;* they have the same electron configuration.

There are a great many monatomic ions that have noble gas configurations; Figure 6.10 shows 24 ions of this type. Note, once again, that ions in a given main group have the same charge (+1 for Group 1, +2 for Group 2, −2 for

Important ideas are worth repeating

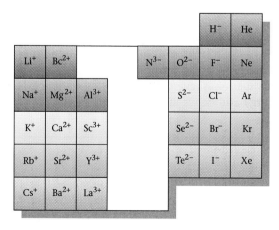

**Figure 6.10**
Species (cations, anions, or atoms) with noble gas structures. The color coding indicates species that are *isoelectronic* (same electronic structure).

Group 16, −1 for Group 17). This explains, in part, the chemical similarity among elements in the same main group. In particular, ionic compounds formed by such elements have similar chemical formulas. For example,

— halides of the alkali metals have the general formula MX, where M = Li, Na, K, . . . and X = F, Cl, Br . . . .
— halides of the alkaline earth metals have the general formula $MX_2$, where M = Mg, Ca, Sr, . . . and X = F, Cl, Br, . . . .
— oxides of the alkaline earth metals have the general formula MO (M = Mg, Ca, Sr, . . .).

## Transition Metal Cations

The transition metals to the right of the scandium subgroup do not form ions with noble gas configurations. To do so, they would have to lose four or more electrons. The energy requirement is too high for that to happen. However, as pointed out in Chapter 2, these metals do form cations with charges of +1, +2, or +3. Applying the principle that, in forming cations, electrons are removed from the sublevel of highest $n$, you can predict correctly that *when transition metal atoms form positive ions, the outer s electrons are lost first.* Consider, for example, the formation of the $Mn^{2+}$ ion from the Mn atom:

$$_{25}Mn \quad [Ar] \, 4s^2 3d^5 \qquad _{25}Mn^{2+} \quad [Ar] \, 3d^5$$

Notice that it is the 4s electrons that are lost rather than the 3d electrons. This is known to be the case because the $Mn^{2+}$ ion has been shown to have five unpaired electrons (the five 3d electrons). If two 3d electrons had been lost, the $Mn^{2+}$ ion would have had only three unpaired electrons.

All the transition metals form cations by a similar process, i.e., loss of outer s electrons. Only after those electrons are lost are electrons removed from the inner d sublevel. Consider, for example, what happens with iron, which, you will recall, forms two different cations. First the 4s electrons are lost to give the $Fe^{2+}$ ion:

*The s electrons are "first in" with the atoms and "first out" with the cations*

$$_{26}Fe \, (Ar \, 4s^2 3d^6) \longrightarrow {}_{26}Fe^{2+} \, (Ar \, 3d^6) + 2e^-$$

Then an electron is removed from the 3d level to form the $Fe^{3+}$ ion:

$$_{26}Fe^{2+} \, (Ar \, 3d^6) \longrightarrow {}_{26}Fe^{3+} \, (Ar \, 3d^5) + e^-$$

In the $Fe^{2+}$ and $Fe^{3+}$ ions, as in all transition metal ions, there are no outer s electrons.

Recall (Figure 6.7) that for fourth-period *atoms,* the 4s level appears to be lower in energy than the 3d. This situation is reversed in the corresponding *ions;* the 4s electrons are higher in energy and hence are removed before the 3d.

Many transition metal ions form brightly colored coordination compounds (Chapter 15).

---

**Example 6.8**    Give the electron configuration of

    (a) $Zn^{2+}$     (b) $Se^{2-}$

*Strategy*    First obtain the electron configuration of the corresponding atom, as in Section 6.4. Then add or remove electrons from sublevels of highest $n$.

## Solution

(a) The electron configuration of Zn ($Z = 30$) is $1s^2 2s^2 2p^6 3s^2 3p^6 4s^2 3d^{10}$

Two electrons are removed from the 4s sublevel to form $Zn^{2+}$:

$$Zn^{2+} \qquad 1s^2 2s^2 2p^6 3s^2 3p^6 3d^{10}$$

(b) The electron configuration of Se ($Z = 34$) is $1s^2 2s^2 2p^6 3s^2 3p^6 4s^2 3d^{10} 4p^4$.

Two electrons are added to the 4p sublevel to form $Se^{2-}$:

$$Se^{2-} \qquad 1s^2 2s^2 2p^6 3s^2 3p^6 4s^2 3d^{10} 4p^6$$

This is the electron configuration of the noble gas krypton ($Z = 36$).

# 6.7  Periodic Trends in the Properties of Atoms

One of the most fundamental principles of chemistry is the periodic law, which states that

***The chemical and physical properties of elements are a periodic function of atomic number.***

To illustrate what this statement means, consider Figure 6.11, which shows how the highest common oxidation number (HCON) of an element changes with atomic number. Notice that it varies periodically from 0 to +7 through the first four periods of the periodic table.

In this section we will consider how the periodic table can be used to correlate properties on an atomic scale. In particular, we will see how atomic radius, ionic radius, ionization energy, and electronegativity vary horizontally and vertically in the periodic table.

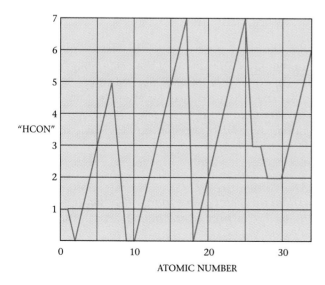

**Figure 6.11**
For the first 25 elements, the highest oxidation number varies periodically from 0 to +7.

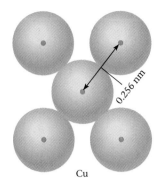

Cu

Atomic radius = $\dfrac{0.256 \text{ nm}}{2}$ = 0.128 nm

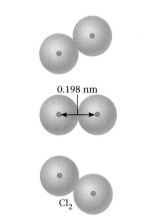

0.198 nm

$Cl_2$

Atomic radius = $\dfrac{0.198 \text{ nm}}{2}$ = 0.099 nm

**Figure 6.12**
Atomic radii are determined by assuming that atoms in closest contact in an element touch one another. The atomic radius is taken to be one half of the closest internuclear distance.

## Atomic Radius

Strictly speaking, the "size" of an atom is a rather nebulous concept. The electron cloud surrounding the nucleus does not have a sharp boundary. However, a quantity called the **atomic radius** can be defined and measured, assuming a spherical atom. Ordinarily, the atomic radius is taken to be one half the distance of closest approach between atoms in an elemental substance (Fig. 6.12).

The atomic radii of the main-group elements are shown at the top of Figure 6.13. Notice that, in general, atomic radii

— decrease across a period from left to right in the periodic table
— increase down a group in the periodic table

It is possible to explain these trends in terms of the electron configurations of the corresponding atoms. Consider first the increase in radius observed as we move down the table, let us say among the alkali metals (Group 1). All these elements have a single s electron outside a filled level or filled p sublevel. Electrons in these inner levels are much closer to the nucleus than the outer s electron and hence effectively shield it from the positive charge of the nucleus. To a first approximation, each inner electron cancels the charge of one proton in the nucleus, so the outer s electron is attracted by a net positive charge of +1. In this sense, it has the properties of an electron in the hydrogen atom. Since the average distance of the electron from the hydrogen nucleus increases with the principle quantum number, the radius increases moving from Li (2s electron) to Na (3s electron) and so on down the group.

The decrease in atomic radius moving across the periodic table can be explained in a similar manner. Consider, for example, the third period, where electrons are being added to the third principal energy level. The added electrons should be relatively poor shields for each other since they are all at about the same distance from the nucleus. Only the ten electrons in inner, filled levels ($\mathbf{n} = 1$, $\mathbf{n} = 2$) are expected to shield the outer electrons from the nucleus. This means that the *"effective nuclear charge"* should increase steadily moving across the period. As effective nuclear charge increases, outer electrons are pulled in more tightly, and atomic radius decreases.

## Ionic Radius

The radii of cations and anions derived from atoms of the main-group elements are shown at the bottom of Figure 6.13. The trends referred to previously for atomic radii are clearly visible here as well. Notice, for example, that ionic radius increases moving down a group in the periodic table. Moreover, the radii of both cations (left) and anions (right) decrease from left to right across a period.

Comparing the radii of cations and anions with those of the atoms from which they are derived (Fig. 6.13):

— *positive ions are smaller than the metal atoms from which they are formed.* The $Na^+$ ion has a radius, 0.095 nm, only a little more than half that of the Na atom, 0.186 nm.
— *negative ions are larger than the nonmetal atoms from which they are formed.* The radius of the $Cl^-$ ion, 0.181 nm, is nearly twice that of the Cl atom, 0.099 nm.

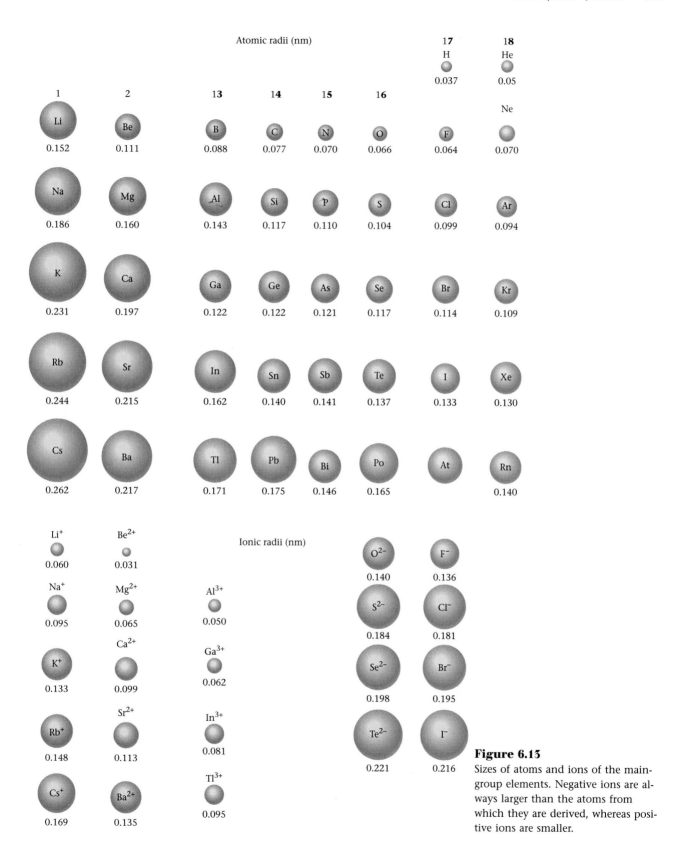

**Figure 6.13**

Sizes of atoms and ions of the main-group elements. Negative ions are always larger than the atoms from which they are derived, whereas positive ions are smaller.

As a result of these effects, anions in general are larger than cations. Compare, for example, the $Cl^-$ ions (radius = 0.181 nm) with the $Na^+$ ion (radius = 0.095 nm). This means that in sodium chloride, and indeed in the vast majority of all ionic compounds, most of the space in the crystal lattice is taken up by anions.

The differences in radii between atoms and ions can be explained quite simply. A cation is smaller than the corresponding metal atom because the excess of protons in the ion draws the outer electrons in closer to the nucleus. In contrast, an extra electron in an anion adds to the repulsion between outer electrons, making a negative ion larger than the corresponding nonmetal atom.

---

**Example 6.9**  Using only the periodic table, arrange each of the following sets of atoms and ions in order of increasing size.

(a) Mg, Al, Ca      (b) S, Cl, $S^{2-}$      (c) Fe, $Fe^{2+}$, $Fe^{3+}$

*Strategy*  Recall that radius decreases across a period and increases moving down a group. An atom is larger than the corresponding cation but smaller than the corresponding anion.

*Solution*

(a) Compare the other atoms with Mg. Al, to the right, is smaller than Mg. Ca, below Mg, is larger. The predicted order is   Al < Mg < Ca.

(b) Compare with the S atom. Cl, to the right, is smaller. The $S^{2-}$ anion is larger than the S atom. The predicted order is   Cl < S < $S^{2-}$.

(c) Compare with the $Fe^{2+}$ ion. The Fe atom, with no charge, is larger. The $Fe^{3+}$ ion, with a +3 charge, is smaller. The predicted order is   $Fe^{3+}$ < $Fe^{2+}$ < Fe.

---

## Ionization Energy

Ionization energy is a measure of how difficult it is to remove an electron from a gaseous atom. Energy must always be *absorbed* to bring about ionization, so ionization energies are always *positive* quantities.

The (first) ionization energy is the energy change for the removal of the outermost electron from a gaseous atom to form a +1 ion:

$$M(g) \longrightarrow M^+(g) + e^- \qquad \Delta E_1 = \text{first ionization energy}$$

The more difficult it is to remove electrons, the larger the ionization energy.

Ionization energies of the main-group elements are listed in Figure 6.14. Notice that ionization energy

— increases across the periodic table from left to right
— decreases moving down the periodic table

Comparing Figures 6.13 and 6.14 shows an inverse correlation between ionization energy and atomic radius. The smaller the atom, the more tightly its electrons are held to the positively charged nucleus and the more difficult they are to remove. Conversely, in a large atom such as that of a Group 1 metal, the

|  | 1 | 2 |  | 13 | 14 | 15 | 16 | 17 | 18 |
|---|---|---|---|---|---|---|---|---|---|
|  |  |  |  |  |  |  |  | H 1312 | He 2372 |
| Ionization energy increases | Li 520 | Be 900 |  | B 801 | C 1086 | N 1402 | O 1314 | F 1681 | Ne 2081 |
|  | Na 496 | Mg 738 |  | Al 578 | Si 786 | P 1012 | S 1000 | Cl 1251 | Ar 1520 |
|  | K 419 | Ca 590 |  | Ga 579 | Ge 762 | As 944 | Se 941 | Br 1140 | Kr 1351 |
|  | Rb 403 | Sr 550 |  | In 558 | Sn 709 | Sb 832 | Te 869 | I 1009 | Xe 1170 |
|  | Cs 376 | Ba 503 |  | Tl 589 | Pb 716 | Bi 703 | Po 812 | At | Rn 1037 |

**Figure 6.14**

First ionization energies of the main-group elements, in kilojoules per mole. In general, ionization energy decreases moving down in the periodic table and increases going across, although there are several exceptions.

electron is relatively far from the nucleus, so less energy has to be supplied to remove it from the atom.

---

**Example 6.10**   Consider the three elements B, C, and Al. Using only the periodic table, predict which of the three elements has

(a) the largest atomic radius; the smallest atomic radius.
(b) the largest ionization energy; the smallest ionization energy.

**Strategy**   Since these three elements form a block,

B  C

Al

in the periodic table, it is convenient to compare both carbon and aluminum with boron. Recall the trends for atomic radius and ionization energy.

**Solution**

(a) B is larger than C but smaller than Al, so   Al must be the largest atom and C the smallest.

(b) B has a smaller ionization energy than C, but a larger ionization energy than Al. Hence   Al has the smallest ionization energy and C the largest.

---

If you look carefully at Figure 6.14, you will note a few exceptions to the general trends referred to above and illustrated in Example 6.10. For example, the ionization energy of B (801 kJ/mol) is *less* than that of Be (900 kJ/mol). This happens because the electron removed from the boron atom comes from the 2p as opposed to the 2s sublevel for beryllium. Since 2p is higher in energy than 2s, it is not too surprising that less energy is required to remove an electron from that sublevel.

TABLE 6.6    **Electronegativity Values**

| H 2.2 | | | | | | |
|------|------|------|------|------|------|------|
| Li 1.0 | Be 1.6 | B 2.0 | C 2.5 | N 3.0 | O 3.5 | F 4.0 |
| Na 0.9 | Mg 1.3 | Al 1.6 | Si 1.9 | P 2.2 | S 2.6 | Cl 3.2 |
| K 0.8 | Ca 1.0 | Sc 1.4 | Ge 2.0 | As 2.2 | Se 2.5 | Br 3.0 |
| Rb 0.8 | Sr 0.9 | Y 1.2 | Sn 1.9 | Sb 2.0 | Te 2.1 | I 2.7 |
| Cs 0.8 | Ba 0.9 | | | | | |

## Electronegativity

*Electronegativity is a positive quantity*

The ionization energy of an atom is a measure of its tendency to lose electrons; the larger the ionization energy, the more difficult it is to remove an electron. There are several different ways of comparing the tendencies of different atoms to gain electrons. The most useful of these for our purposes is the **electronegativity,** which measures the ability of an atom to attract to itself the electron pair forming a covalent bond.

The greater the electronegativity of an atom, the greater its affinity for electrons. Table 6.6 shows a scale of electronegativities first proposed by Linus Pauling. Here, each element is assigned a number, ranging from 4.0 for the most electronegative element, fluorine, to 0.8 for cesium, the least electronegative. Among the main-group elements, electronegativity increases moving from left to right in the periodic table (Groups 1–17). Ordinarily, it decreases moving down a group. You will find Table 6.6 very helpful when we discuss covalent bonding in Chapter 7.

---

## CHEMISTRY
### *Beyond the Classroom*

## The Aurora

*What awesome sights are these northern lights, whatever be their meaning*
*As they dance and prance their ritual dance, shifting, fading, gleaming, . . .*
—Robert H. Eather in "Majestic Lights,"
1980, American Geophysical Union

Far and away the most spectacular atmospheric phenomenon is the light display called the aurora (*aurora borealis* or "northern lights" in the northern hemisphere; *aurora australis* in the southern hemisphere). Still pictures (Figure 6.A) can never capture their full glory; the rays of light pulsate constantly, alternately waxing and waning. Usually the aurora is brightly colored, most often red, green, or blue. The lights can appear as high as 1000 km in the night sky; their greatest intensity is typically found 100 to 150 km above the Earth's surface.

The aurora borealis is concentrated in an oval-shaped area near the Earth's north magnetic pole. Occasionally it becomes visible much farther south, which explains why it was studied extensively by Benjamin Franklin in the United States and John Dalton in England. If you've never seen an aurora (most people haven't), you could travel to Barrow, Alaska, where there is a display every night of the year. (Take a sweater along, because the average nighttime temperature in Barrow is somewhere below −20°C). If you live in New York City, you have about one chance in 20 of seeing the aurora on a clear, moonless night when there is a power outage. Rural New England (e.g., Storrs, CT) where the bright lights are dimmer would be a better choice (for seeing the aurora, that is). To further improve your chances, note that northern lights are seen most often

— around the equinoxes in March and September, between 9 PM and midnight.
— a couple of days after a major solar disturbance ("sun spots").

Hundreds of years ago, native Americans interpreted the aurora to be the spirits of the dead fighting one another in the sky; the red glow was supposed to reflect the blood spilled in these conflicts. Today, scientists have a more mundane explanation. We attribute the colors of the aurora to the visible spectrum produced when electrons and protons from the sun interact with and energize atoms, molecules, and ions in the upper atmosphere. Two types of spectra are observed (Figure 6.B):

**Figure 6.A**
The aurora borealis (Alaska)

— *line spectra,* very similar to those discussed in this chapter. These result from electron transitions in atoms, mostly O and N, or to a lesser extent, in monatomic ions such as $O^+$ or $N^+$. The two brightest lines in the auroral spectrum, at 557.7 nm (green) and 630.0 nm (red), are both due to atomic oxygen.
— *band spectra,* where the emission of light is spread over a range of perhaps 10 to 100 nm in a particular region of the spectrum. These arise from electron transitions in molecules ($O_2$, $N_2$) or molecular ions ($O_2^+$, $N_2^+$). The blue color sometimes seen in auroras is associated with spectral bands around 400 nm in $N_2^+$ and $N_2$.

Curiously, neither of the two lines referred to above, at 557.7 nm and 630.0 nm, is observed in the "normal" spectrum of oxygen, as obtained with laboratory instruments at the Earth's surface. There, excited oxygen atoms lose their energy, not by emission of light, but by collision with other atoms. In the upper atmosphere, atoms are very far apart (the total pressure at 100 km is only $10^{-8}$ atm). Consequently, they seldom collide with one another, so excited oxygen atoms stay around long enough to emit visible light.

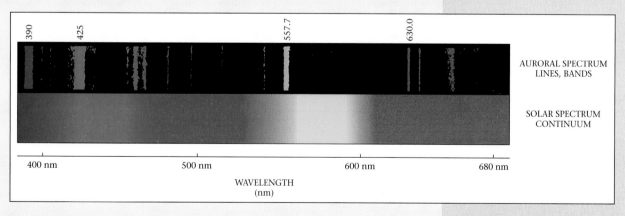

**Figure 6.B**
The red and green colors characteristic of the aurora are due to lines in the spectrum of atomic oxygen at 630.0 nm and 557.7 nm. (From *Majestic Lights* by Robert H. Eather, American Geophysical Union)

# CHAPTER HIGHLIGHTS

## Key Concepts

1. Relate wavelength, frequency, and energy
   (Examples 6.1, 6.2; Problems 1–10, 63, 67)
2. Use the Bohr model to identify lines in the hydrogen spectrum
   (Example 6.3; Problems 13–20)
3. Identify the quantum numbers and capacities corresponding to energy levels, sublevels, and orbitals
   (Example 6.4; Problems 21–32, 41, 42)
4. Write electron configurations of atoms and ions
   (Examples 6.5, 6.6, 6.8; Problems 33–40, 53, 54)
5. Write orbital diagrams for atoms and ions
   (Example 6.7; Problems 43–52, 55, 56)
6. Identify periodic trends in radii, ionization energy, and electronegativity
   (Examples 6.9, 6.10; Problems 57–64)

## Key Equations

| | | |
|---|---|---|
| Frequency–wavelength | $\lambda \nu = c = 2.998 \times 10^8$ m/s | |
| Energy–frequency | $E = h\nu = hc/\lambda$ | $h = 6.626 \times 10^{-34}$ J $\cdot$ s |
| Bohr model | $E_\mathbf{n} = -R_H/\mathbf{n}^2$ | $R_H = 2.180 \times 10^{-18}$ J |

## Key Terms

| | | |
|---|---|---|
| atomic orbital (s, p, d, f) | excited state | photon |
| atomic radius | frequency | principal energy level |
| electron configuration | ground state | quantum no. ($\mathbf{n}$, $\ell$, $\mathbf{m}_\ell$, $\mathbf{m}_s$) |
| —abbreviated | ionic radius | spectrum |
| electron spin | ionization energy | sublevel (s, p, d, f) |
| electronegativity | main-group element | transition metal |
| energy level | orbital diagram | wavelength |

## Summary Problem

Crystals of hydrated nickel (II) chloride, $NiCl_2 \cdot 6H_2O$ (Charles D. Winters)

Consider the element nickel ($Z = 28$).

(a) There is a line in the nickel spectrum at 212.59 nm. In what region of the spectrum (ultraviolet, visible, or infrared) is this line found? What is the frequency of the line? What is the energy difference between the two levels responsible for this line in kJ/mol?

(b) The ionization energy of nickel from the ground state is 757 kJ/mol. Assume that the transition in (a) is from the ground state to an excited state. If that is the case, calculate the ionization energy from the excited state.

(c) Give the electron configuration of the Ni atom; the $Ni^{3+}$ ion.

(d) Give the orbital diagram (beyond argon) for Ni and $Ni^{3+}$.

(e) How many unpaired electrons are there in the Ni atom? in the $Ni^{3+}$ ion?

(f) How many s electrons are there in the Ni atom? in the $Ni^{3+}$ ion?

(g) Rank Ni, $Ni^{2+}$, and $Ni^{3+}$ in order of increasing size.

### Answers

(a) ultraviolet; $1.410 \times 10^{15}$ $s^{-1}$; 562.7 kJ/mol

(b) 194 kJ/mol

(c) $1s^2\ 2s^2\ 2p^6\ 3s^2\ 3p^6\ 4s^2\ 3d^8$; $1s^2\ 2s^2\ 2p^6\ 3s^2\ 3p^6\ 3d^7$

(d)                4s                   3d

   Ni:      (↑↓)    (↑↓) (↑↓) (↑↓) (↑ ) (↑ )

   $Ni^{3+}$:  ( )     (↑↓) (↑↓) (↑ ) (↑ ) (↑ )

(e) 2; 3       (f) 8; 6      (g) $Ni^{3+} < Ni^{2+} < Ni$

# Questions & Problems

## Energy, Wavelength, and Frequency

**\*1.** Compare the energies and wavelengths of two photons, one with a low frequency, the other with a high frequency.

**\*2.** Compare the wavelength of an electron in the ground state with that in an excited state. In which state is the wavelength longer?

**3.** A photon of green light has a frequency of $5.82 \times 10^{14}$ $s^{-1}$. Calculate

  **(a)** the wavelength.

  **(b)** the energy in joules/photon.

  **(c)** the energy in kJ/mol.

**4.** Magnetic Resonance Imaging (MRI) is a powerful diagnostic tool used in medicine. The imagers used in hospitals operate with a wavelength of $7.50 \times 10^8$ nm. Calculate

  **(a)** the frequency in MHz (1 MHz = $10^6$ Hz).

  **(b)** the energy in joules/photon.

  **(c)** the energy in kJ/mol.

**5.** Carbon dioxide absorbs energy at a frequency of $2.001 \times 10^{13}$ $s^{-1}$.

  **(a)** Calculate the wavelength of this absorption in nanometers.

  **(b)** In what spectral range does the absorption occur?

  **(c)** What is the energy difference in kJ/mol?

**6.** The ionization energy of rubidium is 403 kJ/mol. Do x-rays with a wavelength of 78 nm have sufficient energy to ionize rubidium?

**7.** Two states differ in energy by 675 kJ/mol. When an electron moves from the higher to the lower of these states, calculate

  **(a)** $\lambda$ in nm.

  **(b)** $\nu$ in $s^{-1}$.

**8.** Two states differ in energy by $2.88 \times 10^{-18}$ J. When an electron moves from the higher to the lower state, calculate

  **(a)** $\lambda$ in nm and m.

  **(b)** $\nu$ in Hz.

**9.** An x-ray has a wavelength of 4.00 nm. Calculate

  **(a)** the frequency associated with this radiation.

  **(b)** the energy in joules associated with this radiation.

**10.** Microwave ovens heat food by the energy given off by the microwaves. What is the energy associated with a microwave that has a length of $5.00 \times 10^6$ nm?

## Bohr Model of the Hydrogen Atom

**11.** Consider the following transitions

    **(1)** $n = 3$ to $n = 1$

    **(2)** $n = 2$ to $n = 3$

    **(3)** $n = 4$ to $n = 3$

    **(4)** $n = 3$ to $n = 5$

  **(a)** Which one involves the ground state?

  **(b)** Which one absorbs the most energy?

  **(c)** Which one emits the most energy?

**12.** Consider the transitions in Question 11. For which of the transitions is energy absorbed? For which is energy emitted?

**13.** Calculate $E_n$ for $n = 1, 2, 3$, and 4 ($R_H = 2.180 \times 10^{-18}$ J). Make a one-dimensional graph showing energy, at different values of $n$, increasing vertically. On this graph, indicate by vertical arrows transitions in the

  **(a)** Lyman series ($n_{lo} = 1$).

  **(b)** Balmer series ($n_{lo} = 2$).

**14.** According to the Bohr model, the radius of a circular orbit is given by the equation

$$r \text{ (in nm)} = 0.0529\ n^2$$

Draw successive orbits for the hydrogen electron at $n = 1, 2, 3$, and 4. Indicate by arrows transitions between orbits that lead to lines in the

  **(a)** Lyman series ($n_{lo} = 1$).

  **(b)** Balmer series ($n_{lo} = 2$).

**15.** For the Pfund series, $n_{lo} = 5$. Calculate the wavelength in nanometers of a transition from $n = 6$ to $n = 5$.

**16.** The Brackett series lines in the atomic spectrum of hydrogen result from transitions from $n > 4$ to $n = 4$. Calculate the wavelength, in nm, of a line in this series resulting from the $n = 6$ to $n = 4$ transition.

**17.** A line in the Lyman series ($n_{lo} = 1$) occurs at 97.23 nm. Calculate $n_{hi}$ for the transition associated with this line.

**18.** In the Paschen series, $n_{lo} = 3$. Calculate the longest wavelength possible for a transition in this series.

## Quantum Mechanical Model of the Hydrogen Atom

**\*19.** What is the principal difference between the Bohr model and the quantum mechanical model of the hydrogen atom?

**\*20.** Explain in your words the physical meaning of the two graphs in Figure 6.4.

## Energy Levels, Sublevels, and Quantum Numbers

**\*21.** What are the possible values for $m_\ell$ for
   **(a)** $\ell = 1$   **(b)** $\ell = d$   **(c)** $n = 3$ (all sublevels)

**\*22.** What are the possible values for $m_\ell$ for
   **(a)** $\ell = 2$   **(b)** $\ell = f$   **(c)** $n = 2$ (all sublevels)

**\*23.** For the following pairs of orbitals, indicate which is higher in energy in a potassium atom.
   **(a)** 2p or 3p   **(b)** 3s or 3d
   **(c)** 4s or 3d   **(d)** 5s or 4f

**\*24.** For the following pairs of orbitals, indicate which is lower in energy in a rubidium atom.
   **(a)** 3d or 4d   **(b)** 2s or 2p
   **(c)** 5s or 4d   **(d)** 4d or 5f

**\*25.** What type of electron orbital (i.e., s, p, d, or f) is designated by
   **(a)** $n = 1, \ell = 0, m_\ell = 0$
   **(b)** $n = 3, \ell = 2, m_\ell = -1$
   **(c)** $n = 4, \ell = 3, m_\ell = 3$

**\*26.** What type of electron orbital (i.e., s, p, d, or f) is designated by
   **(a)** $n = 2, \ell = 1, m_\ell = -1$
   **(b)** $n = 5, \ell = 2, m_\ell = 2$
   **(c)** $n = 6, \ell = 3, m_\ell = 2$

**\*27.** Give the number of orbitals in
   **(a)** the principal level $n = 2$.
   **(b)** a 3p sublevel.
   **(c)** a d sublevel.

**\*28.** State the total capacity for electrons in
   **(a)** the principal level $n = 3$.
   **(b)** a 3d sublevel.
   **(c)** an f sublevel.

**\*29.** State the relationship, if any, between
   **(a)** $n$ and $m_\ell$   **(b)** $\ell$ and $m_\ell$   **(c)** $m_\ell$ and $m_s$

**\*30.** What is the
   **(a)** minimum $n$ value for $\ell$ designated as p?
   **(b)** letter used to designate the sublevel with $\ell = 3$?
   **(c)** number of orbitals in a sublevel where $\ell$ is designated as d?
   **(d)** number of different sublevels when $n = 2$?

**\*31.** Given the following sets of electron quantum numbers, indicate those that could not occur and explain your answer.
   **(a)** $1, 1, 0, +\frac{1}{2}$   **(b)** $2, 1, 2, +\frac{1}{2}$   **(c)** $3, 0, 0, -\frac{1}{2}$
   **(d)** $3, 2, 1, \frac{1}{2}$   **(e)** $4, 2, -2, 0$

**\*32.** Given the following sets of electron quantum numbers, indicate those that could not occur and explain your answer.
   **(a)** $1, 0, 0, -\frac{1}{2}$   **(b)** $3, 2, -2, +\frac{1}{2}$   **(c)** $4, 0, 2, +\frac{1}{2}$
   **(d)** $2, 2, 1, -\frac{1}{2}$   **(e)** $3, 1, 1, +\frac{1}{2}$

**\*33.** Write the electron configuration (ground state) for
   **(a)** C   **(b)** Cl   **(c)** Co   **(d)** Cd   **(e)** Cs

**\*34.** Write the electron configuration (ground state) for
   **(a)** S   **(b)** Se   **(c)** Sb   **(d)** Sc   **(e)** Si

**\*35.** Write the abbreviated electron configuration (ground state) for
   **(a)** K   **(b)** Br   **(c)** Y   **(d)** W   **(e)** Pb

**\*36.** Write the abbreviated electron configuration (ground state) for
   **(a)** In   **(b)** I   **(c)** Ir   **(d)** Sr   **(e)** Al

**\*37.** Give the symbol of the element of lowest atomic number whose ground state has
   **(a)** a p electron.
   **(b)** four f electrons.
   **(c)** a completed d sublevel.
   **(d)** six s electrons.

**\*38.** Give the symbol of the element of lowest atomic number whose ground state has
   **(a)** a completed f sublevel.
   **(b)** 20 p electrons.
   **(c)** two 4d electrons.
   **(d)** five 5p electrons.

**\*39.** What fraction of the total number of electrons is in p sublevels in
   **(a)** Al   **(b)** Au   **(c)** Ag

**\*40.** What fraction of the total number of electrons is in d sublevels in
   **(a)** Ti   **(b)** Tl   **(c)** Te

**\*41.** Which of the following electron configurations are for atoms in the ground state? In the excited state? Which are impossible?
   **(a)** $1s^2 2s^1 3s^1$   **(b)** $2s^2 2p^2$   **(c)** $1s^2 2s^2 2p^3$
   **(d)** $1s^2 1p^1 2s^2$   **(e)** $1s^2 2s^2 2p^6 3s^2 3p^6 3d^1$
   **(f)** $1s^2 2s^2 2p^6 3s^2 3p^6 3d^1 3f^1$

**\*42.** Which of the following electron configurations are for atoms in the ground state? In the excited state? Which are impossible?
   **(a)** $1s^2 2s^2 3s^2$   **(b)** $1s^2 2p^3$   **(c)** $1s^2 2s^3 2p^5$
   **(d)** $1s^1 2s^2 2p^7$   **(e)** $2s^2 2p^6 3s^1$   **(f)** $1s^2 2s^2 2p^6 3s^2 3d^1$

## Orbital Diagrams; Hund's Rule

**\*43.** Give the orbital diagram of
   **(a)** Li   **(b)** P   **(c)** Fe   **(d)** F

**\*44.** Give the orbital diagram of
   **(a)** N   **(b)** Ni   **(c)** Na   **(d)** Ne

**\*45.** Give the symbol of the atom with the following orbital diagram.

|  | 1s | 2s | 2p | 3s | 3p |
|---|---|---|---|---|---|
| **(a)** | (↑↓) | (↑↓) | (↑↓)(↑↓)(↑↓) | (↑↓) | ( )( )( ) |
| **(b)** | (↑↓) | (↑↓) | (↑↓)(↑↓)(↑↓) | (↑↓) | (↑ )(↑ )(↑ ) |
| **(c)** | (↑↓) | (↑↓) | (↑↓)(↑ )(↑ ) | ( ) | ( )( )( ) |

**\*46.** Give the symbol of the atom with the following orbital diagram beyond argon.

|  | 4s | 3d | 4p |
|---|---|---|---|
| **(a)** | (↑↓) | (↑↓)(↑↓)(↑↓)(↑ )(↑ ) | ( )( )( ) |
| **(b)** | (↑↓) | (↑↓)(↑↓)(↑ )(↑ )(↑ ) | ( )( )( ) |
| **(c)** | (↑↓) | (↑↓)(↑↓)(↑↓)(↑↓)(↑↓) | (↑ )(↑ )( ) |

**\*47.** Give the symbols of
  **(a)** all the elements that have filled 4p sublevels.
  **(b)** all the nonmetals in the 5th period that have no unpaired electrons.
  **(c)** all the metalloids that have filled 3s sublevels.
  **(d)** all the elements in the second period that have one unpaired electron.

**\*48.** Give the symbols of
  **(a)** all the elements in which all the 4d orbitals are half full.
  **(b)** all the nonmetals in period 2 that have two or more unpaired electrons.
  **(c)** all the elements in group 2 in which the 4p sublevel is full.
  **(d)** all the metalloids that have paired 5p electrons.

**\*49.** Give the number of unpaired electrons in an atom of
  **(a)** phosphorus   **(b)** potassium
  **(c)** plutonium (Pu)

**\*50.** Give the number of unpaired electrons in an atom of
  **(a)** mercury   **(b)** manganese   **(c)** magnesium

**\*51.** Give the symbol of the main-group metals in period 4 (transition metals are not included) with the following number of unpaired electrons per atom.
  **(a)** 0   **(b)** 1   **(c)** 2   **(d)** 3

**\*52.** In what main group(s) of the periodic table do element(s) have the following number of filled p orbitals in the outermost principal level?
  **(a)** 0   **(b)** 1   **(c)** 2   **(d)** 3

### Electron Arrangement in Ions

**\*53.** Write the ground-state electron configuration for
  **(a)** Na, $Na^+$   **(b)** S, $S^{2-}$
  **(c)** Sc, $Sc^{3+}$   **(d)** $Mn^{2+}$, $Mn^{4+}$

**\*54.** Write the ground-state electron configuration for
  **(a)** Mg, $Mg^{2+}$   **(b)** N, $N^{3-}$
  **(c)** O, $O^-$   **(d)** $Fe^{2+}$, $Fe^{3+}$

**\*55.** How many unpaired electrons are there in the following ions?
  **(a)** $Ti^{4+}$   **(b)** $Ni^{2+}$   **(c)** $Br^-$   **(d)** $Hg^{2+}$

**\*56.** How many unpaired electrons are there in the following ions?
  **(a)** $V^{3+}$   **(b)** $Sn^{4+}$   **(c)** $I^-$   **(d)** $W^{4+}$

### Trends in the Periodic Table

**\*57.** Arrange the elements Sr, Sn, Te in order of
  **(a)** increasing atomic radius.
  **(b)** increasing first ionization energy.
  **(c)** increasing electronegativity.

**\*58.** Arrange the elements Na, Si, S in order of
  **(a)** decreasing atomic radius.
  **(b)** decreasing first ionization energy.
  **(c)** decreasing electronegativity.

**\*59.** Which of the four atoms Rb, Sr, Sb, or Cs
  **(a)** has the smallest atomic radius?
  **(b)** has the lowest ionization energy?
  **(c)** is the least electronegative?

**\*60.** Which of the four atoms Na, P, Cl, or K
  **(a)** has the largest atomic radius?
  **(b)** has the highest ionization energy?
  **(c)** is the most electronegative?

**\*61.** Select the larger member of each pair.
  **(a)** Ca and $Ca^{2+}$   **(b)** F and $F^-$
  **(c)** Sr and $Sr^{2+}$   **(d)** $Cu^+$ and $Cu^{2+}$

**\*62.** Select the smaller member of each pair.
  **(a)** V and $V^{3+}$   **(b)** Se and $Se^{2-}$
  **(c)** As and $As^{3-}$   **(d)** $Mn^{2+}$ and $Mn^{4+}$

**\*63.** List the following species in order of decreasing radius.
  **(a)** Na, Mg, K, $Mg^{2+}$   **(b)** N, P, As, $As^{3-}$

**\*64.** List the following species in order of increasing radius.
  **(a)** Ni, $Ni^{2+}$, $Ni^{3+}$   **(b)** Br, $Br^-$, $I^-$

### Unclassified

**65.** A light bulb radiates 8.5% of the energy supplied to it as visible light. If the wavelength of the visible light is assumed to be 565 nm, how many photons per second are emitted by a 25-W light bulb? (1 W = 1 J/s)

**\*66.** Write the abbreviated electron configuration for promethium. What actinide would it most resemble chemically?

**67.** A carbon dioxide laser produces radiation of wavelength 10.6 micrometers (1 micrometer = $10^{-6}$ m). If the laser produces about one joule of energy per pulse, how many photons are produced per pulse?

**\*68.** Explain the difference between
  **(a)** the Bohr model of the atom and the quantum mechanical model.
  **(b)** wavelength and frequency.
  **(c)** the geometries of the three different p orbitals.

**\*69.** Explain in your own words what is meant by
  **(a)** the Pauli exclusion principle.
  **(b)** Hund's rule.
  **(c)** a line in an atomic spectrum.
  **(d)** the principal quantum number.

**\*70.** Indicate whether each of the following statements is true or false. If false, correct the statement.
  **(a)** An electron transition from $n = 3$ to $n = 1$ gives off energy.

**(b)** Light emitted by an $n = 4$ to $n = 2$ transition will have a longer wavelength than that from an $n = 5$ to $n = 2$ transition.

**(c)** A sublevel of $\ell = 3$ has a capacity of ten electrons.

**(d)** An atom of Group 13 has three unpaired electrons.

**\*71.** Criticize or comment upon the following statements:

**(a)** The energy of a photon is inversely proportional to wavelength.

**(b)** The energy of the hydrogen electron is inversely proportional to the quantum number $\ell$.

**(c)** Electrons start to enter the fifth principal level as soon as the fourth level is full.

**\*72.** No currently known elements contain electrons in g ($\ell = 4$) orbitals in the ground state. If an element is discovered that has electrons in the g orbital, what is the lowest value for $n$ in which these g orbitals could exist? What are the possible values of $m_\ell$? How many electrons could a set of g orbitals hold?

**\*73.** Explain why

**(a)** positive ions are smaller than their corresponding atoms.

**(b)** scandium, a transition metal, forms an ion with a noble gas structure.

**(c)** electronegativity increases across a period in the periodic table.

**\*74.** Indicate whether each of the following is true or false.

**(a)** Effective nuclear charge increases when one goes down a group.

**(b)** Group 13 elements have three outer electrons.

**(c)** Energy has to be absorbed when an electron is removed from an atom.

**\*75.** Name and give the symbol for the element with the characteristic given below:

**(a)** Electron configuration is $1s^2 2s^2 2p^6 3s^2 3p^5$.

**(b)** Lowest ionization energy in Group 14.

**(c)** Its 2+ ion has the configuration $[_{18}Ar]3d^5$.

**(d)** Alkali metal with the smallest atomic radius.

**(e)** Largest ionization energy in the fourth period.

## Challenge Problems

**76.** The energy of any one-electron species in its $n$th state ($n$ = principal quantum number) is given by $E = -BZ^2/n^2$, where $Z$ is the charge on the nucleus and $B$ is $2.180 \times 10^{-18}$ J. Find the ionization energy of the $Li^{2+}$ ion in its first excited state in kilojoules per mole.

**77.** In 1885 Johann Balmer, a numerologist, derived the following relation for the wavelength of lines in the visible spectrum of hydrogen.

$$\lambda = 364.5n^2/(n^2 - 4)$$

where $\lambda$ is in nanometers and $n$ is an integer that can be 3, 4, 5, . . . . Show that this relation follows from the Bohr equation and the equation using the Rydberg constant. Note that in the Balmer series, the electron is returning to the $n = 2$ level.

**\*78.** Suppose the rules for assigning quantum numbers were as follows:

$$n = 1, 2, 3, \ldots$$
$$\ell = 0, 1, 2, \ldots, n$$
$$m_\ell = 0, 1, 2, \ldots, \ell + 1$$
$$m_s = +\tfrac{1}{2} \text{ or } -\tfrac{1}{2}$$

State the values of $\ell$, $m_\ell$ and $m_s$ for the electrons when $n = 1$ and $n = 2$. Give the electron configuration for an atom with eight electrons.

**\*79.** Suppose that the spin quantum number could have the values $\tfrac{1}{2}$, 0, $-\tfrac{1}{2}$. Assuming that the rules governing the values of the other quantum numbers and the order of filling sublevels were unchanged.

**(a)** what would be the electron capacity of an s sublevel? A p sublevel? A d sublevel?

**(b)** how many electrons could fit in the $n = 3$ level?

**(c)** what would be the electron configuration of the element with atomic number 8? 17?

**80.** In the photoelectric effect, electrons are ejected from a metal surface when light strikes it. A certain minimum energy, $E_{min}$, is required to eject an electron. Any energy absorbed beyond that minimum gives kinetic energy to the electron. It is found that when light at a wavelength of 540 nm falls on a cesium surface, an electron is ejected with a kinetic energy of $2.60 \times 10^{-20}$ J. When the wavelength is 400 nm, the kinetic energy is $1.54 \times 10^{-19}$ J.

**(a)** Calculate $E_{min}$ for cesium in joules.

**(b)** Calculate the longest wavelength, in nanometers, that will eject electrons from cesium.

No one has ever isolated a stable compound of helium, neon, or argon. The heavier noble gases, especially xenon, are more reactive. (see p. 201) (John D. Cunningham/Visuals Unlimited)

# Covalent Bonding

## 7

What immortal hand or eye

Could frame thy fearful symmetry?

—WILLIAM BLAKE

*The Tiger*

Earlier, we referred to the forces that hold nonmetal atoms to one another, covalent bonds. These bonds consist of an electron pair shared between two atoms. To represent the covalent bond in the $H_2$ molecule, two structures can be written:

$$H:H \quad \text{or} \quad H—H$$

These structures can be misleading if they are taken to mean that the two electrons are fixed in position between the two nuclei. A more accurate picture of the electron density in $H_2$ is shown in Figure 7.1. At a given instant, the two electrons may be located at any of various points about the two nuclei. However, they are more likely to be found between the nuclei than at the far ends of the molecule.

You may well wonder why the sharing of an electron pair between two nuclei leads to increased stability. Why, for example, is the $H_2$ molecule more stable than two isolated hydrogen atoms? There are two somewhat different ways of explaining this situation.

1. From a classical electrostatic point of view, locating two electrons between the two protons of the $H_2$ molecule lowers the energy of the system. The attractive energies between oppositely charged particles (electron-proton) exceed the repulsive energies between particles of like charge (electron-electron, proton-proton).
2. From a quantum mechanical point of view, the two atomic orbitals of the hydrogen atoms overlap to form a new bonding orbital. Putting two electrons of opposite spin in this orbital lowers the energy of the system.

This chapter is devoted to the covalent bond as it exists in molecules and polyatomic ions. We consider

— the distribution of outer level *(valence)* electrons in species where atoms are joined by covalent bonds. These distributions are most simply described by *Lewis structures* (Section 7.1)
— molecular geometries. The so-called *VSEPR model* can be used to predict the angles between covalent bonds formed by a central atom (Section 7.2).
— the polarity of covalent bonds and the molecules they form (Section 7.3). Most bonds and many molecules are polar in the sense that they have a positive and a negative pole.
— The distribution of valence electrons among *atomic orbitals*, using the "valence bond" approach (Section 7.4).

**Figure 7.1**
Electron density in $H_2$. The depth of color is proportional to the probability of finding an electron in a particular region.

# 7.1  Lewis Structures; The Octet Rule

The idea of the covalent bond was first suggested by the American physical chemist G. N. Lewis in 1916. He pointed out that the electron configuration of the noble gases appears to be a particularly stable one. Noble-gas atoms are themselves extremely unreactive. Lewis suggested that *atoms, by sharing electrons to form an electron-pair bond, can acquire a stable, noble-gas structure.* Consider, for example, two hydrogen atoms, each with one electron. The process by which they combine to form an $H_2$ molecule can be shown as

*Noble gas structures are stable in molecules as they are in atoms and ions*

$$H\cdot + H\cdot \longrightarrow \left( H (\!:\!) H \right)$$

using dots to represent electrons; the circles emphasize that the pair of electrons in the covalent bond can be considered to occupy the 1s orbital of either hydrogen atom. In that sense, each atom in the $H_2$ molecule has the electronic structure of the noble gas helium, with the electron configuration $1s^2$.

This idea is readily extended to other simple molecules containing nonmetal atoms. An example is the $F_2$ molecule. You will recall that a fluorine atom has the electron configuration $1s^2 2s^2 2p^5$. It has seven electrons in its outermost principal energy level ($\mathbf{n} = 2$). These are referred to as **valence electrons.** If these are shown as dots about the symbol of the element, the fluorine atom can be represented as

$$:\!\ddot{F}\cdot$$

The combination of two fluorine atoms leads to the structure

$$:\!\ddot{F}\cdot + \cdot\ddot{F}: \longrightarrow \left( :\!\ddot{F} (\!:\!) \ddot{F}: \right)$$

This "electron-dot" structure shows that each fluorine atom "owns" six valence electrons outright and shares two others. Putting it another way, each F atom is surrounded by eight valence electrons. By this model, both F atoms achieve the electron configuration $1s^2 2s^2 2p^6$, which is that of the noble gas neon. This, according to Lewis, explains why the $F_2$ molecule is stable and why F atoms combine to form $F_2$ rather than $F_3$, $F_4$, . . . .

*Shared electrons are counted for both atoms*

The structures (without the circles) are referred to as **Lewis structures.** In writing Lewis structures, only the valence electrons written above are shown, since they are the ones that participate in covalent bonding. For the main-group elements, the only ones dealt with here, the number of valence electrons is equal to the last digit of the group number in the periodic table (Table 7.1). Notice

### TABLE 7.1  Lewis Structures of Atoms

| Group | 1 | 2 | 13 | 14 | 15 | 16 | 17 | 18 |
|---|---|---|---|---|---|---|---|---|
| No. of valence $e^-$ | 1 | 2 | 3 | 4 | 5 | 6 | 7 | 8 |
| Lewis structure | H· | ·Be· | ·Ḃ· | ·Ċ· | ·P̈· | ·Ö· | :F̈· | :Ne: |

that elements in a given main group have the same number of valence electrons. This explains why such elements behave similarly when they react to form covalently bonded species.

In the Lewis structure of a molecule or polyatomic ion, valence electrons ordinarily occur in pairs. There are two kinds of electron pairs.

1. A pair of electrons shared between two atoms is a **covalent bond,** ordinarily shown as a straight line between bonded atoms.
2. An **unshared pair** of electrons, belonging entirely to one atom, is shown as a pair of dots on that atom.

The Lewis structures for the species $OH^-$, $H_2O$, $NH_3$, and $NH_4^+$ are

Notice that in each case the oxygen or nitrogen atom is surrounded by eight valence electrons. In each species, a single electron pair is shared between two bonded atoms. These bonds are called **single bonds.** There is one single bond in the $OH^-$ ion, two in the $H_2O$ molecule, three in $NH_3$, and four in $NH_4^+$. There are three unshared pairs in the hydroxide ion, two in the water molecule, one in the ammonia molecule, and none in the ammonium ion.

Bonded atoms can share more than one electron pair. A **double bond** occurs when bonded atoms share two electron pairs; in a **triple bond,** three pairs of electrons are shared. In ethylene ($C_2H_4$) and acetylene ($C_2H_2$), the carbon atoms are linked by a double bond and triple bond, respectively. Using two parallel lines to represent a double bond and three for a triple bond, we write the structures of these molecules as

*Forming a multiple bond "saves" electrons because bonding pairs are counted for both atoms*

Note that each carbon is surrounded by eight valence electrons and each hydrogen by two. *The atoms most commonly joined by double bonds are C, O, N and S.*

These examples illustrate the principle that atoms in covalently bonded species tend to have noble-gas electronic structures. This generalization is often referred to as the **octet rule.** Nonmetals, except for hydrogen, achieve a noble-gas structure by sharing in an "octet" of electrons (eight). Hydrogen atoms, in molecules or polyatomic ions, are surrounded by a *duet* of electrons (two).

*Most molecules follow the octet rule*

## Writing Lewis Structures

For very simple species, Lewis structures can often be written by inspection. Usually, though, you will save time by following the steps listed below.

1. *Count the number of valence electrons.* For a molecule, simply sum up the valence electrons of the atoms present. For a polyatomic anion, one elec-

tron is added for each unit of negative charge. For a polyatomic cation, a number of electrons equal to the positive charge must be subtracted.

**2. Draw a skeleton structure for the species, joining atoms by single bonds.** In some cases, only one arrangement of atoms is possible; in others, there are two or more alternative structures. Most of the molecules and polyatomic ions dealt with in this chapter consist of a *central atom* bonded to two or more *terminal atoms,* located at the outer edges of the molecule or ion. For such species (e.g., $NH_4^+$, $SO_2$, $CCl_4$), it is relatively easy to derive the skeleton structure. **The central atom is usually the one written first in the formula** (N in $NH_4^+$, S in $SO_2$, C in $CCl_4$); **put this in the center of the molecule or ion. Terminal atoms are most often hydrogen, oxygen, or a halogen; bond these atoms to the central atom.**

**3. Determine the number of valence electrons still available for distribution.** To do this, deduct two valence electrons for each single bond written in Step 2.

**4. Determine the number of valence electrons required to fill out an octet for each atom** (except H) **in the skeleton.** Remember that shared electrons are counted for both atoms.

   (a) If the number of electrons available (Step 3) is equal to the number required (Step 4), distribute the available electrons as unshared pairs, satisfying the octet rule for each atom.

   (b) If the number of electrons available (Step 3) is less than the number required (Step 4), the skeleton structure must be modified by changing single to multiple bonds. If you are two electrons short, convert a single bond to a double bond; if there is a deficiency of four electrons, convert a single bond to a triple bond (or two single bonds to double bonds).

*It would be nice if all shortages could be solved so simply*

The application of these steps and some further guiding principles are shown in Example 7.1.

---

**Example 7.1**   Draw Lewis structures of

   (a) the hypochlorite ion, $OCl^-$.      (b) methyl alcohol, $CH_4O$.

*Strategy*   Follow the steps listed above. For the $OCl^-$ ion, only one skeleton is possible; for methyl alcohol, keep in mind that the carbon atom ordinarily forms four bonds. Hydrogen, since it forms only one bond, must be a terminal atom.

*Solution*

   (a)

> (1) The number of valence electrons is 6 (from O) + 7 (from Cl) + 1 (from the −1 charge) = 14.
> (2) The skeleton structure is $[O—Cl]^-$.
> (3) The number of electrons available for distribution is 14 (originally) − 2 (used in skeleton) = 12.
> (4) The number of electrons required to give each atom an octet is 6 (for O) + 6 (for Cl) = 12.

   Since the number of available electrons is the same as the number required, the skeleton structure is correct; there are no multiple bonds. The Lewis structure is

Products that contain the hypochlorite ($OCl^-$) ion.

(b)

(1) The number of valence electrons is 4 (from C) + 4 (1 from each H) + 6 (from O) = 14.

(2) Since carbon forms four bonds and H must always be a terminal atom, the only reasonable skeleton is

$$
\begin{array}{c}
\text{H} \\
| \\
\text{H—C—O—H} \\
| \\
\text{H}
\end{array}
$$

(3) The number of electrons available for distribution = 14 − 10 = 4.

(4) The number of electrons needed to satisfy the octet rule is 4 (for O). Again, the number of electrons available is equal to the number required. The Lewis structure is

$$
\begin{array}{c}
\text{H} \\
| \\
\text{H—C—\ddot{O}—H} \\
| \\
\text{H}
\end{array}
$$

---

**Example 7.2**    Draw Lewis structures for

(a) $SO_2$    (b) $N_2$

***Strategy***    Follow the four steps for writing Lewis structures.

***Solution***

(a)

(1) There are 18 valence electrons. Both sulfur (two atoms) and oxygen (one atom) are in Group **16**; $3 \times 6 = 18$

(2) The skeleton structure, with sulfur as the central atom, is

$$\text{O—S—O}$$

(3) There are $18 − 4 = 14$ electrons available for distribution.

(4) Sixteen electrons are required to give each atom an octet (four for S and six for each O). There is a deficiency of two electrons. This means that a single bond in the skeleton must be converted to a double bond. The Lewis structure of $SO_2$ is

(b)

(1) There are ten valence electrons; nitrogen is in Group **15**.

(2) The skeleton structure is

$$\text{N—N}$$

(3) There are $10 − 2 = 8$ electrons available for distribution.

(4) Each N needs six electrons for an octet, so 12 electrons are needed. This means that there is a deficiency of $12 − 8 = 4$ electrons. Convert the single bond between the two N atoms to a triple bond. The Lewis structure is

$$: \text{N} \equiv \text{N} :$$

Liquid nitrogen. (Charles D. Winters)

## Resonance Forms

In certain cases, the Lewis structure does not adequately describe the properties of the ion or molecule that it represents. Consider, for example, the $SO_2$ structure derived in Example 7.2. This structure implies that there are two different kinds of sulfur-to-oxygen bonds in $SO_2$. One of these appears to be a single bond, the other a double bond. Yet experiment shows that there is only one kind of bond in the molecule.

One way to explain this situation is to assume that each of the bonds in $SO_2$ is intermediate between a single and a double bond. To express this concept, two structures are written:

The double-headed arrow is used to separate resonance structures

with the understanding that the true structure is intermediate between them. These are referred to as *resonance forms*. The concept of **resonance** is invoked whenever a single Lewis structure does not adequately reflect the properties of a substance.

Another species for which it is necessary to invoke the idea of resonance is the nitrate ion. Here three equivalent structures can be written

to explain the experimental observation that the three nitrogen-to-oxygen bonds in the $NO_3^-$ ion are identical in all respects.

We will encounter other examples of molecules and ions whose properties can be interpreted in terms of resonance. In all such species,

**1.** Resonance forms do not imply different kinds of molecules with electrons shifting eternally between them. There is only one type of $SO_2$ molecule; its structure is intermediate between those of the two resonance forms drawn for sulfur dioxide.

**2.** Resonance can be anticipated when it is possible to write two or more Lewis structures that are about equally plausible. In the case of the nitrate ion, the three structures we have written are equivalent. One could, in principle, write many other structures, but none of them would put eight electrons around each atom.

**3.** Resonance forms differ only in the distribution of electrons, not in the arrangement of atoms. The structure

could not be a resonance form of the nitrate ion, since the atoms are arranged in a quite different way.

**Example 7.3**   Write three resonance forms for $SO_3$.

***Strategy***   Write a Lewis structure for $SO_3$ following the four steps for writing Lewis structures. Then write all the resonance forms by changing the position of the multiple bond. Do *not* change the skeleton structure.

***Solution***   The Lewis structure and all the resonance forms for $SO_3$ are

## Formal Charge

Often, it is possible to write two different Lewis structures for a molecule differing in the arrangement of atoms, i.e.,

$$A—A—B \quad \text{or} \quad A—B—A$$

Sometimes both structures represent real compounds, which are *isomers* of one another (Chap. 15, 22). More often, only one structure exists in nature. For example, methanol ($CH_4O$) has the structure

In contrast, the structure

does not correspond to any real compound even though it obeys the octet rule.

There are several ways to choose the more plausible of two structures differing in their arrangement of atoms. As pointed out in Example 7.1, the fact that carbon almost always forms four bonds leads to the correct structure for methanol. Another approach involves a concept called **formal charge**, which can be applied to any atom within a Lewis structure. The formal charge is the difference between the number of valence electrons in the free atom and the number assigned to that atom in the Lewis structure. The "assigned" electrons include

— all the unshared electrons owned by that atom
— one half of the bonding electrons shared by that atom

Mathematically, the equation for the formal charge, $C_f$, is

$$C_f = E_v - (E_u + \tfrac{1}{2}E_b)$$

where $E_v$ = no. of valence $e^-$ in the free atom = last digit of group number in the periodic table
$E_u$ = no. of unshared $e^-$ owned by the atom in the Lewis structure and
$E_b$ = no. of bonding electrons shared by the atom in the Lewis structure.

Thus, for a phosphorus atom (Group **15**), which in a Lewis structure is surrounded by one unshared electron pair and three single bonds,

$$C_f = 5 - (2 + 3) = 0$$

Empirically, it is found that the more likely Lewis structure is the one in which

— the formal charges are as close to zero as possible.
— any negative formal charge is located on a strongly electronegative atom.

To show how this works, let us assign formal charges to the carbon and oxygen atoms in the two structures discussed on p. 180.

$$C_f \text{ of C} = 4 - \tfrac{1}{2}(8) = 0$$
$$C_f \text{ of O} = 6 - [4 + \tfrac{1}{2}(4)] = 0$$

$$C_f \text{ of C} = 4 - [2 + \tfrac{1}{2}(6)] = -1$$
$$C_f \text{ of O} = 6 - [2 + \tfrac{1}{2}(6)] = +1$$

The second structure is frowned upon because the formal charges are not zero; moreover, the negative formal charge is on the *less* electronegative atom (recall from Table 6.6 that EN of C = 2.5, EN of O = 3.5).

The concept of formal charge has a much wider applicability than this short discussion might imply. In particular, it can be used to predict situations where conventional Lewis structures, written in accordance with the octet rule, may be incorrect (Table 7.2).

## Exceptions to the Octet Rule

Although most of the molecules and polyatomic ions referred to in general chemistry follow the octet rule, there are some familiar species that do not. Among these are molecules containing an odd number of valence electrons. Nitric oxide, NO, and nitrogen dioxide, $NO_2$, fall in this category:

NO    no. of valence electrons = 5 + 6 = 11

$NO_2$    no. of valence electrons = 5 + 6(2) = 17

For such *odd electron* species (sometimes called free radicals) it is impossible to write Lewis structures in which each atom obeys the octet rule. In the NO

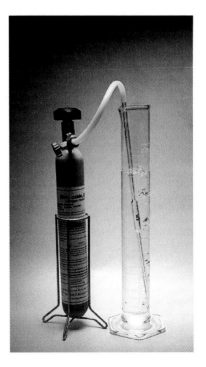

NO is a colorless gas which reacts with oxygen in the air forming a brown mixture of $NO_2$ and $N_2O_4$.

**TABLE 7.2  Possible Structures for $BeF_2$ and $BF_3$**

| Structure I | $C_f$ | Structure II | $C_f$ |
|---|---|---|---|
| $:\!\ddot{F}\!=\!\ddot{Be}\!=\!\ddot{F}\!:$ | Be = −2<br>F = +1 | $:\!\ddot{F}\!-\!\ddot{Be}\!-\!\ddot{F}\!:$ | Be = 0<br>F = 0 |
| (BF₃ structure I) | B = −1<br>F = +1,0,0 | (BF₃ structure II) | B = 0<br>F = 0 |

**Figure 7.2**
Oxygen is attracted into a magnetic field and can actually be suspended between the poles of an electromagnet. Both the paramagnetism and the blue color are due to the unpaired electrons in the $O_2$ molecule.

molecule, the unpaired electron is put on the nitrogen atom, giving both atoms a formal charge of zero:

$$\cdot \ddot{N} = \ddot{O} :$$

In $NO_2$, the "best" structure one can write again puts the unpaired electron on the nitrogen atom:

Elementary oxygen, like NO and $NO_2$, is paramagnetic (Fig. 7.2). Experimental evidence suggests that the $O_2$ molecule contains two unpaired electrons *and* a double bond. It is impossible to write a conventional Lewis structure for $O_2$ that has these two characteristics. A more sophisticated model of bonding, using molecular orbitals (Appendix 5) is required to explain the properties of oxygen.

There are a few species in which the central atom "violates" the octet rule in the sense that it is surrounded by two or three electron pairs rather than four. Examples include the fluorides of beryllium and boron, $BeF_2$ and $BF_3$. Although one could write multiple bonded structures for these molecules in accordance with the octet rule (Table 7.2), experimental evidence suggests the structures

$$: \ddot{F} - Be - \ddot{F} : \quad \text{and}$$

in which the central atom is surrounded by four and six valence electrons, respectively, rather than eight.

## Expanded Octets

In most molecules, the central atom is surrounded by 8 electrons. Rarely, it is surrounded by 4 ($BeF_2$) or 6 ($BF_3$). Occasionally, the number is 10 ($PCl_5$) or 12 ($SF_6$)

The largest class of molecules to "violate" the octet rule consists of species in which the central atom is surrounded by more than four pairs of valence electrons. Typical molecules of this type are phosphorus pentachloride, $PCl_5$, and sulfur hexafluoride, $SF_6$. The Lewis structures of these molecules are

As you can see, the central atoms in these molecules have "expanded octets." In $PCl_5$, the phosphorus atom is surrounded by ten valence electrons (five shared pairs); in $SF_6$, there are 12 valence electrons (six shared pairs) around the sulfur atom.

In molecules of this type, the terminal atoms are most often halogens (F, Cl, Br, I); in a few molecules, oxygen is a terminal atom. The central atom is a nonmetal in the third, fourth, or fifth period of the periodic table. Most frequently, it is one of the following elements:

|            | Group 15 | Group 16 | Group 17 | Group 18 |
|------------|----------|----------|----------|----------|
| 3rd period | P        | S        | Cl       |          |
| 4th period | As       | Se       | Br       | Kr       |
| 5th period | Sb       | Te       | I        | Xe       |

All these atoms have d orbitals available for bonding (3d, 4d, 5d). These are the orbitals in which the "extra" pairs of electrons are located in such species as $PCl_5$ and $SF_6$ (Section 7.4). Since there is no 2d sublevel, C, N, and O never form expanded octets.

Sometimes, as with $PCl_5$ and $SF_6$, it is clear from the formula that the central atom has an expanded octet. Often, however, it is by no means obvious that this is the case. At first glance, formulas such as $ClF_3$ or $XeF_4$ look completely straightforward. However, when you try to draw the Lewis structure it becomes clear that an expanded octet is involved. The number of electrons available after the skeleton is drawn is *greater* than the number required to give each atom an octet. When that happens, **distribute the extra electrons (two or four) around the central atom as unshared pairs.**

---

**Example 7.4**    Draw the Lewis structure of $XeF_4$.

**Strategy**    Follow the usual four-step sequence. If there is an electron surplus, add the extra electrons to the central xenon atom as unshared pairs.

**Solution**

(1) The number of valence electrons is 8 (from Xe) + 28 (from four F atoms) = 36.
(2) The skeleton is

$$
\begin{array}{c}
\text{F} \\
| \\
\text{F—Xe—F} \\
| \\
\text{F}
\end{array}
$$

(3) The number of electrons available for distribution is 36 − 8 = 28.
(4) Each fluorine atom needs six electrons to complete its octet; the number of electrons required = 4(6) = 24.

There is a surplus of electrons; 28 are available and only 24 are required to fill out octets. The four extra electrons are added to the central Xe atom. The Lewis structure for $XeF_4$ is

The "extra" electrons go to the central atom

In this molecule there are six pairs of electrons around the xenon atom.

---

# Chemistry: *The Human Side*

**G. N. Lewis**
(1875–1946)
(Dr. Glenn Seaborg, University of California, Lawrence Berkeley Laboratory)

Born in Massachusetts, G. N. Lewis grew up in Nebraska, then came back east to obtain his B.S. (1896) and Ph.D. (1899) at Harvard. Although he stayed on for a few years as an instructor, Lewis seems never to have been happy at Harvard. A precocious student and an intellectual rebel, he was repelled by the highly traditional atmosphere that prevailed in the chemistry department there in his time. Many years later, he refused an honorary degree from his alma mater.

After leaving Harvard, Lewis made his reputation at M.I.T., where he was promoted to full professor in only four years. In 1912, he moved across the country to the University of California at Berkeley as Dean of the College of Chemistry and department chairman. He remained there for the rest of his life. Under his guidance, the chemistry department at Berkeley became perhaps the most prestigious in the country. Among the faculty and graduate students that he attracted were five future Nobel Prize winners: Harold Urey in 1934, William Giauque in 1949, Glenn Seaborg in 1951, Willard Libby in 1960, and Melvin Calvin in 1961.

In administering the chemistry department at Berkeley, Lewis demanded excellence in both research and teaching. Virtually the entire staff was involved in the general chemistry program; at one time eight full professors carried freshman sections.

Lewis' interest in chemical bonding and structure dated from 1902. In attempting to explain "valence" to a class at Harvard, he devised an atomic model to rationalize the octet rule. His model was deficient in many respects; for one thing, Lewis visualized cubic atoms with electrons located at the corners. Perhaps this explains why his ideas of atomic structure were not published until 1916. In that year, Lewis conceived of the electron-pair bond, perhaps his greatest single contribution to chemistry. At that time, it was widely believed that all bonds were ionic; Lewis's ideas were rejected by many well-known organic chemists.

In 1923, Lewis published a classic book (later reprinted by Dover Publications) entitled *Valence and the Structure of Atoms and Molecules*. Here, in Lewis' characteristically lucid style, we find many of the basic principles of covalent bonding discussed in this chapter. Included are electron-dot structures, the octet rule, and the concept of electronegativity. Here too is the Lewis definition of acids and bases (Chap. 15). That same year, Lewis published with Merle Randall a text called *Thermodynamics and the Free Energy of Chemical Substances*. Today, a revised edition of that text is still widely used in graduate courses in chemistry.

The years from 1923 to 1938 were relatively unproductive for G. N. Lewis insofar as his own research was concerned. The applications of the electron-pair bond came largely in the areas of organic and quantum chemistry; in neither of these fields did Lewis feel at home. In the early 1930s, he published a series of relatively minor papers dealing with the properties of deuterium. Then, in 1939, he began to publish in the field of photochemistry. Of approximately 20 papers in this area, several were of fundamental importance, comparable in quality to the best work of his early years. Retired officially in 1945, Lewis died a year later while carrying out an experiment on fluorescence.

Lewis certainly deserved a Noble prize, but he never received one

## 7.2 Molecular Geometry

The geometry of a diatomic molecule such as $Cl_2$ or HCl can be described very simply. Since two points define a straight line, the molecule must be linear.

$$Cl—Cl \qquad H—Cl$$

With molecules containing three or more atoms, the geometry is not so obvious. Here, the angles between bonds, called **bond angles,** must be considered. For example, a molecule of the type $YX_2$, where Y represents the central atom and X an atom bonded to it, can be

— *linear,* with a bond angle of 180°:     X—Y—X

— *bent,* with a bond angle less than 180°:

$$\overset{\displaystyle Y}{\underset{\displaystyle X \qquad X}{\diagup \quad \diagdown}}$$

The major features of molecular geometry can be predicted on the basis of a quite simple principle—electron-pair repulsion. This principle is the essence of the *valence-shell electron-pair repulsion (VSEPR) model,* first suggested by N. V. Sidgwick and H. M. Powell in 1940. It was developed and expanded later by R. J. Gillespie and R. S. Nyholm. According to VSEPR theory, *the valence electron pairs surrounding an atom repel one another, so the orbitals containing those electron pairs are oriented to be as far apart as possible.*

In this section we apply this model to predict the geometry of some rather simple molecules and polyatomic ions. In all these species, a central atom is surrounded by from two to six pairs of electrons.

## Ideal Geometries with Two to Six Electron Pairs on the Central Atom

We begin by considering species in which a central atom, A, is surrounded by from two to six electron pairs, all of which are used to form single bonds with terminal atoms, X. These species have the general formulas $AX_2, AX_3, \ldots, AX_6$. It is understood that there are no unshared pairs around atom A.

To see how the bonding orbitals surrounding the central atom are oriented with respect to one another, consider Figure 7.3, which shows the positions taken naturally by two to six balloons tied together at the center. The balloons, like the orbitals they represent, arrange themselves to be as far from one another as possible.

Figure 7.4 (p. 186) shows the geometries predicted by the VSEPR model for molecules of the types $AX_2$ to $AX_6$. The geometries for two and three electron pairs are those associated with species in which the central atom has less than an octet of electrons. Molecules of this type include $BeF_2$ and $BF_3$, which have the Lewis structures

$$: \overset{..}{\underset{..}{F}} - Be - \overset{..}{\underset{..}{F}} :$$
$$180°$$

$$: \overset{..}{\underset{..}{F}} \diagdown \quad \diagup \overset{..}{\underset{..}{F}} :$$
$$B \;) \, 120°$$
$$: \overset{..}{\underset{..}{F}} :$$

Two electron pairs are as far apart as possible when they are directed at 180° to one another. This gives $BeF_2$ a **linear** structure. The three electron pairs around

**Figure 7.3**
The balloon models illustrate the geometries predicted by VSEPR theory.

**Figure 7.4**

Geometries to be expected from the VSEPR model for the molecules with two to six electron pairs around a central atom. (Compare with Figure 7.3.)

| Species type | Orientation of electron pairs | Predicted bond angles | Example | Ball and stick model |
|---|---|---|---|---|
| $AX_2$ | Linear | 180° | $BeF_2$ | 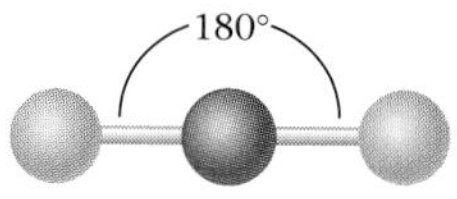 |
| $AX_3$ | Triangular Planar | 120° | $BF_3$ | 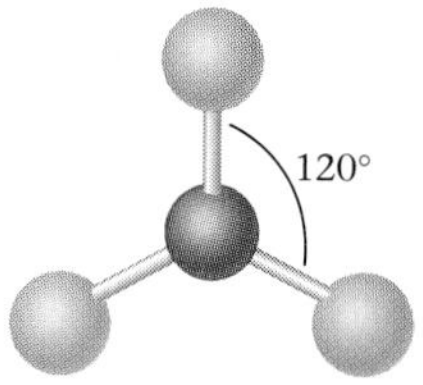 |
| $AX_4$ | Tetrahedron | 109.5° | $CH_4$ | 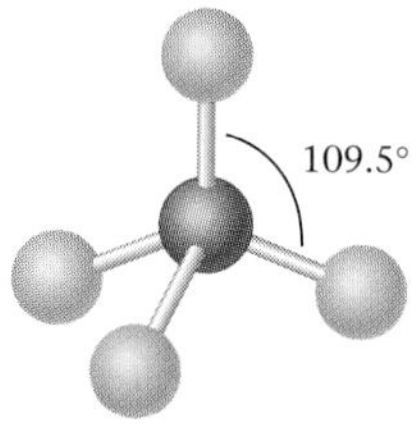 |
| $AX_5$ | Triangular Bipyramid | 90°<br>120°<br>180° | $PCl_5$ | 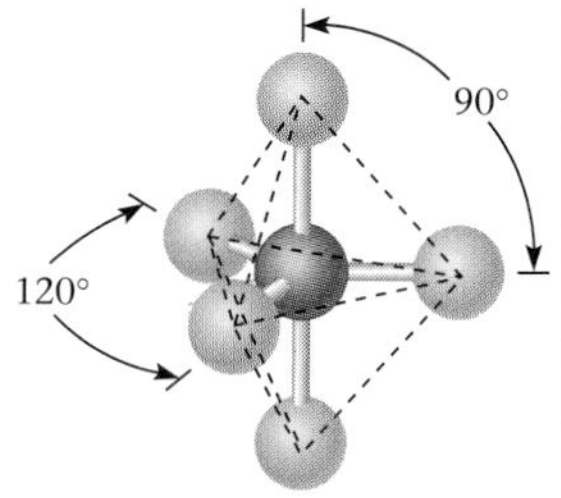 |
| $AX_6$ | Octahedron | 90°<br>180° | $SF_6$ | 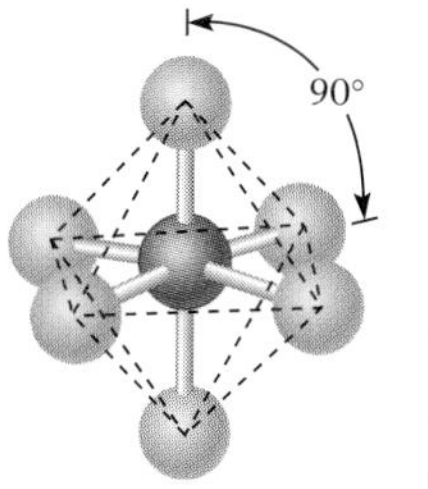 |

the boron atom in $BF_3$ are directed toward the corners of an equilateral triangle; the bond angles are 120°. We describe this geometry as **triangular planar.**

In species that follow the octet rule, the central atom is surrounded by four electron pairs. If each of these pairs forms a single bond with a terminal atom, a molecule of the type $AX_4$ results. The four bonds are directed toward the corners of a regular **tetrahedron.** All the bond angles are 109.5°, the tetrahedral angle. This geometry is found in many polyatomic ions such as $NH_4^+$ and $SO_4^{2-}$ and in a wide variety of organic molecules, the simplest of which is methane, $CH_4$.

This puts the four electron pairs as far apart as possible

Molecules of the type $AX_5$ and $AX_6$ require that the central atom have an expanded octet. The geometries of these molecules are shown at the bottom of Figure 7.4. In $PF_5$, the five bonding pairs are directed toward the corners of a **triangular bipyramid,** a figure formed when two triangular pyramids are fused together, base-to-base. Three of the fluorine atoms are located at the corners of an equilateral triangle with the phosphorus atom at the center; the other two fluorine atoms are directly above and below the P atom. In $SF_6$, the six bonds are directed toward the corners of a regular **octahedron.** An octahedron can be formed by fusing two square pyramids base-to-base. Four of the fluorine atoms in $SF_6$ are located at the corners of a square with the S atom at the center; one fluorine atom is directly above the S atom, another directly below it.

## Effect of Unshared Pairs on Molecular Geometry

In many molecules and polyatomic ions, one or more of the electron pairs around the central atom is unshared. The VSEPR model is readily extended to predict the geometries of these species. In general,

1. The electron-pair geometry is approximately the same as that observed when only single bonds are involved. The bond angles are either equal to the ideal values listed in Figure 7.4 or a little less than these values.
2. The *molecular geometry* is quite different when one or more unshared pairs are present. In describing molecular geometry, we refer only to the positions of the bonded atoms. These positions can be determined experimentally; positions of unshared pairs cannot be established by experiment. Hence, the locations of unshared pairs are not specified in describing molecular geometry.

With these principles in mind, consider the $NH_3$ molecule:

$$\ddot{N} \overset{|}{\underset{H \ H \ H}{}}$$

The apparent orientation of the four electron pairs around the N atom in $NH_3$ is shown at the top of Figure 7.5. Notice that, as in $CH_4$, the four pairs are directed toward the corners of a regular tetrahedron. The diagram at the bottom of Figure 7.5 shows the positions of the atoms in $NH_3$. The nitrogen atom is located above the center of an equilateral triangle formed by the three hydrogen atoms. The molecular geometry of the $NH_3$ molecule is described as a **triangular pyramid.** The nitrogen atom is at the apex of the pyramid, while the three hydrogen atoms form its triangular base. The molecule is three-dimensional, as the word "pyramid" implies.

The development we have just gone through for $NH_3$ is readily extended to the water molecule, $H_2O$. Here, the Lewis structure shows that the central oxygen atom is surrounded by two single bonds and two unshared pairs:

$$\overset{..}{\underset{H \ \overset{..}{} \ H}{O}}$$

The diagram at the top of Figure 7.6 emphasizes that the four electron pairs are oriented tetrahedrally. At the bottom, the positions of the atoms are shown. Clearly they are not in a straight line; the $H_2O$ molecule is **bent.**

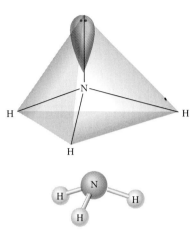

**Figure 7.5**
Two ways of showing the geometry of the $NH_3$ molecule. The orientation of the electron pairs, including the unshared pair (blue lobe), is shown at the top. The orientation of the atoms is shown at the bottom. The nitrogen atom is located directly above the center of the equilateral triangle formed by the three hydrogen atoms. The $NH_3$ molecule is described as a triangular pyramid.

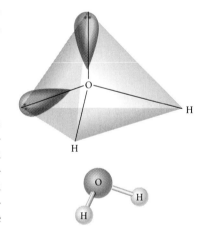

**Figure 7.6**
Two ways of showing the geometry of the $H_2O$ molecule. At the top, the two unshared pairs are shown. As you can see from the drawing at the bottom, $H_2O$ is a bent molecule. The bond angle, 105°, is a little smaller than the tetrahedral angle, 109.5°. The unshared pairs spread out over a larger volume than that occupied by the bonding pairs.

TABLE 7.3   **Geometries with 2, 3, or 4 Electron Pairs Around a Central Atom**

| No. of Terminal Atoms (X) + Unshared Pairs (E) | Species Type | Ideal Bond Angles* | Molecular Geometry | Examples |
|---|---|---|---|---|
| 2 | $AX_2$ | 180° | linear | $BeF_2$, $CO_2$ |
| 3 | $AX_3$ | 120° | triangular planar | $BF_3$, $SO_3$ |
|   | $AX_2E$ | 120°* | bent | $GeF_2$, $SO_2$ |
| 4 | $AX_4$ | 109.5° | tetrahedron | $CH_4$ |
|   | $AX_3E$ | 109.5°* | triangular pyramid | $NH_3$ |
|   | $AX_2E_2$ | 109.5°* | bent | $H_2O$ |

*In these species, the observed bond angle is ordinarily somewhat less than the ideal value.

Experiments show that the bond angles in $NH_3$ and $H_2O$ are slightly less than the ideal value of 109.5°. In $NH_3$ (three single bonds, one unshared pair around N), the bond angle is 107°. In $H_2O$ (two single bonds, two unshared pairs around O), the bond angle is about 105°.

These effects can be explained in a rather simple way. An unshaired pair is attracted by one nucleus, that of the atom to which it belongs. In contrast, a bonding pair is attracted by two nuclei, those of the two atoms it joins. Hence the electron cloud of an unshared pair is expected to spread out over a larger volume than that of a bonding pair. In $NH_3$, this tends to force the bonding pairs closer to one another, thereby reducing the bond angle. Where there are two unshared pairs, as in $H_2O$, this effect is more pronounced. In general, the VSEPR model predicts that unshared electron pairs will occupy slightly more space than bonding pairs.

*Unshared pairs reduce bond angles below ideal values*

Table 7.3 summarizes the molecular geometries of species in which a central atom is surrounded by two, three, or four electron pairs. The table is organized in terms of the number of terminal atoms, X, and unshared pairs, E, surrounding the central atom, A.

---

**Example 7.5**   Predict the geometries of the following molecules.

(a) $BeH_2$   (b) $OF_2$   (c) $PF_3$

**Strategy**   Draw the Lewis structure as a first step. Then decide what type ($AX_2$, $AX_3$, etc.) the molecule is, focusing on the central atom. Remember, X represents a terminal atom, E an unshared pair of electrons.

**Solution**

(a)   The Lewis structure for $BeH_2$ is

$$H\!-\!Be\!-\!H$$

The central atom has no unshared electron pairs and two bonded atoms. The molecule is of type $AX_2$.   It is linear with a bond angle of 180°.

(b)  The Lewis structure of $OF_2$ is

$$\ddot{O}$$
$$:\ddot{F} \quad \ddot{F}:$$

The central oxygen atom has two unshared pairs and two bonds. The molecule is of type $AX_2E_2$ (like $H_2O$); it should be bent with a bond angle somewhat

less than the ideal value of 109.5°.  The observed value is 103°.

(c)  The Lewis structure of $PF_3$ is

$$\ddot{P}$$
$$:\ddot{F} \quad :\ddot{F}: \quad \ddot{F}:$$

The central phosphorus atom has one unshared pair and three bonded atoms. The molecule is of type $AX_3E$; it should be a triangular pyramid (like $NH_3$) with

a bond angle somewhat less than 109.5°  (actually, 104°).

---

In many expanded-octet molecules, one or more of the electron pairs around the central atom is unshared. Recall, for example, the Lewis structure of xenon tetrafluoride, $XeF_4$,

$$:\ddot{F} \qquad \ddot{F}:$$
$$Xe$$
$$:\ddot{F} \,_{(AX_4E_2)}\, \ddot{F}:$$

Here there are six electron pairs around the xenon atom; four of these are covalent bonds to fluorine and the other two pairs are unshared. This molecule is classified as $AX_4E_2$.

Geometries of molecules such as these can be predicted by the VSEPR model. The results are shown in Figure 7.7, p. 190. The structures listed include those of all types of molecules having five or six electron pairs around the central atom, one or more of which may be unshared.

## Multiple Bonds

The VSEPR model is readily extended to species in which double or triple bonds are present. Here, a simple principle applies: *Insofar as molecular geometry is concerned, a multiple bond behaves like a single electron pair.* This makes sense. The four electrons in a double bond, or the six electrons in a triple bond, must be located between the two atoms, as are the two electrons in a single bond. This means that the electron pairs in a multiple bond must occupy the same region of space as those in a single bond. Hence, the "extra" electron pairs in a multiple bond have no effect upon geometry.

To illustrate this principle, consider the $CO_2$ molecule. Its Lewis structure is

$$:\ddot{O}=C=\ddot{O}:$$

The central atom, carbon, has two double bonds and no unshared pairs. For purposes of determining molecular geometry, we pretend that the double bonds are single bonds, ignoring the "extra" bonding pairs. The bonds are directed to be as far apart as possible, giving a 180° O—C—O bond angle. The $CO_2$ molecule, like $BeF_2$, is linear:

Dry Ice, $CO_2(s)$.

**Figure 7.7**

Geometries of molecules with expanded octets. The green spheres represent terminal atoms. The open ellipses represent unshared electron pairs.

**5 ELECTRON PAIRS**

| Species type | Structure | Description | Example | Bond angles |
|---|---|---|---|---|
| $AX_5$ | 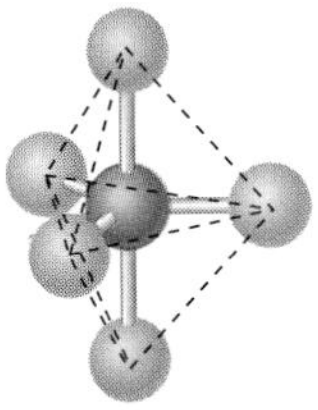 | Triangular bipyramidal | $PF_5$ | 90°, 120°, 180° |
| $AX_4E$ | 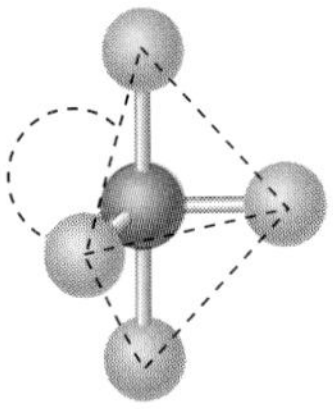 | See-saw | $SF_4$ | 90°, 120°, 180° |
| $AX_3E_2$ | 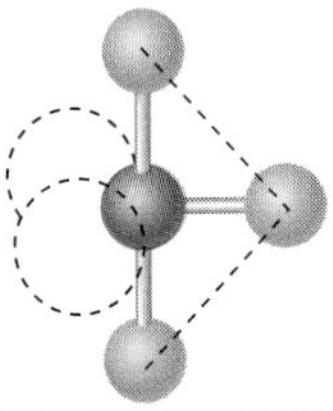 | T-shaped | $ClF_3$ | 90°, 180° |
| $AX_2E_3$ | 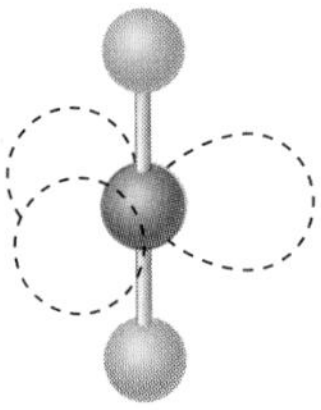 | Linear | $XeF_2$ | 180° |

**6 ELECTRON PAIRS**

| | | | | |
|---|---|---|---|---|
| $AX_6$ | 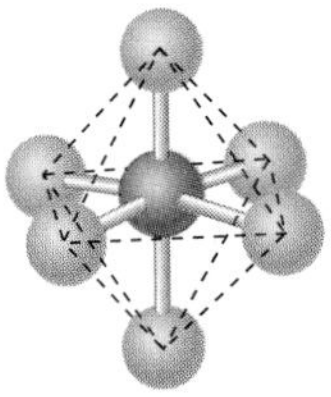 | Octahedral | $SF_6$ | 90°, 180° |
| $AX_5E$ | 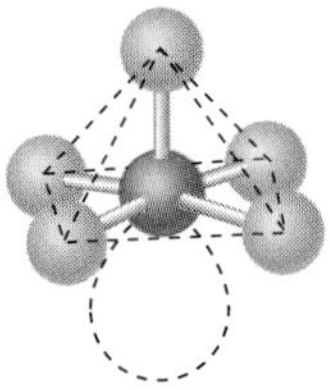 | Square pyramidal | $ClF_5$ | 90°, 180° |
| $AX_4E_2$ | 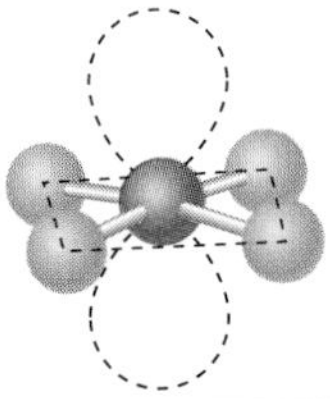 | Square planar | $XeF_4$ | 90°, 180° |

$$F\!-\!Be\!-\!F \qquad O\!=\!C\!=\!O$$
$$180° \qquad\qquad 180°$$

This principle can be restated in a somewhat different way for molecules in which there is a single central atom. The geometry of such a molecule depends only upon

— *the number of terminal atoms, X, bonded to the central atom, irrespective of whether the bonds are single, double, or triple.*
— *the number of unshared pairs, E, around the central atom*

This means that Table 7.3 can be used in the usual way to predict the geometry of a species containing multiple bonds.

---

**Example 7.6**    Predict the geometries of the $ClO_3{}^-$ ion, the $NO_3{}^-$ ion, and the $N_2O$ molecule, which have the Lewis structures

(a) $\left[ :\!\overset{..}{\underset{..}{O}}\!-\!\overset{..}{\underset{\underset{:\!\overset{..}{\underset{..}{O}}\!:}{|}}{Cl}}\!-\!\overset{..}{\underset{..}{O}}\!: \right]^{-}$ 　(b) $\left[ :\!\overset{..}{\underset{..}{O}}\!-\!\overset{\overset{:\!\overset{..}{\underset{..}{O}}\!:}{\|}}{N}\!-\!\overset{..}{\underset{..}{O}}\!: \right]^{-}$ 　(c) $:\!N\!=\!\overset{..}{N}\!=\!\overset{..}{\underset{..}{O}}\!:$

*Strategy*    Classify each species as $AX_mE_n$ and use Table 7.3.

*Solution*

(a) The central atom, chlorine, is bonded to three oxygen atoms; it has one un-

shared pair. The $ClO_3{}^-$ ion is of the type $AX_3E$. It is a　triangular pyramid;　the ideal bond angle is 109.5°.

(b) The central atom, nitrogen, is bonded to three oxygen atoms; it has no un-shared pairs. The $NO_3{}^-$ ion is of the type $AX_3$. It has the geometry of an equilateral triangle, with the nitrogen atom at the center; the bond angle is 120°. The ion

is　triangular planar.

(c) The central nitrogen atom is bonded to two other atoms with no unshared

pairs. The molecule, type $AX_2$, is　linear,　with a bond angle of 180°.

> $AX_3E_2$  means that the central atom A is bonded to three X atoms and has two unshared electron pairs

---

The VSEPR model applies equally well to molecules in which there is no single central atom. Consider the acetylene molecule, $C_2H_2$. Recall that here the two carbon atoms are joined by a triple bond:

$$H\!-\!C\!\equiv\!C\!-\!H$$

Each carbon atom behaves as if it were surrounded by two electron pairs. Both of the bond angles ($H\!-\!C\!\equiv\!C$ and $C\!\equiv\!C\!-\!H$) are 180°. The molecule is linear; the four atoms are in a straight line. The two "extra" electron pairs in the triple bond do not affect the geometry of the molecule.

In ethylene, $C_2H_4$, there is a double bond between the two carbon atoms. The molecule has the geometry to be expected if each carbon atom had only three pairs of electrons around it.

$$\underset{H}{\overset{H}{\diagdown}}C\!=\!C\underset{H}{\overset{H}{\diagup}}$$

The six atoms are located in a plane with bond angles of approximately 120°.

Synthesis and combustion of acetylene, $C_2H_2$.

## 7.5 Polarity of Molecules

Covalent bonds and molecules held together by such bonds may be

— *polar.* As a result of an unsymmetrical distribution of electrons, the bond or molecule contains a positive and a negative pole and is therefore a *dipole.*
— *nonpolar.* A symmetrical distribution of electrons leads to a bond or molecule with no positive or negative poles.

### Polar and Nonpolar Covalent Bonds

The two electrons in the $H_2$ molecule are shared equally by the two nuclei. Stated another way, a bonding electron is as likely to be found in the vicinity of one nucleus as another. Bonds of this type are described as nonpolar. **Nonpolar bonds** are formed whenever the two atoms joined are identical, as in $H_2$ and $F_2$.

*All molecules, except those of elements, have polar bonds*

In the HF molecule, the distribution of the bonding electrons is somewhat different from that found in $H_2$ or $F_2$. Here, the density of the electron cloud is greater about the fluorine atom. The bonding electrons, on the average, are shifted toward fluorine and away from the hydrogen (atom Y in Fig. 7.8). Bonds in which the electron density is unsymmetrical are referred to as **polar bonds.**

Atoms of two different elements always differ at least slightly in their electronegativity (recall Table 6.6, p. 166). Hence, covalent bonds between unlike atoms are always polar. Consider, for example, the H—F bond. Since fluorine has a higher electronegativity (4.0) than does hydrogen (2.2), bonding electrons are displaced toward the fluorine atom. The H—F bond is polar, with a partial negative charge at the fluorine atom and a partial positive charge at the hydrogen atom.

The extent of polarity of a covalent bond is related to the difference in electronegativities of the bonded atoms. If this difference is large, as in HF ($\Delta$EN = 1.8), the bond is strongly polar. Where the difference is small, as in H—C ($\Delta$EN = 0.3), the bond is only slightly polar.

### Polarity of Molecules

A polar molecule is one that contains positive and negative poles. There is a partial positive charge (positive pole) at one point in the molecule and a partial negative charge (negative pole) at a different point. As shown in Figure 7.9, polar molecules orient themselves in the presence of an electric field. The positive pole in the molecule tends to align with the external negative charge, and the negative pole with the external positive charge. In contrast, there are no positive and negative poles in a nonpolar molecule. In an electric field, nonpolar molecules, such as $H_2$, show no preferred orientation.

If a molecule is diatomic, it is easy to decide whether it is polar or nonpolar. A diatomic molecule has only one kind of bond; hence, the polarity of the molecule is the same as the polarity of the bond. Hydrogen and fluorine ($H_2$, $F_2$) are nonpolar because the bonded atoms are identical and the bond is nonpolar. Hydrogen fluoride, HF, on the other hand, has a polar bond, so the molecule is polar. The bonding electrons spend more time near the fluorine atom so that there is a negative pole at that end and a positive pole at the hydrogen end. This is sometimes indicated by writing

$$H \longmapsto F$$

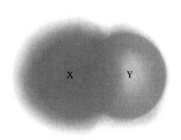

**Figure 7.8**
If two atoms, X and Y, differ in electronegativity, the bond between them is polar. The electron cloud associated with the bonding electrons is concentrated around the more electronegative atom, in this case, X.

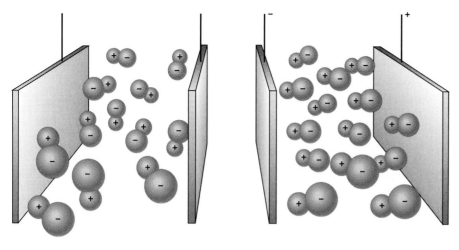

**Figure 7.9**
Orientation of polar molecules in an electric field. In the absence of the field, polar molecules are randomly oriented. In an electric field, polar molecules such as HF line up as shown ("Field on"). Nonpolar molecules such as $H_2$ do not line up.

Field off                                   Field on

The arrow points toward the negative end of the polar bond (F atom); the plus sign is at the positive end (H atom). The HF molecule is called a *dipole;* it contains positive and negative poles.

If a molecule contains more than two atoms it is not so easy to decide whether it is polar or nonpolar. In this case, not only bond polarity but also molecular geometry determines the polarity of the molecule. To illustrate what is involved, consider the molecules shown in Figure 7.10.

**1.** In $BeF_2$ there are two polar Be—F bonds; in both bonds, the electron density is concentrated around the more electronegative fluorine atom. However, since the $BeF_2$ molecule is linear, the two Be$+\!\!\longrightarrow$F dipoles are in opposite directions and cancel one another. The molecule has no net dipole and hence is nonpolar. From a slightly different point of view, in $BeF_2$ the centers of positive and negative charge coincide with each other, at the Be atom. There is no way that a $BeF_2$ molecule can line up in an electric field.

The following molecules are nonpolar: $A_2$, $AX_2$ (linear), $AX_3$ (triangular planar), $AX_4$ (tetrahedral)

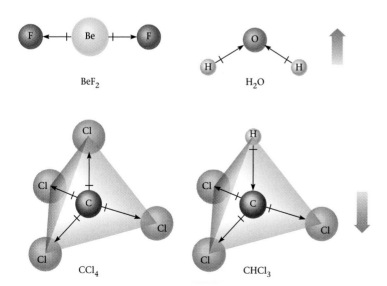

**Figure 7.10**
Polarity of molecules. In all these molecules, the bonds are polar, as indicated by the $+\!\!\longrightarrow$ notation. However, in $BeF_2$ and $CCl_4$, the bond dipoles cancel, and the molecule is nonpolar. In $H_2O$ and $CHCl_3$, on the other hand, there is a net dipole, and the molecule is polar. (The broad arrow beside the molecule points to the negative pole.)

**2.** Since oxygen is more electronegative than hydrogen (3.5 versus 2.2) an O—H bond is polar, with the electron density higher around the oxygen atom. In the bent $H_2O$ molecule, the two $H{+}{\longrightarrow}O$ dipoles do not cancel each other. Instead, they add to give the $H_2O$ molecule a net dipole. The center of negative charge is located at the O atom; this is the negative pole of the molecule. The center of positive charge is located midway between the two H atoms; the positive pole of the molecule is at that point. The $H_2O$ molecule is polar. It tends to line up in an electric field with the oxygen atom oriented toward the positive electrode.

**3.** Carbon tetrachloride, $CCl_4$, is another molecule that, like $BeF_2$, is nonpolar despite the presence of polar bonds. Each of its four bonds is a dipole, $C{+}{\longrightarrow}Cl$. However, because the four bonds are arranged symmetrically around the carbon atom, they cancel. As a result, the molecule has no net dipole; it is nonpolar. If one of the Cl atoms in $CCl_4$ is replaced by hydrogen, the situation changes. In the $CHCl_3$ molecule, the $H{+}{\longrightarrow}C$ dipole does not cancel with the three $C{+}{\longrightarrow}Cl$ dipoles. Hence $CHCl_3$ is polar.

There are two criteria for determining the polarity of a molecule; bond polarity and molecular geometry. *If the polar A—X bonds in a molecule $AX_mE_n$ are arranged symmetrically around the central atom A, the molecule is nonpolar.*

---

**Example 7.7**    Determine whether each of the following is polar or nonpolar:

(a) $SO_2$    (b) $BF_3$    (c) $CO_2$

***Strategy***    Write the Lewis structure of the molecule and classify it as $AX_mE_n$. Using Table 7.3 or Figure 7.7, decide upon the geometry of the molecule. Then decide whether the dipoles cancel, in which case it is nonpolar.

*Solution*

(a) The Lewis structure of $SO_2$ is shown on p. 178; it is of the type $AX_2E$. The molecule is bent, so the dipoles do not cancel. The $SO_2$ molecule is polar.

(b) The Lewis structure of $BF_3$ is shown on p. 182; it is of the type $AX_3$. The molecule is an equilateral triangle with the boron atom at the center. The three polar bonds cancel one another; $BF_3$ is nonpolar.

(c) The Lewis structure of $CO_2$ is shown on p. 189; it is of the type $AX_2$. The molecule is linear, so the two $C{+}{\longrightarrow}O$ dipoles cancel each other; $CO_2$ is nonpolar.

---

## 7.4  Atomic Orbitals; Hybridization

In the 1930s a theoretical treatment of the covalent bond was developed by, among others, Linus Pauling, then at the California Institute of Technology. The *atomic orbital* or **valence bond** model won him the Nobel Prize in chemistry in 1954. Eight years later, Pauling won the Nobel Peace Prize for his efforts to stop nuclear testing.

According to this model, a covalent bond consists of a pair of electrons of opposed spin within an atomic orbital. For example, a hydrogen atom forms a

Pauling, a versatile chemist with contributions in many areas, died in 1994

covalent bond by accepting an electron from another atom to complete its 1s orbital. Using orbital diagrams, we could write

|                            | 1s       |
|----------------------------|----------|
| isolated H atom            | (↑)      |
| H atom in a stable molecule| (↑↓)     |

The second electron, shown in color, is contributed by another atom. This could be another H atom in $H_2$, an F atom in HF, a C atom in $CH_4$, and so on.

This simple model is readily extended to other atoms. The fluorine atom (electron configuration $1s^2 2s^2 2p^5$) has a half-filled p orbital:

|               | 1s   | 2s   |      | 2p        |      |
|---------------|------|------|------|-----------|------|
| isolated F atom | (↑↓) | (↑↓) | (↑↓) | (↑↓) | (↑ ) |

By accepting an electron from another atom, F can complete this 2p orbital:

|                          | 1s   | 2s   |      | 2p        |      |
|--------------------------|------|------|------|-----------|------|
| F atom in HF, $F_2$,. . . | (↑↓) | (↑↓) | (↑↓) | (↑↓) | (↑↓) |

According to this model, it would seem that in order for an atom to form a covalent bond, it must have an unpaired electron. Indeed, the number of bonds formed by an atom should be determined by its number of unpaired electrons. Since hydrogen has an unpaired electron, an H atom should form one covalent bond, as indeed it does. The same holds for the F atom, which forms only one bond. Noble gas atoms, such as He and Ne, which have no unpaired electrons, should not form bonds at all; they don't.

When this simple idea is extended beyond hydrogen, the halogens, and the noble gases, problems arise. Consider, for example, the three atoms Be ($Z = 4$), B ($Z = 5$), and C ($Z = 6$):

|         | 1s   | 2s   |      | 2p   |      |
|---------|------|------|------|------|------|
| Be atom | (↑↓) | (↑↓) | ( )  | ( )  | ( )  |
| B atom  | (↑↓) | (↑↓) | (↑)  | ( )  | ( )  |
| C atom  | (↑↓) | (↑↓) | (↑)  | (↑)  | ( )  |

Notice that the beryllium atom has no unpaired electrons, the boron atom has one, and the carbon atom two. Simple valence bond theory would predict that Be, like He, should not form covalent bonds. A boron atom should form one bond, carbon two. Experience tells us that these predictions are wrong. Beryllium forms two bonds in $BeF_2$; B forms three bonds in $BF_3$. Carbon ordinarily forms four bonds, not two.

To explain these and other discrepancies, simple valence bond theory must be modified. It is necessary to invoke a new kind of atomic orbital, called a **hybrid orbital.**

## Hybrid Orbitals: sp, sp², sp³, sp³d, sp³d²

The formation of the $BeF_2$ molecule can be explained by assuming that, as two fluorine atoms approach, the atomic orbitals of the beryllium atom undergo a significant change. Specifically, the 2s orbital is mixed or *hybridized* with a 2p orbital to form two new **sp hybrid orbitals.**

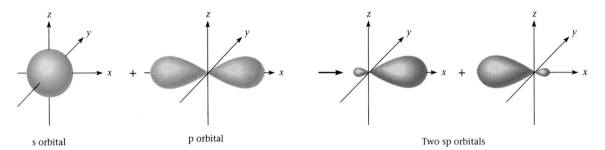

s orbital          p orbital                    Two sp orbitals

**Figure 7.11**
Formation of sp hybrid orbitals. The mixing of an s orbital with a p orbital gives two sp hybrid orbitals.

one s atomic orbital + one p atomic orbital ⟶ *two* sp hybrid orbitals

Notice (Fig. 7.11) that **the number of hybrid orbitals formed is equal to the number of atomic orbitals mixed.** This is always true in hybridization of orbitals. Moreover, the energies of the hybrid orbitals are intermediate between those of the atomic orbitals from which they are derived.

In the $BeF_2$ molecule, there are two electron-pair bonds. These electron pairs are located in the two sp hybrid orbitals just described. In each orbital

— one electron is a valence electron contributed by beryllium; remember that Be has two such electrons (·Be·)
— one electron is a valence electron contributed by a fluorine atom.

The "orbital diagram" for the beryllium atom in the $BeF_2$ molecule can be written as

|  | 1s | 2s |  | 2p |  |
|---|---|---|---|---|---|
| Be in $BeF_2$ | (↑↓) | (↑↓) | (↑↓) | ( ) | ( ) |

(The blue arrows indicate the two electrons supplied by the fluorine atoms. The horizontal lines enclose the hybrid orbitals involved in bond formation).

A similar argument can be used to explain why boron forms three bonds and carbon four.

In the case of boron:

one s atomic orbital + two p atomic orbitals ⟶ *three* $sp^2$ hybrid orbitals

With carbon:

one s atomic orbital + three p atomic orbitals ⟶ *four* $sp^3$ hybrid orbitals

Thus we have

|  | 1s | 2s |  | 2p |  |  |
|---|---|---|---|---|---|---|
| B in $BF_3$ | (↑↓) | (↑↓) | (↑↓) | (↑↓) | ( ) | |
| C in $CH_4$ | (↑↓) | (↑↓) | (↑↓) | (↑↓) | (↑↓) | |

(Remember that boron has three valence electrons and carbon has four; each orbital contains one of these electrons along with another, shown in blue, contributed by a F or H atom).

TABLE 7.4   **Hybrid Orbitals and Their Geometries**

| Number of Electron Pairs | Atomic Orbitals | Hybrid Orbitals | Orientation | Examples |
|---|---|---|---|---|
| 2 | s, p | sp | linear | $BeF_2$, $CO_2$ |
| 3 | s, two p | $sp^2$ | triangular planar | $BF_3$, $SO_3$ |
| 4 | s, three p | $sp^3$ | tetrahedron | $CH_4$, $NH_3$, $H_2O$ |
| 5 | s, three p, d | $sp^3d$ | triangular bipyramid | $PCl_5$, $SF_4$, $ClF_3$ |
| 6 | s, three p, two d | $sp^3d^2$ | octahedron | $SF_6$, $ClF_5$, $XeF_4$ |

You will recall that the bond angles in $NH_3$ and $H_2O$ are very close to that in $CH_4$. This suggests that the four electron pairs surrounding the central atom in $NH_3$ and $H_2O$, like those in $CH_4$, occupy $sp^3$ hybrid orbitals. In $NH_3$, three of these orbitals are filled by bonding electrons, the other by the unshared pair on the nitrogen atom. In $H_2O$, two of the $sp^3$ orbitals of the oxygen atom contain bonding electron pairs; the other two contain unshared pairs. The situation in $NH_3$ and $H_2O$ is not unique. In general, we find that *unshared as well as shared electron pairs can be located in hybrid orbitals.*

The "extra" electron pairs in an expanded octet are accommodated by using d orbitals. The phosphorus atom (five valence electrons) in $PCl_5$ and the sulfur atom (six valence electrons) in $SF_6$ make use of 3d as well as 3s and 3p orbitals:

> Hybridization: sp ($AX_2$), $sp^2$ ($AX_3$, $AX_2E$), $sp^3$ ($AX_4$, $AX_3E$, $AX_2E_2$)

|  | | 3s | 3p | | | | 3d | | | |
|---|---|---|---|---|---|---|---|---|---|---|
| P atom in $PCl_5$ | [$_{10}$Ne] | (↑↓) | (↑↓) (↑↓) (↑↓) | | | (↑↓) | ( ) ( ) ( ) ( ) | | | |
| S atom in $SF_6$ | [$_{10}$Ne] | (↑↓) | (↑↓) (↑↓) (↑↓) | | | (↑↓) (↑↓) | ( ) ( ) ( ) | | | |

The orbitals used by the five pairs of bonding electrons surrounding the phosphorus atom in $PCl_5$ are **$sp^3d$ hybrid orbitals.**

one s orbital + three p orbitals + one d orbital ⟶ *five* $sp^3d$ hybrid orbitals

Similarly, in $SF_6$, the six pairs of bonding electrons are located in **$sp^3d^2$ hybrid orbitals:**

one s orbital + three p orbitals + two d orbitals ⟶ *six* $sp^3d^2$ hybrid orbitals

Table 7.4 gives the orientation in space of hybrid orbitals, as found mathematically by quantum mechanics. Notice that *the geometries are exactly as predicted by VSEPR theory.* In each case, the several hybrid orbitals are directed to be as far apart as possible; recall Figure 7.4, p. 186.

---

**Example 7.8**   Give the hybridization of

(a) carbon in $CF_4$   (b) phosphorus in $PF_3$   (c) sulfur in $SF_4$

***Strategy***   Draw the Lewis structure for the molecule and determine the number of electron pairs (single bonds or unshared pairs) around the central atom. The possible hybridizations are sp (two pairs), $sp^2$ (three pairs), $sp^3$ (four pairs), $sp^3d$ (five pairs), and $sp^3d^2$ (six pairs).

*Solution*

(a) For $CF_4$, the Lewis structure is

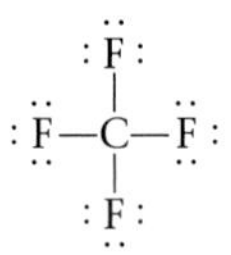

$$\ddot{F}$$

There are four bonds and no unshared electron pairs around C, the central atom. The hybridization is $sp^3$.

(b) The Lewis structure of $PF_3$ is

There are three bonds and one unshared pair, making a total of four pairs. The hybridization is $sp^3$.

(c) Sulfur tetrafluoride has the Lewis structure

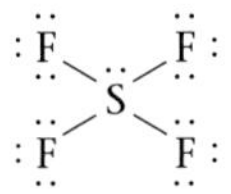

There are five electron pairs around sulfur: four bonds and one unshared pair. The hybridization is $sp^3d$.

## Multiple Bonds

In Section 7.2, we saw that insofar as geometry is concerned, a multiple bond acts as if it were a single bond. In other words, the "extra" electron pairs in a double or triple bond have no effect upon the geometry of the molecule. This behavior is related to hybridization. *The extra electron pairs in a multiple bond (one pair in a double bond, two pairs in a triple bond) are not located in hybrid orbitals.*

To illustrate this rule, consider the ethylene ($C_2H_4$) and acetylene ($C_2H_2$) molecules. You will recall that the bond angles in these molecules are 120° for ethylene and 180° for acetylene. This implies $sp^2$ hybridization in $C_2H_4$ and $sp$ hybridization in $C_2H_2$ (see Table 7.4). Using blue lines to represent hybridized electron pairs,

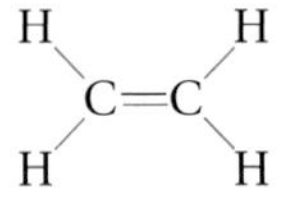

$$\begin{array}{cc} H & H \\ \diagdown & \diagup \\ C= & C \\ \diagup & \diagdown \\ H & H \end{array} \qquad H-C{\equiv}C-H$$

ethylene 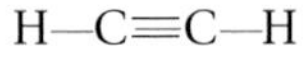                acetylene

In both cases, only one of the electron pairs in the multiple bond is hybridized.

---

**Example 7.9**    What is the hybridization of the nitrogen atom in

(a) $N_2$    (b) $NO_3^-$

**Strategy** The first step is to draw Lewis structures. To find the hybridization of the nitrogen atom, include

— unshared electron pairs
— electron pairs forming single bonds
— one and only one of the electron pairs in a multiple bond

*Solution*

(a) $:N{\equiv}N:$

For each N atom, hybridize one unshared pair and one of the electron pairs in the triple bond; the hybridization is sp.

(b) $\left[ :\overset{..}{\underset{..}{O}}-N-\overset{..}{\underset{..}{O}}: \right]^{-}$
$\quad\quad\quad\underset{:O:}{\overset{\|}{}}$

Three of the four electron pairs around the nitrogen atom are hybridized; the hybridization is sp².

---

## Sigma and Pi Bonds

We have noted that the extra electron pairs in a multiple bond are not hybridized and have no effect upon molecular geometry. At this point, you may well wonder what happened to those electrons. Where are they in molecules like $C_2H_4$ and $C_2H_2$?

To answer this question, it is necessary to consider the shape or spatial distribution of the orbitals filled by bonding electrons in molecules. From this point of view, we can distinguish between two types of bonding orbitals. The first of these, and by far the more common, is called a **sigma** ($\sigma$) bonding orbital. Hybrid orbitals are of this type. Their shape is similar to that of a p orbital in an isolated atom. A sigma bonding orbital consists of a single lobe in which the electron density is concentrated in the region directly between the two bonded atoms. The bonds shown in blue in the structures written for $C_2H_4$ and $C_2H_2$ on p. 198 are of this type.

The unhybridized bonding orbitals associated with multiple bonds have a quite different shape. They are referred to as **pi** ($\pi$) bonding orbitals. A pi orbital consists of two lobes, one above the bond axis, the other below it. Along the bond axis itself, the electron density is zero. There is one such orbital in the $C_2H_4$ molecule, two in $C_2H_2$.

The geometries of the bonding orbitals in ethylene and acetylene are shown in Figure 7.12. Sigma orbitals are shown in blue, pi orbitals in red. Notice that

— the nature of the $\pi$ bond in $C_2H_4$ prohibits free rotation about the carbon-carbon bond. For that to happen the electron density of the $\pi$ orbital would have to be high in areas outside the plane of the paper. In effect, the pi bond is fixed in position, so the $C_2H_4$ molecule is planar.
— the two $\pi$ bonding orbitals in $C_2H_2$, both of which consist of two lobes, are oriented at right angles to one another. This is difficult to show in a two-dimensional figure. In effect, the four lobes of the two $\pi$ bonds (red) are wrapped around the central $\sigma$ bond (blue) like a bun around a hot dog.

Sigma and pi orbitals, like s and p, differ in shape; each orbital can hold 2 e⁻

**Figure 7.12**
Bonding in ethylene, $C_2H_4$, and acetylene, $C_2H_2$. The sigma bond backbone is shown in blue; the pi bonds (one in $C_2H_4$, two in $C_2H_2$) are shown in red. Note that a pi bonding orbital consists of two lobes.

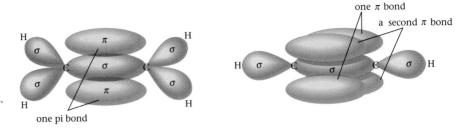

In general, to find the number of $\sigma$ and $\pi$ bonds in a species, remember that

— all single bonds are sigma bonds.
— one of the electron pairs in a multiple bond is a sigma bond; the others are pi bonds.

---

**Example 7.10**    Give the number of pi and sigma bonds in

(a) $N_2$    (b) $NO_3^-$

**Strategy**    Refer back to Example 7.9 for the Lewis structures. Then apply the two rules just cited.

**Solution**

(a) In $N_2$ there is a triple bond consisting of    one $\sigma$ bond and two $\pi$ bonds.

(b) In $NO_3^-$, there is one double bond and two single bonds; this translates to
three $\sigma$ bonds and one $\pi$ bond.

---

## CHEMISTRY

### *Beyond the Classroom*

## The Noble Gases

The modern periodic table contains six relatively unreactive gases that were unknown to Mendeleev: the noble gases that make up Group **18** at the far right of the table. The first of these elements to be isolated was argon, which makes up about 0.9% of air. The physicist Lord Rayleigh found that the density of "atmospheric nitrogen," obtained by removing $O_2$, $CO_2$, and $H_2O$ from air, was slightly greater than that of chemically pure $N_2$ ($\mathcal{M} = 28.02$ g/mol). Following up on that observation, Sir William Ramsey separated argon ($\mathcal{M} = 39.95$ g/mol) from air. Over a three-year period between 1895 and 1898, this remarkable Scotsman, who never took a formal course in chemistry, isolated three more noble gases: Ne, Kr, and Xe. In effect, Ramsey added a whole new group to the periodic table.

Helium, the first member of the group, was detected in the spectrum of the sun in 1868. Because of its low density ($\frac{1}{7}$ that of air), helium is used in all kinds of balloons and in synthetic atmospheres to make breathing easier for people suffering from emphysema. Research laboratories use helium as a liquid coolant to achieve very low temperatures (bp He = $-269°C$). Argon and, more recently, krypton are used to provide an inert atmosphere in light bulbs, thereby extending the life of the tungsten fil ament. In "neon signs," a high voltage is passed through a glass tube containing neon at very low pressures. The red glow emitted corresponds to an intense line at 640 nm in the neon spectrum.

Until about 30 years ago, these elements were referred to as "inert gases"; they were believed to be entirely unreactive toward other substances. In 1962 Neil Bartlett, a 29-year-old chemist at the University of British Columbia, shook up the world of chemistry by preparing the first noble gas compound. In the course of his research on platinum-fluorine compounds, he isolated a reddish solid that he showed to be $O_2^+$ $(PtF_6^-)$. Bartlett realized that the ionization energy of Xe (1170 kJ/mol) is virtually identical to that of the $O_2$ molecule (1165 kJ/mol). This encouraged him to attempt to make the analogous compound $XePtF_6$. His success opened up a new era in noble gas chemisty.

The most stable binary compounds of xenon are the three fluorides, $XeF_2$, $XeF_4$, and $XeF_6$. Xenon difluoride can be prepared quite simply by exposing a 1:1 mol mixture of xenon and fluorine to ulraviolet light; colorless crystals of $XeF_2$ (mp = 129°C) form slowly.

$$Xe(g) + F_2(g) \longrightarrow XeF_2(s)$$

The higher fluorides are prepared using excess fluorine (Figure 7.A). All these compounds are stable in dry air at room temperature. However, they react with water to form compounds in which one or more of the fluorine atoms has been replaced by oxygen. Thus, xenon hexafluoride reacts rapidly with water to give the trioxide

$$XeF_6(s) + 3H_2O(l) \longrightarrow XeO_3(s) + 6HF(g)$$

Xenon trioxide is highly unstable; it detonates if warmed above room temperature.

The chemisty of xenon is much more extensive than that of any other noble gas. Only one binary compound of krypton, $KrF_2$, has been prepared. It is a colorless solid that decomposes at room temperature. The chemistry of radon is difficult to study because all its isotopes are radioactive. Indeed, the radiation given off is so intense that it decomposes any reagent added to radon in an attempt to bring about a reaction.

**Figure 7.A**
Crystals of xenon tetrafluoride, $XeF_4$. (Argonne National Laboratory)

# CHAPTER HIGHLIGHTS

## *Key Concepts*

1. Draw Lewis structures for molecules and polyatomic ions
   (Examples 7.1, 7.2, 7.4; Problems 1–20, 75)
2. Write resonance forms
   (Example 7.3; Problems 21–26)
3. Use Table 7.3 and Figure 7.7, applying VSEPR theory, to predict molecular geometry
   (Examples 7.5, 7.6; Problems 31–46)
4. Having derived the geometry of a molecule or polyatomic ion, predict whether it will be a dipole
   (Example 7.7; Problems 47–52)
5. State the hybridization of an atom in a covalently bonded species
   (Examples 7.8, 7.9; Problems 53–66)
6. State the number of pi and sigma bonds in a molecule or polyatomic ion
   (Example 7.10; Problems 67–72)

## Key Terms

| | | |
|---|---|---|
| bond | dipole | resonance |
| —double | expanded octet | tetrahedron |
| —nonpolar | formal charge | triangular bipyramid |
| —pi, sigma | hybrid orbital | triangular planar |
| —polar | Lewis structure | triangular pyramid |
| —single | octahedron | unshared pair |
| —triple | octet rule | valence electron |

## Summary Problem

Consider the species $NO_2^-$, $SOCl_2$, $SCl_4$, and $SF_6$.

(a) Draw the Lewis structures of these species.
(b) Draw the resonance structures for $NO_2^-$.
(c) Describe the geometries of these species, including bond angles.
(d) Classify each species as being polar or nonpolar.
(e) Give the hybridization of the central atom in each species.
(f) State the number of sigma and pi bonds in each species.
(g) $SOCl_2$ can have two Lewis structures. One has single bonds for both chlorine and oxygen. The other has single bonds for chlorine and a double bond for oxygen. Draw both Lewis structures, calculate the formal charges, and state which is the more likely structure.

### Answers

(a)

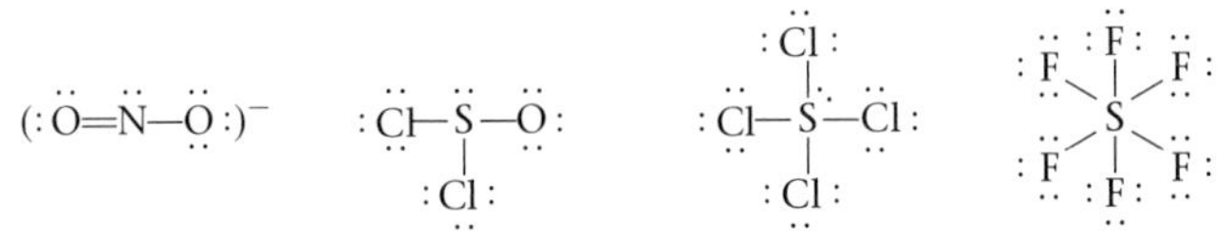

(b) $(:\ddot{O}-N=\ddot{O}:)^- \longleftrightarrow (:\ddot{O}=N-\ddot{O}:)^-$

(c) $NO_2^-$; bent, 120°          $SOCl_2$; triangular pyramid, 109°
$SCl_4$; see-saw, 90°, 120°, 180°     $SF_6$; octahedral; 90°, 180°

(d) All are polar except $SF_6$.
(e) $sp^2$ in $NO_2^-$; $sp^3$ in $SOCl_2$; $sp^3d$ in $SCl_4$; $sp^3d^2$ in $SF_6$
(f) $NO_2^-$; $2\sigma$, $1\pi$   $SCl_4$; $4\sigma$   $SOCl_2$; $3\sigma$   $SF_6$; $6\sigma$

(g)

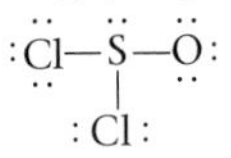

$C_f$: S = 1; Cl = 0; O = −1

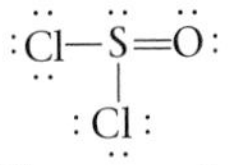

$C_f$: S = 0; Cl = 0; O = 0

The second structure seems more likely.

## Questions & Problems

### Lewis Structures

**\*1.** Write the Lewis structures for the following molecules and polyatomic ions. In each case, the first atom is the central atom.

   **(a)** $CBr_4$   **(b)** $NCl_3$   **(c)** $NO^+$   **(d)** $ClO_2^-$

**\*2.** Follow the directions of Question 1 for
   **(a)** $NHO_2$   **(b)** $KrF_2$   **(c)** $COCl_2$   **(d)** $NO_3^-$

**\*3.** Follow the directions of Question 1 for
   **(a)** $PCl_4^+$   **(b)** $ClF_2^+$   **(c)** $I_3^-$   **(d)** $SiF_4$

*4. Follow the directions of Question 1 for
  **(a)** $IO_3^-$  **(b)** $SeF_6$  **(c)** $BrF_3$  **(d)** $CN^-$
*5. Follow the directions of Question 1 for
  **(a)** CO  **(b)** $NH_3$  **(c)** $XeO_2F_2$  **(d)** $ICl_4^-$
*6. Follow the directions of Question 1 for
  **(a)** $AsH_3$  **(b)** $BF_3$  **(c)** $SO_3^{2-}$  **(d)** $XeF_2$
*7. Oxalic acid, $H_2C_2O_4$, is a poisonous compound found in rhubarb leaves. Draw the Lewis structure for oxalic acid. There is a single bond between the two carbon atoms, each hydrogen atom is bonded to an oxygen atom, and each carbon is bonded to two oxygen atoms.
*8. Radioastronomers have detected the isoformyl ion, $HOC^+$, in outer space. Write the Lewis structure for this ion.
*9. Draw Lewis structures for the following species. (The skeleton is indicated by the way the molecule is written.)

  **(a)** $H_3C-CN$  **(b)** $H_3C-C \overset{\displaystyle O}{\underset{\displaystyle H}{\big<}}$  **(c)** $F_2C-CCl_2$

*10. Follow the directions of Question 9 for the following species.
  **(a)** $Cl_2CO$  **(b)** $H_2C-CH_2$  **(c)** $(HO)_2-S-O$
*11. Peroxyacetylnitrate (PAN) is the substance in smog that makes your eyes water. Its skeletal structure is

$$H_3C-CO_3-NO_2$$

It has one O—O bond and three O—N bonds. Draw its Lewis structure.
*12. Formic acid is the irritating substance that gets on your skin when an ant bites. Its formula is HCOOH. Carbon is the central atom and it has no O—O bonds. Write its Lewis structure.
*13. Two different molecules have the formula $C_2H_2Cl_2$. Draw a Lewis structure for each molecule. (All the H and Cl atoms are bonded to carbon. The two carbon atoms are bonded to each other.)
*14. There are two compounds with the formula $C_3H_6O$. Draw a Lewis structure for each compound.
*15. Give the formula of a molecule that you would expect to have the same Lewis structure as
  **(a)** $OH^-$  **(b)** $O_2^{2-}$  **(c)** $CN^-$  **(d)** $SO_4^{2-}$
*16. Give the formula of a polyatomic ion that you would expect to have the same Lewis structure as
  **(a)** $Cl_2$  **(b)** $H_2SO_4$  **(c)** $CH_4$  **(d)** $GeCl_4$
*17. Write a Lewis structure for
  **(a)** $ClO_2^-$  **(b)** $OCN^-$
  **(c)** $NFCl_2$  **(d)** $P_2O_7^{4-}$ (no O—O bonds)
*18. Write a Lewis structure for
  **(a)** $BCl_4^-$  **(b)** $HPO_4^{2-}$
  **(c)** $ClO^-$  **(d)** $S_2O_3^{2-}$
*19. Write reasonable Lewis structures for the following species, none of which follow the octet rule.
  **(a)** $BeH_2$  **(b)** $CO^-$  **(c)** $SO_2^-$  **(d)** $CH_3$
*20. Write reasonable Lewis structures for the following species, none of which follow the octet rule.
  **(a)** $BF_3$  **(b)** NO  **(c)** $CO^+$  **(d)** $ClO_3$

## Resonance Forms and Formal Charge

*21. Draw resonance structures for
  **(a)** $NO_2^-$  **(b)** NNO  **(c)** $HCO_2^-$
*22. Draw resonance structures for
  **(a)** $Cl-NO_2$  **(b)** $H_2C-N-N$  **(c)** $SO_3$
*23. The Lewis structure for hydrazoic acid may be written as

  **(a)** Draw two other resonance forms for this molecule.
  **(b)** Is

another form of hydrazoic acid? Explain.
*24. The oxalate ion, $C_2O_4^{2-}$, has the skeleton structure

  **(a)** Complete the Lewis structure of this ion.
  **(b)** Draw three resonance forms for $C_2O_4^{2-}$, equivalent to the Lewis structure drawn in (a).
  **(c)** Is

$$\left[\ddot{O}=\ddot{C}-\ddot{O}-C-\ddot{O}:\right]^{2-}$$
$$\underset{:\ddot{O}:}{\big|}$$

a resonance form of the oxalate ion?
*25. Borazine, $B_3N_3H_6$, has the skeleton

Draw the resonance forms of the molecule.
*26. The skeleton structure for disulfur dinitride, $S_2N_2$, is

Draw the resonance forms of this molecule.
*27. What is the formal charge on the indicated atom in each of the following species?
  **(a)** the central nitrogen atom in $N_3^-$
  **(b)** xenon in $XeF_6$
  **(c)** bromine in $BrCl_3$
*28. Follow the directions in Question 27 for
  **(a)** oxygen in HOF
  **(b)** nitrogen in $NO_2^-$

**\*29.** Based on the concept of formal charge, choose the more likely skeleton in each of the following cases:
    **(a)** HCN or HNC    **(b)** NOCl or ONCl

**\*30.** Which of the two Lewis structures of $H_2SO_3$ is more likely, based on formal charges?

$$H-\overset{..}{\underset{..}{O}}-\overset{..}{\underset{\underset{:\overset{..}{O}:}{\|}}{S}}-\overset{..}{\underset{..}{O}}-H \qquad H-\overset{..}{\underset{..}{O}}-\underset{\underset{:\overset{..}{\underset{..}{O}}:}{|}}{S}-\overset{..}{\underset{..}{O}}-H$$

## Molecular Geometry

**\*31.** Predict the geometry of the following species:
    **(a)** $SO_2$    **(b)** $BeCl_2$    **(c)** $SeCl_4$    **(d)** $PCl_5$

**\*32.** Predict the geometry of the following species:
    **(a)** $SCO$    **(b)** $IBr_2^-$    **(c)** $NO_3^-$    **(d)** $RnF_4$

**\*33.** Predict the geometry of the following species:
    **(a)** $NNO$    **(b)** $ONCl$    **(c)** $NH_4^+$    **(d)** $O_3^{2-}$

**\*34.** Predict the geometry of the following species:
    **(a)** $KrF_2$    **(b)** $NH_2Cl$    **(c)** $CH_2Br_2$    **(d)** $SCN^-$

**\*35.** Predict the geometry of the following species:
    **(a)** $ClO_4^-$    **(b)** $TeBr_4$    **(c)** $IF_3$    **(d)** $SeF_6$

**\*36.** Predict the geometry of the following species:
    **(a)** $ClO_2^-$    **(b)** $CO_3^{2-}$    **(c)** $SF_5Cl$    **(d)** $XeF_2$

**\*37.** Describe the geometry of the species in which there are, around the central atom
    **(a)** four single bonds, two unshared pairs of electrons
    **(b)** five single bonds
    **(c)** two single bonds, one unshared pair of electrons

**\*38.** Describe the geometry of the species in which there are, around the central atom
    **(a)** three single bonds, two unshared pairs of electrons
    **(b)** two single bonds, two unshared pairs of electrons
    **(c)** five single bonds, one unshared pair of electrons

**\*39.** Give all the ideal bond angles (109.5°, 120°, or 180°) in the following molecules. (The skeleton does not imply geometry.)
    **(a)** $O=C=O$
    **(b)** $H-B-H$ with $H$ below $B$

    **(c)** $H-O-N$ with $=O$ above and $O$ below

    **(d)** $H-\overset{H}{\underset{H}{C}}-C=O$

**\*40.** Follow the instructions for Question 39 for
    **(a)** $Cl-S-Cl$
    **(b)** $F-Xe-F$
    **(c)** $H-\overset{H}{\underset{H}{C}}-\overset{}{\underset{O}{C}}-N\overset{H}{\underset{H}{{}}}$
    **(d)** $H-\overset{}{\underset{H}{C}}=\overset{}{\underset{H}{C}}-C\equiv N$

**\*41.** An objectionable component of smoggy air is acetylperoxide, which has the skeleton structure

$$H_3C-\overset{}{\underset{\underset{O}{\|}}{C}}-O-O-\overset{}{\underset{\underset{O}{\|}}{C}}-CH_3$$

    **(a)** Draw the Lewis structure of this compound.
    **(b)** Indicate all the bond angles.

**\*42.** Peroxypropionyl nitrate (PPN) is an eye irritant found in smog. Its skeleton structure is

$$H_3C-\overset{H\ O}{\underset{H}{\overset{|\ \|}{C}}}-C-O-O-N-O$$

    **(a)** Draw the Lewis structure of PPN.
    **(b)** Indicate all the bond angles.

**\*43.** Consider the following molecules: $SiH_4$, $PH_3$, $H_2S$. In each case, a central atom is surrounded by four electron pairs. In which of these molecules would you expect the bond angle to be less than 109.5°? Explain your reasoning.

**\*44.** In each of the following molecules, a central atom is surrounded by a total of three atoms or unshared electron pairs: $SnCl_2$, $BCl_3$, $SO_2$. In which of these molecules would you expect the bond angle to be less than 120°? Explain your reasoning.

**\*45.** Vitamin C has the following structure. Estimate the value of the numbered angles (unshared electron pairs are not shown).

**\*46.** The thymine molecule is one of the four bases in DNA. Estimate the approximate values of the indicated bond angles. (Unshared electron pairs are not shown.)

## Molecular Polarity

**\*47.** Which of the species with octets in Question 31 are dipoles?

**\*48.** Which of the species with octets in Question 32 are dipoles?

**\*49.** Which of the species with octets in Question 33 are dipoles?

**\*50.** Which of the species with octets in Question 34 are dipoles?

**\*51.** There are two different molecules with the formula $N_2F_2$:

$$\underset{N=N}{\overset{F}{\diagdown}}\underset{}{\overset{F}{\diagup}} \quad \text{and} \quad \underset{\underset{F}{\diagdown}N=N}{\overset{F}{\diagup}}$$

Is either molecule polar? Explain.

**\*52.** There are three compounds with the formula $C_2H_2Cl_2$:

$$\underset{H}{\overset{Cl}{\diagdown}}C=C\underset{H}{\overset{Cl}{\diagup}} \quad \underset{H}{\overset{Cl}{\diagdown}}C=C\underset{Cl}{\overset{H}{\diagup}} \quad \text{and} \quad \underset{H}{\overset{H}{\diagdown}}C=C\underset{Cl}{\overset{Cl}{\diagup}}$$

Which of these molecules are polar? Nonpolar?

## Hybridization

**\*53.** Give the hybridization of the central atom in each species in Question 31.

**\*54.** Give the hybridization of the central atom in each species in Question 32.

**\*55.** Give the hybridization of the central atom in each species in Question 33.

**\*56.** Give the hybridization of the central atom in each species in Question 34.

**\*57.** Give the hybridization of the central atom in each species in Question 35.

**\*58.** Give the hybridization of the central atom in each species in Question 36.

**\*59.** In each of the following polyatomic ions, the central atom has an expanded octet. Determine the number of electron pairs around the central atom and the hybridization in

    **(a)** $XeF_3^-$    **(b)** $SiCl_6^{2-}$    **(c)** $PCl_4^-$

**\*60.** Follow the directions of Question 59 for the following polyatomic ions.

    **(a)** $ClF_2^-$    **(b)** $GeCl_4^{2-}$    **(c)** $ICl_4^-$

**\*61.** Give the hybridization of the B and N atoms in borazine.

$$\underset{\underset{H}{\overset{\displaystyle|}{N}}}{\underset{\underset{}{\underset{}{}}}{\overset{\displaystyle\underset{B}{\overset{|}{H}}}{}}}$$

H—N, B, N—H ; H—B, N, B—H ; central ring structure of borazine

**\*62.** Give the hybridization of each atom (except H) in the solvent dimethysulfoxide. (Unshared electron pairs are not shown.)

$$H_3C—\underset{\underset{O}{\overset{\|}{}}}{S}—CH_3$$

**\*63.** What is the hybridization of carbon in

    **(a)** $CH_3Cl$     **(b)** $\left[\underset{\underset{O}{\overset{\|}{}}}{O—C—O}\right]^{2-}$

    **(c)** $O=C=O$     **(d)** $H—\underset{\underset{O}{\overset{\|}{}}}{C}—OH$

**\*64.** What is the hybridization of nitrogen in

    **(a)** $\left[\underset{\underset{O}{\overset{\|}{}}}{O—N—O}\right]^-$   **(b)** $H—\underset{\underset{Cl}{\overset{\displaystyle\ddot{}}{}}}{\overset{\displaystyle\ddots}{N}}—H$

    **(c)** $\ddot{N}\equiv\ddot{N}$     **(d)** $N\equiv N—O$

**\*65.** Give the hybridization of the central atom (underlined in red)

    **(a)** $(H_2N)_2\underline{C}O$   **(b)** $F_2\underline{C}O$   **(c)** $\underline{N}HF_2$

**\*66.** Give the hybridization of the central atom (underlined in red)

    **(a)** $HO\underline{I}O_2$   **(b)** $\underline{O}F_2$   **(c)** $O\underline{P}Cl_3$

## Sigma and Pi Bonds

**\*67.** Give the number of sigma and pi bonds in each species in Question 63.

**\*68.** Give the number of sigma and pi bonds in each species in Question 64.

**\*69.** Give the number of sigma and pi bonds in the molecule in Question 61.

**\*70.** Give the number of sigma and pi bonds in the molecule in Question 62.

**\*71.** Give the formula of an ion or molecule in which an atom of

    **(a)** N forms three bonds using $sp^3$ hybrid orbitals.

    **(b)** N forms two pi bonds and one sigma bond.

    **(c)** N forms two sigma bonds and one pi bond.

**\*72.** Give the formula of an ion or molecule in which an atom of

    **(a)** O forms a sigma and a pi bond.

    **(b)** C forms four bonds, in three of which it uses $sp^2$ hybrid orbitals.

    **(c)** Xe forms two bonds using $sp^3d$ hybrid orbitals.

## Unclassified

**\*73.** A possible structure for $BeCl_2$, which would obey the octet rule, is

$$Cl=Be=Cl$$

Assign formal charges to each atom in the structure. Explain why this structure seems less plausible than one with only single bonds.

**\*74.** Two possible structures for the HOCN molecule are

$$H—O—C\equiv N \quad \text{and} \quad H—O=C=N$$

Assign formal charges to each atom in both structures. Which structure do you think is more likely?

**\*75.** Consider the dichromate ion. It has no metal-metal or oxygen-oxygen bonds. Write a Lewis structure for the dichromate ion. Consider chromium to have six valence electrons.

**\*76.** Explain the meaning of the following terms:

    **(a)** expanded octet

    **(b)** resonance

    **(c)** unshared electron pair

    **(d)** odd-electron species

**\*77.** In which of the following molecules does the sulfur have an expanded octet?

(a) $SO_2$    (b) $SF_4$    (c) $SO_2Cl_2$    (d) $SF_6$

**\*78.** Consider acetyl salicylic acid, better known as aspirin. Its structure is

(a) How many sigma and pi bonds are there in aspirin?

(b) What are the approximate values of the angles marked (in blue) A, B, and C?

(c) What is the hybridization of each atom marked (in red) 1, 2, and 3?

**\*79.** Complete the following table:

| Species | Atoms Around Central Atom A | Unshared Pairs Around A | Geometry | Hybridization | Polarity (Assume all X atoms the same) |
|---|---|---|---|---|---|
| $AX_2E_2$ | _____ | _____ | _____ | _____ | _____ |
| _____ | 3 | 0 | _____ | _____ | _____ |
| $AX_4E_2$ | _____ | _____ | _____ | _____ | _____ |
| _____ | _____ | _____ | triangular bipyramid | _____ | _____ |

## Challenge Problems

**80.** A compound of chlorine and fluorine, $ClF_x$, reacts at about 75°C with uranium to produce uranium hexafluoride and chlorine fluoride, ClF. A certain amount of uranium produced 5.63 g of uranium hexafluoride and 457 mL of chlorine fluoride at 75°C and 3.00 atm. What is $x$? Describe the geometry, polarity, and bond angles of the compound and the hybridization of chlorine. How many sigma and pi bonds are there?

**\*81.** Draw the Lewis structure and describe the geometry of the hydrazine molecule, $N_2H_4$. Would you expect this molecule to be polar?

**\*82.** Consider the polyatomic ion $IO_6^{5-}$. How many pairs of electrons are there around the central iodine atom? What is its hybridization? Describe the geometry of the ion.

**\*83.** It is possible to write a simple Lewis structure for the $SO_4^{2-}$ ion, involving only single bonds, that follows the octet rule. However, Linus Pauling and others have suggested an alternative structure, involving double bonds, in which the sulfur atom is surrounded by six electron pairs.

(a) Draw the two Lewis structures.

(b) What geometries are predicted for the two structures?

(c) What is the hybridization of sulfur in each case?

(d) What are the formal charges of the atoms in the two structures?

**\*84.** Phosphoryl chloride, $POCl_3$, has the skeleton structure

$$\begin{array}{c} O \\ | \\ Cl-P-Cl \\ | \\ Cl \end{array}$$

Write

(a) a Lewis structure for $POCl_3$ following the octet rule. Calculate the formal charges in this structure.

(b) a Lewis structure in which all the formal charges are zero. (The octet rule need not be followed.)

To maintain good health, eat five servings of fruits and vegetables daily, exercise regularly (p. 231) and don't grow old. (Charles D. Winters)

# Thermochemistry

# 8

**S**ome say the world will end in fire,

Some say in ice.

From what I've tasted of desire

I hold with those who favor fire.

**—ROBERT FROST**

*Fire and Ice*

This chapter deals with energy and heat, two terms used widely by both the general public and scientists. Energy, in the vernacular, is equated with pep and vitality. Heat conjures images of blast furnaces and sweltering summer days. Scientifically, these terms have quite different meanings. *Energy* can be defined as the capacity to do work. *Heat* is a particular form of energy that is transferred from a body at a high temperature to one at a lower temperature when they are brought into contact with each other. Two centuries ago, heat was believed to be a material fluid ("caloric"); we still use the phrase "heat flow" to refer to heat transfer or to heat effects in general.

**Thermochemistry** refers to the study of the heat flow that accompanies chemical reactions. Our discussion of this subject will focus upon

— the basic principles of heat flow (Section 8.1)
— the experimental measurement of the magnitude and direction of heat flow, known as *calorimetry* (Section 8.2)
— the concept of enthalpy (heat content) and *enthalpy change, $\Delta H$* (Section 8.3)
— the calculation of $\Delta H$ for reactions, using *thermochemical equations* (Section 8.4) and *enthalpies of formation* (Section 8.5)
— heat effects in the breaking and formation of covalent bonds (Section 8.6)
— the relation between heat and other forms of energy, as expressed by the first law of thermodynamics (Section 8.7).

## 8.1 Principles of Heat Flow

In any discussion of heat flow, it is important to distinguish between system and surroundings. The **system** is that part of the universe upon which attention is focused. In a very simple case (Fig. 8.1), it might be a sample of water in contact with a hot plate. The **surroundings,** which exchange energy with the system, comprise in principle the rest of the universe. For all practical purposes, however, they include only those materials in close contact with the system. In Figure 8.1, the surroundings would consist of the hot plate, the beaker holding the water sample, and the air around it.

*The universe is a big place*

When a chemical reaction takes place, we consider the substances involved, reactants and products, to be the system. The surroundings include the vessel in which the reaction takes place (test tube, beaker, and so on) and the air or other material in thermal contact with the reaction system.

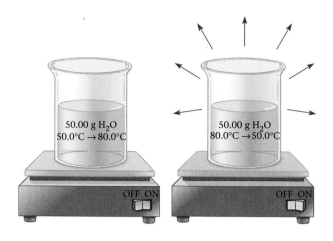

**Figure 8.1**

When the system (50.00 g of $H_2O$) absorbs heat from the surroundings (hot plate), its temperature increases from 50.0°C to 80.0°C. When the hot plate is turned off, the system gives off heat to the surrounding air and its temperature drops.

## State Properties

The **state** of a system is described by giving its composition, temperature, and pressure. The system at the left of Figure 8.1 consists of

$$50.0 \text{ g of } H_2O(l) \text{ at } 50.0°C \text{ and } 1 \text{ atm}$$

When this system is heated, its state changes, perhaps to one described as

$$50.0 \text{ g of } H_2O(l) \text{ at } 80.0°C \text{ and } 1 \text{ atm}$$

Certain quantities, called **state properties,** depend only upon the state of the system, not upon the way the system reached that state. Putting it another way, if $X$ is a state property, then

*The distance between two cities depends upon path so it isn't a state property*

$$\Delta X = X_{final} - X_{initial}$$

That is, the change in $X$ is the difference between its values in final and initial states. Most of the quantities that you are familiar with are state properties; volume is a common example. You may be surprised to learn, however, that heat flow is *not* a state property; its magnitude depends upon how a process is carried out (Section 8.7).

## Direction and Sign of Heat Flow

Consider again the setup in Figure 8.1. If the hot plate is turned on, there is a flow of heat from the surroundings into the system, 50.0 g of water. This situation is described by stating that the heat flow, $q$, for the system is a positive quantity.

***q* is + when heat flows into the system from the surroundings**

Usually, when heat flows into a system, its temperature rises. In this case, the temperature of the 50.0-g water sample might increase from 50.0 to 80.0°C. When the hot plate in Figure 8.1 is shut off, the hot water gives off heat to the surrounding air. In this case, $q$ for the system is a negative quantity.

***q* is − when heat flows out of the system into the surroundings**

Direction of heat flow.

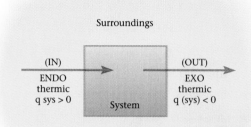

endo = into system; exo = out of system

Here, as is usually the case, the temperature of the system drops when heat flows out of it into the surroundings. The 50.0-g water sample might cool from 80.0°C back to 50.0°C.

This same reasoning can be applied to a reaction where the system consists of the reaction mixture (products and reactants). We can distinguish between

— *an endothermic process (q > 0), in which heat flows from the surroundings into the reaction system.* An example is the melting of ice:

$$H_2O(s) \longrightarrow H_2O(l) \qquad q > 0$$

The melting of ice *absorbs* heat from the surroundings, which might be the water in a glass of iced tea. The temperature of the surroundings drops, perhaps from 25 to 0°C, as they give up heat to the system.

— *an exothermic process (q < 0), in which heat flows from the reaction system into the surroundings.* A familiar example is the combustion of methane gas:

$$CH_4(g) + 2O_2(g) \longrightarrow CO_2(g) + 2H_2O(l) \qquad q < 0$$

This reaction *evolves* heat to the surroundings, which might be the air around a Bunsen burner in the laboratory or a potato being baked in a gas oven. In either case, the effect of the heat transfer is to raise the temperature of the surroundings.

## Magnitude of Heat Flow

In any process, we are interested not only in the direction of heat flow but also in its magnitude. The magnitude of $q$ is ordinarily cited in joules (J) or kilojoules (kJ).

$$1 \text{ kJ} = 10^3 \text{ J}$$

The SI unit of heat, the *joule* (Chapter 1), is named for James Joule (1818–1889), who carried out very precise thermometric measurements that established the first law of thermodynamics (Section 8.7).

Phosphorus reacting with chlorine to give $PCl_3$ is an exothermic reaction.

Most of the remainder of this chapter is devoted to a discussion of the magnitude of the heat flow in chemical reactions or phase changes. Here, however, we will focus upon a simpler process in which the only effect of the heat flow is to change the temperature of a system. In general, the relationship between the magnitude of the heat flow, $q$, and the temperature change, $\Delta t$, is given by the equation

$$q = C \times \Delta t \qquad (\Delta t = t_{final} - t_{initial})$$

The quantity $C$ appearing in this equation is known as the **heat capacity** of the system. It represents the amount of heat required to raise the temperature of the system 1°C and has the units J/°C.

For a pure substance of mass $m$, the expression for $q$ can be written as

$$q = m \times c \times \Delta t$$

The quantity $c$ is known as the **specific heat;** it is defined as the amount of heat required to raise the temperature of one gram of a substance 1°C.

Specific heat, like density or melting point, is an intensive property which can be used to identify a substance. Water has an unusually large specific heat, **4.18 J/g · °C.** This explains why swimming is not a popular pastime in northern Minnesota in May. Even if the air temperature rises to 90°F, the water temperature will remain below 60°F. Metals have a relatively low specific heat (Table 8.1). When you warm water in a stainless steel saucepan, nearly all of the heat is absorbed by the water, very little by the steel.

(a)

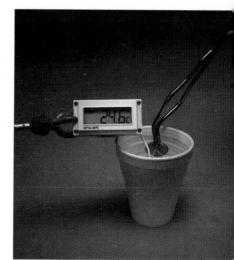
(b)

(a) The heated metal is immersed in a cup containing a measured amount of water at a known temperature. (b) The temperature of the water rises. The specific heat of the metal can be calculated from the data obtained in this experiment.

---

**Example 8.1**   How much heat is given off by a 50.0-g sample of copper when it cools from 80.0 to 50.0°C?

**Strategy**   Calculate $q$ by first determining $\Delta t$ and then substituting into the equation $q = m \times c \times \Delta t$. The specific heat of copper is given in Table 8.1.

**Solution**   The temperature change is

$$\Delta t = t_{final} - t_{initial} = 50.0°C - 80.0° = -30.0°C$$

$$q = 50.0 \text{ g} \times 0.382 \frac{\text{J}}{\text{g°C}} \times (-30.0°C) = \boxed{-573 \text{ J}}$$

The negative sign indicates that heat flows from copper to the surroundings.

---

**TABLE 8.1   Specific Heats of a Few Common Substances**

| | c (J/g · °C) | | c (J/g · °C) |
|---|---|---|---|
| $Br_2(l)$ | 0.474 | $Cu(s)$ | 0.382 |
| $Cl_2(g)$ | 0.478 | $Fe(s)$ | 0.446 |
| $C_2H_5OH(l)$ | 2.43 | $H_2O(g)$ | 1.87 |
| $C_6H_6(l)$ | 1.72 | $H_2O(l)$ | 4.18 |
| $CO_2(g)$ | 0.843 | $NaCl(s)$ | 0.866 |

## 8.2 Measurement of Heat Flow; Calorimetry

To measure the heat flow in a reaction, it is carried out in a device known as a **calorimeter.** The apparatus contains water and/or other materials of known heat capacity. The walls of the calorimeter are insulated so that there is no exchange of heat with the air outside the calorimeter. It follows that the only heat flow is between the reaction system and the calorimeter. The heat flow for the reaction system is equal in magnitude but opposite in sign to that for the calorimeter:

$$q_{reaction} = -q_{calorimeter}$$

Notice that if the reaction is exothermic ($q_{reaction} < 0$), $q_{calorimeter}$ must be positive; i.e., heat flows from the reaction mixture into the calorimeter. Conversely, if the reaction is endothermic, the calorimeter gives up heat to the reaction mixture.

The preceding equation is basic to calorimetric measurements. It allows you to calculate the amount of heat absorbed or evolved in a reaction if you know the heat capacity, $C_{cal}$, and the temperature change, $\Delta t$, of the calorimeter.

$$q_{reaction} = -C_{cal} \times \Delta t$$

*The reaction mixture is the system; the calorimeter is the surroundings*

### Coffee-Cup Calorimeter

Figure 8.2 shows a simple calorimeter used in the general chemistry laboratory. It consists of a polystyrene foam cup partially filled with water. The cup has a tightly fitting cover through which an accurate thermometer is inserted. Since polystyrene foam is a good insulator, there is very little heat flow through the walls of the cup. Essentially all the heat evolved by a reaction taking place within the calorimeter is absorbed by the water. This means that, to a good degree of approximation, the heat capacity of the coffee-cup calorimeter is that of the water:

$$C_{cal} = m_{water} \times c_{water} = m_{water} \times 4.18 \frac{J}{g \cdot °C}$$

and hence

$$q_{reaction} = -m_{water} \times 4.18 \frac{J}{g \cdot °C} \times \Delta t$$

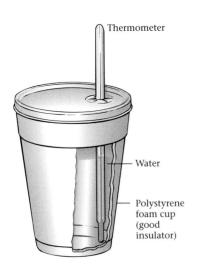

Thermometer

Water

Polystyrene foam cup (good insulator)

**Figure 8.2**
Coffee-cup calorimeter. The heat given off by a reaction is absorbed by the water. If you know the mass of the water, its specific heat (4.18 J/g · °C), and the temperature change as read on the thermometer, you can calculate the heat flow, *q*, for the reaction.

---

**Example 8.2**    When 1.00 g of ammonium nitrate, $NH_4NO_3$, dissolves in 50.0 g of water in a coffee-cup calorimeter, the following reaction occurs:

$$NH_4NO_3(s) \longrightarrow NH_4{}^+(aq) + NO_3{}^-(aq)$$

and the temperature of the water drops from 25.00 to 23.32°C. Assuming that all the heat absorbed by the reaction comes from the water, calculate *q* for the reaction system.

**Strategy**    Apply the relation $q_{reaction} = -m_{water} \times 4.18 \frac{J}{g \cdot °C} \times \Delta t$

**Solution**

$$q_{reaction} = -50.0 \text{ g} \times 4.18 \frac{J}{g \cdot °C} \times (23.32°C - 25.00°C) = \boxed{+351 \text{ J}}$$

Notice that *q* is positive; this is an endothermic reaction.

---

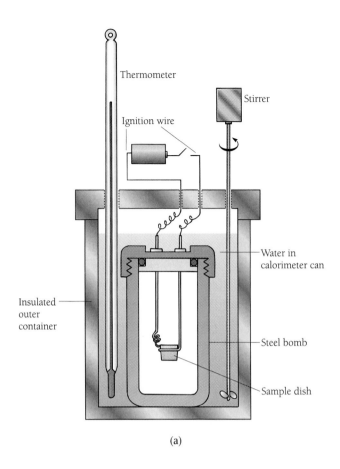

(a)

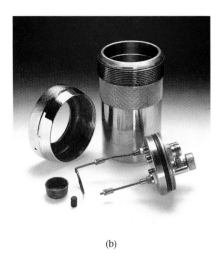

(b)

**Figure 8.3**
(a) Bomb calorimeter. The heat flow, $q$, for the reaction is calculated from the temperature change multiplied by the heat capacity of the calorimeter, which is determined in a preliminary experiment. (b) Parts of a bomb calorimeter.

## Bomb Calorimeter

A coffee-cup calorimeter is suitable for measuring heat flow for reactions in solution. However, it cannot be used for reactions involving gases, which would escape from the cup. Neither would it be appropriate for reactions in which the products reach high temperatures. The bomb calorimeter, shown in Figure 8.3, is more versatile. To use it, a weighed sample of the reactant(s) is added to the heavy-walled steel container called a "bomb." This is then sealed and lowered into a metal vessel that fits snugly within the insulating walls of the calorimeter. An amount of water sufficient to cover the bomb is added, and the entire apparatus is closed. The initial temperature is measured precisely. The reaction is then started, perhaps by electrical ignition. In an exothermic reaction, the hot products give off heat to the walls of the bomb and to the water. The final temperature is taken to be the highest value read on the thermometer.

To analyze the heat flow in a bomb calorimeter, the basic equation

$$q_{reaction} = -q_{calorimeter} = -C_{cal} \times \Delta t$$

is used. Since heat is absorbed by both the water and the metal parts of the calorimeter, $C_{cal}$ must be found experimentally. This is ordinarily done by carrying out a reaction with known $q$ in the bomb calorimeter, measuring $\Delta t$, and calculating $C_{cal}$. Suppose, for example, that you carried out a reaction known to evolve 93.3 kJ of heat and found that the temperature of the calorimeter rose from 20.00 to 30.00°C. It follows that

$C_{cal}$ depends upon the amount of water around the bomb

$$C_{cal} = \frac{93.3 \text{ kJ}}{10.00°C} = 9.33 \text{ kJ/°C}$$

Knowing the heat capacity of the calorimeter, the heat flow for any reaction taking place within the calorimeter can be calculated (Example 8.3).

---

**Example 8.3**   The reaction between hydrogen and chlorine, $H_2(g) + Cl_2(g) \rightarrow 2HCl(g)$, can be studied in a bomb calorimeter. It is found that when a 1.00-g sample of $H_2$ reacts completely, the temperature rises from 20.00 to 29.82°C. Taking the heat capacity of the calorimeter to be 9.33 kJ/°C, calculate the amount of heat evolved in the reaction.

***Strategy***   Use the equation: $q_{reaction} = -C_{cal} \times \Delta t$

***Solution***

$$q_{reaction} = -9.33 \frac{\text{kJ}}{°C} \times (29.82°C - 20.00°C) = \boxed{-91.6 \text{ kJ}}$$

---

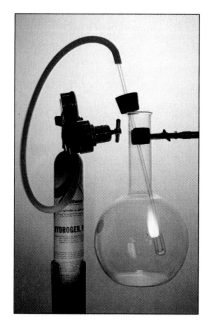

Hydrogen, a colorless gas, reacts with chlorine, a pale yellow gas, to form colorless hydrogen chloride gas.

## 8.3   Enthalpy

We have referred several times to the "heat flow for the reaction system," symbolized as $q_{reaction}$. At this point, you may well find this concept a bit nebulous and wonder if it could be made more concrete by relating $q_{reaction}$ to some property of reactants and products. This can indeed be done; the situation is particularly simple for reactions taking place at constant pressure. Under that condition, the heat flow for the reaction system is equal to the difference in **enthalpy** (*H*) between products and reactants. That is,

$$q_{reaction} \text{ at constant pressure} = \Delta H = H_{products} - H_{reactants}$$

Enthalpy is a type of chemical energy, sometimes referred to as "heat content." Reactions that occur in the laboratory in an open container or in the world around us take place at a constant pressure, that of the atmosphere. For such reactions, the equation just written is valid, making enthalpy a very useful quantity.

Figure 8.4a shows the enthalpy relations for an exothermic reaction such as

$$CH_4(g) + 2O_2(g) \longrightarrow CO_2(g) + 2H_2O(l) \qquad \Delta H < 0$$

Here, the products, 1 mol of $CO_2(g)$ and 2 mol of $H_2O(l)$, have a lower enthalpy than the reactants, 1 mol of $CH_4(g)$ and 2 mol of $O_2(g)$. The decrease in enthalpy

**Figure 8.4**
In an exothermic reaction (a), the products have a lower enthalpy than the reactants; thus, $\Delta H$ is negative, and heat is given off to the surroundings. In an endothermic reaction (b), the products have a higher enthalpy than the reactants so $\Delta H$ is positive, and heat is absorbed from the surroundings.

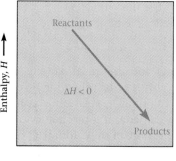

(a)

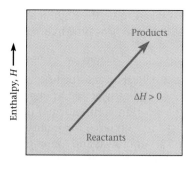

(b)

is the source of the heat evolved to the surroundings. Figure 8.4b shows the situation for an endothermic process such as

$$H_2O(s) \longrightarrow H_2O(l) \qquad\qquad \Delta H > 0$$

Since liquid water has a higher enthalpy than ice, heat must be transferred from the surroundings to melt the ice.

In general, the following relations apply for reactions taking place at constant pressure.

exothermic reaction:   $q = \Delta H < 0$    $H_{products} < H_{reactants}$

endothermic reaction:   $q = \Delta H > 0$    $H_{products} > H_{reactants}$

*When wood burns, $\Delta H$ is negative; when ice melts, $\Delta H$ is positive*

The enthalpy of a substance, like its volume, is a state property. A sample of one gram of liquid water at 25.00°C and 1 atm has a fixed enthalpy, $H$. In practice, no attempt is made to determine absolute values of enthalpy. Instead, scientists deal with changes in enthalpy, which are readily determined. For the process

$$1.00 \text{ g } H_2O(l, 25.00°C, 1 \text{ atm}) \longrightarrow 1.00 \text{ g } H_2O(l, 26.00°C, 1 \text{ atm})$$

$\Delta H$ is 4.18 J because the specific heat of water is 4.18 J/g · °C.

## 8.4   Thermochemical Equations

A chemical equation that shows the enthalpy relation between products and reactants is called a **thermochemical equation.** This type of equation contains, at the right of the balanced chemical equation, the appropriate value and sign for $\Delta H$.

To see where a thermochemical equation comes from, consider again the process by which ammonium nitrate dissolves in water:

$$NH_4NO_3(s) \longrightarrow NH_4{}^+(aq) + NO_3{}^-(aq)$$

Recall (Example 8.2) that a simple experiment with a coffee-cup calorimeter shows that when one gram of $NH_4NO_3$ dissolves, $q_{reaction} = 351$ J. Since the calorimeter is open to the atmosphere, the pressure is constant, and

$$\Delta H \text{ for dissolving } 1.00 \text{ g of } NH_4NO_3 = 351 \text{ J} = 0.351 \text{ kJ}$$

When one mole (80.05 g) of $NH_4NO_3$ dissolves, $\Delta H$ should be eighty times as great:

$$\Delta H \text{ for dissolving } 1.00 \text{ mol of } NH_4NO_3 = 0.351 \frac{\text{kJ}}{\text{g}} \times 80.05 \text{ g} = 28.1 \text{ kJ}$$

The thermochemical equation for this reaction must then be (Fig. 8.5 p. 216)

$$NH_4NO_3(s) \longrightarrow NH_4{}^+(aq) + NO_3{}^-(aq) \qquad \Delta H = 28.1 \text{ kJ}$$

By an entirely analogous procedure, the thermochemical equation for the formation of HCl from the elements (Example 8.3) is found to be

$$H_2(g) + Cl_2(g) \longrightarrow 2HCl(g) \qquad \Delta H = -185 \text{ kJ}$$

In other words, 185 kJ of heat is evolved when two moles of HCl are formed from $H_2$ and $Cl_2$.

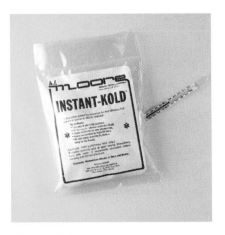  

**Figure 8.5**

When the seal separating the compartment in a "cold pack" is broken, the following endothermic reaction occurs: $NH_4NO_3(s) \longrightarrow NH_4^+(aq) + NO_3^-(aq)$ $\Delta H = +28.1$ kJ and the temperature, as read on the thermometer, drops. (Marna G. Clarke)

This explains why burns from steam are more painful than those from boiling water

These thermochemical equations are typical of those used throughout this text. It is important to realize that

— the sign of $\Delta H$ indicates whether the reaction, when carried out at constant pressure, is endothermic (positive $\Delta H$) or exothermic (negative $\Delta H$).
— in interpreting a thermochemical equation, the coefficients represent numbers of moles ($\Delta H$ is $-185$ kJ when **1 mol** $H_2$ + **1 mol** $Cl_2 \rightarrow$ **2 mol** HCl).
— the phases (physical states) of all species must be specified, using the symbols (*s*), (*l*), (*g*), or (*aq*). The enthalpy of one mole of $H_2O(g)$ at 25°C is 44 kJ larger than that of one mole of $H_2O(l)$; the difference, which represents the heat of vaporization of water, is clearly significant.
— the value quoted for $\Delta H$ applies when products and reactants are at the same temperature, ordinarily taken to be 25°C unless specified otherwise.

## Rules of Thermochemistry

To make effective use of thermochemical equations, three basic rules of thermochemistry are applied.

**1. *The magnitude of $\Delta H$ is directly proportional to the amount of reactant or product.*** This is a common-sense rule, consistent with experience. The amount of heat that must be absorbed to boil a sample of water is directly proportional to its mass. In another case, the more gasoline you burn in your car's engine, the more energy you produce.

This rule allows you to find $\Delta H$ corresponding to any desired amount of reactant or product. To do this, you follow the conversion-factor approach used in Chapter 3 with ordinary chemical equations. Consider, for example,

$$H_2(g) + Cl_2(g) \longrightarrow 2HCl(g) \qquad \Delta H = -185 \text{ kJ}$$

The mass relationships in this equation are such that one mole of hydrogen, one mole of chlorine, and two moles of HCl are chemically equivalent to one another. The thermochemical equation adds the enthalpy relation and hence leads to conversion factors such as

$$\frac{-185 \text{ kJ}}{1 \text{ mol } Cl_2} \qquad \frac{1 \text{ mol } H_2}{-185 \text{ kJ}} \qquad \frac{-185 \text{ kJ}}{2 \text{ mol } HCl}$$

---

**Example 8.4** Consider the thermochemical equation for the formation of two moles of HCl from the elements. Calculate $\Delta H$ when

(a) 1.00 mol of HCl is formed.  (b) 1.00 g of $Cl_2$ reacts.

**Strategy** Use the conversion factor approach. In (a), only one conversion is required, from moles of HCl to $\Delta H$ in kilojoules. In (b), you first have to convert to moles of $Cl_2$ and then to $\Delta H$ in kilojoules.

**Solution**

(a) $\Delta H = 1.00 \text{ mol HCl} \times \dfrac{-185 \text{ kJ}}{2 \text{ mol HCl}} = \boxed{-92.5 \text{ kJ}}$

This means that the thermochemical equation for the formation of one mole of HCl would be

$$\tfrac{1}{2}H_2(g) + \tfrac{1}{2}Cl_2(g) \longrightarrow HCl(g) \qquad \Delta H = -92.5 \text{ kJ}$$

(b) $\Delta H = 1.00 \text{ g } Cl_2 \times \dfrac{1 \text{ mol } Cl_2}{70.90 \text{ g } Cl_2} \times \dfrac{-185 \text{ kJ}}{1 \text{ mol } Cl_2} = \boxed{-2.61 \text{ kJ}}$

---

This relation between $\Delta H$ and amounts of substances is equally useful in dealing with chemical reactions or phase changes.

---

**Example 8.5** Using a coffee-cup calorimeter, it is found that when an ice cube weighing 24.6 g melts, it absorbs 8.19 kJ of heat. Calculate $\Delta H$ for the phase change represented by the thermochemical equation $H_2O(s) \rightarrow H_2O(l)$ $\Delta H = ?$

**Strategy** The data given lead directly to the conversion factor

$$\dfrac{8.19 \text{ kJ}}{24.6 \text{ g ice}}$$

You need to know $\Delta H$ when one mole (18.02 g) of ice melts; two conversions are required.

**Solution**

$$\Delta H = 1 \text{ mol } H_2O \times \dfrac{18.02 \text{ g } H_2O}{1 \text{ mol } H_2O} \times \dfrac{8.19 \text{ kJ}}{24.6 \text{ g } H_2O} = \boxed{6.00 \text{ kJ}}$$

$\Delta H$ per gram of ice is 6.00 kJ/18.02 g = 0.333 kJ/g

This calculation shows that 6.00 kJ of heat must be absorbed to melt one mole of ice:

$$H_2O(s) \longrightarrow H_2O(l) \qquad \Delta H = +6.00 \text{ kJ}$$

---

The heat absorbed when a solid melts $(s \rightarrow l)$ is referred to as the **heat of fusion;** that absorbed when a liquid vaporizes $(l \rightarrow g)$ is called the **heat of vaporization.** Heats of fusion ($\Delta H_{fus}$) and vaporization ($\Delta H_{vap}$) are most often expressed in kilojoules per mole (kJ/mol). Values for several different substances are given in Table 8.2.

**2. $\Delta H$ for a reaction is equal in magnitude but opposite in sign to $\Delta H$ for the reverse reaction.** Another way to state this rule is to say that the amount of heat evolved in a reaction is exactly equal to the amount of heat absorbed in the reverse reaction. This again is a common-sense rule. If 6.00 kJ of heat is absorbed when a mole of ice melts,

$$H_2O(s) \longrightarrow H_2O(l) \qquad\qquad \Delta H = +6.00 \text{ kJ}$$

TABLE 8.2 $\Delta H$(kJ/mol) for Phase Changes

| Substance | | mp(°C) | $\Delta H_{fus}$* | bp(°C) | $\Delta H_{vap}$* |
|---|---|---|---|---|---|
| Benzene | $C_6H_6$ | 5 | 9.84 | 80 | 30.8 |
| Bromine | $Br_2$ | −7 | 10.8 | 59 | 29.6 |
| Mercury | Hg | −39 | 2.33 | 357 | 59.4 |
| Naphthalene | $C_{10}H_8$ | 80 | 19.3 | 218 | 43.3 |
| Water | $H_2O$ | 0 | 6.00 | 100 | 40.7 |

*Values of $\Delta H_{fus}$ are given at the melting point, values of $\Delta H_{vap}$ at the boiling point. The heat of vaporization of water decreases from 44.9 kJ/mol at 0°C to 44.0 kJ/mol at 25°C to 40.7 kJ/mol at 100°C.

then 6.00 kJ of heat should be evolved when a mole of liquid water freezes.

$$H_2O(l) \longrightarrow H_2O(s) \qquad\qquad \Delta H = -6.00 \text{ kJ}$$

**Example 8.6**   Given

$$H_2(g) + \tfrac{1}{2}O_2(g) \longrightarrow H_2O(l) \qquad\qquad \Delta H = -285.8 \text{ kJ}$$

calculate $\Delta H$ for the equation

$$2H_2O(l) \longrightarrow 2H_2(g) + O_2(g)$$

**Strategy**   Note that for the required equation the coefficients are twice as great as in the given equation; the equation is also reversed. Apply Rules 1 and 2 in succession.

**Solution**   Applying Rule 1:

$$2H_2(g) + O_2(g) \longrightarrow 2H_2O(l) \qquad \Delta H = 2(-285.8 \text{ kJ}) = -571.6 \text{ kJ}$$

Applying Rule 2:

$$2H_2O(l) \longrightarrow 2H_2(g) + O_2(g) \qquad \Delta H = \boxed{+571.6 \text{ kJ}}$$

**3. The value of $\Delta H$ for a reaction is the same whether it occurs in one step or in a series of steps** (Figure 8.6). If a thermochemical equation can be expressed as the sum of two or more equations,

$$\text{equation} = \text{equation (1)} + \text{equation (2)} + \cdots$$

then $\Delta H$ for the overall equation is the sum of the $\Delta H$'s for the individual equations:

$$\Delta H = \Delta H_1 + \Delta H_2 + \cdots$$

This relationship is referred to as **Hess' law,** after Germaine Hess, professor of chemistry at the University of St. Petersburg, who deduced it in 1840. Hess' law is a direct consequence of the fact that enthalpy is a state property, dependent only upon initial and final states. This means that, in Figure 8.6, $\Delta H$ *must* equal the sum of $\Delta H_1$ and $\Delta H_2$, because the final and initial states are the same for the two processes.

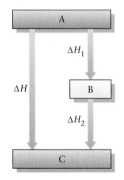

**Figure 8.6**
$\Delta H$ is independent of path since $H$ is a state property

Hess' law is very convenient for obtaining values of $\Delta H$ for reactions that are difficult to carry out in a calorimeter. Consider, for example, the formation of the toxic gas carbon monoxide from the elements

$$C(s) + \tfrac{1}{2}O_2(g) \longrightarrow CO(g)$$

It is difficult, essentially impossible, to measure $\Delta H$ for this reaction because when carbon burns the major product is always carbon dioxide, $CO_2$. It is possible, however, to calculate $\Delta H$, using thermochemical data for two other reactions that are readily carried out in the laboratory (Example 8.7).

---

**Example 8.7**   Given

$$(1) \ \ C(s) + O_2(g) \longrightarrow CO_2(g) \qquad \Delta H = -393.5 \text{ kJ}$$

$$(2) \ \ 2CO(g) + O_2(g) \longrightarrow 2CO_2(g) \qquad \Delta H = -566.0 \text{ kJ}$$

calculate $\Delta H$ for the reaction

$$C(s) + \tfrac{1}{2}O_2(g) \longrightarrow CO(g) \qquad \Delta H = \ ?$$

***Strategy***   The "trick" here is to work with the given information until you arrive at two equations that will add to give the equation you want ($C + \tfrac{1}{2}O_2 \rightarrow CO$). To do this, focus on CO which, unlike $CO_2$ and $O_2$, appears in only one thermochemical equation. Notice that you want *one* mole (not two) of CO on the *right* side (not the left side) of the equation.

***Solution***   To get one mole of CO on the right side, reverse Equation (2) and divide the coefficients by two. Applying Rule 1 and Rule 2 in succession,

$$CO_2(g) \longrightarrow CO(g) + \tfrac{1}{2}O_2(g) \qquad \Delta H = +566.0 \text{ kJ}/2 = +283.0 \text{ kJ}$$

Now, add Equation (1) and simplify:

$$CO_2(g) \longrightarrow CO(g) + \tfrac{1}{2}O_2(g) \qquad \Delta H = +283.0 \text{ kJ}$$

$$(1) \ \underline{\ \ C(s) + O_2(g) \longrightarrow CO_2(g) \qquad\qquad\qquad \Delta H = -393.5 \text{ kJ}\ }$$

$$C(s) + \tfrac{1}{2}O_2(g) \longrightarrow CO(g) \qquad\qquad \Delta H = \ -110.5 \text{ kJ}$$

Note that thermochemical equations can be added in exactly the same manner as algebraic equations; in this case, $1CO_2$ and $\tfrac{1}{2}O_2$ "cancelled" when the equations were added.

---

Carbon burns in oxygen to form carbon dioxide. (Charles Steele)

Summarizing the rules of thermochemistry:

**1.** $\Delta H$ is directly proportional to the amount of reactant or product.
**2.** $\Delta H$ changes sign when a reaction is reversed.
**3.** $\Delta H$ for a reaction has the same value regardless of the number of steps.

## 8.5   Enthalpies of Formation

We have now written several thermochemical equations. In each case, we have cited the corresponding value of $\Delta H$. Literally thousands of such equations would be needed to list the $\Delta H$ values for all the reactions that have been studied. Clearly, there has to be some more concise way of recording data of this sort. These data should be in a form that can easily be used to calculate $\Delta H$ for any reaction. It turns out that there is a simple way to do this, using quantities known as enthalpies of formation.

## Meaning of $\Delta H_f^\circ$

Often called "heat of formation"

The **standard molar enthalpy of formation** of a compound, $\Delta H_f^\circ$, is equal to the enthalpy change when one mole of the compound is formed at a constant pressure of 1 atm and a fixed temperature, ordinarily 25°C, from the elements in their stable states at that pressure and temperature. From the equations

$$Ag(s, 25°C) + \tfrac{1}{2}Cl_2(g, 25°C, 1 \text{ atm}) \longrightarrow AgCl(s, 25°C) \qquad \Delta H = -127.1 \text{ kJ}$$

$$\tfrac{1}{2}N_2(g, 1 \text{ atm}, 25°C) + O_2(g, 1 \text{ atm}, 25°C) \longrightarrow NO_2(g, 1 \text{ atm}, 25°C) \quad \Delta H = +33.2 \text{ kJ}$$

it follows that

$$\Delta H_f^\circ \text{ AgCl}(s) = -127.1 \text{ kJ/mol} \qquad \Delta H_f^\circ \text{ NO}_2(g) = +33.2 \text{ kJ/mol}$$

Enthalpies of formation are listed for a variety of compounds in Table 8.3. Notice that, with a few exceptions, enthalpies of formation are negative quantities. This means that the formation of a compound from the elements is ordinarily exothermic. Conversely, when a compound decomposes to the elements, heat usually must be absorbed.

You will note from Table 8.3 that there are no entries for elemental species such as $Br_2(l)$ and $O_2(g)$. This is a consequence of the way in which enthalpies of formation are defined. In effect, the enthalpy of formation of an element in its stable state at 25°C and 1 atm is taken to be zero. That is,

$$\Delta H_f^\circ \text{ Br}_2(l) = \Delta H_f^\circ \text{ O}_2(g) = 0$$

## Calculation of $\Delta H^\circ$

Enthalpies of formation can be used to calculate $\Delta H^\circ$ for a reaction. To do this, apply this general rule:

*The standard enthalpy change, $\Delta H^\circ$, for a given thermochemical equation is equal to the sum of the standard enthalpies of formation of the product compounds minus the sum of the standard enthalpies of formation of the reactant compounds.*

Using the symbol $\Sigma$ to represent "the sum of,"

$$\Delta H^\circ = \sum \Delta H_f^\circ \text{ products} - \sum \Delta H_f^\circ \text{ reactants}$$

This is a very useful equation, one we will use again and again

In taking these summations, you must take into account the coefficients of products and reactants in the thermochemical equation. For the general reaction

$$aA(g) + bB(g) \longrightarrow cC(g) + dD(g)$$

where *a*, *b*, *c*, and *d* are the coefficients of substances A, B, C, and D,

$$\sum \Delta H_f^\circ \text{ products} = c\Delta H_f^\circ \text{ C}(g) + d\Delta H_f^\circ \text{ D}(g)$$

$$\sum \Delta H_f^\circ \text{ reactants} = a\Delta H_f^\circ \text{ A}(g) + b\Delta H_f^\circ \text{ B}(g)$$

Hence,

$$\Delta H^\circ = c\Delta H_f^\circ \text{ C}(g) + d\Delta H_f^\circ \text{ D}(g) - [a\Delta H_f^\circ \text{ A}(g) + b\Delta H_f^\circ \text{ B}(g)]$$

## TABLE 8.3 Enthalpies of Formation, $\Delta H_f^{\circ}$ (kJ/mol), of Compounds at 25°C, 1 atm

| | | | | | | | |
|---|---|---|---|---|---|---|---|
| AgBr(s) | −100.4 | CaCl₂(s) | −795.8 | H₂O(g) | −241.8 | NH₄NO₃(s) | −365.6 |
| AgCl(s) | −127.1 | CaCO₃(s) | −1206.9 | H₂O(l) | −285.8 | NO(g) | +90.2 |
| AgI(s) | −61.8 | CaO(s) | −635.1 | H₂O₂(l) | −187.8 | NO₂(g) | +33.2 |
| AgNO₃(s) | −124.4 | Ca(OH)₂(s) | −986.1 | H₂S(g) | −20.6 | N₂O₄(g) | +9.2 |
| Ag₂O(s) | −31.0 | CaSO₄(s) | −1434.1 | H₂SO₄(l) | −814.0 | NaCl(s) | −411.2 |
| Al₂O₃(s) | −1675.7 | CdCl₂(s) | −391.5 | HgO(s) | −90.8 | NaF(s) | −573.6 |
| BaCl₂(s) | −858.6 | CdO(s) | −258.2 | KBr(s) | −393.8 | NaOH(s) | −425.6 |
| BaCO₃(s) | −1216.3 | Cr₂O₃(s) | −1139.7 | KCl(s) | −436.7 | NiO(s) | −239.7 |
| BaO(s) | −553.5 | CuO(s) | −157.3 | KClO₃(s) | −397.7 | PbBr₂(s) | −278.7 |
| BaSO₄(s) | −1473.2 | Cu₂O(s) | −168.6 | KClO₄(s) | −432.8 | PbCl₂(s) | −359.4 |
| CCl₄(l) | −135.4 | CuS(s) | −53.1 | KNO₃(s) | −494.6 | PbO(s) | −219.0 |
| CHCl₃(l) | −134.5 | Cu₂S(s) | −79.5 | MgCl₂(s) | −641.3 | PbO₂(s) | −277.4 |
| CH₄(g) | −74.8 | CuSO₄(s) | −771.4 | MgCO₃(s) | −1095.8 | PCl₃(g) | −287.0 |
| C₂H₂(g) | +226.7 | Fe(OH)₃(s) | −823.0 | MgO(s) | −601.7 | PCl₅(g) | −374.9 |
| C₂H₄(g) | +52.3 | Fe₂O₃(s) | −824.2 | Mg(OH)₂(s) | −924.5 | SiO₂(s) | −910.9 |
| C₂H₆(g) | −84.7 | Fe₃O₄(s) | −1118.4 | MgSO₄(s) | −1284.9 | SnO₂(s) | −580.7 |
| C₃H₈(g) | −103.8 | HBr(g) | −36.4 | MnO(s) | −385.2 | SO₂(g) | −296.8 |
| CH₃OH(l) | −238.7 | HCl(g) | −92.3 | MnO₂(s) | −520.0 | SO₃(g) | −395.7 |
| C₂H₅OH(l) | −277.7 | HF(g) | −271.1 | NH₃(g) | −46.1 | ZnI₂(s) | −208.0 |
| CO(g) | −110.5 | HI(g) | +26.5 | N₂H₄(l) | +50.6 | ZnO(s) | −348.3 |
| CO₂(g) | −393.5 | HNO₃(l) | −174.1 | NH₄Cl(s) | −314.4 | ZnS(s) | −206.0 |

Strictly speaking, $\Delta H°$ calculated from enthalpies of formation listed in Table 8.3 represents the enthalpy change at 25°C and 1 atm. Actually, $\Delta H$ is independent of pressure and varies relatively little with temperature, changing by perhaps 1 to 10 kJ per 100°C.

---

**Example 8.8** Calculate $\Delta H°$ for the combustion of one mole of propane, $C_3H_8$, according to the equation

$$C_3H_8(g) + 5O_2(g) \longrightarrow 3CO_2(g) + 4H_2O(l)$$

***Strategy*** Use the general reaction $\Delta H° = \Sigma \Delta H_f^{\circ}$ products $- \Sigma \Delta H_f^{\circ}$ reactants. Take enthalpies of formation from Table 8.3. Remember that the values in the table are for one mole of compound; if there are $n$ moles in the equation, multiply the molar enthalpy of formation by $n$.

***Solution*** Expressing $\Delta H°$ in terms of enthalpies of formation,

$$\Delta H° = 3\Delta H_f^{\circ}\, CO_2(g) + 4\Delta H_f^{\circ}\, H_2O(l) - [\Delta H_f^{\circ}\, C_3H_8(g) + 5\Delta H_f^{\circ}\, O_2(g)]$$

Taking the enthalpy of formation of $O_2(g)$ to be zero and substituting values for the other substances from Table 8.3,

$$\Delta H° = 3 \text{ mol} \left(-393.5 \frac{kJ}{mol}\right) + 4 \text{ mol} \left(-285.8 \frac{kJ}{mol}\right) - 1 \text{ mol} \left(-103.8 \frac{kJ}{mol}\right)$$

$$= -2219.9 \text{ kJ}$$

---

The relation between $\Delta H°$ and enthalpies of formation is perhaps used more often than any other in thermochemistry. Its validity depends upon the fact

Liquid propane in a tank can be attached to a barbecue grill to serve as fuel.

**Figure 8.7**

When a piece of burning magnesium ribbon is used as a fuse, a finely divided mixture of aluminum powder and iron (III) oxide undergoes an exothermic reaction: $2Al(s) + Fe_2O_3(s) \longrightarrow 2Fe(s) + Al_2O_3(s)$ $\Delta H° = -851.5$ kJ. Enough heat is generated to produce molten iron.

that enthalpy is a state property. For a reaction, $\Delta H°$ can be obtained by imagining that the reaction takes place in two steps, one involving the heats of formation of the products, the other the heats of formation of the reactants. To illustrate, consider the thermite reaction, once used to weld rails (Fig. 8.7):

$$2Al(s) + Fe_2O_3(s) \longrightarrow 2Fe(s) + Al_2O_3(s)$$

This can be considered to be the sum of two different reactions:

$$\textbf{(1)} \quad 2Al(s) + \tfrac{3}{2}O_2(g) \longrightarrow Al_2O_3(s) \qquad \Delta H_1°$$

$$\textbf{(2)} \quad Fe_2O_3(s) \longrightarrow 2Fe(s) + \tfrac{3}{2}O_2(g) \qquad \Delta H_2°$$

$$\textbf{(3)} \quad 2Al(s) + Fe_2O_3(s) \longrightarrow 2Fe(s) + Al_2O_3(s) \qquad \Delta H_3°$$

From the definition of enthalpy of formation,

$$\Delta H_1° = \Delta H_f° \, Al_2O_3(s) \qquad \Delta H_2° = -\Delta H_f° \, Fe_2O_3(s)$$

Applying Hess' law,

$$\Delta H°_3 = \Delta H_1° + \Delta H_2°$$

$$= \Delta H_f° \, Al_2O_3(s) - \Delta H_f° \, Fe_2O_3(s)$$

which is exactly the expression obtained by applying the relation

$$\Delta H° = \sum \Delta H_f° \text{ products} - \sum \Delta H_f° \text{ reactants}$$

directly to the equation

$$2Al(s) + Fe_2O_3(s) \longrightarrow 2Fe(s) + Al_2O_3(s)$$

It's a great deal easier to use the general relation than to go through the tedious, stepwise analysis using Hess' law.

The relation between $\Delta H°$ and $\Delta H_f°$ can also be used to calculate the heat of formation of a substance from an experimentally determined heat of reaction. Example 8.9 illustrates the calculations involved.

**Example 8.9**  The thermochemical equation for the combustion of benzene, $C_6H_6$, is

$$C_6H_6(l) + \tfrac{15}{2}O_2(g) \longrightarrow 6CO_2(g) + 3H_2O(l) \qquad \Delta H° = -3267.4 \text{ kJ}$$

Using Table 8.3 to obtain heats of formation for $CO_2$ and $H_2O$, calculate the heat of formation of benzene.

**Strategy**  Work with the general relation $\Delta H° = \Sigma \Delta H_f°$ products $- \Sigma \Delta H_f°$ reactants. All the heats of formation are known except that of benzene; $\Delta H°$ is also known. With only one unknown, the equation is readily solved for $\Delta H_f° \ C_6H_6$.

**Solution**  Taking the heat of formation of $O_2(g)$ to be zero, the relation for $\Delta H°$ is

$$\Delta H° = 6\Delta H_f° \ CO_2(g) + 3\Delta H_f° \ H_2O(l) - \Delta H_f° \ C_6H_6(l)$$

Substituting values for $\Delta H°$, $\Delta H_f° \ CO_2$, and $\Delta H_f° \ H_2O(l)$, we have

$$-3267.4 \text{ kJ} = 6 \text{ mol} \left(-393.5 \ \frac{\text{kJ}}{\text{mol}}\right) + 3 \text{ mol} \left(-285.8 \ \frac{\text{kJ}}{\text{mol}}\right)$$
$$- \ 1 \text{ mol} \ [\Delta H_f° \ C_6H_6(l)]$$

Solving,

$$\Delta H_f° \ C_6H_6(l) = +49.0 \text{ kJ/mol}$$

## Enthalpies of Formation of Ions in Solution

It is possible to set up a table, very much like Table 8.3, for enthalpies of formation of ions in water solution. There is, however, one problem. The enthalpy of formation of an individual ion cannot be measured. In any reaction involving ions, at least two of them are present, as required by the principle of electrical neutrality. Consider, for example, the reaction that occurs when HCl is added to water:

$$HCl(g) \longrightarrow H^+(aq) + Cl^-(aq)$$

Here, two different ions are formed, $H^+$ and $Cl^-$. The same situation applies in all other cases. To get around this dilemma, the enthalpy of formation of the $H^+$ ion is arbitrarily taken to be zero:

*You can't write a balanced equation with a single ion*

$$\Delta H_f° \ H^+(aq) = 0$$

Doing this establishes enthalpies of formation for other ions. Take, for instance, the $Cl^-$ ion. For the reaction that occurs when HCl is added to water, $\Delta H°$ is found to be $-74.9$ kJ. That is,

$$HCl(g) \longrightarrow H^+(aq) + Cl^-(aq) \qquad \Delta H° = -74.9 \text{ kJ}$$

Applying the relation between $\Delta H°$ and enthalpies of formation,

$$-74.9 \text{ kJ} = \Delta H_f° \ H^+(aq) + \Delta H_f° \ Cl^-(aq) - \Delta H_f° \ HCl(g)$$

Taking the enthalpy of formation of $H^+(aq)$ to be zero and that of $HCl(g)$ to be $-92.3$ kJ (Table 8.3) gives

$$-74.9 \text{ kJ} = 0 + \Delta H_f° \ Cl^-(aq) + 92.3 \text{ kJ}$$

or

$$\Delta H_f° \ Cl^-(aq) = -74.9 \text{ kJ} - 92.3 \text{ kJ} = -167.2 \text{ kJ}$$

TABLE 8.4 **Enthalpies of Formation, $\Delta H_f^\circ$ (kJ/mol), of Aqueous Ions at 25°C, 1 $M$**

| Cations | | | | Anions | | | |
|---|---|---|---|---|---|---|---|
| $Ag^+(aq)$ | $+105.6$ | $Hg^{2+}(aq)$ | $+171.1$ | $Br^-(aq)$ | $-121.6$ | $HPO_4^{2-}(aq)$ | $-1292.1$ |
| $Al^{3+}(aq)$ | $-531.0$ | $K^+(aq)$ | $-252.4$ | $CO_3^{2-}(aq)$ | $-677.1$ | $HSO_4^-(aq)$ | $-887.3$ |
| $Ba^{2+}(aq)$ | $-537.6$ | $Mg^{2+}(aq)$ | $-466.8$ | $Cl^-(aq)$ | $-167.2$ | $I^-(aq)$ | $-55.2$ |
| $Ca^{2+}(aq)$ | $-542.8$ | $Mn^{2+}(aq)$ | $-220.8$ | $ClO_3^-(aq)$ | $-104.0$ | $MnO_4^-(aq)$ | $-541.4$ |
| $Cd^{2+}(aq)$ | $-75.9$ | $Na^+(aq)$ | $-240.1$ | $ClO_4^-(aq)$ | $-129.3$ | $NO_2^-(aq)$ | $-104.6$ |
| $Cu^+(aq)$ | $+71.7$ | $NH_4^+(aq)$ | $-132.5$ | $CrO_4^{2-}(aq)$ | $-881.2$ | $NO_3^-(aq)$ | $-205.0$ |
| $Cu^{2+}(aq)$ | $+64.8$ | $Ni^{2+}(aq)$ | $-54.0$ | $Cr_2O_7^{2-}(aq)$ | $-1490.3$ | $OH^-(aq)$ | $-230.0$ |
| $Fe^{2+}(aq)$ | $-89.1$ | $Pb^{2+}(aq)$ | $-1.7$ | $F^-(aq)$ | $-332.6$ | $PO_4^{3-}(aq)$ | $-1277.4$ |
| $Fe^{3+}(aq)$ | $-48.5$ | $Sn^{2+}(aq)$ | $-8.8$ | $HCO_3^-(aq)$ | $-692.0$ | $S^{2-}(aq)$ | $+33.1$ |
| $H^+(aq)$ | $0.0$ | $Zn^{2+}(aq)$ | $-153.9$ | $H_2PO_4^-(aq)$ | $-1296.3$ | $SO_4^{2-}(aq)$ | $-909.3$ |

Enthalpies of formation for other ions in water solution can be established in a similar way. Table 8.4 lists values for several common ions. This table can be used, much like Table 8.3, to calculate $\Delta H$ for reactions in solution. To do that, use the general relation

$$\Delta H^\circ = \sum \Delta H_f^\circ \text{ products} - \sum \Delta H_f^\circ \text{ reactants}$$

realizing that $\Delta H_f^\circ$ for $H^+(aq)$ is zero, as is $\Delta H_f^\circ$ for an element in its stable state.

---

**Example 8.10** When hydrochloric acid is added to a solution of sodium carbonate, carbon dioxide gas is formed (Fig. 8.8). The equation for the reaction is

$$2H^+(aq) + CO_3^{2-}(aq) \longrightarrow CO_2(g) + H_2O(l)$$

Calculate $\Delta H^\circ$ for this thermochemical equation.

***Strategy*** Apply the general relation between $\Delta H^\circ$ and enthalpies of formation. Obtain enthalpies of formation from Tables 8.3 and 8.4. Remember that $\Delta H_f^\circ$ $H^+(aq) = 0$.

*Solution*

$$\Delta H^\circ = \Delta H_f^\circ \, CO_2(g) + \Delta H_f^\circ \, H_2O(l) - [2\Delta H_f^\circ \, H^+(aq) + \Delta H_f^\circ \, CO_3^{2-}(aq)]$$

Taking $\Delta H_f^\circ$ $H^+(aq)$ to be zero and obtaining the other enthalpies of formation from Tables 8.3 and 8.4,

$$\Delta H^\circ = 1 \text{ mol} \left(-393.5 \, \frac{kJ}{mol}\right) + 1 \text{ mol} \left(-285.8 \, \frac{kJ}{mol}\right) - 1 \text{ mol} \left(-677.1 \, \frac{kJ}{mol}\right)$$

$$= \boxed{-2.2 \text{ kJ}}$$

---

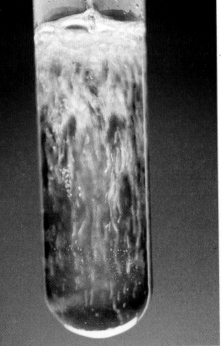

**Figure 8.8**
Hydrochloric acid added to sodium carbonate solution produces $CO_2$ gas.

## 8.6 Bond Energy

For many reactions, $\Delta H$ is a large negative number; the reaction gives off a lot of heat. In other cases, $\Delta H$ is positive; heat must be absorbed for the reaction to occur. You may well wonder why the enthalpy change should vary so widely

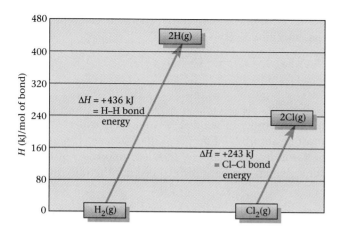

**Figure 8.9**
Bond energies in $H_2$ and $Cl_2$. The H—H bond energy is greater than the Cl—Cl bond energy ($+436$ kJ/mol versus $+243$ kJ/mol). This means that the bond in $H_2$ is stronger than that in $Cl_2$.

from one reaction to another. Is there some basic property of the molecules involved in the reaction that determines the sign and magnitude of $\Delta H$?

These questions can be answered on a molecular level in terms of a quantity known as bond energy. (More properly, but less commonly, it is called bond enthalpy). ***The bond energy is defined as $\Delta H$ when one mole of bonds is broken in the gaseous state.*** From the equations (see also Fig. 8.9)

$$H_2(g) \longrightarrow 2H(g) \qquad \Delta H = +436 \text{ kJ}$$

$$Cl_2(g) \longrightarrow 2Cl(g) \qquad \Delta H = +243 \text{ kJ}$$

What is $\Delta H_f^\circ$ for H(g)? Cl(g)?

it follows that the H—H bond energy is $+436$ kJ/mol, while that for Cl—Cl is 243 kJ/mol. In both reactions, one mole of bonds (H—H and Cl—Cl) is broken.

Bond energies for a variety of single and multiple bonds are listed in Table 8.5. Note that bond energy is always a positive quantity; energy is always ab-

TABLE 8.5   **Bond Energies**

| Single Bond Energies (kJ/mol) | | | | | | | | |
|---|---|---|---|---|---|---|---|---|
| | **H** | **C** | **N** | **O** | **S** | **F** | **Cl** | **Br** | **I** |
| **H** | 436 | 414 | 389 | 464 | 339 | 565 | 431 | 368 | 297 |
| **C** | | 347 | 293 | 351 | 259 | 485 | 331 | 276 | 218 |
| **N** | | | 159 | 222 | — | 272 | 201 | 243 | — |
| **O** | | | | 138 | — | 184 | 205 | 201 | 201 |
| **S** | | | | | 226 | 285 | 255 | 213 | — |
| **F** | | | | | | 153 | 255 | 255 | 277 |
| **Cl** | | | | | | | 243 | 218 | 209 |
| **Br** | | | | | | | | 193 | 180 |
| **I** | | | | | | | | | 151 |

| Multiple Bond Energies (kJ/mol) | | | | | |
|---|---|---|---|---|---|
| C=C | 612 | N=N | 418 | C≡C | 820 |
| C=N | 615 | N=O | 607 | C≡N | 890 |
| C=O | 715 | O=O | 498 | C≡O | 1075 |
| C=S | 477 | S=O | 498 | N≡N | 941 |

sorbed when chemical bonds are broken. Conversely, energy is given off when bonds are formed from gaseous atoms. Thus,

$$H(g) + Cl(g) \longrightarrow HCl(g) \qquad \Delta H = -431 \text{ kJ}$$

$$H(g) + F(g) \longrightarrow HF(g) \qquad \Delta H = -565 \text{ kJ}$$

Using bond energies, it is possible to explain why certain reactions are endothermic and others are exothermic. In general, a reaction is expected to be endothermic (i.e., heat must be absorbed) if *the bonds in the reactants are stronger than those in the products*. Consider, for example, the decomposition of hydrogen fluoride to the elements, which can be shown to be endothermic from the data in Table 8.3.

$$2HF(g) \longrightarrow H_2(g) + F_2(g) \qquad \Delta H = -2\Delta H_f^\circ \text{ HF} = +542.2 \text{ kJ}$$

The bonds in HF (565 kJ/mol) are stronger than the average of those in $H_2$ and $F_2$: (436 kJ/mol + 153 kJ/mol)/2 = 295 kJ/mol.

—*there are more bonds in the reactants than in the products*. For the reaction

$$2H_2O(g) \longrightarrow 2H_2(g) + O_2(g) \qquad \Delta H = -2\Delta H_f^\circ \text{ H}_2\text{O}(g) = +483.6 \text{ kJ}$$

There are four moles of O—H bonds in two moles of $H_2O$ as compared with only three moles of bonds in the products (two mol H—H bonds, one mole O=O bonds).

The converse of this statement is also true. A reaction is expected to be exothermic if strong bonds are formed at the expense of weak bonds.

You will note from Table 8.5 that the bond energy is larger for a multiple bond than for a single bond between the same two atoms. Thus

$$\text{bond energy C—C} = 347 \text{ kJ/mol}$$

$$\text{bond energy C}{=}\text{C} = 612 \text{ kJ/mol}$$

$$\text{bond energy C}{\equiv}\text{C} = 820 \text{ kJ/mol}$$

This effect is a reasonable one; the greater the number of bonding electrons, the more difficult it should be to break the bond between two atoms.

## 8.7    The First Law of Thermodynamics

So far in this chapter, our discussion has focused upon thermochemistry, the study of the heat effects in chemical reactions. Thermochemistry is a branch of *thermodynamics,* which deals with all kinds of energy effects in all kinds of processes. Thermodynamics distinguishes between two types of energy. One of these is heat ($q$); the other is **work,** represented by the symbol $w$. The thermodynamic definition of work is quite different from its colloquial meaning. Quite simply, **work includes all forms of energy except heat.**

The law of conservation of energy states that energy ($E$) can neither be created nor destroyed; it can only be transferred between system and surroundings. That is,

$$\Delta E_{\text{system}} = -\Delta E_{\text{surroundings}}$$

The first law of thermodynamics goes a step further. Taking account of the fact that there are two kinds of energy, heat and work, the first law states:

The higher the bond energy, the stronger the bond

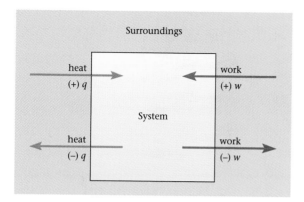

**Figure 8.10**

Heat and work are positive when they enter a system and negative when they leave.

***In any process, the total change in energy of the system, ΔE, is equal to the sum of the heat absorbed q, and the work, w, done on the system.***

> If work is done on the system or heat is added to it, its energy increases

$$\Delta E = q + w$$

In applying the first law, note (Fig. 8.10) that $q$ and $w$ are positive when heat or work enters the system from the surroundings. If the transfer is in the opposite direction, from system to surroundings, $q$ and $w$ are negative.

---

**Example 8.11**   Calculate $\Delta E$ of a gas for a process in which the gas

(a) absorbs 20 J of heat and does 12 J of work by expanding.
(b) evolves 30 J of heat and has 52 J of work done on it as it contracts.

***Strategy***   First decide upon the signs (+ or −) of $q$ and $w$. Then substitute into the equation for $\Delta E$.

*Solution*

(a) $q = +20$ J; $w = -12$ J, since the gas does work on the surroundings.

$$\Delta E = +20\,\text{J} - 12\,\text{J} = \boxed{+8\,\text{J}}$$

(b) $q = -30$ J; $w = +52$ J, since work is done on the gas by the surroundings.

$$\Delta E = -30\,\text{J} + 52\,\text{J} = \boxed{+22\,\text{J}}$$

---

## Expansion Work

When a chemical reaction is carried out directly in the laboratory, only one type of work is ordinarily involved: work of expansion (or contraction) done by a gas against a restraining pressure. Typically, the gases involved are products or reactants in the reaction. The "restraining pressure" is simply the constant pressure of the atmosphere.

To illustrate expansion work, consider what happens when a mole of liquid water is vaporized at 100°C and 1 atm (Fig. 8.11). Work is defined as the product of force ($f$) times the distance ($d$) through which the force moves

$$w = -f \times d$$

(The minus sign arises because when expansion occurs, as is the case here, work

**Figure 8.11**
In general, the expansion work for a system working against a constant pressure, $P$, is $-P\Delta V$, where $\Delta V$ is the change in volume. In this case, where one mole of liquid water is vaporized at 100°C and 1 atm, $w$ turns out to be $-30.6\,L \cdot atm$ or $-3.10\,kJ$.

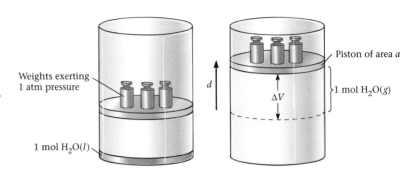

is done by the system on the surroundings). Since pressure ($P$) is force per unit area ($a$), it follows that $f = P \times a$, so

$$w = -P \times a \times d$$

But, as you can see from the figure, the product of area times distance is simply the volume change for the process, $\Delta V$. Hence, the general expression for expansion work occurring against a constant external pressure $P$ is

$$w = -P\Delta V$$

To evaluate the expansion work for the process shown in Figure 8.11,

$$H_2O(l,\ 1\ atm,\ 100°C) \longrightarrow H_2O(g,\ 1\ atm,\ 100°C)$$

note that the volume change is

$$\Delta V = V(H_2O,\ g,\ 1\ atm,\ 100°C) - V(H_2O,\ l,\ 1\ atm,\ 100°C)$$

But, in general, the volume of a liquid or solid is much, much smaller than that of a gas at the same temperature and pressure. So the second term on the right side of this equation can be neglected, and we have

$$\Delta V \approx V\ (H_2O,\ g) = \frac{nRT}{P}$$

$$-P\Delta V = -nRT$$

To calculate the work in kilojoules, use the value of $R$ in $J/mol \cdot K$ (recall Table 5.1, p. 116).

$$w = -nRT = -1\ mol\ \times\ \frac{8.31\ J}{mol \cdot K} \times \frac{1\ kJ}{10^3\ J} \times 373\ K = -3.10\ kJ$$

The calculation we have just gone through is typical of those involved in evaluating the expansion work in a phase change or chemical reaction. In general, we can say that

$$w = -P\Delta V = -\Delta n_g \cdot RT$$

where $\Delta n_g$ is the change in the number of moles of gas in the process and $R = 8.31\ J/mol \cdot K$.

## Constant Volume, Constant Pressure Processes

The quantities $q$ and $w$, unlike $E$ and $H$, are *not* state properties. Their value depends upon the path by which a process is carried out. To illustrate this point, let us evaluate the heat and work effects associated with processes carried out at constant volume versus constant pressure.

**1. Constant volume.** In a bomb calorimeter or any other sealed container, there is no change in volume upon reaction (unless, of course, the container explodes!). It follows that there is no expansion work associated with a constant-volume process:

$$w_v = 0$$

(The subscript v is used to indicate constant volume). Hence, from the first law,

$$q_v = \Delta E$$

**2. Constant pressure.** For a process carried out in a coffee-cup calorimeter or any other container open to the atmosphere such as a beaker or test tube (or other container), the expansion work is given by the expression just derived:

$$w_p = -P\Delta V = -\Delta n_g RT$$

Ordinarily, there is at least a small change in volume, so $w_p$ has a finite value other than zero, and $w_p \neq w_v$.

To find the heat flow for a constant-pressure process, we again apply the first law:

$$q_p = \Delta E - w_p$$

$$q_p = \Delta E + \Delta n_g RT$$

Comparing this expression with that for the heat flow at constant volume ($q_v = \Delta E$), you can see that $q_p$ and $q_v$ are *not* the same. Indeed,

$$q_p = q_v + \Delta n_g RT$$

The heat flow measured in a bomb calorimeter is $\Delta E$; in a coffeecup calorimeter, it's $\Delta H$.

(a)

---

**Example 8.12**   Calculate the difference in kilojoules between $q_p$ and $q_v$ for the decomposition of one mole of ammonium chloride at 25°C.

$$NH_4Cl(s) \longrightarrow NH_3(g) + HCl(g)$$

**Strategy**   Evaluate the quantity $\Delta n_g RT$, where $\Delta n_g$ is the change in the number of moles of gas (products − reactants) and $R = 8.31$ J/mol · K.

**Solution**

$$\Delta n_g = 2 \text{ mol} - 0 = 2 \text{ mol}$$

$$\Delta n_g RT = 2 \text{ mol} \times \frac{8.31 \text{ J}}{\text{mol} \cdot \text{K}} \times 298 \text{ K} = 4.95 \times 10^3 \text{ J} = 4.95 \text{ kJ}$$

Other calculations of this type are shown in Table 8.6.

---

(b)

### TABLE 8.6   **Comparison of $q_p(\Delta H)$ and $q_v(\Delta E)$ at 25°C**

| | $\Delta n_g$ | $q_v$ | $\Delta n_g RT$ | $q_p$ |
|---|---|---|---|---|
| $N_2O_4(g) \longrightarrow 2NO_2(g)$ | +1 | +54.7 kJ | +2.5 kJ | +57.2 kJ |
| $PCl_3(g) + Cl_2(g) \longrightarrow PCl_5(g)$ | −1 | −85.4 kJ | −2.5 kJ | −87.9 kJ |
| $N_2(g) + 3H_2(g) \longrightarrow 2NH_3(g)$ | −2 | −87.2 kJ | −5.0 kJ | −92.2 kJ |
| $2HI(g) \longrightarrow H_2(g) + I_2(g)$ | 0 | +7.0 kJ | 0 | +7.0 kJ |

(a) Pieces of Dry Ice are put inside a plastic bag. (b) Some of the Dry Ice, $CO_2(s)$, has sublimed into $CO_2(g)$. The pressure in the bag is constant, but the volume changes.

### $\Delta H$ and $\Delta E$

The equation

$$q_p = q_v + \Delta n_g RT$$

can be used to relate the change in enthalpy, $\Delta H$, to the change in energy, $\Delta E$. As you have just seen, $q_v = \Delta E$; earlier, we pointed out that $q_p = \Delta H$. It follows that

$$\Delta H = \Delta E + \Delta n_g RT$$

Since $\Delta n_g RT$ is ordinarily a small quantity, it follows that *the values of $\Delta H$ and $\Delta E$ for a reaction are close to one another.* For the reaction considered in Example 8.12, recall that $\Delta n_g RT = +4.95$ kJ; $\Delta H$ for this reaction, calculated from enthalpies of formation, is $+176.0$ kJ. Thus

$$\Delta E = \Delta H - \Delta n_g RT = +176.0 \text{ kJ} - 4.95 \text{ kJ} = +171.0 \text{ kJ}$$

The two quantities $\Delta H$ and $\Delta E$ differ in this case by less than 3%.

---

## CHEMISTRY

### *Beyond the Classroom*

## Energy Balance in the Human Body

All of us require energy to maintain life processes and do the muscular work associated with such activities as studying, writing, walking, or jogging. The fuel used to produce that energy is the food we eat. In this discussion, we focus upon "energy input" or "energy output" and the balance between them.

### *Energy Input*

Energy values of food can be estimated on the basis of the content of carbohydrate, protein, and fat:

| | |
|---|---|
| carbohydrate: | 17 kJ (4.0 kcal) of energy per gram* |
| protein: | 17 kJ (4.0 kcal) of energy per gram |
| fat: | 38 kJ (9.0 kcal) of energy per gram |

Alcohol, which doesn't fit into any of these categories, furnishes about 29 kJ (7.0 kcal) per gram.

Using these guidelines, it is possible to come up with the data given in Table 8.A, which lists approximate energy values for standard portions of different types of foods. With a little practice, you can use the table to estimate your energy input within ± 10%.

### *Energy Output*

Energy output can be divided, more or less arbitrarily, into two categories.

**1. Metabolic energy.** This is the largest item in most people's energy budget; it comprises the energy necessary to maintain life. Metabolic rates are ordinarily estimated based on a person's sex, weight, and age. For a young adult female, the metabolic rate is about 100 kJ per day per kilogram of body weight. For a 20-year-old woman weighing 120 lb,

---

*In most nutrition texts, energy is expressed in kilocalories rather than kilojoules, although that situation is changing. Remember that 1 kcal = 4.18 kJ; 1 kg = 2.20 lb.

### TABLE 8.A  Energy Values of Food Portions*

|  | kJ | kcal |
|---|---|---|
| 8-oz glass (250 g) skim milk | 330 | 80 |
| whole milk | 710 | 170 |
| beer | 420 | 100 |
| 1 portion vegetable | 100 | 25 |
| 1 portion fruit | 170 | 40 |
| 1 portion bread or starchy vegetable | 300 | 70 |
| 1 oz (30 g) lean meat | 230 | 55 |
| medium-fat meat | 320 | 77 |
| high-fat meat | 420 | 100 |
| 1 teaspoon (5 g) fat or oil | 190 | 45 |
| 1 teaspoon (5 g) sugar | 80 | 20 |

*Adapted from E. N. Whitney and E. M. Hamilton, *Understanding Nutrition*, 3rd ed. West Publishing Co., St. Paul, 1984.

$$\text{metabolic rate} = 120 \text{ lb} \times \frac{1 \text{kg}}{2.20 \text{ lb}} \times 100 \frac{\text{kJ}}{\text{d} \cdot \text{kg}}$$

$$= 5500 \text{ kJ/d} \ (1300 \text{ kcal/day})$$

For a young adult male of the same weight, the figure is about 10% higher.

**2. Muscular activity.** Energy is consumed whenever your muscles contract, whether in writing, walking, playing tennis, or even sitting in class. Generally speaking, energy expenditure for muscular activity ranges from 50% to 100% of metabolic energy, depending upon lifestyle. If the 20-year-old woman referred to earlier is a student who does nothing more strenuous than lift textbooks (and that not too often), the energy spent on muscular activity might be

$$0.50 \times 5500 \text{ kJ} = 2800 \frac{\text{kJ}}{\text{d}}$$

Her total energy output per day would be 5500 kJ + 2800 kJ = 8300 kJ (2000 kcal).

## Energy Balance

To maintain constant weight, your daily energy input, as calculated from the foods you eat, should be about 700 kJ (170 kcal) greater than output. The difference allows for the fact that about 40 g of protein is required to maintain body tissues and fluids. If the excess of input over output is greater than 700 kJ/day, the unused food (carbo-

### TABLE 8.B  Energy Consumed by Various Types of Exercise*

| kJ/hour | kcal/hour |  |
|---|---|---|
| 1000–1250 | 240–300 | walking (3 mph), bowling, golf (pulling cart) |
| 1250–1500 | 300–360 | volleyball, calisthenics, golf (carrying clubs) |
| 1500–1750 | 360–420 | ice skating, roller skating |
| 1750–2000 | 420–480 | tennis (singles), water skiing |
| 2000–2500 | 480–600 | jogging (5 mph), downhill skiing, mountain climbing |
| > 2500 | > 600 | running (6 mph), handball, squash |

*Adapted from Jane Brody: *Jane Brody's Nutrition Book: A Lifetime Guide to Good Eating for Better Health and Weight Control by the Personal Health Columnist for the New York Times*. Norton, New York, 1981.

Skiing at 10 mph requires approximately 600 kcal/hour or 2500 kJ/hour. (Mark Junak/Tony Stone Images)

hydrate, protein, or fat) is converted to fatty tissue and stored as such in the body. Fatty tissue consists of about 85% fat and 15% water; its energy value is

$$0.85 \times 9.0 \frac{\text{kcal}}{\text{g}} \times \frac{454 \text{ g}}{1 \text{ lb}} = 3500 \text{ kcal/lb}$$

This means that to lose one pound of fat, energy input from foods must be decreased by about 3500 kcal. To lose weight at a sensible rate of one pound per week, it is necessary to cut down by 500 kcal (2100 kJ) per day.

It's also possible, of course, to lose weight by increasing muscular activity. Table 8.B shows the amount of energy consumed per hour with various types of exercise. In principle, you can lose a pound a week by climbing mountains for an hour each day, provided you're not already doing that (most people aren't). Alternatively, you could spend $1\frac{1}{2}$ hours a day ice skating or water skiing, depending on the weather.

# CHAPTER HIGHLIGHTS

## *Key Concepts*

1. Relate heat flow to specific heat, $m$, and $\Delta t$.
   (Example 8.1; Problems 1–4, 65, 66)
2. Calculate $q$ for a reaction from calorimetric data.
   (Examples 8.2, 8.3; Problems 5–14)
3. Use a thermochemical equation to relate $\Delta H$ to mass of reactant or product and/or to $\Delta H$ of the reverse reaction.
   (Examples 8.4–8.6; Problems 17–26, 63)
4. Use Hess' law to calculate $\Delta H$.
   (Example 8.7; Problems 29–34, 62)
5. Relate $\Delta H$ to enthalpies of formation.
   (Examples 8.8–8.10; Problems 37–50)
6. Calculate $\Delta E$, $q$ or $w$ from the first law.
   (Example 8.11; Problems 51–56)
7. Relate $q_p$ to $q_v$ (or $\Delta H$ to $\Delta E$).
   (Example 8.12; Problems 57–60)

## *Key Equations*

| | |
|---|---|
| Heat flow | $q = C \times \Delta t = m \times c \times \Delta t$ |
| Calorimetry | $q_{\text{reaction}} = -C_{\text{cal}} \times \Delta t$ |
| Enthalpy of formation | $\Delta H° = \sum \Delta H_f^{\circ} \text{ products} - \sum \Delta H_f^{\circ} \text{ reactants}$ |

# Key Terms

| | | |
|---|---|---|
| bond energy | exothermic | mole |
| calorimeter | heat | specific heat |
| —bomb | —capacity | state property |
| —coffee-cup | —of fusion | surroundings |
| endothermic | —of vaporization | system |
| enthalpy | Hess' law | work |
| enthalpy of formation | joule | |

## Summary Problem

Dimethylhydrazine, $N_2H_2(CH_3)_2$, is used as a rocket fuel. When it reacts with oxygen, the thermochemical equation for the reaction is

$$\tfrac{1}{2}N_2H_2(CH_3)_2(l) + 2O_2(g) \longrightarrow CO_2(g) + 2H_2O(g) + \tfrac{1}{2}N_2(g) \qquad \Delta H = -898.1 \text{ kJ}$$

(a) Calculate $\Delta H$ for the following thermochemical equations:
   (1) $N_2H_2(CH_3)_2(l) + 4O_2(g) \longrightarrow 2CO_2(g) + 4H_2O(g) + N_2(g)$
   (2) $CO_2(g) + 2H_2O(g) + \tfrac{1}{2}N_2(g) \longrightarrow \tfrac{1}{2}N_2H_2(CH_3)_2(l) + 2O_2(g)$

(b) The heat of vaporization of $H_2O(l)$ is 44.0 kJ/mol. Calculate $\Delta H$ for the equation

$$\tfrac{1}{2}N_2H_2(CH_3)_2(l) + 2O_2(g) \longrightarrow CO_2(g) + 2H_2O(l) + \tfrac{1}{2}N_2(g)$$

(c) The heats of formation of $CO_2$ and $H_2O(g)$ are $-393.5$ and $-241.8$ kJ/mol, respectively. Calculate $\Delta H_f^\circ$ of $N_2H_2(CH_3)_2(l)$.

(d) How much heat is evolved when 1.00 g of $N_2H_2(CH_3)_2(l)$ is burned in an open container?

(e) The temperature of a bomb calorimeter increases by 2.15°C when it absorbs 10.0 kJ of heat. Calculate $C_{calorimeter}$.

(f) When 0.500 g of $N_2H_2(CH_3)_2(l)$ is burned at 25.00°C in the bomb calorimeter described in (e), what is the expected temperature of the calorimeter after the experiment?

(g) For the reaction, $\Delta H = q_p = -898.1$ kJ. Calculate $\Delta E = q_v$ at 25°C.

### Answers

(a) $-1796.2$ kJ; $+898.1$ kJ    (b) $-986.1$ kJ    (c) $+42.0$ kJ/mol
(d) $\Delta H = -29.9$ kJ    (e) 4.65 kJ/°C    (f) 28.22°C    (g) $-901.8$ kJ

## Questions & Problems

### Specific Heat and Calorimetry

**1.** Titanium is a metal used in jet engines. Its specific heat is 0.523 J/g · °C. How much heat must be absorbed to raise the temperature of 5.88 g of titanium from 22.00°C to 27.34°C?

**2.** Stainless steel accessories in cars are usually plated with chromium to give them a shiny surface and to prevent rusting. Chromium has a specific heat of 0.449 J/g · °C. If 5.00 g of chromium at 23.00°C absorbs 62.5 J of heat, what is the final temperature of the chromium?

**3.** The specific heat of aluminum is 0.902 J/g · °C. How much heat is absorbed by an aluminum pie tin with a mass of 473 g to raise its temperature from room temperature (23.00°C) to oven temperature (375°F)?

**4.** When a 10.0-g block of gold absorbs 2.80 joules, the temperature of the block rises from 23.50°C to 25.67°C. What is the specific heat of gold?

**5.** Calcium chloride is a compound frequently found in first-aid hot packs. It gives off heat when dissolved in water. When 1.50 g of $CaCl_2$ is dissolved in 150.0 g of water, the temperature of the water rises from 20.50°C to 22.25°C. Assume that all the heat is absorbed by the water (specific heat = 4.18 J/g · °C).

    **(a)** Write a balanced equation for the solution process.
    **(b)** What is $q$ for the solution process?
    **(c)** Is the solution process exothermic or endothermic?
    **(d)** How much heat is absorbed by the water if one mole of calcium chloride is dissolved?

**6.** Sodium chloride is added in cooking to enhance the flavor of food. When 15.25 g of NaCl dissolves in 85.0 g of water, the temperature of the water drops from 22.00°C to 19.14°C. Assume that all the heat is absorbed by the water (specific heat = 4.18 J/g · °C).

    **(a)** Write a balanced equation for the solution process.
    **(b)** What is $q$ for the solution process?
    **(c)** Is the process exothermic or endothermic?
    **(d)** What is $q$ when one mole of sodium chloride dissolves?

**7.** Fructose is a sugar commonly found in fruit. A sample of fructose, $C_6H_{12}O_6$, weighing 4.50 g is burned in a bomb calorimeter. The heat capacity of the calorimeter is $2.115 \times 10^4$ J/°C. The temperature in the calorimeter rises from 23.49°C to 27.71°C. Calculate $q$ for the combustion of one mole of fructose.

**8.** Ethyl ether was commonly used as an anesthetic. It is, however, highly flammable. When one mL of ethyl ether, $C_4H_{10}O(l)$, $(d = 0.714$ g/mL) is burned in a bomb calorimeter, the temperature rises from 24.75°C to 28.33°C. The calorimeter heat capacity is 7.346 kJ/°C. Calculate $q$ for the combustion of one mole of ethyl ether.

**9.** When 2.500 g of caffeine ($C_8H_{10}N_4O_2$) burns in a bomb calorimeter, the temperature in the bomb rises from 22.55°C to 27.14°C. Under these conditions, $4.96 \times 10^3$ kJ of heat is evolved per mole of caffeine burned. Calculate the calorimeter heat capacity in J/°C.

**10.** A 700.0-mg sample of urea, $(NH_2)_2CO$, gave off 7.366 kJ of heat when burned in a bomb calorimeter. The temperature of the bomb rose 4.157°C. What is the calorimeter heat capacity in J/°C?

**11.** Methanol, $CH_3OH$, is used as a fuel in some cars. When one mole of methanol burns, 726 kJ of heat is evolved. Five mL of methanol ($d = 0.796$ g/mL) is burned in a bomb calorimeter; heat capacity = 8342 J/°C. If the bomb is initially at 22.73°C, what is the final temperature?

**12.** Naphthalene, $C_{10}H_8$, is the compound present in moth balls. When one mole of naphthalene is burned, $5.15 \times 10^3$ kJ of heat is evolved. A sample of naphthalene burned in a bomb calorimeter (heat capacity = 1.27 kJ/°C) increases the temperature in the calorimeter by 2.49°C. How many mg of naphthalene were burned?

**13.** Salicylic acid, $C_7H_6O_3$, is one of the starting materials in the manufacture of aspirin. When 1.00 g of salicylic acid burns in a bomb calorimeter, the temperature rises from 23.11°C to 28.91°C. The temperature in the bomb calorimeter increases by 2.48°C when the calorimeter absorbs 9.37 kJ. How much heat is given off when one mole of salicylic acid is burned?

**14.** A 3.500 g sample of benzoic acid ($C_7H_6O_2$) gives off 9.258 kJ of heat when burned in a bomb calorimeter. The bomb calorimeter increases its temperature by 1.61°C when 6.23 kJ of heat are absorbed. What is the final temperature of the calorimeter when one mole of benzoic acid is burned and the initial temperature in the bomb is 20.50°C?

## Rules of Thermochemistry

**\*15.** Using Table 8.2, write the thermochemical equations for the following:
    **(a)** mercury thawing
    **(b)** bromine vaporizing
    **(c)** naphthalene freezing

**\*16.** Using Table 8.2, write thermochemical equations for the following:
    **(a)** mercury condensing
    **(b)** benzene vaporizing
    **(c)** bromine freezing

**17.** The combustion of one mole of liquid carbon disulfide in oxygen produces carbon dioxide and sulfur dioxide gases and liberates 1075 kJ of heat.
    **(a)** Write the thermochemical equation for the combustion of carbon disulfide.
    **(b)** Is the reaction exothermic or endothermic?
    **(c)** Draw an energy diagram showing the path of this reaction. (Figure 8.4 is an example of an energy diagram showing the path of a reaction.)
    **(d)** What is $\Delta H$ when 5.00 g of carbon disulfide is burned?
    **(e)** How many grams of carbon disulfide must be burned to liberate ten kilojoules of heat?

**18.** When one mole of hydrogen gas, graphite (solid carbon), and nitrogen gas are combined, hydrogen cyanide gas is formed and 270.3 kJ of heat is absorbed.
    **(a)** Write a thermochemical equation for this reaction.
    **(b)** Is the reaction exothermic or endothermic?
    **(c)** Draw an energy diagram showing the path of this reaction. (Figure 8.4 is an example of an energy diagram showing the path of a reaction.)
    **(d)** What is $\Delta H$ when 1.00 g of $HCN(g)$ is formed?
    **(e)** How many grams of graphite are used up when 20.00 kJ of heat are absorbed?

**19.** When calcium chloride dissolves, the following takes place:

$$CaCl_2(s) \longrightarrow Ca^{2+}(aq) + 2Cl^-(aq) \qquad \Delta H = -81.4 \text{ kJ}$$

    **(a)** Calculate $\Delta H$ when one mole of calcium chloride precipitates from solution.
    **(b)** What is $\Delta H$ when 2.00 g of calcium chloride precipitates?

**20.** When aluminum metal is put in a solution with zinc ions, the following takes place:

$$2Al(s) + 3Zn^{2+}(aq) \longrightarrow 3Zn(s) + 2Al^{3+}(aq)$$
$$\Delta H = -600.3 \text{ kJ}$$

**(a)** Calculate $\Delta H$ when one mole of zinc reacts with aluminum ions.

**(b)** What is $\Delta H$ when 1.00 g of zinc reacts?

**21.** Ethyl mercaptan, $C_2H_5SH$, is responsible for the characteristic odor of natural gas. It is added to natural gas (which is odorless) so leaks in gas lines can be detected. When 1.00 g of ethyl mercaptan burns in air, water, carbon dioxide gas, sulfur dioxide gas, and 30.21 kJ of energy are produced.

**(a)** Write a balanced thermochemical equation for the reaction.

**(b)** What is $\Delta H$ for the formation of 10.00 L of carbon dioxide at 25°C and 1.00 atm?

**22.** Nitroglycerine, $C_3H_5(NO_3)_3(l)$, is an explosive most often used in mine or quarry blasting. It is a powerful explosive because four gases ($N_2$, $O_2$, $CO_2$, and steam) are formed when nitroglycerine is detonated. In addition, 6.26 kJ of heat is given off per gram of nitroglycerine detonated.

**(a)** Write a balanced thermochemical equation for the reaction.

**(b)** What is $\Delta H$ when 5.00 moles of products are formed?

**23.** A typical fat in the body is glyceryl trioleate, $C_{57}H_{104}O_6$. When metabolized in the body, it combines with oxygen to produce carbon dioxide, water, and $3.022 \times 10^4$ kJ of heat per mole of fat.

**(a)** Write a balanced thermochemical equation for the metabolism of fat.

**(b)** How many grams of fat would have to be burned to heat 100.0 mL of water ($d = 1.00$ g/mL) from 22.00°C to 26.00°C? The specific heat of water is 4.18 J/g · °C.

**24.** Using the metabolism of fat reaction in Problem 23, how many kilojoules of energy must be evolved in the form of heat if you want to get rid of two pounds of this fat by combustion? How many calories is this? How many nutritional calories (1 nutritional calorie = $1 \times 10^3$ calories)?

**25.** Which requires the absorption of a greater amount of heat—vaporizing 100.0 g of bromine or boiling 100.0 g of water? (Use Table 8.2.)

**26.** Which evolves more heat—freezing 100.0 g of water or condensing 100.0 g of naphthalene vapor to liquid?

### Hess' Law

**27.** Using Table 8.2, write a thermochemical equation for
**(a)** vaporizing solid naphthalene.
**(b)** freezing bromine gas.

**28.** Using Table 8.2, write a thermochemical equation for
**(a)** freezing mercury vapor.
**(b)** evaporating ice to steam.

**29.** A student wants to change 12.50 g of water at room temperature (22.30°C) to steam at 100.0°C. To do this, she must follow the step-wise process outlined below. Do the following calculations with her. The specific heat of water is 4.18 J/g · °C.

**(a)** Calculate $\Delta H$ for

$$H_2O(l, 22.30°C) \longrightarrow H_2O(l, 100.0°C)$$

**(b)** Calculate $\Delta H$ using Table 8.2 for

$$H_2O(l, 100.0°C) \longrightarrow H_2O(g, 100.0°C)$$

**(c)** Using Hess' law, calculate $\Delta H$ for

$$H_2O(l, 22.30°C) \longrightarrow H_2O(g, 100.0°C)$$

**30.** Follow the step-wise process outlined in Problem 29 to calculate the amount of heat involved in changing 10.00 g of liquid benzene at 65.0°C to vapor at 80.0°C. Use Tables 8.1 and 8.2 for the specific heat, boiling point, and heat of vaporization of benzene.

**31.** Dichloromethane is used as a solvent in industrial applications that once required chloroform, a known carcinogen. It is obtained by first reacting methane gas with chlorine.

$$CH_4(g) + Cl_2(g) \longrightarrow CH_3Cl(g) + HCl(g) \qquad \Delta H = -98.3 \text{ kJ}$$

The product, $CH_3Cl$, is again made to react with chlorine gas according to the equation

$$CH_3Cl(g) + Cl_2(g) \longrightarrow CH_2Cl_2(g) + HCl(g) \qquad \Delta H = -104 \text{ kJ}$$

Write the thermochemical equation for the reaction between methane gas and chlorine to produce dichloromethane.

**32.** Given the following thermochemical equations

| | |
|---|---|
| $2H_2(g) + O_2(g) \longrightarrow 2H_2O(l)$ | $\Delta H = -571.6$ kJ |
| $N_2O_5(g) + H_2O(l) \longrightarrow 2HNO_3(l)$ | $\Delta H = -73.7$ kJ |
| $\frac{1}{2}N_2(g) + \frac{3}{2}O_2(g) + \frac{1}{2}H_2(g) \longrightarrow HNO_3(l)$ | $\Delta H = -174.1$ kJ |

calculate $\Delta H$ for the decomposition of one mole of dinitrogen pentaoxide to its elements in their stable state at 25°C and 1 atm.

**33.** Given the following thermochemical equations

| | |
|---|---|
| $N_2(g) + 3H_2(g) \longrightarrow 2NH_3(g)$ | $\Delta H = -92.2$ kJ |
| $C(s) + 2H_2(g) \longrightarrow CH_4(g)$ | $\Delta H = -74.7$ kJ |
| $2C(s) + H_2(g) + N_2(g) \longrightarrow 2HCN(g)$ | $\Delta H = 270.3$ kJ |

write the thermochemical equation for the reaction between methane, $CH_4$, and ammonia to form hydrogen cyanide and hydrogen gases.

**34.** Given the following thermochemical equations

| | |
|---|---|
| $2KCl(s) + 3O_2(g) \longrightarrow 2KClO_3(s)$ | $\Delta H = 78.0$ kJ |
| $P_4(s) + 6Cl_2(g) \longrightarrow 4PCl_3(g)$ | $\Delta H = -1148.0$ kJ |
| $P_4(s) + 2O_2(g) + 6Cl_2(g) \longrightarrow 4POCl_3(g)$ | $\Delta H = -2168.8$ kJ |

calculate $\Delta H$ for the reaction

$$KClO_3(s) + 3PCl_3(g) \longrightarrow 3POCl_3 + KCl(s)$$

## $\Delta H°$ and Heats of Formation

*35. Write thermochemical equations for the formation of one mole of the following compounds from the elements in their stable states at 25°C and 1 atm.

(a) carbon tetrachloride (*l*)
(b) magnesium hydroxide (*s*)
(c) calcium sulfate (*s*)
(d) hydrogen iodide (*g*)

*36. Write thermochemical equations for the decomposition of one mole of the following compounds into the elements in their stable states at 25°C and 1 atm.

(a) potassium chlorate (*s*)      (b) hydrazine (*l*)
(c) dinitrogen tetraoxide (*g*)   (d) silver chloride (*s*)

37. Given

$$2HF(g) \longrightarrow H_2(g) + F_2(g) \qquad \Delta H° = 542.2 \text{ kJ}$$

(a) Determine the heat of formation of $HF(g)$.
(b) Calculate $\Delta H°$ for the formation of 12.00 g of HF.

38. Given

$$2BaCO_3(s) \longrightarrow 2Ba(s) + 2C(s) + 3O_2(g) \quad \Delta H° = 2432.6 \text{ kJ}$$

(a) Determine the heat of formation of $BaCO_3$.
(b) How many grams of $BaCO_3$ are formed if 100.0 kJ are involved?

39. A first-aid hot pack is made up of two compartments: one has solid calcium chloride, the other has water. When heat is required, you are instructed to twist the hot pack. This breaks the diaphragm separating both compartments. Calcium chloride dissolves in water, giving calcium ions, chloride ions, and heat. How much heat is evolved if the hot pack contains five grams of calcium chloride? (Use Tables 8.3 and 8.4.)

40. When ammonia reacts with oxygen, nitrogen oxide gas and steam are formed. How much heat is evolved or absorbed when one liter of ammonia gas at 23.0°C and 765 mm Hg reacts?

41. Use Tables 8.3 and 8.4 to obtain $\Delta H°$ for the following thermochemical equations:

(a) $Br_2(l) + H_2(g) + 2OH^-(aq) \longrightarrow 2Br^-(aq) + 2H_2O(l)$
(b) $3Fe^{2+}(aq) + 4H_2O(l) + CrO_4^{2-}(aq) \longrightarrow$
$\quad Cr(OH)_3(s) + 5OH^-(aq) + 3Fe^{3+}(aq)$
$\quad (\Delta H_f°$ for $Cr(OH)_3$ is $-1064.0$ kJ/mol.)
(c) $2NO(g) + 3MnO_2(s) + 4H^+(aq) \longrightarrow$
$\quad 3Mn^{2+}(aq) + 2NO_3^-(aq) + 2H_2O(l)$

42. Use Tables 8.3 and 8.4 to obtain $\Delta H°$ for the following thermochemical equations:

(a) $Br_2(l) + 2I^-(aq) \longrightarrow 2Br^-(aq) + I_2(s)$
(b) $Pb(s) + Sn^{2+}(aq) \longrightarrow Pb^{2+}(aq) + Sn(s)$
(c) $Ni^{2+}(aq) + SO_2(g) + 2H_2O(l) \longrightarrow$
$\quad SO_4^{2-}(aq) + 4H^+(aq) + Ni(s)$

43. Use the appropriate tables to calculate $\Delta H°$ for
(a) the reaction of one mole of nitrogen and steam to form ammonia and oxygen.
(b) the reaction between carbon dioxide and steam to form one mole of ethyl alcohol, $C_2H_5OH$, and oxygen.

44. Use the appropriate tables to calculate $\Delta H°$ for
(a) the reaction between two moles of nitrogen and water to form ammonia and oxygen.
(b) the reaction between chloride ions, hydrogen ions, and one mole of permanganate ions yielding chlorine gas, solid manganese dioxide, and water.

45. When one mole of ethylene gas, $C_2H_4$, is burned in oxygen and hydrogen chloride, the products are liquid ethylene chloride, $C_2H_4Cl_2$, and liquid water; 318.7 kJ of heat is evolved.
(a) Write a thermochemical equation for the reaction.
(b) Using Table 8.3, calculate the standard heat of formation of ethylene chloride.

46. When one mole of calcium carbonate reacts with ammonia, solid calcium cyanamide, $CaCN_2$, and liquid water are formed. The reaction absorbs 90.1 kJ of heat.
(a) Write a balanced thermochemical equation for the reaction.
(b) Using Table 8.3, calculate $\Delta H_f°$ for calcium cyanamide.

47. Use the appropriate tables to determine the heat of formation of $Cr_2O_3(s)$ given $\Delta H° = -1854.9$ kJ for the reaction

$$8H^+(aq) + Cr_2O_7^{2-}(aq) + 2Al(s) \longrightarrow$$
$$2Al^{3+}(aq) + Cr_2O_3(s) + 4H_2O(l)$$

48. Use the appropriate tables to determine the heat of formation of $Ni^{2+}(aq)$ given $\Delta H° = -1776.8$ kJ for the reaction

$$5Ni(s) + 12H^+(aq) + 2ClO_3^-(aq) \longrightarrow$$
$$5Ni^{2+}(aq) + Cl_2(g) + 6H_2O(l)$$

49. Glucose, $C_6H_{12}O_6(s)$ ($\Delta H_f° = -1275.2$ kJ/mol) is converted to ethyl alcohol, $C_2H_5OH(l)$, and carbon dioxide in the fermentation of grape juice. What quantity of heat is liberated when a liter of wine containing 12.0% ethyl alcohol by volume ($d = 0.789$ g/cm$^3$) is produced by the fermentation of grape juice?

50. When ammonia reacts with dinitrogen oxide gas ($\Delta H_f° = 82.05$ kJ/mol), liquid water and nitrogen gas are formed. How much heat is liberated or absorbed by the reaction that produces 275 mL of nitrogen gas at 25°C and 777 mm Hg?

## First Law of Thermodynamics

*51. Draw a cylinder with a movable piston containing six molecules of a liquid. A pressure of 1 atm is exerted on the piston. Next draw the same cylinder after the liquid has been vaporized. A pressure of one atmosphere is still exerted on the piston. Is work done on the system or by the system?

*52. Redraw the cylinder in Question 51 after work has been done on the system.

53. Calculate the work done (joules) by a chemical reaction if the volume increases from 2.00 L to 2.50 L against a constant pressure of 1.5 atm. (1 L · atm = 101 J)

54. Calculate the work done (joules) by a chemical reaction if the volume contracts from 1.75 L to 1.00 L against a constant pressure of 0.90 atm. (1 L · atm = 101 J)

55. Calculate
(a) $q$ when a system does 65 J of work and its energy

decreases by 88 J.

**(b)** $\Delta E$ for a gas that releases 42 J of heat and has 115 J of work done on it.

**56.** Find

**(a)** $\Delta E$ when a gas absorbs 22 J of heat and has 15 J of work done on it.

**(b)** $q$ when 83 J of work are done on a system and its energy is increased by 68 J.

**57.** For the vaporization of one mole of water at 100°C, determine

**(a)** $q_p = \Delta H$ **(b)** $q_v$ **(c)** the expansion work, $w_p$

**58.** Determine the difference in kilojoules at 25°C between $q_p$ and $q_v$ for

$$2CO(g) + O_2(g) \longrightarrow 2CO_2(g)$$

**59.** Consider the combustion of propane, $C_3H_8$, the fuel that is commonly used in portable gas barbeque grills. The products of combustion are carbon dioxide and liquid water.

**(a)** Write a thermochemical equation for the combustion of one mole of propane.

**(b)** How much heat would be evolved if 1.00 g of propane were burned in a bomb calorimeter at 25°C?

**60.** Consider the combustion of one mole of acetylene, $C_2H_2$, which yields carbon dioxide and liquid water.

**(a)** Using Table 8.3, calculate $\Delta H$.

**(b)** Calculate $\Delta E$ at 25°C.

### Unclassified

**61.** An 8-oz glass of beer has about 100 nutritional calories (1 nutritrional calorie = 1 kilocalorie). Approximately how many minutes of walking are required to burn up the energy taken in by a glass of beer? Walking uses up about 250 kcal/hour.

**62.** Given the following reactions

$$N_2H_4(l) + O_2(g) \longrightarrow N_2(g) + 2H_2O(g) \quad \Delta H° = -534.2 \text{ kJ}$$

$$H_2(g) + \tfrac{1}{2}O_2(g) \longrightarrow H_2O(g) \quad \Delta H° = -241.8 \text{ kJ}$$

calculate the heat of formation of hydrazine.

**63.** In World War II, the Germans made use of unusable airplane parts by grinding them up into powdered aluminum. This was made to react with ammonium nitrate to produce powerful bombs. The products of this reaction were nitrogen gas, steam, and aluminum oxide. If 5.00 kg of ammonium nitrate is mixed with 5.00 kg of powdered aluminum, how much heat is generated?

**64.** One way to lose weight is to exercise. Walking briskly at 4.0 mph for an hour consumes 1700 kJ of energy. How many miles would you have to walk to lose one pound of body fat? (One gram of body fat is equivalent to 32 kJ of energy.)

**65.** Brass has a density of 8.25 g/cm$^3$ and a specific heat of 0.362 J/g · °C. A cube of brass 19.00 mm on an edge is heated in a Bunsen-burner flame to a temperature of 95.0°C. It is then immersed into 20.0 mL of water ($d$ = 1.00 g/mL, $c$ = 4.18 J/g · °C) at 22.0°C in an insulated container. Assuming

no heat loss, what is the final temperature of the water?

**66.** Some solutes have large heats of solution, and care should be taken in preparing solutions of these substances. The heat evolved when sodium hydroxide dissolves is 44.5 kJ/mol. What is the final temperature of the water, originally at 20.0°C, used to prepare 750.0 cm$^3$ of 6.00 $M$ NaOH solution? Assume that all the heat is absorbed by 750.0 g of water, specific heat = 4.18 J/g · °C.

**67.** A 1.00-L sample of a gaseous mixture at 0°C and 1.00 atm (STP) evolves, upon complete combustion at constant pressure, 75.65 kJ of heat. If the gas is a mixture of ethane ($C_2H_6$) and propane ($C_3H_8$), what is the mole fraction of ethane in the mixture?

**68.** Microwave ovens convert radiation to energy. A microwave oven uses radiation with a wavelength of 12.5 cm. Assuming that all the energy from the radiation is converted to heat without loss, how many moles of photons are required to raise the temperature of a cup of water (350.0 g, specific heat = 4.18 J/g · °C) from 23.0°C to 99.0°C?

### Challenge Problems

**69.** On a hot day, you take a six-pack of soda on a picnic, cooling it with ice. Each empty (aluminum) can weighs 12.5 g and contains 12.0 oz of soda. The specific heat of aluminum is 0.902 J/g · °C; take that of soda to be 4.10 J/g · °C.

**(a)** How much heat must be absorbed from the six-pack to lower the temperature from 25.0 to 5.0°C?

**(b)** How much ice must be melted to absorb this amount of heat? ($\Delta H_{fus}$ of ice is given in Table 8.2.)

**70.** A cafeteria sets out glasses of tea at room temperature. The customer adds ice. Assuming that the customer wants to have some ice left when the tea cools at 0°C, what fraction of the total volume of the glass should be left empty for adding ice? Make any reasonable assumptions needed to work this problem.

**71.** The thermite reaction was once used to weld rails:

$$2Al(s) + Fe_2O_3(s) \longrightarrow Al_2O_3(s) + 2Fe(s)$$

**(a)** Calculate $\Delta H$ for this reaction using heat of formation data.

**(b)** Take the specific heats of $Al_2O_3$ and Fe to be 0.77 and 0.45 J/g·°C, respectively. Calculate the temperature to which the products of this reaction will be raised, starting at room temperature, by the heat given off in the reaction.

**(c)** Will the reaction produce molten iron (mp Fe = 1535°C, $\Delta H_{fus}$ = 270 J/g)?

**72.** A sample of sucrose, $C_{12}H_{22}O_{11}$, is contaminated by sodium chloride. When the contaminated sample is burned in a bomb calorimeter, sodium chloride does not burn. What is the percentage of sucrose in the sample if a temperature increase of 1.67°C is observed when 3.000 g of the sample is burned in the calorimeter? Sucrose gives off $5.64 \times 10^3$ kJ/mol when burned. The heat capacity of the calorimeter is 22.51 kJ/°C.

In about 95% of the minerals in the earth's crust, silicon atoms are bonded tetrahedrally to oxygen (p. 264). These include quartz, several forms of which (clear, smoky, yellow, rose) appear here. (Paul Silverman/Fundamental Photos, New York)

# 9 Liquids and Solids

See plastic nature working to this end,
The single atoms each to other tend.
Attract, attracted to, the next in place
Form'd and impell'd its neighbor to embrace.

—ALEXANDER POPE

In Chapter 5 we pointed out that at ordinary temperatures and pressures all gases follow the ideal gas law. Unfortunately, there is no simple "equation of state" that can be written to correlate the properties of liquids or solids. There are two reasons for this difference in behavior.

**1.** Molecules are much closer to one another in liquids and solids. In the gas state, particles are typically separated by ten molecular diameters or more; in liquids or solids, they touch one another. This explains why liquids and solids have densities so much larger than those of gases. At 100°C and 1 atm, $H_2O(l)$ has a density of 0.95 g/mL; that of $H_2O(g)$ under the same conditions is only 0.00059 g/mL. For the same reason, liquids and solids are much less compressible than gases. When the pressure on liquid water is increased from 1 to 2 atm, the volume decreases by 0.0045%; the same change in pressure reduces the volume of an ideal gas by 50%.

**2.** Intermolecular forces, which are essentially negligible with gases, play a much more important role in liquids and solids. Among the effects of these forces is the phenomenon of *surface tension,* shown by liquids. Molecules at the surface are attracted inward by unbalanced intermolecular forces; as a result, liquids tend to form spherical drops, for which the ratio of surface area to volume is as small as possible (Figure 9.1). Water, in which there is relatively strong hydrogen bonding, has a high surface tension. That of most organic liquids is lower, which explains why toluene tends to "wet" or spread out on solid surfaces more readily than water. The wetting ability of water can be increased by adding a soap or detergent, which drastically lowers the surface tension.

**Figure 9.1**

Surface tension causes water to bead on the polished surface of a car.

(Richard Megna/FUNDAMENTAL PHOTOGRAPHS, New York)

Our discussion of liquids and solids will focus upon

— the relationships among particle structure, interparticle forces, and physical properties. Section 9.1 explores these relationships for molecular substances; Section 9.2 extends the discussion to nonmolecular solids (network covalent, ionic, and metallic). Section 9.3 is devoted to the crystal structure of ionic and metallic solids.

— phase equilibria. Section 9.4 considers several different phenomena related to gas-liquid equilibria, including vapor pressure, boiling point behavior, and critical properties. Section 9.5 deals with *phase diagrams,* which describe all three types of phase equilibria (gas-liquid, gas-solid, and liquid-solid).

Bromine, sulfur, and sugar are molecular substances. (Richard Megna/Fundamental Photographs, New York)

# 9.1 Molecular Substances; Intermolecular Forces

As a class, substances composed of molecules tend to be

**1.** *Nonconductors of electricity when pure.* Since molecules are uncharged, they cannot carry an electric current. In most cases, (e.g., iodine, $I_2$, and ethyl alcohol, $C_2H_5OH$), water solutions of molecular substances are nonconductors. A few polar molecules, including HCl, react with water to form ions:

$$HCl(g) \longrightarrow H^+(aq) + Cl^-(aq)$$

and hence produce a conducting water solution.

**2.** *Insoluble in water but soluble in nonpolar solvents such as $CCl_4$ or toluene.* Iodine is typical of most molecular substances; it is only slightly soluble in water (0.0013 mol/L at 25°C), much more soluble in toluene (0.5 mol/L). A few molecular substances, including ethyl alcohol, are very soluble in water. As you will see later in this section, such substances have intermolecular forces similar to those in water.

**3.** *Low melting and boiling.* Many molecular substances are gases at 25°C and 1 atm (for example, $N_2$, $O_2$, and $CO_2$), which means that they have boiling points below 25°C. Others (such as $H_2O$ and $CCl_4$) are liquids with melting (freezing) points below room temperature. Of the molecular substances that are solids at ordinary temperatures, most are low melting (mp $I_2$ = 114°C, mp naphthalene = 80°C). The upper limit for melting and boiling points of most molecular substances is about 300°C.

The generally low melting and boiling points of molecular substances reflect the fact that the forces between molecules (**intermolecular forces**) are weak. To melt or boil a molecular substance, the molecules must be set free from one another. This requires only that enough energy be supplied to overcome the weak attractive forces between molecules. The strong covalent bonds within molecules remain intact when a molecular substance melts or boils.

The boiling points of different molecular substances are directly related to the strength of the intermolecular forces involved. *The stronger the intermolecular forces, the higher the boiling point of the substance.* In the remainder of this section, we examine the nature of the three different types of intermolecular forces: *dispersion forces, dipole forces,* and *hydrogen bonds.*

It's much easier to separate two molecules than to separate two atoms within a molecule

## Dispersion (London) Forces

The most common type of intermolecular force, found in all molecular substances, is referred to as a **dispersion force.** It is basically electrical in nature, involving an attraction between temporary or *induced* dipoles in adjacent molecules. To understand the origin of dispersion forces, consider Figure 9.2.

**Figure 9.2**
Temporary dipoles in adjacent $H_2$ molecules line up to create an electrical attractive force known as the dispersion force. Deeply shaded areas indicate regions where the electron cloud is momentarily concentrated.

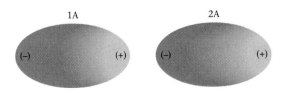

TABLE 9.1   **Effect of Molar Mass on Boiling Points of Molecular Substances**

| Noble Gases* | | | Halogens | | | Hydrocarbons | | |
|---|---|---|---|---|---|---|---|---|
| | $\mathcal{M}$ (g/mol) | bp (°C) | | $\mathcal{M}$ (g/mol) | bp (°C) | | $\mathcal{M}$ (g/mol) | bp (°C) |
| He | 4 | −269 | $F_2$ | 38 | −188 | $CH_4$ | 16 | −161 |
| Ne | 20 | −246 | $Cl_2$ | 71 | −34 | $C_2H_6$ | 30 | −88 |
| Ar | 40 | −186 | $Br_2$ | 160 | 59 | $C_3H_8$ | 44 | −42 |
| Kr | 84 | −152 | $I_2$ | 254 | 184 | $n$-$C_4H_{10}$ | 58 | 0 |

*Strictly speaking, the noble gases are "atomic" rather than molecular. However, like molecules, the noble gas atoms are attracted to one another by dispersion forces.

On the average, electrons in a nonpolar molecule, such as $H_2$, are as close to one nucleus as to the other. However, at a given instant, the electron cloud may be concentrated at one end of the molecule (position 1A in Fig. 9.2). This momentary concentration of the electron cloud on one side of the molecule creates a temporary dipole in $H_2$. One side of the molecule, shown in deeper color in Figure 9.2, acquires a partial negative charge; the other side has a partial positive charge of equal magnitude.

This temporary dipole induces a similar dipole (an induced dipole) in an adjacent molecule. When the electron cloud in the first molecule is at 1A, the electrons in the second molecule are attracted to 2A. These temporary dipoles, both in the same direction, lead to an attractive force between the molecules. This is the dispersion force.

All molecules have dispersion forces. The strength of these forces depends upon two factors:

— The number of electrons in the atoms that make up the molecule.
— The ease with which electrons are *dispersed* to form temporary dipoles.

Both these factors increase with increasing molecular size. Large molecules are made up of more and/or larger atoms. The outer electrons in larger atoms are relatively far from the nucleus and are easier to disperse than the electrons in small atoms. In general, molecular size and molar mass parallel one another. Thus *as molar mass increases, dispersion forces become stronger, and the boiling point of nonpolar molecular substances increases* (Table 9.1).

Dispersion forces are weak in $H_2$, much stronger in $I_2$

## Dipole Forces

**Dipole forces** arise as a result of the interaction between polar molecules. The nature of this interaction is illustrated in Figure 9.3, p. 242, which shows the orientation of polar molecules, such as ICl, in a crystal. Adjacent molecules line up so that the negative pole of one molecule (small Cl atom) is as close as possible to the positive pole (large I atom) of its neighbor. Under these conditions, there is an electrical attractive force, referred to as a dipole force, between adjacent polar molecules.

When iodine chloride is heated to 27°C, the rather weak dipole forces are unable to keep the molecules rigidly aligned, and the solid melts. Dipole forces

**Figure 9.3**
Dipole forces in the ICl crystal. The (+) and (−) indicate fractional charges on the I and Cl atoms in the polar molecules. The existence of these partial charges causes the molecules to line up in the pattern shown. Adjacent molecules are attracted to each other by dipole forces between the (+) pole of one molecule and the (−) pole of the other.

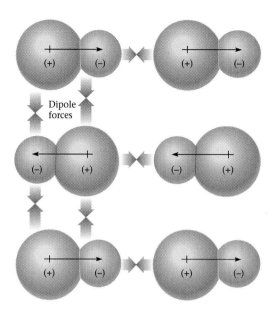

are still important in the liquid state, since the polar molecules remain close to one another. Only in the gas, where the molecules are far apart, do the effects of dipole forces become negligible. Hence, boiling points as well as melting points of polar compounds such as ICl are somewhat higher than those of nonpolar substances of comparable molar mass. This effect is shown in Table 9.2.

**Example 9.1**    Explain, in terms of intermolecular forces, why

(a) the boiling point of $O_2$ (−183°C) is higher than that of $N_2$ (−196°C).
(b) the boiling point of NO (−151°C) is higher than that of either $O_2$ or $N_2$.

***Strategy***    Determine whether the molecule is polar or nonpolar and identify the intermolecular forces present. Remember, dispersion forces are always present and increase with molar mass.

***Solution***

(a) Only dispersion forces are involved with these nonpolar molecules. Since the molar mass of $O_2$ is greater (32.0 g/mol versus 28.0 g/mol for $N_2$), its   dispersion forces are stronger,   making its boiling point higher.

(b) Dispersion forces in NO (30 g/mol) are comparable in strength to those in $O_2$ and $N_2$. The   polar NO molecule shows   an additional type of intermolecular force not present in $N_2$ or $O_2$;   the dipole force.   As a result, its boiling point is the highest of the three substances.

## Hydrogen Bonds

*Usually, dispersion forces are stronger than dipole forces*

Ordinarily, polarity has a relatively small effect on boiling point. In the series HCl ⟶ HBr ⟶ HI, boiling point increases steadily with molar mass, even

TABLE 9.2 **Boiling Points of Nonpolar versus Polar Substances**

| Nonpolar | | | Polar | | |
|---|---|---|---|---|---|
| Formula | $\mathcal{M}$ (g/mol) | bp (°C) | Formula | $\mathcal{M}$ (g/mol) | bp (°C) |
| $N_2$ | 28 | −196 | CO | 28 | −192 |
| $SiH_4$ | 32 | −112 | $PH_3$ | 34 | −88 |
| $GeH_4$ | 77 | −90 | $AsH_3$ | 78 | −62 |
| $Br_2$ | 160 | 59 | ICl | 162 | 97 |

though polarity decreases moving from HCl to HI. However, when hydrogen is bonded to a small, highly electronegative atom (N, O, F), polarity has a much greater effect upon boiling point. Hydrogen fluoride, HF, despite its low molar mass (20 g/mol), has the highest boiling point of all the hydrogen halides. Water ($\mathcal{M}$ = 18 g/mol) and ammonia ($\mathcal{M}$ = 17 g/mol) also have abnormally high boiling points (Table 9.3). In these cases, the effect of polarity reverses the normal trend expected from molar mass alone.

The unusually high boiling points of HF, $H_2O$, and $NH_3$ result from an unusually strong type of dipole force called a **hydrogen bond.** The hydrogen bond is a force exerted between an H atom bonded to an F, O, or N atom in one molecule and an unshared pair on the F, O, or N atom of a neighboring molecule:

$$: X—H \text{ ---- } : X—H \qquad X = N, O, \text{ or } F$$

↑
hydrogen bond

There are two reasons that hydrogen bonds are stronger than ordinary dipole forces:

**1.** The difference in electronegativity between hydrogen (2.2) and fluorine (4.0), oxygen (3.5), or nitrogen (3.0) is quite large. It causes the bonding electrons in molecules such as HF, $H_2O$, and $NH_3$ to be primarily associated with the more electronegative atom (F, O, or N). So the hydrogen atom, insofar as its interaction with a neighboring molecule is concerned, behaves almost like a bare proton.

**2.** The small size of the hydrogen atom allows the unshared pair of an F, O, or N atom of one molecule to approach the H atom in another very closely. It is significant that hydrogen bonding occurs only with these three nonmetals, all of which have small atomic radii.

TABLE 9.3 **Effect of Hydrogen Bonding on Boiling Point**

| | bp(°C) | | bp(°C) | | bp(°C) |
|---|---|---|---|---|---|
| $NH_3$ | −33 | $H_2O$ | 100 | HF | 19 |
| $PH_3$ | −88 | $H_2S$ | −60 | HCl | −85 |
| $AsH_3$ | −63 | $H_2Se$ | −42 | HBr | −67 |
| $SbH_3$ | −18 | $H_2Te$ | −2 | HI | −35 |

*Molecules in color show hydrogen bonding.

Hydrogen bonding is very important in proteins (Ch. 22)

Hydrogen bonds can exist in many molecules other than HF, $H_2O$, and $NH_3$. The basic requirement is simply that hydrogen be bonded to a fluorine, oxygen, or nitrogen atom with at least one unshared pair. Consider, for example, the two compounds whose condensed structural formulas are

$$CH_3CH_2CH_2 - \overset{\overset{\displaystyle H}{|}}{N} - H \qquad\qquad CH_3 - \overset{\overset{\displaystyle CH_3}{|}}{N} - CH_3$$

propylamine                  trimethylamine

Propylamine, in which there are hydrogen atoms bonded to nitrogen, can show hydrogen bonding; it is a liquid with a normal boiling point of 49°C. Trimethylamine, which like propylamine has the molecular formula $C_3H_9N$, cannot hydrogen bond to itself; at 25°C and 1 atm, it is a gas.

---

**Example 9.2**    Would you expect to find hydrogen bonds in

(a) ethyl alcohol?

$$H - \overset{\overset{\displaystyle H}{|}}{\underset{\underset{\displaystyle H}{|}}{C}} - \overset{\overset{\displaystyle H}{|}}{\underset{\underset{\displaystyle H}{|}}{C}} - \overset{..}{\underset{..}{O}} - H$$

(b) dimethyl ether?

$$H - \overset{\overset{\displaystyle H}{|}}{\underset{\underset{\displaystyle H}{|}}{C}} - \overset{..}{\underset{..}{O}} - \overset{\overset{\displaystyle H}{|}}{\underset{\underset{\displaystyle H}{|}}{C}} - H$$

(c) hydrazine, $N_2H_4$?

*Strategy*    For hydrogen bonding to occur, hydrogen must be bonded to F, O, or N. In (c), draw the Lewis structure first.

*Solution*

(a) There should be hydrogen bonds in ethyl alcohol, since an H atom is bonded to oxygen.

(b) Since all the hydrogen atoms are bonded to carbon in dimethyl ether, there should be no hydrogen bonds.

(c) The Lewis structure of hydrazine is

$$H - \overset{..}{N} - \overset{..}{N} - H$$
$$\underset{\underset{\displaystyle H}{|}}{\phantom{H - }} \underset{\underset{\displaystyle H}{|}}{\phantom{N}}$$

Since hydrogen is bonded to nitrogen, hydrogen bonding can occur between neighboring $N_2H_4$ molecules. The boiling point of hydrazine (114°C) is much higher than that of molecular $O_2$ ($-183$°C), which has the same molar mass.

---

If ice were more dense than water, skating would be a lot less popular

Water has many unusual properties in addition to its high boiling point. In particular, water, in contrast to nearly all other substances, expands upon freezing. Ice is one of the very few solids that has a density less than that of the liquid from which it is formed ($d$ ice at 0°C = 0.917 g/cm$^3$; $d$ water at 0°C = 1.000 g/cm$^3$). This behavior is an indirect result of hydrogen bonding. When water freezes to ice, an open hexagonal pattern of molecules results (Fig. 9.4). Each oxygen atom in an ice crystal is bonded to four hydrogens. Two of these

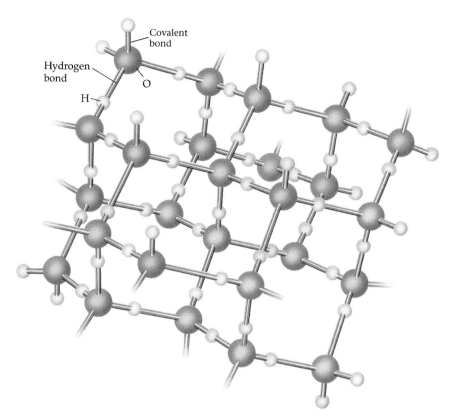

**Figure 9.4**
In ice, the water molecules are arranged in an open hexagonal pattern that gives ice its low density. Each oxygen atom (red) is bonded covalently to two hydrogen atoms (off-white) and forms hydrogen bonds with two other hydrogen atoms.

are attached by ordinary covalent bonds at a distance of 0.099 nm. The other two form hydrogen bonds 0.177 nm in length. The large proportion of "empty space" in the ice structure explains why ice is less dense than water.

---

**Example 9.3**   What types of intermolecular forces are present in $H_2$? $CCl_4$? SCO? $NH_3$?

**Strategy**   Determine whether the molecules are polar or nonpolar; only polar molecules show dipole forces. Check the Lewis structures for H atoms bonded to F, O, or N. All molecules have dispersion forces.

**Solution**   The $H_2$ and $CCl_4$ molecules are nonpolar (Chap. 7). Thus, only dispersion forces are present among neighboring molecules. Both SCO and $NH_3$ are polar molecules:

$$: \overset{\cdot\cdot}{\underset{}{S}} = C = \overset{\cdot\cdot}{\underset{}{O}} :   \qquad  \underset{H \;\; \underset{H}{|} \;\; H}{\overset{\cdot\cdot}{N}}$$

There are dipole forces as well as dispersion forces with SCO molecules. In $NH_3$, there are hydrogen bonds and dispersion forces.

---

We have now discussed three types of intermolecular forces: dispersion forces, dipole forces, and hydrogen bonds. You should bear in mind that all these forces are relatively weak compared with ordinary covalent bonds. Consider, for example, the situation in $H_2O$. The total intermolecular attractive energy in ice is about 50 kJ/mol; the O—H bond energy is an order of magnitude greater, about 464 kJ/mol. This explains why it is a lot easier to melt ice or boil water than it is to decompose water into the elements. Even at a temperature of 1000°C and 1 atm, only about one $H_2O$ molecule in a billion decomposes to hydrogen and oxygen atoms.

## 9.2  Network Covalent, Ionic, and Metallic Solids

Virtually all substances that are gases or liquids at 25°C and 1 atm are molecular. In contrast, there are at least three types of nonmolecular solids (Fig. 9.5). These are

— **network covalent solids,** in which atoms are joined by a continuous network of covalent bonds. The entire crystal, in effect, consists of one huge molecule.
— **ionic solids,** held together by strong electrical forces (ionic bonds) between oppositely charged ions adjacent to one another
— **metallic solids,** in which the structural units are electrons ($e^-$) and cations ($M^+$)

### Network Covalent Solids

As a class, solids of this type are

1. *High melting, often with melting points about 1000°C.* To melt the solid, covalent bonds between atoms must be broken. In this respect, solids of this type differ markedly from molecular solids, which have much lower melting points.
2. *Insoluble in all common solvents.* For solution to occur, covalent bonds throughout the solid would have to be broken.
3. *Poor electrical conductors.* In most network covalent substances (graphite is an exception), there are no mobile electrons to carry a current.

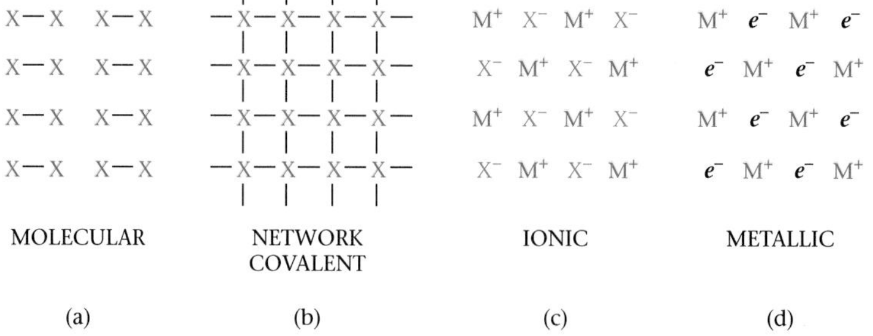

**Figure 9.5**
Schematic diagrams of four types of substances (see text discussion). $M^+$ represents a cation, $X^-$ an anion, X a nonmetal atom, and $e^-$ an electron. Covalent bonds are shown by straight lines joining atoms.

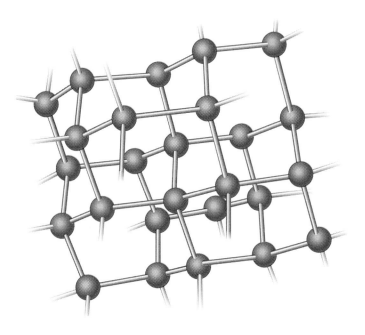

**Figure 9.6**
Diamond has a three-dimensional structure in which each carbon atom is surrounded tetrahedrally by four other carbon atoms.

### Graphite and Diamond

Several nonmetallic elements have a network covalent structure. The most important of these is carbon, which has two different crystalline forms of the network covalent type. Both graphite and diamond have high melting points, above 3500°C. However, the bonding patterns in the two solids are quite different.

In diamond, each carbon atom forms single bonds with four other carbon atoms arranged tetrahedrally around it. The hybridization in diamond is $sp^3$. The three-dimensional covalent bonding contributes to diamond's unusual hardness. Diamond is one of the hardest substances known; it is used in cutting tools and quality grindstones (Fig. 9.6).

Graphite is planar, with the carbon atoms arranged in a hexagonal pattern. Each carbon atom is bonded to three others, two by single bonds, one by a double bond. The hybridization is $sp^2$. The forces between adjacent layers in graphite are of the dispersion type and are quite weak. A "lead" pencil really contains a graphite rod, thin layers of which rub off onto the paper as you write (Fig. 9.7).

An Afro-American engineer, Elija McCoy (1843–1927) was the first to use graphite as an industrial lubricant

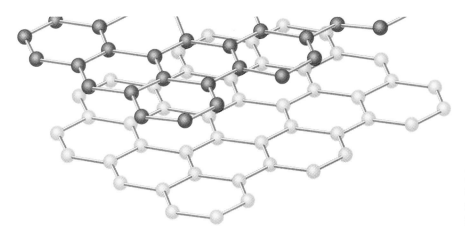

**Figure 9.7**
Graphite has a two-dimensional layer structure with weak dispersion forces between the layers.

At room temperature and atmospheric pressure, graphite is the stable form of carbon. Diamond, in principle, should slowly transform to graphite under ordinary conditions. Fortunately, for the owners of diamond rings, this transition occurs at zero rate unless the diamond is heated to about 1500°C, at which temperature the conversion occurs rapidly. For understandable reasons, no one has ever become very excited over the commercial possibilities of this process. The more difficult task of converting graphite to diamond has aroused much greater enthusiasm.

Diamond has a higher density than graphite (3.51 versus 2.26 g/cm³); as the more dense phase, it should be favored by high pressures. Theoretically, at 25°C and 15,000 atm graphite should turn to diamond. However, under those conditions the reaction has a negligible rate. At higher temperatures it goes faster, but the required pressure goes up too; at 2000°C, a pressure of about 100,000 atm is needed. In 1954, scientists at the General Electric laboratories were able to achieve these high temperatures and pressures and converted graphite carbon to diamond. Nowadays, all industrial diamonds are synthetic.

*Pure research paid off handsomely here*

### Quartz

Among the many compounds with a network covalent structure is quartz, the most common form of $SiO_2$ and the major component of sand. In quartz, each silicon atom bonds tetrahedrally to four oxygen atoms. Each oxygen atom bonds to two silicons, thus linking adjacent tetrahedra to one another (Fig. 9.8). Notice that the network of covalent bonds extends throughout the entire crystal. Unlike most pure solids, quartz does not melt sharply to a liquid. Instead, it turns to a viscous mass over a wide temperature range, first softening at about 1400°C. The viscous fluid probably contains long —Si—O—Si—O— chains, with enough bonds broken to allow flow.

Ordinary glass is made by heating a mixture of sand ($SiO_2$) limestone ($CaCO_3$), and soda ash ($Na_2CO_3$) to the melting point. Carbon dioxide is given off from the hot mass, which contains the oxides of silicon, sodium, and cal-

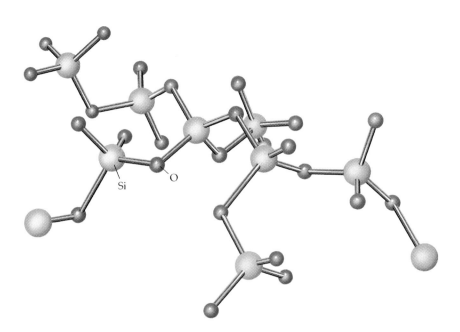

**Figure 9.8**

Crystal structure of quartz, $SiO_2$. Every silicon atom (off-white) is linked tetrahedrally to four oxygen atoms (red).

cium in a mole ratio of about $7:1:1$. The "soft" glass produced this way softens at about 600°C and has a wider range of high viscosity than quartz. Hard glass, called Pyrex or Kimax, is made from a melt of silicon, boron, aluminum, sodium, and potassium oxides. It is much superior to soft glass in withstanding chemical attack and thermal shock. Hard glass softens at about 800°C and is readily worked in the flame of a gas-oxygen torch. Nearly all the glassware used in chemical equipment nowadays is made of hard glass.

## Ionic Solids

An ionic solid consists of cations and anions (e.g., $Na^+$, $Cl^-$). No simple, discrete molecules are present in NaCl or other ionic compounds; rather, the ions are held in a regular, repeating arrangement by strong **ionic bonds,** electrostatic interactions between oppositely charged ions. Because of this structure, ionic solids show the following properties:

1. *Ionic solids are nonvolatile and high-melting* (typically at 600°C to 2000°C). Ionic bonds must be broken to melt the solid, separating oppositely charged ions from each other. Only at high temperatures do the ions acquire enough kinetic energy for this to happen.
2. *Ionic solids do not conduct electricity because the charged ions are fixed in position* (recall Figure 2.9, p. 40). They become good conductors, however, when melted or dissolved in water. In both cases, in the melt or solution, the ions (such as $Na^+$ and $Cl^-$) are free to move through the liquid and thus can conduct an electric current.
3. *Many, but not all, ionic compounds* (for example, NaCl but not $CaCO_3$) *are soluble in water, a polar solvent.* In contrast, ionic compounds are insoluble in nonpolar solvents such as benzene ($C_6H_6$) or carbon tetrachloride ($CCl_4$).

Charged particles must move to carry a current

The relative strengths of different ionic bonds can be estimated from **Coulomb's law,** which gives the electrical energy of interaction between a cation and anion in contact with one another:

$$E = \frac{k \times Q_1 \times Q_2}{d}$$

Here, $Q_1$ and $Q_2$ are the charges of anion and cation, and $d$, the distance between the centers of the two ions, is the sum of the ionic radii:

$$d = r_{cation} + r_{anion}$$

The quantity $k$ is a constant whose magnitude need not concern us here. Since the cation and anion have opposite charges, $E$ is a negative quantity. This makes sense; energy is evolved when two oppositely charged ions, originally far apart with $E = 0$, approach one another closely. Conversely, energy has to be absorbed to separate the ions from each other.

From Coulomb's law, the strength of the ionic bond should depend upon two factors:

1. *The charges of the ions.* The bond in CaO ($+2$, $-2$ ions) is considerably stronger than that in NaCl ($+1$, $-1$ ions). This explains why the melting point of calcium oxide (2927°C) is so much higher than that of sodium chloride (801°C).

The steel sphere, aluminum cylinder, gold chain, and copper wire, are all metals and have metallic bonds.

The band theory of metals is described in Appendix 5

**2.** *The size of the ions.* The ionic bond in NaCl (mp = 801°C) is somewhat stronger than that in KBr (mp = 734°C) because the internuclear distance is smaller in NaCl:

$$d_{NaCl} = r_{Na+} + r_{Cl-} = 0.095 \text{ nm} + 0.181 \text{ nm} = 0.276 \text{ nm}$$

$$d_{KBr} = r_{K+} + r_{Br-} = 0.133 \text{ nm} + 0.195 \text{ nm} = 0.328 \text{ nm}$$

## Metals

Figure 9.5d (p. 246) illustrates a simple model of bonding in metals known as the **electron-sea model.** The metallic crystal is pictured as an array of positive ions, e.g., $Na^+$, $Mg^{2+}$. These are anchored in position, like buoys in a mobile "sea" of electrons. These electrons are not attached to any particular positive ion, but rather can wander through the crystal. The electron-sea model explains many of the characteristic properties of metals:

**1.** *High electrical conductivity.* The presence of large numbers of relatively mobile electrons explains why metals have electrical conductivities several hundred times greater than those of typical nonmetals. Silver is the best electrical conductor but is too expensive for general use. Copper, with a conductivity close to that of silver, is the metal most commonly used for electrical wiring. Although a much poorer conductor than copper, mercury is used in many electrical devices, such as "silent" light switches, where a liquid conductor is required.

**2.** *High thermal conductivity.* Heat is carried through metals by collisions between electrons, which occur frequently. Saucepans used for cooking commonly contain aluminum, copper, or stainless steel; their handles are made of a nonmetallic material that is a good thermal insulator.

**3.** *Ductility and malleability.* Most metals are ductile (capable of being drawn out into a wire) and malleable (capable of being hammered into thin sheets; Fig. 9.9). In a metal, the electrons act like a flexible glue holding the atomic nuclei together. As a result, metal crystals can be deformed without shattering.

**4.** *Luster.* Polished metal surfaces reflect light. Most metals have a silvery white "metallic" color because they reflect light of all wavelengths. Since electrons are not restricted to a particular bond, they can absorb and re-emit light over a wide wavelength range. Gold and copper absorb some light in the blue region of the visible spectrum and so appear yellow (gold) or red (copper).

**5.** *Insolubility in water and other common solvents.* No metals dissolve in water; electrons cannot go into solution, and cations cannot dissolve by themselves. The only liquid metal, mercury, dissolves many metals, forming solutions called *amalgams.* An Ag–Sn–Hg amalgam is used in filling teeth.

In general, the melting points of metals cover a wide range, from −39°C for mercury to 3410°C for tungsten. This variation in melting point corresponds to a similar variation in the strength of the metallic bond. Generally speaking, the lowest-melting metals are those that form +1 cations, like sodium (mp = 98°C) and potassium (mp = 64°C).

Much of what has been said about the four structural types of solids in Sections 9.1 and 9.2 is summarized in Table 9.4 and Example 9.4

**Figure 9.9**
The blacksmith hammers iron into sheets, taking advantage of its malleability. (Linda Dufurrern/Grant Heilman, Inc.)

**TABLE 9.4    Structures and Properties of Types of Substances**

| Type | Structural Particles | Forces Within Particles | Forces Between Particles | Properties | Examples |
|---|---|---|---|---|---|
| Molecular | molecules | | | | |
| (a) nonpolar | | covalent bond | dispersion | low mp, bp; often gas or liquid at 25°C; nonconductors; insoluble in water, soluble in organic solvents | $H_2$ $CCl_4$ |
| (b) polar | | covalent bond | dispersion, dipole, H bond | similar to nonpolar but generally higher mp and bp, more likely to be water-soluble | HCl $NH_3$ |
| Network covalent | atoms | — | covalent bond | hard, very high-melting solids; nonconductors; insoluble in common solvents | C $SiO_2$ |
| Ionic | ions | — | ionic bond | high mp; conductors in molten state or water solution; often soluble in water, insoluble in organic solvents | NaCl MgO $CaCO_3$ |
| Metallic | cations, mobile electrons | — | metallic bond | variable mp; good conductors in solid; insoluble in common solvents | Na Fe |

**Example 9.4**    A certain substance is a liquid at room temperature and is insoluble in water. Suggest which of the four types of basic structural units is present in the substance and list additional experiments that could be carried out to confirm your prediction.

***Strategy***    Use Table 9.4, which shows the general properties of the four different types of substances. Note that a given property (e.g., high electrical conductivity) may be shown by more than one type of substance.

***Solution***    The fact that the substance is a liquid suggests that it is probably molecular. The fact that it is insoluble in water agrees with this classification. To confirm that it is molecular, measure the conductivity, which should be zero. This test eliminates mercury, a liquid metal.

## 9.3   Crystal Structures

Solids tend to crystallize in definite geometric forms that often can be seen by the naked eye. In ordinary table salt, cubic crystals of NaCl are clearly visible. Large, beautifully formed crystals of such minerals as fluorite, $CaF_2$, are found in nature. It is possible to observe distinct crystal forms of many metals under a microscope.

Crystals have definite geometric forms because the atoms or ions present are arranged in a definite, three-dimensional pattern. The nature of this pattern can be deduced by a technique known as x-ray diffraction. The basic informa-

Naturally occurring crystals. They are (clockwise from upper right): calcite, fluorite, garnet, beryl, and barite (center).

tion that comes out of such studies has to do with the dimensions and geometric form of the **unit cell,** the smallest structural unit that, repeated over and over again in three dimensions, generates the crystal.

## Metals

Three of the simpler unit cells found in metals, shown in Figure 9.10, are the following:

1. **Simple cubic cell. (SC)** This is a cube that consists of eight atoms whose centers are located at the corners of the cell. Atoms at adjacent corners of the cube touch one another.
2. **Face-centered cubic cell (FCC).** Here there is an atom at each corner of the cube and one in the center of each of the six faces of the cube. In this structure, atoms at the corners of the cube do not touch one another; they are forced slightly apart. Instead, contact occurs along a face diagonal. The atom at the center of each face touches atoms at opposite corners of the face.
3. **Body-centered cubic cell (BCC).** This is a cube with atoms at each corner and one in the center of the cube. Here again, corner atoms do not touch each other. Instead, contact occurs along the body diagonal; the atom at the center of the cube touches atoms at opposite corners.

If the nature and dimensions of the unit cell of a metallic crystal are known, the atomic radius of the metal can be calculated. It should be evident from Figure 9.11 and the Pythagorean theorem

$$a^2 = b^2 + c^2$$

(*a* is the hypotenuse of a right triangle; *b* and *c* are the other two sides) that the radii (*r*) and sides (*s*) of the unit cells are related:

Simple cubic          Body-centered cubic          Face-centered cubic

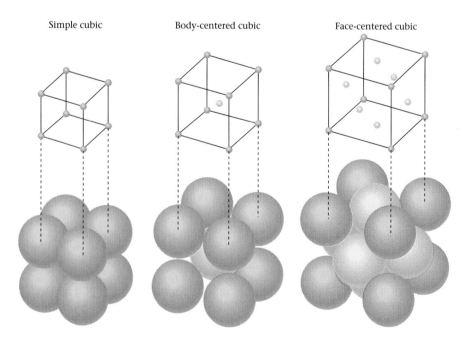

**Figure 9.10**

Three types of unit cells. In each case, there is an atom at each of the eight corners of the cube. In the body-centered lattice, there is an additional atom in the center of the cube. In the face-centered lattice, there is an atom in the center of each of the six faces.

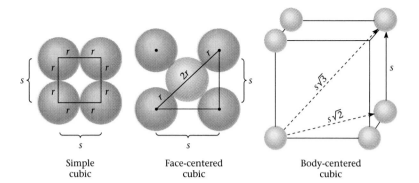

**Figure 9.11**
Relation between atomic radius, *r*, and length of edge, *s*, for cubic cells. Note that:
—in the FCC cell, the face diagonal has a length of 4*r*, hence, $(4r)^2 = s^2 + s^2 = 2s^2$; $4r = s\sqrt{2}$
—in the BCC cell, the body diagonal has a length of 4*r*. It is also the hypotenuse of a right triangle whose other sides are a face diagonal $(s\sqrt{2})$ and an edge (*s*), hence, $(4r)^2 = 2s^2 + s^2 = 3s^2$; $4r = s\sqrt{3}$

Simple cubic    Face-centered cubic    Body-centered cubic

$$\text{simple cubic cell:} \quad 2r = s$$
$$\text{face-centered cubic cell:} \quad 4r = s\sqrt{2}$$
$$\text{body-centered cubic cell:} \quad 4r = s\sqrt{3}$$

---

**Example 9.5**   Silver crystallizes with a face-centered cubic (FCC) unit cell 0.407 nm on an edge. Calculate the atomic radius of silver.

***Strategy***   Find the appropriate relation between the radius and the length of a side of an FCC cell. Substitute 0.407 nm for *s*.

***Solution***   The equation is

$$4r = s\sqrt{2}$$

Substituting *s* = 0.407 nm and solving for *r*:

$$r = \frac{0.407 \text{ nm} \times \sqrt{2}}{4} = \frac{0.407 \text{ nm} \times 1.41}{4} = \boxed{0.144 \text{ nm}}$$

---

When the atoms in a crystal are all of the same size, as with metals, they tend to pack closely together. From this point of view, the simple cubic structure is very unstable, since nearly half of its total volume consists of empty space. A body-centered cubic structure has less waste space; about 20 metals have this type of unit cell. A still more efficient way of packing spheres of the same size is the face-centered cubic structure, where the fraction of empty space is only 0.26. About 40 different metals have a structure based on either the face-centered cubic cell or a close relative in which the packing is equally efficient (hexagonal closest-packed structure).

*Golf balls and oranges pack naturally in an FCC structure*

## Ionic Crystals

The geometry of ionic crystals, in which there are two different kinds of ions, is more difficult to describe than that of metals. However, in many cases the packing can be visualized in terms of the unit cells described above. Lithium chloride, LiCl, is a case in point. Here, the larger $Cl^-$ ions form a face-centered

**Figure 9.12**

Three types of lattices in ionic crystals. In LiCl, the Cl$^-$ ions are in contact with each other, forming a face-centered cubic lattice. In NaCl, the Cl$^-$ ions are forced slightly apart by the larger Na$^+$ ions. In CsCl, the structure is quite different; the large Cs$^+$ ion at the center touches Cl$^-$ ions at each corner of the cube.

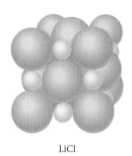

LiCl

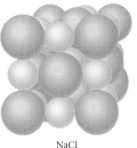

NaCl

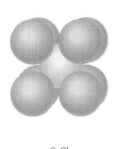

CsCl

cubic lattice (Fig. 9.12). The smaller Li$^+$ ions fit into "holes" between the Cl$^-$ ions. This puts a Li$^+$ ion at the center of each edge of the cube.

In the sodium chloride crystal, the Na$^+$ ion is slightly too large to fit into holes in a face-centered lattice of Cl$^-$ ions (Fig. 9.12). As a result, the Cl$^-$ ions are pushed slightly apart so that they are no longer touching, and only Na$^+$ ions are in contact with Cl$^-$ ions. However, the relative positions of positive and negative ions remain the same as in LiCl: each anion is surrounded by six cations and each cation by six anions.

NaCl is FCC in both Na$^+$ and Cl$^-$ ions; CsCl is BCC in both Cs$^+$ and Cl$^-$ ions

The structures of LiCl and NaCl are typical of all the alkali halides (Group 1 cation, group **17** anion) except those of cesium. Because of the large size of the Cs$^+$ ion, CsCl crystallizes in a quite different structure. Here, each Cs$^+$ ion is located at the center of a simple cube outlined by Cl$^-$ ions. The Cs$^+$ ion at the center touches all the Cl$^-$ ions at the corners; the Cl$^-$ ions do not touch each other. As you can see, each Cs$^+$ ion is surrounded by eight Cl$^-$ ions, while each Cl$^-$ ion is surrounded by eight Cs$^+$ ions.

# Chemistry: *The Human Side*

The crystal structures discussed in this section were determined by a powerful technique known as x-ray diffraction (Figure 9.A). By studying the pattern produced when the scattered rays strike a target, it is possible to deduce the geometry of the unit cell. With molecular crystals, one can go a step further, identifying the geometry and composition of the molecule. A father-son team of two English physicists, William H. and William L. Bragg, won the Nobel Prize in 1915 for pioneer work in this area.

The Braggs and their successors strongly sought out talented women scientists to develop the new field of x-ray crystallography. Foremost among these women was Dorothy Crowfoot Hodgkin, who spent almost all of her professional career at Oxford University in England. As the years passed she unraveled the structures of successively more complex natural products. These included penicillin, which she studied between 1942 and 1949, vitamin B-12 (1948–1957), and her greatest triumph, insulin, which she worked on for more than 30 years. Among her students at Oxford was Margaret Thatcher, the future prime minister. In 1964, Dorothy Hodgkin became the third (and, so far, the last) woman to win the Nobel Prize in chemistry.*

Hodgkin's accomplishments were all the more remarkable when you consider some of the obstacles she had to overcome. At the age of 24, she developed rheumatoid arthritis, a crippling disease of the immune system. Gradually she lost the use of her

*To learn more about Dorothy Hodgkin and other women Nobel laureates, we recommend the book *Nobel Prize Women in Science* by Sharon Bertsch McGrayne, published in 1993.

hands; ultimately she was confined to a wheelchair. Dorothy Hodgkin succeeded because she combined a first-rate intellect with an almost infinite capacity for hard work. Beyond that, she was one of those rare individuals who inspire loyalty by taking genuine pleasure in the successes of other people. A colleague referred to her as "the gentle genius."

Dorothy's husband, Thomas Hodgkin, was a scholar in his own right with an interest in the history of Africa. Apparently realizing that hers was the greater talent, he acted as a "house-husband" for their three children so she would have more time to devote to research. One wonders how Dorothy and Thomas Hodgkin reacted to the 1964 headline in a London tabloid, "British Wife Wins Nobel Prize."

**Dorothy Hodgkin**
(1910–1994)
(UPI/Bettman Archive)

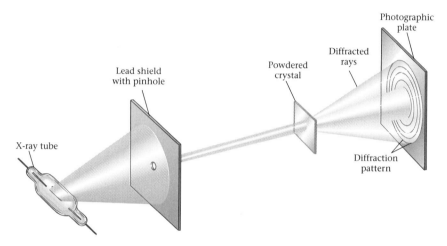

**Figure 9.A**
Knowing the angles and intensities at which x-rays are diffracted by a crystal, it is possible to calculate the distances between layers of atoms.

## 9.4 Liquid-Vapor Equilibrium

All of us are familiar with the process of **vaporization,** in which a liquid is converted to a vapor. In an open container, *evaporation* continues until all the liquid is gone. If the container is closed, the situation is quite different. At first, the movement of molecules is primarily in one direction, from liquid to vapor. Here, however, the vapor molecules cannot escape from the container. Some of them collide with the surface and re-enter the liquid. As time passes, and the concentration of molecules in the vapor increases, so does the rate of condensation. When the rate of condensation becomes equal to the rate of vaporization, the liquid and vapor are in a state of **dynamic equilibrium:**

$$\text{liquid} \rightleftharpoons \text{vapor}$$

The double arrow implies that the forward and reverse processes are occurring at the same rate, which is characteristic of a dynamic equilibrium.

### Vapor Pressure

Once equilibrium between liquid and vapor is reached, the number of molecules per unit volume in the vapor does not change with time. This means that *the pressure exerted by the vapor over the liquid remains constant.* The pressure of vapor in equilibrium with a liquid is called the **vapor pressure.** This quantity is a characteristic property of a given liquid at a particular temperature. It varies from one liquid to another, depending upon the strength of the intermolecular forces. At 25°C, the vapor pressure of water is 24 mm Hg; that of ether, in which intermolecular forces are weaker, is 537 mm Hg.

It is important to realize that *so long as both liquid and vapor are present, the pressure exerted by the vapor is independent of the volume of the container.* If a small amount of liquid is introduced into a closed container, some of it will vaporize, establishing its equilibrium vapor pressure. The greater the volume of the container, the greater will be the amount of liquid that vaporizes to establish that pressure. Only if all the liquid vaporizes will the pressure drop below the equilibrium value.

---

**Example 9.6**    Consider a sample of 1.00 g of water, which has an equilibrium vapor pressure of 24 mm Hg at 25°C.

(a) If this sample is introduced into a 10.0-L flask, how much liquid water will remain when equilibrium is established?
(b) What would be the minimum volume of a flask in which the 1.00-g sample would be completely vaporized?
(c) Suppose the 1.00-g sample is introduced into a 60.0-L flask. What will be the pressure exerted by the vapor?

**Strategy**    In each case, use the ideal gas law. In (a), calculate the mass of $H_2O(g)$ required to establish a vapor pressure of 24 mm Hg in a 10.0-L flask at 25°C; the mass of $H_2O(l)$ left can then be found by subtraction. In (b), calculate the volume occupied by 1.00 g of $H_2O(g)$ at 25°C and 24 mm Hg. In (c), calculate the pressure exerted by 1.00 g of $H_2O(g)$ at 25°C in a 60.0-L flask.

*Solution*

(a) $n_{H_2O(g)} = \dfrac{PV}{RT} = \dfrac{(24/760 \text{ atm})(10.0 \text{ L})}{(0.0821 \text{ L} \cdot \text{atm/mol} \cdot \text{K})(298 \text{ K})} = 0.013 \text{ mol}$

$m_{H_2O(g)} = 0.013 \text{ mol} \times \dfrac{18.02 \text{ g}}{1 \text{ mol}} = 0.23 \text{ g}$

$m_{H_2O(l)} = 1.00 \text{ g} - 0.23 \text{ g} = \boxed{0.77 \text{ g}}$

How would these answers change if 2.00 g of water were used? Ans: (a) 1.77 g (b) 86 L (c) 24 mm Hg

(b) $V = \dfrac{nRT}{P} = \dfrac{(1.00/18.02 \text{ mol})(0.0821 \text{ L} \cdot \text{atm/mol} \cdot \text{K})(298 \text{ K})}{(24/760 \text{ atm})} = \boxed{43 \text{ L}}$

(c) $P = \dfrac{nRT}{V} = \dfrac{(1.00/18.02 \text{ mol})(0.0821 \text{ L} \cdot \text{atm/mol} \cdot \text{K})(298 \text{ K})}{60.0 \text{ L}} = 0.0226 \text{ atm}$

$= 0.0226 \text{ atm} \times \dfrac{760 \text{ mm Hg}}{1 \text{ atm}} = \boxed{17.2 \text{ mm Hg}}$

This is less than the equilibrium vapor pressure of water, which makes sense. Since 1.00 g of $H_2O(l)$ vaporizes completely in a 43-L flask, there is none left to establish equilibrium in a 60.0-L flask , and the pressure drops below its equilibrium value.

---

## Vapor Pressure vs. Temperature

The vapor pressure of a liquid always increases as temperature rises. Water evaporates more readily on a hot, dry day. Stoppers in bottles of volatile liquids such as ether or gasoline pop out when the temperature rises.

The vapor pressure of water, which is 24 mm Hg at 25°C, becomes 92 mm Hg at 50°C and 1 atm (760 mm Hg) at 100°C. The data for water are plotted at the top of Figure 9.13. As you can see, the graph of vapor pressure versus temperature is not a straight line, as it would be if pressure were plotted versus temperature for an ideal gas. Instead, the slope increases steadily as temperature rises, reflecting the fact that more molecules vaporize at higher temperatures. At 100°C, the concentration of $H_2O$ molecules in the vapor in equilibrium with liquid is 25 times as great as at 25°C.

In working with the relationship between two variables, such as vapor pressure and temperature, scientists prefer to deal with linear (straight-line) functions. Straight-line graphs are easier to construct and to interpret. In this case, it is possible to obtain a linear function by making a simple shift in variables. Instead of plotting vapor pressure ($P$) versus temperature ($T$), we plot the *natural logarithm** of the vapor pressure (ln $P$) versus the reciprocal of the absolute temperature ($1/T$). Such a plot for water is shown at the bottom of Figure 9.13; as with all other liquids, a plot of ln $P$ versus $1/T$ is a straight line.

The general equation of a straight line is

$$y = mx + A \quad (m = \text{slope}, A = y\text{-intercept})$$

The $y$-coordinate in Figure 9.13b is ln $P$, while the $x$-coordinate is $1/T$. The slope, which is a negative quantity, turns out to be $-\Delta H_{vap}/R$, where $\Delta H_{vap}$ is the molar heat of vaporization and $R$ is the gas constant, in the proper units. Hence, the equation of the straight line in Figure 9.13b is

$$\ln P = \frac{-\Delta H_{vap}}{RT} + A$$

For many purposes, it is convenient to have a two-point relation between the vapor pressures ($P_2$, $P_1$) at two different temperatures ($T_2$, $T_1$). Such a relation is obtained by first writing separate equations at each temperature:

$$\text{at } T_2: \quad \ln P_2 = A - \frac{\Delta H_{vap}}{RT_2}$$

$$\text{at } T_1: \quad \ln P_1 = A - \frac{\Delta H_{vap}}{RT_1}$$

Upon subtraction, the constant $A$ is eliminated, and we obtain:

$$\ln P_2 - \ln P_1 = -\frac{\Delta H_{vap}}{R}\left[\frac{1}{T_2} - \frac{1}{T_1}\right] = \frac{\Delta H_{vap}}{R}\left[\frac{1}{T_1} - \frac{1}{T_2}\right]$$

or alternatively

$$\ln \frac{P_2}{P_1} = \frac{\Delta H_{vap}}{R}\left(\frac{T_2 - T_1}{T_2 T_1}\right)$$

This equation is known as the **Clausius-Clapeyron equation.** Rudolph Clausius was a prestigious nineteenth-century German scientist. B. P. E. Clapeyron, a French engineer, first proposed a modified version of the equation in 1834.

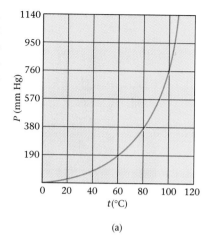

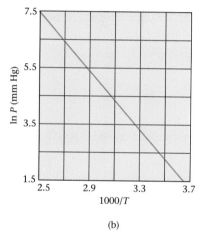

(a)

(b)

**Figure 9.13**
A plot of vapor pressure versus temperature for water (or any other liquid) is a curve with a steadily increasing slope. On the other hand, a plot of the logarithm of vapor pressure versus the reciprocal of the absolute temperature is a straight line.

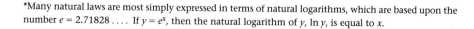

*Many natural laws are most simply expressed in terms of natural logarithms, which are based upon the number $e = 2.71828\ldots$. If $y = e^x$, then the natural logarithm of $y$, ln $y$, is equal to $x$.

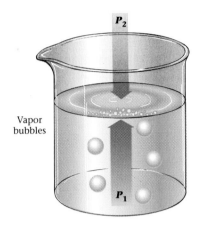

**Figure 9.14**
A liquid boils when its vapor pressure ($P_1$) slightly exceeds the pressure above it ($P_2$).

In using the Clausius-Clapeyron equation, the units of $\Delta H_{vap}$ and $R$ must be consistent. If $\Delta H$ is expressed in joules, then $R$ must be expressed in joules per mole per kelvin. Recall Table 5.1, p. 116 that

$$R = 8.31 \text{ J/mol} \cdot \text{K}$$

---

**Example 9.7** A certain organic compound has a vapor pressure of 91 mm Hg at 40°C. Calculate its vapor pressure at 25°C, taking the heat of vaporization to be $5.31 \times 10^4$ J/mol.

***Strategy*** It is convenient to use the subscript 2 for the higher temperature and pressure. Substitute into the Clausius-Clapeyron equation, solving for $P_1$. Remember to express temperature in K and take $R = 8.31$ J/mol · K.

***Solution***

$$\ln \frac{P_2}{P_1} = \ln \frac{91 \text{ mm Hg}}{P_1} = \frac{5.31 \times 10^4 \text{ J/mol}}{8.31 \text{ J/mol} \cdot \text{K}} \left[ \frac{1}{298} - \frac{1}{313} \right] = 1.03$$

Taking inverse logs,

$$\frac{91 \text{ mm Hg}}{P_1} = e^{1.03} = 2.8$$

(*e* is the base of natural logarithms; on many calculators, to find the number whose natural logarithm is 1.03, you first enter 1.03, then press in succession the "INV" and "ln *x*" keys).
Solving for $P_1$,

$$P_1 = \frac{91 \text{ mm Hg}}{2.8} = \boxed{32 \text{ mm Hg}}$$

This value is reasonable in the sense that lowering the temperature should reduce the vapor pressure.

---

## Boiling Point

When a liquid is heated in an open container, bubbles form, usually at the bottom, where heat is applied. The first small bubbles are air, driven out of solution by the increase in temperature. Eventually, at a certain temperature, large vapor bubbles form throughout the liquid. These vapor bubbles rise to the surface, where they break. When this happens, the liquid is said to be boiling. For a pure liquid, the temperature remains constant throughout the boiling process.

The temperature at which a liquid boils depends upon the pressure above it. To understand why this is the case, consider Figure 9.14. This shows vapor bubbles rising in a boiling liquid. For a vapor bubble to form, the pressure within it, $P_1$, must be at least equal to the pressure above it, $P_2$. Since $P_1$ is simply the vapor pressure of the liquid, it follows that ***a liquid boils at a temperature at which its vapor pressure is equal to the pressure above its surface.*** If this pressure is 1 atm (760 mm Hg), the temperature is referred to as the normal boiling point. (When the term "boiling point" is used without qualification, normal boiling point is implied.) The normal boiling point of water is 100°C; its vapor pressure is 760 mm Hg at that temperature.

The vapor pressure of a substance at its normal bp is 760 mm Hg

As you might expect, the boiling point of a liquid can be reduced by lowering the pressure above it. Water can be made to boil at 25°C by evacuating the space above it. When a pressure of 24 mm Hg, the equilibrium vapor pressure at 25°C, is reached, the water starts to boil. Chemists often take advantage of this effect in purifying a high-boiling compound that might decompose or oxidize at its normal boiling point. They boil it at a reduced temperature under vacuum and condense the vapor.

If you have been fortunate enough to camp in the high Sierras or the Rockies, you may have noticed that it takes longer at high altitudes to cook foods in boiling water. The reduced pressure lowers the temperature at which water boils in an open container and thus slows down the physical and chemical changes that take place when foods like potatoes or eggs are cooked. In principle, this problem can be solved by using a pressure cooker. In that device, the pressure that develops is high enough to raise the boiling point of water above 100°C. Pressure cookers are indeed used in places like Cheyenne, Wyoming (elevation 1848 m), but not by mountain climbers, who have to carry all their equipment on their backs.

Or in the Presidential Range of New Hampshire (WLM)

## Critical Temperature and Pressure

Consider an experiment in which liquid carbon dioxide is introduced into an otherwise evacuated glass tube, which is then sealed (Fig. 9.15). At 0°C, the pressure above the liquid is 34 atm, the equilibrium vapor pressure of $CO_2(l)$ at that temperature. As the tube is heated, some of the liquid is converted to vapor, and the pressure rises, to 44 atm at 10°C and 56 atm at 20°C. Nothing spectacular

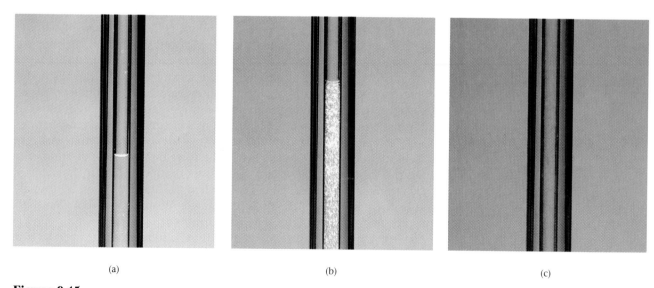

(a)                                    (b)                                    (c)

**Figure 9.15**
When liquid carbon dioxide under pressure is heated, some of it vaporizes (a); vapor bubbles form in the liquid (b). Suddenly, at 31°C, the critical temperature of $CO_2$, the meniscus disappears (c). Above that temperature, only one phase is present, regardless of the applied pressure. (Marna G. Clarke)

TABLE 9.5  **Critical Temperatures (°C)**

| Permanent Gases | | Condensable Gases | | Liquids | |
|---|---|---|---|---|---|
| Helium | −268 | Carbon dioxide | 31 | Ethyl ether | 194 |
| Hydrogen | −240 | Ethane | 32 | Ethyl alcohol | 243 |
| Nitrogen | −147 | Propane | 97 | Benzene | 289 |
| Argon | −122 | Ammonia | 132 | Bromine | 311 |
| Oxygen | −119 | Chlorine | 144 | Water | 374 |
| Methane | −82 | Sulfur dioxide | 158 | | |

happens (unless there happens to be a weak spot in the tube) until 31°C is reached, where the vapor pressure is 73 atm. Suddenly, as the temperature goes above 31°C, the meniscus between the liquid and vapor disappears! The tube now contains only one phase.

It is impossible to have liquid carbon dioxide at temperatures above 31°C, no matter how much pressure is applied. Even at pressures as high as 1000 atm, carbon dioxide gas does not liquefy at 35 or 40°C. This behavior is typical of all substances. There is a temperature, called the **critical temperature,** above which the liquid phase of a pure substance cannot exist. The pressure that must be applied to cause condensation at that temperature is called the **critical pressure.** Quite simply, the critical pressure is the vapor pressure of the liquid at the critical temperature.

Table 9.5 lists the critical temperatures of several common substances. The species in the column at the left all have critical temperatures below 25°C. They are often referred to as "permanent gases." Applying pressure at room temperature will not condense a permanent gas. It must be cooled as well. When you see a truck labeled "liquid nitrogen" on the highway, you can be sure that the cargo trailer is refrigerated to at least −147°C, the critical temperature of $N_2$.

"Permanent gases" are most often stored and sold in steel cylinders under high pressures, often 150 atm or greater. When the valve on a cylinder of $N_2$ or $O_2$ is opened, gas escapes, and the pressure drops accordingly. The substances listed in the center column of Table 9.5, all of which have critical temperatures above 25°C, are handled quite differently. They are available commercially as liquids in high-pressure cylinders. When the valve on a cylinder of propane is opened, the gas that escapes is replaced by vaporization of liquid. The pressure quickly returns to its original value. Only when the liquid is completely vaporized does the pressure drop as gas is withdrawn. This indicates that almost all of the propane is gone, and it's time to recharge the tank.

*This can be done with organic solvents, but they tend to leave a residue*

Above the critical temperature and pressure, a substance is referred to as a supercritical fluid. Such fluids have unusual solvent properties that have led to many practical applications. Water at the critical point has a density of about 0.3 g/mL and a heat capacity that approaches infinity. More important, oxygen and most organic molecules are completely soluble in supercritical water. This may offer a safe, inexpensive method of disposing of hazardous organic waste by oxidation. In a quite different area, supercritical $CO_2$ is now used to extract caffeine from coffee and nicotine from tobacco.

# 9.5 Phase Diagrams

In the preceding section, we discussed several features of the equilibrium between a liquid and its vapor. For a pure substance, at least two other types of phase equilibria need to be considered. One is the equilibrium between a solid and its vapor, the other between solid and liquid at the melting (freezing) point. Many of the important relations in all these equilibria can be shown in a **phase diagram.** A phase diagram is a graph that shows the pressures and temperatures at which different phases are in equilibrium with each other. The phase diagram of water is shown in Figure 9.16. This figure, which covers a wide range of temperatures and pressures, is not drawn to scale.

To understand what a phase diagram implies, consider first the three lines AB, AC, and AD in Figure 9.16. Each of these lines shows the pressures and temperatures at which two adjacent phases are at equilibrium.

**1.** Line AB is a portion of the vapor pressure–temperature curve of liquid water. At any temperature and pressure along this line, liquid water is in equilibrium with water vapor. At point A on the curve, these two phases are in equilibrium at 0°C and about 5 mm Hg (more exactly, 0.01°C and 4.56 mm Hg). At B, corresponding to 100°C, the pressure exerted by the vapor in equilibrium with liquid water is 1 atm; this is the normal boiling point of water. The extension of line AB beyond point B gives the equilibrium vapor pressure of the liquid above the normal boiling point. The line ends at 374°C, the critical temperature of water, where the pressure is 218 atm.

**2.** Line AC represents the vapor pressure curve of ice. At any point along this line, such as point A (0°C, 5 mm Hg) or point C, which might represent −3°C and 3 mm Hg, ice and vapor are in equilibrium with each other.

**3.** Line AD gives the temperatures and applied pressures at which liquid water is in equilibrium with ice.

Point A on a phase diagram is the only one at which all three phases, liquid, solid, and vapor, are in equilibrium with each other. It is called the **triple**

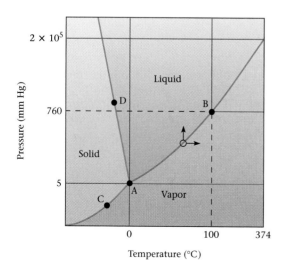

**Figure 9.16**

Phase diagram for water (not to scale). The triple point is at 0.01°C, 4.56 mm Hg; the critical point is at 374°C, $1.66 \times 10^5$ mm Hg (218 atm).

**point.** For water, the triple point temperature is 0.01°C. At this temperature, liquid water and ice have the same vapor pressure, 4.56 mm Hg.

In the three areas of the phase diagram labeled "solid," "liquid," and "vapor," only one phase is present. To understand this, consider what happens to an equilibrium mixture of two phases when the pressure or temperature is changed. Suppose we start at the point on AB indicated by an open circle. Here liquid water and vapor are in equilibrium with each other, let us say at 70°C and 234 mm Hg. If the pressure on this mixture is increased, condensation occurs. The phase diagram confirms this; increasing the pressure at 70°C *(vertical arrow)* puts us in the liquid region. In another experiment, the temperature might be increased at a constant pressure. This should cause the the liquid to vaporize. The phase diagram shows that this is indeed what happens. An increase in temperature *(horizontal arrow)* shifts us to the vapor region.

---

**Example 9.8**    Consider a sample of H$_2$O at point D in Figure 9.16

(a) What phase(s) is (are) present?
(b) If the temperature of the sample were reduced at constant pressure, what would happen?
(c) How would you convert the sample to vapor without changing the temperature?

***Strategy***    Use the phase diagram in Figure 9.16. Note that *P* increases moving up vertically; *T* increases moving to the right.

*Solution*

(a) Point D is on the solid-liquid equilibrium line.   Ice and liquid water are

present.

(b) Move to the left to reduce *T*. This penetrates the solid area, which implies that

the   solid freezes   completely.

(c)   Reduce the pressure to below the triple point value, perhaps to 4 mm Hg.

---

## Sublimation

The process by which a solid changes directly to vapor without passing through the liquid phase is called *sublimation*. A solid can sublime only at temperatures below the triple point; above that temperature it will melt to liquid (Fig. 9.16). At temperatures below the triple point, a solid can be made to sublime by reducing the pressure of the vapor above it to less than the equilibrium value. To illustrate what this means, consider the conditions under which ice sublimes. This happens on a cold, dry, winter day when the temperature is below 0°C and the pressure of water vapor in the air is less than the equilibrium value (4.5 mm Hg at 0°C). The rate of sublimation can be increased by evacuating the space above the ice. This is how foods are freeze-dried. The food is frozen, put into a vacuum chamber, and evacuated to a pressure of 1 mm Hg or less. The ice crystals formed upon freezing sublime, which leaves a product whose mass is only a fraction of that of the original food.

By the reverse process, snow is formed in the atmosphere

Iodine sublimes more readily than ice because its triple-point pressure, 90 mm Hg, is much higher. Sublimation occurs upon heating (Fig. 9.17) below the triple-point temperature, 114°C. If the triple point is exceeded, the solid melts. Solid carbon dioxide (Dry Ice) has a triple-point pressure above 1 atm (5.2 atm at −57°C). Liquid carbon dioxide cannot be made by heating dry ice in an open container. Solid $CO_2$ always passes directly to vapor at 1 atm pressure.

## Melting Point

For a pure substance, the **melting point** is identical to the **freezing point.** It represents the temperature at which solid and liquid phases are in equilibrium. Melting points are usually measured in an open container, i.e., at atmospheric pressure. For most substances, the melting point at 1 atm (the "normal" melting point) is virtually identical with the triple-point temperature. For water, the difference is only 0.01°C.

Although the effect of pressure upon melting point is very small, its direction is still important. To decide whether the melting point will be increased or decreased by compression, a simple principle is applied. *An increase in pressure favors the formation of the more dense phase.*

Two types of behavior are shown in Figure 9.18.

**1.** *The solid is the more dense phase* (Fig. 9.18a). The solid-liquid equilibrium line is inclined to the right, shifting away from the *y*-axis as it rises. At higher pressures, the solid becomes stable at temperatures above the normal melting point. In other words, the melting point is raised by an increase in pressure. This behavior is shown by most substances.

**2.** *The liquid is the more dense phase* (Fig. 9.18b). The liquid-solid line is inclined to the left, toward the *y*-axis. An increase in pressure favors the formation of liquid; that is, the melting point is decreased by raising the pressure. Water is one of the few substances that behaves this way; ice is less dense than liquid water. The effect is exaggerated for emphasis in Figure 9.18b. Actually, an increase in pressure of 134 atm is required to lower the melting point of ice by 1°C.

**Figure 9.17**
Solid iodine passes directly to vapor at any temperature below the triple point, 114°C. The vapor condenses back to solid on a cold surface, as in the upper tube, which is filled with Dry Ice. (Marna G. Clarke)

This is generally true; recall diamond vs. graphite

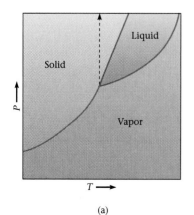

(a)

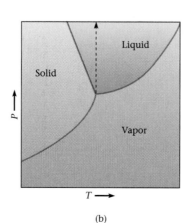

(b)

**Figure 9.18**
Effect of pressure on the melting point of a solid. (a) When the solid is the more dense phase, an increase in pressure converts liquid to solid; the melting point increases. (b) If the liquid is the more dense phase, an increase in pressure converts solid to liquid and the melting point decreases.

# Silicon and the Silicates

Among the most important of the network covalent structures are those of the element silicon and the silicate minerals, in which silicon atoms are bonded to oxygen.

### Elemental Silicon

A crystal of pure silicon has the diamond structure and is an electrical nonconductor. A simplified, two-dimensional model of the bonding in elementary silicon is shown at the left of Figure 9.B. The valence electrons are used in the four covalent bonds each Si atom forms with its neighbors. Conductivity increases dramatically when small amounts of arsenic (Group 15) or boron (Group 13) are introduced into the crystal. As little as 0.0001% of these elements present in Si produces semiconductor devices used in transistors or solar cells.

Notice from Figure 9.B that when an As atom with five valence electrons substitutes for a Si atom (four valence electrons) a mobile "free" electron is formed. This electron can move through the crystal under the influence of an electric field. This gives an n-type semiconductor (current carried by the flow of negative charge). If an atom of boron with three valence electrons is substituted for silicon, a quite different situation applies. An electron deficiency or "positive hole" is created in the lattice. In an electric field an electron moves from a neighboring atom to fill the hole, giving rise to what is known as a p-type semiconductor. In effect, positive holes move through the lattice.

### Silicates

More than 90% of the rocks and minerals found in the Earth's crust are classed as silicates; they contain silicon atoms bonded tetrahedrally to four oxygen atoms. Silicates are essentially ionic; the negative ion is an oxoanion in which silicon is the central atom. The anion may be a very simple species (e.g., $SiO_4^{4-}$) or a more complex species of the network covalent type. The negative charge of the silicon oxoanion is balanced by positive charges on the cations of one or more metals (e.g., $Na^+$, $Mg^{2+}$, $Al^{3+}$) that fit into the crystal lattice.

The simplest silicates are those containing the tetrahedral $SiO_4^{4-}$ anion, found in the semiprecious stone zircon, $ZrSiO_4$, and the garnets, of which $Ca_3Cr_2(SiO_4)_3$ is typical. Most silicate minerals have network covalent structures in which $SiO_4^{4-}$ tetrahedra are linked together in one, two, or three dimensions. In the structure at the left of Figure 9.C, tetrahedra are linked together to form what amounts to a one-dimensional infinite chain. This occurs in fibrous minerals such as diopside, $CaSiO_3 \cdot MgSiO_3$. The best known of the fibrous silicates are the asbestos minerals, with a structure similar

**Figure 9.B**
Semiconductors derived from silicon. In n-type semiconductors, the impurity atoms furnish mobile electrons to the crystal. In p-type semiconductors there is a deficiency of electrons because the impurity atoms have three rather than four valence electrons.

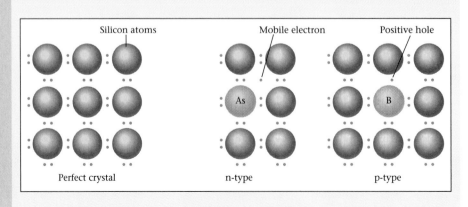

Silicon atoms      Mobile electron      Positive hole

Perfect crystal      n-type      p-type

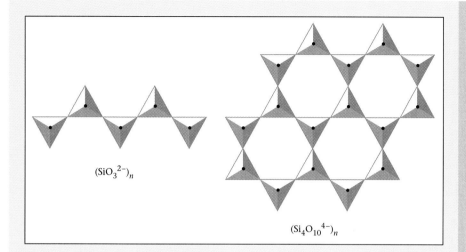

to diopside except that two chains are linked together to form a double strand. As-bestos has been used since ancient Roman times for its insulating and fire-resistant properties. Today, asbestos is being removed from private homes and public buildings because of the potential health hazards associated with exposure to the material. Work-ers in the asbestos industry have abnormally high rates of lung disease, including lung cancer. Presumably, asbestos fibers become embedded in lung tissue, causing an irrita-tion that may lead to the formation of tumors many years later.

The silicate structure shown at the right of Figure 9.C is typical of layer minerals such as talc, $Mg_3(OH)_2Si_4O_{10}$. Here, $SiO_4^{4-}$ tetrahedra are linked together in two di-mensions to give an infinite layer. The sheets are held together only by weak disper-sion forces, so they easily slide past one another. As a result, talcum powder has a slip-pery feeling, much like graphite. Mica is also a layer silicate with a somewhat more complex structure in which cations are located between the layers. The attraction be-tween oppositely charged ions makes sheets of mica more difficult to separate. The shiny luster of mica is exploited by adding small amounts of it to paint to give a sil-very, metallic effect.

The simplest three-dimensional silicate network is that described previously for silicon dioxide, $SiO_2$. Silicon dioxide occurs in slightly impure form as rose quartz, smoky quartz, and amethyst. Flint and obsidian, used centuries ago by Native Ameri-cans for arrowheads, are poorly crystallized forms of $SiO_2$.

Another class of three-dimensional silicates are the materials known as zeolites (Fig. 9.D). These contain relatively large cavities or tunnels in which $H_2O$ molecules or cations such as $Na^+$ or $Ca^{2+}$ may be trapped. A variety of zeolites occur naturally. They can also be made in the laboratory, in which case the dimensions of the "holes" can be made to order.

Zeolites are widely used in home water softeners. When hard water, containing $Ca^{2+}$ (or $Mg^{2+}$) ions, flows through a column packed with a zeolite, NaZ, the follow-ing reaction occurs:

$$Ca^{2+}(aq) + 2NaZ(s) \longrightarrow CaZ_2(s) + 2Na^+(aq)$$

Sodium ions migrate out of the cavities; $Ca^{2+}$ ions move in. The effect is to replace $Ca^{2+}$ ions in the water by $Na^+$ ions. In this way, the objectionable properties of hard water including precipitation of soap, and formation of boiler scale, are removed.

**Figure 9.D**

Structure of a natural zeolite with the simplest formula NaAlSiO$_4$. Na$^+$ ions are held loosely in the anionic lattice, which consists of Al, Si, and O atoms. When water containing Ca$^{2+}$ ions passes through the zeolite, Ca$^{2+}$ ions enter the lattice while Na$^+$ ions take their place in solutions.

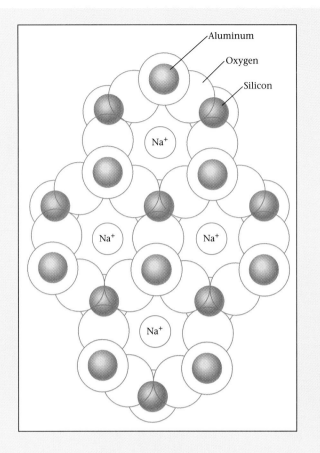

Talc crystal. (Charles D. Winters)

# CHAPTER HIGHLIGHTS

## *Key Concepts*

1. Identify intermolecular forces in different substances and evaluate their effects on boiling point.
   (Examples 9.1–9.3; Problem 1–14)
2. Classify substances as molecular, network covalent, ionic, or metallic on the basis of their properties.
   (Example 9.4; Problems 15–24, 63, 64)
3. Relate atomic radius (*r*) to side (*s*) for different unit cells.
   (Example 9.5; Problems 25–32)
4. Use the ideal gas law to determine whether a liquid will completely vaporize.
   (Example 9.6; Problems 35–38, 68)
5. Use the Clausius-Clapeyron equation to relate the vapor pressure of a liquid to temperature.
   (Example 9.7; Problems 39–46)
6. Use a phase diagram to determine the phase(s) present at a given *T* and *P*.
   (Example 9.8; Problems 49–56)

## *Key Equations*

| | |
|---|---|
| Vapor pressure–temperature | $\ln \dfrac{P_2}{P_1} = \dfrac{\Delta H}{R}\left[\dfrac{1}{T_1} - \dfrac{1}{T_2}\right]$ |
| Unit cell | $2r = s$  (SC)   $4r = s\sqrt{2}$ (FCC)   $4r = s\sqrt{3}$ (BCC) |

## *Key Terms*

| | | |
|---|---|---|
| boiling point | dipole force | phase diagram |
| critical temperature | dispersion force | sublimation |
| cubic cell | electron-sea model | triple point |
| —body-centered (BCC) | hydrogen bond | unit cell |
| —face-centered (FCC) | melting point | vapor pressure |
| —simple | network covalent | |

## *Summary Problems*

Consider propylamine, an organic compound of structural formula   $H_3C—CH_2—CH_2$
                                                                                            $|$
                                                                                            $NH_2$

(a) What kind of intermolecular forces would you expect to find in propylamine?

(b) How would you expect the normal boiling point of propylamine to compare with that of nitrogen? of methylamine ($CH_3NH_2$)? of ammonium chloride? of silicon dioxide ($SiO_2$)?

(c) Would you expect propylamine to conduct electricity? How would its solubility in water compare with that of nitrogen gas? of silicon dioxide?

(d) Give the symbol of a solid nonmetal that is a better conductor than propylamine. Give the formula of a solid compound that, unlike propylamine, has a network structure.

(e) The normal boiling point of propylamine is 48.7°C, its normal freezing point is −83°C, and its critical temperature is 234°C. Draw a rough sketch of the phase diagram of propylamine, assuming that the solid is more dense than the liquid.

(f) If you want to purify propylamine by sublimation, at what approximate temperature would you operate?

(g) A sample of propylamine vapor at 250°C and 1 atm is cooled at constant pressure. At what temperature does the liquid first appear?

(h) Taking the normal boiling point of propylamine to be 48.7°C and its heat of vaporiztion to be 30.7 kJ/mol, estimate its normal vapor pressure at 25.0°C. Estimate the temperature at which its vapor pressure is 0.500 atm.

(i) A sample of 1.00 mL of propylamine ($d = 0.719$ g/mL) is placed in a one-liter flask at 22°C, where its vapor pressure is 269 mm Hg. Show by calculation whether there is any liquid left in the flask when equilibrium is established.

### *Answers*

(a) dispersion, H-bonds

(b) higher than $N_2$ and $CH_3NH_2$; lower than $NH_4Cl$ and $SiO_2$

(c) no; higher than $N_2$ and $SiO_2$

(d) C(graphite) or Si; $SiO_2$, NaCl

(e) See Figure 9.18a for the general configuration. Put in the appropriate temperatures and pressures.

(f) below about −83°C

(g) 48.7°C

(h) 305 mm Hg; 30.5°C

(i) no

## Questions & Problems

### Intermolecular Forces

**\*1.** Arrange the following in order of increasing boiling point.

(a) $CH_4$   (b) $SiH_4$   (c) $SnH_4$   (d) $GeH_4$

**\*2.** Arrange the following in order of decreasing boiling point.

(a) $CH_4$   (b) $CCl_4$   (c) $CF_4$   (d) $CBr_4$

**\*3.** Which of the following would you expect to show dispersion forces? dipole forces?

(a) $Cl_2$   (b) $CH_2Cl_2$   (c) $CCl_4$   (d) $HCl$

**\*4.** Which of the following would you expect to show dispersion forces? dipole forces?

(a) $O_2$   (b) $CO$   (c) $CO_2$   (d) $H_2CO$

**\*5.** Which of the following would show hydrogen bonding?

(a) $CCl_2F_2$   (b) $CH_3F$
(c) $CH_3NH_2$   (d) $H_3C—CO—CH_3$

**\*6.** Which of the following would show hydrogen bonding?

(a) $[H—F—F]^+$   (b) $CH_3CN$
(c) $HO—OH$   (d) $H_3C—O—CH_3$

**\*7.** Explain in terms of forces between structural units why

(a) methane, $CH_4$, has a lower boiling point than octane, $C_8H_{18}$.

(b) HF has a higher boiling point than HI.

(c) ICl has a higher melting point than $Cl_2$.

(d) $N_2$ has a lower boiling point than $O_2$.

**\*8.** Explain in terms of forces between structural units why

(a) iodine has a lower melting point than NaI.

(b) $NH_3$ has a higher boiling point than $CH_4$.

(c) $H_2O$ has a higher boiling point than $H_2Te$.

(d) Formic acid, $H—\overset{\overset{\displaystyle O}{\|}}{C}—OH$, has a lower boiling point

than benzoic acid, $C_6H_5—\overset{\overset{\displaystyle O}{\|}}{C}—OH$

**\*9.** In which of the following processes is it necessary to break covalent bonds, as opposed to simply overcoming intermolecular forces?

(a) melting ice

(b) subliming iodine crystals

(c) decomposing $CO_2$ to C and $O_2$

(d) dissolving HCl gas in water to make hydrochloric acid

**\*10.** In which of the following processes is it necessary to break covalent bonds, as opposed to simply overcoming intermolecular forces?

(a) decomposing hydrogen peroxide ($H_2O_2$) to water and oxygen gas

(b) melting mothballs made of naphthalene

(c) vaporizing chloroform, $CHCl_3$

(d) changing ozone, $O_3$, to oxygen gas, $O_2$

**\*11.** For each of the following pairs, choose the member with the lower boiling point. Explain your reason in each case.

(a) $NO_2$ or $SO_2$   (b) NaCl or HCl
(c) Xe or Ne   (d) $NH_3$ or $AsH_3$

**\*12.** Follow the directions for Question 11 for the following substances.

(a) $CH_4$ or $CCl_4$   (b) $H_2$ or $Br_2$
(c) LiF or HF   (d) $C_2H_6$ or $C_{10}H_{22}$

**\*13.** What are the strongest attractive forces that must be overcome to

(a) boil $H_2Se$?

(b) vaporize water?

(c) dissolve $Br_2$ in trichloroethane, $C_2H_3Cl_3$?

(d) dissolve potassium chloride in water?

**\*14.** What are the strongest attractive forces that must be overcome to

(a) melt table salt, NaCl?

(b) sublime Dry Ice, $CO_2$?

(c) boil ammonia, $NH_3$?

(d) vaporize acetone, $H_3C—CO—CH_3$?

### Types of Substances

**\*15.** Classify each of the following solids as metallic, network covalent, ionic, or molecular.

(a) It is insoluble in water, does not conduct electricity, and has a high melting point (above 500°C).

(b) It is insoluble in water and conducts electricity when melted.

(c) It is insoluble in water and conducts electricity.

**\*16.** Classify each of the following solids as metallic, network covalent, ionic, or molecular.

(a) It dissolves in water, conducts electricity when present in aqueous solution, and melts above 100°C.

(b) It is insoluble in water, melts below 100°C, and does not conduct electricity either as a solid, dissolved in water, or molten.

(c) It is malleable and conducts electricity.

**\*17.** Of the four general types of solids, which one(s)

(a) generally have high melting points?

(b) are generally insoluble in water?

(c) conduct electricity when molten?

**\*18.** Of the four general types of solids, which one(s)

(a) generally have low boiling points?

(b) are shiny and malleable?

(c) are generally soluble in polar solvents?

**\*19.** Classify each of the following species as molecular, network covalent, ionic, or metallic.

(a) $HI(g)$   (b) $CH_3COOH$   (c) C (graphite)
(d) W   (e) NaI

**\*20.** Classify each of the following species as molecular, network covalent, ionic, or metallic.

(a) sand   (b) Ca(s)   (c) ICl
(d) C (diamond)   (e) $CaCl_2$

**\*21.** Give the formula of a solid containing carbon that is

(a) ionic   (b) molecular
(c) metallic   (d) network covalent

**\*22.** Give the formula of a solid containing oxygen that is
  **(a)** ionic  **(b)** polar molecular
  **(c)** network covalent  **(d)** nonpolar molecular
**\*23.** Describe the structural units in
  **(a)** $SiO_2$  **(b)** $SO_2$
  **(c)** $KO_2$  **(d)** K
**\*24.** Describe the structural units in
  **(a)** $CH_2Cl_2$  **(b)** $Al_2O_3$
  **(c)** Al  **(d)** C (diamond)

## Crystal Structure

**25.** Sodium has an atomic radius of 0.186 nm. The edge of its cubic unit cell is 0.430 nm. What is the geometry of the sodium unit cell?

**26.** Barium has an atomic radius of 0.217 nm. The volume of its cubic unit cell is 0.127 $nm^3$. What is the geometry of the barium unit cell?

**27.** Lead crystallizes with a face-centered cubic unit cell. Its atomic radius is 0.175 nm. What is the length of a side of the cell?

**28.** Aluminum crystallizes with a face-centered cubic unit cell. The volume of a unit cell is 0.0664 $nm^3$. What is the atomic radius of aluminum?

**29.** Sodium iodide has a unit cell similar to that of sodium chloride (Figure 9.12). The ionic radii of $Na^+$ and $I^-$ are 0.095 nm and 0.216 nm, respectively. How long is
  **(a)** one side of the cube?
  **(b)** the face diagonal of the cube?

**30.** In the LiCl structure shown in Figure 9.12, the chloride ions form a face-centered cubic unit cell 0.513 nm on an edge. The ionic radius of $Cl^-$ is 0.181 nm.
  **(a)** Along a cell edge, how much space is there between the $Cl^-$ ions?
  **(b)** Would an $Na^+$ ion ($r$ = 0.095 nm) fit into this space? a $K^+$ ion ($r$ = 0.133 nm)?

**31.** For a cell of the CsCl type (Figure 9.12), how is the length of one side of the cell, $s$, related to the sum of the radii of the ions, $r_{cation} + r_{anion}$?

**32.** Consider the CsCl cell (Figure 9.12). The ionic radii of $Cs^+$ and $Cl^-$ are 0.169 and 0.181 nm, respectively. What is the length of
  **(a)** the body diagonal?
  **(b)** the side of the cell?

**\*33.** Consider the sodium chloride unit cell shown in Figure 9.12. Looking only at the front face (five large $Cl^-$ ions, four small $Na^+$ ions),
  **(a)** how many cubes share each of the $Na^+$ ions in this face?
  **(b)** how many cubes share each of the $Cl^-$ ions in this face?

**\*34.** Consider the CsCl unit shown in Figure 9.12. How many $Cs^+$ ions are there per unit cell? How many $Cl^-$ ions? (Note that each $Cl^-$ ion is shared by eight cubes.)

## Liquid–Vapor Equilibrium

**35.** Methyl alcohol can be used as a fuel instead of, or combined with, gasoline. A sample of methyl alcohol, $CH_3OH$, in a flask of constant volume exerts a pressure of 254 mm Hg at 57°C. The flask is slowly cooled.
  **(a)** Assuming no condensation, use the ideal gas law to calculate the pressure of the vapor at 35°C and at 45°C.
  **(b)** Compare your answers in (a) with the equilibrium vapor pressures of methyl alcohol: 203 mm Hg at 35°C; 325 mm Hg at 45°C.
  **(c)** On the basis of your answers to (a) and (b), predict the pressure exerted by the methyl alcohol in the flask at 35°C and at 45°C.
  **(d)** What physical states of methyl alcohol are present in the flask at 35°C? at 45°C?

**36.** The vapor pressure of naphthalene, $C_{10}H_8$, at 25°C is 0.300 mm Hg.
  **(a)** How many mg of naphthalene will sublime into an evacuated 500.0-mL flask?
  **(b)** If 0.500 mg of naphthalene is used, what will the final pressure be? What physical state(s) of naphthalene is (are) in the flask?
  **(c)** If 3.00 mg of naphthalene are used, what will the final pressure be? What physical state(s) of naphthalene is (are) in the flask?

**37.** Chloroform, $CHCl_3$, was once used as an anesthetic. It is the material put in handkerchiefs to render victims in spy movies unconscious. It has a vapor pressure of 186 mm Hg at 23°C. A 25.00-mL sample ($d$ = 1.498 g/mL) is sealed in a 250.0-mL flask.
  **(a)** Will there be any liquid left in the flask at equilibrium at 23°C?
  **(b)** What is the maximum volume that the flask can have if equilibrium is to be established between liquid and vapor?
  **(c)** If the flask has a volume of 50.0 L, what is the pressure of chloroform in the flask?

**38.** A humidifier is used in a bedroom kept at 22°C. The bedroom's volume is $4.0 \times 10^4$ L. Assume that the air is originally dry and that no moisture leaves the room while the humidifier is operating.
  **(a)** If the humidifier has a capacity of 3.0 gallons of water, will there be enough liquid to saturate the room with water vapor (vp of water at 22°C = 19.83 mm Hg; $d$ of water = 1.00 $g/cm^3$)?
  **(b)** What is the final pressure of water vapor in the room when the humidifier has vaporized two thirds of its water supply?

**39.** Carbon disulfide, $CS_2$, normally boils at 46.5°C and has a vapor pressure of 400.0 mm Hg at 28.0°C. Estimate
  **(a)** its heat of vaporization ($\Delta H_{vap}$).
  **(b)** its vapor pressure at 40.0°C.

**40.** Dichloromethane, $CH_2Cl_2$, is an organic solvent used to "decaffeinate" coffee beans. Its vapor pressure is 329.6 mm Hg at 20.0°C and 0.953 atm at 30.0°C. Estimate its
  **(a)** heat of vaporization.
  **(b)** normal boiling point.

**41.** Glacier National Park in Montana is a favorite vacation stop for backpackers. It is about 4100 ft above sea level, with an atmospheric pressure of 681 mm Hg. At what temperature does water boil at Glacier Park? ($\Delta H_{vap}$ = 40.7 kJ/mol)

**42.** Mt. McKinley in Alaska has an altitude of 20,320 ft. Water boils at 77°C atop Mt. McKinley. What is the normal atmospheric pressure on the mountain?

**43.** When water boils inside a pressure cooker, its vapor pressure is $1.500 \times 10^3$ mm Hg. Use the Clausius-Clapeyron equation to calculate the temperature of the water. Take the heat of vaporization to be 40.7 kJ/mol.

**44.** Mercury is used extensively in studying the effect of pressure on the volume of gases. It is fairly inert at the surface, and most gases are insoluble in it. Should we worry about the contribution of the vapor pressure of mercury to the pressure of gases above it? Use the following data to determine the vapor pressure of mercury at 0°C and at 100°C: $\Delta H_{vap}$ of Hg = 59.4 kJ/mol; normal boiling point = 357°C.

**45.** The data below give the vapor pressure of octane, a major component of gasoline:

| vp (mm Hg) | 10 | 40 | 100 | 400 |
|---|---|---|---|---|
| $t$(°C) | 19.2 | 45.1 | 65.7 | 104.0 |

Plot ln vp versus 1/$T$ (ln $P = A - \dfrac{\Delta H_{vap}}{RT}$) and use your graph to estimate the heat of vaporization of octane.

**46.** Consider the following data for the vapor pressure of diethyl ether, a widely used anesthetic in the early days of surgery.

| vp (mm Hg) | 146 | 231 | 355 | 531 |
|---|---|---|---|---|
| $t$(°C) | −5 | 5 | 15 | 25 |

Plot ln vp versus 1/$T$ (ln $P = A - \dfrac{\Delta H_{vap}}{RT}$) and use your graph to estimate the heat of vaporization of diethyl ether.

**\*47.** The critical point of xenon is 16.6°C and 58 atm. Liquid xenon has a vapor pressure of 4 atm at −84.5°C. Which of the following statements must be true?

  **(a)** Xe is a gas at −84.5°C and 1 atm.
  **(b)** A tank of Xe at 25°C can have a pressure of 58 atm.
  **(c)** Xe gas cooled to 10°C and 65 atm will condense.
  **(d)** The normal boiling point of Xe is higher than −84.5°C.

**\*48.** The normal boiling point of ammonia is −34°C. At 27°C, its vapor pressure is 12 atm. Which of the following statements about ammonia must be true?

  **(a)** A tank of ammonia at 27°C that has a pressure of 10 atm must contain liquid ammonia.
  **(b)** A tank of ammonia at 27°C cannot have a pressure of 15 atm.
  **(c)** The critical temperature of ammonia must be 27°C.

### Phase Diagrams

**\*49.** Referring to Figure 9.16, state what phase(s) is (are) present at

  **(a)** 50°C, 5 mm Hg    **(b)** 100°C, 1000 mm Hg
  **(c)** −10°C, 20 mm Hg

**(b)** 100°C, 1000 mm Hg

**\*50.** Referring to Figure 9.16, state what phase(s) is (are) present at

  **(a)** 10°C, 380 mm Hg
  **(b)** 200°C, 50 mm Hg
  **(c)** 375°C, $3 \times 10^5$ mm Hg

**\*51.** Given the following data about oxygen:

   normal boiling point = −183°C
   triple point = −219°C at 2 mm Hg
   normal melting point = −218°C
   critical point = −119°C at 50 atm

  **(a)** Construct an appropriate phase diagram for oxygen.
  **(b)** Is solid oxygen denser than liquid oxygen?
  **(c)** Estimate the vapor pressure of oxygen at −200°C.

**\*52.** Given the following data about benzene:

   normal boiling point = 80°C
   triple point = 6°C at 36 mm Hg
   $d_{solid} > d_{liquid}$
   critical point = 289°C at 48 atm

  **(a)** Construct an appropriate phase diagram for benzene.
  **(b)** Estimate the vapor pressure of benzene at 50°C.

**\*53.** Use the phase diagram constructed for Problem 51. What phase changes occur (if any) when

  **(a)** at a constant temperature of −225°C, the pressure is increased from 10 mm Hg to 500 mm Hg.
  **(b)** at a constant pressure of 380 mm Hg, the temperature is increased from −200°C to −175°C?
  **(c)** at a constant temperature of −190°C, the pressure is increased from 1 mm Hg to 100 mm Hg?

**\*54.** Use the phase diagram constructed for Problem 52. What phase changes occur (if any) when

  **(a)** at a constant pressure of 300 mm Hg, the temperature is increased from 4°C to 90°C?
  **(b)** at a constant temperature of 25°C, the pressure is increased from 740 mm Hg to 790 mm Hg?
  **(c)** at a constant pressure of 20 mm Hg, the temperature is decreased from 10°C to 0°C?

**\*55.** A pure substance X has the following properties: mp = 90°C, increasing slightly as pressure increases; normal bp = 120°C; liquid vp = 65 mm Hg at 100°C, 20 mm Hg at the triple point.

  **(a)** Draw a phase diagram for X.
  **(b)** Label solid, liquid, and vapor regions of the diagram.
  **(c)** What changes occur if, at a constant pressure of 100 mm Hg, the temperature is raised from 100°C to 150°C?

**\*56.** A pure substance A has a liquid vapor pressure of 320 mm Hg at 125°C, 800 mm Hg at 150°C, and 60 mm Hg at the triple point, 85°C. The melting point of A decreases slightly as pressure increases.

  **(a)** Sketch a phase diagram for A.
  **(b)** From the phase diagram, estimate the boiling point.
  **(c)** What changes occur when, at a constant pressure of 320 mm Hg, the temperature drops from 150°C to 100°C?

## Unclassified

**\*57.** Represent pictorially, using ten molecules:
  **(a)** water freezing
  **(b)** water vaporizing
  **(c)** water being electrolyzed into hydrogen and oxygen

**\*58.** Which of the following statements is (are) true?
  **(a)** The critical temperature must be reached to change liquid to gas.
  **(b)** To melt a solid at constant pressure, the temperature must be above the triple point.
  **(c)** $CHF_3$ can be expected to have a higher boiling point than $CHCl_3$ because $CHF_3$ has hydrogen bonding.
  **(d)** One metal crystallizes in a body-centered cubic cell and another in a face-centered cubic cell of the same volume. The two atomic radii are related by the factor $\sqrt{1.5}$.

**\*59.** Criticize or comment upon each of the following statements.
  **(a)** Vapor pressure remains constant regardless of volume.
  **(b)** The only forces that affect boiling point are dispersion forces.
  **(c)** The strength of the covalent bonds within a molecule has no effect on the melting point of the molecular substance.
  **(d)** A compound at its critical temperature is always a gas regardless of pressure.

**\*60.** Differentiate between
  **(a)** a covalent bond and a hydrogen bond.
  **(b)** melting and vaporizing.
  **(c)** normal boiling point and a boiling point.
  **(d)** a molecular compound and a network covalent compound.

**\*61.** Differentiate between
  **(a)** the triple point and the critical point.
  **(b)** a metal and an ionic solid.
  **(c)** a phase diagram and a vapor-pressure curve.
  **(d)** volume effect and temperature effect on vapor pressure.

**62.** The density of liquid mercury at 20°C is 13.6 g/cm³, and its vapor pressure is $1.2 \times 10^{-3}$ mm Hg.
  **(a)** What volume (cm³) is occupied by one mole of Hg(*l*) at 20°C.
  **(b)** What volume (cm³) is occupied by one mole of Hg(*g*) at 20°C and the equilibrium vapor pressure?
  **(c)** The atomic radius of Hg is 0.155 nm. Calculate the volume (cm³) of one mole of Hg atoms ($V = 4\pi r^3/3$).
  **(d)** From your answers to (a), (b), and (c), calculate the percentage of the total volume occupied by the atoms in Hg(*l*) and Hg(*g*) at 20°C and $1.2 \times 10^{-3}$ mm Hg.

**\*63.** Elemental boron is almost as hard as diamond. It is insoluble in water, does not conduct electricity at room temperature, and melts at 2300°C. What type of solid is boron?

**\*64.** Four shiny solids are labeled A, B, C, and D. Given the following information about the solids, deduce the identity of A, B, C, and D. (1) The solids are a graphite rod, a silver bar, a lump of "fool's gold" (iron sulfide), and iodine crystals. (2) B, C, and D are insoluble in water. A is slightly soluble. (3) Only C can be hammered into a sheet. (4) C and D conduct electricity as solids, B conducts when melted. A does not conduct as a solid or when melted or dissolved in water.

**65.** An experiment is performed to determine the vapor pressure of formic acid. A 30.0-L volume of helium gas at 20.0°C is passed through 10.00 g of liquid formic acid (HCOOH) at 20.0°C. After the experiment, 7.50 g of liquid formic acid remains. Assume that the helium gas becomes saturated with formic acid vapor and that the total gas volume and temperature remain constant. What is the vapor pressure of formic acid at 20.0°C?

## Challenge Problems

**66.** Iron crystallizes in a body-centered unit cell. Its atomic radius is 0.124 nm. Its density is 7.86 g/cm³. Body-centered crystals have two atoms per cell. Using this information, estimate Avogadro's number.

**67.** A flask with a volume of 10.0 L contains 0.400 g of hydrogen gas and 3.20 g of oxygen gas. The mixture is ignited and the reaction

$$2H_2(g) + O_2(g) \longrightarrow 2H_2O$$

goes to completion. The mixture is cooled to 27°C. Assuming 100% yield,
  **(a)** what physical state(s) of water is (are) present in the flask?
  **(b)** what is the final pressure in the flask?

**68.** Trichloroethane, $C_2H_3Cl_3$, is the active ingredient in aerosols that claim to "stain proof" men's ties. Trichloroethane has a vapor pressure of 100.0 mm Hg at 20.0°C and boils at 74.1°C. An uncovered cup ($\frac{1}{2}$ pint) of trichloroethane ($d = 1.325$ g/mL) is kept in an 18-ft³ refrigerator at 39°F. What percentage of the trichloroethane is left as a liquid when equilibrium is established?

**69.** It has been suggested that the pressure exerted on a skate blade is sufficient to melt the ice beneath it and form a thin film of water, which makes it easier for the blade to slide over the ice. Assume that a skater weighs 120 lb and that the blade has an area of 0.10 in². Calculate the pressure exerted on the blade (1 atm = 15 lb/in²). From information in the text, calculate the decrease in melting point at this pressure. Comment on the plausibility of this explanation and suggest another mechanism by which the water film might be formed.

**70.** As shown in Figure 9.12, $Li^+$ ions fit into a closely packed array of $Cl^-$ ions, but $Na^+$ ions do not. What is the value of the $r_{cation}/r_{anion}$ ratio at which a cation just fits into a structure of this type?

**\*71.** When the temperature drops from 20°C to 10°C, the pressure of a cylinder of compressed $N_2$ drops by 3.4%. The same temperature change decreases the pressure of a propane ($C_3H_8$) cylinder by 42%. Explain the difference in behavior.

Red and silver maples produce about half as much syrup (p. 293) as sugar maples, whose leaves are shown here. Birches, oaks and telephone poles are successively less productive. (Visuals Unlimited/Richard Thorn)

# 10 Solutions

When water turns ice does it remember one time it was water?
When ice turns back into water does it remember it was ice?

—CARL SANDBURG

*Metamorphoses*

# CHAPTER OUTLINE

**10.1 Concentration Units**      **10.3 Colligative Properties**

**10.2 Principles of Solubility**

In the course of a day, you use or make solutions many times. Your morning cup of coffee is a solution of solids (sugar and coffee) in a liquid (water). The gasoline you fill your gas tank with is a solution of several different liquid hydrocarbons. The soda you drink at a study break is a solution of a gas (carbon dioxide) in a liquid (water).

A solution is a homogeneous mixture of a *solute* (substance being dissolved) distributed through a *solvent* (substance doing the dissolving). Solutions exist in any of the three physical states: gas, liquid, or solid. Air, the most common gaseous solution, is a mixture of nitrogen, oxygen, and lesser amounts of other gases. Many metal alloys are solid solutions. An example is the U.S. "nickel" coin (25% Ni, 75% Cu). The most familiar solutions are those in the liquid state, especially ones in which water is the solvent. Aqueous solutions are most important for our purposes in chemistry and will be emphasized in this chapter.

This chapter covers several of the physical aspects of solutions, including

— methods of expressing solution concentrations by specifying the relative amounts of solute and solvent (Section 10.1)
— factors affecting solubility, including the nature of the solute and the solvent, the temperature, and the pressure (Section 10.2)
— the effect of solutes upon such solvent properties as vapor pressure, freezing point, and boiling point (Section 10.3)

## 10.1 Concentration Units

Several different methods are used to express relative amounts of solute and solvent in a solution. Two concentration units, **molarity** and **mole fraction,** were referred to in previous chapters. Two others, **mass percent** and **molality,** are considered here for the first time.

### Mass Percent; Parts per Million

The mass percent of solute in solution is expressed quite simply:

$$\text{mass percent of solute} = \frac{\text{mass solute}}{\text{total mass solution}} \times 100\%$$

In a solution prepared by dissolving 24 g of NaCl in 152 g of water,

$$\text{mass percent of NaCl} = \frac{24 \text{ g}}{24 \text{ g} + 152 \text{ g}} \times 100\% = \frac{24}{176} \times 100\% = 14\%$$

When the amount of solute is very small, as with trace impurities in water, concentration is often expressed in **parts per million (ppm):**

$$\text{ppm solute} = \frac{\text{mass solute}}{\text{total mass solution}} \times 10^6$$

Parts per billion is defined similarly

Comparing the defining equations for mass percent and parts per million, it should be clear that

$$\text{ppm} = \text{mass percent} \times 10^4$$

In the United States and Canada, drinking water cannot contain more than $5 \times 10^{-4}$ mg of mercury per gram of sample. In parts per million that would be

$$\text{ppm Hg} = \frac{5 \times 10^{-4} \text{ mg Hg}}{1 \times 10^3 \text{ mg}} \times 10^6 = 0.5$$

## Mole Fraction (*X*)

Recall from Chapter 5 the defining equation for mole fraction (*X*) of a component A:

$$X_A = \frac{\text{moles A}}{\text{total moles}} = \frac{n_A}{n_{\text{tot}}}$$

The mole fractions of all components of a solution (A, B, . . .) must add to unity:

$$X_A + X_B + \cdots = 1$$

---

**Example 10.1**    What are the mole fractions of $CH_3OH$ and $H_2O$ in a solution prepared by dissolving 1.20 g of methyl alcohol in 16.8 g of water?

***Strategy***    First (1), convert grams to moles for both components. Then (2), using the defining equation, calculate the mole fraction of $CH_3OH$. Finally (3), obtain the mole fraction of $H_2O$, most simply by using the relation: $X_{CH_3OH} + X_{H_2O} = 1$.

*Solution*

(1) $n_{CH_3OH} = 1.20 \text{ g } CH_3OH \times \dfrac{1 \text{ mol } CH_3OH}{32.04 \text{ g } CH_3OH} = 0.0375 \text{ mol } CH_3OH$

$n_{H_2O} = 16.8 \text{ g } H_2O \times \dfrac{1 \text{ mol } H_2O}{18.02 \text{ g } H_2O} = 0.932 \text{ mol } H_2O$

(2) $X_{CH_3OH} = \dfrac{n_{CH_3OH}}{n_{CH_3OH} + n_{H_2O}} = \dfrac{0.0375}{0.0375 + 0.932} = \boxed{0.0387}$

(3) $X_{H_2O} = 1 - X_{CH_3OH} = 1 - 0.0387 = \boxed{0.9613}$

96% of the molecules in this solution are $H_2O$; the rest are $CH_3OH$

---

## Molality (*m*)

The concentration unit molality, given the symbol *m*, is defined as the number of moles of solute per kilogram (1000 g) of *solvent*.

$$\text{molality } (m) = \frac{\text{moles solute}}{\text{kilograms solvent}}$$

The molality of a solution is readily calculated if the masses of solute and solvent are known (Example 10.2).

**Example 10.2**   A solution used for intravenous feeding contains 4.80 g of glucose, $C_6H_{12}O_6$, in 90.0 g of water. What is the molality of glucose?

***Strategy***   First (1), calculate the number of moles of glucose ($\mathcal{M}$ = 180.16 g/mol), then (2), the number of kilograms of water. Finally (3), use the defining equation to calculate molality.

*Solution*

(1) $n_{\text{glucose}} = 4.80 \text{ g glucose} \times \dfrac{1 \text{ mol glucose}}{180.16 \text{ g glucose}} = 0.0266 \text{ mol glucose}$

(2) $\text{kg water} = 90.0 \text{ g water} \times \dfrac{1 \text{ kg water}}{1000 \text{ g water}} = 0.0900 \text{ kg water}$

(3) $\text{molality} = \dfrac{0.0266 \text{ mol glucose}}{0.0900 \text{ kg water}} = \boxed{0.296 \ m}$

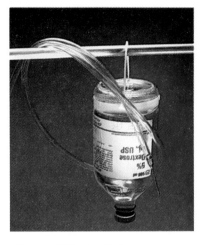

Glucose, $C_6H_{12}O_6$, is the main ingredient of intravenous liquids. (Charles Steele)

## Molarity (*M*)

In Chapter 4, molarity was the concentration unit of choice in dealing with solution stoichiometry. You will recall that molarity is defined as

$$\text{molarity } (M) = \frac{\text{moles solute}}{\text{liters solution}}$$

Molality and molarity are concentration units; morality is something else

A solution can be prepared to a specified molarity by weighing out the calculated mass of solute and dissolving in enough solvent to form the desired volume of solution. Alternatively, you can start with a more concentrated solution and dilute with water to give a solution of the desired molarity (Fig. 10.1). The calculations here are straightforward if you keep a simple point in mind: Adding solvent cannot change the number of moles of solute, B, i.e., $n_B$.

$$n_B \text{ in concentrated solution} = n_B \text{ in dilute solution}$$

In both solutions, the number of moles of solute can be found by multiplying the molarity, [B], by the volume in liters, $V$. Hence,

$$[B]_c V_c = [B]_d V_d$$

**Figure 10.1**

To prepare one liter of 0.100 *M* $CuSO_4$ solution, pipet 50.0 mL of 2.00 *M* $CuSO_4$ into a volumetric flask (left). Then add water in stages (center, right) to bring the level to the narrow mark on the neck of the flask, corresponding to 1.00 L. (Marna G. Clarke)

where the subscripts c and d stand for concentrated and dilute solutions, respectively.

---

**Example 10.3** How would you prepare 1.00 L of 0.100 $M$ $CuSO_4$ starting with 2.00 $M$ $CuSO_4$?

***Strategy*** This question might be restated as: What volume of 2.00 $M$ $CuSO_4$ should be diluted with water to give 1.00 L of 0.100 $M$ $CuSO_4$? To do this, you use the equation $[CuSO_4]_c V_c = [CuSO_4]_d V_d$ to calculate $V_c$, the volume of the more concentrated solution.

***Solution*** Solving for $V_c$

$$V_c = \frac{[CuSO_4]_d \times V_d}{[CuSO_4]_c} = \frac{0.100\ M \times 1.00\ L}{2.00\ M} = 0.0500\ L = 50.0\ mL$$

Measure out 50.0 mL of 2.00 $M$ $CuSO_4$ and dilute with enough water (about 950 mL) to form 1.00 L of 0.100 $M$ solution.

---

*It's easier to dilute a concentrated solution than to start from "scratch"*

The advantage of preparing solutions by the method illustrated in Example 10.3 and Figure 10.1 is that only volume measurements are involved. If you wander into the general chemistry storeroom, you're likely to find concentrated "stock" solutions of various chemicals. Storeroom personnel prepare the more dilute solutions that you use in the laboratory on the basis of calculations like those shown in the example.

## Conversions Between Concentration Units

Frequently it is necessary to convert from one concentration unit to another. This problem arises, for example, in making up solutions of hydrochloric acid. Typically, the analysis or assay that appears on the label (Fig. 10.2) does not give the molarity or molality of the acid. Instead, it lists the mass percent of solute and the density of the solution.

Conversions between concentration units are relatively straightforward provided you *first decide upon a fixed amount of solution*. The amount chosen depends upon the unit in which concentration is originally expressed.

*Complex problems can usually be solved if you know where to start*

| When the Original Concentration Is | Start With |
|---|---|
| Mass percent | 100 g solution |
| Molarity ($M$) | 1.00 L solution |
| Molality ($m$) | 1000 g solvent |
| Mole fraction ($X$) | 1 mol (solute + solvent) |

To illustrate the approach used, let us consider how the molarity of HCl can be calculated from the data shown on the label (Fig. 10.2).

---

**Example 10.4** Calculate the molarity of a concentrated solution of hydrochloric acid that is 37.7% by mass HCl; the solution has a density of 1.19 g/mL.

***Strategy*** Start with 100.0 g of solution. First (1), knowing the percent by mass of HCl (37.7%) and its molar mass (36.46 g/mol), you can readily calculate the number of

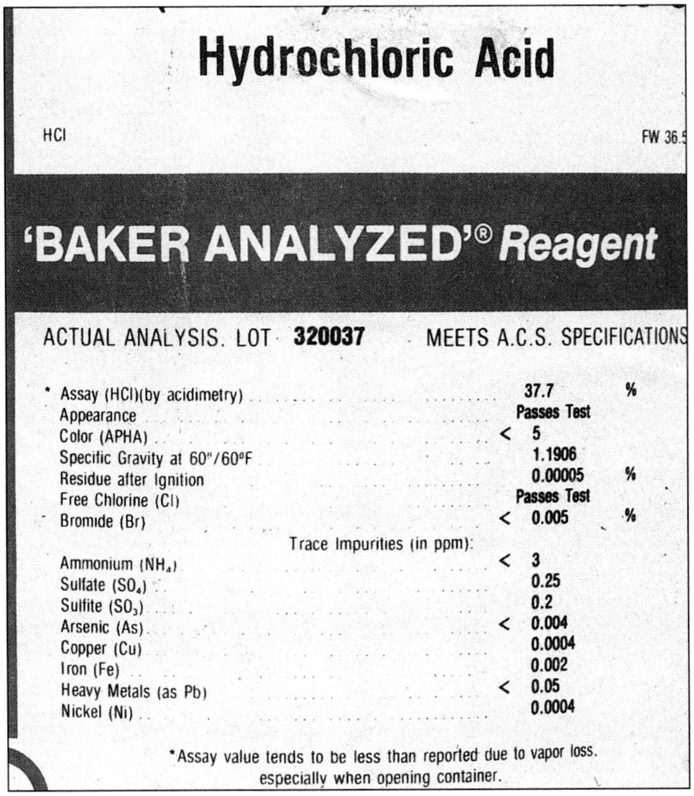

**Hydrochloric Acid**

HCl

FW 36.5

**'BAKER ANALYZED'® Reagent**

ACTUAL ANALYSIS. LOT **320037** · MEETS A.C.S. SPECIFICATIONS

| | | |
|---|---|---|
| * Assay (HCl)(by acidimetry) | 37.7 | % |
| Appearance | Passes Test | |
| Color (APHA) | < 5 | |
| Specific Gravity at 60°/60°F | 1.1906 | |
| Residue after Ignition | 0.00005 | % |
| Free Chlorine (Cl) | Passes Test | |
| Bromide (Br) | < 0.005 | % |
| Trace Impurities (in ppm): | | |
| Ammonium (NH₄) | < 3 | |
| Sulfate (SO₄) | 0.25 | |
| Sulfite (SO₃) | 0.2 | |
| Arsenic (As) | < 0.004 | |
| Copper (Cu) | 0.0004 | |
| Iron (Fe) | 0.002 | |
| Heavy Metals (as Pb) | < 0.05 | |
| Nickel (Ni) | 0.0004 | |

*Assay value tends to be less than reported due to vapor loss, especially when opening container.

**Figure 10.2**
The label on a bottle of concentrated hydrochloric acid typically gives the mass percent of HCl ("assay") and the density (or "specific gravity") of the solution. Given that information, it is possible to calculate the molality, molarity, and mole fraction of HCl.
(Marna G. Clarke)

moles of HCl in 100.0 g of solution. Then (2), knowing the density of the solution, you should be able to calculate the volume occupied by 100.0 g of solution. Finally (3), knowing both the number of moles and the volume in liters, use the defining equation to calculate the molarity of HCl.

*Solution*

(1) $n_{HCl} = 100.0 \text{ g soln.} \times \dfrac{37.7 \text{ g HCl}}{100.0 \text{ g soln.}} \times \dfrac{1 \text{ mol HCl}}{36.46 \text{ g HCl}} = 1.03 \text{ mol HCl}$

(2) $V = 100.0 \text{ g} \times \dfrac{1 \text{ mL}}{1.19 \text{ g}} \times \dfrac{1 \text{ L}}{1000 \text{ mL}} = 0.0840 \text{ L}$

(3) $[HCl] = 1.03 \text{ mol}/0.0840 \text{ L} = \boxed{12.3 \ M}$

This same approach can be used to convert between molarity and molality (Example 10.5).

**Example 10.5**   A 1.13 *M* solution of KOH has a density of 1.05 g/mL. Calculate its molality.

*Strategy*   Start with one liter of solution, which contains 1.13 mol of KOH. Then find, in order, (1) the total mass of solution, knowing the density, (2) the mass of KOH, knowing the molarity, and (3) the mass of water. Finally (4), use the defining equation to calculate molality.

*Solution*

(1) total mass of 1 L of soln. = $1.00 \text{ L} \times \dfrac{1000 \text{ mL}}{1 \text{ L}} \times \dfrac{1.05 \text{ g}}{1 \text{ mL}} = 1050$ g

(2) mass of KOH in 1 L of soln. = $1.13 \text{ mol KOH} \times \dfrac{56.11 \text{ g KOH}}{1 \text{ mol KOH}} = 63.4$ g KOH

(3) mass of water in 1 L of soln. = 1050 g − 63.4 g = 987 g

(4) molality = $\dfrac{1.13 \text{ mol KOH}}{0.987 \text{ kg water}}$ = $\boxed{1.14 \; m}$

---

You will notice from Example 10.5 that the molarity and molality of the KOH solutions are very close to one another (1.13 *M*, 1.14 *m*). This is generally true of *dilute water solutions;* one liter of a dilute aqueous solution contains approximately one kilogram of water. For concentrated or nonaqueous solutions, molarity and molality ordinarily differ considerably from each other.

## 10.2   Principles of Solubility

The extent to which a solute dissolves in a particular solvent depends upon several factors. The most important of these are

— the nature of solvent and solute particles and the interactions between them
— the temperature at which the solution is formed
— the pressure of a gaseous solute

In this section we consider in turn the effect of each of these factors upon solubility.

### Solute-Solvent Interactions

In discussing solubility, it is sometimes stated that "like dissolves like." A more meaningful way to express this idea is to say that two substances with intermolecular forces of about the same type and magnitude are likely to be very soluble in one another. To illustrate, consider the hydrocarbons pentane, $C_5H_{12}$, and hexane, $C_6H_{14}$, which are completely miscible with each other. Molecules of these nonpolar substances are held together by dispersion forces of about the same magnitude. A pentane molecule experiences little or no change in intermolecular forces when it goes into solution in hexane.

Most nonpolar substances have very small water solubilities. Petroleum, a mixture of hydrocarbons, spreads out in a thin film on the surface of a body of water rather than dissolving. The mole fraction of pentane, $C_5H_{12}$, in a saturated water solution is only 0.0001. These low solubilities are readily understood in terms of the structure of liquid water. To dissolve appreciable amounts of pentane in water, it would be necessary to break the hydrogen bonds holding $H_2O$ molecules together. There is no comparable attractive force between $C_5H_{12}$ and $H_2O$ to supply the energy required to break into the water structure.

Of the relatively few organic compounds that dissolve readily in water, most contain —OH groups. Three familiar examples are methyl alcohol (methanol), ethyl alcohol (ethanol), and ethylene glycol, all of which are infinitely soluble in water.

Gasoline and water, shown here as two distinct and separable layers, are insoluble in each other. (Charles D. Winters)

H—C—OH     H—C—C—OH     H—C—C—H

methyl alcohol     ethyl alcohol     ethylene glycol

In these compounds, as in water, the principal intermolecular forces are hydrogen bonds. When a substance like methyl alcohol dissolves in water, it forms hydrogen bonds with $H_2O$ molecules. These hydrogen bonds, joining a $CH_3OH$ molecule to an $H_2O$ molecule, are about as strong as those in the pure substances.

Not all organic compounds that contain —OH groups are soluble in water (Table 10.1). As molar mass increases, the polar —OH group represents an increasingly smaller portion of the molecule. At the same time, the nonpolar hydrocarbon portion becomes larger. As a result, solubility decreases with molar mass. Butanol, $CH_3CH_2CH_2CH_2OH$, is much less soluble in water than methanol, $CH_3OH$. The hydrocarbon portion, shaded in red, is much larger in butanol than in methanol.

The solubility (or insolubility) of different vitamins is of concern in nutrition. Molecules of vitamins B and C contain several —OH groups that can form hydrogen bonds with water (Fig. 10.3 p. 280). As a result, they are water-soluble, readily excreted by the body, and must be consumed daily. In contrast, vitamins A, D, E, and K, whose molecules are relatively nonpolar, are water-insoluble. These vitamins are not so readily excreted; they tend to stay behind in fatty tissues. This means that the body can draw on its reservoir of vitamins A, D, E, and K to deal with sporadic deficiencies. Conversely, megadoses of these vitamins can lead to very high, possibly toxic, concentrations in the body.

*The presence of H bonds has a profound effect on solubility*

As we noted in Chapter 4, the solubility of ionic compounds in water varies tremendously from one solid to another. The extent to which solution occurs depends upon a balance between two forces, both electrical in nature:

1. The force of attraction between $H_2O$ molecules and the ions, which tends to bring the solid into solution. If this factor predominates, the compound is very soluble in water, as is the case with NaCl, NaOH, and many other ionic solids.
2. The force of attraction between oppositely charged ions, which tends to keep them in the solid state. If this is the major factor, the water solubility is very

### TABLE 10.1  Solubilities of Alcohols in Water

| Substance | Formula | Solubility (g solute/L $H_2O$) |
|---|---|---|
| Methanol | $CH_3OH$ | Completely soluble |
| Ethanol | $CH_3CH_2OH$ | Completely soluble |
| Propanol | $CH_3CH_2CH_2OH$ | Completely soluble |
| Butanol | $CH_3CH_2CH_2CH_2OH$ | 74 |
| Pentanol | $CH_3CH_2CH_2CH_2CH_2OH$ | 27 |
| Hexanol | $CH_3CH_2CH_2CH_2CH_2CH_2OH$ | 6.0 |
| Heptanol | $CH_3CH_2CH_2CH_2CH_2CH_2CH_2OH$ | 1.7 |

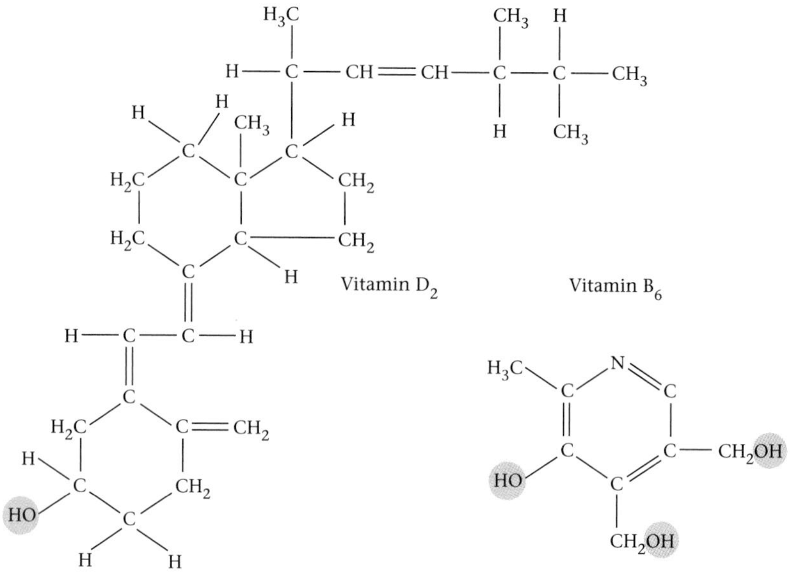

**Figure 10.3**
Molecular structures for Vitamins $D_2$ and $B_6$. Polar groups are in color. Vitamin $D_2$ is water insoluble; Vitamin $B_6$ is water-soluble.

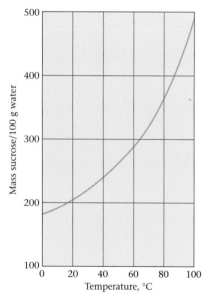

Vitamin $D_2$

Vitamin $B_6$

low. The fact that $CaCO_3$ and $BaSO_4$ are almost insoluble in water implies that interionic attractive forces predominate with these ionic solids.

## Effect of Temperature upon Solubility

When an excess of a solid such as sucrose, $C_{12}H_{22}O_{11}$, (commonly called sugar) is shaken with water, it forms a saturated solution. An equilibrium is established between molecules in the solid state and in solution

$$C_{12}H_{22}O_{11}(s) \rightleftharpoons C_{12}H_{22}O_{11}(aq)$$

At 20°C, the saturated solution contains 204 g of sucrose per 100 g of water.

A similar type of equilibrium is established when a gas such as oxygen is bubbled through water.

$$O_2(g) \rightleftharpoons O_2(aq)$$

At 20°C and 1 atm, 0.00138 mol of $O_2$ dissolves per liter of water.

The effect of a temperature change upon solubility equilibria such as these can be predicted by applying a simple principle. ***An increase in temperature always shifts the position of an equilibrium so as to favor an endothermic process.*** This means that if the solution process absorbs heat, ($\Delta H_{soln.} > 0$), an increase in temperature increases the solubility. Conversely, if the solution process is exothermic ($\Delta H < 0$), an increase in temperature decreases the solubility.

Dissolving a solid in a liquid is usually an endothermic process; heat must be absorbed to break down the crystal lattice.

$$\text{solid} + \text{liquid} \rightleftharpoons \text{solution} \qquad \Delta H_{soln.} > 0$$

Consistent with this effect, the solubilities of solids usually increase as the temperature rises. More sugar dissolves in hot coffee than in cold coffee (Figure 10.4).

**Figure 10.4**
The solubility of sugar, $C_{12}H_{22}O_{11}$, in water increases exponentially with temperature.

*Mass sucrose/100 g water* (y-axis: 100, 200, 300, 400, 500)

*Temperature, °C* (x-axis: 0, 20, 40, 60, 80, 100)

**Figure 10.5**
The column of solid sodium acetate was formed by slowly pouring a super-saturated solution over a seed crystal. (Charles D. Winters)

**Figure 10.6**
Rock candy is formed by the crystallization of sugar from a saturated solution that is cooled slowly. (Marna G. Clarke)

A warm glass rod placed in a glass of ginger ale increases the temperature of the carbonated solution. The increase in temperature decreases the solubility of $CO_2$ in water, causing the $CO_2$ to bubble out of solution. (Charles D. Winters)

Cooling a saturated solution usually causes a solid to crystallize out of solution, since the solubility is smaller at the lower temperature. Sometimes, however, this doesn't happen; the excess solute stays in solution upon cooling, giving what is known as a **supersaturated solution** (Figure 10.5). This happens when a saturated solution of sucrose at 80°C (362 g of $C_{12}H_{22}O_{11}$/100 g of water) is cooled carefully to 20°C, where the solubility is 204 g/100 g of water. The excess sucrose stays in solution until a small seed crystal of sucrose is added, whereupon crystallization quickly takes place. The excess solute (362 g − 204 g = 158 g of $C_{12}H_{22}O_{11}$/100 g of water) comes out of solution, establishing equilibrium between the saturated solution and the sucrose crystals.

The crystallization of excess solute is a common problem in the preparation of candies and in the storage of jam and honey. From these supersaturated solutions, sugar separates either as tiny crystals, causing the "graininess" in fudge, or as large crystals, which often appear in honey kept for a long time (Fig. 10.6).

When a gas condenses to a liquid, heat is always evolved. By the same token, heat is usually evolved when a gas dissolves in a liquid:

$$\text{gas} + \text{liquid} \rightleftharpoons \text{solution} \qquad \Delta H_{\text{soln.}} < 0$$

This means that the reverse process (gas coming out of solution) is endothermic. Hence, it is favored by an increase in temperature; typically, gases become less soluble as the temperature rises. This rule is followed by all gases in water. You have probably noticed this effect when heating water in an open pan or beaker. Bubbles of air are driven out of the water by an increase in temperature.

**Figure 10.7**
The solubility of $O_2(g)$ in water decreases as temperature rises (a) and increases as pressure increases (b). In (a), the pressure is held constant at 1 atm; in (b), the temperature is held constant at 25°C.

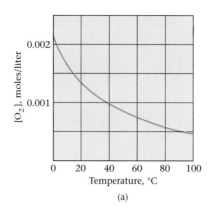

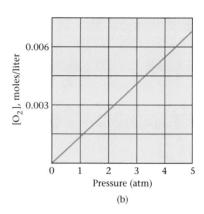

(a)  (b)

The reduced solubility of oxygen in water at high temperatures (Fig. 10.7a) may explain why trout congregate at the bottom of deep pools on hot summer days when the surface water is depleted of dissolved oxygen.

## Effect of Pressure upon Solubility

Pressure has a major effect on solubility only for gas-liquid systems. At a given temperature, a rise in pressure increases the solubility of a gas. Indeed, at low to moderate pressures, gas solubility is directly proportional to pressure (Figure 10.7b).

$$C_g = kP_g$$

where $P_g$ is the partial pressure of the gas over the solution, $C_g$ is its concentration in the solution, and $k$ is a constant characteristic of the particular gas-liquid system. This relation is called **Henry's law** after its discoverer, William Henry (1775–1836), a friend of John Dalton.

Henry's law arises because increasing the pressure raises the concentration of molecules in the gas phase. To balance this change and maintain equilibrium, more gas molecules enter the solution, increasing their concentration in the liquid phase.

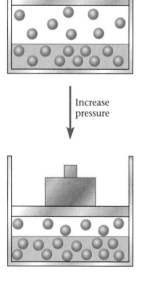

Increase
pressure

An increase in the pressure of a gas above a solution increases the solubility of the gas. This is an illustration of Henry's law.

---

**Example 10.6**   The solubility of pure nitrogen in blood at body temperature, 37°C, and one atmosphere, is $6.2 \times 10^{-4}$ M. If a diver breathes air ($X\ N_2 = 0.78$) at a depth where the total pressure is 2.5 atm, calculate the concentration of nitrogen in his blood.

**Strategy**   Use Henry's law in any problem involving gas solubility and pressure. Perhaps the simplest approach here is to use the data at one atmosphere to calculate $k$. Then apply Henry's law to calculate $C_g$ at the higher pressure. Note that it is the *partial* pressure of $N_2$ that is required; use the relation $P_{N_2} = X_{N_2} \times P_{tot}$ to find it.

**Solution**   At 1 atm:

$$k = \frac{\text{concentration of } N_2}{\text{pressure of } N_2} = \frac{6.2 \times 10^{-4}\ M}{1.00\ \text{atm}} = 6.2 \times 10^{-4}\ M/\text{atm}$$

At the higher pressure:

$$P_{N_2} = X_{N_2} \times P_{tot} = 0.78 \times 2.5 \text{ atm} = 2.0 \text{ atm}$$

$$[N_2] = kP_{N_2} = 6.2 \times 10^{-4} \frac{M}{\text{atm}} \times 2.0 \text{ atm} = \boxed{1.2 \times 10^{-3} \; M}$$

The influence of partial pressure on gas solubility is used in making carbonated beverages such as beer, sparkling wines, and many soft drinks. These beverages are bottled under pressures of $CO_2$ as high as 4 atm. When the bottle or can is opened, the pressure above the liquid drops to 1 atm, and the carbon dioxide bubbles rapidly out of solution. Pressurized containers for shaving cream, whipped cream, and cheese spreads work on a similar principle. Pressing a valve reduces the pressure on the dissolved gas, causing it to rush from solution, carrying liquid with it as a foam.

Another consequence of the effect of pressure on gas solubility is the painful, sometimes fatal, affliction known as the "bends." This occurs when a person goes rapidly from deep water (high pressure) to the surface (lower pressure). The rapid decompression causes air, dissolved in blood and other body fluids, to bubble out of solution. These bubbles impair blood circulation and affect nerve impulses. To minimize these effects, deep-sea divers and aquanauts breathe a helium-oxygen mixture rather than compressed air (nitrogen-oxygen). Helium is only about one-third as soluble as nitrogen, and hence much less gas comes out of solution upon decompression.

## 10.3 Colligative Properties

The properties of a solution differ considerably from those of the pure solvent. Those solution properties that depend primarily upon the *concentration of solute particles* rather than their nature are called **colligative properties.** Such properties include vapor pressure lowering, osmotic pressure, boiling point elevation, and freezing point depression. This section considers the relations between colligative properties and solute concentration, starting with **nonelectrolytes,** which exist in solution as molecules, and then going on (briefly) to electrolytes, which form ions in solution.

The relationships among colligative properties and solute concentration are best regarded as limiting laws. They are approached more closely as the solution becomes more dilute. In practice, the relationships discussed here are valid, for nonelectrolytes, to within a few percent at concentration as high as 1 *M*. At higher concentrations, solute-solute interactions lead to larger deviations.

### Vapor Pressure Lowering (Nonelectrolytes)

You may have noticed that concentrated aqueous solutions evaporate more slowly than does pure water. This reflects the fact that the vapor pressure of water over the solution is less than that of pure water (Fig. 10.8).

Vapor pressure lowering is a true colligative property; that is, it is independent of the nature of the solute but directly proportional to its concentration. For example, the vapor pressure of water above a 0.10 *M* solution of either glucose or sucrose at 0°C is the same, about 0.008 mm Hg less than that of pure

Scuba divers have to worry about this

● Solvent molecules
● Solute molecules

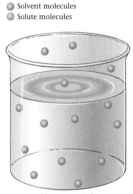

$X_{solvent} = 1.0$

$X_{solvent} = 0.5$

**Figure 10.8**

Raoult's law. Adding a solute lowers the concentration of solvent molecules in the liquid phase. To maintain equilibrium, the concentration of solvent molecules in the gas phase must decrease, thereby lowering the solvent vapor pressure.

water. In 0.30 *M* solution, the vapor pressure lowering is almost exactly three times as great, 0.025 mm Hg.

The relationship between solvent vapor pressure and concentration is ordinarily expressed as

$$P_1 = X_1 P_1°$$

In this equation, $P_1$ is the vapor pressure of solvent over the solution, $P_1°$ is the vapor pressure of the pure solvent at the same temperature, and $X_1$ is the mole fraction of solvent. Note that since $X_1$ in a solution must be less than 1, $P_1$ must be less than $P_1°$. This relationship is called **Raoult's law;** Francois Raoult (1830–1901) carried out a large number of careful experiments on vapor pressures and freezing point lowering.

To obtain a direct expression for vapor pressure lowering, note that $X_1 = 1 - X_2$, where $X_2$ is the mole fraction of solute. Substituting $1 - X_2$ for $X_1$ in Raoult's law,

$$P_1 = (1 - X_2)P_1°$$

Rearranging,

$$P_1° - P_1 = X_2 P_1°$$

*Vapor pressure lowering is directly proportional to solute mole fraction*

The quantity $(P_1° - P_1)$ is the vapor pressure lowering ($\Delta P$). It is the difference between the solvent vapor pressure in the pure solvent and in solution.

$$\Delta P = X_2 P_1°$$

---

**Example 10.7**   A solution contains 102 g of sugar, $C_{12}H_{22}O_{11}$, in 375 g of water. Calculate the vapor pressure lowering at 25°C (vp of pure water = 23.76 mm Hg).

***Strategy***   First (1), calculate the number of moles of sugar ($\mathcal{M} = 342.30$ g/mol) and water ($\mathcal{M} = 18.02$ g/mol). That information allows you to calculate (2) the mole fraction of sugar. Finally (3), use Raoult's law to find the vapor pressure lowering.

*Solution*

(1) $n_{sugar} = 102 \text{ g sugar} \times \dfrac{1 \text{ mol sugar}}{342.30 \text{ g sugar}} = 0.298 \text{ mol sugar}$

$n_{water} = 375 \text{ g water} \times \dfrac{1 \text{ mol water}}{18.02 \text{ g water}} = 20.8 \text{ mol water}$

(2) $X_{sugar} = \dfrac{0.298}{0.298 + 20.8} = \dfrac{0.298}{21.1} = 0.0141$

(3) $\Delta P = X_{sugar} \times P°_{H_2O} = 0.0141 \times 23.76 \text{ mm Hg} = \boxed{0.335 \text{ mm Hg}}$

---

One interesting effect of vapor pressure lowering is shown at the left of Figure 10.9. Here, we start with two beakers, one containing pure water and the other containing a sugar solution. These are placed next to each other, under a bell jar (Fig. 10.9a). As time passes, the liquid level in the beaker containing the solution rises. The level of pure water in the other beaker falls. Eventually, by evaporation and condensation, all the water is transferred to the solution (Fig. 10.9b). At the end of the experiment, the beaker that contained pure water is

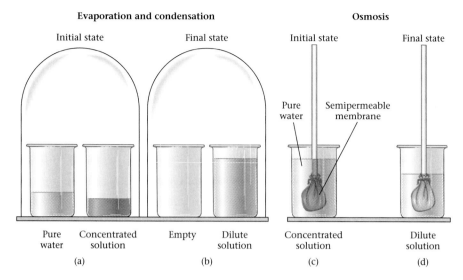

Evaporation and condensation

Initial state  Final state

Pure water  Concentrated solution

Empty  Dilute solution

(a)  (b)

Osmosis

Initial state  Final state

Pure water  Semipermeable membrane

Concentrated solution

Dilute solution

(c)  (d)

**Figure 10.9**
Water tends to move spontaneously from a region where its vapor pressure is high to a region where it is low. In a → b, movement of water molecules occurs through the air trapped under the bell jar. In c → d, water molecules move by osmosis through a semi-permeable membrane. The driving force is the same in the two cases, although the mechanism differs.

empty. The driving force behind this process is the difference in vapor pressure of water in the two beakers. *Water moves from a region in which its vapor pressure is high* (pure water) *to one in which its vapor pressure is low* (sugar solution). This is a general tendency, followed by all liquids, and is responsible for a variety of natural processes, including osmosis.

## Osmotic Pressure (Nonelectrolytes)

The apparatus shown in Figure 10.9c and d can be used to achieve a result similar to that found in the bell jar experiment. Here, a sugar solution is separated from water by a "semipermeable" membrane. This may be an animal bladder, a slice of vegetable tissue, or a piece of parchment. The membrane, by a mechanism that is not well understood, allows water molecules to pass through it, but not sugar molecules. Here, as before, water moves from a region where its vapor pressure is high (pure water) to a region where it is low (sugar solution). This process, taking place through a membrane permeable only to the solvent, is called **osmosis.** As a result of osmosis, the water level rises in the tube and drops in the beaker (Fig. 10.9d).

The osmotic pressure, $\pi$, is equal to the external pressure, $P$, just sufficient to prevent osmosis (Fig. 10.10 p. 286). If $P$ is less than $\pi$, osmosis takes place in the normal way, and water moves through the membrane into the solution (Fig. 10.10a). By making the external pressure large enough, it is possible to reverse this process (Fig. 10.10b). When $P > \pi$, water molecules move through the membrane from the solution to pure water. This process, called **reverse osmosis,** is used to obtain fresh water from seawater in arid regions of the world, including Saudi Arabia.

Osmotic pressure, like vapor pressure lowering, is a colligative property. For any nonelectrolyte B, $\pi$ is directly proportional to molarity, [B]. The equation relating these two quantities is very similar to the ideal gas law:

$$\pi = \frac{n_B RT}{V} = [B]RT$$

**Figure 10.10**
Osmosis can be prevented by applying to the solution a pressure P that just balances the osmotic pressure, π. If P < π, normal osmosis occurs. If P > π, water flows in the opposite direction, producing reverse osmosis, used to obtain fresh water from sea water.

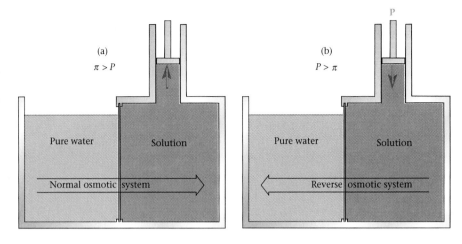

(a) π > P

(b) P > π

Pure water    Solution

Normal osmotic system

Pure water    Solution

Reverse osmotic system

where R is the gas law constant, 0.0821 L · atm/mol · K, and T is the Kelvin temperature. Even in dilute solution, the osmotic pressure is quite large. Consider, for example, a 0.10 M solution at 25°C:

$$\pi = (0.10 \text{ mol/L}) \left(0.0821 \frac{\text{L} \cdot \text{atm}}{\text{mol} \cdot \text{K}}\right) (298 \text{ K}) = 2.4 \text{ atm}$$

A pressure of 2.4 atm is equivalent to a column of water 25 m (more than 80 ft) high.

Dishwasher's hands get wrinkled too

If a cucumber is placed in a concentrated brine solution, it shrinks and assumes the wrinkled skin of a pickle. The skin of the cucumber acts as a semipermeable membrane. The water solution inside the cucumber is more dilute than the solution surrounding it. As a result, water flows out of the cucumber into the brine (Fig. 10.11).

When a dried prune is placed in water, the skin also acts as a semipermeable membrane. This time the solution inside the prune is more concentrated than the water, so that water flows into the prune, making the prune less wrinkled.

Nutrient solutions used in intravenous feeding must be *isotonic* with blood; that is, they must have the same osmotic pressure as blood. If the solution is too dilute, its osmotic pressure will be less than that of the fluids inside blood

**Figure 10.11**
When a cucumber is pickled, water moves out of the cucumber by osmosis into the concentrated brine solution. A prune placed in pure water swells as water moves into the prune, again by osmosis. (Marna G. Clarke)

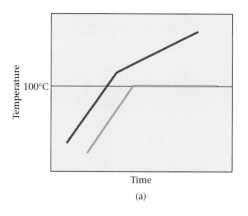

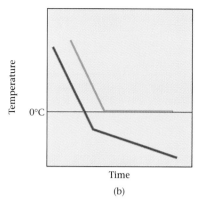

**Figure 10.12**

Heating (a) and cooling (b) curves for pure water (red) and an aqueous solution (green). For pure water, the temperature remains constant during boiling or freezing. For the solution, the temperature changes steadily during the phase change because water is being removed, increasing the concentration of solute.

cells; in that case, water will flow into the cell until it bursts. Conversely, if the nutrient solution has too high a concentration of solutes, water will flow out of the cell until it shrivels and dies.

## Boiling Point Elevation and Freezing Point Lowering (Nonelectrolytes)

When a solution of a nonvolatile solute is heated, it does not begin to boil until the temperature exceeds the boiling point of the solvent. The difference in temperature is called the **boiling point elevation,** $\Delta T_b$.

$$\Delta T_b = T_b - T_b^{\circ}$$

where $T_b$ and $T_b^{\circ}$ are the boiling points of the solution and the pure solvent, respectively. As boiling continues, pure solvent distils off, the concentration of solute increases, and the boiling point continues to rise (Fig. 10.12).

When a solution is cooled, it does not begin to freeze until a temperature below the freezing point of the pure solvent is reached. The **freezing point lowering,** $\Delta T_f$, is defined to be a positive quantity:

$$\Delta T_f = T_f^{\circ} - T_f$$

where $T_f^{\circ}$, the freezing point of the solvent, lies above $T_f$, the freezing point of the solution. As freezing takes place, pure solvent freezes out, the concentration of solute increases, and the freezing point continues to drop.

Boiling point elevation is a direct result of vapor pressure lowering. At any given temperature, a solution of a nonvolatile solute* has a vapor pressure *lower* than that of the pure solvent. Hence, a *higher* temperature must be reached before the solution boils; that is, before its vapor pressure becomes equal to the external pressure. Figure 10.13 illustrates this reasoning graphically.

The freezing point lowering, like the boiling point elevation, is a direct result of the lowering of the solvent vapor pressure by the solute. Notice from Figure 10.13 that the freezing point of the solution is the temperature at which the solvent in solution has the same vapor pressure as the pure solid solvent. This implies that it is pure solvent (e.g., ice) that separates when the solution freezes.

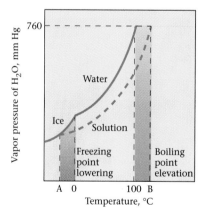

**Figure 10.13**

Since a nonvolatile solute lowers the vapor pressure of a solvent, the boiling point of a solution will be higher and the freezing point lower than the corresponding values for the pure solvent. Water solutions freeze *below* 0°C at point A and boil *above* 100°C at point B.

---

*Volatile solutes ordinarily lower the boiling point because they contribute to the total vapor pressure of the solution.

TABLE 10.2 **Molal Freezing Point and Boiling Point Constants**

| Solvent | fp (°C) | $k_f$ (°C/$m$) | bp (°C) | $k_b$ (°C/$m$) |
|---|---|---|---|---|
| Water | 0.00 | 1.86 | 100.00 | 0.52 |
| Acetic acid | 16.66 | 3.90 | 117.90 | 2.53 |
| Benzene | 5.50 | 5.10 | 80.10 | 2.53 |
| Cyclohexane | 6.50 | 20.2 | 80.72 | 2.75 |
| Camphor | 178.40 | 40.0 | 207.42 | 5.61 |
| p-Dichlorobenzene | 53.1 | 7.1 | 174.1 | 6.2 |
| Naphthalene | 80.29 | 6.94 | 217.96 | 5.80 |

Solutes raise the boiling point and lower the freezing point

Boiling point elevation and freezing point lowering, like vapor pressure lowering, are colligative properties. They are directly proportional to solute concentration, generally expressed as molality, *m*. The relevant equations are

$$\Delta T_b = k_b \text{ (molality)}$$

$$\Delta T_f = k_f \text{ (molality)}$$

The proportionality constants in these equations, $k_b$ and $k_f$, are called the *molal boiling point constant* and the *molal freezing point constant,* respectively. Their magnitudes depend upon the nature of the solvent (Table 10.2). Note that when the solvent is water,

$$k_b = 0.52°C/m \qquad k_f = 1.86°C/m$$

**Example 10.8** An antifreeze solution is prepared containing 50.0 cm³ of ethylene glycol, $C_2H_6O_2$ ($d$ = 1.12 g/cm³), in 50.0 g of water. Calculate the freezing point of this "50-50" mixture.

***Strategy*** First (1) calculate the number of moles of $C_2H_6O_2$ ($\mathcal{M}$ = 62.07 g/mol). Then (2) apply the defining equation to calculate the molality. Finally (3), use the equation $\Delta T_f = (1.86°C/m) \times$ molality to find the freezing point lowering.

***Solution***

(1) $n_{C_2H_6O_2} = 50.0 \text{ cm}^3 \times \dfrac{1.12 \text{ g}}{1 \text{ cm}^3} \times \dfrac{1 \text{ mol}}{62.04 \text{ g}} = 0.903 \text{ mol}$

(2) molality $= \dfrac{\text{moles of } C_2H_6O_2}{\text{kg of water}} = \dfrac{0.903 \text{ mol}}{0.0500 \text{ kg}} = 18.1 \; m$

(3) $\Delta T_f = k_f \text{ (molality)} = 1.86°C/m \times 18.1 \; m = 33.7°C$

The freezing point of the solution is 33.7°C below that of pure water (0°C). Hence the solution should freeze at  $-33.7°C$. Actually, the freezing point is somewhat lower, about $-37°C$ ($-35°F$). The deviation reminds us that the equation used here, $\Delta T_f = k_f$ (molality) is a limiting law.

Commercial-grade ethylene glycol is added to a car's radiator to prevent the water from freezing in the winter and boiling over in the summer. (Union Carbide)

You take advantage of freezing point lowering when you add antifreeze to your automobile radiator in winter. Ethylene glycol is the solute commonly used in so-called "permanent" antifreeze. It has a high boiling point (197°C), is virtually nonvolatile at 100°C, and raises the boiling point of water. Hence, antifreeze that contains ethylene glycol does not boil away in summer driving.

When a water solution is cooled to the freezing point, ice crystallizes out. If the ice is removed, as by filtration, the concentration of solute is raised. This is the way "ice beer" is made; it contains a relatively high concentration of alcohol.

## Determination of Molar Masses of Nonelectrolytes from Colligative Properties

Colligative properties, particularly freezing point depression, can be used to determine molar masses of a wide variety of nonelectrolytes. The approach used is illustrated in Example 10.9.

---

**Example 10.9**   A student dissolves 1.50 g of a newly prepared compound in 75.0 g of cyclohexane. She measures the freezing point of the solution to be 2.70°C; that of pure cyclohexane is 6.50°C. Cyclohexane has a $k_f$ of 20.2°C/$m$. Using these data, calculate the molar mass of the compound.

*Strategy*   First (1), knowing the freezing point of the solution and that of the pure solvent, calculate $\Delta T_f$. Then (2) use the equation $\Delta T_f = k_f$ (molality) to find the molality. Finally (3), obtain the molar mass using the defining equation for molality.

*Solution*

(1)  $\Delta T_f = T_f^\circ - T_f = 6.50°C - 2.70°C = 3.80°C$

(2)  $\Delta T_f = k_f$ (molality)

$$\text{molality} = \frac{\Delta T_f}{k_f} = \frac{3.80°C}{20.2°C/m} = 0.188 \ m$$

(3)  $\text{molality} = \dfrac{\text{moles solute}}{\text{kilograms solvent}} = \dfrac{\text{grams solute}/\mathcal{M}}{\text{kilograms solvent}}$

Solving for the molar mass:

$$\mathcal{M} = \frac{\text{grams solute}}{(\text{molality})(\text{kilograms solvent})}$$

All the quantities on the right side of this equation are known; 1.50 g of solute is dissolved in 75.0 g (0.0750 kg) of solvent. The molality was calculated to be 0.188 $m$. So

$$\mathcal{M} = \frac{1.50 \text{ g}}{\left(\dfrac{0.188 \text{ mol}}{\text{kg solvent}}\right) \times 0.0750 \text{ kg solvent}} = \boxed{106 \text{ g/mol}}$$

---

In carrying out a molar mass determination by freezing point depression, we must choose a solvent in which the solute is readily soluble. Usually, several such solvents are available. Of these, we tend to pick one that has the largest $k_f$. This makes $\Delta T_f$ large and thus reduces the percent error in the freezing point measurement. From this point of view, cyclohexane or other organic solvents are better choices than water, since their $k_f$ values are larger.

Molar masses can also be determined using other colligative properties. Osmotic pressure measurements are often used, particularly for solutes of high molar mass, where the concentration is likely to be quite low. The advantage of using osmotic pressure is that the effect is relatively large. Consider, for example, a 0.0010 $M$ aqueous solution, for which

$$\pi \text{ at } 25°C = 0.024 \text{ atm} = 18 \text{ mm Hg}$$

$$\Delta T_f \approx 1.86 \times 10^{-3}°C$$

$$\Delta T_b \approx 5.2 \times 10^{-4}°C$$

A pressure of 18 mm Hg can be measured relatively accurately; temperature differences of the order of 0.001°C are essentially impossible to measure accurately.

---

**Example 10.10**    A solution contains 1.0 g of hemoglobin dissolved in enough water to form 0.100 L of solution. The osmotic pressure at 20°C is found to be 2.75 mm Hg. Calculate

(a) the molarity of hemoglobin
(b) the molar mass of hemoglobin.

*Strategy*    The molarity of hemoglobin, [Hem], is readily calculated from the relation $\pi = [\text{Hem}]RT$. Then the molar mass can be obtained from the defining equation for molarity.

*Solution*

(a) $[\text{Hem}] = \pi/RT$

$$\pi = (2.75/760) \text{ atm} \qquad R = 0.0821 \text{ L} \cdot \text{atm/mol} \cdot \text{K} \qquad T = 293 \text{ K}$$

$$[\text{Hem}] = \frac{(2.75/760) \text{ atm}}{(0.0821 \text{ L} \cdot \text{atm/mol} \cdot \text{K})(293 \text{ K})} = \boxed{1.50 \times 10^{-4} \text{ mol/L}}$$

(b) From the defining equation for molarity,

$$\text{molarity} = \frac{\text{moles solute}}{\text{liters solution}} = \frac{\text{grams solute}/\mathcal{M}}{\text{liters solution}}$$

Solving for the molar mass,

$$\mathcal{M} = \frac{\text{grams solute}}{(\text{molarity})(\text{liters solution})}$$

Recall that there is 1.0 g of hemoglobin in 0.100 L of solution. Hence,

$$\mathcal{M} = \frac{1.0 \text{ g}}{(1.50 \times 10^{-4} \text{ mol/L})(0.100 \text{ L})} = \boxed{6.7 \times 10^4 \text{ g/mol}}$$

---

*Molar masses of polymers like polyethylene can be measured this way*

## Colligative Properties of Electrolytes

As noted earlier, colligative properties of solutions are directly proportional to the concentration of solute *particles*. On this basis, it is reasonable to suppose that, at a given concentration, an electrolyte should have a greater effect upon these properties than does a nonelectrolyte. When one mole of a nonelectrolyte such as glucose dissolves in water, one mole of solute molecules is obtained. On the other hand, one mole of the electrolyte NaCl yields two moles of ions (1 mol of $Na^+$, 1 mol of $Cl^-$). With $CaCl_2$, three moles of ions are produced per mole of solute (1 mol of $Ca^{2+}$, 2 mol of $Cl^-$).

Truck applying salt, NaCl, on a snow-packed road.

This reasoning is confirmed experimentally. Compare, for example, the vapor pressure lowerings for 1 *M* solutions of glucose, sodium chloride, and calcium chloride at 25°C.

|  | Glucose | NaCl | CaCl$_2$ |
|---|---|---|---|
| $\Delta P$ | 0.42 mm Hg | 0.77 mm Hg | 1.3 mm Hg |

With many electrolytes, $\Delta P$ is so large that the solid, when exposed to moist air, picks up water *(deliquesces)*. This occurs with calcium chloride, whose saturated solution has a vapor pressure only 30% that of pure water. If dry CaCl$_2$ is exposed to air in which the relative humidity is greater than 30%, it absorbs water and forms a saturated solution. Deliquescence continues until the vapor pressure of the solution becomes equal to that of the water in the air.

The freezing points of electrolyte solutions, like their vapor pressures, are lower than those of nonelectrolytes at the same concentration. Sodium chloride and calcium chloride are used to lower the melting point of ice on highways; their aqueous solutions can have freezing points as low as $-21$ and $-55$°C, respectively. The equation for the freezing point lowering of an electrolyte is similar to that for nonelectrolytes, except for the introduction of a multiplier, *i*, called the *Van't Hoff factor*. For aqueous solutions of electrolytes,

$$\Delta T_f = 1.86°C/m \times \text{molality} \times i$$

Similar equations apply for other colligative properties.

$$\Delta T_b = 0.52°C/m \times \text{molality} \times i$$

$$\pi = \text{molarity} \times R \times T \times i$$

If we assume that the ions of an electrolyte behave independently, *i* should be equal to **the number of moles of ions per mole of electrolyte.** Thus, *i* should be 2 for NaCl and MgSO$_4$, 3 for CaCl$_2$ and Na$_2$SO$_4$, and so on.

Potassium chloride is sold for home use because it's kinder to the environment

---

**Example 10.11**  Estimate the freezing points of 0.20 *m* solutions of

(a) KNO$_3$    (b) Cr(NO$_3$)$_3$

Assume that *i* is the number of moles of ions formed per mole of electrolyte.

***Strategy***  Find *i* and apply the equation $\Delta T_f = 1.86°C/m \times \text{molality} \times i$.

***Solution***

(a) One mole of KNO$_3$ forms two moles of ions:

$$KNO_3(s) \longrightarrow K^+(aq) + NO_3^-(aq)$$

Hence, *i* should be 2, and we have

$$\Delta T_f = (1.86°C)(0.20)(2) = 0.74°C \qquad \boxed{T_f = -0.74°C}$$

(b) For Cr(NO$_3$)$_3$, $i = 4$:

$$Cr(NO_3)_3(s) \longrightarrow Cr^{3+}(aq) + 3NO_3^-(aq)$$

$$\Delta T_f = (1.86°C)(0.20)(4) = 1.5°C \qquad \boxed{T_f = -1.5°C}$$

TABLE 10.3 **Freezing Point Lowerings of Solutions**

| Molality | $\Delta T_f$ Observed (°C) | | $i$ (Calc from $\Delta T_f$) | |
| | NaCl | MgSO$_4$ | NaCl | MgSO$_4$ |
| --- | --- | --- | --- | --- |
| 0.005 | 0.0182 | 0.0160 | 1.96 | 1.72 |
| 0.01 | 0.0360 | 0.0285 | 1.94 | 1.53 |
| 0.02 | 0.0714 | 0.0534 | 1.92 | 1.44 |
| 0.05 | 0.176 | 0.121 | 1.89 | 1.30 |
| 0.10 | 0.348 | 0.225 | 1.87 | 1.21 |
| 0.20 | 0.685 | 0.418 | 1.84 | 1.12 |
| 0.50 | 1.68 | 0.995 | 1.81 | 1.07 |

*For a nonelectrolyte, $i = 1$*

The data in Table 10.3 suggest that the situation is not as simple as this discussion implies. The observed freezing point lowerings of NaCl and MgSO$_4$ are smaller than would be predicted with $i = 2$. For example, 0.50 $m$ solutions of NaCl and MgSO$_4$ freeze at $-1.68$ and $-0.995$°C, respetively; the predicted freezing point is $-1.86$°C. Only in very dilute solution does the multiplier $i$ approach the predicted value of 2.

This behavior is generally typical of electrolytes. Their colligative properties deviate considerably from "ideal" values, even at concentrations below 1 $m$. There are at least a couple of reasons for this effect.

**1.** Because of electrostatic attraction, an ion in solution tends to surround itself with more ions of opposite than of like charge (Fig. 10.14). The existence of this *ionic atmosphere,* first proposed by Peter Debye, a Dutch physical chemist, in 1923, prevents ions from acting as completely independent solute particles. The result is to make an ion somewhat less effective than a nonelectrolyte molecule in its influence on colligative properties.

**2.** Oppositely charged ions may interact strongly enough to form a discrete species called an *ion pair.* This effect is essentially nonexistent with electrolytes such as NaCl, in which the ions have low charges ($+1$, $-1$). However, with MgSO$_4$ ($+2$, $-2$ ions), ion pairing plays a major role. Even at concentrations as low as 0.1 $m$, there are more MgSO$_4$ ion pairs than "free" Mg$^{2+}$ and SO$_4^{2-}$ ions.

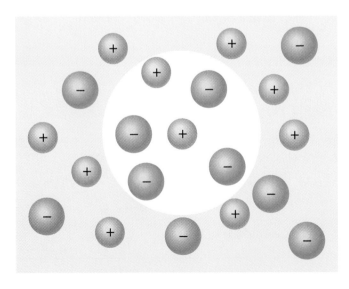

**Figure 10.14**
Ion atmosphere. An ion, on the average, is surrounded by more ions of opposite charge than of like charge.

# Maple Syrup

The collection of maple sap and its conversion to syrup or sugar illustrate many of the principles covered in this chapter. Moreover, in northern New England, making maple syrup is an interesting way to spend the month of March (Fig. 10.A), which separates midwinter from "mud season."

The driving force behind the flow of maple sap is by no means obvious. The calculated osmotic pressure of sap, a 2% solution of sucrose ($\mathcal{M}$ = 342 g/mol) is

$$\pi = \frac{20/342 \text{ mol}}{1 \text{ L}} \times 0.0821 \, \frac{\text{L} \cdot \text{atm}}{\text{mol} \cdot \text{K}} \times 280 \text{ K} = 1.3 \text{ atm}$$

This is sufficient to push water to a height of about 45 ft; many maple trees are taller than that. Besides, a maple tree continues to bleed sap for several days after it has been cut down. An alternative theory suggests that sap is forced out of the tree by bubbles of $CO_2(g)$, produced by respiration. When the temperature drops at night, the carbon dioxide goes into solution, and the flow of sap ceases. This would explain the high sensitivity of sap flow to temperature; the aqueous solubility of carbon dioxide doubles when the temperature falls by 15°C.

If you want to make your own maple syrup on a small scale, perhaps 10 to 20 L per season, there are a few principles to keep in mind.

**1.** Make sure the trees you tap are *maples;* hemlocks would be a particularly poor choice. Identify the trees to be tapped in the fall, before the leaves fall. If possible, select sugar maples, which produce about one liter of maple syrup per tree per season. Other types of maples are less productive.

**2.** It takes 20 to 40 L of sap to yield one liter of maple syrup. To remove the water, you could freeze the sap, as the colonists did 200 years ago. The ice that forms is pure water (p. 288); by discarding it, you increase the concentration of sugar in the remaining solution. Large-scale operators today use reverse osmosis (p. 285) to remove about half of the water. The remainder must be boiled off. The characteristic flavor of maple syrup is caused by compounds formed upon heating, such as

acetol          cyclotene          vanillin

**Figure 10.A**
WLM collecting sap from red maple trees.

It's best to boil off the water outdoors. If you do this in the kitchen, you may not be able to open the doors and windows for a couple of weeks. Wood tends to swell when it absorbs a few hundred liters of water.

**3.** When the concentration of sugar reaches 66%, you are at the maple syrup stage. The calculated boiling point elevation (660 g of sugar, 340 g of water)

$$\Delta T_b = 0.52°C \times \frac{660/342}{0.340} = 3.0°C$$

is somewhat less than the observed value, about 4°C. The temperature rises very rapidly around 104°C (Fig. 10.B), which makes thermometry the method of choice for detecting the "end point." Shortly before that point, add a drop or two of vegetable oil to prevent foaming; perhaps this is the source of the phrase "spreading oil on troubled waters."

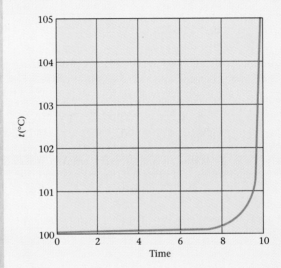

**Figure 10.B**
It seems to take forever to boil down maple sap until you reach 101°C. Then the temperature rises rapidly to the "end point," 104°C. Further *careful* heating produces maple sugar.

# CHAPTER HIGHLIGHTS

## *Key Concepts*

1. Calculate the concentration of a solute (mass percent, mole fraction, molality, molarity) given mass and/or density data
   (Examples 10.1, 10.2, 10.4, 10.5; Problem 1–10, 17, 69)
2. Make dilution calculations
   (Example 10.3; Problems 11, 12, 15, 16)
3. Apply Henry's law to relate gas solubility to partial pressure
   (Example 10.6; Problems 27–30)
4. Apply Raoult's law to calculate the vapor pressure of a solution
   (Example 10.7; Problems 31–34)
5. Relate the freezing point, boiling point, or osmotic pressure of a solution to solute concentration
   (Examples 10.8–10.11; Problems 35–58, 70)

## *Key Equations*

| | |
|---|---|
| Henry's law | $C_g = kP_g$ |
| Raoult's law | $P_1 = X_1 P_1^\circ$ |
| Osmotic pressure | $\pi = MRT$ |
| Boiling point, freezing point | $\Delta T_b = k_b \times m \times i \qquad \Delta T_f = k_f \times m \times i$ |
| | ($i$ = no. of moles of particles per mole of solute) |

## *Key Terms*

boiling point
colligative property
electrolyte
mass percent
melting point
molality

molar mass
molarity
mole fraction
nonelectrolyte
osmosis
osmotic pressure

partial pressure
saturated solution
solvent
supersaturated solution
vapor pressure

## *Summary Problem*

Consider naphthalene, $C_{10}H_8$ ($\mathcal{M}$ = 128.2 g/mol), the active ingredient of moth balls. A solution of naphthalene is prepared by mixing 25.0 g of naphthalene with 0.750 L of carbon disulfide, $CS_2$ ($d$ = 1.263 g/mL). Assume that the volume remains 0.750 L when the solution is prepared.

(a) What is the mass percent of naphthalene in the solution?
(b) What is the concentration of naphthelene in parts per million?
(c) What is the density of the solution?
(d) What is the molarity of the solution?
(e) What is the molality of the solution?
(f) The vapor pressure of pure $CS_2$ at 25°C is 358 mm Hg. Assume that the vapor pressure exerted by naphthalene at 25°C is negligible. (It is 1 mm Hg at 50°C.) What is the vapor pressure of the solution at this temperature?
(g) What is the osmotic pressure of the solution at 25°C?
(h) The normal boiling point of $CS_2$ is 46.13°C ($k_b$ = 2.34°C/m). What is the normal boiling point of the solution?
(i) Vasopressin is a hormone secreted by the pituitary gland to regulate the rate at which water is reabsorbed by the kidneys. To determine its molar mass, 10.00 g of vasopressin is dissolved in 50.00 g of naphthalene ($k_f$ = 6.94°C/m). The freezing point of the mixture is determined to be 79.01°C; that of pure naphthalene is 80.29°C. What is the molar mass of vasopressin?

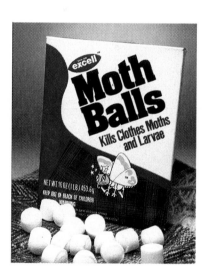

Naphthalene is the major ingredient of moth balls.

### *Answers*

(a)  2.57%   (b)  $2.57 \times 10^4$ ppm   (c)  1.30 g/mL   (d)  0.260 *M*
(e)  0.206 *m*   (f)  352 mm Hg   (g)  6.36 atm   (h)  46.61°C
(i)  $1.08 \times 10^3$ g/mol

# Questions & Problems

## Concentrations of Solutions

**1.** Acetone, $C_3H_6O$, is the active ingredient in nail polish removers. A solution is made up by adding 15.0 mL of acetone ($d = 0.792$ g/mL) to 25.0 mL of isopropyl alcohol, $C_3H_8O$ ($d = 0.7854$ g/mL). Assuming that volumes are additive, calculate

    **(a)** the mass percent of acetone in the solution.
    **(b)** the volume percent of isopropyl alcohol in the solution.
    **(c)** the mole fraction of acetone in the solution.

**2.** A solution is prepared by dissolving 2.86 g of potassium permanganate in 75.0 mL of water ($d = 1.00$ g/mL). Calculate

    **(a)** the mass percent of potassium permanganate in the solution.
    **(b)** the mole fraction of water in the solution.

**3.** For a solution of acetic acid ($CH_3COOH$) to be called "vinegar," it must contain 5.00% acetic acid by mass. If a vinegar is made up only of acetic acid and water, what is the molarity of acetic acid in the vinegar? The density of vinegar is 1.006 g/mL.

**4.** The "proof" of an alcoholic beverage is twice the volume percent of ethyl alcohol, $C_2H_5OH$, in solution. For an 80-proof (two significant figures) rum, what is the molality of the ethyl alcohol? Take the densities of ethyl alcohol and water to be 0.789 g/mL and 1.00 g/mL, respectively.

**5.** Lead is a poisonous metal that especially affects children because they retain a larger fraction of lead than adults do. Lead levels of 0.250 ppm in a child cause delayed cognitive development. How many moles of lead present in 1.00 g of a child's blood would 0.250 ppm represent?

**6.** A 100.0-g sample of water from a river is found to contain $12.6 \times 10^{-4}$ mg of mercury. What is the concentration of mercury in the river in ppm?

**7.** Complete the following table for aqueous solutions of sodium dichromate:

| | Mass of Solute | Volume of Solution | Molarity |
|---|---|---|---|
| (a) | 12.7 g | 225 mL | ——— |
| (b) | ——— | 386 mL | 2.87 $M$ |
| (c) | 3.88 g | ——— | 0.693 $M$ |

**8.** Complete the following table for aqueous solutions of silver nitrate:

| | Mass of Solute | Volume of Solution | Molarity |
|---|---|---|---|
| (a) | ——— | 500.0 mL | 3.875 $M$ |
| (b) | 1.288 g | ——— | 0.2239 $M$ |
| (c) | 6.474 g | 275 mL | ——— |

**9.** Complete the following table for aqueous solutions of sucrose, $C_{12}H_{22}O_{11}$:

| | Molality | Mass Percent of Solvent | Ppm Solute | Mole Fraction of Solvent |
|---|---|---|---|---|
| (a) | 3.894 | ——— | ——— | ——— |
| (b) | ——— | 12.0 | ——— | ——— |
| (c) | ——— | ——— | 2972 | ——— |
| (d) | ——— | ——— | ——— | 0.749 |

**10.** Complete the following table for aqueous solutions of oxalic acid, $H_2C_2O_4$:

| | Molality | Mass Percent of Solvent | Ppm Solute | Mole Fraction of Solvent |
|---|---|---|---|---|
| (a) | ——— | ——— | ——— | 0.634 |
| (b) | ——— | ——— | 437 | ——— |
| (c) | ——— | 78.0 | ——— | ——— |
| (d) | 1.810 | ——— | ——— | ——— |

**11.** Describe how you would prepare 375 mL of 0.295 $M$ copper(II) sulfate solution starting with

    **(a)** solid copper(II) sulfate.
    **(b)** 1.00 $M$ copper(II) sulfate solution.

**12.** Describe how you would prepare 1.00 L of 0.600 $M$ sodium hydroxide solution starting with

    **(a)** solid sodium hydroxide.
    **(b)** 6.00 $M$ sodium hydroxide solution.

**\*13.** One mole of $CaCl_2$ is represented as ⬚⚬, where ⬚ represents Ca and ⚬ represents Cl. Complete the picture below showing only the calcium and chloride ions. The water molecules need not be shown. What is the molarity of $Ca^{2+}$? of $Cl^-$?

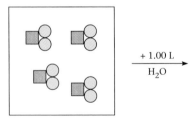

**\*14.** One mole of $Na_2S$ is represented as ⬚⚬, where ⬚ represents Na and ⚬ represents S. Complete the picture showing only the sodium and sulfide ions. The water molecules need not be shown.

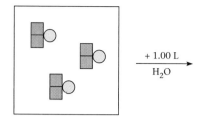

What is the molarity of $Na^+$? of $S^{2-}$?

**15.** A solution is prepared by diluting 325 mL of 0.188 $M$ aluminum sulfate solution with water to a final volume of 0.750 L. Calculate

(a) the molarity of aluminum sulfate, aluminum ions, and sulfate ions.

(b) the number of moles of aluminum ions present in the original solution.

**16.** A solution is prepared by diluting 0.250 L of 1.65 $M$ nickel(II) nitrate solution with water to a final volume of 1.25 L.

(a) What are the molarities of nickel(II) nitrate, nickel(II) ion, and nitrate in the diluted solution?

(b) How many grams of nickel(II) nitrate were dissolved to give the original solution?

**17.** The concentrated nitric acid available in the laboratory is 71.0% nitric acid by mass. Its density is 1.418 g/mL.

(a) What is the molarity of concentrated nitric acid?

(b) How would you prepare 2.00 L of 6.00 $M$ $HNO_3$ from the concentrated solution?

**18.** A bottle of commercial concentrated aqueous ammonia is labeled 29.89% $NH_3$ by mass; density = 0.8960 g/mL.

(a) What is the molarity of the ammonia solution?

(b) If 100.0 mL of the commercial ammonia is diluted with water to make 2.50 L of solution, what is the molarity of the diluted solution?

**19.** Complete the following table for aqueous solutions of iron(III) chloride:

| | Density (g/mL) | Molarity | Molality | Mass Percent of Solute |
|---|---|---|---|---|
| (a) | 1.080 | 0.6203 | —— | —— |
| (b) | 1.230 | —— | 1.982 | —— |
| (c) | 1.350 | —— | —— | 34.47 |

**20.** Complete the following table for aqueous solutions of potassium iodide:

| | Density (g/mL) | Molarity | Molality | Mass Percent of Solute |
|---|---|---|---|---|
| (a) | 1.028 | —— | —— | 4.00 |
| (b) | 1.227 | —— | 2.117 | —— |
| (c) | 1.546 | 4.654 | —— | —— |

## Solubilities

**\*21.** Choose the member of each set that you would expect to be more soluble in water. Explain your answer.

(a) chloromethane, $CH_3Cl$, or methanol, $CH_3OH$

(b) lithium chloride or ethyl chloride, $C_2H_5Cl$

(c) naphthalene, $C_{10}H_8$, or hydrogen peroxide, $H$—$O$—$O$—$H$

(d) chloroform, $CHCl_3$, or hydrogen chloride.

**\*22.** Choose the member of each set that you would expect to be more soluble in water. Explain your answer.

(a) nitrogen triiodide or potassium iodide

(b) ammonia or methane

(c) silicon dioxide or sodium hydroxide

(d) methyl alcohol, $CH_3OH$, or methyl ether, $H_3C$—$O$—$CH_3$

**23.** Consider the process by which calcium carbonate dissolves in water:

$$CaCO_3(s) \longrightarrow Ca^{2+}(aq) + CO_3{}^{2-}(aq)$$

(a) Using data from tables in Chapter 8, calculate $\Delta H$ for this reaction.

(b) Would you expect the solubility of $CaCO_3$ to increase or decrease if the temperature is increased?

**24.** Consider the process by which ammonium nitrate dissolves in water.

(a) Write an equation to represent the solution process.

(b) Calculate $\Delta H$ for the reaction in (a), using tables in Chapter 8.

(c) Discuss the effect of temperature on the solubility of ammonium nitrate in water.

**25.** Describe the following solutions as saturated, unsaturated, or supersaturated. Use Figure 10.4.

(a) 90 g of sucrose in 50 g of water at 20°C.

(b) 58 g of sucrose in 20 g of water at 20°C.

(c) 12 g of sucrose in 35 g of water at 80°C.

**26.** Use Figure 10.4 to answer the following questions:

(a) How many grams of sucrose can be dissolved in 80 g of water at 40°C to make a saturated solution?

(b) A supersaturated solution is prepared by dissolving 750 g of sucrose in 125 g of boiling water (100°C). The solution is then slowly cooled to 25°C. One milligram of sucrose is added. How many grams of sucrose crystallize out?

**\*27.** A certain gaseous solute dissolves in water, evolving 12.0 kJ of heat. Its solubility at 25°C and 4.00 atm is 0.0200 $M$. Would you expect the solubility to be greater or less than 0.0200 $M$ at

(a) 5°C and 6 atm?  (b) 50°C and 2 atm?

(c) 20°C and 4 atm?  (d) 25°C and 1 atm?

**\*28.** Dissolving a gaseous solute in water is an exothermic process. Its solubility at 20°C and 1.00 atm is $1.00 \times 10^{-3}$ $M$. To increase the solubility of the gas in water,

(a) the temperature must be (greater or less) than 20°C.

(b) the pressure must be (greater or less) than 1.00 atm.

**29.** The Henry's law constant for the solubility of oxygen in water is $3.30 \times 10^{-4}$ *M*/atm at 12°C and $2.85 \times 10^{-4}$ *M*/atm at 22°C. Air is 21 mol% oxygen.

(a) How many grams of oxygen can be dissolved in one liter of a trout stream at 12°C (54°F) at an air pressure of 1.00 atm?

(b) How many grams of oxygen can be dissolved per liter in the same trout stream at 22°C (72°F) at the same pressure as in (a)?

(c) A nuclear power plant is responsible for the stream's increase in temperature. What percentage of dissolved oxygen is lost by this increase in the stream's temperature?

**30.** A carbonated beverage is made by saturating water with carbon dioxide at 0°C and a pressure of 3.0 atm. The bottle is then opened at room temperature (25°C) and comes to equilibrium with air in the room containing $CO_2$ ($P_{CO_2} = 3.4 \times 10^{-4}$ atm).

(a) What is the concentration of carbon dioxide in the bottle before it is opened?

(b) What is the concentration of carbon dioxide in the bottle after it has been opened and has come to equilibrium with the air?

The Henry's law constant for the solubility of $CO_2$ in water is 0.0769 *M*/atm at 0°C and 0.0313 *M*/atm at 25°C.

## Colligative Properties

**31.** Calculate the vapor pressure of water over each of the following sucrose ($C_{12}H_{22}O_{11}$) solutions at 24°C (vp pure water = 22.38 mm Hg).

(a) $X_{sucrose} = 0.150$

(b) % sucrose by mass = 22.5%

(c) 1.88 *m* sucrose

**32.** Repeat the calculations called for in Problem 31 at 96°C (vp pure water = 657.6 mm Hg).

**33.** The vapor pressure at 20°C of pure benzene, $C_6H_6$, is 74.7 mm Hg. How many grams of naphthalene ($C_{10}H_8$) must be added to 30.00 g of benzene so that the vapor pressure of benzene over the solution is 70.00 mm Hg?

**34.** How would you prepare 1.00 L of an aqueous glucose ($C_6H_{12}O_6$) solution ($d = 1.05$ g/mL) with a vapor pressure of 16.00 mm Hg at 22°C (vp pure water = 19.83 mm Hg)?

**35.** Calculate the osmotic pressure at 25°C in the following solutions of urea $CO(NH_2)_2$:

(a) 0.250 *M* urea

(b) 10.0 g of urea dissolved in enough water to make 555 mL of solution

(c) 1.50% by mass (density of the solution = 1.03 g/mL)

**36.** Lysozyme is an enzyme that cleaves bacterial cell walls. A sample of lysozyme extracted from egg whites has a molar mass of 13,930 g/mol. If 20.0 mg of this enzyme is dissolved in water to make 225 mL of solution at 23°C, what is the osmotic pressure of the solution?

**37.** Reverse osmosis is one way of obtaining pure water from sea water. If the sea water contains 3.0 g of NaCl (assume $i = 2$) per liter of solution, how much pressure is required to reverse the osmotic process at 25°C?

**38.** Adrenaline is the hormone responsible for supplying cells with extra glucose in times of stress. A 375-mL aqueous solution has 2.50 g of adrenaline. Its osmotic pressure at 20°C is measured to be $6.70 \times 10^2$ mm Hg. What is the molar mass of adrenaline?

**39.** Calculate the freezing point and normal boiling point of each of the following solutions:

(a) 32.0 g of propylene glycol, $C_3H_8O_2$, in 250.0 mL of water ($d = 1.00$ g/cm$^3$)

(b) 43.0 mL of methanol, $CH_3OH$ ($d = 0.792$ g/mL) in 500.0 g of water ($d = 1.00$ g/cm$^3$)

(c) 17.6% by mass glycerine, $C_3H_8O_3$, in water.

**40.** How many grams of the following nonelectrolytes would have to be dissolved in 100.0 g of benzene (see Table 10.2) to increase the boiling point by 1.0°C? To decrease the freezing point by 1.5°C?

(a) citric acid, $C_6H_8O_7$

(b) caffeine, $C_8H_{10}N_4O_2$

**41.** An automobile radiator is filled with an antifreeze solution prepared by mixing equal volumes of ethylene glycol, $C_2H_6O_2$ ($d = 1.12$ g/cm$^3$), and water ($d = 1.00$ g/cm$^3$). Estimate the freezing point of the mixture. Will this mixture protect automobile engines in Connecticut if the lowest temperature expected is −20°F?

**42.** What are the freezing point and normal boiling point of a solution made by adding 35.0 mL of isopropyl alcohol, $C_3H_7OH$, to 95.0 mL of water? The densities of isopropyl alcohol and water are 0.785 g/cm$^3$ and 1.00 g/cm$^3$, respectively.

**43.** When 13.66 g of lactic acid, $C_3H_6O_3$, is mixed with 115 g of stearic acid, the mixture freezes at 63.5°C. The freezing point of pure stearic acid is 69.4°C. What is the freezing point constant of stearic acid?

**44.** The Rast method uses camphor ($C_{10}H_{16}O$) as a solvent for determining the molar mass of a compound. When 2.50 g of cortisone acetate is dissolved in 50.00 g of camphor ($k_f = 40.0$°C/m), the freezing point of the mixture is determined to be 173.44°C; that of pure camphor is 178.40°C. What is the molar mass of cortisone acetate?

**45.** A compound contains 42.9% C, 2.4% H, 16.6% N, and 38.1% O. The addition of 3.16 g of this compound to 75.0 mL of cyclohexane ($d = 0.779$ g/cm$^3$) gives a solution with a freezing point at 0.0°C. Using Table 10.2, determine the molecular formula of the compound.

**46.** Lauryl alcohol is obtained from the coconut and is an ingredient in many shampoos. Its simplest formula is $C_{12}H_{26}O$. A solution of 5.00 g of lauryl alcohol in 100.0 g of benzene boils at 80.78°C. Using Table 10.2, find the molecular formula of lauryl alcohol.

**47.** It has been shown that antioxidants play a role in protecting organisms against certain cancers. Analysis of β-carotene, an antioxidant and source of vitamin A, shows that it is made up of carbon and hydrogen atoms only. It is 89.5% C by mass. When 50.0 mg of β-carotene is dissolved in 5.00 g of cyclohexane ($k_f$ = 20.2°C/m), the freezing point depression of the solution is 0.376°C. What are the simplest and molecular formulas of β-carotene?

**48.** Pure phenol has a boiling point of 182.0°C. Its boiling point constant, $k_b$, is 3.56°C/m. A sample of phenol is contaminated by benzoic acid, $C_7H_6O_2$. The boiling point of the contaminated sample is 182.6°C. How pure is the sample? (Express your answer as mass percent of phenol.)

**49.** Aqueous solutions introduced into the blood stream by injection must have the same osmotic pressure as blood, i.e., they must be "isotonic" with blood. At 25°C, the average osmotic pressure of blood is 7.7 atm. What is the molarity of a glucose solution isotonic with blood?

**50.** Refer to Problem 49 and determine the concentration of an isotonic saline solution (NaCl in $H_2O$). Recall that NaCl is an electrolyte; assume complete conversion to $Na^+$ and $Cl^-$ ions.

**51.** The molar mass of a type of hemoglobin was determined by osmotic pressure measurement. A student measured an osmotic pressure of 4.60 mm Hg for a solution at 20°C containing 3.27 g of hemoglobin in 0.200 L of solution. What is the molar mass of hemoglobin?

**52.** A biochemist isolated a new protein and determined its molar mass by osmotic pressure measurements. She used 0.270 g of the protein in 50.0 mL of solution and observed an osmotic pressure of 3.86 mm Hg for this solution at 25°C. What should she report as the molar mass of the new protein?

**\*53.** Arrange 0.10 *m* solutions of the following solutes in order of decreasing freezing point and boiling point.
  **(a)** $C_2H_6O_2$    **(b)** $Fe(NO_3)_3$
  **(c)** $Cr_2(SO_4)_3$    **(d)** $K_2Cr_2O_7$

**\*54.** Estimate the freezing and boiling points at 1 atm of 0.10 *m* solutions of
  **(a)** NaBr    **(b)** $AlCl_3$    **(c)** $(NH_4)_2CO_3$

**\*55.** The freezing point of 0.20 *m* HF is −0.38°C. Is HF primarily nonionized in this solution (HF molecules), or is it dissociated to $H^+$ and $F^-$ ions?

**\*56.** The freezing point of 0.10 *M* $KHSO_3$ is −0.38°C. Which of the following equations best represents what happens when $KHSO_3$ dissolves in water?
  **(a)** $KHSO_3(s) \longrightarrow KHSO_3(aq)$
  **(b)** $KHSO_3(s) \longrightarrow K^+(aq) + HSO_3^-(aq)$
  **(c)** $KHSO_3(s) \longrightarrow K^+(aq) + SO_3^{2-}(aq) + H^+(aq)$

**57.** Assume that 30.0 L of maple sap yields one kilogram of maple syrup (66% sucrose, $C_{12}H_{22}O_{11}$). What is the molality of the sucrose solution after one third of the water content of the sap has been removed?

**58.** What is the freezing point of maple syrup (66% $C_{12}H_{22}O_{11}$)?

## Unclassified

**\*59.** Explain why
  **(a)** the freezing point of 0.10 *m* $CaCl_2$ is lower than the freezing point of 0.10 *m* $CaSO_4$.
  **(b)** the solubility of solids in water usually increases as the temperature increases.
  **(c)** pressure must be applied to cause reverse osmosis to occur.

**\*60.** Explain why
  **(a)** 0.10 *M* $BaCl_2$ has a higher osmotic pressure than 0.10 *M* glucose.
  **(b)** the solubility of gases in water decreases as temperature increases.
  **(c)** molarity and molality are nearly the same in dilute aqueous solution.

**\*61.** In your own words, explain
  **(a)** why sea water has a lower freezing point than fresh water.
  **(b)** why we believe that vapor pressure lowering is a colligative property.
  **(c)** why one often obtains a "grainy" product when making fudge (a supersaturated sugar solution).
  **(d)** what causes the "bends" in divers.

**\*62.** In your own words, explain
  **(a)** why the concentrations of solutions used for intravenous feeding must be controlled carefully.
  **(b)** why fish in a lake (and fishermen) seek deep, shaded places during summer afternoons.
  **(c)** why champagne "fizzes" in a glass.
  **(d)** the difference between molarity and molality.

**\*63.** Criticize the following statements.
  **(a)** A saturated solution is always a concentrated solution.
  **(b)** The water solubility of a solid always decreases with a drop in temperature.
  **(c)** For all aqueous solutions, molarity and molality are equal.
  **(d)** The freezing point depression of a 0.10 *m* $CaCl_2$ solution is twice that of a 0.10 *m* KCl solution.
  **(e)** A 0.10 *m* sucrose solution and a 0.10 *m* NaCl solution have the same osmotic pressure.

**\*64.** Explain, in your own words,
  **(a)** how to determine experimentally whether a pure substance is an electrolyte or a nonelectrolyte.
  **(b)** why a cold glass of beer goes "flat" upon warming.
  **(c)** why the molality of a solute is ordinarily larger than its mole fraction.
  **(d)** why the boiling point is raised by the presence of a solute.

**65.** A glycerol ($C_3H_8O_3$) solution that is 40.0% glycerol by weight has a density of 1.103 g/cm$^3$ at 15°C. Calculate its
  **(a)** molarity
  **(b)** molality
  **(c)** vapor pressure (vp $H_2O$ at 15°C = 12.81 mm Hg)
  **(d)** freezing point

**66.** An aqueous solution made up of 12.0 g of sodium sulfate in 100.0 mL of solution has a density of 1.11 g/cm$^3$ at 25°C. Calculate its
   (a) molarity
   (b) molality
   (c) osmotic pressure at 25°C (take i = 3)
   (d) normal boiling point

## Challenge Problems

**67.** What is the density of an aqueous solution of potassium nitrate that has a normal boiling point of 103.0°C and an osmotic pressure of 122 atm at 25°C?

**68.** A solution contains 158.2 g of KOH per liter; its density is 1.13 g/cm$^3$. A lab technician wants to prepare 0.250 molal KOH, starting with 100.0 mL of this solution. How much water or solid KOH should be added to the 100.0-mL portion?

**69.** Show that the following relation is generally valid for any solution:

$$\text{molality} = \frac{\text{molarity}}{d - \dfrac{\mathcal{M} \, (\text{molarity})}{1000}}$$

where $d$ is solution density (g/cm$^3$) and $\mathcal{M}$ is the molar mass of the solute. Using this equation, explain why molality approaches molarity in dilute solution when water is the solvent, but not with other solvents.

**70.** The water-soluble nonelectrolyte X has a molar mass of 410 g/mol. A 0.100-g mixture containing this substance and sugar ($\mathcal{M}$ = 342 g/mol) is added to 1.00 g of water to give a solution freezing at −0.500°C. Estimate the mass percent of X in the mixture.

**71.** A martini, weighing about 5.0 oz (142 g), contains 30.0% alcohol by mass. About 15% of the alcohol in the martini passes directly into the blood stream (7.0 L for an adult). Estimate the concentration of alcohol in the blood (g/cm$^3$) of a person who drinks two martinis before dinner. (A concentration of 0.0010 g/cm$^3$ or more is frequently considered indicative of intoxication in a "normal" adult.)

**72.** When water is added to a mixture of aluminum metal and sodium hydroxide, hydrogen gas is produced. This is the reaction used in commercial drain cleaners:

$$2Al(s) + 6H_2O(l) + 2OH^-(aq) \longrightarrow 2Al(OH)_4^-(aq) + 3H_2(g)$$

A sufficient amount of water is added to 49.92 g of NaOH to make 0.600 L of solution; 41.28 g of Al is added to this solution, and hydrogen gas if formed.
   (a) Calculate the molarity of the initial NaOH solution.
   (b) How many moles of hydrogen were formed?
   (c) The hydrogen was collected over water at 25°C and 758.6 mm Hg. The vapor pressure of water at this temperature is 23.8 mm Hg. What volume of hydrogen was generated?

**73.** It is found experimentally that the volume of a gas that dissolves in a given amount of water is independent of the pressure of the gas; that is, if 5 cm$^3$ of a gas dissolves in 100 g of water at 1 atm pressure, 5 cm$^3$ will dissolve at a pressure of 2 atm, 5 atm, 10 atm, . . . . Show that this relationship follows logically from Henry's law and the ideal gas law.

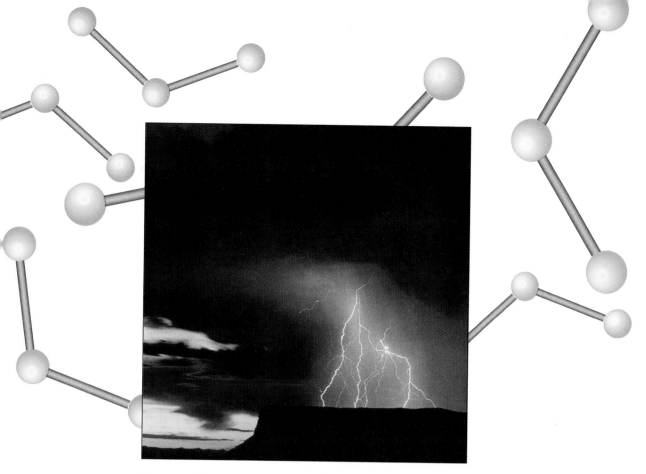

Ozone, produced here by an electrical discharge, has both detrimental and beneficial effects on human health, depending upon where it is found in the atmosphere. (p. 328) (© 1992 Greg Gawlowski/Dembinsky Photo Assoc.)

# Rate of Reaction

## 11

**N**ot every collision,

not every punctilious trajectory

by which billiard-ball complexes

arrive at their calculable meeting

places leads to reaction

Men (and women) are not

as different from molecules

as they think

—ROALD HOFFMANN

*Men and Molecules*

For a chemical reaction to be feasible, it must occur at a reasonable rate. Consequently, it is important to be able to control the rate of reaction. Most often, this means making it occur more rapidly. When you carry out a reaction in the general chemistry laboratory, you want it to take place quickly. A research chemist trying to synthesize a new drug has the same objective. Sometimes, though, it is desirable to reduce the rate of reaction. The aging process, a complex series of biological oxidations, believed to involve "free radicals" such as

$$\cdot \ddot{O}\!\!-\!\!H \qquad \text{and} \qquad \cdot \ddot{O}\!\!-\!\!\ddot{O}\colon$$

*At least all of us over the age of 10*

is one we would all like to slow down.

This chapter sets forth the principles of **chemical kinetics,** the study of reaction rates. The main emphasis is upon those factors that influence rate. These include

— the concentrations of reactants (Sections 11.2, 11.3)
— the process by which the reaction takes place (Section 11.4)
— the presence of a catalyst (Section 11.5)
— the temperature (Section 11.6)
— the reaction mechanism (Section 11.7)

## 11.1 Meaning of Reaction Rate

To discuss reaction rate meaningfully, it must be defined precisely. *The rate of reaction is a positive quantity that expresses how the concentration of a reactant or product changes with time.* To illustrate what this means, consider the reaction

$$N_2O_5(g) \longrightarrow 2NO_2(g) + \tfrac{1}{2}O_2(g)$$

As you can see from Figure 11.1, the concentration of $N_2O_5$ decreases with time; the concentrations of $NO_2$ and $O_2$ increase. Because these species have different coefficients in the balanced equation, their concentrations do not change at the same rate. When *one* mole of $N_2O_5$ decomposes, *two* moles of $NO_2$ and *one-half* mole of $O_2$ are formed. This means that

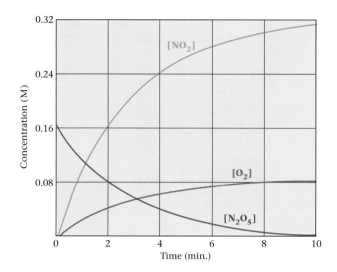

**Figure 11.1**

For the reaction: $N_2O_5(g) \longrightarrow 2NO_2(g) + \frac{1}{2}O_2(g)$, the concentration of $NO_2$ and $O_2$ increase with time, whereas that of $N_2O_5$ decreases. The reaction rate is defined as

$$-\Delta[N_2O_5]/\Delta t = \Delta[NO_2]/2\Delta t = 2\Delta[O_2]/\Delta t$$

$$-\Delta[N_2O_5] = \frac{\Delta[NO_2]}{2} = \frac{\Delta[O_2]}{\frac{1}{2}}$$

where $\Delta[\ ]$ refers to the change in concentration in moles per liter. The minus sign in front of the $N_2O_5$ term takes account of the fact that $[N_2O_5]$ decreases as the reaction takes place; the numbers in the denominator of the terms on the right $(2, \frac{1}{2})$ are the coefficients of these species in the balanced equation. The rate of reaction can now be defined by dividing by the change in time, $\Delta t$:

$$\text{rate} = \frac{-\Delta[N_2O_5]}{\Delta t} = \frac{\Delta[NO_2]}{2\Delta t} = \frac{\Delta[O_2]}{\frac{1}{2}\Delta t}$$

More generally, for the reaction

$$aA + bB \longrightarrow cC + dD$$

where A, B, C, and D represent substances in the gas phase ($g$) or in aqueous solution ($aq$), and $a, b, c, d$ are their coefficients in the balanced equation:

$$\text{rate} = \frac{-\Delta[A]}{a\Delta t} = \frac{-\Delta[B]}{b\Delta t} = \frac{\Delta[C]}{c\Delta t} = \frac{\Delta[D]}{d\Delta t}$$

This is the defining equation for rate

To illustrate the use of this expression, suppose that for the formation of ammonia:

$$N_2(g) + 3H_2(g) \longrightarrow 2NH_3(g)$$

molecular nitrogen is disappearing at the rate of 0.10 mol/L per minute, i.e., $\Delta[N_2]/\Delta t = -0.10$ mol/L · min. From the coefficients of the balanced equation, we see that the concentration of $H_2$ must be decreasing three times as fast: $\Delta[H_2]/\Delta t = -0.30$ mol/L · min. By the same token, the concentration of $NH_3$ must be increasing at the rate of $2 \times 0.10$ mol/L · min: $\Delta[NH_3]/\Delta t = 0.20$ mol/L · min. It follows that

$$\text{rate} = \frac{-\Delta[N_2]}{\Delta t} = \frac{-\Delta[H_2]}{3\Delta t} = \frac{\Delta[NH_3]}{2\Delta t} = \frac{0.10 \text{ mol}}{L \cdot \text{min}}$$

By defining rate this way, it is independent of which species we focus upon, $N_2$, $H_2$, or $NH_3$.

Notice that reaction rate has the units of concentration divided by time. We will always express concentration in moles per liter. Time, on the other hand, can be expressed in seconds, minutes, hours, . . . . A rate of 0.10 mol/L · min corresponds to

$$0.10 \frac{mol}{L \cdot min} \times \frac{1\ min}{60\ s} = 1.7 \times 10^{-3} \frac{mol}{L \cdot s}$$

$$\text{or} \quad 0.10 \frac{mol}{L \cdot min} \times \frac{60\ min}{1\ h} = 6.0 \frac{mol}{L \cdot h}$$

### Measurement of Rate

For the reaction

$$N_2O_5(g) \longrightarrow 2NO_2(g) + \tfrac{1}{2}O_2(g)$$

the rate could be determined by measuring

— the absorption of visible light by the $NO_2$ formed; this species has a reddish-brown color, whereas $N_2O_5$ and $O_2$ are colorless.
— the change in pressure that results from the increase in the number of moles of gas (1 mol reactant → $2\tfrac{1}{2}$ mol product).

The graphs shown in Figure 11.1 were plotted from data obtained by measurements of this type.

To find the rate of decomposition of $N_2O_5$, it is convenient to use Figure 11.2, which is a magnified version of a portion of Figure 11.1. If a tangent is drawn to the curve of concentration versus time, its slope at that point must equal $\Delta[N_2O_5]\Delta t$. But since the reaction rate is $-\Delta[N_2O_5]/\Delta t$, it follows that

instantaneous rate = −slope of tangent

From Figure 11.2 it appears that the slope of the tangent at $t = 2$ min is −0.028 mol/L · min. Hence,

rate at 2 min = −(−0.028 mol/L · min) = 0.028 mol/L · min

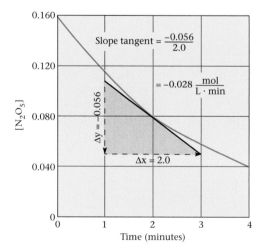

**Figure 11.2**
To determine the rate of a reaction, plot concentration versus time and take the tangent to the curve at the desired point. For the reaction $N_2O_5(g) \longrightarrow 2NO_2(g) + \tfrac{1}{2}O_2(g)$, it appears that the reaction rate at $[N_2O_5] = 0.080\ M$ is 0.028 mol/L · min.

# 11.2  Reaction Rate and Concentration

Ordinarily, reaction rate is directly related to reactant concentration. The higher the concentration of starting materials, the more rapidly a reaction takes place. Pure hydrogen peroxide, in which the concentration of $H_2O_2$ molecules is about 40 mol/L, is an extremely dangerous substance to work with. In the presence of trace impurities, it decomposes explosively

$$H_2O_2(l) \longrightarrow H_2O(g) + \tfrac{1}{2}\,O_2(g)$$

at a rate too rapid to measure. The hydrogen peroxide you buy in a drugstore is a dilute aqueous solution in which $[H_2O_2] \approx 1M$. At this relatively low concentration, decomposition is so slow that the solution is stable for several months.

The dependence of reaction rate upon concentration is readily explained. Ordinarily, *reactions occur as the result of collisions between reactant molecules.* The higher the concentration of molecules, the greater the number of collisions in unit time and hence the faster the reaction. As reactants are consumed, their concentrations drop, collisions occur less frequently, and reaction rate decreases. This explains the common observation that reaction rate drops off with time, eventually going to zero when all the reactants are consumed.

## Rate Expression and Rate Constant

The dependence of reaction rate upon concentration is readily determined for the decomposition of $N_2O_5$. Figure 11.3 shows what happens when reaction rate is plotted versus $[N_2O_5]$. As you would expect, rate increases as concentration increases, going from zero when $[N_2O_5] = 0$ to about 0.06 mol/L · min when $[N_2O_5] = 0.16\ M$. Moreover, as you can see from the figure, the plot of rate versus concentration is a straight line through the origin, which means that rate must be directly proportional to concentration.

$$\text{rate} = k[N_2O_5]$$

This equation is referred to as the **rate expression** for the decomposition of $N_2O_5$. It tells how the rate of the reaction

$$N_2O_5(g) \longrightarrow 2NO_2(g) + \tfrac{1}{2}\,O_2(g)$$

depends upon the concentration of reactant. The proportionality constant $k$ is called a **rate constant.** It is independent of the other quantities in the equation.

Rate depends upon concentration but rate constant does not

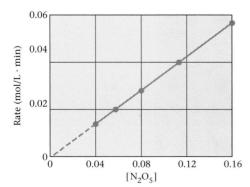

**Figure 11.3**
For the decomposition of $N_2O_5$, a plot of rate versus concentration of $N_2O_5$ is a straight line. The line, if extrapolated, passes through the origin. This means that rate is directly proportional to concentration; that is, rate = $k[N_2O_5]$.

As we will see shortly, the rate expression can take various forms, depending upon the nature of the reaction. It can be quite simple, as in the $N_2O_5$ decomposition, or exceedingly complex.

### Order of Reaction Involving a Single Reactant

Rate expressions have been determined by experiment for a large number of reactions. For the process

$$A \longrightarrow \text{products}$$

the rate expression has the general form

$$\text{rate} = k[A]^m$$

The power to which the concentration of reactant A is raised in the rate expression is called the **order of the reaction,** $m$. If $m$ is 0, the reaction is said to be "zero-order." If $m = 1$, the reaction is "first-order"; if $m = 2$, it is "second-order," and so on.

The order of a reaction must be determined experimentally; *it cannot be deduced from the coefficients in the balanced equation.* This must be true because there is only one reaction order, but there are many different ways in which the equation for the reaction can be balanced. For example, although we wrote

$$N_2O_5(g) \longrightarrow 2NO_2(g) + \tfrac{1}{2} O_2(g)$$

to describe the decomposition of $N_2O_5$, it could have been written

$$2N_2O_5(g) \longrightarrow 4NO_2(g) + O_2(g)$$

The reaction is still first-order no matter how the equation is written.

One way to find the order of a reaction is to measure the **initial rate** (i.e., the rate at $t = 0$) as a function of the concentration of reactant. Suppose, for example, that we make up two different reaction mixtures differing only in the concentration of reactant A. We now measure the rates at the beginning of re-

Another approach is to measure concentration as a function of time (Section 11.3)

The rate of reaction of zinc with aqueous sulfuric acid depends on the concentration of $H_2SO_4$. The beaker with the dilute solution (left) reacts slowly, whereas the beaker with the concentrated solution (right) reacts rapidly. (Charles Steele)

action, before the concentration of A has decreased appreciably. This gives two different initial rates (rate$_1$, rate$_2$) corresponding to two different starting concentrations of A, $[A]_1$ and $[A]_2$. From the rate expression,

$$\text{rate}_2 = k[A]_2{}^m \qquad \text{rate}_1 = k[A]_1{}^m$$

Dividing the second rate by the first,

$$\frac{\text{rate}_2}{\text{rate}_1} = \frac{[A]_2{}^m}{[A]_1{}^m} = \left(\frac{[A]_2}{[A]_1}\right)^m$$

Since all the quantities in this equation are known except $m$, the reaction order can be calculated (Example 11.1).

---

**Example 11.1**  The initial rate of decomposition of acetaldehyde, $CH_3CHO$, at 600°C

$$CH_3CHO(g) \longrightarrow CH_4(g) + CO(g)$$

was measured at a series of concentrations with the following results:

| $[CH_3CHO]$ | 0.10 $M$ | 0.20 $M$ | 0.30 $M$ | 0.40 $M$ |
|---|---|---|---|---|
| rate (mol/L · s) | 0.085 | 0.34 | 0.76 | 1.4 |

Using these data, determine the reaction order, that is, determine the value of $m$ in the equation: rate $= k[CH_3CHO]^m$.

**Strategy**  Choose the first two concentrations, 0.10 $M$ and 0.20 $M$. Calculate the ratio of the rates, the ratio of the concentrations, and finally the order of reaction, using the general relation derived above.

**Solution**

$$\frac{\text{rate}_2}{\text{rate}_1} = \frac{0.34 \text{ mol/L} \cdot \text{s}}{0.085 \text{ mol/L} \cdot \text{s}} = 4.0$$

$$\frac{[CH_3CHO]_2}{[CH_3CHO]_1} = \frac{0.20 \text{ } M}{0.10 \text{ } M} = 2.0$$

Hence, the general relation becomes $4.0 = (2.0)^m$. Clearly, $m = 2$; the reaction is second-order.

---

Once the order of the reaction is known, the rate constant is readily calculated. Consider, for example, the decomposition of acetaldehyde, where we have shown that the rate expression is

$$\text{rate} = k[CH_3CHO]^2$$

There are two variables in this equation, rate and concentration, and two constants, $k$ and reaction order

The data in Example 11.1 show that the rate at 600°C is 0.085 mol/L · s when the concentration is 0.10 mol/L. It follows that

$$k = \frac{\text{rate}}{[CH_3CHO]^2} = \frac{0.085 \text{ mol/L} \cdot \text{s}}{(0.10 \text{ mol/L})^2} = 8.5 \text{ L/mol} \cdot \text{s}$$

The same value of $k$ would be obtained, within experimental error, using any other data pair.

Having established the value of $k$ and the reaction order, the rate is readily calculated at any concentration. Again, using the decomposition of acetaldehyde as an example, we have established that

$$\text{rate} = 8.5 \, \frac{\text{L}}{\text{mol} \cdot \text{s}} \, [CH_3CHO]^2$$

If the concentration of acetaldehyde were 0.50 $M$,

$$\text{rate} = 8.5 \, \frac{\text{L}}{\text{mol} \cdot \text{s}} \, (0.50 \, \text{mol/L})^2 = 2.1 \, \text{mol/L} \cdot \text{s}$$

## Order of Reaction with More Than One Reactant

Many reactions (indeed, most reactions) involve more than one reactant. For a reaction between two species A and B,

$$a\text{A} + b\text{B} \longrightarrow \text{products}$$

the general form of the rate expression is

$$\text{rate} = k[\text{A}]^m \times [\text{B}]^n$$

Here $m$ is referred to as "the order of the reaction with respect to A." Similarly, $n$ is "the order of the reaction with respect to B." The **overall order** of the reaction is the sum of the exponents, $m + n$. If $m = 1$, $n = 2$, then the reaction is first-order in A, second-order in B, and third-order overall.

When more than one reactant is involved, the order can be determined by holding the initial concentration of one reactant constant while varying that of the other reactant. From rates measured under these conditions, it is possible to deduce the order of the reaction with respect to the reactant whose initial concentration is varied.

To see how this is done, consider the reaction between A and B referred to above. Suppose we run two different experiments in which the initial concentrations of A differ ($[\text{A}]_1$, $[\text{A}]_2$) but that of B is held constant at $[\text{B}]$. Then

$$\text{rate}_1 = k[\text{A}]_1^m \times [\text{B}]^n \qquad \text{rate}_2 = k[\text{A}]_2^m \times [\text{B}]^n$$

Dividing the second equation by the first:

$$\frac{\text{rate}_2}{\text{rate}_1} = \frac{k[\text{A}]_2^m \times [\text{B}]^n}{k[\text{A}]_1^m \times [\text{B}]^n} = \frac{[\text{A}]_2^m}{[\text{A}]_1^m} = \left(\frac{[\text{A}]_2}{[\text{A}]_1}\right)^m$$

Knowing the two rates and the ratio of the two concentrations, we can readily find the value of $m$.

---

**Example 11.2**    Consider the reaction at 55°C

$$(CH_3)_3CBr(aq) + OH^-(aq) \longrightarrow (CH_3)_3COH(aq) + Br^-(aq)$$

A series of experiments is carried out with the following results:

|  | Expt. 1 | Expt. 2 | Expt. 3 | Expt. 4 | Expt. 5 |
|---|---|---|---|---|---|
| $[(CH_3)_3CBr]$ | 0.50 | 1.0 | 1.5 | 1.0 | 1.0 |
| $[OH^-]$ | 0.050 | 0.050 | 0.050 | 0.10 | 0.20 |
| rate (mol/L · s) | 0.0050 | 0.010 | 0.015 | 0.010 | 0.010 |

*These are initial rates*

Find the order of the reaction with respect to both $(CH_3)_3CBr$ and $OH^-$.

***Strategy***   To find the order of the reaction with respect to $(CH_3)_3CBr$, choose two experiments, perhaps 1 and 3, where $[OH^-]$ is constant. A similar approach can be used to find $n$; compare experiments 2 and 5, where $[(CH_3)_3CBr]$ is constant.

***Solution***

(1) $\dfrac{rate_3}{rate_1} = \left(\dfrac{[(CH_3)_3CBr]_3}{[(CH_3)_3CBr]_1}\right)^m \qquad \dfrac{0.015}{0.0050} = \left(\dfrac{1.5}{0.50}\right)^m \qquad 3 = 3^m$

Clearly, $m = 1$.

(2) $\dfrac{rate_5}{rate_2} = \left(\dfrac{[OH^-]_5}{[OH^-]_2}\right)^n \qquad \dfrac{0.010}{0.010} = \left(\dfrac{0.20}{0.050}\right)^n \qquad 1 = 4^n$

In this case, $n = 0$; the rate is independent of the concentration of $OH^-$.

Any number raised to the power zero equals one

# 11.3  Reactant Concentration and Time

The rate expression

$$rate = k[N_2O_5]$$

shows how the rate of decomposition of $N_2O_5$ changes with concentration. From a practical standpoint, however, it is more important to know the relation between concentration and *time* rather than between concentration and rate. Suppose, for example, you are studying the decomposition of dinitrogen pentaoxide. Most likely, you would want to know how much $N_2O_5$ is left after 5 min, 1 h, or several days. An equation relating *rate* to concentration does not answer that purpose.

Using calculus, it is possible to develop **integrated rate equations** relating reactant concentration to time. We now examine several such equations, starting with first-order reactions.

## First-Order Reactions

For the decomposition of $N_2O_5$ and other first-order reactions of the type

$$A \longrightarrow products \qquad rate = k[A]$$

it can be shown by using calculus that the relationship between concentration and time is[*]

$$\ln \frac{[A]_o}{[A]} = kt$$

where $[A]_o$ is the original concentration of reactant, $[A]$ is its concentration at time $t$, $k$ is the first-order rate constant, and the abbreviation "ln" refers to the natural logarithm.

Since $\ln a/b = \ln a - \ln b$, the first-order equation can be written in the form

$$\ln [A]_o - \ln [A] = kt$$

---

[*]Throughout Section 11.3, balanced chemical equations are written in such a way that the coefficient of the reactant is 1. In general, if the coefficient of the reactant is $a$, where $a$ may be 2 or 3 or . . . , then $k$ in each integrated rate equation must be replaced by the product $ak$. (See Problem 78.)

## Figure 11.4

The rate constant for a first-order reaction can be determined from the slope of a plot of ln[A] versus time. From the graph of data for the reaction

$$N_2O_5(g) \longrightarrow 2NO_2(g) + \tfrac{1}{2}O_2(g)$$

at 67°C, it appears that the first-order rate constant is about 0.35/min.

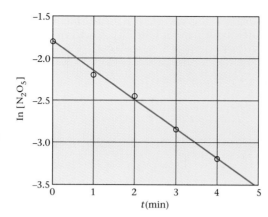

Solving for ln [A],

$$\ln [A] = \ln [A]_o - kt$$

Comparing this equation to the general equation of a straight line,

$$y = b + mx \qquad (b = y\text{-intercept, } m = \text{slope})$$

it is clear that a plot of ln [A] versus $t$ should be a straight line with a $y$-intercept of ln [A]$_o$ and a slope of $-k$. This is indeed the case, as you can see from Figure 11.4 where we have plotted ln [N$_2$O$_5$] versus $t$ for the decomposition of N$_2$O$_5$. Drawing the best straight line through the points and taking the slope based upon the two points on the $y$- and $x$-axes,

$$\text{slope} = \frac{-3.5 - (-1.8)}{4.9 - 0} = -0.35$$

It follows that the rate constant is 0.35/min; the integrated first-order equation for the decomposition of N$_2$O$_5$ is

$$\ln \frac{[N_2O_5]_o}{[N_2O_5]} = \frac{0.35}{\min} t \qquad (\text{at } 67°C)$$

---

**Example 11.3**    For the decomposition of N$_2$O$_5$ at 67°C, where $k$ = 0.35/min, calculate

    (a) the concentration after 4.0 min, starting at 0.160 $M$.
    (b) the time required for the concentration to drop from 0.160 to 0.100 $M$.
    (b) the time required for half a sample of N$_2$O$_5$ to decompose.

*Strategy*    In each case, the equation ln [N$_2$O$_5$]$_o$/[N$_2$O$_5$] = (0.35/min)$t$ is used. In (a) and (b), two of the three variables, [N$_2$O$_5$]$_o$, [N$_2$O$_5$], and $t$, are known; the other is readily calculated. In (c), you should be able to find the ratio [N$_2$O$_5$]$_o$/[N$_2$O$_5$]; knowing that ratio, the time can be calculated.

*Solution*

If ln $x = y$, then $x = e^y$

    (a) Substituting in the integrated first-order equation,

$$\ln \frac{0.160\ M}{[N_2O_5]} = \frac{0.35}{\min} (4.0 \text{ min}) = 1.4$$

Taking inverse logarithms,

$$\frac{0.160\ M}{[N_2O_5]} = e^{1.4} = 4.0 \qquad [N_2O_5] = \frac{0.160\ M}{4.0} = \boxed{0.040\ M}$$

(b) Solving the concentration-time relation for $t$,

$$t = \frac{1}{k} \ln \frac{[A]_o}{[A]} = \frac{1}{0.35/\text{min}} \ln \frac{0.160M}{0.100M} = \frac{\ln 1.60}{0.35/\text{min}} = 0.47 \text{ min}/0.35 = \boxed{1.3 \text{ min}}$$

(c) When half of the sample has decomposed,

$$[N_2O_5] = [N_2O_5]_o/2 \qquad [N_2O_5]_o = 2[N_2O_5] \qquad [N_2O_5]_o/[N_2O_5] = 2$$

Using the equation in (b),

$$t = \frac{1}{k} \ln 2 = \frac{0.693}{k}$$

With $k = 0.35/\text{min}$, we have

$$t = \frac{0.693}{0.35} = \boxed{2.0 \text{ min}}$$

---

The analysis of Example 11.3c reveals an important feature of a first-order reaction: ***The time required for one half of a reactant to decompose via a first-order reaction has a fixed value, independent of concentration.*** This quantity, called the **half-life,** is given by the expression

$$t_{1/2} = \frac{\ln 2}{k} = \frac{0.693}{k} \qquad \text{first-order reaction}$$

where $k$ is the rate constant. For the decomposition of $N_2O_5$, where $k = 0.35/\text{min}$, $t_{1/2} = 2.0$ min. Thus, every 2 minutes, one half of a sample of $N_2O_5$ decomposes (Table 11.1).

Notice that, for a first-order reaction, the rate constant has the units of reciprocal time, e.g., $\text{min}^{-1}$. This suggests a simple physical interpretation of $k$; it is the fraction of reactant decomposing in unit time. For a first-order reaction in which

$$k = 0.010/\text{min} = 0.010 \text{ min}^{-1}$$

we can say that 0.010 (i.e., 1.0%) of the sample decomposes per minute.

Perhaps the most important first-order reaction is that of radioactive decay, introduced in Chapter 2. Letting $X$ be the amount of a radioactive isotope present at time $t$:

$$\text{rate} = kX$$

and

$$\ln \frac{X_0}{X} = kt$$

where $X_0$ is the initial amount. The amount of a radioactive isotope can be expressed in terms of moles, grams, or number of atoms.

**TABLE 11.1   Decomposition of $N_2O_5$ at 67°C ($t_{1/2} = 2.0$ min)**

| $t$ (min) | 0.0 | 2.0 | 4.0 | 6.0 | 8.0 |
|---|---|---|---|---|---|
| $[N_2O_5]$ | 0.160 | 0.080 | 0.040 | 0.020 | 0.010 |
| fraction of $N_2O_5$ decomposed | 0 | $\frac{1}{2}$ | $\frac{3}{4}$ | $\frac{7}{8}$ | $\frac{15}{16}$ |
| fraction of $N_2O_5$ left | 1 | $\frac{1}{2}$ | $\frac{1}{4}$ | $\frac{1}{8}$ | $\frac{1}{16}$ |
| number of half-lives | 0 | 1 | 2 | 3 | 4 |

**Example 11.4**   Plutonium-240, produced in nuclear reactors, has a half-life of $6.58 \times 10^3$ years. Calculate

(a) the first-order rate constant for the decay of plutonium-240.
(b) the fraction of a sample that will remain after 100 $(1.00 \times 10^2)$ years.

**Strategy**   This is a standard type of first-order rate calculation. Use the equation $k = 0.693/t_{1/2}$ to calculate $k$ and then the equation $\ln X_o/X = kt$ to find the fraction left, $X/X_o$.

**Solution**

(a) $k = 0.693/(6.58 \times 10^3 \text{ yr}) =$   $1.05 \times 10^{-4}/\text{yr}$

(b) $\ln \dfrac{X_o}{X} = 1.05 \times 10^{-4} \text{ yr}^{-1} \times 100 \text{ yr} = 1.05 \times 10^{-2}$

Taking the inverse log, $X_o/X = 1.01$.

Hence, $X/X_o = 1/1.01 =$   $0.99$. That is 99% remains.

This problem illustrates the difficulty inherent in storing or disposing of a long-lived radioactive species. Such isotopes cannot be released into the environment in the hope that they will decompose rapidly. In the case of plutonium-240, its radiation level would be virtually unchanged a century from now.

Half-lives can be interpreted in terms of the level of radiation of the corresponding isotopes. Since uranium has a very long half-life $(4.5 \times 10^9 \text{ yr})$, it gives off radiation very slowly. At the opposite extreme is fermium-258, which decays with a half-life of $3.8 \times 10^{-4}$ s. You would expect the rate of decay to be quite high. Within a second all the radiation from fermium-258 is gone. Species such as this produce very high radiation during their brief existence.

## Zero- and Second-Order Reactions

For a **zero-order reaction,**

$$A \longrightarrow \text{products} \qquad \text{rate} = k[A]^0 = k$$

since any quantity raised to the zero power, including [A], is equal to one. In other words, the rate of a zero-order reaction is constant, independent of concentration. As you might expect from our earlier discussion, zero-order reactions are relatively rare. Most of them take place at solid surfaces, where the rate is independent of concentration in the gas phase. A typical example is the thermal decomposition of hydrogen iodide on gold:

$$\text{HI}(g) \xrightarrow{\text{Au}} \tfrac{1}{2} \text{H}_2(g) + \tfrac{1}{2} \text{I}_2(g)$$

When the gold surface is completely covered with HI molecules, increasing the concentration of HI($g$) has no effect on reaction rate.

It can readily be shown (Problem 75) that the concentration-time relation for a zero-order reaction is

In a zero order reaction, all the reactant is consumed in a finite time

$$[A] = -[A]_o - kt$$

Comparing this equation to that for a straight line,

$$y = b + mx \qquad (b = y\text{-intercept}, \ m = \text{slope})$$

it should be clear that a plot of [A] versus $t$ should be a straight line with a slope of $-k$. Putting it another way, if a plot of concentration versus time is linear, the reaction must be zero-order; the rate constant $k$ is numerically equal to the slope of that line but has the opposite sign.

For a **second-order reaction** involving a single reactant, such as acetaldehyde (recall Example 11.1),

$$A \longrightarrow \text{products} \qquad \text{rate} = k[A]^2$$

it is again necessary to resort to calculus* to obtain the concentration-time relationship

$$\frac{1}{[A]} - \frac{1}{[A]_o} = kt$$

where the symbols [A], $[A]_o$, $t$, and $k$ have their usual meanings. For a second-order reaction, a plot of $1/[A]$ versus $t$ should be linear.

The characteristics of zero-, first-, and second-order reactions are summarized in Table 11.2. To determine reaction order, the properties in either of the last two columns at the right of the table can be used (Example 11.5).

---

**Example 11.5**   The following data were obtained for the gas-phase decomposition of hydrogen iodide:

| time (h) | 0 | 2 | 4 | 6 |
|---|---|---|---|---|
| [HI] | 1.00 | 0.50 | 0.33 | 0.25 |

Is this reaction zero-, first-, or second-order in HI?

*Strategy*   Note from the last column of Table 11.2 that

— if the reaction is zero-order, a plot of [A] versus $t$ should be linear.
— if the reaction is first-order, a plot of ln [A] versus $t$ should be linear.
— if the reaction is second-order, a plot of $1/[A]$ versus $t$ should be linear.

Using these data, prepare these plots and determine which one is a straight line.

---

TABLE 11.2   **Characteristics of Zero-, First-, and Second-Order Reactions of the Form A($g$) → products; [A], $[A]_o$ = conc. A at $t$ and $t = 0$, respectively**

| Order | Rate Expression | Conc.-Time Relation | Half-Life | Linear Plot |
|---|---|---|---|---|
| 0 | rate $= k$ | $[A]_o - [A] = kt$ | $[A]_o/2k$ | [A] vs. $t$ |
| 1 | rate $= k[A]$ | $\ln \frac{[A]_o}{[A]} = kt$ | $0.693/k$ | ln [A] vs. $t$ |
| 2 | rate $= k[A]^2$ | $\frac{1}{[A]} - \frac{1}{[A]_o} = kt$ | $1/k[A]_o$ | $\frac{1}{[A]}$ vs. $t$ |

---

*If you are taking a course in calculus, you may be surprised to learn how useful it can be in the real world (e.g., chemistry). The general rate expressions for zero-, first-, and second-order reactions are

$$-d[A]/dt = k \qquad -d[A]/dt = k[A] \qquad -d[A]/dt = k[A]^2$$

Integrating these questions from 0 to $t$ and from $[A]_o$ to [A], you should be able to derive the equations for zero-, first-, and second-order reactions.

**Figure 11.5**
One way to determine reaction order is to search for a linear relation between some function of concentration and time (Table 11.2). Since a plot of 1/[HI] versus *t* is linear, the decomposition of HI must be second-order.

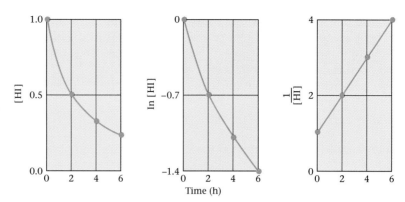

*Solution*   It is useful to prepare a table listing [HI], ln [HI], and 1/[HI] as a function of time.

| *t*(h) | [HI] | ln [HI] | 1/[HI] |
|--------|------|---------|--------|
| 0 | 1.00 | 0 | 1.0 |
| 2 | 0.50 | −0.69 | 2.0 |
| 4 | 0.33 | −1.10 | 3.0 |
| 6 | 0.25 | −1.39 | 4.0 |

If it is not obvious from this table that the only linear plot is that of 1/[HI] versus time,

that point should be clear from Figure 11.5.   This is a second-order reaction.

---

## 11.4   Models for Reaction Rate

So far in this chapter we have approached reaction rate from an experimental point of view, describing what happens in the laboratory or the world around us. Now we change emphasis and try to explain why certain reactions occur rapidly while other take place slowly. To do this, we look at a couple of models that chemists have developed to predict rate constants for reactions.

### Collision Model; Activation Energy

Consider the reaction

$$CO(g) + NO_2(g) \longrightarrow CO_2(g) + NO(g)$$

Above 600 K, this reaction takes place as a direct result of collisions between CO and $NO_2$ molecules. When the concentration of CO doubles (Figure 11.6), the number of these collisions in a given time increases by a factor of two; doubling the concentration of $NO_2$ has the same effect. Assuming that reaction rate is directly proportional to the collision rate, the following relation should hold:

$$\text{reaction rate} = k[CO] \times [NO_2]$$

Experimentally, this prediction is confirmed; the reaction is first-order in both carbon monoxide and nitrogen dioxide.

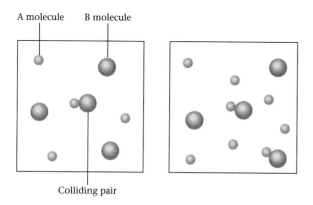

A molecule    B molecule

Colliding pair

**Figure 11.6**

Consider the reaction A + B ⟶ products, where reaction occurs as the result of collision. If we double the number of A molecules (square at right), there will be twice as many collisions per unit time, so the rate will be doubled. In general, the rate will be directly proportional to the concentration of A. The same reasoning applies to B, so the rate law is rate = $k$[A] × [B].

There is a restriction on this simple model for the CO-NO$_2$ reaction. According to the kinetic theory of gases, for a reaction mixture at 700 K and concentrations of 0.10 *M*, every CO molecule should collide with about $10^9$ NO$_2$ molecules in one second. If every collision were effective, the reaction should be over in a fraction of a second. In reality, this does not happen; under these conditions, the half-life is about 10 s. This implies that not every CO-NO$_2$ collision leads to reaction.

There would be an explosion

There are a couple of reasons why collision between reactant molecules does not always lead to reaction. For one thing, the molecules have to be properly oriented with respect to one another when they collide. Suppose, for example, that the carbon atom of a CO molecule strikes the nitrogen atom of an NO$_2$ molecule (Fig. 11.7). This is very unlikely to result in the transfer of an oxygen atom from NO$_2$ to CO, which is required for reaction to occur.

A more important factor that reduces the number of effective collisions has to do with the kinetic energy of reactant molecules. These molecules are held

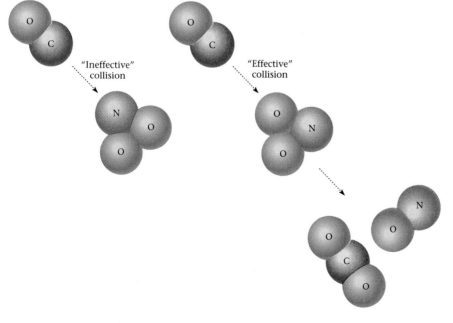

"Ineffective" collision

"Effective" collision

**Figure 11.7**

For a collision to result in reaction, the molecules must be properly oriented. For the reaction CO(*g*) + NO$_2$(*g*) ⟶ CO$_2$(*g*) + NO(*g*), the carbon atom of the CO molecule must strike an oxygen atom of the NO$_2$ molecule, forming CO$_2$ as one product, NO as the other.

together by strong chemical bonds. Only if the colliding molecules are moving very rapidly will the kinetic energy be large enough to supply the energy required to break these bonds. Molecules with small kinetic energies bounce off one another without reacting. As a result, only a small fraction of collisions are effective.

For every reaction, there is a certain minimum energy that molecules must possess for collision to be effective. This is referred to as the **activation energy.** It has the symbol $E_a$ and is expressed in kilojoules. For the reaction between one mole of CO and one mole of $NO_2$, $E_a$ is 134 kJ. The colliding molecules (CO and $NO_2$) must have a total kinetic energy of at least 134 kJ/mol if they are to react. The activation energy for a reaction is a positive quantity ($E_a > 0$) whose value depends upon the nature of the reaction.

The collision model of reaction rates just developed can be made quantitative. We can say that the rate constant for a reaction, $k$, is a product of three factors:

$$k = p \times Z \times f$$

where

— **$p$**, called a *steric factor*, takes into account the fact that only certain orientations of colliding molecules are likely to lead to reaction (recall Figure 11.7). Unfortunately, the collision model cannot predict the value of $p$; about all we can say is that it is likely to be less than 1, sometimes much less.
— **$Z$**, the *collision number*, which gives the number of molecular collisions occurring in unit time at unit concentrations of reactants. This quantity can be calculated quite accurately from the kinetic theory of gases, but we will not describe that calculation.
— **$f$**, the *fraction of collisions—in which the energy of the colliding molecules is equal to or greater than $E_a$*. It can be shown that

$$f = e^{-E_a/RT}$$

where $e$ is the base of natural logarithms, $R$ is the gas constant, and $T$ is the absolute temperature in K.

Substituting this expression for f into the equation written above for $k$, we obtain the basic equation for the collision model:

$$k = p \times Z \times e^{-E_a/RT}$$

This equation tells us, among other things, that *the larger the value of $E_a$, the smaller the rate constant.* Thus,

$$\text{if } E_a = 0, \ e^{-E_a/RT} = e^0 = 1$$

$$E_a = RT, \ e^{-E_a/RT} = e^{-1} = 0.37$$

$$E_a = 2RT, \ e^{-E_a/RT} = e^{-2} = 0.14$$

and so on. This makes sense; the larger the activation energy, the smaller the fraction of molecules having enough energy to react upon collision, so the slower the rate of reaction.

Table 11.3 compares observed rate constants for several reactions with those predicted by collision theory, arbitrarily taking $p = 1$. As you might expect, the calculated $k$'s are too high, suggesting that the steric factor is indeed less than 1.

**TABLE 11.3  Observed and Calculated Rate Constants for Second-Order Gas Phase Reactions**

| Reaction | $k$ Observed | Collision Model | Transition-State Model |
|---|---|---|---|
| $NO + O_3 \longrightarrow NO_2 + O_2$ | $6.3 \times 10^7$ | $4.0 \times 10^9$ | $3.2 \times 10^7$ |
| $NO + Cl_2 \longrightarrow NOCl + Cl$ | $5.2$ | $130$ | $1.6$ |
| $NO_2 + CO \longrightarrow NO + CO_2$ | $1.2 \times 10^{-4}$ | $6.4 \times 10^{-4}$ | $1.0 \times 10^{-4}$ |
| $2NO_2 \longrightarrow 2NO + O_2$ | $5.0 \times 10^{-3}$ | $1.0 \times 10^{-1}$ | $1.2 \times 10^{-2}$ |

## Transition-State Model; Activation Energy Diagrams

Figure 11.8 is an energy diagram for the CO-$NO_2$ reaction. Reactants, CO and $NO_2$, are shown at the left. Products, $CO_2$ and NO, are at the right. They have an energy 226 kJ less than that of the reactants; $\Delta H$ for the reaction is $-226$ kJ. In the center of the figure is an intermediate called an *activated complex*. This is an unstable, high-energy species that must be formed before the reaction can occur. It has an energy 134 kJ greater than that of the reactants and 360 kJ greater than that of the products.

The exact nature of the activated complex is difficult to determine. For this reaction, the activated complex might be a "pseudomolecule" made up of CO and $NO_2$ molecules in close contact. The path of the reaction might be more or less as follows:

$$O{\equiv}C + O{-}N{\diagdown}_O \longrightarrow O{\equiv}C{\cdot}{\cdot}O{\cdot}{\cdot}N{\diagdown}_O \longrightarrow O{=}C{=}O + N{=}O$$

reactants          activated complex          products

The dotted lines stand for "partial bonds" in the activated complex. The N—O bond in the $NO_2$ molecule has been partially broken. A new bond between carbon and oxygen has started to form.

The idea of the activated complex was developed by, among others, Henry Eyring at Princeton in the 1930s. It forms the basis of the *transition-state model* for reaction rate, which assumes that the activated complex

In the activated complex, electrons are often in excited states

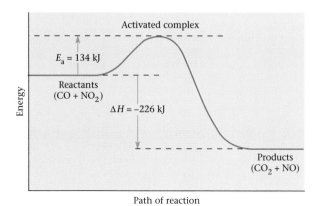

**Figure 11.8**
During the reaction initiation, 134 kJ—the activation energy, $E_a$—must be furnished to the reactants for every mole of CO that reacts. This energy activates each CO-$NO_2$ complex to the point at which reaction can occur.

— is in equilibrium, at low concentrations, with the reactants.

— may either decompose to products by "climbing" over the energy barrier or, alternatively, revert back to reactants.

With these assumptions, it is possible to develop an expression for the rate constant $k$. Although the expression is too complex to discuss here, we show some of the results in Table 11.3.

The transition-state model is generally somewhat more accurate than the collision model (at least with $p = 1$). Another advantage is that it explains why the activation energy is ordinarily much smaller than the bond energies in the reactant molecules. Consider, for example, the reaction

$$CO(g) + NO_2(g) \longrightarrow CO_2(g) + NO(g)$$

where $E_a = 134$ kJ. This is considerably smaller than any of the bond energies in the reactants:

$$C\equiv O \quad 1075 \text{ kJ} \qquad N\equiv O \quad 607 \text{ kJ} \qquad N—O \quad 222 \text{ kJ}$$

Forming the activated complex shown on p. 317 requires the absorption of relatively little energy, because it requires only the weakening of reactant bonds rather than their rupture.

# Chemistry: *The Human Side*

Chemical kinetics, the subject of this chapter, evolved from an art into a science in the first half of the twentieth century. One person more than any other was responsible for that change: Henry Eyring. Born in a Mormon settlement in Mexico, Henry emigrated to Arizona with his parents in 1912, when the Mexican revolution broke out. Many years later, he became a United States citizen, appearing before a judge who had just naturalized Albert Einstein.

Eyring's first exposure to kinetics came as a young chemistry instructor at the University of Wisconsin in the 1920s. There he worked on the thermal decomposition of $N_2O_5$ (Section 11.2). In 1931 Eyring came to Princeton, where he developed the transition-state model of reaction rates. Using this model and making reasonable guesses about the nature of the activated complex, Eyring calculated rate constants that agreed well with experiment. After 15 productive years at Princeton, Eyring returned to his Mormon roots at the University of Utah in Salt Lake City. While serving as dean of the graduate school, he found time to do ground-breaking research on the kinetics of life processes, including the mechanism of aging.

Henry Eyring's approach to research is best described in his own words.

> *I perceive myself as rather uninhibited, with a certain mathematical facility and more interest in the broad aspects of a problem than the delicate nuances. I am more interested in discovering what is over the next rise than in assiduously cultivating the beautiful garden close at hand.*

Henry's enthusiasm for chemistry spilled over into just about every other aspect of his life. Each year, he challenged his students to a 50-yard dash held at the stadium of the University of Utah. He continued these races until he was in his mid-70s. Henry never won, but on one occasion he did make the CBS evening news.

**Henry Eyring**
(1901–1981)

# 11.5 Catalysis

**A catalyst is a substance that increases the rate of a reaction without being consumed by it.** It ordinarily does this by changing the reaction path to one with a lower activation energy. Consider, for example, the decomposition of $N_2O$:

$$N_2O(g) \longrightarrow N_2(g) + \tfrac{1}{2}O_2(g)$$

For the direct reaction, taking place in the gas phase, the activation energy is 250 kJ/mol. On a gold surface, which acts as a catalyst for this reaction, the activation energy is only 120 kJ/mol (Fig. 11.9). Hence, the reaction occurs more rapidly on a gold surface.

## Heterogeneous Catalysis

A *heterogeneous catalyst* is one that is in a different phase from the reaction mixture. Most commonly, the catalyst is a solid that increases the rate of a gas-phase or liquid-phase reaction. An example is the reaction previously cited:

$$N_2O(g) \xrightarrow{\text{Au}} N_2(g) + \tfrac{1}{2}O_2(g)$$

In the catalyzed decomposition, $N_2O$ is chemically adsorbed on the surface of the solid. A chemical bond is formed between the oxygen atom of an $N_2O$ molecule and a gold atom on the surface. This weakens the bond joining nitrogen to oxygen, making it easier for the $N_2O$ molecule to break apart. Symbolically, this process can be shown as

$$N\equiv N-O(g) + Au(s) \longrightarrow N\equiv N\text{---}O\text{---}Au(s) \longrightarrow N\equiv N(g) + O(g) + Au(s)$$

where the broken lines represent weak covalent bonds.

Perhaps the most familiar example of heterogeneous catalysis is the series of reactions that occur in the catalytic converter of an automobile (Fig. 11.10, p. 320). Typically this device contains 1 to 3 g of platinum metal mixed with rhodium. The platinum catalyzes the oxidation of carbon monoxide and unburned hydrocarbons such as benzene, $C_6H_6$:

$$2CO(g) + O_2(g) \xrightarrow{\text{Pt}} 2CO_2(g)$$

$$C_6H_6(g) + \tfrac{15}{2}O_2(g) \xrightarrow{\text{Pt}} 6CO_2(g) + 3H_2O(l)$$

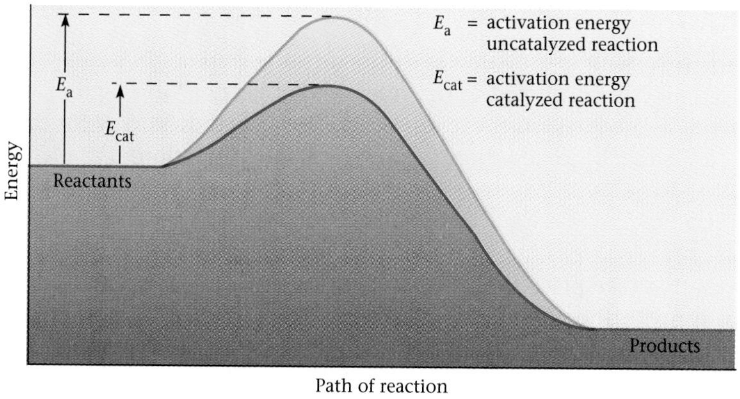

Many industrial processes use heterogeneous catalysts

$E_a$ = activation energy uncatalyzed reaction

$E_{cat}$ = activation energy catalyzed reaction

**Figure 11.9**
By changing the path by which a reaction occurs, a catalyst can lower the activation energy that is required and so speed up the reaction.

**Figure 11.10**

Catalytic converters contain a "three-way" catalyst designed to convert CO to $CO_2$, unburned hydrocarbons to $CO_2$ and $H_2O$, and NO to $N_2$. The active components of the catalysts are the precious metals platinum and rhodium; palladium is sometimes used as well.

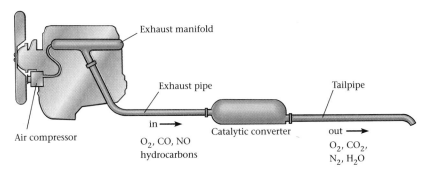

The rhodium acts as a catalyst to convert oxides of nitrogen to the free element:

$$CO(g) + NO(g) \xrightarrow{\text{Rh}} CO_2(g) + \tfrac{1}{2}N_2(g)$$

One problem with heterogeneous catalysis is that the solid catalyst is easily "poisoned." Foreign materials deposited on the catalytic surface during the reaction reduce or even destroy its effectiveness. A major reason for using unleaded gasoline is that lead metal poisons the Pt-Rh mixture in the catalytic converter.

## Homogeneous Catalysis

A *homogeneous catalyst* is one that is present in the same phase as the reactants. It speeds up the reaction by forming a reactive intermediate that decomposes to give products. In this way, the catalyst provides an alternative process of lower activation energy.

An example of a reaction that is subject to homogeneous catalysis is the decomposition of hydrogen peroxide in aqueous solution:

$$2H_2O_2(aq) \longrightarrow 2H_2O + O_2(g)$$

Under ordinary conditions, this reaction occurs very slowly. However, if a solution of sodium iodide, NaI, is added, reaction occurs almost immediately; you can see the bubbles of oxygen forming.

The catalyzed decomposition of hydrogen peroxide is believed to take place by a two-step path:

$$\begin{array}{l} \text{Step 1: } H_2O_2(aq) + I^-(aq) \longrightarrow H_2O + IO^-(aq) \\ \underline{\text{Step 2: } H_2O_2(aq) + IO^-(aq) \longrightarrow H_2O + O_2(g) + I^-(aq)} \\ \qquad\qquad 2H_2O_2(aq) \longrightarrow 2H_2O + O_2(g) \end{array}$$

Notice that the end result is the same as in the direct reaction.

The $I^-$ ions are not consumed in the reaction. For every $I^-$ ion used up in the first step, one is produced in the second step. The activation energy for this two-step process is much smaller than for the uncatalyzed reaction.

## Enzymes

Many reactions that take place slowly under ordinary conditions occur readily in living organisms in the presence of catalysts called **enzymes.** Enzymes are protein molecules of high molar mass. An example of an enzyme-catalyzed reaction is the decomposition of hydrogen peroxide:

$$2H_2O_2(aq) \longrightarrow 2H_2O + O_2(g)$$

In blood or tissues, this reaction is catalyzed by an enzyme called catalase. When 3% hydrogen peroxide is used to treat a fresh cut or wound, oxygen gas is given off rapidly (Fig. 11.11). The function of catalase in the body is to prevent the buildup of hydrogen peroxide, a powerful oxidizing agent.

Many enzymes are extremely specific. For example, the enzyme maltase catalyzes the hydrolysis of maltose:

$$\underset{\text{maltose}}{C_{12}H_{22}O_{11}(aq)} + H_2O \xrightarrow{\text{maltase}} \underset{\text{glucose}}{2C_6H_{12}O_6(aq)}$$

This is the only function of maltase, but it is one that no other enzyme can perform. There are many such digestive enzymes required for the metabolism of carbohydrates, proteins, and fats. It has been estimated that without enzymes, it would take upwards of 50 years to digest a meal.

*That would make dieting a lot easier*

Enzymes, like all other catalysts, lower the activation energy for reaction. They can be enormously effective; it is not uncommon for the rate constant to increase by a factor of $10^{12}$ or more. However, from a commercial standpoint, enzymes have some drawbacks. A particular enzyme operates best over a narrow range of temperature. An increase in temperature frequently deactivates an enzyme by causing the molecule to "unfold," changing its characteristic shape. Recently, chemists have discovered that this effect can be prevented if the enzyme is immobilized by bonding to a solid support. Among the solids that have been used are synthetic polymers, porous glass, and even stainless steel. The development of immobilized enzymes has led to a host of new products. The sweetener aspartame, used in many diet soft drinks, is made using the enzyme aspartase in immobilized form.

## 11.6  Reaction Rate and Temperature

The rates of most reactions increase as the temperature rises. A person in a hurry to prepare dinner applies this principle in using a pressure cooker to raise the temperature for cooking potatoes, apples, or a pot roast (not all at the same time, we trust). By storing the leftovers in a refrigerator, the chemical reactions re-

**Figure 11.11**

An enzyme present in the liver is a catalyst in the rapid decomposition of $H_2O_2$ into water and oxygen gas. Hydrogen peroxide without the catalyst still decomposes, but at a much slower rate.
(Richard Megna/FUNDAMENTAL PHOTOGRAPHS, New York)

**Figure 11.12**

When $T$ is increased to $T + 10$, the fraction of molecules with very high energies increases sharply. Hence, many more molecules possess the activation energy, $E_a$, and reaction occurs more rapidly. If $E_a$ is of the order of 50 kJ/mol, an increase in temperature of 10°C approximately doubles the number of molecules having energy $E_a$ or greater and thus doubles the reaction rate.

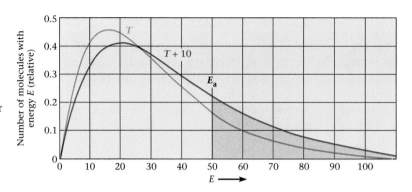

Hibernating animals lower their body temperature, slowing down life processes

sponsible for food spoilage are slowed down. As a general and very approximate rule, it is often stated that an increase in temperature of 10°C doubles the reaction rate. If this rule holds, foods should cook twice as fast in a pressure cooker at 110°C as in an open saucepan and deteriorate four times as rapidly at room temperature (25°C) as they do in a refrigerator at 5°C.

The effect of temperature on reaction rate can be explained in terms of the kinetic theory of gases. Recall from Chapter 5 that raising the temperature greatly increases the fraction of molecules having very high kinetic energies. These are the molecules that are most likely to react when they collide. The higher the temperature, the larger the fraction of molecules that can provide the activation energy required for reaction. This effect is apparent from Figure 11.12, where the distribution of kinetic energies is shown at two different temperatures. Notice that the fraction of molecules having a kinetic energy equal to or greater than the activation energy $E_a$ (shaded area) is considerably larger at the higher temperature. Hence the fraction of effective collisions increases; this is the major factor causing reaction rate to increase with temperature.

## The Arrhenius Equation

You will recall from Section 11.4 that the collision model yields the following expression for the rate constant:

$$k = p \times Z \times e^{-E_a/RT}$$

The steric factor, $p$, is presumably temperature-independent. The collision number, $Z$, is relatively insensitive to temperature. For example, when the temperature increases from 500 to 600 K, $Z$ increases by less than 10%. Hence, to a good degree of approximation we can write, as far as temperature dependence is concerned:

$$k = Ae^{-E_a/RT}$$

where $A$ is a constant. Taking the natural logarithm of both sides of this equation,

$$\ln k = \ln A - E_a/RT$$

This equation was first shown to be valid by Svante Arrhenius (Chapter 4) in 1889. It is commonly referred to as the **Arrhenius equation.**

Comparing the Arrhenius equation to the general equation for a straight line,

$$y = b + mx$$

it should be clear that a plot of $\ln k$ ("$y$") versus $1/T$ ("$x$") should be linear. The slope of the straight line, $m$, is equal to $-E_a/R$. Figure 11.13 shows such a plot for the CO-NO$_2$ reaction. The slope appears to be about $-1.61 \times 10^4$ K. Hence,

$$\frac{-E_a}{R} = -1.61 \times 10^4 \text{ K}$$

$$E_a = 8.31\frac{\text{J}}{\text{mol} \cdot \text{K}}(1.61 \times 10^4 \text{K}) = 1.34 \times 10^5 \text{ J/mol} = 134 \text{ kJ/mol}$$

## "Two-Point" Equation Relating $k$ and $T$

The Arrhenius equation can be expressed in a different form by following the procedure used with the Clausius-Clapeyron equation in Chapter 9. At two different temperatures, $T_2$ and $T_1$,

$$\ln k_2 = \ln A - \frac{E_a}{RT_2}$$

$$\ln k_1 = \ln A - \frac{E_a}{RT_1}$$

Subtracting the second equation from the first,

$$\ln k_2 - \ln k_1 = \frac{-E_a}{R}\left[\frac{1}{T_2} - \frac{1}{T_1}\right]$$

or

$$\ln \frac{k_2}{k_1} = \frac{E_a}{R}\left[\frac{1}{T_1} - \frac{1}{T_2}\right] = \frac{E_a}{R}\left[\frac{T_2 - T_1}{T_2 T_1}\right]$$

Taking $R = 8.31$ J/mol $\cdot$ K, the activation energy is expressed in joules per mole.

Antimony powder reacts more rapidly with bromine at high temperatures. The reaction is

$$2Sb(s) + 3Br_2(l) \longrightarrow 2SbBr_3(s).$$

(J. Morgenthaler)

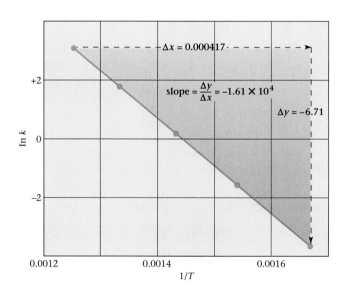

**Figure 11.13**
A plot of $\ln k$ versus $1/T$ ($k$ = rate constant, $T$ = Kelvin temperature) is a straight line. From the slope of this line, the activation energy can be determined: $E_a = -R \times$ slope. For the CO-NO$_2$ reaction shown here,

$$E_a = -(8.31 \text{ J/mol} \cdot \text{K})(-1.61 \times 10^4 \text{K})$$
$$= 1.34 \times 10^5 \text{ J/mol} = 134 \text{ kJ/mol}.$$

**Example 11.6** For a certain reaction, the rate constant doubles when the temperature increases from 15 to 25°C. Calculate

(a) the activation energy, $E_a$.
(b) the rate constant at 100°C, taking $k$ at 25°C to be $1.2 \times 10^{-2}$ L/mol · s.

***Strategy*** Use the two-point form of the Arrhenius equation. In (a), the two temperatures and the ratio $k_2/k_1$ are known, so $E_a$ is readily calculated. That value of $E_a$ is used in (b) to calculate $k_2$, knowing $k_1$, $T_2$, and $T_1$. Remember to express temperatures in K.

*Solution*

(a) $k_2/k_1 = 2.00$     $\ln k_2/k_1 = \ln 2.00 = 0.693$     $T_2 = 298$ K     $T_1 = 288$ K
Substituting into the two-point form of the Arrhenius equation,

With $R = 8.31$ J/mol · K, $E_a$ must be expressed in J/mol

$$0.693 = \frac{E_a}{8.31 \text{ J/mol} \cdot \text{K}} \left[ \frac{1}{288 \text{ K}} - \frac{1}{298 \text{ K}} \right]$$

Solving,

$$E_a = \boxed{49 \text{ kJ/mol}}$$

(b) $\ln \dfrac{k_2}{k_1} = \dfrac{49{,}000 \text{ J/mol}}{8.31 \text{ J/mol} \cdot \text{K}} \left[ \dfrac{1}{298 \text{ K}} - \dfrac{1}{373 \text{ K}} \right] = 3.98$     $k_2/k_1 = e^{3.98} = 53$

Taking $k_1 = 1.2 \times 10^{-2}$ L/mol · s

$$k_2 = 53(1.2 \times 10^{-2} \text{ L/mol} \cdot \text{s}) = \boxed{0.64 \text{ L/mol} \cdot \text{s}}$$

## 11.7 Reaction Mechanisms

A **reaction mechanism** is a description of a path, or a sequence of steps, by which a reaction occurs at the molecular level. In the simplest case, only a single step is involved. This is a collision between two reactant molecules. This is the "mechanism" for the reaction of CO with $NO_2$ at high temperatures, above about 600 K:

Reaction mechanisms frequently change with T, sometimes with P

$$CO(g) + NO_2(g) \longrightarrow NO(g) + CO_2(g)$$

At low temperatures the reaction between carbon monoxide and nitrogen dioxide takes place by a quite different mechanism. Two steps are involved:

$$\begin{array}{r} NO_2(g) + NO_2(g) \longrightarrow NO_3(g) + NO(g) \\ \underline{CO(g) + NO_3(g) \longrightarrow CO_2(g) + NO_2(g)} \\ CO(g) + NO_2(g) \longrightarrow NO(g) + CO_2(g) \end{array}$$

Notice that the overall reaction, obtained by summing the individual steps, is identical with that for the one-step process. The rate expressions are quite different, however:

high temperatures:     rate = $k[CO] \times [NO_2]$

low temperatures:     rate = $k[NO_2]^2$

In general, the nature of the rate expression and hence ***the reaction order depend upon the mechanism by which the reaction takes place.***

## Elementary Steps

The individual steps that constitute a reaction mechanism are referred to as **elementary steps.** These steps may be *unimolecular*

$$A \longrightarrow B + C \qquad \text{rate} = k[A]$$

*bimolecular*

$$A + B \longrightarrow C + D \qquad \text{rate} = k[A] \times [B]$$

or, in rare cases, *termolecular*

$$A + B + C \longrightarrow D + E \qquad \text{rate} = k[A] \times [B] \times [C]$$

(Color is used to distinguish elementary steps from overall reactions.)

Notice from the rate expressions just written that *the rate of an elementary step is equal to a rate constant k multiplied by the concentration of each reactant molecule.* This rule is readily explained. Consider, for example, a step in which two molecules, A and B, collide effectively with each other to form C and D. As pointed out earlier, the rate of collision and hence the rate of reaction will be directly proportional to the concentration of each reactant.

## Slow Steps

Often, one step in a mechanism is much slower than any other. If this is the case, *the slow step is rate-determining.* That is, the rate of the overall reaction can be taken to be that of the slow step. Consider, for example, a three-step reaction:

$$
\begin{array}{lll}
\text{Step 1:} & A \longrightarrow B & \text{fast} \\
\text{Step 2:} & B \longrightarrow C & \text{slow} \\
\text{Step 3:} & \underline{C \longrightarrow D} & \text{fast} \\
 & A \longrightarrow D &
\end{array}
$$

The rate at which A is converted to D (the overall reaction) is approximately equal to the rate of conversion of B to C (the slow step).

To understand the rationale behind this rule, and its limitations, consider an analogous situation. Suppose three people (A, B, and C) are assigned to grade general chemistry examinations that contain three questions. On the average, A spends 10 s grading Question 1 and B spends 15 s grading Question 2. In contrast, C, the ultimate procrastinator, takes 5 min to grade Question 3. The rate at which exams are graded is

Maybe Questions 1 and 2 were multiple choice

$$\frac{1 \text{ exam}}{10 \text{ s} + 15 \text{ s} + 300 \text{ s}} = \frac{1 \text{ exam}}{325 \text{ s}} = 0.00308 \text{ exam/s}$$

This is approximately equal to the rate of the slow grader:

$$\frac{1 \text{ exam}}{300 \text{ s}} = 0.00333 \text{ exam/s}$$

Extrapolating from general chemistry exams to chemical reactions, we can say that

— the overall rate of the reaction cannot exceed that of the slowest step

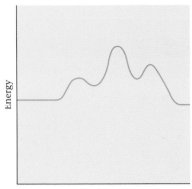

**Figure 11.14**

Activation energy diagram for a reaction that occurs by a three-step mechanism in which the second step is rate-determining.

— if that step is by far the slowest, its rate will be approximately equal to that of the overall reaction

— the slowest step in a mechanism is ordinarily the one with the highest activation energy (Figure 11.14).

## Deducing a Rate Expression from a Proposed Mechanism

The question to be considered here is: *Given a mechanism for a several-step reaction, how can you deduce the rate expression corresponding to that mechanism?* In principle, at least, all you have to do is to apply the rules just cited.

1. ***Find the slowest step and equate the rate of the overall reaction to the rate of that step.***
2. ***Find the rate expression for the slowest step.***

To illustrate this process, consider the two-step mechanism for the low-temperature reaction between CO and $NO_2$.

$$NO_2(g) + NO_2(g) \longrightarrow NO_3(g) + NO(g) \qquad \text{slow}$$
$$\underline{CO(g) + NO_3(g) \longrightarrow CO_2(g) + NO_2(g)} \qquad \text{fast}$$
$$CO(g) + NO_2(g) \longrightarrow CO_2(g) + NO(g)$$

Applying the above rules in order,

$$\text{rate of overall reaction} = \text{rate of 1st step} = k[NO_2]^2$$

This analysis explains why the rate expression for the two-step mechanism is different from that for the direct, one-step reaction.

## Elimination of Intermediates

Sometimes the rate expression obtained by the process just described involves a reactive intermediate, i.e., a species produced in one step of the mechanism and consumed in a later step. Ordinarily, concentrations of such species are too small to be determined experimentally. Hence they must be eliminated from the rate expression if it is to be compared with experiment. The final rate expression usually includes only those species that appear in the balanced equation for the reaction. Sometimes, the concentration of a catalyst is included, but never that of a reactive intermediate.

To illustrate this situation, consider the reaction between nitric oxide and chlorine, which is believed to proceed by a two-step mechanism:

$$\text{Step 1:} \qquad NO(g) + Cl_2(g) \underset{k_{-1}}{\overset{k_1}{\rightleftharpoons}} NOCl_2(g) \qquad \text{fast}$$

$$\text{Step 2:} \qquad \underline{NOCl_2(g) + NO(g) \overset{k_2}{\longrightarrow} 2NOCl(g)} \qquad \text{slow}$$
$$2NO(g) + Cl_2(g) \longrightarrow 2NOCl(g)$$

The first step occurs rapidly and reversibly; a dynamic equilibrium is set up in which the rates of forward and reverse reactions are equal. Since the second step is slow and rate-determining, it follows that

$$\text{rate of overall reaction} = \text{rate of Step 2} = k_2[NOCl_2] \times [NO]$$

The rate expression just written is unsatisfactory in that it cannot be checked against experiment. The species $NOCl_2$ is a reactive intermediate whose concentration is too small to be measured accurately, if at all. To eliminate the $[NOCl_2]$ term from the rate expression, recall that the rates of forward and reverse reactions in Step 1 are equal, which means that

$$k_1[NO] \times [Cl_2] = k_{-1}[NOCl_2]$$

Solving for $[NOCl_2]$ and substituting in this rate expression,

$$[NOCl_2] = \frac{k_1[NO] \times [Cl_2]}{k_{-1}}$$

$$\text{rate of overall reaction} = k_2[NOCl_2] \times [NO] = \frac{k_2k_1[NO]^2 \times [Cl_2]}{k_{-1}}$$

The quotient $k_2k_1/k_{-1}$ is the experimentally observed rate constant for the reaction, which is found to be second-order in NO and first-order in $Cl_2$, as predicted by this mechanism.

> If the concentration of a species can't be measured, it better not appear in the rate expression

---

**Example 11.7**   The decomposition of ozone, $O_3$, to diatomic oxygen, $O_2$, is believed to occur by a two-step mechanism:

Step 1: $\quad\quad O_3(g) \underset{k_{-1}}{\overset{k_1}{\rightleftharpoons}} O_2(g) + O(g) \quad\quad$ fast

Step 2: $\quad\quad \underline{O_3(g) + O(g) \overset{k_2}{\longrightarrow} 2O_2(g)} \quad\quad$ slow

$$2O_3(g) \longrightarrow 3O_2(g)$$

Obtain the rate expression corresponding to this mechanism.

**Strategy**   Write the rate expression for the rate-determining second step. This will involve the unstable intermediate, O. To get rid of [O], use the fact that reactants and products are in equilibrium in Step 1, so forward and reverse reactions occur at the same rate.

*Solution*

(1) Overall rate = rate Step 2 = $k_2[O_3] \times [O]$
(2) $k_1[O_3] = k_{-1}[O_2] \times [O]$

$$[O] = \frac{k_1[O_3]}{k_{-1}[O_2]}$$

(3) Overall rate $= \dfrac{k_1k_2[O_3]^2}{k_{-1}[O_2]} = \dfrac{k[O_3]^2}{[O_2]}$

> The quotient $\dfrac{k_1k_2}{k_{-1}}$ is the observed rate constant, $k$

Notice that the concentration of $O_2$, a product in the reaction, appears in the denominator. The rate is inversely proportional to the concentration of molecular oxygen, a feature that you would never have predicted from the balanced equation for the reaction. As the concentration of $O_2$ builds up, the rate slows down.

---

It is important to point out one of the limitations of mechanism studies. Usually more than one mechanism is compatible with the same experimentally obtained rate expression. To make a choice between alternative mechanisms, other evidence must be considered. A classic example of this situation is the reaction between hydrogen and iodine

$$H_2(g) + I_2(g) \longrightarrow 2HI(g)$$

for which the observed rate expression is

$$\text{rate} = k[H_2] \times [I_2]$$

For many years, it was assumed that the $H_2$-$I_2$ reaction occurs in a single step, a collision between an $H_2$ molecule and an $I_2$ molecule. That would, of course, be compatible with the rate expression above. However, there is now evidence to indicate that a quite different and more complex mechanism is involved (see Problem 62).

## CHEMISTRY

### *Beyond the Classroom*

## The Ozone Story

In recent years, a minor component of the atmosphere, ozone, has received a great deal of attention. Ozone, molecular formula $O_3$, is a pale blue gas with a characteristic odor that can be detected after lightning activity, in the vicinity of electric motors, or near a subway train.

Depending upon its location in the atmosphere, ozone can be a villain or a beleaguered hero. In the lower atmosphere (the *troposphere*), ozone is bad news; it is a major component of photochemical smog, formed by the reaction sequence

$$NO_2(g) \longrightarrow NO(g) + O(g)$$
$$O_2(g) + O(g) \longrightarrow O_3(g)$$
$$\overline{NO_2(g) + O_2(g) \longrightarrow NO(g) + O_3(g)}$$

Ozone is an extremely powerful oxidizing agent, which explains its toxicity to animals and humans. At partial pressures as low as $10^{-7}$ atm, it can cut in half the rate of photosynthesis by plants.

In the upper atmosphere (the *stratosphere*), the situation is quite different. There the partial pressure of ozone goes through a maximum of about $10^{-5}$ atm at an altitude of 30 km. From 95 to 99% of sunlight in the wavelength range 200 to 300 nm is absorbed by ozone in this region, commonly referred to as the "ozone layer." If this ultraviolet radiation were to reach the surface of the earth, it could have several adverse effects. A decrease in ozone concentration of 5% could increase the incidence of skin cancer by 20%. Ultraviolet radiation is also a factor in diseases of the eye, including cataract formation.

Ozone molecules in the stratosphere can decompose by the reaction

$$O_3(g) + O(g) \longrightarrow 2O_2(g)$$

This reaction takes place rather slowly by direct collision between an $O_3$ molecule and an O atom. It can occur more rapidly by a two-step process in which a chlorine atom acts as a catalyst:

$$O_3(g) + Cl(g) \longrightarrow O_2(g) + ClO(g) \qquad k = 5.2 \times 10^9 \text{ L/mol} \cdot \text{s at 220 K}$$
$$\underline{ClO(g) + O(g) \longrightarrow Cl(g) + O_2(g)} \qquad k = 2.6 \times 10^{10} \text{ L/mol} \cdot \text{s at 220 K}$$
$$O_3(g) + O(g) \longrightarrow 2O_2(g)$$

Bromine atoms can substitute for chlorine; indeed, the rate constant for the Br-catalyzed reaction is larger than that for the reaction just cited.

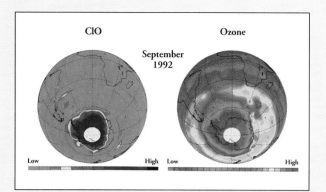 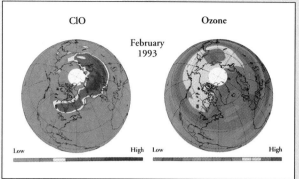

**Figure 11.A**

Chlorine monoxide (ClO) is the dominant form of chlorine that destroys ozone. Note that large abundances of ClO (left) appear where there are very low levels of ozone. The left figures show the northern hemisphere (February 1993) while the right figures show the Antarctic ozone hole (September 1992). (Courtesy of NASA)

The chlorine atoms that catalyze the decomposition of ozone come from chlorofluorocarbons (CFCs) used in many refrigerators and air conditioners. A major culprit is $CF_2Cl_2$, which forms Cl atoms when exposed to ultraviolet radiation at 200 nm:

$$CF_2Cl_2(g) \longrightarrow CF_2Cl(g) + Cl(g)$$

The major source of bromine atoms in the stratosphere are compounds called "halons," which are used to extinguish fires and suppress explosions in military aircraft and in commercial oil and gas production. Chief among these are $CBrF_3$ and $CBrClF_2$.

The catalyzed decomposition of ozone is known to be responsible for the ozone hole (Fig. 11.A) that develops in Antarctica each year in September and October, at the end of winter in the Southern Hemisphere. No ozone is generated during the long, dark Antarctic winter. Meanwhile, a heterogeneous reaction occurring on clouds of ice crystals at $-85°C$ produces species such as $Cl_2$, $Br_2$, and HOCl. When the sun reappears in September, these molecules decompose photochemically to form Cl or Br atoms.

Ozone depletion is by no means restricted to polar regions. In the winter of 1992–1993, ozone levels in the stratosphere over the United States reached record lows, up to 20% below normal. Along the East Coast, these low levels persisted into July, when people (and crops) are more exposed and so more vulnerable to ultraviolet radiation.

In 1987, an international treaty was signed in Montreal to cut back on the use of CFCs and halons. Halon production in the United States was stopped in 1994; the production and use of CFCs is scheduled to end in 1996. As you can imagine, the development of alternatives to these compounds is an active area of research. The best bet to replace CFCs seems to be

$$\begin{array}{ccc} F & H \\ | & | \\ F-C-C-F \\ | & | \\ F & H \end{array}$$

which is now being used in automobile air conditioners. Replacements for halons are more difficult to come by. One possibility is $CIF_3$, in which the bromine atom of $CBrF_3$ is replaced by iodine.

$CF_2Cl_2$ was developed 50 years ago to replace $SO_2$ and $NH_3$ as refrigerants.

# CHAPTER HIGHLIGHTS

## Key Concepts

1. Determine the rate expression (reaction order) from
   —initial rate data
   (Examples 11.1, 11.2; Problems 21–28, 80)
   —concentration-time data, using Table 11.2
   (Example 11.5; Problems 29–32)
   —reaction mechanism
   (Example 11.7; Problems 59–64)
2. Relate concentration to time for a first-order reaction
   (Examples 11.3, 11.4; Problems 33–42, 72, 76)
3. Use the Arrhenius equation to relate rate constant to temperature
   (Example 11.6; Problems 51–58, 73, 74)

## Key Equations

| | | |
|---|---|---|
| Zero-order reaction | rate = $k$ | $[A] = [A]_o - kt$ |
| First-order reaction | rate = $k[A]$ | $\ln [A]_o/[A] = kt$ $\quad$ $t_{1/2} = 0.693/k$ |
| Second-order reaction | rate = $k[A]^2$ | $1/[A] - 1/[A]_o = kt$ |
| Arrhenius equation | $\ln \dfrac{k_2}{k_1} = \dfrac{E_a}{R}\left[\dfrac{1}{T_1} - \dfrac{1}{T_2}\right]$ | |

## Key Terms

| | | |
|---|---|---|
| activation energy | mechanism | rate expression |
| catalyst | order | reaction rate |
| half-life | rate constant | |

## Summary Problem

Hydrogen peroxide decomposes to water and oxygen according to the following reaction:

$$H_2O_2(aq) \longrightarrow H_2O + \tfrac{1}{2}O_2(g)$$

Its rate of decomposition is measured by titrating samples of the solution with potassium permanganate ($KMnO_4$) at certain intervals.

(a) Initial rate determinations at 25°C for the decomposition give the following data

| $[H_2O_2]$ | Initial Rate (mol/L · min) |
|---|---|
| 0.200 | $1.04 \times 10^{-4}$ |
| 0.300 | $1.55 \times 10^{-4}$ |
| 0.500 | $2.59 \times 10^{-4}$ |

What is the order of the reaction? Write the rate equation for the decomposition. Calculate the rate constant and the half-life for the reaction at 25°C.

(b) Hydrogen peroxide is sold commercially as a 30.0% solution. If the solution is kept at room temperature (25°C), how long will it take for the solution to become 10.0% $H_2O_2$?

(c) It has been determined that at 40°C, the rate constant for the reaction is $1.93 \times 10^{-3}$ $min^{-1}$. Calculate the activation energy for the decomposition of $H_2O_2$.

(d) Manufacturers recommend that solutions of hydrogen peroxide be kept in a refrigerator at 4°C. How much longer will it take a 30.0% solution to decompose to 10.0% if the solution is kept in a refrigerator instead of at room temperature?

(e) The rate constant for the catalyzed reaction at 25°C is $2.95 \times 10^8$ $min^{-1}$. What is the half-life of the catalyzed reaction? What is the rate for the catalyzed decomposition of 0.500 $M$ $H_2O_2$? How much faster is the rate of a catalyzed reaction than that of an uncatalyzed reaction at 25°C?

(f) Hydrogen peroxide in basic solution oxidizes iodide ions to iodine. The proposed mechanism for this reaction is

$$H_2O_2(aq) + I^-(aq) \longrightarrow HOI(aq) + OH^-(aq) \qquad \text{slow}$$
$$HOI(aq) + I^-(aq) \rightleftharpoons I_2(aq) + OH^-(aq) \qquad \text{fast}$$

Write the overall redox reaction. Write a rate law consistent with this proposed mechanism.

### Answers

(a) first-order; rate = $k[H_2O_2]$; $5.17 \times 10^{-4}$ $min^{-1}$; 22.3 hours

(b) 35.4 hours

(c) 68 kJ/mol

(d) 249 hours

(e) $2.35 \times 10^{-9}$ min; $1.48 \times 10^8$ mol/L · min; $5.71 \times 10^{11}$ times faster, or 0.500 trillion times faster

(f) $H_2O_2(aq) + 2I^-(aq) \longrightarrow I_2(aq) + 2OH^-(aq)$; rate = $k[H_2O_2] \times [I^-]$

---

## Questions & Problems

### Meaning of Reaction Rate

**\*1.** Express the rate of the reaction

$$2HF(g) + O_2(g) \longrightarrow 2HOF(g)$$

in terms of
  **(a)** $\Delta[HOF]$   **(b)** $\Delta[O_2]$

**\*2.** Express the rate of the reaction

$$2N_2(g) + 5O_2(g) \longrightarrow 2N_2O_5(g)$$

in terms of
  **(a)** $\Delta[N_2O_5]$   **(b)** $\Delta[O_2]$

**3.** Consider the combustion of cyclohexane.

$$C_6H_{12}(l) + 9O_2(g) \longrightarrow 6CO_2(g) + 6H_2O(l)$$

If the cyclohexane is burning at the rate of 0.350 mol/L · s, at what rates are $CO_2$ produced and $O_2$ used up?

**4.** For the reaction

$$5Fe^{2+}(aq) + MnO_4^-\ (aq) + 8H^+(aq) \longrightarrow$$
$$Mn^{2+}(aq) + 5Fe^{3+}(aq) + 4H_2O$$

It was found that at a particular instant, $Mn^{2+}$ was being formed at the rate of 0.213 mol/L · min. At that instant, at what rate is

  **(a)** $MnO_4^-$ being reduced?
  **(b)** water being produced?
  **(c)** $Fe^{2+}(aq)$ being oxidized?

**5.** Ammonia is produced by the reaction between nitrogen and hydrogen gases.

  **(a)** Write a balanced equation using smallest whole-number coefficients for the reaction.
  **(b)** Write an expression for the rate of reaction in terms of $\Delta[NH_3]$.
  **(c)** The concentration of ammonia increases from 0.288 $M$ to 0.736 $M$ in 12.0 min. Calculate the average rate of reaction over this time interval.

**6.** Dinitrogen pentaoxide gas decomposes to nitrogen dioxide and oxygen gases.

  **(a)** Write a balanced equation using smallest whole-number coefficients for the decomposition.

**(b)** Write an expression for the reaction rate in terms of $\Delta[N_2O_5]$.

**(c)** The concentration of $N_2O_5$ decreases by 35.0% in three minutes, starting at 0.500 *M*. Calculate the average rate of reaction over this time interval.

**7.** Experimental data are listed for the hypothetical reaction

$$A \longrightarrow B + C$$

| Time (sec) | 0 | 2 | 4 | 6 | 8 | 10 |
|---|---|---|---|---|---|---|
| [B] | 0 | 0.100 | 0.130 | 0.150 | 0.165 | 0.175 |

Plot these data as in Figure 11.2. Draw a tangent to the curve to find the rate at 4 minutes.

**8.** Using the data and the plot referred to in Question 7,
  **(a)** find the instantaneous rate at 6 minutes.
  **(b)** find the average rate over the 0-to-6 minute interval.

## Rate Expressions

**\*9.** A reaction has two reactants, A and B. What is the order with respect to each reactant and the overall order of the reactions described by the following rate expressions?
  **(a)** rate = $k_1[A][B]^2$     **(b)** rate = $k_2[B]^3$
  **(c)** rate = $k_3[A]$       **(d)** rate = $k_4$

**\*10.** A reaction has two reactants, C and D. What is the order with respect to each reactant and the overall order of the reactions described by the following rate expressions?
  **(a)** rate = $k_1[C]^2$     **(b)** rate = $k_2[C][D]$
  **(c)** rate = $k_3[D]$       **(d)** rate = $k_4[C]^2[D]$

**\*11.** What will the units of the rate constants in Question 9 be if the rate is expressed in mol/L · min?

**\*12.** What will the units of the rate constants in Question 10 be if the rate is expressed in mol/L · s?

**13.** Complete the following table for the first-order reaction

$$A(g) \longrightarrow \text{products}$$

| | [A] | $k(hr^{-1})$ | rate(mol/L · hr) |
|---|---|---|---|
| (a) | 0.236 | 0.830 | ___ |
| (b) | 0.187 | ___ | 6.41 |
| (c) | ___ | 0.0591 | $4.99 \times 10^{-2}$ |

**14.** Complete the following table for the reaction

$$A(g) + B(g) \longrightarrow \text{products}$$

which is first-order in both A and B.

| | [A] | [B] | $k$ (L/mol · min) | rate (mol/L · min) |
|---|---|---|---|---|
| (a) | 0.200 | 0.500 | 0.388 | ___ |
| (b) | ___ | 0.033 | 1.09 | 0.084 |
| (c) | 1.00 | ___ | 0.714 | 0.688 |
| (d) | 0.500 | 0.399 | ___ | 1.73 |

**15.** The reaction

$$NO(g) + \tfrac{1}{2}Br_2(g) \longrightarrow NOBr(g)$$

is second-order with respect to NO and first-order with respect to $Br_2$. The rate constant is $1.33 \times 10^{-3}L^2/mol^2 \cdot min$.
  **(a)** Write the rate expression.
  **(b)** Calculate the rate of the reaction when [NO] = 0.158 *M* and $[Br_2]$ = 2[NO]
  **(c)** At what concentration of $Br_2$ is the rate $2.8 \times 10^{-8}$ mol/L · min and [NO] = 0.039 *M*?

**16.** The decomposition of nitrogen dioxide is a second-order reaction. At 550 K, a 0.400-*M* sample decomposes at the rate of 3.00 mol/L · min.
  **(a)** Write the rate expression.
  **(b)** What is the rate constant at 550 K?
  **(c)** What is the rate of decomposition when $[NO_2]$ = 0.639 *M*?

**17.** The reaction

$$NH_4^+ (aq) + NO_2^- (aq) \longrightarrow N_2(g) + 2H_2O$$

is first-order with respect to both reactants. At 25°C, the rate of the reaction is 0.109 mol/L · hr when $[NH_4^+]$ is 0.450 *M* and $[NO_2^-]$ is 0.225 *M*.
  **(a)** What is the rate constant?
  **(b)** What is the concentration of the ammonium ion when the rate is 0.222 mol/L · hr and the nitrite ion concentration is 0.155 mol/L?
  **(c)** What is the nitrite ion concentration when the rate of the reaction is 0.633 mol/L · hr and $[NH_4^+]$ = $\tfrac{1}{3}[NO_2^-]$?

**18.** The decomposition of ammonia on tungsten at 1100°C is a zero-order reaction. Its rate is $4.17 \times 10^{-6}$ mol/L · s when $[NH_3]$ = 0.296 *M*.
  **(a)** What is $k$?
  **(b)** What is the rate when the concentration of ammonia is doubled?
  **(c)** At what concentration of ammonia will the rate double?

**\*19.** The greatest increase in the reaction rate for the reaction between A and C where rate = $k[A]^{\frac{1}{2}}[B]$ is caused by
  **(a)** doubling [A]     **(b)** halving [B]
  **(c)** halving [A]     **(d)** doubling [A] and [B]

**\*20.** The greatest decrease in the reaction rate for the reaction between V and W where rate = $k[V][W]$ is caused by
  **(a)** doubling [V]       **(b)** doubling [W]
  **(c)** halving [V] and [W]   **(d)** doubling [V] and [W]

## Determination of Reaction Order

**21.** For the decomposition of acetone, the following reaction rate data are obtained:

| Rate (mol/L · min) | 0.0260 | 0.0341 | 0.0380 |
|---|---|---|---|
| $[CH_3COCH_3]$ | 0.0500 | 0.0655 | 0.0731 |

  **(a)** Determine the order of the reaction.

**(b)** Write the rate expression for the decomposition of acetone.

**(c)** Calculate $k$ for the decomposition in the experiment above.

22. The gas-phase decomposition of nitrogen dioxide at 380°C gives the following data:

| | | | |
|---|---|---|---|
| **Rate (mol/L · s)** | 0.0976 | 0.635 | 1.79 |
| **[$NO_2$]** | 0.100 | 0.255 | 0.428 |

**(a)** Determine the order of the reaction.

**(b)** Write the rate expression for the decomposition of $NO_2$.

**(c)** Calculate $k$ for the decomposition in the experiment above.

**23.** When a base is added to an aqueous solution of chlorine dioxide gas, the following reaction occurs:

$$2ClO_2(aq) + 2OH^-(aq) \longrightarrow ClO_3^-(aq) + ClO_2^-(aq) + H_2O$$

The reaction rate can be followed by monitoring the hydroxide concentration. The following data are obtained:

| [$ClO_2$] | [$OH^-$] | Initial Rate (mol/L · s) |
|---|---|---|
| 0.010 | 0.030 | $6.00 \times 10^{-4}$ |
| 0.010 | 0.075 | $1.50 \times 10^{-3}$ |
| 0.055 | 0.030 | $1.82 \times 10^{-2}$ |
| 0.055 | 0.062 | $3.75 \times 10^{-2}$ |

**(a)** What is the order of the reaction with respect to chlorine dioxide, hydroxide ion, and overall?

**(b)** Write the rate expression for the reaction.

**(c)** Calculate $k$ for the reaction.

**(d)** When [$ClO_2$] = 0.25 $M$ and [$OH^-$] = 0.036 $M$, what is the rate of the reaction?

24. The rate of the reaction

$$HgCl_2(aq) + \tfrac{1}{2}C_2O_4^{2-}(aq) \longrightarrow Cl^-(aq) + CO_2(g) + \tfrac{1}{2}Hg_2Cl_2(s)$$

is followed by measuring the number of moles of $Hg_2Cl_2$ that precipitate per liter per second. The following data are obtained:

| [$HgCl_2$] | [$C_2O_4^{2-}$] | Initial Rate (mol/L · min) |
|---|---|---|
| 0.020 | 0.020 | $6.24 \times 10^{-8}$ |
| 0.048 | 0.020 | $1.50 \times 10^{-7}$ |
| 0.020 | 0.033 | $1.70 \times 10^{-7}$ |
| 0.075 | 0.033 | $6.37 \times 10^{-7}$ |

**(a)** What is the order of the reaction with respect to mercury(II) chloride, oxalate ion, and overall?

**(b)** Write the rate expression for the reaction.

**(c)** Calculate $k$ for the reaction.

**(d)** When [$HgCl_2$] = 0.085 $M$ and [$C_2O_4^{2-}$] = 0.068 $M$, what is the rate of the reaction?

**25.** The equation for the reaction between iodide and bromate ions in acidic solution is

$$6I^-(aq) + BrO_3^-(aq) + 6H^+(aq) \longrightarrow 3I_2(aq) + Br^-(aq) + 3H_2O$$

The rate of the reaction is followed by measuring the appearance of $I_2$. The following data are obtained:

| [$I^-$] | [$BrO_3^-$] | [$H^+$] | Initial Rate (mol/L · s) |
|---|---|---|---|
| 0.0020 | 0.0080 | 0.020 | $8.89 \times 10^{-5}$ |
| 0.0040 | 0.0080 | 0.020 | $1.78 \times 10^{-4}$ |
| 0.0020 | 0.0160 | 0.020 | $1.78 \times 10^{-4}$ |
| 0.0020 | 0.0080 | 0.040 | $3.56 \times 10^{-4}$ |
| 0.0015 | 0.0040 | 0.030 | $7.51 \times 10^{-5}$ |

**(a)** What is the order of the reaction with respect to each reactant?

**(b)** Write the rate expression for the reaction.

**(c)** Calculate $k$.

**(d)** What is the hydrogen ion concentration when the rate is $5.00 \times 10^{-4}$ mol/L · s and [$I^-$] = $\tfrac{1}{2}$[$BrO_3^-$] = 0.0075 $M$?

26. The equation for the iodination of acetone in acidic solution is

$$CH_3COCH_3(aq) + I_2(aq) \longrightarrow CH_3COCH_2I(aq) + H^+(aq) + I^-(aq)$$

The rate of the reaction is found to be dependent not only on the concentration of the reactants but also on the hydrogen ion concentration. Hence, the rate expression of this reaction is

$$\text{rate} = k[CH_3COCH_3]^m[I_2]^n[H^+]^p$$

The rate is obtained by following the disappearance of iodine using starch as an indicator. The following data are obtained:

| [$CH_3COCH_3$] | [$H^+$] | [$I_2$] | Initial Rate (mol/L · s) |
|---|---|---|---|
| 0.80 | 0.20 | 0.001 | $4.2 \times 10^{-6}$ |
| 1.6 | 0.20 | 0.001 | $8.2 \times 10^{-6}$ |
| 0.80 | 0.40 | 0.001 | $8.7 \times 10^{-6}$ |
| 0.80 | 0.20 | 0.0005 | $4.3 \times 10^{-6}$ |

**(a)** What is the order of the reaction with respect to each reactant?

**(b)** Write the rate expression for the reaction.

**(c)** Calculate $k$.

**(d)** What is the rate of the reaction when [$H^+$] = 0.933 $M$ and [$CH_3COCH_3$] = 3[$H^+$] = 10[$I^-$]?

**27.** Consider the reaction

$$Cr(H_2O)_6^{3+}(aq) + SCN^-(aq) \longrightarrow Cr(H_2O)_5SCN^{2+}(aq) + H_2O$$

The following data were obtained:

| [$Cr(H_2O)_6^{3+}$] | [$SCN^-$] | Initial Rate (mol/L · min) |
|---|---|---|
| 0.025 | 0.060 | $6.5 \times 10^{-4}$ |
| 0.025 | 0.077 | $8.4 \times 10^{-4}$ |
| 0.042 | 0.077 | $1.4 \times 10^{-3}$ |
| 0.042 | 0.100 | $1.8 \times 10^{-3}$ |

**(a)** Write the rate expression for the reaction.
**(b)** Calculate $k$.
**(c)** What is the rate of the reaction when 15 mg of KSCN is added to 1.50 L of a solution 0.0500 $M$ in $Cr(H_2O)_6^{3+}$?

**28.** In solution at a constant $H^+$ concentration, iodide ions react with hydrogen peroxide to produce iodine.

$$H^+(aq) + I^-(aq) + \tfrac{1}{2}H_2O_2(aq) \longrightarrow \tfrac{1}{2}I_2(aq) + H_2O$$

The reaction rate can be followed by monitoring the appearance of $I_2$. The following data were obtained:

| $[I^-]$ | $[H_2O_2]$ | Initial Rate (mol/L · min) |
|---|---|---|
| 0.015 | 0.030 | 0.0022 |
| 0.035 | 0.030 | 0.0052 |
| 0.055 | 0.030 | 0.0082 |
| 0.035 | 0.050 | 0.0087 |

**(a)** Write the rate expression for the reaction.
**(b)** Calculate $k$.
**(c)** What is the rate of the reaction when 25.0 mL of a 0.100 $M$ solution of KI is added to 25.0 mL of a 10.0% by mass solution of $H_2O_2$ ($d$ = 1.00 g/mL)? Assume volumes are additive.

**29.** Sucrose, $C_{12}H_{22}O_{11}$, decomposes to glucose and fructose, both with molecular formula $C_6H_{12}O_6$, in dilute acidic solution. The following data are obtained for the decomposition of sucrose:

| Time (min) | $[C_{12}H_{22}O_{11}]$ |
|---|---|
| 0 | 0.368 |
| 20 | 0.333 |
| 60 | 0.287 |
| 120 | 0.235 |
| 160 | 0.208 |

Write the rate expression for the reaction.

**30.** Use the following data to determine the order of the decomposition reaction of nitrosyl bromide:

| Time(s) | 0 | 6 | 12 | 18 | 24 |
|---|---|---|---|---|---|
| [NOBr] | 0.0286 | 0.0253 | 0.0229 | 0.0208 | 0.0190 |

**31.** Consider the decomposition of A represented by squares, where each square represents a molecule of A. Is the reaction zero-order, first-order, or second-order?

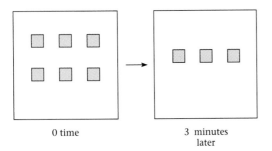

0 time          3 minutes later

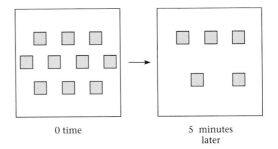

0 time          5 minutes later

**32.** Consider the decomposition of B represented by circles, where each circle represents a molecule of B. Is the reaction zero-order, first-order, or second-order?

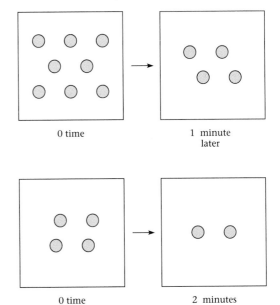

0 time          1 minute later

0 time          2 minutes later

## First-Order Reactions

**33.** Consider the following data for the gas-phase decomposition of ethyl chloride, $C_2H_5Cl$, at 746 K.

| Time (min) | $[C_2H_5Cl]$ |
|---|---|
| 0 | 0.100 |
| 1 | 0.0975 |
| 2 | 0.0951 |
| 3 | 0.0928 |
| 4 | 0.0905 |
| 8 | 0.0819 |
| 16 | 0.0670 |

**(a)** By plotting the data, show that the reaction is first-order.
**(b)** From the graph, determine $k$.
**(c)** Using $k$, find the time it takes to decrease the concentration to 0.0400 $M$.
**(d)** Calculate the rate of the reaction when $[C_2H_5Cl]$ = 0.200 $M$.

**34.** The following data are for the isomerization of cyclopropane, $C_3H_6$, to propylene at 500°C:

| Time (min) | $[C_3H_6]$ |
|:---:|:---:|
| 0 | 0.500 |
| 5 | 0.418 |
| 10 | 0.349 |
| 15 | 0.292 |
| 20 | 0.244 |

**(a)** By plotting the data, show that the reaction is first-order.
**(b)** From the graph, determine $k$.
**(c)** Using $k$, find the time it takes to decrease the concentration from 0.400 $M$ to 0.100 $M$.
**(d)** Calculate the rate of the reaction at 5 minutes when $[C_3H_6] = 0.418$ $M$.

**35.** The first-order rate constant for the decomposition of a certain hormone in water at 25°C is 0.125 $yr^{-1}$.
**(a)** If a 0.0100 $M$ solution of the hormone is stored at 25°C for one month, what will its concentration be at the end of the month?
**(b)** How long will it take for the concentration of the solution to drop from 0.0100 $M$ to 0.0075 $M$?
**(c)** What is the half-life of the hormone?

**36.** In the first-order decomposition of cyclobutane at 750 K, it is found that 15% of the sample has decomposed in 45 seconds. Calculate
**(a)** the rate constant.
**(b)** the half-life.
**(c)** how long it will take for 65% of the cyclobutane to decompose.

**37.** The decomposition of sulfuryl chloride, $SO_2Cl_2$, to sulfur dioxide and chlorine gases is a first-order reaction. It is found that it takes 13.7 hours to decompose a 0.250 $M$ $SO_2Cl_2$ solution to 0.117 $M$.
**(a)** What is the rate constant for the reaction?
**(b)** What is its half-life?
**(c)** What is the rate of decomposition when $[SO_2Cl_2] = 1.00$ $M$?
**(d)** How long will it take to decompose 95.0% of $SO_2Cl_2$?

**38.** The decomposition of dinitrogen pentaoxide to nitrogen dioxide and oxygen is first-order. Its rate constant is 1.91 $hr^{-1}$.
**(a)** What is its half-life?
**(b)** What fraction of dinitrogen pentaoxide is left undecomposed after 45 minutes?
**(c)** How long will it take to decompose 22% of the $N_2O_5$?
**(d)** How fast is a 0.2500 $M$ solution decomposing?

**39.** Copper-64 is one of the metals used to study brain activity. Its decay constant is 0.0541 $hr^{-1}$. If a solution containing 10.0 mg of Cu-64 is used, how many mg of Cu-64 remains after one day?

**40.** Iodine-131 is used to treat tumors in the thyroid. Its half-life is 8.1 days. If a patient is given a sample containing 5.00 mg of I-131, at what time will 1.0% of the isotope remain in her system?

**41.** A sample of sodium-24 chloride contains 0.050 mg of Na-24 to study the sodium balance of an animal. After 24.9 hr, there is 0.016 mg of Na-24 left. What is the half-life of Na-24?

**42.** Argon-41 is used to measure the rate of gas flow. It has a decay constant of $6.3 \times 10^{-3}$ $min^{-1}$.
**(a)** What is its half-life?
**(b)** How long will it take before only 10.0% of the original amount of Ar-41 is left?

## *Zero- and Second-Order Reactions*

**43.** Butadiene, $C_4H_6$, dimerizes (two molecules combine to form a single molecule) to form $C_8H_{12}$. The dimerization is a second-order reaction. If the rate constant for the reaction is 0.84 L/mol · min,
**(a)** how long will it take for the concentration of butadiene to go from 0.200 $M$ to 0.128 $M$?
**(b)** what is the half-life of the reaction if one starts with a 0.100 $M$ solution of butadiene?
**(c)** what is the rate of the reaction when $[C_4H_6] = 0.078$ $M$?

**44.** The second-order reaction for the decomposition of nitrosyl bromide,

$$NOBr(g) \longrightarrow NO(g) + \tfrac{1}{2}Br_2(g)$$

has a half-life of 12.5 seconds at a concentration of 0.100 $M$.
**(a)** Calculate $k$ for the reaction.
**(b)** How long will it take to decompose 80.0% of a 0.0500 $M$ solution of nitrosyl bromide?
**(c)** How fast is a 0.200 $M$ solution of nitrosyl bromide decomposing?

**45.** For the zero-order decomposition of ammonia on tungsten,

$$NH_3(g) \xrightarrow{\text{W}} \tfrac{1}{2}N_2(g) + \tfrac{3}{2}H_2(g)$$

the rate constant is 0.0125 mol/L · min.
**(a)** What is the half-life of a 0.750 $M$ solution of ammonia?
**(b)** How long will it take for the concentration of ammonia to drop from 0.300 $M$ to 0.0600 $M$?

**46.** For the zero-order decomposition of HI on a gold surface,

$$HI(g) \xrightarrow{\text{Au}} \tfrac{1}{2}H_2(g) + \tfrac{1}{2}I_2(g)$$

it takes 1.0 second for the concentration of HI to drop from 0.335 $M$ to 0.285 $M$.
**(a)** What is the rate constant for the reaction?
**(b)** How long will it take for the concentration of a 0.650 $M$ solution to drop to 0.0100 $M$?
**(c)** What is the half-life of a 0.400 $M$ solution?

## Activation Energy, Catalysis, $\Delta H$

**\*47.** Three first-order reactions have the following activation energies:

| Reaction | A | B | C |
|---|---|---|---|
| $E_a$(kJ) | 75 | 136 | 292 |

**(a)** Which reaction is the fastest?
**(b)** Which reaction has the largest half-life?

**\*48.** Consider the data for several systems for the conversion of reactants to products at the same temperature.

| System | $E_a$(kJ) | $\Delta H$(kJ) |
|---|---|---|
| 1 | 75 | −20 |
| 2 | 110 | +50 |
| 3 | 50 | −75 |

**(a)** Which system has the slowest rate?
**(b)** Which system has the smallest rate constant?
**(c)** In each system determine the difference between the energy of the activated complex and that of the products.

**\*49.** For a certain reaction, $E_a$ is 135 kJ and $\Delta H = 45$ kJ. In the presence of a catalyst, the activation energy is 39% of that for the uncatalyzed reaction. Draw a diagram similar to those in Figures 11.8 and 11.9 for the reaction.

**\*50.** The activation energy, $E_a$, for a certain reaction is 128 kJ. The activation energy for the reverse reaction is 95 kJ. In the presence of a catalyst, $E_{acat}$ is 110 kJ. What is $\Delta H$ for the reaction?

## Reaction Rate and Temperature

**51.** The following data were obtained for the thermal decomposition of nitrogen dioxide,

$$2NO_2(g) \longrightarrow 2NO(g) + O_2(g)$$

| $k$(L/mol · s) | 1.57 | 3.59 | 7.73 | 15.8 |
|---|---|---|---|---|
| $t$(°C) | 350 | 375 | 400 | 425 |

Plot these data (ln $k$ versus $1/T$) and find the activation energy for the reaction.

**52.** The following data were obtained for the decomposition of A:

$$2A(g) \longrightarrow products$$

| $k$(L/mol · s) | 0.0300 | 0.725 | 9.96 | 89.02 |
|---|---|---|---|---|
| $t$(°C) | 650 | 750 | 850 | 950 |

Plot these data (ln $k$ versus $1/T$) and find the activation energy for the reaction.

**53.** The uncoiling of deoxyribonucleic acid (DNA) is a first-order reaction. Its activation energy is 420 kJ. At 37°C, the rate constant is $4.90 \times 10^{-4}$ min$^{-1}$.
**(a)** What is the half-life of the uncoiling at 37°C (normal body temperature)?
**(b)** What is the half-life of the uncoiling if the organism has a temperature of 40°C ($\approx 104$°F)?
**(c)** By what factor does the rate of uncoiling increase (per °C) over this temperature interval?

**54.** The activation energy for the reaction involved in the souring of raw milk is 75 kJ. Milk will sour in about eight hours at 21°C (70°F = room temperature). How long will raw milk last in a refrigerator maintained at 5°C? Assume the rate constant to be inversely related to souring time.

**55.** Hypothermia, which can result in death, results when a warm-blooded organism is exposed to severe cold. If the activation energy of a certain enzyme-catalyzed reaction is 98 kJ/mol, by what percentage is the rate of this reaction decreased if the body temperature drops from 98.6°F (37.0°C) to 89.6°F (32.0°C)?

**56.** The chirping rate of a cricket, $X$, in chirps per minute near room temperature is given by

$$X = 7.2t - 32$$

where $t$ is the temperature in °C.
**(a)** Calculate the chirping rates at 23°C and 33°C.
**(b)** Use your answers in (a) to estimate the activation energy for this reaction.
**(c)** What is the percent increase for a 10°C rise in temperature?

**57.** For the reaction

$$C_2H_4(g) + H_2(g) \longrightarrow C_2H_6(g)$$

the activation energy is 181 kJ/mol. The rate constant at 1000°C is $1.6 \times 10^3$ L/mol · s.
**(a)** At what temperature is the rate constant half its value at 1000°C?
**(b)** What is the rate constant at 725°C?

**58.** For the reaction

$$HI(g) + CH_3I(g) \longrightarrow CH_4(g) + I_2(g)$$

the activation energy is $1.4 \times 10^2$ kJ/mol. The rate constant at 227°C is $3.9 \times 10^{-3}$ L/mol · s.
**(a)** What is the rate constant at 310°C?
**(b)** At what temperature is the rate constant $1.0 \times 10^{-3}$ L/mol · s?

## Reaction Mechanisms

**\*59.** Write the rate expression for each of the following elementary steps:
**(a)** $NO_3 + CO \longrightarrow NO_2 + CO_2$
**(b)** $NO + O_2 \longrightarrow NO_3$
**(c)** $2NO_2 \longrightarrow 2NO + O_2$

*60. Write the rate expression for each of the following elementary steps:

(a) $N_2O_2 + H_2 \longrightarrow N_2O + H_2O$

(b) $K + HCl \longrightarrow KCl + H$

(c) $I_2 \longrightarrow 2I$

*61. For the reaction

$$2H_2(g) + 2NO(g) \longrightarrow N_2(g) + 2H_2O(g)$$

the experimental rate expression is rate = $k[NO]^2 \times [H_2]$. The following mechanism is proposed:

| | |
|---|---|
| $2NO \rightleftharpoons N_2O_2$ | (fast) |
| $N_2O_2 + H_2 \longrightarrow H_2O + N_2O$ | (slow) |
| $N_2O + H_2 \longrightarrow N_2 + H_2O$ | (fast) |

Is this mechanism consistent with the rate expression?

*62. For the reaction between hydrogen and iodine,

$$2H_2(g) + I_2(g) \longrightarrow 2HI(g)$$

the experimental rate expression is rate = $k[H_2] \times [I_2]$. Show that this expression is consistent with the mechanism:

| | |
|---|---|
| $I_2(g) \rightleftharpoons 2I(g)$ | (fast) |
| $H_2(g) + I(g) + I(g) \longrightarrow 2HI(g)$ | (slow) |

*63. At low temperatures, the rate law for the reaction

$$CO(g) + NO_2(g) \longrightarrow CO_2(g) + NO(g)$$

is as follows: rate = constant $\times [NO_2]^2$. Which of the following mechanisms is consistent with the rate law?

(a) $CO + NO_2 \longrightarrow CO_2 + NO$

(b) 
| | |
|---|---|
| $2NO_2 \rightleftharpoons N_2O_4$ | (fast) |
| $N_2O_4 + 2CO \longrightarrow 2CO_2 + 2NO$ | (slow) |

(c) 
| | |
|---|---|
| $2NO_2 \longrightarrow NO_3 + NO$ | (slow) |
| $NO_3 + CO \longrightarrow NO_2 + CO_2$ | (fast) |

(d) 
| | |
|---|---|
| $2NO_2 \longrightarrow 2NO + O_2$ | (slow) |
| $O_2 + 2CO \longrightarrow 2CO_2$ | (fast) |

*64. Two mechanisms are proposed for the reaction

$$2NO(g) + O_2(g) \longrightarrow 2NO_2(g)$$

Mechanism 1:
| | |
|---|---|
| $NO + O_2(g) \rightleftharpoons NO_3$ | (fast) |
| $NO_3 + NO \longrightarrow 2NO_2$ | (slow) |

Mechanism 2:
| | |
|---|---|
| $NO + NO \rightleftharpoons N_2O_2$ | (fast) |
| $N_2O_2 + O_2(g) \longrightarrow 2NO_2$ | (slow) |

Show that each of these mechanisms is consistent with the observed rate law: rate = $k[NO]^2 \times [O_2]$.

## Unclassified

*65. Consider the first-order decomposition reaction

$$A \longrightarrow B + C$$

where A (circles) decomposes to B (triangles) and C (squares). Given the following boxes representing the reaction at $t = 2$ minutes, fill in the boxes with products at the indicated time. Estimate the half-life of the reaction.

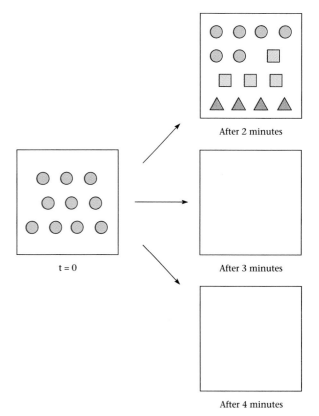

t = 0

After 2 minutes

After 3 minutes

After 4 minutes

*66. Given the following reaction, which is second-order in A and first-order in B:

$$A + B \longrightarrow products$$

(a) Write the rate expression.

(b) Consider the following one-liter vessel, where each square represents a mole of A and each circle represents a mole of B.

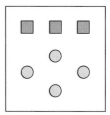

What is the rate of the reaction in terms of $k$?

(c) Fill the similar one-liter vessel shown below with an appropriate number of circles (representing B), assuming the same rate and $k$ as in (b).

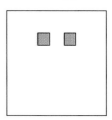

**67.** The following experiments are performed for the first-order reaction

$$A \longrightarrow products$$

Fill in the following blanks. If the answer cannot be calculated with the information given, write NC on the blank(s) provided.

|  | Expt 1 | Expt 2 | Expt 3 | Expt 4 |
|---|---|---|---|---|
| [A] | 0.100 | 0.200 | 0.100 | 0.100 |
| Catalyst | no | no | yes | no |
| Temperature | 25°C | 25°C | 25°C | 30°C |
| $k$ (min$^{-1}$) | 0.5 | _____ | 1 | 0.6 |
| rate (mol/L · min) | _____ | _____ | _____ | _____ |
| $E_a$ (kJ) | 32 | _____ | _____ | _____ |

*68. For the reaction

$$A + B \longrightarrow C$$

the rate expression is rate = $k[A][B]$.

   **(a)** Given three test tubes at the same temperature with different concentrations of A and B, which test tube will have the smallest rate?

     **(1)** 0.10 $M$ A;    0.10 $M$ B
     **(2)** 0.15 $M$ A;    0.15 $M$ B
     **(3)** 0.06 $M$ A;    1.0 $M$ B

   **(b)** If the temperature is increased, describe (using the words *increases, decreases,* or *remains the same*) what happens to the rate and to the values of $k$ and $E_a$.

*69. How does each of the following affect reaction rate?

   **(a)** the passage of time
   **(b)** decreasing container size for a gas-phase reaction
   **(c)** adding a catalyst

*70. In your own words explain why

   **(a)** a decrease in temperature slows the rate of a reaction.

   **(b)** doubling the concentration of a reactant does not always double the rate of the reaction.

   **(c)** a flame lights a cigarette but the cigarette continues to burn after the flame has been removed.

**71.** The decomposition of nitrosyl bromide

$$NOBr(g) \longrightarrow NO(g) + \tfrac{1}{2}Br_2(g)$$

is a second-order reaction. Its rate constant is 0.80 L/mol · s. The heats of formation are 82.2 kJ/mol for NOBr, 90.2 kJ/mol for NO, and 30.9 kJ/mol for $Br_2$. If the initial concentration of NOBr is 0.100 $M$, how much heat is evolved or absorbed per liter of solution in one minute?

**72.** Consider the decomposition of dinitrogen pentaoxide in $CCl_4$:

$$N_2O_5(g) \longrightarrow N_2O_4(g)) + \tfrac{1}{2}O_2(g)$$

The rate constant for this first-order reaction is 0.037/min. A sample of $N_2O_5$ weighing 55.0 g is dissolved in $CCl_4$ and allowed to decompose.

   **(a)** How long will it take to reduce the amount of $N_2O_5$ to 10.0 g?

   **(b)** What total volume of $O_2$ at 35°C and 1.00 atm is produced if the reaction is stopped after 23 minutes?

**73.** How much faster would a reaction proceed at 25°C than at 10°C if the activation energy is $1.00 \times 10^2$ kJ/mol?

**74.** A reaction has an activation energy of 132 kJ. What percent increase would one get for $k$ for every 10°C increase in temperature around room temperature, 25°C?

**75.** Derive the integrated rate law, $[A] = [A]_o - kt$ for a zero-order reaction. (*Hint:* Start with the relation $-\Delta[A] = k\Delta t$.)

**76.** For the first-order thermal decomposition of ozone,

$$O_3(g) \longrightarrow O_2(g) + O(g)$$

$k = 3 \times 10^{-26}$ s$^{-1}$ at 25°C. What is the half-life for this reaction in years? Comment on the likelihood that this reaction contributes to the depletion of the ozone layer.

## Challenge Problems

**77.** The gas-phase reaction between hydrogen and iodine,

$$H_2(g) + I_2(g) \rightleftharpoons 2HI(g)$$

proceeds with a rate constant for the forward reaction at 700°C of 138 L/mol · s and an activation energy of 165 kJ/mol.

   **(a)** Calculate the activation energy of the reverse reaction given that $\Delta H_f^\circ$ for HI is 26.48 kJ/mol and $\Delta H_f^\circ$ for $I_2(g)$ is 62.44 kJ/mol.

   **(b)** Calculate the rate constant for the reverse reaction at 700°C. (Assume A in the equation $k = Ae^{-E_a/RT}$ is the same for both forward and reverse reactions.)

   **(c)** Calculate the rate of the reverse reaction if the concentration of HI is 0.200 $M$. The reverse reaction is second order in HI.

**78.** For a first-order reaction $a$A $\longrightarrow$ products, where $a$ is a number other than 1, the rate is equal to $-\Delta[A]/a\Delta t$ or, in derivative notation, $-d[A]/adt$. Derive the integrated rate law for the first-order decomposition of $a$ moles of reactant.

**79.** Using calculus, derive the equation for

   **(a)** the concentration-time relation for a second-order reaction (see Table 11.2).

   **(b)** the concentration-time relation for a third-order reaction, A $\longrightarrow$ product.

**80.** The following data apply to the reaction

$$A(g) + 3B(g) + 2C(g) \longrightarrow products$$

| [A] | [B] | [C] | rate |
|---|---|---|---|
| 0.20 | 0.40 | 0.10 | $X$ |
| 0.40 | 0.40 | 0.20 | $8X$ |
| 0.20 | 0.20 | 0.20 | $X$ |
| 0.40 | 0.40 | 0.10 | $4X$ |

Determine the rate law for the reaction.

**81.** In a first-order reaction, let us suppose that a quantity, $X$, of a reactant is added at regular intervals of time, $\Delta t$. At first the amount of reactant in the system builds up; eventually, however, it levels off at a "saturation value" given by the expression

$$\text{saturation value} = \frac{X}{1 - 10^{-a}}, \qquad \text{where } a = 0.30 \frac{\Delta t}{t_{\frac{1}{2}}}$$

This analysis applies to prescription drugs when you take a certain amount each day. Suppose you take 0.100 g of a drug three times a day and that the half-life for elimination is 2.0 days. Using this equation, calculate the mass of the drug in the body at saturation. Suppose further that side effects show up when 0.500 g of the drug accumulates in the body. As a pharmacist, what is the maximum dosage you could assign to a patient for an eight-hour period without his suffering side effects?

Chemical plants such as these produce upwards of ten million tons of ammonia each year by the Haber process (p. 361). (Visuals Unlimited/Steve McCutcheon)

# 12 Gaseous Chemical Equilibrium

The cumbrous elements, Earth, Flood, Aire, Fire

Flew upward, spirited various forms,

Each had his place appointed, each his course,

The rest in circuit walles this Universe.

**—JOHN MILTON**

*Paradise Lost*

Among the topics covered in Chapter 9 was the equilibrium between liquid and gaseous water:

$$H_2O(l) \rightleftharpoons H_2O(g)$$

The state of this equilibrium system at a given temperature can be described in a simple way by citing the equilibrium pressure of water vapor: 0.034 atm at 25°C, 1.00 atm at 100°C, and so on.

Chemical reactions involving gases carried out in closed containers resemble in many ways the $H_2O(l)$–$H_2O(g)$ system. The reactions are reversible; reactants are not completely consumed. Instead, an equilibrium mixture containing both products and reactants is obtained. At equilibrium, forward and reverse reactions take place at the same rate. As a result, the amounts of all species at equilibrium remain constant with time.

*They can often be reversed by changing the temperature or pressure*

There is, however, one complicating factor where gaseous chemical equilibria are involved. Ordinarily, more than one gaseous substance is present. We might, for example, have a system

$$aA(g) + bB(g) \rightleftharpoons cC(g) + dD(g)$$

in which the equilibrium mixture contains two gaseous products (C and D) and two reactants (A and B). This means that

— the **partial pressures** of the gases in the mixture ($P_C$, $P_D$, $P_A$, $P_B$) must be specified. Recall from Chapter 5 that the partial pressure of a gas, $P_i$, can be calculated from the ideal gas law:

$$P_i = n_i RT/V$$

— the equilibrium condition cannot be expressed in terms of a single partial pressure, e.g., $P_D$ or $P_A$.

Fortunately, it turns out that there is a relatively simple relationship between the partial pressures of different gases present at equilibrium in a reaction system. This relationship is expressed in terms of a quantity called the **equilibrium constant,** symbol $K$. In this chapter, you will learn how to

— write the expression for $K$ corresponding to any chemical equilibrium (Section 12.2)
— calculate the value for $K$ from experimental data for the equilibrium system (Section 12.3)
— use the value of $K$ to predict the extent to which a reaction will take place (Section 12.4)
— use $K$, along with Le Chatelier's principle, to predict the result of disturbing an equilibrium system (Section 12.5)

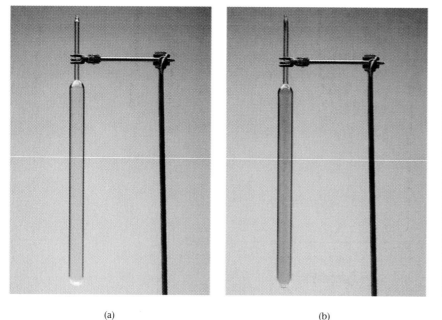

(a)                    (b)                    (c)

**Figure 12.1**

Dinitrogen tetraoxide, $N_2O_4$, is color-less (a); nitrogen dioxide, $NO_2$, is a brownish-red gas (c). The partial pressure of $NO_2$ in an equilibrium mixture of the two gases (b) is directly related to the depth of the brown color.   (Marna G. Clarke)

## 12.1   The $N_2O_4$–$NO_2$ Equilibrium System

Consider what happens when a sample of $N_2O_4$, a colorless gas, is placed in a closed, evacuated container at 100°C. Instantly, a reddish-brown color develops. This color is due to nitrogen dioxide, $NO_2$, formed by decomposition of part of the $N_2O_4$ (Fig. 12.1):

$$N_2O_4(g) \longrightarrow 2NO_2(g)$$

At first, this is the only reaction taking place. As soon as some $NO_2$ is formed, however, the reverse reaction can occur:

$$2NO_2(g) \longrightarrow N_2O_4(g)$$

As time passes, the partial pressure of $N_2O_4$ drops, and the forward reaction slows down. Conversely, the partial pressure of $NO_2$ increases, so the rate of the reverse reaction increases. Soon, these rates become equal. A dynamic equilibrium has been established:

$$N_2O_4(g) \rightleftharpoons 2NO_2(g)$$

**TABLE 12.1   Establishment of Equilibrium in the System $N_2O_4(g) \rightleftharpoons 2NO_2(g)$ (at 100°C)**

| Time(s) | 0 | 20 | 40 | 60 | 80 | 100 |
|---|---|---|---|---|---|---|
| $P_{N_2O_4}$ (atm) | 1.00 | 0.60 | 0.35 | **0.22** | **0.22** | **0.22** |
| $P_{NO_2}$ (atm) | 0.00 | 0.80 | 1.30 | **1.56** | **1.56** | **1.56** |

(Boldface numbers are equilibrium pressures.)

At equilibrium, appreciable amounts of both gases are present. As time passes, their amounts or partial pressures remain constant, as long as the volume of the container and the temperature remain unchanged.

The characteristics just described are typical of all systems at equilibrium. First, *the forward and reverse reactions are taking place at the same rate*. This explains why *the concentrations of species present remain constant with time*. Moreover, these concentrations are independent of the direction from which equilibrium is approached.

The approach to equilibrium in the $N_2O_4$–$NO_2$ system is illustrated by the data in Table 12.1 and Figure 12.2. Originally, only $N_2O_4$ is present; its pressure is 1.00 atm. Since there is no $NO_2$ around, its original pressure is zero. As equilibrium is approached, the *net reaction* is

$$N_2O_4(g) \longrightarrow 2NO_2(g)$$

The amount and hence the partial pressure of $N_2O_4$ drop, rapidly at first, then more slowly. The partial pressure of $NO_2$ increases. Finally, both partial pressures level off and become constant. At equilibrium, at 100°C:

$$P_{N_2O_4} = 0.22 \text{ atm} \qquad P_{NO_2} = 1.56 \text{ atm}$$

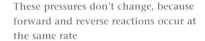
These pressures don't change, because forward and reverse reactions occur at the same rate

There are many ways to approach equilibrium in the $N_2O_4$–$NO_2$ system. Table 12.2 p. 344 gives data for three experiments in which the original conditions are quite different. Experiment 1 is that just described, starting with pure $N_2O_4$. Experiment 2 starts with pure $NO_2$ at a partial pressure of 1.00 atm. As equilibrium is approached, some of the $NO_2$ reacts to form $N_2O_4$:

$$2NO_2(g) \longrightarrow N_2O_4(g)$$

Finally, in Experiment 3, both $N_2O_4$ and $NO_2$ are present originally, each at a partial pressure of 1.00 atm.

Looking at the data in Table 12.2, you might wonder whether these three experiments have anything in common. Specifically, is there any relationship between the equilibrium partial pressures of $NO_2$ and $N_2O_4$ that is valid for all the experiments? It turns out that there is, although it is not an obvious one. The value of the quotient $(P_{NO_2})^2/P_{N_2O_4}$ is the same, about 11, in each case:

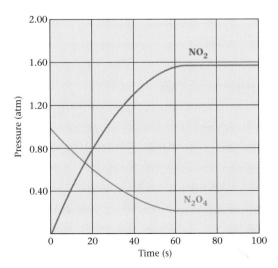

**Figure 12.2**
Approach to equilibrium in the $N_2O_4$–$NO_2$ system. The partial pressure of $N_2O_4$ starts off at 1.00 atm, drops sharply at first, and finally levels off at the equilibrium value of 0.22 atm. Meanwhile, the partial pressure of $NO_2$ rises from zero to its equilibrium value, 1.56 atm.

TABLE 12.2 **Equilibrium Measurements in the $N_2O_4$–$NO_2$ System at 100°C**

|         |          | Original Pressure (atm) | Equilibrium Pressure (atm) |
|---------|----------|-------------------------|----------------------------|
| Expt. 1 | $N_2O_4$ | 1.00                    | 0.22                       |
|         | $NO_2$   | 0.00                    | 1.56                       |
| Expt. 2 | $N_2O_4$ | 0.00                    | 0.07                       |
|         | $NO_2$   | 1.00                    | 0.86                       |
| Expt. 3 | $N_2O_4$ | 1.00                    | 0.42                       |
|         | $NO_2$   | 1.00                    | 2.16                       |

Note that this relationship holds only at equilibrium; initial pressures can have any values

Expt. 1 $\dfrac{(P_{NO_2})^2}{P_{N_2O_4}} = \dfrac{(1.56)^2}{0.22} = 11$

Expt. 2 $\dfrac{(P_{NO_2})^2}{P_{N_2O_4}} = \dfrac{(0.86)^2}{0.07} = 11$

Expt. 3 $\dfrac{(P_{NO_2})^2}{P_{N_2O_4}} = \dfrac{(2.16)^2}{0.42} = 11$

This relationship holds for any equilibrium mixture containing $N_2O_4$ and $NO_2$ at 100°C. More generally, it is found that, at any temperature, the quantity

$$\frac{(P_{NO_2})^2}{P_{N_2O_4}}$$

where $P_{NO_2}$ and $P_{N_2O_4}$ are equilibrium pressures in atmospheres, is a constant, independent of the original composition, the volume of the container, or the total pressure. This constant is referred to as the **equilibrium constant $K$** for the system

$$N_2O_4(g) \rightleftharpoons 2NO_2(g)$$

At 100°C, $K$ for this system at 11; at 150°C, it has a different value, about 110. Any mixture of $NO_2$ and $N_2O_4$ at 100°C will react in such a way that the ratio $(P_{NO_2})^2/P_{N_2O_4}$ becomes equal to 11. At 150°C, reaction occurs until this ratio becomes 110.

## 12.2  The Equilibrium Constant Expression

For every gaseous chemical system, an equilibrium constant expression can be written stating the condition that must be attained at equilibrium. For the general system involving only gases,

$$aA(g) + bB(g) \rightleftharpoons cC(g) + dD(g)$$

where A, B, C, and D represent different substances and *a*, *b*, *c*, and *d* are their coefficients in the balanced equation

$$K = \frac{(P_C)^c \times (P_D)^d}{(P_A)^a \times (P_B)^b}$$

where $P_C$, $P_D$, $P_A$, and $P_B$ are the partial pressures in **atmospheres** of the four gases at equilibrium.* Notice that in the expression for $K$

— the equilibrium partial pressures of **products** (right side of equation) appear in the **numerator.**
— the equilibrium partial pressures of **reactants** (left side of equation) appear in the **denominator.**
— each partial pressure is raised to a **power** equal to its **coefficient** in the balanced equation.

---

**Example 12.1**   Ammonia can be made from hydrogen and nitrogen

$$N_2(g) + 3H_2(g) \rightleftharpoons 2NH_3(g)$$

Write the equilibrium expression for this system.

**Strategy**   Since $NH_3$ is a product, $P_{NH_3}$ appears in the numerator; the partial pressures of $N_2$ and $H_2$ are in the denominator. The exponents are the coefficients in the balanced equation: 2 for $NH_3$, 1 for $N_2$, 3 for $H_2$.

**Solution**   $NH_3$ is a product; $H_2$ and $N_2$ are reactants. The expression is

$$K = \frac{(P_{NH_3})^2}{P_{N_2} \times (P_{H_2})^3}$$

---

## Changing the Chemical Equation

It is important to realize that *the expression for K depends upon the form of the chemical equation written to describe the equilibrium system.* To illustrate what this statement means, consider the $N_2O_4$–$NO_2$ system:

$$N_2O_4(g) \rightleftharpoons 2NO_2(g) \qquad K = \frac{(P_{NO_2})^2}{P_{N_2O_4}}$$

$K$ is meaningless unless accompanied by a chemical equation

Many other equations could be written for this system, for example,

$$\tfrac{1}{2}N_2O_4(g) \rightleftharpoons NO_2(g)$$

In this case, the expression for the equilibrium constant would be

$$K' = \frac{P_{NO_2}}{(P_{N_2O_4})^{1/2}}$$

---

*This equilibrium constant is often given the symbol $K_p$ to emphasize that it involves partial pressures. Other equilibrium expressions for gases are sometimes used, including $K_c$:

$$K_c = \frac{[C]^c \times [D]^d}{[A]^a \times [B]^b}$$

where the brackets represent equilibrium concentrations in moles per liter. The two constants, $K_c$ and $K_p$, are simply related (see Problem 66):

$$K_p = K_c(RT)^{\Delta n_g}$$

where $\Delta n_g$ is the change in the number of moles of gas in the equation, $R$ is the gas law constant, $0.0821 \, L \cdot atm/mol \cdot K$, and $T$ is the Kelvin temperature. Throughout this chapter, we deal exclusively with $K_p$, referring to it simply as $K$.

Comparing the expressions for $K$ and $K'$, it is clear that $K'$ is the square root of $K$. This illustrates a general rule, sometimes referred to as the **coefficient rule,** which states:

*If the coefficients in a balanced equation are multiplied by a factor n, the equilibrium constant is raised to the nth power:*

$$K' = K^n$$

In this particular case, $n = \frac{1}{2}$ because the coefficients were divided by 2. Hence $K'$ is the square root of $K$. If $K = 11$, then $K' = (11)^{1/2} = 3.3$.

Another equation that might be written to describe the $N_2O_4$–$NO_2$ system is

$$2NO_2(g) \rightleftharpoons N_2O_4(g) \qquad K'' = \frac{P_{N_2O_4}}{(P_{NO_2})^2}$$

This chemical equation is simply the reverse of that written originally; $N_2O_4$ and $NO_2$ switch sides of the equation. Notice that $K''$ is the reciprocal of $K$; the numerator and denominator have been inverted. This illustrates the **reciprocal rule:**

*The equilibrium constants for forward and reverse reactions are the reciprocals of each other,*

$$K'' = 1/K$$

If, for example, $K = 11$, then $K'' = 1/11 = 0.091$.

## Adding Chemical Equations

A property of $K$ that you will find very useful in this and succeeding chapters is expressed by the **rule of multiple equilibria,** which states

We will use this relationship frequently in later chapters

*If a reaction can be expressed as the sum of two or more reactions, K for the overall reaction is the product of the equilibrium constants of the individual reactions.*

That is, if

$$\text{Reaction 3} = \text{Reaction 1} + \text{Reaction 2}$$

then

$$K\,(\text{Reaction 3}) = K\,(\text{Reaction 1}) \times K\,(\text{Reaction 2})$$

To illustrate the application of this rule, consider the following reactions at 700°C:

$$SO_2(g) + \tfrac{1}{2}O_2(g) \rightleftharpoons SO_3(g) \qquad\qquad K = 2.2$$

$$NO_2(g) \rightleftharpoons NO(g) + \tfrac{1}{2}O_2(g) \qquad\qquad K = 4.0$$

Adding these equations eliminates $\tfrac{1}{2}O_2$; the result is

$$SO_2(g) + NO_2(g) \rightleftharpoons SO_3(g) + NO(g)$$

For this reaction, $K = 2.2 \times 4.0 = 8.8$.

The validity of this rule can be demonstrated by writing the expression for K for the individual reactions and multiplying:

$$\frac{P_{SO_3}}{P_{SO_2} \times (P_{O_2})^{1/2}} \times \frac{P_{NO} \times (P_{O_2})^{1/2}}{P_{NO_2}} = \frac{P_{SO_3} \times P_{NO}}{P_{SO_2} \times P_{NO_2}}$$

The product expression, at the right, is clearly the equilibrium constant for the overall reaction, as it should be according to the rule of multiple equilibria.

## Heterogeneous Equilibria

In the reactions described so far, all the reactants and products have been gaseous; the equilibrium systems are *homogeneous*. In certain reactions, one or more of the substances involved is a pure liquid or solid; the others are gases. Such a system is *heterogeneous*, since more than one phase is present. Examples include

$$CO_2(g) + H_2(g) \rightleftharpoons CO(g) + H_2O(l)$$

$$I_2(s) \rightleftharpoons I_2(g)$$

Experimentally, it is found that for such systems

— *the position of the equilibrium is independent of the amount of solid or liquid, as long as some is present.*
— *terms for pure liquids or solids need not appear in the expression for K.*

To see why this is the case, consider the data shown in Table 12.3 for the system

$$CO_2(g) + H_2(g) \rightleftharpoons CO(g) + H_2O(l)$$

at 25°C. In Experiments 1 through 4, there is liquid water present, so $P_{H_2O} = 0.03$ atm, the equilibrium vapor pressure of water at 25°C.

Notice that

**1.** In all cases, the quantity

$$K_I = \frac{P_{CO} \times P_{H_2O}}{P_{CO_2} \times P_{H_2}}$$

is constant at $9 \times 10^{-6}$.

TABLE 12.3   **Equilibrium Constant Expressions for** $CO_2(g) + H_2(g) \rightleftharpoons CO(g) + H_2O(l)$

|  | Expt. 1 | Expt. 2 | Expt. 3 | Expt. 4 |
|---|---|---|---|---|
| Mass $H_2O(l)$ | 8 g | 6 g | 4 g | 2 g |
| $P_{H_2O}$ (atm) | $3 \times 10^{-2}$ | $3 \times 10^{-2}$ | $3 \times 10^{-2}$ | $3 \times 10^{-2}$ |
| $K_I$ | $9 \times 10^{-6}$ | $9 \times 10^{-6}$ | $9 \times 10^{-6}$ | $9 \times 10^{-6}$ |
| $K_{II}$ | $3 \times 10^{-4}$ | $3 \times 10^{-4}$ | $3 \times 10^{-4}$ | $3 \times 10^{-4}$ |

$$K_I = \frac{P_{CO} \times P_{H_2O}}{P_{CO_2} \times P_{H_2}} \qquad K_{II} = \frac{P_{CO}}{P_{CO_2} \times P_{H_2}}$$

**2.** In all cases (Experiments 1 through 4), the quantity

$$K_{II} = \frac{P_{CO}}{P_{CO_2} \times P_{H_2}}$$

is constant at $3 \times 10^{-4}$. This must be true because

$$K_{II} = K_I/P_{H_2O}$$

and $P_{H_2O}$ is constant at $3 \times 10^{-2}$ atm. It follows that $K_{II}$ is a valid equilibrium constant for the system

$$CO_2(g) + H_2(g) \rightleftharpoons CO(g) + H_2O(l)$$

Since the expression for $K_{II}$ is simpler than that for $K_I$, it is the equilibrium constant of choice for the heterogeneous system.

A similar argument can be applied to the equilibrium involved in the sublimation of iodine, $I_2(s) \rightleftharpoons I_2(g)$, where $K = P_{I_2}$. At a given temperature, the pressure of iodine vapor is constant, independent of the amount of solid iodine or any other factor.

For the process: $I_2(g) \rightleftharpoons I_2(s)$, $K = 1/P_{I_2}$

---

**Example 12.2** Write the expression for K for

(a) the reduction of black solid copper(II) oxide (1 mol) with hydrogen to form copper metal and steam.
(b) the reaction of 1 mol of steam with red-hot coke (carbon) to form a mixture of hydrogen and carbon monoxide, called water gas.

**Strategy** First write the equation for the equilibrium system. Then write the expression for *K*, deleting pure liquids and solids, and following the usual rules for gases.

*Solution*

If the product were $H_2O(l)$, $K = 1/P_{H_2}$

(a) The reaction is $CuO(s) + H_2(g) \rightleftharpoons Cu(s) + H_2O(g)$.

The equilibrium expression is $K = P_{H_2O}/P_{H_2}$

(b) The reaction is $C(s) + H_2O(g) \rightleftharpoons H_2(g) + CO(g)$

The equilibrium expression is $K = \dfrac{P_{H_2} \times P_{CO}}{P_{H_2O}}$

---

In this and succeeding chapters, a wide variety of different types of equilibria will be covered. They may involve gases, pure liquids or solids, and species in aqueous solution. It will always be true that in the expression for the equilibrium constant

To express *K* for any system, follow these rules

— *gases enter as their partial pressures in atmospheres.*
— *pure liquids or solids do not appear; neither does the solvent for a reaction in dilute solution*
— *species (ions or molecules) in water solution enter as their molar concentrations*

Thus, for the equilibrium attained when zinc metal reacts with acid:

$$Zn(s) + 2H^+(aq) \rightleftharpoons Zn^{2+}(aq) + H_2(g)$$

$$K = \frac{P_{H_2} \times [Zn^{2+}]}{[H^+]^2}$$

We will have more to say in later chapters about equilibria involving species in aqueous solution.

## 12.3   Determination of *K*

Numerical values of equilibrium constants can be calculated if the partial pressures of products and reactants at equilibria are known. Sometimes you will be given equilibrium partial pressures directly (Example 12.3). At other times, you will be given the original partial pressures and the equilibrium partial pressure of one species (Example 12.4). In that case, the calculation of *K* is a bit more difficult.

---

**Example 12.3**   Ammonium chloride is sometimes used as a flux in soldering because it decomposes upon heating:

$$NH_4Cl(s) \rightleftharpoons NH_3(g) + HCl(g)$$

The HCl formed removes oxide films from metals to be soldered. In a certain equilibrium system at 400°C, 12.0 g of NH$_4$Cl is present; the partial pressures of NH$_3$ and HCl are 3.0 atm and 5.0 atm, respectively. Calculate *K* at 400°C.

*Strategy*   First write the expression for *K*. Then substitute equilibrium partial pressures into that expression and solve.

*Solution*   $K = (P_{NH_3}) \times (P_{HCl}) = (3.0) \times (5.0) = \boxed{15}$

---

**Example 12.4**   Consider the equilibrium system

$$2HI(g) \rightleftharpoons H_2(g) + I_2(g)$$

Originally, a system contains only HI at a pressure of 1.00 atm at 520°C. The equilibrium partial pressure of H$_2$ is found to be 0.10 atm. Calculate

   (a) $P_{I_2}$ at equilibrium      (b) $P_{HI}$ at equilibrium      (c) *K*

*Strategy*   In order to solve this problem, you must

— carefully distinguish between *original* and *equilibrium* partial pressures.
— learn how to relate the equilibrium partial pressures of H$_2$, I$_2$ and HI, using the coefficients of the balanced equation.

*Solution*

   (a) Note from the balanced equation that one mole of I$_2$ is formed for every mole of H$_2$. Putting it another way, as reaction proceeds from left to right, the increase in the number of moles of I$_2$ is the same as that for H$_2$:

$$\Delta n_{I_2} = \Delta n_{H_2}$$

But, since $P = nRT/V$, it follows that for an equilibrium system at constant *T* and *V*, pressure is directly proportional to amount in moles, and the above relationship must hold for partial pressures as well:

$$\Delta P_{I_2} = \Delta P_{H_2}$$

Ammonia gas on the left reacts with hydrogen chloride gas on the right. The white smoke is solid NH$_4$Cl.

Since both partial pressures start at zero, the equilibrium partial pressures must be equal. The equilibrium partial pressure of $I_2$, like that of $H_2$, is $\boxed{0.10 \text{ atm.}}$

(b) Two moles of HI are required to form one mole of $H_2$. More generally,

$$\Delta n_{HI} = -2\Delta n_{H_2}$$

where the minus sign takes account of the fact that when the amount of hydrogen increases, the amount of HI decreases. But, since $P$ is directly proportional to $n$,

If $\Delta P_{H_2} = x$, then $\Delta P_{I_2} = x$ and $\Delta P_{HI} = -2x$

$$\Delta P_{HI} = -2\Delta P_{H_2} = -2(0.10 \text{ atm}) = -0.20 \text{ atm}$$

In other words, the partial pressure of HI decreases by 0.20 atm. Since it starts at 1.00 atm, its equilibrium value must be 1.00 atm $-$ 0.20 atm $= \boxed{0.80 \text{ atm.}}$

Summarizing this reasoning in the form of a table:

|  | 2HI($g$) | $\rightleftharpoons$ | $H_2$($g$) | $+$ | $I_2$($g$) |
|---|---|---|---|---|---|
| $P_o$ (atm) | 1.00 |  | 0.00 |  | 0.00 |
| $\Delta P$ (atm) | $-0.20$ |  | $+0.10$ |  | $+0.10$ |
| $P_{eq}$ (atm) | 0.80 |  | 0.10 |  | 0.10 |

($P_o$ = initial pressure; $P_{eq}$ = equilibrium pressure)

Note that the numbers in the center row ($\Delta P$) are related through the coefficients of the balanced equation (2HI, 1$H_2$, 1$I_2$).

(c) $K = \dfrac{P_{H_2} \times P_{I_2}}{(P_{HI})^2} = \dfrac{0.10 \times 0.10}{(0.80)^2} = \boxed{0.016}$

---

Example 12.4 illustrates a principle that you will find very useful in solving equilibrium problems throughout this (and later) chapters. *Changes in partial pressures of reactants and products as a system approaches equilibrium, like changes in molar amounts, are related to one another through the coefficients of the balanced equation.*

## 12.4  Applications of the Equilibrium Constant

Sometimes, knowing only the magnitude of the equilibrium constant, it is possible to decide upon the feasibility of a reaction. Consider, for example, a possible method for "fixing" atmospheric nitrogen—converting it to a compound—by reaction with oxygen:

$$N_2(g) + O_2(g) \rightleftharpoons 2NO(g)$$

$$K = \frac{(P_{NO})^2}{P_{N_2} \times P_{O_2}} = 1 \times 10^{-30} \text{ at } 25°C$$

Since $K$ is so small, the partial pressure of NO in equilibrium with $N_2$ and $O_2$, and hence the amount of NO, must be extremely small, approaching zero. Clearly, this would not be a suitable way to fix nitrogen, at least at 25°C.

An alternate approach to nitrogen fixation involves reacting it with hydrogen:

$$N_2(g) + 3H_2(g) \rightleftharpoons 2NH_3(g)$$

$$K = \frac{(P_{NH_3})^2}{P_{N_2} \times (P_{H_2})^3} = 6 \times 10^5 \text{ at } 25°C$$

Here, the equilibrium system must contain mostly ammonia. A mixture of $N_2$ and $H_2$ should be almost completely converted to $NH_3$ at equilibrium (see, however, the discussion of the Haber process, p. 362).

In general, if $K$ is a very small number, the equilibrium mixture will contain mostly unreacted starting materials; for all practical purposes, the forward reaction does not "go." Conversely, a large $K$ implies a reaction that, at least in principle, is feasible; products should be formed in high yield. Frequently, $K$ has an intermediate value, in which case you must make quantitative calculations concerning the *direction* or *extent* of reaction.

*The concept of equilibrium is most useful when K is neither very large nor very small*

## Direction of Reaction; the Reaction Quotient (*Q*)

Consider the general gas-phase reaction

$$aA(g) + bB(g) \rightleftharpoons cC(g) + dD(g)$$

for which

$$K = \frac{(P_C)^c \times (P_D)^d}{(P_A)^a \times (P_B)^b}$$

As we have seen, for a given system at a particular temperature, the value of $K$ is fixed.

In contrast, the *actual* pressure ratio, $Q$

$$Q = \frac{(P_C)^c \times (P_D)^d}{(P_A)^a \times (P_B)^b} \quad \text{(color is used to distinguish actual from equilibrium pressures)}$$

can have any value at all. If you start with pure reactants (A and B), $P_C$ and $P_D$ are zero, and the value of $Q$ is zero:

$$Q = \frac{0 \times 0}{(P_A)^a \times (P_B)^b} = 0$$

Conversely, starting with only C and D:

$$Q = \frac{(P_C)^c \times (P_D)^d}{0 \times 0} \longrightarrow \infty$$

If all four substances are present, $Q$ can have any value between zero and infinity.

The expression for $Q$, known as the **reaction quotient,** is the same as that for the equilibrium constant, $K$. The difference is that the partial pressures that appear in $Q$ are those that apply at a particular moment, not necessarily when the system is at equilibrium. By comparing the numerical value of $Q$ with that of $K$, it is possible to decide in which direction the system will move to achieve equilibrium.

*As equilibrium is approached, Q approaches K*

**1.** If $Q < K$ the reaction proceeds from left to right:

$$aA(g) + bB(g) \longrightarrow cC(g) + dD(g)$$

In this way, the partial pressures of products increase, while those of reactants decrease. As this happens, the reaction quotient Q increases and eventually at equilibrium becomes equal to K.

**2.** If $Q > K$ the partial pressures of products are "too high" and those of the reactants "too low" to meet the equilibrium condition. Reaction proceeds in the reverse direction:

$$aA(g) + bB(g) \longleftarrow cC(g) + dD(g)$$

increasing the partial pressures of A and B while reducing those of C and D. This lowers Q to its equilibrium value, K.

**3.** If, perchance, $Q = K$, the system is already at equilibrium, and nothing happens.

---

**Example 12.5**    Consider the following system at 100°C:

$$N_2O_4(g) \rightleftharpoons 2NO_2(g) \qquad K = 11$$

Predict the direction in which reaction will occur to reach equilibrium starting with 0.20 mol of $N_2O_4$ and 0.20 mole of $NO_2$ in a 4.0-L container.

***Strategy***    First, calculate the partial pressures of $N_2O_4$ and $NO_2$, using the ideal gas law, as applied to mixtures: $P_i = n_i RT/V$. Then, calculate Q. Finally, compare Q with K to predict the direction of reaction.

*Solution*

1. $P_{N_2O_4} = P_{NO_2} = \dfrac{nRT}{V} = \dfrac{(0.20 \text{ mol})(0.0821 \text{ L} \cdot \text{atm/mol} \cdot \text{K})(373 \text{ K})}{4.0 \text{ L}} = 1.5 \text{ atm}$

2. $Q = \dfrac{(P_{NO_2})^2}{P_{N_2O_4}} = \dfrac{(1.5)^2}{1.5} = 1.5$

3. Since $Q < 11$,   reaction must proceed in the forward direction.   This way, the partial pressure of $NO_2$ will increase, that of $N_2O_4$ will decrease, and Q will become equal to K.

---

## Extent of Reaction; Equilibrium Partial Pressures

The equilibrium constant for a chemical system can be used to calculate the partial pressures of the species present at equilibrium. In the simplest case, one equilibrium pressure can be calculated, knowing all the others. Consider, for example, the system

$$N_2(g) + O_2(g) \rightleftharpoons 2NO(g)$$

$$K = \frac{(P_{NO})^2}{P_{N_2} \times P_{O_2}} = 1 \times 10^{-30} \text{ at } 25°C$$

In air, the partial pressures of $N_2$ and $O_2$ are about 0.78 and 0.21 atm, respectively. The expression for K can be used to calculate the equilibrium partial pressure of NO under these conditions:

$$(P_{NO})^2 = P_{N_2} \times P_{O_2} \times K$$

$$= (0.78)(0.21)(1 \times 10^{-30}) = 1.6 \times 10^{-31}$$

$$P_{NO} = (1.6 \times 10^{-31})^{1/2} = 4 \times 10^{-16} \text{ atm}$$

This is an extremely small pressure, as you might have guessed from the small magnitude of the equilibrium constant.

That's just as well; if $N_2$ and $O_2$ in the atmosphere were converted to NO, we'd be in big trouble

More commonly, $K$ is used to determine the equilibrium partial pressures of all species, reactants, and products, knowing their original pressures. To do this, you follow what amounts to a four-step path.

**1.** *Using the balanced equation for the reaction, write the expression for K.*
**2.** *Express the equilibrium partial pressures of all species in terms of a single unknown, x.* To do this, apply the principle mentioned earlier: *The changes in partial pressures of reactants and products are related through the coefficients of the balanced equation.* To keep track of these values, make an equilibrium table, like the one illustrated in Example 12.6.
**3.** *Substitute the equilibrium terms into the expression for K. This gives an algebraic equation that must be solved for x.*
**4.** *Having found x, refer back to the table and calculate the equilibrium partial pressures of all species.*

---

**Example 12.6**   For the system

$$CO_2(g) + H_2(g) \rightleftharpoons CO(g) + H_2O(g)$$

$K$ is 0.64 at 900 K. Originally, only $CO_2$ and $H_2$ are present, each at a partial pressure of 1.00 atm. What are the equilibrium partial pressures of all species?

**Strategy**   Follow the four-step procedure described above. Be particularly careful about step (2), the most critical one in the process.

*Solution*

(1) $K = \dfrac{P_{CO} \times P_{H_2O}}{P_{CO_2} \times P_{H_2}}$

(2) You must express all equilibrium partial pressures in terms of a single unknown, $x$. A reasonable choice for $x$ is

$x$ = change in partial pressure of CO as the system goes to equilibrium

$= \Delta P_{CO}$

All the coefficients in the balanced equation are the same, 1. It follows that all the partial pressures change by the same amount, $x$.

$$\Delta P_{H_2O} = \Delta P_{CO} = x$$

$$\Delta P_{H_2} = \Delta P_{CO_2} = -x$$

(The minus sign arises because $H_2$ and $CO_2$ are consumed, whereas CO and $H_2O$ are formed.) The equilibrium table becomes

|  | $CO_2(g)$ | + | $H_2(g) \rightleftharpoons CO(g)$ | + | $H_2O(g)$ |
|---|---|---|---|---|---|
| $P_o$ (atm) | 1.00 | | 1.00 | 0 | 0 |
| $\Delta P$ (atm) | $-x$ | | $-x$ | $+x$ | $+x$ |
| $P_{eq}$ (atm) | $1.00 - x$ | | $1.00 - x$ | $x$ | $x$ |

You must distinguish carefully between initial pressures ($P_O$) and equilibrium pressures

(3) Substituting into the expression for $K$,

$$0.64 = \frac{(x)(x)}{(1.00 - x)(1.00 - x)}$$

Taking the square root of both sides,

$$(0.64)^{1/2} = 0.80 = \frac{x}{1.00 - x}$$

Solving for $x$:

$$x = 0.80 - 0.80x \qquad 1.80x = 0.80 \qquad x = 0.44$$

(4) Referring back to the equilibrium table,

$$P_{CO} = P_{H_2O} = x = \boxed{0.44 \text{ atm}}$$

$$P_{CO_2} = P_{H_2} = 1.00 - x = \boxed{0.56 \text{ atm}}$$

To check the validity of the calculation, note that

$$\frac{P_{CO} \times P_{H_2O}}{P_{CO_2} \times P_{H_2}} = \frac{(0.44)^2}{(0.56)^2} = 0.64 = K$$

---

The arithmetic involved in equilibrium calculations can be relatively simple, as it was in Example 12.6. Sometimes it is more complex (Example 12.7). The reasoning involved, however, is the same. It is always helpful to set up an equilibrium table to summarize the analysis of the problem.

---

**Example 12.7**   Consider the system

$$N_2O_4(g) \rightleftharpoons 2NO_2(g) \qquad K = 11 \text{ at } 100°C$$

Starting with pure $N_2O_4$ at a pressure of 1.00 atm, what will be the equilibrium partial pressures of $NO_2$ and $N_2O_4$?

**Strategy**   Again, follow the four-step path. The second step is a bit trickier here because the coefficients in the balanced equation are not equal to one another. The third step is more difficult algebraically, as you will find.

**Solution**

(1) $K = \dfrac{(P_{NO_2})^2}{P_{N_2O_4}}$

(2) Some of the $N_2O_4$ will decompose; let $x$ be the decrease in the partial pressure of $N_2O_4$, i.e., $\Delta P_{N_2O_4} = -x$. The balanced equation shows that one $N_2O_4$ yields two $NO_2$; it follows that the change in partial pressure of $NO_2$ must be $2x$. That is, $\Delta P_{NO_2} = 2x$. The equilibrium table becomes

| | $N_2O_4(g) \rightleftharpoons$ | $2NO_2(g)$ |
|---|---|---|
| $P_o$ (atm) | 1.00 | 0.00 |
| $\Delta P$ (atm) | $-x$ | $+2x$ |
| $P_{eq}$ (atm) | $1.00 - x$ | $2x$ |

(3) $K = \dfrac{(2x)^2}{(1.00 - x)} = \dfrac{4x^2}{(1.00 - x)} = 11$

This time, you cannot solve for $x$ as simply as in Example 12.6; the denominator on the left side is not a perfect square. Use the general method of solving a quadratic equation, discussed in Appendix 3. This involves rearranging to the form

$$ax^2 + bx + c = 0$$

and applying the quadratic formula

$$x = \frac{-b \pm \sqrt{b^2 - 4ac}}{2a}$$

Putting the equilibrium constant expression in the desired form,

$$4x^2 + 11x - 11 = 0$$

So

$$a = 4, \; b = 11, \; c = -11$$

$$x = \frac{-11 \pm \sqrt{121 + 16(11)}}{8} = \frac{-11 \pm \sqrt{297}}{8}$$

For the most part, we'll avoid tedious calculations of this type

Noting that the square root of 297 is 17.2,

$$x = \frac{-11 \pm 17.2}{8} = \frac{6.2}{8} \;\; \text{or} \;\; \frac{-28.2}{8} = 0.78 \;\; \text{or} \;\; -3.52$$

Of the two answers, only 0.78 is plausible; a value of $-3.52$ would imply a negative partial pressure of $NO_2$, which is ridiculous.

(4) $P_{NO_2} = 2(0.78 \text{ atm}) = $ $\boxed{1.56 \text{ atm}}$

$P_{N_2O_4} = 1.00 \text{ atm} - 0.78 \text{ atm} = $ $\boxed{0.22 \text{ atm}}$

In other words, 22% of the $N_2O_4$ originally present is left when equilibrium is established; 78% is converted to $NO_2$. One way to check the validity of this answer is to refer back to Table 12.1, where you find that, sure enough, starting with an $N_2O_4$ pressure of 1.00 atm, the values of the equilibrium partial pressures are those just calculated.

---

You may be curious about what would have happened in Example 12.7 with a different choice of variable. Suppose, for example, $x$ had been chosen to be the *increase* in the partial pressure of $NO_2$. In that case, the partial pressure of $N_2O_4$ at equilibrium would have decreased by $\frac{1}{2}x$, so

$$P_{NO_2} = x \qquad P_{N_2O_4} = 1.00 - \tfrac{1}{2}x$$

This leads to the algebraic equation

$$\frac{x^2}{(1.00 - \tfrac{1}{2}x)} = 11$$

Solving this equation, you should find that $x = 1.56$, so $P_{NO_2} = 1.56$ atm, $P_{N_2O_4} = 1.00 \text{ atm} - \frac{1}{2}(1.56 \text{ atm}) = 0.22$ atm. These are, of course, the same values obtained with different choice of unknown in Example 12.7. In general, it doesn't matter what choice of unknown you make, provided you are consistent in relating equilibrium pressures.

We should point out that the calculations involved in Examples 12.6 and 12.7 assume ideal gas behavior. At the conditions specified (1 atm, relatively high temperatures), this assumption is a good one. However, many industrial gas-phase reactions are carried out at very high pressures (see Table 12.A, p. 362). In that case, intermolecular forces become important, and calculated yields based on ideal gas behavior may be seriously in error.

## 12.5   Effect of Changes in Conditions Upon an Equilibrium System

Once a system has attained equilibrium, it is possible to change the ratio of products to reactants by changing the external conditions. We will consider three ways in which a chemical equilibrium can be disturbed:

**1.** Adding or removing a gaseous reactant or product
**2.** Compressing or expanding the system
**3.** Changing the temperature

You can deduce the direction in which an equilibrium will shift when one of these changes is made by applying the following principle:

Equilibrium systems, like people, resist change

*If a system at equilibrium is disturbed by a change in concentration, pressure, or temperature, the system will, if possible, shift so as to partially counteract the change.*

This law was first stated, in a more complex form, in 1884 by Henri Le Chatelier (1850–1936), a French chemist who had studied a variety of industrial equilibria, including those involved in the production of iron in the blast furnace. The statement is commonly referred to as **Le Chatelier's principle.**

### Adding or Removing a Gaseous Species

According to Le Chatelier's principle, *if a chemical system at equilibrium is disturbed by adding a gaseous\* species (reactant or product), the reaction will proceed in such a direction as to consume part of the added species. Conversely, if a gaseous species is removed, the system shifts to restore part of that species.* In that way, the change in concentration brought about by adding or removing a gaseous species is partially counteracted.

To apply this general rule, consider the reaction

$$N_2O_4(g) \rightleftharpoons 2NO_2(g)$$

Suppose this system has reached equilibrium at a certain temperature. This equilibrium could be disturbed by

— *adding* $N_2O_4$. Here, reaction will occur in the forward direction (left to right). In this way, part of the $N_2O_4$ will be consumed.
— *adding* $NO_2$, which causes the reverse reaction (right to left) to occur, using up part of the $NO_2$ added.
— *removing* $N_2O_4$. Here, reaction occurs in the reverse direction to restore part of the $N_2O_4$.
— *removing* $NO_2$, which causes the forward reaction to occur, restoring part of the $NO_2$ removed.

It is possible to use $K$ to calculate the extent to which reaction occurs when an equilibrium is disturbed by adding or removing a product or reactant. To show how this is done, consider the effect of adding hydrogen iodide to the HI–$H_2$–$I_2$ system (Example 12.8).

$$2HI(g) \rightleftharpoons H_2(g) + I_2(g)$$

---

\*Adding a pure liquid or solid has no effect on the system, as is implied by the discussion of heterogeneous equilibrium, p. 347.

**Example 12.8**   In Example 12.4 you found that this system is in equilibrium at 520°C when $P_{HI} = 0.80$ atm and $P_{H_2} = P_{I_2} = 0.10$ atm. Suppose enough HI is added to raise its pressure temporarily to 1.00 atm. When equilibrium is restored, what are $P_{HI}$, $P_{H_2}$, and $P_{I_2}$?

**Strategy**   This problem is entirely analogous to Examples 12.6 and 12.7; the only real difference is that in this case the "original pressures" are those that prevail immediately after equilibrium is disturbed. Note that

$$Q = \frac{(0.10) \times (0.10)}{(1.00)^2} = 0.010 < K = 0.016$$

so some HI must dissociate to establish equilibrium.

*Solution*

(1) Recall from Example 12.4 that

$$K = \frac{P_{H_2} \times P_{I_2}}{(P_{HI})^2} = 0.016 \text{ at } 520°C$$

(2) The partial pressure of HI must decrease; those of $H_2$ and $I_2$ must increase. Let $x$ be the increase in partial pressure of $H_2$. Then

$$\Delta P_{I_2} = \Delta P_{H_2} = x \qquad \Delta P_{HI} = -2x$$

The equilibrium table becomes

|  | 2HI($g$) $\rightleftharpoons$ H2($g$) | + | I₂($g$) |
|---|---|---|---|
| $P_O$ (atm) | 1.00 | 0.10 | 0.10 |
| $\Delta P$ (atm) | $-2x$ | $+x$ | $+x$ |
| $P_{eq}$ (atm) | $1.00 - 2x$ | $0.10 + x$ | $0.10 + x$ |

(3) Substituting into the expression for $K$,

$$0.016 = \frac{(0.10 + x)^2}{(1.00 - 2x)^2}$$

To solve this equation, take the square root of both sides:

$$0.13 = \frac{0.10 + x}{1.00 - 2x}$$

from which you should find that $x = 0.024$

(4) $P_{H_2} = P_{I_2} = 0.10 + x =$ ⬚ 0.12 atm

$P_{HI} = 1.00 - 2x =$ ⬚ 0.95 atm

Note that the equilibrium partial pressure of HI is intermediate between its value before equilibrium was disturbed (0.80 atm) and that immediately afterward (1.00 atm). This is exactly what Le Chatelier's principle predicts; *part* of the added HI is consumed to re-establish equilibrium.

*The system does not return to its original equilibrium state*

## Compression or Expansion

To understand how a change in pressure can change the position of an equilibrium, consider again the $N_2O_4$–$NO_2$ system:

$$N_2O_4(g) \rightleftharpoons 2NO_2(g)$$

**Figure 12.3**
Effect of compression upon the
$N_2O_4(g) \rightleftharpoons 2NO_2(g)$ system at equilibrium. The immediate effect (mid-
dle cylinder) is to crowd the same
number of moles of gas into a
smaller volume and so increase the
total pressure. This is partially compensated for by the conversion of
some of the $NO_2$ to $N_2O_4$, thereby
reducing the total number of moles
of gas.

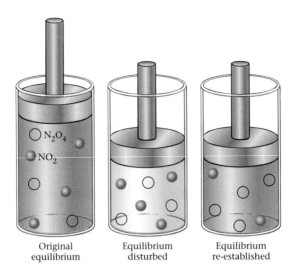

| Original equilibrium | Equilibrium disturbed | Equilibrium re-established |

Suppose the system is compressed by pushing down the piston shown in Figure 12.3. The immediate effect is to increase the gas pressure because the same
number of molecules are crowded into a smaller volume ($P - nRT/V$). According to Le Chatelier's principle, the system will shift so as to partially counteract
this change. There is a simple way in which the gas pressure can be reduced.
Some of the $NO_2$ molecules combine with each other to form $N_2O_4$ (diagram
at right of Figure 12.3). That is, reaction occurs in the reverse direction:

$$N_2O_4(g) \longleftarrow 2NO_2(g)$$

This reduces the number of moles, *n*, and hence the pressure, *P*.

It is possible (but not easy) to calculate from *K* the extent to which $NO_2$ is
converted to $N_2O_4$ when the system is compressed. The results of such calculations are given in Table 12.4. As the pressure is increased from 1.0 to 10.0 atm,
more and more of the $NO_2$ is converted to $N_2O_4$. Notice that the total number
of moles of gas decreases steadily as a result of this conversion.

The analysis we have just gone through for the $N_2O_4$–$NO_2$ system can be
applied to any equilibrium system involving gases:

— *when the system is compressed, increasing the total pressure, reaction takes
place in the direction that decreases the total number of moles of gas.*
— *when the system is expanded, decreasing the total pressure, reaction takes
place in the direction that increases the total number of moles of gas.*

TABLE 12.4 **Effect of Compression on the Equilibrium System**
**$N_2O_4(g) \rightleftharpoons 2NO_2(g)$; $K = 11$ at 100°C**

| $P_{tot}$ (atm) | $n_{NO_2}$ | $n_{N_2O_4}$ | $n_{tot}$ |
|---|---|---|---|
| 1.0 | 0.92 | 0.08 | 1.00 |
| 2.0 | 0.82 | 0.13 | 0.95 |
| 5.0 | 0.64 | 0.22 | 0.86 |
| 10.0 | 0.50 | 0.29 | 0.79 |

TABLE 12.5 **Effect of Pressure upon the Position of Gaseous Equilibria**

| System | $\Delta n_{gas}$ | $P_{tot}$ Increases | $P_{tot}$ Decreases |
|---|---|---|---|
| 1. $N_2O_4(g) \rightleftharpoons 2NO_2(g)$ | $+1$ | $\longleftarrow$ | $\longrightarrow$ |
| 2. $SO_2(g) + \frac{1}{2}O_2(g) \rightleftharpoons SO_3(g)$ | $-\frac{1}{2}$ | $\longrightarrow$ | $\longleftarrow$ |
| 3. $N_2(g) + 3H_2(g) \rightleftharpoons 2NH_3(g)$ | $-2$ | $\longrightarrow$ | $\longleftarrow$ |
| 4. $C(s) + H_2O(g) \rightleftharpoons CO(g) + H_2(g)$ | $+1$ | $\longleftarrow$ | $\longrightarrow$ |
| 5. $N_2(g) + O_2(g) \rightleftharpoons 2NO(g)$ | $0$ | $0$ | $0$ |

The application of this principle to several different systems is shown in Table 12.5. In System 2, the number of moles of gas decreases from $\frac{3}{2}$ to 1 as the reaction goes to the right. Hence, increasing the pressure causes the forward reaction to occur; a decrease in pressure has the reverse effect. Notice that it is the change in the number of moles of *gas* that determines which way the equilibrium shifts (System 4). When there is no change in the number of moles of gas (System 5), a change in pressure has no effect on the position of the equilibrium.

We should emphasize that in applying this principle, it is important to realize that an "increase in pressure" means that the system is compressed. Similarly, a "decrease in pressure" corresponds to expanding the system. There are other ways in which pressure can be changed. One way is to add an unreactive gas such as helium at constant volume. This increases the total number of moles and hence the total pressure. It has no effect, however, upon the position of the equilibrium. In general, the position of an equilibrium is not shifted by adding an unreactive gas, since that does not change the concentrations or partial pressures of reactants or products.

---

**Example 12.9** The pressure on each of the following systems is decreased from 5 atm to 1 atm. Which way does the equilibrium shift?

(a) $2CO_2(g) \rightleftharpoons 2CO(g) + O_2(g)$
(b) $H_2(g) + I_2(g) \rightleftharpoons 2HI(g)$
(c) $H_2(g) + I_2(s) \rightleftharpoons 2HI(g)$

**Strategy** Using the coefficients of the balanced equation, determine what happens to the number of moles of gas. Then apply Le Chatelier's principle.

*Solution*

(a) $\longrightarrow$ ($\Delta n_{gas} = +1$). The immediate effect of the expansion is to decrease the pressure. That is partially compensated for by increasing the number of moles, which increases *P*.

(b) No effect. Same number of moles of gas on both sides.

(c) $\longrightarrow$ (1 mol gas $\longrightarrow$ 2 mol gas). Note that it is the number of moles of gas that is important; solids or liquids don't count.

If the reaction involves no change in the number of moles of gas, pressure has no effect on the equilibrium position

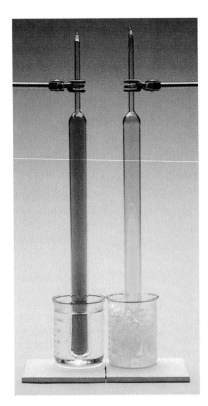

**Figure 12.4**
Effect of temperature on the $N_2O_4$–$NO_2$ system. At 0°C (tube at right), $N_2O_4$, which is colorless, predominates. At 50°C (tube at left), some of the $N_2O_4$ has dissociated to give the deep brown color of $NO_2$.
(Marna G. Clarke)

For a given system, the only way to change $K$ is to change the temperature

## Change in Temperature

Le Chatelier's principle can be used to predict the effect of a change in temperature upon the position of an equilibrium. In general, *an increase in temperature causes the endothermic reaction to occur.* This absorbs heat and so tends to reduce the temperature of the system, partially compensating for the original change.

Applying this principle to the $N_2O_4$–$NO_2$ system,

$$N_2O_4(g) \rightleftharpoons 2NO_2(g) \quad \Delta H° = +57.2 \text{ kJ}$$

it follows that raising the temperature causes the forward reaction to occur, since that reaction absorbs heat. This is confirmed by experiment (Fig. 12.4). At higher temperatures, more $NO_2$ is produced, and the reddish-brown color of that gas becomes more intense.

For the synthesis of ammonia,

$$N_2(g) + 3H_2(g) \rightleftharpoons 2NH_3(g) \quad \Delta H° = -92.2 \text{ kJ}$$

an increase in temperature shifts the equilibrium to the left; some ammonia decomposes to the elements. This reflects the fact that the reverse reaction is endothermic:

$$2NH_3(g) \longrightarrow N_2(g) + 3H_2(g) \quad \Delta H° = +92.2 \text{ kJ}$$

As pointed out earlier, the equilibrium constant of a system changes with temperature. The form of the equation relating $K$ to $T$ is a familiar one, similar to the Clausius-Clapeyron equation (Chapter 9) and the Arrhenius equation (Chapter 11). This one is called the *van't Hoff equation,* honoring Jacobus van't Hoff (1852–1911), who was the first to use the equilibrium constant, $K$. Coincidentally, van't Hoff was a good friend of Arrhenius. The equation is

$$\ln \frac{K_2}{K_1} = \frac{\Delta H°}{R}\left[\frac{1}{T_1} - \frac{1}{T_2}\right]$$

where $K_2$ and $K_1$ are the equilibrium constants at $T_2$ and $T_1$, respectively, $\Delta H°$ is the standard enthalpy change for the forward reaction, and $R$ is the gas constant, 8.31 J/mol · K.

To illustrate how this equation is used, let us apply it to calculate the equilibrium constant at 100°C for the system

$$N_2(g) + 3H_2(g) \rightleftharpoons 2NH_3(g) \quad \Delta H° = -92.2 \text{ kJ}$$

given that $K = 6 \times 10^5$ at 25°C. The relation is

$$\ln \frac{K \text{ at } 100°C}{6 \times 10^5} = \frac{-92,200 \text{ J/mol}}{8.31 \text{ J/mol} \cdot \text{K}}\left[\frac{1}{298 \text{ K}} - \frac{1}{373 \text{ K}}\right] = -7.5$$

Taking inverse logarithms:

$$\frac{K \text{ at } 100°C}{6 \times 10^5} = 6 \times 10^{-4}$$

$$K \text{ at } 100°C = 4 \times 10^2$$

Notice that the equilibrium constant becomes smaller as the temperature increases. In general,

— if the forward reaction is exothermic, as is the case here ($\Delta H° = -92.2$ kJ), $K$ decreases as $T$ increases
— if the forward reaction is endothermic as in the decomposition of $N_2O_4$,

$$N_2O_4(g) \rightleftharpoons 2NO_2(g) \qquad\qquad \Delta H° = +57.2 \text{ kJ}$$

$K$ increases as $T$ increases. For this reaction $K$ is 0.11 at 25°C and 11 at 100°C.

---

**Example 12.10**   Consider the system $I_2(g) \rightleftharpoons 2I(g)$; $\Delta H° = +151$ kJ. Suppose the system is at equilibrium at 1000°C. In which direction will reaction occur if
  (a) I atoms are added?
  (b) the system is compressed?
  (c) the temperature is increased?

**Strategy**   Apply Le Chatelier's principle to predict the effect of (a) adding a product, (b) increasing the pressure, or (c) increasing the temperature.

*Solution*

  (a)   $2I(g) \longrightarrow I_2(g)$

  (b)   $2I(g) \longrightarrow I_2(g)$   This decreases the number of moles of gas.

  (c)   $I_2(g) \longrightarrow 2I(g)$   This reaction absorbs heat.

---

We should emphasize that of the three changes in conditions described in this section

— adding or removing a gaseous species
— compressing or expanding the system
— changing the temperature

*the only one that changes the value of the equilibrium constant is a change in temperature.* In the other two cases, $K$ remains constant.

Equilibrium between liquid bromine and its gas:

$$Br_2(l) \rightleftharpoons Br_2(g)$$

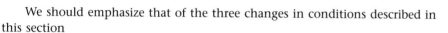

# Industrial Applications of Gaseous Equilibria

A wide variety of materials, both pure substances and mixtures, are made by processes that involve one or more gas-phase reactions. Among these are two of the most important industrial chemicals: ammonia and sulfuric acid.

The processes used to make these chemicals apply many of the principles of chemical equilibrium, discussed in this chapter, and chemical kinetics (Chap. 11).

### Haber Process for Ammonia

Combined ("fixed") nitrogen in the form of protein is essential to all forms of life. There is more than enough elementary nitrogen in the air, about $4 \times 10^{18}$ kg, to meet all our needs. The problem is to convert the element to compounds that can be used by plants to make proteins. At room temperature and atmospheric pressure, $N_2$ does

*continued*

**C H E M I S T R Y**

*Beyond the Classroom*

**Figure 12.A**
Fritz Haber (Professor John Stock, University of Connecticut, Storrs Connecticut)

not react with any nonmetal. However, in 1908 in Germany, Fritz Haber (Figure 12.A) showed that nitrogen does react with hydrogen at high temperatures and pressures to form ammonia in good yield:

$$N_2(g) + 3H_2(g) \rightleftharpoons 2NH_3(g) \qquad \Delta H° = -92.2 \text{ kJ}$$

The Haber process, represented by this equation, is now the main source of fixed nitrogen. Its feasibility depends upon choosing conditions under which nitrogen and hydrogen react rapidly to give a high yield of ammonia. At 25°C and atmospheric pressure, the position of the equilibrium favors the formation of $NH_3(K = 6 \times 10^5)$. Unfortunately, however, the rate of reaction is virtually zero. Equilibrium is reached more rapidly by raising the temperature. However, since the synthesis of ammonia is exothermic, high temperatures reduce $K$ and hence the yield of ammonia. High pressures, on the other hand, have a favorable effect on both the rate of the reaction and the position of the equilibrium (Table 12.A).

**TABLE 12.A    Effect of Temperature and Pressure on the Yield of Ammonia in the Haber Process ($P_{H_2} = 3P_{N_2}$)**

| | | Mole Percent $NH_3$ in Equilibrium Mixture | | | | |
|---|---|---|---|---|---|---|
| °C | $K$ | 10 atm | 50 atm | 100 atm | 300 atm | 1000 atm |
| 200 | 0.4 | 51 | 74 | 82 | 90 | 98 |
| 300 | $4 \times 10^{-3}$ | 15 | 39 | 52 | 71 | 93 |
| 400 | $2 \times 10^{-4}$ | 4 | 15 | 25 | 47 | 80 |
| 500 | $2 \times 10^{-5}$ | 1 | 6 | 11 | 26 | 57 |
| 600 | $3 \times 10^{-6}$ | 0.5 | 2 | 5 | 14 | 31 |

The goal of Haber's research was to find a catalyst to synthesize ammonia at a reasonable rate without going to very high temperatures. Nowadays, the catalyst used is a special mixture of iron, potassium oxide ($K_2O$), and aluminum oxide ($Al_2O_3$). The hydrogen and nitrogen gases must be carefully purified to remove traces of sulfur compounds, which "poison" the catalyst. Reaction takes place at 450°C and at a pressure of 200 to 600 atm. The ammonia formed is passed through a cooling chamber. Since ammonia boils at −33°C, it is condensed out as a liquid, separating it from unreacted nitrogen and hydrogen. The yield is typically less than 50%, so the reactants are recycled to produce more ammonia.

### Contact Process for Sulfuric Acid

Sulfuric acid is made commercially by the three-step **"contact" process:**

**1.** Elemental sulfur is burned in air to form sulfur dioxide:

$$S(s) + O_2(g) \longrightarrow SO_2(g)$$

**2.** Sulfur dioxide is converted to sulfur trioxide by bringing it into "contact" with oxygen on the surface of a solid catalyst:

$$SO_2(g) + \tfrac{1}{2}O_2(g) \rightleftharpoons SO_3(g) \qquad \Delta H° = -98.9 \text{ kJ}$$

This is the key step in the process and the most difficult to carry out. The catalyst used today is an oxide of vanadium, $V_2O_5$. Platinum is equally effective but is more expensive and more easily "poisoned" (made ineffective) by impurities in the gases.

Since the conversion of $SO_2$ to $SO_3$ is exothermic, the equilibrium constant decreases as the temperature rises (Table 12.B). Hence, a low temperature is preferred on equilibrium principles. Too low a temperature, however, reduces the rate to the point at which the reaction becomes impractical. The temperature used in the contact process, 450–600°C represents a compromise between equilibrium and rate considerations.

**TABLE 12.B   Equilibrium Constants for the Reaction**
**$SO_2(g) + \frac{1}{2}O_2(g) \rightleftharpoons SO_3(g)$**

| $t(°C)$ | 200 | 300 | 400 | 500 | 600 |
|---|---|---|---|---|---|
| $K$ | $1.0 \times 10^6$ | $1.3 \times 10^4$ | $5.9 \times 10^2$ | 60 | 10 |

**3.** The sulfur trioxide formed in Step 2 is converted to sulfuric acid by reaction with water:

$$SO_3(g) + H_2O \longrightarrow H_2SO_4(aq)$$

This reaction cannot be carried out directly. If $SO_3$ is bubbled through water, $H_2SO_4$ is formed as a fog of tiny particles that are difficult to condense. Instead, $SO_3$ is absorbed in concentrated sulfuric acid to form an intermediate product, $H_2S_2O_7$, called pyrosulfuric acid. Subsequent dilution with water forms $H_2SO_4$:

$$SO_3(g) + H_2SO_4(l) \longrightarrow H_2S_2O_7(l)$$
$$\underline{H_2S_2O_7(l) + H_2O \longrightarrow 2H_2SO_4(aq)}$$
$$SO_3(g) + H_2O \longrightarrow H_2SO_4(aq)$$

# CHAPTER HIGHLIGHTS

## *Key Concepts*

**1.** Write the expression for $K$ corresponding to a chemical equation
(Examples 12.1, 12.2; Problems 5–12)

**2.** Calculate $K$, knowing
—appropriate $K$'s for other reactions
(Problems 13–18)
—all the equilibrium partial pressures
(Example 12.3; Problems 19–22)
—all the original and one equilibrium partial pressure
(Example 12.4; Problems 23, 24, 62)

**3.** Use the value of $K$ to determine
—the direction of reaction
(Example 12.5; Problems 25–28)
—equilibrium concentrations of all species
(Examples 12.6, 12.7, Problems 31–42, 59, 60, 65, 68–71)

**4.** Apply Le Chatelier's principle to predict the direction in which an equilibrium shifts when conditions are changed
(Examples 12.8–12.10; Problems 43–54)

## Key Equations

| | |
|---|---|
| Expression for $K$ | $K = \dfrac{(P_C)^c \times (P_D)^d}{(P_A)^a \times (P_B)^b}$ |
| Coefficient rule | $K' = K^n$ |
| Reciprocal rule | $K'' = 1/K$ |
| Multiple equilibrium | $K_3 = K_1 \times K_2$ |

## Key Terms

| | | |
|---|---|---|
| endothermic | exothermic | partial pressure |
| equilibrium constant | mole | reaction quotient |

## Summary Problem

Consider phosgene, $COCl_2(g)$. It is a poisonous, suffocating gas used in chemical warfare. It is also an important intermediate in the manufacture of some plastics. It is formed by the reaction between carbon monoxide and chlorine gas. Its $\Delta H_f^\circ$ is $-200.0$ kJ/mol.

(a) Write the chemical equation for the formation of phosgene using simplest whole-number coefficients.

(b) Write the equilibrium expression for the reaction.

(c) At 400°C, after the reaction has reached equilibrium, the partial pressures of $COCl_2$, CO, and $Cl_2$ are 1.29 atm, 0.653 atm, and 0.112 atm, respectively. Calculate $K$.

(d) Initially, a 5.00-L flask contains $Cl_2$, CO, and $COCl_2$ at partial pressures of 0.76 atm, 0.50 atm, and 0.20 atm, respectively. After equilibrium is established at 400°C, the partial pressure of $COCl_2$ is 0.60 atm. Calculate the equilibrium partial pressures of chlorine and carbon monoxide gases.

(e) Initially, a reaction vessel contains $Cl_2$, CO, and $COCl_2$ at 400°C where the partial pressure of $Cl_2$ and CO is 1.00 atm. The partial pressure of phosgene is 0.80 atm. In what direction will the reaction proceed at 400°C? What is the partial pressure of each gas when equilibrium is established? Suppose enough $COCl_2$ is added to double its initial pressure temporarily. In which direction will equilibrium shift? When equilibrium is restored, what is the partial pressure of each gas?

(f) In which direction will this system shift if at equilibrium
  (1) the system is expanded?
  (2) helium gas is added?
  (3) the temperture is increased?

(g) What is the equilibrium constant at $5.00 \times 10^{2}$°C?

### Answers

(a) $Cl_2(g) + CO(g) \longrightarrow COCl_2(g)$

(b) $K = \dfrac{P_{COCl_2}}{P_{CO} \times P_{Cl_2}}$

(c) 17.6

(d) $P_{Cl_2} = 0.36$ atm; $P_{CO} = 0.10$ atm

(e) $\longrightarrow$; $P_{CO} = P_{Cl_2} = 0.29$ atm; $P_{COCl_2} = 1.690$;
  $\longleftarrow$; $P_{CO} = P_{Cl_2} = 0.30$ atm, $P_{COCl_2} = 1.59$ atm

(f) $\longleftarrow$; no change; $\longleftarrow$

(g) 2.22

# Questions & Problems

## Establishment of Equilibrium

**\*1.** The following data are for the system

$$A(g) \rightleftharpoons 2B(g)$$

| Time(s) | 0 | 20 | 40 | 60 | 80 | 100 |
|---|---|---|---|---|---|---|
| $P_A$ (atm) | 1.00 | 0.83 | 0.72 | 0.65 | 0.62 | 0.62 |
| $P_B$ (atm) | 0.00 | 0.34 | 0.56 | 0.70 | 0.76 | 0.76 |

**(a)** How long does it take the system to reach equilibrium?

**(b)** How does the rate of the forward reaction compare with the rate of the reverse reaction after 50 s? after 90 s?

**\*2.** Graph the data given for the system in Question 1. Answer the following questions:

**(a)** Estimate the partial pressure of each gas after 30 seconds.

**(b)** What is the partial pressure of B after 150 s?

**(c)** When will the partial pressure of A become 0.500 atm?

**3.** Complete the table below for the reaction

$$A(g) + 2B(g) \rightleftharpoons 3C(g)$$

| Time(s) | 0 | 1 | 2 | 3 | 4 | 5 | 6 |
|---|---|---|---|---|---|---|---|
| $P_A$ (atm) | 0.800 | 0.717 | — | — | 0.540 | — | 0.533 |
| $P_B$ (atm) | 1.000 | — | — | 0.568 | — | — | — |
| $P_C$ (atm) | 0.000 | — | 0.500 | — | — | 0.800 | — |

**4.** The following data apply to the *unbalanced* equation:

$$A(g) \rightleftharpoons B(g)$$

| Time(s) | 0 | 50 | 100 | 150 | 200 | 250 |
|---|---|---|---|---|---|---|
| $P_A$ (atm) | 2.00 | 1.25 | 0.95 | 0.80 | 0.71 | 0.68 |
| $P_B$ (atm) | 0.10 | 0.60 | 0.80 | 0.90 | 0.96 | 0.98 |

**(a)** Based on the data, balance the equation (simplest whole-number coefficients).

**(b)** Has the system reached equilibrium? Explain.

## Expressions for K

**\*5.** Write equilibrium constant expressions ($K$) for the following reactions:

**(a)** $Xe(g) + 3F_2(g) \rightleftharpoons XeF_6(g)$

**(b)** $CH_4(g) + 2H_2S(g) \rightleftharpoons CS_2(g) + 4H_2(g)$

**(c)** $3CO_2(g) + 4H_2O(g) \rightleftharpoons C_3H_8(g) + 5O_2(g)$

**\*6.** Write equilibrium constant expressions ($K$) for the following reactions:

**(a)** $P_4O_{10}(g) + 6PCl_5(g) \rightleftharpoons 10POCl_3(g)$

**(b)** $2H_2S(g) + 3O_2(g) \rightleftharpoons 2H_2O(g) + 2SO_2(g)$

**(c)** $4NH_3(g) + 5O_2(g) \rightleftharpoons 4NO(g) + 6H_2O(g)$

**\*7.** Write equilibrium constant expressions ($K$) for the following reactions:

**(a)** $2Mg(s) + O_2(g) \rightleftharpoons 2MgO(s)$

**(b)** $HCOOH(aq) \rightleftharpoons H^+(aq) + HCOO^-(aq)$

**(c)** $2Al(s) + 6H^+(aq) \rightleftharpoons 2Al^{3+}(aq) + 3H_2(g)$

**\*8.** Write equilibrium constant expressions ($K$) for the following reactions:

**(a)** $4H^+(aq) + 2Cl^-(aq) + MnO_2(s) \rightleftharpoons$
$\qquad Mn^{2+}(aq) + H_2O + Cl_2(g)$

**(b)** $ZnO(s) + CO(g) \rightleftharpoons Zn(s) + CO_2(g)$

**(c)** $Ni(s) + 4CO(g) \rightleftharpoons Ni(CO)_4(g)$

**\*9.** Given the following descriptions of reversible reactions, write a balanced net ionic equation (simplest whole-number coefficients) and the equilibrium constant expression ($K$) for each.

**(a)** Liquid carbon tetrachloride is in equilibrium with its vapor.

**(b)** Hydrazine gas reacts with chlorine trifluoride gas, producing hydrogen fluoride, nitrogen, and chlorine gases.

**(c)** Carbonate ion in water solution reacts with strong acid at 25°C, giving carbon dioxide and water.

**\*10.** Following the directions in Question 9 for the following reversible reactions.

**(a)** Iodine subliming at equilibrium.

**(b)** Hydrogen gas reacts with elemental sulfur [$S_8(s)$], producing hydrogen sulfide gas.

**(c)** The reaction of ammonia gas with strong acid to form ammonium ion in solution.

**\*11.** Write a chemical equation for an equilibrium system that would lead to the following expressions (a through e) for $K$.

**(a)** $K = \dfrac{(P_{I_2})(P_{F_2})}{(P_{IF})^2}$

**(b)** $K = \dfrac{[Cl^-]^2}{[I^-]^2\, P_{Cl_2}}$

**(c)** $K = \dfrac{(P_{CO_2})^3(P_{H_2O})^4}{(P_{C_3H_8})(P_{O_2})^5}$

**(d)** $K = \dfrac{(P_{NH_3})^2}{(P_{N_2})(P_{H_2})^3}$

**(e)** $K = \dfrac{(P_{H_2O})^2(P_{O_2})}{(P_{H_2O_2})^2}$

**\*12.** Write a chemical equation for an equilibrium system that would lead to the following expressions (a through e) for $K$.

**(a)** $K = \dfrac{(P_{CO})(P_{H_2})^2}{P_{CH_3OH}}$

**(b)** $K = \dfrac{P_{CO_2}}{P_{O_2}}$

**(c)** $K = \dfrac{(P_{SO_3})^2}{(P_{SO_2})^2(P_{O_2})}$

**(d)** $K = \dfrac{[Fe^{3+}]^2[Cl^-]^2}{[Fe^{2+}]^2 P_{Cl_2}}$

**(e)** $K = \dfrac{(P_{HCl})^4(P_{O_2})}{(P_{H_2O})^2(P_{Cl_2})^2}$

## Calculation of K

**13.** At 271°C, $K = 0.0125$ for the reaction

$$3H_2(g) + N_2(g) \rightleftharpoons 2NH_3(g)$$

calculate $K$ at 271°C for
(a) the synthesis of one mole of ammonia.
(b) the decomposition of four moles of ammonia.

**14.** At 1000°C, $K = 442$ for the reaction

$$2NO_2(g) \rightleftharpoons N_2(g) + 2O_2(g)$$

calculate $K$ at 1000°C for
(a) the decomposition of one mole of nitrogen dioxide.
(b) the formation of three moles of oxygen from the decomposition of nitrogen dioxide.

**15.** Given the following reactions and their equilibrium constants,

$$\tfrac{1}{2}Sn(s) + CO_2(g) \rightleftharpoons \tfrac{1}{2}SnO_2(s) + CO(g) \qquad K = 6.45$$

$$CO(g) + H_2O(g) \rightleftharpoons CO_2(g) + H_2(g) \qquad K = 0.034$$

calculate $K$ for the reaction

$$SnO_2(s) + 2H_2(g) \rightleftharpoons Sn(s) + 2H_2O(g)$$

**16.** Given the following reactions and their equilibrium constants,

$$\tfrac{1}{2}C(s) + \tfrac{1}{2}CO_2(g) \rightleftharpoons CO(g) \qquad K = 4.9 \times 10^{-5}$$

$$COCl_2(g) \rightleftharpoons CO(g) + Cl_2(g) \qquad K = 8.8 \times 10^{-13}$$

calculate $K$ for the reaction

$$C(s) + CO_2(g) + 2Cl_2(g) \rightleftharpoons 2COCl_2(g)$$

**17.** Given the following data at a certain temperature,

$$2N_2(g) + O_2(g) \rightleftharpoons 2N_2O(g) \qquad K = 1.2 \times 10^{-35}$$

$$N_2(g) + 2O_2(g) \rightleftharpoons 2NO_2(g) \qquad K = 1.7 \times 10^{-17}$$

$$2N_2O(g) + 3O_2(g) \rightleftharpoons 2N_2O_4(g) \qquad K = 1.2 \times 10^{6}$$

calculate $K$ for the decomposition of one mole of dinitrogen tetraoxide to nitrogen dioxide.

**18.** Given the following data at 25°C,

$$\tfrac{1}{2}N_2(g) + \tfrac{1}{2}O_2 \rightleftharpoons NO(g) \qquad K = 1 \times 10^{-15}$$

$$2NOBr(g) \rightleftharpoons 2NO(g) + Br_2(g) \qquad K = 0.0125$$

calculate $K$ for the formation of one mole of NOBr from its elements in the gaseous state.

**19.** Carbon disulfide gas reacts with hydrogen gas at 1123 K, giving methane and hydrogen sulfide gases. At 1123 K, equilibrium partial pressures are $P_{CH_4} = 0.131$ atm, $P_{H_2S} = 0.084$ atm, $P_{CS_2} = 0.428$ atm, $P_{H_2} = 0.921$ atm.
(a) Write a balanced equation for the reaction of one mole of carbon disulfide gas with hydrogen.
(b) Calculate $K$ for the reaction at 1123 K.

**20.** Ammonia and carbon dioxide gases react to form solid ammonium carbamate, $NH_4CO_2NH_2$. When equilibrium is established at 313 K, the following data are obtained: $P_{NH_3} = 0.179$ atm, $P_{CO_2} = 0.221$ atm, mass of ammonium carbamate = 1.29 g.
(a) Write a balanced equation for the formation of one mole of $NH_4CO_2NH_2$.
(b) Calculate $K$ at 313 K.

**21.** Calculate $K$ for the decomposition of methyl alcohol vapor at 373 K,

$$CH_3OH(g) \rightleftharpoons CO(g) + 2H_2(g)$$

given that at equilibrium there are 0.128 mol of CO, 0.155 mol of $H_2$, and 0.0244 mol of $CH_3OH$ in a 10.0-L flask.

**22.** For the system

$$CH_4(g) + H_2O(g) \rightleftharpoons CO(g) + 3H_2(g)$$

analysis shows that at 1400 K, the concentrations of methane, steam, hydrogen, and carbon monoxide gases are 0.150 M, 0.233 M, 0.259 M, and 0.513 M, respectively. Calculate $K$ at 1400 K.

**23.** When nitrogen oxide, chlorine, and nitrosyl chloride (NOCl) gases with partial pressures of 0.981 atm, 0.483 atm, and 0.100 atm, respectively, are sealed in a flask at 220°C, the following equilibrium is established:

$$2NO(g) + Cl_2(g) \rightleftharpoons 2NOCl(g)$$

Given that the initial partial pressure of NO(g) is reduced by 74.6% at equilibrium, calculate $K$ for the reaction at 220°C.

**24.** Consider the decomposition of nitrosyl bromide, NOBr(g):

$$2NOBr(g) \rightleftharpoons 2NO(g) + Br_2(g)$$

At 220°C in a sealed flask, originally, $P_{NOBr} = 0.434$ atm, $P_{NO} = 0.100$ atm, and $P_{Br_2} = 0.147$ atm. When equilibrium is established at 220°C, it is found that the partial pressure of NO has increased by 137%. Calculate $K$ for the decomposition at 220°C.

## K; Direction and Extent of the Reaction

**25.** For the system

$$PCl_3(g) + Cl_2(g) \rightleftharpoons PCl_5(g)$$

$K$ is $3.8 \times 10^{-2}$ at 300°C. In a 5.0-L flask, a gaseous mixture consists of all three gases with partial pressures as follows: $P_{PCl_3} = 0.28$ atm, $P_{Cl_2} = 0.071$ atm, $P_{PCl_5} = 0.12$ atm.
(a) Is the mixture at equilibrium? Explain.
(b) If it is not at equilibrium, which way will the system shift to establish equilibrium?

**26.** A gaseous reaction mixture contains $SO_2$, $Cl_2$, and $SO_2Cl_2$ in a 2.0-L container with the gases having the following partial pressures: $P_{SO_2} = 0.35$ atm, $P_{SO_2Cl_2} = 0.19$ atm, $P_{Cl_2} = 0.24$ atm. $K = 91$ for the equilibrium system:

$$SO_2(g) + Cl_2(g) \rightleftharpoons SO_2Cl_2(g)$$

**(a)** Is the system at equilibrium? Explain.
**(b)** If it is not at equilibrium, in which direction will the system move to reach equilibrium?

**27.** For the reversible reaction

$$Cl_2(g) + H_2O(g) \rightleftharpoons 2HCl(g) + \tfrac{1}{2}O_2(g)$$

$K = 0.79$ at a certain temperature. Predict the direction in which the system will move to reach equilibrium if one starts with

**(a)** $P_{CL_2} = P_{HCl} = P_{O_2} = 0.44$ atm
**(b)** $P_{HCl} = 0.15$ atm, $P_{H_2O} = 0.27$ atm, $P_{Cl_2} = 0.093$ atm, $P_{O_2} = 0.22$ atm

**28.** For the reaction

$$NO(g) + \tfrac{1}{2}O_2(g) \rightleftharpoons NO_2(g)$$

$K$ at a certain temperature is 1.4. Predict the direction in which the system will move to reach equilibrium if one starts with

**(a)** $P_{O_2} = P_{NO} = 0.75$ atm
**(b)** $P_{NO} = 0.20$ atm, $P_{O_2} = 0.68$ atm, $P_{NO_2} = 0.33$ atm

**29.** For the reaction

$$C(s) + CO_2(g) \rightleftharpoons 2CO(g)$$

$K = 168$ at 1273 K. If one starts with 0.3 atm of $CO_2$ and 12.0 g of C at 1273 K, will the equilibrium mixture contain

**(a)** mostly $CO_2$?
**(b)** mostly CO?
**(c)** roughly equal amounts of $CO_2$ and CO?
**(d)** only C?

**30.** At another temperature, $K$ for the reaction in Question 29 is 0.2. At this temperature and with the same amounts of $CO_2$ and C as in Question 29, which choice will apply?

## K; Equilibrium Pressures

**31.** Carbonyl bromide, $COBr_2$, is prepared by the reaction between $CO(g)$ and $Br_2(g)$.

$$CO(g) + Br_2(g) \rightleftharpoons COBr_2(g)$$

The equilibrium constant for the reaction at 346 K is 0.185. Calculate the total pressure at equilibrium if $P_{COBr_2} = 0.163$ atm and $P_{CO} = 0.734$ atm.

**32.** Consider the decomposition of nitrosyl chloride:

$$NOCl(g) \rightleftharpoons NO(g) + \tfrac{1}{2}CL_2(g)$$

At a certain temperature, $K = 0.82$. Calculate $P_{Cl_2}$ if $P_{NO} = 0.47$ atm and $P_{NOCl} = 0.19$ atm.

**33.** Nitrogen dioxide can be formed by combining nitrogen oxide and oxygen:

$$NO(g) + \tfrac{1}{2}O_2(g) \rightleftharpoons NO_2$$

$K$ is 1.31 at a certain temperature. A 5.0-L flask at equilibrium is determined to have a total pressure of 1.06 atm and oxygen to have a partial pressure of 0.643 atm. Calculate $P_{NO}$ and $P_{NO_2}$ at equilibrium.

**34.** For the reaction

$$NO(g) + H_2(g) \rightleftharpoons \tfrac{1}{2}N_2(g) + H_2O(g)$$

$K = 25.5$ at a certain temperature. When equilibrium is established, the partial pressure of $N_2$ and NO are 0.416 atm and 0.821 atm, respectively. The total pressure is determined to be 2.67 atm. What are the partial pressures of $H_2$ and $H_2O$ at equilibrium?

**35.** Hydrogen cyanide, a highly toxic gas, can be prepared by reducing cyanogen gas, $C_2N_2$, with hydrogen gas:

$$C_2N_2(g) + H_2(g) \rightleftharpoons 2HCN(g)$$

At a certain temperature, the equilibrium constant is 47. What are the partial pressures of all gases at equilibrium if the initial partial pressures of the reactants are 0.500 atm?

**36.** At 800 K, hydrogen iodide gas decomposes into hydrogen and iodine gases.

$$HI(g) \rightleftharpoons \tfrac{1}{2}H_2(g) + \tfrac{1}{2}I_2(g)$$

$K$ is 0.13. What are the partial pressures of all the gases at equilibrium when the initial partial pressure of HI is 0.75 atm?

**37.** The reaction

$$CO(g) + H_2O(g) \rightleftharpoons H_2(g) + CO_2(g)$$

has an equilibrium constant of 1.30 at 650°C. Carbon monoxide and steam both have initial partial pressures of 0.485 atm, while hydrogen and carbon dioxide start with partial pressures of 0.159 atm.

**(a)** Calculate the partial pressure of each gas at equilibrium.
**(b)** Compare the total pressure initially with the total pressure at equilibrium. Would this relation hold for all gaseous systems?

**38.** At 460°C, the reaction

$$NO(g) + SO_3(g) \rightleftharpoons SO_2(g) + NO_2(g)$$

has $K = 0.0118$. What is the partial pressure of $SO_2$ gas at equilibrium if all gases each have an initial pressure of 1.00 atm?

**39.** Solid ammonium iodide decomposes to ammonia and hydrogen iodide gases at sufficiently high temperatures.

$$NH_4I(s) \rightleftharpoons NH_3(g) + HI(g)$$

The equilibrium constant for the decomposition at 673 K is 0.215. Ten grams of ammonium iodide is sealed in a 2.0-L flask and heated to 673 K.

**(a)** What is the total pressure in the flask at equilibrium?
**(b)** How much ammonium iodide reacted?

**40.** Solid ammonium carbamate, $NH_4CO_2NH_2$, decomposes at 25°C to ammonia and carbon dioxide.

$$NH_4CO_2NH_2(s) \rightleftharpoons 2NH_3(g) + CO_2(g)$$

The equilibrium constant for the decomposition at 25°C is

$2.3 \times 10^{-4}$. Ten grams of $NH_4CO_2NH_2$ are sealed in a 5.0-L flask at 25°C and allowed to decompose.
   (a) What is the total pressure in the flask when equilibrium is established?
   (b) What percentage of $NH_4CO_2NH_2$ decomposed?
   (c) Can you state from the data calculated that the decomposition took place slowly?

**41.** The decomposition at 25°C of dinitrogen tetroxide

$$N_2O_4(g) \rightleftharpoons 2NO_2(g)$$

has an equilibrium constant of 0.144. What are the partial pressures at equilibrium of the two gases if a sealed flask initially contains only $N_2O_4$ at a pressure of 0.863 atm?

**42.** Phosphorus pentachloride gas can be produced by reacting phosphorus trichloride and chlorine gases at a certain temperature.

$$PCl_3(g) + Cl_2(g) \rightleftharpoons PCl_5(g)$$

At that temperature, $K = 0.42$. What is the partial pressure of chlorine gas at equilibrium if initial partial pressures are $P_{Cl_2} = 0.500$ atm, $P_{PCl_3} = 1.00$ atm?

## Le Chatelier's Principle

**\*43.** Consider one of the steps in the manufacture of sulfuric acid.

$$2SO_2(g) + O_2(g) \rightleftharpoons 2SO_3(g) \qquad \Delta H = -197.8 \text{ kJ}$$

   (a) Predict whether the forward or reverse reaction will occur when the equilibrium is disturbed by
      (1) adding $SO_3(g)$.
      (2) expanding the system at constant temperature.
      (3) adding argon gas.
      (4) adding $SO_2(g)$.
      (5) increasing the temperature.
   (b) Which of the above factors will increase the value of $K$? Which will decrease it?

**\*44.** For the system

$$2N_2(g) + 6H_2O(l) \rightleftharpoons 4NH_3(g) + 3O_2(g) \qquad \Delta H = 1530.4 \text{ kJ}$$

   (a) How will the amount of ammonia at equilibrium be affected by
      (1) removing $O_2(g)$?
      (2) adding $N_2(g)$?
      (3) compressing the system at constant $T$?
      (4) cooling the system?
      (5) adding helium gas?
   (b) Which of the above factors will increase the value of $K$? Which will decrease it?

**\*45.** Predict the direction in which the following equilibria will shift if the pressure on the system is increased by compression.
   (a) $H_2O(g) + C(s) \rightleftharpoons CO(g) + H_2(g)$
   (b) $SbCl_5(g) \rightleftharpoons SbCl_3(g) + Cl_2(g)$
   (c) $CO(g) + H_2O(g) \rightleftharpoons CO_2(g) + H_2(g)$

**\*46.** Predict the direction in which each of the following equilibria will shift if the pressure on the system is decreased by expansion.
   (a) $Ni(s) + 4CO(g) \rightleftharpoons Ni(CO)_4(g)$
   (b) $ClF_5(g) \rightleftharpoons ClF_3(g) + F_2(g)$
   (c) $HBr(g) \rightleftharpoons \frac{1}{2}H_2(g) + \frac{1}{2}Br_2(g)$

**\*47.** The system

$$3Z(g) + Q(g) \rightleftharpoons 2R(g)$$

is at equilibrium when the partial pressure of Q is 0.44 atm. Sufficient R is added to increase its partial pressure temporarily to 1.5 atm. When equilibrium is re-established, the partial pressure of Q could be which of the following?
   (a) 1.5 atm     (b) 1.2 atm     (c) 0.80 atm
   (d) 0.44 atm     (e) 0.40 atm

**\*48.** For the system in Question 47, the initial partial pressure of Z is 0.60 atm. When equilibrium is re-established, the partial pressure of Z could be which of the following?
   (a) 0.00 atm     (b) 0.20 atm     (c) 0.40 atm
   (d) 0.60 atm     (e) 0.80 atm

**49.** Water gas, a commercial fuel, is made by reacting hot coke with steam.

$$C(s) + H_2O(g) \rightleftharpoons CO(g) + H_2(g)$$

When equilibrium is established at 900 K, the partial pressures of CO, $H_2$, and $H_2O$ are 0.22 atm, 0.63 atm, and 0.31 atm, respectively.
   (a) Calculate $K$ at 900 K.
   (b) Enough steam is added to raise its partial pressure temporarily to 0.50 atm. What are the equilibrium partial pressures of all gases after equilibrium is re-established?

**50.** For the reaction

$$CO_2(g) + H_2(g) \rightleftharpoons CO(g) + H_2O(g)$$

Equilibrium is established at a certain temperature when the partial pressures of CO, $H_2O$, $CO_2$, and $H_2$ are (in atm) 0.27, 0.41, 0.16, 0.68, respectively.
   (a) Calculate $K$.
   (b) If enough steam condenses to water to decrease its partial pressure to 0.30 atm, in which direction will the reaction proceed? What is the partial pressure of steam when equilibrium is re-established?

**51.** For the system

$$2HI(g) \rightleftharpoons H_2(g) + I_2(g) \qquad \Delta H = +9.4 \text{ kJ}$$

$K = 0.016$ at 800 K. What is the equilibrium constant at 1000 K?

**52.** For the system

$$2SO_2(g) + O_2(g) \rightleftharpoons 2SO_3(g)$$

$K = 0.76$ at 627°C. What is the equilibrium constant at 727°C?

**53.** What is the value of $\Delta H°$ for a reaction if $K$ doubles when the temperature increases from 25 to 35°C?

**54.** For a certain reaction, $\Delta H°$ is $-45$ kJ. What is the ratio of the two equilibrium constants when the temperature is doubled from 300 K to 600 K?

### *Unclassified*

**\*55.** The graph below is similar to that of Figure 12.2 If after 100 s have elapsed, the partial pressure of $N_2O_4$ is increased to 1.0 atm, what will the graph for $N_2O_4$ look like beyond 100 s?

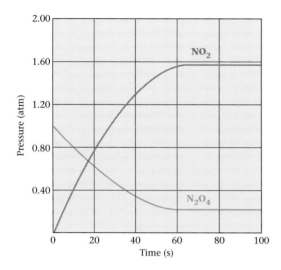

**\*56.** The following figure represents a system at equilibrium. Circles represent the A atoms and squares represent the B atoms. Assume that each molecule corresponds to a partial pressure of one atm.

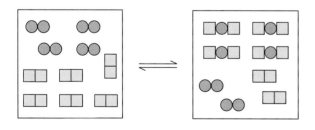

(a) Write a balanced equation to represent the figure.
(b) What is $K$ for the reaction?

**\*57.** The following figure represents initial conditions for the following reaction: $2\ \square\!\!\bigcirc \rightleftharpoons \square\!\!\square + \bigcirc\!\!\bigcirc\ k = 1$
Fill in the blank square with a picture of the system at equilibrium. (*Hint:* You may want to do this by trial and error.)

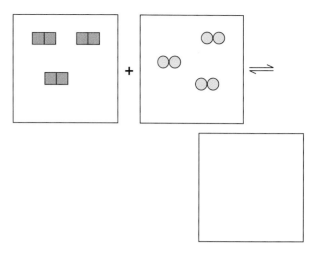

**\*58.** The figures below represent the following reaction at equilibrium at different temperatures.

$$A_2(g) + 3B_2(g) \rightleftharpoons 2AB_3(g)$$

where squares represent atoms of A and circles represent atoms of B. Is the reaction exothermic?

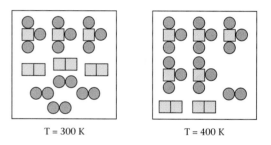

T = 300 K          T = 400 K

**59.** Isopropyl alcohol is the main ingredient in rubbing alcohol. It can decompose into acetone (the main ingredient in nail polish) and hydrogen gas according to the following reaction:

$$C_3H_7OH(g) \rightleftharpoons C_2H_6CO(g) + H_2(g)$$

At 180°C, the equilibrium constant for the decomposition is 0.45. If 20.0 mL ($d = 0.785$ g/mL) of isopropyl alcohol is placed in a 5.00-L vessel and heated to 180°C, what percentage remains at equilibrium?

**\*60.** For the decomposition of $CaCO_3$:

$$CaCO_3(s) \rightleftharpoons CaO(s) + CO_2(g)$$

At a certain temperature, $K = 0.84$. Consider two flasks of equal volume at the same temperature. Flask I has 5.00 g of $CaCO_3$; Flask II has 15.0 g of $CaCO_3$. Answer the following questions with $=$, $<$, $>$, or "can't tell."

(a) The equilibrium partial pressure of $CO_2$:  Flask I _____ Flask II
(b) The amount of $CaCO_3$ at equilibrium:
Flask I _____ Flask II
(c) The rate of decomposition:  Flask I _____ Flask II

**\*61.** An examination question asked students to calculate $K$ for the reaction

$$CH_4(g) + 2H_2S(g) \rightleftharpoons CS_2(g) + 4H_2(g)$$

given that at equilibrium in a 10.0-L vessel, the partial pressures are $P_{CS_2} = 0.42$ atm, $P_{H_2} = 0.17$ atm, $P_{CH_4} = 0.39$ atm, and $P_{H_2S} = 0.26$ atm. Four students gave the following answers, all wrong. Explain the error(s) in their reasoning.

**(a)** $K = \dfrac{(0.39)(0.26)^2}{(0.42)(0.17)^4}$  **(b)** $K = \dfrac{(0.42)(0.68)^4}{(0.39)(0.52)^2}$

**(c)** $K = \dfrac{(0.42) + 4(0.17)}{(0.39) + 2(0.26)}$  **(d)** $K = \dfrac{(0.042)(0.017)^4}{(0.039)(0.026)^2}$

**62.** Consider the equilibrium

$$C(s) + CO_2(g) \rightleftharpoons 2CO(g)$$

When this system is at equilibrium at 700°C in a 2.0-L container, there are 0.10 mol of CO, 0.20 mol of $CO_2$, and 0.40 mol of C present. When the system is cooled to 600°C, an additional 0.040 mol of C(s) forms. Calculate $K$ at 700°C and again at 600°C.

**\*63.** Consider this statement: "The equilibrium constant for a mixture of hydrogen, nitrogen, and ammonia is 3.41." What information is missing from this statement?

**\*64.** Consider this statement: "The equilibrium constant for a reaction at 400 K is 792. It must be a very fast reaction." What is wrong with the statement?

## Challenge Problems

**65.** Ammonia can decompose into its constituent elements according to the reaction

$$2NH_3(g) \rightleftharpoons N_2(g) + 3H_2(g)$$

The equilibrium constant for the decomposition at a certain temperature is 2.5. Calculate the partial pressures of all the gases at equilibrium if ammonia with a pressure of 1.00 atm is sealed in a 3.0-L flask.

**66.** Derive the relationship

$$K = K_c \times (RT)^{\Delta n_g}$$

where $K_c$ is the equilibrium constant using molarities and $\Delta n_g$ is the change in the number of moles of gas in the reaction. (see p. 345). (*Hint:* Recall that $P_A = n_A RT/V$ and $n_A/V = [A]$.)

**67.** Hydrogen iodide gas decomposes to hydrogen gas and iodine gas:

$$2HI(g) \rightleftharpoons H_2(g) + I_2(g)$$

To determine the equilibrium constant of the system, identical one-liter glass bulbs are filled with 3.20 g of HI and maintained at a certain temperature. Each bulb is periodically opened and analyzed for iodine formation by titration with sodium thiosulfate, $Na_2S_2O_3$.

$$I_2(aq) + 2S_2O_3{}^{2-}(aq) \longrightarrow S_4O_6{}^{2-}(aq) + 2I^-(aq)$$

It is determined that when equilibrium is reached, 37.0 mL of 0.200 $M$ $Na_2S_2O_3$ is required to titrate the iodine. What is $K$ at the temperature of the experiment?

**68.** For the system

$$SO_3(g) \rightleftharpoons SO_2(g) + \tfrac{1}{2}O_2(g)$$

at 1000 K, $K = 0.45$. Sulfur trioxide, originally at 1.00 atm pressure, partially dissociates to $SO_2$ and $O_2$ at 1000 K. What is its partial pressure at equilibrium?

**69.** At a certain temperature, the reaction

$$Xe(g) + 2F_2(g) \rightleftharpoons XeF_4(g)$$

gives a 50.0% yield of $XeF_4$, starting with Xe ($P_{Xe} = 0.20$ atm) and $F_2$ ($P_{F_2} = 0.40$ atm). Calculate $K$ at this temperature. What must the initial pressure of $F_2$ be to convert 75.0% of the xenon to $XeF_4$?

**70.** A student studies the equilibrium

$$I_2(g) \rightleftharpoons 2I(g)$$

at a high temperature. He finds that the total pressure at equilibrium is 40% greater than it was originally, when only $I_2$ was present. What is $K$ for this reaction at that temperature?

**71.** Benzaldehyde, a flavoring agent, is obtained by the dehydrogenation of benzyl alcohol.

$$C_6H_5CH_2OH(g) \rightleftharpoons C_6H_5CHO(g) + H_2(g)$$

$K$ for the reaction at 250°C is 0.56. If 1.50 g of benzyl alcohol is placed in a 2.0-L flask and heated to 250°C,

**(a)** what is the partial pressure of the benzaldehyde when equilibrium is established?

**(b)** how many grams of benzyl alcohol remain at equilibrium?

This crystal-clear lake in the Adirondacks seems beautiful unless you're a fisherman or a beaver. (p. 392) (Visuals Unlimited/John D. Cunningham)

# Acids and Bases

## 13

**T**here is nothing in the Universe but alkali and acid,

From which Nature composes all things.

**—OTTO TACHENIUS (1671)**

Among the solution reactions considered in Chapter 4 were those between acids and bases. In this chapter, we take a closer look at the properties of acidic and basic water solutions. In particular, we examine

— the ionization of water and the equilibrium between hydrated $H^+$ ions ($H_3O^+$ ions) and $OH^-$ ions in water solution (Section 13.2)
— the quantities pH and pOH, used to describe the acidity and/or basicity of water solutions (Section 13.3)
— the types of species that act as weak acids and weak bases and the equilibria that apply in their water solutions (Sections 13.4, 13.5)
— the acid-base properties of salt solutions (Section 13.6)

## 13.1 Brønsted-Lowry Acid-Base Model

You will recall that in Chapter 4 we considered acids and bases to be species that supply $H^+$ ions and $OH^-$ ions, respectively, to water. The model of acids and bases used in this chapter is a somewhat more general one developed independently by Johannes Brønsted (1879–1947) in Denmark and Thomas Lowry (1874–1936) in England in 1923. The **Brønsted-Lowry** model considers that

*Sometimes called the Lowry-Brønsted model, at least in England*

— *an acid is a proton ($H^+$ ion) donor*
— *a base is a proton ($H^+$ ion) acceptor*
— *in an acid-base reaction, a proton is transferred from an acid to a base.*

A Brønsted-Lowry acid-base reaction might be represented as

$$HB(aq) + A^-(aq) \rightleftharpoons HA(aq) + B^-(aq)$$

The species shown in red, HB and HA, act as Brønsted-Lowry acids in the forward and reverse reactions, respectively; $A^-$ and $B^-$ (blue) act as Brønsted-Lowry bases. (We will use this color coding consistently throughout the chapter in writing Brønsted-Lowry acid-base equations.)

In connection with the Brønsted-Lowry model, there are some terms that are used frequently.

**1.** The species formed when a proton is removed from an acid is referred to as the **conjugate base** of that acid; $B^-$ is the conjugate base of HB. The species formed when a proton is added to a base is called the **conjugate acid** of that base; HA is the conjugate acid of $A^-$. Thus we have

| Conjugate Acid | Conjugate Base |
|---|---|
| HF | $F^-$ |
| $HSO_4^-$ | $SO_4^{2-}$ |
| $NH_4^+$ | $NH_3$ |

**2.** A species that can either accept or donate a proton is referred to as **amphiprotic.** An example is the $H_2O$ molecule, which can gain a proton to form the hydronium ion, $H_3O^+$, or lose a proton, leaving the hydroxide ion, $OH^-$.

| | | H |
| hydroxide ion | water molecule | hydronium ion |

---

## Example 13.1

(a) What is the conjugate base of $HNO_3$? the conjugate acid of $CN^-$?

(b) The $HCO_3^-$ ion, like the $H_2O$ molecule, is amphiprotic. What is its conjugate base? its conjugate acid?

***Strategy***   To form a conjugate base, remove $H^+$; the effect is to lower the number of hydrogen atoms by one and lower the charge by one unit. Conversely, a conjugate acid is formed by adding $H^+$; this adds a H atom and increases the charge by one unit.

*Solution*

(a)   $NO_3^-$; HCN      (b)   $CO_3^{2-}$; $H_2CO_3$

---

# 13.2   Water Ionization Constant

The acidic and basic properties of aqueous solutions are dependent upon an equilibrium that involves the solvent, water. The reaction involved can be regarded as a Brønsted-Lowry acid-base reaction in which the $H_2O$ molecule shows its amphiprotic nature:

$$H_2O + H_2O \rightleftharpoons H_3O^+(aq) + OH^-(aq)$$

Alternatively, and somewhat more simply, the reaction can be regarded as the ionization of a single $H_2O$ molecule:

$$H_2O \rightleftharpoons H^+(aq) + OH^-(aq)$$

Recall (p. 348, Ch. 12) that in the equilibrium constant expression for reactions in solution

— solutes enter as their molarity, [ ]

— the solvent, $H_2O$ in this case, does not appear: its concentration is essentially the same in all dilute solutions.

Hence, for the ionization of water, the equilibrium constant expression is

$$K_w = [H^+] \times [OH^-]$$

where $K_w$, referred to as the **ionization constant of water,** has a very small value. At 25°C,

$$K_w = 1.0 \times 10^{-14}$$

There are very few $H^+$ and $OH^-$ ions in pure water

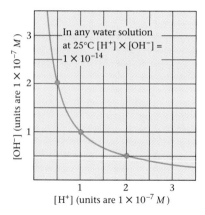

**Figure 13.1**

A graph of [OH⁻] versus [H⁺] looks very much like a graph of gas volume versus pressure. In both cases, the two variables are inversely proportional to one another. When [H⁺] gets larger, [OH⁻] gets smaller.

The concentrations of H⁺ ([H⁺] = [H₃O⁺]) ions and OH⁻ ions in *pure water* at 25°C are readily calculated from $K_w$. Notice from the equation for ionization that these two ions are formed in equal numbers. Hence in pure $H_2O$,

$$[H^+] = [OH^-]$$

$$K_w = [H^+] \times [OH^-] = [H^+]^2 = 1.0 \times 10^{-14}$$

$$[H^+] = 1.0 \times 10^{-7} \, M = [OH^-]$$

An aqueous solution in which [H⁺] *equals* [OH⁻] is called a **neutral** solution. It has a [H⁺] of $1.0 \times 10^{-7} \, M$ at 25°C.

In most water solutions, the concentrations of H⁺ and OH⁻ are not equal. The equation $[H^+] \times [OH^-] = 1.0 \times 10^{-14}$ indicates that these two quantities are inversely proportional to one another (Fig. 13.1). When [H⁺] is very high, [OH⁻] is very low, and vice versa.

---

**Example 13.2**   In a certain tap-water sample, $[H^+] = 3.0 \times 10^{-7} \, M$. What is the concentration of OH⁻?

***Strategy***   Simply substitute into the expression for $K_w$ and solve for [OH⁻].
***Solution***

$$[OH^-] = \frac{K_w}{[H^+]} = \frac{1.0 \times 10^{-14}}{3.0 \times 10^{-7}} = \boxed{3.3 \times 10^{-8} M}$$

---

An aqueous solution in which [H⁺] is greater than [OH⁻] is termed acidic. An aqueous solution in which [OH⁻] is greater than [H⁺] is basic (alkaline). Therefore,

if $[H^+] > 1.0 \times 10^{-7} M$, $[OH^-] < 1.0 \times 10^{-7} \, M$,   solution is acidic

if $[OH^-] > 1.0 \times 10^{-7} M$, $[H^+] < 1.0 \times 10^{-7} \, M$,   solution is basic

## 13.3   pH and pOH

As just pointed out, the acidity or basicity of a solution can be described in terms of its H⁺ concentration. In 1909, Søren Sørensen, a biochemist working at the Carlsberg Brewery in Copenhagen, proposed an alternative method of specifying the acidity of a solution. He defined a term called **pH** (for "power of the hydrogen ion"):

It is simpler to use numbers (pH = 4) than exponents ($[H^+] = 10^{-4} \, M$) to describe acidity

$$pH = -\log_{10}[H^+] = -\log_{10}[H_3O^+]$$

or

$$[H^+] = [H_3O^+] = 10^{-pH}$$

Figure 13.2 shows the relationship between pH and [H⁺]. Notice that, as the defining equation implies, pH increases by one unit when the concentration of H⁺ decreases by a power of 10. Moreover, ***the higher the pH, the less acidic the solution.*** Most aqueous solutions have hydrogen ion concentrations between 1 and $10^{-14} \, M$ and hence have pHs between 0 and 14.

The pH, like [H⁺] or [OH⁻], can be used to differentiate acidic, neutral, and basic solutions. At 25°C,

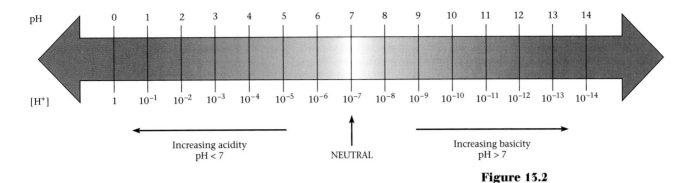

pH

0  1  2  3  4  5  6  7  8  9  10  11  12  13  14

[H⁺]

Increasing acidity
pH < 7

NEUTRAL

Increasing basicity
pH > 7

| if pH < 7.0, | solution is acidic |
| if pH = 7.0, | solution is neutral |
| if pH > 7.0, | solution is basic |

**Figure 13.2**
Acidity is inversely related to pH; the higher the H⁺ ion concentration, the lower the pH. In neutral solution, $[H^+] = [OH^-] = 1.0 \times 10^{-7}\ M$; pH = 7.00 at 25°C.

Table 13.1 shows the pH of some common solutions.

A similar approach is used for the hydroxide ion concentration. The **pOH** of a solution is defined as

$$pOH = -\log_{10}[OH^-]$$

Since $[H^+] \times [OH^-] = 1.0 \times 10^{-14}$ at 25°C, it follows that at this temperature

$$pH + pOH = 14.00$$

Thus, a solution that has a pH of 6.20 must have a pOH of 7.80, and vice versa.

**TABLE 13.1  pH of Some Common Materials**

| | | | |
|---|---|---|---|
| Lemon juice | 2.2–2.4 | Urine, human | 4.8–8.4 |
| Wine | 2.8–3.8 | Cow's milk | 6.3–6.6 |
| Vinegar | 3.0 | Saliva, human | 6.5–7.5 |
| Tomato juice | 4.0 | Drinking water | 5.5–8.0 |
| Beer | 4–5 | Blood, human | 7.3–7.5 |
| Cheese | 4.8–6.4 | Seawater | 8.3 |

Knowing [H⁺], [OH⁻], pH or pOH, you can calculate the other three quantities

**Example 13.3**  Calculate, at 25°C,

(a) the pH and pOH of a lemon juice solution in which [H⁺] is $5.0 \times 10^{-3}\ M$.
(b) the [H⁺] and [OH⁻] of human blood at pH 7.40.

***Strategy***  These calculations can be carried out by substitution into the basic relations

$$pH = -\log_{10}[H^+] \qquad [H^+] \times [OH^-] = 1.0 \times 10^{-14} \qquad pH + pOH = 14.00$$

Use a calculator to find logarithms and inverse logarithms, following the rules for significant figures discussed in Appendix 3.

***Solution***

(a) You should find on your calculator that

$$\log_{10}(5.0 \times 10^{-3}) = -2.30$$

Lemon juice is acidic, has a pH of 2.30, and turns blue litmus red.

Hence pH = −(−2.30) =   2.30    pOH = 14.00 − pH =   11.70

(b) Since the pH is 7.40, $[H^+] = 10^{-7.40}$. To find $[H^+]$, enter −7.40 on your calculator. Then either

— punch the $10^x$ key, if you have one, or
— punch the INV and then the LOG key

Either way, you should find that $[H^+] =$   $4.0 \times 10^{-8}M$.

Knowing $[H^+]$, the concentration of $OH^-$ is calculated as in Example 13.2:

$$[OH^-] = \frac{1.0 \times 10^{-14}}{4.0 \times 10^{-8}} =   2.5 \times 10^{-7}M$$

Since pH > 7 and $[OH^-] > [H^+]$, blood must be (slightly) basic.

## pH of Strong Acids and Strong Bases

As pointed out in Chapter 4, the following acids are strong

**HCl**      **HBr**      **HI**

**HClO₄**      **HNO₃**      **H₂SO₄**

in the sense that they are completely ionized in water. The reaction of the strong acid HCl with water can be represented by the equation (Fig. 13.3)

$$HCl(aq) + H_2O \longrightarrow H_3O^+(aq) + Cl^-(aq)$$

1 *M* HCl has a pH of 0, 1 *M* NaOH a pH of 14

Since this reaction goes to completion, it follows that a 0.10 *M* solution of HCl is 0.10 *M* in both $H_3O^+$ (i.e., $H^+$) and $Cl^-$ ions.

The strong bases

**LiOH**      **NaOH**    **KOH**

**Ca(OH)₂**    **Sr(OH)₂**    **Ba(OH)₂**

are completely ionized in dilute water solution. A 0.10 *M* solution of NaOH is 0.10 *M* in both $Na^+$ and $OH^-$ ions.

The fact that strong acids and bases are completely ionized in water makes it relatively easy to calculate the pH and pOH of their solutions (Example 13.4).

**Figure 13.3**
When HCl is added to water, there is a proton transfer from an HCl to an $H_2O$ molecule, forming a $Cl^-$ ion and an $H_3O^+$ ion. In the reaction, HCl acts as a Brønsted-Lowry acid, $H_2O$ as a Brønsted-Lowry base.

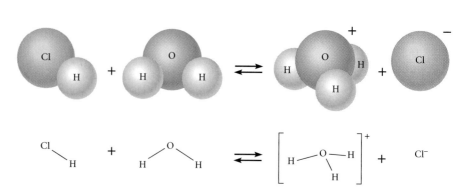

### Example 13.4    What is the pH of

(a) 0.025 $M$ HNO$_3$?
(b) a solution prepared by dissolving 10.0 g of Ba(OH)$_2$ per liter?

***Strategy***    To calculate pH, you can either use the defining equation (a) or the relation between pH and pOH (b). Note that both HNO$_3$ and Ba(OH)$_2$ are completely ionized in water.

*Solution*

(a) Since HNO$_3$ is a strong acid, [H$^+$] = 0.025 $M$

$$pH = -\log_{10}(0.025) = \boxed{1.60}$$

(b) $[Ba(OH)_2] = \dfrac{10.0 \text{ g}}{1 \text{ L}} \times \dfrac{1 \text{ mol}}{171.3 \text{ g}} = 0.0584 \text{ mol/L}$

Since 1 mol of Ba(OH)$_2$ yields 2 mol of OH$^-$ ions,

$$[OH^-] = 2 \times 0.0584 \text{ mol/L} = 0.117 \text{ mol/L}$$

$$pOH = -\log_{10}(0.117) = 0.932 \qquad pH = 14.00 - 0.93 = \boxed{13.07}$$

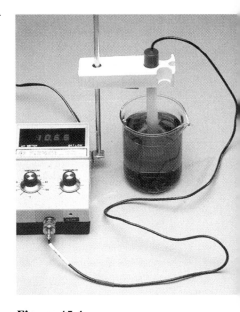

**Figure 13.4**
A pH meter with digital readout. The pH of the solution in the beaker appears to be about 10.65. What do you suppose is in the solution?
(Marna G. Clarke)

## Measuring pH

The pH of a solution can be measured by an instrument called a pH meter. A pH meter translates the H$^+$ ion concentration of a solution into an electrical signal that is converted into either a digital display or a deflection on a meter that reads pH directly (Fig. 13.4). Later, in Chapter 18, we will consider the principle upon which the pH meter works.

A less accurate but more colorful way to measure pH uses a "universal indicator," which is a mixture of *acid-base indicators* that (Fig. 13.5) shows changes

**Figure 13.5**
Universal indicator is deep red in strongly acidic solution (upper left). It changes to yellow and green at pH 6 to 8 and then to deep violet in strongly basic solution (lower right).

**Figure 13.6**
Hydrangeas grown in strongly acidic soil (below pH 5) are blue. In weakly acidic soil (5 < pH < 7), they are mauve. If the soil is neutral or basic, the flowers are rose-pink. (Terry Donnelly/Dembinsky Photo Associates)

in color at different pH values. A similar principle is used with "pH paper." Strips of this paper are coated with a mixture of pH-sensitive dyes; these strips are widely used to test the pH of biological fluids, ground water, and foods. Depending upon the indicators used, a test strip can measure pH over a wide or narrow range.

Home gardeners frequently measure and adjust soil pH in an effort to improve the yield and quality of grass, vegetables, and flowers. Interestingly enough, the colors of many flowers depend upon pH; among these are dahlias, delphiniums, and, in particular, hydrangeas (Fig. 13.6). The first research in this area was carried out by Robert Boyle of gas-law fame, who published a paper on the relation between flower color and acidity in 1664.

## 13.4  Weak Acids and Their Ionization Constants

A wide variety of solutes behave as weak acids; that is, they react reversibly with water to form $H_3O^+$ ions. Using HB to represent a weak acid, its Brønsted-Lowry reaction with water is

$$HB(aq) + H_2O \longrightarrow H_3O^+(aq) + B^-(aq)$$

Typically, this reaction occurs to a very small extent; usually, fewer than 1% of the HB molecules are converted to ions.

Most weak acids fall into one of two categories:

1. **Molecules containing an ionizable hydrogen atom.** This type of weak acid was discussed in Chapter 4. There are literally thousands of molecular weak acids, most of them organic in nature. Among the molecular inorganic acids is nitrous acid:

$$HNO_2(aq) + H_2O \rightleftharpoons H_3O^+(aq) + NO_2^-(aq)$$

*You can assume that all acids other than HCl, HBr, HI, $HNO_3$, $HClO_4$, and $H_2SO_4$ are weak*

2. **Cations.** The ammonium ion, $NH_4^+$, behaves as a weak acid in water; a 0.10 M solution of $NH_4Cl$ has a pH of about 5. The process by which the $NH_4^+$ ion lowers the pH of water can be represented by the (Brønsted-Lowry) equation:

$$NH_4^+(aq) + H_2O \rightleftharpoons H_3O^+(aq) + NH_3(aq)$$

Comparing the equation just written with that cited above for $HNO_2$, you can see that they are very similar. In both cases, a weak acid ($HNO_2$, $NH_4^+$) is converted to its conjugate base ($NO_2^-$, $NH_3$). The fact that the $NH_4^+$ ion has a +1 charge, whereas the $HNO_2$ molecule is neutral, is really irrelevant here.

You may be surprised to learn that most metal cations, except those of Periodic Groups 1 and 2, are weak acids. A 0.10 M solution of $Al_2(SO_4)_3$ has a pH close to 3; you can change the color of hydrangeas from red to blue by adding aluminum salts to soil. At first glance it is not at all obvious how a cation such as $Al^{3+}$ can make a water solution acidic. However, the aluminum cation in water solution is really a hydrated species, $Al(H_2O)_6^{3+}$, in which six water molecules are bonded to the central $Al^{3+}$ ion. This species can transfer a proton to a solvent water molecule to form an $H_3O^+$ ion:

*To get red hydrangeas, add lime*

$$Al(H_2O)_6^{3+}(aq) + H_2O \rightleftharpoons H_3O^+(aq) + Al(H_2O)_5(OH)^{2+}(aq)$$

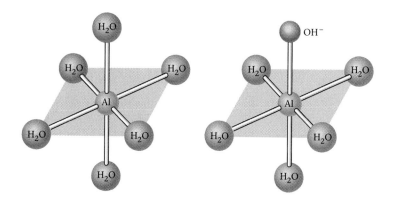

**Figure 13.7**
The $Al^{3+}$ ion is bonded to six $H_2O$ molecules in the $Al(H_2O)_6^{3+}$ ion (left); in the $Al(H_2O)_5(OH)^{2+}$ ion (right), one of the $H_2O$ molecules has been replaced by an $OH^-$ ion.

Figure 13.7 shows the structures of the $Al(H_2O)_6^{3+}$ cation and its conjugate base, $Al(H_2O)_5(OH)^{2+}$.

A very similar equation can be written to explain why solutions of zinc salts are acidic. Here it appears that the hydrated cation in water solution contains four $H_2O$ molecules bonded to a central $Zn^{2+}$ ion. The Brønsted-Lowry equation is

$$Zn(H_2O)_4{}^{2+}(aq) + H_2O \rightleftharpoons H_3O^+(aq) + Zn(H_2O)_3(OH)^+(aq)$$

---

**Example 13.5**   Using the Brønsted-Lowry model, write equations to show why the following species behave as weak acids in water.

(a) acetic acid, $HC_2H_3O_2$, a weak acid found in vinegar
(b) the hydrated iron (II) cation, $Fe(H_2O)_6^{2+}$

**Strategy**   In each case, a proton is transferred to a solvent water molecule, which acts as a Brønsted base. The products are an $H_3O^+$ ion and the conjugate base of the weak acid.

**Solution**

(a) $HC_2H_3O_2(aq) + H_2O \rightleftharpoons H_3O^+(aq) + C_2H_3O_2{}^-(aq)$

(b) $Fe(H_2O)_6{}^{2+}(aq) + H_2O \rightleftharpoons H_3O^+(aq) + Fe(H_2O)_5(OH)^+(aq)$

All transition metal cations behave this way

---

## The Equilibrium Constant for a Weak Acid

In discussing the equilibrium involved when a weak acid is added to water, it is convenient to represent the proton transfer

$$HB(aq) + H_2O \rightleftharpoons H_3O^+(aq) + B^-(aq)$$

as a simple ionization

$$HB(aq) \rightleftharpoons H^+(aq) + B^-(aq)$$

in which case the expression for the equilibrium constant becomes

$$K_a = \frac{[H^+] \times [B^-]}{[HB]}$$

TABLE 13.2 **Ionization Constants of Weak Acids at 25°C**

| Name | Formula | $K_a$ | $pK_a$* |
|------|---------|-------|---------|
| Sulfurous acid | $H_2SO_3$ | $1.7 \times 10^{-2}$ | 1.77 |
| Hexaaquairon(III) ion | $Fe(H_2O)_6^{3+}$ | $6.7 \times 10^{-3}$ | 2.17 |
| Hydrofluoric acid | HF | $6.9 \times 10^{-4}$ | 3.16 |
| Nitrous acid | $HNO_2$ | $6.0 \times 10^{-4}$ | 3.22 |
| Formic acid | $HCHO_2$ | $1.9 \times 10^{-4}$ | 3.72 |
| Benzoic acid | $HC_7H_5O_2$ | $6.6 \times 10^{-5}$ | 4.18 |
| Acetic acid | $HC_2H_3O_2$ | $1.8 \times 10^{-5}$ | 4.74 |
| Hexaaquaaluminum(III) ion | $Al(H_2O)_6^{3+}$ | $1.2 \times 10^{-5}$ | 4.92 |
| Hypochlorous acid | HClO | $2.8 \times 10^{-8}$ | 7.55 |
| Hydrocyanic acid | HCN | $5.8 \times 10^{-10}$ | 9.24 |
| Ammonium ion | $NH_4^+$ | $5.6 \times 10^{-10}$ | 9.25 |
| Tetraaquazinc(II) ion | $Zn(H_2O)_4^{2+}$ | $3.3 \times 10^{-10}$ | 9.48 |
| Diaquasilver(I) ion | $Ag(H_2O)_2^+$ | $1.2 \times 10^{-12}$ | 11.92 |

*$pK_a = -\log_{10} K_a$.

The equilibrium constant $K_a$ is called the *ionization constant* of the weak acid HB. Thus, for the $HNO_2$ molecule and the $NH_4^+$ ion,

$$HNO_2(aq) \rightleftharpoons H^+(aq) + NO_2^-(aq) \qquad K_a = \frac{[H^+] \times [NO_2^-]}{[HNO_2]}$$

$$NH_4^+(aq) \rightleftharpoons H^+ + NH_3(aq) \qquad K_a = \frac{[H^+] \times [NH_3]}{[NH_4^+]}$$

The $K_a$ values of some weak acids (in order of decreasing strength) are given in Table 13.2, along with values of **$pK_a$.**

$$pK_a = -\log_{10}K_a$$

Thus we have

$$HNO_2 \quad K_a = 6.0 \times 10^{-4} \quad pK_a = -\log_{10}(6.0 \times 10^{-4}) = 3.22$$

$$NH_4^+ \quad K_a = 5.6 \times 10^{-10} \quad pK_a = -\log_{10}(5.6 \times 10^{-10}) = 9.25$$

The *weaker* the acid, the *smaller* the value of $K_a$ and the *larger* the value of $pK_a$. The $NH_4^+$ ion is a weaker acid than the $HNO_2$ molecule.

There are several ways to determine the ionization constant, $K_a$ of a weak acid. A simple approach involves measuring $[H^+]$ or pH in a solution prepared by dissolving a known amount of the weak acid to form a given volume of solution.

*Ibuprofen is a somewhat weaker acid than aspirin*

**Example 13.6** Aspirin is a weak organic acid whose molecular formula may be written as $HC_9H_7O_4$. A water solution of aspirin is prepared by dissolving 3.60 g per liter. The pH of this solution is found to be 2.60. Calculate $K_a$ for aspirin.

***Strategy*** The approach used here is very similar to that used in Chapter 12, except that partial pressures are replaced by concentrations in moles per liter. Note that you can readily calculate

— the original concentration of $HC_9H_7O_4$ ($\mathcal{M}$ = 180.15 g/mol)

$$[HC_9H_7O_4]_o = \frac{3.60 \text{ g}}{1 \text{ L}} \times \frac{1 \text{ mol}}{180.15 \text{ g}} = 2.00 \times 10^{-2} \text{ M}$$

— The equilibrium concentration of $H^+$

$$[H^+]_{eq} = 10^{-2.60} = 2.5 \times 10^{-3} \text{ M}$$

Your task is to calculate the acid ionization constant:

$$HC_9H_7O_4(aq) \rightleftharpoons H^+(aq) + C_9H_7O_4^-(aq) \qquad K_a = \frac{[H^+] \times [C_9H_7O_4^-]}{[HC_9H_7O_4]}$$

*Solution*   From the ionization equation it should be clear that *1 mol* of $C_9H_7O_4^-$ is *produced*, and *1 mol* of $HC_9H_7O_4$ is *consumed* for every mole of $H^+$ produced. It follows that

$$\Delta[C_9H_7O_4^-] = \Delta[H^+] \qquad \Delta[HC_9H_7O_4] = -\Delta[H^+]$$

Originally, there is essentially no $H^+$ (ignoring the slight ionization of water). The same holds for the anion $C_9H_7O_4^-$; the only species present originally is the weak acid, $HC_9H_7O_4$, at a concentration of 0.0200 M.

Putting this information together in the form of a table:

| | $HC_9H_7O_4(aq)$ | $\rightleftharpoons$ $H^+(aq)$ | $+$ $C_9H_7O_4^-(aq)$ |
|---|---|---|---|
| [ ]o | 0.0200 | 0.0000 | 0.0000 |
| Δ[ ] | −0.0025 | +0.0025 | +0.0025 |
| [ ]eq | 0.0175 | 0.0025 | 0.0025 |

It is always true that $\Delta[H^+] = \Delta[B^-] = -\Delta[HB]$

(Numbers in color are those given or implied in the statement of the problem; the other numbers are deduced using the ionization equation printed above the table. The symbols [ ]o and [ ]eq refer to original and equilibrium concentrations, respectively.)

All the information needed to calculate $K_a$ is now available,

$$K_a = \frac{(2.5 \times 10^{-3})^2}{0.0175} = \boxed{3.6 \times 10^{-4}}$$

---

In discussing the ionization of a weak acid,

$$HB(aq) \rightleftharpoons H^+(aq) + B^-(aq)$$

we often refer to the **percent ionization:**

$$\% \text{ ionization} = \frac{[H^+]_{eq}}{[HB]_o} \times 100\%$$

For the aspirin solution referred to in Example 13.6,

$$\% \text{ ionization} = \frac{2.5 \times 10^{-3}}{2.0 \times 10^{-2}} \times 100\% = 12\%$$

This is a relatively high percent of ionization; typically, values are of the order of 1% or less (Table 13.3). As you can see from the table, percent ionization at a given concentration correlates directly with acid strength. The weaker the acid, the lower the percent of ionization.

TABLE 13.3   **Percent Ionization of Weak Acids (orig. conc. 1.0 $M$)**

| Weak Acid | $K_a$ | % Ionization |
|---|---|---|
| Lactic acid | $1.4 \times 10^{-4}$ | 1.2 |
| Acetic acid | $1.8 \times 10^{-5}$ | 0.42 |
| Carbonic acid | $4.4 \times 10^{-7}$ | 0.066 |
| Hypochlorous acid | $2.8 \times 10^{-8}$ | 0.017 |
| Ammonium ion | $5.6 \times 10^{-10}$ | 0.0024 |

## Calculation of [H$^+$] in a Water Solution of a Weak Acid

Given the ionization constant of a weak acid and its original concentration, the $H^+$ concentration in solution is readily calculated. The approach used is the inverse of that followed in Example 13.6, where $K_a$ was calculated knowing $[H^+]$; here, $K_a$ is known and $[H^+]$ must be calculated.

**Example 13.7**   Nicotinic acid, $HC_6H_4O_2N$ ($K_a = 1.4 \times 10^{-5}$), is another name for niacin, an important member of the vitamin B group. Determine $[H^+]$ in a solution prepared by dissolving 0.10 mol of nicotinic acid, HNic, in water to form one liter of solution.

*Strategy*   To start with, set up an equilibrium table similar to the one in the previous example. To do that, note that

— the original concentrations of HNic, $H^+$ and $Nic^-$ are 0.10 $M$, 0.00 $M$, and 0.00 $M$, respectively, ignoring, tentatively at least, the $H^+$ ions from the ionization of water.
— the changes in concentration are related by the coefficients of the balanced equation, all of which are 1:

$$\Delta[Nic^-] = \Delta[H^+] \qquad \Delta[HNic] = -\Delta[H^+]$$

Letting $\Delta[H^+] = x$, it follows that $\Delta[Nic^-] = x$; $\Delta[HNic] = -x$. This information should enable you to express the equilibrium concentrations of all species in terms of $x$. The rest is algebra; substitute into the expression for $K_a$ and solve for $x = [H^+]$.

*Solution*   Setting up the table:

|  | HNic(aq) | $\rightleftharpoons$ H$^+$(aq) + | Nic$^-$(aq) |
|---|---|---|---|
| [ ]$_o$ | 0.10 | 0.00 | 0.00 |
| $\Delta$[ ] | $-x$ | $+x$ | $+x$ |
| [ ]$_{eq}$ | $0.10 - x$ | $x$ | $x$ |

*Equilibrium tables always help; be sure to include the equation*

Substituting into the expression for the ionization constant,

$$K_a = \frac{(x)(x)}{0.10 - x} = 1.4 \times 10^{-5}$$

This is a quadratic equation. It could be rearranged to the form $ax^2 + bx + c = 0$ and solved for $x$, using the quadratic formula. Such a procedure is time-consuming and, in this case, unnecessary. Nicotinic acid is a weak acid, only slightly ionized in water. The equilibrium concentration of HNic, $0.10 - x$, is probably only very slightly less than its original concentration, 0.10 M. So let's make the approximation $0.10 - x \approx 0.10$. This simplifies the equation

$$\frac{x^2}{0.10} = 1.4 \times 10^{-5}$$

$$x^2 = 1.4 \times 10^{-6}$$

Taking square roots:

$$x = 1.2 \times 10^{-3} \; M = [H^+]$$

Note that the concentration of $H^+$, 0.0012 M, is

— much smaller than the original concentration of weak acid, 0.10 M. In this case, then, the approximation $0.10 - x \approx 0.10$ is justified. This will usually, but not always, be the case (see Example 13.8).

— much larger than $[H^+]$ in pure water, $1 \times 10^{-7}$ M, justifying the assumption that the ionization of water can be neglected. ***This will always be the case, provided $[H^+]$ from the weak acid is $\geq 10^{-6}$ M.***

In general, the value of $K_a$ is seldom known to be better than $\pm 5\%$. Hence, in the expression

$$K_a = \frac{x^2}{a - x}$$

where $x = [H^+]$ and $a = $ *original* concentration of weak acid, you can neglect the $x$ in the denominator if doing so does not introduce an error of more than 5%. In other words,

$$\text{if } \frac{x}{a} \leq 0.05$$

This is known as the 5% rule

then

$$a - x \approx a.$$

In most of the problems you will work, the approximation $a - x \approx a$ is valid, and you can solve for $[H^+]$ quite simply, as in Example 13.7, where $x = 0.012a$. Sometimes, though, you will find that the calculated $[H^+]$ is greater than 5% of the original concentration of weak acid. In that case, you can solve for $x$ either by using the *quadratic formula* or *the method of successive approximations*.

---

**Example 13.8**   Calculate $[H^+]$ in a 0.100 M solution of nitrous acid, $HNO_2$, for which $K_a = 6.0 \times 10^{-4}$.

***Strategy***   The setup here is identical with that in Example 13.7. However, you will find, upon solving for $x$, that $x > 0.050a$, so the approximation $a - x \approx a$ fails. The simplest way to proceed here is to use the calculated value of $x$ to obtain a better estimate of $[HNO_2]$, then solve again for $[H^+]$. An alternative is to use the quadratic formula. (This is a particularly shrewd choice if you have a calculator that can be programmed to solve quadratic equations.)

***Solution***   Proceeding as in Example 13.7, you arrive at the equation

$$K_a = \frac{x^2}{0.100 - x} = 6.0 \times 10^{-4}$$

Making the same approximation as before, $0.100 - x \approx 0.100$,

$$x^2 = 0.100 \times 6.0 \times 10^{-4} = 6.0 \times 10^{-5}$$

$$x = 7.7 \times 10^{-3} \approx [H^+]$$

To check the validity of the approximation, note that

$$\frac{x}{a} = \frac{7.7 \times 10^{-3}}{1.0 \times 10^{-1}} = 0.077 > 0.050$$

So the approximation is not valid.

To obtain a better value for $x$, you have a choice of two approaches.

**(1)** *The method of successive approximations.* In the first approximation, you ignored the $x$ in the denominator of the equation. This time, you use the approximate value of $x$ just calculated, 0.0077, to find a more exact value for the concentration of $HNO_2$:

$$[HNO_2] = 0.100 - 0.0077 = 0.092 \ M$$

Substituting in the expression for $K_a$,

$$K_a = \frac{x^2}{0.092} = 6.0 \times 10^{-4} \qquad x^2 = 5.5 \times 10^{-5}$$

$$x = \boxed{7.4 \times 10^{-3} M} \approx [H^+]$$

WLM likes this method; CNH leans toward the use of the quadratic formula

This value is closer to the true $[H^+]$, since $0.092 \ M$ is a better approximation for $[HNO_2]$ than was $0.100 \ M$. If you're still not satisfied, you can go one step further. Using $7.4 \times 10^{-3}$ for $x$ instead of $7.7 \times 10^{-3}$, you can recalculate $[HNO_2]$ and solve again for $x$. If you do, you will find that your answer does not change. In other words, "you have gone about as far as you can go."

**(2)** *The quadratic formula.* This gives an exact solution for $x$ but is more time-consuming. Here, rewrite the equation

$$\frac{x^2}{0.100 - x} = 6.0 \times 10^{-4}$$

in the form $ax^2 + bx + c = 0$. Doing this:

$$x^2 + (6.0 \times 10^{-4}x) - (6.0 \times 10^{-5}) = 0$$

thus, $a = 1$; $b = 6.0 \times 10^{-4}$; $c = -6.0 \times 10^{-5}$. Applying the quadratic formula,

$$x = \frac{-b \pm \sqrt{b^2 - 4ac}}{2a}$$

$$= \frac{-6.0 \times 10^{-4} \pm \sqrt{(6.0 \times 10^{-4})^2 + (24.0 \times 10^{-5})}}{2}$$

If you carry out the arithmetic properly, you should get two answers for $x$:

$$x = \boxed{7.4 \times 10^{-3} M} \qquad \text{and} \qquad -8.0 \times 10^{-3} M$$

The second answer is physically ridiculous; the concentration of $H^+$ cannot be a negative quantity. The first answer is the same one obtained by the method of successive approximations.

---

## 13.5 Weak Bases and Their Ionization Constants

Like the weak acids, there are a large number of solutes that act as weak bases. It is convenient to classify weak bases into two groups, molecules and anions.

**1. Molecules.** As pointed out in Chapter 4, there are many molecular weak bases, including the organic compounds known as amines. The simplest weak base is ammonia, whose reversible, Brønsted-Lowry reaction with water is represented by the equation

$$NH_3(aq) + H_2O \rightleftharpoons NH_4^+(aq) + OH^-(aq)$$

This reaction proceeds to only a very small extent. In 0.10 $M$ NH$_3$, the concentrations of NH$_4^+$ and OH$^-$ are only about 0.0013 $M$.

**2. Anions.** *Any anion derived from a weak acid is itself a weak base.* A typical example is the fluoride ion, F$^-$, which is the conjugate base of the weak acid HF. The reaction of the F$^-$ ion with water is

$$F^-(aq) + H_2O \rightleftharpoons HF(aq) + OH^-(aq)$$

The OH$^-$ ions formed make the solution basic. A 0.10 $M$ solution of sodium fluoride, NaF, has a pH of about 8.1.

Notice the similarity between the equation just written and that for the NH$_3$ molecule cited above:

— both weak bases (NH$_3$, F$^-$) accept protons to form the conjugate acid (NH$_4^+$, HF)
— water acts as a Brønsted-Lowry acid in each case, donating a proton to form the OH$^-$ ion

*Only a few HF molecules dissociate, so [OH$^-$] >> [H$^+$]*

---

**Example 13.9**   Write an equation to explain why each of the following species produces a basic water solution.

(a) NO$_2^-$   (b) CO$_3^{2-}$   (c) HCO$_3^-$

***Strategy***   In each case, the weak base reacts reversibly with a water molecule, picking up a proton from it. Two species are formed. One is an OH$^-$ ion, which makes the solution basic. The other product is the conjugate acid of the weak base.

*Solution*

(a) $NO_2^-(aq) + H_2O \rightleftharpoons HNO_2(aq) + OH^-(aq)$
(b) $CO_3^{2-}(aq) + H_2O \rightleftharpoons HCO_3^-(aq) + OH^-(aq)$
(c) $HCO_3^-(aq) + H_2O \rightleftharpoons H_2CO_3(aq) + OH^-(aq)$

---

## The Equilibrium Constant for a Weak Base

For the weak base ammonia

$$NH_3(aq) + H_2O \rightleftharpoons NH_4^+(aq) + OH^-(aq)$$

the equilibrium constant is written in the usual way, omitting the term for solvent water.

$$K_b = \frac{[NH_4^+] \times [OH^-]}{[NH_3]}$$

For an anion B$^-$ that acts as a weak base,

$$B^-(aq) + H_2O \rightleftharpoons HB(aq) + OH^-(aq)$$

a very similar expression can be written:

$$K_b = \frac{[HB] \times [OH^-]}{[B^-]}$$

Baking soda is sodium hydrogen carbonate, NaHCO$_3$. The hydrogen carbonate ion in aqueous solution, HCO$_3^-$(aq), is basic, as shown by the color of the indicator. (Charles D. Winters)

### TABLE 13.4 Ionization Constants for Some Weak Bases at 25°C

| Base | Formula | $K_b$ | $pK_b$* |
|------|---------|-------|---------|
| Phosphate ion | $PO_4^{3-}$ | $2.2 \times 10^{-2}$ | 1.66 |
| Methylamine | $CH_3NH_2$ | $4.2 \times 10^{-4}$ | 3.38 |
| Carbonate ion | $CO_3^{2-}$ | $2.1 \times 10^{-4}$ | 3.68 |
| Ammonia | $NH_3$ | $1.8 \times 10^{-5}$ | 4.74 |
| Cyanide ion | $CN^-$ | $1.7 \times 10^{-5}$ | 4.77 |
| Hypochlorite ion | $ClO^-$ | $3.6 \times 10^{-7}$ | 6.44 |
| Sulfite ion | $SO_3^{2-}$ | $1.7 \times 10^{-7}$ | 6.77 |
| Hydrogen phosphate ion | $HPO_4^{2-}$ | $1.6 \times 10^{-7}$ | 6.80 |
| Hydrogen carbonate ion | $HCO_3^-$ | $2.3 \times 10^{-8}$ | 7.64 |
| Acetate ion | $C_2H_3O_2^-$ | $5.6 \times 10^{-10}$ | 9.25 |
| Aniline | $C_6H_5NH_2$ | $3.8 \times 10^{-10}$ | 9.42 |
| Formate ion | $CHO_2^-$ | $5.3 \times 10^{-11}$ | 10.28 |
| Nitrite ion | $NO_2^-$ | $1.7 \times 10^{-11}$ | 10.77 |
| Fluoride ion | $F^-$ | $1.4 \times 10^{-11}$ | 10.85 |
| Dihydrogen phosphate ion | $H_2PO_4^-$ | $1.4 \times 10^{-12}$ | 11.85 |
| Hydrogen sulfite ion | $HSO_3^-$ | $5.9 \times 10^{-13}$ | 12.23 |

*$pK_b = -\log_{10} K_b$.

Values of $K_b$ are listed in Table 13.4; the larger the value of $K_b$, the stronger the base. Base strength decreases moving down the table. Since $K_b$ for $NH_3$ ($1.8 \times 10^{-5}$) is larger than $K_b$ for the acetate ion, $C_2H_3O_2^-$ ($5.6 \times 10^{-10}$), ammonia is a stronger base than the acetate ion.

## Calculation of [OH⁻] in a Water Solution of a Weak Base

We showed in Section 13.4 how $K_a$ of a weak acid can be used to calculate $[H^+]$ in a solution of that acid. In a very similar way, $K_b$ can be used to find $[OH^-]$ in a solution of a weak base.

---

**Example 13.10**  Calculate the pH of 0.10 $M$ NaF ($K_b$ $F^- = 1.4 \times 10^{-11}$).

***Strategy***  The procedure here is entirely analogous to that followed in Example 13.7. Take $[OH^-] = x$ and set up an equilibrium table, ignoring the $OH^-$ ions present in pure water. Solve for $x$, making the usual approximation. Then calculate pOH and finally pH. Hopefully, you will obtain a pH greater than 7; a solution of sodium fluoride had better be basic!

***Solution***  The equilibrium table is

Note the resemblance to Example 13.7; here you calculate [OH⁻]

| | $F^-(aq)$ + H₂O ⇌ | HF(aq) + | OH⁻(aq) |
|---|---|---|---|
| [ ]₀ | 0.10 | 0.00 | 0.00 |
| Δ[ ] | $-x$ | $+x$ | $+x$ |
| [ ]eq | $0.10 - x$ | $x$ | $x$ |

Substituting into the equilibrium constant expression:

$$K_b = \frac{x^2}{0.10 - x} = 1.4 \times 10^{-11}$$

Assuming $0.10 - x \approx 0.10$ and solving for $x$,

$$x^2 = 1.4 \times 10^{-12} \qquad x = 1.2 \times 10^{-6}$$

Clearly $x \ll 0.10$, so the approximation is justified.

$$[OH^-] = 1.2 \times 10^{-6} \, M$$

$$pOH = -\log_{10}(1.2 \times 10^{-6}) = 5.92$$

$$pH = 14.00 - 5.92 = \boxed{8.08}$$

## Relation Between $K_a$ and $K_b$

Ionization constants of weak bases can be measured in the laboratory by procedures very much like those used for weak acids. In practice, though, it is simpler to take advantage of a simple mathematical relationship between $K_b$ for a weak base and $K_a$ for its conjugate acid. This relationship can be derived by adding together the equations for the ionization of the weak acid HB and the reaction of the weak base $B^-$ with water:

(1) $\quad HB(aq) \rightleftharpoons H^+(aq) + B^-(aq) \qquad\qquad K_I = K_a$ of HB

(2) $\quad B^-(aq) + H_2O(aq) \rightleftharpoons HB(aq) + OH^-(aq) \qquad K_{II} = K_b$ of $B^-$

(3) $\quad H_2O \rightleftharpoons H^+(aq) + OH^-(aq) \qquad\qquad K_{III} = K_w$

Since Equation (1) + Equation (2) = Equation (3), we have, according to the rule of multiple equilibria (Chap. 12),

$$K_I \times K_{II} = K_{III}$$

or

$$(K_a \text{ of HB}) \times (K_b \text{ of B}^-) = K_w = 1.0 \times 10^{-14}$$

---

**Example 13.11**  From the equilibrium constants listed in Tables 13.2 and 13.4, calculate

(a) $K_b$ of $Fe(H_2O)_5(OH)^{2+}$   (b) $K_a$ of $CH_3NH_3^+$

**Strategy**  In each case, identify the conjugate acid or base, find that species in the table, and apply the relation $K_a \times K_b = 1.0 \times 10^{-14}$.

**Solution**

(a) The conjugate acid $Fe(H_2O)_6^{3+}$ is listed in Table 13.2; it has a $K_a$ of $6.7 \times 10^{-3}$. Hence

$$K_b \text{ of } Fe(H_2O)_5(OH)^{2+} = \frac{1.0 \times 10^{-14}}{6.7 \times 10^{-3}} = \boxed{1.5 \times 10^{-12}}$$

(b) Conjugate base: $CH_3NH_2$; $K_b$ of $CH_3NH_2$ from Table 13.4 = $4.2 \times 10^{-4}$. Hence

$$K_a \text{ of } CH_3NH_3^+ = \frac{1.0 \times 10^{-14}}{4.2 \times 10^{-4}} = \boxed{2.4 \times 10^{-11}}$$

---

From the general relation between $K_a$ and $K_b$, it should be clear that these two quantities are inversely related to one another. This makes sense; the stronger the acid, the weaker the conjugate base. Table 13.5 shows this effect and some other interesting features. In particular,

TABLE 13.5    **Relative Strengths of Brønsted-Lowry Acids and Bases**

| $K_a$ | Conjugate Acid | Conjugate Base | $K_b$ |
|---|---|---|---|
| very large | $HClO_4$ | $ClO_4^-$ | very small |
| very large | $HCl$ | $Cl^-$ | very small |
| very large | $HNO_3$ | $NO_3^-$ | very small |
| | **$H_3O^+$** | **$H_2O$** | |
| $6.9 \times 10^{-4}$ | $HF$ | $F^-$ | $1.4 \times 10^{-11}$ |
| $1.8 \times 10^{-5}$ | $HC_2H_3O_2$ | $C_2H_3O_2^-$ | $5.6 \times 10^{-10}$ |
| $1.4 \times 10^{-5}$ | $Al(H_2O)_6^{3+}$ | $Al(H_2O)_5(OH)^{2+}$ | $7.1 \times 10^{-10}$ |
| $4.4 \times 10^{-7}$ | $H_2CO_3$ | $HCO_3^-$ | $2.3 \times 10^{-8}$ |
| $2.8 \times 10^{-8}$ | $HClO$ | $ClO^-$ | $3.6 \times 10^{-7}$ |
| $5.6 \times 10^{-10}$ | $NH_4^+$ | $NH_3$ | $1.8 \times 10^{-5}$ |
| $4.7 \times 10^{-11}$ | $HCO_3^-$ | $CO_3^{2-}$ | $2.1 \times 10^{-4}$ |
| | **$H_2O$** | **$OH^-$** | |
| very small | $C_2H_5OH$ | $C_2H_5O^-$ | very large |
| very small | $OH^-$ | $O^{2-}$ | very large |
| very small | $H_2$ | $H^-$ | very large |

Species shown in deep color are unstable because they react with $H_2O$ molecules

1. Brønsted-Lowry acids (left column) can be divided into three categories:
   (a) strong acids ($HClO_4, \ldots$), which are stronger proton donors than the $H_3O^+$ ion
   (b) weak acids ($HF, \ldots$), which are weaker proton donors than the $H_3O^+$ ion, but stronger than the $H_2O$ molecule
   (c) species such as $C_2H_5OH$, which are weaker proton donors than the $H_2O$ molecule and hence do not form acidic water solutions
2. Brønsted-Lowry bases (right column) can be divided similarly:
   (a) strong bases ($H^-, \ldots$), which are stronger proton acceptors than the $OH^-$ ion
   (b) weak bases ($F^-, \ldots$), which are weaker proton acceptors than the $OH^-$ ion, but stronger than the $H_2O$ molecule
   (c) the anions of strong acids ($ClO_4^-, \ldots$), which are weaker proton acceptors than the $H_2O$ molecule and hence do not form basic water solutions

We should point out that, just as the three strong acids at the top of Table 13.5 are completely converted to $H_3O^+$ ions in aqueous solution:

$$HClO_4(aq) + H_2O \longrightarrow H_3O^+(aq) + ClO_4^-(aq)$$

$$HCl(aq) + H_2O \longrightarrow H_3O^+(aq) + Cl^-(aq)$$

$$HNO_3(aq) + H_2O \longrightarrow H_3O^+(aq) + NO_3^-(aq)$$

the three strong bases at the bottom of the table are completely converted to $OH^-$ ions:

$$H^-(aq) + H_2O \longrightarrow OH^-(aq) + H_2(g)$$

$$O^{2-}(aq) + H_2O \longrightarrow OH^-(aq) + OH^-(aq)$$

$$C_2H_5O^-(aq) + H_2O \longrightarrow OH^-(aq) + C_2H_5OH(aq)$$

*13.6  Acid-Base Properties of Salt Solutions*   **389**


In other words, the species $H^-$, $O^{2-}$, and $C_2H_5O^-$ do not exist in water solution.

# 13.6  Acid-Base Properties of Salt Solutions

*A salt is an ionic solid containing a cation other than $H^+$ and an anion other than $OH^-$ or $O^{2-}$.* When a salt such as NaCl, $K_2CO_3$, or $Al(NO_3)_3$ dissolves in water, the cation and anion separate from one another.

$$NaCl(s) \longrightarrow Na^+(aq) + Cl^-(aq)$$

$$K_2CO_3(s) \longrightarrow 2K^+(aq) + CO_3{}^{2-}(aq)$$

$$Al(NO_3)_3(s) \longrightarrow Al^{3+}(aq) + 3NO_3{}^-(aq)$$

To predict whether a given salt solution will be acidic, basic, or neutral, you consider three factors in turn.

1. *Decide what effect, if any, the cation has on the pH of water.*
2. *Decide what effect, if any, the anion has on the pH of water.*
3. *Combine the two effects to decide upon the behavior of the salt.*

## Cations: Weak Acids or Spectator Ions?

As we pointed out in Section 13.4, certain cations act as weak acids in water solution because of reactions such as

$$NH_4{}^+(aq) + H_2O \rightleftharpoons H_3O^+(aq) + NH_3(aq)$$

$$Zn(H_2O)_4{}^{2+}(aq) + H_2O \rightleftharpoons H_3O^+(aq) + Zn(H_2O)_3(OH)^+(aq)$$

Essentially all transition metal ions behave like $Zn^{2+}$, forming a weakly acidic solution. Among the main-group cations, $Al^{3+}$ and, to a lesser extent, $Mg^{2+}$, act as weak acids.

The cations derived from strong bases, e.g.,

— the alkali metal cations ($Li^+$, $Na^+$, $K^+$, ...)
— the heavier alkaline earth cations ($Ca^{2+}$, $Sr^{2+}$, $Ba^{2+}$)

do not react with water. In that sense, these ions are "spectator" ions and do not affect pH (Table 13.6, p. 390).

*Learn the spectator ions, not the players*

## Anions: Weak Bases or Spectator Ions?

As pointed out in Section 13.5, anions that are the conjugate bases of weak acids act themselves as weak bases in water. They accept a proton from a water molecule, leaving an $OH^-$ ion that makes the solution basic. The reactions of the fluoride and carbonate ions are typical:

$$F^-(aq) + H_2O \rightleftharpoons HF(aq) + OH^-(aq)$$

$$CO_3{}^{2-}(aq) + H_2O \rightleftharpoons HCO_3{}^-(aq) + OH^-(aq)$$

Most anions behave this way, except those derived from six strong acids ($Cl^-$, $Br^-$, $I^-$, $NO_3{}^-$, $ClO_4{}^-$, $SO_4{}^{2-}$). These anions do not react with water; in

| TABLE 13.6  **Acid-Base Properties of Ions\* in Water Solution** | | | | | |
|---|---|---|---|---|---|
| | **Spectator** | | **Basic** | | **Acidic** |
| **Anion** | $Cl^-$  $Br^-$  $I^-$ | $NO_3^-$  $ClO_4^-$  $SO_4^{2-}$ | $C_2H_3O_2^-$  $F^-$  many others | $CO_3^{2-}$  $PO_4^{3-}$ | |
| **Cation** | $Li^+$  $Na^+$  $K^+$ | $Ca^{2+}$  $Sr^{2+}$  $Ba^{2+}$ | | | $NH_4^+$   $Al^{3+}$  $Mg^{2+}$  transition metal ions |

\*Ions are classified as "spectators" unless, in 0.1 *M* solution, they cause a change of at least one pH unit.

that sense they behave as "spectator ions" so far as their effect on pH is concerned (Table 13.6).

## Salts: Acidic, Basic, or Neutral?

If you know how the cation and anion of a salt affect the pH of water, it is a relatively simple matter to decide what the net effect will be (Example 13.12).

---

**Example 13.12**   Consider water solutions of these four salts:

$$NH_4I,\ Zn(NO_3)_3,\ KClO_4,\ Na_3PO_4$$

Which of these solutions are acidic? basic? neutral?

*Strategy*   First, decide what ions are present in the solution. Then classify each cation and anion as acidic, basic, or neutral, using Table 13.6. Finally, consider the combined effects of the two ions in each salt.

*Solution*   To implement this strategy, it is convenient to prepare a table.

| Salt | Cation | Anion | Solution of Salt |
|---|---|---|---|
| $NH_4I$ | $NH_4^+$ (acidic) | $I^-$ (spectator) | acidic |
| $Zn(NO_3)_2$ | $Zn^{2+}$ (acidic) | $NO_3^-$ (spectator) | acidic |
| $KClO_4$ | $K^+$ (spectator) | $ClO_4^-$ (spectator) | neutral |
| $Na_3PO_4$ | $Na^+$ (spectator) | $PO_4^{3-}$ (basic) | basic |

These predictions are confirmed by experiment (Figure 13.8).

---

The procedure described in Example 13.12 does not cover one combination—an acidic cation (e.g., $NH_4^+$) and a basic anion (e.g., $F^-$). Here, $K_a$ of the cation must be compared to $K_b$ of the anion. In general,

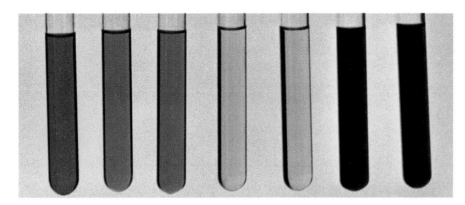

**Figure 13.8**
Colors of salt solutions with universal indicator, which is red in strong acid (far left) and purple in strong base (far right). $NH_4I$ and $Zn(NO_3)_2$ are weakly acidic at about pH 5 (orange); $KClO_4$, like pure water, has a pH of 7 (yellow-green); $Na_3PO_4$ is basic at about pH 12 (purple).

1. **If $K_a > K_b$, the salt is acidic.** An example is $NH_4F$ ($K_a NH_4^+ = 5.6 \times 10^{-10}$; $K_b F^- = 1.4 \times 10^{-11}$), where the pH is about 6.2.
2. **If $K_b > K_a$, the salt is basic.** An example is $NH_4ClO$ ($K_a NH_4^+ = 5.6 \times 10^{-10}$, $K_b ClO^- = 3.6 \times 10^{-7}$) where the pH is 8.4.

A similar argument can be used to decide whether an amphiprotic species such as $HCO_3^-$ or $H_2PO_4^-$ is acidic or basic. In principle, these ions can act as either weak acids or weak bases:

$$HCO_3^-(aq) + H_2O \rightleftharpoons H_3O^+(aq) + CO_3^{2-}(aq) \qquad K_a = 4.7 \times 10^{-11}$$

$$HCO_3^-(aq) + H_2O \rightleftharpoons H_2CO_3(aq) + OH^-(aq) \qquad K_b = 2.3 \times 10^{-8}$$

Since $K_b > K_a$, you would predict, correctly, that the $HCO_3^-$ ion produces a basic water solution. Sodium hydrogen carbonate, $NaHCO_3$ ("bicarbonate of soda") is used as an antacid because its water solution is basic, with a pH of about 8.3. The reverse situation applies with the $H_2PO_4^-$ ion ($K_a = 6.2 \times 10^{-8}$, $K_b = 1.4 \times 10^{-11}$). Since $K_a > K_b$, the $H_2PO_4^-$ ion gives an acidic solution. The compound $NaH_2PO_4$ is weakly acidic, with a solution pH of about 4.7.

Most ions of this type, like $HCO_3^-$, are basic

## Acid Rain

Natural rainfall has a pH of about 5.5. It is slightly acidic because of dissolved carbon dioxide, which reacts with water to form the weak acid $H_2CO_3$. In contrast, the average pH of rainfall in the eastern United States and southeastern Canada is about 4.4, corresponding to a $H^+$ ion concentration more than ten times the normal value. In an extreme case, rain with a pH of 1.5 was recorded in Wheeling, West Virginia (distilled vinegar has a pH of 2.4).

The phenomenon of acid rain results from the presence of two strong acids in polluted air: $H_2SO_4$ and, to a lesser extent, $HNO_3$. Nitric acid comes largely from NO produced by automobile emissions. There are two major sources of $SO_2$, the precursor of sulfuric acid. Coal used in power plants contains up to 4% sulfur, mostly in the form of minerals such as pyrite, $FeS_2$. Combustion forms sulfur dioxide:

$$4FeS_2(s) + 11\ O_2(g) \longrightarrow 2Fe_2O_3(s) + 8SO_2(g)$$

Large quantities of sulfur dioxide are also produced in the metallurgical processes used to treat sulfide ores. A single nickel smelter in Sudbury, Ontario, produces 1% of the global supply of $SO_2$ by the reaction

$$2NiS(s) + 3O_2(g) \longrightarrow 2NiO(s) + 2SO_2(g)$$

Sulfur dioxide formed by reactions such as these is slowly converted to $SO_3$ in the atmosphere

$$SO_2(g) + \tfrac{1}{2}O_2(g) \longrightarrow SO_3(g)$$

Reaction of $SO_3$ with water

$$SO_3(g) + H_2O \longrightarrow H_2SO_4(aq)$$

produces sulfuric acid as tiny droplets high in the atmosphere. These may be carried by prevailing winds as far as 1500 km. The acid rain that falls in the Adirondacks of New York or the Green Mountains of Vermont comes from sulfur dioxide produced by power plants in Ohio and Illinois.

The effects of acid rain are particularly severe in areas where the bedrock is granite or other materials incapable of neutralizing $H^+$ ions. As the concentration of acid builds up in a lake, aquatic life, from algae to brook trout, dies. The end product is a crystal-clear, totally sterile lake (see p. 371).

Acid rain adversely affects trees as well (Figure 13.A). Here it appears that the damage is largely due to the leaching of metal cations from the soil. In particular, $H^+$ ions in acid rain can react with insoluble aluminum compounds in the soil, bringing $Al^{3+}$ ions into solution. The following reaction is typical:

$$Al(OH)_3(s) + 3H^+(aq) \longrightarrow Al^{3+}(aq) + 3H_2O$$

$Al^{3+}$ ions in solution can have a direct toxic effect on the roots of trees. Perhaps more important, $Al^{3+}$ can replace $Ca^{2+}$, an ion that is essential to the growth of all plants.

Strong acids in the atmosphere can also attack building materials such as limestone or marble (calcium carbonate)

$$CaCO_3(s) + 2H^+(aq) \longrightarrow Ca^{2+}(aq) + CO_2(g) + H_2O$$

The relatively soluble calcium compounds formed this way [$CaSO_4$, $Ca(NO_3)_2$] are gradually washed away (Figure 13.B). This process is responsible for the deterioration of the

**Figure 13.A**
Effect of acid rain on evergreens at Camel's Hump, Vermont. The two photographs were taken 15 years apart. (EPA)

Greek ruins on the Acropolis in Athens. These structures have suffered more damage in the twentieth century than in the preceding 2000 years.

To reduce acid rain, sulfur dioxide emissions from power plants and smelters must be brought under control. One way to do this is to spray a water solution of calcium hydroxide directly into the smokestack. This "scrubbing" operation brings about the reaction

$$Ca^{2+}(aq) + 2\ OH^-(aq) + SO_2(g) + \tfrac{1}{2}O_2(g) \longrightarrow CaSO_4(s) + H_2O$$

One problem with this approach is that it produces large quantities of solid calcium sulfate as a waste product. A quite different method is used at the Sudbury smelter referred to earlier. There, the $SO_2$ is separated from the stack gas and converted to sulfuric acid by the contact process. Over $10^6$ tons of $H_2SO_4$, worth $100 million, are produced each year.

**Figure 13.B**
Effect of pollutants on statuary. The photograph on the left was taken about 1910 at Lincoln Cathedral in England; that on the right was taken in 1984.
(Dean and Chapler of Lincoln)

# CHAPTER HIGHLIGHTS

## *Key Concepts*

**1.** Given $[H^+]$, $[OH^-]$ or pH, calculate the other two quantities
(Examples 13.2–13.4; Problems 7–24)
**2.** Using the Brønsted-Lowry model, write an equation to explain why a species acts as a
—weak acid
(Examples 13.1, 13.5; Problems 25, 26)
—weak base
(Examples 13.1, 13.9; Problems 45, 46)
**3.** Given pH and $[HB]_o$, calculate $K_a$
(Example 13.6; Problems 33–36)
**4.** Given $K_a$ and $[HB]_o$, calculate $[H^+]$
(Examples 13.7, 13.8; Problems 37–44, 68)

**5.** Given $K_b$ and $[B^-]_0$, calculate $[OH^-]$
(Example 13.10; Problems 53–56)
**6.** Given $K_a$ or $K_b$, calculate the other quantity
(Example 13.11; Problems 51, 52)
**7.** Predict whether a given salt solution is acidic, basic, or neutral
(Example 13.12; Problems 57–64, 70)

# Key Equations

| | |
|---|---|
| Ionization of water | $K_w = [H^+] \times [OH^-] = 1.0 \times 10^{-14}$ |
| Expressions for $K_a$, $K_b$ | $K_a = \dfrac{[H^+] \times [B^-]}{[HB]}$ |
| | $K_b = \dfrac{[OH^-] \times [HB]}{[B^-]}$ |
| | $K_a \times K_b = K_w$ |

# Key Terms

acid
—strong
—weak
amphiprotic
base
—strong
—weak

conjugate acid
conjugate base
indicator
ionization constant ($K_w$, $K_a$, $K_b$)
molarity
neutral solution
pH

$pK_a$
pOH
percent ionization
polyprotic acid
quadratic formula
salt
successive approximations

# Summary Problem

Consider the salts ammonium nitrate and sodium fluoride.

(a) When aqueous solutions of nitric acid and aqueous ammonia are combined, ammonium nitrate is produced. Sodium fluoride can be prepared by combining aqueous solutions of hydrogen fluoride and sodium hydroxide. Classify each reactant as strong or weak, acid or base.
(b) Using the Brønsted-Lowry model, write equations to explain the acidity or basicity of $HNO_3$, $NH_3$, and HF.
(c) Calculate the pH of 0.20 $M$ solutions of $HNO_3$, NaOH, $NH_4NO_3$, and HF.
(d) What is the conjugate base of $NH_4^+$? What is $K_b$ for the conjugate base? What is the conjugate acid of $F^-$? What is $K_a$ for the conjugate acid?
(e) Classify the salts $NH_4NO_3$ and NaF as acidic or basic. Write net ionic equations to explain your answers.
(f) Would ammonium fluoride, $NH_4F$, be acidic or basic?

### Answers

(a) $HNO_3$—strong acid; NaOH—strong base; $NH_3$—weak base; HF—weak acid
(b) $HNO_3(aq) + H_2O \longrightarrow H_3O^+(aq) + NO_3^-(aq)$
$NH_3(aq) + H_2O \rightleftharpoons NH_4^+(aq) + OH^-(aq)$
$HF(aq) + H_2O \longrightarrow H_3O^+(aq) + F^-(aq)$

(c) $HNO_3 = 0.70$; $NaOH = 13.30$; $NH_4NO_3 = 4.98$; $NaF = 8.22$

(d) Conjugate base of $NH_4^+$ is $NH_3$. $K_b$ for $NH_3 = 1.8 \times 10^{-5}$
Conjugate acid of $F^-$ is HF. $K_a$ for HF $= 6.9 \times 10^{-4}$

(e) $NH_4NO_3$ is acidic; $NH_4^+(aq) + H_2O \rightleftharpoons NH_3(aq) + H_3O^+(aq)$
NaF is basic; $F^-(aq) + H_2O \rightleftharpoons HF(aq) + OH^-(aq)$

(f) ammonium fluoride is acidic.

## Questions & Problems

### Brønsted-Lowry Acid-Base Model

***1.** For each of the following reactions, indicate the Brønsted-Lowry acids and bases. What are the conjugate acid-base pairs?

(a) $H_3O^+(aq) + CN^-(aq) \rightleftharpoons HCN(aq) + H_2O$

(b) $HNO_2(aq) + OH^-(aq) \rightleftharpoons NO_2^-(aq) + H_2O$

(c) $HC_2H_3O_2(aq) + H_2O \rightleftharpoons$
$\qquad C_2H_3O_2^-(aq) + H_3O^+(aq)$

***2.** Follow the directions for Question 1 for the following reactions.

(a) $HClO(aq) + NH_3(aq) \rightleftharpoons NH_4^+(aq) + ClO^-(aq)$

(b) $H_2CO_3(aq) + H_2O \rightleftharpoons H_3O^+(aq) + HCO_3^-(aq)$

(c) $(CH_3)_2NH(aq) + H_2O \rightleftharpoons$
$\qquad (CH_3)_2NH_2^+(aq) + OH^-(aq)$

***3.** According to the Brønsted-Lowry theory, which of the following would you expect to act as an acid? Which as a base?

(a) $NH^{2-}$ (b) HIO (c) $H_3O^+$

***4.** According to the Brønsted-Lowry theory, which of the following would you expect to act as an acid? Which as a base?

(a) $C_2H_3O_2^-$ (b) $CH_3NH_3^+$ (c) HF

**5.** Give the formula of the conjugate acid of

(a) $C_2H_3O_2^-$ (b) $Fe(H_2O)_5(OH)^{2+}$ (c) $SO_3^{2-}$
(d) $(CH_3)_3N$ (e) $HPO_4^{2-}$

***6.** Give the formula of the conjugate base of

(a) $H_2O$ (b) $H_2PO_4^-$ (c) $NH_4^+$
(d) $HS^-$ (e) $Fe(H_2O)_6^{3+}$

### [H⁺], [OH⁻], pH, and pOH

**7.** Find the pH of solutions with the following $[H^+]$. Classify each as acidic or basic.

(a) 0.1 $M$ (b) $1.0 \times 10^{-5}$ $M$
(c) 2.7 $M$ (d) $4.8 \times 10^{-9}$ $M$

**8.** Find the pH of solutions with the following $[H^+]$. Classify each as acidic or basic.

(a) 0.0010 $M$ (b) $8.9 \times 10^{-10}$ $M$
(c) 1.62 $M$ (d) $3.2 \times 10^{-5}$ $M$

**9.** Calculate $[H^+]$ and $[OH^-]$ in solutions with the following pH.

(a) 3.0 (b) 6.23 (c) −0.74 (d) 14.60

**10.** Calculate $[H^+]$ and $[OH^-]$ in solutions with the following pH.

(a) 10.32 (b) 2.59 (c) 0.00 (d) −1.61

**11.** Solution A has pH 8.94. Solution B has $[OH^-] = 4.5 \times 10^{-8}$. Which solution is more basic? Which has the higher pOH?

**12.** Solution I has $[H^+] = 3.6 \times 10^{-8}$ $M$. Solution II has the same $[OH^-]$ as solution I's $[H^+]$. Which solution is more acidic? Which has the higher pH?

**13.** One solution has pH 6.72. What must the pH of another solution be so that $[H^+]$ is twice as large? One half as large?

**14.** Consider two solutions. Solution A has a pH 2 units higher than solution B. What is the ratio of the $H^+$ ion concentrations in the two solutions?

**15.** Gastric juice has a pH of 1.5.

(a) Calculate $[H^+]$.

(b) What is the ratio of $[H^+]$ in gastric juice to that in vinegar, which has a pH of 2.1?

**16.** A laundry detergent has a pH of 11.23. A "mild" shampoo has a pH of 8.71. Calculate the $[OH^-]$ ratio of laundry detergent to "mild" shampoo.

**17.** Find $[H^+]$ and the pH of the following solutions.

(a) 0.200 g of HI dissolved in enough water to make 400.0 mL.

(b) Ten mL of 0.125 $M$ HCl diluted with enough water to make 100.0 mL of solution. How does a tenfold dilution affect the pH?

**18.** Find $[H^+]$ and the pH of the following solutions.

(a) 5.00 g of $HClO_4$ dissolved in water to make 1.75 L of solution.

(b) 25.0 mL of an HCl solution with pH 3.15 diluted fivefold (i.e., diluted in enough water to make 125 mL of solution).

**19.** What is the pH of a solution obtained by adding 3.50 g of HBr to 435 mL of a 0.639 $M$ solution of $HNO_3$? Assume that the HBr addition does not change the volume of the resulting solution.

**20.** What is the pH of a solution obtained by adding 136 mL of 0.277 $M$ HCl to 379 mL of a $HClO_4$ solution with a pH of 0.927?

**21.** Find $[OH^-]$, $[H^+]$, and the pH of the following solutions.

(a) 0.38 $M$ $Ba(OH)_2$.

(b) A solution made by dissolving 12.4 g of NaOH in enough water to make 0.750 L of solution.

22. Find $[OH^-]$, $[H^+]$, and the pH of the following solutions.

(a) 65.0 mL of 0.0239 $M$ $Sr(OH)_2$ diluted with enough water to make 200.0 mL of solution.

(b) A solution made by dissolving 1.69 g of CsOH in enough water to make 300.0 mL of solution.

23. What is the pH of a solution obtained by adding 1.00 g $Ca(OH)_2$ and 37.5 g of KOH to enough water to make 0.500 L of solution?

24. What is the pH of a solution obtained by adding 35.0 mL of 0.406 $M$ RbOH to 118 mL of $Ba(OH)_2$ solution with a pH of 12.93? Assume that volumes are additive.

## Ionization Expressions, Weak Acids

*25. Using the Brønsted-Lowry model, write equations to show why the following species behave as weak acids in water.

(a) $Ni(H_2O)_3OH^+$    (b) $HSO_4^-$    (c) HCN
(d) $Fe(H_2O)_6^{3+}$    (e) $HC_2H_3O_2$

*26. Follow the directions of Question 25 for the following species.

(a) $Zn(H_2O)_3OH^+$    (b) $Al(H_2O)_6^{3+}$    (c) $H_2CO_3$
(d) $HPO_4^{2-}$    (e) $HIO_2$

*27. Write the ionization equation and the $K_a$ expression for each of the following acids.

(a) HF    (b) $HSO_3^-$    (c) $HC_2H_3O_2$

*28. Write the ionization equation and the $K_a$ expression for each of the following acids.

(a) $HS^-$    (b) $PH_4^+$    (c) $H_2C_2O_4$

*29. What are the $pK_a$ values for acids with $K_a$ of

(a) $1.8 \times 10^{-5}$    (b) $5.6 \times 10^{-10}$    (c) $1.9 \times 10^{-4}$

*30. Calculate $K_a$ for the weak acids that have the following $pK_a$ values.

(a) 2.6    (b) 8.3    (c) 10.5

*31. Consider these acids:

| Acid | A | B | C | D |
|------|-----|-----|-----|-----|
| $K_a$ | $2 \times 10^{-2}$ | $4 \times 10^{-6}$ | $8 \times 10^{-3}$ | $5 \times 10^{-4}$ |

(a) Arrange the acids in order of increasing acid strength.

(b) Which acid has the smallest $pK_a$ value?

*32. Consider these acids:

| Acid | A | B | C | D |
|------|------|-----|-----|------|
| $pK_a$ | 12.6 | 1.9 | 4.7 | 13.8 |

(a) Arrange the acids in order of decreasing acid strength.

(b) Which acid has the largest $K_a$ value?

## Equilibrium Calculations, Weak Acids

33. Penicillin, an antibiotic often used to treat bacterial infections, is a weak acid. A solution is made by dissolving 0.250 mol of penicillin to form a liter of solution. The solution has $[H^+] = 2.00 \times 10^{-2}$. What is $K_a$ for penicillin?

34. Butyric acid, $HC_4H_7O_2$, is responsible for the odor of rancid butter and cheese. A solution prepared by dissolving 0.0400 mol $HC_4H_7O_2$ to form 1.00 L of solution has $[H^+] = 7.70 \times 10^{-4}$. What is $K_a$ for butyric acid?

35. Picric acid, $HC_6H_2O_7N_3$, is an intensely yellow acid used in the manufacture of explosives and dyes. A picric acid solution prepared by dissolving 17 g of picric acid in 250.0 mL of solution has a pH of 0.82. What is the $K_a$ for picric acid?

36. Uric acid, $HC_5H_3O_3N_4$, can accumulate in the joints. This accumulation causes severe pain, and the condition is called gout. A uric acid solution prepared by dissolving 0.200 g of uric acid in 1.75 L of solution has a pH of 4.25. What is the $K_a$ for uric acid?

37. Para-amino benzoic acid (PABA), $HC_7H_6NO_2$, is used in some sunscreen agents. Its $K_a$ is $2.2 \times 10^{-5}$. Calculate $[H^+]$ in solutions prepared by adding the following number of moles of PABA to one liter of solution.

(a) 1.25 mol    (b) 0.66 mol

38. Caproic acid, $HC_6H_{11}O_2$, is found in coconut oil and is used in the making of artificial flavors. Its $K_a$ is $1.3 \times 10^{-5}$. Calculate $[H^+]$ in solutions prepared by adding the following number of moles of caproic acid to 2.00 L of solution.

(a) 1.78 mol    (b) 0.450 mol

39. Benzoic acid, $HC_7H_5O_2$ ($K_a = 6.5 \times 10^{-5}$), is present in many berries. For a 1.7 $M$ solution of benzoic acid, calculate

(a) $[H^+]$    (b) $[OH^-]$
(c) pH    (d) percent ionization

40. Phenol ($K_a = 1.3 \times 10^{-10}$) is a weak organic acid used as an antiseptic and in the manufacture of plastics. Follow the directions of Problem 39 for a 0.88 $M$ solution of phenol.

41. Trichloroacetic acid ($K_a = 2.0 \times 10^{-1}$) is sometimes used in the treatment of warts. Calculate the pH of a 3.0 $M$ solution of trichloroacetic acid.

42. Barbituric acid ($K_a = 1.1 \times 10^{-4}$) is used in the manufacture of some sedatives. Calculate the pH of a 1.3 $M$ solution of barbituric acid.

43. Using the $K_a$ values listed in Table 13.2, calculate the pH of a 0.20 $M$ solution of ammonium chloride.

44. Taking $K_a$ of a certain weak acid to be $4.4 \times 10^{-7}$, calculate the pH of a 0.043 $M$ solution.

## Ionization Expressions; Weak Bases

*45. Using the Brønsted-Lowry model, write an equation to show why each of the following species acts as a weak base in water.

(a) $(CH_3)_2NH$    (b) $NO_2^-$    (c) $H_2PO_4^-$
(d) $CO_3^{2-}$    (e) $HS^-$    (f) $HCO_3^-$

*46. Follow the directions of Question 45 for the following species.

(a) $NH_3$    (b) $PO_4^{3-}$    (c) $C_6H_5NH_2$
(d) $CN^-$    (e) $F^-$    (f) $(C_2H_5)_3N$

47. Using the ionization constants listed in Table 13.4, arrange the following 0.1 $M$ aqueous solutions in order of increasing pH.

(a) NaOH    (b) NaCN
(c) $C_6H_5NH_2$    (d) $Ba(OH)_2$

48. Using the ionization constants listed in Tables 13.2 and 13.4, arrange the following 0.1 $M$ aqueous solutions in order of decreasing pH.

(a) $NaNO_2$    (b) $NH_4Cl$    (c) $Na_2CO_3$    (d) HCl

*49. Which of the following are true regarding a 0.10 $M$ solution of a strong base MOH?

(a) $[M^+] = 0.10\ M$       (b) $[MOH] = 0.10\ M$
(c) $[M^+] + [OH^-] = 0.20\ M$
(d) $[H^+] = 0.10\ M$       (e) pH = 14.0

*50. Which of the following are true regarding a 0.10 $M$ solution of a weak base, $B^-$?

(a) $[HB] = 0.10\ M$       (b) $[OH^-] \approx [HB]$
(c) $[B^-] >> [HB]$
(d) $[H^+] = \dfrac{1.0 \times 10^{-14}}{0.10}$       (e) pH = 13.0

## Equilibrium Calculations; Weak Bases

51. Find the value of $K_a$ for the conjugate acid of the following organic bases.

(a) morphine, an addictive opiate; $K_b = 8 \times 10^{-7}$
(b) aniline, an important dye intermediate; $K_b = 3.8 \times 10^{-10}$

52. Find the value of $K_b$ for the conjugate base of the following acids.

(a) formic acid, the irritant in ant bites; $K_a = 1.9 \times 10^{-4}$
(b) gallic acid, present in tea; $K_a = 3.9 \times 10^{-5}$

53. Consider sodium acrylate, $NaC_3H_3O_2$. $K_a$ for acrylic acid is $5.5 \times 10^{-5}$.

(a) Write a balanced net ionic equation for the reaction that makes aqueous solutions of sodium acrylate basic.
(b) Calculate $K_b$ for the reaction in (a).
(c) Find the pH of a solution prepared by dissolving 0.500 g of $NaC_3H_3O_2$ in enough water to make 0.150 L of solution.

54. Use Table 13.4 to determine the pH of a 0.200 $M$ solution of $Na_3PO_4$.

55. Codeine (Cod), a powerful and addictive painkiller, is a weak base. Its ionization in water can be represented by the equation

$$Cod(aq) + H_2O \rightleftharpoons CodH^+(aq) + OH^-(aq)$$

A 0.002 $M$ solution of codeine has a pH of 9.63. Calculate $K_b$ for codeine.

56. The pH of a household cleaning ammonia solution is 11.68. How many grams of ammonia were dissolved in 2.00 L of the solution? Use Table 13.4.

## Salt Solutions

*57. State whether 1 $M$ solutions of the following salts in water would be acidic, basic, or neutral.

(a) $AlCl_3$    (b) $Ba(NO_3)_2$    (c) $NH_4F$
(d) $NaCHO_2$    (e) $KHCO_3$

*58. State whether 1 $M$ solutions of the following salts in water would be acidic, basic, or neutral.

(a) $NH_4Cl$    (b) $NH_4ClO$    (c) $FeI_3$
(d) $KC_2H_3O_2$    (e) $Na_3PO_4$

*59. Write net ionic equations to explain the acidity or basicity of the various salts listed in Question 57.

*60. Write net ionic equations to explain the acidity or basicity of the various salts listed in Question 58.

*61. Arrange the following 0.1 $M$ aqueous solutions in order of increasing pH.

KOH, KF, KCl, $ZnCl_2$, HCl

*62. Arrange the following 0.1 $M$ aqueous solutions in order of decreasing pH.

$Ba(NO_3)_2$, $HNO_3$, $NH_4NO_3$, $Al(NO_3)_3$, NaF

*63. Write formulas for four salts that

(a) contain $K^+$ and are basic.
(b) contain $K^+$ and are neutral.
(c) contain $ClO_4^-$ and are neutral.
(d) contain $ClO_4^-$ and are acidic.

*64. Write formulas for four salts that

(a) contain $Co^{3+}$ and are acidic.
(b) contain $Br^-$ and are neutral.
(c) contain $Sr^{2+}$ and are basic.
(d) contain $Li^+$ and are neutral.

## Unclassified

*65. Each figure below represents an acid solution in equilibrium. Squares represent $H^+$ ions, and circles represent the anions. Water molecules are not shown. Which figure represents a strong acid? Which a weak acid?

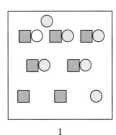     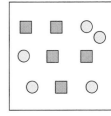

1              2

*66. Each box (p. 598) represents an acid solution at equilibrium. Squares represent $H^+$ ions. Circles represent anions. (Although the anions have different identities in each figure, they are all represented as circles.) Water molecules are not shown. Assume that all solutions have the same volume.

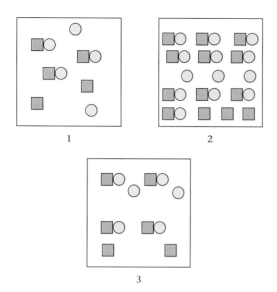

(a) Which figure represents the strongest acid?

(b) Which figure represents the acid with the smallest $K_a$?

(c) Which figure represents the solution with the lowest pH?

*67. If ▢◯ represents the weak acid HA(▢ = $H^+$, ◯ = $A^-$) in Figure a below, fill in Figure b with ▢◯, ▢, and/or ◯ to represent HA as being 10% ionized. (Water molecules are omitted.)

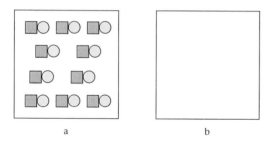

68. There are 324 mg of acetylsalicylic acid ($\mathcal{M}$ = 180.15 g/mol) per aspirin tablet. If two tablets are dissolved to give two ounces ($\frac{1}{16}$ quart) of solution, estimate the pH. $K_a$ of acetylsalicylic acid is $3.6 \times 10^{-4}$.

*69. Give two examples of

(a) an acid stronger than acetic acid, $HC_2H_3O_2$.

(b) a base weaker than KOH.

(c) a salt that dissolves in water to yield a solution with a pH > 7.

(d) a salt that dissolves in water to give a pH = 7.

*70. Using data in Tables 13.2 and 13.4, classify solutions of the following salts as acidic, basic, or neutral.

(a) $NH_4C_2H_3O_2$     (b) $NH_4H_2PO_4$

(c) $AgNO_2$          (d) $CH_3NH_3F$

71. A student is asked to bubble enough ammonia gas through water to make 1.00 L of an aqueous ammonia solution with a pH of 11.21. What volume of ammonia gas at 25°C and 1.00 atm pressure is necessary?

72. Consider the process

$$H_2O \rightleftharpoons H^+(aq) + OH^-(aq) \qquad \Delta H^\circ = 55.8 \text{ kJ}$$

(a) Will the pH of pure water at body temperature (37°C) be 7.0?

(b) If not, calculate the pH of pure water at 37°C.

73. Is a saline (NaCl) solution at 80°C acidic, basic, or neutral?

74. Household bleach is prepared by dissolving chlorine in water:

$$Cl_2(g) + H_2O \rightleftharpoons H^+(aq) + Cl^-(aq) + HOCl(aq)$$

$K_a$ for HOCl is $3.2 \times 10^{-8}$. How much chlorine was dissolved in one liter of water to make the pH of the solution 1.19?

## Challenge Problems

*75. Silver hydroxide, AgOH, is insoluble in water. Describe a simple qualitative experiment that would enable you to determine whether AgOH is a strong or weak base.

76. Using the Tables in Appendix 1, calculate $\Delta H$ for the reaction of

(a) 1.00 L of 0.100 $M$ NaOH with 1.00 L of 0.100 $M$ HCl.

(b) 1.00 L of 0.100 $M$ NaOH with 1.00 L of 0.100 $M$ HF, taking the heat of formation of HF($aq$) to be −320.1 kJ/mol.

77. Show by calculation that when the concentration of a weak acid decreases by a factor of 10, its percent ionization increases by a factor of $10^{\frac{1}{2}}$.

78. What is the freezing point of vinegar, which is an aqueous solution of 5.00% acetic acid, $HC_2H_3O_2$, by mass (d = 1.006 g/cm³)?

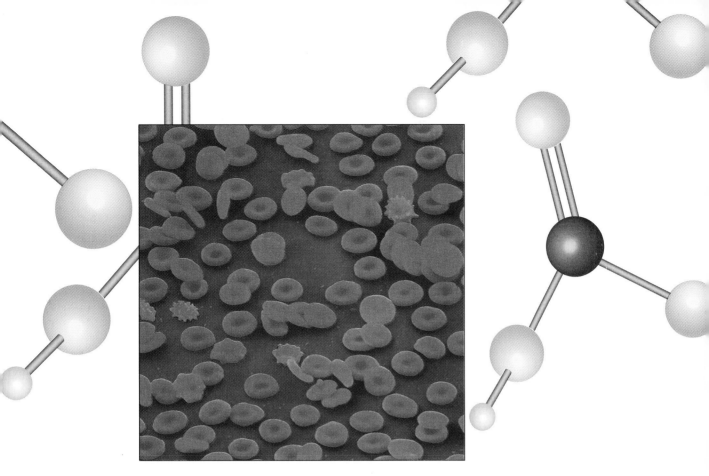

Blood and other body fluids contain enough $HCO_3^-$ ions to neutralize organic acids produced by life processes (p. 418).
(NIBSC/Science Photo Library/Photo Researchers, Inc.)

# Equilibria in Acid–Base Solutions

# 14

No single thing abides; but all things flow
Fragment to fragment clings—the things thus grow
Until we know and name them. By degrees
They melt, and are no more the things we know.
—TITUS LUCRETIUS CARUS (92–52 B.C.)
*No Single Thing Abides*
(Translated by W. H. Mallock)

## CHAPTER OUTLINE

**14.1 Buffers**

**14.2 Acid-Base Indicators**

**14.3 Acid-Base Titrations**

**14.4 Polyprotic Weak Acids**

In Chapter 13, we dealt with the equilibrium established when a single solute, either a weak acid or a weak base, is added to water. This chapter focuses upon the equilibrium established when two different solutes are mixed in water solution. These solutes may be

— *a weak acid, HB, and its conjugate base, B⁻.* Solutions called *buffers* contain roughly equal amounts of these two species. The equilibria involved in buffer solutions are considered in Section 14.1.

— *an acid and a base used in an acid-base titration.* This type of reaction was discussed in Chapter 4. Section 14.3 examines the equilibria involved, the way pH changes during the titration, and the choice of indicator for the titration.

We will also consider equilibria established when an acid-base indicator (Section 14.2) or a polyprotic acid (Section 14.4) is dissolved in water. Here we will be particularly interested in how the concentrations of different species change as a function of pH.

## 14.1 Buffers

Any solution containing appreciable amounts of both a weak acid and its conjugate base

— *is highly resistant to changes in pH brought about by addition of strong acid or strong base.*
— *has a pH close to the $pK_a$ of the weak acid.*

By definition, a buffer is any solution that resists a change in pH

A solution showing these properties is given a special name; it is called a **buffer solution** or simply a **buffer.**

A buffer consisting of a weak acid HB and its conjugate base B⁻ can react with either strong base (OH⁻) or strong acid (H⁺).

$$HB(aq) + OH^-(aq) \longrightarrow B^-(aq) + H_2O$$

$$B^-(aq) + H^+(aq) \longrightarrow HB(aq)$$

These reactions have very large equilibrium constants, as we will see in Section 14.3, and so go virtually to completion. As a result, the added H⁺ or OH⁻ ions are consumed and do not directly affect the pH. This is the principle of buffer action, which explains why a buffered solution is much more resistant to a change in pH than one that is unbuffered (Figure 14.1).

Buffers are widely used to maintain nearly constant pH in a variety of commercial and laboratory products (Fig. 14.2). For these applications and many others, it is essential to be able to determine

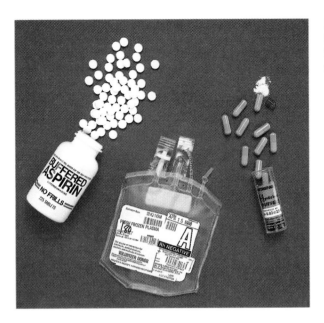

**Figure 14.1**
Many products, including aspirin and blood plasma, are buffered. Buffer tablets are also available in the laboratory (far right) to make up a solution to a specified pH. (Marna G. Clarke)

— the pH of a buffer system made by mixing a weak acid with its conjugate base
— the appropriate buffer system to maintain a desired pH
— the (small) change in pH that occurs when a strong acid or base is added to a buffer.

## Determination of [H⁺] in a Buffer System

The concentration of $H^+$ ion in a buffer can be calculated if you know the concentrations of the weak acid HB and its conjugate base $B^-$. These three quantities are related through the ionization constant of HB:

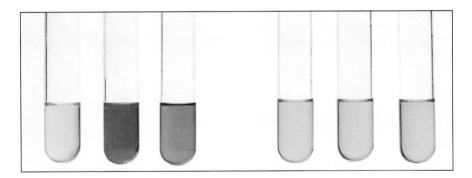

**Figure 14.2**
The three tubes at the left show the effect of adding a few drops of strong acid or strong base to water. The pH changes drastically, giving a pronounced color change with universal indicator. This experiment is repeated in the three tubes at the right, using a buffer of pH 7 instead of water. This time the pH changes only very slightly, and there is no change in color of the indicator. (Marna G. Clarke)

$$HB(aq) \rightleftharpoons H^+(aq) + B^-(aq) \qquad K_a = \frac{[H^+] \times [B^-]}{[HB]}$$

Solving for $[H^+]$,

$$[H^+] = K_a \times \frac{[HB]}{[B^-]} \tag{1}*$$

This is a completely general equation, applicable to all buffer systems. The calculation of $[H^+]$, and hence pH, can be simplified if you keep two points in mind.

**1. You can always assume that equilibrium is established without appreciably changing the original concentrations of either HB or $B^-$.** That is,

$$[HB] = [HB]_o$$

$B^-$ comes from adding a salt of the weak acid such as NaB

$$[B^-] = [B^-]_o$$

**2. Since the two species HB and $B^-$ are present in the same solution, the ratio of their concentrations is also their mole ratio.** That is,

$$\frac{[HB]}{[B^-]} = \frac{\text{no. moles HB}/V}{\text{no. moles B}^-/V} = \frac{n_{HB}}{n_{B^-}} \qquad (n = \text{amount in moles})$$

Hence, Equation 1 can be rewritten as

$$[H^+] = K_a \times \frac{n_{HB}}{n_{B^-}} \tag{2}$$

Frequently, Equation 2 is easier to work with than Equation 1.

---

**Example 14.1**    Lactic acid, $C_3H_6O_3$, is a weak organic acid present in both sour milk and buttermilk. It is also a product of carbohydrate metabolism and is found in the blood after vigorous muscular activity. A buffer is prepared by dissolving 1.00 mol of lactic acid, HLac ($K_a = 1.4 \times 10^{-4}$) and 1.00 mol of sodium lactate, NaLac, in enough water to form 550 mL of solution. Calculate $[H^+]$ and the pH of the buffer.

**Strategy**    You are given $K_a$ and $n_{HLac}$ in the statement of the problem. Since sodium lactate, like all ionic solids, is completely ionized in water, $n_{Lac^-} = n_{NaLac} = 1.00$ mol. Thus, you have all the quantities required to calculate $[H^+]$ by Equation 2.

**Solution**

$$[H^+] = K_a \times \frac{n_{HLac}}{n_{Lac^-}}$$

$$= 1.4 \times 10^{-4} \times \frac{1.00}{1.00} = \boxed{1.4 \times 10^{-4}}$$

$$pH = -\log_{10}(1.4 \times 10^{-4}) = \boxed{3.85}$$

---

*By taking the logarithms of both sides of Equation 1 and multiplying through by $-1$, you can show (Problem 66) that, for any buffer system,

$$pH = pK_a + \log_{10} [B^-]/[HB]$$

This relation, known as the **Henderson-Hasselbach** equation, is often used in biology and biochemistry to calculate the pH of buffers.

Notice in Example 14.1 that the total volume of solution is irrelevant. All that is required to solve for $[H^+]$ is the number of moles of lactic acid and sodium lactate. The pH of the buffer is 3.85 whether the volume of solution is 550 mL, 1 L, or even 10 L.

From a slightly different point of view, we can say that diluting a buffer with water does not change the pH. The mole ratio of HB to $B^-$ stays the same on dilution, so

$$[H^+] = K_a \times \frac{n_{HB}}{n_{B^-}}$$

stays constant. In practice, buffer solutions in the laboratory ordinarily contain 0.1 mol or more of both HB and $B^-$. If the solution is too dilute, it will have very little capacity to react with strong acid or strong base.

A solution 0.001 *M* in both $B^-$ and HB couldn't neutralize much $H^+$ or $OH^-$

## Choosing a Buffer System

Suppose you want to make up a buffer in the laboratory with a specified pH (e.g., 4.0, 7.0, 10.0, ...). Looking at the equation

$$[H^+] = K_a \times \frac{[HB]}{[B^-]} = K_a \times \frac{n_{HB}}{n_{B^-}}$$

it is evident that the pH of the buffer depends upon two factors:

**1. *The ionization constant of the weak acid, $K_a$.*** This has the greatest influence on buffer pH. Since HB and $B^-$ are likely to be present in nearly equal amounts,

$$[H^+] \approx K_a \qquad pH \approx pK_a$$

Thus, to make up a buffer of pH close to 7, you should start with a weak acid–weak base pair in which the ionization constant of the weak acid is about $10^{-7}$.

**2. *The ratio of the concentrations or amounts of HB and $B^-$.*** Small variations in pH can be achieved by adjusting this ratio. To obtain a slightly more acidic buffer, add more weak acid, HB; addition of more weak base, $B^-$, will make the buffer a bit more basic.

---

**Example 14.2**    To prepare a buffer with a pH of 9.00, which of the systems in Table 14.1 p. 404 would you use? What should be the ratio of concentrations, $[HB]/[B^-]$?

***Strategy***    Choose the buffer in which $pK_a$ is closest to 9. Clearly, this is the $NH_4^+$–$NH_3$ buffer. Then use the equation $[H^+] = K_a \times [NH_4^+]/[NH_3]$ to calculate the necessary ratio. Note that since the pH is 9.00, $[H^+] = 1.0 \times 10^{-9} M$.

***Solution***    For the $NH_4^+$–$NH_3$ buffer

$$[H^+] = K_a \times \frac{[NH_4^+]}{[NH_3]}$$

Solving for the ratio,

$$\frac{[NH_4^+]}{[NH_3]} = \frac{[H^+]}{K_a} = \frac{1.0 \times 10^{-9}}{5.6 \times 10^{-10}} = \boxed{1.8}$$

You might add 1.8 mol of $NH_4Cl$ to 1.0 mol of $NH_3$, or 9 mol of $NH_4Cl$ to 5 mol of $NH_3$ or . . . .

---

## TABLE 14.1 Buffer Systems at Different pH Values

| | Buffer System | | | |
|---|---|---|---|---|
| Desired pH | Weak Acid | Weak Base | $K_a$(Weak Acid) | p$K_a$ |
| 4 | lactic acid (HLac) | lactate ion (Lac$^-$) | $1.4 \times 10^{-4}$ | 3.85 |
| 5 | acetic acid (HC$_2$H$_3$O$_2$) | acetate ion (C$_2$H$_3$O$_2^-$) | $1.8 \times 10^{-5}$ | 4.74 |
| 6 | carbonic acid (H$_2$CO$_3$) | hydrogen carbonate ion (HCO$_3^-$) | $4.4 \times 10^{-7}$ | 6.36 |
| 7 | dihydrogen phosphate ion (H$_2$PO$_4^-$) | hydrogen phosphate ion (HPO$_4^{2-}$) | $6.2 \times 10^{-8}$ | 7.21 |
| 8 | hypochlorous acid (HClO) | hypochlorite ion (ClO$^-$) | $2.8 \times 10^{-8}$ | 7.55 |
| 9 | ammonium ion (NH$_4^+$) | ammonia (NH$_3$) | $5.6 \times 10^{-10}$ | 9.25 |
| 10 | hydrogen carbonate ion (HCO$_3^-$) | carbonate ion (CO$_3^{2-}$) | $4.7 \times 10^{-11}$ | 10.32 |

Note that complete neutralization would not produce a buffer; only NH$_4^+$ ions would be present

Most often, buffers are prepared, as implied in Example 14.2, by mixing a weak acid and its conjugate base. However, a somewhat different approach can be followed: partial neutralization of a weak acid or a weak base gives a buffer. To illustrate, suppose that 0.10 mol of a strong acid, perhaps HCl, is added to 0.20 mol of NH$_3$. The following reaction occurs:

$$H^+(aq) + NH_3(aq) \longrightarrow NH_4^+(aq)$$

The limiting reactant, 0.10 mol of H$^+$, is consumed. Hence, 0.10 mol of NH$_4^+$ is formed. ($\Delta n_{NH_4^+} = +0.10$ mol) and 0.10 mol of NH$_3$ is used up ($\Delta n_{NH_3} = -0.10$ mol). We started with no NH$_4^+$ and 0.20 mol of NH$_3$, so after reaction,

$$n_{NH_4^+} = 0.10 \text{ mol} \qquad n_{NH_3} = 0.20 \text{ mol} - 0.10 \text{ mol} = 0.10 \text{ mol}$$

The solution formed contains appreciable amounts of both NH$_4^+$ and NH$_3$; it is a buffer. We will have more to say in Section 14.3 about this way of making a buffer.

### Distribution Curves

Buffer action can be understood in terms of *distribution curves* such as the one shown in Figure 14.3. Here we have plotted, as a function of pH, the fractions of a weak acid HB ($K_a = 1 \times 10^{-5}$) and its conjugate base B$^-$ in a mixture of these two species. These fractions are defined by the relations

$$f(HB) = \frac{[HB]}{[HB] + [B^-]} \qquad f(B^-) = \frac{[B^-]}{[HB] + [B^-]}$$

$$f(HB) + f(B^-) = 1$$

**Figure 14.3**

Distribution curve showing the fractions of weak acid HB ($K_a = 1 \times 10^{-5}$) and weak base B$^-$ as a function of pH. Below pH 4, f$_{HB}$ >> f$_{B^-}$. Above pH 6, f$_{B^-}$ >> f$_{HB}$. Between pH 4 and 6, both species are present in appreciable amounts. At pH 5, F$_{HB}$ = f$_{B^-}$.

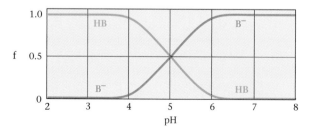

The two quantities $f(HB)$ and $f(B^-)$ are inversely related; if $f(HB) = 0.80$, then $f(B^-) = 0.20$, and vice versa.

Looking at the distribution curve, it should be evident that

— at pH 5.0, $f(HB) = f(B^-) = 0.50$. At this point, where

$$pH = pK_a \text{ of HB}$$

we have a buffer system containing equal amounts of the weak acid HB and its conjugate base $B^-$. This is the point at which the buffer has its maximum capacity to react with both $H^+$ and $OH^-$ ions.

— between pH 4.0 and 6.0, the mixture contains appreciable amounts of both HB and $B^-$, so it can act as a buffer. In this sense, the effective pH range of the buffer system is about two units, one above and one below the $pK_a$ value.

$$pH \text{ range} = pK_a \pm 1$$

— below pH 4.0 or above pH 6.0, the system consists essentially of a single component (HB at low pH, $B^-$ at high pH). Consequently, it cannot act as a buffer. Below pH 4.0, the system has no capacity to absorb strong acid; above pH 6.0, it cannot react with strong base.

## Effect of Added $H^+$ or $OH^-$ on Buffer Systems

The pH of a buffer does change slightly upon addition of moderate amounts of a strong acid or strong base. Addition of $H^+$ ions converts an equal amount of weak base $B^-$ to its conjugate acid HB:

$$H^+(aq) + B^-(aq) \longrightarrow HB(aq)$$

By the same token, addition of $OH^-$ ions converts an equal amount of weak acid to its conjugate base $B^-$:

$$HB(aq) + OH^-(aq) \longrightarrow B^-(aq) + H_2O$$

In either case, the ratio $n_{HB}/n_{B^-}$ changes. This in turn changes the $H^+$ ion concentration and pH of the buffer. The effect ordinarily is small, as illustrated in Example 14.3.

---

**Example 14.3**    Consider the buffer described in Example 14.1, where $n_{HLac} = n_{Lac^-} = 1.00$ mol ($K_a HLac = 1.4 \times 10^{-4}$). You will recall that in this buffer the pH is 3.85. Calculate the pH after addition of

(a) 0.10 mol of HCl     (b) 0.10 mol of NaOH

**Strategy**    First (1), write the chemical equation for the reaction that occurs when strong acid or base is added. Then (2), apply the principles of stoichiometry to find the numbers of moles of weak acid and weak base remaining after reaction. Finally (3), apply the relation $[H^+] = K_a \times n_{HB}/n_{B^-}$ to find $[H^+]$ and then the pH.

*Solution*

(a) (1) The $H^+$ ions from the strong acid HCl convert $Lac^-$ ions to HLac molecules:

$$H^+(aq) + Lac^-(aq) \longrightarrow HLac(aq)$$

To work this problem, you have to combine principles of stoichiometry and equilibrium

(2) Since the reacting mole ratios are 1:1, the addition of 0.10 mol of $H^+$ produces 0.10 mol of HLac and consumes 0.10 mol of $Lac^-$. Originally, there was 1.00 mol of both HLac and $Lac^-$. Hence,

$$n_{HLac} = (1.00 + 0.10) \text{ mol} = 1.10 \text{ mol}$$

$$n_{Lac^-} = (1.00 - 0.10) \text{ mol} = 0.90 \text{ mol}$$

(3) $[H^+] = 1.4 \times 10^{-4} \times \dfrac{n_{HLac}}{n_{Lac^-}} = 1.4 \times 10^{-4} \times \dfrac{1.10}{0.90} = 1.7 \times 10^{-4} \, M$

$$pH = -\log_{10}[H^+] = \boxed{3.77} \quad \text{(about 0.1 unit less than the original pH)}$$

(b) (1) The $OH^-$ ions from the strong base NaOH convert HLac molecules to $Lac^-$ ions:

$$HLac(aq) + OH^-(aq) \longrightarrow Lac^-(aq) + H_2O$$

(2) The reaction produces 0.10 mol of $Lac^-$ and consumes 0.10 mol of HLac:

$$n_{HLac} = (1.00 - 0.10) \text{ mol} = 0.90 \text{ mol}$$

$$n_{Lac^-} = (1.00 + 0.10) \text{ mol} = 1.10 \text{ mol}$$

(3) $[H^+] = 1.4 \times 10^{-4} \times \dfrac{n_{HLac}}{n_{Lac^-}} = 1.4 \times 10^{-4} \times \dfrac{0.90}{1.10} = 1.1 \times 10^{-4} \, M$

$$pH = -\log_{10}[H^+] = \boxed{3.94}$$

Again, the pH changes by less than 0.1 unit. In contrast, addition of 0.10 mol of HCl or NaOH to a liter of pure water would change the pH by 6 units.

---

Example 14.3 shows, among other things, how effectively a buffer "soaks up" $H^+$ or $OH^-$ ions. That can be important. Suppose you are carrying out a reaction whose rate is first order in $H^+$. If the pH increases from 5 to 7, perhaps by the absorption of traces of ammonia from the air, the rate will decrease by a factor of 100. A reaction which should have been complete in three hours will still be going on when you come back ten days later. Small wonder that chemists frequently work with buffered solutions to avoid disasters of that type.

## 14.2   Acid-Base Indicators

As pointed out in Chapter 4, an acid-base indicator is useful in determining the equivalence point of an acid-base titration. In Chapter 13 we saw that, more generally, an acid-base indicator can be used to determine the pH of an aqueous solution. We now consider the equilibrium principle upon which an indicator works.

An acid-base indicator is derived from a weak acid HIn:

$$HIn(aq) \rightleftharpoons H^+(aq) + In^-(aq) \qquad K_a = \dfrac{[H^+] \times [In^-]}{[HIn]}$$

*For litmus, HIn is red; $In^-$ is blue*

where the weak acid HIn and its conjugate base $In^-$ have different colors (Figure 14.4). The color that you see when a drop of indicator solution is added in an acid-base titration depends upon the ratio

$$\dfrac{[HIn]}{[In^-]}$$

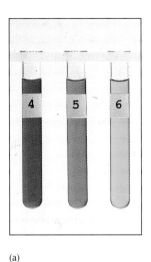
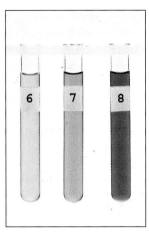
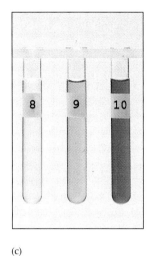

(a)                    (b)                    (c)

**Figure 14.4**
Methyl red (a) goes from red at low pH to orange at about pH 5 to yellow at high pH. Bromthymol blue (b) is yellow at low pH, blue at high pH, and green at about pH 7. Phenolphthalein (c) goes from colorless to pink at about pH 9.   (Marna G. Clarke)

Three cases can be distinguished:

**1.** If $\dfrac{[\text{HIn}]}{[\text{In}^-]} \geqslant 10$

the principal species is HIn; you see the "acid" color, that of the HIn molecule.

**2.** If $\dfrac{[\text{HIn}]}{[\text{In}^-]} \leqslant 0.1$

the principal species is $\text{In}^-$; you see the "base" color, that of the $\text{In}^-$ ion.

**3.** If $\dfrac{[\text{HIn}]}{[\text{In}^-]} \approx 1$

the color observed is intermediate between those of the two species, HIn and $\text{In}^-$.

The expression for the ionization of the indicator molecule, HIn,

$$K_a = \frac{[\text{H}^+] \times [\text{In}^-]}{[\text{HIn}]}$$

can be rearranged to give

$$\frac{[\text{HIn}]}{[\text{In}^-]} = \frac{[\text{H}^+]}{K_a}$$

From this expression, it follows that the ratio $[\text{HIn}]/[\text{In}^-]$ and hence the color of an indicator depends upon two factors.

**1. $[\text{H}^+]$ or the pH of the solution.** At high $[\text{H}^+]$ (low pH), you see the color of the HIn molecule; at low $[\text{H}^+]$ (high pH), the color of the $\text{In}^-$ ion dominates.

**2. $K_a$ of the indicator.** Since the ionization constant $K_a$ varies from one indicator to another, different indicators change colors at different pHs. A color change occurs when

$$[\text{H}^+] \approx K_a \qquad \text{pH} \approx pK_a$$

Table 14.2 p. 408 shows the characteristics of three indicators: methyl red, bromthymol blue, and phenolphthalein. These indicators change colors at pH 5, 7, and 9, respectively.

TABLE 14.2　**Colors and End Points of Indicators**

| | Color [HIn] | Color [In⁻] | $K_a$ | pH at End Point |
|---|---|---|---|---|
| Methyl red | red | yellow | $1 \times 10^{-5}$ | 5 |
| Bromthymol blue | yellow | blue | $1 \times 10^{-7}$ | 7 |
| Phenolphthalein | colorless | pink | $1 \times 10^{-9}$ | 9 |

Since only a drop or two of indicator is used, it does not affect the pH of the solution

## Distribution Curves

As is the case with buffers, a distribution curve can be useful in understanding how an acid-base indicator works. Figure 14.5 shows such a curve for the indicator bromthymol blue, where $K_a$ HIn $= 1 \times 10^{-7}$. Notice that

— at pH 7.0, the two quantities $f$ (HIn) and $f$ (In⁻) are equal to each other. At this point where

$$\text{pH} = \text{p}K_a \text{ of HIn}$$

the color (green) is a 50:50 blend of that of HIn (yellow) and In⁻ (blue).

— between pH 6.0 and 8.0, the system contains appreciable amounts of both HIn and In⁻. There is a smooth, continuous change in color going from pH 6.0 (yellow) to pH 8.0 (blue). In other words, an indicator, like a buffer, has an effective range of about two pH units.

$$\text{pH range} = \text{p}K_a \text{ of HIn} \pm 1$$

— an indicator, like a buffer, is essentially useless outside its pH range. At any pH less than 6.0, bromthymol blue is all in the form of the weak acid HIn, which is yellow. Changing the pH from, let us say, 0 to 6 has no effect on the color.

---

**Example 14.4**　Consider bromthymol blue ($K_a = 1 \times 10^{-7}$). At pH 7.5,

(a) calculate the ratio [In⁻]/[HIn]
(b) calculate the fractions of In⁻ and HIn [$f$ (In⁻) and $f$ (HIn)]
(c) What is the color of the indicator at this point?

*Strategy*　In (a), use the general relation: $K_a$ of HIn $= [H^+] \times [In^-]/[HIn]$ to find the required ratio. In (b), use the defining relations given on p. 404 to find the fractions of the two species. Finally, in (c), use Figure 14.5 to estimate the color at this point.

**Figure 14.5**
Distribution curve for the indicator bromthymol blue, which is yellow below pH 6, blue above pH 8, and green at pH = 6 = p$K_a$ of the indicator.

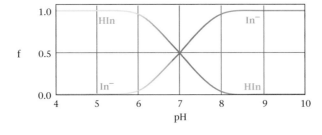

*Solution*

(a) $[In^-]/[HIn] = K_a/[H^+]$

At pH 7.5, $[H^+] = 3 \times 10^{-8}$ *M*, so we have

$$[In^-]/[HIn] = (1 \times 10^{-7})/(3 \times 10^{-8}) = 3 \quad \text{or} \quad \boxed{[In^-] = 3[HIn]}$$

(b) $f(In^-) = \dfrac{[In^-]}{[In^-] + [HIn]}$

Substituting $[In^-] = 3[HIn]$,

$$f(In^-) = \frac{3[HIn]}{3[HIn] + [HIn]} = \frac{3}{4} = 0.75$$

$$f(HIn) = 1 - 0.75 = 0.25$$

Looking at Figure 14.5, it does indeed appear that these   two fractions

are about $\frac{3}{4}$ and $\frac{1}{4}$ at pH 7.5.

(c) At pH 7.5, we are about halfway from 7.0 (green) to the upper end of the range at

pH 8.0 (blue). The color should be   blue-green.   This is reasonable; the color of

the $In^-$ ion should predominate, since there is three times as much of it as HIn.

## 14.3 Acid-Base Titrations

In Chapter 4, we considered the general topic of acid-base titrations. In carry-ing out an acid-base titration, you measure the volume of a standard acidic or basic solution required to just react with a known volume of a base or acid of unknown concentration. The point at which the reaction is complete is referred to as the **equivalence point;** equivalent quantities of acid and base have re-acted. This point is detected using an acid-base indicator. It is important that the point at which the indicator changes color (the **end point**) coincides with the equivalence point.

The discussion in Chapter 4 focused upon the stoichiometry of acid-base titrations. Here, the emphasis is on the equilibrium principles that apply to the acid-base reactions involved. It is convenient to distinguish between titrations involving

— a strong acid (e.g., HCl) and a strong base (e.g., NaOH, $Ba(OH)_2$)
— a weak acid (e.g., $HC_2H_3O_2$) and a strong base (e.g., NaOH)
— a strong acid (e.g. HCl) and a weak base (e.g., $NH_3$)

### Strong Acid–Strong Base

As pointed out in Chapter 13, strong acids ionize completely in water to form $H_3O^+$ ions; strong bases dissolve in water to form $OH^-$ ions. The neutralization reaction that takes place when *any* strong acid reacts with *any* strong base can be represented by a net ionic equation of the Brønsted-Lowry type:

$$H_3O^+(aq) + OH^-(aq) \longrightarrow 2H_2O$$

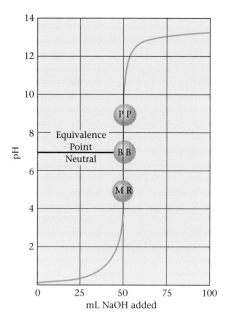

**Figure 14.6**
Titration of 50.00 mL of 1.000 M HCl with 1.000 M NaOH. The solution at the equivalence point is neutral (pH = 7). The pH rises so rapidly near the equivalence point that any of the three indicators methyl red (MR, end point pH = 5), bromthymol blue (BB, end point pH = 7), or phenolphthalein (PP, end point pH = 9) can be used.

or, more simply, substituting $H^+$ ions for $H_3O^+$ ions, as

$$H^+(aq) + OH^-(aq) \longrightarrow H_2O$$

Since the equation just written is the reverse of that for the ionization of water, the equilibrium constant can be calculated by using the reciprocal rule (Chap. 12).

$$K = 1/K_w = 1/(1.0 \times 10^{-14}) = 1.0 \times 10^{14}$$

*K must be large for a titration to work*

The enormous value of $K$ means that for all practical purposes this reaction goes to completion, consuming the limiting reactant, $H^+$ or $OH^-$.

Consider now what happens when HCl, a typical strong acid, is titrated with NaOH. Figure 14.6 shows how the pH changes during the titration. Two features of this curve are of particular importance:

1. At the equivalence point, when all the HCl has been neutralized by NaOH, a solution of NaCl, a neutral salt, is present. The pH at the equivalence point is 7.
2. Near the equivalence point, the pH rises very rapidly. Indeed, the pH may increase by as much as six units (from 4 to 10) when half a drop, $\approx 0.02$ mL, of NaOH is added (Example 14.5).

**Example 14.5**     When 50.00 mL of 1.000 M HCl is titrated with 1.000 M NaOH, the pH increases. Find the pH of the solution after the following volumes of 1.000 M NaOH have been added.

(a) 49.99 mL     (b) 50.01 mL

***Strategy***     This is essentially a stoichiometry problem of the type discussed in Chapter 4. The reaction is $H^+(aq) + OH^-(aq) \longrightarrow H_2O$. The number of moles of $H^+$ in 50.00 mL of 1.000 M HCl is

$$n_{H^+} = 50.00 \times 10^{-3}\ L \times \frac{1.000\ mol\ H^+}{1\ L} = 50.00 \times 10^{-3}\ mol\ of\ H^+$$

In each part of the problem, you start (1) by calculating the number of moles of $OH^-$ added. Then (2), find the number of moles of $H^+$ or $OH^-$ in excess. Finally (3), calculate $[H^+]$ and the pH.

### Solution

(a) (1) $n_{OH^-} = 49.99 \times 10^{-3} \text{ L} \times \dfrac{1.000 \text{ mol } OH^-}{1 \text{ L}} = 49.99 \times 10^{-3} \text{ mol of } OH^-$

(2) Since $H^+$ and $OH^-$ react in a 1:1 mole ratio,

$n_{H^+} \text{ reacted} = 49.99 \times 10^{-3} \text{ mol}$

$n_{H^+} \text{ excess} = 50.00 \times 10^{-3} \text{ mol} - 49.99 \times 10^{-3} \text{ mol}$

$= 0.01 \times 10^{-3} \text{ mol} = 1 \times 10^{-5} \text{ mol}$

(3) The total volume is almost exactly 100 mL:

$[H^+] = \dfrac{1 \times 10^{-5} \text{ mol } H^+}{1 \times 10^{-1} \text{ L}} = 1 \times 10^{-4} M$

$pH = -\log_{10}[H^+] = \boxed{4.0}$

Titrating a solution of HCl with NaOH using phenolphthalein as the indicator.

(b) (1) $n_{OH^-} = 50.01 \times 10^{-3} \text{ L} \times \dfrac{1.000 \text{ mol } OH^-}{1 \text{ L}} = 50.01 \times 10^{-3} \text{ mol of } OH^-$

(2) $n_{H^+} \text{ reacted} = 50.00 \times 10^{-3} \text{ mol}$

$n_{OH^-} \text{ excess} = 50.01 \times 10^{-3} \text{ mol} - 50.00 \times 10^{-3} \text{ mol}$

$= 0.01 \times 10^{-3} \text{ mol} = 1 \times 10^{-5} \text{ mol}$

(3) $[OH^-] = \dfrac{1 \times 10^{-5} \text{ mol } OH^-}{1 \times 10^{-1} \text{ L}} = 1 \times 10^{-4} M$

$pOH = -\log_{10}[OH^-] = 4.0$

$pH = 14.00 - pOH = \boxed{10.0}$

Note that the pH changes from 4 to 10 upon adding $(50.01 - 49.99)$mL = 0.02 mL of titrant.

*pH changes very rapidly near the end point*

From Figure 14.6 and Example 14.5, we conclude that any indicator that changes color between pH 4 and 10 should be satisfactory for a strong acid–strong base titration. Bromthymol blue (end point pH = 7) would work very well, but so would methyl red (end point pH = 5) or phenolphthalein (end point pH = 9).

## Weak Acid–Strong Base

A typical weak acid–strong base titration is that of acetic acid with sodium hydroxide. The net ionic equation for the reaction is

$$HC_2H_3O_2(aq) + OH^-(aq) \longrightarrow C_2H_3O_2^-(aq) + H_2O$$

Notice that this reaction is the reverse of the reaction of the weak base $C_2H_3O_2^-$ (the acetate ion) with water (Chap. 13). It follows from the reciprocal rule that for this reaction,

$$K = \frac{1}{(K_b \; C_2H_3O_2^-)} = \frac{1}{(5.6 \times 10^{-10})} = 1.8 \times 10^9$$

**Figure 14.7**

Titration of 50.00 mL of the weak acid $HC_2H_3O_2$ (1.000 $M$) with 1.000 $M$ NaOH. The solution at the equivalence point is basic (pH = 9.22). Phenolphthalein is a suitable indicator. Methyl red would change color much too early, when only about 33 mL of NaOH had been added. Bromthymol blue would change color slightly too early.

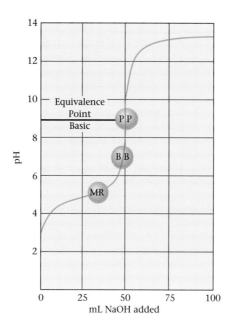

For a titration to work, the reaction must go to completion, which requires $K > 10^4$ (approximately)

Here again, $K$ is a very large number; the reaction of acetic acid with a strong base goes essentially to completion.

Figure 14.7 shows how the pH changes as 50 mL of 1 $M$ $HC_2H_3O_2$ is titrated with 1 $M$ NaOH. Notice that

— the pH starts off above 2; we are dealing with a weak acid, $HC_2H_3O_2$.

— there is a region, centered at the halfway point of the titration, where pH changes very slowly. Here the solution contains equal amounts of unreacted $HC_2H_3O_2$ and its conjugate base, $C_2H_3O_2^-$, formed by reaction with NaOH. As pointed out earlier, such a system acts as a buffer.

— at the equivalence point, there is a solution of sodium acetate, $NaC_2H_3O_2$. Since the acetate ion is a weak base, the pH at the equivalence point must be greater than 7.

**Example 14.6** 50.00 mL of 1.000 $M$ acetic acid, $HC_2H_3O_2$, is titrated with 1.000 $M$ NaOH. Find the pH of the solution after the following volumes of 1.000 $M$ NaOH have been added.

    (a) 0.00 mL        (b) 25.00 mL        (c) 50.00 mL

***Strategy*** In (a), before any base is added, we have a solution of 1 $M$ $HC_2H_3O_2$ ($K_a = 1.8 \times 10^{-5}$). Its pH is calculated like that of any weak acid; recall Example 13.7, p. 382. In (b), we are dealing with a buffer system; the pH is calculated as described in Section 14.1. In (c), we have a solution of a weak base, $C_2H_3O_2^-$ ($K_b = 5.6 \times 10^{-10}$); recall Example 13.10, p. 386.

***Solution***

(a) $[H^+]^2 \approx K_a HC_2H_3O_2 \times [HC_2H_3O_2] = 1.8 \times 10^{-5} \times 1.0 = 1.8 \times 10^{-5}$

$[H^+] = (1.8 \times 10^{-5})^{\frac{1}{2}} = 4.2 \times 10^{-3}$ $M$; pH = 2.38

(b) When 25.00 mL of NaOH has been added, we are halfway to the equivalence point (50.00 mL). It follows that half of the $HC_2H_3O_2$ molecules have been converted to $C_2H_3O_2^-$ ions, so

$$n_{HC_2H_3O_2} = n_{C_2H_3O_2^-}$$

Under these conditions, as pointed out in Section 14.1,

$$pH = pK_a \text{ of } HC_2H_3O_2 = \boxed{4.74}$$

(c) $[OH^-]^2 \approx K_b C_2H_3O_2^- \times [C_2H_3O_2^-]$

To find $[C_2H_3O_2^-]$, note that all of the acetic acid molecules originally present have been converted to acetate ions. The volume of the solution is twice as large as it was originally, 100.00 mL instead of 50.00 mL. It follows that the concentration of acetate ion ($n_{C_2H_3O_2^-}$)/$V$ solution) must be one half of the original concentration of $HC_2H_3O_2$. In other words,

$$[C_2H_3O_2^-] = 1.000 \ M/2 = 0.5000 \ M$$

Hence,

$$[OH^-]^2 = 5.6 \times 10^{-10} \times 0.5000 = 2.8 \times 10^{-10}$$

$$[OH^-] = (2.8 \times 10^{-10})^{\frac{1}{2}} = 1.7 \times 10^{-5} \ M$$

$$pOH = 4.78; \ pH = \boxed{9.22}$$

---

From Figure 14.7 and Example 14.6, it should be clear that the indicator used in this titration must change color at about pH 9. Phenolphthalein (end point pH = 9) is satisfactory. Methyl red (end point pH = 5) is not suitable. If we used methyl red, we would stop the titration much too early, when reaction is only about 65% complete. This situation is typical of *weak acid–strong base titrations*. For such a titration we choose an indicator, such as phenolphthalein, that *changes color above pH 7*.

*pH changes relatively slowly near the end point*

## Strong Acid–Weak Base

The reaction between solutions of hydrochloric acid and ammonia can be represented by the Brønsted-Lowry equation:

$$H_3O^+(aq) + NH_3(aq) \longrightarrow NH_4^+(aq) + H_2O$$

Again, we simplify the equation by replacing $H_3O^+$ with $H^+$:

$$H^+(aq) + NH_3(aq) \longrightarrow NH_4^+(aq)$$

To find the equilibrium constant, note that this equation is the reverse of that for the acid ionization of the $NH_4^+$ ion. Hence

$$K = \frac{1}{K_a NH_4^+} = \frac{1}{5.6 \times 10^{-10}} = 1.8 \times 10^9$$

Since $K$ is so large, the reaction goes virtually to completion.

Figure 14.8 p. 414 shows how pH changes when 50 mL of 1 $M$ $NH_3$ is titrated with 1 $M$ HCl. In many ways, this curve is the "inverse" of that shown in Figure 14.7 for the weak acid–strong base case. In particular

— the original pH is that of 1 $M$ $NH_3$, a weak base ($K_b = 1.8 \times 10^{-5}$). As you would expect, it lies between 7 and 14:

$$[OH^-]^2 \approx 1.8 \times 10^{-5}; \ [OH^-] = 4.2 \times 10^{-3} \ M; \ pOH = 2.38$$

$$pH = 11.62$$

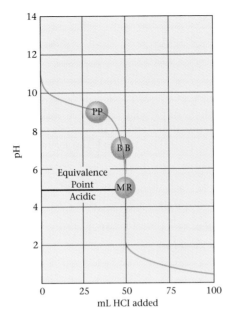

**Figure 14.8**
Titration of 50.00 mL of the weak base $NH_3$ (1.000 $M$) with 1.000 $M$ HCl. The solution at the equivalence point is acidic because of the $NH_4^+$ ion. Methyl red is a suitable indicator; phenolphthalein would change color much too early.

As strong acid is added, the pH drops

— addition of 25 mL of HCl gives a buffer containing equal amounts of $NH_3$ and $NH_4^+$ ($K_a = 5.6 \times 10^{-10}$). Hence

$$pH = pK_a \text{ of } NH_4^+ = 9.25$$

— at the equivalence point, there is a 0.5000 $M$ solution of $NH_4^+$, a weak acid. As you would expect, the pH is on the acid side.

$$[H^+]^2 \approx 0.50 \times 5.6 \times 10^{-10}; \ [H^+] = 1.7 \times 10^{-5} \ M$$

$$pH = 4.77$$

From Figure 14.8, it should be clear that the only suitable indicator listed is methyl red. The other two indicators would change color too early, before the equivalence point. In general, *for the titration of a weak base with a strong acid, the indicator should change color on the acid side of pH 7.*

## Summary

Table 14.3 summarizes our discussion of acid-base titrations. Notice that for these three types of titrations

— *the equations (second column) written to describe the reactions are quite different.* Strong acids and bases are represented by $H^+$ and $OH^-$ ions, respectively; weak acids and weak bases are represented by their chemical formulas.
— *the equilibrium constants (K) for all these reactions are very large, indicating that the reactions go essentially to completion.*
— *the pH at the equivalence point is determined by what species are present.* When only "spectator ions" are present, as in a strong acid–strong base titration, the pH at the equivalence point is 7. Basic anions ($F^-$, $C_2H_3O_2^-$), formed during a weak acid–strong base titration, make the pH at the equivalence

## TABLE 14.3   Characteristics of Acid-Base Titrations

### Strong Acid–Strong Base

| Example | Equation | $K$ | Species Equiv. Pt. | pH Equiv. Pt. | Indicator* |
|---------|----------|-----|--------------------|---------------|------------|
| NaOH–HCl | $H^+(aq) + OH^-(aq) \longrightarrow H_2O$ | $K = 1/K_w$ $1.0 \times 10^{14}$ | $Na^+$, $Cl^-$ | 7.00 | MR, BB, PP |
| Ba(OH)$_2$–HNO$_3$ | $H^+(aq) + OH^-(aq) \longrightarrow H_2O$ | $1.0 \times 10^{14}$ | $Ba^{2+}$, $NO_3^-$ | 7.00 | MR, BB, PP |

### Weak Acid–Strong Base

| Example | Equation | $K$ | Species Equiv. Pt. | pH Equiv. Pt. | Indicator* |
|---------|----------|-----|--------------------|---------------|------------|
| HC$_2$H$_3$O$_2$–NaOH | $HC_2H_3O_2(aq) + OH^-(aq) \longrightarrow$ $C_2H_3O_2^-(aq) + H_2O$ | $K = 1/K_b$ $1.8 \times 10^9$ | $Na^+$, $C_2H_3O_2^-$ | 9.22† | PP |
| HF–KOH | $HF(aq) + OH^-(aq) \longrightarrow$ $F^-(aq) + H_2O$ | $6.9 \times 10^{10}$ | $K^+$, $F^-$ | 9.42† | PP |

### Strong Acid–Weak Base

| Example | Equation | $K$ | Species Equiv. Pt. | pH Equiv. Pt. | Indicator* |
|---------|----------|-----|--------------------|---------------|------------|
| NH$_3$–HCl | $NH_3(aq) + H^+(aq) \longrightarrow NH_4^+(aq)$ | $K = 1/K_a$ $1.8 \times 10^9$ | $NH_4^+$, $Cl^-$ | 4.78† | MR |
| ClO$^-$–HCl | $ClO^-(aq) + H^+(aq) \longrightarrow HClO(aq)$ | $3.6 \times 10^7$ | HClO, $Cl^-$ | 3.93† | MR |

*MR = methyl red (end point pH = 5); BB = bromthymol blue (end point pH = 7); PP = phenolphthalein (end point pH = 9).

†When 1 $M$ acid is titrated with 1 $M$ base.

point greater than 7. Conversely, acidic species ($NH_4^+$, HClO) formed during a strong acid–weak base titration make the pH at the equivalence point less than 7.

**Example 14.7**   Consider the titration of nitrous acid, $HNO_2$, with barium hydroxide.

(a) Write a balanced net ionic equation for the reaction.
(b) Calculate $K$ for the reaction.
(c) Is the solution at the equivalence point acidic, basic, or neutral?
(d) What would be an appropriate indicator for the titration?

*Strategy*   Once you realize that this is a weak acid–strong base titration, the problem unravels; follow the rules cited in Table 14.3.

*Solution*

(a)  $HNO_2(aq) + OH^-(aq) \longrightarrow NO_2^-(aq) + H_2O$

(b) $K = 1/K_b$ $NO_2^-$. From Table 13.4, p. 386, $K_b$ $NO_2^- = 1.7 \times 10^{-11}$.

   $K = 1/(1.7 \times 10^{-11}) = $  $5.9 \times 10^{10}$

(c) At the equivalence point we have a solution of NaNO$_2$, which is  basic  due to the $NO_2^-$ ion.

(d)  phenolphthalein

## 14.4 Polyprotic Weak Acids

Certain weak acids are *polyprotic;* they contain more than one ionizable hydrogen atom. Such acids ionize in steps, with a separate ionization constant for each step. Oxalic acid, a weak organic acid sometimes used to remove bloodstains, is *diprotic:*

$$H_2C_2O_4(aq) \rightleftharpoons H^+(aq) + HC_2O_4^-(aq) \qquad K_{a1} = 5.9 \times 10^{-2}$$

$$HC_2O_4^-(aq) \rightleftharpoons H^+(aq) + C_2O_4^{2-}(aq) \qquad K_{a2} = 5.2 \times 10^{-5}$$

Phosphoric acid, a common ingredient of cola drinks, is *triprotic:*

$$H_3PO_4(aq) \rightleftharpoons H^+(aq) + H_2PO_4^-(aq) \qquad K_{a1} = 7.1 \times 10^{-3}$$

$$H_2PO_4^-(aq) \rightleftharpoons H^+(aq) + HPO_4^{2-}(aq) \qquad K_{a2} = 6.2 \times 10^{-8}$$

$$HPO_4^{2-}(aq) \rightleftharpoons H^+(aq) + PO_4^{3-}(aq) \qquad K_{a3} = 4.5 \times 10^{-13}$$

The behavior of these acids is typical of that of all polyprotic acids in that

— *the anion formed in one step* (e.g., $HC_2O_4^-$, $H_2PO_4^-$) *ionizes in the next step*
— *the ionization constant becomes smaller with each successive step.*

$$K_{a1} > K_{a2} > K_{a3}$$

Putting it another way, **the acids in successive steps become progressively weaker.** This is reasonable; it should be more difficult to remove a positively charged proton, $H^+$, from a negatively charged species like $H_2PO_4^-$ than from a neutral molecule like $H_3PO_4$.

### Calculation of [H⁺] in a Water Solution of the Acid

Ordinarily, successive ionization constants of polyprotic acids decrease by a factor of at least 100 (Table 14.4). In that case, **essentially all the hydrogen ions in the solution come from the first ionization of the acid.** This makes it relatively easy to calculate the pH of a solution of a polyprotic acid.

TABLE 14.4 **Ionization Constants for Some Polyprotic Acids at 25°C**

| Acid | Formula | $K_{a1}$ | $K_{a2}$ | $K_{a3}$ |
|------|---------|----------|----------|----------|
| Sulfuric acid | $H_2SO_4$ | ∞ | $1.0 \times 10^{-2}$ | |
| Carbonic acid* | $H_2CO_3$ | $4.4 \times 10^{-7}$ | $4.7 \times 10^{-11}$ | |
| Citric acid | $H_3C_6H_5O_7$ | $8.7 \times 10^{-4}$ | $1.8 \times 10^{-5}$ | $4.0 \times 10^{-7}$ |
| Oxalic acid | $H_2C_2O_4$ | $5.9 \times 10^{-2}$ | $5.2 \times 10^{-5}$ | |
| Phosphoric acid | $H_3PO_4$ | $7.1 \times 10^{-3}$ | $6.2 \times 10^{-8}$ | $4.5 \times 10^{-13}$ |
| Sulfurous acid | $H_2SO_3$ | $1.7 \times 10^{-2}$ | $6.0 \times 10^{-8}$ | |

*Carbonic acid is a water solution of carbon dioxide:

$$CO_2(g) + H_2O \rightleftharpoons H_2CO_3(aq)$$

The ionization constants listed here are calculated assuming that all the carbon dioxide that dissolves is in the form of $H_2CO_3$.

**Example 14.8**  The distilled water you use in the laboratory is slightly acidic because of dissolved $CO_2$, which reacts to form carbonic acid, $H_2CO_3$. Calculate the pH of a 0.0010 $M$ solution of $H_2CO_3$.

***Strategy***  In principle, there are two different sources of $H^+$ ions from $H_2CO_3$:

$$H_2CO_3(aq) \rightleftharpoons H^+(aq) + HCO_3^-(aq) \qquad K_{a1} = 4.4 \times 10^{-7}$$

$$HCO_3^-(aq) \rightleftharpoons H^+(aq) + CO_3^{2-}(aq) \qquad K_{a2} = 4.7 \times 10^{-11}$$

In practice, essentially all the $H^+$ ions come from the first reaction, since $K_{a1}$ is so much larger than $K_{a2}$. In other words, $H_2CO_3$ can be treated as if it were a weak monoprotic acid.

***Solution***  For the first reaction, $H_2CO_3(aq) \rightleftharpoons H^+(aq) + HCO_3^-(aq)$

$$K_{a1} = \frac{x^2}{0.0010 - x} = 4.4 \times 10^{-7} \qquad (x = [H^+] = [HCO_3^-])$$

*[H$^+$] is calculated for H$_2$CO$_3$, using $K_{a1}$, exactly as it is with HF, using $K_a$*

Making the approximation $0.0010 - x \approx 0.0010$ and solving gives

$$x = (4.4 \times 10^{-10})^{1/2} = 2.1 \times 10^{-5}$$

$$pH = -\log_{10}(2.1 \times 10^{-5}) = 4.68$$

Note that the solution is indeed acidic, with a pH considerably less than 7.
Notice that since:

$$[H^+] = [HCO_3^-] \text{ and } \frac{[H^+] \times [CO_3^{2-}]}{[HCO_3^-]} = K_{a2} = 4.7 \times 10^{-11}$$

It follows that $[CO_3^{2-}] = K_{a2} = 4.7 \times 10^{-11}$

---

The conclusions reached in Example 14.8 are ordinarily valid for any weak diprotic acid, $H_2B$.

— $[H^+]$ is calculated from $K_{a1}$; $[H^+] \approx (K_{a1} \times [H_2B])^{1/2}$
— $[HB^-] \approx [H^+]$
— $[B^{2-}] \approx K_{a2}$

## Distribution Curves

The distribution curve for a polyprotic weak acid resembles that for a monoprotic acid, except that more species are involved. For a weak acid $H_2B$ there are three species: $H_2B$, $HB^-$ and $B^{2-}$. Figure 14.9 shows the distribution curve for the carbonate system, in which the species are $H_2CO_3$, $HCO_3^-$, and $CO_3^{2-}$.

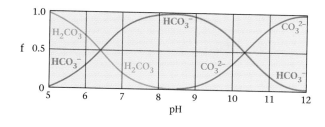

**Figure 14.9**
In this system: (1) $H_2CO_3$ is the major species below pH 6.4 (2) $HCO_3^-$ is the major species between pH 6.4 and 10.3 (3) $CO_3^{2-}$ is the major species above pH 10.3.

Notice that

— there are two buffer systems:

$$H_2CO_3/HCO_3^-, \text{ centered at } pK_{a1} \text{ of } H_2CO_3 = 6.4$$

$$HCO_3^-/CO_3^{2-}, \text{ centered at } pK_{a2} \text{ of } H_2CO_3 = 10.3$$

— the predominant species at various pHs are

$$H_2CO_3 \text{ at low pH (below about 6)}$$

$$HCO_3^- \text{ at intermediate pH (7–10)}$$

$$CO_3^{2-} \text{ at high pH ( >11)}$$

*Adding acid to a solution of NaHCO3 produces H2CO3 (and CO2) when the pH drops below 6*

One other point that may not be obvious from the figure: The pH of a NaHCO$_3$ solution is about 8.3, where [HCO$_3^-$] has its highest value.

One practical application of Figure 14.9 involves predicting the composition of the solid that separates when a water solution is cooled or evaporated. Around pH 8 we would expect to get mostly HCO$_3^-$ ions in the solid (e.g., NaHCO$_3$); above pH 11, CO$_3^{2-}$ ions should predominate (e.g., Na$_2$CO$_3$). The most important carbonate ore is trona, found in dried-up lake beds in the Mojave Desert of California. This compound has the composition NaHCO$_3 \cdot$ 2Na$_2$CO$_3 \cdot$ 2H$_2$O. Looking at Figure 14.9, one would guess that trona must have crystallized out of solution at about pH 10.5, at which point there are roughly two CO$_3^{2-}$ ions for every HCO$_3^-$ ion.

## CHEMISTRY
### *Beyond the Classroom*

## pH Control in the Body

Blood, like many natural fluids, is buffered. Indeed, there are three different buffer systems in blood plasma that hold the pH at about 7.40. By far the most important of these involves H$_2$CO$_3$ (an aqueous solution of CO$_2$), and its conjugate base, the HCO$_3^-$ ion. This system consumes H$^+$ or OH$^-$ ions by the reactions

$$HCO_3^-(aq) + H^+(aq) \longrightarrow H_2CO_3(aq)$$

$$H_2CO_3(aq) + OH^-(aq) \longrightarrow HCO_3^-(aq) + H_2O$$

Since the pH of blood is 7.40, we have

$$[H^+] = 10^{-7.40} = 4.0 \times 10^{-8} \ M$$

The ionization constant of carbonic acid is $4.4 \times 10^{-7}$; it follows that the concentration of HCO$_3^-$ in the blood is more than ten times that of H$_2$CO$_3$:

$$\frac{[HCO_3^-]}{[H_2CO_3]} = \frac{K_a H_2CO_3}{[H^+]} = \frac{4.4 \times 10^{-7}}{4.0 \times 10^{-8}} = 11$$

Consequently, blood has a much greater capacity for absorbing H$^+$ ions than for OH$^-$ ions. It needs this capacity; life processes tend to produce H$^+$ ions rather than OH$^-$ ions. For example, heavy exercise produces lactic acid, a weak organic acid.

If the pH of the blood drops significantly below 7.4, a condition called *acidosis* is created. The nervous system is depressed; fainting and even coma can result. Most often, acidosis is caused by a buildup of carbon dioxide concentration in the blood. You can produce a very mild case of acidosis by holding your breath. A more serious case can be the result of lung diseases such as asthma or emphysema. Diabetics are also susceptible to acidosis because their metabolic processes form organic acids such as

$$CH_3{-}\underset{\underset{O}{\|}}{C}{-}CH_2{-}COOH \qquad \text{acetoacetic acid}$$

If acetoacetic acid is produced in amounts greater than the $HCO_3^-$ ions in the blood can react with, a diabetic coma can result.

*Alkalosis,* which results when the pH of the blood rises significantly above its normal value of 7.4, is much less common than acidosis (fortunately, since blood has a relatively small capacity to absorb $OH^-$ ions). The symptoms of alkalosis are the inverse of those for acidosis; the nervous system is overstimulated, leading to muscle cramps and ultimately convulsions. Most often, alkalosis is caused by rapid or heavy breathing (hyperventilation). This can result from fever, infection, or the action of certain drugs. When a person breathes too deeply, carbon dioxide is expelled in large quantities from the bloodstream, and the pH rises.

# CHAPTER HIGHLIGHTS

## *Key Concepts*

**1.** Calculate the pH of a buffer, before or after adding $H^+$ or $OH^-$
(Examples 14.1, 14.3; Problems 9–12, 15–28, 31–34)
**2.** Choose a buffer to get a specified pH
(Example 14.2; Problems 13, 14, 29, 30)
**3.** Determine the color of an acid-base indicator at a given pH
(Example 14.4; Problems 39, 40)
**4.** Calculate the pH during an acid-base titration
(Examples 14.5, 14.6; Problems 41–46, 59, 64)
**5.** Choose the proper indicator for an acid-base titration
(Example 14.7; Problems 37, 38)
**6.** Calculate $K$ for a given acid-base reaction
(Example 14.7; Problems 5–8)
**7.** Calculate $[H^+]$ and the concentrations of other species in a solution of a weak polyprotic acid
(Example 14.8; Problems 47–52, 67)

## *Key Equations*

The key equation, used over and over again in this chapter, in one form or another, is

$$K_a = \frac{[H^+] \times [B^-]}{[HB]} = [H^+] \times \frac{n_{B^-}}{n_{HB}}$$

# Key Terms

| | | |
|---|---|---|
| acid | buffer | equivalence point |
| —strong | conjugate acid | indicator |
| —weak | conjugate base | pH |
| base | ionization constant | polyprotic acid |
| —strong | distribution curve | titration |
| —weak | | |

# Summary Problem

Consider formic acid, $HCHO_2$, ($K_a = 1.9 \times 10x^{-4}$) and its conjugate base, $CHO_2^-$ ($K_b = 5.3 \times 10^{-11}$)

(a) Write net ionic equations and calculate K for the reactions between
   (1) $HCHO_2$ and NaOH
   (2) $CHO_2^-$ and HCl
   (3) $HCHO_2$ and NaCN ($K_b = 1.7 \times 10^{-5}$)
(b) A buffer is prepared by dissolving 30.0 g of sodium formate, $NaCHO_2$, in one liter of a solution of 0.500 M $HCHO_2$.
   (1) Calculate the pH of the buffer.
   (2) What is the range (in pH units) of this buffer system?
   (3) Calculate the pH of the buffer after 0.100 mol of $HNO_3$ is added.
   (4) Calculate the pH of the buffer after 10.0 g of KOH is added.
   (5) Calculate the pH of the buffer after 0.600 mol of $HNO_3$ is added.
(c) Fifty mL of 1.000 M $HCHO_2$ is titrated with 0.675 M KOH.
   (1) What is the pH of the $HCHO_2$ solution before titration?
   (2) How many mL of KOH are required to reach the equivalence point?
   (3) What is the pH of the solution halfway to neutralization?
   (4) What is the pH of the solution at the equivalence point?
   (5) What indicator could be used for this titration?

## Answers

(a) (1) $HCHO_2(aq) + OH^-(aq) \rightleftharpoons H_2O + CHO_2^-(aq)$      $K = 1.9 \times 10^{10}$
   (2) $CHO_2^-(aq) + H^+(aq) \rightleftharpoons HCHO_2(aq)$      $K = 5.3 \times 10^3$
   (3) $HCHO_2(aq) + CN^-(aq) \rightleftharpoons HCN(aq) + CHO_2^-(aq)$      $K = 3.3 \times 10^5$
(b) (1) 3.67
   (2) pH is between 2.7 and 4.7
   (3) 3.47
   (4) 4.01
   (5) 0.77

(c) (1) 1.86
   (2) 74.1 mL
   (3) 3.72
   (4) 8.66
   (5) phenolphthalein

# Questions & Problems

Equilibrium constants required to solve these problems can be found in the tables in Chapter 13 or in Appendix 1.

## Acid-Base Reactions

**\*1.** Write a net ionic equation for the reaction between solutions of
   **(a)** perchloric acid and barium hydroxide.
   **(b)** ammonia and nitric acid.
   **(c)** nitrous acid and lithium fluoride.
   **(d)** calcium hydroxide and aqueous hydrogen cyanide (hydrocyanic acid).
**\*2.** Follow the direction of Question 1 for
   **(a)** aniline, $C_6H_5NH_2$, and ammonium chloride.
   **(b)** hydrobromic acid and barium hydroxide.
   **(c)** ammonia and hydriodic acid.
   **(d)** hydrochloric acid and strontium hydroxide.

**\*3.** Write a balanced net ionic equation for the reaction of each of the following aqueous solutions with $H^+$ ions.
   **(a)** sodium acetate
   **(b)** cesium hydroxide
   **(c)** potassium sulfite

**\*4.** Write a balanced net ionic equation for the reaction of each of the following aqueous solutions with $OH^-$ ions.
   **(a)** $Fe(H_2O)_6^{3+}$
   **(b)** sulfurous acid
   **(c)** sulfuric acid

**5.** Calculate $K$ for the reactions given in Question 1.

**6.** Calculate $K$ for the reactions given in Question 2.

**7.** Calculate $K$ for the reactions given in Question 3.

**8.** Calculate $K$ for the reactions given in Question 4.

## Buffers

**9.** Calculate $[H^+]$ and pH in a solution in which $[C_2H_3O_2^-]$ is 0.500 $M$ and $[HC_2H_3O_2]$ is
   **(a)** 0.100 $M$     **(b)** 0.2500 $M$
   **(c)** 0.500 $M$     **(d)** 0.750 $M$

**10.** Calculate $[OH^-]$ and pH in a solution in which $[CH_3NH_2]$ is 0.300 $M$ and $[CH_3NH_3^+]$ is
   **(a)** 0.0500 $M$     **(b)** 0.200 $M$
   **(c)** 0.300 $M$     **(d)** 0.500 $M$

**11.** A buffer is prepared by dissolving 0.250 mol of sodium benzoate, $NaC_7H_5O_2$, in 75.0 mL of 0.310 $M$ benzoic acid, $HC_7H_5O_2$. Calculate the pH of this buffer.

**12.** A buffer is prepared by dissolving 0.035 mol of methylamine, $CH_3NH_2$, in 1.32 L of 0.025 $M$ methylammonium chloride, $CH_3NH_3Cl$. Calculate the pH of this buffer.

**\*13.** Consider the weak acids listed in Table 13.2. Which acid-base pair would be best for a buffer at a pH of
   **(a)** 4.2     **(b)** 6.3     **(c)** 9.3

**\*14.** Follow the instructions of Question 13 for a pH of
   **(a)** 3.5     **(b)** 9.0     **(c)** 4.7

**15.** To make a buffer using $HNO_2$ and $NO_2^-$ in which the desired pH = 3.50:
   **(a)** What must be the ratio $[HNO_2]/[NO_2^-]$?
   **(b)** How many moles of $HNO_2$ must be added to a liter of 0.42 $M$ $NaNO_2$ to give this pH?
   **(c)** How many grams of $NaNO_2$ must be added to 750.0 mL of 0.250 $M$ $HNO_2$ to give this pH?

**16.** A methylammonium chloride–methylamine buffer $(CH_3NH_3^+/CH_3NH_2)$ is to be prepared with a pH of 11.00.
   **(a)** What must be the ratio $[CH_3NH_3^+]/[CH_3NH_2]$?
   **(b)** How many moles of methylamine must be added to 0.500 L of 0.680 $M$ $CH_3NH_3^+$ to give this pH?
   **(c)** How many grams of methylamine must be added to 2.00 L of a solution that is 0.74 M in $CH_3NH_3^+$ to give a pH of 11.00?

**17.** A buffer solution is prepared by adding 10.00 g of sodium fluoride and 5.00 g of hydrogen fluoride to enough water to make 1.2 L of solution.

   **(a)** What is the pH of the buffer?
   **(b)** The buffer is diluted by adding enough water to make 2.0 L of solution. What is the pH of the diluted buffer?

**18.** A buffer solution is prepared by adding 50.0 g of both $Na_3PO_4$ and $Na_2HPO_4$ to enough water to make $5.00 \times 10^2$ mL of solution.
   **(a)** What is the pH of the buffer?
   **(b)** If enough water is added to double the volume, what is the pH of the solution?
   **(c)** If 50.00 more grams of both $Na_3PO_4$ and $Na_2HPO_4$ are added, does the pH change?

**19.** A student wants to make an acetic acid–acetate buffer with pH 4.25. She has 0.500 L of vinegar (5.00% by mass acetic acid, $d = 1.006$ g/mL) and potassium acetate on hand. How many grams of potassium acetate are necessary to produce the desired buffer?

**20.** How many grams of ammonium chloride should be added to 475 mL of 0.225 $M$ aqueous ammonia solution to produce a buffer with pH 9.80?

**21.** A 0.171 $M$ solution of a weak acid, HX, has a pH of 4.29. What is the pH of the solution after 0.200 mol of solid KX has been dissolved per liter of solution?

**22.** When 0.416 mol of solid NaX is dissolved in water, 2.00 L of solution with a pH of 8.92 is obtained. What is the pH of the solution after 0.365 mol of the weak acid HX is added?

**23.** A buffer is made up of 1.00 L each of 0.15 $M$ $NaHCO_3$ and 0.20 $M$ $Na_2CO_3$. Calculate
   **(a)** the pH of the buffer.
   **(b)** the pH of the buffer after the addition of 0.0300 mol of HCl.
   **(c)** the pH of the buffer after the addition of 0.0300 mol of KOH per liter.

**24.** A buffer is made up of 0.500 L each of 0.300 $M$ $KH_2PO_4$ and 0.400 $M$ $K_2HPO_4$. Calculate
   **(a)** the pH of the buffer.
   **(b)** the pH of the buffer after the addition of 0.0500 mole of HCl.
   **(c)** the pH of the buffer after the addition of 0.0500 mole of NaOH.

**25.** If enough water is added to the buffer in Question 23 to make the total volume 10.0 L, what do the answers in (a), (b) and (c) become? What conclusions can you reach about the dilution of a buffer?

**26.** Consider the buffer prepared in Question 24. Enough water is added to the buffer to make the total volume 10.0 L. Work questions a through c. What conclusions can you reach about the dilution of a buffer?

**27.** A buffer is prepared in which the ratio $[HCO_3^-]/[CO_3^{2-}]$ is 2.0.
   **(a)** What is the pH of this buffer? ($K_a$ $HCO_3^- = 4.7 \times 10^{-11}$)
   **(b)** Enough strong acid is added to make the pH of the buffer 9.50. What is the ratio $[HCO_3^-]/[CO_3^{2-}]$ at this point?

**28.** A buffer is prepared by using the conjugate acid–base pair $NH_4^+$–$NH_3$. The pH of the prepared buffer is 10.00.

    **(a)** What is the ratio $[NH_4^+]/[NH_3]$?

    **(b)** What does the pH become if 25% of the $NH_4^+$ ions are converted to ammonia?

    **(c)** What does the pH become if 15% of the ammonia molecules are converted to ammonium ions?

**\*29.** Which of the following would form a buffer if added to one liter of 0.30 *M* $Ba(OH)_2$?

    **(a)** 0.30 mol of $HNO_2$     **(b)** 0.60 mol of $HNO_2$

    **(c)** 0.80 mol of $HNO_2$     **(d)** 0.30 mol of $NaNO_2$

    **(e)** 0.30 mol of HCl

    Explain your reasoning in each case.

**\*30.** Which of the following would form a buffer if added to one liter of 0.30 *M* $NH_3$?

    **(a)** 0.30 mol of HCl     **(b)** 0.10 mol of HCl

    **(c)** 0.10 mol of $NH_4Cl$     **(d)** 0.50 mol of HCl

    **(e)** 0.30 mol of NaOH

    Explain your reasoning in each case.

**31.** Calculate the pH of a solution that is prepared by mixing 2.00 g of butyric acid ($HC_4H_7O_2$) with 0.50 g of NaOH in water ($K_a$ butyric acid $= 1.55 \times 10^{-5}$).

**32.** Calculate the pH of a solution prepared by mixing 100.0 mL of 1.20 *M* ethanolamine, $C_2H_5ONH_2$, with 50.0 mL of 1.0 *M* HCl. $K_a$ for $C_2H_5ONH_3^+$ is $3.61 \times 10^{-10}$.

**33.** Calculate the pH of a solution prepared by dissolving 10.00 g of lactic acid ($HC_3H_5O_3$) and 8.00 g of KOH in enough water to make 475 mL of solution.

**34.** Calculate the pH of a solution prepared by dissolving 5.00 g of NaCN in 250.0 mL of 1.50 *M* HCl. Assume that the volume of the solution is 250.0 mL.

**35.** There is a buffer system ($H_2PO_4^-/HPO_4^{2-}$) in blood that helps keep the blood pH at about 7.40. ($K_a$ $H_2PO_4^- = 6.2 \times 10^{-8}$)

    **(a)** Calculate the ratio of $H_2PO_4^-/HPO_4^{2-}$ at the normal pH of blood.

    **(b)** What percentage of the $HPO_4^{2-}$ ions are converted to $H_2PO_4^-$ when the pH goes down to 7.00?

    **(c)** What percentage of $H_2PO_4^-$ ions are converted to $HPO_4^{2-}$ when the pH goes up to 7.50?

**36.** Blood is buffered mainly by the $HCO_3^-/H_2CO_3$ system ($K_a$ $H_2CO_3 = 4.4 \times 10^{-7}$). The normal pH of blood is 7.40.

    **(a)** What is the ratio $[H_2CO_3/HCO_3^-]$

    **(b)** What does the pH become if 10% of the $HCO_3^-$ ions are converted to $H_2CO_3$?

    **(c)** What does the pH become if 10% of the $H_2CO_3$ molecules are converted to $HCO_3^-$ ions?

## Titrations and Indicators

**\*37.** Given three acid-base indicators: methyl orange (end point at pH 4), bromthymol blue (end point at pH 7), and phenolphthalein (end point at pH 9)—which would you select for the following acid-base titrations?

    **(a)** nitrous acid with lithium hydroxide

    **(b)** sodium fluoride with nitric acid

    **(c)** methylamine with hydrochloric acid

    **(d)** sulfuric acid with potassium hydroxide

**\*38.** Given the acid-base indicators in Question 37, select a suitable indicator for the following titrations:

    **(a)** perchloric acid with an aqueous solution of ammonia

    **(b)** hydrobromic acid with strontium hydroxide

    **(c)** sulfurous acid with sodium hydroxide

    **(d)** sodium acetate with hydrochloric acid

**39.** Methyl yellow is an indicator with $K_a = 3.56 \times 10^{-4}$. It is red in acid solution and yellow in alkaline solution. What is its pH range? What would its color be at its $pK_a$?

**40.** Alizarin yellow has a pH range of 10.1 to 12.1. What is its $K_a$? It is yellow at pH 10 and red at pH 12. What is the color of a solution with pH 10.9 that has a few drops of alizarin yellow?

**41.** Twenty-five mL of 0.3000 *M* sodium hydroxide is titrated with 0.5125 *M* nitric acid. Calculate

    **(a)** the volume of nitric acid required to reach the equivalence point.

    **(b)** the pH of the solution before any $HNO_3$ is added.

    **(c)** the pH of the solution halfway to the equivalence point.

    **(d)** the pH of the solution at the equivalence point.

    For parts (c) and (d), assume that volumes are additive.

**42.** Forty mL of 0.2500 *M* hydrochloric acid is titrated with 0.1216 *M* potassium hydroxide. Calculate

    **(a)** the volume of potassium hydroxide required to reach the equivalence point.

    **(b)** the pH of the solution before any KOH is added.

    **(c)** the pH of the solution halfway to the equivalence point.

    **(d)** the pH of the solution at the equivalence point.

    For parts (c) and (d), assume that volumes are additive.

**43.** Consider the titration of sodium acetate ($NaC_2H_3O_2$) with HCl. $K_b$ for $C_2H_3O_2^-$ is $5.6 \times 10^{-10}$. In the titration, 25.00 mL of 0.2000 *M* sodium acetate is titrated with 0.1500 *M* HCl.

    **(a)** Write a balanced net ionic equation for the reaction that takes place during titration.

    **(b)** What are the principal species present before titration begins, halfway to the equivalence point, and at the equivalence point?

    **(c)** What volume of HCl is required to reach the equivalence point?

    **(d)** What is the pH of the solution before titration begins, halfway to the equivalence point, and at the equivalence point? Assume that volumes are additive.

**44.** Consider the titration of formic acid ($HCHO_2$) with potassium hydroxide. $K_a$ for formic acid is $1.9 \times 10^{-4}$. For an experiment, 30.00 mL of 0.4500 *M* formic acid is titrated with 0.6216 *M* KOH.

    **(a)** Write a balanced net ionic equation for the reaction that takes place during titration.

    **(b)** What are the principal species present before titration begins, halfway to the equivalence point, and at the equivalence point?

**(c)** What volume of KOH is required to reach the equivalence point?

**(d)** What is the pH of the solution before titration begins, halfway to the equivalence point, and at the equivalence point? Assume that volumes are additive.

**45.** Five grams of sodium fluoride is dissolved in enough water to make 1.00 L of solution. Twenty-five mL of the prepared solution is titrated with 0.2500 $M$ HBr. Calculate the pH of the solution

   **(a)** before titration
   **(b)** halfway to the equivalence point
   **(c)** at the equivalence point

**46.** Two liters of ammonia at 25°C and 1.00 atm is bubbled into 1.00 L of water. Assume that all the ammonia dissolves and that the volume of the solution is the volume of the water. Fifty mL of the prepared solution is titrated with 0.0893 $M$ HNO$_3$. Calculate the pH of the solution

   **(a)** before titration
   **(b)** halfway to the equivalence point
   **(c)** at the equivalence point

## Polyprotic Acids

Use the $K_a$ values listed in Table 14.4 for polyprotic acids.

**47.** Write the overall chemical equation and calculate $K$ for the complete ionization of $H_3PO_4$.

**48.** Write the overall chemical equation and calculate $K$ for the complete ionization of oxalic acid, $H_2C_2O_4$.

**49.** Calculate the pH of 0.25 $M$ $H_2CO_3$. Estimate [HCO$_3^-$] and [CO$_3^{2-}$].

**50.** Calculate the pH of a 0.40 $M$ solution of citric acid, $H_3C_6H_5O_7$. Estimate [$H_2C_6H_5O_7^-$], [$HC_6H_5O_7^{2-}$], and [$C_6H_5O_7^{3-}$].

**51.** Using the distribution curve for $H_2CO_3$ shown in Figure 14.9, estimate the concentrations of $H_2CO_3$, HCO$_3^-$, and CO$_3^{2-}$ when the pH is 7.00. The total concentration of all three species is 0.200 $M$.

**52.** Using the distribution curve for $H_2CO_3$ shown in Figure 14.9, estimate the pH at which 2[CO$_3^{2-}$] = [HCO$_3^-$].

## Unclassified

**\*53.** Each symbol in the box below represents a mole of a component in one liter of a buffer solution; $\bigcirc$ represents the anion ($X^-$), $\Box\!\bigcirc$ = the weak acid, (HX), $\Box$ = H$^+$, and $\triangle$ = OH$^-$. Water molecules and the few H$^+$ and OH$^-$ from the ionization of HX and $X^-$ are not shown.

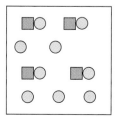

**(a)** Fill in a similar box (representing one liter of the same solution) after 2 moles of H$^+$ (2 $\Box$ ) have been added. Indicate whether the resulting solution is an acid, base, or buffer.

**(b)** Follow the directions of part (a) after 2 moles of OH$^-$ (2 $\triangle$) have been added.

**(c)** Follow the directions of part (a) after 5 moles of OH$^-$ (5 $\triangle$) have been added.

*Hint:* Write the reaction before you draw the results.

**\*54.** The box below contains 10 moles of a weak acid, $\Box\!\bigcirc$ , in a liter of solution. Using the same symbols as in Question 53 ($\bigcirc$ = anion,= $\triangle$ OH$^-$), show what happens upon

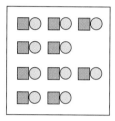

**(a)** the addition of 2 moles of OH$^-$ (2 $\triangle$ ).
**(b)** the addition of 5 moles of OH$^-$ (5 $\triangle$ ).
**(c)** the addition of 10 moles of OH$^-$ (10 $\triangle$ ).
**(d)** the addition of 12 moles of OH$^-$ (12 $\triangle$ ).
Which addition (a through d) represents neutralization halfway to the equivalence point?

**\*55.** The following is the titration curve for the titration of 50.00 mL of a 0.100 $M$ acid with 0.100 $M$ KOH.

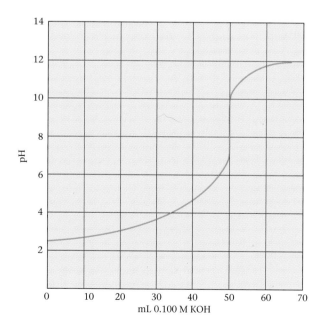

**(a)** Is the acid strong or weak?
**(b)** If the acid is weak, what is its $K_a$?
**(c)** Estimate the pH at the equivalence point.

*56. Carbonic acid, $H_2CO_3$, is a solution of $CO_2$ in water. Explain, using Le Chatelier's principle, the buildup of $CO_2$ in the blood when acidosis occurs.

*57. Explain why

(a) the pH decreases when lactic acid is added to a sodium lactate solution.

(b) the pH of 0.1 $M$ $NH_3$ is less than 13.0.

(c) a buffer resists changes in pH caused by the addition of $H^+$ or $OH^-$.

(d) a solution with a low pH is not necessarily a strong acid solution.

*58. Indicate whether each of the following statements is true or false. If the statement is false, restate it to make it true.

(a) The formate ion ($CHO_2^-$) concentration in 0.10 $M$ $HCHO_2$ is the same as in 0.10 $M$ $NaCHO_2$.

(b) A buffer can be destroyed by adding too much strong acid.

(c) A buffer can be made up by any combination of weak acid and weak base.

(d) Since $K_a$ for $HCO_3^-$ is $4.7 \times 10^{-11}$, $K_b$ for $HCO_3^-$ is $2.1 \times 10^{-4}$.

59. A solution contains 45.00 mL of 0.1250 $M$ HCl. Calculate the pH after 38.00 mL of 0.125 $M$ $Ba(OH)_2$ is added.

*60. Three test tubes labeled A, B, and C all have the same pH. The test tubes are known to contain $1.0 \times 10^{-3}$ $M$ HCl, $6.0 \times 10^{-3}$ $M$ $HCHO_2$, and $4 \times 10^{-2}$ $M$ $C_6H_5NH_3^+$. Describe a procedure for identifying the solutions.

## Challenge Problems

61. Glacial acetic acid is 98% acetic acid by mass ($d = 1.0542$ g/mL). What volume of glacial acetic acid must be added to 75.0 mL of 1.50 $M$ NaOH to give a buffer with a pH of 4.00?

62. Explain why it is not possible to prepare a buffer with a pH of 6.50 using a mixture of $NH_3$ and $NH_4Cl$.

63. Four grams of a monoprotic weak acid is dissolved in water to make 250.0 mL of solution with a pH of 2.56. The solution is divided into two equal parts, A and B. Solution A is titrated with strong base to its equivalence point. Solution B is added to solution A after it is neutralized. The pH of the resulting solution is 4.26. What is the molar mass of the acid?

64. Fifty $cm^3$ of 1.000 $M$ $HC_2H_3O_2$ ($K_a = 1.8 \times 10^{-5}$) is titrated with 1.000 $M$ NaOH. What is the pH of the solution after the following volumes (in $cm^3$) of NaOH have been added?

(a) 0.00      (b) 25.00      (c) 49.90

(d) 50.00     (e) 50.10      (f) 100.00

Use your data to construct a plot similar to that shown in Figure 14.7 (pH versus volume of NaOH added).

65. In a titration of 50.00 mL of 1.00 $M$ $HC_2H_3O_2$ with 1.00 $M$ NaOH, a student used bromcresol green as an indicator ($K_a = 1.0 \times 10^{-5}$). About how many mL of NaOH would it take to reach the end point with this indicator? What would be a better indicator for this titration?

*66. Starting with the relation

$$[H^+] = K_a \frac{[HB]}{[B^-]}$$

derive the Henderson-Hasselbach equation:

$$pH = pK_a + \log_{10} \frac{[B^-]}{[HB]}$$

67. What is the pH of a 0.1500 $M$ $H_2SO_4$ solution if

(a) the ionization of $HSO_4^-$ is ignored?

(b) the ionization of $HSO_4^-$ is taken into account? ($K_a$ for $HSO_4^-$ is $1.1 \times 10^{-2}$.)

Chlorophyll, the green coloring matter in leaves, and hemoglobin, which is responsible for the red color of blood, have very similar structures (p. 442). (William Clay/Tony Stone Images)

# Complex Ions

15

Chromium ions are sensitive to their
Chemical environment.
They make the ruby red.
The emerald green.

The ruby and emerald are similar.
You say red, I say green.

**—ANN RAE JONAS**

*The Causes of Color*

## CHAPTER OUTLINE

**15.1 Composition of Complex Ions**

**15.2 Geometry of Complex Ions**

**15.3 Electronic Structure of Complex Ions**

**15.4 Formation Constants of Complex Ions**

In previous chapters we have referred from time to time to compounds of the transition metals. Many of these have relatively simple formulas such as $CuSO_4$, $CrCl_3$, and $Fe(NO_3)_3$. These compounds are ionic: the transition metal is present as a simple cation ($Cu^{2+}$, $Cr^{3+}$, $Fe^{3+}$). In that sense, they resemble the ionic compounds formed by the main-group metals, such as $CaSO_4$ and $Al(NO_3)_3$.

It has been known for more than a century, however, that transition metals also form a variety of ionic compounds with more complex formulas such as

$$[Cu(NH_3)_4]SO_4 \quad [Cr(NH_3)_6]Cl_3 \quad K_3[Fe(CN)_6]$$

In these so-called *coordination compounds,* the transition metal is present as a complex ion, enclosed within the brackets. In the three compounds listed above, the following complex ions are present:

$$Cu(NH_3)_4{}^{2+} \quad Cr(NH_3)_6{}^{3+} \quad Fe(CN)_6{}^{3-}$$

**$Cu^{2+}$ is a simple cation; $Cu(NH_3)_4{}^{2+}$ is a complex ion**

The charges of these complex ions are balanced by those of simple anions or cations (e.g., $SO_4{}^{2-}$, $3Cl^-$, $3K^+$).

This chapter is devoted to complex ions and the important role they play in inorganic chemistry. We consider in turn

— the compositions of complex ions* (Section 15.1)
— the geometry of complex ions (Section 15.2)
— the electronic structure of the central metal ion (or atom) in a complex (Section 15.3)
— the equilibrium constant for the formation of a complex ion (Section 15.4)

## 15.1 Composition of Complex Ions

When ammonia is added to an aqueous solution of a copper(II) salt, a deep, almost opaque, blue color develops (Fig. 15.1). This color is due to the formation of the $Cu(NH_3)_4{}^{2+}$ ion, in which four $NH_3$ molecules are bonded to a central $Cu^{2+}$ ion. The formation of this species can be represented by the equation

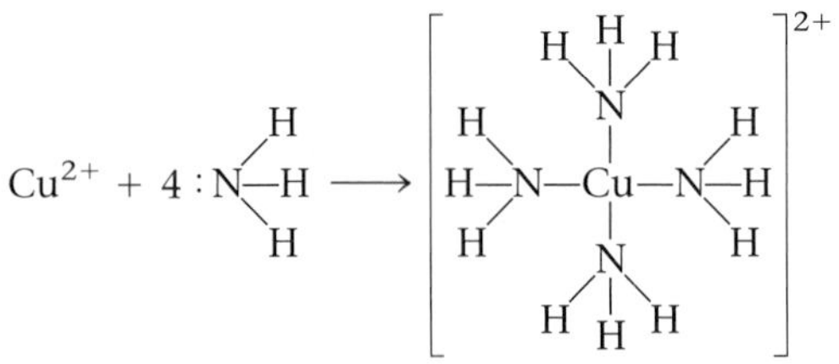

$$Cu^{2+} + 4\,:\!\overset{\textstyle H}{\underset{\textstyle H}{N}}\!\!-\!H \longrightarrow \left[ \begin{array}{c} H\overset{\textstyle H\;\;H}{\underset{\textstyle N}{\phantom{x}}} H \\ H\!-\!N\!-\!Cu\!-\!N\!-\!H \\ H\underset{\textstyle N}{\phantom{x}}H \\ H\;\;H \end{array} \right]^{2+}$$

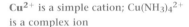

**Figure 15.1**

The $Cu(NH_3)_4{}^{2+}$ ion has an intense blue, almost violet, color, in contrast to the light blue $Cu(H_2O)_4{}^{2+}$ ion.

*The system used to name complex ions is described in Appendix 4.

The nitrogen atom of each $NH_3$ molecule contributes a pair of unshared electrons to form a covalent bond with the $Cu^{2+}$ ion. This bond and others like it, where both electrons are contributed by the same atom, is referred to as a **coordinate** covalent bond.

The $Cu(NH_3)_4^{2+}$ ion is commonly referred to as a **complex ion,** a charged species in which a **central metal ion** (or atom) is bonded to molecules and/or anions referred to collectively as **ligands.** The number of atoms bonded to the central metal is referred to as its **coordination number.** In the $Cu(NH_3)_4^{2+}$ complex ion

— the central metal is copper(II)
— the ligands are $NH_3$ molecules
— the coordination number is four

Complex ions are commonly formed by transition metals, particularly those toward the right of a transition series ($_{24}Cr \longrightarrow _{30}Zn$ in the first transition series). Nontransition metals, including Al, Sn, and Pb, form a more limited number of stable complex ions.

Cations of these metals invariably exist in aqueous solution as complex ions. In a water solution of copper(II) sulfate, the $Cu(H_2O)_4^{2+}$ cation is present. This is converted to $Cu(NH_3)_4^{2+}$ by adding ammonia or to $CuCl_4^{2-}$ by adding hydrochloric acid. Again, zinc(II) can be present in water solution as $Zn(H_2O)_4^{2+}$, $Zn(NH_3)_4^{2+}$, or $Zn(OH)_4^{2-}$.

When a complex ion is formed from a simple cation, the electron pairs required for bond formation come solely from the ligands. Reactions such as these, in which one species donates an electron pair to another, are referred to as **Lewis acid-base reactions.** In particular

*The Lewis acid-base model is the most general of the three we have considered*

— a **Lewis base** is a species that donates a pair of electrons. Ligands such as the $H_2O$ molecule, the $NH_3$ molecule, or the $OH^-$ ion act as Lewis bases in complex ion formation.
— a **Lewis acid** is a species that accepts a pair of electrons. Cations such as $Cu^{2+}$ and $Zn^{2+}$ act as Lewis acids when they form complex ions. Thus we have

$$Cu^{2+}(aq) + 4NH_3(aq) \longrightarrow Cu(NH_3)_4^{2+}(aq)$$
$$\text{Lewis acid} \qquad \text{Lewis base}$$

$$Zn^{2+}(aq) + 4OH^-(aq) \longrightarrow Zn(OH)_4^{2-}(aq)$$
$$\text{Lewis acid} \qquad \text{Lewis base}$$

Notice that

— species that act as Lewis bases (e.g., $H_2O$, $NH_3$, $OH^-$) are also Brønsted bases in the sense that they can accept a proton.
— species that act as Lewis acids (e.g., $Cu^{2+}$, $Zn^{2+}$) need not be Brønsted acids (proton donors). In that sense the Lewis model broadens the concept of an acid.

## Charges of Complexes

The charge of a complex is readily determined by applying a simple principle:

charge of complex = oxid. no. central metal + charges of ligands

The application of this principle is shown in Table 15.1, where we list the formulas of several complexes formed by platinum(II), which shows a coordi-

**TABLE 15.1    Complexes of Pt$^{2+}$ with NH$_3$ and Cl$^-$**

| Complex | Oxid. No. of Pt | Ligands | Charge of Ligands | Charge Complex |
|---|---|---|---|---|
| Pt(NH$_3$)$_4$$^{2+}$ | +2 | 4NH$_3$ | 0 | +2 |
| Pt(NH$_3$)$_3$Cl$^+$ | +2 | 3NH$_3$, 1Cl$^-$ | −1 | +1 |
| Pt(NH$_3$)$_2$Cl$_2$ | +2 | 2NH$_3$, 2Cl$^-$ | −2 | 0 |
| Pt(NH$_3$)Cl$_3$$^-$ | +2 | 1NH$_3$, 3Cl$^-$ | −3 | −1 |
| PtCl$_4$$^{2-}$ | +2 | 4Cl$^-$ | −4 | −2 |

nation number of four. Notice that one of the species, Pt(NH$_3$)$_2$Cl$_2$, is a neutral complex rather than a complex ion; the charges of the two Cl$^-$ ions just cancel that of the central Pt$^{2+}$ ion.

Note that this complex contains two Cl$^-$ ions, not a Cl$_2$ molecule

**Example 15.1**    Consider the Co(H$_2$O)$_4$Cl$_2$$^+$ ion.

(a) What is the oxidation number of cobalt?
(b) What is the formula of the coordination compound containing this cation and the Br$^-$ anion? the SO$_4$$^{2-}$ anion?

*Strategy*    To find the oxidation number, apply the relation

$$\text{charge complex ion} = \text{oxid. no. central atom} + \text{charges of ligands}$$

You know the charge of the complex and those of the ligands. To find the formulas of the coordination compounds, apply the principle of electrical neutrality.

*Solution*
(a) +1 = oxid. no. Co + 2(−1); oxid. no. Co =   +3

(b)  [Co(H$_2$O)$_4$Cl$_2$]Br    [Co(H$_2$O)$_4$Cl$_2$]$_2$SO$_4$   (compare NaBr and Na$_2$SO$_4$)

The brackets enclose the formula of the complex ion.

## Ligands; Chelating Agents

In principle, any molecule or anion with an unshared pair of electrons can donate them to a metal ion to form a coordinate covalent bond. In practice, a ligand usually contains an atom of one of the more electronegative elements (C, N, O, S, F, Cl, Br, I). Several hundred different ligands are known. Those most commonly encountered in general chemistry are NH$_3$ and H$_2$O molecules and Cl$^-$ and OH$^-$ ions.

Some ligands have more than one atom with an unshared pair of electrons and hence can form more than one bond with a central metal atom. Ligands of this type are referred to as **chelating agents;** the complexes formed are referred to as **chelates** (from the Greek *chela*, crab's claw). Two of the most com-

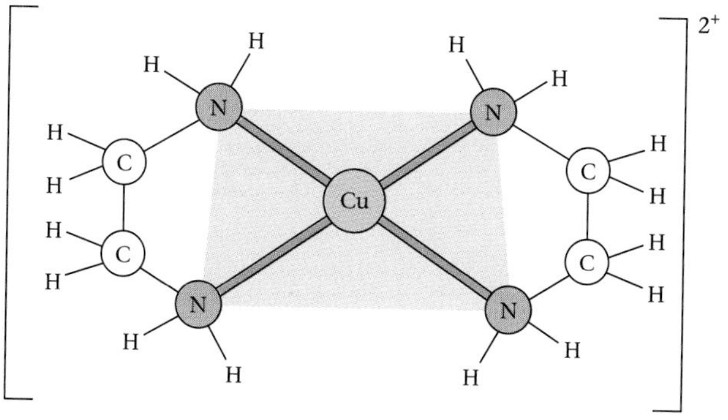

Cu(en)$_2^{2+}$

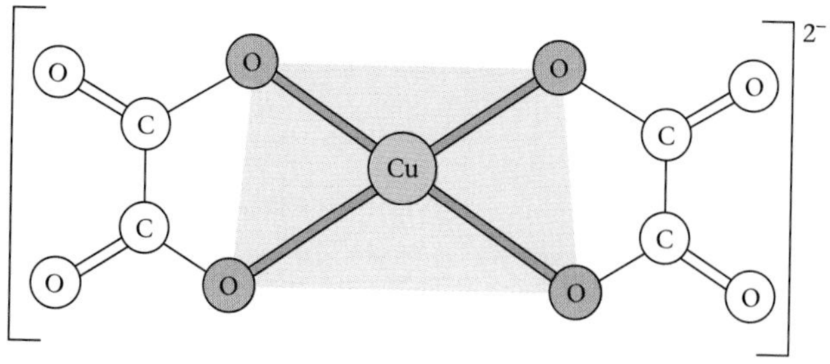

Cu(C$_2$O$_4$)$_2^{2-}$

**Figure 15.2**

Structures of the chelates formed by Cu$^{2+}$ with the ethylenediamine molecule (en) and the oxalate ion (C$_2$O$_4^{2-}$).

mon chelating agents are the oxalate anion (abbr. *ox*) and the ethylenediamine molecule (abbr. *en*), whose Lewis structures are

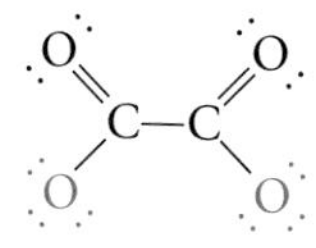

oxalate ion, C$_2$O$_4^{2-}$ 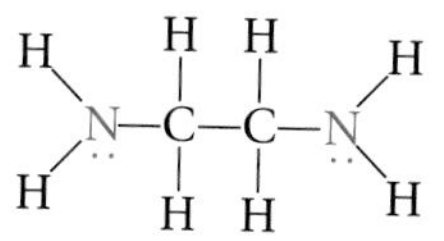          ethylenediamine molecule  H$_2$N—(CH$_2$)$_2$—NH$_2$

(The atoms which form bonds with the central metal are shown in color.)

Figure 15.2 shows the structure of the chelates formed by copper(II) with these ligands. Notice that

— in both of these complex ions, the coordination number of copper(II) is *four* (even though the complex ion contains only two ligands). Copper(II) is bonded to four atoms, two from each chelating agent.
— both the oxalate ion and the ethylenediamine molecule form five-membered rings. Most chelates contain five- or 6-membered rings. Smaller and larger rings are less stable.

## Coordination Number

As shown in Table 15.2, the most common coordination number is 6. A coordination number of 4 is less common. A value of 2 is restricted largely to Cu$^+$, Ag$^+$, and Au$^+$.

Odd coordination numbers are rare

**TABLE 15.2  Coordination Number and Geometry of Complex Ions***

| Metal Ion | Coordination Number | Geometry | Example |
|---|---|---|---|
| Ag$^+$, Au$^+$, Cu$^+$ | 2 | linear | Ag(NH$_3$)$_2^+$ |
| Cu$^{2+}$, Ni$^{2+}$, **Pd$^{2+}$**, **Pt$^{2+}$** | 4 | square planar | Pt(NH$_3$)$_4^{2+}$ |
| Al$^{3+}$, Au$^+$, **Cd$^{2+}$**, Co$^{2+}$, Cu$^+$, Ni$^{2+}$, **Zn$^{2+}$** | 4 | tetrahedral | Zn(NH$_3$)$_4^{2+}$ |
| Al$^{3+}$, Co$^{2+}$, **Co$^{3+}$**, **Cr$^{3+}$**, Cu$^{2+}$, **Fe$^{2+}$**, **Fe$^{3+}$**, Ni$^{2+}$, **Pt$^{4+}$** | 6 | octahedral | Co(NH$_3$)$_6^{3+}$ |

*The most common coordination numbers are indicated by bold type.

A few cations show only one coordination number in their complexes. Thus, $Co^{3+}$ always shows a coordination number of 6, as in

$$Co(NH_3)_6{}^{3+} \qquad Co(NH_3)_4Cl_2{}^{+} \qquad Co(en)_3{}^{3+}$$

Other cations, such as $Al^{3+}$ and $Ni^{2+}$, have variable coordination numbers, depending upon the nature of the ligand.

---

**Example 15.2**    Write formulas (including charges) of all the complex ions formed by cobalt(III) with $NH_3$ and/or *en* molecules as ligands.

***Strategy***    As indicated in Table 15.2, the coordination number of $Co^{3+}$ is 6. This means that $Co^{3+}$ forms six bonds with ligands, *including two with each en molecule.* The complex ions can contain 0 to 3 *en* molecules. Since both ligands are neutral, each complex ion has a charge of +3.

***Solution***    There are four in all:

$$Co(NH_3)_6{}^{3+} \qquad Co(NH_3)_4(en)^{3+} \qquad Co(NH_3)_2(en)_2{}^{3+} \qquad Co(en)_3{}^{3+}$$

---

## 15.2   Geometry of Complex Ions

The physical and chemical properties of complex ions and of the coordination compounds they form depend upon the spatial orientation of ligands around the central metal atom. Here we consider the geometries associated with the coordination numbers: 2, 4, and 6. With that background, we then examine the phenomenon of **geometric isomerism,** where two or more complex ions have the same chemical formula but different properties because of their different geometries.

### Coordination Number = 2

Complex ions in which the central metal forms only two bonds to ligands are *linear;* that is, the two bonds are directed at 180° angles. The structures of $CuCl_2{}^{-}$, $Ag(NH_3)_2{}^{+}$, and $Au(CN)_2{}^{-}$ may be represented as

$$(\text{Cl—Cu—Cl})^{-} \qquad \left[\begin{array}{c} H \\ | \\ H-N-Ag-N-H \\ | \\ H \end{array}\right]^{+} \qquad (N\equiv C-Au-C\equiv N)^{-}$$

### Coordination Number = 4

Four-coordinate metal complexes may have either of two different geometries (Figure 15.3). The four bonds from the central metal may be directed toward the corners of a regular *tetrahedron.* This is what we would expect from VSEPR theory (recall Chapter 7). Two common *tetrahedral* complexes are $Zn(NH_3)_4{}^{2+}$ and $CoCl_4{}^{2-}$.

Square planar complexes, in which the four bonds are directed toward the corners of a square, are more common. Certain complexes of copper(II) and nickel(II) show this geometry; it is characteristic of the complexes of $Pd^{2+}$ and $Pt^{2+}$, including $Pt(NH_3)_4{}^{2+}$.

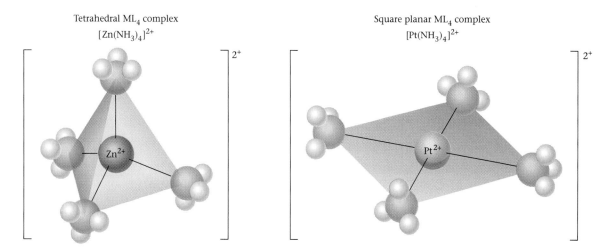

Tetrahedral $ML_4$ complex
$[Zn(NH_3)_4]^{2+}$

Square planar $ML_4$ complex
$[Pt(NH_3)_4]^{2+}$

**Figure 15.3**
Complexes with a coordination number of four can be tetrahedral, as is $Zn(NH_3)_4^{2+}$, or square planar, like $Pt(NH_3)_4^{2+}$.

## Coordination Number = 6

We saw in Chapter 7 that octahedral geometry is characteristic of many molecules (e.g., $SF_6$) in which a central atom is surrounded by six other atoms. All complex ions in which the coordination number is six are *octahedral*. The metal ion (or atom) is at the center of the *octahedron;* the six ligands are at the corners. This geometry is shown in Figure 15.4.

The structure is often shown simply as

It is important to realize that

— the six ligands can be considered to be equidistant from the central metal
— an octahedral complex can be regarded as a derivative of a square planar complex. The two extra ligands are located above and below the square, on a line perpendicular to the square at its center.

## Geometric Isomerism

Two or more species with different physical and chemical properties but the same formula are said to be **isomers** of one another. Complex ions can show many different kinds of isomerism, only one of which is considered here. *Geometric isomers* are ones that differ only in the spatial orientation of ligands around the central metal atom.

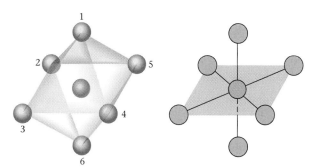

**Figure 15.4**
The drawing at the left shows six ligands at the corners of an octahedron with a metal atom at the center. A simpler way to represent an octahedral complex is shown at the right.

**1. *Square planar.*** There are two compounds with the formula $Pt(NH_3)_2Cl_2$, differing in water solubility, melting point, chemical behavior, and biological activity.* Their structures are

Cl, NH₃    NH₃, Cl
　  Pt　　　　Pt
Cl　 NH₃   Cl　 NH₃
　  *cis*　　　*trans*

The complex in which the two chlorine atoms are at adjacent corners of the square, as close to one another as possible, is called the ***cis*** isomer. It is made by reacting ammonia with the $PtCl_4^{2-}$ ion. In the ***trans*** isomer, made by reacting $Pt(NH_3)_4^{2+}$ with HCl, the two chlorine atoms are at opposite corners of the square, as far apart as possible.

*cis* = close together, *trans* = far apart

Geometric isomerism can occur with any square complex of the type Mabcd, Ma₂bc, or Ma₂b₂, where M refers to the central metal and a, b, c, and d are different ligands. Conversely, geometric isomerism cannot occur with a square complex of the type Ma₄ or Ma₃b. Thus, there are two different square complexes with the formula $Pt(NH_3)_2ClBr$ but only one with the formula $Pt(NH_3)_3Cl^+$.

Br, NH₃    NH₃, Br    ⎡NH₃, Cl⎤⁺
　  Pt　　　　Pt　　　　   Pt
Cl　 NH₃   Cl　 NH₃   ⎣NH₃　 NH₃⎦

　*cis*　　　*trans*

**2. *Octahedral.*** To understand how geometric isomerism can arise in octahedral complexes, refer back to Figure 15.4. Notice that for any given position of a ligand, four positions are equivalent, while a fifth is at a greater distance. Taking position 1 as a point of reference, you can see that groups at 2, 3, 4, and 5 are equidistant from 1; 6 is farther away. In other words, positions 1 and 2, 1 and 3, 1 and 4, 1 and 5 are *cis* to one another; positions 1 and 6 are *trans*. Hence, a complex ion like $Co(NH_3)_4Cl_2^+$ (Fig. 15.5) can exist in two different isomeric forms. In the *cis* isomer, the two $Cl^-$ ions are at adjacent corners of the octahedron, as close together as possible. In the *trans* isomer they are at opposite corners, as far away from one another as possible.

*The *cis* isomer ("cisplatin") is an effective anticancer drug. This reflects the ability of the two Cl atoms to interact with the nitrogen atoms of DNA, a molecule responsible for cell reproduction. The *trans* isomer is ineffective in chemotherapy, presumably because the Cl atoms are too far apart to react with a DNA molecule.

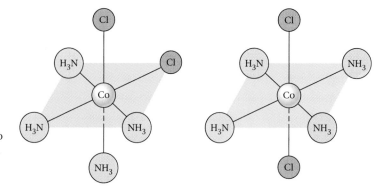

**Figure 15.5**

*Cis* and *trans* isomers of the $Co(NH_3)_4Cl_2^+$ ion. Note that the two $Cl^-$ ions are closer to one another in the *cis* isomer (left) than in the *trans* isomer (right).

**Example 15.3**   How many geometric isomers are possible for the neutral complex [Co(NH$_3$)$_3$Cl$_3$]?

**Strategy**   It's best to approach a problem of this type systematically. You might start by putting two NH$_3$ molecules *trans* to one another and see how many different structures can be obtained by placing the third NH$_3$ molecule at different positions. Then start over with two NH$_3$ molecules *cis* to one another, and follow the same procedure.

*Solution*

(1) Starting with two NH$_3$ molecules *trans* to one another (Fig. 15.6a), it clearly doesn't matter where you put the third NH$_3$ molecule. All the available vacancies are equivalent in the sense that they are *cis* to the two NH$_3$ molecules already in place. Choosing one of these positions arbitrarily for NH$_3$ gives the first isomer (Fig. 15.6b); the Cl$^-$ ions occupy the three remaining positions.

(2) Start again with the two NH$_3$ molecules *cis* to one another (Fig. 15.6c). If you place the third NH$_3$ molecule at either of the other corners of the square, you will simply reproduce the first isomer. (Remember that the symmetry of a regular octahedron requires that the distance across the diagonal of the square be the same as that from "top" to "bottom".) The only way you can get a "new" isomer is to place the third NH$_3$ molecule above or below the square so that all three NH$_3$ molecules are *cis* to one another. Putting that molecule arbitrarily at the top gives the second isomer (Fig. 15.6d). That's all there are.

Unless you're careful, you'll come up with too many isomers

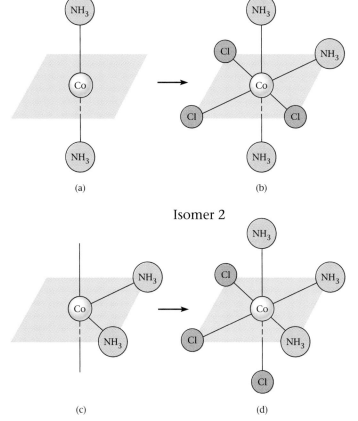

Isomer 1

(a)          (b)

Isomer 2

(c)          (d)

**Figure 15.6**
Isomers of Co(NH$_3$)$_3$Cl$_3$ (Example 15.3). You may be tempted to write down additional structures, but you will find that they are equivalent to one of the two isomers shown here.

**Figure 15.7**
There are two complex ions with the formula $Co(en)_2Cl_2^+$. Compounds containing the *cis* cation are purple; those derived from the *trans* cation are green.

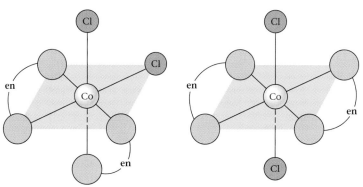

Geometric isomerism can occur in chelated octahedral complexes (Fig. 15.7). Notice that an *ethylenediamine molecule,* here and indeed in all complexes, **can only bridge** cis **positions.** It is not long enough to connect to *trans* positions.

# Chemistry: *The Human Side*

The basic ideas concerning the structure and geometry of complex ions presented in this chapter were developed by one of the most gifted individuals in the history of inorganic chemistry, Alfred Werner. His theory of coordination chemistry was published in 1893 when Werner was 26 years old. In his paper Werner made the revolutionary suggestion that metal ions such as $Co^{3+}$ could show two different kinds of valences. For the compound $Co(NH_3)_6Cl_3$, Werner postulated a central $Co^{3+}$ ion joined by "primary valences" (ionic bonds) to three $Cl^-$ ions and by "secondary valences" (coordinate covalent bonds) to six $NH_3$ molecules. Moreover, he made the inspired guess that the six secondary valences were directed toward the corners of a regular octahedron.

Werner spent the next twenty years obtaining experimental evidence to prove his theory. (At the University of Zürich, there remain several *thousand* samples of coordination compounds prepared by Werner and his students.) He was able to show, for example, that the electrical conductivities in water solution decreased in the order $[Co(NH_3)_6]Cl_3 > [Co(NH_3)_5Cl]Cl_2 > [Co(NH_3)_4Cl_2]Cl$ in much the same way as with simple salts, e.g., $ScCl_3 > CaCl_2 > NaCl$. Another property he studied was isomerism. In 1907 he was able to isolate a second geometric isomer of $[Co(NH_3)_4Cl_2]Cl$, in complete accord with his theory. Six years later, Werner won the Nobel Prize in chemistry.

By all accounts, Werner was a superb lecturer. Sometimes as many as 300 students crowded into a hall with a capacity of 150 to hear him speak. So great was his reputation that students in theology and law came to hear him talk about chemistry. There was, however, a darker side to Werner that few students saw. A young woman badgered by Werner during an oral examination came to see him later to ask whether she had passed; he threw a chair at her. Werner died at 52 of hardening of the arteries, perhaps caused in part by his addiction to alcohol and strong black cigars.

# 15.3  Electronic Structure of Complex Ions

Over the years, several different models have been developed to treat electronic structure and bonding in complex ions. One of the simplest and most useful is the crystal field model, originally proposed by Hans Bethe and J. H. van Vleck. Here the magnetic properties and brilliant colors (Figure 15.8) of coordination compounds are explained in terms of the effect of ligands on the energies of electronic levels in transition metal cations. We will apply this model to octahedral complexes. Before doing so, it may be helpful to review the electronic structure of uncomplexed transition metal cations, originally covered in Chapter 6.

## Transition Metal Cations

Recall (pp. 160–161) that in a simple transition metal cation

1. There are no outer s electrons. Electrons beyond the preceding noble gas are located in an inner d sublevel (3d for the first transition series).
2. Electrons are distributed among the five d orbitals in accordance with Hund's rule (p. 157), giving the maximum number of unpaired electrons.

In transition metal ions, 3d is lower in energy than 4s

To illustrate these rules, consider the $Fe^{2+}$ ion. Since the atomic number of iron is 26, this +2 ion must contain $26 - 2 = 24e^-$. Of these electrons, the first 18 have the argon structure; the remaining six are located in the 3d sublevel. The electron configuration is

$$Fe^{2+} \quad [Ar]3d^6$$

These six electrons are spread over all five orbitals; the orbital diagram is

3d

$$Fe^{2+} \quad [Ar] \; (\uparrow\downarrow)(\uparrow)(\uparrow)(\uparrow)(\uparrow)$$

The $Fe^{2+}$ ion is *paramagnetic,* with four unpaired electrons.

---

**Example 15.4**    For the $Cr^{3+}$ ion, derive the electron configuration, orbital diagram, and number of unpaired electrons.

***Strategy***    First (1), find the total number of electrons ($Z$ Cr = 24). Then (2), find the electron configuration; the first 18 electrons form the argon core, and the remaining electrons enter the 3d sublevel. Finally (3), apply Hund's rule to obtain the orbital diagram.

**Figure 15.8**
Most coordination compounds are brilliantly colored, a property that can be explained readily by the crystal-field model.  (Marna G. Clarke)

*Solution*

(1) no. of electrons = 24 − 3 = 21

(2) electron configuration   [Ar]3d³

3d

(3) orbital diagram (beyond Ar)  (↑)(↑)(↑)( ) ( )    3 unpaired electrons

---

The shapes of the five d orbitals are shown in Figure 15.9. These orbitals are given the symbols

$$d_{z^2} \quad d_{x^2-y^2} \quad d_{xy} \quad d_{yz} \quad d_{xz}$$

*In the uncomplexed transition metal cation, all of these orbitals have the same energy.*

## Octahedral Complexes

As six ligands approach a central metal ion to form an octahedral complex, they change the energies of electrons in the d orbitals. The effect (Figure 15.10) is to split the five d orbitals into two groups of different energy.

**1.** A higher-energy pair, the $d_{x^2-y^2}$ and $d_{z^2}$ orbitals
**2.** A lower-energy trio, the $d_{xy}$, $d_{yz}$, and $d_{xz}$ orbitals

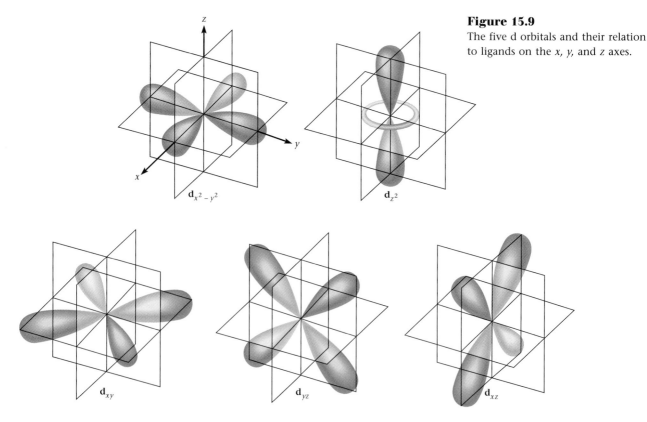

**Figure 15.9**
The five d orbitals and their relation to ligands on the *x, y,* and *z* axes.

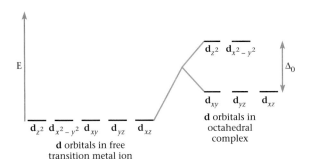

**Figure 15.10**
In a free transition metal ion, all five d orbitals have the same energy. When the ion forms an octahedral complex, two orbitals ($d_{z^2}$, $d_{x^2-y^2}$) have a higher energy than the other three ($d_{xy}$, $d_{yz}$, $d_{xz}$).

The difference in energy between the two groups is called the **crystal field splitting energy** and given the symbol $\Delta_o$ (the subscript "o" stands for "octahedral").

To see why this splitting occurs, consider what happens when six ligands (e.g., $H_2O$, $CN^-$, $NH_3$) approach a central metal cation along the x-, y-, and z-axes (Fig. 15.9). The unshared electron pairs on these ligands repel the electrons in the d orbitals of the cation. The repulsion is greatest for the $d_{x^2-y^2}$ and $d_{z^2}$ orbitals, which have their maximum electron density directly along the x-, y-, and z-axes, respectively. Electrons in the other three orbitals ($d_{xy}$, $d_{yz}$, $d_{xz}$) are less affected because their densities are concentrated between the axes rather than along them.

This gives the $d_{xy}$, $d_{yz}$, and $d_{xz}$ orbitals a lower energy

Depending upon the magnitude of $\Delta_o$, a cation may form either of two different kinds of complexes. This is illustrated for the $Fe^{2+}$ ion in the two diagrams shown in Figure 15.11.

**1.** In the $Fe(CN)_6^{4-}$ ion shown at the left, $\Delta_o$ is so large that the six 3d electrons of the $Fe^{2+}$ ion pair up in the lower-energy orbitals. In this way, the $Fe^{2+}$ ion achieves its most stable (lowest energy) electronic structure. The complex is diamagnetic, with no unpaired electrons.

**2.** In the $Fe(H_2O)_6^{2+}$ ion shown at the right, $\Delta_o$ is relatively small, and the six 3d electrons of the $Fe^{2+}$ ion spread out over all the orbitals as expected by

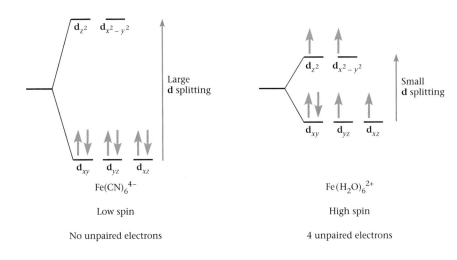

**Figure 15.11**
Depending upon the magnitude of $\Delta_o$, either of two different complexes may be formed by $Fe^{2+}$.

When $\Delta_o$ is small, the electron distribution is the same as in the simple cation; if $\Delta_o$ is large, Hund's rule is not followed

Hund's rule. In other words, the electron distribution in the complex is the same as in the $Fe^{2+}$ ion itself (where $\Delta_o$ is 0!). The $Fe(H_2O)_6^{2+}$ ion is paramagnetic, with four unpaired electrons

More generally, it is possible to distinguish between

— *low spin complexes,* containing the minimum number of unpaired electrons, formed when $\Delta_o$ is large; electrons are concentrated in the lower-energy orbitals.
— *high spin complexes,* containing the maximum number of unpaired electrons, formed when $\Delta_o$ is small; electrons are spread out among the d orbitals, exactly as in the uncomplexed transition metal cation.

For a given cation

— the high-spin complex always contains more unpaired electrons than the low-spin complex and so is more strongly paramagnetic.
— the value of $\Delta_o$ is determined by the nature of the ligand. So-called "strong field" ligands (e.g., $CN^-$), which interact strongly with d-orbital electrons, produce a large $\Delta_o$. Conversely, "weak-field" ligands (e.g., $H_2O$) interact weakly with d-orbital electrons, producing a small $\Delta_o$.

---

**Example 15.5**    Derive the structure of the $Co^{2+}$ ion in the low-spin and high-spin octahedral complexes.

**Strategy**    First, find out how many 3d electrons there are in $Co^{2+}$. Now, consider how those electrons will be distributed if Hund's rule is followed; that corresponds to the high-spin complex. Finally, sprinkle as many electrons as possible into the lower three orbitals; this is the low-spin complex.

**Solution**

(1) no. of electrons = 27 − 2 = 25; the electron configuration is $[Ar]3d^7$.

(2)    (↑)(↑)

    (↑↓)(↑↓)(↑)       high-spin       3 unpaired electrons

(3)    (↑)( )

    (↑↓)(↑↓)(↑↓)       low-spin       1 unpaired electron

Both types of complexes are known. The $Co(CN)_6^{4-}$ ion is of the low-spin type; $Co(H_2O)_6^{2+}$ is a high-spin complex.

---

This model of the electronic structure of complex ions explains why high-spin and low-spin complexes occur only with ions that have four to seven electrons ($d^4$, $d^5$, $d^6$, $d^7$). With three or fewer electrons, only one distribution is possible; the same is true with eight or more electrons.

        ( ) ( )        (↑)(↑)

(↑)(↑)(↑)      (↑↓)(↑↓)(↑↓)

3 electrons        8 electrons

## Color

Most coordination compounds are brightly colored except for those of cations whose d sublevels are completely empty (such as $Sc^{3+}$) or completely filled (such as $Zn^{2+}$). These colors are readily explained, at least qualitatively, by the model we have just described. The energy difference between two sets of d orbitals in a complex is ordinarily equal to that of a photon in the visible region. Hence, by absorbing visible light, an electron may be able to move from the lower energy set of d orbitals to the higher one. This removes some of the component wavelengths of white light, so that the light reflected or transmitted by the complex is colored. The relation between absorbed and transmitted light can be deduced from the "color wheel" shown in Figure 15.12. For example, a complex that absorbs blue-green light appears orange, and vice versa.

The color we see is what is left over after certain wavelengths are absorbed

From the color (absorption spectrum) of a complex ion, it is sometimes possible to deduce the value of $\Delta_o$, the crystal-field splitting energy. The situation is particularly simple in $_{22}Ti^{3+}$, where there is only one 3d electron. Consider, for example, the $Ti(H_2O)_6^{3+}$ ion, which has an intense purple color. This ion absorbs at 510 nm, in the green region. The purple (red-violet) color of a solution of $Ti(H_2O)_6^{3+}$ is what is left over when the green component is subtracted from the visible spectrum. Using the Einstein equation (Chap. 6), it is possible to calculate the energy of a photon with a wavelength of 510 nm:

$$E = \frac{hc}{\lambda} = \frac{(6.626 \times 10^{-34} J \cdot s)(2.998 \times 10^8 \text{ m/s})}{510 \times 10^{-9} \text{ m}} = 3.90 \times 10^{-19} \text{ J}$$

The energy in kilojoules per mol is

$$E = 3.90 \times 10^{-19} J \times \frac{1 \text{ kJ}}{10^3 J} \times \frac{6.022 \times 10^{23}}{1 \text{ mol}} = 2.35 \times 10^2 \text{ kJ/mol}$$

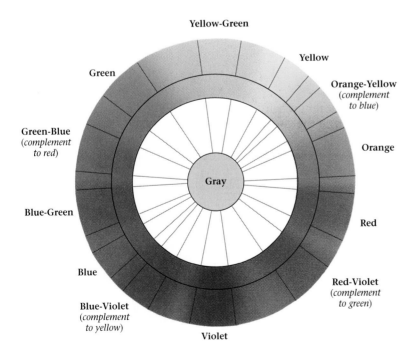

**Figure 15.12**

The color of a complex is complementary to (180° across from) the color of the light absorbed.

**TABLE 15.3 Colors of Complex Ions of Co³⁺**

| Complex | Color Observed | Color Absorbed | Approximate Wavelength (nm) Absorbed |
|---------|----------------|----------------|--------------------------------------|
| $Co(NH_3)_6^{3+}$ | yellow | violet | 430 |
| $Co(NH_3)_5NCS^{2+}$ | orange | blue-green | 470 |
| $Co(NH_3)_5H_2O^{3+}$ | red | green-blue | 500 |
| $Co(NH_3)_5Cl^{2+}$ | purple | yellow-green | 530 |
| *trans*-$Co(NH_3)_4Cl_2^+$ | green | red | 680 |

This is the energy absorbed in raising the 3d electron from a lower to a higher orbital. In other words, in the $Ti(H_2O)_6^{3+}$ ion, the two sets of d orbitals are separated by this amount of energy. The splitting energy, $\Delta_o$, is 235 kJ/mol.

The smaller the value of $\Delta_o$, the longer the wavelength of the light absorbed. This effect is shown in Table 15.3 and Figure 15.13. Notice that when weak-field ligands such as $NCS^-$, $H_2O$, or $Cl^-$ are substituted for $NH_3$, the light absorbed shifts to longer wavelengths (lower energies). On the basis of observations like this, ligands can be arranged in order of decreasing tendency to split the d orbitals. A short version of such a *spectrochemical series* is

$$CN^- > NO_2^- > en > NH_3 > NCS^- > H_2O > F^- > Cl^-$$

$$\text{strong field} \qquad \xrightarrow[\text{decreasing } \Delta_o]{} \qquad \text{weak field}$$

## 15.4 Formation Constants of Complex Ions

The equilibrium constant for the formation of a complex ion is called a **formation constant** (or stability constant) and given the symbol $K_f$. A typical example is

$$Cu^{2+}\ (aq) + 4NH_3(aq) \rightleftharpoons Cu(NH_3)_4^{2+}(aq) \qquad K_f = \frac{[Cu(NH_3)_4^{2+}]}{[Cu^{2+}] \times [NH_3]^4}$$

**Figure 15.13**
Compounds containing the five cations listed in Table 15.3 form a spectrochemical series. They vary in color from yellow [$Co(NH_3)_6$]$Cl_3$ (left) to green [$t$-$Co(NH_3)_4Cl_2$]$Cl$ (right). (Charles D. Winters)

Table 15.4 lists formation constants of complex ions. In each case, $K_f$ applies to the formation of the complex by a reaction of the type just cited. Notice that for each complex ion listed, $K_f$ is a large number, $10^5$ or greater. This means that equilibrium considerations strongly favor complex formation. Consider, for example, the system

TABLE 15.4   **Formation Constants of Complex Ions**

| Complex Ion | $K_f$ | Complex Ion | $K_f$ |
|---|---|---|---|
| $AgCl_2^-$ | $1.8 \times 10^5$ | $CuCl_2^-$ | $1 \times 10^5$ |
| $Ag(CN)_2^-$ | $2 \times 10^{20}$ | $Cu(NH_3)_4^{2+}$ | $2 \times 10^{12}$ |
| $Ag(NH_3)_2^+$ | $1.7 \times 10^7$ | $FeSCN^{2+}$ | $9.2 \times 10^2$ |
| $Ag(S_2O_3)_2^{3-}$ | $1 \times 10^{13}$ | $Fe(CN)_6^{3-}$ | $4 \times 10^{52}$ |
| $Al(OH)_4^-$ | $1 \times 10^{33}$ | $Fe(CN)_6^{4-}$ | $4 \times 10^{45}$ |
| $Cd(CN)_4^{2-}$ | $2 \times 10^{18}$ | $Hg(CN)_4^{2-}$ | $2 \times 10^{41}$ |
| $Cd(NH_3)_4^{2+}$ | $2.8 \times 10^7$ | $Ni(NH_3)_6^{2+}$ | $9 \times 10^8$ |
| $Cd(OH)_4^{2-}$ | $1.2 \times 10^9$ | $PtCl_4^{2-}$ | $1 \times 10^{16}$ |
| $Co(NH_3)_6^{2+}$ | $1 \times 10^5$ | $t\text{-}Pt(NH_3)_2Cl_2$ | $3 \times 10^{28}$ |
| $Co(NH_3)_6^{3+}$ | $1 \times 10^{23}$ | $c\text{-}Pt(NH_3)_2Cl_2$ | $3 \times 10^{29}$ |
| $Co(NH_3)_5Cl^{2+}$ | $2 \times 10^{28}$ | $Zn(CN)_4^{2-}$ | $6 \times 10^{16}$ |
| $Co(NH_3)_5NO_2^{2+}$ | $1 \times 10^{24}$ | $Zn(NH_3)_4^{2+}$ | $3.6 \times 10^8$ |
| | | $Zn(OH)_4^{2-}$ | $3 \times 10^{14}$ |

$$Ag^+(aq) + 2NH_3(aq) \rightleftharpoons Ag(NH_3)_2^+(aq)$$

$$K_f = \frac{[Ag(NH_3)_2^+]}{[Ag^+] \times [NH_3]^2} = 1.7 \times 10^7$$

The large $K_f$ value means that the forward reaction goes virtually to completion. Addition of ammonia to a solution of $AgNO_3$ will convert nearly all the $Ag^+$ ions to $Ag(NH_3)_2^+$.

The stabilities of different complexes of the same cation are directly related to their formation constants; the larger the $K_f$ value, the more stable the complex. For the reaction

$$Ag^+(aq) + 2S_2O_3^{2-}(aq) \rightleftharpoons Ag(S_2O_3)_2^{3-}(aq)$$

$$K_f = \frac{[Ag(S_2O_3)_2^{3-}]}{[Ag^+] \times [S_2O_3^{2-}]^2} = 1 \times 10^{13}$$

This $K_f$ is larger than that for $Ag(NH_3)_2^+$ ($K_f = 1.7 \times 10^7$). Hence, the $Ag(S_2O_3)_2^{3-}$ complex ion is more stable than $Ag(NH_3)_2^+$.

**Example 15.6**   Determine the ratios

(a) $[Ag(NH_3)_2^+]/[Ag^+]$ in 0.10 $M$ $NH_3$
(b) $[Ag(S_2O_3)_2^{3-}]/[Ag^+]$ in 0.10 $M$ $S_2O_3^{2-}$

***Strategy***   In each case, write down the expression for $K_f$ and solve for the desired ratio.

***Solution***

(a)   $K_f = 1.7 \times 10^7 = \dfrac{[Ag(NH_3)_2^+]}{[Ag^+] \times [NH_3]^2}$

$$\frac{[Ag(NH_3)_2^+]}{[Ag^+]} = K_f \times [NH_3]^2 = 1.7 \times 10^7(0.10)^2 = \boxed{1.7 \times 10^5}$$

This means that in 0.10 $M$ $NH_3$ there are 170,000 $Ag(NH_3)_2^+$ complex ions for every $Ag^+$ ion.

(b)  Proceeding as in (a),

$$\frac{[Ag(S_2O_3)_2{}^{3-}]}{[Ag^+]} = K_f \times [S_2O_3{}^{2-}]^2 = 1 \times 10^{13}(0.1)^2 = \boxed{1 \times 10^{11}}$$

Essentially all of the $Ag^+$ ions are complexed

Note that this ratio is much higher than the comparable ratio for the $Ag(NH_3)_2{}^+$ complex. This is really what we mean when we say that the $Ag(S_2O_3)_2{}^{3-}$ complex is "more stable" than $Ag(NH_3)_2{}^+$.

## CHEMISTRY

### *Beyond the Classroom*

## Chelates: Natural and Synthetic

Chelating agents, capable of bonding to metal atoms at more than one position, are abundant in nature. Certain species of soybeans synthesize and secrete organic chelating agents that extract iron from insoluble compounds in the soil, thereby making it available to the plant. Mosses and lichens growing on boulders use a similar process to obtain the metal ions they need for growth.

Many important natural products are chelates in which a central metal atom is bonded into a large organic molecule. In chlorophyll, the green coloring matter of plants, the central atom is magnesium; in the essential vitamin $B_{12}$, it is cobalt. The structure of heme, the pigment responsible for the red color of blood, is shown in Figure 15.A. There is an $Fe^{2+}$ ion at the center of the complex surrounded by four nitrogen atoms at the corners of a square. A fifth coordination position around the iron is occupied by an organic molecule globin, which in combination with heme, gives the protein referred to as hemoglobin.

The sixth coordination position of $Fe^{2+}$ in heme is occupied by a water molecule that can be replaced reversibly by oxygen to give a derivative known as oxyhemoglobin, which has the bright-red color characteristic of arterial blood.

$$hemoglobin + O_2 \rightleftharpoons oxyhemoglobin + H_2O$$

The position of this equilibrium is sensitive to the pressure of oxygen. In the lungs, where the blood is saturated with air ($P_{O_2} = 0.2$ atm), the hemoglobin is almost com-

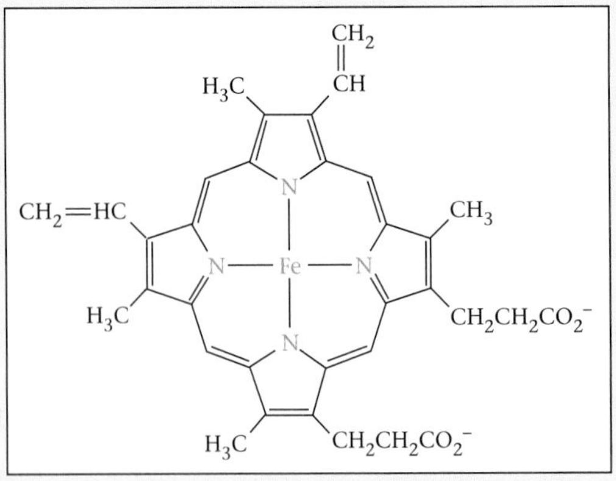

**Figure 15.A**
Structure of heme. In the hemoglobin molecule, the $Fe^{2+}$ ion is at the center of an octahedron, surrounded by four nitrogen atoms, a globin molecule, and a water molecule.

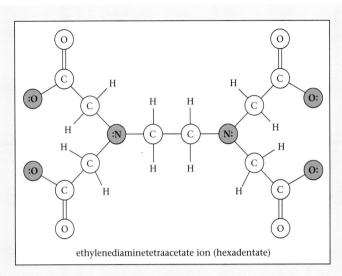

ethylenediaminetetraacetate ion (hexadentate)

**Figure 15.B**

Structure of the EDTA ion; the chelating atoms, six in all, are shown in color.

pletely converted to oxyhemoglobin. In the tissues, the partial pressure of oxygen drops, and the oxyhemoglobin breaks down to release $O_2$ essential for metabolism. In this way, hemoglobin acts as an oxygen carrier, absorbing oxygen in the lungs and liberating it to the tissues.

Unfortunately, hemoglobin forms a complex with carbon monoxide that is considerably more stable than oxyhemoglobin. The equilibrium constant for the reaction

$$\text{hemoglobin} \cdot O_2(aq) + CO(g) \rightleftharpoons \text{hemoglobin} \cdot CO(aq) + O_2(g)$$

is about 200 at body temperature. Consequently, the carbon monoxide complex is formed preferentially in the lungs even at CO concentrations as low as one part per thousand. When this happens, the flow of oxygen to the tissues is cut off, resulting eventually in muscular paralysis and death.

Many chelating agents have been synthesized in the laboratory to take advantage of their ability to tie up (*sequester*) metal cations. Perhaps the best-known synthetic chelating agent is ethylenediaminetetraacetic acid (EDTA), shown in Figure 15.B. An EDTA molecule can bond at as many as six positions, thereby occupying all the octahedral sites surrounding a central metal ion.

EDTA forms 1:1 complexes with a large number of cations, including those of some of the main-group metals. The complex formed by calcium with EDTA is used to treat lead poisoning. When a solution containing the Ca–EDTA complex is given by injection, the calcium is displaced by lead. The more stable Pb–EDTA complex is eliminated in the urine. EDTA has also been used to remove radioactive isotopes of metals, notably plutonium, from body tissues.

You may have noticed Ca–EDTA on the list of ingredients of many prepared foods, ranging from beer to mayonnaise. Here, EDTA acts as a scavenger to pick up traces of metal ions that catalyze the chemical reactions responsible for flavor deterioration, loss of color, or rancidity. Typically, Ca–EDTA is added at a level of 30 to 800 ppm.

Many household products contain EDTA. (Charles D. Winters)

# CHAPTER HIGHLIGHTS

## *Key Concepts*

1. Relate the composition of a complex ion to its charge, the coordination number, and the oxidation number of the central metal.
   (Examples 15.1, 15.2; Problems 1–8)
2. Sketch the geometry of complex ions and identify geometric isomers
   (Example 15.3; Problems 13–20)
3. Give the electron configuration and/or orbital diagram of a transition metal ion
   (Example 15.4; Problems 21–24)
4. Distinguish between high-spin and low-spin complexes
   (Example 15.5; Problems 25–32)
5. Determine the ratio of the concentration of complex ion to that of metal cation, given $K_f$ for the complex.
   (Example 15.6; Problems 37–40)

## *Key Terms*

| | | |
|---|---|---|
| central atom | geometric isomerism | square planar |
| chelate | ligand | splitting energy, $\Delta_o$ |
| *cis-* | octahedral | tetrahedral |
| complex ion | spin | *trans-* |
| coordination number | —high | transition metal |
| formation constant, $K_f$ | —low | |

## *Summary Problem*

Consider the complex $Pt(NH_3)_2(CN)_4$.

(a) Identify the ligands and their charges.
(b) What is the oxidation number of platinum?
(c) Suppose the ammonia molecules were replaced by $OH^-$ ions. Write the formula of the resulting complex.
(d) What is the coordination number of platinum in the compound?
(e) Describe the geometry of the complex.
(f) How many possible geometric isomers are there?
(g) Write the abbreviated electron configuration for the platinum(IV) ion.
(h) Cyanide ligands are ligands with the strongest crystal field (large $\Delta_o$). Show the electron distribution for platinum in this compound. Is this a high-spin or a low-spin complex?
(i) Is the complex paramagnetic or diamagnetic?
(j) Write $K$ for the formation of the complex $PtCl_4^{2-}$.
(k) Calculate $[Pt^{2+}]$ in a solution where $[Cl^-] = [PtCl_4^{2-}] = 0.50\ M$

### *Answers*

(a) $NH_3 = 0$; $CN^- = -1$
(b) $+4$
(c) $[Pt(OH)_2(CN)_4]^{2-}$
(d) 6

(e) octahedral
(f) two
(g) $[Xe]\ 4f^{14}5d^6$
(h) __ __ __  low-spin
    ↑↓ ↑↓ ↑↓

(i) diamagnetic
(j) $K_f = \dfrac{[PtCl_4^{2-}]}{[Pt^{2+}][Cl^-]^4}$
(k) $8 \times 10^{-16}$

## *Questions & Problems*

### Composition of Complex Ions and Coordination Compounds

**\*1.** Consider the complex ion $[Ni(C_2O_4)_2(NH_3)Cl]^{3-}$
  **(a)** Identify the ligands and their charges.
  **(b)** What is the oxidation number of nickel?
  **(c)** What is the formula for the sodium salt of this ion?

**\*2.** Consider the complex ion $[Cr(H_2O)_4Cl_2]^+$.
  **(a)** Identify the ligands and their charges.
  **(b)** What is the oxidation number of chromium?
  **(c)** What is the formula for the sulfate salt of this ion?

**\*3.** Iron(III) can form many complexes. Give the formula and charge of each complex with the following ligands.
  **(a)** four carbonyl molecules (CO) and two chloride ions
  **(b)** two ethylenediamine molecules
  **(c)** two thiocyanate ions ($SCN^-$), two water molecules, and two hydroxide ions

**\*4.** Ruthenium(III) (Ru) can form complex ions and compounds. Give the formula and charge of each complex with the following ligands.
  **(a)** four ammonia molecules
  **(b)** two oxalate ions ($C_2O_4^{2-}$) and two water molecules
  **(c)** one ethylenediamine molecule, two nitrite ions, and two iodide ions.

**\*5.** What is the coordination number of the metal in the following complexes?
  **(a)** $[Co(en)_2(SCN)Cl]^+$   **(b)** $[Ni(CN)_4]^{2-}$
  **(c)** $[Zn(C_2O_4)_2]^{2-}$   **(d)** $[Ag(NH_3)Cl]$

**\*6.** What is the coordination number of the metal in the following complexes?
  **(a)** $[Co(NH_3)_5(SO_4)]^+$   **(b)** $[Zn(en)(C_2O_4)]$
  **(c)** $[Mo(Cl)_6]^{3-}$   **(d)** $[Cu(NH_3)_2]^+$

**\*7.** Refer to Table 15.2 to predict the formula of the complex formed by
  **(a)** $Ag^+$ with $CN^-$   **(b)** $Pt^{2+}$ with $H_2O$
  **(c)** $Cd^{2+}$ with $CN^-$   **(d)** $Fe^{3+}$ with $C_2O_4^{2-}$

**\*8.** Refer to Table 15.2 to predict the formula of the complex formed by
  **(a)** $Pt^{4+}$ with $NH_3$   **(b)** $Zn^{2+}$ with $C_2O_4^{2-}$
  **(c)** $Ag^+$ with $H_2O$   **(d)** $Pd^{2+}$ with $Br^-$

**9.** What is the mass percent of sulfur in the $Ag(SCN)_2^-$ complex ion?

**10.** What is the mass percent of cobalt in the sulfate salt of $[Co(H_2O)_4(OH)_2]^+$?

**11.** There are four iron atoms in each hemoglobin molecule. The mass percent of iron in a hemoglobin molecule is 0.35%. Estimate the molar mass of a hemoglobin molecule.

**12.** Vitamin $B_{12}$ is a coordination compound with cobalt as its central atom. It contains 0.44% cobalt by mass and has a molar mass of $1.3 \times 10^4$ g/mol. How many cobalt atoms are there in a molecule of vitamin $B_{12}$?

### Geometry of Complex Ions

**\*13.** Sketch the geometry of
  **(a)** $Zn(H_2O)_2Cl_2$ (tetrahedral)
  **(b)** *trans*-$Ni(NH_3)_2(CN)_2$
  **(c)** *cis*-$[Ni(C_2O_4)_2(OH)_2]^{4-}$
  **(d)** $Cu(en)_3^{2+}$   **(e)** *trans*-$[Co(NO_2)_4(SCN)_2]^{3-}$

**\*14.** Sketch the geometry of
  **(a)** *trans*-$Cu(Br)_2(H_2O)_4$   **(b)** $Zn(en)_2^{2+}$ (tetrahedral)
  **(c)** *cis*-$[Ni(H_2O)_2(C_2O_4)_2]^{2-}$
  **(d)** $Pt(en)_2^{2+}$ (square planar)   **(e)** $[Au(CN)(NH_3)]$

**\*15.** The compound 1,2-diaminocyclohexane

(abbreviated "dech") is a ligand in the promising anticancer complex *cis*-$Pd(H_2O)_2(dech)^{2+}$. Sketch the geometry of this complex.

**\*16.** The acetylacetonate ion (acac$^-$)

$$\left( \begin{array}{c} \quad\;\; \overset{\displaystyle :O:}{\underset{\displaystyle \|}{}} \quad\quad \overset{\displaystyle :\ddot{O}:}{\underset{\displaystyle |}{}} \\ CH_3\!-\!C\!-\!CH\!=\!C\!-\!CH_3 \end{array} \right)^{-}$$

forms complexes with many metal ions. Sketch the geometry of Fe(acac)$_3$.

**\*17.** Which of the following octahedral complexes show geometric isomerism? If geometric isomers are possible, draw their structures.
  **(a)** $[Ru(C_2O_4)Br_4]^{2-}$   **(b)** $[Ni(en)_2Cl(SCN)]$
  **(c)** $[Co(NH_3)_4Cl_2]$

**\*18.** Follow the directions of Question 17 for
  **(a)** $[Mn(H_2O)_2(NO_2)_4]^{2-}$   **(b)** $[Pt(NH_3)_3(Cl)_3]^+$
  **(c)** $[Al(C_2O_4)(CO)_2Br_2]^-$

**\*19.** Draw all the structural formulas for the octahedral complexes of $[Fe(NH_3)_3Cl_2Br]$.

**\*20.** How many different octahedral complexes of $Cr^{3+}$ can you write using only *en* and/or $H_2O$ as ligands?

### Electronic Structure of Metal Ions

**\*21.** Give the electron configuration for
  **(a)** $Pd^{2+}$   **(b)** $Cr^{3+}$   **(c)** $Ir^{3+}$
  **(d)** $Mn^{4+}$   **(e)** $Ru^{4+}$

**\*22.** Give the electron configuration for
  **(a)** $V^{2+}$   **(b)** $Rh^{2+}$   **(c)** $Pt^{4+}$
  **(d)** $Nb^{2+}$   **(e)** $Mo^{3+}$

**\*23.** Write an orbital diagram and determine the number of unpaired electrons in each species in Question 21.

**\*24.** Write an orbital diagram and determine the number of unpaired electrons in each species in Question 22.

## Electron Distributions and Crystal-Field Energy

**\*25.** Give the electron distribution in low-spin and high-spin complexes of
  **(a)** $Zn^{2+}$    **(b)** $Pt^{4+}$

**\*26.** Follow the directions of Question 25 for
  **(a)** $Cr^{3+}$    **(b)** $Ru^{4+}$

**\*27.** Explain why $Mo^{2+}$ forms high-spin and low-spin octahedral complexes but $Mo^{3+}$ does not.

**\*28.** For complexes of $Cu^{2+}$, only one distribution of electrons is possible. Explain.

**\*29.** Why is $[Co(CN)_6]^{3-}$ diamagnetic, whereas $[Co(H_2O)_6^{3+}]$ is paramagnetic?

**\*30.** $[Fe(NO_2)_6]^{3-}$ is less paramagnetic than $[FeF_6]^{3-}$. Explain.

**\*31.** Give the number of unpaired electrons in octahedral complexes with weak-field ligands for
  **(a)** $Mn^{3+}$    **(b)** $Co^{3+}$    **(c)** $Rh^{3+}$
  **(d)** $Ti^{2+}$    **(e)** $Mo^{2+}$

**\*32.** For the species in Question 31, indicate the number of unpaired electrons with strong-field ligands.

**\*33.** $Ti(NH_3)_6^{3+}$ has a d-orbital electron transition at 399 nm. Find $\Delta_o$ at this wavelength.

**\*34.** $MnF_6^{2-}$ has a crystal field splitting energy, $\Delta_o$, of $2.60 \times 10^2$ kJ/mol. What is the wavelength corresponding to this energy?

**\*35.** A solution of $Fe(CN)_6^{3-}$ appears red. Using Figures 15.12 and 6.2, estimate the wavelength of maximum absorption.

**\*36.** The wavelength of maximum absorption of $Cu(NH_3)_4^{2+}$ is 580 nm (orange-yellow). What color is a solution of $Cu(NH_3)_4^{2+}$?

## Formation Constants of Complex Ions

**37.** Using the data in Table 15.4, calculate the ratio $[Zn^{2+}]/[Zn(OH)_4^{2-}]$ at pH 1.0, 7.0, and 10.0.

**38.** At what concentration of $S_2O_3^{2-}$ is 99% of the $Ag^+$ in a solution converted to $Ag(S_2O_3)_2^{3-}$?

**39.** At what concentration of ammonia is
  **(a)** $[Ag^+] = [Ag(NH_3)_2^+]$?
  **(b)** $[Ni^{2+}] = [Ni(NH_3)_6^{2+}]$?

**40.** At what concentration of cyanide ion is
  **(a)** $[Zn^{2+}] = 10^{-8} \times [Zn(CN)_4^{2-}]$?
  **(b)** $[Fe^{3+}] = 10^{-20} \times [Fe(CN)_6^{3-}]$?

## Unclassified

**\*41.** Explain why
  **(a)** oxalic acid removes rust stains.
  **(b)** there are no geometric isomers of tetrahedral complexes.
  **(c)** cations such as $Co^{2+}$ act as Lewis acids.

**\*42.** Explain why
  **(a)** $C_2O_4^{2-}$ is a chelating agent.
  **(b)** $NH_3$ can be a ligand but $NH_4^+$ is not.
  **(c)** a pale-green solution of nickel(II) turns blue when $NH_3$ is added.

**\*43.** Indicate whether each of the following is true or false. If the statement is false, correct it.

**(a)** The coordination number of iron(III) in $Fe(NH_3)_4(en)^{3+}$ is 5.
**(b)** $Zn^{2+}$ has two unpaired d electrons.
**(c)** $Ni(CN)_6^{4-}$ is expected to absorb at a longer wavelength than $Ni(NH_3)_6^{2+}$.

**\*44.** Indicate whether each of the following statements is true or false. If the statement is false, correct it.
  **(a)** In $[Pt(NH_3)_3Cl_3]Cl$, platinum has an oxidation number of 4 and a coordination number of 6.
  **(b)** Complexes of $Cr^{3+}$ are brightly colored, whereas those of $Zn^{2+}$ are colorless.
  **(c)** Ions with seven or more d electrons cannot form both high- and low-spin octahedral complexes.

**45.** A chemist synthesizes two coordination compounds. One compound decomposes at 280°C, the other at 240°C. Analysis of the compounds gives the same mass percent data: 52.6% Pt, 7.6% N, 1.63% H, and 38.2% Cl. Both compounds contain a +4 central ion.
  **(a)** What is the simplest formula of the compounds?
  **(b)** Draw structural formulas for the complexes present.

**46.** Analysis of a coordination compound gives the following results: 22.0% Co, 31.4% N, 6.78% H, and 39.8% Cl. One mole of the compound dissolves in water to form four moles of ions.
  **(a)** What is the simplest formula of the compound?
  **(b)** Write an equation for its dissolving in water.

**47.** For the system

$$\text{hemoglobin} \cdot O_2(aq) + CO(g) \rightleftharpoons$$
$$\text{hemoglobin} \cdot CO\ (aq) + O_2(g)$$

$K = 2.0 \times 10^2$. What must be the ratio of $P_{CO}/P_{O_2}$ if 5.0% of the hemoglobin in the blood stream is converted to the CO complex?

**\*48.** Oxyhemoglobin is red, whereas hemoglobin is blue. Explain, using Le Chatelier's principle, why venous blood is blue but arterial blood is bright red.

## Challenge Problems

**49.** A child eats 10.0 g of paint containing 5.0% Pb. How many grams of the sodium salt of EDTA, $Na_4(EDTA)$, should he receive to bring the lead into solution as $Pb \cdot EDTA$?

**50.** A certain coordination compound has the simplest formula $PtN_2H_6Cl_2$. It has a molar mass of about 600 g/mol and contains both a complex cation and a complex anion. What is its structure?

**51.** Two coordination compounds decompose at different temperatures but have the same mass percent analysis data: 20.25% Cu, 15.29% C, 7.07% H, 26.86% N, 10.23% S, and 20.39% O. Each contains $Cu^{2+}$.
  **(a)** Determine the simplest formula of the compounds.
  **(b)** Draw the structural formulas of the complex ion in each case.

**52.** In the $Ti(H_2O)_6^{3+}$ ion, the splitting between the d levels, $\Delta_o$, is 55 kcal/mol. What is the color of this ion, assuming that the color results from a transition between upper and lower d levels?

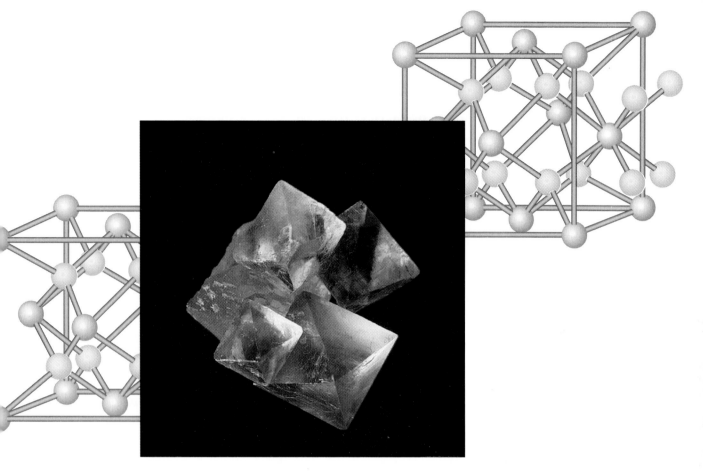

The concentration of F$^-$ ions in equilibrium with calcium fluoride, a common mineral shown above, is 13 ppm; in fluoridated water (p. 460), [F$^-$] ≈ 1 ppm.  (Paul Silverman/Fundamental Photographs, NYC)

# Precipitation Equilibria    16

**S**urrounded by beakers, by strange coils,

By ovens and flasks with twisted necks,

The chemist, fathoming the whims of attractions,

Artfully imposes on them their precise meetings.

—SULLY-PRUDHOMME

*The Naked World*

**(translated by William Dock)**

## CHAPTER OUTLINE

**16.1 Precipitate Formation; Solubility Product Constant ($K_{sp}$)**

**16.2 Dissolving Precipitates**

**16.3 Qualitative Analysis**

**Figure 16.1**
A solution prepared by mixing solutions of $Sr(NO_3)_2$ and $K_2CrO_4$ is in equilibrium with yellow $SrCrO_4(s)$. In such a solution $[Sr^{2+}] \times [CrO_4^{2-}] = 3.6 \times 10^{-5}$.

In this chapter, we consider two types of equilibria, both in water solution:

— that between a precipitate and its ions (Section 16.1), for example,

$$AgCl(s) \rightleftharpoons Ag^+(aq) + Cl^-(aq)$$

— that between a precipitate and a species used to dissolve it (Section 16.2), for example,

$$AgCl(s) + 2NH_3(aq) \rightleftharpoons Ag(NH_3)_2^+(aq) + Cl^-(aq)$$

Equilibria such as these have applications in fields as diverse as geology, medicine, and agriculture. In chemistry, you are most likely to meet up with precipitation equilibria in the laboratory when you carry out experiments in qualitative analysis (Section 16.3).

## 16.1 Precipitate Formation; Solubility Product Constant ($K_{sp}$)

As we saw in Chapter 4, a precipitate forms when a cation from one solution combines with an anion from another solution to form an insoluble solid. Precipitation reactions, like all reactions, reach a position of equilibrium. Suppose, for example, solutions of $Sr(NO_3)_2$ and $K_2CrO_4$ are mixed. In this case, $Sr^{2+}$ ions combine with $CrO_4^{2-}$ ions to form a yellow precipitate of strontium chromate, $SrCrO_4$ (Fig. 16.1). Very quickly, an equilibrium is established between the solid and the corresponding ions in solution:

$$SrCrO_4(s) \rightleftharpoons Sr^{2+}(aq) + CrO_4^{2-}(aq)$$

### $K_{sp}$ Expression

The equilibrium constant expression for the dissolving of $SrCrO_4$ can be written following the rules cited in Chapters 12 and 13. In particular, the solid does not appear in the expression; the concentration of each ion is raised to a power equal to its coefficient in the chemical equation.

$$K_{sp} = [Sr^{2+}] \times [CrO_4^{2-}]$$

Although the solid doesn't appear in $K_{sp}$, it must be present for equilibrium

The symbol $K_{sp}$ represents a particular type of equilibrium constant known as the **solubility product constant.** Like all equilibrium constants, $K_{sp}$ has a fixed value for a given system at a particular temperature. At 25°C, $K_{Sp}$ for $SrCrO_4$ is about $3.6 \times 10^{-5}$; that is,

$$[Sr^{2+}] \times [CrO_4{}^{2-}] = 3.6 \times 10^{-5}$$

This relation says that the product of the two ion concentrations at equilibrium must be $3.6 \times 10^{-5}$, regardless of how equilibrium is established.

---

**Example 16.1** Write expressions for $K_{Sp}$ for

   (a) $PbCl_2$    (b) $Ag_2CrO_4$

***Strategy*** Start by writing the chemical equation for the solution process (solid on the left, ions in solution on the right). Then write the expression for $K_{Sp}$, noting that

— solids do not appear
— the concentration of each ion is raised to a power equal to its coefficient in the chemical equation

*Solution*

   (a) $PbCl_2(s) \rightleftharpoons Pb^{2+}(aq) + 2Cl^-(aq)$

$K_{Sp} = \boxed{[Pb^{2+}] \times [Cl^-]^2}$

   (b) $Ag_2CrO_4(s) \rightleftharpoons 2Ag^+(aq) + CrO_4{}^{2-}(aq)$

$K_{Sp} = \boxed{[Ag^+]^2 \times [CrO_4{}^{2-}]}$

---

## $K_{sp}$ and the Equilibrium Concentrations of Ions

The relation

$$K_{Sp}\ SrCrO_4 = [Sr^{2+}] \times [CrO_4{}^{2-}] = 3.6 \times 10^{-5}$$

can be used to calculate the equilibrium concentration of one ion if you know that of the other. Suppose, for example, the concentration of $CrO_4{}^{2-}$ in a certain solution in equilibrium with $SrCrO_4$ is known to be $2.0 \times 10^{-3}\ M$. It follows that

$$[Sr^{2+}] = \frac{K_{Sp}\ SrCrO_4}{[CrO_4{}^{2-}]} = \frac{3.6 \times 10^{-5}}{2.0 \times 10^{-3}} = 1.8 \times 10^{-2}\ M$$

If in another case, $[Sr^{2+}] = 1.0 \times 10^{-4}\ M$,

$$[CrO_4{}^{2-}] = \frac{K_{Sp}\ SrCrO_4}{[Sr^{2+}]} = \frac{3.6 \times 10^{-5}}{1.0 \times 10^{-4}} = 3.6 \times 10^{-1}\ M$$

Example 16.2 illustrates the same kind of calculation for a different electrolyte; the math is a bit more difficult, but the principle is the same.

---

**Example 16.2** Calcium phosphate, $Ca_3(PO_4)_2$, is a water-insoluble mineral, large quantities of which are used to make commercial fertilizers. Taking its $K_{Sp}$ value from Table 16.1 p. 450, calculate

   (a) the concentration of $PO_4{}^{3-}$ in equilibrium with the solid if $[Ca^{2+}] = 1 \times 10^{-9}\ M$.
   (b) the concentration of $Ca^{2+}$ in equilibrium with the solid if $[PO_4{}^{3-}] = 1 \times 10^{-5}\ M$.

***Strategy*** The first step is to write down the $K_{sp}$ expression:

$$Ca_3(PO_4)_2(s) \rightleftharpoons 3Ca^{2+}(aq) + 2PO_4^{3-}(aq)$$

$$K_{sp} = [Ca^{2+}]^3 \times [PO_4^{3-}]^2 = 1 \times 10^{-33}$$

Now substitute the concentration of one ion and solve for that of the other.

***Solution***

(a) $[PO_4^{3-}]^2 = \dfrac{1 \times 10^{-33}}{[Ca^{2+}]^3} = \dfrac{1 \times 10^{-33}}{(1 \times 10^{-9})^3} = 1 \times 10^{-6}$ $\qquad [PO_4^{3-}] = \boxed{1 \times 10^{-3}\,M}$

(b) $[Ca^{2+}]^3 = \dfrac{1 \times 10^{-33}}{[PO_4^{3-}]^2} = \dfrac{1 \times 10^{-33}}{(1 \times 10^{-5})^2} = 1 \times 10^{-23}$ $\qquad [Ca^{2+}] = \boxed{2 \times 10^{-8}\,M}$

(To find a cube root on your calculator, use the $y^x$ key, where $x = 1/3 = 0.333333\ldots.$)

## $K_{sp}$ and Precipitate Formation

In Chapter 4, we used the solubility rules (Table 4.1) to predict whether a precipitate will form when two solutions are mixed. That type of prediction is limited to the situation where the ions involved are at a concentration of 0.1 $M$ or greater. If the ion concentrations are appreciably less than 0.1 mol/L, a precipitate may not form even though the solid is listed as being "insoluble" in water.

### TABLE 16.1 Solubility Product Constants at 25°C

| | | $K_{sp}$ | | | $K_{sp}$ |
|---|---|---|---|---|---|
| Acetates | $AgC_2H_3O_2$ | $1.9 \times 10^{-3}$ | Hydroxides | $Al(OH)_3$ | $2 \times 10^{-31}$ |
| | | | | $Fe(OH)_2$ | $5 \times 10^{-17}$ |
| Bromides | $AgBr$ | $5 \times 10^{-13}$ | | $Fe(OH)_3$ | $3 \times 10^{-39}$ |
| | $Hg_2Br_2$ | $6 \times 10^{-23}$ | | $Mg(OH)_2$ | $6 \times 10^{-12}$ |
| | $PbBr_2$ | $6.6 \times 10^{-6}$ | | $Tl(OH)_3$ | $2 \times 10^{-44}$ |
| Carbonates | $Ag_2CO_3$ | $8 \times 10^{-12}$ | | $Zn(OH)_2$ | $4 \times 10^{-17}$ |
| | $BaCO_3$ | $2.6 \times 10^{-9}$ | | $Ca(OH)_2$ | $4.0 \times 10^{-6}$ |
| | $CaCO_3$ | $4.9 \times 10^{-9}$ | | | |
| | $MgCO_3$ | $6.8 \times 10^{-6}$ | | | |
| | $SrCO_3$ | $5.6 \times 10^{-10}$ | Iodides | $AgI$ | $1 \times 10^{-16}$ |
| | $PbCO_3$ | $1 \times 10^{-13}$ | | $Hg_2I_2$ | $5 \times 10^{-29}$ |
| | | | | $PbI_2$ | $8.4 \times 10^{-9}$ |
| Chlorides | $AgCl$ | $1.8 \times 10^{-10}$ | | | |
| | $Hg_2Cl_2$ | $1 \times 10^{-18}$ | | | |
| | $PbCl_2$ | $1.7 \times 10^{-5}$ | Phosphates | $Ag_3PO_4$ | $1 \times 10^{-16}$ |
| | | | | $AlPO_4$ | $1 \times 10^{-20}$ |
| Chromates | $Ag_2CrO_4$ | $1 \times 10^{-12}$ | | $Ca_3(PO_4)_2$ | $1 \times 10^{-33}$ |
| | $BaCrO_4$ | $1.2 \times 10^{-10}$ | | $Mg_3(PO_4)_2$ | $1 \times 10^{-24}$ |
| | $PbCrO_4$ | $2 \times 10^{-14}$ | | | |
| | $SrCrO_4$ | $3.6 \times 10^{-5}$ | | | |
| Fluorides | $BaF_2$ | $1.8 \times 10^{-7}$ | Sulfates | $BaSO_4$ | $1.1 \times 10^{-10}$ |
| | $CaF_2$ | $1.5 \times 10^{-10}$ | | $CaSO_4$ | $7.1 \times 10^{-5}$ |
| | $MgF_2$ | $7 \times 10^{-11}$ | | $PbSO_4$ | $1.8 \times 10^{-8}$ |
| | $PbF_2$ | $7.1 \times 10^{-7}$ | | $SrSO_4$ | $3.4 \times 10^{-7}$ |

$K_{sp}$ values can be used to make a more general prediction concerning precipitate formation, regardless of the concentrations of the ions involved. To do this, a quantity called the **ion product, $P$,** is compared with the solubility product constant, $K_{sp}$. The ion product, $P$, is entirely analogous to the reaction quotient, $Q$, discussed in Chapter 12. The form of the expression for $P$ is exactly the same as that for the equilibrium constant, $K_{sp}$. The difference is that the concentrations used to calculate $P$ are the actual values that apply at a particular moment. Those that appear in $K_{sp}$ are equilibrium concentrations. Putting it another way, the value of $P$ is expected to change as a precipitation reaction proceeds, approaching $K_{sp}$ and eventually becoming equal to it.

Three cases can be distinguished (Fig 16.2):

1. If $P > K_{sp}$, the solution contains a higher concentration of ions than it can hold at equilibrium. In other words, the solution is supersaturated. **A precipitate forms,** decreasing the concentrations until the ion product becomes equal to $K_{sp}$ and equilibrium is established.
2. If $P < K_{sp}$, the solution contains a lower concentration of ions than is required for equilibrium with the solid. The solution is unsaturated. **No precipitate forms;** equilibrium is not established.
3. If $P = K_{sp}$, the solution is just saturated with ions and is at the point of precipitation.

$K_{sp}$ is a constant; $P$ can have any value

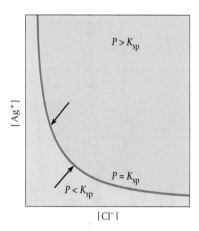

**Figure 16.2**
Silver chloride($s$) is in contact with $Ag^+$ and $Cl^-$ ions in aqueous solution. The product $P$ of the concentration of the ions $[Ag^+] \times [Cl^-]$ is equal to $K_{sp}$ (curved line) when equilibrium exists. If $P > K_{sp}$, $AgCl(s)$ tends to precipitate out until equilibrium is reached. If $P < K_{sp}$, additional solid dissolves.

---

**Example 16.3**   Sodium chromate is added to a solution in which the original concentration of $Sr^{2+}$ is $1.0 \times 10^{-3}$ M. Assuming $[Sr^{2+}]$ stays constant,

(a) will a precipitate of $SrCrO_4$ ($K_{sp} = 3.6 \times 10^{-5}$) form when the concentration of $CrO_4^{2-} = 3.0 \times 10^{-2}$ M?
(b) will a precipitate form when the concentration of $CrO_4^{2-}$ is $5.0 \times 10^{-2}$ M?

**Strategy**   First calculate $P$, using the concentrations quoted. Then compare with $K_{sp}$ and follow the rules just cited.

**Solution**
(a) $P = (1.0 \times 10^{-3}) \times (3.0 \times 10^{-2}) = 3.0 \times 10^{-5}$

Since $P$ is less than $K_{sp}$, $3.6 \times 10^{-5}$,    no precipitate forms.

(b) $P = (1.0 \times 10^{-3}) \times (5.0 \times 10^{-2}) = 5.0 \times 10^{-5}$    $P > K_{sp}$

A precipitate forms,    reducing the concentrations of $Sr^{2+}$ and $CrO_4^{2-}$ until the ion product becomes equal to $3.6 \times 10^{-5}$.

---

The calculations in Example 16.3 were simplified by assuming that the concentration of one ion ($Sr^{2+}$) stayed constant while the other ion ($CrO_4^{2-}$) was added. That assumption will not hold if precipitation comes about through the mixing of two solutions. In that case, the concentrations of the ions present will decrease as volume increases. The concentrations to be used to calculate the ion product $P$ are those obtained after mixing.

The two solutions dilute each other

---

**Example 16.4**   A student mixes 0.200 L of 0.0060 M $Sr(NO_3)_2$ solution with 0.100 L of 0.015 M $K_2CrO_4$ solution to give a final volume of 0.300 L. Will a precipitate of $SrCrO_4$ ($K_{sp} = 3.6 \times 10^{-5}$) form under these conditions?

***Strategy***   This is similar to Example 16.3 with one twist: You must first calculate the concentration of each ion after mixing, taking the volume increase into account. To do this, it's convenient to find the number of moles of each ion and then divide by the volume after mixing, 0.300 L.

***Solution***

$$n_{Sr^{2+}} = 0.200\ L \times 0.0060\ mol/L = 1.2 \times 10^{-3}\ mol$$

$$[Sr^{2+}] = 1.2 \times 10^{-3}\ mol/0.300\ L = 4.0 \times 10^{-3}\ M$$

$$n_{CrO_4^{2-}} = 0.100\ L \times 0.015\ mol/L = 1.5 \times 10^{-3}\ mol$$

$$[CrO_4^{2-}] = 1.5 \times 10^{-3}\ mol/0.300\ L = 5.0 \times 10^{-3}\ M$$

$$P = (4.0 \times 10^{-3})(5.0 \times 10^{-3}) = 2.0 \times 10^{-5}$$

Since $P$ is less than $K_{sp}$ ($3.6 \times 10^{-5}$),   no precipitate forms.

---

The development we have just gone through is the basis of an experimental method of separating ions in solution. Suppose, for example, that we slowly add $SO_4^{2-}$ ions to a solution which is 0.10 $M$ in both $Ba^{2+}$ and $Ca^{2+}$. Barium sulfate, which is much less soluble than calcium sulfate, will start to precipitate when the concentration of $SO_4^{2-}$ is very small, about $1.1 \times 10^{-9}\ M$.

The less soluble solid precipitates first; the more soluble solid stays in solution

$$P = (1.0 \times 10^{-1}) \times (1.1 \times 10^{-9}) = 1.1 \times 10^{-10} = K_{sp}BaSO_4$$

At this point, calcium ions will stay in solution, since

$$P = 1.1 \times 10^{-10} < K_{sp}CaSO_4\ (7.1 \times 10^{-5})$$

As more sulfate ions are added, $BaSO_4$ will continue to precipitate. In contrast, $CaSO_4$ will not precipitate until the concentration of $SO_4^{2-}$ reaches $7.1 \times 10^{-4}\ M$.

$$[SO_4^{2-}] = \frac{K_{sp}CaSO_4}{[Ca^{2+}]} = \frac{7.1 \times 10^{-5}}{1.0 \times 10^{-1}} = 7.1 \times 10^{-4}\ M$$

By this time, essentially all the $Ba^{2+}$ ions (about 99.9998%) will have been removed from solution as $BaSO_4$.

The process just described is referred to as *fractional precipitation*. It is particularly useful in separating cations from one another in qualitative analysis (Section 16.3). The calculations involved are further illustrated in Example 16.10, p. 458.

## $K_{sp}$ and Water Solubility

The relationship between $K_{sp}$ and $s$ depends upon the type of solid

The water solubility of an ionic compound, $s$, in moles per liter, can be calculated from its solubility product constant, $K_{sp}$*. The approach followed is illustrated in Example 16.5.

---

*Experimentally  we usually find that the solubility is slightly greater than that predicted from $K_{sp}$. For example, the measured solubility of $PbI_2$ in water at 25°C is $1.7 \times 10^{-3}\ M$. This compares with a value of $1.4 \times 10^{-3}\ M$ calculated from the $K_{sp}$ of $PbI_2$. The reason for this is that some of the lead in $PbI_2$ goes into solution in the form of species other than $Pb^{2+}$. For example, we can detect ions such as $Pb(OH)^+$ and $PbI^+$ in a water solution of lead iodide.

**Example 16.5** Calculate the water solubility of

(a) $CaCO_3$ ($K_{sp} = 4.9 \times 10^{-9}$) in mol/L.    (b) $BaF_2$ ($K_{sp} = 1.8 \times 10^{-7}$) in g/L.

**Strategy** To relate molar solubility, $s$, to $K_{sp}$, start by writing the equation for the dissolving of the solid in water:

$$CaCO_3(s) \longrightarrow Ca^{2+}(aq) + CO_3{}^{2-}(aq)$$

$$BaF_2(s) \longrightarrow Ba^{2+}(aq) + 2F^-(aq)$$

Now express the equilibrium concentration of each ion in terms of $s$ and substitute into the expression for $K_{sp}$. Knowing the value of $K_{sp}$, solve for $s$. In (b), an additional conversion (moles to grams) is required.

**Solution**

(a) For every mole of $CaCO_3$ that dissolves, 1 mol of $Ca^{2+}$ and 1 mol of $CO_3{}^{2-}$ form.

$$[Ca^{2+}] = s \qquad [CO_3{}^{2-}] = s$$

$$K_{sp} = [Ca^{2+}] \times [CO_3{}^{2-}] = s \times s = s^2$$

$$s = (K_{sp})^{1/2} = (4.9 \times 10^{-9})^{1/2} = \boxed{7.0 \times 10^{-5} \text{ mol/L}}$$

(b) For every mole of $BaF_2$ that dissolves, 1 mol of $Ba^{2+}$ and 2 mol of $F^-$ form.

$$[Ba^{2+}] = s \qquad [F^-] = 2s$$

$$K_{sp} = [Ba^{2+}] \times [F^-]^2 = s \times (2s)^2 = 4s^3$$

$$s = (K_{sp}/4)^{1/3} = (1.8 \times 10^{-7}/4)^{1/3} = 3.6 \times 10^{-3} \text{ mol/L}$$

To find the solubility in grams per liter, note that the molar mass of $BaF_2$ is 175.3 g/mol. So

$$\text{solubility} = \frac{3.6 \times 10^{-3} \text{ mol } BaF_2}{1 \text{ L}} \times \frac{175.3 \text{ g } BaF_2}{1 \text{ mol } BaF_2} = \boxed{0.63 \text{ g } BaF_2/\text{L}}$$

Stalactites (upper) and stalagmites (lower) consist of calcium carbonate. They are formed when a water solution containing $Ca^{2+}$ and $HCO_3{}^-$ ions enters a cave. Carbon dioxide is released, and calcium carbonate precipitates:

$$Ca^{2+}(aq) + 2HCO_3{}^-(aq) \longrightarrow$$
$$CaCO_3(s) + H_2O + CO_2(g).$$

(Visuals Unlimited/Richard Thom)

## $K_{sp}$ and the Common Ion Effect

How would you expect the solubility of $CaCO_3$ in water,

$$CaCO_3(s) \rightleftharpoons Ca^{2+}(aq) + CO_3{}^{2-}(aq)$$

to compare with that in a 0.10 *M* solution of $Na_2CO_3$, which contains the same (**common**) anion, $CO_3{}^{2-}$? A moment's reflection should convince you that the solubility in 0.10 *M* $Na_2CO_3$ must be *less* than that in pure water. Recall (Example 16.5) that in pure water $[CO_3{}^{2-}]$ is only $7 \times 10^{-5}$ *M*. Increasing the concentration of $CO_3{}^{2-}$ to 0.10 *M* should, by Le Chatelier's principle, drive the above equilibrium to the *left*, repressing the solubility of calcium carbonate. This is, indeed, the case (Example 16.6).

A "common" ion comes from two sources such as $CaCO_3$ and $Na_2CO_3$

**Example 16.6** Taking $K_{sp}$ of $CaCO_3$ to be $4.9 \times 10^{-9}$, estimate its solubility (moles per liter) in 0.10 *M* $Na_2CO_3$ solution.

**Strategy** Let $s$ = solubility = no. of moles per liter of $CaCO_3$ that dissolves. Relate $[Ca^{2+}]$ and $[CO_3{}^{2-}]$ to $s$; notice that in this case *there are two different sources of $CO_3{}^{2-}$*. Substitute into the expression for $K_{sp}$ and solve for $s$.

(a)

(b)

**Figure 16.3**
Sodium chloride can be precipitated from its saturated solution (6 *M*) by adding a solution containing either Na⁺ or Cl⁻ ions at a concentration greater than 6 *M*. Both 12 *M* NaOH (a) and 12 *M* HCl(b) will bring about this reaction, illustrating the common ion effect. (Marna G. Clarke)

*Solution*   Every mole of $CaCO_3$ that dissolves forms 1 mol of $Ca^{2+}$ and 1 mol of $CO_3^{2-}$. Since there was no $Ca^{2+}$ present originally, its concentration becomes $s$:

$$[Ca^{2+}] = s$$

Since the concentration of $CO_3^{2-}$ was originally 0.10 *M*, and it increases by $s$ mol/L,

$$[CO_3^{2-}] = 0.10 + s$$

Hence $K_{sp} = 4.9 \times 10^{-9} = [Ca^{2+}] \times [CO_3^{2-}] = s(0.10 + s)$. This is a quadratic equation; as usual, we look for ways to avoid solving it by brute force. A suitable approximation here would appear to be

$$s + 0.10 \approx 0.10$$

The solubility should be much less than 0.10 mol/L, since $CaCO_3$ is quite insoluble. With that assumption,

$$s(0.10) = 4.9 \times 10^{-9}$$

$$s = \frac{4.9 \times 10^{-9}}{1.0 \times 10^{-1}} = \boxed{4.9 \times 10^{-8} \text{ mol/L}}$$

The solubility is indeed much less than 0.10 *M*, so the approximation is justified. More important, the solubility in 0.10 *M* $Na_2CO_3$ is much less than in pure water (Example 16.5):

$$4.9 \times 10^{-8} \ll 7.0 \times 10^{-5}$$

which is the point we set out to prove.

---

The effect illustrated in Example 16.6 is a general one. ***An ionic solid is less soluble in a solution containing a common ion than it is in water*** (Fig. 16.3).

## 16.2   Dissolving Precipitates

Many different methods can be used to bring water-insoluble ionic solids into solution. Most commonly, this is done by adding a reagent to react with either the anion or the cation. The two most useful reagents for this purpose are

— a strong acid, H⁺, used to react with basic anions
— a complexing agent, most often $NH_3$ or OH⁻, added to react with metal cations

### Strong Acid

Water-insoluble metal hydroxides can be brought into solution with a strong acid such as HCl. The reaction with zinc hydroxide is typical:

$$Zn(OH)_2(s) + 2H^+(aq) \longrightarrow Zn^{2+}(aq) + 2H_2O$$

We can imagine that this reaction occurs in two (reversible) steps:

BaSO₄ cannot be dissolved by acid. Explain

dissolving $Zn(OH)_2$ in water:     $Zn(OH)_2(s) \rightleftharpoons Zn^{2+}(aq) + 2OH^-(aq)$

neutralizing OH⁻ ions by H⁺:     $\dfrac{2H^+(aq) + 2OH^-(aq) \rightleftharpoons 2H_2O}{Zn(OH)_2(s) + 2H^+(aq) \rightleftharpoons Zn^{2+}(aq) + 2H_2O}$

The equilibrium constant for the neutralization (Example 16.7) is so large that the overall reaction goes essentially to completion.

**Example 16.7**    Consider $Zn(OH)_2(s) + 2H^+(aq) \rightleftharpoons Zn^{2+}(aq) + 2H_2O$

(a) Determine $K$ for the reaction, applying the rule of multiple equilibria to the two-step process referred to above.

(b) Using $K$, calculate the molar solubility, $s$, of $Zn(OH)_2$ in acid at pH 5.0.

**Strategy**    Start by finding the equilibrium constants $K_1$ and $K_2$ for the two reactions ($K = K_1 \times K_2$). Then write the expression for $K$ and use it to relate $s$ to $[H^+]$.

**Solution**

(a) The first reaction

$$Zn(OH)_2(s) \rightleftharpoons Zn^{2+}(aq) + 2OH^-(aq)$$

is that for the dissolving of $Zn(OH)_2$ in water. Hence $K_1 = K_{sp} \, Zn(OH)_2$. The second reaction

$$2H^+(aq) + 2OH^-(aq) \rightleftharpoons 2H_2O$$

is the *reverse* of that for the ionization of water ($H_2O \rightleftharpoons H^+ + OH^-$) with all the coefficients multiplied by *two*. Applying the reciprocal rule and the coefficient rule (Chapter 12) in succession: $K_2 = 1/K_w^2$.

Hence

$$K = K_1 \times K_2 = \frac{K_{sp} Zn(OH)_2}{K_w^2} = \frac{4 \times 10^{-17}}{(1 \times 10^{-14})^2} = \boxed{4 \times 10^{11}}$$

The fact that the equilibrium constant is huge implies that $Zn(OH)_2$ is extremely soluble in strong acid.

(b) For the reaction $Zn(OH)_2(s) + 2H^+(aq) \rightleftharpoons Zn^{2+}(aq) + 2H_2O$:

$$K = \frac{[Zn^{2+}]}{[H^+]^2}$$

For every mole of $Zn(OH)_2$ that dissolves, a mole of $Zn^{2+}$ is formed, i.e., $[Zn^{2+}] = s$. Hence

$$s = [H^+]^2 \times K$$

At pH 5.0, $[H^+] = 1 \times 10^{-5} \, M$; $s = (1 \times 10^{-5})^2(4 \times 10^{11}) = \boxed{4 \times 10^1 \, mol/L}$

This extremely high concentration, $40 \, M$, means that $Zn(OH)_2$ is completely soluble in acid at pH 5 (or any lower pH).

---

Strong acid can also be used to dissolve many water-insoluble salts in which the anion is a weak base. In particular, $H^+$ ions will dissolve

— *virtually all carbonates* ($CO_3^{2-}$). Here, the product is the weak acid $H_2CO_3$, which then decomposes into $CO_2$ and $H_2O$. The equation for the reaction of $H^+$ ions with $ZnCO_3$ is typical

$$ZnCO_3(s) + 2H^+(aq) \longrightarrow Zn^{2+}(aq) + H_2CO_3(aq)$$

— *many sulfides* ($S^{2-}$). The "driving force" behind this reaction is the formation of the weak acid $H_2S$, much of which evolves as a gas. The reaction in the case of zinc sulfide is

$$ZnS(s) + 2H^+(aq) \longrightarrow Zn^{2+}(aq) + H_2S(aq)$$

---

**Example 16.8**  Write balanced net ionic equations to explain why each of the following precipitates dissolves in strong acid.

  (a) $Cr(OH)_3$     (b) $Ag_2CO_3$     (c) MnS

***Strategy***  Compare the equations given above for dissolving $Zn(OH)_2$, $ZnCO_3$, and ZnS. In each case, the cation is brought into solution; the other product is $H_2O$, $H_2CO_3$, or $H_2S$.

***Solution***

  (a) $Cr(OH)_3(s) + 3H^+(aq) \longrightarrow Cr^{3+}(aq) + 3H_2O$
  (b) $Ag_2CO_3(s) + 2H^+(aq) \longrightarrow 2Ag^+(aq) + H_2CO_3(aq)$
  (c) $MnS(s) + 2H^+(aq) \longrightarrow Mn^{2+}(aq) + H_2S(aq)$

---

## Complex Formation

Ammonia and sodium hydroxide are commonly used to dissolve precipitates containing a cation that forms a stable complex with $NH_3$ or $OH^-$ (Table 16.2). The reactions with zinc hydroxide are typical:

$$Zn(OH)_2(s) + 4NH_3(aq) \longrightarrow Zn(NH_3)_4^{2+}(aq) + 2OH^-(aq)$$

$$Zn(OH)_2(s) + 2OH^-(aq) \longrightarrow Zn(OH)_4^{2-}(aq)$$

$Mg(OH)_2$ cannot be dissolved by $NH_3$ or NaOH. Explain

The equilibrium constant for the solubility reaction is readily calculated. Consider, for example, the reaction by which zinc hydroxide dissolves in ammonia. Again, imagine that the reaction occurs in two steps:

$$Zn(OH)_2(s) \rightleftharpoons Zn^{2+}(aq) + 2OH^-(aq) \qquad K_1 = K_{sp}\ Zn(OH)_2$$

$$\frac{Zn^{2+}(aq) + 4NH_3(aq) \rightleftharpoons Zn(NH_3)_4^{2+}(aq) \qquad K_2 = K_f\ Zn(NH_3)_4^{2+}}{Zn(OH)_2(s) + 4NH_3(aq) \rightleftharpoons Zn(NH_3)_4^{2+}(aq) + 2OH^-(aq) \quad K = K_1 \times K_2}$$

Applying the rule of multiple equilibria,

$$K = K_{sp}\ Zn(OH)_2 \times K_f\ Zn(NH_3)_4^{2+} = (4 \times 10^{-17})(3.6 \times 10^8) = 1 \times 10^{-8}$$

In general, for any reaction of this type

$$K = K_{sp} \times K_f$$

---

**TABLE 16.2  Complexes of Cations with $NH_3$ and $OH^-$**

| Cation | $NH_3$ Complex | $OH^-$ Complex |
|---|---|---|
| $Ag^+$ | $Ag(NH_3)_2^+$ | |
| $Cu^{2+}$ | $Cu(NH_3)_4^{2+}$ (blue) | |
| $Cd^{2+}$ | $Cd(NH_3)_4^{2+}$ | |
| $Sn^{4+}$ | | $Sn(OH)_6^{2-}$ |
| $Sb^{3+}$ | | $Sb(OH)_4^-$ |
| $Al^{3+}$ | | $Al(OH)_4^-$ |
| $Ni^{2+}$ | $Ni(NH_3)_6^{2+}$ (blue) | |
| $Zn^{2+}$ | $Zn(NH_3)_4^{2+}$ | $Zn(OH)_4^{2-}$ |

where $K_{sp}$ is the solubility product constant of the solid and $K_f$ is the formation constant of the complex.

---

**Example 16.9**   Consider the reaction by which silver chloride dissolves in ammonia:

$$AgCl(s) + 2NH_3(aq) \rightleftharpoons Ag(NH_3)_2{}^+(aq) + Cl^-(aq)$$

(a) Taking $K_{sp}$ AgCl $= 1.8 \times 10^{-10}$ and $K_f$ Ag(NH$_3$)$_2{}^+ = 1.7 \times 10^7$, calculate $K$ for the above reaction.
(b) Calculate the solubility (moles/liter) of AgCl in 6.0 $M$ NH$_3$.

**Strategy**   Apply the rule of multiple equilibria to find $K$. Then work with the expression for $K$ to relate the molar solubility, $s$, to [NH$_3$].

**Solution**

(a) $K = K_{sp} \times K_f = (1.8 \times 10^{-10})(1.7 \times 10^7) = \boxed{3.1 \times 10^{-3}}$

(b) $AgCl(s) + 2NH_3(aq) \rightleftharpoons Ag(NH_3)_2{}^+(aq) + Cl^-(aq)$

$$K = \frac{[Ag(NH_3)_2{}^+] \times [Cl^-]}{[NH_3]^2} = \boxed{3.1 \times 10^{-3}}$$

For every mole of AgCl that dissolves, one mole of Ag(NH$_3$)$_2{}^+$ and one mole of Cl$^-$ are formed. In other words, $s = [Ag(NH_3)_2{}^+] = [Cl^-]$
It follows that

$$\frac{s^2}{[NH_3]^2} = K$$

$$s = [NH_3] \times K^{\frac{1}{2}} = (6.0)(3.1 \times 10^{-3})^{\frac{1}{2}} = \boxed{0.33\ M}$$

---

# 16.3   Qualitative Analysis

In the general chemistry laboratory, you will most likely carry out at least one experiment dealing with the qualitative analysis of metal cations. The objective here is to separate and identify the cations present in an "unknown" solution. A scheme of analysis for 22 different cations is shown in Table 16.3. As you can see, the general approach is to use precipitation reactions to divide the ions into

**TABLE 16.3   Cation Groups of Qualitative Analysis**

| Group | Cations | Precipitating Reagent/Conditions |
|---|---|---|
| I | Ag$^+$, Pb$^{2+}$, Hg$_2{}^{2+}$ | 6 $M$ HCl |
| II | Cu$^{2+}$, Bi$^{3+}$, Hg$^{2+}$, Cd$^{2+}$, Sn$^{4+}$, Sb$^{3+}$ | 0.1 $M$ H$_2$S at a pH of 0.5 |
| III | Al$^{3+}$, Cr$^{3+}$, Co$^{2+}$, Fe$^{2+}$, Mn$^{2+}$, Ni$^{2+}$, Zn$^{2+}$ | 0.1 $M$ H$_2$S at a pH of 9 |
| IV | Ba$^{2+}$, Ca$^{2+}$, Mg$^{2+}$; Na$^+$, K$^+$, NH$_4{}^+$ | 0.2 $M$ (NH$_4$)$_2$CO$_3$ at a pH of 9.5. No precipitates with Na$^+$, K$^+$, NH$_4{}^+$; separate tests for identification |

four different groups. The ions within a group are then brought into solution, separated from one another, and identified.

**Group I** contains the only three common cations that form insoluble chlorides: $Ag^+$, $Pb^{2+}$, and $Hg_2^{2+}$. Addition of HCl precipitates AgCl, $PbCl_2$, and $Hg_2Cl_2$, all of which are white solids. To separate the cations, lead chloride is brought into solution in hot water

$Hg_2^{2+}$ is a polyatomic cation; there is a covalent bond between the mercury atoms

$$PbCl_2(s) \longrightarrow Pb^{2+}(aq) + 2Cl^-(aq)$$

while AgCl is dissolved in ammonia through complex ion formation

$$AgCl(s) - 2NH_3(aq) \longrightarrow Ag(NH_3)_2^+(aq) + Cl^-(aq)$$

**Group II** consists of six different cations, all of which form very insoluble sulfides with characteristic colors (Fig 16.4). These compounds are precipitated by adding hydrogen sulfide, a toxic, foul-smelling gas, at a pH of 0.5. At this rather high $H^+$ ion concentration, the equilibrium

$$H_2S(aq) \rightleftharpoons 2H^+(aq) + S^{2-}(aq)$$

lies far to the left. The concentration of $S^{2-}$ is extremely low but sufficient to precipitate the very insoluble Group II sulfides such as CuS and $Bi_2S_3$.

$$Cu^{2+}(aq) + H_2S(aq) \longrightarrow CuS(s) + 2H^+(aq)$$

$$2Bi^{3+}(aq) + 3H_2S(aq) \longrightarrow Bi_2S_3(s) + 6H^+(aq)$$

**Group III** cations form sulfides that are considerably more soluble than those of Group II. Consequently, they do not precipitate at pH 0.5, allowing for their separation from Group II. However, at pH 9, where the concentration of $S^{2-}$ is considerably higher, five Group III cations precipitate as sulfides; $Al^{3+}$ and $Cr^{3+}$ come down as hydroxides in this basic solution (Figure 16.5):

$$Al^{3+}(aq) + 3\ OH^-(aq) \longrightarrow Al(OH)_3(s)$$

$$Cr^{3+}(aq) + 3\ OH^-(aq) \longrightarrow Cr(OH)_3(s)$$

**Figure 16.4**
The Group II sulfides show a variety of colors: CuS, $Bi_2S_3$, and HgS are black, CdS is orange-yellow, $SnS_2$ is pale yellow, and $Sb_2S_3$ is a brilliant red-orange.

---

**Example 16.10**   For the general solution reaction

$$MS(s) + 2H^+(aq) \rightleftharpoons M^{2+}(aq) + H_2S(aq)$$

$K = 1 \times 10^{-1}$ for NiS; $K = 1 \times 10^{-16}$ for CuS. Calculate the molar solubility of NiS and CuS in 0.3 $M$ $H^+$ and 0.1 $M$ $H_2S$, the conditions under which Group II is precipitated.

**Figure 16.5**
Five Group III cations precipitate as sulfides. These are NiS, CoS, and FeS (all of which are black), MnS (light pink), and ZnS (white). Two cations precipitate as hydroxides, $Al(OH)_3$ (white) and $Cr(OH)_3$ (gray-green).

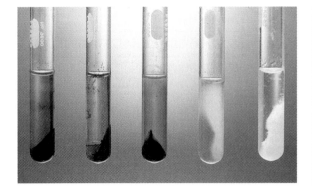

***Strategy***   For every mole of MS that dissolves, one mole of $M^{2+}$ is formed. Hence the molar solubility *s* is equal to $[M^{2+}]$. To calculate *s* from the equilibrium constant, set up the expression for *K*.

*Solution*

$$K = \frac{[M^{2+}] \times [H_2S]}{[H^+]^2} = s \times \frac{(0.1)}{(0.3)^2} = s$$

In other words, for this particular system, $s = K$

Hence for   $NiS, s = 1 \times 10^{-1}\,M$   ; for   $CuS, s = 1 \times 10^{-16}\,M$

Clearly, NiS is appreciably soluble under these conditions, but CuS is not. This explains why CuS precipitates in Group II while $Ni^{2+}$ stays in solution and does not precipitate until $[H^+]$ is lowered in Group III.

---

**Group IV** cations have soluble chlorides and sulfides. However, the alkaline earth cations in this group ($Mg^{2+}$, $Ca^{2+}$, $Ba^{2+}$) can be precipitated as carbonates, thereby separating them from the other three cations in this group. The reaction with $Mg^{2+}$ is typical:

$$Mg^{2+}(aq) + CO_3{}^{2-}(aq) \longrightarrow MgCO_3(s)$$

The alkali metal cations ($Na^+$, $K^+$) are identified by flame tests (Fig. 16.6). The $NH_4{}^+$ ion is heated with strong base to form $NH_3$, the only gas that turns red litmus blue:

$$NH_4{}^+(aq) + OH^-(aq) \longrightarrow NH_3(g) + H_2O$$

The fewer insoluble compounds a cation forms, the closer it is to the end of the qual scheme

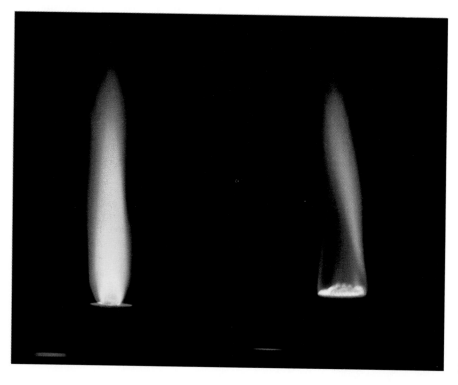

**Figure 16.6**
Flame tests are used for $Na^+$ (yellow) and $K^+$ (violet). A drop of solution is picked up on a platinum loop and immersed in the flame. The test for $K^+$ is best done with a filter that hides the strong $Na^+$ color.

**Example 16.11**   Write balanced net ionic equations for two reactions by which $Al(OH)_3$, precipitated in Group III, can be brought into solution.

***Strategy***   Like any metal hydroxide, $Al(OH)_3$ dissolves in strong acid. Another way to dissolve it is suggested by Table 16.2, which lists $Al(OH)_4^-$ as a stable complex ion.

***Solution***

$$Al(OH)_3(s) + 3H^+(aq) \longrightarrow Al^{3+}(aq) + 3H_2O$$

$$Al(OH)_3(s) + OH^-(aq) \longrightarrow Al(OH)_4^-(aq)$$

Hydroxides like $Al(OH)_3$ that react with either strong acid ($H^+$) or strong base ($OH^-$) are described as being **amphoteric.**

---

# CHEMISTRY

## *Beyond the Classroom*

**Figure 16.A**
Dental products containing $F^-$ ions. (Charles D. Winters)

# Fluoridation of Drinking Water

Many years ago, it was found that children in areas where the drinking water contained relatively high concentrations of $F^-$ ions from natural sources showed a remarkably low incidence of tooth decay (Table 16.A). Data of this sort led to the fluoridation of public water supplies, carried out first in Grand Rapids, Michigan, in 1945. Today about half the communities in the United States and Canada add metal fluorides to drinking water as needed to raise $[F^-]$ to 0.7–1.2 parts per million. Children and adults who grow up in areas with fluoridated water have about 25% fewer cavities.

**TABLE 16.A   Correlation of $[F^-]$ in Drinking Water with Tooth Decay in Children 12–14 Years of Age (1942 Study)**

| Location | $[F^-]$ (ppm) | % of Children with No Decay | Avg. No. Decayed Teeth per Child |
|---|---|---|---|
| Colorado Springs, CO | 2.6 | 28.5 | 2.5 |
| Galesburg, IL | 1.9 | 27.8 | 2.4 |
| East Moline, IL | 1.2 | 20.4 | 3.0 |
| Kewanee, IL | 0.9 | 17.9 | 3.4 |
| Pueblo, CO | 0.6 | 10.6 | 4.1 |
| Lima, OH | 0.3 | 2.2 | 6.5 |
| Elkhart, IN | 0.1 | 1.4 | 8.2 |

On average, an adult takes in about 2.4 mg/day of $F^-$ ions from drinking fluoridated water. This compares with perhaps 0.4 mg/day from dietary sources; tea and seafoods are rich in fluoride. Another source these days is dental products. Most toothpastes contain small amounts of a metal fluoride, typically sodium fluoride (NaF) or tin(II) fluoride, $SnF_2$ (Figure 16.A). Mouthwashes are usually fluoridated, as are the gels dentists use to coat teeth.

The protective action of $F^-$ ions is explained at least in part by a chemical reaction with tooth enamel composed of hydroxyapatite, whose formula may be written as $Ca(OH)_2 \cdot 3Ca_3(PO_4)_2$. The $OH^-$ ions in this compound are readily replaced by $F^-$, which has the same charge as $OH^-$ and is nearly the same size. The product is fluor-

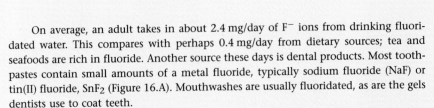

apatite, $CaF_2 \cdot 3Ca_3(PO_4)_2$. Just as $CaF_2$ ($K_{sp} = 1.5 \times 10^{-10}$) is much less soluble than $Ca(OH)_2$ ($K_{sp} = 4.0 \times 10^{-6}$), so fluorapatite is less soluble than hydroxyapatite. Beyond that, fluorapatite is less soluble in acid, since the $F^-$ ion is a much weaker base than $OH^-$:

$$F^-(aq) + H^+(aq) \rightleftharpoons HF(aq) \qquad K = 1.4 \times 10^3$$

$$OH^-(aq) + H^+(aq) \rightleftharpoons H_2O \qquad K = 1.0 \times 10^{14}$$

Tooth decay starts when bacteria in the mouth convert sugar or other carbohydrates into organic acids. These acids attack the enamel of the tooth, exposing the interior to decay.

In some ground water, $[F^-]$ is well above that produced by fluoridation. To understand why this happens, consider the solubility of calcium fluoride, $CaF_2$, a common mineral in many areas (Figure 16.B).

$$s = (K_{sp}/4)^{1/3} = (1.5 \times 10^{-10}/4)^{1/3} = 3.3 \times 10^{-4} \text{ mol/L}$$

This corresponds to a fluoride concentration of 0.013 g/L or 13 ppm:

$$[F^-] = 2s = 6.6 \times 10^{-4} \frac{\text{mol}}{\text{L}} \times 19.0 \frac{\text{g}}{\text{mol}} = 0.013 \frac{\text{g}}{\text{L}} = 13 \text{ ppm}$$

Fortunately, $F^-$ ion concentrations are rarely, if ever, as high as 13 ppm in North America. In general
— if $[F^-] > 10$ ppm, longtime exposure can lead to a crippling bone disease, skeletal fluorosis, which is common in parts of India.
— if $[F^-] > 5$ ppm, children's teeth can develop with mottled enamel. Chalky white patches form, along with yellowish stains.
Taking this into account, the Environmental Protection Agency has set an upper limit of 4 ppm for $F^-$ in drinking water.

**Figure 16.B**
A fluorite crystal ($CaF_2$).

# CHAPTER HIGHLIGHTS

## *Key Concepts*

**1.** Set up the expression for $K_{sp}$ for any ionic solid
(Example 16.1; Problems 1–4)
**2.** Use the value of $K_{sp}$ to
—calculate the concentration of one ion, knowing that of the other
(Example 16.2; Problems 5–8)
—determine whether a precipitate will form
(Examples 16.3, 16.4; Problems 9–14, 47, 52)
—calculate the solubility
(Examples 16.5, 16.6; Problems 15–20)
—determine which ion will precipitate first from a mixture
(Example 16.10; Problems 41, 42, 51)
**3.** Calculate $K$ for
—the dissolving of a metal hydroxide in strong acid
(Example 16.7; Problems 27, 31)
—the dissolving of a precipitate in a complexing agent
(Example 16.9; Problems 28–30, 35, 36, 53)
**4.** Write balanced net ionic equations to explain why a precipitate dissolves in
—strong acid
(Examples 16.8, 16.11; Problems 21, 22)
—$NH_3$ or $OH^-$
(Example 16.11; Problems 23–26)

## Key Terms

| | | |
|---|---|---|
| equilibrium rules | formation constant, $K_f$ | pH |
| —coefficient | ion product, $P$ | precipitate |
| —multiple | molar solubility, $s$ | solubility product constant, $K_{sp}$ |
| —reciprocal | | |

## Summary Problem

Consider zinc hydroxide. It can be formed when solutions of zinc nitrate and sodium hydroxide are mixed. Its $K_{sp}$ is $4 \times 10^{-17}$.

(a) Write the $K_{sp}$ expression for zinc hydroxide.

(b) How many moles of NaOH must be added to 1.00 L of 0.0500 $M$ Zn(NO$_3$)$_2$ to just start the precipitation of zinc hydroxide?

(c) Will a precipitate form if 5.00 mL of 0.0300 $M$ NaOH is added to 100.0 mL of 0.125 $M$ Zn(NO$_3$)$_2$?

(d) How many milligrams of zinc hydroxide are necessary to prepare 1.00 L of a saturated Zn(OH)$_2$ solution?

(e) How many milligrams of zinc hydroxide can be dissolved in a 0.100 $M$ ZnCl$_2$ solution?

(f) Which would precipitate first upon adding 0.100 $M$ NaOH to a 1.00 L solution containing 0.0500 moles of Zn$^{2+}$ and 0.0200 moles of Al$^{3+}$, Zn(OH)$_2$ or Al(OH)$_3$? ($K_{sp}$ Al(OH)$_3$ = $2 \times 10^{-31}$)

(g) Write the equation for dissolving Zn(OH)$_2$ with a strong acid. Calculate $K$ for this reaction.

(h) Write the equation for dissolving Zn(OH)$_2$ with a strong base. Calculate $K$ for this reaction. ($K_f$ Zn(OH)$_4{}^{2-}$ = $3 \times 10^{14}$)

(i) Write the equation for dissolving Zn(OH)$_2$ with an aqueous solution of ammonia. Assume that the ammonia instead of the hydroxide complex is formed. Calculate $K$ for this reaction. ($K_f$ Zn(NH$_3$)$_4{}^{2+}$ = $3.6 \times 10^8$)

(j) How many milligrams of Zn(OH)$_2$ can be dissolved by complex formation in 1.00 L of 0.200 $M$ NaOH?

### Answers

(a) $K_{sp} = [Zn^{2+}][OH^-]^2$     (b) $3 \times 10^{-8}$     (c) yes

(d) 0.2 mg     (e) 0.001 mg     (f) Al(OH)$_3$

(g) Zn(OH)$_2$(s) + 2H$^+$(aq) $\rightleftharpoons$ Zn$^{2+}$(aq) + 2H$_2$O     $K = 4 \times 10^{11}$

(h) Zn(OH)$_2$(s) + 2OH$^-$(aq) $\rightleftharpoons$ Zn(OH)$_4{}^{2-}$(aq)     $K = 0.01$

(i) Zn(OH)$_2$(s) + 4NH$_3$(aq) $\rightleftharpoons$ Zn(NH$_3$)$_4{}^{2+}$(aq)     $K = 1 \times 10^{-8}$

(j) $4 \times 10^1$ mg

## Questions & Problems

### Expression for $K_{sp}$

*1. Write the equilibrium equation and the $K_{sp}$ expression for each of the following.

(a) Ba(C$_2$H$_3$O$_2$)$_2$     (b) Al$_2$(SO$_4$)$_3$

(c) Zr(OH)$_4$     (d) Mn(SO$_4$)$_2$

*2. Write the equilibrium equation and the $K_{sp}$ expression for each of the following.

(a) K$_2$SiF$_6$     (b) PbI$_2$     (c) Ni$_2$S$_3$     (d) Zn$_2$P$_2$O$_7$

*3. Write the equilibrium equations on which the following $K_{sp}$ expressions are based.

(a) $[Ti^{4+}][O^{2-}]^2$     (b) $[K^+]^2[Cr_2O_7{}^{2-}]$

(c) $[Al^{3+}]^2[S^{2-}]^3$     (d) $[Pb][C_2O_4{}^{2-}]$

*4. Write the equilibrium equations on which the following $K_{sp}$ expressions are based.

(a) $[Sr^{2+}][ClO_3{}^-]^2$     (b) $[Co^{2+}]^3[AsO_4{}^{3-}]^2$

(c) $[Eu^{3+}][OH^-]^3$     (d) $[Hg_2{}^{2+}][Cl^-]^2$

## $K_{sp}$ *and Precipitation*

**5.** Complete the following table for concentrations in equilibrium with $Mg_3(PO_4)_2$

|  | $[Mg^{2+}]$ | $[PO_4{}^{3-}]$ |
|---|---|---|
| **(a)** | ——— | $3 \times 10^{-5}$ |
| **(b)** | $2 \times 10^{-3}$ | ——— |
| **(c)** | $2[PO_4{}^{3-}]$ | ——— |
| **(d)** | ——— | $3[Mg^{2+}]$ |

**6.** Complete the following table for concentrations in equilibrium with $PbBr_2$.

|  | $[Pb^{2+}]$ | $[Br^-]$ |
|---|---|---|
| **(a)** | ——— | $0.007$ |
| **(b)** | $0.4$ | ——— |
| **(c)** | ——— | $\frac{1}{2}[Pb^{2+}]$ |
| **(d)** | $4[Br^-]$ | ——— |

**7.** Calculate the concentration of each of the following ions in equilibrium with $2.0 \times 10^{-4}\,M\ OH^-$.
  **(a)** $Mg^{2+}$   **(b)** $Fe^{2+}$   **(c)** $Fe^{3+}$

**8.** Calculate the concentration of each of the following ions in equilibrium with $5.0 \times 10^{-3}\,M\ Ag^+$.
  **(a)** $C_2H_3O_2{}^-$   **(b)** $CrO_4{}^{2-}$   **(c)** $PO_4{}^{3-}$

**9.** Barium nitrate is added to a solution of $0.025\,M$ sodium fluoride.
  **(a)** At what concentration of $Ba^{2+}$ does a precipitate start to form?
  **(b)** Enough barium nitrate is added to make $[Ba^{2+}] = 0.200\,M$. What is $[F^-]$? What percentage of the original fluoride ion has precipitated?

**10.** A solution contains $0.045\,M\ SO_4{}^{2-}$. Calcium chloride is added to precipitate calcium sulfate.
  **(a)** At what concentration of $Ca^{2+}$ does a precipitate first start to form?
  **(b)** Calculate $[Ca^{2+}]$ when 35% of the $SO_4{}^{2-}$ originally in solution remains.

**11.** Before the use of lead in paint was discontinued, lead sulfate was a common pigment in white paint. A solution is prepared by mixing a solution $1 \times 10^{-3}\,M$ in $SO_4{}^{2-}$ with another solution $5 \times 10^{-4}\,M$ in $Pb^{2+}$. Ignoring dilution effects, would you expect a precipitate to form? What should $[Pb^{2+}]$ be to just start precipitation?

**12.** Water from a well is found to contain 3.5 mg of iodide ion per liter of well water. If 1.0 mg of silver nitrate is added to one liter of the well water without changing its volume, will silver iodide precipitate?

**13.** A solution is prepared by mixing 25.00 mL of a $0.075\,M$ solution of iron(II) nitrate with 45.0 mL of NaOH with pH of 8.50.
  **(a)** Will precipitation occur?
  **(b)** Calculate $[Fe^{2+}]$, $[NO_3{}^-]$, $[Na^+]$, and the pH after equilibrium is established.

**14.** Twenty mg of strontium chloride is added to 95.0 mL of a $0.0200\,M$ solution of potassium chromate. Assume no volume changes.
  **(a)** Will precipitation occur?
  **(b)** Calculate $[Sr^{2+}]$, $[Cl^-]$, $[K^+]$, and $[CrO_4{}^{2-}]$ after equilibrium is established.

## *Solubility*

**15.** At 25°C, 27 mg of barium iodate must be added to 100.0 g of water to prepare a saturated solution. Assuming that the volume of the solution is 100.0 mL, what is the $K_{sp}$ for barium iodate?

**16.** At 25°C, 0.16 g of silver dichromate $(Ag_2Cr_2O_7)$ must be dissolved in 100.0 mL to prepare a saturated solution. What is the $K_{sp}$ for $Ag_2Cr_2O_7$?

**17.** Calculate the solubility (in grams per liter) of calcium phosphate in
  **(a)** pure water
  **(b)** $0.0010\,M\ CaCl_2$
  **(c)** $0.020\,M\ K_3PO_4$

**18.** Calculate the solubility (in grams per liter) of magnesium fluoride in
  **(a)** pure water
  **(b)** $0.015\,M\ Mg(NO_3)_2$
  **(c)** $0.0050\,M\ SbF_3$

**19.** One gram of $PbCl_2$ is dissolved in 1.0 L of hot water. When the solution is cooled to 25°C, will some of the $PbCl_2$ crystallize out? If so, how much?

**20.** $K_{sp}$ for silver acetate $(AgC_2H_3O_2)$ at 80°C is estimated to be $2 \times 10^{-2}$. Ten grams of silver acetate is added to 1.0 L of water at 25°C.
  **(a)** Will all the silver acetate dissolve at 25°C?
  **(b)** If the solution (assume the volume to be 1.0 L) is heated to 80°C, will all the silver acetate dissolve?

## *Dissolving Precipitates*

**\*21.** Write net ionic equations for the reactions of each of the following with strong acid.
  **(a)** $FeS$         **(b)** $Bi(OH)_3$
  **(c)** $Cu(NH_3)_4{}^{2+}$   **(d)** $MgCO_3$

**\*22.** Write net ionic equations for the reaction of $H^+$ with
  **(a)** $SrCO_3$     **(b)** $Hg_2Cl_2$
  **(c)** $Co(OH)_3$   **(d)** $Sb(OH)_4{}^-$

**\*23.** Write a net ionic equation for the reaction with ammonia by which
  **(a)** silver chloride dissolves.
  **(b)** aluminum ion forms a precipitate.
  **(c)** copper(II) forms a complex ion.

**\*24.** Write a net ionic equation for the reaction with ammonia by which
    **(a)** $Cu(OH)_2$ dissolves.
    **(b)** $Cd^{2+}$ forms a complex ion.
    **(c)** $Pb^{2+}$ forms a precipitate.
**\*25.** Write a net ionic equation for the reaction with $OH^-$ by which
    **(a)** $Sb^{3+}$ forms a precipitate.
    **(b)** antimony(III) hydroxide dissolves when more $OH^-$ is added.
    **(c)** $Sb^{3+}$ forms a complex ion.
**\*26.** Write a net ionic equation for the reaction with $OH^-$ by which
    **(a)** $Ni^{2+}$ forms a precipitate.
    **(b)** $Sn^{4+}$ forms a complex ion.
    **(c)** $Al(OH)_3$ dissolves.

## Solution Equilibria

**27.** Calculate $K$ for the reaction in which aluminum hydroxide is dissolved by 6 $M$ HCl.
**28.** Calculate $K$ for the reaction in which silver bromide is dissolved by 6 $M$ aqueous ammonia.
**29.** Calculate $K$ for the reaction

$$Al(OH)_3(s) + OH^-(aq) \rightleftharpoons Al(OH)_4^-(aq)$$

$$K_{sp}\ Al(OH)_3 = 2 \times 10^{-31} \qquad K_f\ Al(OH)_4^- = 1 \times 10^{33}$$

**30.** Calculate $K$ for the reaction

$$Cu(OH)_2(s) + 4NH_3(aq) \rightleftharpoons Cu(NH_3)_4^{2+}(aq) + 2\ OH^-(aq)$$

$$K_{sp}\ Cu(OH)_2 = 2 \times 10^{-19} \qquad K_f\ Cu(NH_3)_4^{2+} = 2 \times 10^{12}$$

**31.** Using the value of $K$ calculated in Problem 29, determine the solubility (moles/liter) of $Al(OH)_3$ at pH 11.0.
**32.** Using the value of $K$ calculated in Problem 30, determine the solubility (moles/liter) of $Cu(OH)_2$ in 6.0 $M$ $NH_3$.
**33.** Calculate the molar solubility of silver iodide in 3.0 $M$ aqueous ammonia solution.
**34.** What is the minimum $[H^+]$ that can be used to dissolve 0.1 mol/L of $Mg(OH)_2$?
**35.** For the reaction

$$Zn(OH)_2(s) + 2\ OH^-(aq) \rightleftharpoons Zn(OH)_4^{2-}(aq)$$

    **(a)** Calculate $K$.
    **(b)** Calculate $[OH^-]$ required to dissolve 1.00 g of $Zn(OH)_2$ in 1.00 L of solution.
**36.** For the reaction

$$AgCl(s) + 2NH_3(aq) \rightleftharpoons Ag(NH_3)_2^+(aq) + Cl^-(aq)$$

    **(a)** Calculate $K$.
    **(b)** Calculate $[NH_3]$ required to dissolve 1.00 g of AgCl in 1.00 L of solution.

## Qualitative Analysis

**\*37.** Complete the following table for cations.

| | Cation | Analytical Group | Precipitating Agent | Precipitate Formed |
|---|---|---|---|---|
| **(a)** | $Ag^+$ | ——— | ——— | ——— |
| **(b)** | $Bi^{3+}$ | ——— | ——— | ——— |
| **(c)** | $Co^{2+}$ | ——— | ——— | ——— |
| **(d)** | $Mg^{2+}$ | ——— | ——— | ——— |

**\*38.** Complete the following table.

| | Species | Test/Reagent | Response to Test/Reagent |
|---|---|---|---|
| **(a)** | ——— | flame test | yellow |
| **(b)** | $AgCl(s)$ | ——— | dissolves |
| **(c)** | $MgCO_3(s)$ | $H^+$ | ——— |
| **(d)** | $Cu^{2+}$ | $H_2S$ at pH 0.5 | ——— |

**\*39.** A solution may contain any of several unknown metal ions. Tests for $K^+$, $Na^+$, and $NH_4^+$ are negative. When the solution is treated with dilute HCl, a precipitate forms. The precipitate is separated, and the filtrate is acidified to a pH of 0.5. $H_2S$ is added to the filtrate. No precipitate forms. The solution is then treated with base to a pH of 9. $H_2S$ is again added. No precipitate forms. Addition of ammonium carbonate gives a precipitate. What possible cations are present?
**\*40.** A solution contains several unknown metal ions. All tests done prove negative except that a precipitate forms upon addition of $H_2S$ in basic solution. What group of cations is present?
**\*41.** Consider the following reactions:

$$CdS(s) + 2H^+(aq) \rightleftharpoons Cd^{2+}(aq) + H_2S(aq) \qquad K = 1 \times 10^{-9}$$
$$MnS(s) + 2H^+(aq) \rightleftharpoons Mn^{2+}(aq) + H_2S(aq) \qquad K = 5 \times 10^6$$

Under the following conditions: pH = 0.50 and $[H_2S]$ = 0.10 $M$—which sulfide will precipitate first?

**\*42.** Consider the following reactions:

$$HgS(s) + 2H^+(aq) \rightleftharpoons Hg^{2+}(aq) + H_2S(aq) \qquad K = 1 \times 10^{-32}$$
$$CoS(s) + 2H^+(aq) \rightleftharpoons Co^{2+}(aq) + H_2S(aq) \qquad K = 3 \times 10^{-6}$$

Using the same conditions as in Question 41, which sulfide will precipitate first?

## Unclassified

**\*43.** Shown (p. 465) in the box at the left, is the ionic solid MX, where M cations are represented by squares and X anions are represented by circles. Fill in the box at the right to show what happens to the solid after it completely dissolves in water. Show only ions, not water molecules.

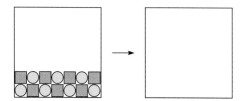

***44.** The box below represents one liter of a saturated solution of the solid $MX_2$ where M cations are represented by squares and X anions are represented by circles.

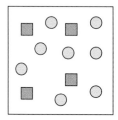

Complete the following three figures by filling one liter boxes to the right of each arrow, showing what happens when water is added to the solid on the left. Represent cations in solution by squares, anions by circles, and any undissolved solid by the symbol 🔲○.

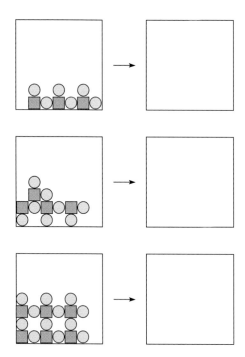

***45.** For the species $MX_2$ referred to in Question 44, show what happens when cations and anions in solution, shown in the filled boxes are mixed (spectator ions are not shown). Assume the empty box has a volume of one liter; use the same symbols as in Question 44.

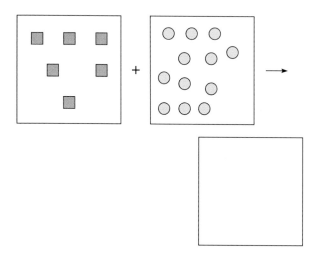

***46.** Predict what effect each of the following has on the position of the equilibrium

$$PbCl_2(s) \rightleftharpoons Pb^{2+}(aq) + 2Cl^-(aq) \qquad \Delta H = 23.4 \text{ kJ}$$

**(a)** addition of $1\,M$ $Pb(NO_3)_2$ solution
**(b)** increase in temperature
**(c)** addition of $Ag^+$ forming AgCl
**(d)** addition of $1\,M$ hydrochloric acid

**47.** A town adds 1.0 ppm of $F^-$ ion to fluoridate its water supply (fluoridation of water reduces the incidence of dental carries). If the concentration of $Ca^{2+}$ in the water is $2.0 \times 10^{-4}\,M$, will a precipitate of $CaF_2$ form when the water is fluoridated?

**48.** Using $K_{sp}$ data to calculate $K$ for the reaction, decide whether this reaction is likely to go to completion:

$$Ag_2CrO_4(s) + 2Cl^-(aq) \longrightarrow 2AgCl(s) + CrO_4{}^{2-}(aq)$$

## Challenge Problems

**49.** Ammonium chloride solutions are slightly acidic, so they are better solvents than water for insoluble hydroxides such as $Mg(OH)_2$. Find the solubility of $Mg(OH)_2$ in moles/liter in $0.2\,M$ $NH_4Cl$ and compare with the solubility in water. *Hint:* Find $K$ for the reaction

$$Mg(OH)_2(s) + 2NH_4{}^+(aq) \longrightarrow Mg^{2+}(aq) + 2NH_3(aq) + 2H_2O$$

**50.** What is the solubility of $CaF_2$ in a buffer solution containing $0.30\,M$ $HCHO_2$ and $0.20\,M$ $NaCHO_2$? (*Hint:* Consider the equation

$$CaF_2(s) + 2H^+(aq) \longrightarrow Ca^{2+}(aq) + 2HF(aq)$$

and solve the equilibrium problem.)

**51.** What is the $I^-$ concentration just as AgCl begins to precipitate when $1.0\,M$ $AgNO_3$ is slowly added to a solution containing $0.020\,M$ $Cl^-$ and $0.020\,M$ $I^-$?

52. The concentrations of various cations in seawater, in moles per liter, are

| Ion | $Na^+$ | $Mg^{2+}$ | $Ca^{2+}$ | $Al^{3+}$ | $Fe^{3+}$ |
|---|---|---|---|---|---|
| **Molarity ($M$)** | 0.46 | 0.056 | 0.01 | $4 \times 10^{-7}$ | $2 \times 10^{-7}$ |

(a) At what $[OH^-]$ does $Mg(OH)_2$ start to precipitate?
(b) At this concentration, will any of the other ions precipitate?
(c) If enough $OH^-$ is added to precipitate 50% of the $Mg^{2+}$, what percentage of each of the other ions will be precipitated?
(d) Under the conditions in (c), what mass of precipitate will be obtained from one liter of seawater?

53. Consider the equilibrium

$$Zn(NH_3)_4{}^{2+}(aq) + 4OH^-(aq) \rightleftharpoons Zn(OH)_4{}^{2-}(aq) + 4NH_3(aq)$$

(a) Calculate $K$ for this reaction.
(b) What is the ratio $[Zn(NH_3)_4{}^{2+}]/[Zn(OH)_4{}^{2-}]$ in a solution $1.0\,M$ in $NH_3$?

54. Using the equilibrium constants in the appendix, calculate $K$ for the reaction

$$Ag(NH_3)_2{}^+(aq) + 2H^+(aq) + Cl^-(aq) \rightleftharpoons AgCl(s) + 2NH_4{}^+(aq).$$

There are lots of different kinds of "greenhouse effects" (p. 488).
(William H. Allen Photography)

# Spontaneity of Reaction

## 17

I am a sleepless

Slowfaring eater,

Maker of rust and rot

In your bastioned fastenings,

I am the crumbler: tomorrow

—CARL SANDBURG

*Under*

The goal of this chapter is to answer a basic question: Will a given reaction occur "by itself" at a particular temperature and pressure, without the exertion of any outside force? In that sense, is the reaction *spontaneous?* This is a critical question in just about every area of science and technology. A synthetic organic chemist looking for a source of acetylene, $C_2H_2$, would like to know whether this compound can be made by heating the elements together. A metallurgist trying to produce titanium metal from $TiO_2$ would like to know what reaction to use; would hydrogen, carbon, or aluminum be feasible reducing agents?

To develop a general criterion for spontaneity, we will apply the principles of *thermodynamics,* the science that deals with heat and energy effects. Three different thermodynamic functions are of value here.

1. Δ*H,* the change in enthalpy (Chap. 8); a *negative* value of Δ*H* tends to make a reaction spontaneous.
2. Δ*S,* the change in entropy (Section 17.2); a *positive* value of Δ*S* tends to make a reaction spontaneous.
3. Δ*G,* the change in free energy (Sections 17.3, 17.4); a reaction at constant temperature and pressure will be *spontaneous* if Δ*G* is *negative,* no ifs, ands, or buts.

Besides serving as a general criterion for spontaneity, the free energy change can be used to

— determine the effect of temperature, pressure, and concentration on reaction spontaneity (Section 17.5)
— calculate the equilibrium constant for a reaction (Section 17.6)
— determine whether coupled reactions will be spontaneous (Section 17.7)

## 17.1 Spontaneous Processes

All of us are familiar with certain spontaneous processes. For example,

— an ice cube melts when added to a glass of water at room temperature

$$H_2O(s) \longrightarrow H_2O(l)$$

— a mixture of hydrogen and oxygen burns if ignited by a spark

$$2H_2(g) + O_2(g) \longrightarrow 2H_2O(l)$$

— an iron (steel) tool exposed to moist air rusts

$$2Fe(s) + \tfrac{3}{2}O_2(g) + 3H_2O(l) \longrightarrow 2Fe(OH)_3(s)$$

In other words, these three reactions are spontaneous at 25°C and 1 atm.

The word "spontaneous" does not imply anything about how rapidly a reaction occurs. Some spontaneous reactions, notably the rusting of iron, are quite slow. Often a reaction that is spontaneous does not occur without some sort of stimulus to get the reaction started. A mixture of hydrogen and oxygen shows no sign of reaction in the absence of a spark or match. Once started, though, a spontaneous reaction continues by itself without further input of energy from the outside.

If a reaction is spontaneous under a given set of conditions, the reverse reaction must be nonspontaneous. For example, water does not decompose to the elements by the reverse of the reaction referred to above.

$$2H_2O(l) \longrightarrow 2H_2(g) + O_2(g) \quad \text{nonspontaneous}$$

However, it is often possible to bring about a nonspontaneous reaction by supplying energy in the form of work. Electrolysis can be used to decompose water to the elements. To do this, electrical energy must be furnished, perhaps from a storage battery.

## The Energy Factor

*Many spontaneous processes proceed with a decrease of energy.* Boulders roll downhill, not uphill. A storage battery discharges when you leave your car's headlights on. Extrapolating to chemical reactions, one might guess that spontaneous reactions would be exothermic ($\Delta H < 0$). A century ago, P. M. Berthelot in Paris and Julius Thomsen in Copenhagen proposed this as a general principle, applicable to all reactions.

It turns out that almost all exothermic chemical reactions are indeed spontaneous at 25°C and 1 atm. Consider, for example, the formation of water from the elements and the rusting of iron:

$$2H_2(g) + O_2(g) \longrightarrow 2H_2O(l) \qquad \Delta H = -571.6 \text{ kJ}$$

$$2Fe(s) + \tfrac{3}{2}O_2(g) + 3H_2O(l) \longrightarrow 2Fe(OH)_3(s) \qquad \Delta H = -780.6 \text{ kJ}$$

For both of these spontaneous reactions, $\Delta H$ is a negative quantity.

On the other hand, this simple rule fails for many familiar phase changes. An example is the melting of ice. This takes place spontaneously at 1 atm above 0°C, even though it is endothermic:

$$H_2O(s) \longrightarrow H_2O(l) \qquad \Delta H = +6.0 \text{ kJ}$$

There is still another basic objection to using the sign of $\Delta H$ as a general criterion for spontaneity. Endothermic reactions that are nonspontaneous at room temperature often become spontaneous when the temperature is raised. Consider, for example, the decomposition of limestone:

$$CaCO_3(s) \longrightarrow CaO(s) + CO_2(g) \qquad \Delta H = +178.3 \text{ kJ}$$

At 25°C and 1 atm, this reaction is nonspontaneous. Witness the existence of the white cliffs of Dover and other limestone deposits over eons of time. How-

A spark is OK, but a continuous input of energy isn't

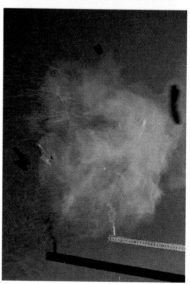

When a hydrogen-filled balloon is ignited from a distance (caution!), it explodes. The reaction $2H_2(g) + O_2(g) \longrightarrow 2H_2O(l)$ is spontaneous.

ever, if the temperature is raised to about 1100 K, the limestone decomposes to give off carbon dioxide gas at 1 atm. In other words, this endothermic reaction becomes spontaneous at high temperatures. This is true despite the fact that $\Delta H$ remains about 178 kJ, nearly independent of temperature.

## The Randomness Factor

Clearly, the direction of a spontaneous change is not always determined by the tendency for a system to go to a state of lower energy. There is another natural tendency that must be taken into account to predict the direction of spontaneity. *Nature tends to move spontaneously from a state of lower probability to one of higher probability.* Or, as G. N. Lewis put it,

> *Each system which is left to its own will, over time, change toward a condition of maximum probability.*

To illustrate what these statements mean, consider a pastime far removed from chemistry: tossing dice. If you've ever shot craps (and maybe even if you haven't), you know that when a pair of dice is thrown, a 7 is much more likely to come up than a 12. Figure 17.1 shows why this is the case. There are six different ways to throw a 7 and only one way to throw a 12. Over time, dice will come up 7 six times as often as 12. A 7 is a state of "high" probability; a 12 is a state of "low" probability.

Now, let's consider a process a bit closer to chemistry (Fig. 17.2). Two different gases, let us say $H_2$ and $N_2$, are originally contained in different glass bulbs, separated by a stopcock. When the stopcock is opened, the two different kinds of molecules distribute themselves evenly between the two bulbs. Eventually, half of the $H_2$ molecules will end up in the left bulb and half in the right; the same holds for the $N_2$ molecules. Each gas achieves its own most probable distribution, independent of the presence of the other gas.

We could explain the results of this experiment the way we did before; the final distribution is clearly much more probable than the initial distribution. There is, however, another useful way of looking at this process. The system has gone from a highly ordered state (all the $H_2$ molecules on the left, all the $N_2$ molecules on the right) to a more disordered, or random, state in which the molecules are distributed evenly between the two bulbs. The same situation holds when marbles rather than molecules are mixed (Fig. 17.3). *In general, nature tends to move spontaneously from more ordered to more random states.*

This statement is quite easy for parents or students to understand. Your room tends to get messy because an ordered room has few options for objects to be moved around (socks on the floor is not an option for an orderly room). The comedian Bill Cosby insists that with an army of 80 two-year-olds he could take over any country in the world, because they have a remarkable ability for disorganization.

**Figure 17.1**

Certain "states" are more probable than others. For example, when you toss a pair of dice, a 7 is much more likely to come up than a 12.

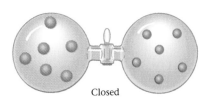

Closed

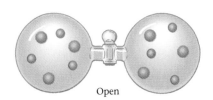

Open

**Figure 17.2**

Different kinds of gas molecules mix spontaneously, going from a more ordered to a more random state.

**Example 17.1**  Choose the state that is more random and thus more probable.

(a) A chessboard before the first move or while a game is in process.
(b) A solved jigsaw puzzle or the unassembled pieces in a box.
(c) Humpty Dumpty after the fall or Humpty Dumpty together again.

## Solution

(a) The chess game in progress is more random. There are many more ways the pieces on the board could be arranged when the game is going on.

(b) A solved jigsaw puzzle is quite orderly. A box has all the pieces at random.

(c) Humpty Dumpty after the fall is more probable. Putting him together again is nonspontaneous; all the king's horses and all the king's men couldn't do it.

**Figure 17.3**
Some states are much more probable than others. If you shake red and black marbles with each other, the random distribution at the left is much more probable than the highly ordered distribution at the right.
(Charles Steele)

## 17.2 Entropy, *S*

The randomness factor can be treated quantitatively in terms of a function called **entropy,** symbol *S*. In general, the more probable a state or the more random the distribution of molecules, the greater the entropy.

Entropy, like enthalpy (Chap. 8) is a state property. That is, the entropy depends only upon the state of a system, not upon its history. The entropy change is determined by the entropies of the final and initial states, not upon the path followed from one state to another.

$$\Delta S = S_{final} - S_{initial}$$

Several factors influence the amount of entropy that a system has in a particular state. In general,

— *a liquid has a higher entropy than the solid from which it is formed.* In a solid, the atoms, molecules, or ions are fixed in position; in the liquid, these particles are free to move past one another. In that sense, the liquid structure is more random, the solid more ordered.

— *a gas has a higher entropy than the liquid from which it is formed.* Upon vaporization, the particles acquire greater freedom to move about. They are distributed throughout the entire container instead of being restricted to a small volume.

— *increasing the temperature of a substance increases its entropy.* Raising the temperature increases the kinetic energy of the molecules (or atoms or ions) and hence their freedom of motion. In the solid, the molecules vibrate with a greater amplitude at higher temperatures. In a liquid or a gas, they move about more rapidly.

These effects are shown in Figure 17.4 p. 473, where the entropy of ammonia, $NH_3$, is plotted versus temperature. Note that the entropy of solid ammonia at 0 K is zero. This reflects the fact that molecules are completely ordered in the solid state at this temperature; there is only one way in which this system can be attained. More generally, the ***third law of thermodynamics*** tells us that *a completely ordered pure crystalline solid has an entropy of zero at 0 K.*

We'll get to the second law shortly

Notice from the figure that an increase in temperature alone has relatively little effect on entropy. The slope of the curve is small in regions where only one phase is present. In contrast, there is a large jump in entropy when the solid melts and an even larger one when the liquid vaporizes. This behavior is typical of all substances; fusion and vaporization are accompanied by relatively large increases in entropy.

**Figure 17.4**
Molar entropy of $NH_3$ as a function of temperature. Note the large increases upon fusion and vaporization.

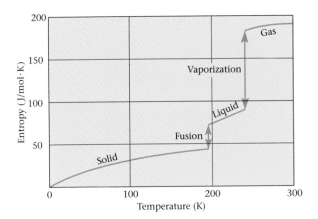

What is the sign of $\Delta S$ for: egg $\longrightarrow$ omelet?

---

**Example 17.2**   Predict whether $\Delta S$ is positive or negative for each of the following processes:

(a) Taking Dry Ice from a freezer where its temperature is $-80°C$ and allowing it to warm to room temperature.
(b) Dissolving bromine in hexane.
(c) Condensing gaseous bromine to liquid bromine.

***Strategy***   Consider the relative disorder of final and initial states; remember that entropy increases in the order solid $<$ liquid $<$ gas.

***Solution***

(a) This process involves an increase in temperature and a phase change from solid to gas:   $\Delta S > 0.$

(b) The solution is more random than the original liquids:   $\Delta S > 0.$

(c) This is a phase change from gas to liquid, from disorder to order,   $\Delta S < 0.$

---

## Standard Molar Entropies

The entropy of a substance, unlike its energy or enthalpy, can be evaluated directly. The details of how this is done are beyond the level of this text, but Figure 17.4, (top of page), shows the results for one substance, ammonia. From such a plot you can read off the **standard molar entropy** at 1 atm pressure and any given temperature, most often 25°C. This quantity is given the symbol $S°$ and has the units of joules per mole per kelvin (J/mol · K). From Figure 17.4, it appears that

$$S° \, NH_3(g) \text{ at } 25°C = 192 \text{ J/mol} \cdot K$$

Standard molar entropies of elements, compounds, and aqueous ions are listed in Table 17.1. Notice that

— *elements have nonzero standard entropies.* This means that in calculating the standard entropy change for a reaction, $\Delta S°$, elements as well as compounds must be taken into account.

Tᴀʙʟᴇ 17.1   **Standard Entropies at 25°C (J/mol · K) of Elements and Compounds at 1 atm, Aqueous Ions at 1 $M$**

| Elements | | | | | | | | | |
|---|---|---|---|---|---|---|---|---|---|
| Ag(s) | 42.6 | Cl$_2$(g) | 223.0 | I$_2$(s) | 116.1 | O$_2$(g) | 205.0 | | |
| Al(s) | 28.3 | Cr(s) | 23.8 | K(s) | 64.2 | Pb(s) | 64.8 | | |
| Ba(s) | 62.8 | Cu(s) | 33.2 | Mg(s) | 32.7 | P$_4$(s) | 164.4 | | |
| Br$_2$(l) | 152.2 | F$_2$(g) | 202.7 | Mn(s) | 32.0 | S(s) | 31.8 | | |
| C(s) | 5.7 | Fe(s) | 27.3 | N$_2$(g) | 191.5 | Si(s) | 18.8 | | |
| Ca(s) | 41.4 | H$_2$(g) | 130.6 | Na(s) | 51.2 | Sn(s) | 51.6 | | |
| Cd(s) | 51.8 | Hg(l) | 76.0 | Ni(s) | 29.9 | Zn(s) | 41.6 | | |

| Compounds | | | | | | | | | |
|---|---|---|---|---|---|---|---|---|---|
| AgBr(s) | 107.1 | CaCl$_2$(s) | 104.6 | H$_2$O(g) | 188.7 | NH$_4$NO$_3$(s) | 151.1 | | |
| AgCl(s) | 96.2 | CaCO$_3$(s) | 92.9 | H$_2$O(l) | 69.9 | NO(g) | 210.7 | | |
| AgI(s) | 115.5 | CaO(s) | 39.8 | H$_2$O$_2$(l) | 109.6 | NO$_2$(g) | 240.0 | | |
| AgNO$_3$(s) | 140.9 | Ca(OH)$_2$(s) | 83.4 | H$_2$S(g) | 205.7 | N$_2$O$_4$(g) | 304.2 | | |
| Ag$_2$O(s) | 121.3 | CaSO$_4$(s) | 106.7 | H$_2$SO$_4$(l) | 156.9 | NaCl(s) | 72.1 | | |
| Al$_2$O$_3$(s) | 50.9 | CdCl$_2$(s) | 115.3 | HgO(s) | 70.3 | NaF(s) | 51.5 | | |
| BaCl$_2$(s) | 123.7 | CdO(s) | 54.8 | KBr(s) | 95.9 | NaOH(s) | 64.5 | | |
| BaCO$_3$(s) | 112.1 | Cr$_2$O$_3$(s) | 81.2 | KCl(s) | 82.6 | NiO(s) | 38.0 | | |
| BaO(s) | 70.4 | CuO(s) | 42.6 | KClO$_3$(s) | 143.1 | PbBr$_2$(s) | 161.5 | | |
| BaSO$_4$(s) | 132.2 | Cu$_2$O(s) | 93.1 | KClO$_4$(s) | 151.0 | PbCl$_2$(s) | 136.0 | | |
| CCl$_4$(l) | 216.4 | CuS(s) | 66.5 | KNO$_3$(s) | 133.0 | PbO(s) | 66.5 | | |
| CHCl$_3$(l) | 201.7 | Cu$_2$S(s) | 120.9 | MgCl$_2$(s) | 89.6 | PbO$_2$(s) | 68.6 | | |
| CH$_4$(g) | 186.2 | CuSO$_4$(s) | 107.6 | MgCO$_3$(s) | 65.7 | PCl$_3$(g) | 311.7 | | |
| C$_2$H$_2$(g) | 200.8 | Fe(OH)$_3$(s) | 106.7 | MgO(s) | 26.9 | PCl$_5$(g) | 364.5 | | |
| C$_2$H$_4$(g) | 219.5 | Fe$_2$O$_3$(s) | 87.4 | Mg(OH)$_2$(s) | 63.2 | SiO$_2$(s) | 41.8 | | |
| C$_2$H$_6$(g) | 229.5 | Fe$_3$O$_4$(s) | 146.4 | MgSO$_4$(s) | 91.6 | SnO$_2$(s) | 52.3 | | |
| C$_3$H$_8$(g) | 269.9 | HBr(g) | 198.6 | MnO(s) | 59.7 | SO$_2$(g) | 248.1 | | |
| CH$_3$OH(l) | 126.8 | HCl(g) | 186.8 | MnO$_2$(s) | 53.0 | SO$_3$(g) | 256.7 | | |
| C$_2$H$_5$OH(l) | 160.7 | HF(g) | 173.7 | NH$_3$(g) | 192.3 | ZnI$_2$(s) | 161.1 | | |
| CO(g) | 197.6 | HI(g) | 206.5 | N$_2$H$_4$(l) | 121.2 | ZnO(s) | 43.6 | | |
| CO$_2$(g) | 213.6 | HNO$_3$(l) | 155.6 | NH$_4$Cl(s) | 94.6 | ZnS(s) | 57.7 | | |

| Cations | | | | Anions | | | |
|---|---|---|---|---|---|---|---|
| Ag$^+$(aq) | 72.7 | Hg$^{2+}$(aq) | −32.2 | Br$^-$(aq) | 82.4 | HPO$_4^{2-}$(aq) | −33.5 |
| Al$^{3+}$(aq) | −321.7 | K$^+$(aq) | 102.5 | CO$_3^{2-}$(aq) | −56.9 | HSO$_4^-$(aq) | 131.8 |
| Ba$^{2+}$(aq) | 9.6 | Mg$^{2+}$(aq) | −138.1 | Cl$^-$(aq) | 56.5 | I$^-$(aq) | 111.3 |
| Ca$^{2+}$(aq) | −53.1 | Mn$^{2+}$(aq) | −73.6 | ClO$_3^-$(aq) | 162.3 | MnO$_4^-$(aq) | 191.2 |
| Cd$^{2+}$(aq) | −73.2 | Na$^+$(aq) | 59.0 | ClO$_4^-$(aq) | 182.0 | NO$_2^-$(aq) | 123.0 |
| Cu$^+$(aq) | 40.6 | NH$_4^+$(aq) | 113.4 | CrO$_4^{2-}$(aq) | 50.2 | NO$_3^-$(aq) | 146.4 |
| Cu$^{2+}$(aq) | −99.6 | Ni$^{2+}$(aq) | −128.9 | Cr$_2$O$_7^{2-}$(aq) | 261.9 | OH$^-$(aq) | −10.8 |
| Fe$^{2+}$(aq) | −137.7 | Pb$^{2+}$(aq) | 10.5 | F$^-$(aq) | −13.8 | PO$_4^{3-}$(aq) | −222 |
| Fe$^{3+}$(aq) | −315.9 | Sn$^{2+}$(aq) | −17.4 | HCO$_3^-$(aq) | 91.2 | S$^{2-}$(aq) | −14.6 |
| H$^+$(aq) | 0.0 | Zn$^{2+}$(aq) | −112.1 | H$_2$PO$_4^-$(aq) | 90.4 | SO$_4^{2-}$(aq) | 20.1 |

— *standard molar entropies of pure substances* (elements and compounds) *are always positive quantities* ($S° > 0$).
— *aqueous ions may have negative $S°$ values.* This is a consequence of the arbitrary way in which ionic entropies are defined, taking

$$S° \; H^+(aq) = 0$$

The fluoride ion has a standard entropy 13.8 units less than that of $H^+$; hence $S°\ F^-(aq) = -13.8$ J/mol K.

You will also notice that gases, as a group, have higher entropies than liquids or solids. Moreover, among substances of similar structure and physical state, entropy usually increases with molar mass. Compare, for example, the hydrocarbons

$$CH_4(g) \qquad S° = 186.2 \text{ J/mol} \cdot \text{K}$$

$$C_2H_6(g) \qquad S° = 229.5 \text{ J/mol} \cdot \text{K}$$

$$C_3H_8(g) \qquad S° = 269.9 \text{ J/mol} \cdot \text{K}$$

As the molecule becomes more complex, there are more ways for the atoms to move about with respect to one another, leading to a higher entropy.

## $\Delta S°$ for Reactions

Table 17.1 can be used to calculate the standard entropy change, $\Delta S°$, for reactions, using the relation

$$\Delta S° = \sum S°_{\text{products}} - \sum S°_{\text{reactants}}$$

In taking these sums, the standard molar entropies are multiplied by the number of moles specified in the balanced chemical equation.

To show how this relation is used, consider the reaction

$$CaCO_3(s) \longrightarrow CaO(s) + CO_2(g)$$

$$\Delta S° = 1 \text{ mol}\left(39.8\,\frac{\text{J}}{\text{mol} \cdot \text{K}}\right) + 1 \text{ mol}\left(213.6\,\frac{\text{J}}{\text{mol} \cdot \text{K}}\right) - 1 \text{ mol}\left(92.9\,\frac{\text{J}}{\text{mol} \cdot \text{K}}\right)$$

$$= 39.8 \text{ J/K} + 213.6 \text{ J/K} - 92.9 \text{ J/K} = +160.5 \text{ J/K}$$

You will notice that $\Delta S°$ for the decomposition of calcium carbonate is a positive quantity. This is reasonable since the gas formed, $CO_2$, has a much higher molar entropy than either of the solids, CaO or $CaCO_3$. As a matter of fact, it is almost always found that *a reaction that results in an increase in the number of moles of gas is accompanied by an increase in entropy. Conversely, if the number of moles of gas decreases, $\Delta S°$ is a negative quantity.* Consider, for example, the reaction

*If $\Delta n_g$ is zero, $\Delta S$ is usually small*

$$2H_2(g) + O_2(g) \longrightarrow 2H_2O(l)$$

$$\Delta S° = 2S°\ H_2O(l) - 2S°\ H_2(g) - S°\ O_2(g)$$

$$= 139.8 \text{ J/K} - 261.2 \text{ J/K} - 205.0 \text{ J/K} = -326.4 \text{ J/K}$$

---

**Example 17.3**    Calculate $\Delta S°$ for the process by which calcium hydroxide dissolves in water: $Ca(OH)_2(s) \longrightarrow Ca^{2+}(aq) + 2OH^-(aq)$.

**Strategy**    Apply the relation

$$\Delta S° = \sum S°_{\text{products}} - \sum S°_{\text{reactants}}$$

Use data in Table 17.1; take account of the coefficients in the balanced equation.

## Solution

$$\Delta S° = S° \; Ca^{2+}(aq) + 2S° \; OH^-(aq) - S° \; Ca(OH)_2(s)$$

$$= -53.1 \text{ J/K} + 2(-10.8 \text{ J/K}) - 83.4 \text{ J/K} = \boxed{-158.1 \text{ J/K}}$$

## The Second Law of Thermodynamics

The relationship between entropy change and spontaneity can be expressed through a basic principle of nature known as the second law of thermodynamics. One way to state this law is to say that, *in a spontaneous process, there is a net increase in entropy, taking into account both system and surroundings.* That is,

$$\Delta S_{universe} = (\Delta S_{system} + \Delta S_{surroundings}) > 0 \qquad \text{spontaneous process}$$

(Recall from Chapter 8 that the system is that portion of the universe upon which attention is focused; the surroundings include everything else.)

  Notice that the second law refers to the total entropy change, involving both system and surroundings. For many spontaneous processes, the entropy change for the system is a *negative* quantity. Consider, for example, the rusting of iron, a spontaneous process:

*In this sense, the universe is running down*

$$2Fe(s) + \tfrac{3}{2}O_2(g) + 3H_2O(l) \longrightarrow 2Fe(OH)_3(s)$$

$\Delta S°$ for this system at 25°C and 1 atm can be calculated from a table of standard entropies; it is found to be $-358.4$ J/K. The negative sign of $\Delta S°$ is entirely consistent with the second law. All the law requires is that the entropy change of the surroundings be greater than 358.4 J/K, so that $\Delta S_{universe} > 0$.

  In principle, the second law can be used to determine whether a reaction is spontaneous. To do that, however, requires calculating the entropy change for the surroundings, which is not easy. We follow a conceptually simpler approach (Section 17.3), which deals only with the thermodynamic properties of chemical *systems*.

## 17.3   Free Energy, *G*

As pointed out earlier, there are two thermodynamic quantities that affect reaction spontaneity. One of these is the enthalpy, *H*; the other is the entropy, *S*. The problem is to put these two quantities together in such a way as to arrive at a single function whose sign will determine whether a reaction is spontaneous. This problem was first solved a century ago by J. Willard Gibbs (p. 477), who introduced a new quantity, now called the **Gibbs free energy** and given the symbol *G*. Gibbs showed that for a reaction taking place at constant pressure and temperature, $\Delta G$ represents that portion of the total energy change that is available (i.e., "free") to do useful work. If, for example, $\Delta G$ for a reaction is $-270$ kJ, it is possible to obtain 270 kJ of useful work from the reaction. Conversely, if $\Delta G$ is $+270$ kJ, at least that much energy in the form of work must be supplied to make the reaction take place.

*A spontaneous process is capable of producing useful work*

  The basic definition of the Gibbs free energy is

$$G = H - TS$$

**Figure 17.5**

For a spontaneous reaction, the free energy of the products is less than that of the reactants: $\Delta G < 0$. For a nonspontaneous reaction, the reverse is true: $\Delta G > 0$.

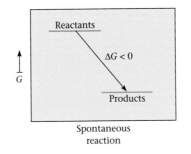

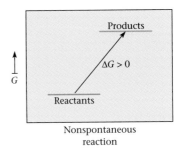

where $T$ is the absolute (Kelvin) temperature. The free energy of a substance, like its enthalpy and entropy, is a state property; its value is determined only by the state of a system, not by how it got there. Putting it another way, $\Delta G$ for a reaction depends only upon the nature of products and reactants, their concentrations or pressures, and the temperature. It does *not* depend upon the path by which the reaction is carried out.

The sign of the free energy change can be used to determine the spontaneity of a reaction carried out at constant temperature and pressure.

1. *If $\Delta G$ is negative, the reaction is spontaneous.*
2. *If $\Delta G$ is positive, the reaction will not occur spontaneously.* Instead, the reverse reaction will occur.
3. *If $\Delta G$ is 0, the system is at equilibrium;* there is no tendency for reaction to occur in either direction.

*A system has its minimum free energy at equilibrium*

Putting it another way, $\Delta G$ is a measure of the driving force of a reaction. ***Reactions, at constant pressure and temperature, go in such a direction as to decrease the free energy of the system.*** This means that the direction in which a reaction takes place is determined by the relative free energies of products and reactants. If the products have a lower free energy than the reactants ($G_{products} < G_{reactants}$), the forward reaction will occur (Fig. 17.5). If the reverse is true ($G_{reactants} < G_{products}$), the reverse reaction is spontaneous. Finally, if $G_{products} = G_{reactants}$, there is no driving force to make the reaction go in either direction.

## Relation Among $\Delta G$, $\Delta H$, and $\Delta S$

From the defining equation for free energy, it follows that at constant temperature

$$\Delta G = \Delta H - T\Delta S$$

where $\Delta G$, $\Delta H$, and $\Delta S$ are the changes in free energy, enthalpy, and entropy for a reaction. This relation, known as the **Gibbs-Helmholtz equation,** is perhaps the most important equation in chemical thermodynamics. As you can see, there are two factors that tend to make $\Delta G$ negative and hence lead to a spontaneous reaction:

1. *A negative value of $\Delta H$.* Exothermic reactions ($\Delta H < 0$) tend to be spontaneous, inasmuch as they contribute to a negative value of $\Delta G$. On the molecular level, this means that there will be a tendency to form "strong" bonds at the expense of "weak" ones.

**2. A positive value of ΔS.** If the entropy change is positive ($\Delta S > 0$), the term $-T\Delta S$ will make a negative contribution to $\Delta G$. Hence, there will be a tendency for a reaction to be spontaneous if the products are less ordered than the reactants.

In many physical changes, the entropy increase is the major driving force. This situation applies when two liquids with similar intermolecular forces, such as benzene ($C_6H_6$) and toluene ($C_7H_8$), are mixed. There is no change in enthalpy, but the entropy increases because the pure substances become diluted when they form a solution.[*]

In certain reactions, $\Delta S$ is nearly zero, and $\Delta H$ is the only important component of the driving force for spontaneity. An example is the synthesis of hydrogen fluoride from the elements:

$$\tfrac{1}{2}H_2(g) + \tfrac{1}{2}F_2(g) \longrightarrow HF(g)$$

For this reaction, $\Delta H$ is a large negative number, $-271.1$ kJ, showing that the bonds in HF are stronger than those in the $H_2$ and $F_2$ molecules. As you might expect for a gaseous reaction in which there is no change in the number of moles, $\Delta S$ is very small, about 0.0070 kJ/K. The free energy change, $\Delta G$, at 1 atm is $-273.2$ kJ at 25°C, almost identical to $\Delta H$. Even at very high temperatures, the difference between $\Delta G$ and $\Delta H$ is small, amounting to only about 14 kJ at 2000 K.

# Chemistry: *The Human Side*

Two theoreticians working in the latter half of the nineteenth century changed the very nature of chemistry by deriving the mathematical laws that govern the behavior of matter undergoing physical or chemical change. One of these was James Clerk Maxwell, whose contributions to kinetic theory were discussed in Chapter 5. The other was J. Willard Gibbs, Professor of Mathematical Physics at Yale from 1871 until his death in 1903.

In 1876 Gibbs published the first portion of a remarkable paper in the *Transactions of the Connecticut Academy of Sciences* entitled "On the Equilibrium of Heterogeneous Substances." When the paper was completed in 1878 (it was 323 pages long), the foundation was laid for the science of chemical thermodynamics. Here, for the first time, the concept of free energy appeared. Included as well were the basic principles of chemical equilibrium (Chap. 12), phase equilibrium (Chap. 9), and the relations governing energy changes in electrical cells (Chap. 18).

If Gibbs had never published another paper, this single contribution would have placed him among the greatest theoreticians in the history of science. Generations of experimental scientists have established their reputations by demonstrating in the laboratory the validity of the relationships that Gibbs derived at his desk. Many of these relationships were rediscovered by others; an example is the Gibbs-Helmholtz equation developed in 1882 by Helmholtz, a prestigious German physiologist and physicist who was completely unaware of Gibbs' work.

**J. Willard Gibbs**
(1839–1903)
(Burndy Library/AIP Emilio Segre
Visual Archives)

*continued*

[*]Formation of a *water* solution is often accompanied by a *decrease* in entropy because of hydrogen bonding or hydration effects that lead to a highly ordered solution structure. Examples include

$$C_2H_5OH(l) \longrightarrow C_2H_5OH(aq) \qquad\qquad \Delta S° = -12.2 \text{ J/K}$$
$$CaCl_2(s) \longrightarrow Ca^{2+}(aq) + 2Cl^-(aq) \qquad\qquad \Delta S° = -44.7 \text{ K/J}$$

They started paying him when he got an offer from Johns Hopkins

J. Willard Gibbs is often cited as an example of the "prophet without honor in his own country." His colleagues in New Haven and elsewhere in the United States seem not to have realized the significance of his work until late in his life. During his first ten years as a professor at Yale he received no salary. In 1920, when he was first proposed for the Hall of Fame of Distinguished Americans at New York University, he received 9 votes out of a possible 100. Not until 1950 was he elected to that body. Even today the name of J. Willard Gibbs is generally unknown among educated Americans outside of those interested in the natural sciences.

Admittedly, Gibbs himself was largely responsible for the fact that for many years his work did not attract the attention it deserved. He made little effort to publicize it; the *Transactions of the Connecticut Academy of Sciences* was hardly the leading scientific journal of its day. Gibbs was one of those rare individuals who seem to have no inner need for recognition by contemporaries. His satisfaction came from solving a problem in his mind; having done so, he was ready to proceed to other problems. His papers are not easy to read; he seldom cites examples to illustrate his abstract reasoning. Frequently, the implications of the laws that he derives are left for the readers to grasp on their own. One of his colleagues at Yale confessed many years later that none of the members of the Connecticut Academy of Sciences understood his paper on thermodynamics; as he put it, "We knew Gibbs and took his contributions on faith."

## 17.4   Standard Free Energy Change, $\Delta G°$

Although the Gibbs-Helmholtz equation is valid under all conditions, we will apply it only under standard conditions, where any gas involved in the reaction is at 1 atm pressure and all species in aqueous solution are at a concentration of 1 $M$. In other words, we will use the equation in the form

$$\Delta G° = \Delta H° - T\Delta S°$$

where $\Delta G°$ is the standard free energy change (1 atm, 1 $M$); $\Delta H°$ is the standard enthalpy change, which can be calculated from heats of formation, $\Delta H_f°$ (listed in Tables 8.3 and 8.4, Chapter 8); and $\Delta S°$ is the standard entropy change (Table 17.1).

### Calculation of $\Delta G°$ at 25°C; Free Energies of Formation

To illustrate the use of the Gibbs-Helmholtz equation

$$\Delta G° = \Delta H° - T\Delta S°$$

we apply it first to find $\Delta G°$ at 25°C (Example 17.4). To do this, the units of $\Delta H°$ and $\Delta S°$ must be consistent. Using $\Delta H°$ in kilojoules, it is necessary to convert $\Delta S°$ from joules per kelvin to *kilojoules per kelvin*.

---

**Example 17.4**   For the reaction $CaSO_4(s) \longrightarrow Ca^{2+}(aq) + SO_4{}^{2-}(aq)$, calculate

(a) $\Delta H°$   (b) $\Delta S°$   (c) $\Delta G°$ at 25°C

*Strategy*   $\Delta H°$ is calculated from Tables 8.3 and 8.4, Chapter 8, using the relation

$$\Delta H° = \sum \Delta H_f° \text{ products} - \sum \Delta H_f° \text{ reactants}$$

$\Delta S°$ is obtained from Table 17.1 using the relation

$$\Delta S° = \sum S°_{\text{products}} - \sum S°_{\text{reactants}}$$

Convert $\Delta S°$ to kJ/K using the relation $1\ \text{J} = 10^{-3}$ kJ. Finally, calculate $\Delta G°$ from the Gibbs-Helmholtz equation with $T = 298$ K.

## Solution

(a) $\Delta H° = \Delta H°_f\ Ca^{2+}(aq) + \Delta H°_f\ SO_4^{2-}(aq) - \Delta H°_f\ CaSO_4(s)$

$\quad = -542.8\ \text{kJ} - 909.3\ \text{kJ} - (-1434.1\ \text{kJ}) = \boxed{-18.0\ \text{kJ}}$

(b) $\Delta S° = S°\ Ca^{2+}(aq) + S°\ SO_4^{2-}(aq) - S°\ CaSO_4(s)$

$\quad = -53.1\ \text{J/K} + 20.1\ \text{J/K} - 106.7\ \text{J/K} = -139.7\ \text{J/K}$

$\quad = -139.7\ \dfrac{\text{J}}{\text{K}} \times \dfrac{10^{-3}\ \text{kJ}}{1\ \text{J}} = \boxed{-0.1397\ \dfrac{\text{kJ}}{\text{K}}}$

(c) $\Delta G° = \Delta H° - T\Delta S°$

$\quad = -18.0\ \text{kJ} - 298\ \text{K} \times (-0.1397\ \text{kJ/K}) = \boxed{+23.6\ \text{kJ}}$

This conversion is necessary to find $\Delta G°$ in kilojoules

Since $\Delta G°$ is positive, this reaction at standard conditions

$$CaSO_4(s) \longrightarrow Ca^{2+}(aq,\ 1\ M) + SO_4^{2-}(aq,\ 1\ M)$$

should not be spontaneous. In other words, calcium sulfate should not dissolve in water to give a 1 *M* solution. This is indeed the case. The solubility of $CaSO_4$ at 25°C is considerably less than 1 mol/L, only about 0.02 mol/L.

---

The Gibbs-Helmholtz equation can be used to calculate the *standard free energy of formation* of a compound. This quantity, $\Delta G°_f$, is analogous to the enthalpy of formation, $\Delta H°_f$. It is defined as the free energy change per mole when a compound is formed from the elements in their stable states at 1 atm.

---

**Example 17.5**   Calculate the standard free energy of formation, $\Delta G°_f$, at 25°C for $CH_4(g)$, using data in Tables 8.3 and 17.1.

***Strategy***   By definition, $\Delta G°_f$ for $CH_4$ is $\Delta G°$ for the reaction $C(s) + 2H_2(g) \longrightarrow CH_4(g)$. The calculations are entirely analogous to those in Example 17.4.

## Solution

$\quad \Delta H° = \Delta H°_f\ CH_4(g) = -74.8\ \text{kJ}$

$\quad \Delta S° = S°\ CH_4(g) - S°\ C(s) - 2S°\ H_2(g)$

$\quad\quad = 186.2\ \text{J/K} - 5.7\ \text{J/K} - 2(130.6\ \text{J/K}) = -80.7\ \text{J/K} = -0.0807\ \text{kJ/K}$

At 25°C, that is, 298 K,

$\quad \Delta G° = \Delta H° - 298\ \Delta S° = -74.8\ \text{kJ} - 298\ \text{K}(-0.0807\ \text{kJ/K}) = -50.7\ \text{kJ}$

Hence, the free energy of formation of $CH_4(g)$ at 25°C is $\boxed{-50.7\ \text{kJ/mol.}}$

---

Tables of standard free energies of formation at 25°C of compounds and ions in solution are given in Appendix 1 (along with standard heats of formation and standard entropies). Notice that, for most compounds, $\Delta G°_f$ is a negative quantity, which means that the compound can be formed spontaneously from the elements. This is true for water:

$$H_2(g) + \tfrac{1}{2}O_2(g) \longrightarrow H_2O(l) \qquad \Delta G_f^\circ \; H_2O(l) = -237.2 \text{ kJ/mol}$$

and, at least in principle, for methane:

$$C(s) + 2H_2(g) \longrightarrow CH_4(g) \qquad \Delta G_f^\circ \; CH_4(g) = -50.7 \text{ kJ/mol}$$

A few compounds, including acetylene, have positive free energies of formation ($\Delta G_f^\circ \; C_2H_2(g) = +209.2$ kJ/mol). These compounds cannot be made from the elements; indeed, they are potentially unstable with respect to the elements. In the case of acetylene, the reaction

$$C_2H_2(g) \longrightarrow 2C(s) + H_2(g) \qquad \Delta G^\circ \text{ at } 25°C = -209.2 \text{ kJ}$$

occurs with explosive violence unless special precautions are taken.

Values of $\Delta G_f^\circ$ can be used to calculate free energy changes for reactions. The relationship here is entirely analogous to that given for enthalpies in Chapter 8:

$$\Delta G^\circ{}_{reaction} = \sum \Delta G_f^\circ \; \text{products} - \sum \Delta G_f^\circ \; \text{reactants}$$

If you calculate $\Delta G^\circ$ in this way, you should keep in mind an important limitation. $\Delta G^\circ{}_{reaction}$ **is valid only at the temperature at which $\Delta G_f^\circ$ data are tabulated, in this case 25°C.** Since $\Delta G^\circ$ varies considerably with temperature, this approach is not even approximately valid at other temperatures.

*To find $\Delta G^\circ$ at temperatures other than 25°C, use the Gibbs-Helmholtz equation*

## Calculation of $\Delta G^\circ$ at Other Temperatures

To a good degree of approximation, the temperature variation of $\Delta H^\circ$ and $\Delta S^\circ$ can be neglected.[*] This means that to apply the Gibbs-Helmholtz equation

$$\Delta G^\circ = \Delta H^\circ - T\Delta S^\circ$$

at temperatures other than 25°C, you need only change the value of $T$. The quantities $\Delta H^\circ$ and $\Delta S^\circ$ can be calculated in the usual way from tables of standard enthalpies and entropies.

---

**Example 17.6**   Calculate $\Delta G^\circ$ for the reaction $Cu(s) + H_2O(g) \longrightarrow$ $CuO(s) + H_2(g)$ at 500 K.

*Strategy*   Calculate $\Delta H^\circ$ using Table 8.3 and $\Delta S^\circ$ using Table 17.1. Then apply the Gibbs-Helmholtz equation with $T = 500$ K.

*Solution*   From Table 8.3

$$\Delta H^\circ = \Delta H_f^\circ \; CuO(s) - \Delta H_f^\circ \; H_2O(g)$$

$$= -157.3 \text{ kJ} + 241.8 \text{ kJ} = +84.5 \text{ kJ}$$

From Table 17.1

$$\Delta S^\circ = S^\circ \; CuO(s) + S^\circ \; H_2(g) - S^\circ \; Cu(s) - S^\circ \; H_2O(g)$$

$$= 42.6 \text{ J/K} + 130.6 \text{ J/K} - 33.2 \text{ J/K} - 188.7 \text{ J/K}$$

$$= -48.7 \text{ J/K} = -0.0487 \text{ kJ/K}$$

---

[*]As far as the Gibbs-Helmholtz equation is concerned, there is another reason for ignoring the temperature dependence of $\Delta H$ and $\Delta S$. These two quantities always change in the same direction as the temperature changes (that is, if $\Delta H$ becomes more positive, so does $\Delta S$). Hence, the two effects tend to cancel each other.

Hence, since $\Delta G° = \Delta H° - T\Delta S°$:

$$\Delta G° = +84.5 \text{ kJ} - 500 \text{ K} \times (-0.0487 \text{ kJ/K}) = \boxed{+108.9 \text{ kJ}}$$

Since $\Delta G°$ is positive, the reaction is not spontaneous. Instead, the reverse reaction occurs at 1 atm and 500 K (227°C). Under these conditions, CuO is reduced to copper in a stream of $H_2$ gas.

From Example 17.6 and the preceding discussion, it should be clear that $\Delta G°$, unlike $\Delta H°$ and $\Delta S°$, is strongly dependent upon temperature. This comes about, of course, because of the $T$ in the Gibbs-Helmholtz equation:

$$\Delta G° = \Delta H° - T\Delta S°$$

Comparing this equation with that of a straight line,

$$y = b + mx$$

it is clear that a plot of $\Delta G°$ versus $T$ should be linear, with a slope of $-\Delta S°$ and a $y$-intercept (at 0 K) of $\Delta H°$ (Figure 17.6).

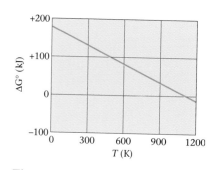

**Figure 17.6**

A plot of $\Delta G°$ vs. $T$ for the decomposition of $CaCO_3$:

$$CaCO_3(s) \longrightarrow CaO(s) + CO_2(g)$$

$\Delta G°$ decreases from +178.3 kJ at 0 K (the value of $\Delta H°$) to −14.3 kJ at 1200 K. The slope of the straight line is $-\Delta S° = -0.1605$ kJ/K.

## 17.5 Effect of Temperature, Pressure, and Concentration on Reaction Spontaneity

A change in reaction conditions can, and often does, change the direction in which a reaction occurs spontaneously. The Gibbs-Helmholtz equation in the form

$$\Delta G° = \Delta H° - T\Delta S°$$

is readily applied to deduce the effect of temperature upon reaction spontaneity. To consider the effect of pressure or concentration, we need an analogous expression for the dependence of $\Delta G$ upon these factors.

### Temperature

When the temperature of a reaction system is increased, the sign of $\Delta G°$, and hence the direction in which the reaction proceeds spontaneously, may or may not change. Whether it does or not depends upon the signs of $\Delta H°$ and $\Delta S°$. The four possible situations, deduced from the Gibbs-Helmholtz equation, are summarized in Figure 17.7.

If $\Delta H°$ and $\Delta S°$ have opposite signs (Fig. 17.7, I and III), it is impossible to reverse the direction of spontaneity by a change in temperature alone. The two terms $\Delta H°$ and $-T\Delta S°$ reinforce one another. Hence, $\Delta G°$ has the same sign at all temperatures. Reactions of this type are rather uncommon. One such reaction is that discussed in Example 17.6:

$$Cu(s) + H_2O(g) \longrightarrow CuO(s) + H_2(g)$$

Here, $\Delta H°$ is +84.5 kJ and $\Delta S°$ is −0.0487 kJ/K. Hence,

$$\Delta G° = \Delta H° - T\Delta S°$$

$$= +84.5 \text{ kJ} + T(0.0487 \text{ kJ/K})$$

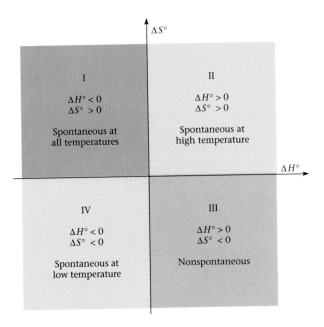

**Figure 17.7**
Spontaneity of reaction depends upon the sign and magnitude of $\Delta H°$ and $T\Delta S°$. Most commonly, $\Delta H°$ and $\Delta S°$ have the same sign and the direction of spontaneity depends upon the temperature.

Copper metal doesn't react with water vapor, period

Clearly, $\Delta G°$ is positive at all temperatures. The reaction cannot take place spontaneously at 1 atm regardless of temperature.

It is more common to find that $\Delta H°$ and $\Delta S°$ have the same sign (Fig. 17.7, II and IV). When this happens, the enthalpy and entropy factors oppose each other. $\Delta G°$ changes sign as temperature increases, and the direction of spontaneity reverses. At low temperatures, $\Delta H°$ predominates, and the exothermic reaction, which may be either the forward or the reverse reaction, occurs. As the temperature rises, the quantity $T\Delta S°$ increases in magnitude and eventually exceeds $\Delta H°$. At high temperatures, the reaction that leads to an increase in entropy occurs. In most cases, 25°C is a "low" temperature, at least at a pressure of 1 atm. This explains why exothermic reactions are usually spontaneous at room temperature and atmospheric pressure.

An example of a reaction for which $\Delta H°$ and $\Delta S°$ have the same sign is the decomposition of calcium carbonate:

$$CaCO_3(s) \longrightarrow CaO(s) + CO_2(g)$$

Here, $\Delta H° = +178.3$ kJ and $\Delta S° = +160.5$ J/K. Hence,

$$\Delta G° = +178.3 \text{ kJ} - T(0.1605 \text{ kJ/K})$$

For this reaction at "low" temperatures, e.g., 298 K,

$$\Delta G° = +178.3 \text{ kJ} - 47.8 \text{ kJ} = +130.5 \text{ kJ}$$

the $\Delta H°$ term predominates, $\Delta G°$ is positive, and the reaction is nonspontaneous at 1 atm. Conversely, at "high" temperatures, e.g., 2000 K,

$$\Delta G° = +178.3 \text{ kJ} - 321.0 \text{ kJ} = -142.7 \text{ kJ}$$

the $T\Delta S°$ term predominates, $\Delta G°$ is negative, and the reaction is spontaneous at 1 atm.

CaO is made commercially by heating $CaCO_3$

---

**Example 17.7** At what temperature does $\Delta G°$ become zero for the reaction

$$CaCO_3(s) \longrightarrow CaO(s) + CO_2(g)$$

*Strategy* This is another application of the Gibbs-Helmholtz equation. Setting $\Delta G° = 0$ gives $\Delta H° = T\Delta S°$; solving for $T$ gives $T = \Delta H°/\Delta S°$. Since you know $\Delta H°$ and $\Delta S°$, $T$ is readily calculated.

*Solution* From the preceding discussion.

$$\Delta H° = +178.3 \text{ kJ} \qquad \Delta S° = 0.1605 \text{ kJ/K}$$

Hence

$$T = \frac{\Delta H°}{\Delta S°} = \frac{178.3 \text{ kJ}}{0.1605 \text{ kJ/K}} = \boxed{1110 \text{ K}} \quad \text{i.e. } 1.110 \times 10^3 \text{ K}$$

At this temperature, the reaction is at equilibrium at 1 atm pressure. If some $CaCO_3$ is placed in a container and heated to 1110 K, the pressure of $CO_2$ developed is 1 atm.

---

The development presented in Example 17.7 is an important one from a practical standpoint. It tells us the temperature at which the direction of spontaneity changes. In this case, one can deduce that calcium carbonate must be heated to about 1110 K ($\approx 840°C$) to decompose it to calcium oxide and carbon dioxide. In another case, for the reaction

$$Fe_2O_3(s) + 3H_2(g) \longrightarrow 2Fe(s) + 3H_2O(g)$$

it turns out that $\Delta G°$, which is positive at low temperatures, becomes zero at about 400°C. This means that to reduce hematite ore, $Fe_2O_3$, to iron in a stream of hydrogen, the system must be heated to 400°C.

## Pressure and Concentration

All of the free energy calculations to this point have involved the standard free energy change, $\Delta G°$. It is possible, however, to write a general relation for the free energy change, $\Delta G$, valid under any conditions:

$P$ and $T$ affect $\Delta G$ but not $\Delta H$; $P$ affects $\Delta S$

$$\Delta G = \Delta G° + RT \ln Q$$

The quantity $Q$ that appears in this equation is the "reaction quotient" referred to in Chapter 12. It has the same mathematical form as the equilibrium constant, $K$; the difference is that the terms that appear in $Q$ are arbitrary, instantaneous pressures or concentrations rather than equilibrium values.

To use this general expression for $\Delta G$, you need to express $T$ in kelvin and $R$ in kilojoules per kelvin ($8.31 \times 10^{-3}$ kJ/K). So far as $Q$ is concerned, the general rules for expressing $K$ are followed:

— gases enter as their partial pressures in atmospheres
— species in aqueous solution enter as their molar concentrations
— pure liquids or solids do not appear; neither does the solvent in a dilute solution

**Example 17.8** Consider the reaction of zinc with strong acid:

$$Zn(s) + 2H^+(aq) \longrightarrow Zn^{2+}(aq) + H_2(g)$$

(a) Write the expression for $Q$.

(b) Calculate $\Delta G°$ at 25°C.

(c) Find $\Delta G$ at 25°C when $P_{H_2} = 750$ mm Hg, $[Zn^{2+}] = 0.10$ M, $[H^+] = 1.0 \times 10^{-4}$ M.

**Strategy** In (a), follow the rules on p. 483 for $Q$. In (b), calculate $\Delta G°$ using the Gibbs-Helmholtz equation, as in Examples 17.4–17.7. Finally (c), determine $\Delta G$ using the relation $\Delta G = \Delta G° + RT \ln Q$. Note that the pressure of $H_2$ must be expressed in atmospheres.

**Solution**

(a) $\quad Q = \dfrac{[Zn^{2+}] \times P_{H_2}}{[H^+]^2}$

(b) $\Delta H° = \Delta H_f° \ Zn^{2+} = -153.9$ kJ

$\Delta S° = S° \ Zn^{2+} + S° \ H_2 - S° \ Zn = -23.1$ J/K $= -0.0231$ kJ/K

$\Delta G° = \Delta H° - T\Delta S° = -153.9$ kJ $+ 298$ K $\times (0.0231$ kJ/K$) = \boxed{-147.0 \text{ kJ}}$

(c) $\Delta G = \Delta G° + RT \ln Q$

$= -147.0$ kJ $+ 298$ K $(8.31 \times 10^{-3}$ kJ/K$) \left[ \ln\dfrac{(0.10)(750/760)}{(1.0 \times 10^{-4})^2} \right]$

$= -147.0$ kJ $+ 39.9$ kJ $= \boxed{-107.1 \text{ kJ}}$

---

As Example 17.8 implies, changes in pressure and/or concentration can have a considerable effect upon $\Delta G$. Sometimes, when $\Delta G°$ is close to zero, changing the pressure from 1 atm to some other value can change the direction of spontaneity. Consider, for example, the reaction at 300°C:

$$NH_4Cl(s) \longrightarrow NH_3(g) + HCl(g) \qquad \Delta G° = +13.0 \text{ kJ}$$

The positive sign of $\Delta G°$ implies that ammonium chloride will not decompose at 300°C to give $NH_3$ and HCl, both at 1 atm pressure. However, notice what happens if $P_{NH_3} = P_{HCl} = 0.10$ atm, in which case $Q = (0.10)^2$:

$\Delta G = \Delta G° + RT \ln(0.10)^2$

$= +13.0$ kJ $+ (8.31 \times 10^{-3})(573)$ kJ $\times \ln 0.010$

$= +13.0$ kJ $- 21.9$ kJ $= -8.9$ kJ

This means that $NH_3$ and HCl *can* be formed, each at 0.10 atm pressure, by heating ammonium chloride to 300°C.

Another example of how a change in concentration can change the direction of spontaneity involves the reaction

$$SrCrO_4(s) \longrightarrow Sr^{2+}(aq) + CrO_4{}^{2-}(aq) \qquad \Delta G° = +25.3 \text{ kJ at } 25°C$$

Remember, when $\Delta G = 0$ the system is at equilibrium

Since $\Delta G°$ is a positive quantity, $SrCrO_4$ does not dissolve spontaneously to form a 1 M solution at 25°C (Fig. 17.8). Suppose, however, the concentrations of $Sr^{2+}$ and $CrO_4{}^{2-}$ are reduced to 0.0010 M:

$$\Delta G = \Delta G° + RT \ln [\text{Sr}^{2+}] \times [\text{CrO}_4{}^{2-}]$$

$$= +23.6 \text{ kJ} + (8.31 \times 10^{-3})(298) \text{ kJ} \times [\ln (1.0 \times 10^{-3})^2]$$

$$= +25.3 \text{ kJ} - 34.2 \text{ kJ} = -8.9 \text{ kJ}$$

Strontium chromate should, and does, dissolve spontaneously to form a 0.0010 *M* water solution. As you might expect, its solubility at 25°C lies between 1 *M* and 0.0010 *M*, at about 0.006 *M*.

$$\Delta G \text{ at } 0.006 \text{ M} = +25.3 \text{ kJ} + RT \ln (0.006)^2 \approx 0$$

**Figure 17.8**

A saturated solution of $SrCrO_4$ contains 0.006 mol/L of $Sr^{2+}$ and $CrO_4{}^{2-}$. For the reaction at 25°C: $SrCrO_4(s) \rightleftharpoons Sr^{2+}$ *(aq,* 0.006 *M)* + $CrO_4{}^{2-}$ *(aq,* 0.006 *M),* $\Delta G = 0$.

# 17.6   The Free Energy Change and The Equilibrium Constant

Throughout this chapter, we have stressed the relation between the free energy change and reaction spontaneity. For a reaction to be spontaneous at standard conditions (1 atm, 1 *M*), $\Delta G°$ must be negative. Another indicator of reaction spontaneity is the equilibrium constant, *K*; if *K* is greater than 1, the reaction is spontaneous at standard conditions. As you might suppose, the two quantities $\Delta G°$ and *K* are related. The nature of that relationship is readily deduced by starting with the general equation

$$\Delta G = \Delta G° + RT \ln Q$$

Suppose now that reaction takes place until equilibrium is established, at which point $Q = K$ and $\Delta G = 0$.

$$0 = \Delta G° + RT \ln K$$

or

$$\Delta G° = -RT \ln K$$

The equation just written is generally applicable to any system. The equilibrium constant may be the *K* referred to in our discussion of gaseous equilibrium (Chap. 12), or any of the solution equilibrium constants ($K_w$, $K_a$, $K_b$, $K_{sp}$, . . . ) discussed in subsequent chapters. Notice that $\Delta G°$ is the *standard* free energy change (gases at *1 atm*, species in solution at *1 M*). That is why, in the expression for *K*, gases enter as their partial pressures in *atmospheres* and ions or molecules in solution as their *molarities*.

---

**Example 17.9**   Using $\Delta G_f°$ tables in Appendix 1, calculate the solubility product constant, $K_{sp}$, for AgBr at 25°C.

**Strategy**   Write out the chemical equation for the solution reaction $AgBr(s) \rightleftharpoons Ag^+(aq) + Br^-(aq)$. Now apply the equation

$$\Delta G° = \sum \Delta G_f° \text{ products} - \sum \Delta G_f° \text{ reactants}$$

to calculate $\Delta G°$. Then use the equation $\Delta G° = -RT \ln K_{sp}$ to find $K_{sp}$.

**Solution**

(1) $\Delta G° = \Delta G_f° \text{ Ag}^+ + \Delta G_f° \text{ Br}^- - \Delta G_f° \text{ AgBr}$

$= 77.1 \text{ kJ} - 104.0 \text{ kJ} + 96.9 \text{ kJ} = 70.0 \text{ kJ}$

*How would you find $K_{sp}$ at 100°C?*

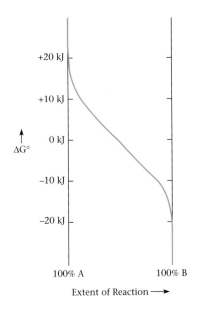

**Figure 17.9**
Dependence of extent of reaction upon the value of $\Delta G°$ for the general system $A(g) \rightleftharpoons B(g)$.

$$(2)\ \ln K_{sp} = \frac{-\Delta G°}{RT} = \frac{-70.0\ kJ}{(8.31 \times 10^{-3}\ kJ/K)(298\ K)} = -28.3$$

$$K_{sp} = \boxed{5 \times 10^{-13}}$$

This is the value listed in Chapter 16 for $K_{sp}$ of AgBr.

---

The relationship $\Delta G° = -RT \ln K$ allows us to relate the standard free energy change to the extent of reaction. Consider, for example, the simple equilibrium system

$$A(g) \rightleftharpoons B(g)$$

at 25°C. Figure 17.9 shows how the extent of reaction varies with the value of $\Delta G°$ for this system. Notice that

— if $\Delta G°$ is greater than about +20 kJ, the equilibrium constant is so small that virtually no A is converted to B. We would say that, for all practical purposes, the reaction does not occur.
— if $\Delta G°$ is less than about −20 kJ, the equilibrium constant is so large that virtually all of A is converted to B. In essence, the reaction goes to completion.
— only if $\Delta G°$ lies between +20 kJ and −20 kJ will the equilibrium mixture contain appreciable amounts of both A and B. In particular, if $\Delta G° = 0$, $K = 1$ and the equilibrium system contains equal amounts of A and B.

## 17.7  Additivity of Free Energy Changes; Coupled Reactions

Free energy changes for reactions, like enthalpy or entropy changes, are additive. That is,

if Reaction 3 = Reaction 1 + Reaction 2

then $\Delta G_3 = \Delta G_1 + \Delta G_2$

This relation can be regarded as the free energy equivalent of Hess' law (Chap. 8). To illustrate its application, consider the synthesis of $CuCl_2$ from the elements:

$$Cu(s) + \tfrac{1}{2}Cl_2(g) \longrightarrow CuCl(s) \qquad \Delta G°\ at\ 25°C = -119.9\ kJ$$
$$\underline{CuCl(s) + \tfrac{1}{2}Cl_2(g) \longrightarrow CuCl_2(s) \qquad \Delta G°\ at\ 25°C = -55.8\ kJ}$$
$$Cu(s) + Cl_2(g) \longrightarrow CuCl_2(s)$$

For the overall reaction, $\Delta G°$ at 25°C = −119.9 kJ − 55.8 kJ = −175.7 kJ.

Since free energy changes are additive, it is often possible to bring about a nonspontaneous reaction by coupling it with a reaction for which $\Delta G°$ is a large negative number. As an example, consider the preparation of iron metal from hematite ore. The reaction

$$Fe_2O_3(s) \longrightarrow 2Fe(s) + \tfrac{3}{2}O_2(g) \qquad \Delta G°\ at\ 25°C = +742.2\ kJ$$

is clearly nonspontaneous; even at temperatures as high as 2000°C, $\Delta G°$ is a positive quantity. Suppose, though, that this reaction is "coupled" with the spontaneous oxidation of carbon monoxide:

$$CO(g) + \tfrac{1}{2}O_2(g) \longrightarrow CO_2(g) \qquad \Delta G° \text{ at } 25°C = -257.1 \text{ kJ}$$

The overall reaction is spontaneous:

$$Fe_2O_3(s) \longrightarrow 2Fe(s) + \tfrac{3}{2}O_2(g)$$

$$\underline{3CO(g) + \tfrac{3}{2}O_2(g) \longrightarrow 3CO_2(g)}$$

$$Fe_2O_3(s) + 3CO(g) \longrightarrow 2Fe(s) + 3CO_2(g)$$

For the coupled reaction at 25°C,

$$\Delta G° = +742.2 \text{ kJ} + 3(-257.1 \text{ kJ}) = -29.1 \text{ kJ}$$

The negative sign of $\Delta G°$ implies that although $Fe_2O_3$ does not spontaneously decompose, it can be converted to iron by reaction with carbon monoxide. This is in fact the reaction used in the blast furnace when iron ore consisting mainly of $Fe_2O_3$ is reduced to iron (Chap. 20).

*Many industrial processes involve coupled reactions*

Coupled reactions are common in human metabolism. Spontaneous processes, such as the oxidation of glucose,

$$C_6H_{12}O_6(aq) + 6 \text{ } O_2(g) \longrightarrow 6CO_2(g) + 6H_2O \qquad \Delta G° = -2870 \text{ kJ at } 25°C$$

ordinarily do not serve directly as a source of energy. Instead these reactions are used to bring about a nonspontaneous reaction:

$$ADP(aq) + HPO_4{}^{2-}(aq) + 2H^+(aq) \longrightarrow ATP(aq) + H_2O$$

$$\Delta G° = +31 \text{ kJ at } 25°C$$

ADP (adenosine diphosphate) and ATP (adenosine triphosphate) are complex organic molecules (Fig. 17.10) that, in essence, differ only by the presence of an

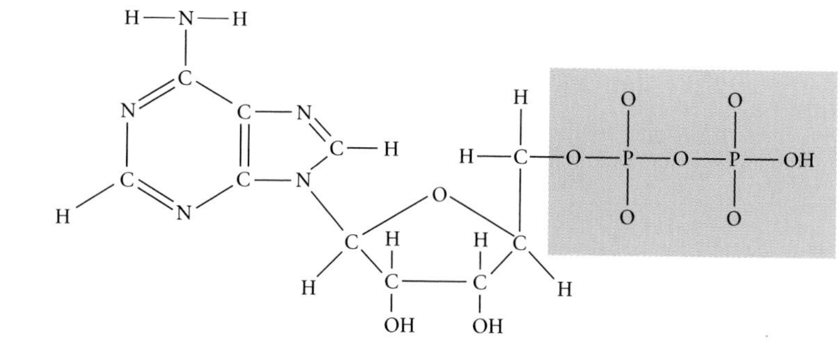

Adenosine diphosphate (ADP)

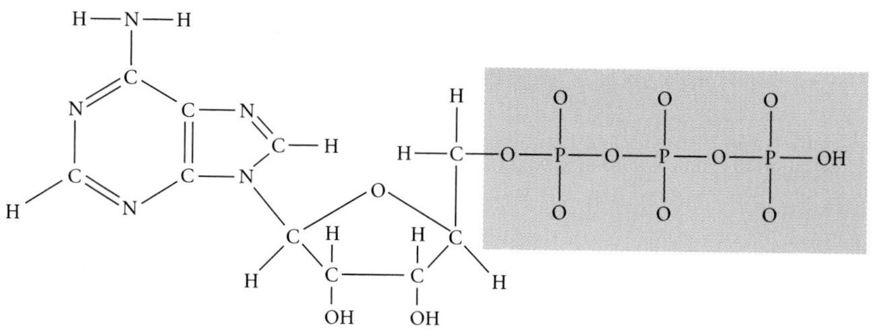

Adenosine triphosphate (ATP)

**Figure 17.10**

Structures of ADP and ATP.

extra phosphate group in ATP. In the coupled reaction with glucose, about 38 mol of ATP are synthesized for every mole of glucose consumed. This gives an overall free energy change for the coupled reaction of

$$-2870 \text{ kJ} + 38(+31 \text{ kJ}) \approx -1700 \text{ kJ}$$

In a very real sense, your body "stores" energy available from the metabolism of foods in the form of ATP. This molecule in turn supplies the energy required for all sorts of biochemical reactions taking place in the body. It does this by reverting to ADP, i.e., by reversing the above reaction. The amount of ATP consumed is amazingly large; a competitive sprinter may hydrolyze as much as 500 g (about 1 lb) of ATP per minute.

*Conversion of ATP to ADP gives you energy in a hurry*

---

**Example 17.10** The lactic acid ($C_3H_6O_3(aq)$, $\Delta G_f^\circ = -559$ kJ) produced in muscle cells after vigorous exercise eventually is absorbed into the blood stream, where it is metabolized back to glucose ($\Delta G_f^\circ = -919$ kJ) in the liver. The reaction is

$$2C_3H_6O_3(aq) \longrightarrow C_6H_{12}O_6(aq)$$

(a) Calculate $\Delta G^\circ$ at 25°C for this reaction, using free energies of formation.
(b) If the hydrolysis of ATP to ADP is coupled with this reaction, how many moles of ATP must react to make the process spontaneous?

**Strategy** Use the relation

$$\Delta G^\circ = \sum \Delta G_f^\circ \text{ products} - \sum \Delta G_f^\circ \text{ reactants}$$

to find $\Delta G^\circ$. To solve part (b), note that 31 kJ of energy is produced per mole of ATP consumed.

**Solution**

(a) $\Delta G^\circ = \Delta G_f^\circ \, C_6H_{12}O_6(aq) - 2\Delta G_f^\circ \, C_3H_6O_3(aq)$

$= -919 \text{ kJ} + 2(559 \text{ kJ}) = \boxed{+199 \text{ kJ}}$

(b) $199 \text{ kJ} \times \dfrac{1 \text{ mol ATP}}{31 \text{ kJ}} \approx \boxed{7 \text{ mol ATP}}$

---

## The Greenhouse Effect

The thermodynamics of our planet is governed in large part by the interaction of the Earth with its atmosphere. If it were not for the presence of two relatively minor components of the atmosphere, water vapor and carbon dioxide, the Earth would be essentially barren of life. Its mean temperature would be about −15°C, roughly 35°C lower than it is today.

Water vapor and carbon dioxide, unlike $N_2$, $O_2$, and Ar, strongly absorb infrared radiation (heat) given off by the earth. In this way, $CO_2$ and $H_2O$ act as an insulating blanket to prevent heat from escaping into outer space. This is often referred to as the **greenhouse effect.**

Of the two gases, water vapor absorbs more infrared radiation than carbon dioxide because its concentration is higher. This property of water vapor accounts for the fact that the temperature drops less on nights when there is a heavy cloud cover. In desert regions, where there is very little water vapor, large variations between day and night temperatures are common.

Although the concentration of water vapor in the atmosphere varies greatly with location, it remains relatively constant over time. In contrast, the concentration of carbon dioxide has increased by more than 20% over the past century, as a result of human activities. Increased combustion of fossil fuels is mainly responsible. Every gram of fossil fuel burned releases about three grams of carbon dioxide into the atmosphere. Part of this $CO_2$ is used by plants in photosynthesis or is absorbed by the oceans, but at least half of it remains. Extensive land clearing, which reduces the amount of carbon dioxide consumed by photosynthesis, is also a factor in raising the $CO_2$ content of the atmosphere. This is one of the adverse effects of the destruction of tropical rain forests for agricultural purposes.

It has been estimated that unless preventive action is taken, increasing $CO_2$ levels could raise the Earth's temperature by 3°C by the year 2100. That may seem like a small amount, but it is worth pointing out that the mean annual temperatures of Burlington, Vermont, and Chicago, Illinois, differ by only about 3°C. Looking at it from another point of view, an increase in temperature of this magnitude could raise sea level by as much as 1 m, flooding many coastal areas, including much of the state of Florida. On a more optimistic note, an increase in $CO_2$ concentration would promote photosynthesis, perhaps increasing the world's food supply.

Recent studies show that average global temperatures have indeed increased by about 0.5°C over the past century (Figure 17.A). One of the effects of this warming trend has been the melting back of glaciers (Figure 17.B). Nobody knows whether these trends reflect increasing concentrations of $CO_2$ and other greenhouse gases or statistical fluctuations. The consensus among atmospheric scientists is that carbon dioxide emissions should be reduced to avoid a worst-case scenario. There are several ways to do this:

— raise fuel efficiency standards for automobiles
— impose a surtax on all carbon-containing fuels
— develop "clean" sources of energy, notably solar

**Figure 17.B**
The Muir glacier in Alaska in 1978 (top) and 1993 (bottom).
(Tom Bean)

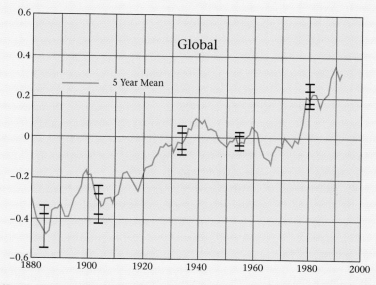

**Note:** Zero is the 1951–80 average temperature. **Source:** NASA Goddard Institute for Space Studies

**Figure 17.A**
Global mean temperature has been on the rise since 1880

# CHAPTER HIGHLIGHTS

## Key Concepts

1. Deduce the sign of $\Delta S$ based on randomness
   (Examples 17.1, 17.2; Problems 5–10)
2. Calculate $\Delta S°$ from tables of standard entropies
   (Example 17.3; Problems 11–16)
3. Calculate $\Delta G°$ from
   —the Gibbs-Helmholtz equation
   (Examples 17.4, 17.6; Problems 17–20, 23, 24)
   —free energies of formation
   (Example 17.5; Problems 21, 22, 25, 26)
4. Calculate the temperature at which $\Delta G° = 0$
   (Example 17.7; Problems 31–44, 75, 76)
5. Calculate $\Delta G$ from $\Delta G°$ and $Q$
   (Example 17.8; Problems 45–50)
6. Relate $\Delta G°$ to $K$
   (Example 17.9; Problems 55–60, 66, 67, 72)
7. Calculate $\Delta G°$ for coupled reactions
   (Example 17.10; Problems 51–57)

## Key Equations

Entropy change $\qquad \Delta S° = \sum S°_{products} - \sum S°_{reactants}$

Free energy change $\qquad \Delta G = \Delta H - T\Delta S$

$$\Delta G° = \sum \Delta G°_f \text{ products} - \sum \Delta G°_f \text{ reactants}$$

$$\Delta G° = -RT \ln K$$

$$\Delta G = \Delta G° + RT \ln Q$$

## Key Terms

| | | |
|---|---|---|
| atmosphere | exothermic | molarity |
| endothermic | free energy | $Q$ |
| enthalpy | —of formation | $\Delta S°$ |
| —of formation | $\Delta G°$ | spontaneous |
| entropy | $\Delta H°$ | system |

## Summary Problem

Consider formic acid, HCOOH. It is given off by many insects when they sting their victims. Its name comes from *formica*, the Latin name for "ant." It is extensively used by the textile industry. It can be obtained by the exposure of methyl alcohol (an alcohol present in gasohol) to air.

$$CH_3OH(aq) + O_2(g) \longrightarrow HCOOH(aq) + H_2O(l)$$

The following data may be useful:

$$CH_3OH(aq): \quad \Delta H_f^\circ = -246 \text{ kJ/mol}$$

$$S^\circ = 133.1 \text{ J/mol} \cdot \text{K}$$

$$HCOOH(aq): \quad \Delta H_f^\circ = -425 \text{ kJ/mol}$$

$$S^\circ = 163 \text{ J/mol} \cdot \text{K}$$

(a)   Calculate $\Delta H^\circ$ and $\Delta S^\circ$ for this process.

(b)   Is the reaction spontaneous at 25°C? at 100°C?

(c)   The heat of vaporization for formic acid is 23.1 kJ/mol. Its normal boiling point is 101°C. Calculate $\Delta S^\circ$ for the reaction $HCOOH(l) \longrightarrow HCOOH(g)$

(d)   What is the standard molar entropy for $HCOOH(g)$ taking $S^\circ$ for $HCOOH(l)$ to be 129 J/mol · K

(e)   Calculate $\Delta G$ at 25°C for the formation of formic acid from methyl alcohol (equation given above) when $[HCOOH] = 0.250 \text{ M}$, $P_{O_2} = 0.75$ atm, and $[CH_3OH] = 0.0100 \text{ M}$.

(f)   Calculate $\Delta G^\circ$ for the ionization of formic acid at 25°C. ($K_a = 1.9 \times 10^{-4}$)

## Answers

(a)   $\Delta H^\circ = -465$ kJ; $\Delta S^\circ = -105$ J/K

(b)   $\Delta G^\circ = -434$ kJ; yes; yes

(c)   61.8 J/K

(d)   191 J/K

(e)   $-425$ kJ

(f)   21.2 kJ

## Questions & Problems

## Spontaneity

*1. Which of the following processes is/are spontaneous?
   (a) building a sand castle
   (b) outlining your chemistry notes
   (c) cleaning up your desk
   (d) wind scattering leaves in a pile

*2. Which of the following processes is/are spontaneous?
   (a) a snowman melting in the sun
   (b) a drop of ink dispersing in water
   (c) building a house of cards
   (d) folding clothes from a laundry basket

*3. Based on your experience, predict which of the following reactions is/are spontaneous.
   (a) $NaCl(s) \longrightarrow NaCl(l)$ at 25°C
   (b) $CO_2(g) \longrightarrow C(s) + O_2(g)$
   (c) $CaCO_3(s) + 2H_2O(l) \longrightarrow Ca(OH)_2(s) + H_2CO_3(aq)$
   (d) $Ag^+(aq) + Cl^-(aq) \longrightarrow AgCl(s)$

*4. Based on your experience, predict which of the following reactions is/are spontaneous.
   (a) $CO_2(s) \longrightarrow CO_2(g)$ at 25°C
   (b) $2NaCl(s) \longrightarrow 2Na(s) + Cl_2(g)$
   (c) $Zn(s) + 2H^+(aq) \longrightarrow Zn^{2+}(aq) + H_2(g)$
   (d) $CH_4(g) + O_2(g) \longrightarrow CO_2(g) + 2H_2O(g)$

## Entropy, $\Delta S^\circ$

*5. Predict the sign of $\Delta S$ for
   (a) boiling water
   (b) sugar, $C_{12}H_{22}O_{11}$, decomposing to carbon and steam
   (c) precipitating lead chloride
   (d) weeding a garden

*6. Predict the sign of $\Delta S$ for
   (a) ice cream melting
   (b) dissolving instant coffee in hot water
   (c) a lake freezing
   (d) a candle burning

*7. Predict the sign of $\Delta S^\circ$ for each of the following reactions.
   (a) $2C(s) + O_2(g) \longrightarrow 2CO(g)$
   (b) $N_2(g) + 2H_2(g) \longrightarrow N_2H_4(g)$
   (c) $2K(s) + Cl_2(g) \longrightarrow 2KCl(s)$
   (d) $2CH_3OH(l) + 3 O_2(g) \longrightarrow 2CO_2(g) + 4H_2O(g)$

*8. Predict the sign of $\Delta S^\circ$ for each of the following reactions.
   (a) $2Fe_2O_3(s) + 3C(s) \longrightarrow 4Fe(s) + 3CO_2(g)$
   (b) $H_2(g) + O_2(g) \longrightarrow H_2O_2(l)$
   (c) $Mg(s) + 2H_2O(l) \longrightarrow Mg(OH)_2(s) + H_2(g)$
   (d) $2NH_3(g) \longrightarrow N_2(g) + 3H_2(g)$

**\*9.** Predict the sign of $\Delta S°$ for each of the following reactions.

**(a)** $Cu(s) + 2H^+(aq) \longrightarrow H_2(g) + Cu^{2+}(aq)$
**(b)** $O_3(g) \longrightarrow O_2(g) + O(g)$
**(c)** $CuSO_4(s) + 5H_2O(l) \longrightarrow CuSO_4 \cdot 5H_2O(s)$

**\*10.** Predict the sign of $\Delta S°$ for each of the following reactions.

**(a)** $H_2(g) + Ni^{2+}(aq) \longrightarrow 2H^+(aq) + Ni(s)$
**(b)** $N_2O_4(g) \longrightarrow 2NO_2(g)$
**(c)** $PCl_3(g) + Cl_2(g) \longrightarrow PCl_5(g)$

**11.** Use Table 17.1 to calculate $\Delta S°$ for each of the following reactions.

**(a)** $MgCl_2(s) + H_2O(l) \longrightarrow MgO(s) + 2HCl(g)$
**(b)** $BaO(s) + CO_2(g) \longrightarrow BaCO_3(s)$
**(c)** $4NO_2(g) + 6H_2O(l) \longrightarrow 4NH_3(g) + 7\ O_2(g)$
**(d)** $2C_2H_6(g) + 7\ O_2(g) \longrightarrow 4CO_2(g) + 6H_2O(l)$

**12.** Use Table 17.1 to calculate $\Delta S°$ for each of the following reactions.

**(a)** $S_8(s) + 8\ O_2(g) \longrightarrow 8SO_2(g)$
**(b)** $Fe_2O_3(s) + 3H_2O(l) \longrightarrow 2Fe(OH)_3(s)$
**(c)** $2ZnS(s) + 3\ O_2(g) \longrightarrow 2ZnO(s) + 2SO_2(g)$

**13.** Use Table 17.1 to calculate $\Delta S°$ for each of the following reactions.

**(a)** $H_2O(l) \longrightarrow H^+(aq) + OH^-(aq)$
**(b)** $2MnO_4^-(aq) + 6I^-(aq) + 8H^+(aq) \longrightarrow$ $2MnO_2(s) + 3I_2(s) + 4H_2O(l)$
**(c)** $ClO_3^-(aq) + 3Zn(s) + 6H^+(aq) \longrightarrow Cl^-(aq) +$ $3Zn^{2+}(aq) + 3H_2O(l)$

**14.** Use Table 17.1 to calculate $\Delta S°$ for each of the following reactions.

**(a)** $2Al^{3+}(aq) + 3Ni(s) \longrightarrow 3Ni^{2+}(aq) + 2Al(s)$
**(b)** $2K(s) + 2H_2O(g) \longrightarrow 2K^+(aq) + 2OH^-(aq) + H_2(g)$
**(c)** $NH_4^+(aq) + OH^-(aq) \longrightarrow NH_3(g) + H_2O(l)$

**15.** Use Table 17.1 to calculate $\Delta S°$ for each of the following reactions.

**(a)** $SO_3(g) + H_2O(l) \longrightarrow 2H^+(aq) + SO_4^{2-}(aq)$
**(b)** $PCl_5(g) + 4H_2O(l) \longrightarrow 6H^+(aq) + 5Cl^-(aq) +$ $H_2PO_4^-(aq)$
**(c)** $SO_4^{2-}(aq) + 4H^+(aq) + Ni(s) \longrightarrow Ni^{2+}(aq) +$ $SO_2(g) + 2H_2O(l)$

**16.** Use Table 17.1 to calculate $\Delta S°$ for each of the following reactions.

**(a)** $2Br^-(aq) + 2H_2O(l) \longrightarrow Br_2(l) + H_2(g) + 2\ OH^-(aq)$
**(b)** $4Fe^{3+}(aq) + 2H_2O(l) \longrightarrow 4Fe^{2+}(aq) + O_2(g) +$ $4H^+(aq)$
**(c)** $ClO_4^-(aq) + H_2O(l) + S^{2-}(aq) \longrightarrow ClO_3^-(aq) +$ $2\ OH^-(aq) + S(s)$

## $\Delta G°$ *and the Gibbs-Helmholtz Equation*

**17.** Calculate $\Delta G°$ at 45°C for reactions for which

**(a)** $\Delta H° = 136$ kJ; $\Delta S° = 39.0$ J/K
**(b)** $\Delta H° = -713$ kJ; $\Delta S° = 118.0$ J/K
**(c)** $\Delta H° = 12.9$ kJ; $\Delta S° = -216.1$ J/K

**18.** Calculate $\Delta G°$ at 45°C for reactions for which

**(a)** $\Delta H° = -649$ kJ; $\Delta S° = -92.5$ J/K
**(b)** $\Delta H° = 384$ kJ; $\Delta S° = -177.0$ J/K
**(c)** $\Delta H° = -506$ kJ; $\Delta S° = 62.1$ J/K

**19.** Calculate $\Delta G°$ at 425 K for each of the reactions in Question 11. State whether the reactions are spontaneous.

**20.** Calculate $\Delta G°$ at 375 K for each of the reactions in Question 12. State whether the reactions are spontaneous.

**21.** Using values for $\Delta G_f°$ given in Appendix 1, calculate $\Delta G°$ at 25°C for each of the reactions in Question 13.

**22.** Follow the directions of Problem 21 for each of the reactions in Question 14.

**23.** Calculate $\Delta G_f°$ at 25°C, using standard entropies and heats of formation for

**(a)** zinc iodide     **(b)** chloroform ($CHCl_3$)
**(c)** silver nitrate

**24.** Follow the directions of Question 23 for the following compounds.

**(a)** ammonium nitrate
**(b)** ethyl alcohol ($C_2H_5OH$)
**(c)** barium carbonate

**25.** It has been proposed that wood alcohol, $CH_3OH$, a relatively inexpensive fuel to produce, be decomposed to produce methane. Methane is a natural gas commonly used for heating homes. Is the decomposition of wood alcohol to methane and oxygen thermodynamically feasible at 25°C and 1 atm?

**26.** A student warned his friends not to swim in a river close to an electric plant. He claimed that the ozone produced by the plant turned the river water to hydrogen peroxide, which would bleach hair. The reaction is

$$O_3(g) + H_2O(l) \longrightarrow H_2O_2(aq) + O_2(g)$$

Show by calculation whether his claim is plausible, assuming that the temperature is 25°C and that all species are at standard concentrations. Take $\Delta G_f°$ $O_3(g)$ at 25°C to be $+163.2$ kJ/mol and $\Delta G_f°$ $H_2O_2(aq)$ to be $-134$ kJ/mol.

**27.** When one mole of ethylene gas, $C_2H_4$, is burned in oxygen and hydrogen chloride gas, the products are liquid ethylene chloride, $C_2H_4Cl_2$, and water. The reaction evolves 318.7 kJ of heat and has a standard free energy change of $-194.4$ kJ at 25°C.

**(a)** Write a thermochemical equation for the reaction.
**(b)** Calculate $\Delta S°$ for the reaction.
**(c)** Calculate $S°$ for $C_2H_4Cl_2$.

**28.** When permanganate ions in aqueous solution react with cobalt metal in strong acid, the equation for the reaction that takes place is as follows:

$$2MnO_4^-(aq) + 16H^+(aq) + 5Co(s) \longrightarrow$$
$$2Mn^{2+}(aq) + 5Co^{2+}(aq) + 8H_2O(l)$$

$$\Delta H° = -2024.6\ kJ \qquad \Delta G° = -1750.9\ kJ\ at\ 25°C$$

**(a)** Calculate $\Delta S°$ for the reaction.
**(b)** Calculate $S°$ for $Co^{2+}$, given $S°$ for Co is 30.04 J/mol · K.

**29.** Phosgene, $COCl_2$, can be formed by the reaction of chloroform, $CHCl_3(l)$, with oxygen:

$$2CHCl_3(l) + O_2(g) \longrightarrow 2COCl_2(g) + 2HCl(g)$$

$$\Delta H° = -353.2 \text{ kJ} \qquad \Delta G° = -452.4 \text{ kJ at } 25°C$$

**(a)** Calculate $\Delta S°$ for the reaction. Is the sign reasonable?
**(b)** Calculate $S°$ for phosgene.
**(c)** Calculate $\Delta H_f°$ for phosgene.

**30.** Consider the reaction

$$2CuCl(s) + 2\ OH^-(aq) \longrightarrow Cu_2O(s) + 2Cl^-(aq) + H_2O(l)$$

$$\Delta H° = -54.3 \text{ kJ} \qquad \Delta S° = -125.1 \text{ J/K}$$

**(a)** Calculate $\Delta G°$ for this reaction at $25°C$.
**(b)** Determine $\Delta H_f°$ for $CuCl(s)$.
**(c)** Calculate $S°$ for $CuCl(s)$.

## *Temperature Dependence of Spontaneity*

**\*31.** Discuss the effect of temperature change upon the spontaneity of the following reactions at 1 atm.
**(a)** $4Fe(s) + 3O_2(g) \longrightarrow 2Fe_2O_3(s)$
$\Delta H° = -1648.4 \text{ kJ} \qquad \Delta S° = -549.4 \text{ J/K}$
**(b)** $N_2O(g) + 2H_2O(g) \longrightarrow NH_4NO_3(s)$
$\Delta H° = 35.95 \text{ kJ} \qquad \Delta S° = -446 \text{ J/K}$
**(c)** $2SO_3(g) \longrightarrow 2SO_2(g) + O_2(g)$
$\Delta H° = 197.8 \text{ kJ} \qquad \Delta S° = 187.8 \text{ J/K}$

**\*32.** Discuss the effect of temperature change upon the spontaneity of the following reactions at 1 atm.
**(a)** $N_2H_4(l) \longrightarrow N_2(g) + 2H_2(g)$
$\Delta H° = -50.6 \text{ kJ} \qquad \Delta S° = -531.5 \text{ J/K}$
**(b)** $2AsF_3(l) \longrightarrow 2As(s) + 3F_2(g)$
$\Delta H° = -1643 \text{ kJ} \qquad \Delta S° = +316 \text{ J/K}$
**(c)** $2PbS(s) + 3O_2(g) \longrightarrow 2PbO(s) + 2SO_2(g)$
$\Delta H° = -830.8 \text{ kJ} \qquad \Delta S° = -168 \text{ J/K}$

**33.** At what temperature does $\Delta G°$ become zero for each of the reactions in Problem 31? Explain the significance of your answers.

**34.** Over what temperature range are the reactions in Problem 32 spontaneous?

**35.** Sodium hydrogen carbonate, commonly called "baking soda," changes to sodium carbonate, "washing soda," when heated. What is the temperature range for spontaneity of the reaction?

$$2NaHCO_3(s) \longrightarrow Na_2CO_3(s) + CO_2(g) + H_2O(g)$$

Additional thermodynamic information that may be useful:

$NaHCO_3$: $\Delta H_f° = -950.8 \text{ kJ/mol} \qquad S° = 101.7 \text{ J/K} \cdot \text{mol}$

$Na_2CO_3$: $\Delta H_f° = -1130.7 \text{ kJ/mol} \qquad S° = 135.0 \text{ J/K} \cdot \text{mol}$

**36.** Oxygen can be made in the laboratory by reacting sodium peroxide and water.

$$2Na_2O_2(s) + 2H_2O(l) \longrightarrow 4NaOH(s) + O_2(g)$$

What is the temperature range for the spontaneity of the reaction? Thermodynamic data for $Na_2O_2(s)$ at $25°C$ and 1 atm are $\Delta H_f° = -510.9 \text{ kJ/mol}$; $S° = 95.0 \text{ J/mol} \cdot \text{K}$.

**37.** Earlier civilizations smelted iron from ore by heating it with charcoal from a wood fire:

$$2Fe_2O_3(s) + 3C(s) \longrightarrow 4Fe(s) + 3CO_2(g)$$

**(a)** Obtain an expression for $\Delta G°$ as a function of temperature. Prepare a table of $\Delta G°$ values at 100-K intervals between 100 and 500 K.
**(b)** Calculate the lowest temperature at which the smelting could be carried out.

**38.** When ammonia gas is mixed with hydrogen chloride gas, a dense white gas is formed that solidifies to ammonium chloride.

$$NH_3(g) + HCl(g) \longrightarrow NH_4Cl(s)$$

**(a)** Using the appropriate tables in Appendix 1, obtain an expression for $\Delta G°$ as a function of temperature. Use it to prepare a table of $\Delta G°$ values at 100-K intervals between 200 K and 800 K.
**(b)** Plot your values from (a) and determine when $\Delta G°$ is 0.

**39.** It is desired to produce tin from its ore, cassiterite, $SnO_2$, at as low a temperature as possible. The ore could be
**(a)** decomposed by heating, producing tin and oxygen.
**(b)** heated with hydrogen gas, producing tin and water vapor.
**(c)** heated with carbon, producing tin and carbon dioxide.
Based solely on thermodynamic principles, which method would you recommend? Show calculations.

**40.** Two possible ways of producing iron from iron ore are
**(a)** $Fe_2O_3(s) + \frac{3}{2}C(s) \longrightarrow 2Fe(s) + \frac{3}{2}CO_2(g)$
**(b)** $Fe_2O_3(s) + 3H_2(g) \longrightarrow 2Fe(s) + 3H_2O(g)$
Which of these reactions would proceed spontaneously at the lower temperature?

**41.** Pencil "lead" is almost pure graphite. Graphite is the elemental form of carbon stable at $25°C$ and 1 atm. Diamond is an allotrope of graphite. Given

diamond: $\Delta H_f° = 1.9 \text{ kJ/mol}$, $S° = 2.4 \text{ J/mol} \cdot \text{K}$

at what temperature are the two forms in equilibrium at 1 atm?

$$C(\text{graphite}) \rightleftharpoons C(\text{diamond})$$

**42.** Sulfur has about 20 different allotropes. The most common are rhombic sulfur (the stable form at $25°C$ and 1 atm) and monoclinic sulfur. They differ in their crystal structures. Given

$S(s, \text{monoclinic})$: $\Delta H_f° = 0.30 \text{ kJ/mol}$, $S° = 0.0326 \text{ kJ/mol} \cdot \text{K}$

at what temperature are the two forms in equilibrium?

**43.** Given the following data for iodine

$I_2(s)$: $S° = 116.1$ J/mol · K

$I_2(g)$: $S° = 260.6$ J/mol · K    $\Delta H_f° = 62.4$ kJ/mol

estimate the temperature at which iodine sublimes at 1 atm.

$$I_2(s) \rightleftharpoons I_2(g)$$

**44.** Given the following data for mercury

$Hg(l)$: $S° = 76.0$ J/mol · K

$Hg(g)$: $S° = 175.0$ J/mol · K    $\Delta H_f° = 61.32$ kJ/mol

estimate the normal boiling point ($\Delta G° = 0$) of Hg.

### Effect of Concentration/Pressure on Spontaneity

**45.** Is the reaction

$$HC_2H_3O_2(aq) \rightleftharpoons H^+(aq) + C_2H_3O_2^-(aq) \qquad \Delta G° = 27.2 \text{ kJ}$$

spontaneous at 25°C

  **(a)** when $[H^+] = [C_2H_3O_2^-] = [HC_2H_3O_2] = 1.0$ $M$?
  **(b)** when $[H^+] = [C_2H_3O_2^-] = 3.0 \times 10^{-3}$ $M$ and $[HC_2H_3O_2] = 1.0$ $M$?

**46.** Show by calculation, using Appendix 1, whether the reaction

$$PbBr_2(s) \rightleftharpoons Pb^{2+}(aq) + 2Br^-(aq)$$

is spontaneous at 25°C

  **(a)** when $[Pb^{2+}] = [Br^-] = 1.0$ $M$
  **(b)** when $[Pb^{2+}] = [Br^-] = 0.0050$ $M$

**47.** For the reaction

$$2H_2O(l) + 2Br^-(aq) \longrightarrow H_2(g) + Br_2(l) + 2\ OH^-(aq)$$

  **(a)** calculate $\Delta G°$ at 25°C.
  **(b)** calculate $\Delta G$ when $P_{H_2} = 0.100$ atm, $[Br^-] = 0.75$ $M$, and the pH of the solution is 9.73.

**48.** For the reaction

$$2NO_3^-(aq) + Cu(s) + 8H^+(aq) \longrightarrow$$
$$2NO(g) + 3Cu^{2+}(aq) + 4H_2O(l)$$

  **(a)** Calculate $\Delta G°$ at 25°C.
  **(d)** Calculate $\Delta G$ when $P_{NO} = 0.300$ atm, $[Cu^{2+}] = 0.100$ $M$, $[NO_3^-] = 0.125$ $M$, and the pH of the solution is 2.56.

**49.** Consider the reaction

$$CaF_2(s) \longrightarrow Ca^{2+}(aq) + 2F^-(aq)$$

  **(a)** Calculate $\Delta G°$ at 25°C. $\Delta G_f°$ for $CaF_2(s) = -1167.3$ kJ/mol
  **(b)** What should the concentrations of $Ca^{2+}$ and $F^-$ be so that $\Delta G = -1.0$ kJ (just spontaneous)? Take $[F^-] = 2[Ca^{2+}]$.

**50.** Consider the reaction

$$2NO(g) + O_2(g) \longrightarrow 2NO_2(g)$$

  **(a)** Calculate $\Delta G°$ at 25°C.
  **(b)** If $P_{NO_2} = 3.00$ atm and $P_{O_2} = 0.001$ atm, what partial pressure should NO have so that the reaction becomes nonspontaneous?

### Additivity of Coupled Reactions

**51.** Natural gas, which is mostly methane, $CH_4$, is a resource that the United States has in abundance. In principle, ethane can be obtained from methane by the reaction

$$2CH_4(g) \longrightarrow C_2H_6(g) + H_2(g)$$

  **(a)** Calculate $\Delta G°$ at 25°C for the reaction. Comment on the feasibility of this reaction at 25°C.
  **(b)** Couple the reaction above with the formation of steam from the elements:

$$H_2(g) + \tfrac{1}{2}O_2(g) \longrightarrow H_2O(g) \qquad \Delta G° = -228.6 \text{ kJ}$$

What is the equation for the overall reaction? Comment on the feasibility of the overall reaction.

**52.** Theoretically, one can obtain zinc from an ore containing zinc sulfide, ZnS, by the reaction

$$ZnS(s) \longrightarrow Zn(s) + S(s)$$

  **(a)** Show by calculation that this reaction is not feasible at 25°C.
  **(b)** Show that by coupling the above reaction with the reaction

$$S(s) + O_2(g) \longrightarrow SO_2(g)$$

the overall reaction, where Zn is obtained by roasting in oxygen, is feasible at 25°C.

**53.** Consider the following reactions at 25°C:

$$C_6H_{12}O_6(aq) + 6\ O_2(g) \longrightarrow 6CO_2(g) + 6H_2O$$
$$\Delta G° = -2870 \text{ kJ}$$

$$ADP(aq) + HPO_4^{2-}(aq) + 2H^+(aq) \longrightarrow ATP(aq) + H_2O$$
$$\Delta G° = 31 \text{ kJ}$$

Write an equation for a coupled reaction between glucose, $C_6H_{12}O_6$, and ADP in which $\Delta G° = -390$ kJ

**54.** How many moles of ATP must be converted to ADP by the reaction

$$ATP(aq) + H_2O \longrightarrow ADP(aq) + HPO_4^{2-}(aq) + 2H^+(aq)$$
$$\Delta G° = -31 \text{ kJ}$$

to bring about a nonspontaneous biochemical reaction in which $\Delta G° = +372$ kJ?

### Free Energy and Equilibrium

**55.** Consider the reaction

$$2HI(g) + Cl_2(g) \longrightarrow 2HCl(g) + I_2(s)$$

using the appropriate tables, calculate
  **(a)** $\Delta G°$ at 25°C      **(b)** $K$ at 25°C

**56.** Consider the reaction

$$N_2O(g) + NO_2(g) \longrightarrow 3NO(g) \qquad K = 4.4 \times 10^{-19}$$

(a) Calculate $\Delta G°$ for the reaction at 25°C.
(b) Calculate $\Delta G_f°$ for $N_2O$.

**57.** For the reaction

$$O_3(g) \longrightarrow O(g) + O_2(g)$$

$\Delta G°$ at 298 K is +68.6 kJ and at 1200 K is −46.3 kJ. Calculate $K$ at both temperatures.

**58.** For the reaction

$$NH_4Cl(s) \rightleftharpoons NH_3(g) + HCl(g)$$

(a) Calculate $K$ at 25°C.
(b) Calculate $\Delta G°$ at 100°C.

**59.** Calculate $K_{sp}$ for magnesium sulfate at 25°C.

**60.** Given that $\Delta H_f°$ for $HF(aq)$ is −320.1 kJ/mol and $S°$ for $HF(aq)$ is 88.7 J/mol · K, find $K_a$ for HF at 25°C. Compare with $K_a$ for HF at body temperature, 37°C.

## Unclassified

*61. Determine whether each of the following statements is true or false. If false, modify it to make it true.
(a) An exothermic reaction is spontaneous.
(b) When $\Delta G°$ is positive, the reaction cannot occur under any conditions.
(c) $\Delta S°$ is positive for a reaction in which there is an increase in the number of moles.
(d) If $\Delta H°$ and $\Delta S°$ are both negative, $\Delta G°$ will be negative.

*62. Which of the following quantities can be taken to be independent of temperature? independent of pressure?
(a) $\Delta H$ for a reaction      (b) $\Delta S$ for a reaction
(c) $\Delta G$ for a reaction      (d) $S$ for a substance

*63. Fill in the blanks:
(a) $\Delta H°$ and $\Delta G°$ become equal at _____K.
(b) $\Delta G°$ and $\Delta G$ are equal when $Q =$ _____.
(c) $S°$ for steam is _____ than $S°$ for water.

*64. Fill in the blanks:
(a) At equilibrium, $\Delta G$ is _____.
(b) For $C_6H_6(l) \rightleftharpoons C_6H_6(g)$, $\Delta H°$ is _____ (+, −, 0).
(c) When a pure solid melts, the temperature at which liquid and solid are in equilibrium and $\Delta G° = 0$ is called _____.

*65. In your own words, explain why
(a) $\Delta S°$ is negative for a reaction in which the number of moles of gas decreases.
(b) we take $\Delta S°$ to be independent of $T$, even though entropy increases with $T$.
(c) a solid has lower entropy than its corresponding liquid.

**66.** At 1200 K, an equilibrium mixture of CO and $CO_2$ gases contains 98.31 mol percent CO and some solid carbon. The total pressure of the mixture is 1.00 atm. For the system

$$CO_2(g) + C(s) \rightleftharpoons 2CO(g)$$

calculate
(a) $P_{CO}$ and $P_{CO_2}$      (b) $K$      (c) $\Delta G°$ at 1200 K

**67.** At 25°C, a 0.200 $M$ solution of a weak acid, HB, has a pH of 4.92. What is $\Delta G°$ for

$$H^+(aq) + B^-(aq) \rightleftharpoons HB(aq)$$

**68.** Some bacteria use light energy to convert carbon dioxide and water to glucose and oxygen:

$$6CO_2(g) + 6H_2O(l) \longrightarrow C_6H_{12}O_6(aq) + 6\,O_2(g)$$
$$\Delta G° = 2870 \text{ kJ at 25°C}$$

Other bacteria, those that do not have light available to them, couple the reaction

$$H_2S(g) + \tfrac{1}{2}O_2(g) \longrightarrow H_2O(l) + S(s)$$

to the glucose synthesis above. Coupling the two reactions, the overall reaction is

$$24H_2S(g) + 6CO_2(g) + 6\,O_2(g) \longrightarrow C_6H_{12}O_6(s) + 18H_2O(l)$$
$$+ 24S(s)$$

Show that the reaction is spontaneous at 25°C.

**69.** It has been proposed that if ammonia, methane, and oxygen gas are combined at 25°C in their standard states, glycine, the simplest of all amino acids, can be formed.

$$2CH_4(g) + NH_3(g) + \tfrac{5}{2}O_2(g) \longrightarrow$$
$$NH_2CH_2COOH(s) + 3H_2O(l)$$

Given $\Delta G_f°$ for glycine = −368.57 kJ/mol,
(a) will this reaction proceed spontaneously at 25°C?
(b) does the equilibrium constant favor the formation of products?
(c) do your calculations for (a) and (b) indicate that glycine is formed as soon as the three gases are combined?

**70.** A student is asked to prepare a 0.030 $M$ aqueous solution of $PbCl_2$.
(a) Is this possible at 25°C? (*Hint:* Is dissolving 0.030 mol of $PbCl_2$ at 25°C possible?)
(b) If the student used water at 100°C, would this be possible?

## Challenge Problems

**71.** The normal boiling point for ethyl alcohol is 78.4°C. $S°$ for $C_2H_5OH(g)$ is 282.7 J/mol · K. At what temperature is the vapor pressure of ethyl alcohol 357 mm Hg?

**72.** $\Delta H_f°$ for iodine gas is 62.4 kJ/mol, while $S°$ is 260.7 J/mol · K. Calculate the equilibrium partial pressures of $I_2(g)$, $H_2(g)$, and $HI(g)$ for the system

$$2HI(g) \rightleftharpoons H_2(g) + I_2(g)$$

at 500°C if the initial partial pressures are all 0.200 atm.

73. The heat of fusion of ice is 333 J/g. For the process

$$H_2O(s) \longrightarrow H_2O(l)$$

determine

(a) $\Delta H°$      (b) $\Delta G°$ at 0°C      (c) $\Delta S°$
(d) $\Delta G°$ at −20°C      (e) $\Delta G°$ at 20°C

74. The overall reaction that occurs when sugar is metabolized is

$$C_{12}H_{22}O_{11}(s) + 12\ O_2(g) \longrightarrow 12CO_2(g) + 11H_2O(l)$$

For this reaction, $\Delta H°$ is −5650 kJ and $\Delta G°$ is −5790 kJ at 25°C.

(a) If 25% of the free energy change is actually converted to useful work, how many kilojoules of work could be obtained when one gram of sugar is metabolized at body temperature, 37°C?
(b) How many grams of sugar would a 120-lb woman have to eat to get the energy to climb the Jungfrau in the Alps, which is 4158 meters high? ($w = 9.79 \times 10^{-3}$ $mh$, where $w$ = work in kilojoules, $m$ is body mass in kilograms, and $h$ is height in meters.)

75. Hydrogen has been suggested as the fuel of the future. One way to store it is to convert it to a compound that can be heated to release the hydrogen. One such compound is calcium hydride, $CaH_2$. This compound has a heat of formation of −186.2 kJ/mol and a standard entropy of 42.0 J/mol · K. What is the minimum temperature to which calcium hydride would have to be heated to produce hydrogen at one atmosphere pressure?

76. When a copper wire is exposed to air at room temperature, it becomes coated with a black oxide, CuO. If the wire is heated above a certain temperature, the black oxide is converted to a red oxide, $Cu_2O$. At a still higher temperature, the oxide coating disappears. Explain these observations in terms of the thermodynamics of the reactions

$$2CuO(s) \longrightarrow Cu_2O(s) + \tfrac{1}{2}O_2(g)$$

$$Cu_2O(s) \longrightarrow 2Cu(s) + \tfrac{1}{2}O_2(g)$$

and estimate the temperatures at which the changes occur.

The Statue of Liberty, erected in 1885 is constructed of sheets of copper which have corroded over the years (p. 523). (Andy Levin/Photo Researchers, Inc.)

# Electrochemistry

## 18

If by fire

Of sooty coal th' empiric Alchymist

Can turn, or holds it possible to turn

Metals of drossiest ore to perfect gold

—JOHN MILTON

lectrochemistry is the study of the interconversion of electrical and chemical energy. This conversion takes place in an electrochemical cell that may be a(n)

— **voltaic cell** (Section 18.1), in which a spontaneous reaction generates electrical energy
— **electrolytic cell** (Section 18.5), in which electrical energy is used to bring about a nonspontaneous reaction

Cathode = reduction; anode = oxidation

The reaction taking place in an electrochemical cell is of the oxidation-reduction type (Chap. 4). At one electrode, called the **cathode,** a **reduction** half-reaction occurs; electrons are consumed. Typical examples include

$$Cu^{2+}(aq) + 2e^- \longrightarrow Cu(s)$$

$$Cl_2(g) + 2e^- \longrightarrow 2Cl^-(aq)$$

At the other electrode, called the **anode,** an **oxidation** half-reaction occurs. Electrons are produced, as in the half-reactions

$$Zn(s) \longrightarrow Zn^{2+}(aq) + 2e^-$$

$$2I^-(aq) \longrightarrow I_2(s) + 2e^-$$

The number of electrons produced at the anode is exactly equal to the number of electrons consumed at the cathode. In any cell, *anions* move to the *anode; cations* move to the *cathode.*

One of the most important characteristics of a cell is its *voltage,* which is a measure of reaction spontaneity. Cell voltages depend upon the nature of the half-reactions occurring at the electrodes (Section 18.2) and upon the concentrations of species involved (Section 18.4). From the voltage measured at standard concentrations, it is possible to calculate the standard free energy change and the equilibrium constant (Section 18.3) of the reaction involved.

The principles discussed in this chapter have a host of practical applications. Whenever you start your car, turn on a flashlight, or take a logarithm on a calculator, you are making use of a voltaic cell. Many of our most important elements, including hydrogen and chlorine, are made in electrolytic cells. These applications, among others, are discussed in Section 18.6.

## 18.1 Voltaic Cells

In principle at least, any spontaneous redox reaction can serve as a source of energy in a voltaic cell. The cell must be designed in such a way that oxidation occurs at one electrode (anode) with reduction at the other electrode (cathode).

The electrons produced at the anode must be transferred to the cathode, where they are consumed. To do this, the electrons move through an external circuit, where they do electrical work.

To understand how a voltaic cell operates, let us start with some simple cells that are readily made in the general chemistry laboratory.

## The Zn–$Cu^{2+}$ Cell (Zn | $Zn^{2+}$ ‖ $Cu^{2+}$ | Cu)

When a piece of zinc is added to a water solution containing $Cu^{2+}$ ions, the following redox reaction takes place:

$$Zn(s) + Cu^{2+}(aq) \longrightarrow Zn^{2+}(aq) + Cu(s)$$

In this reaction, copper metal plates out on the surface of the zinc. The blue color of the aqueous $Cu^{2+}$ ion fades as it is replaced by the colorless aqueous $Zn^{2+}$ ion (Fig. 18.1). Clearly, this redox reaction is spontaneous; it involves electron transfer from a Zn atom to $Cu^{2+}$ ion.

To design a voltaic cell using the Zn–$Cu^{2+}$ reaction as a source of electrical energy, the electron transfer must occur indirectly; that is, the electrons given off by zinc atoms must be made to pass through an external electric circuit before they reduce $Cu^{2+}$ ions to copper atoms. One way to do this is shown in Figure 18.2. The voltaic cell consists of two half-cells:

— a Zn anode dipping into a solution containing $Zn^{2+}$ ions, shown in the beaker at the far right
— a Cu cathode dipping into a solution containing $Cu^{2+}$ ions (blue), shown in the beaker at the center of Figure 18.2.

The "external circuit" consists of a voltmeter with leads (red and black) to the anode and cathode.

Let us trace the flow of electric current through this cell.

**1.** At the zinc *anode,* electrons are produced by the *oxidation* half-reaction

$$Zn(s) \longrightarrow Zn^{2+}(aq) + 2e^-$$

This electrode, which "pumps" electrons into the external circuit, is ordinarily marked as the negative pole of the cell.

**Figure 18.1**
When a strip of zinc is placed in a solution containing $Cu^{2+}$ ions (left), a spontaneous redox reaction occurs. The final result is shown at the right. Copper metal plates out, and the blue color due to $Cu^{2+}$ fades. (Marna G. Clarke)

The Zn electrode must not come in contact with $Cu^{2+}$ ions. Why?

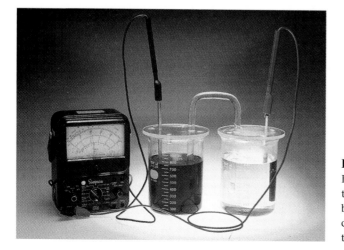

**Figure 18.2**
In this voltaic cell, the following spontaneous redox reaction takes place: $Zn(s) + Cu^{2+}(aq) \longrightarrow Zn^{2+}(aq) + Cu(s)$. The salt bridge allows ions to pass from one solution to the other to complete the circuit. At the same time, it prevents direct contact between Zn atoms and $Cu^{2+}$ ions.

**2.** Electrons generated at the anode move through the external circuit (right to left in Fig. 18.2) to the copper *cathode*. There they are consumed, reducing $Cu^{2+}$ ions present in the solution around the electrode:

$$Cu^{2+}(aq) + 2e^- \longrightarrow Cu(s)$$

The copper electrode, which "pulls" electrons from the external circuit, is considered to be the positive pole of the cell.

**3.** As the above half-reactions proceed, a surplus of positive ions ($Zn^{2+}$) tends to build up around the zinc electrode. The region around the copper electrode tends to become deficient in positive ions as $Cu^{2+}$ ions are consumed. To maintain electrical neutrality, cations must move toward the copper cathode or, alternatively, anions must move toward the zinc anode. In practice, both migrations occur.

In the cell shown in Figure 18.2, movement of ions occurs through a *salt bridge*. In its simplest form, a salt bridge may consist of an inverted U-tube, plugged with glass wool at each end. The tube is filled with a solution of a salt that takes no part in the electrode reactions; potassium nitrate, $KNO_3$, is frequently used. As current is drawn from the cell, $K^+$ ions move from the salt bridge into the cathode half-cell. At the same time, $NO_3^-$ ions move into the anode half-cell. In this way, electrical neutrality is maintained without $Cu^{2+}$ ions coming in contact with the Zn electrode, which would short-circuit the cell.

The cell shown in Figure 18.2 is often abbreviated as

$$Zn \,|\, Zn^{2+} \,\|\, Cu^{2+} \,|\, Cu$$

In this notation,

Anode ‖ cathode

— the **anode** reaction (**oxidation**) is shown at the left. Zn atoms are oxidized to $Zn^{2+}$ ions.
— the salt bridge (or other means of separating the half cells) is indicated by the symbol ‖.
— the **cathode** reaction (**reduction**) is shown at the right. $Cu^{2+}$ ions are reduced to Cu atoms.
— a single vertical line indicates a phase boundary, such as that between a solid electrode and an aqueous solution.

Notice that the anode half-reaction comes first in the cell notation, just as the letter *a* comes before *c*.

## Other Salt Bridge Cells

Cells similar to that shown in Figure 18.2 can be set up for many different spontaneous redox reactions. Consider, for example, the reaction

$$Ni(s) + Cu^{2+}(aq) \longrightarrow Ni^{2+}(aq) + Cu(s)$$

This reaction, like that between Zn and $Cu^{2+}$, can serve as a source of electrical energy in a voltaic cell. The cell is similar to that shown in Figure 18.2 except that, in the anode compartment, a nickel electrode is surrounded by a solution of a nickel(II) salt, such as $NiCl_2$ or $NiSO_4$. The cell notation is $Ni \,|\, Ni^{2+} \,\|\, Cu^{2+} \,|\, Cu$.

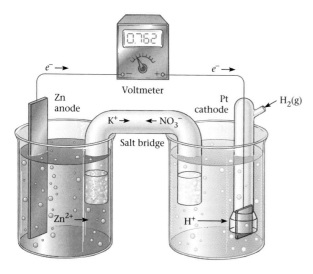

**Figure 18.3**
A voltaic cell in which the reaction $Zn(s) + 2H^+(aq) \longrightarrow$ $Zn^{2+}(aq) + H_2(g)$ occurs. Hydrogen gas is bubbled over a specially prepared platinum electrode that is surrounded by a solution containing $H^+$ ions.

Another spontaneous redox reaction that can serve as a source of electrical energy is that between zinc metal and $H^+$ ions:

$$Zn(s) + 2H^+(aq) \longrightarrow Zn^{2+}(aq) + H_2(g)$$

A voltaic cell using this reaction is similar to the $Zn$-$Cu^{2+}$ cell; the $Zn\,|\,Zn^{2+}$ half-cell and the salt bridge are the same. Since no metal is involved in the cathode half-reaction, an *inert* electrode that conducts an electric current is used. Frequently, the cathode is made out of platinum (Fig. 18.3). Hydrogen gas is bubbled over the cathode, which is surrounded by $H^+$ ions from a solution of HCl.

Nichrome or graphite can also be used

The half-reactions occurring in the cell are

anode:   $Zn(s) \longrightarrow Zn^{2+}(aq) + 2e^-$   (oxidation)

cathode:   $2H^+(aq) + 2e^- \longrightarrow H_2(g)$   (reduction)

The cell notation is $Zn\,|\,Zn^{2+}\,\|\,H^+\,|\,H_2\,|\,Pt$. The symbol Pt is used to indicate the presence of an inert platinum electrode.

---

**Example 18.1**   When chlorine gas is bubbled through a water solution of NaBr, a spontaneous redox reaction occurs:

$$Cl_2(g) + 2Br^-(aq) \longrightarrow 2Cl^-(aq) + Br_2(l)$$

This reaction can serve as a source of electrical energy in the voltaic cell shown in Figure 18.4 p. 502. In this cell,

(a) what is the cathode reaction? the anode reaction?
(b) which way do electrons move in the external circuit?
(c) which way do anions move within the cell? cations?

**Strategy**   Split the reaction into two half-reactions. Remember that oxidation occurs at the anode, reduction at the cathode. Anions move to the anode, cations to the cathode. Electrons are produced at the anode and transferred through the external circuit to the cathode, where they are consumed.

*Solution*

(a) cathode:     $Cl_2(g) + 2e^- \longrightarrow 2Cl^-(aq)$     (reduction)

anode:     $2Br^-(aq) \longrightarrow Br_2(l) + 2e^-$     (oxidation)

(b) From   anode to cathode   (left to right in Figure 18.4).

(c)  Anions move to the anode   (right to left);   cations move to the cathode

(left to right).

**Figure 18.4**
In this voltaic cell, the spontaneous redox reaction is $Cl_2(g) + 2Br^-(aq) \longrightarrow 2Cl^-(aq) + Br_2(l)$. Both electrodes are made of platinum; the one at the left is the anode.

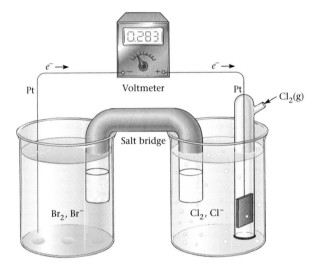

To summarize our discussion of the structure of voltaic cells, note that

— a voltaic cell consists of two half-cells. They are joined by an external electric circuit through which electrons move and a salt bridge through which ions move.

— each half-cell consists of an electrode dipping into a water solution. If a metal participates in the cell reaction, either as a product or a reactant, it is ordinarily used as the electrode; otherwise, an inert electrode such as platinum is used.

— in one half-cell, oxidation occurs at the anode; in the other, reduction takes place at the cathode. The overall cell reaction is the sum of the half-reactions taking place at the anode and cathode.

## 18.2  Standard Voltages

The driving force behind the spontaneous reaction in a voltaic cell is measured by the cell voltage, which is an *intensive* property, independent of the number of electrons passing through the cell. Cell voltage depends upon the nature of the redox reaction and the concentrations of the species involved; for the moment, we'll concentrate upon the first of these factors.

The standard voltage for a given cell is that measured when the current flow is essentially zero, *all ions and molecules in solution are at a concentration of 1 M, and all gases are at a pressure of 1 atm.* To illustrate, consider the Zn-$H^+$ cell referred to earlier. Let us suppose that the half-cells are set up in such a way that the concentrations of $Zn^{2+}$ and $H^+$ are both 1 M and the pressure of $H_2(g)$ is 1 atm. Under these conditions, the cell voltage at very low current flow is +0.762 V. This quantity is referred to as the **standard voltage** and is given the symbol $E°$.

$$Zn(s) + 2H^+(aq, 1\ M) \longrightarrow Zn^{2+}(aq, 1\ M) + H_2(g, 1\ atm) \qquad E° = +0.762\ V$$

## $E°_{ox}$ and $E°_{red}$

Any redox reaction can be split into two half-reactions, an oxidation and a reduction. It is possible to associate standard voltages $E°_{ox}$ (standard oxidation voltage) and $E°_{red}$ (standard reduction voltage) with the oxidation and reduction half-reactions. The standard voltage for the overall reaction, $E°$, is the sum of these two quantities

$$E° = E°_{ox} + E°_{red}$$

To illustrate, consider the reaction between Zn and $H^+$ ions, for which the standard voltage is +0.762 V.

$$+0.762\ V = E°_{ox}(Zn \longrightarrow Zn^{2+}) + E°_{red}(H^+ \longrightarrow H_2)$$

There is no way to measure the standard voltage for a half-reaction; only $E°$ can be measured directly. To obtain values for $E°_{ox}$ and $E°_{red}$, the value zero is arbitrarily assigned to the standard voltage for reduction of $H^+$ ions to $H_2$ gas:

You need two half-cells to measure a voltage

$$2H^+(aq, 1\ M) + 2e^- \longrightarrow H_2(g, 1\ atm) \qquad E°_{red}(H^+ \longrightarrow H_2) = 0.000\ V$$

Using this convention, it follows that the standard voltage for the oxidation of zinc must be +0.762 V; that is,

$$Zn(s) \longrightarrow Zn^{2+}(aq, 1M) + 2e^- \qquad E°_{ox}(Zn \longrightarrow Zn^{2+}) = +0.762\ V$$

As soon as one half-reaction voltage is established, others can be determined. For example, the standard voltage for the Zn-$Cu^{2+}$ cell shown in Figure 18.2 is found to be +1.101 V. Knowing that $E°_{ox}$ for zinc is +0.762 V, it follows that

$$E°_{red}(Cu^{2+} \longrightarrow Cu) = E° - E°_{ox}(Zn \longrightarrow Zn^{2+})$$

$$= 1.101\ V - 0.762\ V = +0.339\ V$$

Standard half-cell voltages are ordinarily obtained from a list of **standard potentials** such as those in Table 18.1, pp. 504–505. The potentials listed are the standard voltages for reduction half-reactions, i.e.,

$$standard\ potential = E°_{red}$$

For example, since the standard potentials listed in the table for $Zn^{2+} \longrightarrow$ Zn and $Cu^{2+} \longrightarrow$ Cu are −0.762 V and +0.339 V, respectively,

$$Zn^{2+}(aq) + 2e^- \longrightarrow Zn(s) \qquad E°_{red} = -0.762\ V$$

$$Cu^{2+}(aq) + 2e^- \longrightarrow Cu(s) \qquad E°_{red} = +0.339\ V$$

TABLE 18.1 **Standard Potentials in Water Solution at 25°C**

| Oxidizing Agent | Reducing Agent | $E°_{red}(V)$ |
|---|---|---|
| **Acidic Solution** | | |
| $Li^+(aq) + e^-$ | $\longrightarrow Li(s)$ | $-3.040$ |
| $K^+(aq) + e^-$ | $\longrightarrow K(s)$ | $-2.936$ |
| $Ba^{2+}(aq) + 2e^-$ | $\longrightarrow Ba(s)$ | $-2.906$ |
| $Ca^{2+}(aq) + 2e^-$ | $\longrightarrow Ca(s)$ | $-2.869$ |
| $Na^+(aq) + e^-$ | $\longrightarrow Na(s)$ | $-2.714$ |
| $Mg^{2+}(aq) + 2e^-$ | $\longrightarrow Mg(s)$ | $-2.357$ |
| $Al^{3+}(aq) + 3e^-$ | $\longrightarrow Al(s)$ | $-1.68$ |
| $Mn^{2+}(aq) + 2e^-$ | $\longrightarrow Mn(s)$ | $-1.182$ |
| $Zn^{2+}(aq) + 2e^-$ | $\longrightarrow Zn(s)$ | $-0.762$ |
| $Cr^{3+}(aq) + 3e^-$ | $\longrightarrow Cr(s)$ | $-0.744$ |
| $Fe^{2+}(aq) + 2e^-$ | $\longrightarrow Fe(s)$ | $-0.409$ |
| $Cr^{3+}(aq) + e^-$ | $\longrightarrow Cr^{2+}(aq)$ | $-0.408$ |
| $Cd^{2+}(aq) + 2e^-$ | $\longrightarrow Cd(s)$ | $-0.402$ |
| $PbSO_4(s) + 2e^-$ | $\longrightarrow Pb(s) + SO_4{}^{2-}(aq)$ | $-0.356$ |
| $Tl^+(aq) + e^-$ | $\longrightarrow Tl(s)$ | $-0.336$ |
| $Co^{2+}(aq) + 2e^-$ | $\longrightarrow Co(s)$ | $-0.282$ |
| $Ni^{2+}(aq) + 2e^-$ | $\longrightarrow Ni(s)$ | $-0.236$ |
| $AgI(s) + e^-$ | $\longrightarrow Ag(s) + I^-(aq)$ | $-0.152$ |
| $Sn^{2+}(aq) + 2e^-$ | $\longrightarrow Sn(s)$ | $-0.141$ |
| $Pb^{2+}(aq) + 2e^-$ | $\longrightarrow Pb(s)$ | $-0.127$ |
| $2H^+(aq) + 2e^-$ | $\longrightarrow H_2(g)$ | $0.000$ |
| $AgBr(s) + e^-$ | $\longrightarrow Ag(s) + Br^-(aq)$ | $0.073$ |
| $S(s) + 2H^+(aq) + 2e^-$ | $\longrightarrow H_2S(aq)$ | $0.144$ |
| $Sn^{4+}(aq) + 2e^-$ | $\longrightarrow Sn^{2+}(aq)$ | $0.154$ |
| $SO_4{}^{2-}(aq) + 4H^+(aq) + 2e^-$ | $\longrightarrow SO_2(g) + 2H_2O$ | $0.155$ |
| $Cu^{2+}(aq) + e^-$ | $\longrightarrow Cu^+(aq)$ | $0.161$ |
| $Cu^{2+}(aq) + 2e^-$ | $\longrightarrow Cu(s)$ | $0.339$ |
| $Cu^+(aq) + e^-$ | $\longrightarrow Cu(s)$ | $0.518$ |
| $I_2(s) + 2e^-$ | $\longrightarrow 2I^-(aq)$ | $0.534$ |
| $Fe^{3+}(aq) + e^-$ | $\longrightarrow Fe^{2+}(aq)$ | $0.769$ |
| $Hg_2{}^{2+}(aq) + 2e^-$ | $\longrightarrow 2Hg(l)$ | $0.796$ |
| $Ag^+(aq) + e^-$ | $\longrightarrow Ag(s)$ | $0.799$ |
| $2Hg^{2+}(aq) + 2e^-$ | $\longrightarrow Hg_2{}^{2+}(aq)$ | $0.908$ |
| $NO_3{}^-(aq) + 4H^+(aq) + 3e^-$ | $\longrightarrow NO(g) + 2H_2O$ | $0.964$ |
| $AuCl_4{}^-(aq) + 3e^-$ | $\longrightarrow Au(s) + 4Cl^-(aq)$ | $1.001$ |
| $Br_2(l) + 2e^-$ | $\longrightarrow 2Br^-(aq)$ | $1.077$ |
| $O_2(g) + 4H^+(aq) + e^-$ | $\longrightarrow 2H_2O$ | $1.229$ |
| $MnO_2(s) + 4H^+(aq) + 2e^-$ | $\longrightarrow Mn^{2+}(aq) + 2H_2O$ | $1.229$ |
| $Cr_2O_7{}^{2-}(aq) + 14H^+(aq) + 6e^-$ | $\longrightarrow 2Cr^{3+}(aq) + 7H_2O$ | $1.33$ |
| $Cl_2(g) + 2e^-$ | $\longrightarrow 2Cl^-(aq)$ | $1.360$ |
| $ClO_3{}^-(aq) + 6H^+(aq) + 5e^-$ | $\longrightarrow \frac{1}{2}Cl_2(g) + 3H_2O$ | $1.458$ |
| $Au^{3+}(aq) + 3e^-$ | $\longrightarrow Au(s)$ | $1.498$ |
| $MnO_4{}^-(aq) + 8H^+(aq) + 5e^-$ | $\longrightarrow Mn^{2+}(aq) + 4H_2O$ | $1.512$ |
| $PbO_2(s) + SO_4{}^{2-}(aq) + 4H^+(aq) + 2e^-$ | $\longrightarrow PbSO_4(s) + 2H_2O$ | $1.687$ |
| $H_2O_2(aq) + 2H^+(aq) + 2e^-$ | $\longrightarrow 2H_2O$ | $1.763$ |
| $Co^{3+}(aq) + e^-$ | $\longrightarrow Co^{2+}(aq)$ | $1.953$ |
| $F_2(g) + 2e^-$ | $\longrightarrow 2F^-(aq)$ | $2.889$ |

R   O = strongest oxidizing agent

O   R = strongest reducing agent

Table 18.1

TABLE 18.1 (*Continued*)

| Oxidizing Agent | Reducing Agent | $E°_{red}(V)$ |
|---|---|---|
| | **Basic Solution** | |
| $Fe(OH)_2(s) + 2e^-$ | $\longrightarrow Fe(s) + 2\,OH^-(aq)$ | $-0.891$ |
| $2H_2O + 2e^-$ | $\longrightarrow H_2(g) + 2\,OH^-(aq)$ | $-0.828$ |
| $Fe(OH)_3(s) + e^-$ | $\longrightarrow Fe(OH)_2(s) + OH^-(aq)$ | $-0.547$ |
| $S(s) + 2e^-$ | $\longrightarrow S^{2-}(aq)$ | $-0.445$ |
| $NO_3^-(aq) + 2H_2O + 3e^-$ | $\longrightarrow NO(g) + 4\,OH^-(aq)$ | $-0.140$ |
| $NO_3^-(aq) + H_2O + 2e^-$ | $\longrightarrow NO_2^-(aq) + 2\,OH^-(aq)$ | $0.004$ |
| $ClO_4^-(aq) + H_2O + 2e^-$ | $\longrightarrow ClO_3^-(aq) + 2\,OH^-(aq)$ | $0.398$ |
| $O_2(g) + 2H_2O + 4e^-$ | $\longrightarrow 4OH^-(aq)$ | $0.401$ |
| $ClO_3^-(aq) + 3H_2O + 6e^-$ | $\longrightarrow Cl^-(aq) + 6\,OH^-(aq)$ | $0.614$ |
| $ClO^-(aq) + H_2O + 2e^-$ | $\longrightarrow Cl^-(aq) + 2\,OH^-(aq)$ | $0.890$ |

*Standard voltages for oxidation half-reactions are obtained by changing the sign of the standard potential listed in Table 18.1.* Thus,

For the system: $X \rightleftharpoons X^+ + e^-$; $E°_{ox}X = -E°_{red}X^+$

$$Zn(s) \longrightarrow Zn^{2+}(aq) + 2e^- \qquad E°_{ox} = +0.762\ V$$

$$Cu(s) \longrightarrow Cu^{2+}(aq) + 2e^- \qquad E°_{ox} = -0.339\ V$$

In general, standard voltages for forward and reverse half-reactions are equal in magnitude but opposite in sign.

## Strength of Oxidizing and Reducing Agents

As pointed out in Chapter 4, an oxidizing agent is a species that can gain electrons; it is a reactant in a reduction half-reaction. Since Table 18.1 lists reduction half-reactions from left to right, it follows that oxidizing agents are located in the left column of the table. All of the species listed in that column ($Li^+, \ldots, F_2$) are, in principle, oxidizing agents. A "strong" oxidizing agent is one that has a strong attraction for electrons and hence *can readily oxidize other species.* In contrast, a "weak" oxidizing agent does not gain electrons readily. It is capable of reacting only with those species that are very easily oxidized.

The strength of an oxidizing agent is directly related to the standard voltage for its reduction, $E°_{red}$. *The more positive $E°_{red}$ is, the stronger the oxidizing agent.* Looking at Table 18.1, you can see that *oxidizing strength increases moving down the left column.* The $Li^+$ ion, at the top of that column, is a very weak oxidizing agent with a large negative reduction voltage ($E°_{red} = -3.040\ V$). In practice, cations of the Group 1 metals ($Li^+$, $Na^+$, $K^+, \ldots$) and the Group 2 metals ($Mg^{2+}$, $Ca^{2+}, \ldots$) never act as oxidizing agents in water solution.

Further down the left column of Table 18.1, the $H^+$ ion ($E°_{red} = 0.000\ V$) is a reasonably strong oxidizing agent. It is capable of oxidizing many metals, including magnesium and zinc:

$$Mg(s) + 2H^+(aq) \longrightarrow Mg^{2+}(aq) + H_2(g) \qquad E° = +2.357\ V$$

$$Zn(s) + 2H^+(aq) \longrightarrow Zn^{2+}(aq) + H_2(g) \qquad E° = +0.762\ V$$

The strongest oxidizing agents are those at the bottom of the left column. Species such as $Cr_2O_7^{2-}$ ($E°_{red} = +1.33\ V$) and $Cl_2$ ($E°_{red} = +1.360\ V$) are com-

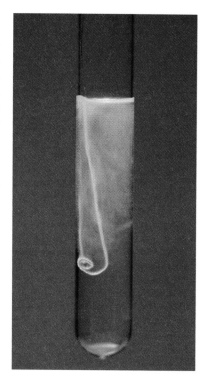

Magnesium metal reacts with $H^+$ ions, giving $Mg^{2+}$ and hydrogen gas.

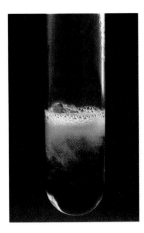

Lithium metal reacts with water to give hydrogen gas, lithium ions, and hydroxide ions.

monly used as oxidizing agents in the laboratory. The fluorine molecule, at the bottom of the left column, should be the strongest of all oxidizing agents. In practice, fluorine is never used as an oxidizing agent in water solution. The $F_2$ molecule takes electrons away from just about anything, including water, often with explosive violence.

The argument we have just gone through can be applied, in reverse, to reducing agents. These species are listed in the column to the right of Table 18.1 (Li, . . . , $F^-$). In principle, at least, all of them can supply electrons to another species in a redox reaction. Their strength as reducing agents is directly related to their $E^\circ_{ox}$ values. ***The more positive $E^\circ_{ox}$ is, the stronger the reducing agent.*** Looking at the values, remembering that $E^\circ_{ox} = -E^\circ_{red}$,

$$Li(s) \longrightarrow Li^+(aq) + e^- \qquad\qquad E^\circ_{ox} = +3.040 \text{ V}$$

$$H_2(g) \longrightarrow 2H^+(aq) + 2e^- \qquad\qquad E^\circ_{ox} = \phantom{+}0.000 \text{ V}$$

$$2F^-(aq) \longrightarrow F_2(g) + 2e^- \qquad\qquad E^\circ_{ox} = -2.889 \text{ V}$$

it should be clear that ***reducing strength decreases moving down the table.*** Actually, the five metals at the top of the right column, (Li, K, Ba, Ca, Na) cannot be used as reducing agents in water solution because they react directly with water, reducing it to hydrogen gas:

$$Li(s) + H_2O \longrightarrow Li^+(aq) + OH^-(aq) + \tfrac{1}{2}H_2(g)$$

$$Ba(s) + 2H_2O \longrightarrow Ba^{2+}(aq) + 2OH^-(aq) + H_2(g)$$

---

**Example 18.2**   Consider the following species in acidic solution: $Cr_2O_7^{2-}$, $NO_3^-$, $Br^-$, Mg, and $Sn^{2+}$. Using Table 18.1,

   (a) classify each of these as an oxidizing and/or reducing agent.
   (b) arrange the oxidizing agents in order of increasing strength.
   (c) do the same with the reducing agents.

***Strategy***   Remember that oxidizing agents are located in the left column of Table 18.1, reducing agents in the right column. Large positive values of $E^\circ_{red}$ and $E^\circ_{ox}$ are associated with strong oxidizing and reducing agents, respectively.

*Solution*

(a) Oxidizing agents:

   $Sn^{2+}$   $(E^\circ_{red} = -0.141 \text{ V})$,   $NO_3^-$   $(E^\circ_{red} = +0.964 \text{ V})$,   $Cr_2O_7^{2-}$   $(E^\circ_{red} = +1.33 \text{ V})$

   Reducing agents:

   Mg   $(E^\circ_{ox} = +2.357 \text{ V})$,   $Sn^{2+}$   $(E^\circ_{ox} = -0.154 \text{ V})$,   $Br^-$   $(E^\circ_{ox} = -1.077 \text{ V})$

   Note that the $Sn^{2+}$ ion can act as either an oxidizing agent, when it is reduced to Sn, or a reducing agent, when it is oxidized to $Sn^{4+}$.

(b) Comparing values of $E^\circ_{red}$:

$$Sn^{2+} < NO_3^- < Cr_2O_7^{2-}$$

   This ranking correlates with the positions of these species in the left column of Table 18.1; $Cr_2O_7^{2-}$ is near the bottom, $Sn^{2+}$ closest to the top.

(c) Comparing values of $E^\circ_{ox}$:

$$Br^- < Sn^{2+} < Mg$$

# Calculation of $E°$ from $E°_{red}$ and $E°_{ox}$

As pointed out earlier, the standard voltage for a redox reaction is the sum of the standard voltages of the two half-reactions, reduction and oxidation; that is,

$$E° = E°_{red} + E°_{ox}$$

This simple relation makes it possible, using Table 18.1, to calculate the standard voltages for more than 3000 different redox reactions.

---

**Example 18.3**   Using Table 18.1, calculate $E°$ for a voltaic cell in which the reaction is

$$2Fe^{3+}(aq) + 2I^-(aq) \longrightarrow 2Fe^{2+}(aq) + I_2(s)$$

*Strategy*   Split the reaction into two half-reactions, a reduction and an oxidation. Using Table 18.1, find the appropriate values for $E°_{red}$ and $E°_{ox}$. Add these to obtain $E°$.

*Solution*   Since the oxidation number of iron drops from +3 to +2, the reduction half-reaction is

$$2Fe^{3+}(aq) + 2e^- \longrightarrow 2Fe^{2+}(aq) \qquad E°_{red} = +0.769 \text{ V}$$

The oxidation number of iodine increases from $-1$ to 0, so the oxidation half-reaction is

$$2I^-(aq) \longrightarrow I_2(s) + 2e^- \qquad E°_{ox} = -0.534 \text{ V}$$

(Notice that the sign of the standard potential is changed, because this is an oxidation.)
   Just as the half-equations are added to obtain the overall equation, the half-reaction voltages are added to obtain the overall voltage.

| | |
|---|---|
| $2Fe^{3+}(aq) + 2e^- \longrightarrow 2Fe^{2+}(aq)$ | $E°_{red} = +0.769$ V |
| $2I^-(aq) \longrightarrow I_2(s) + 2e^-$ | $E°_{ox} = -0.534$ V |
| $2Fe^{3+}(aq) + 2I^-(aq) \longrightarrow 2Fe^{2+}(aq) + I_2(s)$ | $E° = +0.235$ V |

---

Two general points concerning cell voltages are illustrated by Example 18.3.

1. The calculated voltage, $E°$, is always a positive quantity for a reaction taking place in a voltaic cell.
2. The quantities $E°$, $E°_{ox}$, and $E°_{red}$ are independent of how the equation for the cell reaction is written. You *never* multiply the voltage by the coefficients of the balanced equation.

*It would be nice if you could double the voltage by rewriting the equation, but you can't.*

## Spontaneity of Redox Reactions

To determine whether a given redox reaction is spontaneous, apply a simple principle:
   *If the calculated voltage for a redox reaction is a positive quantity, the reaction will be spontaneous. If the calculated voltage is negative, the reaction will not occur.*
   Ordinarily, this principle is applied at standard concentrations (1 atm for gases, 1 $M$ for species in aqueous solution). Here, it is the sign of $E°$ that serves as the criterion for spontaneity. To show how this works, consider the problem

**Figure 18.5**

Nickel metal reacts spontaneously with $Cu^{2+}$ ions, producing Cu metal and $Ni^{2+}$ ions. Copper plates out on the surface of the nickel, and the blue color of $Cu^{2+}$ is replaced by the green color of $Ni^{2+}$. (Marna G. Clarke)

of oxidizing nickel metal to $Ni^{2+}$ ions. This *cannot* be accomplished using $Zn^{2+}$ ions:

$$Ni(s) + Zn^{2+}(aq, 1\ M) \longrightarrow Ni^{2+}(aq, 1\ M) + Zn(s)$$

$$E° = E°_{ox}\ Ni + E°_{red}\ Zn^{2+} = +0.236\ V - 0.762\ V = -0.526\ V$$

Sure enough, if you immerse a piece of nickel in a solution of $ZnSO_4$, nothing happens. Suppose, however, you add a bar of nickel to a solution of $CuSO_4$ (Fig. 18.5). Now the nickel is oxidized through the spontaneous redox reaction:

$$Ni(s) + Cu^{2+}\ (aq, 1\ M) \longrightarrow Ni^{2+}\ (aq, 1M) + Cu(s)$$

$$E° = E°_{ox}\ Ni + E°_{red}\ Cu^{2+} = +0.236\ V + 0.339\ V = +0.575\ V$$

---

**Example 18.4** Using standard potentials listed in Table 18.1, decide whether, at standard concentrations,

(a) Fe(s) will be oxidized to $Fe^{2+}$ by treatment with hydrochloric acid (HCl).
(b) Cu(s) will be oxidized to $Cu^{2+}$ by treatment with hydrochloric acid.
(c) Cu(s) will be oxidized to $Cu^{2+}$ by treatment with nitric acid ($HNO_3$).

***Strategy*** The oxidation half-reaction is given in each case; you must decide upon the nature of the reduction half-reaction. Once that is done, look up the appropriate standard potentials and combine them to find out whether $E°$ is positive or negative.

***Solution***

(a) The oxidation half-reaction is

$$Fe(s) \longrightarrow Fe^{2+}(aq) + 2e^- \qquad\qquad E°_{ox} = +0.409\ V$$

Hydrochloric acid consists of $H^+$ and $Cl^-$ ions. Of these two ions, only $H^+$ is listed in the left column of Table 18.1; the $Cl^-$ ion cannot be reduced. The reduction half-reaction must be

$$2H^+(aq) + 2e^- \longrightarrow H_2(g) \qquad\qquad E°_{red} = 0.000\ V$$

Since the calculated voltage is positive,

$$E° = +0.409\ V + 0.000\ V = +0.409\ V$$

The following redox reaction, obtained by summing the half-reactions,

$$Fe(s) + 2H^+(aq) \longrightarrow Fe^{2+}(aq) + H_2(g)$$

should and does occur (Fig. 18.6).

(b) Proceeding in the same way,

$$\begin{aligned} Cu(s) &\longrightarrow Cu^{2+}(aq) + 2e^- & E°_{ox} &= -0.339\ V \\ 2H^+(aq) + 2e^- &\longrightarrow H_2(g) & E°_{red} &=\ \ 0.000\ V \\ \hline Cu(s) + 2H^+(aq) &\longrightarrow Cu^{2+}(aq) + H_2(g) & E° &= -0.339\ V \end{aligned}$$

As predicted, no reaction occurs when copper is added to 1 *M* hydrochloric acid.

(c) Here, there is another possible oxidizing agent, the $NO_3^-$ ion. Looking at Table 18.1, you should find that $E°_{red}$ for the $NO_3^-$ ion is +0.964 V. It follows that $E°$ is positive:

$$E° = E°_{ox}\ for\ Cu + E°_{red}\ for\ NO_3^- = -0.339\ V + 0.964\ V = +0.625\ V$$

This means that copper is oxidized (to $Cu^{2+}$) by nitric acid, which is reduced to NO or $NO_2$ (Figure 18.7).

**Figure 18.6**

Finely divided iron in the form of steel wool reacts with hydrochloric acid to evolve hydrogen: $Fe(s) + 2H^+(aq) \longrightarrow Fe^{2+}(aq) + H_2(g)$.

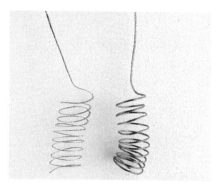
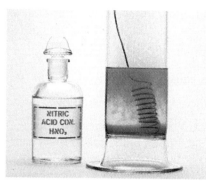
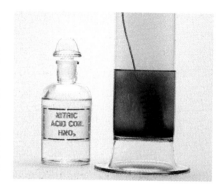

**Figure 18.7**
Copper metal is comparatively inactive, but it reacts with concentrated nitric acid. The brown fumes are $NO_2(g)$, a reduction product of $HNO_3$. The copper is oxidized to $Cu^{2+}$ ions, which impart their color to the solution. (Marna G. Clarke)

# 18.3  Relations Between $E°$, $\Delta G°$, and K

As pointed out previously, the value of the standard cell voltage, $E°$, is a measure of the spontaneity of a cell reaction. In Chapter 17, we showed that the standard free energy change, $\Delta G°$, is a general criterion for reaction spontaneity. As you might suppose, these two quantities are simply related to one another and to the equilibrium constant, $K$, for the cell reaction.

Spontaneous reaction: $E° > 0$, $\Delta G° < 0$, $K = 1$

## $E°$ and $\Delta G°$

It was pointed out in Chapter 17 (p. 475) that the free energy change is a measure of the amount of useful work that can be obtained from a reaction carried out at constant temperature and pressure. The relation between these two quantities is

$$\Delta G = w_{max}$$

where $w_{max}$ is the maximum amount of useful work done *on* the reaction system. For a voltaic cell, the "useful work" produced *by* the cell reaction, $-w_{max}$, is the electrical energy generated. This in turn is the product of the charge, $Q'$, in coulombs, times the voltage, $E$:

$$-w_{max} = Q'E$$

Hence

$$\Delta G = -Q'E$$

We would like to relate $\Delta G$ to the number of moles of electrons taking part in the reaction, $n$. To do this, note that[*]

$$1 \text{ mol } e^- = 9.648 \times 10^4 \text{C}$$

$$Q' = 9.648 \times 10^4 \; \frac{\text{C}}{\text{mol}} \times n \quad (n = \text{no. of moles of } e^- \text{ exchanged in reaction})$$

[*]The charge of an electron is $1.602 \times 10^{-19}$ C. The charge on one mole of electrons must then be $(6.022 \times 10^{23})(1.602 \times 10^{-19} \text{ C}) = 9.648 \times 10^4$ C.

Two voltaic cells that produce "useful work." These cells are called dry cells.

Substituting for $Q'$ in the expression for $\Delta G$:

$$\Delta G = -9.648 \times 10^4 \frac{C}{mol} \times n \times E$$

This equation is usually written in the form

$$\Delta G = -nFE$$

where $F$ is the **Faraday constant,** $9.648 \times 10^4$ C/mol. Noting that 1 J = 1 V × 1 C, it follows that an alternative expression for the Faraday constant is

$$F = 9.648 \times 10^4 \frac{C}{mol} \times \frac{1\ J}{1V \cdot 1C} = 9.648 \times 10^4 \frac{J}{mol \cdot V}$$

At standard concentrations,

$$\Delta G° = -nFE°$$

Notice from this equation that $\Delta G°$ and $E°$ have opposite signs. This is reasonable; a spontaneous reaction is one for which $\Delta G°$ is *negative* but $E°$ is *positive*.

$F = 9.648 \times 10^4$ C/mol $= 9.648 \times 10^4$ J/mol · V

---

**Example 18.5**  Calculate $\Delta G°$ for the reaction

$$Cl_2(g) + 2Br^-(aq) \longrightarrow 2Cl^-(aq) + Br_2(l)$$

using data in Table 18.1

***Strategy***  Split the reaction into two half-reactions; find $E°_{ox}$ and $E°_{red}$ from Table 18.1 and add to obtain $E°$. Then use the relation $\Delta G° = -nFE°$.

*Solution*

(1) The balanced half-equations for reduction and oxidation are

$$
\begin{array}{ll}
Cl_2(g) + 2e^- \longrightarrow 2Cl^-(aq) & E°_{red} = +1.360\ V \\
2Br^-(aq) \longrightarrow Br_2(l) + 2e^- & E°_{ox} = -1.077\ V \\
\hline
Cl_2(g) + 2Br^-(aq) \longrightarrow 2Cl^-(aq) + Br_2(l) & E° = +0.283\ V
\end{array}
$$

(2) Note from the half-equations that $n = 2$:

$$\Delta G° = -2\ mol\ (9.648 \times 10^4\ J/mol \cdot V)(0.283\ V) = -5.46 \times 10^4\ J = \boxed{-54.6\ kJ}$$

Notice that $E°$ is positive, while $\Delta G°$ is negative, indicating a spontaneous reaction. This reaction can serve as a basis for a voltaic cell (Fig. 18.4) or as a way of testing for $Br^-$ ions in solution (Fig. 18.8).

---

The approach used in Example 18.5 to find the number of moles of electrons transferred, *n,* is generally useful.

$$n = \text{coefficient of } e^- \text{ in either half-equation}$$

## $E°$ and $K$

Redox reactions, like all reactions, eventually reach a state of equilibrium. It is possible to calculate the equilibrium constant for a redox reaction from the standard voltage. To do that, we start with the relation obtained in Chapter 17:

$$\Delta G° = -RT \ln K$$

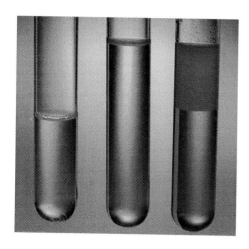

**Figure 18.8**
When a water solution saturated with $Cl_2(g)$ is added to a solution containing $Br^-$ ions (left), a redox reaction occurs. The $Br_2$ formed gives the water solution a light orange color (center). Extraction with a small amount of an organic solvent gives the characteristic reddish-orange color of $Br_2$ (right).

We have just shown that $\Delta G° = -nFE°$. It follows that

$$RT \ln K = nFE°$$

or

$$E° = \frac{RT}{nF} \ln K$$

The quantity $RT/F$ is readily evaluated at 25°C, the temperature at which standard potentials are tabulated.

$$\frac{RT}{F} = \frac{8.31 \text{ J/mol} \cdot \text{K} \times 298 \text{ K}}{9.648 \times 10^4 \text{ J/mol} \cdot \text{V}} = 0.0257 \text{ V}$$

Hence

$$E° = \frac{(0.0257 \text{ V})}{n} \ln K \quad \text{(at 25°C)}$$

Notice that if the standard voltage is positive, ln $K$ is also positive, and $K$ is greater than one. Conversely, if the standard voltage is negative, ln $K$ is also negative, and $K$ is less than one.

---

**Example 18.6**   For the reaction

$$3Ag(s) + NO_3^-(aq) + 4H^+(aq) \longrightarrow 3Ag^+(aq) + NO(g) + 2H_2O$$

calculate the equilibrium constant $K$, using data in Table 18.1.

**Strategy**   Use Table 18.1 to find $E°$. Split the equation into two half-equations to find $n$. Then use the relation $nE° = (0.0257 \text{ V}) \ln K$ to find $K$.

**Solution**

(1) $E° = E°_{red} NO_3^- + E°_{ox} Ag = +0.964 \text{ V} - 0.799 \text{ V} = +0.165 \text{ V}$

(2) Ag is oxidized to $Ag^+$; $NO_3^-$ is reduced to NO. The half-equations are

$$3Ag(s) \longrightarrow 3Ag^+(aq) + 3e^-$$

$$NO_3^-(aq) + 4H^+(aq) + 3e^- \longrightarrow NO(g) + 2H_2O$$

Clearly, $n = 3$.

(3)
$$\ln K = \frac{nE°}{(0.0257 \text{ V})} = \frac{3(+0.165 \text{ V})}{(0.0257 \text{ V})} = 19.3; \quad K = \boxed{2 \times 10^8}$$

---

TABLE 18.2   **Relation Between $E°$, $K$, and $\Delta G°$ ($n = 2$)**

| $E°(V)$ | $K$ | $\Delta G°(kJ)$ | $E°(V)$ | $K$ | $\Delta G°(kJ)$ |
|---|---|---|---|---|---|
| +2.00 | $4 \times 10^{67}$ | −400 | −2.00 | $3 \times 10^{-68}$ | +400 |
| +1.00 | $6 \times 10^{33}$ | −200 | −1.00 | $2 \times 10^{-34}$ | +200 |
| +0.50 | $8 \times 10^{16}$ | −100 | −0.50 | $1 \times 10^{-17}$ | +100 |
| +0.25 | $3 \times 10^{8}$ | −50 | −0.25 | $4 \times 10^{-9}$ | +50 |
| +0.10 | $2 \times 10^{3}$ | −20 | −0.10 | 0.0004 | +20 |
| +0.05 | 50 | −10 | −0.05 | 0.02 | +10 |
| 0.00 | 1 | 0 | | | |

Table 18.2 lists values of $K$ and $\Delta G°$ corresponding to various values of $E°$ with $n = 2$. Notice that

— if $E°$ is greater than about +0.10 V, $K$ is very large, and the reaction goes virtually to completion.
— if $E°$ is smaller than about −0.10 V, $K$ is very small, so the reaction does not proceed to any appreciable extent.

Only if the standard voltage falls within a rather narrow range, say +0.10 to −0.10 V, will the value of $K$ (and that of $\Delta G°$ ) be such that the reaction will produce an equilibrium mixture containing appreciable amounts of both reactants and products.

*Most redox reactions either go to completion or not at all*

## 18.4   Effect of Concentration Upon Voltage

To this point, we have dealt only with "standard" voltages, i.e., voltages when all gases are at 1 atm pressure and all species in aqueous solution are at a concentration of 1 $M$. When the concentration of a reactant or product changes, the voltage changes as well. Qualitatively, the direction in which the voltage shifts is readily predicted when you realize that cell voltage is directly related to reaction spontaneity.

1. Voltage will *increase* if the concentration of a reactant is increased or that of a product is decreased. Either of these changes increases the driving force behind the redox reaction, making it more spontaneous.
2. Voltage will *decrease* if the concentration of a reactant is decreased or that of a product is increased. Either of these changes makes the redox reaction less spontaneous.

When a voltaic cell operates, supplying electrical energy, the concentration of reactants decreases and that of the products increases. As time passes, the voltage drops steadily. Eventually it becomes zero, and we say that the cell is "dead." At that point, the redox reaction taking place within the cell is at equilibrium, and there is no driving force to produce a voltage.

### Nernst Equation

To obtain a quantitative relation between cell voltage and concentration, it is convenient to start with the general expression for the free energy change discussed in Chapter 17:

$$\Delta G = \Delta G° + RT \ln Q$$

Substituting for $\Delta G$ and $\Delta G°$ from the relations obtained in Section 18.3,

$$\Delta G = -nFE \qquad \Delta G° = -nFE°$$

yields
$$-nFE = -nFE° + RT \ln Q.$$

Solving for the cell voltage $E$,

$$E = E° - \frac{RT}{nF} \ln Q$$

This relationship is known as the **Nernst equation,** after Walther Nernst, a brilliant but egocentric colleague of Arrhenius, who first proposed it in 1888. Recalling that, at 25°C, the quantity $RT/F$ is 0.0257 V,

$$E = E° - \frac{(0.0257 \text{ V})}{n} \ln Q$$

If $Q > 1$, the reaction is less spontaneous and $E < E°$

In terms of base-10 logarithms ($\ln x = 2.303 \log_{10} x$),

$$E = E° - \frac{(0.0591 \text{ V})}{n} \log_{10} Q$$

In these equations, $E$ is the cell voltage, $E°$ the standard voltage, $n$ is the number of moles of electrons exchanged in the reaction, and $Q$ is the reaction quotient.

Remember that gases enter $Q$ as their partial pressures in atmospheres. Species in water solution enter as their molar concentrations. Pure liquids and solids do not appear in the expression for $Q$. For example,

$$aA(s) + bB(aq) \longrightarrow cC(aq) + dD(g) \qquad Q = \frac{[C]^c \times (P_D)^d}{[B]^b}$$

---

**Example 18.7** Consider a voltaic cell in which the following reaction occurs:

$$O_2(g) + 4H^+(aq) + 4Br^-(aq) \longrightarrow 2H_2O + 2Br_2(l)$$

Calculate the cell voltage, $E$, when $O_2$ is at 1.0 atm pressure, $[H^+] = [Br^-] = 0.10$ M.

***Strategy*** First, set up the Nernst equation, following the rules for $Q$ listed above. Then calculate $E°$, using the standard potentials in Table 18.1. Finally, using the Nernst equation, calculate $E$.

***Solution***

(1) $Q = \dfrac{1}{(P_{O_2}) \times [H^+]^4 \times [Br^-]^4}$

To find $n$, break the reaction down into two half-reactions:

$$O_2(g) + 4H^+(aq) + 4e^- \longrightarrow 2H_2O$$

$$4Br^-(aq) \longrightarrow 2Br_2(l) + 4e^-$$

Clearly, $n = 4$. The Nernst equation must then be

$$E = E° - \frac{(0.0257 \text{ V})}{4} \ln \frac{1}{(P_{O_2}) \times [H^+]^4 \times [Br^-]^4}$$

(2) $E° = E°_{red} O_2 + E°_{ox} Br^- = +1.229$ V $- 1.077$ V $= +0.152$ V

(3) $E = +0.152 \text{ V} - \dfrac{(0.0257 \text{ V})}{4} \ln \dfrac{1}{1.0(0.10)^4(0.10)^4}$

$\quad = +0.152 \text{ V} - \dfrac{(0.0257 \text{ V})}{4} \ln (1.0 \times 10^8)$

$\quad = +0.152 \text{ V} - \dfrac{(0.0257 \text{ V})(18.4)}{4} = \boxed{+0.034 \text{ V}}$

---

The Nernst equation can also be used to determine the effect of changes in concentration upon the voltage of an individual half-cell, $E^\circ_{red}$ or $E^\circ_{ox}$. Consider, for example, the half-reaction

$$MnO_4^-(aq) + 8H^+(aq) + 5e^- \longrightarrow Mn^{2+}(aq) + 4H_2O \qquad E^\circ_{red} = +1.512 \text{ V}$$

Here the Nernst equation takes the form

$$E_{red} = +1.512 \text{ V} - \dfrac{(0.0257 \text{ V})}{5} \ln \dfrac{[Mn^{2+}]}{[MnO_4^-] \times [H^+]^8}$$

where $E_{red}$ is the observed reduction voltage corresponding to any given concentrations of $Mn^{2+}$, $MnO_4^-$, and $H^+$.

## Use of the Nernst Equation to Determine Ion Concentrations

In chemistry, the most important use of the Nernst equation lies in the experimental determination of the concentration of ions in solution. Suppose you measure the cell voltage $E$ and know the concentration of all but one species in the two half-cells. It should then be possible to calculate the concentration of that species by using the Nernst equation (Example 18.8).

---

**Example 18.8** Consider a voltaic cell in which the reaction is

$$Zn(s) + 2H^+(aq) \longrightarrow Zn^{2+}(aq) + H_2(g)$$

It is found that the cell voltage is +0.460 V when $[Zn^{2+}] = 1.0 \, M$, $P_{H_2} = 1.0$ atm. What must be the concentration of $H^+$ in the $H_2$-$H^+$ half-cell?

*Strategy* This example is handled exactly like Example 18.7, except that in the last step you solve for $[H^+]$ instead of $E$.

*Solution*

This approach is particularly useful at very low concentrations

(1) Setting up the Nernst equation with $n = 2$.

$$E = E^\circ - \dfrac{(0.0257 \text{ V})}{2} \ln \dfrac{[Zn^{2+}] \times (P_{H_2})}{[H^+]^2}$$

(2) $E^\circ = E^\circ_{ox} \, Zn + E^\circ_{red} \, H^+ = +0.762 \text{ V}$

(3) All that remains is to substitute for $E$, $E^\circ_{tot}$, $[Zn^{2+}]$, $P_{H_2}$, and solve for $[H^+]$:

$$+0.460 \text{ V} = +0.762 \text{ V} - \dfrac{0.0257 \text{ V}}{2} \ln \dfrac{1 \times 1}{[H^+]^2}$$

Solving:

$$\ln \frac{1}{[H^+]^2} = \frac{2(+0.460\ V - 0.762\ V)}{-0.0257\ V} = 23.5$$

$$\frac{1}{[H^+]^2} = 1.6 \times 10^{10} \qquad [H^+] = 8 \times 10^{-6}M$$

As Example 18.8 implies, the $Zn\,|\,Zn^{2+}\,\|\,H^+\,|\,H_2\,|\,Pt$ cell can be used to measure the concentration of $H^+$ or pH of a solution. Indeed, cells of this type are commonly used to measure pH; Figure 18.9 shows a schematic diagram of a cell used with a pH meter. The pH meter, referred to in Chapter 13, is actually a high-resistance voltmeter calibrated to read pH rather than voltage. The cell connected to the pH meter consists of two half-cells. One of these is a reference half-cell of known voltage. The other half-cell contains a solution of known pH separated by a thin, fragile *glass electrode* from a solution whose pH is to be determined. The voltage of this cell is a linear function of the pH of the solution in the beaker.

*Specific ion electrodes,* similar in design to the glass electrode, have been developed to analyze for a variety of cations and anions. One of the first to be used extensively was a fluoride ion electrode that is sensitive to $F^-$ at concentrations as low as 0.1 part per million and hence is ideal for monitoring fluoridated water supplies. An electrode that is specific for $Cl^-$ ions is used to diagnose for cystic fibrosis. Attached directly to the skin, it detects the abnormally high concentrations of sodium chloride in sweat that are a characteristic symptom of this disorder. Diagnoses that used to require an hour or more can now be carried out in a few minutes; as a result, large numbers of children can be screened rapidly and routinely.

The general approach illustrated by Example 18.8 is widely used to determine equilibrium constants for solution reactions. The pH meter in particular can be used to determine acid or base ionization constants by measuring the pH of solutions containing known concentrations of weak acids or bases. Specific ion electrodes are readily adapted to the determination of solubility product constants. For example, a chloride ion electrode can be used to find $[Cl^-]$ in equilibrium with $AgCl(s)$ and a known $[Ag^+]$. From that information, $K_{sp}$ of $AgCl$ can be calculated.

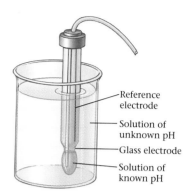

**Figure 18.9**

The pH of a solution can be determined with the aid of a "glass electrode." The voltage between the glass electrode and the reference electrode is directly related to pH. The leads from the electrode are connected to a pH meter of the type discussed in Chapter 13.

Reference electrode
Solution of unknown pH
Glass electrode
Solution of known pH

## 18.5   Electrolytic Cells

In an electrolytic cell, a nonspontaneous redox reaction is made to occur by pumping electrical energy into the system. A generalized diagram for such a cell is shown in Figure 18.10. The storage battery at the left provides a source of direct electric current. From the terminals of the battery, two wires lead to the electrolytic cell. This consists of two electrodes, A and C, dipping into a solution containing ions $M^+$ and $X^-$.

The battery acts as an electron pump, pushing electrons into the *cathode,* C, and removing them from the *anode,* A. To maintain electrical neutrality, some process within the cell must consume electrons at C and liberate them at A. This process, in which a redox reaction is brought about by electrical energy, is called **electrolysis.**

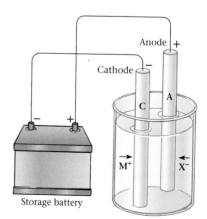

Anode +
Cathode −
C
A
$M^+$
$X^-$
Storage battery

**Figure 18.10**

Schematic diagram of an electrolytic cell.

## Cell Reactions (Water Solution)

As is always the case, a reduction half-reaction occurs at the cathode of an electrolytic cell. This half-reaction may be

— *the reduction of a cation to the corresponding metal.* This commonly occurs with transition metal cations, which are relatively easy to reduce. Examples include

$$Ag^+(aq) + e^- \longrightarrow Ag(s) \qquad\qquad E^\circ_{red} = +0.799 \text{ V}$$

$$Cu^{2+}(aq) + 2e^- \longrightarrow Cu(s) \qquad\qquad E^\circ_{red} = +0.339 \text{ V}$$

This type of half-reaction is characteristic of electroplating processes, in which a metal object serves as the cathode (Figure 18.11)

— *the reduction of a water molecule to hydrogen gas*

$$2H_2O + 2e^- \longrightarrow H_2(g) + 2OH^-(aq) \qquad E^\circ_{red} = -0.828 \text{ V}$$

This half-reaction commonly occurs when the cation in solution is very difficult to reduce. For example, electrolysis of a solution containing $K^+$ ions ($E^\circ_{red} = -2.936$ V) or $Na^+$ ions ($E^\circ_{red} = -2.714$ V) yields hydrogen gas at the cathode.

At the anode of an electrolytic cell, the half-reaction may be

— *the oxidation of an anion to the corresponding nonmetal*

$$2I^-(aq) \longrightarrow I_2(s) + 2e^- \qquad\qquad E^\circ_{ox} = -0.534 \text{ V}$$

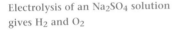
Electrolysis of an $Na_2SO_4$ solution gives $H_2$ and $O_2$

Electrolysis of a water solution of KI gives a saturated solution of iodine at the anode.

— *the oxidation of a water molecule to oxygen gas*

$$2H_2O \longrightarrow O_2(g) + 4H^+(aq) + 4e^- \qquad E^\circ_{ox} = -1.229 \text{ V}$$

This half-reaction occurs when the anion present is virtually impossible to oxidize, as is the case with $F^-$ ($E^\circ_{ox} = -2.889$ V) or $SO_4^{2-}$.

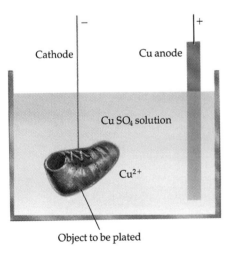

**Figure 18.11**
Copper metal can be plated onto a baby's shoe by electrolysis. The shoe's surface is coated with graphite to make it conduct.

## Quantitative Relationships

There is a simple relationship between the amount of electricity passed through an electrolytic cell and the amounts of substances produced by oxidation or reduction at the electrodes. From the balanced half-equations

$$Ag^+(aq) + e^- \longrightarrow Ag(s)$$

$$Cu^{2+}(aq) + 2e^- \longrightarrow Cu(s)$$

$$Au^{3+}(aq) + 3e^- \longrightarrow Au(s)$$

you can deduce that

$$1 \text{ mol of } e^- \longrightarrow 1 \text{ mol of Ag (107.9 g of Ag)}$$

$$2 \text{ mol of } e^- \longrightarrow 1 \text{ mol of Cu (63.55 g of Cu)}$$

$$3 \text{ mol of } e^- \longrightarrow 1 \text{ mol of Au (197.0 g of Au)}$$

Relations of this type, obtained from balanced half-equations, can be used in many practical calculations involving electrolytic cells. Frequently, the relations between electrical units provided in Table 18.3 will be required as well. The following relationships are particularly useful in electrochemical calculations:

$$\text{no. of moles of } e^- = \text{no. of coulombs}/96480$$

$$\text{no. of coulombs} = \text{no. of amperes} \times \text{no. of seconds}$$

$$\text{no. of joules} = \text{no. of volts} \times \text{no. of coulombs}$$

TABLE 18.3   **Electrical Units**

| Quantity | Unit | Defining Relation | Conversion Factors |
|---|---|---|---|
| Charge | coulomb (C) | $1 \text{ C} = 1 \text{ A} \cdot \text{s} = 1 \text{ J/V}$ | $1 \text{ mol } e^- = 9.648 \times 10^4 \text{ C}$ |
| Current | ampere (A) | $1 \text{ A} = 1 \text{ C/s}$ | |
| Potential | volt (V) | $1 \text{ V} = 1 \text{ J/C}$ | |
| Power | watt (W) | $1 \text{ W} = 1 \text{ J/s}$ | |
| Energy | joule (J) | $1 \text{ J} = 1 \text{ V} \cdot \text{C}$ | $1 \text{ kWh} = 3.600 \times 10^6 \text{ J}$ |

**Example 18.9**   Chromium metal can be electroplated from an acidic solution of $CrO_3$.

(a) How many grams of chromium will be plated by $1.00 \times 10^4$ C?
(b) How long will it take to plate one gram of chromium using a current of 6.00 A?

*Strategy*   The first thing to do is to write the half-equation for the reduction. Following the rules cited in Chapter 4, you should arrive at

$$CrO_3(aq) + 6H^+(aq) + 6e^- \longrightarrow Cr(s) + 3H_2O$$

From this equation, it should be clear that 6 mol of $e^- = 1$ mol of Cr (52.0 g of Cr). That relationship, along with the relations 1 mol $e^- = 9.648 \times 10^4$ C and 1 A = 1 C/s, can be used to calculate the required quantities. Follow a conversion-factor approach.

*Solution*

(a) $1.00 \times 10^4 \text{ C} \times \dfrac{1 \text{ mol } e^-}{9.648 \times 10^4 \text{ C}} \times \dfrac{52.0 \text{ g Cr}}{6 \text{ mol } e^-} = \boxed{0.898 \text{ g of Cr}}$

(b) The most straightforward approach here is to calculate the number of coulombs first and then the time.

(1) $1.00 \text{ g Cr} \times \dfrac{6 \text{ mol } e^-}{52.0 \text{ g Cr}} \times \dfrac{9.648 \times 10^4 \text{ C}}{1 \text{ mol } e^-} = 1.11 \times 10^4 \text{ C}$

(2) Since $1 \text{ A} = 1 \text{ C/s}$,

$$\text{time (s)} = \dfrac{\text{charge (C)}}{\text{current (A)}} = \dfrac{1.11 \times 10^4 \text{ C}}{6.00 \text{ C/s}} = \boxed{1.85 \times 10^3 \text{ s}} \quad \text{(about half an hour)}$$

---

**Example 18.10**   Consider the electroplating of chromium, referred to in Example 18.9. If the applied voltage is 4.5 V, calculate the amount of electrical energy absorbed in plating 1.00 g of Cr, first in joules and then in kilowatt-hours.

*Strategy*   Recall that, in Example 18.9b, you calculated the number of coulombs required to plate one gram of Cr. Multiplying that by the number of volts will give the energy in joules, which can then be converted to kilowatt-hours.

*Solution*

$$\text{energy (J)} = \text{voltage (V)} \times \text{charge (C)} = 4.5 \text{ V} \times 1.1 \times 10^4 \text{ C} = \boxed{5.0 \times 10^4 \text{ J}}$$

$$\text{energy in kWh} = 5.0 \times 10^4 \text{ J} \times \dfrac{1 \text{ kWh}}{3.6 \times 10^6 \text{ J}} = \boxed{1.4 \times 10^{-2} \text{ kWh}}$$

---

*Nobody is 100% efficient*

In working Examples 18.9 and 18.10, we have in effect assumed that the electrolyses were 100% efficient in converting electrical energy into chemical energy. In practice, this is almost never the case. Some electrical energy is wasted in side reactions at the electrodes and in the form of heat. This means that the actual yield of products is less than the "theoretical yield."

# Chemistry: *The Human Side*

The laws of electrolysis were discovered by Michael Faraday, perhaps the most talented experimental scientist of the nineteenth century. Faraday lived his entire life in what is now greater London. The son of a blacksmith, he had no formal education beyond the rudiments of reading, writing, and arithmetic. Apprenticed to a bookbinder at the age of 13, Faraday educated himself by reading virtually every book that came into the shop. One that particularly impressed him was a textbook, *Conversations in Chemistry*, written by Mrs. Jane Marcet. Anxious to escape a life of drudgery as a tradesman, Faraday wrote to Sir Humphry Davy at the Royal Institution, requesting employment. Shortly afterwards, a vacancy arose, and Faraday was hired as a laboratory assistant.

Davy quickly recognized Faraday's talents and as time passed allowed him to work more and more independently. In his years with Davy, Faraday published papers covering almost every field of chemistry. They included studies on the condensation of gases (he was the first to liquefy ammonia), the reaction of silver compounds with ammonia, and the isolation of several organic compounds, the most important of which was benzene. In 1825, Faraday began a series of lectures at the Royal Institution that were brilliantly successful. That same year he succeeded Davy as director of the laboratory. As Faraday's reputation grew, it was said that "Humphrey Davy's greatest discovery was Michael Faraday." Perhaps it was witticisms of this sort that led to an estrangement between master and protégé. Late in his life, Davy opposed Faraday's nomination as a Fellow of the Royal Society and is reputed to have cast the only vote against him.

To Michael Faraday, science was an obsession; one of his biographers describes him as a "work maniac." An observer (Faraday had no students) said of him:

> ...if he had to cross the laboratory for anything, he did not walk, he ran; the quickness of his perception was equalled by the calm rapidity of his movements.

In 1839, he suffered a nervous breakdown, the result of overwork. For much of the rest of his life, Faraday was in poor health. He gradually gave up more and more of his social engagements but continued to do research at the same pace as before.

Faraday developed the laws of electrolysis between 1831 and 1834. In mid-December of 1833, he began a quantitative study of the electrolysis of several metal cations, including $Sn^{2+}$, $Pb^{2+}$, and $Zn^{2+}$. Despite taking a whole day off for Christmas, he managed to complete these experiments, write up the results of three years' work, and get his paper published in the *Philosophic Transactions of the Royal Society* on January 9, 1834. In this paper, Faraday introduced the basic vocabulary of electrochemistry, using for the first time the terms "anode," "cathode," "ion," "electrolyte," and "electrolysis."

**Michael Faraday**
(1791–1867)
(Oesper Collection in the History of Chemistry, University of Cincinnati)

# 18.6 Commercial Cells

To a chemist, electrochemical cells are of interest primarily for the information they yield concerning the spontaneity of redox reactions, the strengths of oxidizing and reducing agents, and the concentrations of trace species in solution. The viewpoint of an engineer is somewhat different; here, applications of electrolytic cells in electroplating and electrosynthesis are of particular importance. To the layman, electrochemistry is important primarily because of commercial voltaic cells, which supply the electrical energy for instruments ranging in size from pacemakers to automobiles.

## Electrolysis of Aqueous NaCl

From a commercial standpoint, the most important electrolysis carried out in water solution is that of sodium chloride (Fig. 18.12). At the anode, $Cl^-$ ions are oxidized to chlorine gas:

$$\text{anode:} \qquad 2Cl^-(aq) \longrightarrow Cl_2(g) + 2e^-$$

At the cathode, the half-reaction involves $H_2O$ molecules, which are easier to reduce ($E^\circ_{red} = -0.828$ V) than $Na^+$ ions ($E^\circ_{red} = -2.714$ V).

$$\text{cathode:} \qquad 2H_2O + 2e^- \longrightarrow H_2(g) + 2OH^-(aq)$$

The overall cell reaction is obtained by summing the half-reactions:

$$2Cl^-(aq) + 2H_2O \longrightarrow Cl_2(g) + H_2(g) + 2OH^-(aq)$$

Chlorine gas bubbles out of solution at the anode. At the cathode, hydrogen gas is formed, and the solution around the electrode becomes strongly basic.

**Figure 18.12**
Schematic diagram for the electrolysis of aqueous NaCl (brine). Migration of ions through the membrane maintains charge balance.

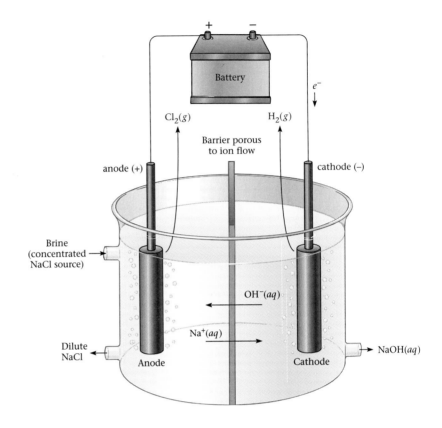

The products of electrolysis have a variety of uses. Chlorine is used to purify drinking water; large quantities of it are consumed in making plastics such as polyvinyl chloride (PVC). Hydrogen, prepared in this and many other industrial processes, is used chiefly in the synthesis of ammonia (Chap. 12). Sodium hydroxide (lye), obtained on evaporation of the electrolyte, is used in processing pulp and paper, in the purification of aluminum ore, in the manufacture of glass and textiles, and for many other purposes.

## Primary (Nonrechargeable) Voltaic Cells

The construction of the ordinary dry cell (Leclanché cell) used in flashlights is shown in Figure 18.13. The zinc wall of the cell is the anode. The graphite rod through the center of the cell is the cathode. The space between the electrodes is filled with a moist paste. This contains $MnO_2$, $ZnCl_2$, and $NH_4Cl$. When the cell operates, the half-reaction at the anode is

$$Zn(s) \longrightarrow Zn^{2+}(aq) + 2e^-$$

At the cathode, manganese dioxide is reduced to species in which Mn is in the +3 oxidation state, such as $Mn_2O_3$:

$$2MnO_2(s) + 2NH_4^+(aq) + 2e^- \longrightarrow Mn_2O_3(s) + 2NH_3(aq) + H_2O$$

The overall reaction occurring in this voltaic cell is

$$Zn(s) + 2MnO_2(s) + 2NH_4^+(aq) \longrightarrow Zn^{2+}(aq) + Mn_2O_3(s) + 2NH_3(aq) + H_2O$$

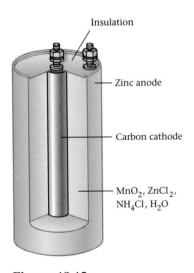

**Figure 18.13**
Section of an ordinary Zn–MnO₂ dry cell. This cell produces 1.5 V and will deliver a current of about half an ampere for six hours.

If too large a current is drawn from a Leclanché cell, the ammonia forms a gaseous insulating layer around the carbon cathode. When this happens, the voltage drops sharply and then returns slowly to its normal value of 1.5 V. This problem can be avoided by using an "alkaline" dry cell, in which the paste between the electrodes contains KOH rather than $NH_4Cl$. In this case the overall cell reaction is simply

A flashlight draws about 1 A and lasts about an hour before "dying"

$$Zn(s) + 2MnO_2(s) \longrightarrow ZnO(s) + Mn_2O_3(s)$$

No gas is produced. The alkaline dry cell, although more expensive than the Leclanché cell, has a longer shelf-life and provides more current.

Another important primary battery is the mercury cell. It usually comes in very small sizes and is used in hearing aids, watches, cameras, and some calculators. The anode of this cell is a zinc-mercury amalgam; the reacting species is zinc. The cathode is a plate made up of mercury(II) oxide, HgO. The electrolyte is a paste containing HgO and sodium or potassium hydroxide. The electrode reactions are

anode:      $Zn(s) + 2OH^-(aq) \longrightarrow Zn(OH)_2(s) + 2e^-$
cathode:     $\underline{HgO(s) + H_2O + 2e^- \longrightarrow Hg(l) + 2\ OH^-(aq)}$

$$Zn(s) + HgO(s) + H_2O \longrightarrow Zn(OH)_2(s) + Hg(l)$$

Notice that the overall reaction does not involve any ions in solution, so there are no concentration changes when current is drawn. As a result, the battery maintains a constant voltage of about 1.3 V throughout its life.

## Storage (Rechargeable) Voltaic Cells

A storage cell, unlike an ordinary dry cell, can be recharged repeatedly. This can be accomplished because the products of the reaction are deposited directly on the electrodes. By passing a current through a storage cell, it is possible to reverse the electrode reactions and restore the cell to its original condition.

A 12-V storage battery can deliver 300 A for a minute or so

The best-known voltaic cell of this type is the lead storage battery. The 12-V battery used in automobiles consists of six voltaic cells of the type shown in Figure 18.14. A group of lead plates, the grills of which are filled with spongy gray lead, forms the anode of the cell. The multiple cathode consists of another group of plates of similar design filled with lead(IV) oxide, $PbO_2$. These two sets

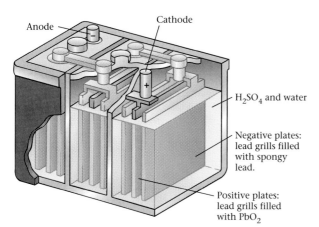

Anode

Cathode

$H_2SO_4$ and water

Negative plates: lead grills filled with spongy lead.

Positive plates: lead grills filled with $PbO_2$

**Figure 18.14**
One cell of a lead storage battery. Three advantages of the lead storage battery are its ability to deliver large amounts of energy for a short time, the ease of recharging, and a nearly constant voltage from full charge to discharge. A disadvantage is its high mass/energy ratio.

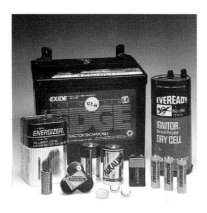

A variety of commercial voltaic cells.
(Charles D. Winters)

of plates alternate through the cell. They are immersed in a water solution of sulfuric acid, $H_2SO_4$, which acts as the electrolyte.

When a lead storage battery is supplying current, the lead in the anode grids is oxidized to $Pb^{2+}$ ions. These immediately react with $SO_4^{2-}$ ions in the electrolyte, precipitating $PbSO_4$ (lead sulfate) on the plates. At the cathode, lead dioxide is reduced to $Pb^{2+}$ ions, which also precipitate as $PbSO_4$:

$$Pb(s) + SO_4^{2-}(aq) \longrightarrow PbSO_4(s) + 2e^-$$
$$PbO_2(s) + 4H^+(aq) + SO_4^{2-}(aq) + 2e^- \longrightarrow PbSO_4(s) + 2H_2O$$
$$\overline{Pb(s) + PbO_2(s) + 4H^+(aq) + 2SO_4^{2-}(aq) \longrightarrow 2PbSO_4(s) + 2H_2O}$$

Deposits of lead sulfate slowly build up on the plates, partially covering and replacing the lead dioxide. As the cell discharges, the concentration of sulfuric acid decreases. For every mole of lead reacting, two moles of $H_2SO_4$ ($4H^+$, $2SO_4^{2-}$) are replaced by two moles of water. The state of charge of a storage battery can be checked by measuring the density of the electrolyte. When the battery is fully charged, the density is in the range of 1.25 to 1.30 $g/cm^3$. A density below 1.20 $g/cm^3$ indicates a low sulfuric acid concentration and hence a partially discharged cell.

A lead storage battery can be recharged and thus restored to its original condition. To do this, a direct current is passed through the cell in the reverse direction. While a storage battery is being recharged, it acts as an electrolytic cell. The overall cell reaction is the reverse of that occurring when the battery discharges:

$$2PbSO_4(s) + 2H_2O \longrightarrow Pb(s) + PbO_2(s) + 4H^+(aq) + 2SO_4^{2-}(aq)$$

The electrical energy required to bring about this nonspontaneous reaction in an automobile is furnished by an alternator equipped with a rectifier to convert alternating to direct current.

As you may have found from experience, lead storage batteries do not endure forever, particularly if they are allowed to stand for some time when discharged. Repeated quick-charging can cause Pb, $PbO_2$, and $PbSO_4$ to flake off the electrodes. This collects as a sludge at the bottom of the battery, often short-circuiting one or more cells. Discharged batteries are also susceptible to freezing, since sulfuric acid concentration is low. If freezing occurs, the electrodes may warp and come in contact with one another.

One of the advantages of the lead storage battery is that its voltage stays constant at 2 V per cell over a wide range of sulfuric acid concentrations. Only when the battery is nearly completely discharged does the voltage drop. It is also true that the cell voltage is virtually independent of temperature. You have trouble starting your car on a cold morning because the conductivity of the electrolyte drops off sharply with temperature; the voltage is still 2 V/cell at $-40°C$.

Another type of rechargeable voltaic cell is the "Nicad" storage battery, used for small appliances, tools, and calculators. The anode in this cell is made of cadmium metal, and the cathode contains nickel (IV) oxide, $NiO_2$. The electrolyte is a concentrated solution of potassium hydroxide. The discharge reactions are

anode: $\quad\quad Cd(s) + 2OH^-(aq) \longrightarrow Cd(OH)_2(s) + 2e^-$

cathode: $\quad \dfrac{NiO_2(s) + 2H_2O + 2e^- \longrightarrow Ni(OH)_2(s) + 2OH^-(aq)}{Cd(s) + NiO_2(s) + 2H_2O \longrightarrow Cd(OH)_2(s) + Ni(OH)_2(s)}$

The insoluble hydroxides of cadmium and nickel deposit on the electrodes. Hence, the half-reactions are readily reversed during recharging. Nicad batteries are more expensive than lead storage batteries for a given amount of electrical energy delivered but also have a longer life.

## Corrosion of Metals

Most metals corrode when exposed to the atmosphere, reacting with oxygen, water vapor, or carbon dioxide. Gold and platinum are among the few metals that retain their shiny appearance indefinitely when exposed to air; these metals are very difficult to oxidize ($E^\circ_{ox}$ Pt = $-1.320$ V, Au = $-1.498$ V).

Aluminum ($E^\circ_{ox}$ = $+1.68$ V) reacts readily with oxygen of the air:

$$4Al(s) + 3O_2(g) \longrightarrow 2Al_2O_3(s)$$

However, the $Al_2O_3$ coating, which is only about $10^{-8}$ m thick, adheres tightly to the surface of the metal. This prevents further corrosion and explains why aluminum cookware does not disintegrate upon exposure to air.

Copper in moist air (Fig. 18.A) slowly acquires a dull green coating. The green material is a 1:1 mole mixture of $Cu(OH)_2$ and $CuCO_3$:

$$2Cu(s) + H_2O(g) + CO_2(g) + O_2(g) \longrightarrow Cu(OH)_2 \cdot CuCO_3(s)$$

Several other elements, including zinc and lead, react similarly. The products, $Zn(OH)_2 \cdot ZnCO_3$ and $Pb(OH)_2 \cdot PbCO_3$, are white and adhere tightly to the metal, preventing further corrosion. In the case of lead, the protective coating dissolves in acetic acid, primarily because $Pb^{2+}$ forms a very stable complex with the acetate ion. It has been suggested that the ancient Romans suffered from lead poisoning because they stored wine (containing some acetic acid) in pottery vessels glazed with lead compounds.

From an economic standpoint, the most important corrosion reaction is that involving iron and steel. About 20% of all the iron produced each year goes to replace products whose usefulness has been destroyed by rust. When a piece of iron is exposed to water containing dissolved oxygen, the half-reactions of oxidation

$$Fe(s) \longrightarrow Fe^{2+}(aq) + 2e^-$$

and reduction

$$\tfrac{1}{2}O_2(g) + H_2O + 2e^- \longrightarrow 2OH^-(aq)$$

occur at different locations. The surface of a piece of corroding iron consists of a series of tiny voltaic cells. At *anodic areas,* iron is oxidized to $Fe^{2+}$ ions; at *cathodic areas,* elementary oxygen is reduced to $OH^-$ ions. Electrons are transferred through the iron, which acts like the external conductor of an ordinary voltaic cell. The electrical circuit is completed by the flow of ions through the water solution or film covering the iron.

Many characteristics of corrosion are most readily explained in terms of an electrochemical mechanism. A perfectly dry metal surface is not attacked by oxygen; iron exposed to dry air does not corrode. This seems plausible if corrosion occurs through a voltaic cell, which requires a water solution through which ions can move to complete the circuit. The fact that corrosion occurs more readily in seawater than in fresh water has a similar explanation. The dissolved salts in seawater supply the ions necessary for the conduction of current.

The existence of discrete cathodic and anodic areas on a piece of corroding iron requires that adjacent surface areas differ from each other chemically. This can happen if there are differences in oxygen concentration along the metal surface, as when

**Figure 18.A**
Copper exposed to air eventually develops a green coating of $Cu(OH)_2 \cdot CuCO_3$. (Andy Levin/Photo Researchers, Inc.)

**Figure 18.B**
Corrosion of iron under a drop of water. The $Fe^{2+}$ ions migrate toward the edge of the drop, where they precipitate as $Fe(OH)_2$, which later forms $Fe(OH)_3$.

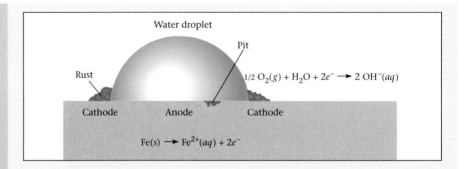

Water droplet

Pit

Rust

$1/2\ O_2(g) + H_2O + 2e^- \longrightarrow 2\ OH^-(aq)$

Cathode    Anode    Cathode

$Fe(s) \longrightarrow Fe^{2+}(aq) + 2e^-$

a drop of water adheres to the surface of a piece of iron exposed to the air (Fig. 18.B). The metal around the edges of the drop is in contact with water containing a high concentration of dissolved oxygen. The water touching the metal beneath the center of the drop is depleted in oxygen, since it is cut off from contact with air. As a result, a small oxygen concentration cell is set up. The area around the edge of the drop, where the oxygen concentration is high, becomes cathodic; oxygen molecules are reduced there. Directly beneath the drop is an anodic area where the iron is oxidized. A particle of dirt on the surface of an iron object can act in much the same way as a drop of water to cut off the supply of oxygen to the area beneath it and thereby establish anodic and cathodic areas. This explains why garden tools left covered with soil are particularly susceptible to rusting.

# CHAPTER HIGHLIGHTS

## *Key Concepts*

1. Draw a diagram for a voltaic cell, labeling electrodes and direction of current flow
   (Example 18.1; Problems 3–6)
2. Use standard potentials (Table 18.1) to
   —rank oxidizing and reducing agents
   (Example 18.2; Problems 7–14)
   —calculate $E°$ and/or reaction spontaneity
   (Examples 18.3, 18.4; Problems 15–32)
3. Relate $E°$ to $\Delta G°$ and $K$
   (Examples 18.5, 18.6; Problems 33–40)
4. Use the Nernst equation to relate voltage to concentration
   (Examples 18.7, 18.8; Problems 41–54, 67)
5. Relate mass to coulombs or joules in electrolysis reactions
   (Examples 18.9, 18.10; Problems 55–60, 65)

## *Key Equations*

Standard voltage    $E° = E°_{ox} + E°_{red}$

$E°, \Delta G°, K$    $E° = \dfrac{\Delta G°}{-nF} = \dfrac{RT \ln K}{nF}$

Nernst equation    $E = E° - \dfrac{RT}{nF} \ln Q$

| ampere | electrolytic cell | reduction |
|--------|-------------------|-----------|
| anode | $\Delta G°$ | salt bridge |
| cathode | oxidation | standard potential |
| coulomb | oxidizing agent | volt |
| $E°$, $E°_{ox}$, $E°_{red}$ | reducing agent | voltaic cell |
| electrolysis | | |

*Summary Problem*

A voltaic cell consists of two half-cells. One of the half-cells contains a platinum electrode surrounded by permanganate and manganese(II) ions. The other half-cell contains a platinum electrode surrounded by chlorate ions and chlorine gas. Assume that the cell reaction, which produces a positive voltage, involves both permanganate and chlorate ions in acidic solution. Take $T = 25°C$.

(a) Write the anode half-equation, the cathode half-equation, and the overall equation for the cell.
(b) Write the cell notation.
(c) Calculate $E°$ for the cell.
(d) For the redox reaction in (a), calculate $K$ and $\Delta G°$.
(e) Calculate the pH of the cell when the cell voltage is 0.023 volts, all ionic species except $H^+$ are at 0.200 $M$, and $P_{Cl_2} = 0.873$ atm.

An electrolytic cell contains a solution of manganese(II) nitrate. Assume that manganese plates out at one electrode and that oxygen gas is evolved at the other electrode.

(f) Write the anode half-reaction, the cathode half-reaction, and the overall equation for the electrolysis.
(g) How long will it take to deposit 5.00 g of manganese, using a current of 6.20 A?
(h) A current of 5.00 A is passed through the cell for 1.00 hr. Starting out with 175 mL of 1.25 $M$ $Mn(NO_3)_2$, what is $[Mn^{2+}]$ after electrolysis? What is the pH of the solution, neglecting the $H^+$ originally present? What is the volume of oxygen given off at 757 mm Hg and 25°C? Assume 100% efficiency and no change in volume during electrolysis.

### Answers

(a) anode:     $Cl_2(g) + 6H_2O \longrightarrow 2ClO_3^-(aq) + 12H^+(aq) + 10e^-$
    cathode:  $MnO_4^-(aq) + 8H^+(aq) + 5e^- \longrightarrow Mn^{2+}(aq) + 4H_2O$
    overall:   $Cl_2(g) + 2MnO_4^-(aq) + 4H^+(aq) \longrightarrow$
                                   $2ClO_3^-(aq) + 2Mn^{2+}(aq) + 2H_2O$

(b) $Pt \mid Cl_2 \mid ClO_3^- \parallel MnO_4^- \mid Mn^{2+} \mid Pt$
(c) 0.054 V
(d) $\Delta G° = -52$ kJ; $K = 1 \times 10^9$
(e) 1.65
(f) anode:     $2H_2O \longrightarrow O_2(g) + 4H^+(aq) + 4e^-$
    cathode:  $Mn^{2+}(aq) + 2e^- \longrightarrow Mn(s)$
    overall:   $2Mn^{2+}(aq) + 2H_2O \longrightarrow 2Mn(s) + O_2(g) + 4H^+(aq)$
(g) 47.2 min
(h) 0.717 $M$; −0.028; 1.15 L

# Questions & Problems

## Voltaic Cells

**\*1.** Write a balanced chemical equation for the overall cell reaction represented as

(a) $Cu \mid Cu^{2+} \parallel I_2 \mid I^- \mid Pt$

(b) $Pt \mid Mn^{2+} \mid MnO_2 \parallel Cl_2 \mid Cl^- \mid Pt$

(c) $Al \mid Al^{3+} \parallel Fe^{2+} \mid Fe$

**\*2.** Write a balanced chemical equation for the overall cell reaction represented as

(a) $Sn \mid Sn^{2+} \parallel S \mid H_2S \mid Pt$

(b) $Pt \mid H_2 \mid H_2O \parallel O_2 \mid OH^- \mid Pt$

(c) $Co \mid Co^{2+} \parallel H_2O_2, H^+ \mid Pt$

**\*3.** Draw a diagram for a salt bridge cell for each of the following reactions. Label the anode and cathode, and indicate the direction of current flow throughout the circuit.

(a) $Pb(s) + 2Fe^{3+}(aq) \longrightarrow Pb^{2+}(aq) + 2Fe^{2+}(aq)$

(b) $4Au(s) + 16Cl^-(aq) + 3O_2(g) + 12H^+(aq) \longrightarrow$ $4AuCl_4^-(aq) + 6H_2O$

(c) $Fe(s) + 2H_2O \longrightarrow Fe(OH)_2(s) + H_2(g)$

**\*4.** Follow the directions for Question 3 for the following:

(a) $3Mg(s) + 2Al^{3+}(aq) \longrightarrow 2Al(s) + 3Mg^{2+}(aq)$

(b) $2NO_3^-(aq) + 3SO_2(g) + 2H_2O \longrightarrow 2NO(g) +$ $3SO_4^{2-}(aq) + 4H^+(aq)$

(c) $3S^{2-}(aq) + 2NO_3^-(aq) + 4H_2O \longrightarrow 2NO(g) +$ $8OH^-(aq) + 3S(s)$

**\*5.** Consider a salt bridge cell in which the anode is an iron rod immersed in a saturated aqueous solution of iron(II) hydroxide. The cathode is a copper strip immersed in a saturated aqueous copper(II) hydroxide solution. Sketch a diagram of the cell, indicating the flow of the current throughout. Write the half-equations for the electrode reactions, the overall equation, and the notation for the cell.

**\*6.** Follow the directions for Question 5 for a salt bridge cell in which the anode is a gold rod immersed in an aqueous solution of NaCl and NaAuCl₄. The cathode is a platinum rod immersed in a solution of sodium bromide and liquid bromine.

## Strength of Oxidizing and Reducing Species

**\*7.** Which species in each pair is the better oxidizing agent?

(a) $NO_3^-$ or $I_2$

(b) $Mn^{2+}$ or $MnO_2$

(c) $Fe(OH)_3$ or $S$

**\*8.** Which species in each pair is the better reducing agent?

(a) Au or Ag

(b) $Br^-$ or $Cl^-$

(c) $OH^-$ or $NO_2^-$

**\*9.** Using Table 18.1, arrange the following reducing agents in order of increasing strength.

NO (acidic)    NO (basic)    $H_2$ (acidic)    Cd    $Cr^{3+}$

**\*10.** Using Table 18.1, arrange the following oxidizing agents in order of increasing strength.

$ClO_3^-$ (acidic)    $ClO_3^-$ (basic)    $SO_4^{2-}$    $H^+$    $O_2$ (acidic)

**\*11.** Consider the following species

$Cr^{2+}$    $Cr^{3+}$    $Cr$    $Cr_2O_7^{2-}$    $H_2O$ (basic)

Classify each species as oxidizing agent, reducing agent, or both. Arrange the oxidizing agents in order of increasing strength. Do the same for the reducing agents.

**\*12.** Follow the directions for Question 11 for the following species:

$Mn^{2+}$    $Mn$    $MnO_2$    $MnO_4^-$    $O_2$ (basic)

**\*13.** Use Table 18.1 to select

(a) a reducing agent that will convert $Sn^{2+}$ to Sn but not $Co^{2+}$ to Co.

(b) a reducing agent to convert $Ag^+$ to Ag but not $Cu^+$ to Cu.

(c) an oxidizing agent that converts $Mn^{2+}$ to $MnO_2$ but not $Mn^{2+}$ to $MnO_4^-$.

**\*14.** Use Table 18.1 to select

(a) an oxidizing agent that converts $ClO_3^-$ to $ClO_4^-$ but not $Cl^-$ to $ClO_3^-$ (basic solution).

(b) a reducing agent capable of converting $Mg^{2+}$ to Mg but not $Ca^{2+}$ to Ca.

(c) an oxidizing agent that converts $I^-$ to $I_2$ but not $Cl^-$ to $Cl_2$.

## Calculation of E°

**15.** Calculate $E°$ for the following voltaic cells:

(a) $Sn(s) + 2Ag^+(aq) \longrightarrow Sn^{2+}(aq) + 2Ag(s)$

(b) $H_2(g) + Hg_2^{2+}(aq) \longrightarrow 2H^+(aq) + 2Hg(l)$

(c) a $Pb$–$PbSO_4$ half cell and a $PbO_2$–$PbSO_4$ half-cell

**16.** Calculate $E°$ for the following voltaic cells:

(a) $2AuCl_4^-(aq) + 3Cu(s) \longrightarrow 2Au(s) + 8Cl^-(aq) +$ $3Cu^{2+}(aq)$

(b) $Fe(s) + Cu^{2+}(aq) \longrightarrow Cu(s) + Fe^{2+}(aq)$

(c) $Ni$–$Ni^{2+}$ half-cell and $Cd$–$Cd^{2+}$ half-cell.

**17.** Using Table 18.1, calculate $E°$ for

(a) the reaction of hydrogen gas with iron(III) ions to produce hydrogen and iron(II) ions.

(b) the reaction between chloride and permanganate ions to produce chlorine gas and manganese(II) ions.

(c) the reaction between hydrogen gas and nitrate ions to produce water and nitrogen oxide gas in basic solution.

**18.** Using Table 18.1, calculate $E°$ for the reaction between

(a) zinc metal and chromium(III) ions to produce zinc ions and chromium.

**(b)** tin and oxygen to produce an acidic solution of tin(II) ions.

**(c)** hydrogen peroxide and nitrogen oxide gas to produce an aqueous solution of nitric acid.

**19.** Calculate $E°$ for the following cells:
    **(a)** $Co|Co^{2+}||Ag^{+}|Ag$
    **(b)** $Zn|Zn^{2+}||Br_2|Br^{-}|Pt$
    **(c)** $C$ (graphite)$|I^{-}|I_2||AuCl_4^{-}|Au$

**20.** Calculate $E°$ for the following cells:
    **(a)** $Tl|Tl^{+}||NO_3^{-}, H^{+}|NO|Pt$
    **(b)** $C$ (graphite)$|H_2S|S||O_2|H^{+}, H_2O|C$ (graphite)
    **(c)** $C$ (graphite)$|H_2|OH^{-}||O_2|OH^{-}|Pt$

**21.** Suppose $E_{red}°$ for $Ni^{2+} \longrightarrow Ni$ was set equal to zero instead of $E_{red}°$ $H^{+} \longrightarrow H_2$. What would be
    **(a)** $E_{red}°$ for $H^{+} \longrightarrow H_2$?
    **(b)** $E_{ox}°$ for $Fe \longrightarrow Fe^{2+}$?
    **(c)** $E°$ for the cell in 19(a)? Compare your answers.

**22.** Suppose $E_{red}°$ for $H^{+} \longrightarrow H_2$ were taken to be 0.500 V instead of 0.000 V. What would be
    **(a)** $E_{ox}°$ for $H_2 \longrightarrow H^{+}$?
    **(b)** $E_{red}°$ for $Br_2 \longrightarrow Br^{-}$?
    **(c)** $E°$ for the cell in 19(a)? Compare your answers.

## Spontaneity and E°

**23.** Which of the following reactions are spontaneous at standard conditions?
    **(a)** $2H_2O + 2Cl_2(g) \longrightarrow O_2(g) + 4H^{+}(aq) + 4Cl^{-}(aq)$
    **(b)** $Zn^{2+}(aq) + 2Fe^{2+}(aq) \longrightarrow Zn(s) + 2Fe^{3+}(aq)$
    **(c)** $Br_2(l) + 2I^{-}(aq) \longrightarrow 2Br^{-}(aq) + I_2(s)$

**24.** Which of the following reactions are spontaneous at standard conditions?
    **(a)** $2NO(g) + 4H_2O + 3Cl_2(g) \longrightarrow 2NO_3^{-}(aq) + 8H^{+}(aq) + 6Cl^{-}(aq)$
    **(b)** $2Au(s) + 8Cl^{-}(aq) + 3Fe^{2+}(aq) \longrightarrow 3Fe(s) + 2AuCl_4^{-}(aq)$
    **(c)** $Cu^{2+}(aq) + H_2(g) \longrightarrow Cu(s) + 2H^{+}(aq)$

**25.** Using Table 18.1, calculate $E°$ and decide whether the following ions will oxidize chlorine gas to chlorate ions in acidic solution at standard conditions.
    **(a)** $Au^{3+}$     **(b)** $NO_3^{-}$     **(c)** $MnO_4^{-}$

**26.** Using Table 18.1, calculate $E°$ and decide whether the following metals will reduce solid silver iodide to silver metal at standard conditions.
    **(a)** Cu     **(b)** Au     **(c)** Na

**27.** Write the equation for the reaction that occurs, if any, when each of the following experiments are performed under standard conditions.
    **(a)** Hydrochloric acid is added to mercury.
    **(b)** Nitric acid is added to mercury.
    **(c)** Iron(III) bromide is added to copper(II) sulfate.

**28.** Write the equation for the reaction that occurs, if any, when each of the following experiments are performed under standard conditions.

**(a)** Chlorine gas is added to an aqueous solution of potassium dichromate.

**(b)** Liquid bromine is added to an aqueous solution of potassium dichromate.

**(c)** Manganese(II) chloride is added to cobalt(II) nitrate in acidic solution.

**29.** Which of the following metals will react with 1 $M$ HBr?
    **(a)** Mg     **(b)** Cu     **(c)** Ag     **(d)** Cr

**30.** Which of the following species will be oxidized by 1 $M$ $HNO_3$?
    **(a)** $Fe^{2+}$     **(b)** Mn     **(c)** Hg     **(d)** $SO_2$

**31.** Using Table 18.1, predict what reactions, if any, will occur when the following are mixed under standard conditions.
    **(a)** Cr, $Ni^{2+}$, $Cl^{-}$     **(b)** $F_2$, $Na^{+}$, $Cl^{-}$
    **(c)** S, $Fe^{3+}$, $NO_3^{-}$

**32.** Predict what reaction, if any, will occur when iodine crystals are added to an acidic aqueous solution of each of the following species at standard conditions.
    **(a)** $MgCl_2$     **(b)** $Ni(NO_3)_2$     **(c)** $H_2S(aq)$

## E°, ΔG°, and K

**33.** Consider a cell reaction at 25°C where $n = 5$. Fill in the following table.

| | $\Delta G°$ | $E°$ | $K$ |
|---|---|---|---|
| (a) | +12 kJ | —— | —— |
| (b) | —— | 0.223 V | —— |
| (c) | —— | —— | $4.6 \times 10^4$ |

**34.** Consider a cell reaction at 25°C where $n = 4$. Fill in the following table.

| | $\Delta G°$ | $E°$ | $K$ |
|---|---|---|---|
| (a) | —— | −0.16 V | —— |
| (b) | −53 kJ | —— | —— |
| (c) | —— | —— | $2.6 \times 10^{-4}$ |

**35.** Calculate $E°$, $\Delta G°$, and $K$ at 25°C for the reaction

$$3Mn^{2+}(aq) + 2H_2O(l) + 2MnO_4^{-}(aq) \longrightarrow 5MnO_2(s) + 4H^{+}(aq)$$

**36.** Calculate $E°$, $\Delta G°$, and $K$ at 25°C for the reaction

$$2NO(g) + NO_3^{-}(aq) + 2OH^{-}(aq) \longrightarrow 3NO_2^{-}(aq) + H_2O$$

**37.** Calculate $\Delta G°$ at 25°C for each of the reactions referred to in Question 15. Assume smallest whole-number coefficients.

**38.** Calculate $\Delta G°$ at 25°C for each of the reactions referred to in Question 16. Assume smallest whole-number coefficients.

**39.** Calculate $K$ at 25°C for each of the reactions referred to in Question 17. Assume smallest whole-number coefficients.

**40.** Calculate $K$ at 25°C for each of the reactions referred to in Question 18. Assume smallest whole-number coefficients.

## Nernst Equation

**41.** Consider a voltaic cell in which the following reaction takes place:

$$2Al(s) + 6H^+(aq) \longrightarrow 3H_2(g) + 2Al^{3+}(aq)$$

**(a)** Calculate $E°$.
**(b)** Write the Nernst equation for the cell.
**(c)** Calculate $E$ under the following conditions: $[H^+] = 1.00 \times 10^{-3}$ $M$, $[Al^{3+}] = 0.100$ $M$, $P_{H_2} = 1.00$ atm.

**42.** Consider a voltaic cell in which the following reaction takes place:

$$Pb(s) + SO_4{}^{2-}(aq) + I_2(s) \longrightarrow PbSO_4(s) + 2I^-(aq)$$

**(a)** Calculate $E°$.
**(b)** Write the Nernst equation for the cell.
**(c)** Calculate $E$ under the following conditions: $[I^-] = 2[SO_4{}^{2-}] = 0.0500$ $M$

**43.** Consider a voltaic cell in which the following reaction takes place:

$$2Fe^{2+}(aq) + H_2O_2(aq) + 2H^+(aq) \longrightarrow 2Fe^{3+}(aq) + 2H_2O$$

**(a)** Calculate $E°$.
**(b)** Write the Nernst equation for the cell.
**(c)** Calculate $E$ under the following conditions: $[Fe^{2+}] = 0.0037$ $M$, $[H_2O_2] = 1.50$ $M$, $[Fe^{3+}] = 0.214$ $M$, pH = 3.16.

**44.** Consider a voltaic cell in which the following reaction takes place:

$$2NO_3{}^-(aq) + 3H_2(g) \longrightarrow 2NO(g) + 2OH^-(aq) + 2H_2O$$

**(a)** Calculate $E°$.
**(b)** Write the Nernst equation for the cell.
**(c)** Calculate $E$ under the following conditions: $[NO_3{}^-] = 0.0255$ $M$, $P_{NO} = 0.866$ atm, $P_{H_2} = 0.325$ atm, pH = 10.67.

**45.** Calculate $E$ for the following cell at 25°C.

$$Cr\,|\,Cr^{3+}(2.0 \times 10^{-3}\ M)\,\|\,Co^{2+}(1.5\ M)\,|\,Co$$

**46.** Calculate $E$ for the following cell at 25°C.

$$Pt\,|\,H_2(P = 0.813\ atm)\,|\,H^+(0.750\ M)\,\|\,S(s)\,|\,H_2S(0.30\ M)\,|\,Pt$$

**47.** Consider the reaction

$$2Au(s) + 8Cl^-(aq) + 3Br_2(l) \longrightarrow 6Br^-(aq) + 2AuCl_4{}^-(aq)$$

Calculate $[Cl^-]$ when the voltage is zero and all the other ionic species have a concentration of 0.100 $M$.

**48.** Consider the reaction

$$O_2(g) + 4H^+(aq) + 4Cl^-(aq) \longrightarrow 2Cl_2(g) + 2H_2O$$

At what pH will the voltage be $-0.200$ V if all species other than $H^+$ are at standard conditions?

**49.** Complete the following cell notation:

$$Cu(s)\,|\,Cu^{2+}(0.020\ M)\,\|\,Ag^+(\ \ )\,|\,Ag(s) \qquad E = +0.15\ V$$

**50.** Complete the following cell notation:

$$Ag(s)\,|\,Br^-(1.0\ M)\,|\,AgBr(s)\,\|\,H^+(2.5\ M)\,|\,H_2(\ \ )\,|\,(Pt)$$
$$E = +0.077\ V$$

**51.** The reaction

$$2H_2O(l) + 2Br_2(l) \longrightarrow O_2(g) + 4H^+(aq) + 4Br^-(aq)$$

is nonspontaneous at standard conditions. If the pH is adjusted to 4.00 while all the other species are kept at standard conditions, is the reaction still nonspontaneous?

**52.** The reaction

$$2Cr^{3+}(aq) + 3Cl_2(g) + 7H_2O \longrightarrow$$
$$Cr_2O_7{}^{2-}(aq) + 6Cl^-(aq) + 14H^+(aq)$$

is spontaneous at standard conditions. If the pH is adjusted to 0.70 while all the other species are kept at standard conditions, is the reaction still spontaneous?

**53.** Consider a cell in which the reaction is

$$2Ag(s) + Cu^{2+}(aq) \longrightarrow 2Ag^+(aq) + Cu(s)$$

**(a)** Calculate $E°$ for this cell.
**(b)** Chloride ions are added to the $Ag\,|\,Ag^+$ half-cell to precipitate AgCl. The measured voltage is $+0.060$ V. Taking $[Cu^{2+}] = 1.0$ $M$, calculate $[Ag^+]$.
**(c)** Taking $[Cl^-]$ in (b) to be 0.10 $M$, calculate $K_{sp}$ of AgCl.

**54.** Consider a cell in which the reaction is

$$Pb(s) + 2H^+(aq) \longrightarrow Pb^{2+}(aq) + H_2(g)$$

**(a)** Calculate $E°$ for this cell.
**(b)** Chloride ions are added to the $Pb\,|\,Pb^{2+}$ half-cell to precipitate $PbCl_2$. The voltage is measured to be $+0.210$ V. Taking $[H^+] = 1.0$ $M$ and $P_{H_2} = 1.0$ atm, calculate $[Pb^{2+}]$.
**(c)** Taking $[Cl^-]$ in (b) to be 0.10 $M$, calculate $K_{sp}$ of $PbCl_2$.

## Electrolytic Cells

**55.** An electrolytic cell produces aluminum from $Al_2O_3$ at the rate of five kilograms a day. Assuming a yield of 100%,
**(a)** how many moles of electrons must pass through the cell in one day?
**(b)** how many amperes are passing through the cell?
**(c)** how many moles of oxygen ($O_2$) are being produced simultaneously?

**56.** The electrolysis of an aqueous solution of NaCl has the overall reaction

$$2H_2O + 2Cl^-(aq) \longrightarrow H_2(g) + Cl_2(g) + 2OH^-(aq)$$

During the electrolysis, 0.415 moles of electrons pass through the cell.
**(a)** How many electrons does this represent?
**(b)** How many coulombs does this represent?
**(c)** What masses of $H_2$ and $Cl_2$ are produced, assuming 100% yield?

**57.** It is desired to silver-plate a thin sheet with the following dimensions: 2.5 in $\times$ 10.0 in $\times$ 0.0500 in. The silver plating is to be 0.0010 in thick.

    **(a)** How many grams of silver ($d = 10.5$ g/cm$^3$) are required?

    **(b)** How long will it take to plate the sheet from $AgNO_3$ using a current of 4.00 A and assuming 100% efficiency?

**58.** A baby's spoon with an area of 6.50 cm$^2$ is plated with gold from AuCN using a current of 0.700 A for 155 minutes.

    **(a)** If the current efficiency is 75.0%, how many grams of gold are plated?

    **(b)** What is the thickness of the gold plate formed ($d = 19.3$ g/cm$^3$)?

**59.** A lead storage battery delivers a current of 2.00 A for one hour at a voltage of 12.0 V. If the battery is run for three hours,

    **(a)** how many grams of lead are converted to $PbSO_4$?

    **(b)** how much electrical energy is produced in kilowatt-hours?

**60.** Calcium can be produced by the electrolysis of CaO. A voltage of 5.0 V is used.

    **(a)** How many joules of electrical energy are required to form 138 g of calcium?

    **(b)** What is the cost of the electrical energy in (a) at the rate of 6.0 cents per kilowatt-hour?

## Unclassified

**\*61.** Explain why

    **(a)** a salt bridge is used in a voltaic cell.

    **(b)** in a salt bridge with $KNO_3$, the $K^+$ ions move from the salt bridge to the cathode.

    **(c)** a lead storage battery won't work if the level of $H_2SO_4$ is low.

**\*62.** Choose the figure below that best represents the results after the electrolysis of water. (Circles represent hydrogen atoms, and squares represent oxygen atoms.)

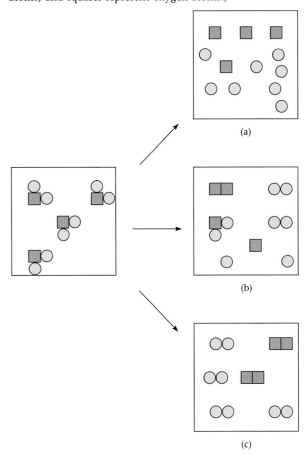

(a)

(b)

(c)

*63. Which of the following changes will increase the voltage of the following cell?

$$Co\,|\,Co^{2+}(0.010\ M)\,\|\,H^+(0.010\ M)\,|\,H_2(0.500\ atm)\,|\,Pt$$

(a) Increase the volume of $CoCl_2$ solution from 100 mL to 300 mL.
(b) Increase $[H^+]$ from 0.010 $M$ to 0.500 $M$.
(c) Increase the pressure of $H_2$ from 0.500 atm to 1 atm.
(d) Increase the mass of the Co electrode from 15 g to 25 g.
(e) Increase $[Co^{2+}]$ from 0.010 $M$ to 0.500 $M$.

*64. For the cell

$$Zn\,|\,Zn^{2+}\,\|\,Cu^{2+}\,|\,Cu$$

$E°$ is 1.10 V. A student prepared the same cell in the lab at standard conditions. Her experimental $E°$ was 1.0 V. A possible explanation for the difference is

(a) A larger volume of $Zn^{2+}$ than $Cu^{2+}$ was used.
(b) The zinc electrode had twice the mass of the copper electrode.
(c) $[Zn^{2+}]$ was less than 1 $M$.
(d) $[Cu^{2+}]$ was less than 1 $M$.
(e) The copper electrode had twice the surface area of the zinc electrode.

65. Hydrogen gas is produced when water is electrolyzed:

$$2H_2O(g) \longrightarrow 2H_2(g) + O_2(g)$$

A balloonist wants to fill a balloon with hydrogen gas. How long must a current of 10.0 A be used in the electrolysis of water to fill the balloon to a volume of 5.00 L and a pressure of 1.20 atm at 25°C?

66. An electrolysis experiment is performed to determine the value of the Faraday constant (number of coulombs per mole of electrons). In this experiment, 28.8 g of gold is plated out from a AuCN solution by running an electrolytic cell for two hours with a current of 2.00 A. What is the experimental value obtained for the Faraday constant?

67. Consider the following reaction at 25°C:

$$O_2(g) + 4H^+(aq) + 4Br^-(aq) \longrightarrow 2H_2O + 2Br_2(l)$$

If $[H^+]$ is adjusted by adding a buffer that is 0.100 $M$ in sodium acetate and 0.100 $M$ in acetic acid, the pressure of oxygen gas is 1.00 atm, and the bromide concentration is 0.100 $M$, what is the calculated cell voltage? ($K_a$ acetic acid = $1.8 \times 10^{-5}$)

68. Given the standard reduction potential for $Zn(OH)_4{}^{2-}$:

$$Zn(OH)_4{}^{2-}(aq) + 2e^- \longrightarrow Zn(s) + 4OH^-(aq)$$
$$E°_{red} = -1.19\ V$$

calculate the formation constant ($K_f$) for the reaction

$$Zn^{2+}(aq) + 4OH^-(aq) \rightleftharpoons Zn(OH)_4{}^{2-}(aq)$$

69. Consider the following reaction carried out at 1000°C:

$$CO(g) + \tfrac{1}{2}O_2(g) \longrightarrow CO_2(g)$$

Assuming that all gases are at 1.00 atm, calculate the voltage produced at the given conditions. (Use Appendix 1 and assume that $\Delta H°$ and $S°$ do not change with an increase in temperature.)

70. Atomic masses can be determined by electrolysis. In one hour, a current of 0.600 A deposits 2.42 g of a certain metal, M, which is present in solution as $M^+$ ions. What is the atomic mass of the metal?

## Challenge Problems

71. In a fully charged lead storage battery, the electrolyte consists of 38% sulfuric acid by mass. The solution has a density of 1.286 g/cm$^3$. Calculate $E$ for the cell. (Assume that all the $H^+$ ions come from the first dissociation of $H_2SO_4$, which is complete; $K_aHSO_4{}^- = 1.0 \times 10^{-2}$.)

72. Consider a voltaic cell in which the following reaction occurs:

$$Zn(s) + Sn^{2+}(aq) \longrightarrow Zn^{2+}(aq) + Sn(s)$$

(a) Calculate $E°$ for the cell.
(b) When the cell operates, what happens to the concentration of $Zn^{2+}$? the concentration of $Sn^{2+}$?
(c) When the cell voltage drops to zero, what is the ratio of the concentration of $Zn^{2+}$ to that of $Sn^{2+}$?
(d) If the concentration of both cations is 1.0 $M$ originally, what are the concentrations when the voltage drops to zero?

73. In biological systems, acetate ion is converted to ethyl alcohol in a two-step process:

$$CH_3COO^-(aq) + 3H^+(aq) + 2e^- \longrightarrow CH_3CHO(aq) + H_2O$$
$$E°' = -0.581\ V$$

$$CH_3CHO(aq) + 2H^+(aq) + 2e^- \longrightarrow C_2H_5OH(aq)$$
$$E°' = -0.197\ V$$

($E°'$ is the standard reduction voltage at 25°C and a pH of 7.00.)

(a) Calculate $\Delta G°'$ for each step and for the overall conversion.
(b) Calculate $E°'$ for the overall conversion.

74. Consider the cell

$$Pt\,|\,H_2\,|\,H^+\,\|\,H^+\,|\,H_2\,|\,Pt$$

In the anode half-cell, hydrogen gas at 1.0 atm is bubbled over a platinum electrode dipping into a solution that has a pH of 7.0. The other half-cell is identical to the first except that the solution around the platinum electrode has a pH of 0.0. What is the cell voltage?

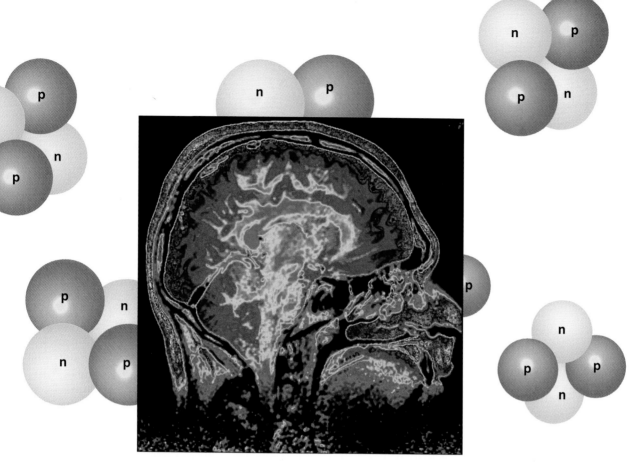

On the average, a person's exposure to radiation from medical procedures such as a PET scan (p. 535), is about equal to that from radon (Table 19.B, p. 550). (FPG International)

# Nuclear Chemistry 19

The soul, perhaps, is a gust of gas
And wrong is a form of right—
But we know that Energy equals Mass
By the Square of the Speed of Light.

—MORRIS BISHOP

$E = MC^2$

Nuclear reactions, involving changes in the composition of atomic nuclei, were introduced in Chapter 2. Recall that nuclei are represented by symbols such as

$$^{12}_{6}C \qquad ^{14}_{6}C$$

Here, the atomic number $Z$ (number of protons in the nucleus) is shown as a left subscript. The mass number, $A$ (number of protons + number of neutrons in the nucleus), appears as a left superscript. Nuclei with the same number of protons but different numbers of neutrons are called *isotopes*. The symbols written above represent two isotopes of the element carbon ($Z = 6$). One isotope has six neutrons in the nucleus and hence has a mass number of $6 + 6 = 12$. The heavier isotope has eight neutrons and hence a mass number of 14.

Nuclear equations are used to describe nuclear reactions such as natural radioactivity (Chapter 2). The quantities appearing in these equations are the nuclear symbols referred to above. In a balanced nuclear equation, the sum of atomic numbers must be the same on both sides; the same is true of mass numbers. Consider, for example, the equation

$$^{14}_{7}N + ^{1}_{0}n \longrightarrow ^{14}_{6}C + ^{1}_{1}H$$

Nuclear reactions usually involve the transmutation of elements

The atomic numbers add to seven on both sides; the mass numbers add to 15.

In this chapter, we will take a closer look at some of the more relevant topics in nuclear chemistry. Many of these topics will be ones that you have heard about on TV or read about in newspapers. Specifically, we will consider

— *radioactive isotopes*, their production in the laboratory, and their use in medicine and other areas (Section 19.1).
— *rate of nuclear reactions* as related to the "age" of inorganic and organic materials (Section 19.2).
— *mass-energy relations*, as expressed by the Einstein equation: $\Delta E = \Delta mc^2$ (Section 19.3).
— *nuclear fission*, a major but controversial source of energy today (Section 19.4).
— *nuclear fusion*, a potentially limitless source of energy for the future (Section 19.5).

## 19.1   Radioactive Nuclei

A radioactive nucleus spontaneously decomposes ("decays") with the evolution of energy. As pointed out in Chapter 2, a few such nuclei occur in nature, accounting for *natural radioactivity*. Many more can be made ("induced") in the laboratory by bombarding stable nuclei with high-energy particles.

## Mode of Decay

Naturally occurring radioactive nuclei can decompose by

— *alpha emission,* in which an ordinary helium nucleus, $^4_2\text{He}$, is given off. Uranium-238 behaves this way:

$$^{238}_{92}\text{U} \longrightarrow {}^4_2\text{He} + {}^{234}_{90}\text{Th}$$

— *beta emission,* which produces an electron, given the symbol $_{-1}^{0}e$. An example of beta emission is

$$^{234}_{90}\text{Th} \longrightarrow {}_{-1}^{0}e + {}^{234}_{91}\text{Pa}$$

— *gamma emission,* which consists of high-energy photons.

Radioactive nuclei produced "artificially" in the laboratory can show $\alpha$-, $\beta$-, or $\gamma$-emission. They can also decompose by

— *positron emission.* A positron is identical to an electron except that it has a charge of $+1$ rather than $-1$. A positron has the symbol $_{1}^{0}e$.

$$^{40}_{19}\text{K} \longrightarrow {}_{1}^{0}e + {}^{40}_{18}\text{Ar}$$

— *K-electron capture,* in which an electron in the innermost energy level ($n = 1$) "falls" into the nucleus.

$$^{82}_{37}\text{Rb} + {}_{-1}^{0}e \longrightarrow {}^{82}_{36}\text{Kr}$$

Notice that the result of K-electron capture is the same as positron emission; mass number remains unchanged while atomic number decreases by one unit. Electron capture is more common with heavy nuclei, presumably because the $n = 1$ level is closer to the nucleus.

---

**Example 19.1**   Promethium ($Z = 61$) is essentially nonexistent in nature; all of its isotopes are radioactive. Write balanced nuclear equations for the decomposition of

*Promethium is by far the rarest of the "rare earths"*

(a) Pm-142 by positron emission; by K-electron capture.
(b) Pm-147 by beta emission.

*Strategy*   Use the principle that atomic and mass numbers must balance to find the symbol of the product nucleus.

*Solution*

(a) $^{142}_{61}\text{Pm} \longrightarrow {}_{1}^{0}e + {}^{142}_{60}\text{Nd}$

$^{142}_{61}\text{Pm} + {}_{-1}^{0}e \longrightarrow {}^{142}_{60}\text{Nd}$

(b) $^{147}_{61}\text{Pm} \longrightarrow {}_{-1}^{0}e + {}^{147}_{62}\text{Sm}$

---

## Bombardment Reactions

During the past 60 years, more than 1500 radioactive isotopes have been prepared in the laboratory. The number of such isotopes per element ranges from one (hydrogen and boron) to 34 (indium). They are all prepared by bombardment reactions in which a stable nucleus is converted to one that is radioactive.

The bombarding particle may be

— a *neutron,* usually of rather low energy, produced in a fission reactor (Section 19.4). A typical reaction of this type is

$$\ce{^{27}_{13}Al + ^{1}_{0}n \longrightarrow ^{28}_{13}Al}$$

The product nucleus, Al-28, is radioactive, decaying by beta emission:

$$\ce{^{28}_{13}Al \longrightarrow ^{28}_{14}Si + ^{0}_{-1}e}$$

— a *charged particle* (electron, positron, α-particle, . . .), which can be accelerated to very high velocities in electric and/or magnetic fields. In this way, the charged particle acquires enough energy to bring about a nuclear reaction, despite electrostatic repulsion with the components of the atom.

The first radioactive isotopes to be made in the laboratory were prepared in 1934 by Irene Curie and her husband, Frederic Joliot. They achieved this by bombarding certain stable isotopes with high-energy alpha particles. One reaction was

$$\ce{^{27}_{13}Al + ^{4}_{2}He \longrightarrow ^{30}_{15}P + ^{1}_{0}n}$$

The product, phosphorus-30, is radioactive, decaying by positron emission:

$$\ce{^{30}_{15}P \longrightarrow ^{30}_{14}Si + ^{0}_{1}e}$$

The names of these elements reflect the names of their discoverers and other nuclear scientists

An interesting application of bombardment reactions is the preparation of the so-called transuranium elements. During the past 50 years, 20 elements with atomic numbers greater than that of uranium (93 through 112) have been synthesized. Much of this work was done by a group at the University of California at Berkeley, under the direction of Glenn Seaborg and, later, Albert Ghiorso. A Russian group at Dubna near Moscow, led by G. N. Flerov, made significant contributions. The heaviest elements (107 through 112) were prepared in the past few years by a German group at Darmstadt under the direction of Peter Armbruster.

Some of the reactions used to prepare transuranium elements are listed in Table 19.1. Neutron bombardment is effective for the lower members of the series (elements 93 through 95), but the yield of product decreases sharply with increasing atomic number. To form the heavier transuranium elements, it is necessary to bombard appropriate targets with high-energy positive ions. By using relatively heavy bombarding particles such as carbon-12, one can achieve a considerable increase in atomic number.

## Applications

A large number of radioactive nuclei have been used both in industry and in many areas of basic and applied research. A few of these are discussed below.

### *Medicine*

Radioactive isotopes are commonly used in cancer therapy, usually to eliminate any malignant cells left after surgery. Cobalt-60 is most often used; γ-rays from this source are focused at small areas where cancer is suspected. Certain types of cancer can be treated internally with radioactive isotopes. If a patient suffering from cancer of the thyroid drinks a solution of NaI containing radioactive iodide ions ($^{131}$I or $^{123}$I), the iodine moves preferentially to the thyroid gland.

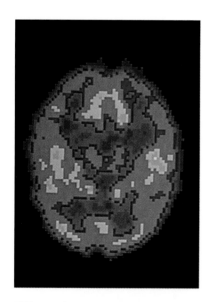

PET scan (positron emission tomography) of a normal human brain. (Hank Morgan/Photo Researchers, Inc.)

## TABLE 19.1 Synthesis of Transuranium Elements

### Neutron Bombardment

| | |
|---|---|
| Neptunium, plutonium | $^{238}_{92}U + ^1_0n \longrightarrow ^{239}_{93}Np + ^0_{-1}e$ |
| | $^{239}_{93}Np \longrightarrow ^{239}_{94}Pu + ^0_{-1}e$ |
| Americium | $^{239}_{94}Pu + 2\,^1_0n \longrightarrow ^{241}_{95}Am + ^0_{-1}e$ |

### Positive Ion Bombardment

| | |
|---|---|
| Curium | $^{239}_{94}Pu + ^4_2He \longrightarrow ^{242}_{96}Cm + ^1_0n$ |
| Californium | $^{242}_{96}Cm + ^4_2He \longrightarrow ^{245}_{98}Cf + ^1_0n$ |
| Rutherfordium* | $^{249}_{98}Cf + ^{12}_6C \longrightarrow ^{257}_{104}Rf + 4\,^1_0n$ |
| Hahnium* | $^{249}_{98}Cf + ^{15}_7N \longrightarrow ^{260}_{105}Ha + 4\,^1_0n$ |
| Seaborgium* | $^{249}_{98}Cf + ^{18}_8O \longrightarrow ^{263}_{106}Sg + 4\,^1_0n$ |

*The names and hence the symbols of elements 104 through 109 are in dispute. The ones used here and throughout this text are those approved by the American Chemical Society. They honor the nuclear scientists Ernest Rutherford (104), Otto Hahn (105), Glenn Seaborg (106), Niels Bohr (107) and Lise Meitner (109).

There, the radiation destroys malignant cells without affecting the rest of the body.

More commonly, radioactive nuclei are used for diagnosis (Table 19.2). Positron emission tomography (PET) is a technique used to study brain disorders. The patient is given a dose of glucose ($C_6H_{12}O_6$) containing a small amount of carbon-11, a positron emitter. The brain is then scanned to detect positron emission from the radioactive, "labeled" glucose. In this way, differences in glucose uptake and metabolism in the brains of normal and abnormal patients are established. For example, PET scans have determined that the brain of a schizophrenic metabolizes only about 20% as much glucose as that of most people.

### Chemistry

Radioactive nuclei are used extensively in chemical analysis. One technique of particular importance is *neutron activation analysis*. This procedure depends upon the phenomenon of induced radioactivity. A sample is bombarded by neutrons, bringing about such reactions as

$$^{84}_{38}Sr + ^1_0n \longrightarrow ^{85}_{38}Sr$$

Ordinarily the element retains its chemical identity, but the isotope formed is radioactive, decaying by gamma emission. The magnitude of the energy change and hence the wavelength of the gamma ray vary from one element to another and so can serve for the qualitative analysis of the sample. The intensity of the radiation depends upon the amount of the element present in the sample; this permits quantitative analysis of the sample. Neutron activation analysis can be used to analyze for 50 different elements in amounts as small as one picogram ($10^{-12}$g).

One application of neutron activation analysis is in the field of archaeology. By measuring the amount of strontium in the bones of prehistoric humans, it is possible to get some idea of their diet. Plants contain considerably more

## TABLE 19.2 Diagnostic Uses of Radioactive Isotopes

| Isotope | Use |
|---|---|
| $^{11}_6C$ | brain scan (PET); see text |
| $^{24}_{11}Na$ | circulatory disorders |
| $^{32}_{15}P$ | detection of eye tumors |
| $^{59}_{26}Fe$ | anemia |
| $^{67}_{31}Ga$ | scan for lung tumors, abscesses |
| $^{75}_{34}Se$ | pancreas scan |
| $^{99}_{43}Tc$ | imaging of brain, liver, kidneys, bone marrow |
| $^{133}_{54}Xe$ | lung imaging |
| $^{201}_{81}Tl$ | heart disorders |

strontium than animals do, so a high strontium content suggests a largely vegetarian diet. Strontium analyses of bones taken from ancient farming communities consistently show a difference by sex; women have higher strontium levels than men. Apparently, in those days, women did most of the farming; men spent a lot of time away from home hunting and eating their kill.

### Commercial Applications

Most smoke alarms (Fig. 19.1) use a radioactive species, typically americium-241. A tiny amount of this isotope is placed in a small ionization chamber; decay of Am-241 ionizes air molecules within the chamber. Under the influence of a potential applied by a battery, these ions move across the chamber, producing an electric current. If smoke particles get into the chamber, the flow of ions is impeded and the current drops. This is detected by electronic circuitry, and an alarm sounds. The alarm also goes off if the battery voltage drops, indicating that it needs to be replaced.

Another potential application of radioactive species is in food preservation (Fig. 19.2). It is well known that gamma rays can kill insects, larvae, and parasites such as trichina that cause trichinosis in pork. Radiation can also inhibit sprouting of onions and potatoes. Perhaps most important from a commercial standpoint, it can extend the shelf lives of many foods for weeks or even months. Since many chemicals used to preserve foods have later been shown to have adverse health effects, irradiation would seem to be an attractive alternative.

**Figure 19.1**
Most smoke detectors use a tiny amount of a radioactive isotope to produce a current flow that drops off sharply in the presence of smoke particles, emitting an alarm in the process. (Marna G. Clarke)

## 19.2 Rate of Radioactive Decay

As pointed out in Chapter 11, radioactive decay is a first-order process. This means that the following equations apply:

$$\text{rate} = kX \tag{1}$$

$$\ln\frac{X_0}{X} = kt \tag{2}$$

$$k = \frac{0.693}{t_{1/2}} \tag{3}$$

**Figure 19.2**
Strawberries irradiated with gamma rays from radioactive isotopes to keep them fresh. (Courtesy of Nordion International, Inc.)

# Chemistry: *The Human Side*

The history of radiochemistry is in no small measure the story of two remarkable women, Marie and Irene Curie, and their husbands, Pierre Curie and Frederic Joliot. Marie Curie was born Maria Sklodowska in 1867 in Warsaw, Poland, then a part of the Russian empire. In 1891 she emigrated to Paris to study at the Sorbonne, where she met and married a French physicist, Pierre Curie. The Curies were associates of Henri Becquerel, the man who discovered that uranium salts are radioactive. They showed that thorium, like uranium, is radioactive and that the amount of radiation emitted is directly proportional to the amount of uranium or thorium in the sample.

In 1898, Marie and Pierre Curie isolated two new radioactive elements, which they named radium and polonium. To obtain a few milligrams of these elements, they started with several tons of pitchblende ore and carried out a long series of tedious separations. Their work was done in a poorly equipped, unheated shed where the temperature reached 6°C (43°F) in winter. Four years later, in 1902, Marie determined the atomic mass of radium to within 0.5%, working with a tiny sample.

In 1903, the Curies received the Nobel Prize in physics (with Becquerel) for the discovery of radioactivity. Three years later, Pierre Curie died at the age of 46, the victim of a tragic accident. He stepped from behind a carriage in a busy Paris street and was run down by a horse-driven truck. That same year, Marie became the first woman instructor at the Sorbonne. In 1911, she won the Nobel Prize in chemistry for the discovery of radium and polonium, thereby becoming the first person to win two Nobel Prizes.

When Europe exploded into war in 1914, scientists largely abandoned their studies to go to the front. Marie Curie, with her daughter Irene, then 17 years old, organized medical units equipped with x-ray machinery. These were used to locate foreign metallic objects in wounded soldiers. Many of the wounds were to the head; French soldiers came out of the trenches without head protection because their government had decided that helmets looked too German. In November of 1918, the Curies celebrated the end of World War I; France was victorious, and Marie's beloved Poland was free again.

In 1921, Irene Curie began research at the Radium Institute. Five years later she married Frederic Joliot, a brilliant young physicist who was also an assistant at the Institute. In 1931, they began a research program in nuclear chemistry that led to several important discoveries and at least one near-miss. The Joliot-Curies were the first to demonstrate induced radioactivity. They also discovered the positron, a particle that scientists had been seeking for many years. They narrowly missed finding another, more fundamental particle, the neutron. That honor went to Chadwick in England. In 1935, Irene Curie and Frederic Joliot received the Nobel Prize in physics. The award came too late for Irene's mother, who had died of leukemia in 1934. Twenty-two years later, Irene Curie-Joliot died of the same disease. Both women acquired leukemia through prolonged exposure to radiation.

Marie and Pierre Curie with daughter Irene, at their home near Paris. (Muller Museum, Philadelphia College of Physicians)

Chadwick, a student of Rutherford, discovered the neutron in 1932

where $k$ is the first-order rate constant, $t_{1/2}$ is the half-life, $X$ is the amount of radioactive species present at time $t$, and $X_0$ is the amount at $t = 0$.

Because of the way in which rate of decay is measured (Figure 19.3 p. 538), it is often described by the **activity** ($A$) of the sample, which expresses the number of atoms decaying in unit time. Equation (1) can be written

$$A = kN \qquad (4)$$

where $A$ is the activity, $k$ the first-order rate constant, and $N$ the number of radioactive nuclei present.

$k$ = rate constant = fraction of atoms decaying in unit time

**Figure 19.3**
Liquid scintillation counter used to detect radiation and measure disintegrations per minute quickly and accurately. (Courtesy of Beckman Instruments)

Activity can be expressed in terms of the number of atoms decaying per second, or becquerels (Bq).

$$1 \text{ Bq} = 1 \text{ atom/s}$$

Alternatively, activity may be cited in disintegrations per minute or, perhaps most commonly, in *curies* (Ci)

$$1 \text{ Ci} = 3.700 \times 10^{10} \text{ atom/s}$$

---

**Example 19.2** The half-life of radium-226 is $1.60 \times 10^3$ y $= 5.05 \times 10^{10}$ s.

(a) Calculate $k$ in $s^{-1}$.
(b) What is the activity in curies of a 1.00-g sample of Ra-226?
(c) What is the mass in grams of a sample of Ra-226 that has an activity of $1.00 \times 10^9$ atom/s?

**Strategy** Use Equation (3) to find $k$. In (b) and (c), use Equation (4). To relate $N$ to mass in grams, note that Avogadro's number ($6.022 \times 10^{23}$) of atoms of Ra-226 weigh 226 g.

**Solution**

(a) $k = 0.693/(5.05 \times 10^{10} \text{ s}) = \boxed{1.37 \times 10^{-11}/s}$

(b) First find $N$ for the 1.00-g sample:

$$N = 1.00 \text{ g} \times \frac{6.022 \times 10^{23} \text{ atoms}}{226 \text{ g}} = 2.66 \times 10^{21} \text{ atoms}$$

Now apply Equation (4):

$$A = (1.37 \times 10^{-11}/s) \times 2.66 \times 10^{21} \text{ atoms} \times \frac{1 \text{ Ci}}{3.700 \times 10^{10} \text{ atom/s}} = \boxed{0.985 \text{ Ci}}$$

(The curie was originally supposed to be the activity of a one-gram sample of radium, the element discovered by the Curies; it isn't quite.)

(c) $N = A/k = \dfrac{1.00 \times 10^9 \text{ atoms/s}}{1.37 \times 10^{-11}/s} = 7.30 \times 10^{19} \text{ atoms}$

$$\text{mass Ra-226} = 7.30 \times 10^{19} \text{ atoms} \times \frac{226 \text{ g}}{6.022 \times 10^{23} \text{ atoms}} = \boxed{0.0274 \text{ g}}$$

---

### Age of Rocks

As pointed out earlier, uranium-238 is radioactive:

$$^{238}_{92}\text{U} \longrightarrow {}^{234}_{90}\text{Th} + {}^4_2\text{He} \qquad t_{1/2} = 4.5 \times 10^9 \text{ yr}$$

The product, thorium-234, is itself radioactive; decay continues until a stable product, lead-206, is formed. The overall, 14-step process can be represented by the nuclear equation

$$^{238}_{92}\text{U} \longrightarrow {}^{206}_{82}\text{Pb} + 8 \, {}^4_2\text{He} + 6 \, {}^{\,0}_{-1}e \qquad t_{1/2} = 4.5 \times 10^9 \text{ yr}$$

Since all the other steps have short half-lives, the slow first step is rate-determining.

By analyzing a uranium-containing rock to determine the relative amounts of U-238 and Pb-206, it is possible to determine the "age" of the rock, i.e., the amount of time elapsed since it solidified. Suppose, for example, that equal numbers of atoms of U-238 and Pb-206 are present in the rock. Assuming that there was no Pb-206 present originally, this means that one half-life, or $4.5 \times 10^9$ years, has passed; the rock is 4.5 billion years old.

Ages of rocks determined by this method range from 3 to $4.5 \times 10^9$ years. The larger number is often taken as an approximate value for the age of the Earth. Analyses of rock samples from the moon indicate ages in the same range. This argues against the once-prevalent idea that the moon was torn from the Earth's surface by a violent event a long time after the Earth solidified.

## Age of Organic Material

During the 1950s, Professor W. F. Libby of the University of Chicago and others worked out a method for determining the age of organic material. It is based upon the decay rate of carbon-14. The method can be applied to objects from a few hundred up to 50,000 years old. It has been used to determine the authenticity of canvases of Renaissance painters and to check the ages of relics left by prehistoric cavemen.

Carbon-14 is produced in the atmosphere by the interaction of neutrons from cosmic radiation with ordinary nitrogen atoms:

$$^{14}_{7}N + ^{1}_{0}n \longrightarrow {}^{14}_{6}C + {}^{1}_{1}H$$

Willard Libby and his equipment for carbon-14 dating. (Oesper Collection in the History of Chemistry/University of Cincinnati)

The carbon-14 formed by this nuclear reaction is eventually incorporated into the carbon dioxide of the air. A steady-state concentration, amounting to about one atom of carbon-14 for every $10^{12}$ atoms of carbon-12, is established in atmospheric $CO_2$. More specifically, the concentration of C-14 is such that a sample containing one gram of carbon has an activity of 13.6 atoms/min*. A living plant, taking in carbon dioxide, has this same activity, as do plant-eating animals or human beings.

When a plant or animal dies, the intake of radioactive carbon stops. Consequently, the radioactive decay of Carbon-14

$$^{14}_{6}C \longrightarrow {}^{14}_{7}N + {}^{0}_{-1}e \qquad t_{1/2} = 5720 \text{ yr}$$

takes over, and the C-14 activity drops. Since the activity of a sample is directly proportional to the amount of C-14, Equation (2) can be rewritten as

$$\ln\frac{A_0}{A} = kt$$

where $A_0$ is the original activity, assumed to be 13.6 atoms/s, and $A$ is the measured activity today; $t$ is the "age" of the sample.

---

**Example 19.3**    A tiny piece of paper taken from the Dead Sea Scrolls, believed to date back to the first century AD, was found to have an activity per gram of carbon of 10.8 atoms/min. Taking $A_0$ to be 13.6 atoms/min, estimate the age of the scrolls.

*Strategy*    First, calculate the rate constant, knowing that the half-life is 5720 y. Then find $t$, using the equation $\ln (A_0/A) = kt$.

---

*Actually, this was the activity in 1950, prior to nuclear testing, which raised the C-14 content considerably. Moreover, it is now known that the C-14 content has varied significantly over the past several thousand years; in very accurate work, a correction must be made for this effect.

The Shroud of Turin had been represented to be the burial cloth of Jesus Christ. (Patrick Mesner/Gamma Liaison)

*Solution*

(1) $k = 0.693/5720 \text{ y} = 1.21 \times 10^{-4}/\text{y}$

(2) $\ln \dfrac{13.6}{10.8} = 0.231 = 1.21 \times 10^{-4}/\text{y} \times t$

$$t = 0.231/(1.21 \times 10^{-4}/\text{y}) = \boxed{1.91 \times 10^3 \text{ y}}$$

---

As you can imagine, it is not easy to determine accurately activities of the order of 10 atoms decaying per minute, about one "event" every six seconds. Elaborate precautions have to be taken to exclude background radiation. Moreover, relatively large samples must be used to increase the counting rate. Recently, a technique has been developed whereby C-14 atoms can be counted very accurately in a specially designed mass spectrometer. This method was used to date the Shroud of Turin, using six samples with a total mass of about 0.1 g. Analysis by an international team of scientists in 1988 showed that the flax used to make the linen of which the Shroud is composed grew in the fourteenth century AD. Clearly, this remarkable burial garment could not have been used for the body of Christ.

## 19.3  Mass–Energy Relations

The energy change accompanying a nuclear reaction can be calculated from the relation

$$\Delta E = c^2 \Delta m$$

where $\Delta m$ is the change in mass*, $\Delta E$ is the change in energy, and $c$ is the speed of light. In an "ordinary" chemical reaction, $\Delta m$ is immeasurably small. In a nuclear reaction, on the other hand, $\Delta m$ is appreciable, amounting to 0.002% or more of the mass of reactants. The change in mass can readily be calculated from a table of nuclear masses (Table 19.3).

To obtain a more useful form of the equation $\Delta E = c^2 \Delta m$, we substitute for $c$ its value in meters per second:

$$c = 3.00 \times 10^8 \text{ m/s}$$

Thus,

$$\Delta E = 9.00 \times 10^{16} \frac{\text{m}^2}{\text{s}^2} \times \Delta m$$

But

$$1 \text{ J} = 1\text{kg} \cdot \frac{\text{m}^2}{\text{s}^2} \; ; \; 1\frac{\text{m}^2}{\text{s}^2} = 1\frac{\text{J}}{\text{kg}}$$

So

$$\Delta E = 9.00 \times 10^{16} \frac{\text{J}}{\text{kg}} \times \Delta m$$

---

*Specifically, $\Delta m$ = mass of products − mass of reactants; $\Delta E$ = energy of products − energy of reactants. In spontaneous nuclear reactions, the products weigh less than the reactants ($\Delta m$ negative). In this case, the energy of the products is less than that of the reactants ($\Delta E$ negative), and energy is evolved to the surroundings.

TABLE 19.3   **Nuclear Masses on the $^{12}$C Scale\***

| | At. No. | Mass No. | Mass (amu) | | At. No. | Mass No. | Mass (amu) |
|---|---|---|---|---|---|---|---|
| *e* | 0 | 0 | 0.00055 | | | | |
| *n* | 0 | 1 | 1.00867 | Br | 35 | 79 | 78.8992 |
| H | 1 | 1 | 1.00728 | | 35 | 81 | 80.8971 |
| | 1 | 2 | 2.01355 | | 35 | 87 | 86.9028 |
| | 1 | 3 | 3.01550 | Rb | 37 | 89 | 88.8913 |
| He | 2 | 3 | 3.01493 | Sr | 38 | 90 | 89.8869 |
| | 2 | 4 | 4.00150 | Mo | 42 | 99 | 98.8846 |
| Li | 3 | 6 | 6.01348 | Ru | 44 | 106 | 105.8832 |
| | 3 | 7 | 7.01436 | Ag | 47 | 109 | 108.8790 |
| Be | 4 | 9 | 9.00999 | Cd | 48 | 109 | 108.8786 |
| | 4 | 10 | 10.01134 | | 48 | 115 | 114.8791 |
| B | 5 | 10 | 10.01019 | Sn | 50 | 120 | 119.8748 |
| | 5 | 11 | 11.00656 | Ce | 58 | 144 | 143.8817 |
| C | 6 | 11 | 11.00814 | | 58 | 146 | 145.8868 |
| | 6 | 12 | 11.99671 | Pr | 59 | 144 | 143.8809 |
| | 6 | 13 | 13.00006 | Sm | 62 | 152 | 151.8857 |
| | 6 | 14 | 13.99995 | Eu | 63 | 157 | 156.8908 |
| O | 8 | 16 | 15.99052 | Er | 68 | 168 | 167.8951 |
| | 8 | 17 | 16.99474 | Hf | 72 | 179 | 178.9065 |
| | 8 | 18 | 17.99477 | W | 74 | 186 | 185.9138 |
| F | 9 | 18 | 17.99601 | Os | 76 | 192 | 191.9197 |
| | 9 | 19 | 18.99346 | Au | 79 | 196 | 195.9231 |
| Na | 11 | 23 | 22.98373 | Hg | 80 | 196 | 195.9219 |
| Mg | 12 | 24 | 23.97845 | Pb | 82 | 206 | 205.9295 |
| | 12 | 25 | 24.97925 | | 82 | 207 | 206.9309 |
| | 12 | 26 | 25.97600 | | 82 | 208 | 207.9316 |
| Al | 13 | 26 | 25.97977 | Po | 84 | 210 | 209.9368 |
| | 13 | 27 | 26.97439 | | 84 | 218 | 217.9628 |
| | 13 | 28 | 27.97477 | Rn | 86 | 222 | 221.9703 |
| Si | 14 | 28 | 27.96924 | Ra | 88 | 226 | 225.9771 |
| S | 16 | 32 | 31.96329 | Th | 90 | 230 | 229.9837 |
| Cl | 17 | 35 | 34.95952 | Pa | 91 | 234 | 233.9934 |
| | 17 | 37 | 36.95657 | U | 92 | 233 | 232.9890 |
| Ar | 18 | 40 | 39.95250 | | 92 | 235 | 234.9934 |
| K | 19 | 39 | 38.95328 | | 92 | 238 | 238.0003 |
| | 19 | 40 | 39.95358 | | 92 | 239 | 239.0038 |
| Ca | 20 | 40 | 39.95162 | Np | 93 | 239 | 239.0019 |
| Ti | 22 | 48 | 47.93588 | Pu | 94 | 239 | 239.0006 |
| Cr | 24 | 52 | 51.92734 | | 94 | 241 | 241.0051 |
| Fe | 26 | 56 | 55.92066 | Am | 95 | 241 | 241.0045 |
| Co | 27 | 59 | 58.91837 | Cm | 96 | 242 | 242.0061 |
| Ni | 28 | 59 | 58.91897 | Bk | 97 | 245 | 245.0129 |
| Zn | 30 | 64 | 63.91268 | Cf | 98 | 248 | 248.0186 |
| | 30 | 72 | 71.91128 | Es | 99 | 251 | 251.0255 |
| Ge | 32 | 76 | 75.90380 | Fm | 100 | 252 | 252.0278 |
| As | 33 | 79 | 78.90288 | | 100 | 254 | 254.0331 |

\*Note that these are *nuclear masses*. The masses of the corresponding atoms can be calculated by adding the masses of each extranuclear electron (0.000549). For example, for an *atom* of $^{4}_{2}$He we have 4.00150 + 2(0.000549) = 4.00260. Similarly, for an atom of $^{12}_{6}$C, 11.99671 + 6(0.000549) = 12.00000.

This equation can be used to calculate the energy change in *joules*, if you know $\Delta m$ in *kilograms*. Ordinarily, $\Delta m$ is expressed in *grams*, while $\Delta E$ is calculated in *kilojoules*. The relationship between $\Delta E$ and $\Delta m$ in these units can be found by using conversion factors:

*This is the most useful form of the mass-energy relation*

$$\Delta E = 9.00 \times 10^{16} \frac{J}{kg} \times \frac{1 \text{ kg}}{10^3 \text{ g}} \times \frac{1 \text{ kJ}}{10^3 \text{ J}} \times \Delta m$$

$$\Delta E = 9.00 \times 10^{10} \frac{kJ}{g} \times \Delta m$$

---

**Example 19.4** For the radioactive decay of radium, $^{226}_{88}\text{Ra} \longrightarrow {}^{222}_{86}\text{Rn} + {}^{4}_{2}\text{He}$, calculate $\Delta E$ in kilojoules when 1.02 g of radium decays.

***Strategy*** Using Table 19.3, calculate $\Delta m$ when one mole of radium decays. Then find $\Delta m$ when 1.02 g of radium decays (1 mol Ra = 226.0 g Ra). Finally, find $\Delta E$ using the relation $\Delta E = 9.00 \times 10^{10} \Delta m$ kJ/g.

*Solution*

(1) Using Table 19.3 for one mole of Ra-226,

$\Delta m$ = mass 1 mol He-4 + mass 1 mol Rn-222 − mass 1 mol Ra-226

$\quad$ = 4.0015 g + 221.9703 g − 225.9771 g

$\quad$ = −0.0053 g/mol Ra

(2) When 1.02 g of radium decays,

$$\Delta m = \frac{-0.0053 \text{ g}}{\text{mol Ra}} \times \frac{1 \text{ mol Ra}}{226.0 \text{ g Ra}} \times 1.02 \text{ g Ra} = -2.4 \times 10^{-5} \text{ g}$$

(3) $\Delta E = 9.00 \times 10^{10} \dfrac{kJ}{g} \times (-2.4 \times 10^{-5} \text{ g}) = \boxed{-2.2 \times 10^6 \text{ kJ}}$

In ordinary chemical reactions, the energy change is of the order of 50 kJ/g or less. The energy change in this nuclear reaction is greater by a factor of

$$2.2 \times 10^6/50 = 4 \times 10^4$$

In other words, about 40,000 times as much energy is evolved.

---

## Nuclear Binding Energy

It is always true that *a nucleus weighs less than the individual protons and neutrons of which it is composed.* Consider, for example, the $^{6}_{3}\text{Li}$ nucleus, which contains three protons and three neutrons. According to Table 19.3, one mole of Li-6 nuclei weighs 6.01348 g. In contrast, the total mass of three moles of neutrons and three moles of protons is

$$3(1.00867 \text{ g}) + 3(1.00728 \text{ g}) = 6.04785 \text{ g}$$

Clearly, one mole of Li-6 weighs *less* than the corresponding protons and neutrons. For the process

$$^{6}_{3}\text{Li} \longrightarrow 3 \, {}^{1}_{1}\text{H} + 3 \, {}^{1}_{0}n$$

$\Delta m$ = 6.04785 g − 6.01348 g = 0.03437 g/mol Li.

The quantity just calculated is referred to as the *mass defect*. The corresponding energy difference is

$$\Delta E = 9.00 \times 10^{10} \text{ kJ} \times 0.03437 \frac{\text{g}}{\text{mol Li}} = 3.09 \times 10^9 \frac{\text{kJ}}{\text{mol Li}}$$

It takes a lot of energy to blow a nucleus apart

This energy is referred to as the **binding energy.** It follows that $3.09 \times 10^9$ kJ of energy would have to be absorbed to decompose one mole of Li-6 nuclei into protons and neutrons.

$$^6_3\text{Li} \longrightarrow 3 \, ^1_0n + 3 \, ^1_1\text{H} \qquad \Delta E = 3.09 \times 10^9 \frac{\text{kJ}}{\text{mol Li}}$$

By the same token, $3.09 \times 10^9$ kJ of energy would be evolved when one mole of Li-6 is formed from protons and neutrons.

In a sense, the binding energy that holds a nucleus together is analogous to the bond energy that holds a molecule together. Binding energies, however, are much larger; they vary from $10^9$ to $10^{11}$ kJ/mol, as compared with $10^2$ to $10^3$ kJ/mol for bond energies (Chap. 8).

---

**Example 19.5**   Calculate the binding energy of Be-9 in kilojoules per mole.

***Strategy***   Use Table 19.3 to find $\Delta m$ for the decomposition of Be-9 into neutrons and protons. Then calculate $\Delta E = 9.00 \times 10^{10}$ kJ/g $\times \Delta m$.

***Solution***

(1) $^9_4\text{Be} \longrightarrow 5 \, ^1_0n + 4 \, ^1_1\text{H}$

(2) $\Delta m = 5(1.00867 \text{ g}) + 4(1.00728 \text{ g}) - 9.00999 \text{ g} = 0.06248 \text{ g/mol Be}$

$$\Delta E = 9.00 \times 10^{10} \frac{\text{kJ}}{\text{g}} \times 0.06248 \frac{\text{g}}{\text{mol Be}} = \boxed{5.62 \times 10^9 \text{ kJ/mol Be}}$$

---

The binding energy of a nucleus is, in a sense, a measure of its stability. The greater the binding energy, the more difficult it would be to decompose the nucleus into protons and neutrons. As you might expect, the binding energy as calculated above increases steadily as the nucleus gets heavier, containing more protons and neutrons. A better measure of the relative stabilities of different nuclei is the ***binding energy per mole of nuclear particles (nucleons).*** This quantity is calculated by dividing the binding energy per mole of nuclei by the number of particles per nucleus. Thus,

Li-6: $\quad 3.09 \times 10^9 \dfrac{\text{kJ}}{\text{mol Li-6}} \times \dfrac{1 \text{ mol Li-6}}{6 \text{ mol nucleons}} = 5.15 \times 10^8 \dfrac{\text{kJ}}{\text{mol}}$

Be-9: $\quad 5.62 \times 10^9 \dfrac{\text{kJ}}{\text{mol Be-9}} \times \dfrac{1 \text{ mol Be-9}}{9 \text{ mol nucleons}} = 6.24 \times 10^8 \dfrac{\text{kJ}}{\text{mol}}$

Figure 19.4 p. 544 shows a plot of this quantity, binding energy/mole of nucleons, versus mass number. Notice that the curve has a broad maximum in the vicinity of mass numbers 50 to 80. Consider what would happen if a heavy nucleus such as $^{235}_{92}\text{U}$ were to split into smaller nuclei with mass numbers near the maximum. This process, referred to as *nuclear fission*, should result in an evolution of energy. The same effect would be obtained if very light nuclei such as $^2_1\text{H}$ were to combine with one another. Indeed, this process, called *nuclear fusion*,

Both very heavy and very light nuclei are relatively unstable; they decompose with the evolution of energy

**Figure 19.4**

The binding energy per nucleon is a measure of nuclear stability. It has its maximum value for nuclei of intermediate mass, falling off for very heavy or very light nuclei. The form of this curve accounts for the fact that both fission and fusion give off large amounts of energy.

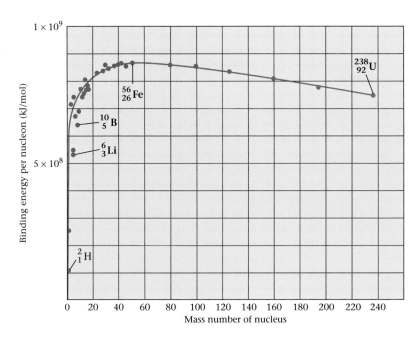

should evolve even more energy, since the binding energy per nucleon increases very sharply at the beginning of the curve.

## 19.4 Nuclear Fission

The process of nuclear fission was discovered more than half a century ago in 1938 by Lise Meitner and Otto Hahn in Germany. With the outbreak of World War II a year later, interest focused on the enormous amount of energy released in the process. At Los Alamos, in the mountains of New Mexico, a group of scientists led by J. Robert Oppenheimer worked feverishly to produce the fission, or "atomic" bomb. Many of the members of this group were exiles from Nazi Germany. They were spurred on by the fear that Hitler would obtain the bomb first. Their work led to the explosion of the first atomic bomb in the New Mexico desert at 5:30 AM on July 16, 1945. Less than a month later (August 6, 1945), the world learned of this new weapon when another bomb was exploded over Hiroshima. This bomb killed 70,000 people and completely devastated an area of 10 square kilometers. Three days later, Nagasaki and its inhabitants met a similar fate. On August 14, Japan surrendered, and World War II was over.

### The Fission Process ($^{235}_{92}U$)

Several isotopes of the heavy elements undergo fission if bombarded by neutrons of high enough energy. In practice, attention has centered upon two particular isotopes, $^{235}_{92}U$ and $^{239}_{94}Pu$. Both of these can be split into fragments by low-energy neutrons.

Our discussion concentrates upon the uranium-235 isotope. It makes up only about 0.7% of naturally occurring uranium. The more abundant isotope, uranium-238, does not undergo fission. The first process used to separate these isotopes, and until recently the only one available, was that of gaseous effusion

(Chap. 5). The volatile compound uranium hexafluoride, $UF_6$, which sublimes at 56°C, is used for this purpose.

**Fission Products**   When a uranium-235 atom undergoes fission, it splits into two unequal fragments and a number of neutrons and beta particles. The fission process is complicated by the fact that different uranium-235 atoms split up in many different ways. For example, while one atom of $^{235}_{92}U$ is splitting to give isotopes of rubidium ($Z = 37$) and cesium ($Z = 55$), another may break up to give isotopes of bromine ($Z = 35$) and lanthanum ($Z = 57$), while still another atom yields isotopes of zinc ($Z = 30$) and samarium ($Z = 62$):

$$^{1}_{0}n + ^{235}_{92}U \begin{cases} \nearrow\ ^{90}_{37}Rb + ^{144}_{55}Cs + 2\ ^{1}_{0}n \\ \longrightarrow\ ^{87}_{35}Br + ^{146}_{57}La + 3\ ^{1}_{0}n \\ \searrow\ ^{72}_{30}Zn + ^{160}_{62}Sm + 4\ ^{1}_{0}n \end{cases}$$

More than 200 isotopes of 35 different elements have been identified among the fission products of uranium-235.

The stable neutron-to-proton ratio near the middle of the periodic table, where the fission products are located, is considerably smaller (~1.2) than that of uranium-235 (1.5). Hence, the immediate products of the fission process contain too many neutrons for stability; they decompose by beta emission. In the case of rubidium-90, three steps are required to reach a stable nucleus:

$$^{90}_{37}Rb \longrightarrow ^{90}_{38}Sr + _{-1}^{0}e \qquad t_{1/2} = 2.8\ \text{min}$$

$$^{90}_{38}Sr \longrightarrow ^{90}_{39}Y + _{-1}^{0}e \qquad t_{1/2} = 29\ \text{yr}$$

$$^{90}_{39}Y \longrightarrow ^{90}_{40}Zr + _{-1}^{0}e \qquad t_{1/2} = 64\ \text{h}$$

The radiation hazard associated with nuclear fallout arises from the formation of radioactive isotopes such as these. One of the most dangerous is strontium-90. In the form of strontium carbonate, $SrCO_3$, it is incorporated into the bones of animals and human beings.

Notice from the fission equations written above that two to four neutrons are produced by fission for every one consumed. Once a few atoms of uranium-235 split, the neutrons produced can bring about the fission of many more uranium-235 atoms. This creates the possibility of a *chain reaction,* whose rate increases exponentially with time. This is precisely what happens in the atomic bomb. The energy evolved in successive fissions escalates to give a tremendous explosion within a few seconds.

For nuclear fission to result in a chain reaction, the sample must be large enough so that most of the neutrons are captured internally. If the samples is too small, most of the neutrons escape, breaking the chain. The *critical mass* of uranium-235 required to maintain a chain reaction in a bomb appears to be about 1 to 10 kg. In the bomb dropped on Hiroshima, the critical mass was achieved by using a conventional explosive to fire one piece of uranium-235 into another.

The Hiroshima bomb was equivalent to 20,000 tons of TNT

**Fission Energy**   The evolution of energy in nuclear fission is directly related to the decrease in mass that takes place. About 80,000,000 kJ of energy is given off for every gram of $^{235}_{92}U$ that reacts. This is about 40 times as great as the energy change for simple nuclear reactions such as radioactive decay. The heat of combustion of coal is only about 30 kJ/g; the energy given off when TNT explodes is still smaller, about 2.8 kJ/g. Putting it another way, the fission of one gram of $^{235}_{92}U$ produces as much energy as the combustion of 2700 kg of coal or the explosion of 30 metric tons ($3 \times 10^4$ kg) of TNT.

## Nuclear Reactors

Even before the first atomic bomb exploded, scientists and political leaders began to speculate on the use of fission as a peacetime energy source. Twenty-five years ago, it was widely believed that nuclear fission would replace fossil fuels (oil, natural gas, coal), at least for the generation of electrical energy. During the 1970s, disillusionment set in with nuclear power in the United States. Economic factors, along with the problem of disposing of nuclear waste, were largely responsible. The newest of the 110 nuclear reactors now operating in the United States was started when Richard Nixon was president (1969–1974). The incident at Three Mile Island (p. 547) in 1979 seemed to be the last straw, stirring fears of nuclear disaster.

However, in the 1990s, there has been a reassessment of the prospects for nuclear power. Mostly, this reflects deficiencies and uncertainties associated with conventional power plants. Coal-fired plants are the major source of acid rain (p. 391). The 1991 war in the Persian Gulf emphasized once again the danger of depending upon oil imported from the Middle East. Finally, the combustion of all fossil fuels is principally responsible for the greenhouse effect. Nuclear energy has none of these shortcomings.

The so-called "light water reactor" (LWR) used in the United States today is shown in Figure 19.5. Fuel rods alternate with control rods in a containment chamber. The *fuel rods* are cylinders that contain fissionable material, uranium dioxide ($UO_2$) pellets, in a zirconium alloy tube. The uranium in these reactors

*All methods of generating large amounts of energy have risks and hazards*

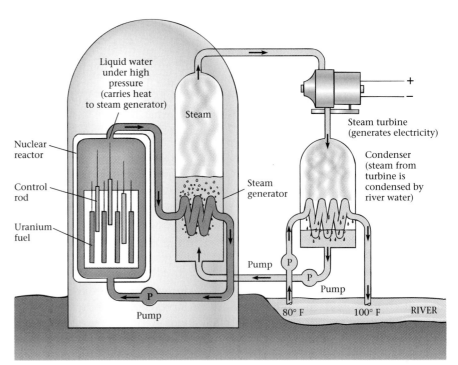

**Figure 19.5**
Nuclear reactor of the pressurized-water type. The control rods are made of a material such as cadmium or boron, which absorb neutrons effectively. The fuel rods contain uranium oxide, usually enriched in U-235.

is "enriched" so that it contains about 3% U-235, the fissionable isotope. The *control rods* are cylinders composed of substances, such as boron and cadmium, that absorb neutrons. Increased absorption of neutrons slows down the chain reaction. By varying the depth of the control rods within the fuel-rod assembly, the speed of the chain reaction can be controlled. Water at a pressure of 140 atm is passed through the reactor to absorb the heat given off by fission. The water, coming out of the reactor core at 320°C, circulates through a closed loop containing a heat exchanger that produces steam at 270°C. This steam is used to drive a turbogenerator that produces electrical energy.

In the light water reactor, the circulating water serves another purpose in addition to heat transfer. It acts to slow down or *moderate* the neutrons given off by fission. This is necessary if the chain reaction is to continue; fast neutrons are not readily absorbed by U-235. Reactors in Canada use "heavy water," $D_2O$, which has an important advantage over $H_2O$. Its moderating properties are such that naturally occurring uranium can be used as a fuel; enrichment in U-235 is not necessary.

**Nuclear Accidents: Three Mile Island and Chernobyl**   To operate a nuclear reactor safely, it is necessary to prevent significant amounts of radioactive fission products from escaping into the environment. This can happen if the temperature of the nuclear fuel rises to the melting point ("meltdown"), leading to an explosion.

The most likely cause of meltdown is a loss of cooling water. If this happens in a light water reactor, fission stops because the unmoderated neutrons move too fast to continue the chain reaction. However, large amounts of heat are given off by the radioactive decay of fission products (recall Example 19.4). This happened at Three Mile Island in Pennsylvania in March of 1979. Through operator error, water was lost from the cooling system. For some time, the fuel rods were uncovered, and there was a real danger of meltdown. Fortunately, this did not happen. Virtually all of the radioactive isotopes released were confined within a complete containment building surrounding the reactor.

A much more severe nuclear accident took place at Chernobyl, near Kiev in the Ukraine in April of 1986. In the absence of a containment building, large amounts of radiation, estimated at 100 million curies, were released to the atmosphere. Much of this was carried westward by prevailing winds to other European countries. There were 31 fatalities from acute radiation exposure at the reactor; 135,000 people had to be permanently evacuated from the Chernobyl area, one of the most fertile farmlands in the world.

**Nuclear Waste**   When a nuclear reactor operates, the fuel rods undergo physical and chemical changes due to the enormous amount of radiation to which they are exposed. Each year, on the average, one fourth of these intensely radioactive rods must be replaced. A typical nuclear reactor produces about 20 metric tons ($2 \times 10^4$ kg) of spent fuel rods per year. Ideally, these rods should be reprocessed to recover uranium and plutonium, producing a relatively small amount of high-level radioactive waste. In 1982, a federal program was established for the disposal of such waste, which is supposed to be buried in an underground site.

Calculations indicate that waste from reprocessed fuel requires 20,000 years to decay to a "safe" level. Unprocessed waste takes 100 times longer. Data like

The nuclear power plant at Chernobyl shortly after the fire and explosion in 1986. (© Peter A. Simon/Phototake)(Novosti Press Agency/Gamma Liaison)

these contribute to the "nimby" syndrome (*not in my back yard*) where nuclear wastes are concerned. The latest estimate is that a repository for high-activity waste may be available by the year 2010, perhaps within a remote mountain in Nevada.

## 19.5  Nuclear Fusion

Recall (Fig. 19.4) that very light nuclei, such as those of hydrogen, are unstable with respect to fusion into heavier isotopes. Indeed, the energy available from nuclear fusion is considerably greater than that given off in the fission of an equal mass of a heavy element (Example 19.6).

---

**Example 19.6**    Calculate $\Delta E$, in kilojoules per gram of reactants, in

(a) a fusion reaction, $_1^2 H + _1^2 H \longrightarrow {}_2^4 He$.

(b) a fission reaction, $_{92}^{235} U \longrightarrow {}_{38}^{90} Sr + {}_{58}^{144} Ce + {}_0^1 n + 4\ {}_{-1}^0 e$

**Strategy**    This problem is entirely analogous to Example 19.4. First, find $\Delta m$ for the equation as written, and then find $\Delta m$ for one gram of reactant. Finally, calculate $\Delta E = 9.00 \times 10^{10}$ kJ/g $\times \Delta m$.

*Solution*

(a) (1) using Table 19.3,

$$\Delta m = 4.00150\ g - 2(2.01355\ g) = -0.02560\ g$$

(2) The total mass of reactant, $_1^2 H$, is 4.027 g:

$$\Delta m = \frac{-0.02560\ g}{4.027\ g\ \text{reactant}} = -6.357 \times 10^{-3}\ g/g\ \text{reactant}$$

(3) $\Delta E = 9.00 \times 10^{10} \dfrac{kJ}{g} \times \dfrac{(-6.357 \times 10^{-3}\ g)}{g\ \text{reactant}} = \dfrac{-5.72 \times 10^8\ kJ}{g\ \text{reactant}}$

(b) (1) $\Delta m = 89.8869\ g + 143.8817\ g + 1.0087\ g + 4(0.00055\ g) - 234.9934\ g$

$= -0.2139\ g$

(2) The total mass of reactant, U-235, is 235.0 g:

$$\Delta m = \frac{-0.2139\ g}{235.0\ g\ \text{reactant}} = 9.102 \times 10^{-4}\ g/g\ \text{reactant}$$

(3) $\Delta E = 9.00 \times 10^{10} \dfrac{kJ}{g} \times \left( \dfrac{-9.102 \times 10^{-4}\ g}{g\ \text{reactant}} \right) = \dfrac{-8.19 \times 10^7\ kJ}{g\ \text{reactant}}$

Comparing the answers to (a) and (b), it appears that the fusion reaction produces about seven times as much energy per gram of reactant ($57.2 \times 10^7$ versus $8.19 \times 10^7$ kJ) as does the fission reaction. This factor varies from about three to ten, depending upon the particular reactions chosen to represent the fusion and fission processes.

---

As an energy source, nuclear fusion possesses several additional advantages over nuclear fission. In particular, light isotopes suitable for fusion are far more abundant than the heavy isotopes required for fission. You can calculate, for example (Problem 66), that the fusion of only $2 \times 10^{-9}\%$ of the deuterium ($^2_1$H) in seawater would meet the total annual energy requirements of the world.

Unfortunately, fusion processes, unlike neutron-induced fission, have very high activation energies. In order to overcome the electrostatic repulsion between two deuterium nuclei and cause them to react, they have to be accelerated to velocities of about $10^6$ m/s, about 10,000 times greater than ordinary molecular velocities at room temperature. The corresponding temperature for fusion, as calculated from kinetic theory, is of the order of $10^9$ °C. In the hydrogen bomb, temperatures of this magnitude were achieved by using a fission reaction to trigger nuclear fusion. If fusion reactions are to be used to generate electricity, it will be necessary to develop equipment in which very high temperatures can be maintained long enough to allow fusion to occur and give off energy. In any conventional container, the reactant nuclei would quickly lose their high kinetic energies by collisions with the walls.

One fusion reaction currently under study is a two-step process involving deuterium and lithium as the basic starting materials:

$$\frac{\begin{array}{r} ^2_1\text{H} + {}^3_1\text{H} \longrightarrow {}^4_2\text{He} + {}^1_0 n \\ {}^6_3\text{Li} + {}^1_0 n \longrightarrow {}^4_2\text{He} + {}^3_1\text{H} \end{array}}{^2_1\text{H} + {}^6_3\text{Li} \longrightarrow 2\ {}^4_2\text{He}}$$

This process is attractive because it has a lower activation energy than other fusion reactions.

Promising results have been obtained with nuclear fusion using magnetic fields to confine the reactant nuclei and prevent them from touching the walls of the container, where they would quickly slow down below the velocity required for fusion. Using 400-ton magnets, it is possible to sustain the reaction for a fraction of a second. To achieve a net evolution of energy, this time must be extended to about one second. A practical fusion reactor would have to produce 20 times as much energy as it consumes. It is likely to take at least 20 years to achieve that goal with magnetic confinement.

Another approach to nuclear fusion is shown in Figure 19.6. Tiny glass pellets (about 0.1 mm in diameter) filled with frozen deuterium and tritium serve as a target. The pellets are illuminated by a powerful laser beam, which delivers $10^{12}$ kilowatts of power in one nanosecond ($10^{-9}$ s). The reaction is the same as with magnetic confinement; unfortunately, at this point energy breakeven seems several years away.

Nuclear fusion has been the "fuel of the future" for at least 40 years

**Figure 19.6**
Laser fusion. (Lawrence Livermore National Laboratory)

## Biological Effects of Radiation

**CHEMISTRY**

*Beyond the Classroom*

The harmful effects of radiation result from its high energy, sufficient to break the chemical bonds holding organic molecules together in body cells. The extent of damage depends mainly upon two factors. These are the amount of radiation absorbed and the type of radiation. The former is commonly expressed in *rads* (radiation *a*bsorbed *dose*). A rad corresponds to the absorption of $10^{-2}$ J of energy per kilogram of tissue:

$$1 \text{ rad} = 10^{-2} \text{ J/kg}$$

The total biological effect of radiation is expressed in *rems* (radiation *e*quivalent for *m*an). The number of rems is found by multiplying the number of rads by an appropriate factor, *n*, for the particular type of radiation:

$$\text{no. rems} = n \text{ (no. rads)}$$

$$n = 1 \text{ for } \beta, \gamma, \text{x-rays}$$

$$n = 10 \text{ for } \alpha \text{ rays, high-energy neutrons}$$

Table 19.A lists some of the effects to be expected when a person is exposed to a single dose of radiation at various levels.

**TABLE 19.A  Effect of Exposure to a Single Dose of Radiation**

| Dose (rems) | Probable Effect |
|---|---|
| 0 to 25 | no observable effect |
| 25 to 50 | small decrease in white blood cell count |
| 50 to 100 | lesions, marked decrease in white blood cells |
| 100 to 200 | nausea, vomiting, loss of hair |
| 200 to 500 | hemorrhaging, ulcers, possible death |
| 500+ | fatal |

Small doses of radiation repeated over long periods of time can have very serious consequences. Many of the early workers in the field of radioactivity developed cancer in this way. Cases are known in which cancers developed as long as 40 years after initial exposure. Studies have shown an abnormally large number of cases of leukemia among the survivors of the atomic bombs dropped on Hiroshima and Nagasaki.

Table 19.B lists average exposures to radiation of people living in the United States. Notice that about two thirds of the radiation exposure comes from natural sources. The level depends upon location. Cosmic radiation is much more intense at high elevations. A person living in Denver is exposed to about 100 millirems/year from this source. This is twice the national average.

**TABLE 19.B  Typical Radiation Exposures in the United States ($1 \text{ millirem} = 10^{-3} \text{ rem}$)**

| Sources | Millirems/yr |
|---|---|
| I. Natural | |
|   A. External to the body | |
|     1. From cosmic radiation | 50 |
|     2. From the Earth | 47 |
|     3. From building materials | 3 |
|   B. Inside the body | |
|     1. Inhalation of air | 5 |
|     2. In human tissues (mostly $^{40}_{19}K$) | 21 |
|   Total from natural sources | **126** |
| II. Man-made | |
|   A. Medical procedures | |
|     1. Diagnostic x-rays | 50 |
|     2. Radiotherapy x-rays, radioisotopes | 10 |
|     3. Internal diagnosis, therapy | 1 |
|   B. Nuclear power industry | 0.2 |
|   C. Luminous watch dials, TV tubes, industrial wastes | 2 |
|   D. Radioactive fallout (nuclear tests) | 4 |
|   Total from man-made sources | **67** |
| TOTAL | **193** |

Radiation listed in Table 19.B as coming from the Earth is mostly in the form of $^{222}_{86}$Rn, a radioactive isotope of radon, which is a decay product of uranium-238. Because it is gaseous and chemically inert, radon seeps through cracks in concrete and masonry from the ground into houses. There its concentration builds up, particularly if the house is tightly insulated.

Inhalation of radon-222 can cause health problems because its decay products, including Po-218 and Po-214, are intensely radioactive and readily adsorbed in lung tissue. The Environmental Protection Agency estimates that radon inhalation causes between 7000 and 30,000 of the 130,000 deaths from lung cancer annually in the United States. To reduce these numbers, EPA recommends that special ventilation devices be used to remove radon from basements if the radiation level exceeds $4 \times 10^{-12}$ Ci/L. It is estimated that at least five million homes in the United States exceed this level and that the measures recommended by EPA would cost homeowners and landlords at least $50 billion. Present U.S. regulations set an upper limit of $20 \times 10^{-12}$ Ci/L for radon exposure; about fifty thousand homes exceed this limit.

A commercially available home-test kit for radon.

# CHAPTER HIGHLIGHTS

## *Key Concepts*

**1.** Write balanced nuclear equations
   (Example 19.1; Problems 1–10)
**2.** Relate activity to rate constant and number of atoms
   (Examples 19.2, 19.3; Problems 11–22, 48, 50, 63)
**3.** Relate rate constant to half-life
   (Examples 19.2, 19.3; Problems 23–30, 47, 61)
**4.** Relate $\Delta m$ to $\Delta E$
   (Examples 19.4, 19.6; Problems 31, 32, 36–47, 49)
**5.** Calculate binding energies
   (Example 19.5; Problems 33–35)

## *Key Equations*

| | |
|---|---|
| Rate of decay | $\ln X_0/X = kt \qquad k = 0.693/t_{1/2} \qquad A = kN$ |
| Mass-energy | $\Delta E = 9.00 \times 10^{10} \dfrac{\text{kJ}}{\text{g}} \times \Delta m$ |

## *Key Terms*

| | | |
|---|---|---|
| activity | curie (Ci) | K-electron capture |
| alpha particle | fission | mass number |
| atomic number | fusion | positron |
| beta particle | half-life | radioactivity |
| binding energy | | |

## Summary Problem

Consider the isotopes of gallium.

(a) Write the nuclear symbol for Ga-67, which is used medically to scan for various tumors. How many protons are there in the nucleus? How many neutrons?

(b) Write the reaction for the decomposition of Ga-67 by K-electron capture.

(c) Write nuclear equations to represent the bombardment of Ga-64 by a neutron and the decay of the product nucleus by beta emission.

(d) Calculate $\Delta E$ in kJ when 1.00 g of Ga-71 (nuclear mass = 70.9476 amu) is bombarded by an $\alpha$-particle producing As-72 (nuclear mass = 71.9493 amu) and neutrons.

(e) What are the mass defect and binding energy of Ga-71?

(f) The half-life of Ga-67 is 77.9 hours. What is the rate constant for Ga-67? What is the activity (in curies) of a $5.00 \times 10^2$-mg sample of Ga-67?

(g) How long will it take Ga-67 to decay and lose 95.0% of its radioactivity?

### Answers

(a) $^{67}_{31}$Ga; 31 protons, 36 neutrons

(b) $^{67}_{31}$Ga $+ \, ^{0}_{-1}e \longrightarrow \, ^{67}_{30}$Zn

(c) $^{64}_{31}$Ga $+ \, ^{1}_{0}n \longrightarrow \, ^{65}_{31}$Ga $\longrightarrow \, ^{0}_{-1}e + \, ^{65}_{32}$Ge

(d) $3.32 \times 10^7$ kJ

(e) mass defect = 0.6249 g; binding energy = $5.62 \times 10^{10}$ kJ

(f) $k = 8.90 \times 10^{-3}$ hr$^{-1}$; $A = 3.00 \times 10^5$ Ci

(g) 337 hr $\approx$ 14 days

## Questions & Problems

### Nuclear Equations

**\*1.** Consider the "new" isotope $^{282}_{115}$X. Compare the product nuclei after K-capture and positron emission.

**\*2.** Follow the directions for Question 1 but compare the product nuclei after beta emission and positron emission.

**\*3.** The blood volume of a patient can be measured by using chromium-51, a positron emitter, administered as a solution of sodium chromate. Write the nuclear equation for the decay of chromium-51.

**\*4.** Phosphorus-32 is used in the treatment of leukemia. It decays by beta emission. Write the nuclear reaction for the decay of P-32.

**\*5.** Write balanced nuclear equations for
   **(a)** the fusion of two C-12 nuclei to give another nucleus and a neutron.
   **(b)** the fission of U-235 to give Ba-140, another nucleus, and an excess of two neutrons.
   **(c)** the alpha decay of Ra-210.

**\*6.** Write balanced nuclear equations for
   **(a)** the K-capture of Ar-37.
   **(b)** the fusion of two C-12 nuclei to give sodium-23 and another particle.
   **(c)** the fission of Pu-239 to give tin-130, another nucleus, and an excess of three neutrons.

**\*7.** Write balanced nuclear equations for the bombardment of
   **(a)** silver-109 with an alpha particle to form a single product.
   **(b)** curium (Cm)-246 with carbon-12 to produce four neutrons and another nucleus.
   **(c)** Mo-96 with deuterium, $^{2}_{1}$H, to produce a neutron and another nucleus.
   **(d)** a nucleus with a neutron to produce a proton and P-31.

**\*8.** Write balanced nuclear reactions for the bombardment of
   **(a)** Al-27 with deuterium, $^{2}_{1}$H, to produce an alpha particle and another nucleus.
   **(b)** californium (Cf)-250 with boron-11 to form four neutrons and another nucleus.
   **(c)** a nucleus with one neutron to form Mo-99.
   **(d)** copper-65 with carbon-12 to give three neutrons and another particle.

**\*9.** Balance the following equations by filling in the blanks.
   **(a)** $^{126}_{56}$Ba $+$ ⎯⎯⎯ $\longrightarrow \, ^{126}_{57}$ ⎯⎯⎯
   **(b)** ⎯⎯⎯ $\longrightarrow \, ^{0}_{1}e + \, ^{125}_{55}$⎯⎯⎯
   **(c)** $^{14}_{7}$N $+$ ⎯⎯⎯ $\longrightarrow \, ^{17}_{8}$O $+ \, ^{1}_{1}$ ⎯⎯⎯
   **(d)** ⎯⎯⎯ $+ \, ^{4}_{2}$⎯⎯⎯ $\longrightarrow \, ^{27}$ ⎯⎯ Si $+ \, ^{1}_{0}n$

*10. Balance the following nuclear equations by filling in the blanks.

(a) _____ $\longrightarrow$ $_{-1}^{0}e$ + $_{26}^{49}$ _____

(b) $_{}^{76}$ _____ Kr + $_{-1}^{0}e$ $\longrightarrow$ _____

(c) _____ + $_{2}^{4}$He $\longrightarrow$ $_{0}^{1}n$ + $_{6}^{12}$ _____

(d) $_{13}^{27}$Al + _____ $\longrightarrow$ $_{2}^{4}$ _____ + $_{11}^{24}$ _____

## Rate of Nuclear Decay

**11.** Food can be preserved by radiation with gamma rays. A $\gamma$-ray source has an activity of 2156 Ci. How many disintegrations are there per minute?

**12.** How many atoms decay in two hours in a 12.0-mCi (millicuries) source?

**13.** A scintillation counter registers emitted radiation caused by the disintegration of nuclides. If each atom of nuclide emits one count, what is the activity of a sample that registers 25,650 (four significant figures) disintegrations in ten minutes?

**14.** A Geiger counter counts 0.070% of all particles emitted by a sample. What is the activity that registers $12.3 \times 10^3$ counts in thirty seconds?

**15.** Krypton-87 has a rate constant of $1.5 \times 10^{-4}\,s^{-1}$. What is the activity of a 10.0-mg sample?

**16.** Yttrium-87 has a rate constant of $2.6 \times 10^{-6}\,s^{-1}$. What is the activity of a 1.00-mg sample?

**17.** Bromine-82 has a half-life of 36 hours. A sample containing Br-82 was found to have an activity of $8.7 \times 10^4$ disintegrations/min. How many grams of Br-82 were present in the sample? Assume that there were no other radioactive nuclides in the sample.

**18.** Lead-210 has a half-life of 20.4 years. A counter registers $7.5 \times 10^3$ disintegrations in three minutes. How many grams of Pb-210 are there?

**19.** Technetium-99 (at. mass = 98.9 amu) is used for bone scans. It has a half-life of 6.0 h. How many disintegrations/s can you expect from 1.00 mg of technetium-99?

**20.** I-131 is used to locate tumors in the thyroid glands. A 1.00-g sample of I-131 (at. mass = 130.9 amu) has an activity of $1.3 \times 10^5$ Ci. Calculate its decay constant and half-life in days.

**21.** Chlorine-36 decays by beta emission. It has a decay constant of $2.3 \times 10^{-6}\,yr^{-1}$. How many $\beta$-particles are emitted in half a minute from a 10.0-mg sample of Cl-36? How many curies does this represent?

**22.** Fluorine-18 has a decay constant of $6.31 \times 10^{-3}\,min^{-1}$. How many counts will one get on a Geiger counter in 30.0 seconds from 1.00 mg of fluorine-18? Assume that the sensitivity of the counter is such that it intercepts 0.10% of the emitted radiation.

**23.** The remains of an ancient cave were unearthed. Analysis of charcoal in the cave gives 12.0 disintegrations/min/g of carbon. The half-life of C-14 is 5720 years. Analysis of a tree cut down when the cave was unearthed showed 22.1 disintegrations/min/g of carbon. How old are the remains in the cave?

**24.** A "new" painting supposed to be by Michelangelo (1475–1564) is authenticated by C-14 dating. The C-14 content ($t_{1/2}$ = 5720 years) of the canvas is 0.972 times that in a living plant. Could the painting have been by Michelangelo?

**25.** The radioactive isotope tritium, H-3, is produced in nature in much the same way as C-14. Its half-life is 12.3 yr. Estimate the age of a sample of Scotch whiskey that has a tritium content 0.62 times that of the water in the area where the whiskey was produced.

**26.** The half-life of tritium, H-3, is 12.3 years. Estimate the ratio of the tritium of water in the area to the tritium in a bottle of wine claimed to be 19 years old.

**27.** What is the approximate age of a rock in which the mass ratio (grams) of U-238 to Pb-206 is 1.30? ($t_{1/2}$ of U-238 = $4.5 \times 10^9$ yr)

**28.** What is the approximate age of a rock in which the mole ratio of U-238 to Pb-206 is 1.30? Take the half-life of U-238 to be $4.5 \times 10^9$ yr.

**29.** One way of dating rocks is to determine the relative amounts of K-40 and Ar-40. The decay of K-40 has a half-life of $1.26 \times 10^9$ yr. Analysis of a certain lunar sample gives the following results in mole ratios.

$$^{40}Ar/^{40}K = 4.13 \qquad t_{1/2}\,^{40}K = 1.26 \times 10^9\ yr$$

$$^{206}Pb/^{238}U = 0.66 \qquad t_{1/2}\,^{238}U = 4.5 \times 10^9\ yr$$

$$^{87}Sr/^{87}Rb = 0.049 \qquad t_{1/2}\,^{87}Rb = 4.8 \times 10^{10}\ yr$$

Using these data, obtain the best possible value for the age of the sample. Can you suggest why the K-Ar method gives a low result?

**30.** The $^{87}$Rb to $^{87}$Sr method of dating rocks was used to analyze lunar samples from the Apollo-15 mission. Estimate the age of the lunar sample in which

(a) the mole ratio of $^{87}$Rb to $^{87}$Sr is 25.0 (see Problem 29).

(b) the mole ratio of $^{87}$Rb to $^{87}$Sr is 20.0.

## Mass Changes

**31.** For the $\alpha$-decay of Rn-222:

(a) Calculate $\Delta m$ in grams/mol of Rn-222.

(b) Calculate $\Delta E$ in kJ when one mole decays; when one gram decays.

**32.** Beryllium-10 decays by beta emission. Calculate $\Delta m$ (in grams) and $\Delta E$ (in kJ) when 5.00 mg decays.

**33.** For Bk-245, calculate

(a) the mass defect.

(b) the binding energy.

**34.** Which has the larger binding energy, Mg-26 or Al-26?

**35.** Which has the larger binding energy, fluorine-19 or oxygen-17?

**36.** Some of the sun's energy comes from the reaction

$$4 \, {}^1_1\text{H} \longrightarrow {}^4_2\text{He} + 2 \, {}^0_1 e$$

Calculate the energy change in this reaction per gram of hydrogen.

**37.** Some rocket engines get energy from the reaction of hydrazine, $N_2H_4(l)$, with oxygen to form nitrogen gas and steam. Calculate the change in mass when one mole of hydrazine burns.

**38.** The sun radiates energy into space at the rate of $3.9 \times 10^{26}$ J/s. Calculate the rate of mass loss by the sun.

**39.** For the fission reaction

$$ {}^1_0 n + {}^{235}_{92}\text{U} \longrightarrow {}^{89}_{37}\text{Rb} + {}^{144}_{58}\text{Ce} + 3 \, {}^{0}_{-1} e + 3 \, {}^1_0 n $$

**(a)** how much energy (in kJ) is given off per gram of ${}^{235}_{92}\text{U}$?

**(b)** how many kg of TNT must be detonated to produce the same amount of energy ($\Delta E = -2.76$ kJ/g.)

**(c)** How many grams of U-235 would have to react to produce the same amount of energy as one metric ton (1000 kg) of TNT?

**40.** Consider the fission reaction

$$ {}^1_0 n + {}^{239}_{94}\text{Pu} \longrightarrow {}^{90}_{38}\text{Sr} + {}^{146}_{58}\text{Ce} + 2 \, {}^{0}_{-1} e + 4 \, {}^1_0 n $$

**(a)** How many grams of ${}^{239}_{94}\text{Pu}$ would have to react to produce 1.00 kJ of energy?

**(b)** How many atoms of ${}^{239}_{94}\text{Pu}$ would have to react to produce 1.00 kJ of energy?

**41.** Show by calculation whether K-capture by Al-26 is spontaneous.

**42.** Show by calculation whether positron emission by Si-28 is spontaneous.

**43.** Consider the fission reaction in which U-235 is bombarded with a neutron. Cerium-144, bromine-87, electrons, and neutrons are produced.

**(a)** Write a balanced nuclear equation for the reaction.

**(b)** Calculate $\Delta E$ when one gram of U-235 undergoes fission.

**(c)** The decomposition of ammonium nitrate, an explosive, evolves 37.0 kJ/mol. How many kilograms of $NH_4NO_3$ are required to produce the same amount of energy as one milligram of U-235?

**44.** An explosion used five tons (1 ton = 2000 lb) of ammonium nitrate ($\Delta E = -37.0$ kJ/mol).

**(a)** How much energy was released by the explosion?

**(b)** How many grams of TNT ($\Delta E = -2.76$ kJ/g) are needed to release the energy calculated in (a)?

**(c)** How many grams of U-235 are needed to obtain the same amount of energy calculated in (a)? (See the equation in Problem 39.)

## Unclassified

**\*45.** Classify the following statements as true or false. If false, correct the statement to make it true.

**(a)** The mass number increases in beta emission.

**(b)** A radioactive species with a large rate constant, $k$, decays very slowly.

**(c)** Fusion gives off less energy per gram of fuel than fission.

**\*46.** Explain how

**(a)** alpha and beta radiation are separated by an electric field.

**(b)** radioactive C-11 can be used as a tracer to study brain disorders.

**(c)** a self-sustaining chain reaction occurs in nuclear fission.

**47.** Iodine-131 is used in the treatment of tumors in the thyroid gland. Its half-life is 8.1 days. If a patient is scheduled to receive 15.0 mg of I-131 monthly (every 30 days), how much I-131 is in his blood stream when he comes for his second dose?

**48.** How many disintegrations per second occur in a basement that is $40 \times 40 \times 10$ feet if the radiation level from radon is the allowed $4 \times 10^{-12}$ Ci/L?

**49.** A positron and an electron have the same mass but opposite charges. When a positron meets an electron, annihilation occurs; that is, both masses are converted entirely into energy in the form of $\gamma$-rays. Calculate

**(a)** the energy emitted by the annihilation of a mole of positrons with electrons.

**(b)** the energy of one gamma ray.

**50.** Smoke detectors contain small amounts of americium-241. Am-241 decays by emitting $\alpha$-particles and has a decay constant of $1.51 \times 10^{-3}$ yr$^{-1}$. If a smoke detector gives off ten disintegrations per second, how many grams of Am-241 are present in the detector?

**51.** The principle behind the home smoke detector is described on p. 536. Americium-241 is present in most of these detectors. It has a decay constant of $1.51 \times 10^{-3}$ yr$^{-1}$. You are urged to check the battery in the detector at least once a year. You are however, never encouraged to check how much Am-241 remains undecayed. Explain why.

**52.** Suppose the $^{14}\text{C}/^{12}\text{C}$ ratio in plants a thousand years ago was 10% higher than it is today. What effect, if any, would this have on the calculated age of an artifact found by the C-14 method to be a thousand years old?

**53.** The amount of oxygen dissolved in a sample of water can be determined by using thallium metal containing a small amount of the isotope Tl-204. When excess thallium is added to oxygen-containing water, the following reaction occurs:

$$ 2\text{Tl}(s) + \tfrac{1}{2}\text{O}_2(g) + \text{H}_2\text{O} \longrightarrow 2\text{Tl}^+(aq) + 2\text{OH}^-(aq) $$

It is found that, after reaction, the activity of a 25.0-mL water sample is 815 counts per minute (cpm), caused by the presence of Tl$^+$-204 ions. The activity of Tl-204 is $5.53 \times 10^5$ cpm per gram of thallium metal. Assuming that $O_2$ is the limiting reactant in the above equation, calculate its concentration in moles per liter.

**54.** A 35-mL sample of 0.050 $M$ $AgNO_3$ is mixed with 35 mL of 0.050 $M$ NaI labeled with I-131. The following reaction occurs.

$$Ag^+(aq) + I^-(aq) \longrightarrow AgI$$

The filtrate is found to have an activity of $2.50 \times 10^3$ counts per minute per milliliter. The 0.050 $M$ NaI solution had an activity of $1.25 \times 10^{10}$ counts per minute per milliliter. Calculate $K_{sp}$ for AgI.

**55.** A 100.0-g sample of water containing tritium, $^3_1H$, emits $4.22 \times 10^3$ beta particles per second. Tritium has a half-life of 12.3 years. What percentage of all the hydrogen atoms in the water sample is tritium?

**56.** Using the half-life of tritium given in Problem 55, calculate the activity in curies of 1.00 mL of $^3_1H_2$ at STP.

**57.** In order to measure the volume of the blood in an animal's circulatory system, the following experiment was performed. A 5.0-mL sample of an aqueous solution containing $1.7 \times 10^5$ counts per minute (cpm) of tritium was injected into the blood stream. After an adequate period of time to allow for the complete circulation of the tritium, a 5.0-mL sample of blood was withdrawn and found to have $1.3 \times 10^3$ cpm on the scintillation counter. What is the volume of the animal's circulatory system, assuming that only a negligible amount of tritium has decayed during the experiment?

**58.** One of the causes of the explosion at Chernobyl may have been the reaction between zirconium, which coated the fuel rods, and steam.

$$Zr(s) + 2H_2O(g) \longrightarrow ZrO_2(s) + 2H_2(g)$$

If half a metric ton of zirconium reacted, what pressure was exerted by the hydrogen gas produced at 55°C in the containment chamber, which had a volume of $2.0 \times 10^4$ L?

**59.** Consider the fission reaction

$$^1_0n + {}^{235}_{92}U \longrightarrow {}^{89}_{37}Rb + {}^{144}_{58}Ce + 3\,{}^{0}_{-1}e + 3\,{}^1_0n$$

How many liters of octane, $C_8H_{18}$, the primary component of gasoline, must be burned to $CO_2(g)$ and $H_2O(g)$ to produce as much energy as the fission of one gram of U-235 fuel? Octane has a density of 0.703 g/mL; its heat of formation is $-249.9$ kJ/mol.

**60.** A chelate of $Cr^{3+}$ and $C_2O_4^{2-}$ is made by a reaction that involves $Na_2CrO_4$ and oxalic acid, $H_2C_2O_4$, a reducing agent. The sodium chromate has an activity of 765 counts per minute per gram, from Cr-51. The oxalic acid has an activity of 512 counts per minute per gram. It is labeled with C-14. Since Cr-51 and C-14 emit different particles during decay, their activities can be counted independently. A sample of the chelate was found to have a Cr-51 count of 314 cpm and a C-14 count of 235 cpm. How many oxalate ions are bound to one $Cr^{3+}$ ion?

**61.** Polonium-210 decays to Pb-206 by alpha emission. Its half-life is 138 days. What volume of helium at 25°C and 1.20 atm would be obtained from a 25.00-g sample of Po-210 left to decay for 75.0 hours?

**62.** Radium-226 decays by alpha emission to radon-222. Suppose that 25.0% of the energy given off by one gram of radium is converted to electrical energy. What is the minimum mass of lithium that would be needed for the voltaic cell Li|Li$^+$||Cu$^{2+}$|Cu at standard conditions to produce the same amount of electrical work ($\Delta G°$)?

## Challenge Problems

**63.** An activity of 20 picocuries ($20 \times 10^{-12}$ Ci) of radon-222 per liter of air in a house constitutes a health hazard to anyone living in the house. Taking the half-life of radon-222 to be 3.82 days, calculate the concentration of radon in air (moles per liter) corresponding to this activity.

**64.** Plutonium-239 decays by the reaction

$$^{239}_{94}Pu \longrightarrow {}^{235}_{92}U + {}^4_2He$$

Its rate constant is $5.5 \times 10^{-11}$/min. In a one-gram sample of Pu-239,

**(a)** how many grams decompose in 25 minutes?

**(b)** how much energy in kilojoules is given off in 25 minutes?

**(c)** what radiation dosage in rems (p. 551) is received by a 75-kg man exposed to a gram of Pu-239 for 25 minutes?

**65.** It is possible to estimate the activation energy for fusion by calculating the energy required to bring two deuterons close enough to one another to form an alpha particle. This energy can be obtained by using Coulomb's law in the form $E = 8.99 \times 10^9 q_1 q_2/r$, where $q_1$ and $q_2$ are the charges of the deuterons ($1.60 \times 10^{-19}$ C), $r$ is the radius of the He nucleus, about $2 \times 10^{-15}$ m, and $E$ is the energy in joules.

**(a)** Estimate $E$ in joules per alpha particle.

**(b)** Using the equation $E = mv^2/2$, estimate the velocity (meters/second) each deuteron must have if a collision between two of them is to supply the activation energy for fusion ($m$ is the mass of the deuteron in kilograms).

**66.** Consider the reaction

$$2\,{}^2_1H \longrightarrow {}^4_2He$$

**(a)** Calculate $\Delta E$ in kilojoules per gram of deuterium fused.

**(b)** How much energy is potentially available from the fusion of all the deuterium in sea water? The percentage of deuterium in water is about 0.0017%. The total mass of water in the oceans is $1.3 \times 10^{24}$ g.

**(c)** What fraction of the deuterium in the oceans would have to be consumed to supply the annual energy requirements of the world ($2.3 \times 10^{17}$ kJ)?

Foods such as these and many others (Table 20.A, p. 575) supply trace amounts of metal cations that are essential to life. (Charles D. Winters)

# 20 Chemistry of the Metals

The fields of Nature long prepared and fallow
    the silent cyclic chemistry
The slow and steady ages plodding, the unoccupied surface
    ripening, the rich ores forming beneath;

                    **–WALT WHITMAN**

                  *Song of the Redwood Tree*

## CHAPTER OUTLINE

**20.1 Metallurgy**

**20.2 Reactions of the Alkali and Alkaline Earth Metals**

**20.3 Redox Chemistry of the Transition Metals**

**20.4 Alloys**

The diagonal line or "stairway" that runs from the left to the lower right of the periodic table separates metals from nonmetals; the 80 or more elements below and to the left of the line are metals. In discussing the descriptive chemistry of the metals, we concentrate upon

— the *main-group metals in Groups 1 and 2* at the far left of the periodic table. These are commonly referred to as the *alkali* metals (Group 1) and *alkaline earth* metals (Group 2). The group names reflect the strongly basic nature of the oxides ($K_2O$, $CaO$, . . .) and hydroxides ($KOH$, $Ca(OH)_2$, . . .) of these elements.

— *the transition metals,* located in the center of the periodic table. There are three series of transition metals, each consisting of ten elements, located in the fourth, fifth, and sixth periods. We focus on a few of the more important transition metals (Fig. 20.1), particularly those toward the right of the first series.

Section 20.1 deals with the processes by which these metals are obtained from their principal ores. Section 20.2 describes the reactions of the alkali and alkaline earth metals, particularly those with hydrogen, oxygen, and water. Section 20.3 considers the redox chemistry of the transition metals, their cations (e.g., $Fe^{2+}$, $Fe^{3+}$) and their oxoanions (e.g., $CrO_4^{2-}$). The last section of the chapter deals with alloys of the metals (Section 20.4).

## 20.1 Metallurgy

The processes by which metals are extracted from their ores fall within the science of metallurgy. As you might expect, the chemical reactions involved depend upon the type of ore (Fig. 20.2 p. 558). We consider some typical processes used to obtain metals from chloride, oxide, sulfide, or "native" ores.

An ore is a natural source from which a metal can be extracted profitably

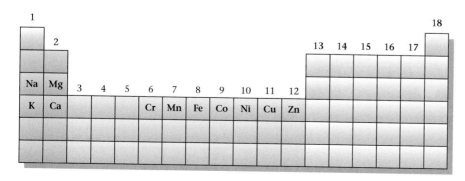

**Figure 20.1**

The metals considered in this chapter are those in Groups 1 and 2 (red) and the transition metals (blue). Symbols are shown for the more common metals.

**Figure 20.2**
Principal ores of the Group 1, Group 2, and transition metals.

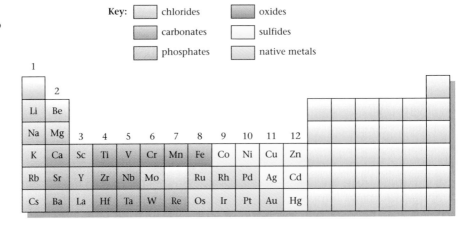

## Chloride Ores: Na from NaCl

Sodium metal is obtained by the electrolysis of molten sodium chloride (Fig. 20.3). The electrode reactions are quite simple:

$$
\begin{aligned}
\text{cathode:} \quad & 2Na^+(l) + 2e^- \longrightarrow 2Na(l) \\
\text{anode:} \quad & \underline{2Cl^-(l) \longrightarrow Cl_2(g) + 2e^-} \\
& 2NaCl(l) \longrightarrow 2Na(l) + Cl_2(g)
\end{aligned}
$$

The cell is operated at about 600°C to keep the electrolyte molten; calcium chloride is added to lower the melting point. About 14 kJ of electrical energy is required to produce one gram of sodium (Example 20.1), which is drawn off as a

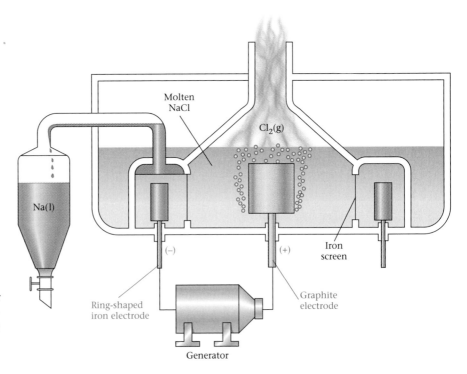

**Figure 20.3**
Electrolysis of molten sodium chloride, containing some $CaCl_2$ to lower the melting point. The iron screen is used to prevent sodium and chlorine from coming in contact with each other.

liquid (mp of Na = 98°C). The chlorine gas produced at the anode is a valuable by-product.

---

**Example 20.1** Taking $\Delta H°$ and $\Delta S°$ for the reaction

$$2NaCl(l) \longrightarrow 2Na(l) + Cl_2(g)$$

to be +820 kJ and +0.180 kJ/K, respectively, calculate

(a) $\Delta G°$ at the electrolysis temperature, 600°C.
(b) the voltage required to carry out the electrolysis.

*Strategy* To find $\Delta G°$, apply the Gibbs-Helmholtz equation, $\Delta G° = \Delta H° - T\Delta S°$ (Chap. 17). To find $E°$, use the relation $\Delta G° = -nFE°$ (Chap. 18).

*Solution*

(a) $\Delta G° = +820$ kJ $- 873$ K(0.180 kJ/K) $= \boxed{+663 \text{ kJ}}$

This is the free energy change for the formation of two moles of sodium (45.98 g); the electrical energy required per gram is 663 kJ/45.98 g = 14.4 kJ/g.

(b) $E° = \dfrac{-\Delta G°}{nF} = \dfrac{-6.63 \times 10^5 \text{ J}}{(2 \text{ mol})(9.648 \times 10^4 \text{ J/mol} \cdot \text{V})} = -3.44$ V

At least $\boxed{3.44 \text{ V}}$ must be applied to carry out the electrolysis.

---

## Oxide Ores: Al from $Al_2O_3$, Fe from $Fe_2O_3$

Oxides of very reactive metals such as calcium or aluminum are reduced by electrolysis. In the case of aluminum, bauxite ore, $Al_2O_3$, is used:

$$2Al_2O_3(l) \longrightarrow 4Al(l) + 3O_2(g)$$

Cryolite, $Na_3AlF_6$, is added to $Al_2O_3$ to produce a mixture melting at about 1000°C. (A mixture of $AlF_3$, NaF, and $CaF_2$ may be substituted for cryolite.) The cell is heated electrically to keep the mixture molten so that ions can move through it, carrying the electric current. About 30 kJ of electrical energy is consumed per gram of aluminum formed. The high energy requirement explains in large part the value of recycling aluminum cans.

The process for obtaining aluminum from bauxite was worked out in 1886 by Charles Hall, a chemistry student at Oberlin College. The problem that Hall faced was to find a way to electrolyze $Al_2O_3$ at a temperature below its melting point of 2000°C. His general approach was to look for ionic compounds in which $Al_2O_3$ would dissolve at a reasonable temperature. After several unsuccessful attempts, Hall found that cryolite was the ideal "solvent." Curiously enough, the same electrolytic process was worked out by Heroult in France, also in 1886. Each young man (both were 22 years old) was entirely unaware of the other's work!

Chemistry majors can be very productive

With less active metals, a chemical reducing agent can be used to reduce a metal cation to the element. The most common reducing agent in metallurgical processes is carbon, in the form of coke or, more exactly, carbon monoxide formed from the coke.

The most important metallurgical process involving carbon is the reduction of hematite ore, which consists largely of iron(III) oxide, $Fe_2O_3$, mixed with sil-

**Figure 20.4**

Blast furnace for the production of pig iron (a). The carbon content of the crude iron is lowered by heating with oxygen in a basic furnace (b), forming steel. (Courtesy of Bethlehem Steel)

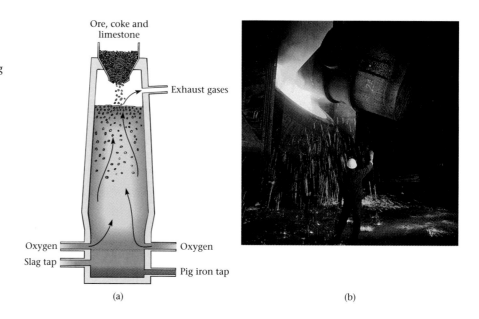

(a)　　　　　　　　　　(b)

icon dioxide, $SiO_2$ (sand). Reduction occurs in a blast furnace (Fig. 20.4a) typically 30 m high and 10 m in diameter. The solid charge admitted at the top of the furnace consists of iron ore, coke, and limestone ($CaCO_3$). To get the process started, a blast of compressed air or pure $O_2$ at 500°C is blown into the furnace through nozzles near the bottom. Several different reactions occur, of which three are most important:

**1. *Conversion of carbon to carbon monoxide.*** In the lower part of the furnace, coke burns to form carbon dioxide, $CO_2$. As the $CO_2$ rises through the solid mixture, it reacts further with the coke to form carbon monoxide, $CO$. The overall reaction is

$$2C(s) + O_2(g) \longrightarrow 2CO(g) \qquad\qquad \Delta H = -221 \text{ kJ}$$

The heat given off by this reaction maintains a high temperature within the furnace.

**2. *Reduction of $Fe^{3+}$ ions to Fe.*** The carbon monoxide reacts with the iron(III) oxide in the ore:

$$Fe_2O_3(s) + 3CO(g) \longrightarrow 2Fe(l) + 3CO_2(g)$$

Molten iron, formed at a temperature of 1600°C, collects at the bottom of the furnace. Four or five times a day, it is drawn off. The daily production of iron from a single blast furnace is about 1000 metric tons. This is enough to make 500 Cadillacs or 1000 Hyundais.

**3. *Formation of slag.*** The limestone added to the furnace decomposes at about 800°C:

$$CaCO_3(s) \longrightarrow CaO(s) + CO_2(g)$$

The calcium oxide formed reacts with impurities in the iron ore to form a glassy material called **slag.** The main reaction is with $SiO_2$ to form calcium silicate, $CaSiO_3$:

$$CaO(s) + SiO_2(s) \longrightarrow CaSiO_3(l)$$

The slag, which is less dense than molten iron, forms a layer on the surface of the metal. This makes it possible to draw off the slag through an opening in the furnace above that used to remove the iron. The slag is used to make cement and as a base in road construction.

The product that comes out of the blast furnace, called "pig iron," is highly impure. On the average, it contains about 4% carbon along with lesser amounts of silicon, manganese, and phosphorus. To make steel from pig iron, the carbon content must be lowered below 2%. Most of the steel produced in the world today is made by the basic oxygen process (Fig. 20.4b). The "converter" is filled with a mixture of about 70% molten iron from the blast furnace, 25% scrap iron or steel, and 5% limestone. Pure oxygen under a pressure of about 10 atm is blown through the molten metal. The reaction

*Where do you suppose the phrase "pig iron" came from?*

$$C(s) + O_2(g) \longrightarrow CO_2(g)$$

occurs rapidly; at the same time, impurities such as silicon are converted to oxides, which react with limestone to form a slag. When the carbon content drops to the desired level, the supply of oxygen is cut off. At this stage, the steel is ready to be poured. The whole process takes from 30 min to 1 h and yields about 200 metric tons of steel in a single "blow."

---

**Example 20.2**  Write balanced equations for the reduction of each of the following oxide ores by carbon monoxide:

   (a) ZnO     (b) $MnO_2$     (c) $Fe_3O_4$

*Strategy*  In each case, the products are solid metal and $CO_2(g)$. The equations are most simply balanced by the process described in Chapter 3.

*Solution*

   (a) $ZnO(s) + CO(g) \longrightarrow Zn(s) + CO_2(g)$
   (b) $MnO_2(s) + 2CO(g) \longrightarrow Mn(s) + 2CO_2(g)$
   (c) $Fe_3O_4(s) + 4CO(g) \longrightarrow 3Fe(s) + 4CO_2(g)$

---

## Sulfide Ores: Cu from $Cu_2S$

Sulfide ores, after preliminary treatment, most often undergo roasting, i.e., heating with air or pure oxygen. With a relatively reactive transition metal such as zinc, the product is the oxide

$$2ZnS(s) + 3\,O_2(g) \longrightarrow 2ZnO(s) + 2SO_2(g)$$

which can then be reduced to the metal with carbon. With sulfides of less reactive metals such as copper or mercury, the free metal is formed directly upon roasting. The reaction with cinnabar, the sulfide ore of mercury, is

$$HgS(s) + O_2(g) \longrightarrow Hg(g) + SO_2(g)$$

Among the several ores of copper, one of the most important is chalcocite, which contains copper(I) sulfide, $Cu_2S$, in highly impure form. Rocky material typically lowers the fraction of copper in the ore to 1% or less. The $Cu_2S$ is concentrated by a process called *flotation* (Fig. 20.5), which raises the fraction of copper to 20–40%. The concentrated ore is then converted to the metal by blow-

*Because Cu can be produced by simple "roasting", it was one of the first metals known*

**Figure 20.5**
Low-grade sulfide ores, including
$Cu_2S$, are often concentrated by flotation. The finely divided sulfide particles are trapped in soap bubbles,
while the rocky waste sinks to the
bottom and is discarded.

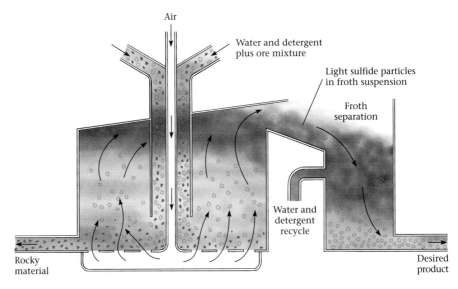

ing air through it at a high temperature, typically above 1000°C. (Pure $O_2$ is often used instead of air.) The overall reaction that occurs is a simple one:

$$Cu_2S(s) + O_2(g) \longrightarrow 2Cu(s) + SO_2(g)$$

The solid produced is called "blister copper." It has an irregular appearance due
to air bubbles that enter the copper while it is still molten. Blister copper is impure, containing small amounts of several other metals.

Copper is purified by electrolysis. The anode, which may weigh as much as
300 kg, is made of blister copper. The electrolyte is 0.5 to 1.0 $M$ $CuSO_4$, adjusted
to a pH of about 0 with sulfuric acid. The cathode is a piece of pure copper,
weighing perhaps 150 kg. The half-reactions are

$$\text{oxidation:} \quad Cu(s, \text{impure}) \longrightarrow Cu^{2+}(aq) + 2e^-$$

$$\text{reduction:} \quad Cu^{2+}(aq) + 2e^- \longrightarrow Cu(s, \text{pure})$$

The overall reaction, obtained by adding these two half-reactions, is

$$Cu(s, \text{impure}) \longrightarrow Cu(s, \text{pure})$$

Thus, the net effect of electrolysis is to transfer copper metal from the impure
blister copper used as one electrode to the pure copper sheet used as the other
electrode. Electrolytic copper is 99.95% pure.

---

**Example 20.3**    A major ore of bismuth is bismuth(III) sulfide, $Bi_2S_3$. Upon roasting in air, it is converted to the corresponding oxide; sulfur dioxide is a by-product.
What volume of $SO_2$, at 25°C and 1.00 atm, is formed from one metric ton ($10^6$ g) of
ore containing 1.25% $Bi_2S_3$?

***Strategy***    First, write a balanced equation for the reaction, which is very similar to
that for ZnS, except that $Zn^{2+}$ is replaced by $Bi^{3+}$. Using the balanced equation, calculate the number of moles of $SO_2$. Finally, use the ideal gas law to calculate the volume
of $SO_2$.

## Solution

(1) The product is $Bi_2O_3(s)$; the equation is

$$2Bi_2S_3(s) + 9O_2(g) \longrightarrow 2Bi_2O_3(s) + 6SO_2(g)$$

(2) $n_{SO_2} = 0.0125 \times 10^6$ g $Bi_2S_3 \times \dfrac{1 \text{ mol } Bi_2S_3}{514.2 \text{ g } Bi_2S_3} \times \dfrac{6 \text{ mol } SO_2}{2 \text{ mol } Bi_2S_3} = 72.9$ mol $SO_2$

(3) $V_{SO_2} = \dfrac{nRT}{P} = \dfrac{(72.9 \text{ mol})(0.0821 \text{ L} \cdot \text{atm/mol} \cdot \text{K})(298 \text{ K})}{1.00 \text{ atm}} = \boxed{1.78 \times 10^3 \text{ L}}$

The roasting of sulfide ores is a major source of the gaseous pollutant $SO_2$, which contributes to acid rain. (Chap. 13).

---

## Native Metals: Au

A few very unreactive metals, notably silver and gold, are found in nature in elemental form, mixed with large amounts of rocky material. For countless centuries, people have extracted gold by taking advantage of its high density (19.3 g/mL). In ancient times, gold-bearing sands were washed over sheepskins, which retained the gold; this is believed to be the source of the "Golden Fleece" of Greek mythology. The forty-niners in California obtained gold by swirling gold-bearing sands with water in a pan. Less dense impurities were washed away, leaving gold nuggets or flakes at the bottom of the pan.

Nowadays, the gold content of ores is much too low for these simple mechanical separation methods to be effective. Instead, the ore is treated with very dilute (0.01 *M*) sodium cyanide solution, through which air is blown. The following redox reaction takes place:

*Gold at 10 ppm can be extracted profitably*

$$4Au(s) + 8CN^-(aq) + O_2(g) + 2H_2O \longrightarrow 4Au(CN)_2{}^-(aq) + 4OH^-(aq)$$

The oxidizing agent is $O_2$, which takes gold to the +1 state. The cyanide ion acts as a complexing ligand, forming the stable $Au(CN)_2{}^-$ ion. Metallic gold is recovered from solution by adding zinc; the gold in the complex ion is reduced to the metal.

$$Zn(s) + 2Au(CN)_2{}^-(aq) \longrightarrow Zn(CN)_4{}^{2-}(aq) + 2Au(s)$$

## 20.2 Reactions of the Alkali and Alkaline Earth Metals

The metals in Groups 1 and 2 are among the most reactive of all elements (Table 20.1 p. 564). Their low ionization energies and high $E°_{ox}$ values explain why they are so readily oxidized to cations, +1 cations for Group 1, +2 for Group 2. The alkali metals and the heavier alkaline earth metals (Ca, Sr, Ba) are commonly stored under dry mineral oil or kerosene to prevent them from reacting with oxygen or water vapor in the air. Magnesium is less reactive; it is commonly available in the form of ribbon or powder. Beryllium, as one would expect from its position in the periodic table, is the least metallic element in these two groups. It is also the least reactive toward water, oxygen, or other nonmetals.

*Not under water!*

### TABLE 20.1 Reactions of the Alkali and Alkaline Earth Metals

| Reactant | Product | Comments |
|---|---|---|
| | **Alkali Metals (M)** | |
| $H_2(g)$ | $MH(s)$ | Upon heating in hydrogen gas |
| $X_2(g)$ | $MX(s)$ | X = F, Cl, Br, I |
| $N_2(g)$ | $M_3N(s)$ | Only Li reacts; product contains $N^{3-}$ ion |
| $S(s)$ | $M_2S(s)$ | Upon heating |
| $O_2(g)$ | $M_2O(s)$ | Li; product contains $O^{2-}$ ion |
| | $M_2O_2(s)$ | Na; product contains $O_2^{2-}$ ion |
| | $MO_2(s)$ | K, Rb, Cs; product contains $O_2^{-}$ ion |
| $H_2O(l)$ | $H_2(g), M^+, OH^-$ | Violent reaction with Na, K |
| | **Alkaline Earth Metals (M)** | |
| $H_2(g)$ | $MH_2(s)$ | All except Be; heating required |
| $X_2(g)$ | $MX_2(s)$ | Any halogen |
| $N_2(g)$ | $M_3N_2(s)$ | All except Be; heating required |
| $S(s)$ | $MS(s)$ | Upon heating |
| $O_2(g)$ | $MO(s)$ | All; product contains $O^{2-}$ ion |
| | $MO_2(s)$ | Ba; product contains $O_2^{2-}$ ion |
| $H_2O(l)$ | $H_2(g), M^{2+}, OH^-$ | Ca, Sr, Ba |

**Example 20.4**   Write balanced equations for the reaction of

(a) sodium with hydrogen.     (b) barium with oxygen.

***Strategy***   The formulas of the products can be deduced from Table 20.1. Note that barium forms two products with oxygen, $BaO$ and $BaO_2$.

***Solution***

(a)   $2Na(s) + H_2(g) \longrightarrow 2NaH(s)$

(b)   $2Ba(s) + O_2(g) \longrightarrow 2BaO(s)$
$Ba(s) + O_2(g) \longrightarrow BaO_2(s)$

## Hydrogen

The compounds formed by the reaction of hydrogen with the alkali and alkaline earth metals contain $H^-$ ions; for example, sodium hydride consists of $Na^+$ and $H^-$ ions. These white crystalline solids are often referred to as saline hydrides because of their physical resemblance to NaCl. Chemically, they behave quite differently from sodium chloride; for example, they react with water to produce hydrogen gas. Typical reactions are

$$NaH(s) + H_2O \longrightarrow H_2(g) + Na^+(aq) + OH^-(aq)$$

$$CaH_2(s) + 2H_2O \longrightarrow 2H_2(g) + Ca^{2+}(aq) + 2OH^-(aq)$$

In this way, saline hydrides can serve as compact, portable sources of hydrogen gas for inflating life rafts and balloons.

**Figure 20.6**
When a *very small* piece of sodium is added to water, it reacts violently:

$Na(s) + H_2O \longrightarrow$
$\quad Na^+(aq) + OH^-(aq) + \frac{1}{2}H_2(g)$

The $OH^-$ ions formed turn the organic dye phenolphthalein from colorless to pink. (Marna G. Clarke)

## Water

The alkali metals react vigorously with water (Fig. 20.6) to evolve hydrogen and form a water solution of the alkali hydroxide. The reaction of sodium is typical:

$$2Na(s) + 2H_2O(l) \longrightarrow 2Na^+(aq) + 2OH^-(aq) + H_2(g) \qquad \Delta H^\circ = -368.6 \text{ kJ}$$

The heat evolved in the reaction frequently causes the hydrogen to ignite.

Among the Group 2 metals, Ca, Sr, and Ba react with water in much the same way as the alkali metals. The reaction with calcium is

$$Ca(s) + 2H_2O \longrightarrow Ca^{2+}(aq) + 2OH^-(aq) + H_2(g)$$

Beryllium does not react with water at all. Magnesium reacts very slowly with boiling water but reacts more readily with steam at high temperatures:

$$Mg(s) + H_2O(g) \longrightarrow MgO(s) + H_2(g)$$

This reaction, like that of sodium with water, produces enough heat to ignite the hydrogen. Firefighters who try to put out a magnesium fire by spraying water on it have discovered this reaction, often with tragic results. The best way to extinguish burning magnesium is to dump dry sand on it.

Milk, although not very photogenic, is a good source of $Ca^{2+}$ ions.

## Oxygen

Note from Table 20.1 that several different products are possible when an alkali or alkaline earth metal reacts with oxygen. The product may be a normal oxide ($O^{2-}$ ion), a peroxide ($O_2^{2-}$ ion), or superoxide ($O_2^-$ ion).

$$\left[ :\ddot{O}: \right]^{2-} \qquad \left[ :\ddot{O}\!-\!\ddot{O}: \right]^{2-} \qquad \left[ :\ddot{O}\!-\!\ddot{O}\cdot \right]^{-}$$

oxide ion          peroxide ion          superoxide ion

The superoxide ion has an unpaired electron; $KO_2$ is paramagnetic

Lithium is the only Group 1 metal that forms the normal oxide in good yield by direct reaction with oxygen. The other Group 1 oxides ($Na_2O$, $K_2O$, $Rb_2O$, $Cs_2O$) must be prepared by other means. In contrast, the Group 2 metals usually react with oxygen to give the normal oxide. Beryllium and magnesium must be heated strongly to give BeO and MgO (Fig. 20.7). Calcium and strontium react more readily to give CaO and SrO. Barium, the most reactive of the Group 2 metals, catches fire when exposed to moist air. The product is a mixture of the normal oxide BaO ($Ba^{2+}$, $O^{2-}$ ions) and the peroxide $BaO_2$ ($Ba^{2+}$, $O_2^{2-}$ ions). The higher the temperature, the greater the fraction of BaO in the mixture.

The oxides of these metals react with water to form hydroxides:

$$Li_2O(s) + H_2O(l) \longrightarrow 2LiOH(s)$$

$$CaO(s) + H_2O(l) \longrightarrow Ca(OH)_2(s)$$

The oxides of the Group 1 and Group 2 metals are sometimes referred to as "basic anhydrides" (bases without water) because of this reaction. The reaction with CaO is referred to as the "slaking" of lime; it gives off 65 kJ of heat per mole of $Ca(OH)_2$ formed. A similar reaction with MgO takes place slowly to form $Mg(OH)_2$, the antacid commonly referred to as "milk of magnesia."

When sodium burns in air, the principal product is yellowish sodium peroxide, $Na_2O_2$:

$$2Na(s) + O_2(g) \longrightarrow Na_2O_2(s)$$

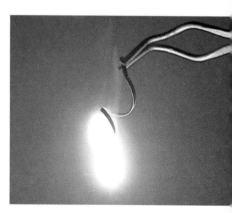

**Figure 20.7**
A piece of magnesium ribbon bursts into flame when heated in air; the product of the reaction is MgO(s).

Addition of sodium peroxide to water gives hydrogen peroxide, $H_2O_2$:

$$Na_2O_2(s) + 2H_2O \longrightarrow 2Na^+(aq) + 2OH^-(aq) + H_2O_2(aq)$$

Through this reaction, sodium peroxide finds use as a bleaching agent in the pulp and paper industry.

The heavier alkali metals (K, Rb, Cs) form the superoxide when they burn in air. For example,

$$K(s) + O_2(g) \longrightarrow KO_2(s)$$

Potassium superoxide is used in self-contained breathing devices for firefighters and miners. It reacts with the moisture in exhaled air to generate oxygen:

$$4KO_2(s) + 2H_2O(g) \longrightarrow 3O_2(g) + 4KOH(s)$$

The carbon dioxide in the exhaled air is removed by reaction with the KOH formed:

$$KOH(s) + CO_2(g) \longrightarrow KHCO_3(s)$$

A person using a mask charged with $KO_2$ can rebreathe the same air for an extended period of time. This allows that person to enter an area where there are poisonous gases or oxygen-deficient air.

---

**Example 20.5**  Consider the compounds strontium hydride, radium peroxide, and cesium superoxide.

(a) Give the formulas of these compounds.
(b) Write equations for the formation of these compounds from the elements.
(c) Write equations for the reactions of strontium hydride and radium peroxide with water.

***Strategy***  The formulas can be deduced from the charges of the ions ($Sr^{2+}$, $Ra^{2+}$, $Cs^+$; $H^-$, $O_2^{2-}$, $O_2^-$). If you know the formulas, the equations are readily written. In part (c), note that

— hydrides on reaction with water give $H_2(g)$ and a solution of the metal hydroxide
— peroxides on reaction with water give $H_2O_2(aq)$ and a solution of the metal hydroxide

*Solution*

$RaO_2$ is a peroxide; $CsO_2$ is a superoxide. Explain

(a) $SrH_2$, $RaO_2$, $CsO_2$
(b) $Sr(s) + H_2(g) \longrightarrow SrH_2(s)$
   $Ra(s) + O_2(g) \longrightarrow RaO_2(s)$
   $Cs(s) + O_2(g) \longrightarrow CsO_2(s)$
(c) $SrH_2(s) + 2H_2O \longrightarrow 2H_2(g) + Sr^{2+}(aq) + 2OH^-(aq)$
   $RaO_2(s) + 2H_2O \longrightarrow H_2O_2(aq) + Ra^{2+}(aq) + 2OH^-(aq)$

---

## 20.3  Redox Chemistry of the Transition Metals

The transition metals, unlike those in Groups 1 and 2, typically show several different oxidation numbers in their compounds. This tends to make their redox chemistry more complex (and more colorful). Only in the lower oxidation states (+1, +2, +3) are the transition metals present as cations (e.g., $Ag^+$, $Zn^{2+}$, $Fe^{3+}$). In higher oxidation states (+4 to +7) a transition metal is covalently bonded to a nonmetal atom, most often oxygen.

## Reactions of the Transition Metals with Oxygen

Table 20.2 lists the formulas of the oxides formed when the more common transition metals react with oxygen. With one exception ($Cu_2O$), the transition metal is present as a +2 and/or a +3 ion. In $Co_3O_4$ and in other oxides of general formula $M_3O_4$, there are two different types of cations: +2 and +3. To be specific, there are twice as many +3 as +2 cations; we might show the composition of $Co_3O_4$ as

$$1\ Co^{2+} : 2\ Co^{3+}\ ;\ 4\ O^{2-}$$

We should point out that many oxides of the transition metals beyond those listed in Table 20.2 can be prepared indirectly. For example, although silver does not react directly with oxygen, silver(I) oxide, $Ag_2O$, can be made by treating a solution of a silver salt with strong base.

$$2Ag^+(aq) + 2OH^-(aq) \longrightarrow Ag_2O(s) + H_2O$$

In another case, cobalt(II) oxide can be prepared by heating the carbonate in the absence of air.

$$CoCO_3(s) \longrightarrow CoO(s) + CO_2(g)$$

Many oxides of the transition metals can be *nonstoichiometric*; the atom ratios of the elements differ slightly from whole-number values. Stoichiometric nickel(II) oxide, in which the atom ratio is exactly 1:1 (i.e., $Ni_{1.00}O_{1.00}$), is a green, nonconducting solid. If this compound is heated to 1200°C in pure oxygen, it turns black and undergoes a change in composition to something close to $Ni_{0.97}O_{1.00}$. This nonstoichiometric solid was one of the early semiconductors; it has the structure shown in Figure 20.8. A few $Ni^{2+}$ ions are missing from the crystal lattice; to maintain electrical neutrality, two $Ni^{2+}$ ions are converted to $Ni^{3+}$ ions.

Nonstoichiometry is relatively common among "mixed" metal oxides, in which more than one metal is present. In 1986 it was discovered that certain compounds of this type showed the phenomenon of *superconductivity*; upon cooling to about 100 K, their electrical resistance drops to zero (Figure 20.9 p. 568). A typical formula here is $YBa_2Cu_3O_x$, where $x$ varies from 6.5 to 7.2, depending upon the method of preparation of the solid.

## Reaction of Transition Metals With Acids

Any metal with a positive standard oxidation voltage, $E^\circ_{ox}$, can be oxidized by the $H^+$ ions present in a 1 $M$ solution of a strong acid. All the transition metals in the left column of Table 20.3 react spontaneously with dilute solutions of such strong acids as HCl, HBr, and $H_2SO_4$. The products are hydrogen gas and a cation of the transition metal. A typical reaction is that of nickel:

$$Ni(s) + 2H^+(aq) \longrightarrow Ni^{2+}(aq) + H_2(g)$$

The end result of this reaction is shown in Figure 20.10, p. 569.

TABLE 20.2 **Products of Reactions of the Transition Metals with Oxygen**

| Metal | Product |
|-------|---------|
| Cr | $Cr_2O_3$ |
| Mn | $Mn_3O_4$ |
| Fe | $Fe_2O_3$, $Fe_3O_4$ |
| Co | $Co_3O_4$ |
| Ni | $NiO$ |
| Cu | $Cu_2O$, $CuO$ |
| Zn | $ZnO$ |
| Ag | — |
| Cd | $CdO$ |
| Au | — |
| Hg | $HgO$ |

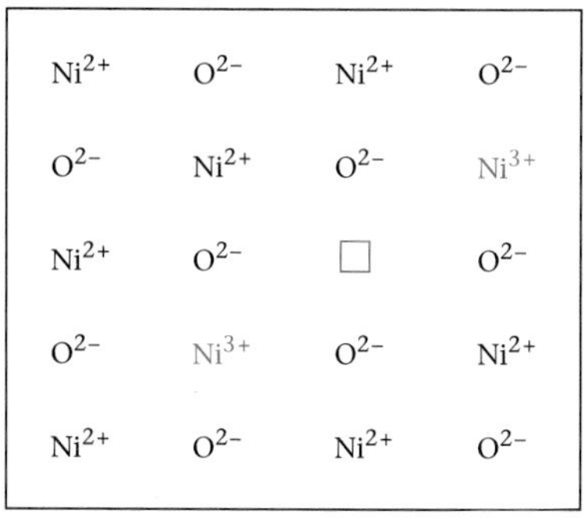

| | | | |
|---|---|---|---|
| $Ni^{2+}$ | $O^{2-}$ | $Ni^{2+}$ | $O^{2-}$ |
| $O^{2-}$ | $Ni^{2+}$ | $O^{2-}$ | $Ni^{3+}$ |
| $Ni^{2+}$ | $O^{2-}$ | $\square$ | $O^{2-}$ |
| $O^{2-}$ | $Ni^{3+}$ | $O^{2-}$ | $Ni^{2+}$ |
| $Ni^{2+}$ | $O^{2-}$ | $Ni^{2+}$ | $O^{2-}$ |

**Figure 20.8**
In nonstoichiometric nickel (II) oxide, a few $Ni^{2+}$ ions are missing from the crystal lattice. For every $Ni^{2+}$ ion removed, two others are converted to $Ni^{3+}$ ions.

**Figure 20.9**
A pellet of superconducting material, previously cooled to 77 K with liquid nitrogen, floats above a magnet.
(Courtesy of Edmund Scientific)

With metals that can form more than one cation, such as iron, the product upon reaction with $H^+$ in the absence of air is ordinarily the cation of lower charge, e.g., $Fe^{2+}$:

$$Fe(s) + 2H^+(aq) \longrightarrow Fe^{2+}(aq) + H_2(g)$$

Metals with negative values of $E_{ox}^\circ$, listed at the right of Table 20.3, are too inactive to react with hydrochloric acid. The $H^+$ ion is not a strong enough oxidizing agent to convert a metal such as copper ($E_{ox}^\circ = -0.339$ V) to a cation. However, copper can be oxidized by nitric acid. Here, the oxidizing agent is the nitrate ion, $NO_3^-$, which may be reduced to $NO_2$ or $NO$:

$$3Cu(s) + 8H^+(aq) + 2NO_3^-(aq) \longrightarrow 3Cu^{2+}(aq) + 2NO(g) + 4H_2O$$

$$E^\circ = E_{ox}^\circ \ Cu + E_{red}^\circ \ NO_3^- = -0.339V + 0.964V = +0.625V$$

TABLE 20.3    **Ease of Oxidation of Transition Metals**

| Metal | | Cation | $E_{ox}^\circ$ (V) | Metal | | Cation | $E_{ox}^\circ$ (V) |
|---|---|---|---|---|---|---|---|
| Mn | $\longrightarrow$ | $Mn^{2+}$ | +1.182 | Cu | $\longrightarrow$ | $Cu^{2+}$ | −0.339 |
| Cr | $\longrightarrow$ | $Cr^{2+}$ | +0.912 | Ag | $\longrightarrow$ | $Ag^+$ | −0.799 |
| Zn | $\longrightarrow$ | $Zn^{2+}$ | +0.762 | Hg | $\longrightarrow$ | $Hg^{2+}$ | −0.852 |
| Fe | $\longrightarrow$ | $Fe^{2+}$ | +0.409 | Au | $\longrightarrow$ | $Au^{3+}$ | −1.498 |
| Cd | $\longrightarrow$ | $Cd^{2+}$ | +0.402 | | | | |
| Co | $\longrightarrow$ | $Co^{2+}$ | +0.282 | | | | |
| Ni | $\longrightarrow$ | $Ni^{2+}$ | +0.236 | | | | |

**Example 20.6**    Write balanced equations for the reactions, if any, at standard concentrations of

  (a) chromium with hydrochloric acid.      (b) silver with nitric acid.

***Strategy***    Use standard potentials to decide whether reaction will occur. Note that with HCl only the $H^+$ ion can be reduced (to $H_2$); in $HNO_3$, the $NO_3^-$ ion can be reduced to NO ($E_{red}^\circ = +0.964$ V).

**Figure 20.10**
Nickel reacts slowly with hydrochloric acid to form $H_2(g)$ and $Ni^{2+}$ ions in solution. Evaporation of the solution formed gives green crystals of $NiCl_2 \cdot 6H_2O$. (Charles D. Winters)

## Solution

(a)
$$Cr(s) \longrightarrow Cr^{2+}(aq) + 2e^- \qquad E^{\circ}_{ox} = +0.912 \text{ V}$$
$$2H^+(aq) + 2e^- \longrightarrow H_2(g) \qquad E^{\circ}_{red} = 0.000 \text{ V}$$

Upon exposure to air, $Cr^{2+}$ is oxidized to $Cr^{3+}$

The reaction is

$$Cr(s) + 2H^+(aq) \longrightarrow Cr^{2+}(aq) + H_2(g) \qquad E^{\circ} = +0.912 \text{ V}$$

If chromium metal is added to hydrochloric acid in the absence of air, it slowly reacts, forming blue $Cr^{2+}$ and bubbles of hydrogen gas.

(b)
$$Ag(s) \longrightarrow Ag^+(aq) \qquad E^{\circ}_{ox} = -0.799 \text{ V}$$
$$NO_3^-(aq) \longrightarrow NO(g) \qquad E^{\circ}_{red} = +0.964 \text{ V}$$

Since $E^{\circ}$ is a positive quantity, $+0.165$ V, a redox reaction should occur. The balanced half-equations are

oxidation:     $Ag(s) \longrightarrow Ag^+(aq) + e^-$

reduction:     $NO_3^-(aq) + 4H^+(aq) + 3e^- \longrightarrow NO(g) + 2H_2O$

To obtain the final balanced equation, multiply the first half-equation by 3 and add to the second. The result is

$$3Ag(s) + NO_3^-(aq) + 4H^+(aq) \longrightarrow 3Ag^+(aq) + NO(g) + 2H_2O$$

---

Although gold is not oxidized by nitric acid, it can be brought into solution in *aqua regia*, a 3:1 mixture by volume of 12 *M* HCl and 16 *M* $HNO_3$:

$$Au(s) + 4H^+(aq) + 4Cl^-(aq) + NO_3^-(aq) \longrightarrow AuCl_4^-(aq) + NO(g) + 2H_2O$$

The nitrate ion of the nitric acid acts as the oxidizing agent. The function of the hydrochloric acid is to furnish $Cl^-$ ions to form the very stable complex ion $AuCl_4^-$.

## Equilibria Between Different Cations of a Transition Metal

Several transition metals form more than one cation. For example, chromium forms $Cr^{2+}$ and $Cr^{3+}$; copper forms $Cu^+$ and $Cu^{2+}$. Table 20.4, p. 570, lists values of $E^{\circ}_{red}$ for several such systems. Using the data in this table and in

| TABLE 20.4 | **Ease of Reduction of Transition Metal Cations** | | | | |
|---|---|---|---|---|---|
| Chromium | $Cr^{3+}$ | $\xrightarrow{-0.408\ V}$ | $Cr^{2+}$ | $\xrightarrow{-0.912\ V}$ | $Cr$ |
| Manganese | $Mn^{3+}$ | $\xrightarrow{+1.559\ V}$ | $Mn^{2+}$ | $\xrightarrow{-1.182\ V}$ | $Mn$ |
| Iron | $Fe^{3+}$ | $\xrightarrow{+0.769\ V}$ | $Fe^{2+}$ | $\xrightarrow{-0.409\ V}$ | $Fe$ |
| Cobalt | $Co^{3+}$ | $\xrightarrow{+1.953\ V}$ | $Co^{2+}$ | $\xrightarrow{-0.282\ V}$ | $Co$ |
| Copper | $Cu^{2+}$ | $\xrightarrow{+0.161\ V}$ | $Cu^{+}$ | $\xrightarrow{+0.518\ V}$ | $Cu$ |
| Gold | $Au^{3+}$ | $\xrightarrow{+1.400\ V}$ | $Au^{+}$ | $\xrightarrow{+1.695\ V}$ | $Au$ |
| Mercury | $Hg^{2+}$ | $\xrightarrow{+0.908\ V}$ | $Hg_2^{2+}$ | $\xrightarrow{+0.796\ V}$ | $Hg$ |

Table 18.1 (p. 507), it is possible to decide upon the relative stabilities of different transition metal cations in water solution.

Cations for which $E^{\circ}_{red}$ is a large, positive number are readily reduced and hence tend to be unstable in water solution. A case in point is the $Mn^{3+}$ ion, which reacts spontaneously with water:

$$2Mn^{3+}(aq) + H_2O \longrightarrow 2Mn^{2+}(aq) + \tfrac{1}{2}O_2(g) + 2H^+(aq)$$

$$E^{\circ} = E^{\circ}_{red}\ Mn^{3+} + E^{\circ}_{ox}\ H_2O = +1.559\ V - 1.229\ V = +0.330\ V$$

As a result of this reaction, $Mn^{3+}$ cations are never found in water solution. Manganese(III) occurs only in insoluble oxides and hydroxides such as $Mn_2O_3$ and $MnO(OH)$.

The cations in the center column of Table 20.4 ($Cr^{2+}$, $Mn^{2+}$, ...) are in an intermediate oxidation state. They can either be oxidized to a cation of higher charge ($Cr^{2+} \longrightarrow Cr^{3+}$) or reduced to the metal ($Cr^{2+} \longrightarrow Cr$). With certain cations of this type, these two half-reactions occur simultaneously. Consider, for example, the $Cu^+$ ion. In water solution, copper(I) salts **disproportionate,** undergoing reduction (to copper metal) and oxidation (to $Cu^{2+}$) at the same time:

$$2Cu^+(aq) \longrightarrow Cu(s) + Cu^{2+}(aq); \quad E^{\circ} = +0.518\ V - 0.161\ V = +0.357\ V$$

As a result, the only stable copper(I) species are insoluble compounds such as CuCN or complex ions such as $Cu(CN)_2^-$.

The $Cu^+$ ion is one of the few species in the center column of Table 20.4 that disproportionates in water, undergoing simultaneous reduction to the metal and oxidation to a cation of higher charge. However, cations of this type may be unstable for a quite different reason. Water ordinarily contains dissolved air; the $O_2$ in air may oxidize the cation. When a blue solution of chromium(II) salt is exposed to air, the color quickly changes to violet or green as the $Cr^{3+}$ ion is formed by the reaction

$$2Cr^{2+}(aq) \longrightarrow 2Cr^{3+}(aq) + 2e^- \qquad\qquad E^{\circ}_{ox} = +0.408\ V$$

$$\underline{\tfrac{1}{2}O_2(g) + 2H^+(aq) + 2e^- \longrightarrow H_2O \qquad\qquad E^{\circ}_{red} = +1.229\ V}$$

$$2Cr^{2+}(aq) + \tfrac{1}{2}O_2(g) + 2H^+(aq) \longrightarrow 2Cr^{3+}(aq) + H_2O \qquad E^{\circ} = +1.637\ V$$

As a result of this reaction, chromium(II) salts are difficult to prepare and even more difficult to store.

*$Co^{3+}$ is also rare, except in complexes*

The $Fe^{2+}$ ion ($E^\circ_{ox} = -0.769$ V) is much more stable toward oxidation than $Cr^{2+}$. However, iron(II) salts in water solution are slowly converted to iron(III) by dissolved oxygen. In acidic solution, the reaction is

$$2Fe^{2+}(aq) + \tfrac{1}{2}O_2(g) + 2H^+(aq) \longrightarrow 2Fe^{3+}(aq) + H_2O$$

$$E^\circ = E^\circ_{ox}\, Fe^{2+} + E^\circ_{red}\, O_2 = -0.769 \text{ V} + 1.229 \text{ V} = +0.460 \text{ V}$$

A similar reaction takes place in basic solution. Iron(II) hydroxide is pure white when first precipitated, but in the presence of air it turns first green and then brown as it is oxidized by $O_2$:

$$2Fe(OH)_2(s) + \tfrac{1}{2}O_2(g) + H_2O \longrightarrow 2Fe(OH)_3(s)$$

---

**Example 20.7**   Using Table 20.4, find

(a) three different cations, in addition to $Mn^{3+}$, that react with $H_2O$ to form $O_2(g)$ ($E^\circ_{ox}\, H_2O = -1.229$ V).
(b) another cation, in addition to $Cu^+$, that disproportionates in water.
(c) two other cations, in addition to $Cr^{2+}$ and $Fe^{2+}$, that are oxidized by $O_2(g)$ dissolved in water ($E^\circ_{red}\, O_2 = +1.229$ V)

**Strategy**   In each case, look for a reaction in which $E^\circ$ is a positive quantity. In (a), $E^\circ_{red}$ for the cation must exceed $+1.229$ V. In (b), $E^\circ_{red} + E^\circ_{ox}$ for the cation must be a positive quantity. In (c), the cation must have an $E^\circ_{ox}$ value no smaller than $-1.229$ V.

*Solution*

(a)   $Co^{3+}$   ($E^\circ_{red} = +1.953$ V),   $Au^{3+}$   ($E^\circ_{red} = +1.400$ V) and   $Au^+$   ($E^\circ_{red} = +1.695$ V)

(b)   $Au^+$   ($E^\circ_{red} + E^\circ_{ox} = +1.695$ V $- 1.400$ V $= 0.295$ V)

(c)   $Cu^+$   ($E^\circ_{ox} = -0.161$ V) and   $Hg_2^{2+}$   ($E^\circ_{ox} = -0.908$ V)

---

## Oxoanions of the Transition Metals ($CrO_4^{2-}$, $Cr_2O_7^{2-}$, $MnO_4^-$)

Chromium in the +6 state forms two different oxoanions, the yellow chromate ion, $CrO_4^{2-}$, and the red dichromate ion, $Cr_2O_7^{2-}$ (Fig. 20.11). The chromate ion is stable in basic or neutral solution; in acid, it is converted to the dichromate ion:

$$2CrO_4^{2-}(aq) + 2H^+(aq) \rightleftharpoons Cr_2O_7^{2-}(aq) + H_2O \qquad K = 3 \times 10^{14}$$

yellow                                            red

The dichromate ion in acidic solution is a powerful oxidizing agent,

$$Cr_2O_7^{2-}(aq) + 14H^+(aq) + 6e^- \longrightarrow 2Cr^{3+}(aq) + 7H_2O \qquad E^\circ_{red} = +1.33 \text{ V}$$

As you might expect from the half-equation for its reduction, the oxidizing strength of the dichromate ion decreases as the concentration of $H^+$ decreases (increasing pH).

The $Cr_2O_7^{2-}$ ion can act as an oxidizing agent in the solid state as well as in water solution. In particular, it can oxidize the $NH_4^+$ ion to molecular ni-

**Figure 20.11**
Oxidation states of chromium. The first two containers (green and violet) contain +3 chromium. The $Cr(H_2O)_4Cl_2^+$ ion is green; $Cr(H_2O)_6^{3+}$ is violet. The two bottles at the right contain +6 chromium; the $CrO_4^{2-}$ ion is yellow, while the $Cr_2O_7^{2-}$ ion is red. (Charles D. Winters)

**Figure 20.12**
The compound ammonium dichromate, $(NH_4)_2Cr_2O_7$, has a reddish color owing to the presence of the $Cr_2O_7^{2-}$ ion. When ignited, it decomposes to give finely divided $Cr_2O_3$ (which is green), nitrogen gas, and water vapor. (Charles D. Winters)

trogen. When a pile of ammonium dichromate is ignited, a spectacular reaction occurs (Fig. 20.12).

*This makes a great lecture demonstration*

$$(NH_4)_2Cr_2O_7(s) \longrightarrow N_2(g) + 4H_2O(g) + \underset{\text{green}}{Cr_2O_3(s)}$$
$$\underset{\text{red}}{\phantom{(NH_4)_2Cr_2O_7(s)}}$$

The ammonium dichromate resembles a tiny volcano as it burns, emitting hot gases, sparks, and a voluminous green dust of chromium(III) oxide.

Chromates and dichromates are gradually disappearing from chemistry teaching laboratories because of concern about their toxicity. Long-term exposure of industrial workers to dust containing chromates has, in a few cases, been implicated in lung cancer. More commonly, repeated contact with chromate salts leads to skin disorders; a few people are extremely allergic to $CrO_4^{2-}$ and $Cr_2O_7^{2-}$ ions, breaking into a rash upon first exposure.

The permanganate ion, $MnO_4^-$, has an intense purple color (Fig. 20.13), easily visible even in very dilute solution.* Crystals of solid potassium permanganate, $KMnO_4$, have a deep purple, almost black color. This compound is used to treat such diverse ailments as "athlete's foot" and rattlesnake bites. These applications depend upon the fact that the $MnO_4^-$ ion is a very powerful oxidizing agent. This is especially true in acidic solution, where $MnO_4^-$ is reduced to $Mn^{2+}$:

$$MnO_4^-(aq) + 8H^+(aq) + 5e^- \longrightarrow Mn^{2+}(aq) + 4H_2O \qquad E^\circ_{red} = +1.512 \text{ V}$$

In basic solution, $MnO_4^-$ is reduced to $MnO_2$, with a considerably smaller value of $E^\circ_{red}$:

$$MnO_4^-(aq) + 2H_2O + 3e^- \longrightarrow MnO_2(s) + 4OH^-(aq) \qquad E^\circ_{red} = +0.596 \text{ V}$$

However, even in basic solution, $MnO_4^-$ can oxidize water:

$$4MnO_4^-(aq) + 2H_2O \longrightarrow 4MnO_2(s) + 3O_2(g) + 4OH^-(aq)$$

$$E^\circ = E^\circ_{red} \text{ } MnO_4^- + E^\circ_{ox} \text{ } H_2O = +0.596 \text{ V} - 0.401 \text{ V} = +0.195 \text{ V}$$

This reaction accounts for the fact that laboratory solutions of $KMnO_4$ slowly decompose, producing a brownish solid ($MnO_2$) and gas bubbles ($O_2$).

KMnO₄

**Figure 20.13**
The $MnO_4^-$ ion has an intense purple color. (Charles D. Winters)

*The purple color of old bottles exposed to the sun for a long time is due to $MnO_4^-$ ions. These are formed when ultraviolet light oxidizes manganese compounds in the glass.

## 20.4 Alloys

An important characteristic of metallic elements is their ability to form alloys. An **alloy** is a material with metallic properties that contains two or more elements, at least one of which is a metal. Solid alloys are ordinarily prepared by melting the elements together, stirring the molten mixture until it is homogeneous, and allowing it to cool. Many alloys, notably bronze, brass, and pewter, have been made for centuries by this method.

Structurally, two metals A and B can form three different types of alloys.

**1. Solid solutions.** Here A and B are completely soluble in one another in the solid as well as the liquid phase. The solid alloy formed upon cooling is homogeneous. It is referred to as a *substitutional* solid solution; atoms of A can substitute for B in any proportions.

Systems of this type are relatively rare; complete solubility in the solid requires that the two metals have the same type of unit cell (Chap. 9) and very similar atomic radii. Monel, a Cu–Ni alloy, has such a structure. Both copper and nickel crystallize in the face-centered cubic system; their atomic radii are 0.128 nm and 0.124 nm, respectively.

More commonly, two elements show limited solubility in the solid state. Iron can dissolve a maximum of 2% carbon. The carbon atoms, which are much smaller than iron (at. rad. C = 0.077 nm; at. rad. Fe = 0.126 nm) form an *interstitial* solid solution. They fit into holes in the face-centered cubic structure of iron, stable between 900 and 1400°C. The phase formed, called martensite, makes high-carbon steel hard and brittle.

**2. Heterogeneous mixtures.** Here, the two metals A and B are completely insoluble in each other in the solid phase. When the melt is cooled, the first solid that separates may be either A or B, depending upon the composition of the melt. At a lower temperature, an intimate but heterogeneous mixture of tiny crystals of A and B forms. Solder, a Sn–Pb alloy, approximates this behavior. If you look at solder under the microscope, you will see a mixture of nearly pure crystals of the two metals.

**3. Intermetallic compounds,** in which two elements A and B are present in a fixed atom ratio, e.g., AB, $AB_2$. Iron and carbon form a compound called cementite, which has the composition $Fe_3C$. Copper and zinc form a series of different compounds, of which the simplest is CuZn. This 1:1 compound has a body-centered cubic structure very similar to that of CsCl (Chap. 9).

The properties of alloys are quite different from those of the component metals. The effect of alloying is ordinarily to

— *lower the melting point.* An alloy usually (but not always!) melts at a lower temperature than the pure metals. An extreme example is Wood's metal (Table 20.5), which melts at 70°C. Wood's metal is used in fusible plugs that melt to set off automatic sprinkler systems.

— *increase the hardness.* A small amount of copper is present in sterling silver, which is much harder than pure silver. Gold is alloyed with silver and copper to form a metal hard enough to be used in jewelry. The lead plates used in storage batteries contain small amounts of antimony to prevent them from bending under stress.

Solder melts at a lower temperature than either Sn or Pb

TABLE 20.5 **Some Commercially Important Alloys**

| Common Name | Composition, Element (Mass Percent) | Uses |
|---|---|---|
| Alnico | Fe(50), Al(20), Ni(20), Co(10) | magnets |
| Aluminum bronze | Cu(90), Al(10) | crankcases, hinges |
| Brass | Cu(67–90), Zn(10–33) | plumbing, hardware |
| Bronze | Cu(70–95), Zn(1–25), Sn(1–18) | bearings, bells, medals |
| Cast iron | Fe(96–97), C(3–4) | castings |
| Coinage, U.S. | Cu(75), Ni(25) | 5¢, 10¢, 25¢, 50¢ coins |
| Dental amalgam | Hg(50), Ag(35), Sn(15) | dental fillings |
| Duriron | Fe(84), Si(14), C(1), Mn(1) | pipes, kettles, condensers |
| German silver | Cu(60), Zn(25), Ni(15) | teapots, jugs, faucets |
| Gold, 18 carat | Au(75), Ag(10–20), Cu(5–15) | jewelry |
| Gold, 10 carat | Au(42), Ag(12–20), Cu(38–46) | jewelry |
| Gunmetal | Cu(88), Sn(10), Zn(2) | gun barrels, machine parts |
| Monel | Ni(60–70), Cu(25–35), Fe, Mn | instruments, machine parts |
| Nichrome | Ni(60), Fe(25), Cr(15) | electrical resistance wire |
| Pewter | Sn(70–95), Sb(5–15), Pb(0–15) | tableware |
| Silver solder | Ag(63), Cu(30), Zn(7) | high-melting solder |
| Solder | Pb(67), Sn(33) | joining metals |
| Spiegeleisen | Fe(80–95), Mn(5–20), C(0–1) | safes, armor plate, rails |
| Stainless steel | Fe(73–79), Cr(14–18), Ni(7–9) | instruments, sinks |
| Steel | Fe(98–99.5), C(0.5–2) | structural metal |
| Sterling silver | Ag(92.5), Cu(7.5) | tableware, jewelry |
| Vitallium | Co(62), Cr(26), Mo, Ni | artificial hip, knee joints |
| White gold | Au(90), Pd(10) | jewelry |
| Wood's metal | Bi(50), Pb(25), Sn(13), Cd(12) | automatic sprinkler systems |

— *lower the electrical and thermal conductivity.* Copper used in electrical wiring must be extremely pure; as little as 0.03% of arsenic can lower its conductivity by 15%. Sometimes we take advantage of this effect. High-resistance nichrome wire (Ni, Cr) is used in the heating elements of hair dryers and electric toasters.

## CHEMISTRY

### *Beyond the Classroom*

## Essential Metals in Nutrition

Four of the main-group cations are essential in human nutrition (Table 20.A). Of these, the most important is $Ca^{2+}$. About 90% of the calcium in the body is found in bones and teeth, largely in the form of hydroxyapatite, $Ca(OH)_2 \cdot 3Ca_3(PO_4)_2$. Calcium ions in bones and teeth exchange readily with those in the blood; about 0.6 g of $Ca^{2+}$ enters and leaves your bones every day. In a normal adult this exchange is in balance, but in elderly people, particularly women, there is sometimes a net loss of bone calcium, leading to the disease known as osteoporosis. It is generally supposed that osteoporosis is related to $Ca^{2+}$ deficiency in early adult years, although the evidence is far from conclusive.

Of the ten trace elements known to be essential to human nutrition, seven are transition metals. For the most part, transition metals in biochemical compounds are present as complex ions, chelated by organic ligands. You will recall (Chap. 15) that hemoglobin has such a structure with $Fe^{2+}$ as the central ion of the complex. The $Co^{3+}$ ion occupies a similar position in vitamin $B_{12}$, octahedrally coordinated to organic molecules somewhat similar to those in hemoglobin.

### TABLE 20.A  **Essential Metal Ions**

#### Major Species (Main-Group Cations)

| Ion | Amount in Body | Daily Requirement | Function in Body | Rich Sources* |
|-----|---------------|-------------------|------------------|---------------|
| $Na^+$ | 63 g | ~2 g | principal cation *outside* cell fluid | table salt, many processed and preserved foods |
| $K^+$ | 150 g | ~2 g | principal cation *within* cell fluid | fruits, nuts, fish, instant coffee, wheat bran |
| $Mg^{2+}$ | 21 g | 0.35 g | activates enzymes for body processes | chocolate, nuts, instant coffee, wheat bran |
| $Ca^{2+}$ | 1160 g | 0.8 g | bone and tooth formation | milk, cheese, broccoli, canned salmon (with bones) |

#### Trace Species (Transition Metal Cations)

| Ion | Amount in Body | Daily Requirement | Function in Body | Rich Sources* |
|-----|---------------|-------------------|------------------|---------------|
| $Fe^{2+}, Fe^{3+}$ | 5000 mg | 15 mg | component of hemoglobin, myoglobin | liver, meat, clams, spinach |
| $Zn^{2+}$ | 3000 mg | 15 mg | component of many enzymes, hormones | oysters, crab, meat, nuts |
| $Cu^{2+}, Cu^+$ | 100 mg | 3 mg | iron metabolism, component of enzymes | liver, lobster, cherries |
| $Mn^{2+}$ | 15 mg | 3 mg | metabolism of carbohydrates, lipids | beet greens, nuts, blueberries |
| $Mo(IV,V,VI)$ | 10 mg | 0.3 mg | Fe, N metabolism; component of enzymes | legumes, green vegetables |
| $Cr^{3+}$ | 5 mg | 0.1 mg | glucose metabolism; affects action of insulin | corn, clams, whole grains |
| $Co^{2+}, Co^{3+}$ | 1 mg | 0.005 mg | component of vitamin $B_{12}$ | liver, shellfish |

*Other, more mundane, sources can be found in any nutrition textbook.

As you can judge from Table 20.A, transition metal cations are frequently found in enzymes. The $Zn^{2+}$ ion alone is known to be a component of at least 70 different enzymes. One of these, referred to as "alcohol dehydrogenase" is concentrated in the liver, where it acts to break down alcohols. Another zinc-containing enzyme is involved in the normal functioning of oil glands in the skin, which accounts for the use of $Zn^{2+}$ compounds in the treatment of acne.

Although $Zn^{2+}$ is essential to human nutrition, compounds of the two elements below zinc in the periodic table, Cd and Hg, are extremely toxic. This reflects the fact that $Cd^{2+}$ and $Hg^{2+}$, in contrast to $Zn^{2+}$, form very stable complexes with ligands containing sulfur atoms. As a result, these two cations react with and thereby deactivate enzymes containing —SH groups.

Over the years, compounds of the transition metals have been used for a variety of medicinal purposes. Calomel, $Hg_2Cl_2$, was the wonder drug of the early nineteenth century, prescribed for everything from constipation to pneumonia. It probably killed more patients than it cured. Quite recently, gold salts have been found effective in the treatment of rheumatoid arthritis. On a more prosaic level, iron(II) compounds such as $FeSO_4$ are used in treating anemia.

# CHAPTER HIGHLIGHTS

## Key Concepts

1. Calculate $\Delta G°$ from thermodynamic data
   (Example 20.1; Problems 5, 6, 37, 46)
2. Write balanced equations to represent
   — metallurgical processes
   (Examples 20.2, 20.3; Problems 1–4, 7, 8, 40)
   — reactions of Group 1 and Group 2 metals
   (Examples 20.4, 20.5; Problems 15, 16)
   — redox reactions of transition metals
   (Example 20.6; Problems 19–22)
3. Determine $E°$ and reaction spontaneity from standard potentials
   (Example 20.7; Problems 25–30, 41)

## Key Terms

| | | |
|---|---|---|
| alkali metal | electrolysis | peroxide |
| alkaline earth metal | main-group metal | reduction |
| alloy | metallurgy | superoxide |
| disproportionation | oxidation | transition metal |
| $E°_{ox}, E°_{red}$ | oxoanion | |

## Summary Problem

Consider the alkaline earth strontium and the transition metal manganese.

(a) Write a balanced equation for the reaction of strontium with oxygen; with water.
(b) Give the formulas of strontium hydride, strontium nitride, and strontium sulfide.
(c) Electrolysis of strontium chloride gives strontium metal and chlorine gas. What mass of the chloride must be electrolyzed to form one kilogram of strontium? One liter of $Cl_2(g)$ at STP?
(d) Give the formulas of three manganese compounds containing manganese in different oxidation states.
(e) Write a balanced equation for the reaction of manganese metal with hydrochloric acid; $Mn^{3+}$ with water; Mn with $O_2$.

(f) Write a balanced equation for the reduction of the principal ore of manganese, pyrolusite, $MnO_2$, with CO.

(g) For the reaction $MnO(s) + H_2(g) \longrightarrow Mn(s) + H_2O(g)$, $\Delta H° = +143$ kJ, $\Delta S° = +0.0297$ kJ/K, would it be feasible to reduce MnO to the metal by heating with hydrogen?

(h) Given that $E°_{red}$ $Mn^{3+} \longrightarrow Mn^{2+} = +1.559$ V, $E°_{red}$ $Mn^{2+} \longrightarrow Mn = -1.182$ V, and $E°_{red}$ $O_2(g) \longrightarrow H_2O = +1.229$ V, show by calculation whether $Mn^{2+}$ in water solution will disproportionate; whether it will be oxidized to $Mn^{3+}$ by dissolved oxygen; whether it will be reduced by hydrogen gas to the metal.

(i) Given

$$MnO_4{}^-(aq) + 2H_2O + 3e^- \longrightarrow MnO_2(s) + 4OH^-(aq), \quad E°_{red} = +0.59 \text{ V}$$

write a balanced equation for the reaction of $MnO_4{}^-$ in basic solution with $Fe^{2+}$ to form $Fe(OH)_3$. Calculate $E_{red}$ for the $MnO_4{}^-$ ion at pH 9.0, taking conc. $MnO_4{}^- = 1$ M.

## Answers

(a) $2Sr(s) + O_2(g) \longrightarrow 2SrO(s)$
$Sr(s) + 2H_2O \longrightarrow Sr^{2+}(aq) + 2OH^-(aq) + H_2(g)$

(b) $SrH_2$; $Sr_3N_2$; SrS

(c) 1.809 kg; 7.07 g

(d) $MnCl_2$, $Mn_2O_3$; $KMnO_4$

(e) $Mn(s) + 2H^+(aq) \longrightarrow Mn^{2+}(aq) + H_2(g)$
$2Mn^{3+}(aq) + H_2O \longrightarrow 2Mn^{2+}(aq) + \frac{1}{2}O_2(g) + 2H^+(aq)$
$3Mn(s) + 2O_2(g) \longrightarrow Mn_3O_4(s)$

(f) $MnO_2(s) + 2CO(g) \longrightarrow Mn(s) + 2CO_2(g)$

(g) no; $T_{calc} \approx 4800$ K

(h) no; in all cases, $E°$ is negative

(i) $MnO_4{}^-(aq) + 3Fe^{2+}(aq) + 5OH^-(aq) + 2H_2O \longrightarrow$
$MnO_2(s) + 3Fe(OH)_3(s) \qquad E_{red} = +0.98$ V

## Questions & Problems

### Metallurgy

**1.** Write a balanced equation to represent the electrolysis of molten sodium chloride. What volume of $Cl_2$ at STP is formed at the anode when 1.00 g of sodium is formed at the cathode?

**2.** Write a balanced equation to represent the electrolysis of aluminum oxide. If 2.00 L of $O_2$ at 25°C and 751 mm Hg is formed at the anode, what mass of Al is formed at the cathode?

**\*3.** Write a balanced equation to represent
(a) the roasting of nickel(II) sulfide to form nickel(II) oxide.
(b) the reduction of nickel(II) oxide by carbon monoxide.

**\*4.** Write a balanced equation to represent the roasting of copper(I) sulfide to form "blister copper."

**5.** Show by calculation whether the reaction in Problem 3(b) is spontaneous at 25°C and 1 atm.

**6.** Calculate $\Delta G°$ at 200°C for the reaction in Problem 4.

**\*7.** Write a balanced equation for the reaction that occurs when

(a) finely divided gold is treated with $CN^-(aq)$ in the presence of $O_2(g)$.
(b) finely divided zinc metal is added to the solution formed in (a).

**\*8.** Write a balanced equation for the reaction that occurs when

(a) iron(III) oxide is reduced with carbon monoxide.
(b) the excess carbon in pig iron is removed by the basic oxygen process.

**9.** How many cubic feet of air (assume 21% by volume of oxygen in air) at 25°C and 1.00 atm are required to react with coke to form the CO needed to convert one metric ton of hematite ore (92% $Fe_2O_3$) to iron?

**10.** Zinc is produced by electrolytic refining. The electrolytic process, which is similar to that for copper, can be represented by the two half-reactions

$$Zn(\text{impure}, s) \longrightarrow Zn^{2+} + 2e^-$$

$$Zn^{2+} + 2e^- \longrightarrow Zn(\text{pure}, s)$$

For this process, a voltage of 3.0 V is used. How many kilowatt hours are needed to produce one metric ton of pure zinc?

**11.** When 2.876 g of a certain metal sulfide is roasted in air, 2.368 g of the metal oxide are formed. If the metal has an oxidation number of +2, what is its molar mass?

**12.** Chalcopyrite, $CuFeS_2$, is an important source of copper. A typical chalcopyrite ore contains about 0.75% Cu. What volume of sulfur dioxide at 25°C and 1.00 atm pressure is produced when one boxcar load ($4.00 \times 10^3$ ft$^3$) of chalcopyrite ore ($d = 2.6$ g/cm$^3$) is roasted? Assume all the sulfur in the ore is converted to $SO_2$ and no other source of sulfur is present.

### Reactions of Alkali Metals and Alkaline Earth Metals

**\*13.** Give the formula and name of the compound formed by strontium with
   **(a)** nitrogen    **(b)** bromine
   **(c)** water       **(d)** oxygen

**\*14.** Give the formula and name of the compound formed by potassium with
   **(a)** nitrogen    **(b)** iodine    **(c)** water
   **(d)** hydrogen    **(e)** sulfur

**\*15.** Write a balanced equation and give the names of the products for the reaction of
   **(a)** magnesium with chlorine.
   **(b)** barium peroxide with water.
   **(c)** lithium with sulfur.
   **(d)** sodium with water.

**\*16.** Write a balanced equation and give the names of the products for the reaction of
   **(a)** sodium peroxide and water.
   **(b)** calcium and oxygen.
   **(c)** rubidium and oxygen.
   **(d)** strontium hydride and water.

**17.** To inflate a life raft with hydrogen to a volume of 25.0 L at 25°C and 1.10 atm, what mass of calcium hydride must react with water?

**18.** What mass of $KO_2$ is required to remove 90.0% of the $CO_2$ from a sample of 1.00 L of exhaled air (37°C, 1.00 atm) containing 5.00 mole percent $CO_2$?

### Redox Chemistry of Transition Metals

**\*19.** Write a balanced equation to show
   **(a)** the reaction of chromate ion with strong acid.
   **(b)** the oxidation of water to oxygen gas by permanganate ion in basic solution.
   **(c)** the reduction half-reaction of chromate ion to chromium(III) hydroxide in basic solution.

**\*20.** Write a balanced equation to show
   **(a)** the formation of gas bubbles when cobalt reacts with hydrochloric acid.
   **(b)** the reaction of copper with nitric acid.
   **(c)** the reduction half-reaction of dichromate ion to $Cr^{3+}$ in acidic solution.

**\*21.** Write a balanced redox equation for the reaction of mercury with aqua regia, assuming the products include $HgCl_4^{2-}$ and $NO_2(g)$.

**\*22.** Write a balanced redox equation for the reaction of cadmium with aqua regia, assuming the products include $CdCl_4^{2-}$ and $NO(g)$.

**\*23.** Balance the following redox equations.
   **(a)** $Cu(s) + NO_3^-(aq) \longrightarrow Cu^{2+}(aq) + NO_2(g)$ (acidic)
   **(b)** $Cr(OH)_3(s) + ClO^-(aq) \longrightarrow$
   $$CrO_4^{2-}(aq) + Cl^-(aq) \text{ (basic)}$$

**\*24.** Balance the following redox equations.
   **(a)** $Fe(s) + NO_3^-(aq) \longrightarrow Fe^{3+}(aq) + NO_2(g)$ (acidic)
   **(b)** $Cr(OH)_3(s) + O_2(g) \longrightarrow CrO_4^{2-}(aq)$ (basic)

**25.** Show by calculation which of the following metals will react with hydrochloric acid (standard concentrations).
   **(a)** Cd    **(b)** Cr    **(c)** Co
   **(d)** Ag    **(e)** Au

**26.** Show by calculation which of the metals in Problem 25 will react with nitric acid to form NO (standard concentrations).

**27.** Of the cations listed in Table 20.4, show by calculation which one (besides $Cu^+$) will disproportionate at standard conditions.

**28.** Using Table 18.1 (Chap. 18) calculate $E°$ for
   **(a)** $2Co^{3+}(aq) + H_2O \longrightarrow$
   $$2Co^{2+}(aq) + \tfrac{1}{2}O_2(g) + 2H^+(aq)$$
   **(b)** $2Cr^{2+}(s) + I_2(s) \longrightarrow 2Cr^{3+}(aq) + 2I^-(aq)$

**29.** Using Table 20.4, calculate, for the disproportionation of $Fe^{2+}$.
   **(a)** the equilibrium constant, $K$.
   **(b)** the concentration of $Fe^{3+}$ in equilibrium with 0.10 $M$ $Fe^{2+}$.

**30.** Using Table 20.4, calculate, for the disproportionation of $Au^+$.
   **(a)** $K$.
   **(b)** the concentration of $Au^+$ in equilibrium with 0.10 $M$ $Au^{3+}$.

**31.** Referring to Table 20.5, calculate
   **(a)** the mole percent of copper in the U.S. dime.
   **(b)** the number of atoms of silver in 1.00 g of sterling silver.

**\*32.** Give the name(s) of
   **(a)** a low-melting alloy containing bismuth.
   **(b)** an alloy containing two metals in Group **14**.
   **(c)** two different alloys of manganese.

### Unclassified

**33.** A sample of sodium liberates 2.73 L hydrogen at 752 mm Hg and 22°C when it is added to a large amount of water. How much sodium is used?

**34.** A self-contained breathing apparatus contains 248 g of potassium superoxide. A firefighter exhales 116 L of air at 37°C and 748 mm Hg. The volume percent of water in exhaled air is 6.2. What mass of potassium superoxide is left after the water in the exhaled air reacts with it?

**35.** Taking $K_{sp}$ $PbCl_2 = 1.7 \times 10^{-5}$ and assuming $[Cl^-] = 0.20$ $M$, calculate the concentration of $Pb^{2+}$ at equilibrium.

**36.** The equilibrium constant for the reaction

$$2CrO_4{}^{2-}(aq) + 2H^+(aq) \rightleftharpoons Cr_2O_7{}^{2-}(aq) + H_2O$$

is $3 \times 10^{14}$. What must the pH be so that the concentrations of chromate and dichromate ion are both 0.10 $M$?

**37.** Using data in Appendix 1, estimate the temperature at which $Fe_2O_3$ can be reduced to iron, using hydrogen gas as a reducing agent (assume $H_2O(g)$ is the other product).

**38.** A 0.500-g sample of zinc–copper alloy was treated with dilute hydrochloric acid. The hydrogen gas evolved was collected by water displacement at 27°C and a total pressure of 755 mm Hg. The volume of the water displaced by the gas is 105.7 mL. What is the percent composition, by mass, of the alloy? (Vapor pressure of $H_2O$ at 27°C is 26.74 mm Hg.) Assume only the zinc reacts.

**39.** One type of stainless steel contains 22% nickel by mass. How much nickel sulfide ore, NiS, is required to produce one metric ton of stainless steel?

**40.** Silver is obtained in much the same manner as gold, using NaCN solution and $O_2$. Describe with appropriate equations the extraction of silver from argentite ore, $Ag_2S$. (The products are $SO_2$ and $Ag(CN)_2{}^-$, which is reduced with zinc.)

**41.** Iron(II) can be oxidized to iron(III) by permanganate ion in acidic solution. The permanganate ion is reduced to manganese(II) ion.

    **(a)** Write the oxidation half-reaction, the reduction half-reaction, and the overall redox equation.

    **(b)** Calculate $E°$ for the reaction.

    **(c)** Calculate the percent Fe in an ore if a 0.3500-g sample is dissolved and the $Fe^{2+}$ formed requires for titration 55.63 mL of a 0.0200 $M$ solution of $KMnO_4$.

**42.** Of the cations listed in the center column of Table 20.4, which one is the

    **(a)** strongest reducing agent?

    **(b)** strongest oxidizing agent?

    **(c)** weakest reducing agent?

    **(d)** weakest oxidizing agent?

## *Challenge Problems*

**43.** A sample of 20.00 g of barium reacts with oxygen to form 22.38 g of a mixture of barium oxide and barium peroxide. Determine the composition of the mixture.

**44.** Rust, which you can take to be $Fe(OH)_3$, can be dissolved by treating it with oxalic acid. An acid-base reaction occurs, and a complex ion is formed.

    **(a)** Write a balanced equation for the reaction.

    **(b)** What volume of 0.10 $M$ $H_2C_2O_4$ would be required to remove a rust stain weighing 1.0 g?

**45.** A 0.500-g sample of steel is analyzed for manganese. The sample is dissolved in acid and the manganese is oxidized to permanganate ion. A measured excess of $Fe^{2+}$ is added to reduce $MnO_4{}^-$ to $Mn^{2+}$. The excess $Fe^{2+}$ is determined by titration with $K_2Cr_2O_7$. If 75.00 mL of 0.125 $M$ $FeSO_4$ is added and the excess requires 13.50 mL of 0.100 $M$ $K_2Cr_2O_7$ to oxidize $Fe^{2+}$, calculate the percent by mass of Mn in the sample.

**46.** Calculate the temperature in °C at which the equilibrium constant ($K$) for the following reaction is 1.00.

$$MnO_2(s) \longrightarrow Mn(s) + O_2(g)$$

**47.** A solution of potassium dichromate is made basic with sodium hydroxide; the color changes from red to yellow. Addition of silver nitrate to the yellow solution gives a precipitate. This precipitate dissolves in concentrated ammonia but re-forms when nitric acid is added. Write balanced net ionic equations for all the reactions in this sequence.

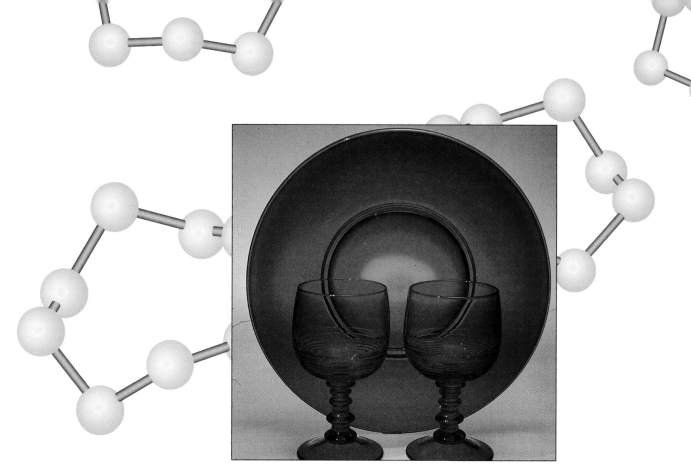

Selenium compounds impart a red color to glass; moreover, they play an important role in nutrition and perhaps in cancer prevention. (p. 605) (Charles D. Winters)

# 21 Chemistry of the Nonmetals

**F**or what can so fire us,

Enrapture, inspire us,

As Oxygen? What so delicious to quaff?

It is so animating,

And so titillating,

E'en grey-beards turn freshy, dance, caper, and

laugh.

—JOHN SHIELD

*Oxygen Gas*

## CHAPTER OUTLINE

Approximately 17 elements are classified as nonmetals; they lie above and to the right of the "stairway" that runs diagonally across the periodic table (Fig. 21.1). As the word "nonmetal" implies, these elements do not show metallic properties; in the solid state they are brittle as opposed to ductile, insulators rather than conductors. Most of the nonmetals, particularly those in Groups 15 to 17 of the periodic table, are molecular in nature (e.g., $N_2$, $O_2$, $F_2$). The noble gases (Group 18) consist of individual atoms attracted to each other by weak dispersion forces. Carbon and silicon in Group 14 have a network covalent structure.

As indicated in Figure 21.1, this chapter concentrates upon the more common and/or more reactive nonmetals, namely

— nitrogen and phosphorus in Group 15
— oxygen and sulfur in Group 16
— the halogens (F, Cl, Br, I) in Group 17

We will consider

— the properties of these elements and methods of preparing them (Section 21.1)
— their hydrogen compounds (Section 21.2)
— their oxides (Section 21.3)
— their oxoacids and oxoanions (Section 21.4)

## 21.1 The Elements and their Preparation

Table 21.1 lists some of the properties of the eight nonmetals considered in this chapter. Notice that all of these elements are molecular; those of low molar mass ($N_2$, $O_2$, $F_2$, $Cl_2$) are gases at room temperature and atmospheric pressure (Fig.

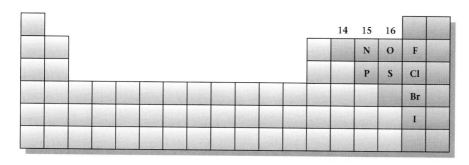

**Figure 21.1**

The 18 nonmetals are shown in color; the symbols indicate those elements whose chemistry is discussed in this chapter.

TABLE 21.1  **Properties of Nonmetallic Elements**

| | Nitrogen | Phosphorus | Oxygen | Sulfur | Fluorine | Chlorine | Bromine | Iodine |
|---|---|---|---|---|---|---|---|---|
| Outer electron configuration | $2s^2 2p^3$ | $3s^2 3p^3$ | $2s^2 2p^4$ | $3s^2 3p^4$ | $2s^2 2p^5$ | $3s^2 3p^5$ | $4s^2 4p^5$ | $5s^2 5p^5$ |
| Molecular formula | $N_2$ | $P_4$ | $O_2$ | $S_8$ | $F_2$ | $Cl_2$ | $Br_2$ | $I_2$ |
| Molar mass (g/mol) | 28 | 124 | 32 | 257 | 38 | 71 | 160 | 254 |
| State (25°C, 1 atm) | gas | solid | gas | solid | gas | gas | liquid | solid |
| Melting point (°C) | −210 | 44 | −218 | 119 | −220 | −101 | −7 | 114 |
| Boiling point (°C) | −196 | 280 | −183 | 444 | −188 | −34 | 59 | 184 |
| Bond energy* (kJ/mol) | 941 | 200 | 498 | 266 | 153 | 243 | 193 | 151 |
| $E^\circ_{red}$ | | | | | +2.889 V | +1.360 V | +1.077 V | +0.534 V |

*In the element (triple bond in $N_2$, double bond in $O_2$.

21.2). Stronger dispersion forces cause the nonmetals of higher molar mass to be either liquids ($Br_2$) or solids ($I_2$, $P_4$, $S_8$).

## Chemical Reactivity

Of the eight nonmetals listed in Table 21.1, **nitrogen** is by far the least reactive. Its inertness is due to the strength of the triple bond holding the $N_2$ molecule together (B.E. $N{\equiv}N$ = 941 kJ/mol). This same factor explains why virtually all chemical explosives are compounds of nitrogen (e.g., nitroglycerine, trinitrotoluene, ammonium nitrate, lead azide). These compounds detonate exothermically to form molecular nitrogen. The reaction with lead azide is

$$Pb(N_3)_2(s) \longrightarrow Pb(s) + 3N_2(g) \qquad \Delta H^\circ = -476 \text{ kJ}$$

Sodium azide, $NaN_3$, decomposes more smoothly; it is used in automobile airbags that inflate upon collision.

**Figure 21.2**
Flasks containing $Cl_2$, $Br_2$, and $I_2$ show a gradation in color from greenish-yellow through deep red to violet. The colors shown for bromine and iodine are those of the vapors in equilibrium with $Br_2(l)$ and $I_2(s)$.
(Marna G. Clarke)

*Fluorine* is the most reactive of all elements, in part because of the weakness of the F—F bond (B.E. F—F = 153 kJ/mol), but mostly because it is such a powerful oxidizing agent ($E°_{red}$ = +2.889 V). Fluorine combines with every element in the periodic table except He, Ne, and Ar. With a few metals, it forms a surface film of metal fluoride, which adheres tightly enough to prevent further reaction. This is the case with nickel, where the product is $NiF_2$. Fluorine gas is ordinarily stored in containers made of a nickel alloy, such as stainless steel (Fe, Cr, Ni) or Monel (Ni, Cu). Fluorine also reacts with many compounds including water. This means that reactions of fluorine with other species cannot be carried out in water solution.

*Chlorine* is somewhat less reactive than fluorine. Although it reacts with nearly all metals (Fig. 21.3), heating is often required. This reflects the relatively strong bond in the $Cl_2$ molecule (B.E. Cl—Cl = 243 kJ/mol). Chlorine disproportionates in water, forming $Cl^-$ ions (oxid. no. Cl = −1) and HClO molecules (oxid. no. Cl = +1).

A species disproportionates when it is oxidized and reduced at the same time

$$Cl_2(g) + H_2O \rightleftharpoons Cl^-(aq) + H^+(aq) + HClO(aq)$$

The hypochlorous acid, HClO, formed by this reaction is a powerful oxidizing agent ($E°_{red}$ = +1.630 V); it kills bacteria, apparently by destroying certain enzymes essential to their metabolism. The taste and odor that we associate with "chlorinated water" is actually due to compounds such as $CH_3NHCl$, produced by the action of hypochlorous acid on bacteria.

---

**Example 21.1**  For the reaction $Cl_2(g) + H_2O \rightleftharpoons Cl^-(aq) + H^+(aq) + HClO(aq)$

(a) Write the expression for the equilibrium constant K.
(b) Given that $K = 2.7 \times 10^{-5}$, calculate the concentration of HClO in equilibrium with $Cl_2(g)$ at 1.0 atm.

*Strategy*  To answer (a), note that, in the expression for K, gases enter as their partial pressures in atmospheres, species in aqueous solution as their molarities; water does not appear, since it is the solvent. To answer (b), note that $H^+$ ions, $Cl^-$ ions, and HClO molecules are formed in equimolar amounts.

*Solution*

(a) $K = \dfrac{[HClO] \times [H^+] \times [Cl^-]}{P_{Cl_2}}$

(b) Let x = [HClO]. Since $H^+$, $Cl^-$, and HClO all have coefficients of 1 in the balanced equation, $[H^+] = [Cl^-] = [HClO] = x$. Substituting in the expression for K:

$$K = \frac{[HClO] \times [H^+] \times [Cl^-]}{P_{Cl_2}} ; \quad 2.7 \times 10^{-5} = \frac{x^3}{1.0}$$

Solving,

$$x = (2.7 \times 10^{-5})^{1/3} = \boxed{0.030 \ M}$$

In other words, the concentration of hypochlorous acid in a solution formed by bubbling chlorine gas through water should be about 0.03 mol/L.

---

The oxidizing power of the halogens makes them hazardous to work with. Fluorine is the most dangerous, but it is very unlikely that you will ever come across it in a teaching laboratory. You are most likely to encounter chlorine as

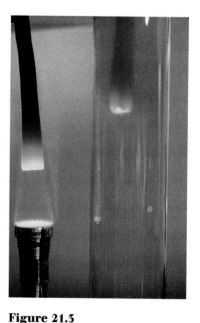

**Figure 21.3**
When a heated piece of copper foil is plunged into a cylinder containing chlorine gas, it reacts vigorously, giving off sparks. The equation for the reaction is:
$Cu(s) + Cl_2(g) \longrightarrow CuCl_2(s)$.

its saturated water solution, called "chlorine water." Remember that the pressure of chlorine gas over this solution (if it is freshly prepared) is 1 atm and that chlorine was used as a poison gas in World War I. Use small quantities of chlorine water and don't breathe the vapors. Bromine, although not as strong an oxidizing agent as chlorine, can cause severe burns if it comes in contact with your skin, particularly if it gets under your fingernails.

Of the four halogens, iodine is the weakest oxidizing agent. Many years ago, "tincture of iodine," a 10% solution of $I_2$ in alcohol, was widely used as an antiseptic. Today, hospitals use a product called "povidone-iodine," a quite powerful iodine-containing antiseptic and disinfectant, which can be diluted with water to the desired strength. These applications of molecular iodine should not delude you into thinking that the solid is harmless. On the contrary, if $I_2(s)$ is allowed to remain in contact with your skin, it can cause painful burns that are slow to heal.

> When the senior author (WLM) was young

## Occurrence and Preparation

Of the eight nonmetals considered here, three (nitrogen, oxygen, and sulfur) occur in nature in elemental form. *Nitrogen* and *oxygen* are obtained from air, where their mole fractions are 0.7808 and 0.2095, respectively. When liquid air at $-200°C$ (73 K) is allowed to warm, the first substance that boils off is nitrogen (bp $N_2$ = 77 K). After most of the nitrogen has been removed, further warming gives oxygen (bp $O_2$ = 90 K). About $2 \times 10^{10}$ kg of $O_2$ and lesser amounts of $N_2$ are produced annually in the United States from liquid air.

At the close of the Civil War in 1865, oil prospectors in Louisiana discovered (to their disgust) elemental *sulfur* in the caprock of vast salt domes up to 20 km$^2$ in area. The sulfur lies 60 to 600 m below the surface of the earth. The process used to mine sulfur is named after its inventor, Herman Frasch, an American chemical engineer (born in Germany). A diagram of the Frasch process is shown in Figure 21.4. The sulfur is heated to its melting point (119°C) by pumping superheated water at 165°C down one of three concentric pipes. Compressed air is used to bring the sulfur to the surface. The air and sulfur form a frothy

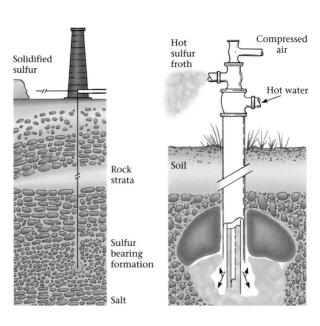

**Figure 21.4**

Frasch process for mining sulfur. Superheated water at 165°C is sent down through the outer pipe to form a pool of molten sulfur (mp = 119°C) at the base. Compressed air, pumped down the inner pipe, brings the sulfur to the surface. Sulfur deposits are often 100 m or more beneath the Earth's surface, covered with quicksand and rock.

mixture that rises through the middle pipe. Upon cooling, the sulfur solidifies, filling huge vats that may be 0.5 km long. The sulfur obtained in this way has a purity approaching 99.9%.

The *halogens* are far too reactive to occur in nature as the free elements. Instead, they are found as anions:

— $F^-$ in the mineral calcium fluoride, $CaF_2$ (fluorite)
— $Cl^-$ in huge underground deposits of sodium chloride, $NaCl$ (rock salt), underlying parts of Oklahoma, Texas, and Kansas
— $Br^-(aq)$ and $I^-(aq)$ in brine wells in Arkansas (conc. $Br^- = 0.05\ M$) and Michigan (conc. $I^- = 0.001\ M$), respectively

$Br_2$ and $I_2$ can also be obtained from seawater

The fluoride and chloride ions are very difficult to oxidize ($E^\circ_{ox}\ F^- = -2.889$ V; $E^\circ_{ox}\ Cl^- = -1.360$ V). Hence, the elements fluorine and chlorine are ordinarily prepared by electrolytic oxidation, using a high voltage. As pointed out in Chapter 18, chlorine is prepared by the electrolysis of aqueous sodium chloride:

$$2Cl^-(aq) + 2H_2O \longrightarrow Cl_2(g) + H_2(g) + 2OH^-(aq)$$

The process used to prepare fluorine was developed by Henri Moissan in Paris more than a century ago; it won him one of the early (1906) Nobel Prizes in chemistry. The electrolyte is a mixture of HF and KF in a 2:1 mole ratio. At 100°C, fluorine is generated by the decomposition of hydrogen fluoride:

$$2HF(l) \longrightarrow H_2(g) + F_2(g)$$

Potassium fluoride furnishes the ions required to carry the electric current.

Since bromide and iodide ions are easier to oxidize ($E^\circ_{ox}\ Br^- = -1.077$ V; $E^\circ_{ox}\ I^- = -0.534$ V), bromine and iodine can be prepared by chemical oxidation. Commonly, the oxidizing agent is chlorine gas ($E^\circ_{red} = +1.360$ V):

$$Cl_2(g) + 2Br^-(aq) \longrightarrow 2Cl^-(aq) + Br_2(l)$$

$$Cl_2(g) + 2I^-(aq) \longrightarrow 2Cl^-(aq) + I_2(s)$$

In the general chemistry laboratory, these same reactions are often used to test for $Br^-$ and $I^-$ ions (Fig. 21.5).

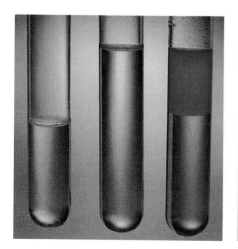

(a)

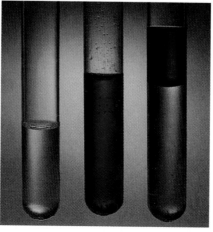

(b)

**Figure 21.5**
The reaction of chlorine with $Br^-$ and $I^-$ ions is shown in (a) and (b), respectively. The anions themselves (test tubes at left) are colorless. Oxidation by chlorine yields the free halogens, $Br_2$ and $I_2$. which are colored in water solution (center tubes) and more intensely colored in an organic solvent (upper layer in the tubes at right).

## Allotropy

Two or more different structural forms of an element in the same phase (gas, liquid, or solid) are referred to as ***allotropes.*** One of the simplest forms of allotropy is shown by *oxygen,* which can exist in the gas phase of either $O_2$ or $O_3$ (ozone). Commercially, ozone is prepared by passing $O_2$ gas through a high-voltage ($10^4$ V) electric discharge. At atmospheric pressure, the reaction

$$2O_3(g) \longrightarrow 3O_2(g) \qquad \Delta H° = -285.4 \text{ kJ} \quad \Delta S° = +137.5 \text{ J/K}$$

is thermodynamically spontaneous at all temperatures. Kinetically, however, ozone stays around long enough to find some application as a substitute for chlorine in disinfecting municipal water supplies.

*Phosphorus* forms several allotropes in the solid state, of which white and red phosphorus are the most common.

**1. *White phosphorus*** consists of $P_4$ molecules with the structure shown in Figure 21.6. It is a soft, waxy substance with a low melting point (44°C) and boiling point (280°C). Like most molecular substances, white phosphorus is readily soluble in such nonpolar solvents as $CCl_4$. The chemical reactivity of white phosphorus is so great that it is stored under water to protect it from $O_2$. A piece of $P_4$ exposed to air in a dark room glows because of the light given off upon oxidation (Fig. 21.7). White phosphorus is extremely toxic. As little as 0.1 g taken internally can be fatal. Direct contact with the skin produces painful burns.

**2. *Red phosphorus*** is the form in which the element is usually found in the laboratory. This allotrope has properties quite different from those of white phosphorus. It is much higher-melting (mp = 590°C at 43 atm) and insoluble in common solvents. The low volatility of red phosphorus makes it much less toxic than the white form. It is also less reactive and must be heated to 250°C to burn in air. These properties are consistent with the structure of red phosphorus, which is known to be network covalent (Fig. 21.6). This allotrope can be made by heating white phosphorus in the absence of air to about 300°C.

Before the undesirable properties of white phosphorus were known, it was used in matches. Today, two different kinds of matches are available, neither of which contains white phosphorus. The heads of "strike-anywhere" matches contain a mixture of a sulfide of phosphorus, $P_4S_3$, potassium chlorate, $KClO_3$, and

White phosphorus is used in napalm, a really nasty chemical weapon

**Figure 21.6**
At the top of the figure are shown the two most common allotropes of phosphorus, the white and the red forms. Below are their structures. White phosphorus is molecular, formula $P_4$. Red phosphorus has a network structure, shown here in simplified form.

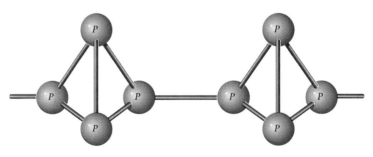

White phosphorus                               Red phosphorus

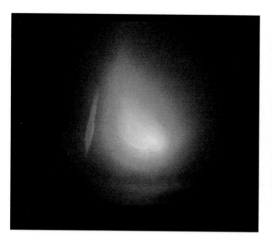
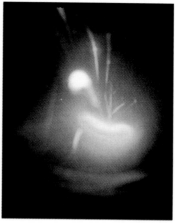

**Figure 21.7**
When white phosphorus is exposed to oxygen gas, it first glows (phosphorescence) and then bursts into flame. The reaction is
$$P_4(s) + 5O_2(g) \longrightarrow P_4O_{10}(s).$$

powdered glass. When struck against a rough surface, the mixture ignites. Safety matches contain sulfur and potassium chlorate; the special surface against which they are struck contains red phosphorus and powdered glass. Friction sets off a reaction between red phosphorus and $KClO_3$.

---

**Example 21.2**  For the allotropic conversion P(white) $\longrightarrow$ P(red), $\Delta H°$ is −17.6 kJ and $\Delta S°$ is −18.3 J/K. At what temperature are the two allotropes in equilibrium at 1 atm?

**Strategy**  A system is at equilibrium at 1 atm when $\Delta G° = 0$. To find the temperature at which that happens, use the Gibbs-Helmholtz equation: $\Delta G° = \Delta H° − T\Delta S°$.

**Solution**  Setting $\Delta G° = 0$, it follows that $\Delta H° = T\Delta S°$. Hence

$$T = \frac{\Delta H°}{\Delta S°} = \frac{-17.6 \text{ kJ}}{-0.0183 \text{ kJ/K}} = 962 \text{ K} \approx \boxed{690°C}$$

At any temperature below about 700°C, red phosphorus is the stable allotrope.

---

In the solid state, **sulfur** can have more than 20 different allotropic forms. You will no doubt be relieved to learn that we will talk about only two of these, rhombic and monoclinic sulfur (Fig. 21.8). Both consist of ring molecules of formula $S_8$. They differ only in the way that molecules are packed in the solid, reflected in different crystal structures.

At room temperature, rhombic sulfur is the stable allotrope. However, since the process

$$S_8(\text{rhombic}) \longrightarrow S_8(\text{monoclinic}) \qquad \Delta H° = +2.6 \text{ kJ}$$

is endothermic, the monoclinic form is stable at high temperatures. The two forms are in equilibrium at 96°C; if rhombic sulfur is heated to that temperature, it slowly converts to monoclinic sulfur. More commonly, the monoclinic allotrope is prepared by freezing liquid sulfur at the melting point (119°C) and then cooling quickly to room temperature. Typically, at 25°C, monoclinic crystals stay around for a day or more before converting to the rhombic form.

The free-flowing, pale yellow liquid formed when sulfur melts contains $S_8$ molecules. However, upon heating to 160°C, a striking change occurs. The liquid becomes so viscous that it cannot be poured readily. At the same time its color changes to a deep reddish-brown. These effects reflect a change in mole-

**Figure 21.8**
Naturally occurring crystals of monoclinic sulfur (above) and rhombic sulfur (below). ((a) Gregory G. Dimijian/Photo Researchers, Inc. (b) Kip Peticolas/Fundamental Photographs, Inc.)

cular structure. The $S_8$ rings break apart and then link to one another to form long chains such as

Under these conditions, sulfur is a polymer

Liquid sulfur between 160 and 250°C contains a high proportion of such chains. They vary in length from eight to perhaps a million atoms. The chains become tangled, producing a highly viscous liquid. The deep color is due to the absorption of light by the unpaired electrons at the ends of the chains.

If liquid sulfur at 200°C is quickly poured into water, a rubbery mass results. This is referred to as "plastic sulfur." It consists of long-chain molecules that did not have time to rearrange to the $S_8$ molecules stable at room temperature. Within a few hours, the plastic sulfur loses its elasticity as it converts to rhombic crystals.

## 21.2 Hydrogen Compounds of Nonmetals

Table 21.2 lists some of the more important nonmetal hydrides (the hydrogen compounds of carbon are discussed in Chap. 22). The physical states listed are those observed at 25°C and 1 atm. The remainder of this section is devoted to a discussion of the chemical properties of the compounds shown in boldface in the table.

### Ammonia, $NH_3$

Ammonia is one of the most important industrial chemicals; more than ten million tons of $NH_3$ are produced annually in the United States. You will recall (Chap. 12) that it is made by the Haber process

$$N_2(g) + 3H_2(g) \rightleftharpoons 2NH_3(g) \qquad 450°C, 200-600 \text{ atm, solid catalyst}$$

Ammonia is used to make fertilizers and a host of different nitrogen compounds, notably nitric acid, $HNO_3$.

The $NH_3$ molecule acts as a *Brønsted-Lowry base* in water, accepting a proton from a water molecule:

$$NH_3(aq) + H_2O \rightleftharpoons NH_4^+(aq) + OH^-(aq)$$

TABLE 21.2 **Hydrogen Compounds of the Nonmetals**

| Group 15 | Group 16 | Group 17 |
|---|---|---|
| **Ammonia, $NH_3(g)$** | Water, $H_2O(l)$ | **Hydrogen fluoride, $HF(g)$** |
| Hydrazine, $N_2H_4(l)$ | **Hydrogen peroxide, $H_2O_2(l)$** | |
| Hydrazoic acid, $HN_3(l)$ | | |
| Phosphine, $PH_3(g)$ | **Hydrogen sulfide, $H_2S(g)$** | **Hydrogen chloride, $HCl(g)$** |
| Diphosphine, $P_2H_4(l)$ | | |
| | | Hydrogen bromide, $HBr(g)$ |
| | | Hydrogen iodide, $HI(g)$ |

Ammonia can also act as a *Lewis base* when it reacts with a metal cation to form a complex ion

$$2NH_3(aq) + Ag^+(aq) \longrightarrow Ag(NH_3)_2^+(aq)$$

The $NH_3$ molecule donates an electron pair to $Ag^+$

or with an electron-deficient compound such as boron trifluoride:

Lewis base    Lewis acid

Ammonia is often used to precipitate insoluble metal hydroxides such as $Al(OH)_3$. The $OH^-$ ions formed when ammonia reacts with water precipitate the cation from solution as the hydroxide. The overall equation for the reaction is

$$Al^{3+}(aq) + 3NH_3(aq) + 3H_2O \longrightarrow Al(OH)_3(s) + 3NH_4^+(aq)$$

Nitrogen cannot have an oxidation number lower than $-3$, which means that when $NH_3$ takes part in a redox reaction, it always acts as a *reducing agent.* Ammonia may be oxidized to elementary nitrogen or to a compound of nitrogen. An important redox reaction of ammonia is that with hypochlorite ion:

$$2NH_3(aq) + ClO^-(aq) \longrightarrow N_2H_4(aq) + Cl^-(aq) + H_2O$$

Hydrazine, $N_2H_4$, is made commercially by this process. Certain by-products of this reaction, notably $NH_2Cl$ and $NHCl_2$, are both toxic and explosive, so solutions of household bleach and ammonia should never be mixed with one another.

## Hydrogen Sulfide, $H_2S$

In water solution, hydrogen sulfide acts as a *Brønsted-Lowry acid;* it can donate a proton to a water molecule:

$$H_2S(aq) + H_2O \rightleftharpoons HS^-(aq) + H_3O^+(aq)$$

Like ammonia, hydrogen sulfide can act as a precipitating agent toward metal cations (recall the discussion of qualitative analysis in Chap. 16). The reaction with $Cd^{2+}$ is typical:

$$Cd^{2+}(aq) + H_2S(aq) \longrightarrow CdS(s) + 2H^+(aq)$$

Like ammonia, hydrogen sulfide (oxid. no. $S = -2$) can act only as a reducing agent when it takes part in redox reactions. Most often the $H_2S$ is oxidized to elementary sulfur, as in the reaction

$$2H_2S(aq) + O_2(g) \longrightarrow 2S(s) + 2H_2O$$

The sulfur formed is often very finely dispersed, which explains why aqueous solutions of hydrogen sulfide in contact with air have a milky appearance.

If you've worked with $H_2S$ in the laboratory, you won't soon forget its rotten-egg odor. In a sense, it's fortunate that hydrogen sulfide has such a distinctive odor. The gas is highly toxic, as poisonous as HCN. At a concentration of 10 parts per million, $H_2S$ can cause headaches and nausea; at 100 ppm it can be fatal.

**Example 21.3** When a solution containing $Cu^{2+}$ is treated with hydrogen sulfide, a black precipitate forms. When another portion of the solution is treated with ammonia, a blue precipitate forms. This precipitate dissolves in excess ammonia to form a deep blue solution containing the $Cu(NH_3)_4^{2+}$ ion. Write balanced net ionic equations to explain these observations.

**Strategy** Before you can write the equations, you must identify the products. The black precipitate is CuS. The blue precipitate must be copper(II) hydroxide, $Cu(OH)_2$. The identity of the final product is given: $Cu(NH_3)_4^{2+}$. The equations are readily written if you know the formulas of the products.

**Solution**

$$Cu^{2+}(aq) + H_2S(aq) \longrightarrow CuS(s) + 2H^+(aq)$$

$$Cu^{2+}(aq) + 2NH_3(aq) + 2H_2O \longrightarrow Cu(OH)_2(s) + 2NH_4^+(aq)$$

$$Cu(OH)_2(s) + 4NH_3(aq) \longrightarrow Cu(NH_3)_4^{2+}(aq) + 2OH^-(aq)$$

## Hydrogen Peroxide

In hydrogen peroxide, oxygen has an oxidation number of $-1$, intermediate between the extremes for the element, 0 and $-2$. This means that $H_2O_2$ can act as either an oxidizing agent, in which case it is reduced to $H_2O$, or as a reducing agent, when it is oxidized to $O_2$. In practice, hydrogen peroxide is an extremely strong oxidizing agent:

$$H_2O_2(aq) + 2H^+(aq) + 2e^- \longrightarrow 2H_2O \qquad\qquad E^\circ_{red} = +1.763 \text{ V}$$

but a very weak reducing agent:

$$H_2O_2(aq) \longrightarrow O_2(g) + 2H^+(aq) + 2e^- \qquad\qquad E^\circ_{ox} = -0.695 \text{ V}$$

Hydrogen peroxide tends to decompose in water, which explains why its solutions soon lose their oxidizing power. The reaction involved is *disproportionation*, combining the two half-reactions referred to above:

Half of the $H_2O_2$ molecules are reduced to $H_2O$; half are oxidized to $O_2$

$$\begin{array}{ll} H_2O_2(aq) + 2H^+(aq) + 2e^- \longrightarrow 2H_2O & E^\circ_{red} = +1.763 \text{ V} \\ H_2O_2(aq) \longrightarrow O_2(g) + 2H^+(aq) + 2e^- & E^\circ_{ox} = -0.695 \text{ V} \\ \hline 2H_2O_2(aq) \longrightarrow O_2(g) + 2H_2O & E^\circ = +1.068 \text{ V} \end{array}$$

This reaction is catalyzed by a wide variety of materials, including $I^-$ ions, $MnO_2$, and metal surfaces (Pt, Ag), and even by traces of $OH^-$ ions dissolved from glass.

You are most likely to come across hydrogen peroxide as its water solution. Solutions available in a drugstore for medicinal use contain 3 to 6 percent by mass of $H_2O_2$. Industrially, concentrations as high as 86% $H_2O_2$ are available. All water solutions of hydrogen peroxide contain stabilizers to prevent disproportionation from taking place during storage.

**Example 21.4** Using Table 18.1, p. 504, decide whether hydrogen peroxide will react with (oxidize or reduce) the following ions in acid solution at standard concentrations.

    (a) $I^-$   (b) $Sn^{2+}$   (c) $Co^{2+}$

**Strategy**    For a redox reaction to occur spontaneously, either

$$E^\circ_{ox}\ X + E^\circ_{red}\ H_2O_2 > 0$$

or

$$E^\circ_{red}\ X + E^\circ_{ox}\ H_2O_2 > 0$$

where $X$ is one of the ions listed and $E^\circ_{red}\ H_2O_2 = +1.763$ V; $E^\circ_{ox}\ H_2O_2 = -0.695$ V.

**Solution**

(a) $E^\circ_{red}\ H_2O_2 + E^\circ_{ox}\ I^- = +1.763$ V $- 0.534$ V $= +1.229$ V

> Hydrogen peroxide oxidizes $I^-$ ions to $I_2$.

(b) $E^\circ_{red}\ H_2O_2 + E^\circ_{ox}\ Sn^{2+} = +1.763$ V $- 0.154$ V $= +1.609$ V

$E^\circ_{ox}\ H_2O_2 + E^\circ_{red}\ Sn^{2+} = -0.695$ V $- 0.141$ V $= -0.836$ V

> Hydrogen peroxide oxidizes $Sn^{2+}$ to $Sn^{4+}$.

(c) $Co^{2+}$ can be reduced to Co ($E^\circ_{red} = -0.282$ V) or oxidized to $Co^{3+}$ ($E^\circ_{ox} = -1.953$ V). Neither gives a positive combination with $H_2O_2$;   no reaction.

## Hydrogen Fluoride and Hydrogen Chloride

The most common hydrogen halides are HF (annual U.S. production = $3 \times 10^8$ kg) and HCl ($3 \times 10^9$ kg/yr). They are most familiar as water solutions, referred to as hydrofluoric acid and hydrochloric acid, respectively. Recall (Chap. 13) that hydrofluoric acid is weak, incompletely dissociated in water, whereas HCl is a strong acid.

$$HF(aq) \rightleftharpoons H^+(aq) + F^-(aq) \qquad\qquad K_a = 6.9 \times 10^{-4}$$

$$HCl(aq) \longrightarrow H^+(aq) + Cl^-(aq) \qquad\qquad K_a \longrightarrow \infty$$

Hydrofluoric and hydrochloric acids undergo very similar reactions with bases such as $OH^-$ or $CO_3^{2-}$ ions. The equations for these reactions look somewhat different because of the difference in acid strength. Thus, for the reaction of hydrochloric acid with a solution of sodium hydroxide, the equation is simply

$$H^+(aq) + OH^-(aq) \longrightarrow H_2O$$

With hydrofluoric acid, the equation is

$$HF(aq) + OH^-(aq) \longrightarrow H_2O + F^-(aq)$$

HF appears in this equation because hydrofluoric acid is weak, containing many more HF molecules than $H^+$ ions. Similarly, for the reaction of these two acids with a solution of sodium carbonate:

hydrochloric acid: $2H^+(aq) + CO_3^{2-}(aq) \longrightarrow CO_2(g) + H_2O$

hydrofluoric acid: $2HF(aq) + CO_3^{2-}(aq) \longrightarrow CO_2(g) + H_2O + 2F^-(aq)$

Concentrated hydrofluoric acid reacts with glass, which can be considered to be a mixture of $SiO_2$ and ionic silicates such as calcium silicate, $CaSiO_3$:

> Hydrochloric acid contains $H^+$ and $Cl^-$ ions; hydrofluoric acid contains mostly HF molecules

$$SiO_2(s) + 4HF(aq) \longrightarrow SiF_4(g) + 2H_2O$$

$$CaSiO_3(s) + 6HF(aq) \longrightarrow SiF_4(g) + CaF_2(s) + 3H_2O$$

As you might guess, HF solutions are never stored in glass bottles; plastic is used instead. Hydrofluoric acid is sometimes used to etch glass. The glass object is first covered with a thin protective coating of wax or plastic. Then the coating is removed from the area to be etched and the glass is exposed to the HF solution. Thermometer stems and burets can be etched or light bulbs frosted in this way.

Hydrogen fluoride is a very unpleasant chemical to work with. If spilled on the skin, it removes $Ca^{2+}$ ions from the tissues, forming insoluble $CaF_2$. A white patch forms that is agonizingly painful to the touch. To make matters worse, HF is a local anesthetic, so a person may be unaware of what's happening until it's too late.

Rest assured that you will never have occasion to use hydrofluoric acid in the general chemistry laboratory. Neither are you likely to come in contact with HBr or HI, both of which are relatively expensive. Hydrochloric acid, on the other hand, is a "workhorse" chemical of the teaching laboratory. It is commonly available as "dilute HCl" (6 M) or "concentrated HCl" (12 M). You will use it as a source of $H^+$ ions for such purposes as

— dissolving insoluble carbonates or hydroxides

$$Ag_2CO_3(s) + 2H^+(aq) \longrightarrow 2Ag^+(aq) + CO_2(g) + H_2O$$

$$Zn(OH)_2(s) + 2H^+(aq) \longrightarrow Zn^{2+}(aq) + 2H_2O$$

— converting a weak base such as $NH_3$ to its conjugate acid

$$NH_3(aq) + H^+(aq) \longrightarrow NH_4^+(aq)$$

— generating $H_2(g)$ by reaction with a metal

$$Zn(s) + 2H^+(aq) \longrightarrow Zn^{2+}(aq) + H_2(g)$$

---

**Example 21.5**  Write balanced equations to explain why

(a) aluminum hydroxide dissolves in hydrochloric acid.
(b) carbon dioxide gas is evolved when calcium carbonate is treated with hydrochloric acid.
(c) carbon dioxide gas is evolved when calcium carbonate is treated with hydrofluoric acid.

**Strategy**  In each case, decide upon the nature of the products. You can reason by analogy with the equations cited in the text.

**Solution**

(a)  $Al(OH)_3(s) + 3H^+(aq) \longrightarrow Al^{3+}(aq) + 3H_2O$

(b)  $CaCO_3(s) + 2H^+(aq) \longrightarrow Ca^{2+}(aq) + CO_2(g) + H_2O$

(c)  $CaCO_3(s) + 2HF(aq) \longrightarrow Ca^{2+}(aq) + CO_2(g) + H_2O + 2F^-(aq)$

## 21.3 Oxygen Compounds of Nonmetals

Table 21.3 lists some of the more familiar nonmetal oxides. Curiously enough, only five of the 21 compounds shown are thermodynamically stable at 25°C and 1 atm ($P_4O_{10}$, $P_4O_6$, $SO_3$, $SO_2$, $I_2O_5$). The others, including all of the oxides of nitrogen and chlorine, have positive free energies of formation at 25°C and 1 atm. For example, $\Delta G_f^\circ$ $NO_2(g)$ = +51.3 kJ; $\Delta G_f^\circ$ $ClO_2(g)$ = +120.6 kJ. Kinetically, these compounds stay around long enough to have an extensive chemistry. Nitrogen dioxide, a reddish-brown gas, is a major factor in the formation of photochemical smog (Fig. 21.9). Chlorine dioxide, a yellow gas, is widely used as an industrial bleach and water purifier, even though it tends to explode at partial pressures higher than 50 mm Hg.

All such compounds are potentially unstable

TABLE 21.3   **Nonmetal Oxides***

| Group 15 | Group 16 | Group 17 |
|---|---|---|
| $N_2O_5(s)$, $N_2O_4(g)$, $NO_2(g)$ | | $OF_2(g)$, $O_2F_2(g)$ |
| $N_2O_3(d)$, $NO(g)$, $N_2O(g)$ | | |
| $P_4O_{10}(s)$, $P_4O_6(s)$ | $SO_3(l)$, $SO_2(g)$ | $Cl_2O_7(l)$, $Cl_2O_6(l)$ |
| | | $ClO_2(g)$, $Cl_2O(g)$ |
| | | $BrO_2(d)$, $Br_2O(d)$ |
| | | $I_2O_5(s)$, $I_4O_9(s)$, $I_2O_4(s)$ |

*The states listed are those observed at 25°C and 1 atm. Compounds that decompose below 25°C are listed as *(d)*. Oxides shown in boldface are discussed in the text.

### Molecular Structures of Nonmetal Oxides

The Lewis structures of the oxides of nitrogen are shown in Figure 21.10, p. 594. Two of these species, NO and $NO_2$, are paramagnetic, with one unpaired electron. When nitrogen dioxide is cooled, it dimerizes; the unpaired electrons combine to form a single bond between the two nitrogen atoms:

**Figure 21.9**
The brown haze covering a city is pollution caused by $NO_2$. (Walter Hodges/Tony Stone Images)

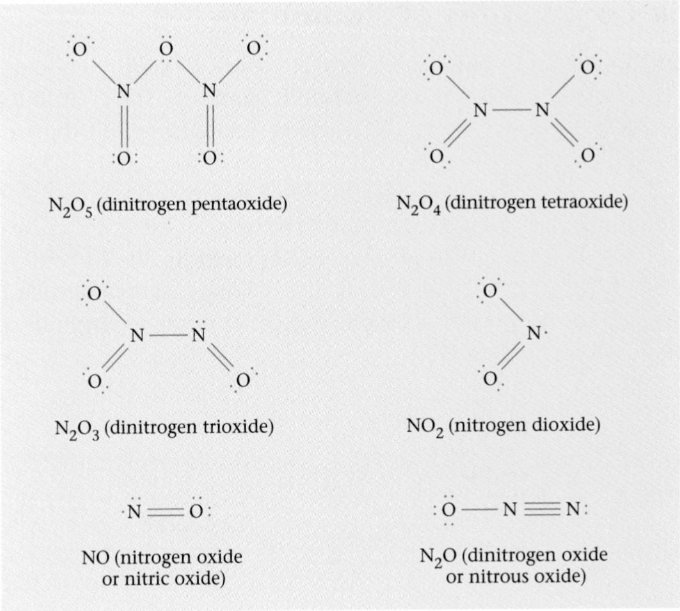

**Figure 21.10**
Lewis structures of the oxides of nitrogen. Many other resonance forms are possible.

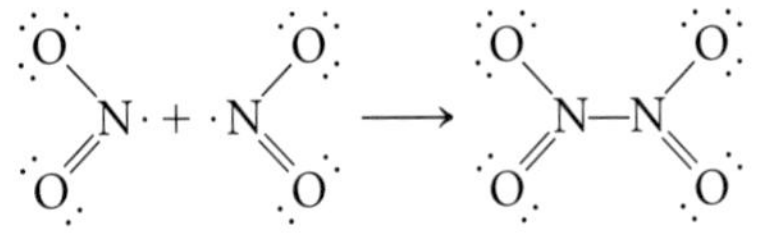

N₂O₅ (dinitrogen pentaoxide)  N₂O₄ (dinitrogen tetraoxide)

N₂O₃ (dinitrogen trioxide)  NO₂ (nitrogen dioxide)

NO (nitrogen oxide or nitric oxide)  N₂O (dinitrogen oxide or nitrous oxide)

$$2NO_2(g) \rightleftharpoons N_2O_4(g)$$

A similar reaction occurs when an equimolar mixture of NO and NO₂ is cooled. Two odd electrons, one from each molecule, pair off to form an N—N bond:

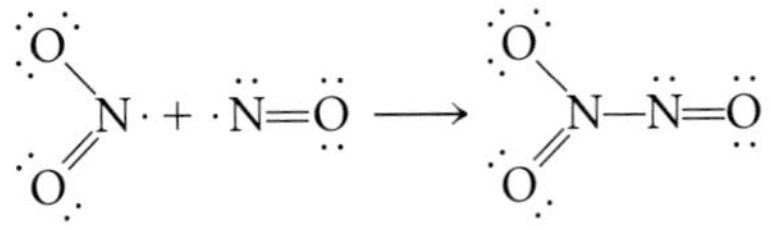

$$NO_2(g) + NO(g) \rightleftharpoons N_2O_3(g)$$

At $-20°C$, dinitrogen trioxide separates from the mixture as a blue liquid.

Perhaps the best-known oxide of nitrogen is $N_2O$, commonly called nitrous oxide or "laughing gas." Nitrous oxide is frequently used as an anesthetic, particularly in dentistry. It is also the propellant gas used in whipped cream containers; $N_2O$ is nontoxic, virtually tasteless, and quite soluble in vegetable oils. The $N_2O$ molecule, like all those in Figure 21.10, can be represented as a resonance hybrid.

---

**Example 21.6**  Consider the $N_2O$ molecule shown in Figure 21.10.

(a) Draw another resonance form of $N_2O$.
(b) What is the bond angle in $N_2O$?
(c) Is the $N_2O$ molecule polar or nonpolar?

**Strategy** Recall the discussion in Chapter 7, Sections 7.1 to 7.3, where the principles of resonance, molecular geometry, and polarity were considered.

*Solution*

(a)  $:\ddot{O}=N=\ddot{N}:$  or  $:O\equiv N-\ddot{\ddot{N}}:$

(b) In any of the resonance forms, the central nitrogen atom, insofar as geometry is concerned, would behave as if it were surrounded by two electron pairs.

The bond angle is 180°; the molecule is linear, like $BeF_2$.

(c) Polar (unsymmetrical).

---

The structures of $SO_2$ and $SO_3$ were referred to in Chapter 7. These molecules are often cited as examples of resonance; sulfur trioxide, for example, has three equivalent resonance structures:

As you might expect, the $SO_3$ molecule is nonpolar, with 120° bond angles.

Of the two oxides of phosphorus, $P_4O_6$ and $P_4O_{10}$, the latter is the more stable; it is formed when white phosphorus burns in air:

$$P_4(s) + 5O_2(g) \longrightarrow P_4O_{10}(s)$$

Note from Figure 21.11 that in both oxides, as in $P_4$ itself, the four phosphorus atoms are at the corners of a tetrahedron. The $P_4O_6$ molecule can be visualized as derived from $P_4$ by inserting an oxygen atom between each pair of phosphorus atoms. In $P_4O_{10}$, an extra oxygen atom is bonded to each phosphorus.

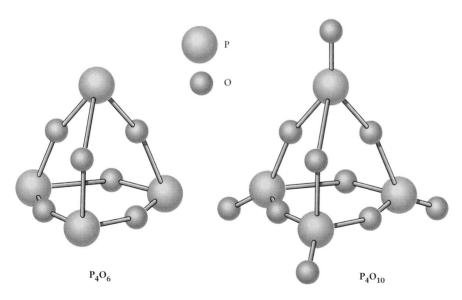

P

O

$P_4O_6$

$P_4O_{10}$

**Figure 21.11**
Structures of the oxides of phosphorus, simplest formulas $P_2O_3$ and $P_2O_5$.

## Reactions of Nonmetal Oxides with Water

Many nonmetal oxides react with water to form acids. Compounds that behave in this way are referred to as **acid anhydrides.** Looking at the reaction

$$SO_3(g) + H_2O(l) \longrightarrow H_2SO_4(l)$$

you can see that sulfur trioxide is the acid anhydride of sulfuric acid. Notice that, in this reaction, the nonmetal does not change oxidation number; sulfur is in the +6 state in both $SO_3$ and $H_2SO_4$. Other acid anhydrides include $N_2O_5$ and $N_2O_3$:

What is the acid derived from $CO_2$?
Ans. $H_2CO_3$

+5 nitrogen:  $N_2O_5(s) + H_2O(l) \longrightarrow 2HNO_3(l)$

+3 nitrogen:  $N_2O_3(g) + H_2O(l) \longrightarrow 2HNO_2(aq)$

The products here are nitric acid, $HNO_3$, and an aqueous solution of nitrous acid, $HNO_2$.

One of the most important reactions of this type involves the +5 oxide of phosphorus, $P_4O_{10}$. Here the product is phosphoric acid, $H_3PO_4$:

$$P_4O_{10}(s) + 6H_2O(l) \longrightarrow 4H_3PO_4(s)$$

This reaction is used to prepare high-purity phosphoric acid and salts of that acid for use in food products. Phosphoric acid, $H_3PO_4$, is added in small amounts to soft drinks to give them a tart taste. It is present to the extent of about 0.05 mass percent in colas, 0.01 mass percent in root beers.

---

**Example 21.7**    Give the formula of the acid anhydride of

(a) $H_3PO_3$    (b) $HClO$    (c) $H_2SO_3$

**Strategy**    Referring to Table 21.3, find an oxide in which the central nonmetal atom has the same oxidation number as in the acid.

*Solution*

(a) +3 P: $H_3PO_3$ and    $P_4O_6$

(b) +1 Cl: $HClO$ and    $Cl_2O$

(c) +4 S: $H_2SO_3$ and    $SO_2$

---

## 21.4    Oxoacids and Oxoanions

Table 21.4 lists some of the more important oxoacids of the nonmetals. In all these compounds, the ionizable hydrogen atoms are bonded to oxygen, not to the central nonmetal atom. Dissociation of one or more protons from the oxoacid gives the corresponding oxoanion (Fig. 21.12).

In this section, we discuss the principles that allow you to predict the relative acid strengths of oxoacids such as those listed in Table 21.4. Then we consider their strengths as oxidizing and/or reducing agents. Finally, we take a closer look at the chemistry of three important oxoacids: $HNO_3$, $H_2SO_4$, and $H_3PO_4$.

TABLE 21.4    **Oxoacids of the Nonmetals**

| Group 15 | Group 16 | Group 17 |
|---|---|---|
| $HNO_3$, $HNO_2$* $H_3PO_4$, $H_3PO_3$ | $H_2SO_4$, $H_2SO_3$* | $HClO_4$, $HClO_3$*, $HClO_2$*, $HClO$* $HBrO_4$*, $HBrO_3$*, $HBrO$* $HIO_4$, $H_5IO_6$, $HIO_3$, $HIO$* |

*These compounds cannot be isolated from water solution.

## Acid Strength

The ionization constants of the oxoacids of the halogens are listed in Table 21.5. Notice that the value of the ionization constant, $K_a$, increases with

— *increasing oxidation number of the central atom* ($HClO < HClO_2 < HClO_3 < HClO_4$)
— *increasing electronegativity of the central atom* ($HIO < HBrO < HClO$)

These trends are general ones, observed with other oxoacids of the nonmetals. Recall, for example, that nitric acid, $HNO_3$ (oxid. no. N = +5), is a strong acid, completely ionized in water. In contrast, nitrous acid, $HNO_2$ (oxid. no. N = +3) is a weak acid ($K_a = 6.0 \times 10^{-4}$). The electronegativity effect shows up with the strengths of the oxoacids of sulfur and selenium:

$$K_{a1} \ H_2SO_3 = 1.7 \times 10^{-2} \qquad K_{a1} \ H_2SeO_3 = 2.7 \times 10^{-3} \ (\text{E.N. S} = 2.6, \text{Se} = 2.5)$$

Trends in acid strength can be explained in terms of molecular structure. In an oxoacid molecule, the hydrogen atom that dissociates is bonded to oxygen, which in turn is bonded to a nonmetal atom, X. The ionization in water of an oxoacid H—O—X can be represented as

$$\text{H—O—X}(aq) \Longrightarrow \text{H}^+(aq) + \text{XO}^-(aq)$$

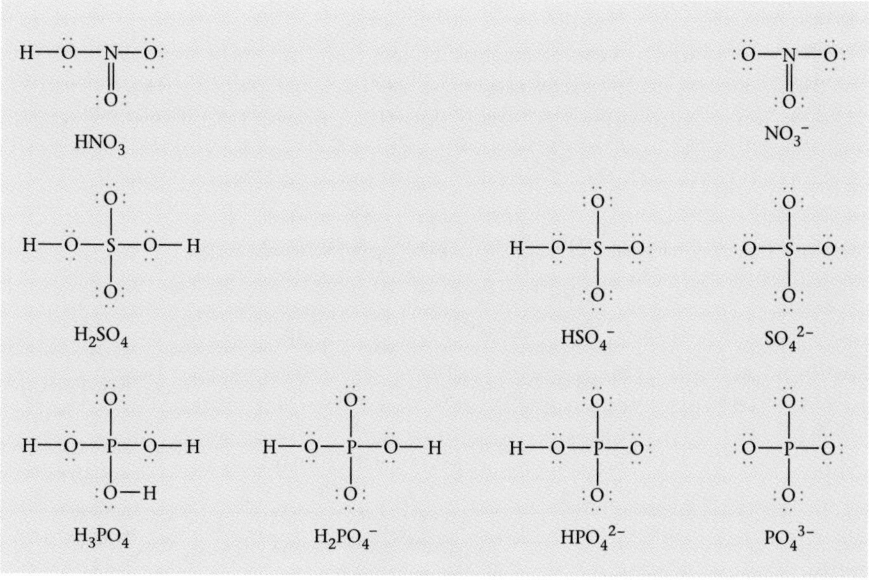

**Figure 21.12**
Lewis structures of the oxoacids $HNO_3$, $H_2SO_4$, $H_3PO_4$, and the oxoanions derived from them.

TABLE 21.5 **Ionization Constants of Oxoacids of the Halogens**

| Oxid. State | | $K_a$ | | $K_a$ | | $K_a$ |
|---|---|---|---|---|---|---|
| +7 | $HClO_4$ | $\sim 10^7$ | $HBrO_4$ | $\sim 10^6$ | $HIO_4^*$ | $1.4 \times 10^1$ |
| +5 | $HClO_3$ | $\sim 10^3$ | $HBrO_3$ | 3.0 | $HIO_3$ | $1.6 \times 10^{-1}$ |
| +3 | $HClO_2$ | $1.0 \times 10^{-2}$ | | | | |
| +1 | $HClO$ | $2.8 \times 10^{-8}$ | $HBrO$ | $2.6 \times 10^{-9}$ | $HIO$ | $2.4 \times 10^{-11}$ |

*Estimated; in water solution the stable species is $H_5IO_6$, whose first dissociation constant is $5 \times 10^{-4}$.

For a proton, with its +1 charge, to separate from the molecule, the electron density around the oxygen should be as low as possible. This will weaken the O—H bond and favor ionization. The electron density around the oxygen atom is decreased when

— *X is a highly electronegative atom such as Cl.* This draws electrons away from the oxygen atom and makes hypochlorous acid stronger than hypoiodous acid.
— *Additional, strongly electronegative oxygen atoms are bonded to X.* These tend to draw electrons away from the oxygen atom bonded to H. Thus, we would predict that the ease of dissociation of a proton, and hence $K_a$, should increase in the following order, from left to right:

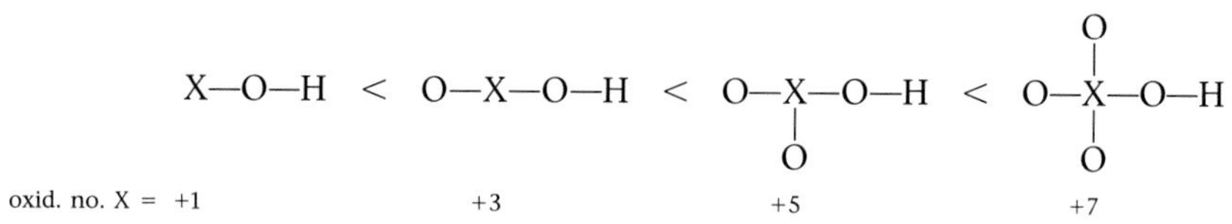

$$X-O-H \quad < \quad \underset{O}{O-X-O-H} \quad < \quad \underset{\underset{O}{|}}{\overset{}{O-X-O-H}} \quad < \quad \underset{\underset{O}{|}}{\overset{\overset{O}{|}}{O-X-O-H}}$$

oxid. no. X = +1        +3        +5        +7

**Example 21.8**  Consider sulfurous acid, $H_2SO_3$.

(a) Show its Lewis structure and that of the $HSO_3^-$ and $SO_3^{2-}$ ions.
(b) How would its acid strength compare with that of $H_2SO_4$? $H_2TeO_3$?

**Strategy**  The structure can be obtained by removing an oxygen atom from $H_2SO_4$ (Fig. 21.12). Relative acid strengths can be predicted on the basis of the electronegativity and oxidation number of the central nonmetal atom, following the rules cited above.

**Solution**

(a)

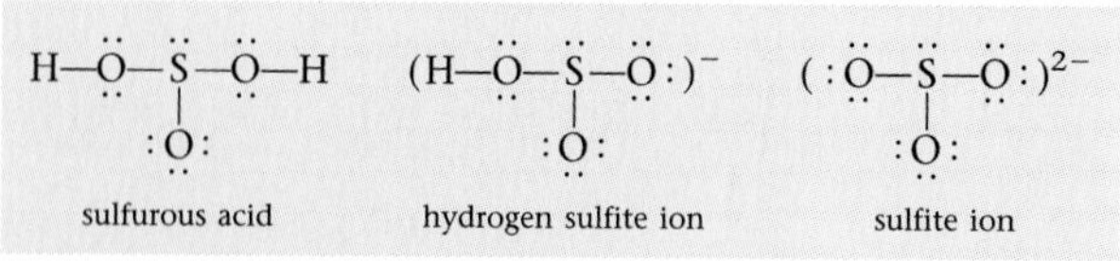

     sulfurous acid      hydrogen sulfite ion      sulfite ion

(b)  $H_2SO_3 < H_2SO_4$      (oxid. no. S = +4, +6)

    $H_2SO_3 > H_2TeO_3$      (S more electronegative than Te)

## Oxidizing and Reducing Strength

Many of the reactions of oxoacids and oxoanions involve oxidation and reduction. There are certain general principles that apply here, regardless of the particular species involved.

**1. *A species in which a nonmetal is in its highest oxidation state can act only as an oxidizing agent, never as a reducing agent.*** Consider, for example, the $ClO_4^-$ ion, in which chlorine is in its highest oxidation state, +7. In any redox reaction in which this ion takes part, chlorine must be reduced to a lower oxidation state. When that happens, the $ClO_4^-$ ions act as an oxidizing agent, taking electrons away from something else. The same argument applies to

*Many oxoanions are very powerful oxidizing agents*

— the $SO_4^{2-}$ ion (highest oxid. no. S = +6)
— the $NO_3^-$ ion (highest oxid. no. N = +5)

Note that, in general, *the highest oxidation number of a nonmetal is given by the second digit of its group number* (**17** for Cl, **16** for S, **15** for N).

**2. *A species in which a nonmetal is in an intermediate oxidation state can act as either an oxidizing agent or a reducing agent.*** Consider, for example, the $ClO_3^-$ ion (oxid. no. Cl = +5). It can be oxidized to the perchlorate ion, in which case $ClO_3^-$ acts as a reducing agent:

$$ClO_3^-(aq) + H_2O \longrightarrow ClO_4^-(aq) + 2H^+(aq) + 2e^- \qquad E_{ox}^\circ = -1.226 \text{ V}$$

Alternatively, the $ClO_3^-$ ion can be reduced, perhaps to a $Cl^-$ ion. When that occurs, $ClO_3^-$ acts as an oxidizing agent:

$$ClO_3^-(aq) + 6H^+(aq) + 6e^- \longrightarrow Cl^-(aq) + 3H_2O \qquad E_{red}^\circ = +1.442 \text{ V}$$

Notice that the $ClO_3^-$ ion is a much stronger oxidizing agent ($E_{red}^\circ = +1.442$ V) than reducing agent ($E_{ox}^\circ = -1.226$ V). This is generally true of oxoanions and oxoacids in an intermediate oxidation state, at least in acidic solution. Compare, for example

HClO      $E_{red}^\circ$ (to $Cl_2$) = +1.630 V; $E_{ox}^\circ$ (to $HClO_2$) = −1.157 V

HNO₂      $E_{red}^\circ$ (to NO) = +1.036 V; $E_{ox}^\circ$ (to $NO_2$) = −1.056 V

**3. *Sometimes, with a species such as $ClO_3^-$, oxidation and reduction occur together, resulting in disproportionation:***

$$4ClO_3^-(aq) \longrightarrow 3ClO_4^-(aq) + Cl^-(aq) \qquad E^\circ = +0.216 \text{ V}$$

In general, a *species in an intermediate oxidation state is expected to disproportionate if the sum $E_{ox}^\circ + E_{red}^\circ$ is a positive number.*

**4. *The oxidizing strength of an oxoacid or oxoanion is greatest at high [$H^+$] (low pH). Conversely, its reducing strength is greatest at low [$H^+$] (high pH).***

This principle has a simple explanation. Looking back at the half-equations written on p. 599, you can see that

— when $ClO_3^-$ acts as an oxidizing agent, the $H^+$ ion is a reactant, so increasing its concentration makes the process more spontaneous.
— when $ClO_3^-$ acts as a reducing agent, the $H^+$ ion is a product; to make the process more spontaneous, [$H^+$] should be lowered.

---

**Example 21.9**  Calculate $E_{red}$ and $E_{ox}$ for the $ClO_3^-$ ion in neutral solution, at pH 7.00, assuming all other species are at standard concentration ($E^\circ_{red} = +1.442$ V; $E^\circ_{ox} = -1.226$ V). Will the $ClO_3^-$ ion disproportionate at pH 7.00?

**Strategy**  First (1) set up the Nernst equation for the reduction half-reaction and calculate $E_{red}$. (It's convenient here to use the base-10 form of the Nernst equation, since pH is involved.) Then (2) repeat the calculation for the oxidation half-reaction, finding $E_{ox}$. Finally (3), add $E_{red} + E_{ox}$; if the sum is positive, disproportionation should occur.

**Solution**

(1) $ClO_3^-(aq) + 6H^+(aq) + 6e^- \longrightarrow Cl^-(aq) + 3H_2O$

$$E_{red} = +1.442 \text{ V} - \frac{0.0592}{6} \log \frac{1}{[H^+]^6}$$

$\log 1/[H^+]^6 = -6 \log [H^+]$

$$= +1.442 \text{ V} + 0.0592 \log [H^+]$$

$$= +1.442 \text{ V} - 0.0592(\text{pH}) = \boxed{+1.028 \text{ V}}$$

As expected, decreasing the concentration of $H^+$ makes the half-reaction less spontaneous and hence makes $E_{red}$ less positive.

(2) $ClO_3^-(aq) + H_2O \longrightarrow ClO_4^-(aq) + 2H^+(aq) + 2e^-$

$$E_{ox} = -1.226 \text{ V} - \frac{0.0592}{2} \log [H^+]^2$$

$$= -1.226 \text{ V} - 0.0592 \log [H^+]$$

$$= -1.226 \text{ V} + 0.0592(\text{pH}) = \boxed{-0.812 \text{ V}}$$

(3) $E = +1.028 \text{ V} - 0.812 \text{ V} = +0.216 \text{ V}$; disproportionation should occur.

Notice that $E$ is the same as $E^\circ$, +0.216 V. You could have predicted that (and saved a lot of work!), since the overall equation for disproportionation

$$4ClO_3^-(aq) \longrightarrow 3ClO_4^-(aq) + Cl^-(aq)$$

does not involve $H^+$ or $OH^-$ ions.

---

## Nitric Acid, HNO₃

Commercially, nitric acid is made by a three-step process developed by the German physical chemist Wilhelm Ostwald (1853–1932). The starting material is ammonia, which is burned in an excess of air at 900°C, using a platinum-rhodium catalyst:

$$4NH_3(g) + 5O_2(g) \longrightarrow 4NO(g) + 6H_2O(g)$$

The gaseous mixture formed is cooled and mixed with more air to convert NO to $NO_2$:

$$2NO(g) + O_2(g) \longrightarrow 2NO_2(g)$$

Finally, nitrogen dioxide is bubbled through water to produce nitric acid:

$$3NO_2(g) + H_2O(l) \longrightarrow NO(g) + 2HNO_3(aq)$$

Nitric acid is a strong acid, completely ionized to $H^+$ and $NO_3^-$ ions in dilute water solution:

$$HNO_3(aq) \longrightarrow H^+(aq) + NO_3^-(aq)$$

Many of the reactions of nitric acid are those associated with all strong acids. For example, dilute (6 M) nitric acid can be used to dissolve aluminum hydroxide

$$Al(OH)_3(s) + 3H^+(aq) \longrightarrow Al^{3+}(aq) + 3H_2O$$

or to generate carbon dioxide gas from calcium carbonate:

$$CaCO_3(s) + 2H^+(aq) \longrightarrow Ca^{2+}(aq) + CO_2(g) + H_2O$$

Referring back to Example 21.5, you will find that these equations are identical with those written for the reactions of hydrochloric acid with $Al(OH)_3$ and $CaCO_3$. It is the $H^+$ ion that reacts in either case: $Cl^-$ and $NO_3^-$ ions take no part in the reactions and hence do not appear in the equation.

Concentrated (16 M) nitric acid is a strong oxidizing agent; here the nitrate ion is reduced to nitrogen dioxide. This happens when 16 M $HNO_3$ reacts with copper metal (Fig. 21.13):

$$Cu(s) + 4H^+(aq) + 2NO_3^-(aq) \longrightarrow Cu^{2+}(aq) + 2NO_2(g) + 2H_2O$$

Dilute nitric acid (6 M) is a weaker oxidizing agent than 16 M $HNO_3$. It also gives a wider variety of reduction products, depending upon the nature of the reducing agent. With inactive metals such as copper ($E°_{ox} = -0.339$ V), the major product is usually NO (oxid. no. N = +2):

$$3Cu(s) + 2NO_3^-(aq) + 8H^+(aq) \longrightarrow 3Cu^{2+}(aq) + 2NO(g) + 4H_2O$$

With very dilute acid and a strong reducing agent such as zinc ($E°_{ox} = +0.762$ V) reduction may go all the way to the $NH_4^+$ ion (oxid. no. N = −3):

$$4Zn(s) + NO_3^-(aq) + 10H^+(aq) \longrightarrow 4Zn^{2+}(aq) + NH_4^+(aq) + 3H_2O$$

As you would expect, the oxidizing strength of the $NO_3^-$ ion drops off sharply as pH increases; the reduction voltage (to NO) is +0.964 V at pH 0,

You can always smell $NO_2$ over 16 M $HNO_3$

**Figure 21.13**

Copper metal is comparatively inactive, but it reacts with concentrated nitric acid. The brown fumes are $NO_2(g)$, a reduction product of $HNO_3$. The copper is oxidized to $Cu^{2+}$ ions, which impart their color to the solution. (Marna G. Clarke)

+0.412 V at pH 7, and −0.140 V at pH 14. The nitrate ion is a very weak oxidizing agent in basic solution.

---

**Example 21.10**    Write a balanced net ionic equation for the reaction of nitric acid with insoluble copper(II) sulfide; the products include $Cu^{2+}$, $S(s)$, and $NO_2(g)$.

***Strategy***    Follow the general procedure for writing and balancing redox equations in Chapter 4 and reviewed in Chapter 18. Note that since nitric acid is strong, it should be represented as $H^+$ and $NO_3^-$ ions.

*Solution*

(1) The "skeleton" half-equations are

$$oxidation: \quad CuS(s) \longrightarrow S(s)$$

$$reduction: \quad NO_3^-(aq) \longrightarrow NO_2(g)$$

(2) The balanced half-equations are

$$oxidation: \quad CuS(s) \longrightarrow Cu^{2+}(aq) + S(s) + 2e^-$$

$$reduction: \quad NO_3^-(aq) + 2H^+(aq) + e^- \longrightarrow NO_2(g) + H_2O$$

(3) Multiplying the reduction half-equation by two and adding to the oxidation half-equation, you should obtain, after simplification,

$$CuS(s) + 2NO_3^-(aq) + 4H^+(aq) \longrightarrow Cu^{2+}(aq) + S(s) + 2NO_2(g) + 2H_2O$$

---

Concentrated nitric acid (16 *M*) is colorless when pure. In sunlight, it turns yellow (Fig. 21.14) because it decomposes to $NO_2(g)$:

$$4HNO_3(aq) \longrightarrow 4NO_2(g) + 2H_2O + O_2(g)$$

The yellow color that appears on your skin if it comes in contact with nitric acid has quite a different explanation. Nitric acid reacts with proteins to give a yellow material called xanthoprotein.

**Figure 21.14**
A water solution of nitric acid slowly turns yellow because of the $NO_2(g)$ formed by decomposition. Nitric acid also reacts with proteins (casein in milk [test tube] and albumen in eggs) to give a characteristic yellow color. (Marna G. Clarke and Charles D. Winters)

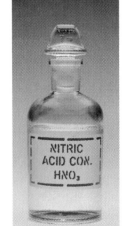
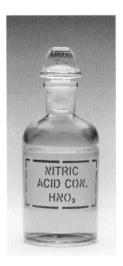
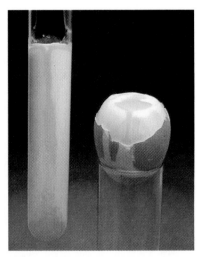

## Sulfuric Acid, $H_2SO_4$

You will recall (Chapter 12) that sulfuric acid is made by the *contact process,* using elementary sulfur, oxygen and water. The overall reaction is

$$S(s) + \tfrac{3}{2}O_2(g) + H_2O(l) \longrightarrow H_2SO_4(aq)$$

Sulfuric acid is a strong acid, completely ionized to $H^+$ and $HSO_4^-$ ions in dilute water solution. The $HSO_4^-$ ion ionizes further to give $H^+$ and $SO_4^{2-}$ ions:

$$H_2SO_4(aq) \longrightarrow H^+(aq) + HSO_4^-(aq) \qquad K_a \longrightarrow \infty$$

$$HSO_4^-(aq) \rightleftharpoons H^+(aq) + SO_4^{2-}(aq) \qquad K_a = 1.0 \times 10^{-2}$$

The ionization constant of the $HSO_4^-$ ion is relatively large. This explains why in writing equations for the reactions of sulfuric acid, we often consider it to consist of $2H^+(aq) + SO_4^{2-}(aq)$.

Sulfuric acid is a relatively weak oxidizing agent; most often it is reduced to sulfur dioxide:

$$SO_4^{2-}(aq) + 4H^+(aq) + 2e^- \longrightarrow SO_2(g) + 2H_2O \qquad E^\circ_{red} = +0.155 \text{ V}$$

This is the case when copper metal is oxidized by hot concentrated sulfuric acid:

$$Cu(s) + 4H^+(aq) + SO_4^{2-}(aq) \longrightarrow Cu^{2+}(aq) + 2H_2O + SO_2(g)$$

When dilute sulfuric acid (3 *M*) reacts with metals, it is ordinarily the $H^+$ ion rather than the $SO_4^{2-}$ ion that is reduced. For example, zinc reacts with dilute sulfuric acid to form hydrogen gas:

$$Zn(s) + 2H^+(aq) \longrightarrow Zn^{2+}(aq) + H_2(g)$$

Concentrated sulfuric acid, in addition to being an acid and an oxidizing agent, is also a dehydrating agent. Small amounts of water can be removed from organic liquids such as gasoline by extraction with sulfuric acid. Sometimes it is even possible to remove the elements of water from a compound by treating it with 18 *M* $H_2SO_4$. This happens with table sugar, $C_{12}H_{22}O_{11}$; the product is a black char that is mostly carbon (Fig. 21.15).

$$C_{12}H_{22}O_{11}(s) \longrightarrow 12C(s) + 11H_2O(l)$$

When concentrated sulfuric acid dissolves in water, a great deal of heat is given off, nearly 100 kJ per mole of $H_2SO_4$. Sometimes enough heat is evolved to bring the solution to the boiling point. To prevent this and to avoid splattering, the acid should be added slowly to water, with constant stirring. If it comes in contact with the skin, concentrated sulfuric acid can cause painful chemical burns. Never add water to concentrated $H_2SO_4$!

## Phosphoric Acid, $H_3PO_4$

Phosphoric acid is a weak *triprotic* acid:

$$H_3PO_4(aq) \rightleftharpoons H^+(aq) + H_2PO_4^-(aq) \qquad K_1 = 7.1 \times 10^{-3}$$

$$H_2PO_4^-(aq) \rightleftharpoons H^+(aq) + HPO_4^{2-}(aq) \qquad K_2 = 6.2 \times 10^{-8}$$

$$HPO_4^{2-}(aq) \rightleftharpoons H^+(aq) + PO_4^{3-}(aq) \qquad K_3 = 4.5 \times 10^{-13}$$

**Figure 21.15**
When sulfuric acid is added to sugar (sucrose), an exothermic reaction occurs. The elements of water are removed from the sugar, $C_{12}H_{22}O_{11}$, leaving a mass of black carbon.

**Figure 21.16**

In water solution, phosphorus in the +5 state can exist as $H_3PO_4$, $H_2PO_4^-$, $HPO_4^{2-}$, or $PO_4^{3-}$, depending upon the pH. The $H_3PO_4$ molecule dominates in strongly acidic solution; the $PO_4^{3-}$ ion is the principal species in strongly basic solution. The $H_2PO_4^-$ ion has its maximum concentration at pH 5, the $HPO_4^{2-}$ ion at pH 10.

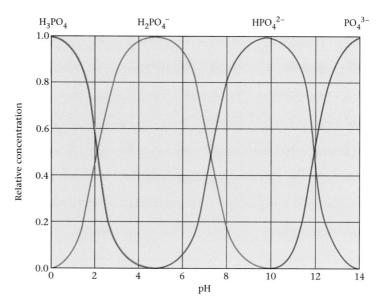

In a water solution containing +5 phosphorus, four different phosphorus-containing species can be present, depending upon pH: $H_3PO_4$, $H_2PO_4^-$, $HPO_4^{2-}$ and $PO_4^{3-}$. The *distribution curve* shown in Figure 21.16 demonstrates how the relative amounts of these species change with pH. Notice that

— in strongly acidic solution, at pH < 2, the molecule $H_3PO_4$ is the dominant species.
— in weakly acidic solution (2 < pH < 7), the principal species is the dihydrogen phosphate ion, $H_2PO_4^-$. The sodium salt of this anion, $NaH_2PO_4$, has a pH of about 5.
— in weakly basic solution (7 < pH < 12), the principal species is the monohydrogen phosphate ion, $HPO_4^{2-}$. The compound $Na_2HPO_4$ has a pH of about 10.
— in strongly basic solution (pH > 12), the dominant species is the phosphate ion, $PO_4^{3-}$.

**Example 21.11** Taking $K_a\ H_3PO_4 = 7.1 \times 10^{-3}$ and $K_a\ H_2PO_4^- = 6.2 \times 10^{-8}$, calculate, at pH 5.0, the ratio

(a) $[H_2PO_4^-]/[H_3PO_4]$     (b) $[H_2PO_4^-]/[HPO_4^{2-}]$

**Strategy**  In each case, set up the expression for $K_a$, substitute $[H^+] = 1 \times 10^{-5}$, and solve for the desired ratio.

**Solution**

(a) $K_a\ H_3PO_4 = \dfrac{[H^+] \times [H_2PO_4^-]}{[H_3PO_4]} = 7.1 \times 10^{-3}$

$[H_2PO_4^-]/[H_3PO_4] = $ $(7.1 \times 10^{-3})/(1 \times 10^{-5}) = 700$

(b) $K_a$ $H_2PO_4^- = \dfrac{[H^+] \times [HPO_4^{2-}]}{[H_2PO_4^-]} = 6.2 \times 10^{-8}$

$[H_2PO_4^-]/[HPO_4^{2-}] = [H^+]/(6.2 \times 10^{-8}) = \quad (1 \times 10^{-5})/(6.2 \times 10^{-8}) = 200$

Clearly, there is much more $H_2PO_4^-$ than either $H_3PO_4$ or $HPO_4^{2-}$ at pH 5.0. That is, of course, what Figure 21.16 tells us.

Some common products that contain phosphate.

Of the three compounds $NaH_2PO_4$, $Na_2HPO_4$, and $Na_3PO_4$, two are used as cleaning agents: $NaH_2PO_4$ in acid-type cleaners, $Na_3PO_4$ in strongly basic cleaners. The principal use of $Na_2HPO_4$ is in the manufacture of cheese; J. L. Kraft discovered 75 years ago that this compound is an excellent emulsifying agent. To this day no one quite understands why this happens.

A great many "phosphates" are used in commercial fertilizers. Perhaps the most important of these is calcium dihydrogen phosphate, $Ca(H_2PO_4)_2$. In relatively pure form, this compound is known as "triple superphosphate of lime." A 1:2 mol mixture of $Ca(H_2PO_4)_2$ and gypsum, $CaSO_4 \cdot 2H_2O$, is commonly referred to as "superphosphate of lime."

## Selenium

Of the 17 nonmetals in the periodic table, only one has not been mentioned to this point: selenium in Group **16**, directly below sulfur in the table. Selenium was discovered in 1817 by J. J. Berzelius, the great Swedish analytical chemist.

Selenium is a relatively rare element (Table 21.A), ranking sixty-fifth in abundance. It has several allotropes, one of which has the molecular formula $Se_8$, analogous to sulfur. The element is ordinarily obtained as a by-product in the metallurgy of copper or lead. About 300 metric tons ($3 \times 10^5$ kg) of Se are produced annually in the United States; in contrast, the annual production of sulfur is about $10^{10}$ kg.

Selenium has long been used as an additive in glassmaking. Particles of colloidally dispersed selenium give the ruby-red color seen in stained glass windows and traffic lights (Figure 21.A). Very small amounts of a sulfide of selenium are used in certain dandruff shampoos (e.g., Selsun Blue).

The principal use of selenium today takes advantage of its *photoconductivity;* the electrical conductivity of selenium increases by a factor of 1000 when it is exposed to light. Modern photocopiers use a plate in which a thin film of photoconductive material, usually selenium, is deposited on an aluminum base. A high potential is placed on the plate, which is then exposed to a light and dark image pattern, obtained by illuminating the article to be copied. In the light areas, the electrostatic charge on the plate is reduced as a result of a photoconductive discharge; the dark areas retain their original charge. An image is formed on the plate when toner particles (black or col-

# CHEMISTRY

## *Beyond the Classroom*

**Figure 21.A**

Glass is made red by adding a compound of selenium. (John D. Cunningham/Visuals Unlimited)

| TABLE 21.A  **Properties of Selenium** | | | |
|---|---|---|---|
| Outer Electron Configuration | $4s^2 4p^4$ | Melting point | 217° |
| Abundance | $5 \times 10^{-6}\%$ | Boiling point | 685° |
| Principal ores | $Cu_2Se$, $Ag_2Se$, PbSe | Oxid. states | $-2$, $+4$, $+6$ |

ored) are attracted to the high-charge areas. The image is then transferred to a piece of plain paper by charging it sufficiently to pull the toner particles off the plate.

The chemistry of selenium strongly resembles that of sulfur. Compare, for example, the formulas of silver selenide and lead selenide listed in Table 21.A with those of the corresponding sulfides, $Ag_2S$, and $PbS$. As you might guess, the formula of the principal hydrogen compound of selenium is $H_2Se$. Like $H_2S$, it is a poisonous, evil-smelling gas.

### Physiological Properties

It has long been known that high concentrations of selenium are toxic. The cattle disease known picturesquely as "blind staggers" arises from grazing on grass growing in soil with a high selenium content. It is now known, however, that selenium is an essential element in human nutrition. A selenium-containing enzyme, glutathione peroxidase, in combination with vitamin E, destroys harmful free radicals in the body.

There is considerable evidence to suggest that selenium compounds are anticarcinogens. For one thing, tests with laboratory animals show that the incidence and size of malignant tumors are reduced when a solution containing $Na_2SeO_3$ is injected at the part per million level. Beyond that evidence, statistical studies show an inverse correlation between selenium levels in the soil and the incidence of certain types of cancer.

A few years ago, it was shown that selenium dietary supplements can diminish some of the symptoms of people with AIDS. Recently (August 1994), it was suggested that the organic acid

$$H-Se-\underset{\underset{H}{|}}{\overset{\overset{H}{|}}{C}}-\underset{\underset{NH_2}{|}}{\overset{\overset{H}{|}}{C}}-C\overset{\displaystyle O}{\underset{\displaystyle OH}{\diagup}}\qquad \text{selenocysteine}$$

may be incorporated into a protein that represses HIV (*h*uman *i*mmunodeficiency *v*irus).

# CHAPTER HIGHLIGHTS

## Key Concepts

**1.** Carry out equilibrium calculations for solution reactions
(Examples 21.1, 21.11; Problems 47–52)
**2.** Apply the Gibbs-Helmholtz equation
(Example 21.2; Problems 53–58)
**3.** Write balanced equations for solution reactions
(Examples 21.3, 21.5, 21.10; Problems 15–26, 69)
**4.** Draw Lewis structures for compounds of the nonmetals
(Examples 21.6, 21.8; Problems 31–38)
**5.** Relate oxoacids to acid anhydrides and compare their acid strengths
(Examples 21.7, 21.8; Problems 7, 8, 67)
**6.** Carry out electrochemical calculations involving $E°$, the Nernst equation, and/or electrolysis
(Examples 21.4, 21.8; Problems 59–66)

| | | |
|---|---|---|
| acid | $E°_{ox}$, $E°_{red}$ | oxidizing agent |
| acid anhydride | electronegativity | oxoacid |
| allotrope | Lewis structure | oxoanion |
| base | nonmetal | reducing agent |
| disproportionation | oxidation number | resonance |
| distribution curve | | |

*Summary Problem*

Consider bromine, whose chemistry is quite similar to that of chlorine.

(a) Write equations for the reaction of bromine with $I^-$ ions; for the reaction of $Br_2$ with water (disproportionation).

(b) Write equations for the preparation of bromine by the electrolysis of aqueous NaBr, by the reaction of chlorine with bromide ions in aqueous solution.

(c) Write equations for the reaction of hydrobromic acid with $OH^-$ ions; with $CO_3^{2-}$ ions; with ammonia.

(d) Consider the species $Br^-$, $Br_2$, $BrO^-$, $BrO_4^-$. In a redox reaction, which of these species can act only as an oxidizing agent? only as a reducing agent? Which can act as either oxidizing or reducing agents?

(e) When aqueous sodium bromide is heated with a concentrated solution of sulfuric acid, the products include liquid bromine and sulfur dioxide. Write a balanced equation for the redox reaction involved.

(f) Write Lewis structures for the following species: $Br^-$, $BrO^-$, $BrO_4^-$. What is the bond angle in the $BrO_4^-$ ion? For which of these ions is the conjugate acid the weakest?

(g) Give the formula of the acid anhydride of HBrO. Knowing that $K_a$ of HBrO is $2.6 \times 10^{-9}$, calculate the ratio $[HBrO]/[BrO^-]$ at pH 10.00.

(h) For the reduction of HBrO to $Br^-$, what is the change in voltage when the pH increases by one unit?

### Answers

(a) $Br_2(l) + 2I^-(aq) \longrightarrow 2Br^-(aq) + I_2(s)$
$Br_2(l) + H_2O \longrightarrow HBrO(aq) + H^+(aq) + Br^-(aq)$

(b) $2Br^-(aq) + 2H_2O \longrightarrow Br_2(l) + H_2(g) + 2OH^-(aq)$
$2Br^-(aq) + Cl_2(aq) \longrightarrow Br_2(l) + 2Cl^-(aq)$

(c) $H^+(aq) + OH^-(aq) \longrightarrow H_2O$; $2H^+(aq) + CO_3^{2-}(aq) \longrightarrow CO_2(g) + H_2O$
$H^+(aq) + NH_3(aq) \longrightarrow NH_4^+(aq)$

(d) $BrO_4^-$; $Br^-$; $Br_2$; $BrO^-$

(e) $2Br^-(aq) + SO_4^{2-}(aq) + 4H^+(aq) \longrightarrow Br_2(l) + SO_2(g) + 2H_2O$

(f) $[:\!\overset{..}{\underset{..}{Br}}\!:]^-$    $[:\!\overset{..}{\underset{..}{Br}}\!-\!\overset{..}{\underset{..}{O}}\!:]^-$    $[:\!\overset{..}{\underset{..}{O}}\!-\!\overset{\overset{\displaystyle :\overset{..}{O}:}{|}}{\underset{\underset{\displaystyle :\overset{..}{O}:}{|}}{Br}}\!-\!\overset{..}{\underset{..}{O}}\!:]^-$    $109.5°$; $BrO^-$

(g) $Br_2O$; 0.038    (h) $-0.0296$ V

## Questions & Problems

### Formulas, Equations, and Reactions

**\*1.** Name the following species.
  **(a)** $HIO_4$  **(b)** $BrO_2^-$  **(c)** $HIO$  **(d)** $NaClO_3$

**\*2.** Name the following compounds.
  **(a)** $HBrO_3$  **(b)** $KIO$  **(c)** $NaClO_2$  **(d)** $NaBrO_4$

**\*3.** Write the formula for each of the following compounds.
  **(a)** chloric acid  **(b)** periodic acid
  **(c)** hypobromous acid  **(d)** hydriodic acid

**\*4.** Write the formula for each of the following compounds.
  **(a)** potassium bromite  **(b)** calcium bromide
  **(c)** sodium periodate  **(d)** magnesium hypochlorite

**\*5.** Write the formula of a compound of each of the following elements that *cannot* act as an oxidizing agent.
  **(a)** N  **(b)** S  **(c)** Cl

**\*6.** Write the formula of an oxoanion of each of the following elements that *cannot* act as a reducing agent.
  **(a)** N  **(b)** S  **(c)** Cl

**\*7.** Give the formula of the acid anhydride of
  **(a)** $HNO_3$  **(b)** $HNO_2$  **(c)** $H_2SO_4$

**\*8.** Write the formula of the acid formed when each of the following anhydrides reacts with water.
  **(a)** $SO_2$  **(b)** $Cl_2O$  **(c)** $P_4O_6$

**\*9.** Write the formulas of the following compounds.
  **(a)** ammonia  **(b)** laughing gas
  **(c)** hydrogen peroxide  **(d)** sulfur trioxide

**\*10.** Write the formula for the following compounds.
  **(a)** sodium azide  **(b)** sulfurous acid
  **(c)** hydrazine  **(d)** sodium dihydrogen phosphate

**\*11.** Write the formula of a compound of hydrogen with
  **(a)** nitrogen, which is a gas at 25°C and 1 atm.
  **(b)** phosphorus, which is a liquid at 25°C and 1 atm.
  **(c)** oxygen, which contains an O—O bond.

**\*12.** Write the formula of a compound of hydrogen with
  **(a)** sulfur
  **(b)** nitrogen, which is a liquid at 25°C and 1 atm.
  **(c)** phosphorus, which is a poisonous gas at 25°C and 1 atm.

**\*13.** Give the formula of
  **(a)** an anion in which S has an oxidation number of $-2$.
  **(b)** two anions in which S has an oxidation number of $+4$.
  **(c)** two different acids of sulfur.

**\*14.** Give the formula of a compound of nitrogen that is
  **(a)** a weak base.  **(b)** a strong acid.
  **(c)** a weak acid.  **(d)** capable of oxidizing copper.

**\*15.** Write a balanced net ionic equation for
  **(a)** the electrolytic decomposition of hydrogen fluoride.
  **(b)** the oxidation of iodide ion to iodine by hydrogen peroxide in acidic solution. Hydrogen peroxide is reduced to water.

**\*16.** Write a balanced net ionic equation for
  **(a)** the oxidation of iodide to iodine by sulfate ion in acidic solution. Sulfur dioxide gas is also produced.
  **(b)** The preparation of iodine from an iodide salt and chlorine gas.

**\*17.** Write a balanced net ionic equation for the disproportionation reaction
  **(a)** of iodine to give iodate and iodide ions in basic solution.
  **(b)** of chlorine gas to chloride and perchlorate ions in basic solution.

**\*18.** Write a balanced net ionic equation for the disproportionation reaction of
  **(a)** hypochlorous acid to chlorine gas and chlorous acid in acidic solution.
  **(b)** chlorate ion to perchlorate and chlorite ions.

**\*19.** Complete and balance the following equation. If no reaction occurs, write NR.
  **(a)** $Cl_2(g) + I^-(aq) \longrightarrow$  **(b)** $F_2(g) + Br^-(aq) \longrightarrow$
  **(c)** $I_2(s) + Cl^-(aq) \longrightarrow$  **(d)** $Br_2(l) + I^-(aq) \longrightarrow$

**\*20.** Complete and balance the following equations. If no reaction occurs, write NR.
  **(a)** $Cl_2(g) + Br^-(aq) \longrightarrow$  **(b)** $I_2(s) + Cl^-(aq) \longrightarrow$
  **(c)** $I_2(s) + Br^-(aq) \longrightarrow$  **(d)** $Br_2(l) + Cl^-(aq) \longrightarrow$

**\*21.** Write a balanced equation for the preparation of
  **(a)** $F_2$ from HF.  **(b)** $Br_2$ from NaBr.
  **(c)** $NH_4^+$ from $NH_3$.

**\*22.** Write a balanced equation for the preparation of
  **(a)** $N_2$ from $Pb(N_3)_2$.  **(b)** $O_2$ from $O_3$.
  **(c)** S from $H_2S$.

**\*23.** Write a balanced equation for the reaction of ammonia with
  **(a)** $Cu^{2+}$  **(b)** $H^+$  **(c)** $Al^{3+}$

**\*24.** Write a balanced equation for the reaction of hydrogen sulfide with
  **(a)** $Cd^{2+}$  **(b)** $OH^-$  **(c)** $O_2(g)$

**\*25.** Write a balanced net ionic equation for the reaction of nitric acid with
  **(a)** a solution of $Ca(OH)_2$
  **(b)** $Ag(s)$; assume the nitrate ion is reduced to $NO_2(g)$.
  **(c)** $Cd(s)$; assume the nitrate ion is reduced to $N_2(g)$.

**\*26.** Write a balanced net ionic equation for the reaction of sulfuric acid with
  **(a)** $CaCO_3(s)$.
  **(b)** a solution of NaOH.
  **(c)** Cu; assume the $SO_4^{2-}$ ion is reduced to $SO_2$.

### Allotropy

**\*27.** Which of the following elements show allotropy?
  **(a)** F  **(b)** O  **(c)** S  **(d)** Cl

*28. Which of the following elements show allotropy?
   (a) Br   (b) N   (c) P   (d) I
*29. Describe the structural difference between
   (a) ozone and ordinary oxygen.
   (b) white phosphorus and red phosphorus.
   (c) liquid sulfur at 120° and 180°.
*30. Describe how
   (a) ozone can be prepared from $O_2$.
   (b) monoclinic sulfur can be prepared from rhombic sulfur.
   (c) red phosphorus can be made from white phosphorus.

## Molecular Structure

*31. Give the Lewis structure of
   (a) $NO_2$   (b) NO   (c) $SO_2$   (d) $SO_3$
*32. Give the Lewis structure of
   (a) $Cl_2O$   (b) $N_2O$   (c) $P_4$   (d) $N_2$
*33. Which of the molecules in Question 31 are polar?
*34. Which of the molecules in Question 32 are polar?
*35. Give the Lewis structure of
   (a) $HNO_3$   (b) $H_2SO_4$   (c) $H_3PO_4$
*36. Give the Lewis structures of the conjugate bases of the species in Question 35.
*37. Give the Lewis structure of
   (a) the strongest oxoacid of bromine.
   (b) a hydride of nitrogen in which there is an —N—N— bond.
   (c) an acid added to cola drinks.
*38. Give the Lewis structure of
   (a) an oxide of nitrogen in the +5 state.
   (b) the strongest oxoacid of nitrogen.
   (c) a tetrahedral oxoanion of sulfur.

## Stoichiometry

39. The average concentration of bromine (as bromide) in seawater is 65 ppm. Calculate
   (a) the volume of seawater ($d = 64.0$ lb/ft$^3$) in cubic feet required to produce one kilogram of liquid bromine.
   (b) the volume of chlorine gas in liters, measured at 20°C and 762 mm Hg, required to react with this volume of seawater.
40. A 425-gallon tank is filled with water containing 175 g of sodium iodide. How many liters of chlorine gas at 758 mm Hg and 25°C will be required to oxidize all the iodide to iodine?
41. Iodine can be prepared by allowing an aqueous solution of hydrogen iodide to react with manganese dioxide, $MnO_2$. The reaction is

$$2I^-(aq) + 4H^+(aq) + MnO_2(s) \longrightarrow Mn^{2+}(aq) + 2H_2O + I_2(s)$$

If an excess of hydrogen iodide is added to 0.200 g $MnO_2$, how many grams of iodine are obtained, assuming 100% yield?

42. When a solution of hydrogen bromide is prepared, 1.283 L of HBr gas at 25°C and 0.974 atm is bubbled into 250.0 mL of water. Assuming all the HBr dissolves with no volume change, what is the molarity of the hydrobromic acid solution produced?
43. When ammonium nitrate explodes, nitrogen, steam, and oxygen gas are produced. If the explosion is carried out by heating one kilogram of ammonium nitrate sealed in a rigid bomb with a volume of one liter, what is the total pressure produced by the gases before the bomb ruptures? Assume that the reaction goes to completion and that the final temperature is 500°C. (3 significant figures)
44. Sulfur dioxide can be removed from the smokestack emissions of power plants by reacting it with hydrogen sulfide, producing sulfur and water. What volume of hydrogen sulfide at 27°C and 755 mm Hg is required to remove the sulfur dioxide produced by a power plant that burns one metric ton of coal containing 5.0% sulfur by mass? How many grams of sulfur are produced by the reaction of $H_2S$ with $SO_2$?
45. A 1.500-g sample containing sodium nitrate was heated to form $NaNO_2$ and $O_2$. The oxygen evolved was collected over water at 23°C and 752 mm Hg; its volume was 125.0 mL. Calculate the percentage of $NaNO_3$ in the sample. The vapor pressure of water at 23°C is 21.07 mm Hg.
46. Chlorine can remove the foul smell of $H_2S$ in water. The reaction is

$$H_2S(aq) + Cl_2(aq) \longrightarrow 2H^+(aq) + 2Cl^-(aq) + S(s)$$

If the contaminated water has 5.0 ppm hydrogen sulfide by mass, what volume of chlorine gas at STP is required to remove all the $H_2S$ from $1.00 \times 10^3$ gallons of water ($d = 1.00$ g/mL)? What is the pH of the solution after treatment with chlorine?

## Equilibria

47. The equilibrium constant at 25°C for the reaction

$$Br_2(l) + H_2O \rightleftharpoons H^+(aq) + Br^-(aq) + HBrO(aq)$$

is $1.2 \times 10^{-9}$. This is the system present in a bottle of "bromine water." Assuming that HBrO does not ionize appreciably, what is the pH of the bromine water?
48. Calculate the pH and the equilibrium concentration of HClO in a 0.10 $M$ solution of hypochlorous acid. $K_a$ HClO = $2.8 \times 10^{-8}$.
49. At equilibrium, a gas mixture has a partial pressure of 0.7324 atm for HBr and $2.80 \times 10^{-3}$ atm for both hydrogen and bromine gases. What is $K$ for the formation of two moles of HBr from $H_2$ and $Br_2$?
50. Given

$$HF(aq) \rightleftharpoons H^+(aq) + F^-(aq) \qquad K_a = 6.9 \times 10^{-4}$$

$$HF(aq) + F^-(aq) \rightleftharpoons HF_2^-(aq) \qquad K = 2.7$$

Calculate K for the reaction

$$2HF(aq) \rightleftharpoons H^+(aq) + HF_2^-(aq)$$

**51.** What is the concentration of fluoride ion in a water solution saturated with $BaF_2$, $K_{sp} = 1.8 \times 10^{-7}$?

**52.** Calculate the solubility in grams per 100 mL of $BaF_2$ in 0.10 $M$ $BaCl_2$ solution.

## Thermodynamics

**53.** Determine whether the following redox reaction is spontaneous at 25°C:

$$2KIO_3(s) + Cl_2(g) \longrightarrow 2KClO_3(s) + I_2(s)$$

Use data in Appendix 1 and the following information: $\Delta H_f^\circ$ $\Delta KIO_3(s) = -501.4$ kJ/mol, $S^\circ$ $KIO_3(s) = 151.5$ J/mol · K. What is the lowest temperature at which the reaction is spontaneous?

**54.** Follow the directions for Problem 53 for the reaction

$$2KBrO_3(s) + Cl_2(g) \longrightarrow 2KClO_3(s) + Br_2(l)$$

The following thermodynamic data may be useful:

$$\Delta H_f^\circ KBrO_3 = -360.2 \text{ kJ/mol}; \; S^\circ KBrO_3 = 149.2 \text{ J/mol · K}$$

**55.** Consider the equilibrium system

$$HF(aq) \rightleftharpoons H^+(aq) + F^-(aq)$$

Given $\Delta H_f^\circ$ $HF(aq) = -320.1$ kJ/mol;

$$\Delta H_f^\circ F^-(aq) = -332.6 \text{ kJ/mol}; \; S^\circ \, F^-(aq) = -13.8 \text{ J/mol · K};$$
$$K_a \, HF = 6.9 \times 10^{-4} \text{ at } 25°C$$

calculate $S^\circ$ for $HF(aq)$.

**56.** Applying the Tables in Appendix 1 to

$$4HCl(g) + O_2(g) \longrightarrow 2Cl_2(g) + 2H_2O(l)$$

determine

**(a)** whether the reaction is spontaneous at 25°C.

**(b)** $K$ for the reaction at 25°C.

**57.** Consider the reaction

$$4NH_3(g) + 5O_2(g) \longrightarrow 4NO(g) + 6H_2O(g)$$

**(a)** Calculate $\Delta H^\circ$ for this reaction. Is it exothermic or endothermic?

**(b)** Would you expect $\Delta S^\circ$ to be positive or negative? Calculate $\Delta S^\circ$.

**(c)** Is the reaction spontaneous at 25°C and 1 atm?

**(d)** At what temperature, if any, is the reaction at equilibrium at 1 atm pressure?

**58.** Data is given in Appendix 1 for white phosphorus, $P_4(s)$. $P_4(g)$ has the following thermodynamic values: $\Delta H_f^\circ = 58.9$ kJ/mol, $S^\circ = 280.0$ J/K · mol. What is the temperature at which white phosphorus sublimes at 1 atm pressure?

## Electrochemistry

**59.** In the electrolysis of a KI solution, using 5.00 V, how much electrical energy in kilojoules is consumed when one mole of $I_2$ is formed?

**60.** If an electrolytic cell producing fluorine uses a current of $7.00 \times 10^3$ A (at 10.0 V), how many grams of fluorine gas can be produced in two days (assuming that the cell operates continuously at 95% efficiency)?

**61.** Sodium hypochlorite is produced by the electrolysis of cold sodium chloride solution. How long must a cell operate to produce $1.500 \times 10^3$ L of 5.00% NaClO by mass if the cell current is $2.00 \times 10^3$ A? Assume that the density of the solution is 1.00 g/cm$^3$.

**62.** Sodium perchlorate is produced by the electrolysis of sodium chlorate. If a current of $1.50 \times 10^3$ A passes through an electrolytic cell, how many kilograms of sodium perchlorate are produced in an eight-hour run?

**63.** Taking $E_{ox}^\circ$ $H_2O_2 = -0.695$ V, determine which of the following species will be reduced by hydrogen peroxide (use Table 18.1 to find $E_{red}^\circ$ values).

**(a)** $Cr_2O_7^{2-}$  **(b)** $Fe^{2+}$  **(c)** $I_2$  **(d)** $Br_2$

**64.** Taking $E_{red}^\circ$ $H_2O_2 = +1.763$ V, determine which of the following species will be oxidized by hydrogen peroxide (use Table 18.1 to find $E_{ox}^\circ$ values).

**(a)** $Co^{2+}$  **(b)** $Cl^-$  **(c)** $Fe^{2+}$  **(d)** $Sn^{2+}$

**65.** Consider the reduction of nitrate ion in acidic solution to nitrogen oxide ($E_{red}^\circ = 0.964$ V) by sulfur dioxide which is oxidized to sulfate ion ($E_{red}^\circ = 0.155$ V). Calculate the voltage of a cell involving this reaction in which all the gases have pressures of 1.00 atm, all the ionic species (except $H^+$) are at 0.100 $M$, and the pH is 4.30.

**66.** For the reaction in Problem 65 if gas pressures are at 1.00 atm and ionic species are at 0.100 $M$ (except $H^+$), at what pH will the voltage be 1.000 V?

## Unclassified

**\*67.** Choose the strongest acid from each group.

**(a)** HClO, HBrO, HIO  **(b)** HIO, HIO$_3$, HIO$_4$

**(c)** HIO, HBrO$_2$, HBrO$_4$

**\*68.** What intermolecular forces are present in the following?

**(a)** $Cl_2$  **(b)** HBr  **(c)** HF

**(d)** $HClO_4$  **(e)** $MgI_2$

**\*69.** Write a balanced equation for the reaction of hydrofluoric acid with $SiO_2$. What volume of 2.0 $M$ HF is required to react with one gram of silicon dioxide?

**\*70.** State the oxidation number of N in

**(a)** $NO_2^-$  **(b)** $NO_2$  **(c)** $HNO_3$  **(d)** $NH_4^+$

**\*71.** The density of sulfur vapor at one atmosphere pressure and 973 K is 0.8012 g/L. What is the molecular formula of the vapor?

**\*72.** Give the formula of a substance discussed in this chapter that is used

**(a)** to disinfect water.  **(b)** in safety matches.

**(c)** to prepare hydrazine.  **(d)** to etch glass.

**\*73.** Why does concentrated nitric acid often have a yellow color even though pure HNO$_3$ is colorless?

*74. Explain why
    (a) acid strength increases as the oxidation number of the central nonmetal atom increases.
    (b) nitrogen dioxide is paramagnetic.
    (c) the oxidizing strength of an oxoanion is inversely related to pH.
    (d) sugar turns black when treated with concentrated sulfuric acid.

## Challenge Problems

75. Suppose you wish to calculate the mass of sulfuric acid that can be obtained from an underground deposit of sulfur $1.00 \text{ km}^2$ in area. What additional information do you need to make this calculation?

76. The reaction

$$4HF(aq) + SiO_2(aq) \longrightarrow SiF_4(aq) + 2H_2O$$

can be used to release gold that is distributed in certain quartz ($SiO_2$) veins of hydrothermal origin. If the quartz contains $1.0 \times 10^{-3}\%$ Au by weight and the gold has a market value of \$425 per troy ounce, would the process be economically feasible if commercial HF (50% by weight, $d = 1.17 \text{ g/cm}^3$) costs 75¢ a liter? (1 troy ounce = 31.1 g.)

77. The amount of sodium hypochlorite in a bleach solution can be determined by using a given volume of bleach to oxidize excess iodide ion to iodine; $ClO^-$ is reduced to $Cl^-$. The amount of iodine produced by the redox reaction is determined by titration with sodium thiosulfate, $Na_2S_2O_3$; $I_2$ is reduced to $I^-$. The sodium thiosulfate is oxidized to sodium tetrathionate, $Na_2S_4O_6$. In this analysis, potassium iodide was added in excess to 5.00 mL of bleach ($d = 1.00 \text{ g/cm}^3$). If 25.00 mL of $0.0700 \ M \ Na_2S_2O_3$ was required to reduce all the iodine produced by the bleach back to iodide, what is the mass percent of NaClO in the bleach?

78. What is the minimum amount of sodium azide that can be added to an automobile airbag to give a volume of 20.0 L upon inflation? Make any reasonable assumptions required to obtain an answer, but state what these assumptions are.

Proteins, which make up about 15% of the human body, are polymeric substances with structures related to, but more complex than, synthetic polymers such as nylon (p. 634). (Mark Kozlowski/FPG International)

# 22 Organic Chemistry

**W**hat is life? It is the flash of a firefly in the night.

It is the breath of a buffalo in the wintertime.

It is the little shadow which runs across grass

And loses itself in the sunset.

**—DYING WORDS OF CROWFOOT**

(*1890*)

Organic chemistry deals with the compounds of carbon, of which there are literally millions. More than 90% of all known compounds contain carbon atoms. There is a simple explanation for this remarkable fact. Carbon atoms bond to one another to a far greater extent than do atoms of any other element. Carbon atoms may link together to form chains or rings.

The bonds may be single (one electron pair), double (two electron pairs), or triple (three electron pairs).

There are a wide variety of different organic compounds that have quite different structures and properties. However, all these substances have certain features in common:

1. *Organic compounds are ordinarily molecular rather than ionic.* Most of the compounds we discuss consist of small, discrete molecules. Many of them are gases or liquids at room temperature.

2. *Each carbon atom forms a total of four covalent bonds.* This is illustrated by the structures written above. A particular carbon atom may form four single bonds, two single bonds and a double bond, two double bonds, or one single bond and a triple bond. One way or another, though, the bonds add up to four.

3. *Carbon atoms may be bonded to each other or to other nonmetal atoms, most often hydrogen, a halogen, oxygen, or nitrogen.* In most organic compounds

— a hydrogen or halogen atom (F, Cl, Br, I) forms one covalent bond, —H, —X
— an oxygen atom forms two covalent bonds, —O— or =O
— a nitrogen atom forms three covalent bonds, —N—, =N—, or ≡N

In this chapter, we consider

— the simplest type of organic compound, called a **hydrocarbon,** which contains only two kinds of atoms, hydrogen and carbon. Hydrocarbons can be classified as alkanes (Section 22.1), alkenes and alkynes (Section 22.2), and aromatics (Section 22.3).

*Carbon always follows the octet rule*

— organic compounds containing oxygen atoms in addition to carbon and hydrogen (Section 22.4)

— organic compounds of very high molar mass, known as polymers (Section 22.5)

## 22.1   Saturated Hydrocarbons: Alkanes

One large and structurally simple class of hydrocarbons includes those substances in which all the carbon-carbon bonds are single bonds. These are called *saturated* hydrocarbons, or **alkanes.** In the alkanes the carbon atoms are bonded to each other in chains, which may be long or short, straight or branched.

The simplest alkanes are methane ($CH_4$), ethane ($C_2H_6$), and propane ($C_3H_8$):

$$
\begin{array}{ccc}
\begin{array}{c}
\text{H} \\
| \\
\text{H}-\text{C}-\text{H} \\
| \\
\text{H}
\end{array}
&
\begin{array}{c}
\text{H}\quad\text{H} \\
|\quad\; | \\
\text{H}-\text{C}-\text{C}-\text{H} \\
|\quad\; | \\
\text{H}\quad\text{H}
\end{array}
&
\begin{array}{c}
\text{H}\quad\text{H}\quad\text{H} \\
|\quad\; |\quad\; | \\
\text{H}-\text{C}-\text{C}-\text{C}-\text{H} \\
|\quad\; |\quad\; | \\
\text{H}\quad\text{H}\quad\text{H}
\end{array} \\
\text{methane} & \text{ethane} & \text{propane}
\end{array}
$$

Around the carbon atoms in these molecules and indeed in any saturated hydrocarbon, there are four single bonds involving $sp^3$ hybrid orbitals. As would be expected from the VSEPR model, these bonds are directed toward the corners of a regular tetrahedron. The bond angles are approximately 109.5°, the tetrahedral angle. This means that in propane ($C_3H_8$) and in the higher alkanes, the carbon atoms are arranged in a "zigzag" pattern (Figure 22.1).

Two different alkanes are known with the molecular formula $C_4H_{10}$. In one of these, called butane, the four carbon atoms are linked in a "straight chain." In the other, called 2-methylpropane, there is a "branched chain." The longest continuous chain in the molecule contains three carbon atoms; there is a $CH_3$ branch from the central carbon atom. The geometries of these molecules are shown in Figure 22.2. The structures are

$$
\begin{array}{cc}
\begin{array}{c}
\text{H}\quad\text{H}\quad\text{H}\quad\text{H} \\
|\quad\; |\quad\; |\quad\; | \\
\text{H}-\text{C}-\text{C}-\text{C}-\text{C}-\text{H} \\
|\quad\; |\quad\; |\quad\; | \\
\text{H}\quad\text{H}\quad\text{H}\quad\text{H}
\end{array}
&
\begin{array}{c}
\text{H}\quad\text{CH}_3\;\;\text{H} \\
|\qquad|\qquad| \\
\text{H}-\text{C}-\text{C}-\;\;-\text{C}-\text{H} \\
|\qquad|\qquad| \\
\text{H}\quad\;\;\text{H}\quad\;\;\text{H}
\end{array} \\
\text{butane} & \text{2-methylpropane}
\end{array}
$$

In structural isomers, the atoms are bonded in different patterns

Compounds having the same molecular formula but different molecular structures are called **structural isomers.** Butane and 2-methylpropane are re-

**Figure 22.1**
Ball-and-stick models of methane, ethane, and propane. The bond angles in all these compounds are above 109.5°, the tetrahedral angle.
(Charles Steele)

Methane

Ethane

Propane

ferred to as structural isomers of $C_4H_{10}$. They are two distinct compounds with their own characteristic physical and chemical properties.

Butane

---

**Example 22.1**   Draw structures for the isomers of $C_5H_{12}$.

*Strategy*   Start with the straight-chain structure, stringing all five carbon atoms one after the other. Then work with structures containing four carbon atoms in a chain with one branch; find all the nonequivalent structures of this type. Continue this process using a three-carbon chain, which is the shortest one that can be drawn for $C_5H_{12}$.

*Solution*   For simplicity, we show only the carbon atoms; there is an H atom attached to each bond extending from a C atom.

five-C chain:   −C−C−C−C−C−   isomer (I)

four-C chain:   −C−C−C−C−   isomer (II)
                       −C−

three-C chain:   −C−C−C−   isomer (III)

Working with only pencil and paper, you might be tempted to draw other structures, such as

−C−C−C−C−        −C−C−C−C−
−C−                           −C−

However, a few moments' reflection (or access to a molecular model kit) should convince you that these are in fact equivalent to structures written previously. In particular, the first one, like I, has a five-carbon chain in which no carbon atom is attached to more than two other carbons. The second structure, like II, has a four-carbon chain with one carbon atom bonded to three other carbons. Structures I, II, and III represent the three possible isomers of $C_5H_{12}$; there are no others.

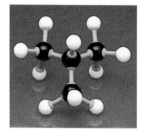

2-methylpropane

**Figure 22.2**
Ball-and-stick models of butane and 2-methylpropane, the isomers of $C_4H_{10}$.  (Charles Steele)

If you want a really tedious job, try drawing the 366,319 structural isomers, of $C_{20}H_{42}$

**Figure 22.3**
The barbecue grill is fueled by "bottled gas," a mixture of liquid propane ($C_3H_8$) and liquid butane ($C_4H_{10}$).
(David R Frazier/Photolibrary)

---

Natural gas, transmitted around the United States and Canada by pipeline, consists largely of methane (80 to 90%), with smaller amounts of $C_2H_6$, $C_3H_8$, and $C_4H_{10}$. Cylinders of "bottled gas," used with campstoves, barbecue grills, and the like, contain liquid propane ($C_3H_8$) and butane ($C_4H_{10}$) (Fig. 22.3). The pressure remains constant as long as any liquid is present, then drops abruptly to zero, indicating that it's time for a recharge.

The higher alkanes are most often obtained from petroleum, a dark brown, viscous liquid dispersed through porous rock deposits. Distillation of petroleum gives a series of fractions of different boiling points (Fig. 22.4). The most important of these is gasoline; distillation of a liter of petroleum gives about 250

**Figure 22.4**
Petroleum refinery. (Courtesy of Ashland Oil, Inc.)

mL of "straight-run" gasoline. It is possible to double the yield of gasoline by converting higher- or lower-boiling fractions to hydrocarbons in the gasoline range ($C_5$ to $C_{12}$).

## Nomenclature

As organic chemistry developed, it became apparent that some systematic way of naming compounds was needed. About 50 years ago, the International Union of Pure and Applied Chemistry (IUPAC) devised a system that could be used for all organic compounds. To illustrate this system, we will show how it works with alkanes.

For straight-chain alkanes such as

$$CH_3{-}CH_2{-}CH_3 \qquad CH_3{-}CH_2{-}CH_2{-}CH_3$$
propane          butane

the IUPAC name consists of a single word. These names, for up to eight carbon atoms, are listed in Table 22.1.

TABLE 22.1    **Nomenclature of Alkanes**

| Straight-Chain Alkanes | | Alkyl Groups | |
|---|---|---|---|
| methane | $CH_4$ | methyl | $CH_3{-}$ |
| ethane | $CH_3CH_3$ | ethyl | $CH_3{-}CH_2{-}$ |
| propane | $CH_3CH_2CH_3$ | propyl | $CH_3{-}CH_2{-}CH_2{-}$ |
| butane | $CH_3(CH_2)_2CH_3$ | | |
| pentane | $CH_3(CH_2)_3CH_3$ | isopropyl | $CH_3{-}\overset{\overset{\textstyle H}{\textstyle \vert}}{\underset{\underset{\textstyle CH_3}{\textstyle \vert}}{C}}{-}$ |
| hexane | $CH_3(CH_2)_4CH_3$ | | |
| heptane | $CH_3(CH_2)_5CH_3$ | butyl | $CH_3{-}CH_2{-}CH_2{-}CH_2{-}$ |
| octane | $CH_3(CH_2)_6CH_3$ | | |

With alkanes containing a **branched chain,** such as

$$CH_3{-}\overset{\overset{\textstyle H}{\textstyle \vert}}{\underset{\underset{\textstyle CH_3}{\textstyle \vert}}{C}}{-}CH_3$$
2-methylpropane

the name is more complex. A branched-chain alkane such as 2-methylpropane can be considered to be derived from a straight-chain alkane by replacing one or more hydrogen atoms by alkyl groups. The name consists of two parts:

— *a suffix that identifies the parent straight-chain alkane.* To find the suffix, count the number of carbon atoms in the longest continuous chain. For a three-carbon chain, the suffix is *propane;* for a four-carbon chain it is *butane,* and so on.

— *a prefix that identifies the branching alkyl group (Table 22.1) and indicates by a number the carbon atom where branching occurs.* In 2-methylpropane, referred to above, the methyl group is located at the second carbon from the end of the chain:

$$
\overset{\displaystyle 1 \quad 2 \quad 3}{C\!-\!\underset{|}{C}\!-\!C}
$$

Following this system, the IUPAC names of the isomers of pentane are

$$CH_3\!-\!CH_2\!-\!CH_2\!-\!CH_2\!-\!CH_3$$

pentane

$$CH_3\!-\!\underset{\underset{\displaystyle CH_3}{|}}{\overset{\overset{\displaystyle H}{|}}{C}}\!-\!CH_2\!-\!CH_3$$

2-methylbutane

$$CH_3\!-\!\underset{\underset{\displaystyle CH_3}{|}}{\overset{\overset{\displaystyle CH_3}{|}}{C}}\!-\!CH_3$$

2,2-dimethylpropane

Structural isomers have different names; if they don't, they're not isomers

Notice that

— *if the same alkyl group is at two branches, the prefix "di-" is used (2,2-dimethylpropane).* If there were three methyl branches, we would write trimethyl, and so on.
— *the number in the name is made as small as possible.* Thus, we refer to 2-methylbutane, numbering the chain from the left, rather than from the right.

$$
\overset{\displaystyle 1 \quad 2 \quad 3 \quad 4}{C\!-\!\underset{|}{C}\!-\!C\!-\!C}
$$

---

**Example 22.2**   Assign IUPAC names to the following:

(a) $CH_3\!-\!\underset{\underset{\displaystyle CH_3}{|}}{\overset{\overset{\displaystyle CH_3}{|}}{C}}\!-\!CH_2\!-\!CH_3$    (b) $CH_3\!-\!CH_2\!-\!\underset{\underset{\displaystyle CH_2}{\underset{\underset{\displaystyle CH_3}{|}}{|}}}{\overset{\overset{\displaystyle H}{|}}{C}}\!-\!CH_2\!-\!CH_3$

***Strategy***   Find the longest chain and use the proper suffix to identify it. Then find the branching alkyl group; locate the carbon atom where branching occurs. Number the carbon atoms so as to give the lower number for the prefix identifying the alkyl group.

*Solution*

(a) The longest chain contains four carbon atoms (butane). There are two $CH_3$ (methyl) groups branching at the second carbon from the end of the chain.

The correct name is    2,2-dimethylbutane.

(b) The longest chain, however you count it, contains five carbon atoms. There is a $CH_3\!-\!CH_2$ branch at the number three carbon, whichever end of the chain you start

from. The IUPAC name is    3-ethylpentane.

---

# 22.2   Unsaturated Hydrocarbons: Alkenes and Alkynes

In an unsaturated hydrocarbon, at least one of the carbon-carbon bonds in the molecule is a multiple bond. As a result, there are fewer hydrogen atoms in an unsaturated hydrocarbon than there are in a saturated one with the same number of carbons. There are many types of unsaturated hydrocarbons, only two of which are discussed here:

— *alkenes,* in which there is one carbon-carbon double bond in the molecule:

$$\text{C=C}$$

— *alkynes,* in which there is one carbon-carbon triple bond in the molecule:

$$-C\equiv C-$$

## Alkenes

The simplest alkene is ethylene, $C_2H_4$ (systematic name, ethene).

$$\begin{array}{cc} H & H \\ & \diagdown\diagup \\ & C=C \\ & \diagup\diagdown \\ H & H \end{array}$$

Ethylene is produced in larger amounts than any other organic compound; about $1.7 \times 10^7$ metric tons of $C_2H_4$ are manufactured in the United States each year. Ethylene is used primarily as a starting material for the preparation of other organic compounds, including ethyl alcohol (p. 625), ethylene glycol (p. 625), and, in particular, polyethylene (p. 630). Smaller amounts of ethylene are used to ripen fruit, picked while still unripe to avoid spoilage.

*That's how we get red tomatoes that taste like sawdust*

You may recall that we discussed the bonding in ethylene in Chapter 7. The double bond in ethylene and other alkenes consists of a sigma bond and a pi bond. The ethylene molecule is planar. There is no rotation about the double bond, since that would require "breaking" the pi bond. The bond angles in ethylene are approximately 120°, corresponding to $sp^2$ hybridization about each carbon atom. The geometries of ethylene and the next member of the alkene series, $C_3H_6$, are shown in Figure 22.5.

Ethylene and other alkenes are considerably more reactive than the corresponding alkanes. In particular, they undergo addition reactions, in which a small molecule such as $H_2$, HBr, or $Br_2$ "adds" across the double bond. The reaction with bromine is typical (Figure 22.6):

$$\begin{array}{cc} & H \quad H \\ & | \quad\ | \\ H-C=C-H + Br-Br \longrightarrow H-C-C-H \\ & | \quad\ | \\ & Br \quad Br \end{array}$$

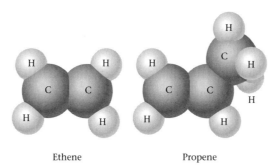

Ethene          Propene

**Figure 22.5**
Space-filling models of ethene and propene. The ethene molecule is planar. Propene contains three carbon atoms, two of which are joined by a double bond.

The systematic names of alkenes are derived from those of the corresponding alkanes with the same number of carbon atoms per molecule. There are two modifications.

— the ending *-ane* is replaced by *-ene*.

$$CH_3—CH_3 \qquad CH_2=CH_2$$
ethane         ethene

— where necessary, a number is used to designate the double-bonded carbon; the number is made as small as possible.

1   2   3   4      1   2   3   4

$$CH_2=CH—CH_2—CH_3 \qquad CH_3—CH=CH—CH_3$$
1-butene            2-butene

$$CH_2=C—CH_2—CH_3 \qquad CH_3—C=C—CH_3$$
$$\qquad \underset{CH_3}{|} \qquad\qquad\qquad \underset{H_3C \;\; H}{|\quad\;\; |}$$
2-methyl-1-butene        2-methyl-2-butene

You may be surprised to learn that there are actually two different 2-butenes, differing from each other in molecular geometry.

cis-2-butene          trans-2-butene
bp = 4°C          bp = 1°C

These compounds are **geometric isomers.** In the *cis-* isomer, the two $CH_3$ groups (or the two H atoms) are as close to one another as possible. In the *trans-* isomer, the two identical groups are farther apart. The two forms exist because there is no free rotation about the carbon-carbon double bond. Here, as with complex ions (Chap. 15), geometry is responsible for *cis-trans* isomerism; the order in which the atoms are bonded to each other is the same in both isomers.

Geometric isomerism is common among alkenes. Indeed, it occurs with all alkenes *except those in which two identical atoms or groups are attached to one of the double-bonded carbons.* Thus, although 2-butene has *cis-* and *trans-* isomers, 1-butene does not because two H atoms are bonded to carbon-1.

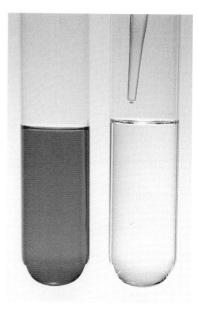

**Figure 22.6**
A solution of bromine in carbon tetrachloride is red. Add a few drops of an alkene and the color disappears. (Charles D. Winters)

---

**Example 22.3**   Draw all the isomers of $C_2H_2Cl_2$ in which two of the H atoms of ethylene are replaced by Cl atoms.

**Strategy**   Draw the C=C skeleton. Attach both chlorine atoms to the same carbon atom; there is only one such compound. Then attach a chlorine atom to each of the carbon atoms. In this case, there are *cis-* and *trans-* isomers.

**Solution**   The isomers of $C_2H_2Cl_2$ are

1,1-dichloroethene      cis-1,2-dichloroethene      trans-1,2-dichloroethene

Name the compound:    H    Pa
         C=C
     Ma    H

Ans: transparent

*Cis-* and *trans-* isomers differ from one another in their physical and, to a lesser extent, their chemical properties. They may also differ in their physiological behavior. For example, the compound *cis-9-tricosene*

$$CH_3-(CH_2)_7 \quad\quad (CH_2)_{12}-CH_3$$
$$C=C$$
$$H \quad\quad\quad H$$

is a sex attractant secreted by the female housefly. The *trans-* isomer is totally ineffective in this area.

## Alkynes

The IUPAC names of alkynes are derived from those of the corresponding alkenes by replacing the suffix *-ene* with *-yne*. Thus,

$$H-C{\equiv}C-H \quad\quad H-C{\equiv}C-CH_3$$
ethyne             propyne

$$H-C{\equiv}C-CH_2-CH_3 \quad\quad CH_3-C{\equiv}C-CH_3$$
1-butyne                2-butyne

The most common alkyne by far is the first member of the series, commonly called acetylene. Recall from Chapter 7 that the $C_2H_2$ molecule is linear, with 180° bond angles. The triple bond consists of a sigma bond and two pi bonds; each carbon atom is sp-hybridized. The geometries of acetylene and the next member of the series, $C_3H_4$, are shown in Figure 22.7.

Acetylene can be made in the laboratory by allowing water to drop on calcium carbide.

$$CaC_2(s) + H_2O(l) \longrightarrow C_2H_2(g) + CaO(s)$$

The gas produced in this way has a garlic-like odor due to traces of phosphine, $PH_3$, formed from $Ca_3P_2$ in the impure calcium carbide. Commercially, acetylene is made by heating methane to about 1500°C in the absence of air:

$$2CH_4(g) \longrightarrow C_2H_2(g) + 3H_2(g)$$

Thermodynamically, acetylene is unstable with respect to decomposition to the elements:

$$C_2H_2(g) \longrightarrow 2C(s) + H_2(g) \quad\quad\quad \Delta G° = -209.2 \text{ kJ at } 25°C$$

At high pressures, this reaction can occur explosively. For that reason, cylinders of acetylene do not contain the pure gas. Instead the cylinder is packed with an inert, porous material that holds a solution of acetylene gas in acetone.

You are probably most familiar with acetylene as a gaseous fuel used in welding and cutting metals (Fig. 22.8). When mixed with pure oxygen in a torch, acetylene burns at temperatures above 2000°C. The heat comes from the reaction

$$C_2H_2(g) + \tfrac{5}{2}O_2(g) \longrightarrow 2CO_2(g) + H_2O(l) \quad\quad\quad \Delta H = -1300 \text{ kJ}$$

Also miner's headlamps

The reaction gives off a brilliant white light, which served as a source of illumination in the headlights of early automobiles.

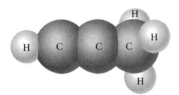

Ethyne (Acetylene)

Propyne (Methylacetylene)

**Figure 22.7**
In acetylene and methylacetylene, two carbon atoms are linked by a triple bond. Both molecules contain four atoms in a straight line.

# 22.3   Aromatic Hydrocarbons and Their Derivatives

Aromatic hydrocarbons, sometimes referred to as *arenes*, can be considered to be derived from benzene, $C_6H_6$. Benzene is a transparent, volatile liquid (bp = 80°C) that was discovered by Michael Faraday in 1825. Its formula, $C_6H_6$, suggests a high degree of unsaturation, yet its properties are quite different from those of alkenes or alkynes.

The atomic orbital (valence bond) approach regards benzene as a resonance hybrid of the two structures

**Figure 22.8**
Acetylene (ethyne) is used in welding. (Charles D. Winters)

This model is consistent with many of the properties of benzene. The molecule is a planar hexagon with bond angles of 120°. The hybridization of each carbon is $sp^2$. However, this structure is misleading in one respect. Chemically, benzene does not behave as if there were double bonds present. In particular, benzene does not undergo addition reactions like the one shown in Figure 22.6.

A more satisfactory model of the electron distribution in benzene, based upon molecular orbital theory (Appendix 5) assumes that

— each carbon atom forms three sigma bonds, one to a hydrogen atom and two to adjacent carbon atoms
— the three electron pairs remaining are spread symmetrically over the entire molecule to form "delocalized" pi bonds.

This model is often represented by the structure

where it is understood that

— there is a carbon atom at each corner of the hexagon
— there is an H atom bonded to each carbon atom
— the circle in the center of the molecule represents the six delocalized electrons

At one time, benzene was widely used as a solvent, both commercially and in research and teaching laboratories. Its use for that purpose has largely been abandoned because of its toxicity. Chronic exposure to benzene vapor leads to various blood disorders and, in extreme cases, aplastic anemia and leukemia. It appears that the culprits here are oxidation products of the aromatic ring, formed in an attempt to solubilize benzene and thus eliminate it from the body.

There are three pi bonds in benzene

## Derivatives of Benzene

Monosubstituted benzenes are ordinarily named as derivatives of benzene.

| chlorobenzene | nitrobenzene | aminobenzene (aniline) | hydroxybenzene (phenol) | methylbenzene (toluene) |

The last three compounds listed are always referred to by their common names, shown in red. Phenol was the first commercial antiseptic; its introduction into hospitals in the 1870s led to a dramatic decrease in deaths from postoperative infections. Its use for this purpose has long since been abandoned because phenol burns exposed tissue, but many modern antiseptics are phenol derivatives. Toluene has largely replaced benzene as a solvent because it is much less toxic. Oxidation of toluene in the body gives benzoic acid, $C_6H_5COOH$, which is readily eliminated and has none of the toxic properties of oxidation products of benzene.

When there are two groups attached to the benzene ring, three isomers are possible. These are designated by the prefixes *ortho-*, *meta-*, and *para-*, often abbreviated as *o-*, *m-*, and *p-*.

*o*-dichlorobenzene          *m*-dichlorobenzene          *p*-dichlorobenzene

Numbers can also be used; these three compounds may be referred to as 1,2-dichlorobenzene, 1,3-dichlorobenzene, and 1,4-dichlorobenzene, respectively. As always, the numbers used are as small as possible.

## Condensed Ring Structures

In another type of aromatic hydrocarbon, two or more benzene rings are fused together. Naphthalene, the solid that gives mothballs their aromatic odor, is the simplest compound of this type. Fusion of three benzene rings gives two different isomers, anthracene and phenanthrene.

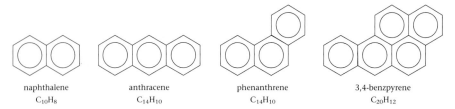

| naphthalene $C_{10}H_8$ | anthracene $C_{14}H_{10}$ | phenanthrene $C_{14}H_{10}$ | 3,4-benzpyrene $C_{20}H_{12}$ |

It is important to realize that in these structures there are no hydrogen atoms bonded to the carbons at the juncture of two rings. The eight hydrogen atoms in naphthalene are located as shown at the top of p. 623.

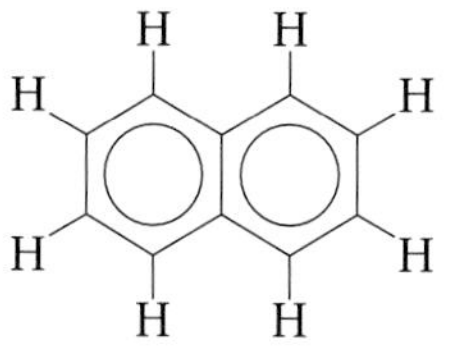

Certain compounds of this type are potent carcinogens. One of the most dangerous is 3,4-benzpyrene, which has been detected in cigarette smoke. It is believed to be a cause of lung cancer, to which heavy smokers are susceptible.

Many natural products contain fused rings. One of the most important compounds of this type is cholesterol, which has the structure

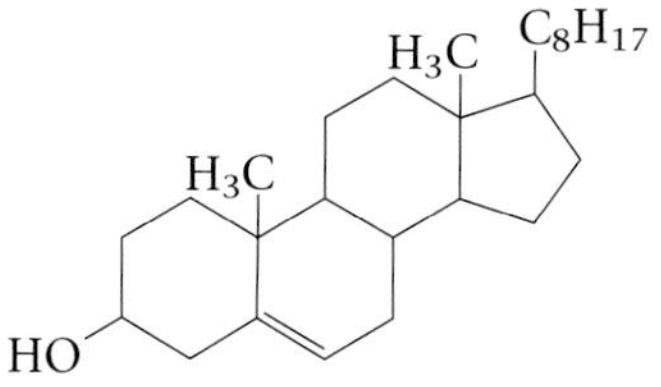

cholesterol

(This molecule has only one pi bond.)

Several derivatives of naphthalene are carcinogenic

Your body contains approximately 150 g of cholesterol. Every day, about one gram of cholesterol is synthesized in the liver; another half gram is taken in food. Cholesterol is the starting material used by the body to synthesize many important steroids, including the sex hormones testosterone (male) and estradiol (female). If there is too high a concentration of cholesterol in blood serum, it can precipitate in plaque-like deposits on the inner walls of arteries. The resultant reduction in the diameter of blood vessels can lead to a heart attack, stroke, or kidney failure.

## 22.4  Functional Groups

Many organic molecules can be considered to be derived from hydrocarbons by substituting a **functional group** for a hydrogen atom. The functional group can be a nonmetal atom or small group of atoms that is bonded to carbon. Table 22.2 p. 624 lists the types of functional groups commonly found in organic compounds. In this section, we will discuss four classes of compounds: alcohols, carboxylic acids, esters, and amines.

### Alcohols

An alcohol is derived from a hydrocarbon by replacing one or more H atoms by —OH groups. Alcohols are named by substituting the suffix -ol for the -ane suffix of the corresponding alkane. The first four members of the series are (common names shown in red)

$CH_3OH$         $CH_3—CH_2OH$         $CH_3—CH_2—CH_2OH$         $CH_3—\underset{\underset{\displaystyle OH}{|}}{CH}—CH_3$

methanol
(methyl alcohol)

ethanol
(ethyl alcohol)

1-propanol
(propyl alcohol)

2-propanol
(isopropyl alcohol)

TABLE 22.2   **Common Functional Groups**

| Group | Class | Example | Name* |
|---|---|---|---|
| —F, —Cl, —Br, —I | halides | $C_2H_5Cl$ | choloroethane (ethyl chloride) |
| **—OH** | **alcohols** | $C_2H_5OH$ | ethanol (ethyl alcohol) |
| —O— | ethers | $CH_3$—O—$CH_3$ | dimethyl ether |
| $\overset{O}{\overset{\|}{-C-H}}$ | aldehydes | $CH_3-\overset{O}{\overset{\|}{C}}-H$ | ethanal (acetaldehyde) |
| $\overset{O}{\overset{\|}{-C-}}$ | ketones | $CH_3-\overset{O}{\overset{\|}{C}}-CH_3$ | propanone (acetone) |
| $\overset{O}{\overset{\|}{-C}}-\textbf{OH}$ | **carboxylic acids** | $CH_3-\overset{O}{\overset{\|}{C}}-OH$ | ethanoic acid (acetic acid) |
| $\overset{O}{\overset{\|}{-C}}-O-$ | **esters** | $CH_3-\overset{O}{\overset{\|}{C}}-OCH_3$ | methyl methanoate (methylacetate) |
| $-\overset{\|}{N}-$ | **amines** | $CH_3NH_2$ | aminomethane (methylamine) |
| $-\overset{O}{\overset{\|}{C}}-\overset{H}{\overset{\|}{N}}-$ | amides | $CH_3-\overset{O}{\overset{\|}{C}}-NH_2$ | ethanamide (acetamide) |

*Common names are shown in red.

Notice that 1-propanol and 2-propanol are structural isomers. In 1-propanol, the —OH group is bonded to a terminal carbon atom; in 2-propanol, it is bonded to the central carbon atom.

About $3 \times 10^9$ kg of ***methanol*** are produced annually in the United States from water gas, a mixture of carbon monoxide and hydrogen:

$$CO(g) + 2H_2(g) \xrightarrow[\text{250 atm, 350°C}]{\text{ZnO, Cr}_2\text{O}_3} CH_3OH(g)$$

Methanol is also formed as a by-product when charcoal is made by heating wood in the absence of air. For this reason, methanol is sometimes called wood alcohol. Methanol is used in jet fuels and as a solvent, gasoline additive, and starting material for several industrial syntheses. Methanol is a deadly poison; ingestion of as little as 25 mL can be fatal. The antidote for methanol poisoning is a solution of sodium hydrogen carbonate, $NaHCO_3$.

***Ethanol,*** the most common alcohol, can be prepared by the fermentation of grains or sugar (Fig. 22.9). It is the active ingredient of alcoholic beverages. There it is present in various concentrations (4 to 8% in beer, 12 to 15% in wine, and 40% or more in distilled spirits). The "proof" of an alcoholic beverage is twice the volume percentage of ethanol. Thus, an 86-proof bourbon whiskey contains 43% ethanol. The taste of alcoholic beverages is due to impurities; ethanol itself is tasteless and colorless.

Industrial ethanol is made by the reaction of ethylene with water, using sulfuric acid as a catalyst.

**Figure 22.9**
Wine (far right) is produced from the glucose in grape juice (left) by fermentation. The reaction is $C_6H_{12}O_6(aq) \longrightarrow 2C_2H_5OH(aq) + 2CO_2(g)$. The purpose of the bubble chamber in the fermentation jug (center) is to allow the carbon dioxide to escape but prevent oxygen from entering and oxidizing ethanol to acetic acid. (Courtesy of Prof. James M. Bobbit, Twenty Mile Vineyard)

Frequently, ethanol is "denatured" by adding small quantities of methanol or benzene. This avoids the high federal tax on beverage alcohol; denatured alcohol is poisonous.

Certain alcohols contain two or more —OH groups per molecule. Perhaps the most familiar compounds of this type are ethylene glycol and glycerol:

ethylene glycol        glycerol

Ethylene glycol is widely used as an antifreeze. Glycerol is formed as a by-product in making soaps and detergents. It is a viscous, sweet-tasting liquid, used in making drugs, antibiotics, plastics, and explosives (nitroglycerin).

## Carboxylic Acids

**Carboxylic acids** are derived from hydrocarbons by replacing one or more H atoms by a carboxyl group, —C—OH, often abbreviated —COOH. The systematic names of these compounds are obtained by adding the suffix *-oic* to the stem of the name of the corresponding alkanes. In practice, these names are seldom used for the first two members of the series, which are commonly referred to as formic acid and acetic acid.

$$\underset{\substack{\|\\ O}}{H-C-OH} \qquad \underset{\substack{\|\\ O}}{CH_3-C-OH}$$

methanoic acid       ethanoic acid
(formic acid)        (acetic acid)

Homemade wine often contains some vinegar

Acetic acid is the active ingredient of vinegar, responsible for its sour taste. "White" vinegar is made by adding pure acetic acid to water, forming a 5% solution. "Brown" or "apple cider" vinegar is made from apple juice; the ethanol formed by fermentation is oxidized to acetic acid:

$$C_2H_5OH(aq) + O_2(g) \longrightarrow CH_3COOH(aq) + H_2O$$

The most important chemical property of carboxylic acids is implied by their name; they act as weak acids in water solution.

$$RCOOH(aq) \rightleftharpoons H^+(aq) + RCOO^-(aq)$$

Carboxylic acids vary considerably in acid strength. Acetic acid has an ionization constant of $1.8 \times 10^{-5}$; that of formic acid is about ten times as great, $1.9 \times 10^{-4}$. Among the strongest acids of this type is trichloroacetic acid ($K_a = 0.20$); a 0.10 $M$ solution of $Cl_3C-COOH$ is about 73% ionized. Trichloroacetic acid is an ingredient of over-the-counter preparations used to treat canker sores and remove warts.

Treatment of a carboxylic acid with the strong base NaOH forms the sodium salt of the acid. With acetic acid, the acid-base reaction is

$$\underset{\substack{\|\\ O}}{CH_3-C-OH}(aq) + OH^-(aq) \longrightarrow \underset{\substack{\|\\ O}}{CH_3-C-O^-}(aq) + H_2O$$

Evaporation gives the salt sodium acetate, which contains $Na^+$ and $CH_3COO^-$ ions. Many of the salts of carboxylic acids have important uses. Sodium and calcium propionate ($Na^+$ or $Ca^{2+}$ ions, $CH_3CH_2COO^-$ ions) are added to bread, cake, and cheese to inhibit the growth of mold. *Soaps* are sodium salts of long-chain carboxylic acids such as stearic acid:

$$\underset{\substack{\|\\ O}}{CH_3(CH_2)_{16}-C-OH} \qquad Na^+, \underset{\substack{\|\\ O}}{CH_3(CH_2)_{16}-C-O^-}$$

stearic acid           sodium stearate, a soap

The cleaning action of soap reflects the nature of the long-chain carboxylate anion. The long hydrocarbon group is a good solvent for greases and oils, while the ionic $COO^-$ group gives high water-solubility.

A wide variety of carboxylic acids occur in nature, giving a sour or tart taste to foods (Table 22.3). Certain drugs, both prescription and over-the-counter, contain organic acids. Two of the most popular products of this type are aspirin and ibuprofen (Advil and Nuprin, for example).

aspirin                    ibuprofen

TABLE 22.3  **Some Naturally Occurring Organic Acids**

| Name | | Source |
|---|---|---|
| acetic acid | $CH_3$—COOH | vinegar |
| citric acid | HOOC—$CH_2$—$\overset{\overset{\displaystyle OH}{\mid}}{\underset{\underset{\displaystyle COOH}{\mid}}{C}}$—$CH_2$—COOH | citrus fruits |
| lactic acid | $CH_3$—$\underset{\underset{\displaystyle OH}{\mid}}{CH}$—COOH | sour milk |
| malic acid | HOOC—$CH_2$—$\underset{\underset{\displaystyle OH}{\mid}}{CH}$—COOH | apples, watermelons, grape juice, wine |
| oxalic acid | HOOC—COOH | rhubarb, spinach, tomatoes |
| quinic acid | | cranberries |
| tartaric acid | HOOC—$\underset{\underset{\displaystyle OH}{\mid}}{CH}$—$\underset{\underset{\displaystyle OH}{\mid}}{CH}$—COOH | grape juice, wine |

Because these compounds are acidic, they can cause stomach irritation unless taken with food or water.

## Esters

The reaction between a carboxylic acid and an alcohol forms an **ester,** containing the functional group $-\overset{\overset{\displaystyle O}{\|}}{C}-O-$, often abbreviated —COO—. The reac-

Esters in common household items.
(Charles D. Winters)

tion between acetic acid and methyl alcohol is typical:

$$CH_3—OH(aq) + HO—\overset{\overset{\displaystyle O}{\|}}{C}—CH_3(aq) \longrightarrow CH_3—O—\overset{\overset{\displaystyle O}{\|}}{C}—CH_3(aq) + H_2O$$

methyl alcohol  |  acetic acid  |  methyl acetate

The common name of an ester consists of two words. The first word (methyl, ethyl, . . .) is the name of the alkyl group of the alcohol. The second word (formate, acetate, . . .) is the name of the acid with the *-ic* suffix replaced by *-ate*. Thus ethyl butyrate (Table 22.4) is made from ethyl alcohol and butyric acid.

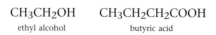

$$CH_3CH_2OH \qquad CH_3CH_2CH_2COOH$$
ethyl alcohol  |  butyric acid

Most esters have pleasant odors

| TABLE 22.4 | **Properties of Esters** | |
|---|---|---|
| **Ester** | **Structure** | **Odor, Flavor** |
| ethyl formate | $CH_3CH_2$—O—C—H<br>$\quad\quad\quad\quad\quad\|\|$<br>$\quad\quad\quad\quad\quad O$ | rum |
| isobutyl formate | $(CH_3)_2$—$CHCH_2$—O—C—H<br>$\quad\quad\quad\quad\quad\quad\quad\quad\|\|$<br>$\quad\quad\quad\quad\quad\quad\quad\quad O$ | raspberry |
| methyl butyrate | $CH_3$—O—C—$(CH_2)_2CH_3$<br>$\quad\quad\quad\|\|$<br>$\quad\quad\quad O$ | apple |
| ethyl butyrate | $CH_3CH_2$—O—C—$(CH_2)_2CH_3$<br>$\quad\quad\quad\quad\quad\|\|$<br>$\quad\quad\quad\quad\quad O$ | pineapple |
| isopentyl acetate | $(CH_3)_2$—CH—$(CH_2)_2$—O—C—$CH_3$<br>$\quad\quad\quad\quad\quad\quad\quad\quad\quad\quad\|\|$<br>$\quad\quad\quad\quad\quad\quad\quad\quad\quad\quad O$ | banana |
| octyl acetate | $CH_3$—$(CH_2)_7$—O—C—$CH_3$<br>$\quad\quad\quad\quad\quad\quad\|\|$<br>$\quad\quad\quad\quad\quad\quad O$ | orange |
| pentyl propionate | $CH_3$—$(CH_2)_4$—O—C—$CH_2CH_3$<br>$\quad\quad\quad\quad\quad\quad\|\|$<br>$\quad\quad\quad\quad\quad\quad O$ | apricot |

**Example 22.4**   Show the structure of

(a) the three-carbon alcohol with an —OH group at the end of the chain.
(b) the three-carbon carboxylic acid.
(c) the ester formed when these two compounds react.

***Strategy***   To work (a) and (b), start by drawing the parent carbon chain. Put the functional group in the specified position; fill out the structural formula with hydrogen atoms. In (c), remember that an —OH group comes from the acid, an H atom from the alcohol.

***Solution***

(a)   $CH_3CH_2CH_2$—OH     (b)   $CH_3CH_2$—C—OH
$\quad\quad\quad\quad\quad\quad\quad\quad\quad\quad\quad\quad\quad\quad\quad\quad\quad\quad\|\|$
$\quad\quad\quad\quad\quad\quad\quad\quad\quad\quad\quad\quad\quad\quad\quad\quad\quad\quad O$

(c)   $CH_3CH_2CH_2$—O—C—$CH_2CH_3$
$\quad\quad\quad\quad\quad\quad\quad\quad\quad\|\|$
$\quad\quad\quad\quad\quad\quad\quad\quad\quad O$

Animal fats and vegetable oils are esters of long-chain carboxylic acids with glycerol. A typical fat molecule might have the structure

$$CH_3(CH_2)_{14}\text{—COO—}CH_2$$
$$CH_3(CH_2)_7CH\text{=}CH(CH_2)_7\text{—COO—}CH$$
$$CH_3(CH_2)_{16}\text{—COO—}CH_2$$

The three carboxylic acids from which this fat is derived are

— palmitic acid, $CH_3(CH_2)_{14}COOH$

— oleic acid, $CH_3(CH_2)_7CH=CH(CH_2)_7COOH$
— stearic acid, $CH_3(CH_2)_{16}COOH$

These acids are typical of those found in fats. Some "fatty acids" are *saturated*, such as palmitic and stearic acid; the hydrocarbon chain contains no multiple bonds. Others, such as oleic acid, are *unsaturated*; there are one or more carbon-carbon multiple bonds in the molecule.

So-called saturated fats contain relatively few carbon-carbon multiple bonds. They are found primarily in animal products such as lard and butter. Unsaturated fats contain a higher proportion of multiple bonds. They are liquids at room temperature and are found in such vegetable products as corn oil and cottonseed oil. Recently, there has been a trend toward the use of unsaturated as opposed to saturated fats. It appears that saturated fats raise the level of cholesterol in the blood, perhaps contributing to the risk of heart attacks and other circulatory disorders.

*The evidence here is somewhat ambiguous*

---

**Example 22.5**    Classify the following as alcohols, carboxylic acids, or esters. More than one functional group may be present in the molecule.

(a) HO—C—C—C—OH    (b) [structures]

*Strategy*    Look for the following functional groups:

—OH     —C—OH     —C—O—
          ‖          ‖
          O          O

alcohol     acid       ester

*Solution*    (a)    alcohol, carboxylic acid        (b)    alcohol, ester

---

Brazil: Ripe coffee beans await the hand pickers. (Johnathan Wilkins/Photo Researchers, Inc.)

## Amines

As pointed out in Chapter 4, amines are an important class of molecular weak bases. Generally speaking, amines derived from alkanes tend to be somewhat stronger bases than ammonia; in contrast, aromatic amines are ordinarily weaker than ammonia.

|  | $NH_3$ | $CH_3NH_2$ | $C_6H_5NH_2$ |
|---|---|---|---|
|  | ammonia | methylamine | aniline |
| $pK_b =$ | 4.74 | 3.33 | 9.42 |

*Alkaloids* such as caffeine, coniine, and morphine (Fig. 22.10) form an important class of naturally occurring amines. Caffeine occurs in tea leaves, coffee beans, and cola nuts. Morphine is obtained from unripe opium poppy seed pods. Coniine, extracted from hemlock, is the alkaloid that killed Socrates. He was sentenced to death because of unconventional teaching methods; teacher evaluations in ancient Greece had teeth in them.

Opium poppy flower and unripe seed capsules. The capsule has a milky, gummy substance that contains the alkaloids morphine, heroin, and codeine. (Scott Camazine/Photo Researchers, Inc.)

**Figure 22.10**
Structural formulas of three alkaloids. Other amines in this category include nicotine, cocaine, quinine, and strychnine.

Caffeine          Coniine          Morphine

## 22.5   Synthetic Organic Polymers

More industrial chemists work with polymers than any other class of materials

Since about 1930, a wide variety of polymers have been synthesized by chemists. A **polymer** is made up of a large number of small molecular units called *monomers*, combined chemically. There are two different kinds of synthetic polymers:

— *addition polymers,* in which monomer units add directly to one another; typically, only one kind of monomer is involved.
— *condensation polymers,* in which monomer units combine by splitting out (condensing) a simple molecule such as $H_2O$. Typically, two different monomers react with one another to form a condensation polymer.

### Addition Polymers

Table 22.5 lists some of the more familiar synthetic addition polymers. Notice that each of these is derived from a monomer containing a carbon-carbon double bond. Upon polymerization, the double bond is converted to a single bond

$$\overset{|}{\underset{/}{C}}=\overset{|}{\underset{\backslash}{C}} \longrightarrow -\overset{|}{\underset{|}{C}}-\overset{|}{\underset{|}{C}}-$$

and successive monomer units add to one another.

Perhaps the most familiar addition polymer is **polyethylene,** a solid derived from the monomer ethylene. We might represent the polymerization process as

$$n\left(\begin{array}{c} H \\ | \\ C \\ | \\ H \end{array}=\begin{array}{c} H \\ | \\ C \\ | \\ H \end{array}\right) \longrightarrow \left(\begin{array}{cc} H & H \\ | & | \\ -C-C- \\ | & | \\ H & H \end{array}\right)_n$$

ethylene                    polyethylene

where $n$ is a very large number, of the order of 2000.

Depending upon the conditions of polymerization, the product may be

— *branched* polyethylene, which has a structure of the type

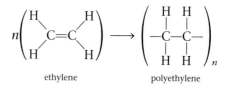

Low-density (top) and high-density (bottom) polyethylene. (Charles D. Winters)

**TABLE 22.5    Some Common Addition Polymers**

| Monomer | Name | Polymer | Uses |
|---|---|---|---|
| ethylene | ethylene | polyethylene | bags, coatings, toys |
| propylene | propylene | polypropylene | beakers, milk cartons |
| vinyl chloride | vinyl chloride | polyvinyl chloride, PVC | floor tile, raincoats, pipe, phonograph records |
| acrylonitrile | acrylonitrile | polyacrylonitrile, PAN | rugs; Orlon and Acrilan are copolymers with other monomers. |
| styrene | styrene | polystyrene | cast articles using a transparent plastic |
| methyl methacrylate | methyl methacrylate | Plexiglas, Lucite, acrylic resins | high-quality transparent objects, latex paints |
| tetrafluoroethylene | tetrafluoroethylene | Teflon | gaskets, insulation, bearings, pan coatings |

Here, neighboring chains are arranged in a somewhat random fashion, producing a soft, flexible solid (Fig. 22.11). The plastic bags at the vegetable counters of supermarkets are made of this material.

— *linear* polyethylene, which consists almost entirely of unbranched chains:

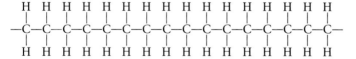

Neighboring chains in linear polyethylene line up nearly parallel to each other. This gives a polymer that approaches a crystalline material. It is used for bottles, toys, and other semirigid objects.

In **polyvinyl chloride,** there is another factor that can complicate the polymer structure. Vinyl chloride, in contrast to ethylene, is unsymmetrical. We might refer to the $CH_2$ group in vinyl chloride as the "head" of the molecule and the CHCl group as the "tail":

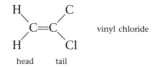

vinyl chloride

**Figure 22.11**
The polyethylene bottle at the left is made of pliable, branched polyethylene. The one at the right is made of semirigid, linear polyethylene. (Marna G. Clarke)

In principle at least, vinyl chloride molecules can add to one another to give three different types of polymers:

**1.** A *head-to-tail* polymer, in which there is a Cl atom on every other C atom in the chain:

$$
\begin{array}{cccccccccc}
\text{H} & \text{H} & \text{H} & \text{H} & \text{H} & \text{H} & \text{H} & \text{H} & \text{H} & \text{H} \\
| & | & | & | & | & | & | & | & | & | \\
-\text{C}-\text{C}-\text{C}-\text{C}-\text{C}-\text{C}-\text{C}-\text{C}-\text{C}-\text{C}- \\
| & | & | & | & | & | & | & | & | & | \\
\text{H} & \text{Cl} & \text{H} & \text{Cl} & \text{H} & \text{Cl} & \text{H} & \text{Cl} & \text{H} & \text{Cl}
\end{array}
$$

PVC is a stiff, rugged, cheap polymer

**2.** A *head-to-head, tail-to-tail* polymer, in which Cl atoms occur in pairs on adjacent carbon atoms in the chain:

$$
\begin{array}{cccccccccc}
\text{H} & \text{H} & \text{H} & \text{H} & \text{H} & \text{H} & \text{H} & \text{H} & \text{H} & \text{H} \\
| & | & | & | & | & | & | & | & | & | \\
-\text{C}-\text{C}-\text{C}-\text{C}-\text{C}-\text{C}-\text{C}-\text{C}-\text{C}-\text{C}- \\
| & | & | & | & | & | & | & | & | & | \\
\text{H} & \text{Cl} & \text{Cl} & \text{H} & \text{H} & \text{Cl} & \text{Cl} & \text{H} & \text{H} & \text{Cl}
\end{array}
$$

**3.** A *random* polymer:

$$
\begin{array}{cccccccccc}
\text{H} & \text{H} & \text{H} & \text{H} & \text{H} & \text{H} & \text{H} & \text{H} & \text{H} & \text{H} \\
| & | & | & | & | & | & | & | & | & | \\
-\text{C}-\text{C}-\text{C}-\text{C}-\text{C}-\text{C}-\text{C}-\text{C}-\text{C}-\text{C}- \\
| & | & | & | & | & | & | & | & | & | \\
\text{H} & \text{Cl} & \text{H} & \text{Cl} & \text{Cl} & \text{H} & \text{H} & \text{Cl} & \text{H} & \text{Cl}
\end{array}
$$

Similar arrangements are possible with polymers made from other asymmetrical monomers (Example 22.6). In practice, the addition polymer is usually of the head-to-tail type. This is the case with polyvinyl chloride and polypropylene.

---

**Example 22.6**    Sketch a polymer derived from propene:

$$
\begin{array}{ccc}
\text{H} & & \text{H} \\
& \diagdown \quad \diagup & \\
& \text{C}=\text{C} & \\
& \diagup \quad \diagdown & \\
\text{H} & & \text{CH}_3
\end{array}
$$

assuming it to be a

(a) head-to-tail polymer.          (b) head-to-head, tail-to-tail polymer

***Strategy***    In the head-to-tail polymer, the two different groups (H and $CH_3$ in this case) alternate along the carbon chain. In the other type of polymer, different groups alternate in pairs (two H atoms, then two $CH_3$ groups, . . .).

***Solution***

$$
\text{(a)} \quad
\begin{array}{cccccccccc}
\text{H} & \text{H} & \text{H} & \text{H} & \text{H} & \text{H} & \text{H} & \text{H} & \text{H} & \text{H} \\
| & | & | & | & | & | & | & | & | & | \\
-\text{C}-\text{C}-\text{C}-\text{C}-\text{C}-\text{C}-\text{C}-\text{C}-\text{C}-\text{C}- \\
| & | & | & | & | & | & | & | & | & | \\
\text{H} & \text{CH}_3 & \text{H} & \text{CH}_3 & \text{H} & \text{CH}_3 & \text{H} & \text{CH}_3 & \text{H} & \text{CH}_3
\end{array}
$$

$$
\text{(b)} \quad
\begin{array}{cccccccccc}
\text{H} & \text{H} & \text{H} & \text{H} & \text{H} & \text{H} & \text{H} & \text{H} & \text{H} & \text{H} \\
| & | & | & | & | & | & | & | & | & | \\
-\text{C}-\text{C}-\text{C}-\text{C}-\text{C}-\text{C}-\text{C}-\text{C}-\text{C}-\text{C}- \\
| & | & | & | & | & | & | & | & | & | \\
\text{H} & \text{CH}_3 & \text{CH}_3 & \text{H} & \text{H} & \text{CH}_3 & \text{CH}_3 & \text{H} & \text{H} & \text{CH}_3
\end{array}
$$

---

## Condensation Polymers

In order to produce a condensation polymer, ***the molecules involved must have functional groups at both ends of the molecule.*** When an alcohol with two —OH groups, HO—R—OH, reacts with a dicarboxylic acid, HOOC—R'—COOH, a **polyester** is formed. The first step in the process is the formation of a simple ester that has a reactive group at both ends of the molecule.

$$\text{HO—R—OH} + \text{HO—}\underset{\underset{O}{\|}}{C}\text{—R'—}\underset{\underset{O}{\|}}{C}\text{—OH} \longrightarrow$$

dihydroxy        dicarboxylic
alcohol           acid

$$\text{HO—R—O—}\underset{\underset{O}{\|}}{C}\text{—R'—}\underset{\underset{O}{\|}}{C}\text{—OH} + \boxed{H_2O}$$

ester with active
end groups

Thread made from polyester. (Charles D. Winters)

The —COOH group at one end of the ester molecule can react with another alcohol molecule. The —OH group at the other end can react with an acid molecule. This process can continue, leading eventually to a long-chain polymer containing 500 or more ester groups. The general structure of the polyester can be represented as

$$\text{—}\underset{\underset{O}{\|}}{C}\text{—R'—}\underset{\underset{O}{\|}}{C}\text{—O—R—O—}\underset{\underset{O}{\|}}{C}\text{—R'—}\underset{\underset{O}{\|}}{C}\text{—O—R—O—}$$

section of a polyester molecule

A thin polyester film was used to cover the wings and pilot compartment of the *Gossamer Albatross,* the first human-powered aircraft to cross the English Channel (Fig. 22.12).

One of the most familiar polyesters is Dacron, in which the monomers are ethylene glycol and terephthalic acid:

$$\text{HO—CH}_2\text{—CH}_2\text{—OH} \qquad \text{HO—}\underset{\underset{O}{\|}}{C}\text{—}\bigcirc\text{—}\underset{\underset{O}{\|}}{C}\text{—OH}$$

ethylene glycol                terephthalic acid

**Figure 22.12**
The *Gossamer Albatross,* the first human-powered craft to fly across the English Channel (1979). Lightweight, durable polyesters cover the wings and body. (UPI/Corbis-Bettman)

**Example 22.7**

(a) Show the structure of the ester formed when one molecule of ethylene glycol reacts with one molecule of terephthalic acid.
(b) Draw the structure of a section of the Dacron (Mylar) polymer.

***Strategy*** In (a), split out $H_2O$ between the alcohol and acid molecules. In (b), continue the esterification process at both ends of the molecule obtained in (a).

***Solution***

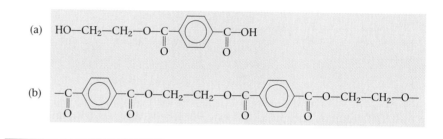

Another type of condensation polymer is a **polyamide** formed when a diamine reacts with a dicarboxylic acid:

**Figure 22.13**
Nylon can be made by the reaction between hexamethylenediamine, $H_2N—(CH_2)_6—NH_2$ and adipic acid, $HOOC—(CH_2)_4—COOH$. It forms at the interface between the two reagents. (Charles D. Winters)

Condensation can continue to form a long-chain polymer such as nylon-66, which has the structure

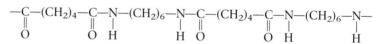

This polymer was first made in 1935 by Wallace Carothers at DuPont; it is readily prepared in the laboratory by bringing the two monomers into contact with one another (Fig. 22.13).

# CHEMISTRY

## *Beyond the Classroom*

## Proteins

Natural polymers produced by plants and animals are essential to all forms of life. In many ways, their structures resemble those of the synthetic polymers discussed in Section 22.5, except that natural polymers are usually more complex. Consider, in particular, the substances known as proteins, which make up about 15% of the human body. Like nylon, proteins are polymers of high molar mass ($6 \times 10^3$ to $1 \times 10^6$ g/mol) in which monomer units are joined to one another by the linkage referred to as the *amide* or *peptide* group. However, proteins are more complex than nylon in the sense that they can contain as many as 20 different monomers, known collectively as α-*amino acids*, which have the structure

$$R-\underset{\underset{NH_2}{|}}{\overset{\overset{H}{|}}{C}}-\overset{\overset{O}{\parallel}}{C}\diagdown_{OH} \qquad (R = H, \text{ alkyl group, } \ldots)$$

A few of the simpler α-amino acids are shown in Table 22.A.

### TABLE 22.A   Some α-Amino Acids

| Acid | R Group | Acid | R Group |
|---|---|---|---|
| glycine | $-H$ | valine | $(CH_3)_2CH-$ |
| alanine | $-CH_3$ | leucine | $(CH_3)_2CH-CH_2-$ |
| serine | $-CH_2OH$ | aspartic acid | $HOOC-CH_2-$ |
| cysteine | $-CH_2SH$ | glutamic acid | $HOOC-CH_2-CH_2-$ |

A protein polymer is formed from α-amino acids by splitting out an $H_2O$ molecule between a —COOH group in one acid and the —NH$_2$ group in another:

$$H-\underset{\underset{H}{|}}{\overset{\overset{H}{|}}{N}}-\underset{\underset{R_1}{|}}{\overset{\overset{O}{\parallel}}{C}}-OH + H-\underset{\underset{H}{|}}{\overset{\overset{H}{|}}{N}}-\underset{\underset{R_2}{|}}{\overset{\overset{O}{\parallel}}{C}}-OH \longrightarrow H-\underset{\underset{H}{|}}{\overset{\overset{H}{|}}{N}}-\underset{\underset{R_1}{|}}{\overset{\overset{O}{\parallel}}{C}}-\underset{\underset{H}{|}}{\overset{\overset{H}{|}}{N}}-\underset{\underset{R_2}{|}}{\overset{\overset{O}{\parallel}}{C}}-OH + H_2O$$

The product of this condensation has reactive groups at both ends of the molecule (—NH$_2$ at one end, —COOH at the other). Hence, condensation can continue, giving the long-chain polymers called proteins. The general structure of a protein can be represented as shown below; its 3-dimensional form is shown in Fig. 22.A.

$$-N-\underset{R_1}{C}-C-N-\underset{R_2}{C}-C-N-\underset{R_3}{C}-C-N-\underset{R_4}{C}-C-$$

Peptide linkages are shown in color.

In the human digestive system, proteins from foods are broken down into α-amino acids. These are then reassembled in cells with the aid of DNA (*deoxyribonucleic acid*) to form other proteins required by the body. In principle at least, there are a huge number of possible proteins. Consider, for example, the proteins formed by linking together 50 amino acid units. Since there are 20 possibilities for each of these units, we have

$$20^{50} = 1 \times 10^{65}$$

possible proteins containing 50 amino acid residues. In practice, the number of proteins is considerably smaller; your body contains about $10^5$ different proteins.

In the body, proteins are built up by a series of reactions that in general produce a specific sequence of amino acids. Even tiny errors in this sequence may have serious effects. Among the genetic diseases known to be caused by improper sequencing are hemophilia, sickle-cell anemia, and albinism. Sickle-cell anemia is caused by the substitution of *one* valine unit for a glutamic acid unit in a chain containing 146 monomers.

*continued*

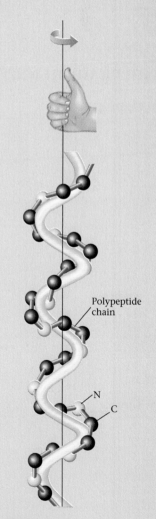

Polypeptide chain

N

C

**Figure 22.A**
α-helix structure of protein

The way in which protein chains are oriented in three dimensions is determined in large part by hydrogen bonding. Oxygen atoms in C=O groups can interact with hydrogen atoms in nearby N—H groups to form these bonds. One structure that maximizes hydrogen bonding is the α-helix, originally proposed by Linus Pauling and Robert Corey in 1948 (Figure 22.A).

Proteins are often classified according to their biological function. As pointed out in Chapter 11, the proteins known as enzymes act as *catalysts* for a wide variety of biochemical reactions. Hemoglobin (Chapter 15) is an example of a *transport* protein; it carries oxygen through the blood stream. The relatively simple protein insulin (51 amino acid units, molar mass = 5777 g/mol) is a *hormone*; it regulates the metabolism of glucose in the body. *Structural* proteins include collagen, found in tendons and ligaments. An example of a *protective* protein is fibrinogen, which is involved in blood clotting. When you cut yourself, fibrinogen is converted by a complex biochemical reaction into fibrin, an insoluble mass of protein fibers that seals the cut and stops the flow of blood.

# CHAPTER HIGHLIGHTS

## Key Concepts

**1.** Draw
—structural isomers
(Example 22.1; Problems 15–24, 53)
—geometric isomers
(Example 22.3; Problems 25–28)
**2.** Name hydrocarbons
(Example 22.2; Problems 1–8)
**3.** Write structural formulas for alcohols, carboxylic acids, and esters
(Examples 22.4, 22.5; Problems 11–14, 29–32)
**4.** Draw structures for
—addition polymers
(Example 22.6; Problems 33–38)
—condensation polymers
(Example 22.7; Problems 39–42, 55)

## Key Terms

| | | |
|---|---|---|
| alcohol | functional group | polymer |
| alkane | geometric isomer | —addition |
| alkene | —*cis-* | —condensation |
| alkyne | —*trans-* | —head-to-head |
| aromatic | o-, m-, p- | —head-to-tail |
| carboxylic acid | polyamide | saturated hydrocarbon |
| ester | polyester | structural isomer |
| | | unsaturated hydrocarbon |

## Summary Problem

Consider the alkane that has the structure

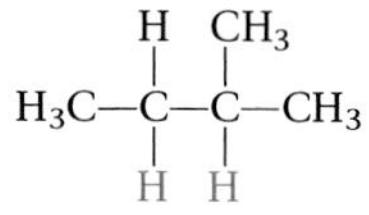

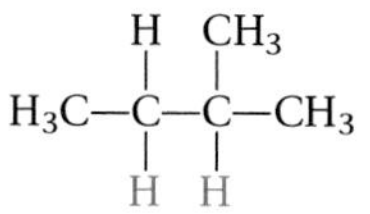

H₃C—C—C—CH₃ (with H CH₃ above, H H below)

(a) Name this alkane.
(b) Draw the structures of all the alkanes isomeric with it.
(c) Draw the structure of the alkene obtained by removing the two hydrogen atoms shown in red. What is the name of this alkene?
(d) Does the alkene referred to in (c) show structural isomerism? geometric isomerism?
(e) Sketch a portion of the head-to-tail addition polymer derived from the alkene in (c).
(f) Draw the structures of the two alcohols formed by replacing an H atom shown in red with an —OH group.
(g) Give the molecular formula of the ester formed when one of the alcohols in (f) reacts with formic acid.

## Answers

(a) 2-methylbutane

(b)

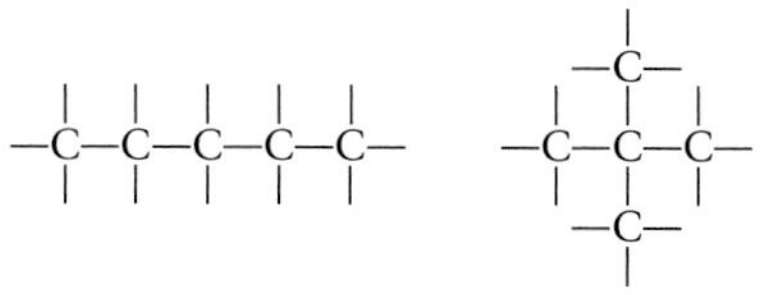

(c) 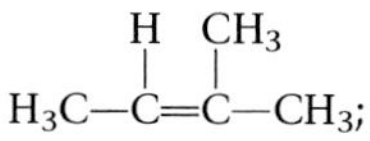 H₃C—C=C—CH₃; 2-methyl-2-butene    (d) yes; no

(e)

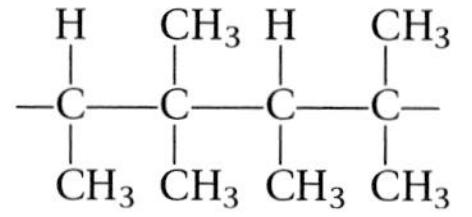

(f) 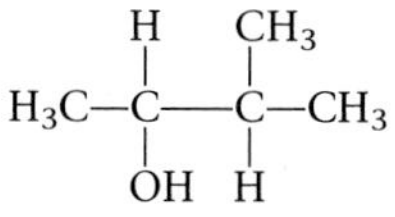 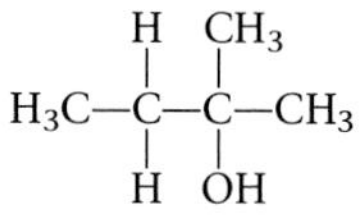 H₃C—C—C—CH₃   H₃C—C—C—CH₃    (g) $C_6H_{12}O_2$

## Questions & Problems

### Nomenclature

*1. Name the following alkanes.

(a) 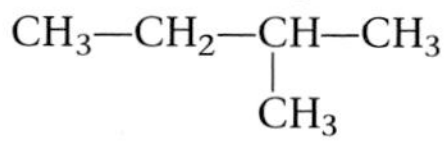 CH₃—CH₂—CH—CH₃
|
CH₃

(b) 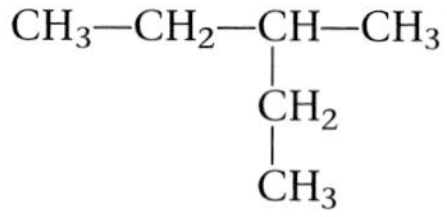 CH₃—CH₂—CH—CH₃
|
CH₂
|
CH₃

(c) 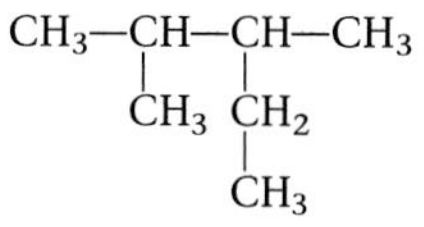 CH₃—CH—CH—CH₃
|      |
CH₃  CH₂
|
CH₃

(d) 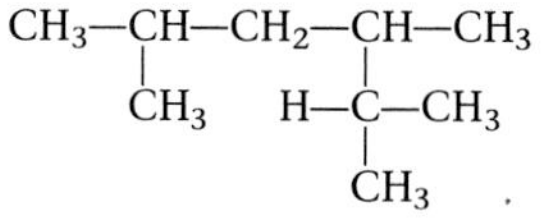 CH₃—CH—CH₂—CH—CH₃
|              |
CH₃      H—C—CH₃
|
CH₃

**\*2.** Name the following alkanes.

**(a)** CH₃—(CH₂)₅—CH—CH₃
            |
            CH₃

**(b)** (CH₃)₄C

**(c)** CH₃—CH—CH₂—C(CH₃)₃
            |
            CH₃

**(d)** 
            H
            |
    CH₃—C—(CH₂)₂—CH—CH₃
            |            |
            CH₂      CH₃
            |
            CH₃

**\*3.** Write structures for the following alkanes.
**(a)** 3-ethylpentane    **(b)** 2,2-dimethylbutane
**(c)** 2-methyl-3-ethylheptane
**(d)** 2,3-dimethylpentane

**\*4.** Write structures for the following alkanes.
**(a)** 2,2,4-trimethylpentane   **(b)** 2,2-dimethylpropane
**(c)** 4-isopropyloctane    **(d)** 2,3,4-trimethylheptane

**\*5.** The following names are incorrect; draw a reasonable structure for the alkane and give the proper IUPAC name.
**(a)** 5-isopropyloctane   **(b)** 2-ethylpropane
**(c)** 1,2-dimethylpropane

**\*6.** Follow the directions of Problem 5 for the following names.
**(a)** 2-dimethylbutane   **(b)** 4-methylpentane
**(c)** 2-ethylpropane

**\*7.** Name the following alkenes.
**(a)** CH₂=C—(CH₃)₂
**(b)** (CH₃)₂—C=C—(CH₃)₂
**(c)** CH₃—CH=CH—CH₂CH₃
**(d)** CH₃—C=CH₂
            |
            CH₂
            |
            CH₃

**\*8.** Write structures for the following alkynes.
**(a)** 2-pentyne    **(b)** 4-methyl-2-pentyne
**(c)** 2-methyl-3-hexyne   **(d)** 3,3-dimethyl-1-butyne

**\*9.** Name the following compounds, using the *ortho-, meta-, para-* notation.

**(a)** [benzene ring with Br and Br substituents]
**(b)** [benzene ring with NO₂ and NO₂ substituents]
**(c)** [benzene ring with I and I substituents]

**\*10.** Name the following compounds using numbers (1, 2, . . .).

**(a)** [benzene ring with Cl and Cl]
**(b)** [benzene ring with Cl, Cl, Cl]
**(c)** [benzene ring with Cl, Cl, Cl]

**\*11.** Give the IUPAC and common names of
**(a)** CH₃OH   **(b)** CH₃COOH   **(c)** CH₃CH₂CH₂OH

**\*12.** Give the IUPAC and common names of
**(a)** HCOOH    **(b)** CH₃CH₂OH
**(c)** CH₃—CH—CH₃
              |
              OH

**\*13.** Show the structure of
**(a)** methyl formate.    **(b)** ethyl propionate.
**(c)** propyl acetate.    **(d)** isopropyl acetate.

**\*14.** Name the following esters.

**(a)** CH₃—O—C—CH₃
                ‖
                O

**(b)** H—C—O—CH₂CH₃
            ‖
            O

**(c)** CH₃—O—C—CH₂CH₃
                ‖
                O

## Structural Isomerism

**\*15.** Draw the structural isomers of the alkane C₆H₁₄.

**\*16.** Draw the structural isomers of the alkene C₄H₈.

**\*17.** Draw the structural isomers of C₄H₉Cl in which one hydrogen atom of a C₄H₁₀ molecule has been replaced by chlorine.

**\*18.** Draw the structural isomers of C₃H₆Cl₂ in which two of the hydrogen atoms of C₃H₈ have been replaced by chlorine atoms.

**\*19.** There are three compounds with the formula C₆H₄ClBr in which two of the hydrogen atoms of the benzene molecule have been replaced by halogen atoms. Draw structures for these compounds.

**\*20.** There are three compounds with the formula C₆H₃Cl₃ in which three of the hydrogen atoms of the benzene molecule have been replaced by chlorine atoms. Draw structures for these compounds.

**\*21.** Write structures for all the structural isomers of double-bonded compounds with the molecular formula C₅H₁₀.

**\*22.** Write structures for all the structural isomers of compounds with the molecular formula C₄H₆ClBr in which Cl and Br are bonded to a double-bonded carbon.

**\*23.** Draw structures for all the alcohols with molecular formula C₅H₁₂O.

**\*24.** Draw structures for all the saturated carboxylic acids with four carbon atoms per molecule.

## Geometric Isomerism

**\*25.** Of the compounds in Problem 21, which ones show geometric isomerism? Draw the *cis-* and *trans-* isomers.

**\*26.** Of the compounds in Problem 22, which ones show geometric isomerism? Draw the *cis-* and *trans-* isomers.

**\*27.** Maleic acid and fumaric acid are the *cis-* and *trans-* isomers, respectively, of C₂H₂(COOH)₂, a dicarboxylic acid. Draw and label their structures.

**\*28.** For which of the following is geometric isomerism possible?

(a) $(CH_3)_2C=CCl_2$   (b) $CH_3ClC=CCH_3Cl$

(c) $CH_3BrC=CCH_3Cl$

## Functional Groups

**\*29.** Classify each of the following as a carboxylic acid, ester, and/or alcohol.

(a) $HO-CH_2-CH_2-CH_2-OH$

(b)

(c) $CH_3-\overset{\overset{H}{|}}{\underset{\underset{OH}{|}}{C}}-COOH$

**\*30.** Classify each of the following as a carboxylic acid, ester, and/or alcohol.

(a) $CH_3-(CH_2)_3-OH$

(b) $CH_3-CH_2-\overset{}{\underset{\underset{O}{\|}}{C}}-O-CH_2-CH_3$

(c) $CH_3-CH_2-O-\overset{}{\underset{\underset{O}{\|}}{C}}-(CH_2)_6-COOH$

**\*31.** Give the structure of

(a) a four-carbon straight-chain alcohol with an $-OH$ group not at the end of the chain.

(b) a five-carbon straight-chain carboxylic acid.

(c) the ester formed when these two compounds react.

**\*32.** Give the structure of

(a) a three-carbon alcohol with the $-OH$ group on the center carbon.

(b) a four-carbon branched-chain carboxylic acid.

(c) the ester formed when these two compounds react.

## Addition Polymers

**33.** Consider a polymer made from tetrachloroethylene, $C_2Cl_4$.

(a) Draw a portion of the polymer chain.

(b) What is the molar mass of the polymer if it contains $3.2 \times 10^3$ tetrachloroethylene molecules?

(c) What are the mass percents of C and Cl in the polymer?

**34.** Consider Teflon, the polymer made from tetrafluoroethylene.

(a) Draw a portion of the Teflon molecule.

(b) Calculate the molar mass of a Teflon molecule that contains $5.0 \times 10^4$ $CF_2$ units.

(c) What are the mass percents of C and F in Teflon?

**\*35.** Sketch a portion of the acrylonitrile polymer (Table 22.5), assuming it is a

(a) head-to-tail polymer.

(b) head-to-head, tail-to-tail polymer.

**\*36.** Styrene, , forms a head-to-tail addition polymer. Sketch a portion of a polystyrene molecule.

**\*37.** The polymer whose structure is shown below is made from two different monomers. Identify the monomers.

**\*38.** Show the structure of the monomer used to make the following addition polymers.

## Condensation Polymers

**\*39.** A rather simple polymer can be made from ethylene glycol, $HO-CH_2-CH_2-OH$, and oxalic acid, $HO-\overset{}{\underset{\underset{O}{\|}}{C}}-\overset{}{\underset{\underset{O}{\|}}{C}}-OH$. Sketch a portion of the polymer chain obtained from these monomers.

**\*40.** Lexan is a very rugged polyester in which the monomers can be taken to be carbonic acid, $HO-\overset{}{\underset{\underset{O}{\|}}{C}}-OH$, and

Sketch a section of the Lexan chain.

**\*41.** The following condensation polymer is made from a single monomer. Identify the monomer.

**\*42.** Identify the monomers from which the following condensation polymers are made.

(a)

(b)

## Unclassified

**\*43.** Which of the following are expected to show bond angles of 109.5°? 120°? 180°?

(a) $H_3C-CH_3$   (b) $H_3C-\overset{\overset{H}{|}}{C}=CH_2$

(c) $H_3C-C\equiv C-CH_3$

**\*44.** State at least one use for each of the following hydrocarbons.

    **(a)** $C_3H_8$   **(b)** $C_2H_4$   **(c)** $C_2H_2$   **(d)** toluene

**\*45.** What does the "circle" represent in the structural formula of benzene?

**\*46.** Correct each of the following statements to make it true.

    **(a)** Gasoline contains mostly alkenes and alkynes.

    **(b)** Benzene is used in chemotherapy.

    **(c)** Calcium carbide reacts with water to form ethene.

    **(d)** Alkanes show geometric isomerism.

    **(e)** Cholesterol is a condensed-ring hydrocarbon.

**\*47.** Explain how the following compounds are prepared.

    **(a)** methanol   **(b)** ethanol

    **(c)** acetic acid   **(d)** ethyl acetate

**48.** Calculate $[H^+]$ and the pH of a 0.10 $M$ solution of chloroacetic acid ($K_a = 1.5 \times 10^{-3}$); use successive approximations.

**\*49.** Explain what is meant by

    **(a)** a saturated fat.   **(b)** a soap.   **(c)** the "proof" of an alcoholic beverage.   **(d)** denatured alcohol.

**\*50.** Which of the following monomers could form an addition polymer? a condensation polymer?

    **(a)** $C_2H_6$            **(b)** $C_2H_4$

    **(c)** $HO-CH_2-CH_2-OH$   **(d)** $HO-CH_2-CH_3$

**\*51.** Draw the structures of the monomers that could be used to make the following polymers

    **(a)**

**(b)**

**(c)**

## Challenge Problems

**\*52.** The structure of cholesterol is given in the text. What is its molecular formula?

**\*53.** Draw structures for all the alcohols with molecular formula $C_6H_{14}O$.

**54.** What mass of propane must be burned to bring one quart of water in a saucepan at 25°C to the boiling point? Use data in Appendix 1 and elsewhere and make any reasonable assumptions.

**\*55.** Glycerol, $C_3H_5(OH)_3$, and orthophthalic acid,

form a cross-linked polymer in which adjacent polymer chains are linked together; this polymer is used in floor coverings and dentures.

    **(a)** Write the structural formula for a portion of the polymer chain.

    **(b)** Use your answer in (a) to show how cross-linking can occur between the polymer chains to form a water-insoluble, network covalent solid.

# Constants and Reference Data

## Fundamental Constants

| | |
|---|---|
| Acceleration of gravity (standard) | $9.8066 \text{ m/s}^2$ |
| Atomic mass unit (amu) | $1.6605 \times 10^{-24} \text{ g}$ |
| Avogadro constant | $6.0221 \times 10^{23}/\text{mol}$ |
| Electronic charge | $1.6022 \times 10^{-19} \text{ C}$ |
| Electronic mass | $9.1094 \times 10^{-28} \text{ g}$ |
| Faraday constant | $9.6485 \times 10^{4} \text{ C/mol}$ |
| Gas constant | $0.082058 \text{ L·atm/mol·K}$ |
| | $8.3145 \text{ J/mol·K}$ |
| Planck constant | $6.6261 \times 10^{-34} \text{ J·s}$ |
| Rydberg constant | $2.1799 \times 10^{-18} \text{ J}$ |
| Velocity of light | $2.9979 \times 10^{8} \text{ m/s}$ |
| $\pi$ | $3.1416$ |
| $e$ | $2.7183$ |
| $\ln x$ | $2.3026 \log_{10} x$ |

## Vapor Pressure of Water (mm Hg)

| T(°C) | vp | T(°C) | vp | T(°C) | vp | T(°C) | vp |
|---|---|---|---|---|---|---|---|
| 0 | 4.58 | 21 | 18.65 | 35 | 42.2 | 92 | 567.0 |
| 5 | 6.54 | 22 | 19.83 | 40 | 55.3 | 94 | 610.9 |
| 10 | 9.21 | 23 | 21.07 | 45 | 71.9 | 96 | 657.6 |
| 12 | 10.52 | 24 | 22.38 | 50 | 92.5 | 98 | 707.3 |
| 14 | 11.99 | 25 | 23.76 | 55 | 118.0 | 100 | 760.0 |
| 16 | 13.63 | 26 | 25.21 | 60 | 149.4 | 102 | 815.9 |
| 17 | 14.53 | 27 | 26.74 | 65 | 187.5 | 104 | 875.1 |
| 18 | 15.48 | 28 | 28.35 | 70 | 233.7 | 106 | 937.9 |
| 19 | 16.48 | 29 | 30.04 | 80 | 355.1 | 108 | 1004.4 |
| 20 | 17.54 | 30 | 31.82 | 90 | 525.8 | 110 | 1074.6 |

# Thermodynamic Data

| | $\Delta H_f^\circ$ (kJ/mol) | $S^\circ$ (kJ/K · mol) | $\Delta G_f^\circ$ (kJ/mol) at 25°C | | $\Delta H_f^\circ$ (kJ/mol) | $S^\circ$ (kJ/K · mol) | $\Delta G_f^\circ$ (kJ/mol) at 25°C |
|---|---|---|---|---|---|---|---|
| Ag(s) | 0.0 | +0.0426 | 0.0 | Cd(s) | 0.0 | +0.0518 | 0.0 |
| $Ag^+(aq)$ | +105.6 | +0.0727 | +77.1 | $Cd^{2+}(aq)$ | −75.9 | −0.0732 | −77.6 |
| AgBr(s) | −100.4 | +0.1071 | −96.9 | $CdCl_2(s)$ | −391.5 | +0.1153 | −344.0 |
| AgCl(s) | −127.1 | +0.0962 | −109.8 | CdO(s) | −258.2 | +0.0548 | −228.4 |
| AgI(s) | −61.8 | +0.1155 | −66.2 | $Cl_2(g)$ | 0.0 | +0.2230 | 0.0 |
| $AgNO_3(s)$ | −124.4 | +0.1409 | −33.4 | $Cl^-(aq)$ | −167.2 | +0.0565 | −131.2 |
| $Ag_2O(s)$ | −31.0 | +0.1213 | −11.2 | $ClO_3^-(aq)$ | −104.0 | +0.1623 | −8.0 |
| Al(s) | 0.0 | +0.0283 | 0.0 | $ClO_4^-(aq)$ | −129.3 | +0.1820 | −8.5 |
| $Al^{3+}(aq)$ | −531.0 | −0.3217 | −485.0 | Cr(s) | 0.0 | +0.0238 | 0.0 |
| $Al_2O_3(s)$ | −1675.7 | +0.0509 | −1582.3 | $CrO_4^{2-}(aq)$ | −881.2 | +0.0502 | −727.8 |
| Ba(s) | 0.0 | +0.0628 | 0.0 | $Cr_2O_3(s)$ | −1139.7 | +0.0812 | −1058.1 |
| $Ba^{2+}(aq)$ | −537.6 | +0.0096 | −560.8 | $Cr_2O_7^{2-}(aq)$ | −1490.3 | +0.2619 | −1301.1 |
| $BaCl_2(s)$ | −858.6 | +0.1237 | −810.4 | Cu(s) | 0.0 | +0.0332 | 0.0 |
| $BaCO_3(s)$ | −1216.3 | +0.1121 | −1137.6 | $Cu^+(aq)$ | +71.7 | +0.0406 | +50.0 |
| BaO(s) | −553.5 | +0.0704 | −525.1 | $Cu^{2+}(aq)$ | +64.8 | −0.0996 | +65.5 |
| $BaSO_4(s)$ | −1473.2 | +0.1322 | −1362.3 | CuO(s) | −157.3 | +0.0426 | −129.7 |
| $Br_2(l)$ | 0.0 | +0.1522 | 0.0 | $Cu_2O(s)$ | −168.6 | +0.0931 | −146.0 |
| $Br^-(aq)$ | −121.6 | +0.0824 | −104.0 | CuS(s) | −53.1 | +0.0665 | −53.6 |
| C(s) | 0.0 | +0.0057 | 0.0 | $Cu_2S(s)$ | −79.5 | +0.1209 | −86.2 |
| $CCl_4(l)$ | −135.4 | +0.2164 | −65.3 | $CuSO_4(s)$ | −771.4 | +0.1076 | −661.9 |
| $CHCl_3(l)$ | −134.5 | +0.2017 | −73.7 | $F_2(g)$ | 0.0 | +0.2027 | 0.0 |
| $CH_4(g)$ | −74.8 | +0.1862 | −50.7 | $F^-(aq)$ | −332.6 | −0.0138 | −278.8 |
| $C_2H_2(g)$ | +226.7 | +0.2008 | +209.2 | Fe(s) | 0.0 | +0.0273 | 0.0 |
| $C_2H_4(g)$ | +52.3 | +0.2195 | +68.1 | $Fe^{2+}(aq)$ | −89.1 | −0.1377 | −78.9 |
| $C_2H_6(g)$ | −84.7 | +0.2295 | −32.9 | $Fe^{3+}(aq)$ | −48.5 | −0.3159 | −4.7 |
| $C_3H_8(g)$ | −103.8 | +0.2699 | −23.5 | $Fe(OH)_3(s)$ | −823.0 | +0.1067 | −696.6 |
| $CH_3OH(l)$ | −238.7 | +0.1268 | −166.3 | $Fe_2O_3(s)$ | −824.2 | +0.0874 | −742.2 |
| $C_2H_5OH(l)$ | −277.7 | +0.1607 | −174.9 | $Fe_3O_4(s)$ | −1118.4 | +0.1464 | −1015.5 |
| CO(g) | −110.5 | +0.1976 | −137.2 | $H_2(g)$ | 0.0 | +0.1306 | 0.0 |
| $CO_2(g)$ | −393.5 | +0.2136 | −394.4 | $H^+(aq)$ | 0.0 | 0.0000 | 0.0 |
| $CO_3^{2-}(aq)$ | −677.1 | −0.0569 | −527.8 | HBr(g) | −36.4 | +0.1986 | −53.4 |
| Ca(s) | 0.0 | +0.0414 | 0.0 | HCl(g) | −92.3 | +0.1868 | −95.3 |
| $Ca^{2+}(aq)$ | −542.8 | −0.0531 | −553.6 | $HCO_3^-(aq)$ | −692.0 | +0.0912 | −586.8 |
| $CaCl_2(s)$ | −795.8 | +0.1046 | −748.1 | HF(g) | −271.1 | +0.1737 | −273.2 |
| $CaCO_3(s)$ | −1206.9 | +0.0929 | −1128.8 | HI(g) | +26.5 | +0.2065 | +1.7 |
| CaO(s) | −635.1 | +0.0398 | −604.0 | $HNO_3(l)$ | −174.1 | +0.1556 | −80.8 |
| $Ca(OH)_2(s)$ | −986.1 | +0.0834 | −898.5 | $H_2O(g)$ | −241.8 | +0.1887 | −228.6 |
| $CaSO_4(s)$ | −1434.1 | +0.1067 | −1321.8 | $H_2O(l)$ | −285.8 | +0.0699 | −237.2 |

(*continued*)

# Thermodynamic Data *(continued)*

| | $\Delta H_f^\circ$ (kJ/mol) | $S^\circ$ (kJ/K · mol) | $\Delta G_f^\circ$ (kJ/mol) at 25°C | | $\Delta H_f^\circ$ (kJ/mol) | $S^\circ$ (kJ/K · mol) | $\Delta G_f^\circ$ (kJ/mol) at 25°C |
|---|---|---|---|---|---|---|---|
| $H_2O_2(l)$ | −187.8 | +0.1096 | −120.4 | $NO_2^-(aq)$ | −104.6 | +0.1230 | −32.2 |
| $H_2PO_4^-(aq)$ | −1296.3 | +0.0904 | −1130.3 | $NO_3^-(aq)$ | −205.0 | +0.1464 | −108.7 |
| $HPO_4^{2-}(aq)$ | −1292.1 | −0.0335 | −1089.2 | $N_2O_4(g)$ | +9.2 | +0.3042 | +97.9 |
| $H_2S(g)$ | −20.6 | +0.2057 | −33.6 | $Na(s)$ | 0.0 | +0.0512 | 0.0 |
| $H_2SO_4(l)$ | −814.0 | +0.1569 | −690.1 | $Na^+(aq)$ | −240.1 | +0.0590 | −261.9 |
| $HSO_4^-(aq)$ | −887.3 | +0.1318 | −755.9 | $NaCl(s)$ | −411.2 | +0.0721 | −384.2 |
| $Hg(l)$ | 0.0 | +0.0760 | 0.0 | $NaF(s)$ | −573.6 | +0.0515 | −543.5 |
| $Hg^{2+}(aq)$ | +171.1 | −0.0322 | +164.4 | $NaOH(s)$ | −425.6 | +0.0645 | −379.5 |
| $HgO(s)$ | −90.8 | +0.0703 | −58.6 | $Ni(s)$ | 0.0 | +0.0299 | 0.0 |
| $I_2(s)$ | 0.0 | +0.1161 | 0.0 | $Ni^{2+}(aq)$ | −54.0 | −0.1289 | −45.6 |
| $I^-(aq)$ | −55.2 | +0.1113 | −51.6 | $NiO(s)$ | −239.7 | +0.0380 | −211.7 |
| $K(s)$ | 0.0 | +0.0642 | 0.0 | $O_2(g)$ | 0.0 | +0.2050 | 0.0 |
| $K^+(aq)$ | −252.4 | +0.1025 | −283.3 | $OH^-(aq)$ | −230.0 | −0.0108 | −157.2 |
| $KBr(s)$ | −393.8 | +0.0959 | −380.7 | $P_4(s)$ | 0.0 | +0.1644 | 0.0 |
| $KCl(s)$ | −436.7 | +0.0826 | −409.1 | $PCl_3(g)$ | −287.0 | +0.3117 | −267.8 |
| $KClO_3(s)$ | −397.7 | +0.1431 | −296.3 | $PCl_5(g)$ | −374.9 | +0.3645 | −305.0 |
| $KClO_4(s)$ | −432.8 | +0.1510 | −303.2 | $PO_4^{3-}(aq)$ | −1277.4 | −0.222 | −1018.7 |
| $KNO_3(s)$ | −494.6 | +0.1330 | −394.9 | $Pb(s)$ | 0.0 | +0.0648 | 0.0 |
| $Mg(s)$ | 0.0 | +0.0327 | 0.0 | $Pb^{2+}(aq)$ | −1.7 | +0.0105 | −24.4 |
| $Mg^{2+}(aq)$ | −466.8 | −0.1381 | −454.8 | $PbBr_2(s)$ | −278.7 | +0.1615 | −261.9 |
| $MgCl_2(s)$ | −641.3 | +0.0896 | −591.8 | $PbCl_2(s)$ | −359.4 | +0.1360 | −314.1 |
| $MgCO_3(s)$ | −1095.8 | +0.0657 | −1012.1 | $PbO(s)$ | −219.0 | +0.0665 | −188.9 |
| $MgO(s)$ | −601.7 | +0.0269 | −569.4 | $PbO_2(s)$ | −277.4 | +0.0686 | −217.4 |
| $Mg(OH)_2(s)$ | −924.5 | +0.0632 | −833.6 | $S(s)$ | 0.0 | +0.0318 | 0.0 |
| $MgSO_4(s)$ | −1284.9 | +0.0916 | −1170.7 | $S^{2-}(aq)$ | +33.1 | −0.0146 | +85.8 |
| $Mn(s)$ | 0.0 | +0.0320 | 0.0 | $SO_2(g)$ | −296.8 | +0.2481 | −300.2 |
| $Mn^{2+}(aq)$ | −220.8 | −0.0736 | −228.1 | $SO_3(g)$ | −395.7 | +0.2567 | −371.1 |
| $MnO(s)$ | −385.2 | +0.0597 | −362.9 | $SO_4^{2-}(aq)$ | −909.3 | +0.0201 | −744.5 |
| $MnO_2(s)$ | −520.0 | +0.0530 | −465.2 | $Si(s)$ | 0.0 | +0.0188 | 0.0 |
| $MnO_4^-(aq)$ | −541.4 | +0.1912 | −447.2 | $SiO_2(s)$ | −910.9 | +0.0418 | −856.7 |
| $N_2(g)$ | 0.0 | +0.1915 | 0.0 | $Sn(s)$ | 0.0 | +0.0516 | 0.0 |
| $NH_3(g)$ | −46.1 | +0.1923 | −16.5 | $Sn^{2+}(aq)$ | −8.8 | −0.0174 | −27.2 |
| $NH_4^+(aq)$ | −132.5 | +0.1134 | −79.3 | $SnO_2(s)$ | −580.7 | +0.0523 | −519.6 |
| $NH_4Cl(s)$ | −314.4 | +0.0946 | −203.0 | $Zn(s)$ | 0.0 | +0.0416 | 0.0 |
| $NH_4NO_3(s)$ | −365.6 | +0.1511 | −184.0 | $Zn^{2+}(aq)$ | −153.9 | −0.1121 | −147.1 |
| $N_2H_4(l)$ | +50.6 | +0.1212 | +149.2 | $ZnI_2(s)$ | −208.0 | +0.1611 | −209.0 |
| $NO(g)$ | +90.2 | +0.2107 | +86.6 | $ZnO(s)$ | −348.3 | +0.0436 | −318.3 |
| $NO_2(g)$ | +33.2 | +0.2400 | +51.3 | $ZnS(s)$ | −206.0 | +0.0577 | −201.3 |

# Equilibrium Constants

## Ionization Constants, Weak acids, $K_a$

| | | | | | |
|---|---|---|---|---|---|
| $H_3AsO_4$ | $5.7 \times 10^{-3}$ | $HNO_2$ | $6.0 \times 10^{-4}$ | $N_2H_5^+$ | $1.0 \times 10^{-8}$ |
| $H_2AsO_4^-$ | $1.8 \times 10^{-7}$ | $H_3PO_4$ | $7.1 \times 10^{-3}$ | $Al(H_2O)_6^{3+}$ | $1.2 \times 10^{-5}$ |
| $HAsO_4^{2-}$ | $2.5 \times 10^{-12}$ | $H_2PO_4^-$ | $6.2 \times 10^{-8}$ | $Ag(H_2O)_2^+$ | $1.2 \times 10^{-12}$ |
| $HBrO$ | $2.6 \times 10^{-9}$ | $HPO_4^{2-}$ | $4.5 \times 10^{-13}$ | $Ca(H_2O)_6^{2+}$ | $2.2 \times 10^{-13}$ |
| $HCHO_2$ | $1.9 \times 10^{-4}$ | $H_2S$ | $1.0 \times 10^{-7}$ | $Cd(H_2O)_4^{2+}$ | $4.0 \times 10^{-10}$ |
| $HC_2H_3O_2$ | $1.8 \times 10^{-5}$ | $H_2SO_3$ | $1.7 \times 10^{-2}$ | $Fe(H_2O)_6^{3+}$ | $6.7 \times 10^{-3}$ |
| $HCN$ | $5.8 \times 10^{-10}$ | $HSO_3^-$ | $6.0 \times 10^{-8}$ | $Fe(H_2O)_6^{2+}$ | $1.7 \times 10^{-7}$ |
| $H_2CO_3$ | $4.4 \times 10^{-7}$ | $HSO_4^-$ | $1.0 \times 10^{-2}$ | $Mg(H_2O)_6^{2+}$ | $3.7 \times 10^{-12}$ |
| $HCO_3^-$ | $4.7 \times 10^{-11}$ | $H_2Se$ | $1.5 \times 10^{-4}$ | $Mn(H_2O)_6^{2+}$ | $2.8 \times 10^{-11}$ |
| $HClO_2$ | $1.0 \times 10^{-2}$ | $H_2SeO_3$ | $2.7 \times 10^{-3}$ | $Ni(H_2O)_6^{2+}$ | $2.2 \times 10^{-10}$ |
| $HClO$ | $2.8 \times 10^{-8}$ | $HSeO_3^-$ | $5.0 \times 10^{-8}$ | $Pb(H_2O)_6^{2+}$ | $6.7 \times 10^{-7}$ |
| $HF$ | $6.9 \times 10^{-4}$ | $CH_3NH_3^+$ | $2.4 \times 10^{-11}$ | $Sc(H_2O)_6^{3+}$ | $1.1 \times 10^{-4}$ |
| $HIO$ | $2.4 \times 10^{-11}$ | $NH_4^+$ | $5.6 \times 10^{-10}$ | $Zn(H_2O)_4^{2+}$ | $3.3 \times 10^{-10}$ |
| $HN_3$ | $2.4 \times 10^{-5}$ | | | | |

## Ionization Constants, Weak Bases, $K_b$

| | | | | | |
|---|---|---|---|---|---|
| $AsO_4^{3-}$ | $4.0 \times 10^{-3}$ | $N_3^-$ | $4.2 \times 10^{-10}$ | $HSeO_3^-$ | $3.7 \times 10^{-12}$ |
| $HAsO_4^{2-}$ | $5.6 \times 10^{-8}$ | $NH_3$ | $1.8 \times 10^{-5}$ | $AlOH^{2+}$ | $8.3 \times 10^{-10}$ |
| $H_2AsO_4^-$ | $1.8 \times 10^{-12}$ | $N_2H_4$ | $1.0 \times 10^{-6}$ | $AgOH$ | $8.3 \times 10^{-3}$ |
| $BrO^-$ | $3.8 \times 10^{-6}$ | $NO_2^-$ | $1.7 \times 10^{-11}$ | $CaOH^+$ | $4.5 \times 10^{-2}$ |
| $CH_3NH_2$ | $4.2 \times 10^{-4}$ | $PO_4^{3-}$ | $2.2 \times 10^{-2}$ | $CdOH^+$ | $2.5 \times 10^{-5}$ |
| $CHO_2^-$ | $5.3 \times 10^{-11}$ | $HPO_4^{2-}$ | $1.6 \times 10^{-7}$ | $FeOH^{2+}$ | $1.5 \times 10^{-12}$ |
| $C_2H_3O_2^-$ | $5.6 \times 10^{-10}$ | $H_2PO_4^-$ | $1.4 \times 10^{-12}$ | $FeOH^+$ | $5.9 \times 10^{-8}$ |
| $CN^-$ | $1.7 \times 10^{-5}$ | $HS^-$ | $1.0 \times 10^{-7}$ | $MgOH^+$ | $2.7 \times 10^{-3}$ |
| $CO_3^{2-}$ | $2.1 \times 10^{-4}$ | $SO_3^{2-}$ | $1.7 \times 10^{-7}$ | $MnOH^+$ | $3.6 \times 10^{-4}$ |
| $HCO_3^-$ | $2.3 \times 10^{-8}$ | $HSO_3^-$ | $5.9 \times 10^{-13}$ | $NiOH^+$ | $4.5 \times 10^{-5}$ |
| $ClO_2^-$ | $1.0 \times 10^{-12}$ | $SO_4^{2-}$ | $1.0 \times 10^{-12}$ | $PbOH^+$ | $1.5 \times 10^{-8}$ |
| $ClO^-$ | $3.6 \times 10^{-7}$ | $HSe^-$ | $6.7 \times 10^{-11}$ | $ScOH^{2+}$ | $9.1 \times 10^{-11}$ |
| $F^-$ | $1.4 \times 10^{-11}$ | $SeO_3^{2-}$ | $2.0 \times 10^{-7}$ | $ZnOH^+$ | $3.0 \times 10^{-5}$ |
| $IO^-$ | $4.2 \times 10^{-4}$ | | | | |

## Formation Constants of Complex Ions, $K_f$

| | | | | | |
|---|---|---|---|---|---|
| $AgBr_2^-$ | $2 \times 10^7$ | $Cu(CN)_2^-$ | $1 \times 10^{24}$ | $PdBr_4^{2-}$ | $6 \times 10^{13}$ |
| $AgCl_2^-$ | $1.8 \times 10^5$ | $Cu(NH_3)_4^{2+}$ | $2 \times 10^{12}$ | $PtBr_4^{2-}$ | $6 \times 10^{17}$ |
| $Ag(CN)_2^-$ | $2 \times 10^{20}$ | $FeSCN^{2+}$ | $9.2 \times 10^2$ | $PtCl_4^{2-}$ | $1 \times 10^{16}$ |
| $AgI_2^-$ | $5 \times 10^{10}$ | $Fe(CN)_6^{3-}$ | $4 \times 10^{52}$ | $Pt(CN)_4^{2-}$ | $1 \times 10^{41}$ |
| $Ag(NH_3)_2^+$ | $1.7 \times 10^7$ | $Fe(CN)_6^{4-}$ | $4 \times 10^{45}$ | $c\text{-}Pt(NH_3)_2Cl_2$ | $3 \times 10^{29}$ |
| $Al(OH)_4^-$ | $1 \times 10^{33}$ | $HgBr_4^{2-}$ | $1 \times 10^{21}$ | $t\text{-}Pt(NH_3)_2Cl_2$ | $3 \times 10^{28}$ |
| $Cd(CN)_4^{2-}$ | $2 \times 10^{18}$ | $HgCl_4^{2-}$ | $1 \times 10^{15}$ | $c\text{-}Pt(NH_3)_2(H_2O)_2^{2+}$ | $4 \times 10^{23}$ |
| $CdI_4^{2-}$ | $4.0 \times 10^5$ | $Hg(CN)_4^{2-}$ | $2 \times 10^{41}$ | $c\text{-}Pt(NH_3)_2I_2$ | $2 \times 10^{33}$ |
| $Cd(NH_3)_4^{2+}$ | $2.8 \times 10^7$ | $HgI_4^{2-}$ | $1 \times 10^{30}$ | $t\text{-}Pt(NH_3)_2I_2$ | $5 \times 10^{32}$ |
| $Cd(OH)_4^{2-}$ | $1.2 \times 10^9$ | $Hg(NH_3)_4^{2+}$ | $2 \times 10^{19}$ | $c\text{-}Pt(NH_3)_2(OH)_2$ | $1 \times 10^{39}$ |
| $Co(NH_3)_6^{2+}$ | $1 \times 10^5$ | $Ni(CN)_4^{2-}$ | $1 \times 10^{30}$ | $Pt(NH_3)_4^{2+}$ | $2 \times 10^{35}$ |
| $Co(NH_3)_6^{3+}$ | $1 \times 10^{23}$ | $Ni(NH_3)_6^{2+}$ | $9 \times 10^8$ | $Zn(CN)_4^{2-}$ | $6 \times 10^{16}$ |
| $Co(NH_3)_5Cl^{2+}$ | $2 \times 10^{28}$ | $PbI_4^{2-}$ | $1.7 \times 10^4$ | $Zn(NH_3)_4^{2+}$ | $3.6 \times 10^8$ |
| $Co(NH_3)_5NO_2^{2+}$ | $1 \times 10^{24}$ | $Pb(OH)_3^-$ | $8 \times 10^{13}$ | $Zn(OH)_4^{2-}$ | $3 \times 10^{14}$ |

## Solubility Product Constants, $K_{sp}$

| | | | | | |
|---|---|---|---|---|---|
| $AgBr$ | $5 \times 10^{-13}$ | $Co(OH)_2$ | $2 \times 10^{-16}$ | $NiCO_3$ | $1.4 \times 10^{-7}$ |
| $AgC_2H_3O_2$ | $1.9 \times 10^{-3}$ | $Co_3(PO_4)_2$ | $1 \times 10^{-35}$ | $Ni_3(PO_4)_2$ | $1 \times 10^{-32}$ |
| $Ag_2CO_3$ | $8 \times 10^{-12}$ | $CuCl$ | $1.7 \times 10^{-7}$ | $NiS$ | $1 \times 10^{-21}$ |
| $AgCl$ | $1.8 \times 10^{-10}$ | $CuBr$ | $6.3 \times 10^{-9}$ | $PbBr_2$ | $6.6 \times 10^{-6}$ |
| $Ag_2CrO_4$ | $1 \times 10^{-12}$ | $CuI$ | $1.2 \times 10^{-12}$ | $PbCO_3$ | $1 \times 10^{-13}$ |
| $AgI$ | $1 \times 10^{-16}$ | $Cu_3(PO_4)_2$ | $1 \times 10^{-37}$ | $PbCl_2$ | $1.7 \times 10^{-5}$ |
| $Ag_3PO_4$ | $1 \times 10^{-16}$ | $CuS$ | $1 \times 10^{-36}$ | $PbCrO_4$ | $2 \times 10^{-14}$ |
| $Ag_2S$ | $1 \times 10^{-49}$ | $Cu_2S$ | $1 \times 10^{-48}$ | $PbF_2$ | $7.1 \times 10^{-7}$ |
| $AgSCN$ | $1.0 \times 10^{-12}$ | $FeCO_3$ | $3.1 \times 10^{-11}$ | $PbI_2$ | $8.4 \times 10^{-9}$ |
| $AlF_3$ | $1 \times 10^{-18}$ | $Fe(OH)_2$ | $5 \times 10^{-17}$ | $Pb(OH)_2$ | $1 \times 10^{-20}$ |
| $Al(OH)_3$ | $2 \times 10^{-31}$ | $Fe(OH)_3$ | $3 \times 10^{-39}$ | $PbS$ | $1 \times 10^{-28}$ |
| $AlPO_4$ | $1 \times 10^{-20}$ | $FeS$ | $2 \times 10^{-19}$ | $Pb(SCN)_2$ | $2.1 \times 10^{-5}$ |
| $BaCO_3$ | $2.6 \times 10^{-9}$ | $GaF_3$ | $2 \times 10^{-16}$ | $PbSO_4$ | $1.8 \times 10^{-8}$ |
| $BaCrO_4$ | $1.2 \times 10^{-10}$ | $Ga(OH)_3$ | $1 \times 10^{-35}$ | $Sc(OH)_3$ | $2 \times 10^{-31}$ |
| $BaF_2$ | $1.8 \times 10^{-7}$ | $GaPO_4$ | $1 \times 10^{-21}$ | $SnS$ | $3 \times 10^{-28}$ |
| $BaSO_4$ | $1.1 \times 10^{-10}$ | $Hg_2Br_2$ | $6 \times 10^{-23}$ | $SrCO_3$ | $5.6 \times 10^{-10}$ |
| $Bi_2S_3$ | $1 \times 10^{-99}$ | $Hg_2Cl_2$ | $1 \times 10^{-18}$ | $SrCrO_4$ | $3.6 \times 10^{-5}$ |
| $CaCO_3$ | $4.9 \times 10^{-9}$ | $Hg_2I_2$ | $5 \times 10^{-29}$ | $SrF_2$ | $4.3 \times 10^{-9}$ |
| $CaF_2$ | $1.5 \times 10^{-10}$ | $HgS$ | $1 \times 10^{-52}$ | $SrSO_4$ | $3.4 \times 10^{-7}$ |
| $Ca(OH)_2$ | $4.0 \times 10^{-6}$ | $Li_2CO_3$ | $8.2 \times 10^{-4}$ | $TlCl$ | $1.9 \times 10^{-4}$ |
| $Ca_3(PO_4)_2$ | $1 \times 10^{-33}$ | $MgCO_3$ | $6.8 \times 10^{-6}$ | $TlBr$ | $3.7 \times 10^{-6}$ |
| $CaSO_4$ | $7.1 \times 10^{-5}$ | $MgF_2$ | $7 \times 10^{-11}$ | $TlI$ | $5.6 \times 10^{-8}$ |
| $CdCO_3$ | $6 \times 10^{-12}$ | $Mg(OH)_2$ | $6 \times 10^{-12}$ | $Tl(OH)_3$ | $2 \times 10^{-44}$ |
| $Cd(OH)_2$ | $5 \times 10^{-15}$ | $Mg_3(PO_4)_2$ | $1 \times 10^{-24}$ | $Tl_2S$ | $1 \times 10^{-20}$ |
| $Cd_3(PO_4)_2$ | $1 \times 10^{-33}$ | $Mn(OH)_2$ | $2 \times 10^{-13}$ | $ZnCO_3$ | $1.1 \times 10^{-10}$ |
| $CdS$ | $1 \times 10^{-29}$ | $MnS$ | $5 \times 10^{-14}$ | $Zn(OH)_2$ | $4 \times 10^{-17}$ |

# APPENDIX 2

# Properties of the Elements

| Element | At. No. | mp(°C) | bp(°C) | E.N. | Ion. Ener. (kJ/mol) | At. Rad. (nm) | Ion. Rad. (nm) |
|---|---|---|---|---|---|---|---|
| H | 1 | −259 | −253 | 2.2 | 1312 | 0.037 | (−1)0.208 |
| He | 2 | −272 | −269 | | 2372 | 0.05 | |
| Li | 3 | 186 | 1326 | 1.0 | 520 | 0.152 | (+1)0.060 |
| Be | 4 | 1283 | 2970 | 1.6 | 900 | 0.111 | (+2)0.031 |
| B | 5 | 2300 | 2550 | 2.0 | 801 | 0.088 | |
| C | 6 | 3570 | subl. | 2.5 | 1086 | 0.077 | |
| N | 7 | −210 | −196 | 3.0 | 1402 | 0.070 | |
| O | 8 | −218 | −183 | 3.5 | 1314 | 0.066 | (−2)0.140 |
| F | 9 | −220 | −188 | 4.0 | 1681 | 0.064 | (−1)0.136 |
| Ne | 10 | −249 | −246 | | 2081 | 0.070 | |
| Na | 11 | 98 | 889 | 0.9 | 496 | 0.186 | (+1)0.095 |
| Mg | 12 | 650 | 1120 | 1.3 | 738 | 0.160 | (+2)0.065 |
| Al | 13 | 660 | 2327 | 1.6 | 578 | 0.143 | (+3)0.050 |
| Si | 14 | 1414 | 2355 | 1.9 | 786 | 0.117 | |
| P | 15 | 44 | 280 | 2.2 | 1012 | 0.110 | |
| S | 16 | 119 | 444 | 2.6 | 1000 | 0.104 | (−2)0.184 |
| Cl | 17 | −101 | −34 | 3.2 | 1251 | 0.099 | (−1)0.181 |
| Ar | 18 | −189 | −186 | | 1520 | 0.094 | |
| K | 19 | 64 | 774 | 0.8 | 419 | 0.231 | (+1)0.133 |
| Ca | 20 | 845 | 1420 | 1.0 | 590 | 0.197 | (+2)0.099 |
| Sc | 21 | 1541 | 2831 | 1.4 | 631 | 0.160 | (+3)0.081 |
| Ti | 22 | 1660 | 3287 | 1.5 | 658 | 0.146 | |
| V | 23 | 1890 | 3380 | 1.6 | 650 | 0.131 | |
| Cr | 24 | 1857 | 2672 | 1.6 | 653 | 0.125 | (+3)0.064 |
| Mn | 25 | 1244 | 1962 | 1.5 | 717 | 0.129 | (+2)0.080 |
| Fe | 26 | 1535 | 2750 | 1.8 | 759 | 0.126 | (+2)0.075 |
| Co | 27 | 1495 | 2870 | 1.9 | 758 | 0.125 | (+2)0.072 |
| Ni | 28 | 1453 | 2732 | 1.9 | 737 | 0.124 | (+2)0.069 |
| Cu | 29 | 1083 | 2567 | 1.9 | 746 | 0.128 | (+1)0.096 |
| Zn | 30 | 420 | 907 | 1.6 | 906 | 0.133 | (+2)0.074 |
| Ga | 31 | 30 | 2403 | 1.6 | 579 | 0.122 | (+3)0.062 |
| Ge | 32 | 937 | 2830 | 2.0 | 762 | 0.122 | |
| As | 33 | 814 | subl. | 2.2 | 944 | 0.121 | |
| Se | 34 | 217 | 685 | 2.5 | 941 | 0.117 | (−2)0.198 |
| Br | 35 | −7 | 59 | 3.0 | 1140 | 0.114 | (−1)0.195 |
| Kr | 36 | −157 | −152 | | 1351 | 0.109 | |
| Rb | 37 | 39 | 688 | 0.8 | 403 | 0.244 | (+1)0.148 |

| Element | At. No. | mp(°C) | bp(°C) | E.N. | Ion. Ener. (kJ/mol) | At. Rad. (nm) | Ion. Rad. (nm) |
|---------|---------|--------|--------|------|---------------------|---------------|----------------|
| Sr | 38 | 770 | 1380 | 0.9 | 550 | 0.215 | (+2)0.113 |
| Y | 39 | 1509 | 2930 | 1.2 | 616 | 0.180 | (+3)0.093 |
| Zr | 40 | 1852 | 3580 | 1.4 | 660 | 0.157 | |
| Nb | 41 | 2468 | 5127 | 1.6 | 664 | 0.143 | |
| Mo | 42 | 2610 | 5560 | 1.8 | 685 | 0.136 | |
| Tc | 43 | 2200 | 4700 | 1.9 | 702 | 0.136 | |
| Ru | 44 | 2430 | 3700 | 2.2 | 711 | 0.133 | |
| Rh | 45 | 1966 | 3700 | 2.2 | 720 | 0.134 | |
| Pd | 46 | 1550 | 3170 | 2.2 | 805 | 0.138 | |
| Ag | 47 | 961 | 2210 | 1.9 | 731 | 0.144 | (+1)0.126 |
| Cd | 48 | 321 | 767 | 1.7 | 868 | 0.149 | (+2)0.097 |
| In | 49 | 157 | 2000 | 1.7 | 558 | 0.162 | (+3)0.081 |
| Sn | 50 | 232 | 2270 | 1.9 | 709 | 0.140 | |
| Sb | 51 | 631 | 1380 | 2.0 | 832 | 0.141 | |
| Te | 52 | 450 | 990 | 2.1 | 869 | 0.137 | (−2)0.221 |
| I | 53 | 114 | 184 | 2.7 | 1009 | 0.133 | (−1)0.216 |
| Xe | 54 | −112 | −107 | | 1170 | 0.130 | |
| Cs | 55 | 28 | 690 | 0.8 | 376 | 0.262 | (+1)0.169 |
| Ba | 56 | 725 | 1640 | 0.9 | 503 | 0.217 | (+2)0.135 |
| La | 57 | 920 | 3469 | 1.1 | 538 | 0.187 | (+3)0.115 |
| Ce | 58 | 795 | 3468 | 1.1 | 528 | 0.182 | (+3)0.101 |
| Pr | 59 | 935 | 3127 | 1.1 | 523 | 0.182 | (+3)0.100 |
| Nd | 60 | 1024 | 3027 | 1.1 | 530 | 0.182 | (+3)0.099 |
| Pm | 61 | 1027 | 2727 | 1.1 | 536 | 0.181 | |
| Sm | 62 | 1072 | 1900 | 1.1 | 543 | 0.180 | |
| Eu | 63 | 826 | 1439 | 1.1 | 547 | 0.204 | (+2)0.097 |
| Gd | 64 | 1312 | 3000 | 1.1 | 592 | 0.179 | (+3)0.096 |
| Tb | 65 | 1356 | 2800 | 1.1 | 564 | 0.177 | (+3)0.095 |
| Dy | 66 | 1407 | 2600 | 1.1 | 572 | 0.177 | (+3)0.094 |
| Ho | 67 | 1461 | 2600 | 1.1 | 581 | 0.176 | (+3)0.093 |
| Er | 68 | 1497 | 2900 | 1.1 | 589 | 0.175 | (+3)0.092 |
| Tm | 69 | 1356 | 2800 | 1.1 | 597 | 0.174 | (+3)0.091 |
| Yb | 70 | 824 | 1427 | 1.1 | 603 | 0.193 | (+3)0.089 |
| Lu | 71 | 1652 | 3327 | 1.1 | 524 | 0.174 | (+3)0.089 |
| Hf | 72 | 2225 | 5200 | 1.3 | 654 | 0.157 | |
| Ta | 73 | 2980 | 5425 | 1.5 | 761 | 0.143 | |
| W | 74 | 3410 | 5930 | 1.7 | 770 | 0.137 | |
| Re | 75 | 3180 | 5885 | 1.9 | 760 | 0.137 | |
| Os | 76 | 2727 | 4100 | 2.2 | 840 | 0.134 | |
| Ir | 77 | 2448 | 4500 | 2.2 | 880 | 0.135 | |
| Pt | 78 | 1769 | 4530 | 2.2 | 870 | 0.138 | |
| Au | 79 | 1063 | 2966 | 2.4 | 890 | 0.144 | (+1)0.137 |
| Hg | 80 | −39 | 357 | 1.9 | 1007 | 0.155 | (+2)0.110 |

*(continued)*

| Element | At. No. | mp(°C) | bp(°C) | E.N. | Ion. Ener. (kJ/mol) | At. Rad. (nm) | Ion. Rad. (nm) |
|---------|---------|--------|--------|------|---------------------|---------------|----------------|
| Tl | 81 | 304 | 1457 | 1.8 | 589 | 0.171 | (+3)0.095 |
| Pb | 82 | 328 | 1750 | 1.9 | 716 | 0.175 | |
| Bi | 83 | 271 | 1560 | 1.9 | 703 | 0.146 | |
| Po | 84 | 254 | 962 | 2.0 | 812 | 0.165 | |
| At | 85 | 302 | 334 | 2.2 | | | |
| Rn | 86 | −71 | −62 | | 1037 | 0.14 | |
| Fr | 87 | 27 | 677 | 0.7 | | | |
| Ra | 88 | 700 | 1140 | 0.9 | 509 | 0.220 | |
| Ac | 89 | 1050 | 3200 | 1.1 | 490 | 0.20 | |
| Th | 90 | 1750 | 4790 | 1.3 | 590 | 0.180 | |
| Pa | 91 | 1600 | 4200 | 1.4 | 570 | | |
| U | 92 | 1132 | 3818 | 1.4 | 590 | 0.14 | |

# Review of Mathematics

The mathematics you will use in general chemistry is relatively simple. You will, however, be expected to

— make calculations involving exponential numbers, such as $6.022 \times 10^{23}$ or $1.6 \times 10^{-10}$.
— work with logarithms or inverse logarithms, particularly in problems involving pH:

$$\text{pH} = -\log_{10} (\text{conc. H}^+)$$

This appendix reviews each of these topics. It will be helpful to start with electronic calculators, which are very useful for all kinds of calculations in general chemistry.

## Electronic Calculators

A simple "scientific calculator" selling for $20 or less is entirely adequate for general chemistry. Like any calculator, it will allow you to carry out such simple operations as addition, subtraction, multiplication, and division. Beyond that, make sure that the scientific calculator you buy can be used to

— enter and perform operations on numbers expressed in exponential (scientific) notation.
— find a base-10 or natural logarithm or inverse logarithm (number corresponding to a given logarithm).
— raise a number to any power, $n$, or extract the $n$th root of a number.

   The first thing you should do after buying a calculator is to learn how to use it. Read the instruction manual and work with the calculator until you become familiar with it. To get started, try carrying out the following operations:

(a) $2.2 \times 6.1 = 13.42$
(b) $8.1/2.7 = 3$
(c) $(64)^{1/2} = 8$
(d) $(27)^{1/3} = 3$
(e) $3^4 = 81$

(In d and e, you will need to use the $\boxed{\mathbf{y^x}}$ or $\boxed{\mathbf{x^y}}$ key. Refer to your instruction manual for the sequence of operations, which differs depending upon the brand of calculator.)

(f) $\dfrac{16 \times 9}{3 \times 8} = 6$

(This is carried out as a single operation; you do *not* solve for intermediate answers.)
   In working with a calculator, you should be aware of one of its limitations.

It does not indicate the number of significant figures in the answer. Consider, for example, the operations in (a) and (b) on p. A.9. Assume that 2.2, 6.1, 8.1, and 2.7 represent experimentally measured quantities. Following the rules for significant figures (Chap. 1), the answers should be 13 and 3.0, in that order, *not* 13.42 and 3, the numbers appearing on the calculator. In another case, if you are asked to obtain the reciprocal of 3.68, your answer should be

$$1/3.68 = 0.272$$

*not* 0.2717391 . . . , or whatever other number appears on your calculator.

## Exponential Notation

Chemists deal frequently with very large or very small numbers. In one gram of the element carbon there are

50,140,000,000,000,000,000,000 atoms of carbon

At the opposite extreme, the mass of a single atom is

0.00000000000000000000001994 g

Numbers such as these are very awkward to work with. For example, neither of the numbers just written could be entered directly on a calculator. Operations involving very large or very small numbers can be simplified by using **exponential (scientific)** notation. To express a number in exponential notation, write it in the form

$$C \times 10^n$$

where $C$ is a number between 1 and 10 (for example, 1, 2.62, 5.8) and $n$ is a positive or negative integer such as 1, $-1$, $-3$. To find $n$, count the number of places that the decimal point must be moved to give the coefficient, $C$. If the decimal point must be moved to the *left*, $n$ is a *positive* integer; if it must be moved to the *right*, $n$ is a *negative* integer. Thus,

$$26.23 = 2.623 \times 10^1 \quad \text{(decimal point moved 1 place to left)}$$
$$5609 = 5.609 \times 10^3 \quad \text{(decimal point moved 3 places to left)}$$
$$0.0918 = 9.18 \times 10^{-2} \quad \text{(decimal point moved 2 places to right)}$$

## Multiplication and Division

A major advantage of exponential notation is that it simplifies the processes of multiplication and division. To *multiply, add exponents:*

$$10^1 \times 10^2 = 10^{1+2} = 10^3 \qquad 10^6 \times 10^{-4} = 10^{6+(-4)} = 10^2$$

To *divide, subtract* exponents:

$$10^3/10^2 = 10^{3-2} = 10^1 \qquad 10^{-3}/10^6 = 10^{-3-6} = 10^{-9}$$

It often happens that multiplication or division yields an answer that is not in standard exponential notation. For example,

$$(5.0 \times 10^4) \times (6.0 \times 10^3) = (5.0 \times 6.0) \times 10^4 \times 10^3 = 30 \times 10^7$$

The product is not in standard exponential notation since the coefficient, 30, does not lie between 1 and 10. To correct this situation, rewrite the coefficient as $3.0 \times 10^1$ and then add exponents:

$$30 \times 10^7 = (3.0 \times 10^1) \times 10^7 = 3.0 \times 10^8$$

In another case,

$$0.526 \times 10^3 = (5.26 \times 10^{-1}) \times 10^3 = 5.26 \times 10^2$$

## Exponential Notation on the Calculator

On all scientific calculators, it is possible to enter numbers in exponential notation. The method used depends upon the brand of calculator. Most often, it involves using a key labeled $\boxed{\textbf{EXP}}$ , $\boxed{\textbf{EE}}$ , or $\boxed{\textbf{EEX}}$ . Check your instruction manual for the procedure to be followed. To make sure you understand it, try entering the following numbers:

$$2.4 \times 10^6 \qquad 3.16 \times 10^{-8} \qquad 6.2 \times 10^{-16}$$

Multiplication, division, and many other operations can be carried out directly on your calculator. Try the following exercises, using your calculator:

(a) $(6.0 \times 10^2) \times (4.2 \times 10^{-4}) = ?$
(b) $\dfrac{6.0 \times 10^2}{4.2 \times 10^{-4}} = ?$
(c) $(2.50 \times 10^{-9})^{1/2} = ?$
(d) $3.6 \times 10^{-4} + 4 \times 10^{-5} = ?$

The answers, expressed in exponential notation and following the rules of significant figures, are as follows: (a) $2.5 \times 10^{-1}$ (b) $1.4 \times 10^6$ (c) $5.00 \times 10^{-5}$ (d) $4.0 \times 10^{-4}$.

## Logarithms and Inverse Logarithms

The logarithm of a number **n** to the base **m** is defined as the power to which **m** must be raised to give the number **n**. Thus,

$$\text{if } m^x = n, \text{ then } \log_m n = x$$

In other words, a *logarithm* is an *exponent,* which may be a whole number (e.g., 1, −1) but more often is not (e.g., 1.500, −0.301).

In general chemistry, you will encounter two kinds of logarithms.

**1.** *Common logarithms,* where the base is 10, ordinarily denoted as $\log_{10}$. If $10^x = n$, then $\log_{10} n = x$. Examples include the following:

$$\log_{10} 100 = 2.00 \qquad (\text{since } 10^2 = 100)$$

$$\log_{10} 1 = 0.000 \qquad (\text{since } 10^0 = 1)$$

$$\log_{10} 0.001 = -3.000 \qquad (\text{since } 10^{-3} = 0.001)$$

**2.** *Natural logarithms,* where the base is the quantity **e** = 2.718. . . . Many of the equations used in general chemistry are expressed most simply in terms of natural logarithms, denoted as ln. If $e^x = n$, then $\ln n = x$.

$$\ln 100 = 4.606 \qquad (\text{i.e., } 100 = e^{4.606})$$

$$\ln 1 = 0 \qquad (\text{i.e., } 1 = e^{0})$$

$$\ln 0.001 = -6.908 \qquad (\text{i.e., } 0.001 = e^{-6.908})$$

Notice that

(a) **$\log_{10} 1 = \ln 1 = 0$.** The logarithm of 1 to any base is zero, since any number raised to the zero power is 1. That is

$$10^{0} = e^{0} = 2^{0} = \cdots = n^{0} = 1$$

(b) **Numbers larger than 1 have a positive logarithm; numbers smaller than 1 have a negative logarithm. For example,**

$$\log_{10} 100 = 2 \qquad \ln 100 = 4.606$$

$$\log_{10} 10 = 1 \qquad \ln 10 = 2.303$$

$$\log_{10} 0.1 = -1 \qquad \ln 0.1 = -2.303$$

$$\log_{10} 0.01 = -2 \qquad \ln 0.01 = -4.606$$

(c) The common and natural logarithms of a number are related by the expression

**$\ln n = 2.303 \log_{10} n$**

That is, the natural logarithm of a number is 2.303 . . . times its base-10 logarithm.

(d) Finally, if $x \leq 0$, both $\log x$ and $\ln x$ are undefined.

An **inverse** logarithm (antilogarithm) is the number corresponding to a given logarithm. In general,

if $m^{x} = n$, then inverse $\log_{m} x = n$

For example:

$$10^{2} = 100 \qquad \text{inverse } \log_{10} 2 = 100$$

$$10^{-3} = 0.001 \qquad \text{inverse } \log_{10} (-3) = 0.001$$

In other words, the numbers whose base-10 logarithms are 2 and $-3$ are 100 and 0.001, respectively. The same reasoning applies to natural logarithms.

$$e^{4.606} = 100 \qquad \text{inverse } \ln 4.606 = 100$$

$$e^{-6.908} = 0.001 \qquad \text{inverse } \ln -6.908 = 0.001$$

The numbers whose natural logarithms are 4.606 and $-6.908$ are 100 and 0.001, respectively. Notice that if the inverse logarithm is positive, the corresponding number is larger than 1; if it is negative, the number is smaller than 1.

## Finding Logarithms on a Calculator

To obtain a base-10 logarithm using a calculator, all you need do is enter the number and press the $\boxed{\textbf{LOG}}$ key. This way you should find that

$$\log_{10} 2.00 = 0.301 \ldots$$

$$\log_{10} 0.526 = -0.279 \ldots$$

Similarly, to find a natural logarithm, you enter the number and press the $\boxed{\textbf{LN X}}$ key.

$$\ln 2.00 = 0.693 \ldots$$

$$\ln 0.526 = -0.642 \ldots$$

To find the logarithm of an exponential number, you enter the number in exponential form and take the logarithm in the usual way. This way you should find that

$$\log_{10} 2.00 \times 10^3 = 3.301 \ldots \qquad \ln 2.00 \times 10^3 = 7.601 \ldots$$

$$\log_{10} 5.3 \times 10^{-12} = -11.28 \ldots \qquad \ln 5.3 \times 10^{-12} = -25.96 \ldots$$

The base-10 logarithm of an exponential number can be found in a somewhat different way by applying the relation

$$\log_{10} (C \times 10^n) = n + \log_{10} C$$

Thus

$$\log_{10} 2.00 \times 10^3 = 3 + \log_{10} 2.00 = 3 + 0.301 = 3.301$$

$$\log_{10} 5.3 \times 10^{-12} = -12 + \log_{10} 5.3 = -12 + 0.72 = -11.28$$

## Finding Inverse Logarithms on a Calculator

The method used to find inverse logarithms depends upon the type of calculator. On certain calculators, you enter the number and then press, in succession, the $\boxed{\textbf{INV}}$ and either $\boxed{\textbf{LOG}}$ or $\boxed{\textbf{LN X}}$ keys. With other calculators, you press the $\boxed{\textbf{10}^x}$ or $\boxed{\textbf{e}^x}$ key. Either way, you should find that

$$\text{inverse } \log_{10} 1.632 = 42.8 \ldots \qquad \text{inverse } \ln 1.632 = 5.11 \ldots$$

$$\text{inverse } \log_{10} - 8.82 = 1.5 \times 10^{-9} \qquad \text{inverse } \ln -8.82 = 1.5 \times 10^{-4}$$

## Significant Figures in Logarithms and Inverse Logarithms

For base-10 logarithms, the rules governing significant figures are as follows:

**1.** In taking the logarithm of a number, retain after the decimal point in the log as many digits as there are significant figures in the number. (This part of the logarithm is often referred to as the *mantissa;* digits that precede the decimal point comprise the *characteristic* of the logarithm.) To illustrate this rule, consider the following:

$$\log_{10} 2.00 = 0.301 \qquad \log_{10} (2.00 \times 10^3) = 3.301$$

$$\log_{10} 2.0 = 0.30 \qquad \log_{10} (2.0 \times 10^1) = 1.30$$

$$\log_{10} 2 = 0.3 \qquad \log_{10} (2 \times 10^{-3}) = 0.3 - 3 = -2.7$$

**2.** In taking the inverse logarithm of a number, retain as many significant figures as there are after the decimal point in the number. Thus,

$$\text{inverse } \log_{10} 0.301 = 2.00 \qquad \text{inverse } \log_{10} 3.301 = 2.00 \times 10^3$$

$$\text{inverse } \log_{10} 0.30 = 2.0 \qquad \text{inverse } \log_{10} 1.30 = 2.0 \times 10^1$$

$$\text{inverse } \log_{10} 0.3 = 2 \qquad \text{inverse } \log_{10} (-2.7) = 2 \times 10^{-3}$$

These rules take into account the fact that, as mentioned earlier,

$$\log_{10} (C \times 10^n) = n + \log_{10} C$$

The digits that appear before (to the left of) the decimal point specify the value of $n$, i.e., the power of 10 involved in the expression. In that sense, they are not experimentally significant. In contrast, the digits that appear after (to the right of) the decimal point specify the value of the logarithm of $C$; the number of such digits reflects the uncertainty in $C$. Thus,

$$\log_{10} 209 = 2.320 \qquad \text{(3 sig. fig.)}$$

$$\log_{10} 209.0 = 2.3201 \qquad \text{(4 sig. fig.)}$$

The rules for significant figures involving natural logarithms and inverse logarithms are somewhat more complex than those for base-10 logs. However, for simplicity, we will assume that the rules listed above apply here as well. Thus,

$$\ln 209 = 5.342 \qquad \text{(3 sig. fig.)}$$

$$\ln 209.0 = 5.3423 \qquad \text{(4 sig. fig.)}$$

## Operations Involving Logarithms

Since logarithms are exponents, the rules governing the use of exponents apply here as well. The rules that follow are valid for all types of logarithms, regardless of the base. We illustrate the rules with natural logarithms, since that is where you are most likely to use them in working with this text.

**Multiplication: $\ln(xy) = \ln x + \ln y$**

Example: $\ln(2.50 \times 1.25) = \ln 2.50 + \ln 1.25 = 0.916 + 0.223 = 1.139$

**Division: $\ln(x/y) = \ln x - \ln y$**

Example: $\ln(2.50/1.25) = 0.916 - 0.223 = 0.693$

**Raising to a Power: $\ln(x^n) = n \ln x$**

Example: $\ln(2.00)^4 = 4 \ln 2.00 = 4(0.693) = 2.772$

**Extracting a Root: $\ln(x^{1/n}) = \dfrac{1}{n} \ln x$**

Example: $\ln(2.00)^{1/3} = \dfrac{\ln 2.00}{3} = \dfrac{0.693}{3} = 0.231$

**Taking a Reciprocal: $\ln(1/x) = -\ln x$**

Example: $\ln(1/2.00) = -\ln 2.00 = -0.693$

# Naming Complex Ions

To name a complex ion, it is necessary to show

— the number and identity of each ligand attached to the central metal ion
— the identity and oxidation number of the central metal ion
— whether the complex is a cation or an anion

To accomplish this, a simple set of rules is followed.

**1.** The names of anionic ligands are obtained by substituting the suffix *-o* for the normal ending. Examples include

$Cl^-$     chlor*o*      $SO_4^{2-}$     sulfat*o*

$OH^-$     hydrox*o*      $CO_3^{2-}$     carbonat*o*

Ordinarily, the names of molecular ligands are unchanged. Two important exceptions are

$H_2O$     aqua      $NH_3$     ammine

**2.** The number of ligands of a particular type is ordinarily indicated by the Greek prefixes *di, tri, tetra, penta, hexa:*

$Cu(H_2O)_4^{2+}$     *tetra*aquacopper(II)

$Cr(NH_3)_6^{3+}$     *hexa*amminechromium(III)

If the name of the ligand is itself complex (e.g., ethylenediamine), the number of such ligands is indicated by the prefixes *bis, tris, . . .* The name of the ligand is enclosed in parentheses:

$Cr(en)_3^{3+}$     *tris*(ethylenediamine)chromium(III)

**3.** If more than one type of ligand is present, they are named in alphabetical order (without regard for prefixes):

$Cu(NH_3)_2(H_2O)_2^{2+}$     di*ammine*di*aqua*copper(II)

$Cr(NH_3)_5Cl^{2+}$     penta*ammine*chloro*chromium(III)

**4.** As you can deduce from the preceding examples, the oxidation number of the central metal ion is indicated by a Roman numeral written at the end of the name.

**5.** If the complex is an anion, the suffix *-ate* is inserted between the name of the metal and the oxidation number:

$Zn(OH)_4^{2-}$     tetrahydroxozinc*ate*(II)

Coordination compounds are named in much the same way as simple ionic compounds. The cation is named first, followed by the anion. Examples include

$[Cr(NH_3)_4Cl_2]NO_3$     tetraamminedichlorochromium(III) nitrate

$K_2[PtCl_6]$     potassium hexachloroplatinate(IV)

$[Co(NH_3)_2(en)_2]Br_3$     diamminebis(ethylenediamine)cobalt(III) bromide

# Molecular Orbitals

In Chapter 7, we used valence bond theory to explain bonding in molecules. It accounts, at least qualitatively, for the stability of the covalent bond in terms of the overlap of atomic orbitals. By invoking hybridization, valence bond theory can account for the molecular geometries predicted by electron-pair repulsion. Where Lewis structures are inadequate, as in $SO_2$, the concept of resonance allows us to explain the observed properties.

A major weakness of valence bond theory has been its inability to predict the magnetic properties of molecules. We mentioned this problem in Chapter 7 with regard to the $O_2$ molecule which is paramagnetic, even though it has an even number (12) of valence electrons. The octet rule, or valence bond theory, would predict that all of the electrons in $O_2$ should be paired, which would make it diamagnetic.

This discrepancy between experiment and theory (and many others) can be explained in terms of an alternative model of covalent bonding, the molecular orbital (MO) approach. Molecular orbital theory treats bonds in terms of orbitals characteristic of the molecule as a whole. To apply this approach, we carry out three basic operations.

**1.** The atomic orbitals of atoms are combined to give a new set of molecular orbitals characteristic of the molecule as a whole. Here, *the number of molecular orbitals formed is equal to the number of atomic orbitals combined.* When two H atoms combine to form $H_2$, two s orbitals, one from each atom, yield two molecular orbitals. In another case, six p orbitals, three from each atom, give a total of six molecular orbitals.

**2.** The molecular orbitals are arranged in order of increasing energy. The relative energies of these orbitals are ordinarily deduced from experiment. Spectra and magnetic properties of molecules are used.

**3.** The *valence electrons* in a molecule are distributed among the available molecular orbitals. The process followed is much like that used with electrons in atoms. In particular, we find the following:

(a) *Each molecular orbital can hold a maximum of two electrons.* When an orbital is filled the two electrons have opposed spins, in accordance with the Pauli principle (Chapter 7).

(b) *Electrons go into the lowest energy molecular orbital available.* A higher orbital starts to fill only when each orbital below it has its quota of two electrons.

(c) *Hund's rule is obeyed.* When two orbitals of equal energy are available to two electrons, one electron goes into each, giving two half-filled orbitals.

## Diatomic Molecules of the Elements

To illustrate molecular orbital theory, we apply it to the diatomic molecules of the elements in the first two periods of the periodic table.

# Hydrogen and Helium (Combination of 1s Orbitals)

Molecular orbital (MO) theory predicts that two 1s orbitals will combine to give two molecular orbitals. One of these has an energy lower than that of the atomic orbitals from which it is formed (Fig. 1). Placing electrons in this orbital gives a species that is more stable than the isolated atoms. For that reason the lower molecular orbital in Figure 1 is called a **bonding orbital.** The other molecular orbital has a higher energy than the corresponding atomic orbitals. Electrons entering it are in an unstable, higher energy state. It is referred to as an **antibonding orbital.**

The electron density in these molecular orbitals is shown at the right of Figure 1. Notice that the bonding orbital has a higher density between the nuclei. This accounts for its stability. In the antibonding orbital, the chance of finding the electron between the nuclei is very small. The electron density is concentrated at the far ends of the "molecule." This means that the nuclei are less shielded from each other than they are in the isolated atoms.

The electron density in both molecular orbitals is symmetrical about the axis between the two nuclei. This means that both of these are sigma orbitals. In MO notation, the 1s bonding orbital is designated as $\sigma_{1s}$. The antibonding orbital is given the symbol $\sigma_{1s}^{*}$. An asterisk designates an antibonding orbital.

In the $H_2$ molecule, there are two 1s electrons. They fill the $\sigma_{1s}$ orbital, giving a single bond. In the $He_2$ molecule, there would be four electrons, two from each atom. These would fill the bonding and antibonding orbitals. As a result, the number of bonds (the *bond order*) in $He_2$ is zero. The general relation is

$$\text{no. of bonds} = \text{bond order} = \frac{B - AB}{2}$$

where B is the number of electrons in bonding orbitals and AB is the number of electrons in antibonding orbitals. In $H_2$, B = 2 and AB = 0, so we have one bond. In $He_2$, B = AB = 2, so the number of bonds is zero. The $He_2$ molecule should not and does not exist.

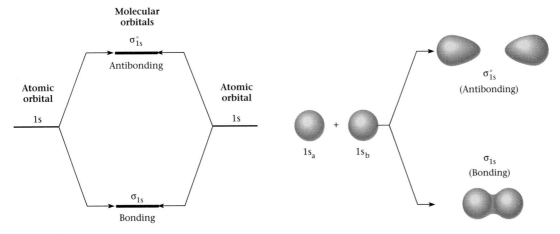

**Figure 1**
Molecular orbital formation. Two molecular orbitals are formed by combining two 1s atomic orbitals.

## Second Period Elements (Combination of 2s and 2p Orbitals)

Among the diatomic molecules of the second period elements are three familiar ones, $N_2$, $O_2$, and $F_2$. The molecules $Li_2$, $B_2$, and $C_2$ are less common but have been observed and studied in the gas phase. In contrast, the molecules $Be_2$ and $Ne_2$ are either highly unstable or nonexistent. Let us see what molecular orbital theory predicts about the structure and stability of these molecules. We start by considering how the atomic orbitals containing the valence electrons (2s and 2p) are used to form molecular orbitals.

Combining two 2s atomic orbitals, one from each atom, gives two molecular orbitals. These are very similar to the ones discussed above. They are designated as $\sigma_{2s}$(sigma, bonding, 2s) and $\sigma_{2s}^*$(sigma, antibonding, 2s).

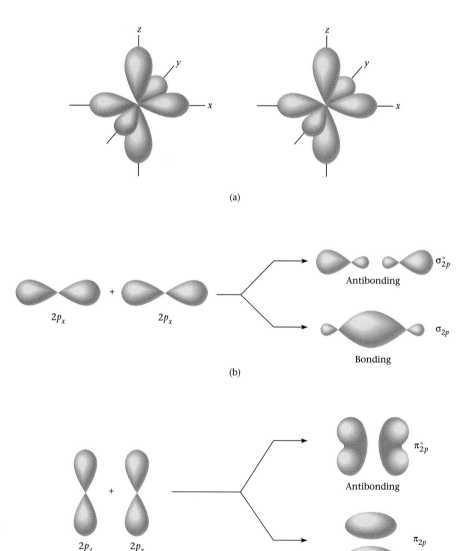

**Figure 2**

(See text, p. A.19). When p orbitals from two different atoms (*a*) overlap, there are two quite different possibilities. If they overlap head-to-head (*b*) two sigma molecular orbitals are produced. If, on the other hand, they overlap side-to-side (*c*) two pi molecular obitals result.

Consider now what happens to the 2p orbitals. In an isolated atom, there are three such orbitals, oriented at right angles to each other. We call these atomic orbitals $p_x$, $p_y$, and $p_z$ (top of Figure 2). When two $p_x$ atomic orbitals, one from each atom, overlap head-to-head, they form two sigma orbitals. These are a bonding orbital, $\sigma_{2p}$, and an antibonding orbital, $\sigma_{2p}^*$. The situation is quite different when the $p_z$ orbitals overlap. Since they are oriented parallel to one another, they overlap side-to-side (Fig. 2c). The two molecular orbitals formed in this case are pi orbitals; one is a bonding orbital, $\pi_{2p}$, the other a nonbonding orbital, $\pi_{2p}^*$. In an entirely similar way, the $p_y$ orbitals of the two atoms interact to form another pair of pi molecular orbitals, $\pi_{2p}$ and $\pi_{2p}^*$ (these orbitals are not shown in Fig. 2).

The relative energies of the molecular orbitals available for occupancy by the valence electrons of the second period elements are shown in Figure 3. This order applies at least through $N_2$*.

To obtain the MO structure of the diatomic molecules of the elements in the second period, we fill the available molecular orbitals in order of increasing energy. The results are shown in Table 1 p. A.20. Note the agreement between MO theory and the properties of these molecules. In particular, the number of unpaired electrons predicted agrees with experiment. There is also a general correlation between the predicted bond order:

$$\text{bond order} = \frac{B - AB}{2}$$

and the bond energy. We would expect a double bond ($C_2$ or $O_2$) to be stronger than a single bond ($Li_2$, $B_2$, $F_2$). A triple bond, as in $N_2$, should be still stronger.

A major triumph of MO theory is its ability to explain the properties of $O_2$. It explains how the molecule can have a double bond and, at the same time, have two unpaired electrons.

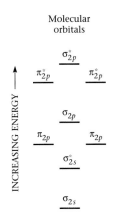

Molecular orbitals

INCREASING ENERGY →

**Figure 3**
Relative energies, so far as filling order is concerned, for the molecular orbitals formed by combining 2s and 2p atomic orbitals.

---

**Example**   Using MO theory, predict the electronic structure, bond order, and number of unpaired electrons in the peroxide ion, $O_2^{2-}$.

**Strategy**   First, find the number of valence electrons. Then construct an orbital diagram, filling the available molecular orbitals (Figure 3) in order of increasing energy.

**Solution**   Recall that oxygen is in Group **16** of the periodic table.

$$\text{no. valence e}^- = 2(6) + 2 = 14$$

The orbital diagram is

|  | $\sigma_{2s}$ | $\sigma_{2s}^*$ | $\pi_{2p}$ | $\pi_{2p}$ | $\sigma_{2p}$ | $\pi_{2p}^*$ | $\pi_{2p}^*$ | $\sigma_{2p}^*$ |
|---|---|---|---|---|---|---|---|---|
| $O_2^{2-}$ | (↑↓) | (↑↓) | (↑↓) | (↑↓) | (↑↓) | (↑↓) | (↑↓) | ( ) |

There are eight electrons in bonding orbitals, six in antibonding orbitals.

$$\text{bond order} = \frac{8 - 6}{2} = 1$$

There are no unpaired electrons. These conclusions are in agreement with the Lewis structure of the peroxide ion: $\left( :\ddot{O}-\ddot{O}: \right)^{2-}$

---

*It appears that beyond $N_2$ the $\sigma_{2p}$ orbital lies below the two $\pi_{2p}$ orbitals. This does not affect the filling order because, in $O_2$, all of these orbitals are filled.

TABLE 1   **Predicted and Observed Properties of Diatomic Molecules of Second Period Elements**

**Occupancy of Orbitals**

| | $\sigma_{2s}$ | $\sigma_{2s}^*$ | $\pi_{2p}$ | $\pi_{2p}$ | $\sigma_{2p}$ | $\pi_{2p}^*$ | $\pi_{2p}^*$ | $\sigma_{2p}^*$ |
|---|---|---|---|---|---|---|---|---|
| $Li_2$ | (↑↓) | ( ) | ( ) | ( ) | ( ) | ( ) | ( ) | ( ) |
| $Be_2$ | (↑↓) | (↑↓) | ( ) | ( ) | ( ) | ( ) | ( ) | ( ) |
| $B_2$ | (↑↓) | (↑↓) | (↑) | (↑) | ( ) | ( ) | ( ) | ( ) |
| $C_2$ | (↑↓) | (↑↓) | (↑↓) | (↑↓) | ( ) | ( ) | ( ) | ( ) |
| $N_2$ | (↑↓) | (↑↓) | (↑↓) | (↑↓) | (↑↓) | ( ) | ( ) | ( ) |
| $O_2$ | (↑↓) | (↑↓) | (↑↓) | (↑↓) | (↑↓) | (↑) | (↑) | ( ) |
| $F_2$ | (↑↓) | (↑↓) | (↑↓) | (↑↓) | (↑↓) | (↑↓) | (↑↓) | ( ) |
| $Ne_2$ | (↑↓) | (↑↓) | (↑↓) | (↑↓) | (↑↓) | (↑↓) | (↑↓) | (↑↓) |

| | **Predicted Properties** | | **Observed Properties** | |
|---|---|---|---|---|
| | Number of Unpaired $e^-$ | Bond order | Number of Unpaired $e^-$ | Bond Energy (kJ/mol) |
| $Li_2$ | 0 | 1 | 0 | 105 |
| $Be_2$ | 0 | 0 | 0 | unstable |
| $B_2$ | 2 | 1 | 2 | 289 |
| $C_2$ | 0 | 2 | 0 | 628 |
| $N_2$ | 0 | 3 | 0 | 941 |
| $O_2$ | 2 | 2 | 2 | 494 |
| $F_2$ | 0 | 1 | 0 | 153 |
| $Ne_2$ | 0 | 0 | 0 | nonexistent |

# Polyatomic Molecules; Delocalized π Electrons

The bonding in molecules containing more than two atoms can also be described in terms of molecular orbitals. We will not attempt to do this here; the energy level structure is considerably more complex than the one we have considered. However, one point is worth mentioning. In polyatomic species, *a pi molecular orbital can be spread over the entire molecule* rather than being concentrated between two atoms.

This principle can be applied to species such as the nitrate ion, whose Lewis structure is:

$$:\ddot{O}—N—\ddot{O}:$$
$$\|$$
$$:\ddot{O}:$$

Valence bond theory (Chapter 7) explains the fact that the three N—O bonds are identical by invoking the idea of resonance, with three contributing structures (p. 179). MO theory, on the other hand, considers that the skeleton of the nitrate ion is established by the three sigma bonds while the electron pair in the pi orbital is *delocalized,* shared by all of the atoms in the molecule. According to MO theory, a similar interpretation applies with all of the "resonance hybrids" described in Chapter 7, including $SO_2$, $SO_3$ and $CO_3^{2-}$.

Another species in which delocalized pi orbitals play an important role is benzene, $C_6H_6$. There are 30 valence electrons in the molecule, 24 of which are required to form the sigma bond framework:

The remaining six electrons are located in three $\pi$ orbitals, which according to MO theory extend over the entire molecule. Figure 4 is one way of representing this structure; more commonly it is shown simply as

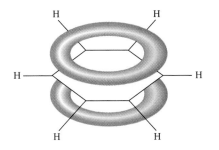

**Figure 4**
In benzene, three electron pairs are not localized on particular carbon atoms. Instead, they are spread out over two lobes of the shape shown, one above the plane of the benzene ring and the other below it.

You may recall from Chapter 9 that in the network covalent structure of graphite (p. 247), there are a large number of pi bonds. The corresponding pi orbitals are delocalized over the entire structure; the electrons are not tied down to a particular pair of atoms. This is reminiscent of the "electron sea" model of metallic bonding, which may explain why graphite has an electrical conductivity approaching that of metals. Interestingly, graphite is a good conductor in two dimensions but not in the third dimension, perpendicular to the planes of atoms.

## Metals; Band Theory

In Chapter 9, we considered a simple picture of metallic bonding, the electron-sea model. The molecular orbital approach leads to a refinement of this model known as *band theory*. Here, a crystal of a metal is considered to be one huge molecule. Valence electrons of the metal are fed into delocalized molecular orbitals, formed in the usual way from atomic orbitals. A huge number of these MOs are grouped into an energy band; the energy separation between adjacent MOs is extremely small.

For purposes of illustration, consider a lithium crystal weighing one gram, which contains roughly $10^{23}$ atoms. Each Li atom has a half-filled 2s atomic orbital (elect. conf. Li = $1s^2 2s^1$). When these atomic orbitals combine, they form an equal number, $10^{23}$, of molecular orbitals. These orbitals are spread over an energy band covering about 100 kJ/mol. It follows that the spacing between adjacent MOs is of the order of

$$\frac{100 \text{ kJ/mol}}{10^{23}} = 10^{-21} \text{ kJ/mol}$$

Since each lithium atom has one valence electron and each molecular orbital can hold two electrons, it follows that the lower half of the valence band (shown in color in Figure 5) is filled with electrons. The upper half of the band is empty. Electrons near the top of the filled MOs can readily jump to empty MOs only an infinitesimal distance above them. This is what happens when an

**Figure 5**

In both lithium and beryllium metal, there are vacant MOs only slightly higher in energy than filled MOs. This is the basic requirement for metallic conductivity.

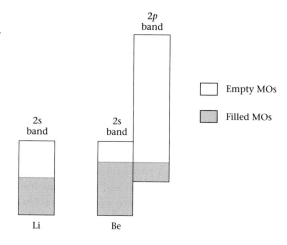

electrical field is applied to the crystal; the movement of electrons through delocalized MOs accounts for the electrical conductivity of lithium metal.

The situation in beryllium metal is more complex. Here, we might expect all of the 2s molecular orbitals to be filled since beryllium has the electron configuration $1s^2 2s^2$. However, in a crystal of beryllium, the 2p MO band overlaps the 2s (Figure 5). This means that, once again, there are vacant MOs which differ only infinitesimally in energy from filled MOs below them. This is indeed the basic requirement for electron conductivity; it is characteristic of all metals including lithium and beryllium.

Materials in which there is a substantial difference in energy between occupied and vacant MOs are poor electron conductors. Diamond, where the gap between the filled valence band and the empty conduction band is 500 kJ/mol, is an insulator. Silicon and germanium, where the gaps are 100 kJ/mol and 60 kJ/mol respectively, are semiconductors.

# Answers to Even-Numbered & Challenge Questions & Problems

## Chapter 1

2. **(a)** element **(b)** mixture
**(c)** compound **(d)** mixture
4. **(a)** heterogeneous mixture
**(b)** solution **(c)** solution
6. **(a)** mass—lb **(b)** pressure—mm Hg
**(c)** energy—BTU, kilowatt-hr, or kJ
8. **(a)** $<$ **(b)** $>$ **(c)** $=$
10. $126°F$
12. $216.6 K$
14. **(a)** $1 kg\text{-}m/s^2$ **(b)** $1.01325 \times 10^5 kg/m\text{-}s^2$
**(c)** $1 \times 10^{-6} m^3$
16. **(a)** $1.28 \times 10^6 Pa$ **(b)** $63 kJ$
18. **(a)** 6 **(b)** 5 **(c)** 4
**(d)** 4 **(e)** ambiguous (2, 3, or 4)
20. $7 cm^3$
22. **(a)** 4 **(b)** 3 **(c)** 3 **(d)** 4
24. **(a)** $132.5 g$ **(b)** $17.3 cm$
**(c)** $169 lb$ **(d)** $4.4 \times 10^3 oz$
26. **(a)** $6.743 \times 10^4 Å$ **(b)** $2.655 \times 10^{-4} in$
**(c)** $4.191 \times 10^{-9} mi$
28. **(a)** $7.7 \times 10^3 grains$ **(b)** $0.819 lb$
30. $5.9 \times 10^{12} mi$
32. $1.85 \times 10^{-3}$
34. $2.8 mL$
36. $3 \times 10^2 garages$
38. $1.07 g/mL$
40. $9.2 g/cm^3$
42. $1.20 \times 10^2 mL$
44. $2.7 \times 10^5 L$
46. **(a)** $49.1 g$ **(b)** $50.4 g$ **(c)** $2.6 g$
48. **(a)** physical **(b)** chemical
**(c)** physical **(d)** physical
50. **(a)** density is mass/volume of a substance; solubility is mass of solute/mass of solution
**(b)** elements make up compounds
**(c)** A solution is a homogeneous mixture.
52. $\$2.6 \times 10^2$

54. $6.5 \times 10^{-4} in$
56. $7.95 g/mL$
57. $320°F$
58. $1.2 km^2$
59. $18.3 cm$
60. $8.1 \times 10^{-3} g$

## Chapter 2

2. A compound always has the same elements in the same mass ratio.
4. **(a)** Conservation of Mass
**(b)** none **(c)** none
6. (4)
8. In both experiments the product is made up of 60.0% Mg and 40.0% O.
10. (c)
12. J. J. Thomson—cathode ray tube experiment (Fig. 2.3)
14. $86 p^+$; $136 n$
16. N-14 has one less neutron than N-15; $^{14}_7N$; $^{15}_7N$
18. **(a)** $^{22}_{11}Na$ **(b)** $^{21}_{10}Ne$
**(c)** $^{23}_{11}Na$–isotope of Na-21
20. **(a)** $3 p^+$ **(b)** $4 n$
**(c)** $3 e^-$ **(d)** $4n, 3 p^+, 2 e^-$
22. $^{85}_{37}Rb$    0    37    48    37
$^{27}_{13}Al^{3+}$    3    13    14    10
$^{19}_9F^-$    $-1$    9    10    10
24. **(a)** $14 p^+, 14 e^-$ **(b)** $7 p^+, 10 e^-$
**(c)** $46 p^+, 46 e^-$ **(d)** $1 p^+, 2 e^-$
26. **(a)** carbon **(b)** cobalt **(c)** cadmium
**(d)** chlorine **(e)** copper
28. **(a)** nonmetal **(b)** metal **(c)** metal
**(d)** nonmetal **(e)** metal
30. **(a)** 2 **(b)** 8 **(c)** 8 **(d)** 18 **(e)** 18
32. $^{241}_{95}Am \longrightarrow ^4_2He + ^{237}_{93}Np$
34. **(a)** $^{237}_{93}Np \longrightarrow ^4_2He + ^{233}_{91}Pa$
**(b)** $^{90}_{38}Sr \longrightarrow ^0_{-1}e + ^{90}_{39}Y$
**(c)** $^{37}_{18}Ar \longrightarrow ^0_{-1}e + ^{37}_{19}K$
**(d)** $^{218}_{85}At \longrightarrow ^4_2He + ^{214}_{83}Bi$
36. **(a)** Ra–210 **(b)** Fr–206
38. Np–237; Pa–233; U–233; Th–229
40. **(a)** Si–28 **(b)** F–19 **(c)** Na–23
42. Rn–228, P–34, Po–209
44. **(a)** $C_2H_7N$ **(b)** $C_3H_8O$
46. **(a)** $CH_4$ **(b)** $H_2O_2$ **(c)** $N_2O_5$
**(d)** $BF_3$ **(e)** $Se_2Cl_2$
48. **(a)** nitrogen trifluoride
**(b)** phosphorus trichloride **(c)** phosphine
**(d)** iodine heptafluoride **(e)** silicon carbide

50. **(a)** $MgBr_2$, $Mg_3N_2$   **(b)** $CoS$, $Co_2S_3$
52. **(a)** $Al_2(SO_4)_3$   **(b)** $NaC_2H_3O_2$
    **(c)** $RbBr$   **(d)** $Na_3N$   **(e)** $NH_4Cl$
54. **(a)** iron(III) carbonate   **(b)** calcium sulfate
    **(c)** copper(I) sulfide   **(d)** lead(IV) oxide
    **(e)** sulfuric acid
56. $HClO$, barium nitrite, gold(III) sulfide, $NO$,
    $Ni(IO_4)_2$, disulfur dichloride
58.

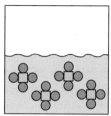

60. **(a)** True only if cation has same charge as anion.
    **(b)** Ionic formula, $SrBr_2$ is not a molecule.
    **(c)** True only for H–1.
    **(d)** True only for anions.
62. **(a)** Br   **(b)** Se   **(c)** Ni   **(d)** B
63. **(a)** ratio of C in ethane to C in ethene/g H is
    $3:2$.
    **(b)** $CH_3$, $CH_2$; $C_2H_6$, $C_2H_4$
64. $3.71\ g/cm^3$; Lots of space between atoms
65. $1.4963 \times 10^{-23}\ g$
66. **(a)** $2.5 \times 10^{24}$ molecules   **(b)** $2.3 \times 10^{-20}$
    **(c)** $\approx 2.9 \times 10^2$ molecules

## Chapter 3

2. **(a)** 9.980   **(b)** 0.5332   **(c)** 0.3703
4. 87.62 amu
6. 150.9 amu
8. 9.9%—Cr-53; 2.2%—Cr-54
10. **(a)** two—HCl-35 and HCl-37   **(b)** 36 and 38
    **(c)**

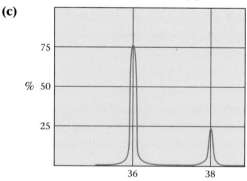

12. **(a)** $5.93 \times 10^{14}\ g$   **(b)** $7.65 \times 10^{-22}\ mol$
14. **(a)** $9.786 \times 10^{-11}\ g$   **(b)** $2.9 \times 10^{21}$ atoms
16. **(a)** 0.016681 mol   **(b)** $1.005 \times 10^{22}$ atoms
    **(c)** $8.840 \times 10^{23}$

18. **(a)** 160   **(b)** 180
    **(c)** $1.084 \times 10^{25}$   **(d)** $9.635 \times 10^{24}$
20. **(a)** 256.6 g/mol   **(b)** 145.2 g/mol
    **(c)** 176.1 g/mol
22. **(a)** $3.120 \times 10^4\ mol$   **(b)** 0.001030 mol
    **(c)** 0.0996 mol
24. **(a)** $5.80 \times 10^2\ g$   **(b)** 427 g
    **(c)** $2.73 \times 10^3\ g$
26. **(a)** 0.2204 mol; $1.327 \times 10^{23}$ molecules; $2.654 \times 10^{23}$ O atoms   **(b)** 30.74 g; $2.982 \times 10^{23}$ molecules; $5.964 \times 10^{23}$ O atoms   **(c)** $6.2 \times 10^3\ g$; $1.0 \times 10^2\ mol$; $1.2 \times 10^{26}$ O atoms
    **(d)** 0.088 g; 0.0014 mol; $8.5 \times 10^{20}$ molecules
28. 5.030% Be; 10.04% Al; 31.35% Si; 53.58% O
30. 201.9 g
32. 40.0%
34. 79.1% C; 9.79% H; 11.1% O
36. $BH_3$
38. **(a)** $C_8H_{20}Pb$   **(b)** $C_9H_8O_4$   **(c)** $C_6H_{10}S_2O$
40. $C_{13}H_{18}O_2$
42. $C_2HClBrF_3$
44. $CH_4N$; $C_2H_8N_2$
46. 52.68%; 16
48.

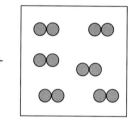

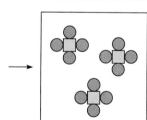

$A_3 + 6B_2 \longrightarrow 3AB_4$

50. **(a)** $3Br_2(l) + I_2(s) \longrightarrow 2IBr_3(g)$
    **(b)** $3H_2(g) + N_2(g) \longrightarrow 2NH_3(g)$
    **(c)** $2BF_3(s) + 3H_2O(g) \longrightarrow B_2O_3(s) + 6HF(g)$
52. **(a)** $4Na(s) + O_2(g) \longrightarrow 2Na_2O(s)$
    **(b)** $2Ca(s) + O_2(g) \longrightarrow 2CaO(s)$
    **(c)** $4Al(s) + 3O_2(g) \longrightarrow 2Al_2O_3(s)$
    **(d)** $2Sr(s) + O_2(g) \longrightarrow 2SrO(s)$
    **(e)** $4Co(s) + 3O_2(g) \longrightarrow 2Co_2O_3(s)$
54. **(a)** $Fe_2O_3(s) + 3H_2(g) \longrightarrow 2Fe(s) + 3H_2O(l)$
    **(b)** $UO_2(s) + 4HF(l) \longrightarrow UF_4(s) + 2H_2O(l)$
    **(c)** $C_3H_8(g) + 5O_2(g) \longrightarrow 3CO_2(g) + 4H_2O(l)$
    **(d)** $XeF_4(g) + 2H_2O(l) \longrightarrow Xe(g) + O_2(g) +$
    $4HF(g)$
    **(e)** $2KClO_3(s) \longrightarrow 2KCl(s) + 3O_2(g)$

**56. (a)** 14.73   **(b)** 19.37
　　**(c)** 0.183   **(d)** 1.84
**58. (a)** 628.6 g   **(b)** 71.05 mol
　　**(c)** 17.96 g   **(d)** 66.86 g
**60. (a)** $3Al(s) + 3NH_4ClO_4(s) \longrightarrow$
　　　　$Al_2O_3(s) + AlCl_3(s) + 3NO(g) + 6H_2O(g)$
　　**(b)** 0.4795 mol
　　**(c)** 3.471 g $Al_2O_3$; 4.539 g $AlCl_3$; 3.065 g NO;
　　3.681 g $H_2O$
**62. (a)** 2.94 L   **(b)** 154 g
**64.** 68.0 g
**66. (a)** $3X + Y \longrightarrow X_3Y$   **(b)** 10 mol X; 5 mol Y
　　**(c)** 3 mol $X_3Y$; 1 mol X; 2 mol Y
**68. (a)** $H_2(g) + Br_2(l) \longrightarrow 2HBr(g)$   **(b)** $H_2$
　　**(c)** 12.9 mol   **(d)** 1.55 mol $Br_2$
**70. (a)** $TiCl_4(s) + O_2(g) \longrightarrow TiO_2(s) + 2Cl_2(g)$
　　**(b)** $5.7 \times 10^2$ g
**72. (a)** $3Al(s) + 3NH_4ClO_4(s) \longrightarrow$
　　　　$Al_2O_3(s) + AlCl_3(s) + 3NO(g) + 6H_2O(g)$
　　**(b)** 3.53 g   **(c)** 71.8%   **(d)** 4.86 g
**74.** 29.5 g $C_7H_6O_3$; 16.4 g $C_4H_6O_3$
**76. (a)** 0.72 g   **(b)** Fe   **(c)** $O_2$
　　**(d)** 3.00 g   **(e)** $Fe_2O_3$
**78.** 9.61 mL
**80.** $NH_4NO_3$
**81.** 893 g/mol
**82.** $6.01 \times 10^{23}$
**83.** 3.657 g CaO; 2.973 g $Ca_3N_2$
**84.** 34.8%
**85. (a)** $V_2O_3$; $V_2O_5$   **(b)** 2.271 g
**86.** 28%

# Chapter 4

**2. (a)** $HgCl_2$—soluble   **(b)** $Ce(NO_3)_3$—soluble
　　**(c)** $Al_2(SO_4)_3$—soluble
　　**(d)** $Ca(OH)_2$—slightly soluble
**4. (a)** Add $H_2SO_4$   **(b)** Add $Al(NO_3)_3$
　　**(c)** Add $Na_2CO_3$
**6. (a)** $Ni^{2+}(aq) + 2OH^-(aq) \longrightarrow Ni(OH)_2(s)$
　　**(b)** $Mg^{2+}(aq) + 2OH^-(aq) \longrightarrow Mg(OH)_2(s)$
**8. (a)** $Pb^{2+}(aq) + 2Cl^-(aq) \longrightarrow PbCl_2(s)$
　　**(b)** $Ba^{2+}(aq) + SO_4^{2-}(aq) \longrightarrow BaSO_4(s)$;
　　$Co^{2+}(aq) + 2OH^-(aq) \longrightarrow Co(OH)_2(s)$
　　**(c)** No reaction
　　**(d)** $2Co^{3+}(aq) + 3CO_3^{2-}(aq) \longrightarrow Co_2(CO_3)_3(s)$
　　**(e)** No reaction
**10. (a)** No reaction
　　**(b)** $Ca^{2+}(aq) + CO_3^{2-}(aq) \longrightarrow CaCO_3(s)$
　　**(c)** $Pb^{2+}(aq) + S^{2-}(aq) \longrightarrow PbS(s)$
　　**(d)** $Fe^{3+}(aq) + 3OH^-(aq) \longrightarrow Fe(OH)_3(s)$
**12.** (1) – (a)   (2) – (a)   (3) – (b)
**14. (a)** weak   **(b)** nonelectrolyte
　　**(c)** strong   **(d)** very weak

**16. (a)** $H^+(aq)$   **(b)** $H^+(aq)$
　　**(c)** $HC_3H_5O_2(aq)$   **(d)** $H_2SO_3(aq)$
　　**(e)** $HC_3H_5O_3(aq)$
**18. (a)** $OH^-(aq)$   **(b)** $C_6H_5NH_2(aq)$
　　**(c)** $OH^-(aq)$   **(d)** $C_5H_5N(aq)$
**20. (a)** weak acid   **(b)** weak base
　　**(c)** strong base   **(d)** strong acid
**22. (a)** $HNO_2(aq) + OH^-(aq) \longrightarrow H_2O + NO_2^-(aq)$
　　**(b)** $OH^-(aq) + H^+(aq) \longrightarrow H_2O$
　　**(c)** $C_6H_5NH_2(aq) + H^+(aq) \longrightarrow C_6H_5NH_3^+(aq)$
**24. (a)** correct
　　**(b)** $H^+(aq) + CH_3NH_2(aq) \longrightarrow CH_3NH_3^+(aq)$
　　**(c)** $H^+(aq) + NH_3(aq) \longrightarrow NH_4^+(aq)$
　　**(d)** correct
　　**(e)** $HF(aq) + OH^-(aq) \longrightarrow H_2O + F^-(aq)$
**26. (a)** N = +5; O = −2   **(b)** S = +6; O = −2
　　**(c)** Na = +1; O = −1   **(d)** C = +3; O = −2
**28. (a)** H = +1; I = +3; O = −2
　　**(b)** Na = +1; Mo = +6; O = −2
　　**(c)** Fe = +3; O = −2
　　**(d)** N = +3; O = −2; F = −1
　　**(e)** K = +1; O = $-\frac{1}{2}$
**30. (a)** reduction   **(b)** reduction
　　**(c)** oxidation   **(d)** oxidation
**32. (a)** $O_2(g) + 4e^- + 2H_2O \longrightarrow 4OH^-(aq)$
　　**(b)** $MnO_4^-(aq) + e^- \longrightarrow MnO_4^{2-}(aq)$
　　**(c)** $2Cr^{3+}(aq) + 7H_2O \longrightarrow$
　　　　$Cr_2O_7^{2-}(aq) + 6e^- + 14H^+(aq)$
　　**(d)** $NH_4^+(aq) + 3H_2O \longrightarrow$
　　　　$NO_3^-(aq) + 8e^- + 10H^+(aq)$
**34. (a)** reduction: $ClO^-(aq) + 2e^- + H_2O \longrightarrow$
　　　　$Cl^-(aq) + 2OH^-(aq)$
　　**(b)** reduction: $NO_3^-(aq) + 3e^- + 4H^+(aq) \longrightarrow$
　　　　$NO(g) + 2H_2O$
　　**(c)** oxidation: $2Ni(OH)_2(s) + 2OH^-(aq) \longrightarrow$
　　　　$Ni_2O_3(s) + 2e^- + 3H_2O$
　　**(d)** oxidation: $Mn^{2+}(aq) + 2H_2O \longrightarrow$
　　　　$MnO_2(s) + 2e^- + 4H^+(aq)$
**36. (a)** $Te(s) \longrightarrow TeO_2(s) + 4e^-$
　　species oxidized and reducing agent: Te
　　$NO_3^-(aq) + 3e^- \longrightarrow NO(g)$
　　species reduced and oxidizing agent: $NO_3^-$
　　**(b)** $Sn^{2+}(aq) \longrightarrow Sn^{4+}(aq) + 2e^-$
　　species oxidized and reducing agent: $Sn^{2+}$
　　$Cr_2O_7^{2-}(aq) + 6e^- \longrightarrow 2Cr^{3+}(aq)$
　　species reduced and oxidizing agent: $Cr_2O_7^{2-}$
**38. (a)** $3Te(s) + 4NO_3^-(aq) + 4H^+(aq) \longrightarrow$
　　　　$3TeO_2(s) + 4NO(g) + 2H_2O$
　　**(b)** $Cr_2O_7^{2-}(aq) + 3Sn^{2+}(aq) + 14H^+(aq) \longrightarrow$
　　　　$2Cr^{3+}(aq) + 3Sn^{4+}(aq) + 7H_2O$
**40. (a)** $P_4(s) + 6H_2O \longrightarrow$
　　　　$2PH_3(g) + 2HPO_3^{2-}(aq) + 4H^+(aq)$
　　**(b)** $2MnO_4^-(aq) + 5C_2H_5OH(aq) + 6H^+(aq) \longrightarrow$
　　　　$2Mn^{2+}(aq) + 5C_2H_4O(aq) + 8H_2O$

**(c)** $5H_3AsO_3(aq) + 2BiO_3^-(aq) + 2H^+(aq) \longrightarrow$
$5H_3AsO_4(aq) + 2Bi(s) + H_2O$

**(d)** $2CrO_4^{2-}(aq) + 3HSO_3^-(aq) + 7H^+(aq) \longrightarrow$
$2Cr^{3+}(aq) + 3SO_4^{2-}(aq) + 5H_2O$

**42.** **(a)** $2S_2O_3^{2-}(aq) + I_2(aq) \longrightarrow$
$S_4O_6^{2-}(aq) + 2I^-(aq)$

**(b)** $4Zn(s) + NO_3^-(aq) + 7OH^-(aq) + 6H_2O \longrightarrow$
$4Zn(OH)_4^{2-}(aq) + NH_3(aq)$

**(c)** $3ClO^-(aq) + 2CrO_2^-(aq) + 2OH^-(aq) \longrightarrow$
$3Cl^-(aq) + 2CrO_4^{2-}(aq) + H_2O$

**(d)** $2Al(s) + 6H_2O + 2OH^-(aq) \longrightarrow$
$2Al(OH)_4^-(aq) + 3H_2(g)$

**44.** **(a)** $P_4(s) + 10HClO(aq) + 6H_2O \longrightarrow$
$4H_3PO_4(aq) + 10Cl^-(aq) + 10H^+(aq)$

**(b)** $3Te(s) + 4NO_3^-(aq) + 4H^+(aq) \longrightarrow$
$3TeO_2(s) + 4NO(g) + 2H_2O$

**(c)** $3Br_2(aq) + I^-(aq) + 3H_2O \longrightarrow$
$6Br^-(aq) + IO_3^-(aq) + 6H^+(aq)$

**46.** All three

**48.** **(a)** 72.0 g NaOH in enough water to make
0.300 L of solution

**(b)** 153 g $NH_4NO_3$ in enough water to make
2.55 L of solution

**50.** **(a)** 0.190 L    **(b)** 29.3 mL    **(c)** $2.6 \times 10^2$ g
**(d)** Add enough water to 0.310 L to make
0.500 L of solution.

**52.** **(a)** 0.256 mol    **(b)** 0.171 mol
**(c)** 0.342 mol    **(d)** 0.171 mol

**54.** **(a)** 32.1 mL    **(b)** 363 mL    **(c)** 40.4 mL

**56.** **(a)** $2Al^{3+}(aq) + 3CO_3^{2-}(aq) \longrightarrow Al_2(CO_3)_3(s)$
**(b)** 0.147 $M$
**(c)** 0.429 g

**58.** 16.6 mL

**60.** **(a)** 16.0 mL    **(b)** 63.5 mL    **(c)** 238 mL

**62.** **(a)** $I_2(s) + 2S_2O_3^{2-}(aq) \longrightarrow$
$2I^-(aq) + S_4O_6^{2-}(aq)$
**(b)** 632 mL

**64.** **(a)** $4Zn(s) + NO_3^-(aq) + 6H_2O + 7OH^-(aq) \longrightarrow$
$4Zn(OH)_4^{2-}(aq) + NH_3(aq)$
**(b)** 134 mL

**66.** 984 mL

**68.** 87.6%

**70.** 172 mL

**72.** 0.0941 $M$

**74.** Yes

**76.** **(a)** $Au(s) + 4Cl^-(aq) + 4H^+(aq) + NO_3^-(aq) \longrightarrow$
$AuCl_4^-(aq) + NO(g) + 2H_2O$
**(b)** $4HCl:1HNO_3$
**(c)** 42 mL HCl, 7.9 mL $HNO_3$

**78.** 93.9%

**79.** 0.794 g; yes

**80.** 0.30 L

**81.** 1.09 g Cu; 3.09 g Ag

**82.** 0.0980 $M$ $Fe^{2+}$; 0.0364 $M$ $Fe^{3+}$

# Chapter 5

**2.** $5.1 \times 10^3$ L; $3.81 \times 10^3$ mol; $3.00 \times 10^2$ K

**4.**

| 432 mm Hg | 0.568 atm | 57.6 kPa |
|---|---|---|
| $1.01 \times 10^3$ mm Hg | 1.33 atm | 135 kPa |
| 564 mm Hg | 0.742 atm | 75.2 kPa |

**6.**

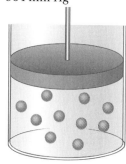

before:

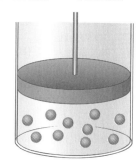

after: **(a)** piston goes down

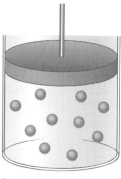

**(b)** piston goes up

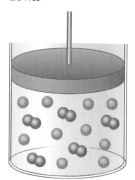

**(c)** piston goes up

**8.** 819 mm Hg

**10.** **(a)** 102 K    **(b)** 203 K

**12.** 20.3 psi; 5.6 psi

**14.** 5.13 $cm^3$

**16.** 0.254 mol

**18.** $P = 32$ atm; It will explode.

**20.** $1.8 \times 10^2$ g

**22.**

| 493 mm Hg | 3.75 L | 36°C | 0.0959 mol | 2.69 g |
|---|---|---|---|---|
| 1.28 atm | 6.39 L | 199 K | 0.500 mol | 14.0 g |
| 125 kPa | 38.2 L | 99°C | 1.54 mol | 43.2 g |
| 0.324 atm | 2.98 L | 125°C | 0.0295 mol | 0.827 g |

**24.** **(a)** Same pressure in both tanks
**(b)** Pressure in Tank A twice that in tank B

**26.** **(a)** 3.60 g/L    **(b)** 3.73 g/L    **(c)** 0.731 g/L

**28.** $1.63 \times 10^3 : 1$

**30.** **(a)** 98.8 g/mol    **(b)** $COCl_2$

**32.** **(a)** 9.20 g/mol    **(b)** 0.288 : 1

**34.** F; sulfur hexafluoride

**36.** **(a)** $2H_2S(g) + 3O_2(g) \longrightarrow 2SO_2(g) + 2H_2O(g)$
**(b)** 18.0 L

**38.** 2.83 L

**40.** 3.84 L
**42.** 3.04 atm, HCl
**44.** 15.6 L
**46.** **(a)** $P_B = 2.00$ atm; $P_A + P_B = 4.00$ atm
**(b)**

**(c)** $P_A = 1.00$ atm; $P_B = 1.00$ atm; $P_A + P_B = 2.00$ atm; Total pressure is half that in (a)
**48.** 3.12 atm
**50.** 0.2682 : 1
**52.** **(a)** $CO_2 < O_2 < CO < Ne$
**(b)** $Ne < CO < O_2 < CO_2$
**54.** $5.20 \times 10^4$ K
**56.** **(a)** CO **(b)** increase $T$; decrease $P$
**58.** **(a)** $1.7 \times 10^2$ g/L
**(b)** From Ideal gas law: $1.31 \times 10^2$ g/L
**60.** >10.3 m
**62.**

**(a)** 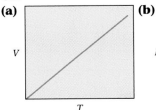 **(b)**

**(c)** **(d)**

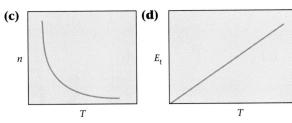

**64.** **(a)** 1.00 **(b)** 0.6869 **(c)** 1.00 **(d)** 1.00
**66.** $O_2$
**68.** 0.572 g/L; 28.0 g/mol
**70.** 3.0 ft
**71.** 0.0456 L-atm/mol-°R
**72.** 78.7%
**73.** 6.62 m
**74.** 0.897 atm
**75.** $\dfrac{V_A}{V} = \dfrac{n_a}{n}$; mol fraction ≠ mass fraction

## Chapter 6

**2.** ground state
**4.** **(a)** $4.00 \times 10^2$ MHz **(b)** $2.65 \times 10^{-25}$ J
**(c)** $1.60 \times 10^{-4}$ kJ/mol

**6.** yes
**8.** **(a)** 69.0 nm **(b)** $4.35 \times 10^{15}$ Hz
**10.** $3.97 \times 10^{-23}$ J
**12.** energy absorbed for (2) and (4); energy emitted for (1) and (3)
**14.** concentric circles of increasing radius:
**(a)** transition from higher levels to $n = 1$
**(b)** transition from higher levels to $n = 2$
**16.** $2.624 \times 10^3$ nm
**18.** 1875 nm
**20.** The probability of finding the hydrogen electron increases as the distance from the nucleus decreases.
**22.** **(a)** −2, −1, 0, 1, 2
**(b)** −3, −2, −1, 0, 1, 2, 3 **(c)** 0; −1, 0, 1
**24.** **(a)** 3d **(b)** 2s **(c)** 5s **(d)** 4d
**26.** **(a)** p **(b)** d **(c)** f
**28.** **(a)** 18 **(b)** 10 **(c)** 14
**30.** **(a)** 2 **(b)** f **(c)** 5 **(d)** 2
**32.** **(c)** can't occur; $m_\ell$ must be 0
**(d)** can't occur; $\ell$ can only be 0 or 1
**34.** **(a)** $1s^2\, 2s^2\, 2p^6\, 3s^2\, 3p^4$
**(b)** $1s^2\, 2s^2\, 2p^6\, 3s^2\, 3p^6\, 4s^2\, 3d^{10}\, 4p^4$
**(c)** $1s^2\, 2s^2\, 2p^6\, 3s^2\, 3p^6\, 4s^2\, 3d^{10}\, 4p^6\, 5s^2\, 4d^{10}\, 5p^3$
**(d)** $1s^2\, 2s^2\, 2p^6\, 3s^2\, 3p^6\, 4s^2\, 3d^1$
**(e)** $1s^2\, 2s^2\, 2p^6\, 3s^2\, 3p^2$
**36.** **(a)** $[Kr]\, 5s^2\, 4d^{10}\, 5p^1$ **(b)** $[Kr]\, 5s^2\, 4d^{10}\, 5p^5$
**(c)** $[Xe]\, 6s^2\, 4f^{14}\, 5d^7$ **(d)** $[Kr]\, 5s^2$
**(e)** $[Ne]\, 3s^2\, 3p^1$
**38.** **(a)** Yb **(b)** Sn **(c)** Zr **(d)** I
**40.** **(a)** 2/22 **(b)** 30/81 **(c)** 20/52
**42.** **(a)** excited **(b)** excited **(c)** impossible
**(d)** impossible **(e)** excited **(f)** excited
**44.**

| | 1s | 2s | 2p | 3s |
|---|---|---|---|---|
| **(a)** N: | (↑↓) | (↑↓) | (↑ )(↑ )(↑ ) | |
| **(b)** Ni: | (↑↓) | (↑↓) | (↑↓)(↑↓)(↑↓) | (↑↓) |

| 3p | 4s | 3d |
|---|---|---|
| (↑↓)(↑↓)(↑↓) | (↑↓) | (↑↓)(↑↓)(↑↓)(↑ )(↑ ) |

| | 1s | 2s | 2p | 3s |
|---|---|---|---|---|
| **(c)** Na: | (↑↓) | (↑↓) | (↑↓)(↑↓)(↑↓) | (↑ ) |
| **(d)** Ne: | (↑↓) | (↑↓) | (↑↓)(↑↓)(↑↓) | |

**46.** **(a)** Ni **(b)** Co **(c)** Ge
**48.** **(a)** Tc **(b)** C, N, O
**(c)** Sr, Ba, Ra **(d)** Te, Po, At
**50.** **(a)** 0 **(b)** 5 **(c)** 0
**52.** **(a)** Groups 1, 2, 13, 14, 15 **(b)** Group 16
**(c)** Group 17 **(d)** Group 18
**54.** **(a)** Mg: $1s^2\, 2s^2\, 2p^6\, 3s^2$  $Mg^{2+}: 1s^2\, 2s^2\, 2p^6$
**(b)** N: $1s^2\, 2s^2\, 2p^3$  $N^{3-}: 1s^2\, 2s^2\, 2p^6$
**(c)** O: $1s^2\, 2s^2\, 2p^4$  $O^-: 1s^2\, 2s^2\, 2p^5$
**(d)** $Fe^{2+}: 1s^2\, 2s^2\, 2p^6\, 3s^2\, 3p^6\, 3d^6$
$Fe^{3+}: 1s^2\, 2s^2\, 2p^6\, 3s^2\, 3p^6\, 3d^5$
**56.** **(a)** 2 **(b)** 0 **(c)** 0 **(d)** 2

**58. (a)** Na > Si > S    **(b)** S > Si > Na
**(c)** S > Si > Na
**60. (a)** K    **(b)** Cl    **(c)** Cl
**62. (a)** $V^{3+}$    **(b)** Se    **(c)** As    **(d)** $Mn^{4+}$
**64. (a)** $Ni^{3+} < Ni^{2+} < Ni$    **(b)** $Br < Br^- < I^-$
**66. (a)** [Xe] $6s^2\,4f^5$    **(b)** Np (neptunium)
**68. (a)** Bohr model specifies position; quantum
mechanical model deals with probability
**(b)** See Fig. 6.1
**(c)** each orbital at 90° to each other
**70. (a)** true    **(b)** true    **(c)** false; $14e^-$
**(d)** false; 1 unpaired $e^-$
**72.** 5; 4, 3, 2, 1, 0, −1, −2, −3, −4; 18
**74. (a)** false, stays the same    **(b)** true    **(c)** true
**76.** +2954 kJ/mol

**77.** $\Delta E = 2.180 \times 10^{-18}\,J\left(\dfrac{1}{4} - \dfrac{1}{n^2}\right) =$

$2.180 \times 10^{-18}\,J\left(\dfrac{n^2 - 4}{4n^2}\right)$

$\lambda = \dfrac{hc}{\Delta E}$

$= \dfrac{(6.626 \times 10^{-34}\,J \cdot s)(2.998 \times 10^{17}\,nm)(4n^2)}{(2.180 \times 10^{-18}\,J)(n^2 - 4)}$

$= \dfrac{364.5}{n^2 - 4}\,n^2\,nm$

**78.** $1s^4\,1p^4$

**79. (a)** s sublevel   $m_\ell = 0$ = $3\,e^-$

p sublevel   $m_\ell = -1, 0, 1$ = $9\,e^-$

d sublevel   $m_\ell = -2, -1, 0, 1, 2$ = $15\,e^-$

**(b)** n = 3; $\ell$ = 0, 1, 2; total electrons = 27
**(c)** $1s^3\,2s^3\,2p^2$   $1s^3\,2s^3\,2p^9\,3s^2$
**80. (a)** $3.42 \times 10^{-19}\,J$    **(b)** 581 nm

# Chapter 7

**2. (a)** H—N—O with double bond O below
**(b)** :F̈—Kr—F̈:

**(c)** :C̈l—C=O with :C̈l: below
**(d)** ( :Ö=N—Ö: with :Ö: below )⁻

**4. (a)** ( :Ö—I—Ö: with :Ö: above )⁻
**(b)** :F̈—Se—F̈: with F above and below
**(c)** :F̈—Br—F̈: with :F̈: below
**(d)** (:C≡N:)⁻

**6. (a)** H—Äs—H with H below
**(b)** :F̈—B—F̈: with :F̈: below
**(c)** ( :Ö—S—Ö: with :Ö: below )²⁻
**(d)** :F̈—Xe—F̈:

**8.** ( H—O≡C: )⁺

**10. (a)** :C̈l \ C=Ö: / :C̈l
**(b)** H—C=C—H with H, H above and H below
**(c)** H—Ö—S—Ö—H with :Ö: below

**12.** :O: double bonded, H—C—Ö—H

**14.** H—C—C—C=Ö: with H,H,H and H below   or   H—C—C—C—H with H,H and :Ö: H

**16. (a)** $O_2^{2-}$    **(b)** $H_2PO_4^-$
**(c)** $NH_4^+$    **(d)** $SO_4^{2-}$

**18. (a)** ( :C̈l—B—C̈l: with :Cl: above and :Cl: below )⁻
**(b)** ( :Ö—P—Ö—H with :O: above and :O: below )²⁻
**(c)** (:C̈l—Ö:)⁻
**(d)** ( :S: above, :Ö—S—Ö: with :Ö: below )²⁻

**20. (a)** :F̈—B—F̈: with :F̈: below
**(b)** ·N=Ö:
**(c)** (·C≡O:)⁺
**(d)** :Ö—C̈l—Ö: with :Ö: below

**22. (a)** :C̈l—N with O: above and O below ⟷ :C̈l—N with O: above and O: below ⟷ :C̈l=N with O: above and O: below
**(b)** H \ C=N=N: / H  ⟷  H \ C—N≡N: / H

**(c)**

$$:\ddot{O}-S=\ddot{O}: \longleftrightarrow :\ddot{O}=S-\ddot{O}: \longleftrightarrow :\ddot{O}-S-\ddot{O}:$$
$$\qquad |\qquad\qquad\qquad |\qquad\qquad\qquad \|$$
$$\quad:\ddot{O}:\qquad\qquad\quad:\ddot{O}:\qquad\qquad\quad:\ddot{O}:$$

**24. (a)**

$$\left(\begin{array}{c}:\ddot{O}\qquad\ddot{O}:\\ \diagdown\quad/\\ C-C\\ /\quad\diagdown\\ :O\qquad O:\end{array}\right)^{2-}$$

**(b)**

$$\left(\begin{array}{c}\ddot{O}:\qquad\ddot{O}:\\ \diagdown\quad/\\ C-C\\ /\quad\diagdown\\ :\ddot{O}:\qquad:\ddot{O}:\end{array}\right)^{2-} \longleftrightarrow \left(\begin{array}{c}\ddot{O}:\qquad\ddot{O}:\\ \diagdown\quad/\\ C-C\\ /\quad\diagdown\\ :\ddot{O}:\qquad:\ddot{O}:\end{array}\right)^{2-} \longleftrightarrow$$

$$\left(\begin{array}{c}:\ddot{O}:\qquad:\ddot{O}:\\ \diagdown\quad/\\ C-C\\ /\quad\diagdown\\ :O\qquad O:\end{array}\right)^{2-}$$

**(c)** No

**26.**

$$\begin{array}{ccc} S & S & S \\ \diagup\diagdown & \diagup\diagdown & \diagup\diagdown \\ N\quad N: & :N\quad N: & :N\quad N: \\ \diagdown\diagup & \diagdown\diagup & \diagdown\diagup \\ S & S & S \end{array} \longleftrightarrow \cdots \longleftrightarrow \cdots \longleftrightarrow$$

$$\begin{array}{c} S \\ \diagup\diagdown \\ :N\quad :N: \\ \diagdown\diagup \\ S \end{array}$$

**28. (a)** 0   **(b)** 0   **(c)** 0

**30.**  $H-\ddot{O}-\overset{\displaystyle\ddot{S}}{\underset{\displaystyle\ddot{O}:}{S}}-\ddot{O}-H$

**32. (a)** linear                 **(b)** linear
**(c)** triangular planar      **(d)** square planar
**34. (a)** linear                 **(b)** triangular pyramid
**(c)** tetrahedron      **(d)** linear
**36. (a)** bent                 **(b)** triangular planar
**(c)** octahedral      **(d)** linear
**38. (a)** T-shaped      **(b)** bent
**(c)** square pyramid
**40. (a)** 109.5°                 **(b)** 180°
**(c)** 109.5°, 120°, 109.5°      **(d)** 120°, 120°, 180°
**42. (a)**

$$H-\overset{\displaystyle H}{\underset{\displaystyle H}{C}}-\overset{\displaystyle H}{\underset{\displaystyle H}{C}}-\overset{\displaystyle :\ddot{O}:}{C}-\ddot{O}-\ddot{O}-N\diagup^{\ddot{O}\cdot}_{\diagdown\ddot{O}\cdot}$$

**(b)** around $CH_3$, $CH_2$ − 109.5°
around C=O − 120°
around central O atoms − 109.5°
around N − 120°
**44.** $SnCl_2$, and $SO_2$; unshared pair occupies a larger volume

**46.**  1 = 120°      2 = 109.5°      3 = 109.5°
**48.**  a
**50.**  b, c, d
**52.**  1st and 3rd are polar
**54. (a)** sp      **(b)** $sp^3d$   **(c)** $sp^2$      **(d)** $sp^3d^2$
**56. (a)** $sp^3d$      **(b)** $sp^3$   **(c)** $sp^3$      **(d)** sp
**58. (a)** $sp^3$      **(b)** $sp^2$   **(c)** $sp^3d^2$      **(d)** $sp^3d$
**60. (a)** $5e^-$ pairs; $sp^3d$
**(b)** $5e^-$ pairs; $sp^3d$
**(c)** $6e^-$ pairs; $sp^3d^2$
**62.**  C − $sp^3$; S − $sp^2$; O − $sp^2$
**64. (a)** $sp^2$      **(b)** $sp^3$   **(c)** sp      **(d)** sp
**66. (a)** $sp^3$      **(b)** $sp^3$   **(c)** $sp^3$
**68. (a)** 3σ, 1π      **(b)** 3σ
**(c)** 1σ, 2π      **(d)** 2σ, 2π
**70.**  9σ, 1π
**72. (a)** $CH_2O$      **(b)** $CH_2O$      **(c)** $XeF_2$
**74.**  First structure: H = 0; O = 0; C = 0; N = 0
Second structure: H = 0; O = 1; C = 0; N = −1
First structure more likely
**76. (a)** more than eight valence electrons around the atom
**(b)** molecular structures with the same skeleton differing only in electron distribution
**(c)** electron pairs assigned only to one atom
**(d)** odd number of valence electrons (3, 5, 7…) around the atom
**78. (a)** 5π, 21σ      **(b)** 109.5°, 120°, 109.5°
**(c)** $sp^2$, $sp^3$, $sp^2$
**80.**  x = 3; T-shaped; polar; $sp^3d$; 90°, 180°; 3 sigma
**81.**

$$\begin{array}{cc} H & H \\ \diagdown & \diagup \\ N-N \\ \diagup & \diagdown \\ H & H \end{array}$$   bent; 109.5°; polar

**82.**  6; octahedron; $sp^3d^2$
**83. (a)**

$$\left(\begin{array}{c}:\ddot{O}:\\ |\\ :\ddot{O}-S-\ddot{O}:\\ |\\ :\ddot{O}:\end{array}\right)^{2-} \left(\begin{array}{c}:\ddot{O}:\\ |\\ :\ddot{O}-S-\ddot{O}:\\ \|\\ :\ddot{O}:\end{array}\right)^{2-}$$

**(b)** tetrahedral for both
**(c)** $sp^3$ for both
**(d)** 1st structure: O = −1, S = 2
2nd structure S = 0, O = −1, −1, 0, 0

**84.**

$$:\ddot{Cl}-\overset{\displaystyle:\ddot{O}:}{\underset{\displaystyle:\ddot{Cl}:}{P}}-\ddot{Cl}:$$   P = +1; Cl = 0; O = −1

$$:\ddot{Cl}-\overset{\displaystyle:O:}{\underset{\displaystyle:\ddot{Cl}:}{P}}-\ddot{Cl}:$$   P = 0; Cl = 0; O = 0

## Chapter 8

**2.** 50.8°C
**4.** 0.129 J/g-°C
**6.** **(a)** NaCl $(s) \longrightarrow$ Na$^+(aq)$ + Cl$^-(aq)$
    **(b)** 1.02 kJ   **(c)** endothermic   **(d)** 3.91 kJ
**8.** $-2.73 \times 10^3$ kJ
**10.** 1.772 kJ/°C
**12.** 78.6 mg
**14.** 104.0°C
**16.** **(a)** Hg$(g) \longrightarrow$ Hg$(l)$   $\Delta H = -59.4$ kJ
    **(b)** C$_6$H$_6(l) \longrightarrow$ C$_6$H$_6(g)$   $\Delta H = 30.8$ kJ
    **(c)** Br$_2(l) \longrightarrow$ Br$_2(s)$   $\Delta H = -10.8$ kJ
**18.** **(a)** H$_2(g)$ + 2C$(s)$ + N$_2(g) \longrightarrow$ 2HCN$(g)$
                            $\Delta H = 270.3$ kJ
    **(b)** endothermic
    **(c)**

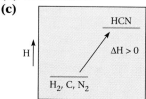

    **(d)** 5.01 kJ   **(e)** 1.777 g C
**20.** **(a)** 200.1 kJ   **(b)** 3.06 kJ
**22.** **(a)** 2 C$_3$H$_5$(NO$_3$)$_3(l) \longrightarrow$ 3 N$_2(g)$ + $+\frac{1}{2}$ O$_2(g)$ +
      6 CO$_2(g)$ + 5H$_2$O$(g)$   $\Delta H° = -2.84 \times 10^3$ kJ
    **(b)** $-979$ kJ
**24.** $3.10 \times 10^4$ kJ; $7.42 \times 10^3$ nutritional calories
**26.** C$_{10}$H$_8$
**28.** **(a)** Hg$(g) \longrightarrow$ Hg$(s)$   $\Delta H = -61.7$ kJ
    **(b)** H$_2$O$(s) \longrightarrow$ H$_2$O$(g)$   $\Delta H = 46.7$ kJ
**30.** 4.20 kJ
**32.** $-11.3$ kJ
**34.** $-804.6$ kJ
**36.** **(a)** KClO$_3(s) \longrightarrow$ K$(s)$ + $\frac{1}{2}$Cl$_2(g)$ + $\frac{3}{2}$O$_2(g)$
                            $\Delta H° = 397.7$ kJ
    **(b)** N$_2$H$_4(l) \longrightarrow$ N$_2(g)$ + 2H$_2(g)$
                            $\Delta H° = -50.6$ kJ
    **(c)** N$_2$O$_4(g) \longrightarrow$ N$_2(g)$ + 2 O$_2(g)$
                            $\Delta H° = -9.2$ kJ
    **(d)** AgCl$(s) \longrightarrow$ Ag$(s)$ + $\frac{1}{2}$Cl$_2(g)$
                            $\Delta H° = 127.1$ kJ
**38.** **(a)** $-1216.3$ kJ/mol   **(b)** 16.22 g
**40.** $\Delta H = -9.38$ kJ
**42.** **(a)** $-132.8$ kJ   **(b)** 7.1 kJ   **(c)** 13.1 kJ
**44.** **(a)** 1530 kJ   **(b)** $-48.6$ kJ
**46.** **(a)** CaCO$_3(s)$ + 2NH$_3(g) \longrightarrow$
          CaCN$_2(s)$ + 3H$_2$O$(l)$   $\Delta H° = 90.1$ kJ
    **(b)** $-351.6$ kJ/mol
**48.** $-54.0$ kJ/mol
**50.** 2.90 kJ liberated
**52.** piston drops
**54.** 68 J
**56.** **(a)** 37 J   **(b)** $-15$ J

**58.** $-2.48$ kJ
**60.** **(a)** $-1299.5$ kJ   **(b)** $-1295.8$ kJ
**62.** 50.6 kJ/mol
**64.** 34 mi
**66.** 83.9°C
**68.** $1.16 \times 10^5$ mol
**69.** **(a)** 168 kJ   **(b)** 505 g
**70.** 22%
**71.** **(a)** $-851.5$ kJ   **(b)** $6.6 \times 10^3$ °C   **(c)** yes
**72.** 76.0%

## Chapter 9

**2.** CBr$_4$ > CCl$_4$ > CF$_4$ > CH$_4$
**4.** dispersion forces: all    dipole forces: b, d
**6.** a, c
**8.** **(a)** NaI is ionic   **(b)** NH$_3$ has H-bonds
    **(c)** H$_2$O has H-bonds
    **(d)** Benzoic acid has larger dispersion forces
**10.** a, d
**12.** **(a)** CH$_4$–smaller dispersion force
    **(b)** H$_2$–smaller dispersion force
    **(c)** HF–LiF is ionic
    **(d)** C$_2$H$_6$–smaller dispersion force
**14.** **(a)** ionic bonds   **(b)** dispersion forces
    **(c)** H-bonds    **(d)** dispersion forces
**16.** **(a)** ionic   **(b)** molecular   **(c)** metallic
**18.** **(a)** molecule   **(b)** metal
    **(c)** polar molecule and ionic
**20.** **(a)** network covalent   **(b)** metal
    **(c)** molecular    **(d)** network covalent
    **(e)** ionic
**22.** **(a)** Na$_2$O   **(b)** CO   **(c)** SiO$_2$   **(d)** CO$_2$
**24.** **(a)** molecule    **(b)** ions
    **(c)** cations, mobile electrons   **(d)** atoms
**26.** body–centered cubic
**28.** 0.143 nm
**30.** **(a)** 0.151 nm   **(b)** no for both
**32.** **(a)** 0.700 nm   **(b)** 0.404 nm
**34.** 1; 1
**36.** **(a)** 1.03 mg   **(b)** 0.145 mm Hg; only gas
    **(c)** 0.300 mm Hg; solid and gas
**38.** **(a)** yes   **(b)** 19.83 mm Hg
**40.** **(a)** 58.1 kJ/mol   **(b)** 30.5°C
**42.** 321 mm Hg
**44.** at 0°C: $2.7 \times 10^{-4}$ mm Hg
    at 100°C: 0.31 mm Hg
**46.** 29 kJ/mol
**48.** **(a)** false   **(b)** true   **(c)** false
**50.** **(a)** liquid   **(b)** vapor   **(c)** vapor

**52. (a)**

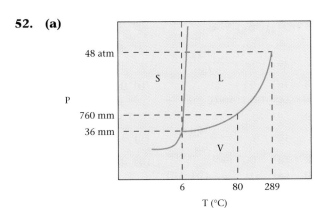

**(b)** ≈ 250 mm Hg

**54. (a)** solid to liquid to gas   **(b)** nothing
**(c)** vapor to solid

**56. (a)**

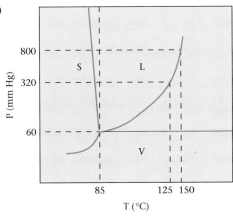

**(b)** ≈ 145°C   **(c)** condensation

**58.** d
**60. (a)** Covalent bond between atoms within a molecule; H-bond between atoms in two different molecules.
**(b)** Melting changes a solid to liquid; vaporizing changes a liquid to gas.
**(c)** Normal boiling point is the temperature where vapor pressure is 760 mm Hg. Boiling point is the temperature where vapor pressure is pressure above the liquid.
**(d)** Molecular compound has a molecule as its structural unit; network covalent is one large molecule.

**62. (a)** $14.8\ cm^3$   **(b)** $1.5 \times 10^{10}\ cm^3$
**(c)** $9.39\ cm^3$
**(d)** 63.4% in liquid; $6.3 \times 10^{-8}\%$ in gas

**64.** A = iodine crystals   B = iron sulfide
C = silver bar   D = graphite rod

**66.** $6.05 \times 10^{23}$ atoms/mol
**67. (a)** liquid and vapor   **(b)** 26.7 mm Hg
**68.** 41%
**69.** 80 atm; 0.60°C—heat conduction is more likely

**70.** $\dfrac{r_{cation}}{r_{anion}} = 0.414$

**71.** $P_{C_3H_8}$ is the vapor pressure of propane which decreases exponentially with $T$. $P_{N_2}$ is gas pressure which decreases linearly with $T$.

# Chapter 10

**2. (a)** 3.67%   **(b)** 0.996
**4.** 11 m
**6.** $1.26 \times 10^{-2}$ ppm
**8. (a)** 329.2 g   **(b)** 33.86 mL   **(c)** 0.139 M

**10.**

|     | $m$ | Mass % solv | ppm solute | $X_{solvent}$ |
| --- | --- | --- | --- | --- |
| **(a)** | 32.1 | 25.7 | $7.43 \times 10^5$ | 0.634 |
| **(b)** | $4.85 \times 10^{-3}$ | 99.9563 | $4.37 \times 10^2$ | 0.9999126 |
| **(c)** | 3.13 | 78.0 | $2.20 \times 10^5$ | 0.947 |
| **(d)** | 1.810 | 85.98 | $1.402 \times 10^5$ | 0.9684 |

**12. (a)** Dissolve 24.0 g NaOH in 1.00 L of solution
**(b)** Dilute 100.0 mL to 1.00 L of solution

**14.**

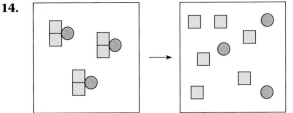

$Na^+ = 6\ M$; $S^{2-} = 3\ M$

**16. (a)** $0.330\ M$; $0.330\ M$; $0.660\ M$   **(b)** 75.4 g
**18. (a)** $15.72\ M$   **(b)** $0.629\ M$

**20.**

|     | density (g/mL) | $M$ | $m$ | Mass % solute |
| --- | --- | --- | --- | --- |
| **(a)** | 1.028 | 0.248 | 0.251 | 4.00 |
| **(b)** | 1.227 | 1.923 | 2.117 | 26.00 |
| **(c)** | 1.546 | 4.654 | 6.02 | 49.97 |

**22. (a)** KI—ionic   **(b)** $NH_3$—H-bonds with $H_2O$
**(c)** NaOH—ionic
**(d)** methyl alcohol—H-bonds with $H_2O$

**24. (a)** $NH_4NO_3(s) \longrightarrow NH_4^+(aq) + NO_3^-(aq)$
**(b)** 28.1
**(c)** solubility should increase as $T$ increases

**26. (a)** 190 g   **(b)** 485 g
**28. (a)** less   **(b)** greater
**30. (a)** $0.23\ M$   **(b)** $1.1 \times 10^{-5}\ M$
**32. (a)** 559 mm Hg   **(b)** 647.7 mm Hg
**(c)** 636.0 mm Hg

**34.** Dissolve 740 g in 310 g of water.
**36.** $1.55 \times 10^{-4}$ atm
**38.** 182 g/mol
**40.** **(a)** 7.6 g; 5.7 g   **(b)** 7.7 g; 5.7 g
**42.** $-8.95°C$; 102.5°C
**44.** 403 g/mol
**46.** $C_{12}H_{26}O$
**48.** 98% phenol
**50.** 0.16 $M$
**52.** $2.60 \times 10^4$ g/mol
**54.** freezing points:
   **(a)** $-0.37°C$   **(b)** $-0.74°C$   **(c)** $-0.56°C$
   boiling points:
   **(a)** 100.10°C   **(b)** 100.21°C   **(c)** 100.16°C
**56.** b
**58.** $-11°C$
**60.** **(a)** $i = 3$ for $BaCl_2$ and only 1 for glucose.
   **(b)** $\Delta H$ for the solution of gases is usually exothermic.
   **(c)** $1\ L \approx 1$ kg for water; dilute solutions have more water and approach that approximation.
**62.** **(a)** $\pi$ of solution must equal that of blood
   **(b)** more $O_2$ in cooler water (for fish not fishermen)
   **(c)** $P$ decreases; solubility of $CO_2$ decreases
   **(d)** $M$ = moles solute/L solution; $m$ = moles solute/kg solvent
**64.** **(a)** test conductivity
   **(b)** increase in $T$ decreases solubility of $CO_2$
   **(c)** number of kg of solvent is most often less than total moles
   **(d)** vapor pressure is lowered
**66.** **(a)** 0.845 $M$   **(b)** 0.85 $m$
   **(c)** 62.0 atm   **(d)** 101.3°C
**67.** 1.1 g/mL
**68.** add $1.03 \times 10^3$ g of water
**69.** $m = \dfrac{n \text{ solute}}{\text{kg solvent}}$; in 1 L of solution, $n$ solute = $M$

   $\text{kg solvent} = \dfrac{\text{mass solution (g)} - \text{mass solute (g)}}{1000}$

   $= \dfrac{(1000 \times d) - M(\mathcal{M})}{1000} = d - \dfrac{M(\mathcal{M})}{1000}$

   $m = \dfrac{M}{d - \dfrac{M(\mathcal{M})}{1000}}$

   In dilute solution, $m \longrightarrow M/d$;
   for water, $d = 1.00$ g/mL
**70.** 49%
**71.** 0.0018 g/cm$^3$—intoxicated
**72.** **(a)** 2.08 $M$   **(b)** 1.872 mol   **(c)** 47.4 L
**73.** $V_{gas} = \dfrac{n_{gas} \times RT}{P_{gas}}$; $n_{gas} = k \times P_{gas}$; $V_{gas} = kRT$

# Chapter 11

**2.** **(a)** rate = $\dfrac{\Delta[N_2O_5]}{2\Delta t}$   **(b)** rate = $\dfrac{-\Delta[O_2]}{5\Delta t}$
**4.** **(a)** 0.213 mol/L-min   **(b)** 0.852 mol/L-min
   **(c)** 1.06 mol/L-min
**6.** **(a)** $2N_2O_5(g) \longrightarrow 4NO_2(g) + O_2(g)$
   **(b)** rate = $\dfrac{-\Delta[N_2O_5]}{2\Delta t}$   **(c)** 0.0292 mol/L-min
**8.** **(a)** 0.009 mol/L-min   **(b)** 0.025 mol/L-min
**10.** **(a)** 2, 0, 2   **(b)** 1, 1, 2
   **(c)** 0, 1, 1   **(d)** 2, 1, 3
**12.** **(a)** L/mol-s   **(b)** L/mol-s
   **(c)** s$^{-1}$   **(d)** $L^2/mol^2 \cdot s$
**14.** **(a)** 0.0388 mol/L-min   **(b)** 2.3 mol/L
   **(c)** 0.964 mol/L   **(d)** 8.67 L/mol-min
**16.** **(a)** rate = $k[NO_2]^2$   **(b)** 18.8 L/mol-min
   **(c)** 7.68 mol/L-min
**18.** **(a)** $4.17 \times 10^{-6}$ mol/L-s
   **(b)** $4.17 \times 10^{-6}$ mol/L-s
   **(c)** rate independent of $[NH_3]$
**20.** c
**22.** **(a)** 2   **(b)** rate = $k[NO_2]^2$   **(c)** 9.76 L/mol-s
**24.** **(a)** 1 for $HgCl_2$, 2 for $C_2O_4^{2-}$, 3 overall
   **(b)** rate = $k[HgCl_2][C_2O_4^{2-}]^2$
   **(c)** $7.8 \times 10^{-3}$ $L^2/mol^2 \cdot min$
   **(d)** $3.1 \times 10^{-6}$ mol/L-min
**26.** **(a)** 1st order in $CH_3COCH_3$, 1st order in $[H^+]$, zero order in $[I_2]$
   **(b)** rate = $k[CH_3COCH_3][H^+]$
   **(c)** $2.6 \times 10^{-5}$ L/mol-s
   **(d)** $6.8 \times 10^{-5}$ mol/L-s
**28.** **(a)** rate = $k[I^-][H_2O_2]$   **(b)** 4.9 L/mol-min
   **(c)** 0.36 mol/L-min
**30.** 2nd order
**32.** 2nd order
**34.** **(a)** linear plot obtained for $\ln[C_3H_6]$ vs. $t$
   **(b)** 0.036 min$^{-1}$   **(c)** 39 min
   **(d)** 0.015 mol/L-min
**36.** **(a)** 0.0036 s$^{-1}$   **(b)** $1.9 \times 10^2$ s
   **(c)** $2.9 \times 10^2$ s
**38.** **(a)** 0.363 hr   **(b)** 24%
   **(c)** 7.8 min   **(d)** 0.478 mol/L-hr
**40.** 54 days
**42.** **(a)** $1.1 \times 10^2$ min   **(b)** 6.1 hr
**44.** **(a)** 0.800 L/mol-s   **(b)** $1.00 \times 10^2$ s
   **(c)** 0.0320 mol/L-s
**46.** **(a)** 0.050 mol/L-s   **(b)** 13 s   **(c)** 4.0 s
**48.** **(a)** System 2   **(b)** System 2
   **(c)** 95 kJ, 60 kJ, 125 kJ
**50.** 33 kJ
**52.** $2.5 \times 10^2$ kJ
**54.** 47 hr
**56.** **(a)** $X_{23°C} = 134$; $X_{33°C} = 206$   **(b)** 32 kJ
   **(c)** 54%

**58.** **(a)** 0.47 L/mol–s    **(b)** $2.1 \times 10^2$ °C
**60.** **(a)** rate = $k[N_2O_2][H_2]$
 **(b)** rate = $k[K][HCl]$    **(c)** rate = $k[I_2]$
**62.** rate = $k_2[H_2][I]^2$; $[I]^2 = \dfrac{k_1}{k_{-1}}[I_2]$;

 rate = $\dfrac{k_2 k_1}{k_{-1}}[H_2][I_2]$

**64.** Mechanism 1: rate = $k_1[NO_3][NO]$;

 $[NO_3] = \dfrac{k_2}{k_{-2}}[O_2][NO]$; rate = $\dfrac{k_2 k_1}{k_{-2}}[NO]^2[O_2]$
 Mechanism 2: rate = $k_1[N_2O_2][O_2]$;

 $[N_2O_2] = \dfrac{k_2}{k_{-2}}[NO]^2$; rate = $\dfrac{k_2 k_1}{k_{-2}}[NO]^2[O_2]$

**66.** **(a)** rate = $k[A]^2[B]$    **(b)** $36k$
 **(c)**

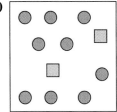

**68.** **(a)** test tube 1
 **(b)** rate and $k$ increase; $E_a$ remains the same
**70.** **(a)** fewer molecules have required $E_a$
 **(b)** works only for 1st order reactions
 **(c)** flame supplies $E_a$
**72.** **(a)** 46 min    **(b)** 3.7 L
**74.** $4.6 \times 10^2$%
**76.** $7 \times 10^{17}$ years; Depletion occurred a lot faster.
**77.** **(a)** 174 kJ    **(b)** 46 L/mol-s    **(c)** 1.8 mol/L-s
**78.** $\dfrac{-d[A]}{a\,dt} = k[A]$; $\dfrac{-d[A]}{[A]} = k\,a\,dt$; $\ln\dfrac{[A_o]}{[A]} = a\,k\,t$

**79.** **(a)** $\dfrac{-d[A]}{dt} = k[A]^2$; $\dfrac{-d[A]}{[A]^2} = k\,dt$; $\dfrac{1}{[A]} - \dfrac{1}{[A_o]} = k\,t$

 **(b)** $\dfrac{-d[A]}{dt} = k[A]^3$; $\dfrac{-d[A]}{[A]^3} = k\,dt$;

 $\dfrac{1}{2[A]^2} - \dfrac{1}{2[A_o]^2} = k\,t$; $\dfrac{1}{[A]^2} - \dfrac{1}{[A_o]^2} = 2\,k\,t$

**80.** rate = $k\,[A]^2[B][C]$
**81.** 0.90 g; 54 mg no more than three times a day

# Chapter 12

**2.** **(a)** $P_A \approx 0.75$ atm; $P_B \approx 0.50$ atm
 **(b)** 0.76 atm
 **(c)** never
**4.** **(a)** $3\,A(g) \longrightarrow 2\,B(g)$
 **(b)** not yet—$P_A$ continues to decrease

**6.** **(a)** $K = \dfrac{(P_{POCl_3})^{10}}{(P_{P_4O_{10}})(P_{PCl_5})^6}$

 **(b)** $K = \dfrac{(P_{H_2O})^2(P_{SO_2})^2}{(P_{H_2S})^2(P_{O_2})^3}$

 **(c)** $K = \dfrac{(P_{NO})^4(P_{H_2O})^6}{(P_{NH_3})^4(P_{O_2})^5}$

**8.** **(a)** $K = \dfrac{[Mn^{2+}]\,(P_{Cl_2})}{[H^+]^4[Cl^-]^2}$    **(b)** $K = \dfrac{P_{CO_2}}{P_{CO}}$

 **(c)** $K = \dfrac{P_{Ni(CO)_4}}{(P_{CO})^4}$

**10.** **(a)** $I_2(s) \rightleftharpoons I_2(g)$    $K = P_{I_2}$
 **(b)** $8\,H_2(g) + S_8(s) \rightleftharpoons 8\,H_2S(g)$

 $K = \dfrac{(P_{H_2S})^8}{(P_{H_2})^8}$

 **(c)** $NH_3(g) + H^+(aq) \rightleftharpoons NH_4^+(aq)$

 $K = \dfrac{[NH_4^+]}{[H^+](P_{NH_3})}$

**12.** **(a)** $CH_3OH(g) \rightleftharpoons CO(g) + 2H_2(g)$
 **(b)** $O_2(g) + C(s) \rightleftharpoons CO_2(g)$
 **(c)** $2SO_2(g) + O_2(g) \rightleftharpoons 2SO_3(g)$
 **(d)** $2Fe^{2+}(aq) + Cl_2(g) \longrightarrow 2Fe^{3+}(aq) + 2Cl^-(aq)$
 **(e)** $2H_2O(g) + 2Cl_2(g) \rightleftharpoons 4HCl(g) + O_2(g)$
**14.** **(a)** 21.0    **(b)** $9.29 \times 10^3$
**16.** $3.1 \times 10^{15}$
**18.** $9 \times 10^{-15}$
**20.** **(a)** $2NH_3(g) + CO_2(g) \rightleftharpoons NH_4CO_2NH_2(s)$
 **(b)** 141
**22.** $3.39 \times 10^3$
**24.** 0.138
**26.** **(a)** no; $Q \neq K$    **(b)** $\longrightarrow$
**28.** **(a)** $\longrightarrow$    **(b)** $\longleftarrow$
**30.** c
**32.** 0.11 atm
**34.** $P_{H_2} = 0.0427$ atm; $P_{H_2O} = 1.39$ atm
**36.** $P_{I_2} = P_{H_2} = 0.078$ atm; $P_{HI} = 0.60$ atm
**38.** $P_{SO_2} = P_{NO_2} = 0.20$ atm
**40.** **(a)** 0.12 atm    **(b)** 6.2%    **(c)** No
**42.** 0.37 atm
**44.** **(a)** increase; increase; decrease; decrease; no effect
 **(b)** none; (4)
**46.** **(a)** $\longleftarrow$    **(b)** $\longrightarrow$    **(c)** no effect
**48.** e
**50.** **(a)** 1.0    **(b)** $\longrightarrow$; 0.32 atm
**52.** 0.054
**54.** $1.2 \times 10^{-4}$
**56.** **(a)** $A_2(g) + 2B_2(g) \rightleftharpoons 2AB_2(g)$
 **(b)** 2
**58.** no
**60.** **(a)** =    **(b)** <    **(c)** can't tell
**62.** $K_{700°C} = 2.0$; $K_{600°C} = 0.06$
**64.** $K$ does not indicate the rate of a reaction.
**65.** $P_{N_2} = 0.33$ atm; $P_{H_2} = 0.99$ atm; $P_{NH_3} = 0.34$ atm

**66.** $K = \dfrac{(P_C)^c (P_D)^d}{(P_A)^a (P_B)^b} = \dfrac{([C]RT)^c ([D]RT)^d}{([A]RT)^a ([B]RT)^b}$

$= \dfrac{[C]^c [D]^d}{[A]^a [B]^b} (RT)^{(c+d)-(a+b)}$

$\Delta n_g = (c+d) - (a+b) \qquad K = K_c (RT)^{\Delta n_g}$

**67.** 0.0442

**68.** 0.52 atm

**69.** 25; 0.65 atm

**70.** 1.1

**71.** 0.22 atm; 0.4 g

# Chapter 13

**2.** **(a)** acids: HClO, $NH_4^+$; bases: $NH_3$, $ClO^-$; pairs $HClO/ClO^-$ and $NH_4^+/NH_3$

**(b)** acids: $H_3O^+$, $H_2CO_3$; bases: $HCO_3^-$, $H_2O$; pairs $H_3O^+/H_2O$ and $H_2CO_3/HCO_3^-$

**(c)** acids: $H_2O$, $(CH_3)_2NH_2^+$; bases: $OH^-$, $(CH_3)_2NH$; pairs $H_2O/OH^-$ and $(CH_3)_2NH_2^+/(CH_3)_2NH$

**4.** acid: b, c; base: a

**6.** **(a)** $OH^-$     **(b)** $HPO_4^{2-}$     **(c)** $NH_3$
    **(d)** $S^{2-}$     **(e)** $Fe(H_2O)_5OH^{2+}$

**8.** **(a)** 3.00 (A)     **(b)** 9.05 (B)
    **(c)** $-0.210$ (A)     **(d)** 4.49 (A)

**10.** **(a)** $4.8 \times 10^{-11}\ M$; $2.1 \times 10^{-4}\ M$
    **(b)** $2.6 \times 10^{-3}\ M$; $3.9 \times 10^{-12}\ M$
    **(c)** $1.0\ M$; $1.0 \times 10^{-14}\ M$
    **(d)** $41\ M$; $2.5 \times 10^{-16}\ M$

**12.** solution II; solution I

**14.** B is 100 times greater

**16.** $3.3 \times 10^2$

**18.** **(a)** $2.84 \times 10^{-2}\ M$; 1.546
    **(b)** $1.4 \times 10^{-4}\ M$; 3.85

**20.** 0.796

**22.** **(a)** $0.0155\ M$; $6.4 \times 10^{-13}\ M$; 12.19
    **(b)** $0.0376\ M$; $2.7 \times 10^{-13}\ M$; 12.57

**24.** 13.19

**26.** **(a)** $Zn(H_2O)_3OH^+(aq) + H_2O \rightleftharpoons$
    $Zn(H_2O)_2(OH)_2(aq) + H_3O^+(aq)$

**(b)** $Al(H_2O)_6^{3+}(aq) + H_2O \rightleftharpoons$
    $Al(H_2O)_5OH^{2+}(aq) + H_3O^+(aq)$

**(c)** $H_2CO_3(aq) + H_2O \rightleftharpoons$
    $H_3O^+(aq) + HCO_3^-(aq)$

**(d)** $HPO_4^{2-}(aq) + H_2O \rightleftharpoons$
    $PO_4^{3-}(aq) + H_3O^+(aq)$

**(e)** $HIO_2(aq) + H_2O \rightleftharpoons H_3O^+(aq) + IO_2^-(aq)$

**28.** **(a)** $HS^-(aq) \rightleftharpoons H^+(aq) + S^{2-}(aq)$

$K_a = \dfrac{[H^+][S^{2-}]}{[HS^-]}$

**(b)** $PH_4^+(aq) \rightleftharpoons H^+(aq) + PH_3(aq)$

$K_a = \dfrac{[H^+][PH_3]}{[PH_4^+]}$

**(c)** $H_2C_2O_4(aq) \rightleftharpoons HC_2O_4^-(aq) + H^+(aq)$

$K_a = \dfrac{[H^+][HC_2O_4^-]}{[H_2C_2O_4]}$

**30.** **(a)** $3 \times 10^{-3}$     **(b)** $5 \times 10^{-9}$     **(c)** $3 \times 10^{-11}$

**32.** **(a)** B>C>A>D     **(b)** B

**34.** $1.51 \times 10^{-5}$

**36.** $5.1 \times 10^{-6}$

**38.** **(a)** $0.0034\ M$     **(b)** $0.0017\ M$

**40.** **(a)** $1.1 \times 10^{-5}\ M$     **(b)** $9.1 \times 10^{-10}\ M$
    **(c)** 4.96     **(d)** $1.2 \times 10^{-3}\%$

**42.** 1.92

**44.** 3.85

**46.** **(a)** $NH_3(aq) + H_2O \rightleftharpoons NH_4^+(aq) + OH^-(aq)$

**(b)** $PO_4^{3-}(aq) + H_2O \rightleftharpoons HPO_4^{2-}(aq) + OH^-(aq)$

**(c)** $C_6H_5NH_2(aq) + H_2O \rightleftharpoons C_6H_5NH_3^+(aq) + OH^-(aq)$

**(d)** $CN^-(aq) + H_2O \rightleftharpoons HCN(aq) + OH^-(aq)$

**(e)** $F^-(aq) + H_2O \rightleftharpoons HF(aq) + OH^-(aq)$

**(f)** $(C_2H_5)_3N(aq) + H_2O \rightleftharpoons (C_2H_5)_3NH^+(aq) + OH^-(aq)$

**48.** $Na_2CO_3 > NaNO_2 > NH_4Cl > HCl$

**50.** b, c

**52.** **(a)** $5.3 \times 10^{-11}$     **(b)** $2.6 \times 10^{-10}$

**54.** 12.75

**56.** 44 g

**58.** **(a)** acidic     **(b)** basic     **(c)** acidic
    **(d)** basic     **(e)** basic

**60.** **(a)** $NH_4^+(aq) \rightleftharpoons NH_3(aq) + H^+(aq)$
    **(b)** $NH_4^+(aq) \rightleftharpoons NH_3(aq) + H^+(aq)$

$K_a = 5.6 \times 10^{-10}$

$ClO^-(aq) + H_2O \rightleftharpoons OH^-(aq) + HClO(aq)$

$K_b = 3.6 \times 10^{-7}$

**(c)** $Fe(H_2O)_6^{3+}(aq) \rightleftharpoons Fe(H_2O)_5OH^{2+}(aq) + H^+(aq)$

**(d)** $C_2H_3O_2^-(aq) + H_2O \rightleftharpoons HC_2H_3O_2(aq) + OH^-(aq)$

**(e)** $PO_4^{3-}(aq) + H_2O \rightleftharpoons HPO_4^{2-}(aq) + OH^-(aq)$

**62.** $NaF > Ba(NO_3)_2 > NH_4NO_3 > Al(NO_3)_3 > HNO_3$

**64.** **(a)** $Co(NO_3)_3$, $CoCl_3$, $Co_2(SO_4)_3$, $Co(ClO_4)_3$
    **(b)** $NaBr$, $KBr$, $LiBr$, $BaBr_2$
    **(c)** $SrF_2$, $Sr(CN)_2$, $Sr(NO_2)_2$, $SrSO_3$
    **(d)** $LiCl$, $LiBr$, $LiI$, $LiNO_3$

**66.** **(a)** Figure 1     **(b)** Figure 2     **(c)** Figure 2

**68.** 2.33

**70.** **(a)** neutral     **(b)** acidic
    **(c)** basic     **(d)** acidic

**72.** **(a)** No     **(b)** 6.80

**74.** 4.6 g

**75.** Test pH of $AgNO_3$ in water. If acidic, then AgOH is a weak base.

**76.** **(a)** $-5.58$ kJ     **(b)** $-6.83$ kJ

**77.** $\%\ \text{ion} = \dfrac{[H^+]}{[HA]_0} \times 100$; $[H^+]^2 \approx K_a \times [HA]_0$;

$[H^+] \approx K_a^{1/2} \times [HA]_o^{1/2}$; % ion $= \dfrac{K_a^{1/2}}{[HA]_o^{1/2}} \times 100$;

% ion is inversely proportional to $[HA]_o^{1/2}$

**78.** $-1.64°C$

# Chapter 14

**2. (a)** $C_6H_5NH_2(aq) + NH_4^+(aq) \rightleftharpoons$
$\qquad\qquad\qquad C_6H_5NH_3^+(aq) + NH_3(aq)$
**(b)** $H^+(aq) + OH^-(aq) \rightleftharpoons H_2O$
**(c)** $NH_3(aq) + H^+(aq) \rightleftharpoons NH_4^+(aq)$
**(d)** $H^+(aq) + OH^-(aq) \rightleftharpoons H_2O$
**4. (a)** $Fe(H_2O)_6^{3+}(aq) + OH^-(aq) \rightleftharpoons$
$\qquad\qquad\qquad Fe(H_2O)_5OH^{2+}(aq) + H_2O$
**(b)** $H_2SO_3(aq) + OH^-(aq) \rightleftharpoons HSO_3^-(aq) + H_2O$
**(c)** $H^+(aq) + OH^-(aq) \rightleftharpoons H_2O$
**6. (a)** $2.1 \times 10^{-5}$  **(b)** $1.0 \times 10^{14}$
**(c)** $1.8 \times 10^9$  **(d)** $1.0 \times 10^{14}$
**8. (a)** $6.7 \times 10^{11}$  **(b)** $1.7 \times 10^{12}$
**(c)** $1.0 \times 10^{14}$
**10. (a)** $2.5 \times 10^{-3} M$; 11.40
**(b)** $6.3 \times 10^{-4} M$; 10.80
**(c)** $4.2 \times 10^{-4} M$; 10.62
**(d)** $2.5 \times 10^{-4} M$; 10.40
**12.** 10.65
**14. (a)** $HCHO_2/CHO_2^-$  **(b)** $HCN/CN^-$
**(c)** $HC_2H_3O_2/C_2H_3O_2^-$
**16. (a)** 0.42  **(b)** 0.82 mol  **(c)** $1.1 \times 10^2$ g
**18. (a)** 12.28  **(b)** 12.28
**(c)** no; mole ratio stays the same
**20.** 1.6 g
**22.** 4.58
**24. (a)** 7.33  **(b)** 7.08  **(c)** 7.61
**26. (a)** 7.33  **(b)** 7.08  **(c)** 7.61
Dilution does not change the pH of the buffer but changes its buffering capacity per liter.
**28. (a)** 0.18  **(b)** 10.14  **(c)** 9.66
**30.** b, c
**32.** 9.59
**34.** $-0.038$
**36. (a)** 0.090  **(b)** 7.03  **(c)** 7.45
**38. (a)** methyl orange  **(b)** any
**(c)** phenolphthalein  **(d)** methyl orange
**40.** $8 \times 10^{-12}$; orange
**42. (a)** 82.24 mL  **(b)** 0.602
**(c)** 1.21  **(d)** 7.00
**44. (a)** $HCHO_2(aq) + OH^-(aq) \rightleftharpoons CHO_2^-(aq) + H_2O$
**(b)** $HCHO_2$; $HCHO_2$, $CHO_2^-$, $K^+$; $CHO_2^-$, $K^+$
**(c)** 21.72 mL  **(d)** 2.03; 3.72; 8.57
**46. (a)** 11.08  **(b)** 9.25  **(c)** 5.31
**48.** $H_2C_2O_4(aq) \rightleftharpoons 2\,H^+(aq) + C_2O_4^{2-}(aq)$
$\qquad\qquad\qquad\qquad\qquad K = 3.1 \times 10^{-6}$

**50.** 1.72; $\approx 1.9 \times 10^{-2}$; $\approx 1.8 \times 10^{-5}$; $\approx 3.8 \times 10^{-10}$
**52.** 10

**54.**

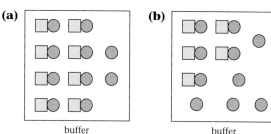

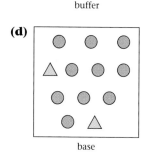

half-neutralization occurs at b

**56.** $H^+(aq) + HCO_3^-(aq) \rightleftharpoons H_2CO_3(aq) \rightleftharpoons$
$\qquad\qquad\qquad\qquad\qquad CO_2(g) + H_2O$
An increase in $H^+$ ions shifts the equilibrium to the right forming $CO_2$.
**58. (a)** False; $[CHO_2^-]$ is 0.1 $M$ only in 0.1 $M$ $NaCHO_2$
**(b)** True
**(c)** False; A buffer can be made up by combining a weak acid and its conjugate base in approximately equal concentrations.
**(d)** False; $K_b$ for $HCO_3^-$ is $2.3 \times 10^{-8}$; $K_b$ for $CO_3^{2-}$ is $2.1 \times 10^{-4}$
**60.** Titrate with NaOH with phenolphthalein as indicator. The test tube with $C_6H_5NH_3^+$ will require the most NaOH for neutralization. The test tube with HCl the least.
**61.** 43 mL
**62.** $\dfrac{[NH_4^+]}{[NH_3]} = 5.7 \times 10^2$; too little $NH_3$
**63.** $1.1 \times 10^2$ g/mol
**64. (a)** 2.37  **(b)** 4.74  **(c)** 7.44
**(d)** 9.22  **(e)** 11.00  **(f)** 13.52
**65.** $\approx 30$ mL; phenolphthalein
**66.** $-\log[H^+] = -\log K_a - \log\left(\dfrac{[HB]}{[B^-]}\right)$;
$pH = pK_a - \log\dfrac{[HB]}{[B^-]} = pK_a + \log\dfrac{[B^-]}{[HB]}$
**67. (a)** 0.8239  **(b)** 0.80

# Chapter 15

**2.** **(a)** $H_2O$:0; Cl: $-1$
 **(b)** $+3$
 **(c)** $[Cr(H_2O)_4Cl_2]_2SO_4$
**4.** **(a)** $[Ru(NH_3)_4]^{3+}$
 **(b)** $[Ru(C_2O_4)_2(H_2O)_2]^-$
 **(c)** $[Ru(en)(NO_2)_2I_2]^-$
**6.** **(a)** 6   **(b)** 4   **(c)** 6   **(d)** 2
**8.** **(a)** $[Pt(NH_3)_6]^{4+}$   **(b)** $[Zn(C_2O_4)_2]^{2-}$
 **(c)** $[Ag(H_2O)_2]^+$   **(d)** $[Pd(Br)_4]^{2-}$
**10.** 27.66%
**12.** one
**14.** **(a)**

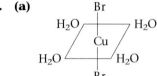

 **(b)**

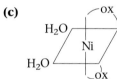

 **(c)**                 **(d)**

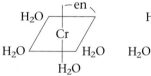

 **(e)** $H_3N$—Au—CN

**16.**

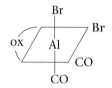

 where R =

**18.** **(a)**

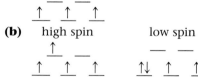

 **(b)**

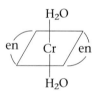

**(c)**

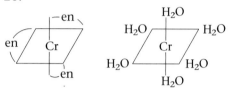

**20.**

**22.** **(a)** $1s^2 2s^2 2p^6 3s^2 3p^6 3d^3$
 **(b)** $1s^2 2s^2 2p^6 3s^2 3p^6 4s^2 3d^{10} 4p^6 4d^7$
 **(c)** $[Xe]4f^{14}5d^6$
 **(d)** $1s^2 2s^2 2p^6 3s^2 3p^6 4s^2 3d^{10} 4p^6 4d^3$
 **(e)** $1s^2 2s^2 2p^6 3s^2 3p^6 4s^2 3d^{10} 4p^6 4d^3$
**24.** **(a)** [Ar] (↑ ) (↑ ) (↑ ) (  ) (  ); 3
 **(b)** [Kr] (↑↓) (↑↓) (↑ ) (↑ ) (↑ ); 3
 **(c)** [Xe] $4f^{14}$ (↑↓) (↑ ) (↑ ) (↑ ) (↑ ); 4
 **(d)** [Kr] (↑ ) (↑ ) (↑ ) (  ) (  ); 3
 **(e)** [Kr] (↑ ) (↑ ) (↑ ) (  ) (  ); 3
**26.** **(a)** (single complex)

 ─ ─ ─
 ↑  ↑  ↑

 **(b)**     high spin          low spin

 ↑  ─
 ↑  ↑  ↑      ↑↓ ↑  ↑

**28.** $Cu^{2+}$ has an outer electron configuration of $3d^9$. For outer configurations of $3d^8$ to $3d^{10}$ only one distribution is possible.
**30.** $[Fe(NO_2)_6]^{3-}$ is low spin with one unpaired electron while $[FeF_6]^{3-}$ is high spin with 5 unpaired electrons.
**32.** **(a)** 2   **(b)** 0   **(c)** 0   **(d)** 2   **(e)** 2

**34.** $4.60 \times 10^2$ nm
**36.** blue
**38.** $3 \times 10^{-6} M$
**40.** **(a)** $6 \times 10^{-3} M$  **(b)** $4 \times 10^{-6} M$
**42.** **(a)** $C_2O_4^{2-}$ has 2 unshared electron pairs that can bond.
 **(b)** $NH_3$ has an unshared electron pair, $NH_4^+$ does not.
 **(c)** $Ni^{2+}$ forms complex ions with $NH_3$.
**44.** **(a)** True  **(b)** True
 **(c)** False; eight or more d electrons
**46.** **(a)** $CoN_6H_{18}Cl_3$
 **(b)** $[Co(NH_3)_6]Cl_3(s) \rightleftharpoons$
$Co(NH_3)_6^{3+}(aq) + 3Cl^-(aq)$
**48.** Lungs saturated with oxygen so the shift is to oxyhemoglobin formation. Tissues low in oxygen, so in venous blood, the shift is to hemoglobin formation.
**49.** 0.92 g
**50.** $[Pt(NH_3)_4][PtCl_4]$ or $[Pt(NH_3)_3Cl][Pt(NH_3)Cl_3]$
**51.** **(a)** $CuC_4H_{22}N_6SO_4$
 **(b)**

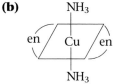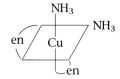

**52.** red-violet

# Chapter 16

**2.** **(a)** $K_2SiF_6(s) \rightleftharpoons 2K^+(aq) + SiF_6^{2-}(aq)$
$K_{sp} = [K^+]^2[SiF_6^{2-}]$
 **(b)** $PbI_2(s) \rightleftharpoons 2I^-(aq) + Pb^{2+}(aq)$
$K_{sp} = [I^-]^2[Pb^{2+}]$
 **(c)** $Ni_2S_3(s) \rightleftharpoons 2Ni^{3+}(aq) + 3S^{2-}(aq)$
$K_{sp} = [Ni^{3+}]^2[S^{2-}]^3$
$Zn_2P_2O_7(s) \rightleftharpoons 2Zn^{2+}(aq) + P_2O_7^{4-}(aq)$
$K_{sp} = [Zn^{2+}]^2[P_2O_7^{4-}]$
**4.** **(a)** $Sr(ClO_3)_2(s) \rightleftharpoons Sr^{2+}(aq) + 2ClO_3^-(aq)$
 **(b)** $Co_3(AsO_4)_2(s) \rightleftharpoons 3Co^{2+}(aq) + 2AsO_4^{3-}(aq)$
 **(c)** $Eu(OH)_3(s) \rightleftharpoons Eu^{3+}(aq) + 3OH^-(aq)$
 **(d)** $Hg_2Cl_2(s) \rightleftharpoons Hg_2^{2+}(aq) + 2Cl^-(aq)$
**6.** **(a)** $0.1 M$  **(b)** $4 \times 10^{-3} M$
 **(c)** $3.0 \times 10^{-2} M$  **(d)** $1.2 \times 10^{-2} M$
**8.** **(a)** $0.38 M$  **(b)** $4 \times 10^{-8} M$
 **(c)** $8 \times 10^{-10} M$
**10.** **(a)** $1.6 \times 10^{-3} M$  **(b)** $4.5 \times 10^{-3} M$
**12.** yes
**14.** **(a)** no
 **(b)** $[Sr^{2+}] = 1.33 \times 10^{-3} M$; $[Cl^-] = 2.66 \times 10^{-3} M$; $[K^+] = 0.0400 M$; $[CrO_4^{2-}] = 0.0200 M$
**16.** $2.0 \times 10^{-7}$

**18.** **(a)** 0.02 g/L  **(b)** 0.002 g/L  **(c)** $2 \times 10^{-5}$ g/L
**20.** **(a)** no  **(b)** yes
**22.** **(a)** $SrCO_3(s) + 2H^+(aq) \longrightarrow Sr^{2+}(aq) + CO_2(g) + H_2O$
 **(b)** no reaction
 **(c)** $Co(OH)_3 + 3H^+(aq) \longrightarrow Co^{3+}(aq) + 3H_2O$
 **(d)** $Sb(OH)_4^-(aq) + 4H^+(aq) \longrightarrow Sb^{3+}(aq) + 4H_2O$
**24.** **(a)** $Cu(OH)_2(s) + 4NH_3(aq) \longrightarrow Cu(NH_3)_4^{2+}(aq) + 2OH^-(aq)$
 **(b)** $Cd^{2+}(aq) + 4NH_3(aq) \longrightarrow Cd(NH_3)_4^{2+}(aq)$
 **(c)** $Pb^{2+}(aq) + 2NH_3(aq) + 2H_2O \longrightarrow Pb(OH)_2(s) + 2NH_4^+(aq)$
**26.** **(a)** $Ni^{2+}(aq) + 2OH^-(aq) \longrightarrow Ni(OH)_2(s)$
 **(b)** $Sn^{4+}(aq) + 6OH^-(aq) \longrightarrow Sn(OH)_6^{2-}(aq)$
 **(c)** $Al(OH)_3(s) + OH^-(aq) \longrightarrow Al(OH)_4^-(aq)$
**28.** $8 \times 10^{-6}$
**30.** $4 \times 10^{-7}$
**32.** 0.05 mol/L
**34.** $1 \times 10^{-9} M$
**36.** **(a)** $3.1 \times 10^{-3}$  **(b)** $0.13 M$
**38.** **(a)** $Na^+$  **(b)** $NH_3$
 **(c)** dissolves  **(d)** CuS precipitate
**40.** Group III
**42.** HgS

**44.** **(a)** 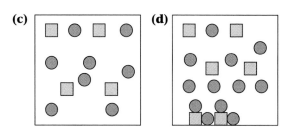 **(b)**

 **(c)** **(d)**

**46.** **(a)** $\longleftarrow$  **(b)** $\longrightarrow$  **(c)** $\longrightarrow$  **(d)** $\longleftarrow$
**48.** $3 \times 10^7$; Likely to go to completion.
**49.** in water: $1 \times 10^{-4}$ mol/L
 in $NH_4Cl$: 0.06 mol/L
**50.** $1.9 \times 10^{-4}$ mol/L
**51.** $1 \times 10^{-8} M$
**52.** **(a)** $1 \times 10^{-5} M$  **(b)** $Al(OH)_3$, $Fe(OH)_3$
 **(c)** virtually all  **(d)** 1.6 g
**53.** **(a)** $8 \times 10^5$  **(b)** $4 \times 10^3$
**54.** $1.0 \times 10^{21}$

# Chapter 17

2. a, b
4. a, c, d
6. **(a)** +   **(b)** +   **(c)** −   **(d)** +
8. **(a)** +   **(b)** −   **(c)** +   **(d)** +
10. **(a)** −   **(b)** +   **(c)** −
12. **(a)** 90.4 J/K   **(b)** −83.7 J/K   **(c)** −147.0 J/K
14. **(a)** 223.6 J/K   **(b)** −191.8 J/K
    **(c)** 159.6 J/K
16. **(a)** −43.4 J/K   **(b)** 778.0 J/K   **(c)** −64.8 J/K
18. **(a)** $-6.20 \times 10^2$ kJ   **(b)** $4.40 \times 10^2$ kJ
    **(c)** −526 kJ
20. **(a)** −2408.3 kJ; spontaneous
    **(b)** 67.0 kJ; nonspontaneous
    **(c)** −823.1 kJ; spontaneous
22. **(a)** 833.2 kJ   **(b)** −423.8 kJ   **(c)** −17.2 kJ
24. **(a)** −184 kJ/mol   **(b)** −175 kJ/mol
    **(c)** −1138 kJ/mol
26. yes
28. **(a)** −0.918 kJ/K   **(b)** −0.159 kJ/mol-K
30. **(a)** −17.0 kJ   **(b)** −137.2 kJ/mol
    **(c)** 0.2114 kJ/mol-K
32. **(a)** $T$ dependent, spontaneous at low $T$
    **(b)** $T$ independent, spontaneous at any $T$
    **(c)** $T$ dependent, spontaneous at low $T$
34. **(a)** spontaneous at $T$ below 95.2 K
    **(b)** no $T$ effect
    **(c)** spontaneous at $T$ below 4900 K
36. Spontaneous at all $T$
38. **(a)**

| $\Delta G°$ | −119.1 | −90.6 | −62.2 | −33.8 | −5.3 | 23.2 | 51.6 |
|---|---|---|---|---|---|---|---|
| $T$ | 200 | 300 | 400 | 500 | 600 | 700 | 800 |

    **(b)** $6.2 \times 10^2$ K
40. b
42. $4 \times 10^2$ K
44. 619 K
46. **(a)** $\Delta G° = 29.5$ kJ; nonspontaneous
    **(b)** $\Delta G° = -9.9$ kJ; spontaneous
48. **(a)** −361.7 kJ   **(b)** −258 kJ
50. **(a)** −70.6 kJ   **(b)** $P_{NO} < 6.0 \times 10^{-5}$ atm
52. **(a)** 201.3 kJ; reaction is nonspontaneous
    **(b)** $ZnS(s) + O_2(g) \longrightarrow ZnS(s) + SO_2(g)$
    $\Delta G° = -98.9$ kJ
    The reaction is spontaneous.
54. 12 moles
56. **(a)** 104.7 kJ   **(b)** 103.8 kJ/mol
58. **(a)** $1 \times 10^{-16}$   **(b)** 69.9 kJ
60. $7 \times 10^{-4}$ at 25°C; $6 \times 10^{-4}$ at 37°C
62. **(a)** independent of $T$ and $P$
    **(b)** independent of $T$   **(c)** neither
    **(d)** neither

64. **(a)** zero   **(b)** +   **(c)** melting point
66. **(a)** $P_{CO} = 0.983$ atm   $P_{CO_2} = 0.0169$ atm
    **(b)** 57.2   **(c)** −40.4 kJ
68. $\Delta G° = -2016$ kJ
70. **(a)** no   **(b)** yes, just barely
71. 61°C
72. $P_{H_2} = P_{I_2} = 0.06$ atm   $P_{HI} = 0.48$ atm
73. **(a)** 6.00 kJ   **(b)** 0
    **(c)** 0.0220 kJ/K   **(d)** 0.43 kJ   **(e)** −0.45 kJ
74. **(a)** 4.2 kJ   **(b)** $\approx 5.2 \times 10^2$ g
75. 1430 K
76. **(1)** $\Delta G° = 146.0 - 0.1104\,T$; becomes spontaneous above 1050°C
    **(2)** $\Delta G° = 168.6 - 0.0758\,T$; becomes spontaneous above 1950°C

# Chapter 18

2. **(a)** $Sn(s) + S(s) + 2H^+(aq) \longrightarrow Sn^{2+}(aq) + H_2S(g)$
   **(b)** $2H_2(g) + O_2(g) \longrightarrow 2H_2O$
   **(c)** $Co(s) + H_2O_2(aq) + 2H^+(aq) \longrightarrow$
   $\qquad\qquad\qquad\qquad\qquad Co^{2+}(aq) + 2H_2O$
4. **(a)** Mg anode, Al cathode; $e^-$ move from Mg to Al. Anions move to Mg, cations to Al.
   **(b)** Pt both anode and cathode. Anode has $SO_2$ gas and $SO_4^{2-}$; cathode has nitrate ions and NO gas. $e^-$ move from anode to cathode.
   **(c)** Pt both anode and cathode. $e^-$ move from Pt immersed in sulfide ions to Pt immersed in nitrate ions and NO gas.
6. anode:        $Au(s) + 4\,Cl^-(aq) \longrightarrow$
                 $\qquad\qquad\qquad AuCl_4^-(aq) + 3\,e^-$
   cathode:      $Br_2(l) + 2\,e^- \longrightarrow 2\,Br^-(aq)$
   overall:      $2\,Au(s) + 8\,Cl^-(aq) + 3\,Br_2(l) \longrightarrow$
                 $\qquad\qquad 2\,AuCl_4^-(aq) + 6\,Br^-(aq)$
   cell notation:   $Au|AuCl_4^-\|Br_2|Br^-|Pt$
   Electrons move from Au anode to Pt cathode.
   Anions move to Au, cations to Pt.
8. **(a)** Ag   **(b)** $Br^-$   **(c)** $NO_2^-$
10. $H^+ < SO_4^{2-} < ClO_3^-$ (basic) $< O_2$ (acidic) $<$
    $\qquad\qquad\qquad\qquad\qquad\qquad ClO_3^-$ (acidic)
12. $Mn^{2+}$: oxidizing and reducing agent
    Mn: reducing agent   $MnO_2$: oxidizing agent
    $MnO_4^-$: oxidizing agent
    $O_2$(basic): oxidizing agent
    oxidizing agents: $Mn^{2+} < O_2$(basic) $< MnO_2 <$
    $MnO_4^-$   reducing agents: $Mn^{2+} < Mn$
14. **(a)** $O_2$ (basic)   **(b)** Na
    **(c)** $Fe^{3+}$ through $Cr_2O_7^{2-}$
16. **(a)** 0.662 V   **(b)** 0.748 V   **(c)** 0.166 V
18. **(a)** 0.018 V   **(b)** 1.370 V   **(c)** 0.799 V
20. **(a)** 1.300 V   **(b)** 1.085 V   **(c)** 1.229 V
22. **(a)** −0.500 V   **(b)** +1.577 V
    **(c)** 1.081 V; same

**24.** a, c

**26.** **(a)** $-0.491$ V; no     **(b)** $-1.650$ V; no
**(c)** 2.562 V; yes

**28.** **(a)** no reaction     **(b)** no reaction
**(c)** no reaction

**30.** a, b, c, d

**32.** **(a)** no reaction     **(b)** no reaction
**(c)** $I_2(s) + H_2S(aq) \longrightarrow 2\,I^-(aq) + S(s) + 2\,H^+(aq)$

**34.**

|  | $\Delta G°$ | $E°$ | $K$ |
|---|---|---|---|
| **(a)** | 62 kJ | $-0.16$ V | $1 \times 10^{-11}$ |
| **(b)** | $-53$ kJ | 0.14 V | $2 \times 10^9$ |
| **(c)** | 20.4 kJ | $-0.053$ V | $2.6 \times 10^{-4}$ |

**36.** $\Delta G° = -83.9$ kJ     $E° = 0.435$ V
$K = 5.2 \times 10^{14}$

**38.** **(a)** $-383$ kJ     **(b)** $-144$ kJ     **(c)** $-32.0$ kJ

**40.** **(a)** 67     **(b)** $4.0 \times 10^{92}$     **(c)** $1 \times 10^{81}$

**42.** **(a)** 0.890 V

**(b)** $E = 0.890 - \dfrac{0.0257}{2} \ln \dfrac{[I^-]^2}{[SO_4^{2-}]}$

**(c)** 0.920 V

**44.** **(a)** 0.688 V

**(b)** $E = 0.688 - \dfrac{0.0257}{6} \ln \dfrac{(P_{NO})^2[OH^-]^2}{[NO_3^-]^2(P_{H_2})^3}$

**(c)** 0.709 V

**46.** 0.157 V

**48.** 1.2

**50.** $P_{H_2} = 5.2 \times 10^{-5}$ atm

**52.** yes

**54.** **(a)** 0.127 V     **(b)** $1.6 \times 10^{-3}\,M$
**(c)** $1.6 \times 10^{-5}$

**56.** **(a)** $2.50 \times 10^{23}\,e^-$     **(b)** $4.00 \times 10^4$ C
**(c)** 14.7 g $Cl_2$; 0.418 g $H_2$

**58.** **(a)** 9.97 g Au     **(b)** 0.0795 cm

**60.** **(a)** $3.3 \times 10^6$ J     **(b)** 5.5 cents

**62.** c

**64.** d

**66.** $9.85 \times 10^4$ C/mol $e^-$

**68.** $3 \times 10^{14}$

**70.** 108 amu

**71.** 2.007 V

**72.** **(a)** 0.621 V     **(b)** increases; decreases
**(c)** $1 \times 10^{21}$
**(d)** $[Zn^{2+}] = 2.0\,M$; $[Sn^{2+}] = 2 \times 10^{-21}\,M$

**73.** **(a)** $1.12 \times 10^5$ J; $0.38 \times 10^5$ J; $1.50 \times 10^5$ J
**(b)** $-0.389$ V

**74.** 0.414 V

# Chapter 19

**2.** $\beta$-emission: $^{282}_{116}Y$     positron emission: $^{282}_{114}Z$
The products have the same mass number but different atomic number.

**4.** $^{32}_{15}P \longrightarrow {}^{\phantom{-}0}_{-1}e + {}^{32}_{16}S$

**6.** **(a)** $^{37}_{18}Ar + {}^{\phantom{-}0}_{-1}e \longrightarrow {}^{37}_{17}Cl$

**(b)** $^{12}_{6}C + {}^{12}_{6}C \longrightarrow {}^{23}_{11}Na + {}^{1}_{1}H$

**(c)** $^{239}_{94}Pu + {}^{1}_{0}n \longrightarrow {}^{130}_{50}Sn + 4\,{}^{1}_{0}n + {}^{106}_{44}Ru$

**8.** **(a)** $^{27}_{13}Al + {}^{2}_{1}H \longrightarrow {}^{4}_{2}He + {}^{25}_{12}Mg$

**(b)** $^{250}_{98}Cf + {}^{11}_{5}B \longrightarrow 4\,{}^{1}_{0}n + {}^{257}_{103}Lr$

**(c)** $^{98}_{42}Mo + {}^{1}_{0}n \longrightarrow {}^{99}_{42}Mo$

**(d)** $^{65}_{29}Cu + {}^{12}_{6}C \longrightarrow 3\,{}^{1}_{0}n + {}^{74}_{35}Br$

**10.** **(a)** $^{49}_{25}Mn \longrightarrow {}^{\phantom{-}0}_{-1}e + {}^{49}_{26}Fe$

**(b)** $^{76}_{36}Kr + {}^{\phantom{-}0}_{-1}e \longrightarrow {}^{76}_{35}Br$

**(c)** $^{9}_{4}Be + {}^{4}_{2}He \longrightarrow {}^{1}_{0}n + {}^{12}_{6}C$

**(d)** $^{27}_{13}Al + {}^{1}_{0}n \longrightarrow {}^{4}_{2}He + {}^{24}_{11}Na$

**12.** $3.20 \times 10^{12}$ atoms

**14.** $1.6 \times 10^{-5}$ Ci

**16.** $4.9 \times 10^2$ Ci

**18.** $1.3 \times 10^{-11}$ g

**20.** 0.086 day$^{-1}$; 8.0 days

**22.** $1.1 \times 10^{14}$ counts

**24.** No; painting was done $\approx 1760$

**26.** 2.9

**28.** $3.8 \times 10^9$ years

**30.** **(a)** $2.8 \times 10^9$ years
**(b)** $3.5 \times 10^9$ years

**32.** $\Delta m = -3.0 \times 10^{-7}$ g;     $\Delta E = -2.7 \times 10^4$ kJ

**34.** Mg-26

**36.** $-5.92 \times 10^8$ kJ/g

**38.** $4.3 \times 10^{12}$ g/s

**40.** **(a)** $1.33 \times 10^{-8}$ g
**(b)** $3.35 \times 10^{13}$ atoms

**42.** $\Delta m = 0.00608 > 0$ nonspontaneous

**44.** **(a)** $\Delta E = -2.10 \times 10^6$ kJ
**(b)** 761 kg     **(c)** 27.2 mg

**46.** **(a)** $\alpha$ and $\beta$ rays have opposite charges.
**(b)** See text-p. 535
**(c)** See text-p. 545

**48.** $7 \times 10^4$ disintegrations/s

**50.** $8.36 \times 10^{-11}$ g

**52.** $X_0$ would be higher; $t > 1000$ years

**54.** $1.0 \times 10^{-16}$

**56.** 2.59 Ci

**58.** 15 atm

**60.** 2 oxalate ions/chromium ion

**62.** 11 kg

**63.** $5.8 \times 10^{-19}$ mol/L

**64.** **(a)** $1.4 \times 10^{-9}$ g     **(b)** $3.0 \times 10^{-3}$ kJ
**(c)** 40 rems

**65.** **(a)** $1 \times 10^{-13}$ J     **(b)** $6 \times 10^6$ m/s

**66.** **(a)** $-5.72 \times 10^8$ kJ/g     **(b)** $1.3 \times 10^{28}$ kJ
**(c)** $1.8 \times 10^{-11}$

# Chapter 20

2. $2Al_2O_3(l) \longrightarrow 4Al(l) + 3 O_2(g)$; 2.91 g
4. $Cu_2S(s) + O_2(g) \longrightarrow 2Cu(s) + SO_2(g)$
6. $-211.9$ kJ
8. (a) $Fe_2O_3(s) + 3CO(g) \longrightarrow 2Fe(l) + 3CO_2(g)$
   (b) $C(s) + O_2(g) \longrightarrow CO_2(g)$
10. $2.5 \times 10^3$ kWh
12. $1.7 \times 10^6$ L
14. (a) potassium nitride, $K_3N$
   (b) potassium iodide, $KI$
   (c) potassium hydroxide, $KOH$
   (d) potassium hydride, $KH$
   (e) potassium sulfide, $K_2S$
16. (a) $Na_2O_2(s) + 2H_2O \longrightarrow$
   $$2Na^+(aq) + 2OH^-(aq) + H_2O_2(aq)$$
   sodium and hydroxide ions, hydrogen peroxide
   (b) $2Ca(s) + O_2(g) \longrightarrow 2CaO(s)$; calcium oxide
   (c) $Rb(s) + O_2(g) \longrightarrow RbO_2(s)$;
   rubidium superoxide
   (d) $SrH_2(s) + 2H_2O \longrightarrow$
   $$Sr^{2+}(aq) + 2OH^-(aq) + 2H_2(g)$$
   strontium and hydroxide ions, hydrogen gas
18. 0.126 g
20. (a) $Co(s) + 2H^+(aq) \longrightarrow Co^{2+}(aq) + H_2(g)$
   (b) $3Cu(s) + 2NO_3^-(aq) + 8H^+(aq) \longrightarrow$
   $$3Cu^{2+}(aq) + 2NO(g) + 4H_2O$$
   (c) $Cr_2O_7^{2-}(aq) + 6e^- + 14H^+(aq) \longrightarrow$
   $$2Cr^{3+}(aq) + 7H_2O$$
22. $3Cd(s) + 12Cl^-(aq) + 2NO_3^-(aq) + 8H^+(aq) \longrightarrow$
   $$3CdCl_4^{2-}(aq) + 2NO(g) + 4H_2O$$
24. (a) $Fe(s) + 3NO_3^-(aq) + 6H^+(aq) \longrightarrow$
   $$Fe^{3+}(aq) + 3NO_2(g) + 3H_2O$$
   (b) $4Cr(OH)_3(s) + 3 O_2(g) + 8OH^-(aq) \longrightarrow$
   $$4CrO_4^{2-}(aq) + 10H_2O$$
26. (a) Cd ($E° = 1.366$ V)    (b) Cr ($E° = 1.708$ V)
   (c) Co ($E° = 1.246$ V)    (d) Ag ($E° = 0.165$ V)
28. (a) 0.724 V    (b) 0.942 V
30. (a) $9 \times 10^9$    (b) $2 \times 10^{-4}$ $M$
32. (a) Wood's metal
   (b) Wood's metal, solder, pewter
   (c) Duriron, monel, spiegeleisen
34. 208 g
36. 6.7
38. 53.8% Zn, 46.2% Cu
40. $2Ag_2S(s) + 8CN^-(aq) + 3 O_2(g) + 2H_2O \longrightarrow$
   $$4Ag(CN)_2^-(aq) + 2SO_2(g) + 4OH^-(aq)$$
   $Ag(CN)_2^-(aq) + Zn(s) \longrightarrow$
   $$Zn(CN)_4^{2-}(aq) + 2Ag(s)$$
42. (a) $Cr^{2+}$    (b) $Au^+$    (c) $Co^{2+}$    (d) $Mn^{2+}$
43. 2% $BaO_2$
44. (a) $Fe(OH)_3(s) + 3H_2C_2O_4(aq) \longrightarrow$
   $$Fe(C_2O_4)_3^{3-}(aq) + 3H_2O + 3H^+(aq)$$
   (b) 0.28 L
45. 2.80%

46. $2.83 \times 10^3$ K
47. $Cr_2O_7^{2-}(aq) + 2OH^-(aq) \longrightarrow$
   $$2CrO_4^{2-}(aq) + H_2O$$
   $2Ag^+(aq) + CrO_4^{2-}(aq) \longrightarrow Ag_2CrO_4(s)$
   $Ag_2CrO_4(s) + 4 NH_3(aq) \longrightarrow$
   $$2Ag(NH_3)_2^+(aq) + CrO_4^{2-}(aq)$$
   $2Ag(NH_3)_2^+(aq) + 4H^+(aq) + CrO_4^{2-}(aq) \longrightarrow$
   $$Ag_2CrO_4(s) + 4NH_4^+(aq)$$

# Chapter 21

2. (a) bromic acid          (b) potassium hypoiodite
   (c) sodium chlorite      (d) sodium perbromate
4. (a) $KBrO_2$    (b) $CaBr_2$
   (c) $NaIO_4$    (d) $Mg(ClO)_2$
6. (a) $NO_3^-$    (b) $SO_4^{2-}$    (c) $ClO_4^-$
8. (a) $H_2SO_3$    (b) $HClO$    (c) $H_3PO_3$
10. (a) $NaN_3$    (b) $H_2SO_3$
   (c) $N_2H_4$    (d) $NaH_2PO_4$
12. (a) $H_2S$    (b) $N_2H_4$    (c) $PH_3$
14. (a) $NH_3$, $N_2H_4$    (b) $HNO_3$
   (c) $HNO_2$    (d) $HNO_3$
16. (a) $2I^-(aq) + SO_4^{2-}(aq) + 4H^+(aq) \longrightarrow$
   $$I_2(s) + SO_2(g) + 2H_2O$$
   (b) $2I^-(aq) + Cl_2(g) \longrightarrow I_2(s) + 2Cl^-(aq)$
18. (a) $3HClO(aq) \longrightarrow Cl_2(g) + HClO_2(g) + H_2O$
   (b) $2ClO_3^-(aq) \longrightarrow ClO_4^-(aq) + ClO_2^-(aq)$
20. (a) $Cl_2(g) + 2Br^-(aq) \longrightarrow 2Cl^-(aq) + Br_2(l)$
   (b) NR    (c) NR    (d) NR
22. (a) $Pb(N_3)_2(s) \longrightarrow 3N_2(g) + Pb(s)$
   (b) $2 O_3(g) \longrightarrow 3 O_2(g)$
   (c) $2H_2S(g) + O_2(g) \longrightarrow 2S(s) + 2H_2O$
24. (a) $Cd^{2+}(aq) + H_2S(aq) \longrightarrow CdS(s) + 2H^+(aq)$
   (b) $H_2S(aq) + OH^-(aq) \longrightarrow H_2O + HS^-(aq)$
   (c) $2H_2S(aq) + O_2(g) \longrightarrow 2H_2O + 2S(s)$
26. (a) $2H^+(aq) + CaCO_3(s) \longrightarrow$
   $$CO_2(g) + H_2O + Ca^{2+}(aq)$$
   (b) $H^+(aq) + OH^-(aq) \longrightarrow H_2O$
   (c) $Cu(s) + 4H^+(aq) + SO_4^{2-}(aq) \longrightarrow$
   $$Cu^{2+}(aq) + 2H_2O + SO_2(g)$$
28. P
30. (a) Pass $O_2$ through $10^4$ V of electric discharge.
   (b) Freeze liquid sulfur at 119°C and cool quickly to room temperature.
   (c) Heat white phosphorus in the absence of air to about 300°C.

32. (a) $:\ddot{Cl}-\ddot{O}-\ddot{Cl}:$    (b) $:\ddot{O}-N\equiv N:$

   (c)
   (d) $:N\equiv N:$

**34.** a and b

**36.** **(a)** $\left(\begin{array}{c} :\ddot{O}-N=\ddot{O}: \\ \phantom{xxx}:\overset{..}{\underset{..}{O}}: \end{array}\right)^{-}$ 　**(b)** $\left(\begin{array}{c} :\overset{..}{\underset{..}{O}}: \\ H-\overset{..}{\underset{..}{O}}-S-\overset{..}{\underset{..}{O}}: \\ :\overset{..}{\underset{..}{O}}: \end{array}\right)^{-}$

**(c)** $\left(\begin{array}{c} :\overset{..}{\underset{..}{O}}: \\ H-\overset{..}{\underset{..}{O}}-P-\overset{..}{\underset{..}{O}}-H \\ :\overset{..}{\underset{..}{O}}: \end{array}\right)^{-}$

**38.** **(a)** $:\overset{..}{\underset{..}{O}}-N-\overset{..}{\underset{..}{O}}-N-\overset{..}{\underset{..}{O}}:$ , $\overset{\|}{\underset{:\overset{..}{\underset{..}{O}}:}{}} \quad \overset{\|}{\underset{:\overset{..}{\underset{..}{O}}:}{}}$ 　**(b)** $H-\overset{..}{\underset{..}{O}}-N=\overset{..}{\underset{..}{O}}:$ , $:\overset{..}{\underset{..}{O}}:$

**(c)** $\left(\begin{array}{c} :\overset{..}{\underset{..}{O}}: \\ :\overset{..}{\underset{..}{O}}-S-\overset{..}{\underset{..}{O}}: \\ :\overset{..}{\underset{..}{O}}: \end{array}\right)^{2-}$

**40.** 14.3 L
**42.** 0.204 $M$
**44.** $7.7 \times 10^4$ L; $1.5 \times 10^5$ g
**46.** 12 L; 3.53
**48.** 4.28, 0.10 $M$
**50.** $1.9 \times 10^{-3}$
**52.** 0.012 g/100 mL
**54.** yes; 0 K
**56.** **(a)** yes　　**(b)** $2 \times 10^{16}$
**58.** 510 K
**60.** $2.26 \times 10^5$ g
**62.** 27.4 kg
**64.** b, c, d
**66.** 4.6
**68.** **(a)** dispersion
**(b)** dispersion, dipole
**(c)** dispersion, H— bonds
**(d)** dispersion, H— bonds
**(e)** no intermolecular forces; not a molecule
**70.** **(a)** +3　　**(b)** +4　　**(c)** +5　　**(d)** −3
**72.** **(a)** HClO　　**(b)** S, $KClO_3$
**(c)** $NH_3$, NaClO　　**(d)** HF
**74.** **(a)** see text p. 598
**(b)** has an unpaired electron
**(c)** $H^+$ is a reactant.
**(d)** C is a product.
**75.** density of sulfur; depth of deposit, purity of S
**76.** no
**77.** 1.30%
**78.** Assume reaction is: $NaN_3(s) \longrightarrow Na(s) + \frac{3}{2}N_2(g)$
Assume 25°C, 1 atm pressure
mass of $NaN_3$ is 35 g

# Chapter 22

**2.** **(a)** 2-methyloctane　　**(b)** 2,2-dimethylpropane
**(c)** 2,2,4-trimethylpentane
**(d)** 2,5-dimethylheptane
**4.** **(a)**
$$CH_3-\underset{\underset{CH_3}{|}}{\overset{\overset{CH_3}{|}}{C}}-CH_2-\underset{\underset{CH_3}{|}}{\overset{\overset{H}{|}}{C}}-CH_3$$
**(b)**
$$CH_3-\underset{\underset{CH_3}{|}}{\overset{\overset{CH_3}{|}}{C}}-CH_3$$
**(c)**
$$CH_3-(CH_2)_2-\underset{\underset{\underset{H}{|}}{CH_3-C-CH_3}}{\overset{\overset{H}{|}}{C}}-(CH_2)_3-CH_3$$
**(d)**
$$CH_3-\underset{\underset{H}{|}}{\overset{\overset{CH_3}{|}}{C}}-\underset{\underset{CH_3}{|}}{\overset{\overset{H}{|}}{C}}-\underset{\underset{H}{|}}{\overset{\overset{CH_3}{|}}{C}}-(CH_2)_2-CH_3$$
**6.** **(a)**
$$CH_3-\underset{\underset{CH_3}{|}}{\overset{\overset{CH_3}{|}}{C}}-CH_2-CH_3$$
2,2-dimethylbutane
**(b)** $CH_3-(CH_2)_2-\underset{\underset{CH_3}{|}}{CH}-CH_3$
2-methylpentane
**(c)** $CH_3-\underset{\underset{\underset{CH_3}{|}}{\underset{CH_2}{|}}}{CH}-CH_3$
2-methylbutane
**8.** **(a)** $CH_3-C\equiv C-CH_2-CH_3$
**(b)** $CH_3-C\equiv C-\underset{\underset{CH_3}{|}}{CH}-CH_3$
**(c)**
$$CH_3-\underset{\underset{}{}}{\overset{\overset{CH_3}{|}}{CH}}-C\equiv C-CH_2-CH_3$$
**(d)**
$$H-C\equiv C-\underset{\underset{CH_3}{|}}{\overset{\overset{CH_3}{|}}{C}}-CH_3$$
**10.** **(a)** 1,3-dichlorobenzene
**(b)** 1,2,4-trichlorobenzene
**(c)** 1,3,5-trichlorobenzene
**12.** **(a)** methanoic acid; formic acid
**(b)** ethanol; ethyl alcohol
**(c)** 2-propanol; isopropyl alcohol

**14. (a)** methyl acetate    **(b)** ethyl formate
**(c)** methyl propionate

**16.**

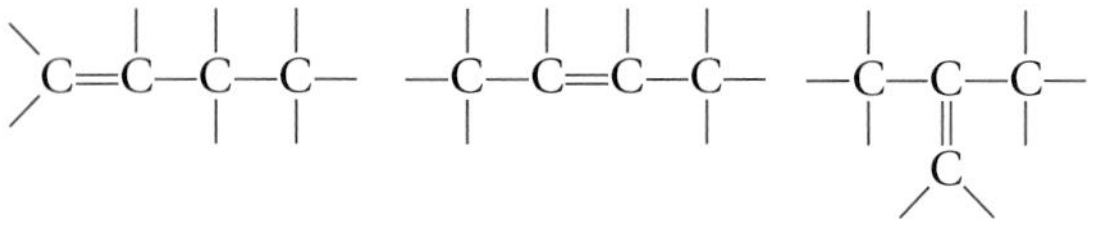

$$\text{C=C-C-C-}\qquad\text{-C-C=C-C-}\qquad\text{-C-C-C-} \atop \text{C}$$

**18.**

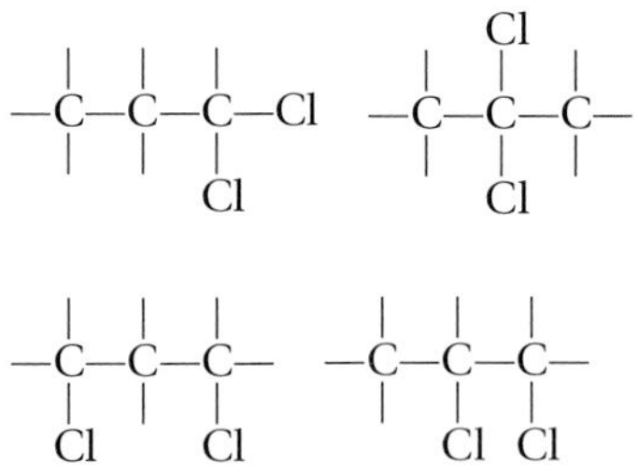

$$\text{-C-C-C-Cl}\qquad\text{-C-C-C-} \atop \text{Cl}\qquad\qquad\text{Cl}$$

$$\text{-C-C-C-}\qquad\text{-C-C-C-} \atop \text{Cl}\quad\text{Cl}\qquad\text{Cl Cl}$$

**20.**

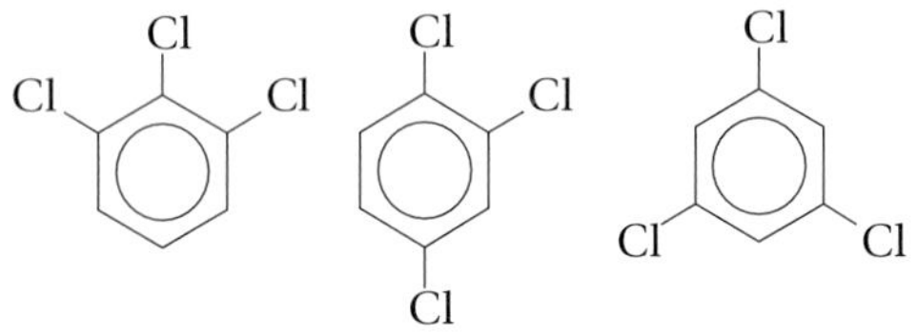

$$\text{trichlorobenzene isomers}$$

**22.**

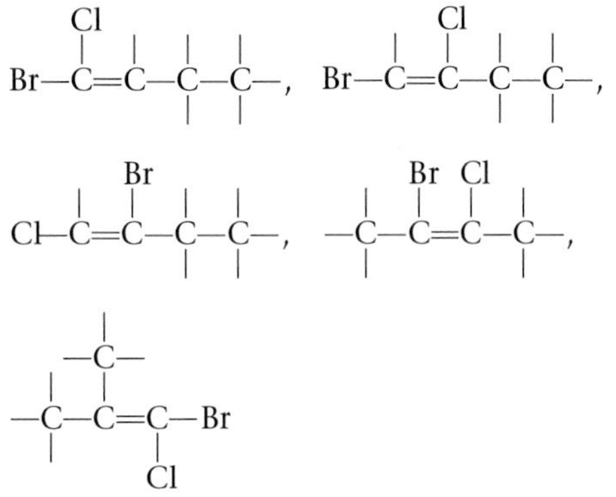

$$\text{Br-C=C-C-C-,}\quad\text{Br-C=C-C-C-,}$$

$$\text{Cl-C=C-C-C-,}\quad\text{-C-C=C-C-,}$$

$$\text{-C-C=C-Br} \atop \text{Cl}$$

**24.**

$$\text{-C-C-C-C-OH,}\quad\text{-C-C-C-} \atop \text{O}\qquad\qquad\text{HO-C=O}$$

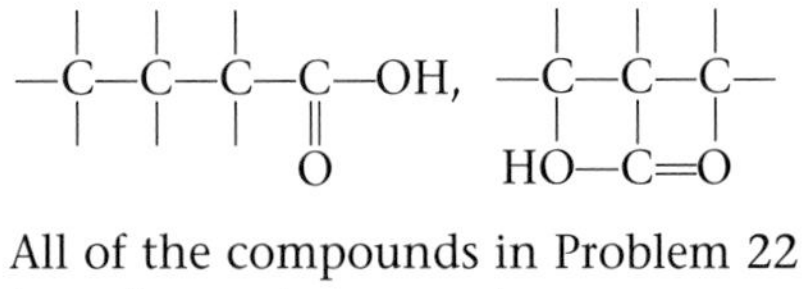

**26.** All of the compounds in Problem 22 show cis-
trans isomerism except

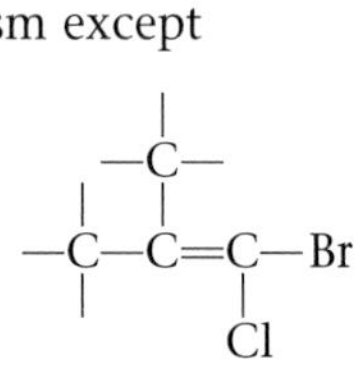

$$\text{-C-C=C-Br} \atop \text{Cl}$$

where there are two —$CH_3$ groups attached to
the same carbon

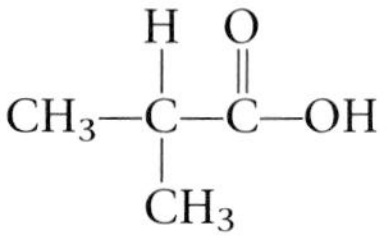

**28.** b, c
**30. (a)** alcohol    **(b)** ester    **(c)** ester, acid

**32. (a)**

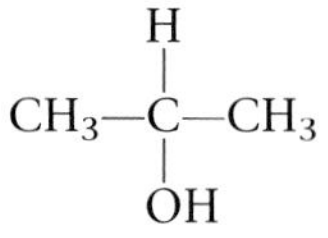

$$CH_3\text{-C-}CH_3 \atop OH$$

**(b)**

$$CH_3\text{-C-C-OH} \atop CH_3$$

**(c)**

$$CH_3\text{-C-C-O-C-}CH_3$$

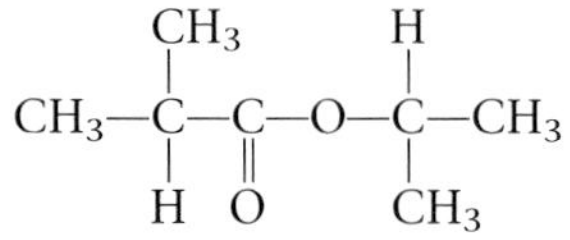

**34. (a)**

$$\text{-C-C-C-C-}$$ with F groups

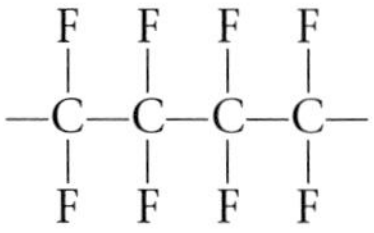

**(b)** $2.5 \times 10^6$ g/mol
**(c)** 24.02% C, 75.98% F

**36.**

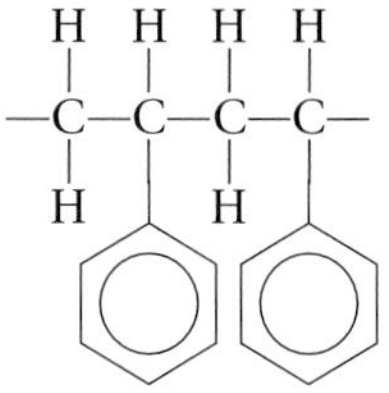

$$\text{-C-C-C-C-}$$

**38. (a)** $H_2C=CHF$    **(b)** 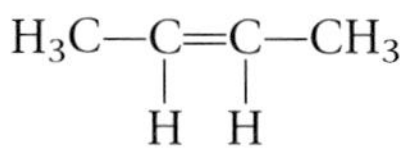 $H_3C\text{-C=C-}CH_3$

**40.**

$$\text{-O-C-O-}\bigcirc\text{-C-}\bigcirc\text{-O-} \atop O$$ with $CH_3$ groups

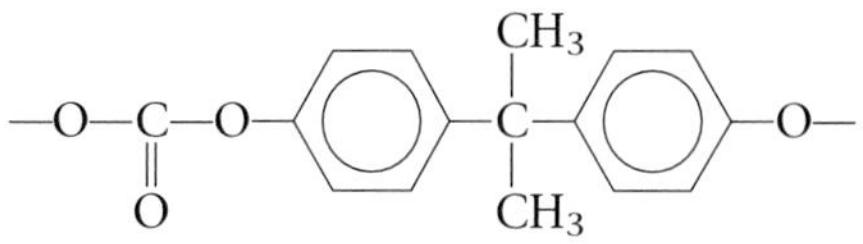

**42. (a)** $H_2N\text{-}CH_2\text{-}CH_2\text{-}NH_2$ and
$HOOC\text{-}CH_2\text{-}COOH$

**(b)**

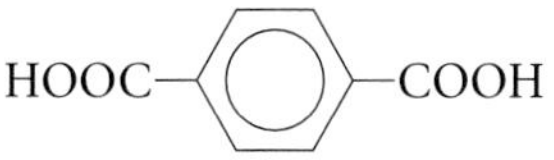

$$HOOC\text{-}\bigcirc\text{-}COOH$$

and

$$HO\text{-}CH_2\text{-C-OH} \atop CH_3$$

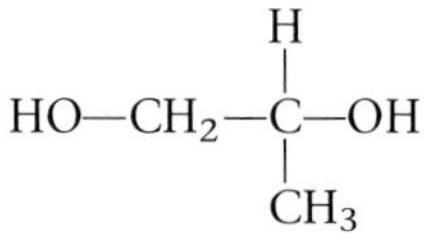

**44. (a)** present in "bottled gas."
**(b)** starting material for organic compounds like
ethyl alcohol.
**(c)** fuel used in welding.
**(d)** nontoxic solvent; benzene replacement
**46. (a)** Gasoline contains mostly alkanes.
**(b)** Benzene is a carcinogen (cancer-causing).
**(c)** $CaC_2$ reacts with $H_2O$ to form acetylene.
**(d)** Alkenes show geometric isomerism.
**(e)** Cholesterol is a fused-ring compound.

**48.** 1.94

**50.** **(a)** neither          **(b)** addition
**(c)** condensation     **(d)** neither

**52.** $C_{27}H_{46}O$

**53.**

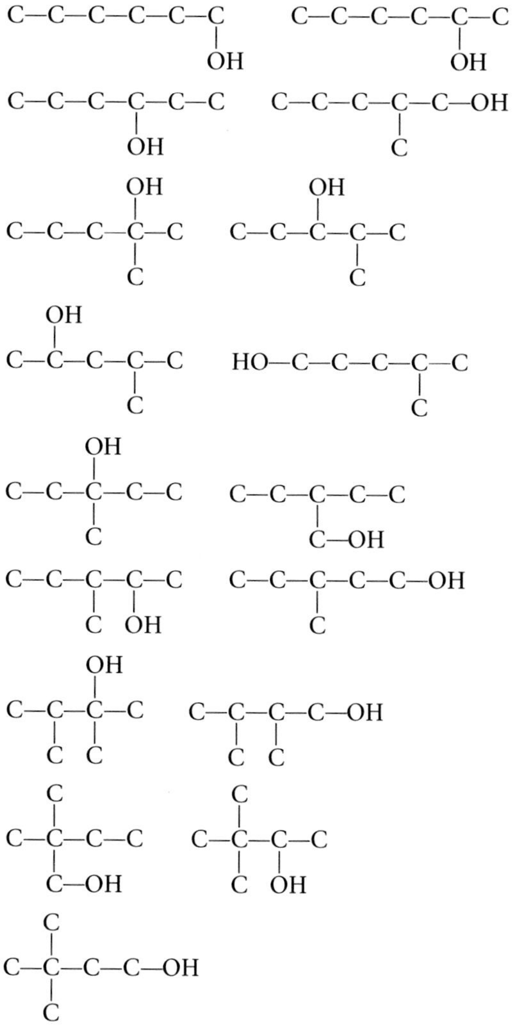

C—C—C—C—C—C
           |
          OH

C—C—C—C—C—C
              |
              OH

C—C—C—C—C—C
        |
        OH

C—C—C—C—C—OH
        |
        C

      OH
      |
C—C—C—C—C
        |
        C

      OH
      |
C—C—C—C—C
    |
    C

  OH
  |
C—C—C—C—C
    |
    C

HO—C—C—C—C—C
             |
             C

    OH
    |
C—C—C—C—C
    |
    C

C—C—C—C—C
        |
        C—OH

C—C—C—C—C
      |  |
      C  OH

C—C—C—C—C—OH
        |
        C

    OH
    |
C—C—C—C
   |  |
   C  C

C—C—C—C—OH
    |  |
    C  C

   C
   |
C—C—C—C
   |
   C—OH

   C
   |
C—C—C—C
   |  |
   C  OH

   C
   |
C—C—C—C—OH
   |
   C

**54.** approximately 6 g (Assume $H_2O(l)$ is a product; neglect heat capacity of pan.)

**55.** **(a)**

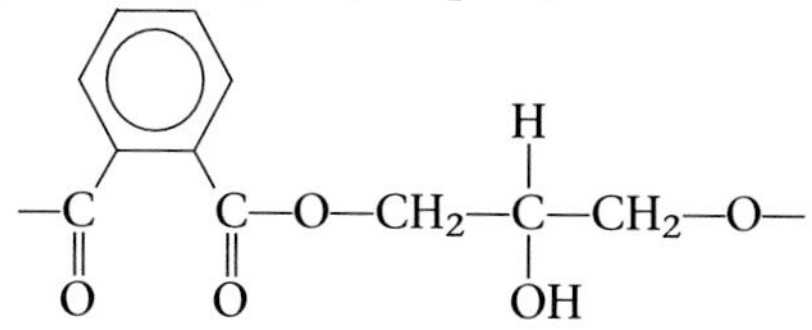

**(b)** orthophthalic acid condenses with OH groups in 2 adjacent chains.

# Index/ Glossary

*Note:* Boldface terms are defined in the context used in the text. Italic page numbers indicate figures; t indicates table; q indicates an end-of-chapter question.

A (mass number), 31

**Abbreviated electron configuration** Brief notation in which only those electrons beyond the preceding noble gas are shown. The abbreviated electron configuration of the Fe atom is [Ar]4s$^2$3d$^6$, 143, 153

Absolute temperature, 9, 113

Absolute zero, 9

**Abundance (isotopic)** Mole percent of an isotope in a natural sample of an element, 55, 56
  of chlorine, 55, *56*
  of chromium, 75q
  of magnesium, 75q
  of oxygen, 75q
  of silicon, 75q
  of strontium, 75q
  of sulfur, 56
  use in calculating atomic mass, 56

Acetal, 295

Acetaldehyde (ethanal), 307, 624t

Acetamide, 624t

Acetate ion, 42t, 386t, 390t, 404t

Acetic acid (ethanoic acid), 288t, 379, 380t, 388t, 404t, 626
  reaction with strong base, 411
  titration with strong base, 411–413, *412*

Acetic acid, glacial, 424q

Acetic anhydride, 79q

Acetoacetic acid, 419

Acetol, 293

Acetominophen, 76q

Acetone (propanone), 76q, 103, 624t

Acetylacetonate ion, 445q

Acetylene (ethyne), 191, 620, *620*
  bonding in, 176, 620
  decomposition of, 480, 620
  hybridization of, 198
  pi and sigma bonds, 200, *200*
  preparation of, 620
  welding, 620, 621

Acetylperoxide, 204q

Acetylsalicylic acid, 60

**Acid** A substance that dissolves in water to produce a solution in which [H$^+$] is greater than 10$^{-7}$ *M*, 86, 87, 372–398
  Arrhenius, 86, 87
  Brønsted-Lowry, 372, 373, 388, 388t
  carboxylic, 625–627
  conjugate, 372
  Lewis, 427
  organic, 625–627
  polyprotic, 416–418
  reacting species, 88, 89, 89t
  strong, 87, 87t, 376, 377
    reaction with strong bases, 88
    reaction with weak bases, 89, 90
  weak, 87, 378–384
    reaction with strong bases, 89

**Acid anhydride** A nonmetal oxide that reacts with water to form an acidic solution, 596

**Acid ionization constant** The equilibrium constant for the ionization of a weak acid, 379, 380, 380t

Acid-base equilibria, 399–424

**Acid-base indicator** A weak acid that changes color with pH; the weak acid molecule and its conjugate base have different colors, 86, 377, *377*, 378, 406–409, 408t

Acid-base reaction, 86–90, 90t
  writing of equation for, 88–90

**Acid-base titration** Procedure used to determine the concentration of an acid or base. The volume of a solution of an acid (or base) of known molarity required to react with a known volume of base (or acid) is measured, *100*, 409–415, 415t
  strong acid-strong base, 409–411, *410*
  strong acid-weak base, 413, 414, *414*
  weak acid-strong base, 411–413, *412*

Acid rain, 2, 391–393, *392, 393*

Acid strength, 87
  of oxoacids, 597, 598

**Acidic ion** Ion that forms H$^+$ ions in water. The ammonium ion is acidic because of the reaction NH$_4^+$(aq) $\rightleftharpoons$ H$^+$(aq) + NH$_3$(aq), 378, 379, 389

**Acidic solution** An aqueous solution with a pH less than 7, *87*, 374, 375
  balancing redox equations in, 94, 95
  end point in titration, 415t
  salts forming, 389
  standard potentials in, 504t, 505t

Acidosis, 418

Acne, 575

Acre, 24q

Acrylonitrile, 631t

**Actinides** Elements 89 (Ac) through 102 (No) in the periodic table, 154

Actinium, A.8

**Activated complex** A species, formed by collision of energetic particles, that can react to form products, 317, 318

**Activation energy** The minimum energy that must be possessed by a pair of molecules if collision is to result in reaction, 314–317
  Arrhenius equation, 322–324
  catalyzed reaction, *319*
  diagrams, *318, 326*
  stepwise reaction, *326*

**Activity, nuclear** Rate of radioactive decay; number of atoms decaying per unit time, 537, 538

Actual yield. *See* experimental yield

Addition, uncertainties in, 14

**Addition polymer** Polymer produced by a monomer, usually a derivative of ethylene, adding to itself; no other product is formed, 630–632

Adenosine diphosphate (ADP), *487*

Adenosine triphosphate (ATP), *487*

ADP. *See* adenosine diphosphate

Adrenaline, 298q

Age
  of organic matter, 539, 540
  of rocks, 538, 539, 553q

AIDS, 47, 505

Air, composition of, 130t
  fractional distillation, 584
  pollution, 391–393

Airbags, 582, 611q

Albertus Magnus, 20

Albinism, 635

Alchemy, 10

**Alcohol(s)** A compound containing an OH group attached to a hydrocarbon chain. The simplest example is methanol, CH$_3$OH, 36, 623–625
  nomenclature, 623
  solubility in water, 279t

Alcohol dehydrogenase, 575

Alcoholic beverages, 624
  ethanol in, 624
  intoxication from, 103
  proof of, 296q

Aldehyde, 624t

Alizarin yellow, 422q

**Alkali metal(s)** A metal in Group 1 of the periodic table, 33, 557, 563–566
  chemical properties of, 563–566, *564*, 564t
  electron configurations of, 155, 155t
  oxidation number of, 92
  oxygen compounds of, 565, 566
  reaction of, with halogens, 564t
  reaction of, with hydrogen, 564, 564t
  reaction of, with nitrogen, 564t
  reaction of, with oxygen, 564t, 565, 566
  reaction of, with sulfur, 564t
  reaction of, with water, *506*, 564, 564t, 565

Alkaline, 374

Alkaline dry cell, 521

**Alkaline earth metal(s)** A metal in Group 2 of the periodic table, 33, 557, 563–566
  chemical properties of, 563–566, 564t, *565*
  electron configurations of, 155, 155t
  oxidation number of, 92